Coral Reefs: An Ecosystem in Transition

Zvy Dubinsky • Noga Stambler
Editors

Coral Reefs: An Ecosystem in Transition

Editors
Zvy Dubinsky
Bar Ilan University
The Mina & Everard Goodman Faculty
of Life Sciences
52900 Ramat Gan
Israel
dubinz@mail.biu.ac.il

Noga Stambler
Bar Ilan University
The Mina and Everard Goodman Faculty
of Life Sciences
52900 Ramat Gan
Israel
stambln@mail.biu.ac.il; drnogas@gmail.com

ISBN 978-94-007-0113-7   e-ISBN 978-94-007-0114-4
DOI 10.1007/978-94-007-0114-4
Springer Dordrecht Heidelberg London New York

© Springer Science+Business Media B.V. 2011
© US government copyright 2011 for Chapter 20
No part of this work may be reproduced, stored in a retrieval system, or transmitted in any form or by any means, electronic, mechanical, photocopying, microfilming, recording or otherwise, without written permission from the Publisher, with the exception of any material supplied specifically for the purpose of being entered and executed on a computer system, for exclusive use by the purchaser of the work.

Printed on acid-free paper

Springer is part of Springer Science+Business Media (www.springer.com)

Preface

In the wake of the alarming decline in the vitality of coral reefs worldwide, and its resulting catastrophic effects on the biodiversity of associated biota, it is timely and topical to revisit, revise and refresh our views of the main processes related to corals, reefs, and their myriads of associated denizens. Global warming, ocean acidification, increasing UV doses, raising sea level, anthropogenic eutrophication, pollution, overfishing, and shoreline development all lead to increase in the spread of coral disease, the frequency and intensity of devastating bleaching events, and the irreversible decline of some 70% of the world's reefs. Coral reefs are one of the peaks of life on Earth, unmatched in diversity and beauty for us to admire, understand, cherish, and make sure to preserve them for generations to come as a precious part of Mankind's heritage.

Leading authorities, established and young, have contributed their up-to-date summaries and evaluations of developments in their respective fields of expertise. The resulting book covers and integrates in one volume materials scattered in hundreds of research articles, in most cases focusing on narrow and specialized aspects of coral science. Besides the latest developments in coral evolution and physiology, chapters are devoted to novel frontiers in coral reef research such as the molecular biology of corals and their symbiotic algae, remote sensing of reef systems, ecology of coral disease spread, effects of various scenarios of global climate change, ocean acidification effects of increasing CO_2 levels on coral calcification, and damaged coral reef remediation. In addition to the extensive coverage of the above, key issues regarding the coral organism and the reef ecosystem, such as calcification, reproduction, modeling, algae, reef invertebrates, competition, and fish, are reevaluated in the light of new research tools and emerging new insights.

In all chapters novel theories as well as challenges to established paradigms are raised, evaluated, and discussed. Chapters are offering attempts of synthesis among aspects and organizational levels.

The volume is an essential asset in any marine laboratory and in all university libraries where there is a biology program.

This volume is humbly dedicated to the memory of Len Muscatine, friend and mentor, from whom we all learned so much about corals and excellence in science.

Zvy Dubinsky and Noga Stambler

Contents

Part I History and Perspective

Coral Research: Past Efforts and Future Horizons 3
Robert H. Richmond and Eric Wolanski

Part II Geology and Evolution

The Paleoecology of Coral Reefs 13
John M. Pandolfi

Remote Sensing of Coral Reef Processes 25
Eric J. Hochberg

Coral Taxonomy and Evolution 37
John (Charlie) E.N. Veron

The Coral Triangle 47
John (Charlie) E.N. Veron, Lyndon M. DeVantier, Emre Turak, Alison L. Green,
Stuart Kininmonth, M. Stafford-Smith, and N. Peterson

Part III Coral Biology: Symbiosis, Photosynthesis and Calcification

Sexual Reproduction of Scleractinian Corals 59
Peter L. Harrison

Zooxanthellae: The Yellow Symbionts Inside Animals 87
Noga Stambler

Light as a Source of Information and Energy in Zooxanthellate Corals 107
Zvy Dubinsky and Paul Falkowski

Coral Calcification, Cells to Reefs 119
Denis Allemand, Éric Tambutté, Didier Zoccola, and Sylvie Tambutté

Coral Calcification Under Ocean Acidification and Global Change 151
Jonathan Erez, Stéphanie Reynaud, Jacob Silverman, Kenneth Schneider,
and Denis Allemand

**Simulating and Quantifying the Environmental Influence
on Coral Colony Growth and Form** 177
Jaap A. Kaandorp, Maxim Filatov, and Nol Chindapol

Physiological Adaptation to Symbiosis in Cnidarians 187
Paola Furla, Sophie Richier, and Denis Allemand

Part IV The Coral Reef Ecosystem: Bacteria, Zooplankton, Algae, Invertebrates, Fishes and Model

Biogeochemistry of Nutrients 199
Marlin J. Atkinson

The Role of Dissolved Organic Nitrogen (DON) in Coral Biology and Reef Ecology 207
Yoshimi Suzuki and Beatriz E. Casareto

The Role of Plankton in Coral Trophodynamics 215
Christine Ferrier-Pagès, Mia Hoogenboom, and Fanny Houlbrèque

Fish or Germs? Microbial Dynamics Associated with Changing Trophic Structures on Coral Reefs 231
Elizabeth A. Dinsdale and Forest Rohwer

Coral Reef Algae 241
Peggy Fong and Valerie J. Paul

Invertebrates and Their Roles in Coral Reef Ecosystems 273
Peter W. Glynn and Ian C. Enochs

Coral Reef Fishes: Opportunities, Challenges and Concerns 327
W. Linn Montgomery

Competition Among Sessile Organisms on Coral Reefs 347
Nanette E. Chadwick and Kathleen M. Morrow

Scaling Up Models of the Dynamics of Coral Reef Ecosystems: An Approach for Science-Based Management of Global Change 373
Jesús Ernesto Arias-González, Craig Johnson, Rob M. Seymour, Pascal Perez, and Porfirio Aliño

Part V Disturbances

The Impact of Climate Change on Coral Reef Ecosystems 391
Ove Hoegh-Guldberg

Coral Bleaching: Causes and Mechanisms 405
Michael P. Lesser

The Potential for Temperature Acclimatisation of Reef Corals in the Face of Climate Change 421
Barbara E. Brown and Andrew R. Cossins

Reef Bioerosion: Agents and Processes 435
Aline Tribollet and Stjepko Golubic

Microbial Diseases of Corals: Pathology and Ecology 451
Eugene Rosenberg and Ariel Kushmaro

Coral Reef Diseases in the Atlantic-Caribbean 465
Ernesto Weil and Caroline S. Rogers

**Factors Determining the Resilience of Coral Reefs
to Eutrophication: A Review and Conceptual Model** 493
Katharina E. Fabricius

Part VI Conservation and Management

The Resilience of Coral Reefs and Its Implications for Reef Management 509
Peter J. Mumby and Robert S. Steneck

Index 521

Part I
History and Perspective

Coral Research: Past Efforts and Future Horizons

Robert H. Richmond and Eric Wolanski

Abstract Modern coral reefs had their origins in the Triassic Period, and over the past 65 million years, have expanded and contracted due to a variety of extrinsic factors such as sea level and climatic changes. As humans evolved, so did a new era for coral reefs: that of exposure to anthropogenic stressors on top of the already persistent natural events such as hurricanes/typhoons, volcanic eruptions and *Acanthaster plancii* outbreaks. Early studies of corals and coral reefs focused on taxonomy, ecology and physiology. The establishment of reef-based marine laboratories and technological advances in SCUBA enabled a rapid expansion of *in situ* studies and the ability to compare reef processes and changes over both space and time. The documented declines in the state of coral reefs worldwide have shifted the focus of many research programs from basic to applied and management-directed research. Expanded studies of reproductive processes, animal-algal symbioses, molecular genetics, ecotoxicology, connectivity, ecological modelling, calcification, cellular biology and biochemistry have allowed researchers to better understand how coral reefs function and how best to respond to a variety of stressors. Coral reef research has necessarily become multidisciplinary in nature, embracing the social and economic as well as the biophysical sciences. In the face of increasing effects of stressors tied to local activities such as poor land-use practices, and mounting concerns tied to global climate change, the bridging of science to policy development, management and conservation is critical if there is to be a legacy of vital reefs left for future generations to enjoy.

Keywords Pollution • climate change • management • biophysical research • modeling

R.H. Richmond (✉)
Kewalo Marine Laboratory, University of Hawaii, 41 Ahui Street, Honolulu, HI 96813, USA
e-mail: Richmond@Hawaii.edu

E. Wolanski
James Cook University, Townsville, Queensland, 4811, Australia
e-mail: eric.wolanski@jcu.edu.au

1 Introduction

Modern coral reefs had their origins in the Triassic Period, and over the past 65 million years, have expanded and contracted due to a variety of extrinsic factors such as sea level and climatic changes. As humans evolved, so did a new era for coral reefs: that of exposure to anthropogenic stressors on top of the already persistent natural events such as hurricanes/typhoons, volcanic eruptions and *Acanthaster plancii* outbreaks (Richmond 1993).

Human interactions with coral reefs date back thousands of years, and there is ample evidence of interests in and appreciation for coral reefs and their related resources passed down in oral traditions prior to the advent of the peer-reviewed literature. One can only ponder how the review process worked to determine which knowledge would be accepted and handed-down in the oral history versions of *Science* and *Nature*, yet such oral and hieroglyphic records of traditional ecological knowledge (TEK) exist in a variety of cultures. Marine protected areas have been in existence in many Pacific islands and atolls for as many generations as oral history can track (centuries, at very least), and many traditional practices that reflect a solid basis in biology are still in use today in islands such as Yap, Federated States of Micronesia, Palau, Melanesia and by aboriginal populations in Australia (Johannes 1981, 1997; Cinner et al. 2005; Richmond et al. 2007). These traditional practices reflect an understanding of spawning behavior in fishes and invertebrates tied to annual events and lunar cycles, current patterns and connectivity (as reflected by traditional navigation in sailing canoes), and carrying capacity (sustainability) as evidenced by specific limits put on resource exploitation. The Hawaiian creation chant (Kumulipo), which predates written records in Hawaii, references the coral polyp, demonstrating that corals and related organisms were a central element of the associated human cultures (Beckwith 1951).

For the purposes of this chapter, our focus will commence with early studies documented in the literature, the directions and trends that have occurred over time, and we will provide an overview of future directions, concerns, and needs.

2 Early Coral Reef Research

The first few decades of place-based coral reef research focused on fundamental questions of ecology, taxonomy, and physiology. As the number of marine laboratories and field stations increased, so did the amount of data in accessible peer-reviewed journals, allowing comparative studies over space and time, including on the effects of both natural and anthropogenic disturbance. A number of excellent reviews exist that summarize past coral research, for which there is no need for repetition (Fautin 2002; Kinzie 2002; Mather 2002; Salvat 2002); however, we present the following brief overview to help set the stage for examining future horizons. Disciplines of coral reef research reflect an evolution that parallels technological advances in both the field and the laboratory, notably SCUBA and the establishment of marine laboratories for extended *in situ* studies (Table 1). Research began in the areas of taxonomy, ecology, and

Table 1 Key coral reef research institutes and marine laboratories

Laboratory	Location	Affiliation	Comments
Australian Institute of Marine Science	Townsville, Australia	Australian Government	Established by the Australian Institute of Marine Science Act 1972
Akajima Marine Laboratory	Okinawa, Japan Aka Island	Private/Independent	Established in 1989 as a private research institution
Bocas del Toro Research Station	Panama, Caribbean	Smithsonian Tropical Research Institute	Began laboratory research operations on site in 2002
Carrie Bow Cay	Belize, Central America	Smithsonian Institution	Opened in 1972 on the Belize Barrier reef
Centre Scientifique de Monaco	Monaco	Monaco Government	Opened in 1960, has focused on climate change effects since 1989
Discovery Bay Marine Laboratory	Jamaica	University of the West Indies	Founded in 1965, has been a site for seminal work in the Caribbean
Galeta Marine Laboratory	Caribbean coast, Panama	Smithsonian Tropical Research Institute	Opened in 1964, provides access to coastal Caribbean fringing reefs
Hawaii Institute of Marine Biology	Oahu, Hawaii Coconut Island	University of Hawaii at Manoa	Opened as the Hawaii Marine Lab in 1951, became HIMB in 1965
Heron Island Research Station	Great Barrier Reef (southern)	University of Queensland	Established in the 1950s, new facilities opened 2009
Inter-University Institute	Eilat, Israel	All of Israel's Universities	Opened in 1968, serves all of Israel's universities and international researchers
Kewalo Marine Laboratory	Oahu, Hawaii	University of Hawaii at Manoa	In an urbanized setting, close to main campus, opened in 1972
Marine Biological Station	Ghardaqa, Red Sea, Egypt	University of Egypt	Opened in 1929, unique facility and access to Red Sea with important history
Mid-Pacific Marine Laboratory	Enewetak, Marshall Islands	University of Hawaii at Manoa	Initially used for visits, resident staff researched from 1980; Closed in 1986
Mote Marine Laboratory	Summerland Key, Florida	Independent	Established in 1955, includes research and education as core activities
Naos Marine Laboratory	Panama City, Pacific Ocean	Smithsonian Tropical Research Institute	Opened in 1968, provided research access to eastern Pacific coral reefs. Now includes a modern molecular lab
One Tree Island	Great Barrier Reef, Australia	University of Sidney	Research expedition used in 1965, the first building was constructed in 1969
Palao Tropical Research Station	Koror, Palau	Japan	Center for coral reef research 1934–1943
Palau International Coral Reef Center	Koror, Palau	Government of Palau	Opened in 2001; Common Agenda for Cooperation between Japan and the US
Phuket Marine Biological Center	Phuket, Thailand	Government of Thailand	Established by agreement between Thai and Danish governments in 1966
Rosentiel School Mar & Atm Sci	Miami, Florida	University of Miami	Base for many coral reef studies in the Atlantic, Caribbean and Pacific
Sesoko Marine Science Center	Okinawa, Japan	University of the Ryukyus	International research facility established in 1972
Tortugas Marine Laboratory	Loggerhead Key, Florida	Carnegie Institute	1905–1939; First Tropical Marine Lab in the Western Hemisphere
University of Guam Marine Lab	Guam, Mariana Islands	University of Guam	Established in 1970, serves Guam and Micronesia
West Indies Laboratory	St. Croix, U.S. Virgin Islands	Farleigh Dickinson University	Closed following Hurricane Hugo in 1989

geology and expanded the frontiers of investigations into the areas of animal-algal symbioses (from CZAR – the "contribution of zooxanthellae to animal respiration" made famous by Muscatine, Trench, Falkowski, Dubinski and others), reproductive biology/larval ecology, population genetics, biochemistry, cellular biology, ecotoxicology and evolutionary biology.

The taxonomy of corals was first developed in museums far from reefs (e.g., the British Museum), often from samples collected by various expeditions and merchant vessels anchoring in tropical waters. Geologists and specifically paleontologists were among those who developed the framework for species identification based on morphological characteristics including septal and collumellar patterns and structures. Cladistics, morphometrics, and numerical taxonomy established the present system of coral identification, with notable revisions occurring over the past 2 decades. When morphological variation along ecological gradients of light, sediment load, and water motion became apparent, some previously distinguished species groups became lumped. With the addition of genetic tools and reproductive studies of hybridization, additional taxonomic revisions have occurred and more are expected.

Great advances in coral reef studies occurred as a result of focused shipboard expeditions, including those by Charles Darwin (the second voyage of the Beagle, 1831–1836), the U.S. Exploring Expedition, 1844–1874, following which James Dana wrote his book *Corals and Coral Islands*, the Challenger Expedition (1872–1876), the Symbios Expedition aboard the R/V Alpha Helix in 1971, and the second International coral reef symposium held aboard the M.V. Marco Polo on the Great Barrier Reef (GBR) of Australia in 1973. Notable efforts resulting in seminal coral reef research include studies performed at the Palao Tropical Research Station by Japanese researchers Atoda and Kawaguti, in Hawaii by Edmondson, in Jamaica by T. and N. Goreau, by Yonge on the Great Barrier Reef of Australia, Boschma in Bermuda, and Vaughan in the Dry Tortugas.

The international coral reef symposia (ICRS) and associated proceedings also reflect both the expanding wealth of knowledge on coral reefs as well as trends including the notable transition from basic research to management applications. The most recent (2008) 11th ICRS included a total of 2,176 presentations (986 oral and 1,190 posters), with the management sessions being the largest of the 26 sessions including 75 oral and 142 poster presentations. Attendees came from 78 countries. This meeting was an order of magnitude larger than the second ICRS, which was held aboard the R/V Marco Polo in 1974 with 264 scientists and 121 papers. It is also worth noting that coral reef research and its applications have been truly international endeavors, with substantial contributions from individuals from Asia, Africa, Australia, the Caribbean, Europe, the Middle East, the Pacific Islands, and North and South America (see Table 1). Coral reef science, like truth, transcends politics, and many researchers and managers from countries whose politicians have been at odds, have exemplified the value of science as an area of common ground, collaboration, and understanding.

3 Present Areas of Research and Future Directions

Population Biology – Reproduction and recruitment of coral reef organisms has been an area of great advancement over the past 3 decades, with an ever-expanding database on reproductive mode, timing, and geographical variations. In scleractinian corals, species have been identified as being either hermaphroditic (possessing both male and female characteristics simultaneously or sequentially) or gonochoric (separate sexes) and as either brooders, in which planula larvae develop internally within the colony or on the colony surface in some soft corals, or spawners which release their eggs and sperm into the water column with subsequent external fertilization and development. Synchronous reproductive timing, particularly mass spawning events, was first documented in the seminal paper by researchers from James Cook University (Harrison et al. 1984), and opened the field for additional studies that have demonstrated some regional patterns as well as variability within sympatric populations and among geographically separated populations. Research has also demonstrated selectivity of larvae in response to putative metamorphic inducers associated with conspecifics or associated flora and fauna.

The applications of this basic knowledge on reproduction and recruitment in a variety of coral reef organisms are critical to a broader understanding of how stressors affect these essential processes responsible for maintaining reef populations. Specifically, reduced water and substratum quality have been found to affect reproduction and recruitment success, and hence are now targets for management and mitigation efforts. Studies of the population biology of the corallivorous starfish *Acanthaster plancii* also helped guide the feasibility (or lack thereof) of predator control, and it was geologists studying the evidence in sediment cores that determined that massive crown-of-thorns starfish outbreaks like those observed since 1960, had not occurred historically; infestations did occur before 1960, but they had a much lesser impact on corals (Devantier and Done 2007). Birkeland (1982) further demonstrated a link between terrestrial runoff, nutrients, and *Acanthaster* outbreaks.

Coral reef fishes have also been the subject of research on reproduction and recruitment, and the identification of site-specific spawning aggregations as well as environmental cues for fish larval recruitment back to reefs. Such studies provided the basis for developing scientific criteria for use in developing marine protected area (MPA) networks, including size, distance, and placement based on current patterns and reef connectivity.

Animal-algal symbiosis has been an area of great interest and recent advances. From the seminal work by Boschma (1925) on the basis of the symbiosis, to light-enhanced calcification studies by T. and N. Goreau (1959) and the discovery of different clades of zooxanthellae (Rowan and Knowlton 1995; Baker and Rowan 1997), researchers are working to better understand the coral bleaching phenomenon, and predict adaptive responses. While previous research has provided the foundation for understanding the role of zooxanthellae in coral nutrition and growth, recent mass-bleaching events have increased interest in the physiological and biochemical interactions between algal symbionts and their animal hosts. Multiple stressors have been found to initiate coral bleaching, and several underlying mechanisms have been indentified including symbiophagy (Downs et al. 2009).

Currents and Connectivity – Marine protected areas (MPAs) are among the tools that can help address declining coral reef ecosystem health, resilience, structure, and function. The size, location and pattern of networks require an understanding of the state of source areas, corridors of larval transport and recipient reef characteristics. Research progress has been made in addressing questions of genetic linkages, oceanic and local marine current patterns, larval retention versus long-range dispersal, and the effects of extreme events (El Niño Southern Oscillation events) on reef connectivity. Satellite imagery, *in situ* oceanographic instrumentation, and modeling have been important areas of research, tool development, and predictions that bridge science to management applications. Marine Spatial Planning (MSP; Ehler and Douvere 2009) is a growing area of policy applications that relies on both connectivity and use of pattern data. Such an integrated and proactive approach is critical to managing human activities as conflicting uses and demands increase.

Algal Ecology and Competition – Coral reefs are composed of far more elements than corals, with a variety of plants, invertebrates, vertebrates, and bacteria often providing the greatest biomass and diversity. Algae, including calcified crustose coralline algae (CCA), fleshy, filamentous, and blue-greens (cyanobacteria), are major components of coral reefs and affect their ecology in many ways (Fong and Paul this book). Crustose coralline species have been found to induce larval recruitment of some coral species, although it is not known if the putative chemical inducers are produced by the algae or bacterial associates. These algae are also instrumental in cementing reefs through accretion. Invasive algae are also a concern on many reefs, e.g., those in Hawaii, where species of *Gracilaria* and *Kappaphycus* brought in for aquaculture have overtaken reefs, smothering corals, and changing water motion characteristics and the chemistry of the substratum boundary layer. Fleshy algae can trap sediments on reefs resulting in chronic reductions in recruitment and survivorship of fish, corals, and other benthic organisms due to wave-induced resuspension events (Wolanski et al. 2003). Alternate stable states where coral reef substrata become overgrown with fleshy algae often occur after damage events, and these may persist for decades in the absence of intervention.

Biogeochemistry and Calcification – As concerns surrounding greenhouse gases, specifically CO_2, came to the forefront of climate change science, the role of corals and other calcifying reef organisms in mediating atmospheric levels became an area of great interest. While a great deal of variability exists among coral reefs globally, there is a general consensus that CO_2 is produced during coral calcification based on the chemical equations, and those reefs that do sequester carbon have a high biomass of fleshy algae. On an annual basis, it appears that "healthy" coral reefs are close to carbon neutral. The ecological, economic, and cultural values of coral reefs appear to outweigh any values associated with global CO_2 sequestration. The looming concern tied to global CO_2 levels and those of other greenhouse gases are tied to changes in ocean aragonite saturation state and pH (see Erez et al., this book, Hoegh-Guldberg this book).

Coral Reefs and Climate Change – The effects of climate change have become more evident in both the field and the scientific literature, and there is an increased focus on two particular issues: increased seawater temperatures and decreased ocean pH, often referred to as ocean acidification. Global warming attributed to "greenhouse gases" has been manifested by mass coral bleaching events and field data already show that corals are exhibiting decreased calcification and growth rates that correspond with reductions in seawater pH. A number of researchers and modelers have argued that reduced oceanic pH and aragonite saturation state present the greatest threat to the future of coral reefs (Kleypas et al. 2006; Veron 2008). Experiments on corals have shown that elevated global CO_2 levels approaching 450 ppm will indeed reduce calcification rates (Fine and Tchernov 2007; Jokiel et al. 2008). Data on coral growth from the Great Barrier Reef have already demonstrated reduced calcification rates over the past decade (De'ath et al. 2009). The effects of lowered pH from an increase in atmospheric CO_2 are predicted to be substantial as early as 2040. While decadal predictions may be debatable, the effects of pH on aragonite saturation levels are not.

Considering the numerous threats facing coral reefs in space and time, the impacts of global climate change will be the most difficult to address, as these require international cooperation at unprecedented levels of commitment and sacrifices by the present generation to insure that resources exist for the future generations. Additionally, populations with the smallest contribution to climate change are among those at the greatest risk from resource losses and sea level changes.

Ecosystem Responses to Stress – While there are numerous sources of stress affecting coral reefs, several priority areas have been identified by researchers, managers, policymakers, and stakeholders, and include land-based sources of pollution, overexploitation of fisheries resources, and the effects of global climate change. Traditional coral reef assessment and monitoring efforts have focused on population-related parameters including coral cover, fish/invertebrate abundance, and species diversity, with mortality as the major metric of change, that is, the loss of individuals, species, or percent cover. Using a human health analogy, death is a very crude estimator of stress and does not allow for management intervention in a timely or effective manner. For that reason, techniques allowing for the detection of stress at sublethal levels and which identify cause-and-effect relationships are critical for addressing the disturbing trend of coral reef decline documented in the recent literature and through coordinated efforts such as the global status and trends series (Hughes and Connell 1999; Wilkinson 2004; Pandolfi et al. 2005).

Land-based Sources of Pollution – Coral reef researchers and managers have been paying more attention to activities within watershed adjacent to coral reef ecosystems, with an understanding that land-based sources of pollution are among the most serious sources of stress to coastal reefs. Watershed discharges have been documented to reach reefs over 100 km from shore on the Great Barrier Reef of Australia, affecting the key parameters of water and substratum quality, and hence, biological parameters of coral reproduction and recruitment (Wolanski et al. 2004; Richmond et al. 2007). Such discharges result in multiple stressors including physical stress from sediments (burial and light attenuation), chemical stress from a variety of xenobiotics (e.g., polycyclic aromatic hydrocarbons – PAHs, pesticides, heavy metals, pharmaceuticals and endocrine disruptors). In such multistressor systems, it becomes critically important to determine cause-and-effect relationships at sublethal levels, when management intervention has the greatest potential for positive outcomes. Emerging technologies including the use of molecular biomarkers of exposure, such as the upregulation of specific proteins (e.g., cytochromes P450, multi-xenobiotic resistance protein, superoxide dismutases and ubiquitin) indicative of specific classes of stressors, can be used to guide management activities and help measure their efficacy (Downs et al. 2006).

Coral Diseases – A number of diseases have been found to affect corals, and these may be caused by fungi (Aspergillosis), bacteria (white plague), or consortia of microbial associates (black band disease). The causes of some other syndromes are presently unknown (e.g., yellow band disease), with new techniques and international efforts being applied to determine the pathologies and options to prevent these from spreading (Bruno et al. 2003). A Coral Health and Disease Consortium was formed in 2002 with the subsequent development of an International Registry of Coral Pathology (see http://www.coral.noaa.gov/coral_disease/cdhc.shtml). The potential for the transmission of diseases among affected corals and sites has raised concerns, and models from medicine are being applied to develop appropriate protocols, such as disinfecting equipment, wet suits and other possible vectors when divers move among coral reef sites. Molecular techniques including DNA sequencing are being applied to help identify and understand the etiology of coral diseases, with the goal of prevention and intervention. Diseases are one manifestation of stress in corals, and efforts to address physical and chemical stressors may help reduce the impacts of pathogens and other biological stressors on coral reef ecosystems.

4 Future Horizons

New Tools – New tools support both field research (mixed gas diving) and laboratory studies (genomics and proteomics). Coral reef ecosystem modeling (using either agent-based models or deterministic, analytical models such as the HOME model) is allowing rapid advances in our understanding of population dynamics of coral reef organisms, trophic relationships, and numerous reef processes (see also Arias-Gonzalez et al., this book). Such models also provide a tool to quantify coral reef responses to human-induced stressors within the context of natural variability and in an integrated manner that includes the functioning of the adjacent ecosystems that comprise the coral reef, such as adjoining watersheds, mangroves, seagrass beds, and the coastal ocean. The results of modeling studies identify the importance of synergies between the components of coral reef ecosystems that compete internally for space and resources (e.g., corals, coralline algae, macroalgae, *Halimeda*, crown-of-thorns starfish, and herbivorous and carnivorous fish), the physical parameters of the surrounding waters (temperature, pH, sediments, turbidity and nutrients), and the external forcing elements, including land-use in upstream watersheds and the effects of climate change (ocean acidification and more frequent and intense warming events associated with the El Niño-La Nina fluctuations; Pandolfi et al. 2005; Hoegh-Guldberg et al. 2007; Baker et al. 2008; De'ath et al. 2009; Semesi et al. 2009).

Bridging Science to Management and Policy – The need for management-directed science is pressing. The Great Barrier Reef Marine Park Authority (GBRMPA) Reef Outlook Report (2009) was the first of its type written to assess management performance in an accountable and transparent manner. It identified the following issues in order of priority as threatening the ecological resilience of the Great Barrier Reef: climate change, continued declining water quality from catchment runoff, loss of coastal habitats from development, and impacts from commercial and traditional fishing on threatened species (see Mumby and Steneck, this book). The overall outlook for coral reefs worldwide is poor, especially in view of climate change (Fig. 1) and it may not be possible to avert widespread and catastrophic damage. The same threats occur, at varying levels, for most coral reefs worldwide. The notable shift in research from purely academic to management-directed studies does not discount the critical importance of basic research, but rather, demonstrates the clear understanding by the coral research community that these incredible ecosystems are in dire straits worldwide, and science is an essential element in guiding management responses to insure a legacy of sound, functional reefs for future generations.

While the validity of science is confirmed through the peer-review process and publication, the management value of data often depends on a different metric: will they hold up in court? Most scientists would agree that the courtroom is not the ideal place to "do science," yet there are times when that is exactly what is needed to advance coral reef management and conservation actions and initiatives. Additionally, the scientific and management perspectives of "uncertainty" are quite different, with the former practitioners usually accepting only 95–99% certainty (p = 0.05 or 0.001, respectively), whereas a manager may be comfortable making a permit decision with a 70% degree of certainty if the default by the deadline is permitting a potentially damaging activity.

Whether science is presented in the courtroom, at public hearings, policy forums, in the classroom or at scientific symposia, communication skills are essential to effective outcomes. More effort is clearly needed in helping scientists effectively and appropriately explain their work, and provide context that makes their important findings relevant to the broader community. Several organizations including SeaWeb (www.seaweb.org), the Communication Partnership for Science and the Sea (COMPASS; www.compassonline.org), and many scientific societies provide formal media and communication training that can help scientists move beyond outputs (publications and workshops) to outcomes (increased coral cover and fish populations on reefs).

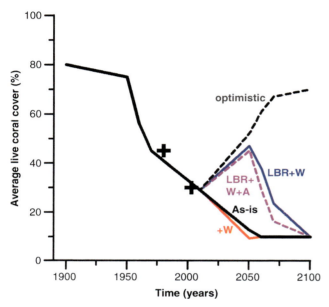

Fig. 1 Time-series plot of the predicted average coral cover of 261 reefs in the central region of the Great Barrier Reef between Bowen and Lizard Island, using the HOME ecohydrology model (Wolanski et al. 2004; Wolanski and De'ath 2005) for the five scenarios: As-is – present land-use; +W – present land-use plus global warming (IPCC scenario A2); LBR + W: land-use activities halving the outflow of nutrients and fine sediment, plus global warming; LBR + W + A: scenario LBR + W plus ocean acidification; "Optimistic" – effective management reductions in local stressors such as land-based sources of pollution in combination with substantial reductions in global climate change effects. The ecology element of the model is determined by herbivorous fish, algae, corals and crown-of-thorns starfish (COTS). The ecosystem is forced by (1) fine sediment from runoff, (2) nutrients and turbidity increasing as a result of land use and transient river floods, (3) occasional tropical cyclones (hurricanes), and (4) bleaching events in the summer months attributable to climate change (warming and ocean acidification) (Field data points (+) from Bruno and Selig (2007)). Coral cover could be less than predicted if flood or hurricane frequency and/or intensity increase, and if the coral death rate in the model from bleaching from future warming events is greater than that experienced in the two GBR coral mass bleaching events that occurred during the last 10 years. This mortality was much less than that observed in other places such as the Seychelles in the 1998 global coral bleaching event. The model does not include biological adaptation, for which there are no data

Coral Reef Valuation –The loss of ecosystem services provided by coral reefs has serious economic consequences as well as ecological and cultural ones, and thus the ability to put a value on coral reefs and related resources is essential for sound policy development and guidance. This is indicative of the broader and growing importance of the social science disciplines as contributors to the knowledge base established by biologists, physical oceanographers, and geologists. Coral reefs have economic and cultural value, and there is a need to quantify their monetary values if policy-makers are to take coral reef conservation seriously. This is paradigm shift for biophysical and socio-economic sciences on land and at sea because historically and practically, any economic development that contributes to Gross Domestic Product (GDP) has been seen as a good thing, regardless of sustainability and long-term effects on the quality of human life. The paradigm shift is occurring because natural and social capital are now recognized as the limiting factors to sustainable human well-being (Millennium Ecosystem Assessment 2005; Costanza et al. 1997, 2008a). In this paradigm shift, coral reef biophysicists and social scientists focus on prioritizing developments that improve the quality of human life within the carrying capacity of the supporting ecosystems, taking into account the ecological footprint of the built capital and maximizing the desirable social capital, which includes culture, equity, social networks, and participatory democracy. A measurement of that could be the Genuine Progress Indicator (GPI), or something similar, instead of the GDP (Costanza 2009; Costanza et al. 2009). There have been partially successful attempts to generate such a paradigm shift for coastal wetlands using their value for hurricane protection (Costanza et al. 2008b), but a similar paradigm shift for coral reefs in the western world is still needed although it did exist, and it partially still exists, in a few traditional communities in Micronesia and elsewhere (Richmond et al. 2007). Coral reef valuation is an area that is still developing and advances are critical to sound policy development and implementation that address the needs of both present and future generations.

Integration of Disciplines – As the knowledge based on coral reefs continues to expand, so has the need for specialization. Using the human health model, documented declines in coral reef state, resilience and persistence require more expertise than can be provided by a general practitioner, e.g., the coral reef ecologist, and require the attention of geochemists, physiologists, geneticists, ecotoxicologists, ichthyologists, algologists, statisticians, modelers, and physical oceanographers. Moreover, there is a need for the inclusion of the social sciences for the application of knowledge and to develop implementation strategies, educators to inform stakeholders, economists to develop a realistic framework for policy-makers, managers to address regulatory issues and enforcement, and environmental lawyers able to ensure the judicial process works when needed. At present, coral reef assessment and monitoring programs still largely depend on mortality as the metric of change, that is, the loss of species and reductions in coral cover or fish population size. Simply put, mortality is an ineffective metric of stress, when the desired outcome is health. Using the human health analogy, it is better to determine and treat the effects of stress or injury prior to death. Research tools developed in the biomedical sciences are now being used to diagnose and treat environmental health disorders (Downs et al. 2005). Corals have been found to have steroidal-based metabolic functions, and respond to exposures to a variety of toxicants including endocrine disruptors. Interactions between coral reef scientists and medical schools/practitioners are already bearing fruit and biomedical research is a ripe area for additional collaborations.

Future Scenarios – Figure 1 shows five possible scenarios, only one of which is acceptable. While the actual dates may be arguable, the trajectories are not, based on the best available science. It is abundantly clear that the only hope for coral reefs, and the people who depend on these magnificent ecosystems for their ecological, economic, and cultural services, as well as outright beauty, is a combination of immediate and effective local and global actions. While human communities living adjacent to coral reefs can undertake the local actions, the predicted outcomes will eventually end up in catastrophic losses unless distant populations address the issue of global climate change. It is understandable for someone far removed from coral reef ecosystems to lack concern, but this is unfortunately unacceptable if there is to be a viable and vital legacy for future generations. The hope lies in the "optimistic" curve that through good science, policy, education, collaboration, and implementation, there is an option that coral reefs and human kind can both co-exist far into the future.

Acknowledgments The authors gratefully acknowledge research support for elements of this chapter from NOAA/CSCOR, and a Pew Fellowship in Marine Conservation to RHR. This chapter is dedicated to the memory of Prof. Kiyoshi Yamazato, a pioneer in coral reef research and its applications, and an advocate for international cooperation and collaboration in supporting coral reef conservation.

References

Baker AC, Rowan R (1997) Diversity of symbiotic dinoflagellates (zooxanthellae) in scleractinian corals of the Caribbean and Eastern Pacific. Proc 8th Int Coral Reef Symp 2:1301–1306

Baker AC, Glynn PW, Riegl B (2008) Climate change and coral reef bleaching: an ecological assessment of long-term impacts, recovery trends and future outlook. Estuar Coast Shelf Sci 80:435–471

Beckwith MW (1951) The Kumulipo. A Hawaiian creation chant University of Chicago Press, Chicago p 257

Birkeland CE (1982) Terrestrial runoff as a cause of outbreaks of *Acanthaster plancii* (Echinodermata: Asteroidea). Mar Biol 69:175–185

Boschma H (1925) The nature of the association between Anthozoa and zooxanthellae. Proc Natl Acad Sci U S A 11:65–67

Bruno JF, Selig ER (2007) Regional decline of coral cover in the Indo-Pacific: timing, extent, and subregional comparisons. PLoS ONE 2(8):e711. doi:10.1371/journal.pone.0000711

Bruno JF, Petes LE, Harvell CD, Hettinger A (2003) Nutrient enrichment can increase the severity of coral diseases. Ecol Lett 6:1056–1061

Cinner JE, Marnane MJ, McClanahan TR (2005) Conservation and community benefits from traditional coral reef management at Ahus Island, Papua New Guinea. Conserv Biol 19:1714–1723

Costanza R (2009) A new development model for a 'Full' world. Development 52:369–376

Costanza R, D'Arge R, de Groot R, Farber S, Grasso M, Hannon B, Naeem S, Limburg K, Paruelo J, O'Neill RV, Raskin R, Sutton P, van den Belt M (1997) The value of the world's ecosystem services and natural capital. Nature 387:253–260

Costanza R, Fisher B, Ali S, Beer C, Bond L, Boumans R, Danigelis NL, Dickinson J, Elliott C, Farley J, Elliott Gayer D, MacDonald Glenn L, Hudspeth TR, Mahoney DF, McCahill L, McIntosh B, Reed B, Abu Turab Rizvi S, Rizzo DM, Simpatico T, Snapp R (2008a) An integrative approach to quality of life measurement, research, and policy. Surv Perspect Integr Environ Soc 1:11–15

Costanza R, Pérez-Maqueo O, Martinez ML, Sutton P, Anderson SJ, Mulder K (2008b) The value of coastal wetlands for hurricane protection. Ambio 37:241–248

Costanza R, Hart M, Posner S, Talberth J (2009) Beyond GDP: the need for new measures of progress. The Pardee Papers 4, Boston University Press, Boston

De'ath G, Lough JM, Fabricius KE (2009) Declining coral calcification on the Great Barrier Reef. Science 323:116–119

DeVantier LM, Done TJ (2007) Inferring past outbreaks of the Crown-of-Thorns Seastar from scar patterns on coral heads, chap. 4. In: Aronson R (ed) Geological approaches to coral reef ecology. Ecolog Stud 192:85–125. Springer, New York, p 439

Downs CA, Woodley CM, Richmond RH, Lanning LL, Owen R (2005) Shifting the paradigm for coral-reef 'Health' assessment. Mar Pollut Bull 51:486–494

Downs CA, Richmond RH, Mendiola WJ, Rougee L, Ostrander G (2006) Cellular physiological effects of the M.V. Kyowa Violet grounding on the hard coral Porites lobata. Environ Toxicol Chem 25:3171–3180

Downs C, Kramarsky-Winter E, Martinez J, Kushmaro A, Woodley CM, Loya Y, Ostrander GK (2009) Symbiophagy as a cellular mechanism for coral bleaching. Autophagy 5(2):211–216

Ehler C, Douvere F (2009). Marine spatial planning: a step-by-step approach toward ecosystem-based management, p 99. Intergovernmental Oceanographic Commission and Man and the Biosphere Programme. IOC Manual and Guides No. 53, ICAM Dossier No. 6. UNESCO, Paris

Fautin DG (2002) Beyond Darwin: coral reef research in the twentieth century, pp 446–449. In: Benson KR, Rehbock PF (eds) Oceanographic history, The Pacific and beyond. University of Washington Press, Washington, DC, p 556

Fine M, Tchernov D (2007) Scleractinian coral species survive and recover from decalcification. Science 315:1811

Goreau TF (1959) The physiology of skeleton formation in corals. I. A method for measuring the rate of calcium deposition by corals under different conditions. Biol Bull 116:59–75

Goreau TF, Goreau N (1959) The physiology of skeleton formation in corals. II. Calcium deposition by hermatypic corals under various conditions in the reef. Biol Bull 117:239–250

Harrison PL, Babcock RC, Bull GD et al (1984) Mass spawning in tropical reef corals. Science 223:1186–1189

Hoegh-Guldberg O, Mumby PJ, Hooten AJ, Steneck RS, Greenfield P, Gomez E, Harvell CD, Sale PF, Edwards AJ, Caldeira K, Knowlton N, Eakin CM, Iglesias-Prieto R, Muthiga N, Bradbury RH, Dubi A, Hatziolos ME (2007) Coral Reefs under rapid climate change and ocean acidification. Science 318:1737–1742

Hughes TP, Connell J (1999) Multiple stressors on coral reefs: A long-term perspective. Limnol Oceanogr 44:932–940

Johannes RE (1981) Words of the Lagoon: fishing and Marine Lore in the Palau District of Micronesia. University of California Press, Berkeley, p 245

Johannes RE (1997) Traditional coral-reef fisheries management. In: Birkeland C (ed) Life and death of coral reefs. Chapman and Hall, New York, pp 380–385, 536

Jokiel P, Rodgers KS, Kuffner IB et al (2008) Ocean acidification and calcifying reef organisms: a mesocosm investigation. Coral Reefs 27:473–483

Kinzie RA (2002) Caribbean contributions to coral reef science, pp 450–457. In: Benson KR, Rehbock PF (eds) Oceanographic history, The Pacific and beyond. University of Washington Press, Washington, DC, p 556

Kleypas JA, Feely RA, Fabry VJ, Langdon C, Sabine CL, Robbins LL (2006) Impacts of ocean acidification on coral reefs and other marine calcifiers: a guide for future research. Report of a workshop, 18–20 April 2005. St. Petersburg, FL, sponsored by NSF, NOAA, and the US Geological Survey. http://www.ucar.edu/communications/Final_acidification.pdf

Mather P (2002). From steady state to stochastic systems: the revolution in coral reef biology, pp. 458–467 In: Benson KR, Rehbock PF (eds) Oceanographic history, The Pacific and beyond. University of Washington Press, Washington, DC, p 556

Millennium Ecosystem Assessment (2005) Ecosystems and human well-being: synthesis. Island Press

Pandolfi JM, Jackson JBC, Baron N, Bradbury RH, Guzman HM, Hughes TP, Kappel CV, Micheli F, Ogden JC, Possingham HP, Sala E (2005) Are US coral reefs on the slippery slope to slime? Science 307(5716):1725–1726

Richmond RH (1993) Coral reefs: present problems and future concerns resulting from anthropogenic disturbance. Am Zool 33:524–536

Richmond RH, Rongo T, Golbuu Y, Victor S, Idechong N, Davis G, Kostka W, Neth L, Hamnett M, Wolanski E (2007) Watersheds and coral reefs: Conservation science, policy and implementation. Bioscience 57:598–607

Rowan R, Knowlton N (1995) Intraspecific diversity and ecological zonation in coral-algal symbiosis. Proc Natl Acad Sci U S A 92:2850–2854

Salvat B (2002) Coral reefs, science and politics: relationships and criteria for decisions over two centuries – a French case history, pp. 468–478. In: Benson KR, Rehbock PF (eds) Oceanographic history, The Pacific and beyond. University of Washington Press, Washington, DC, p 556

Semesi S, Kangwe J, Bjork M (2009) Alterations in seawater pH and CO2 affect calcification and photosynthesis in the tropical coralline alga, Hydrolithon sp. (Rhodophyta). Estuar Coast Shelf Sci 84:337–341

Veron JEN (2008) A reef in time. Harvard University Press, Cambridge

Wilkinson C (2004) Status of coral reefs of the world: 2004, vol 1 and 2. Australian Institute of Marine Science Townsville, 557

Wolanski E, De'ath G (2005) Predicting the present and future human impact on the Great Barrier Reef. Estuar Coast Shelf Sci 64:504–508

Wolanski E, Richmond R, McCook L, Sweatman H (2003) Mud, marine snow and coral reefs. Am Sci 91:44–51

Wolanski E, Richmond RH, McCook L (2004) A model of the effects of land-based, human activities on the health of coral reefs in the Great Barrier Reef and in Fouha Bay, Guam, Micronesia. J Mar Syst 46:133–144

Part II
Geology and Evolution

The Paleoecology of Coral Reefs

John M. Pandolfi

Abstract Reefs are one of the oldest ecosystems in the world, and coral reefs have had a rich and varied history over hundreds of millions of years. The long-term history of living reef organisms provides an essential window in which to view a number of fundamental evolutionary and ecological processes over extended time frames not available to modern ecology over years or decades. Many of the constituents of modern reefs are calcifying organisms that leave a record of their presence in the fossil record. Thus, coral reef paleoecology has been undertaken on tropical ecosystems worldwide with applications in ecology, evolution, biogeography, extinction risk, conservation and management, and global change biology. Because many reef organisms secrete their calcareous skeletons at or near isotopic equilibrium with ambient seawater, they have also been used to reconstruct environmental conditions over long time frames. The examination of ecological and evolutionary change in the context of environmental variability provides an ideal framework for understanding coral reef paleoecology and placing the modern biodiversity crisis in an historical context.

Keywords Coral reefs • paleoecology • reef management • ecology • evolution • global change • biodiversity • evolutionary turnover • biogeography • conservation biology

1 Introduction

To those not familiar with the nuances of time and geology, mining the literature on the fossil record of extant groups of marine organisms can seem a daunting pursuit. In many cases, this has led to a reduced appreciation of the utility of placing modern empirical studies on the ecology of living organisms into a historically informed context. As in all facets of life, ignoring the lessons of history is done at one's own peril. In this chapter, I briefly review the salient aspects of coral reef paleoecology that help to place modern studies into a historical framework. I mostly concentrate on the ecology of coral reefs of the Quaternary Period (over the past 1.8 million years – myr), but where it becomes useful some examples are drawn from reefs that lived during much older intervals.

1.1 What Is Paleoecology?

Paleoecology is the study of the distribution and abundance of organisms based on their remains from the fossil record. What constitutes a fossil has been much debated, but in general fossils are the remains of organisms and their activity that have been preserved within the sedimentary or rock record.

Predictive ecology is a central but elusive goal for experimental ecology on modern ecosystems, and this can hold even more so when confronting the vagaries of the fossil record. The fossil record has the advantage of recording the state of the ecosystem during repetitive time periods and thus under various combinations of environmental factors; however, processes must always be inferred from "natural experiments" where certain conditions have fortuitously been constant while a single or small number of factors vary. In this way, patterns derived from the fossil record can be used constructively for inferring potential processes acting over extended time periods.

1.2 A Brief History of Reefs

While the definition of just what is a "reef" has a long and tortuous history of debate in the scientific literature (Hatcher 1997), there is no dispute about the importance of the reef ecosystems of the coastlines, continental shelves, and ocean provinces of the tropical realm. Hatcher (1997) defines reefs

J.M. Pandolfi (✉)
ARC Centre of Excellence for Coral Reef Studies, Centre for Marine Science, School of Biological Sciences, University of Queensland, 4072St. Lucia, Queensland, 4072, Australia
e-mail: j.pandolfi@uq.edu.au

as "…a marine limestone structure built by calcium-carbonate secreting organisms, which, with its associated water volumes supports a diverse community of predominantly tropical affinities, at a higher density of biomass than the surrounding ocean…" (p. S78). Reefs in their many forms are found throughout the fossil record and represent some of the earliest structure-forming ecosystems on Earth. Since the explosion of metazoans in the Cambrian around 540 million years ago (mya), many groups of organisms have formed "reef-like" features on the seafloor making reef communities difficult to singularly characterize (Wood 1999). *Coral* reefs are one of the oldest reef systems on Earth. Following the greatest extinction of all time, the Permo-Triassic event (251 mya), scleractinian corals, bivalve molluscs, and crustose coralline algae have dominated the construction of wave-resistant organic carbonate structures on the planet.

1.3 The Past, The Present, and The Future

All students of geology are indoctrinated with Hutton's famous principle that "the present is the key to the past," the "Law of Uniformitarianism." However, it has become increasingly clear over the past couple of decades that the past can also provide key insights into modern processes, especially in predicting future likely ecosystem states. Studies of the paleoecology of reefs have provided some interesting and useful data with which to place the modern biodiversity crisis in some perspective, identifying the degree to which reefs have been subject to the shifting baseline syndrome (Pauly 1995; Pandolfi and Jackson 2006), and continue to offer ecological insight into the potential future health of modern ecosystems in the light of predicted environmental change.

2 Constraints and Influences over Coral Reef Development

2.1 Local Controls

The growth of living coral reefs is intimately tied to environmental conditions. Local controls over reef growth include wave energy, water quality, turbidity, salinity, tidal regime, and light. In the Caribbean Sea, clearly defined zones that characterize both modern and Pleistocene assemblages could reliably be predicted based on their spatial distribution with respect to wave energy (Geister 1977). Early efforts to understand the relationship between coral growth and local environmental parameters were almost exclusively concentrated on fossil sequences (Adey 1975, 1978; Geister 1977; Chappell 1980) and have been confirmed with more quantitative studies on both modern (Done 1982) and fossil (Pandolfi and Jackson 2001) taxa. Ecophenotypic plasticity in response to wave energy and light was long thought to be the main driver of differential reef growth but much earlier work focusing on environmental controls (Graus and Macintyre 1976, 1982; Connell 1978) turned out to have a largely genetic component (Knowlton et al. 1992), and the full implications for the fossil record have yet to be investigated (Pandolfi et al. 2002; Budd and Pandolfi 2004; Pandolfi and Budd 2008).

2.2 Regional and Global: Secular

Physical controls on reef building and decline, over geologic timescales, include variations in seawater chemistry and cyclic changes (at varying time scales) in sea level, sea-surface temperatures, and global levels of atmospheric CO_2. These factors are necessarily interrelated. Regional and global variation in sea-surface temperatures (SST), CO_2, and sea level have continued to exert substantial influences over coral reef growth during the Cenozoic, all the way through to the Holocene period (Montaggioni 2005). However, a long-term database on reef development throughout the past 540 million years (my) of the Phanerozoic Eon shows that, whereas long-term climate change is influential in the biotic composition of reefs, neither climate nor chemical changes in the sea are correlated with the waxing and waning of reefs (reviewed in Kiessling 2009).

2.3 Latitudinal Range Limits

A large number of factors can contribute to latitudinal variability in coral species composition and community structure: water temperature, aragonite saturation, light availability, currents and larval dispersal, competition between corals and other biota including macroalgae, reduced coral growth rates, and failure of coral reproduction or recruitment (summarized in Harriott and Banks 2002). Living reef-building corals are generally confined to the tropical oceans between 30°N and 30°S latitudes, and these limits often characterized reefs in the past (Copper 1994; Kiessling 2001). In the Silurian period (443–417 mya), tropical reefs closely matched the geographic distribution of their modern counterparts (Copper 1994). But over the broad swath of the Phanerozoic, there is no strict correlation between the geographic distribution of reef growth and tropical placement (Kiessling 2001, 2005).

2.4 Biotic Factors

Paleontologists have used ancient reefs as model ecosystems in the study of ecological succession, or the orderly changes in communities of a specific place over a period of time (Masse and Montaggioni 2001). For ecologists, one of the great limitations of assessing the frequency or validity of the processes of succession is that the complete successional sequence might take longer than the period of time over which most ecosystems are studied within ecological research programs. Since succession is by its very nature a process that occurs over time, it is not surprising that paleontologists have investigated the degree to which it can be understood from evidence in the fossil record.

Four major phases of succession in eight ancient reefs ranging in age from the Early Ordovician (488 mya) to the Late Cretaceous (65.5 mya) have been established: stabilization, colonization, diversification, and domination (Walker and Alberstadt 1975). The stabilization stage involves the initial colonization of the sea floor that results in the establishment of a firm substrate. The colonization stage is characterized by encrusters and frame builders, those organisms capable of colonizing a hard substrate and that begin to build three-dimensional structures above the sediment–water interface. Some authors consider the first two of these stages to be the same, equivalent to the pioneering stage of ecologists studying succession in living forest communities. The third stage in the succession of fossil reefs is referred to as the diversification stage, where maximum diversity of the reef community is developed. Diversity then decreases as a "domination" stage sets in whereby a single functional entity characterizes the reef community. As the communities ascend through the first three stages of succession, species diversity, degree of stratification and pattern diversity, niche specialization, and the complexity of food chains all increase. These successional sequences typically occur through significant intervals of time and are the result of the gradual alteration of the submarine substratum by individual species, and the elaboration of energy-flow pathways as the community proceeds through time. Shorter-term successional episodes have also been observed where occasional environmental perturbations destroy the pre-existing community. Here, subsequent community development involves a rapid biologically induced sequence of communities that culminate in the climax community that was established prior to the perturbation. Just as in the longer-term successional sequences, increases in major functional and structural attributes of the community accompany these shorter-term successional sequences.

There are, of course, a number of potential problems associated with the identification of succession in the fossil record. In Walker and Alberstadt's (1975) model, an increase in diversity may come about just as easily from an increase in the zonation of the reef (habitat heterogeneity) as to intrinsic biological changes. In the domination phase, the accompanying decrease in diversity may reflect a community that caps a reef after it dies, as opposed to a decrease during the life of the reef. However, in the fossil record of the Quaternary period (the past 1.8 Myr), these difficulties can often be overcome through precise age dating and detailed understanding of the sea-level history of the reefs. Quaternary sequences of coral reefs are particularly well suited for understanding community development through time (Mesolella 1967; Geister 1977; Crame 1980).

In Kenya, "obligatory succession" in the Pleistocene consisted of an early assemblage of sediment-tolerant corals (dominated by massive or doming corals) that was replaced by predominantly branching and platy/encrusting corals (Crame 1980). This kind of succession was mainly confined to the earliest portions of reef development on a bare substrate. Patterns through longer successional phases were varied, but under certain conditions massive or domed shape corals might replace the branching assemblages. However, clearly defined zones were rare – most temporal changes in reef species associations were random or unstructured.

2.5 Autecology of Reef Organisms

Autecological studies have also provided glimpses into biological processes occurring in the past. For example, Kaufman (1981) found galls on the branches of Pleistocene *Acropora* colonies indicating the antiquity of threespot damselfish (*Eupomacentrus planifrons*) territories on Caribbean coral reefs. While the presence of modern processes on past reefs should not be surprising, it is important for our predictive capacity of modern reefs to understand the temporal framework over which such processes have operated, how susceptible they might be to varying disturbance regimes and environmental change, and how their role in the maintenance of species diversity varied through time. The effects of some processes might be long-term or involve significant lag times, so their real significance might only be borne out through studies conducted over broad temporal scales.

The principal features of fossil coral reefs, like their modern relatives, lie in their ability to build three-dimensional structures above their surrounding substrate. While we are concentrating on "coral" reef paleoecology for this chapter, it is important to realize that reefs have not always been built by corals, and that corals have come and gone in their function as dominant frame builders throughout the geological record. In the Cretaceous Period (144–65 mya), rudist bivalves once formed lush and diverse reefs and may have even competitively displaced corals from their role as reef framework builders. However, others dispute this prominent role and believe that rudists were confined to nonreefal, mobile sedimentary settings (Gili et al. 1995). Stromatolites have built wave-resistant structures through the entire Phanerozoic Eon (543 mya to present).

Within the corals themselves, there are vast amounts of untapped information on autecology that may help us uncover the nature of biodiversity crises in the sea. For example, Kiessling et al. (2005) found abundant massive colonies of *Haimesiastraea Conferta,* occurring in the Paleocene of the Lefipán Formation of Argentina. This coral made it through the Cretaceous-Tertiary mass extinction event, presumably as a result of a broad ecological niche and a wide geographic range. Further work on the characteristics of extinction-resistant reef taxa will aid in our ability to predict response of living coral reefs to environmental change.

Attention has been given to the analysis of growth rates of living corals from the colony level all the way down to the microstructural skeletal elements (Dullo 2005). Commonly, linear extension rates are compared, but other measures, including density and biomass, can be used when linear extension does not apply or is an inadequate descriptor of coral growth. Fossil coral growth rate studies are not as common, but have been undertaken on some Caribbean corals (Pandolfi et al. 2002; Dullo 2005; Johnson and Perez 2006). There is clearly more work needing completion to understand environmental controls over reef accretion and variation in coral growth rates over multiple environmental and anthropogenic regimes (Guzman et al. 2008).

3 Reef Paleoproductivity

Primary productivity in marine ecosystems is governed largely by nutrient availability and for coral reefs the balance between carbonate production and bioerosion may teeter on small changes in nutrient availability. This is because many of the organisms that compete with corals thrive better in high nutrient situations – those atypical of most oligotrophic coral reefs. However, small changes in nutrient availability have been implicated in the turn on and demise of coral reefs throughout the geological history of reefs (Hallock and Schlager 1986; Wood 1993). Paleoproductivity is a difficult parameter to measure in fossil sequences and its detection in fossil reefs has met with only limited success (Edinger and Risk 1996).

4 Biotic Interactions

Coral reefs are renowned for their biodiversity as well as the numerous vertical and horizontal links that characterize that biodiversity. These processes include predation, competition, herbivory, and commensalism and can be traced back through the fossil record, providing important information on the evolutionary ecology of the reef ecosystem. Symbiosis is almost a defining characteristic of tropical reef communities. It almost certainly has been so for much of the geological history of the modern scleractinian reef-building corals (Stanley and Swart 1995; Muscatine et al. 2005). Despite the increased interest in corals as "holobionts," we have very little evidence of nonalgal symbiosis (endolithic algae and fungi, Bacteria, Archaea, and viruses) (Wegley et al. 2007) in the fossil record of coral reef organisms. Commensalism is also an age-old occurrence in corals, with first evidence between Tabulate corals and endosymbiont trace fossils in the Ordovician Period (490–443 mya) (Tapanila 2004).

The origins of herbivory and history of grazing in marine fishes was derived from analyses of functional morphospace (Bellwood 2003). The Cenozoic witnessed the appearance and proliferation of herbivory and grazing by marine fishes – opening the door for radical modification of benthic marine communities.

Competition among Pleistocene corals from the *Montastraea "annularis"* species group resulted in competitive release when one of the species went extinct (Pandolfi et al. 2002). *Montastraea nancyi* (Pandolfi 2007), with an organ-pipe growth form, dominated leeward shallow water habitats from the Caribbean Sea. It became extinct sometime between 82 and 3 ka (thousand years ago) (Pandolfi et al. 2001); in modern reefs, columnar *M. annularis* s.s. is dominant over other species in the complex in shallow water, whereas it was subordinate to *M. nancyi* prior to the latter's demise. Ecological release resulted in the dominance of *M. annularis* and a decrease in the diameter of its columns – toward the ecology and morphology of its extinct counterpart.

Detecting the influences of density dependence on processes in the marine realm is very difficult in the fossil record, first and foremost because population size is never really known. However, there are still clear instances of shifts in dominance of reef frame builders in the geologic past. One of the classic instances involved the alternation in dominance between the scleractinian corals and the rudist bivalves in the Cretaceous Period. Some workers believe that corals and rudist bivalves were locked in a competitive battle for dominance on Cretaceous reefs (Kauffman and Johnson 1988), but other authors refute that claim and interpret distributional trends on the basis of differential response to environmental change (Ross and Skelton 1993). Regardless, post-Cretaceous reefs were again dominated by the scleractinian corals, as the rudists never recovered from the Cretaceous–Tertiary (K–T) extinction event.

5 Paleo Community Ecology

Much of our knowledge about the long-term ecological dynamics of modern corals comes from studies of their counterparts in the Quaternary fossil record, that is the past two million years of Earth history. One of the most important questions in ecology is: how are species, and hence biodiversity, maintained in ecological communities? This question has always been elusive, at least in part because so much of ecology is confined to studies of living species at limited temporal scales. Recent studies of the paleoecology of corals use the recent past history (hundreds to hundreds of thousands of years) of living species to investigate ecological dynamics over meaningful time periods, to "re-run the tape" (Savage et al. 2000) for community assembly. Contrary to many shorter-term studies, which have evoked relative chaos in the community structure of living corals, persistence appears to characterize Pleistocene reef corals from both the Indo-Pacific (Pandolfi 1996, 1999) and the Caribbean (Pandolfi and Jackson 2006). Evidence for stability in coral community composition over broad time scales has also been found in both the older Pleistocene development of the Great Barrier Reef (Webster and Davies 2003) and the Caribbean (Klaus and Budd 2003), and the younger Holocene deposits of Belize and Panamá (Aronson and Precht 1997; Aronson et al. 2004). These results challenge conventional ecological views of coral reefs (derived mainly from studies at restricted spatial and temporal scales) as mainly disturbance-influenced ecosystems that contain species that are ephemeral in space and time (summarized in Karlson 2002). In fact, temporal persistence in community structure has been detected in a number of ecosystems throughout many intervals of geological time (DiMichele et al. 2004). Studies conducted on small spatial and temporal scales show the greatest degrees of variability in coral composition and those completed at the largest scales within biogeographic provinces show intermediate variance (Pandolfi 2002). However, studies conducted at intermediate spatial and temporal scales show high degrees of order in coral composition (Fig. 1) (Pandolfi 2002).

In the study of Pleistocene coral assemblages from the Huon Peninsula in Papua New Guinea (Pandolfi 1996), differences in reef coral community composition during successive high stands of sea level were greater among sites of the same age than among reefs of different ages, even though global changes in sea level, atmospheric CO_2 concentration, tropical benthic habitat area, and temperature varied at each high sea-level stand (Pandolfi 1999). One question this study raises is what communities might have looked like during low sea-level stands. Were they the same and so faunal persistence merely reflects "habitat tracking" (*sensu* Brett et al. 2007): "lateral migration of species or biofacies in response to shifting environments" p. 231) or were they different and the recurrent high-stand communities represent multiple niche assembly of similar communities? Tager et al. (2010) investigated this problem and showed that not only were low-stand communities distinct in coral species composition from their high-stand counterparts, but that successive low-stand coral assemblages, in contrast to the high-stand assemblages, showed directional change in coral community composition through time. Thus, it appears that simple habitat tracking does not provide a reasonable explanation for the ecological processes involved in sustaining similar ecological communities over successive intervals of re-assembly of high-stand reefs.

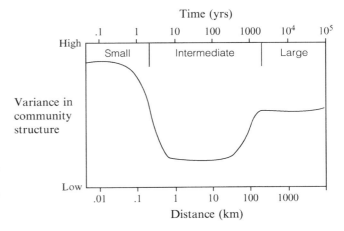

Fig. 1 Three-phase model for variability in coral community structure with respect to spatial and temporal scale of study. Studies conducted on small spatial and temporal scales show the greatest degrees of variability in coral composition and those completed at the largest scales within biogeographic provinces show intermediate variance. However, studies conducted at intermediate spatial and temporal scales show high degrees of order in coral composition (From Pandolfi 2002)

Perhaps, one of the most significant advances of the past 50 years of ecological research has been our ability to formulate null models to help explain species distribution patterns (Gotelli and Graves 1996). One of the most influential of these has been the formulation of the Unified Theory of Biodiversity and Biogeography (Hubbell 1997, 2001), which predicts species abundance distribution using a small number of demographic variables. While simulations over time characterize tests of the theory, there are still relatively few studies that utilize the fossil record (Olszewski and Erwin 2004; McGill et al. 2005). Earlier studies of fossil corals (Pandolfi 1996, 1999) informed the theory to accommodate high levels of similarity in community structure through time [see (Chave 2004) for discussion of neutral theory maturation and fossil corals].

Niche theory has also been discussed in the context of the ecology of ancient reefs. Watkins (2000) interpreted patterns in morphology related to feeding (corallite size and spacing) in associations of Silurian (425 mya) tabulate corals as niche-partitioning, which was also used to explain the ecological dynamics of Pleistocene scleractinian corals (Pandolfi et al. 2002).

6 Global Change: Environmental Influences on Species Distribution Patterns

6.1 Reef Growth and Global Change

Some of the physical changes that are projected to occur in the coming century (Solomon et al. 2007) have occurred repeatedly throughout both the past two million years of the Quaternary period and in the more distant past (e.g. sea-level and temperature rise), while others have not (e.g. rates of CO_2 rise) (Pandolfi and Greenstein 2007). Reef coral communities in the distant past rebounded from decimation by climatic events that affected the global marine biota. Recovery intervals varied from 4 to 100 my, during which time framework building organisms were largely absent from reef ecosystems (Newell 1971). More recently, Quaternary coral reef development either proceeded undeterred throughout climatic changes or recovered so quickly as to leave no record of their demise (Pandolfi et al. 2006; Pandolfi and Greenstein 2007). One major difference between Quaternary reefs and those confronting climate changes in the coming decades is that today's reefs have been preconditioned by human impacts such that the frequency of anthropogenic disturbance might have decreased their resilience to perturbations (Hughes et al. 2003; Pandolfi et al. 2006). Another is that the rates and magnitude of CO_2 concentration change appear to be unprecedented over millions of years.

6.2 Range Expansions/Contractions

One of the most critical challenges facing ecologists today is to understand the changing geographic distribution of species in response to current and predicted global warming. Range movements in living coral communities have begun to be documented (Marsh 1993; Vargas-Angel et al. 2003; Precht and Aronson 2004). However, it is not yet clear whether such shifts represent ephemeral or more permanent change. The fossil record can be used to assess the effect of climate change on coral communities over a temporal scale unavailable to studies conducted solely on modern communities, to determine how labile reef coral species are in the face of sustained changes in climate.

Recent work in coastal Western Australia has provided preliminary data on how such range movements might affect the long-term ecological dynamics of coral reef ecosystems. Pleistocene coral reefs were established throughout the length of the Western Australia coastline for several reasons. First, sea-surface temperatures were at least 2°C warmer than today (Kendrick et al. 1991). Second, the Leeuwin Current, which bathes the coast of Western Australia with warm, relatively low salinity water derived from the western central Pacific via the Indonesian throughflow, was present and may have been more intense as a result of the southward migration of the west wind drift and subtropical convergence during Late Pleistocene time (Kendrick et al. 1991; McGowran et al. 1997; Li et al. 1999). Wells and Wells (1994) used transfer functions derived from planktic foraminifera to show that the Leeuwin Current was intensified during the last interglacial maximum (Marine Isotopic Substage 5e). Third, higher sea levels pushed warmer water down from the Indonesian throughflow facilitating the extension of the warm-water Leeuwin Current along the western Australia coastline. Hence, the gradient in community composition of Late Pleistocene reef corals exposed along the west Australian coast was not as strong as now occurs in adjacent modern reef coral communities, which show a pronounced gradient in coral composition over their latitudinal range (Greenstein and Pandolfi 2008).

Comparison of reef coral community composition between adjacent modern and fossil reefs along 10° of this environmental gradient revealed that coral taxa expanded their latitudinal ranges during warmer Late Pleistocene time compared with today. Tropical-adapted taxa contracted their ranges north since Late Pleistocene time as temperatures cooled, emplacing two biogeographic provinces in a region in which a single province had existed previously. Beta diversity values for adjacent communities also reflect this change. Modern reefs show a distinct peak in beta diversity in the middle of the region; the peak is not matched by Pleistocene reefs. Beta diversity is correlated with distance only for comparisons between modern reefs in the north and the fossil assemblages, further supporting change in distribution of the biogeographic provinces in the study area. Coral taxa present in modern communities clearly expanded and contracted their geographic ranges in response to climate change during the Pleistocene.

One interpretation from this work is that those taxa that distinguish Pleistocene from modern reefs are predicted to migrate south in response to future climate change, and potentially persist in "temperature refugia" as tropical reef communities farther north decline. However, sea-level changes and ocean acidification as well as anthropogenic impacts on habitat degradation need to be incorporated into such predictions.

7 Diversity Through Time: Evolutionary Ecology and Biotic Turnover

The analysis of taxonomic abundance data embedded within a detailed and precise environmental context is enabling paleontologists to rigorously explore the dynamics and underlying processes of ecological and evolutionary change in deep time (Klaus and Budd 2003; Jackson and Erwin 2006). To ensure success, paleontologists now work with other geologists to obtain extensive new collections of fossils tied to detailed and independent records of stratigraphic correlation and changes in climate, oceanography, tectonic events, and other aspects of the physical–chemical environment. Analysis of the history of diversity and evolutionary turnover of fossil corals provides context for understanding both the present standing diversity of modern corals and the rates and magnitudes of any future changes in coral diversity.

7.1 Cenozoic Patterns

Perhaps one of the best examples of the insights gained through such interdisciplinary synergy is the response of tropical marine communities to the oceanographic events associated with the rise of the Isthmus of Panama three to five million years ago (Jackson and Johnson 2001; O'Dea et al. 2007). Here, extinction occurred one to two million years after the environmental event, while the collapse of coastal upwelling and primary productivity corresponded with an upsurge in carbonate production in the tropical western Atlantic (Collins et al. 1996). Knowledge of the environmental events and corresponding development of tropical marine ecosystems in the Indo-West Pacific (IWP) is embryonic when compared with the detailed view of tropical ecosystem development for the Caribbean and the eastern Pacific (Jackson et al. 1996). However, the rise of the modern IWP biota is thought to have occurred in the critical interval spanning the transition from the Late Oligocene to the Early Miocene (Wilson 2002) during large fluctuations in global ice volume (Zachos et al. 2001), regional changes in tectonics, and major shifts in the genesis of carbonate build-ups. It has also been proposed (Renema et al. 2008; Williams and Duda 2008) that increased basin complexity, driven by tectonics, resulted in increased habitat availability and heterogeneity in tropical Neogene faunas, driving higher species origination rates. The unraveling of the interplay between biotic turnover, ecological change, and environmental history is fundamental to understanding the likely ecological response of modern tropical marine biotas to predicted global change.

Paleoecological insights into the timing and antiquity of modern-day biodiversity can also feed into genetic studies and conservation strategies. Many genetic studies on modern-day corals and other tropical marine organisms have focused on Pleistocene sea-level changes as speciation "pumps." However, new data from a number of recent molecular studies, as well as calibration curves and new fossil finds are revealing that Indo-Pacific marine diversity is much older, commonly extending millions of years into the geological past (Renema et al. 2008; Williams and Duda 2008).

Finally, investigations into the relationship between reef growth and biodiversity through time highlight the capacity for coral reefs to engage in reef growth as biodiversity loss occurs at local and regional scales on living coral reefs (Johnson et al. 2008). Modern ecologists can only speculate on the relationship between coral biodiversity and potential for reef growth. In fact, rates of reef growth are broadly similar among regions where coral diversity may vary up to tenfold (Montaggioni 2005), leading to the question of just how important diversity might be for reef development. Johnson et al. (2008) found that reef development did not correlate with coral diversity within the tropical western Atlantic over the past 28 my. Moreover, the largest reef tracts formed after extinction had reduced diversity by as much as 50%. Similar to the spatial patterns found on many modern reefs (Riegl 2001), these temporal patterns suggest that high coral diversity is not essential for the growth and persistence of coral reefs (see review in Kiessling 2009).

7.2 Deep Time

A complement to the specimen-based and genetic studies noted above are a considerable number of studies that have tracked the biodiversity of corals through various intervals of deep time, including the 540 my of the Phanerozoic Eon (summarized in Kiessling 2009). A significant number of challenges face coral paleontologists in estimating broad patterns of global diversity through time. Besides taxonomic uncertainties, there are mineralogical differences among the three major coral groups (Rugosa, Tabulata, Scleractinia). However, even though chemical alteration occurs frequently, especially within the scleractinian corals, who build their skeletons from the unstable aragonite, corals form a major group of fossilized organisms throughout the history of life. There is little evidence to suggest that the diversity of Paleozoic rugose and tabulate corals, which likely built their skeletons from calcite, was preferentially enriched by skeletal mineralogy in relation to their Mesozoic and Cenozoic scleractinian counterparts. Challenges for future work include quantifying patterns of biotic change in carefully dated and correlated stratigraphic sections in the context of multiscaled environmental change in order to test whether patterns of faunal turnover are random or clumped in time, and whether they correlate with tectonic or climatic events.

8 The History of Modern Biogeographic Patterns

Through study of the fossil record of the major carbonate-producing organisms, the biogeographic dynamics of reefs and their constituent organisms can be understood. For example, Miocene (24–5 mya) to Pleistocene (last 2 mya) interchange of molluscan species between biogeographic provinces in the subtropical equatorial Atlantic resulted in the enrichment of the regional and global species pool and the spread of adaptations reflecting intense competition and predation (Vermeij 2005).

Unlocking the history of modern geographic patterns of biodiversity relies on two fundamental data sets – the fossil record and molecular phylogeography. These data sets often appear in conflict, but recent work in the Indo-Australian Archipelago (IAA) has shown remarkable congruence that indicates a series of marine biodiversity "hotspots" in different places over the past 50 my (Fig. 2) (Renema et al. 2008). These "hopping hotspots" have a birth, life, and death that is closely tied to regional (tectonic) and global (paleoclimatic) environmental variation. Data were derived from alpha-diversity expressed in the fossil record of corals, large benthic foraminifera, mangroves, and molluscs, and from molecular studies of fish and molluscs. These combined paleontological/molecular data sets show that the modern biodiversity hotspot found in the present Indo-Australian Archipelago is not so much unique in its biodiversity as its present position (see also Harzhauser et al. 2007). The movement of the Earth's tectonic plates is strongly associated with hotspot history (Renema et al. 2008). Understanding the history of biodiversity hotspots informs us that there always have been, and probably always will be, hotspots; it is vital to understand why they form, what drives them, and how they maintain their integrity. For this, our only recourse is the fossil record.

One of the major controversies surrounding Indo-Pacific coral biodiversity hotspots has concerned the mechanism by which they are formed – in situ as a "center-of-origin" (e.g. Briggs 2003), as a "museum" where species accumulate (Jablonski et al. 2006), and how they arrive – rafting (Jokiel 1984), and the role of dispersal and vicariance (Pandolfi 1992); see Rosen (1988) for an earlier review. All of these theories and mechanisms can be informed through a richer understanding of the Indo-Pacific fossil record, and

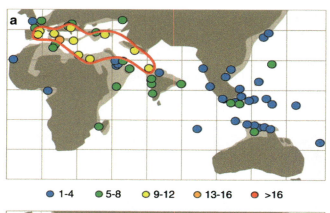
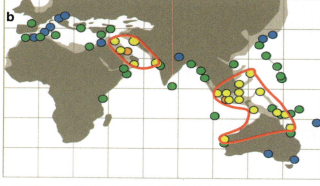
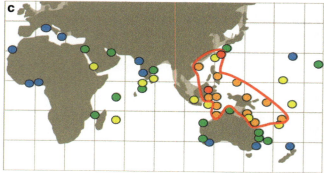

Fig. 2 Generic (alpha)-diversity of large benthic foraminifera in (**a**) the late Middle Eocene (42–39 ma), (**b**) the Early Miocene (23–16 ma), and (**c**) the Recent. Solid lines delimit the West Tethys, Arabian, and IAA biodiversity hotspots. Note the relocation of hotspots across the globe (From Renema et al. 2008)

future research in this area should prove to be especially insightful.

9 Reef Paleoecology, Historical Ecology, and Conservation Biology

The diversity, frequency, and scale of human impacts on coral reefs are increasing to the extent that reefs are threatened globally (Wilkinson 2008). Until recently, the direct and indirect effects of overfishing and pollution from agriculture and land development have been the major drivers of massive and accelerating decreases in abundance of coral reef species (Moberg and Folke 1999; Abram et al. 2001; Jackson et al. 2001; Gardner et al. 2003; Hughes et al. 2003; Pandolfi et al. 2003). These human impacts and the increased fragmentation of coral reef habitat are unprecedented and have the possibility to undermine reef resilience (Bellwood et al. 2004), raising the likelihood that modern coral reefs might be much more susceptible to current and future climate change than is suggested by their geologic history (Hughes et al. 2003; Bellwood et al. 2004).

One of the many pressing issues in conservation science is what was natural in the world before humans impacted it. By combining ecologists, oceanographers, archeologists, and social scientists, it has been possible to document global-scale changes in the sea (Jackson et al. 2001). The work documents in broad terms the history of human alterations in the sea, identifying overfishing as the first and farthest-reaching anthropogenic cause of decline in coastal marine ecosystems.

Very recent work has sought to understand human impacts by developing time-series data archives that can be used to evaluate trends in the global decline of coral reefs since the arrival of humans (Pandolfi et al. 2003). The approach has been to use a number of different kinds of data during several time periods to examine the recent past history and present condition of coral reefs to provide a natural baseline for community ecology and coral growth rates. Archeological sites provide insight into the relationship between the development of civilization and its evolving impact on coastal marine resources over thousands of years. Historical records such as those found in ships logs, and publications of early naturalists and European colonialists provide a moving window of the natural history and inferred ecology of many coral reef inhabitants over a centennial time scale. Fisheries records and modern ecological surveys can be used in association with remote sensing data (going back the last 20 or 30 years) to provide a detailed picture of changing environments and biodiversity as human population and consumption, as well as economic globalization, have accelerated during the past several decades. Taken together, these databases provide a holistic view of changing environments and ecology on coral reefs that includes the onset of human disturbances and against which the acquisition of present-day data can be evaluated.

Study of the historical ecology of coral reefs, relying on the past history of corals and other components of the reef ecosystem documented the history of ecosystem change among tropical reef sites worldwide (Pandolfi et al. 2003, 2005). These findings showed the importance of pre-industrial factors to reef health, and point to the immense importance of understanding historical events when attempting to tease out factors that have or may influence present coral reef biodiversity (Pandolfi et al. 2003). At 14 sites worldwide, early and effective overfishing appears to have been the major culprit in reef decline (Pandolfi et al. 2003). By considering these global "ecological trajectories" in the light of potential responses to projected global climate change, the history of human exploitation can be linked with global environmental parameters (Hughes et al. 2003). Knowledge of past ecosystem states provides at the minimum an understanding of what was natural and may even aid in assessing the success of management toward particular conservation goals.

An understanding of the history of reef coral communities can aid conservation initiatives by supplying information on natural baselines in the sea that are immune to the "shifting baseline syndrome" (Pauly 1995), where each new generation defines what it perceives to be "natural" or "pristine." For example, in Barbados, coral community composition was very similar during four separate reef-building episodes between 220 and 104 ka (Pandolfi and Jackson 2006), and community structure is now very different in modern degraded habitats (Lewis 1984; Tomascik and Sander 1987). Similar results have been obtained from sedimentary cores from Belize and Panamá, where recently observed declines in the abundance of *Acropora* show no historical precedence (Aronson and Precht 1997; Aronson et al. 2004), and from coral death assemblages along the Florida reef tract (Greenstein et al. 1998) and the nearshore Great Barrier Reef (Perry et al. 2008; Roff 2010). Perspectives on baselines that do not include this longer-term data will be shifted to degraded habitats.

10 Proxies for Environmental Change

Corals and other reef inhabitants have long been studied as proxies for environmental change in earth history. Early work (Wells 1963; Runcorn 1966) uncovered a way to determine that the number of days in the year decreased through the Phanerozoic (review in Hughes 1985). More recent work has attempted to correlate skeletal trace element and isotopic composition to ambient seawater. Such variables as sea-surface temperatures, salinity, turbidity, and even paleo pH (Hönisch et al. 2004) are currently being ascertained to reconstruct the environmental history of both nearshore and oceanic coral reefs. And to document changing physical environments, coring of reef corals provides a proxy for sea-surface temperatures, rainfall, and river discharge (McCulloch et al. 2003) from the geological past to the present. These studies have been extended back into the Quaternary, with work completed on both Pleistocene (Tudhope et al. 2001) and Holocene (Gagan et al. 1998) reef corals.

11 Summary

Coral reefs have a rich and illustrious history, going back to hundreds of millions of years. Living coral reefs have ancient counterparts that provide an almost continuous time series on scales ranging from millions to hundreds of years ago. Thus, the history of coral reefs can be reconstructed by examination of their skeletal remains and sedimentary record. Study of these deposits is the purview of coral reef paleoecology and much information has been gained, information that can be readily applied to a host of biological (and geological) problems and issues. Paleoecological study of coral reefs has informed modern ecology, environmental science, conservation biology, and climate change. In this chapter, we explore the rich history of contemporary coral reefs by summarizing the major paleoecological approaches applied to understanding patterns and processes over multiple time scales. Coral reef paleoecology is a vast and vibrant field with applications to a variety of natural history, evolutionary, ecological, environmental, and conservation fields. A comprehensive review could follow each of the major subjects I cover here, but this chapter is intended to provide a flavor for what has been and can be done in the paleoecological study of fossil coral reefs.

References

Abram NJ, Webster JM, Davies PJ, Dullo WC (2001) Biological response of coral reefs to sea surface temperature variation: evidence from the raised Holocene reefs of Kikai-jima (Ryukyu Islands, Japan). Coral Reefs 20:221–234

Adey WH (1975) The algal ridges and coral reefs of St. Croix their structure and Holocene development. Atoll Res Bull 187:1–67

Adey WH (1978) Coral reef morphogenesis: a multidimensional model. Science 202:831–837

Aronson RB, Precht WF (1997) Stasis, biological disturbance, and community structure of a Holocene coral reef. Paleobiology 23:326–346

Aronson RB, MacIntyre IG, Wapnick CM, O'Neill MW (2004) Phase shifts, alternative states, and the unprecedented convergence of two reef systems. Ecology 85:1876–1891

Bellwood DR (2003) Origins and escalation of herbivory in fishes: a functional perspective. Paleobiology 29:71–83

Bellwood DR, Hughes TP, Folke C, Nystrom M (2004) Confronting the coral reef crisis. Nature 429:827–833

Brett CE, Hendy AJW, Bartholomew AJ, Bonelli JR, McLaughlin PI (2007) Response of shallow marine biotas to sea-level fluctuations: a review of faunal replacement and the process of habitat tracking. Palaios 22:228–244

Briggs JC (2003) Marine centres of origin as evolutionary engines. J Biogeogr 30:1–18

Budd AF, Pandolfi JM (2004) Overlapping species boundaries and hybridization within the *Montastraea* 'annular' reef coral complex in the Pleistocene of the Bahama Islands. Paleobiology 30:396–425

Chappell J (1980) Coral morphology, diversity and reef growth. Nature 286:249–252

Chave J (2004) Neutral theory and community ecology. Ecol Lett 7:241–253

Collins LS, Budd AF, Coates AG (1996) Earliest evolution associated with closure of the tropical American seaway. Proc Natl Acad Sci U S A 93:6069–6072

Connell JH (1978) Diversity in tropical rain forests and coral reefs. Science 199:1302–1310

Copper P (1994) Ancient reef ecosystem expansion and collapse. Coral Reefs 13:3–11

Crame JA (1980) Succession and diversity in the Pleistocene coral reefs of the Kenya coast. Palaeontology (Oxford) 23:1–37

DiMichele WA, Behrensmeyer AK, Olszewski TD, Labandeira CC, Pandolfi JM, Wing SL, Bobe R (2004) Long-term stasis in ecological assemblages: evidence from the fossil record. Ann Rev Ecol Evol Systemat 35:285–322

Done TJ (1982) Patterns in the distribution of coral communities across the central Great Barrier Reef. Coral Reefs 1:95–107

Dullo WC (2005) Coral growth and reef growth: a brief review. Facies 51:43–58

Edinger EN, Risk MJ (1996) Sponge borehole size as a relative measure of bioerosion and paleoproductivity. Lethaia 29:275–286

Gagan MK, Ayliffe LK, Hopley D, Cali JA, Mortimer GE, Chappell J, McCulloch MT, Head MJ (1998) Temperature and surface-ocean water balance of the mid-Holocene tropical Western Pacific. Science 279:1014–1018

Gardner TA, Cote IM, Gill JA, Grant A, Watkinson AR (2003) Long-term region-wide declines in Caribbean corals. Science 301:958–960

Geister J (1977) The influence of wave exposure on the ecological zonation of Caribbean coral reefs. In: Procedings of the third international coral reef symposium, vol. 1, Miami, 1977, pp 23–29

Gili E, Masse JP, Skelton PW (1995) Rudists as gregarious sediment-dwellers, not reef-builders, on Cretaceous carbonate platforms. Palaeogeogr Palaeoclimatol Palaeoecol 118:3–4

Gotelli NJ, Graves GR (1996) Null models in ecology. Smithsonian Institution Press, Washington, DC

Graus RR, Macintyre IG (1976) Light control of growth form in colonial reef corals: computer simulation. Science 193:895–897

Graus RR, Macintyre IG (1982) Variations in the growth forms of the reef coral *Montastraea annularis* (Ellis and Solander): a quantitative evaluation of growth response to light distribution using computer simulation. In: Rutzler K, Macintyre IG (eds) The Atlantic barrier reef ecosystem at Carrie Bow Cay, Belize. 1. Structure and communities. Smithsonian Institution Press, Washington, DC, pp 441–464

Greenstein BJ, Curran HA, Pandolfi JM (1998) Shifting ecological baselines and the demise of *Acropora cervicornis* in the western North Atlantic and Caribbean Province: a Pleistocene perspective. Coral Reefs 17:249–261

Greenstein BJ, Pandolfi JM (2008) Escaping the heat: range shifts of reef coral taxa in coastal Western Australia. Global Change Biol 14:513–528

Guzman HM, Cipriani R, Jackson JBC (2008) Historical decline in coral reef growth after the Panama Canal. Ambio 37:342–346

Hallock P, Schlager W (1986) Nutrient excess and the demise of coral reefs and carbonate platforms. Palaios 1:389–398

Harriott VJ, Banks SA (2002) Latitudinal variation in coral communities in eastern Australia: a qualitative biophysical model of factors regulating coral reefs. Coral Reefs 21:83–94

Harzhauser M, Kroh A, Mandic O, Piller WE, Göhlich U, Reuter M, Berning B (2007) Biogeographic responses to geodynamics: a key study all around the Oligo-Miocene Tethyan Seaway. Zool Anz 246:241–256

Hatcher BG (1997) Coral reef ecosystems: how much greater is the whole than the sum of the parts? Coral Reefs 16:S77–S91

Hönisch B, Hemming NG, Grottoli AG, Amat A, Hanson GN, Bijma J (2004) Assessing scleractinian corals as recorders for paleo-pH: empirical calibration and vital effects. Geochim Cosmochim Acta 68:3675–3685

Hubbell SP (1997) A unified theory of biogeography and relative species abundance and its application to tropical rain forests and coral reefs. Coral Reefs 16(Suppl):S9–S21

Hubbell SP (2001) The unified neutral theory of biodiversity and biogeography. Princeton University Press, Princeton

Hughes TP, Baird AH, Bellwood DR. Card M, Connolly SR, Folke C, Grosberg R, Hoegh-Guldberg O, Jackson JBC, Kleypas J, Lough JM, Marshall P, Nystrom M, Palumbi SR, Pandolfi JM, Rosen B, Roughgarden J (2003) Climate change, human impacts, and the resilience of coral reefs. Science 301:929–933

Hughes WW (1985) Planetary rotation and invertebrate skeletal patterns: prospects for extant taxa. Geophys Surv 7:169–183

Jablonski D, Roy K, Valentine JW (2006) Out of the tropics: evolutionary dynamics of the latitudinal diversity gradient. Science 314: 102–106

Jackson JBC, Johnson KG (2001) Paleoecology – measuring past biodiversity. Science 293:2401–2404

Jackson JBC, Erwin DH (2006) What can we learn about ecology and evolution from the fossil record? Trends Ecol Evol 21:322–328

Jackson JBC, Budd AF, Coates AG (eds) (1996) Evolution and environment in Tropical America. University of Chicago Press, Chicago

Jackson JBC, Kirby MX, Berger WH, Bjorndal KA, Botsford LW, Bourque BJ, Bradbury RH, Cooke R, Erlandson J, Estes JA, Hughes TP, Kidwell S, Lange CB, Lenihan HS, Pandolfi JM, Peterson CH, Steneck RS, Tegner MJ, Warner RR (2001) Historical overfishing and the recent collapse of coastal ecosystems. Science 293:629–638

Johnson KG, Jackson JBC, Budd AF (2008) Caribbean reef development was independent of coral diversity over 28 million years. Science 319:1521–1523

Johnson KG, Perez ME (2006) Skeletal extension rates of Cenozoic Caribbean reef corals. Palaios 21:262–271

Jokiel PL (1984) Long distance dispersal of reef corals by rafting. Coral Reefs 3:113–116

Karlson RH (2002) Dynamics of coral communities. Kluwer, Dordrecht

Kauffman EG, Johnson CC (1988) The morphological and ecological evolution of Middle and Upper Cretaceous reef-building Rudistids. Palaios 3:194–216

Kaufman L (1981) There was biological disturbance on Pleistocene coral reefs. Paleobiology 7:527–532

Kendrick G, Wyrwoll K, Szabo B (1991) Pliocene-Pleistocene coastal events and history along the western margin of Australia. Quatern Sci Rev 10:419–439

Kiessling W (2001) Paleoclimatic significance of Phanerozoic reefs. Geology 29:751–754

Kiessling W (2005) Habitat effects and sampling bias on Phanerozoic reef distribution. Facies 51:24–32

Kiessling W (2009) Geological and biologic controls on the evolution of reefs. Ann Rev Ecol Evol Systemat 40:173–192

Kiessling W, Aberhan M, Aragon E, Scasso R, Medina F, Kriwet J, Fracchia D (2005) Massive corals in Paleocene siliciclastic sediments of Chubut (Argentina). Facies 51:233–241

Klaus JS, Budd AF (2003) Comparison of Caribbean coral reef communities before and after Plio-Pleistocene faunal turnover: analyses of two Dominican Republic reef sequences. Palaios 18:3–21

Knowlton N, Weil E, Weigt LA, Guzman HM (1992) Sibling species in *Montastraea annularis*, coral bleaching, and the coral climate record. Science 255:330–333

Lewis JB (1984) The *Acropora* inheritance: a reinterpretation of the development of fringing reefs in Barbados, West Indies. Coral Reefs 3:117–122

Li Q, James N, Bone Y, McGowran B (1999) Palaeoceanographic significance of recent foraminiferal biofacies on the southern shelf of Western Australia: a preliminary study. Palaeogeogr Palaeoclimatol Palaeoecol 147:101–120

Marsh LM (1993) The occurrence and growth of Acropora in extratropical waters off Perth, Western Australia. In: Proceedings of the seventh international coral reef symposium, vol. 7, guam, 1997, pp 1233–1238

Masse JP, Montaggioni LF (2001) Growth history of shallow-water carbonates: control of accommodation on ecological and depositional processes. Int J Earth Sci 90:452–469

McCulloch M, Fallon S, Wyndham T, Hendy E, Lough J, Barnes D (2003) Coral record of increased sediment flux to the inner Great Barrier Reef since European settlement. Nature 421:727–730

McGill BJ, Hadly EA, Maurer BA (2005) Community inertia of quaternary small mammal assemblages in North America. Proc Natl Acad Sci U S A 102:16701–16706

McGowran B, Li QY, Cann J, Padley D, McKirdy D, Shafik S (1997) Biogeographic impact of the Leeuwin current in south-ern Australia since the late middle Eocene. Palaeogeogr Palaeoclimatol Palaeoecol 136:19–40

Mesolella K (1967) Zonation of uplifted Pleistocene coral reefs on Barbados West Indies. Science 156:638–640

Moberg F, Folke C (1999) Ecological goods and services of coral reef ecosystems. Ecol Econ 29:215–233

Montaggioni LF (2005) History of Indo-Pacific coral reef systems since the last glaciation: development patterns and controlling factors. Earth Sci Rev 71:1–75

Muscatine L, Goiran C, Land L, Jaubert J, Cuif J-P, Allemand D (2005) Stable isotopes (δ13C and δ15N) of organic matrix from coral skeleton. Proc Natl Acad Sci U S A 102:1525–1530

Newell ND (1971) An outline history of tropical organic reefs. Am Mus Novit 2465:1–37

O'Dea A, Jackson JBC, Fortunato H, Smith JT, D'Croz L, Johnson KG, Todd JA (2007) Environmental change preceded Caribbean extinction by 2 million years. Proc Natl Acad Sci 104:5501–5506

Olszewski TD, Erwin DH (2004) Dynamic response of Permian brachiopod communities to long-term environmental change. Nature 428:738–741

Pandolfi JM (1992) Successive isolation rather than evolutionary centres for the origination of Indo-Pacific reef corals. J Biogeogr 19:593–609

Pandolfi JM (1996) Limited membership in Pleistocene reef coral assemblages from the Huon Peninsula, Papua New Guinea: constancy during global change. Paleobiology 22:152–176

Pandolfi JM (1999) Response of Pleistocene coral reefs to environmental change over long temporal scales. Am Zool 39:113–130

Pandolfi JM (2002) Coral community dynamics at multiple scales. Coral Reefs 21:13–23

Pandolfi JM (2007) A new, extinct Pleistocene reef coral from the *Montastraea annularis* species complex. J Paleontol 81:472–482

Pandolfi JM, Budd AF (2008) Morphology and ecological zonation of Caribbean reef corals: the *Montastraea annularis* species complex. Mar Ecol Progr Ser 369:89–102

Pandolfi JM, Greenstein BJ (2007) Using the past to understand the future: palaeoecology of coral reefs. In: Johnson JE, Marshall PA (eds) Climate change and the Great Barrier Reef. Great Barrier Reef Marine Park Authority and the Australian Greenhouse Office, Townsville, Australia, pp 717–744

Pandolfi JM, Jackson JBC (2001) Community structure of Pleistocene coral reefs of Curacao, Netherlands Antilles. Ecol Monogr 71:49–67

Pandolfi JM, Jackson JBC (2006) Ecological persistence interrupted in Caribbean coral reefs. Ecol Lett 9:818–826

Pandolfi JM, Jackson JBC, Geister J (2001) Geologically sudden natural extinction of two widespread Late Pleistocene Caribbean reef corals. In: Jackson JBC, Lidgard S, McKinney FK (eds) Evolutionary patterns: growth. Form and tempo in the fossil record. University of Chicago Press, Chicago, pp 120–158

Pandolfi JM, Lovelock CE, Budd AF (2002) Character release following extinction in a Caribbean reef coral species complex. Evolution 53:479–501

Pandolfi JM, Bradbury RH, Sala E, Hughes TP, Bjorndal KA, Cooke RG, McArdle D, McClenachan L, Newman MJH, Paredes G, Warner RR, Jackson JBC (2003) Global trajectories of the long-term decline of coral reef ecosystems. Science 301:955–958

Pandolfi JM, Jackson JBC, Baron N, Bradbury RH, Guzman HM, Hughes TP, Kappel CV, Micheli F, Ogden JC, Possingham HP, Sala E (2005) Are U.S. coral reefs on the slippery slope to slime? Science 307:1725–1726

Pandolfi JM, Tudhope A, Burr G, Chappell J, Edinger E, Frey M, Steneck R, Sharma C, Yeates A, Jennions M, Lescinsky H, Newton A (2006) Mass mortality following disturbance in Holocene coral reefs from Papua New Guinea. Geology 34:949–952

Pauly D (1995) Anecdotes and the shifting baseline syndrome of fisheries. Trends Ecol Evol 10:430–430

Perry CT, Smithers SG, Palmer SE, Larcombe P, Johnson KG (2008) 1200 year paleoecological record of coral community development from the terrigenous inner shelf of the Great Barrier Reef. Geology 36:691–694

Precht WF, Aronson RB (2004) Climate flickers and range shifts of reef corals. Front Ecol Environ 2:307–314

Renema W, Bellwood DR, Braga JC, Bromfield K, Hall R, Johnson KG, Lunt P, Meyer CP, McMonagle LB, Morley RJ, O'Dea A, Todd JA, Wesselingh FP, Wilson MEJ, Pandolfi JM (2008) Hopping hotspots: global shifts in marine biodiversity. Science 321:654–657

Riegl B (2001) Inhibition of reef framework by frequent disturbance: examples from the Arabian Gulf, South Africa, and the Cayman Islands. Palaeogeogr Palaeoclimatol Palaeoecol 175:79–101

Roff G (2010) Historical ecology of coral communities from the inshore Great Barrier Reef. Ph.D. thesis, University of Queensland, Brisbane

Rosen B (1988) Progress, problems and patterns in the biogeography of reef corals and other tropical marine organisms. Helgol Mar Res 42:269–301

Ross DJ, Skelton PW (1993) Rudist formations of the Cretaceous: a palaeoecological, sedimentological and stratigraphical review. In: Wright VP (ed) Sedimentology review, vol 1. Blackwell Science, London, pp 73–91

Runcorn SK (1966) Corals as paleontological clocks. Sci Am 215:26–33

Savage M, Sawhill B, Askenazi M (2000) Community dynamics: what happens when we rerun the tape? J Theor Biol 205:515–526

Solomon S, Qin D, Manning M, Chen Z, Marquis M, Averyt KB, Tignor M, Miller HL (eds) (2007) IPCC, 2007. Climate change 2007: the physical science basis. contribution of Working Group I to the fourth assessment report of the intergovernmental panel on climate change. Cambridge University Press, Cambridge

Stanley GD Jr, Swart PK (1995) Evolution of the coral-zooxanthellae symbiosis during the Triassic: a geochemical approach. Paleobiology 21:179–199

Tager D, Webster JW, Potts DC, Renema W, Braga JC, Pandolfi JM (2010) Community dynamics of Pleistocene coral reefs during alternative climatic regimes. Ecology 91:191–200

Tapanila L (2004) The earliest Helicosalpinx from Canada and the global expansion of commensalism in Late Ordovician sarcinulid corals (Tabulata). Palaeogeogr Palaeoclimatol Palaeoecol 215:99–110

Tomascik T, Sander F (1987) Effects of eutrophication on reef building corals. Part II. Structure of scleractinian coral communities on inshore fringing reefs, Barbados, WI. Mar Biol 94:53–75

Tudhope AW, Chilcott CP, McCulloch MT, Cook ER, Chappell J, Ellam RM, Lea DW, Lough JM, Shimmield GB (2001) Variability in the El Nino-Southern Oscillation through a glacial-interglacial cycle. Science 291:1511–1517

Vargas-Angel B, Thomas JD, Hokel SM (2003) High-latitude *Acropora cervicornis* thickets off Fort Lauderdale, Florida, USA. Coral Reefs 22:465–473

Vermeij GJ (2005) One-way traffic in the western Atlantic: causes and consequences of Miocene to early Pleistocene molluscan invasions in Florida and the Caribbean. Paleobiology 31:624–642

Walker K, Alberstadt LP (1975) Ecological succession as an aspect of structure in fossil communities. Paleobiology 1:238–257

Watkins R (2000) Corallite size and spacing as an aspect of niche-partitioning in tabulate corals of Silurian reefs, Racine formation, North America. Lethaia 33:55–63

Webster JM, Davies PJ (2003) Coral variation in two deep drill cores: significance for the Pleistocene development of the Great Barrier Reef. Sediment Geol 159:61–80

Wegley L, Edwards R, Rodriguez-Brito B, Liu H, Rohwer F (2007) Metagenomic analysis of the microbial community associated with the coral *Porites astreoides*. Environ Microbiol 9:2707–2719

Wells JW (1963) Coral growth and geochronometry. Nature 197:948–950

Wells P, Wells G (1994) Large-scale reorganization of ocean currents offshore Western Australia during the late Quaternary. Mar Micropaleontol 24:157–186

Wilkinson C (ed) (2008) Status of coral reefs of the world. Global Coral Reef Monitoring Network and Reef and Rainforest Research Centre, Townsville, Australia

Williams ST, Duda TF (2008) Did tectonic activity stimulate Oligo-Miocene speciation in the Indo-West Pacific? Evolution 62:1618–1634

Wilson MEJ (2002) Cenozoic carbonates in Southeast Asia: implications for equatorial carbonate development. Sediment Geol 147:295–428

Wood R (1993) Nutrients, predation and the history of reef-building. Palaios 8:526–543

Wood R (1999) Reef evolution. Oxford University Press, Oxford

Zachos JC, Shackleton NJ, Revenaugh JS, Palike H, Flower BP (2001) Climate response to orbital forcing across the Oligocene-Miocene boundary. Science 292:274–278

Remote Sensing of Coral Reef Processes

Eric J. Hochberg

Abstract Digital remote sensing of coral reefs dates to the first Landsat mission of the mid-1970s. Early studies utilized moderate-spatial-resolution satellite broadband multispectral image data and focused on reef geomorphology. Technological advances have since led to development of airborne narrow-band hyperspectral sensors, airborne hydrographic lidar systems, and commercial high-spatial-resolution satellite broadband multispectral imagers. High quality remote sensing data have become widely available, and this has spurred investigations by the reef science and management communities into the technology. Studies utilizing remote sensing data range from predictions of reef fish diversity to multidecadal assessment of reef habitat decline. Fundamental issues remain, both in remote sensing science and in specific coral reef applications. Nevertheless, investigators are increasingly turning to remote sensing for the unique perspective it affords of reef systems.

Keywords Coral reef remote sensing • reef community structure • reef biogeochemistry • reef optical properties

1 Introduction

Quantification of benthic community structure and distribution is fundamental to understanding coral reef ecosystem processes. Community structure determines relative reef status (Connell 1997), and different community-types tend to exhibit modal rates of productivity and calcification (Kinsey 1985). Reef-dwelling organisms rely on different community-types at various stages of their life histories (Chabanet et al. 1997; Light and Jones 1997; Miller et al. 2000). Reef community structure is highly heterogeneous on spatial

E.J. Hochberg (✉)
National Coral Reef Institute, Nova Southeastern University
Oceanographic Center, 8000 North Ocean Drive,
Dania Beach, FL 33004, USA
e-mail: eric.hochberg@nova.edu

scales of centimeters to hundreds of meters, but relative to phytoplankton and macrophyte communities, it is inherently stable on time scales of months to years (Buddemeier and Smith 1999). Although reef communities have evolved and persisted in an environment fraught with natural destructive processes, there is ever-increasing concern that direct and indirect anthropogenic forces are devastating reefs throughout the world (Smith and Buddemeier 1992; Kleypas et al. 1999; Hoegh-Guldberg et al. 2007).

This concern over the decline in reef status has led to development of numerous reports on the status of coral reef ecosystems, both regionally and globally (Waddell and Clarke 2008; Wilkinson 2008). These reports are syntheses of local and regional monitoring efforts that utilize in situ observations. There are three common methods for determining benthic community structure and distribution on coral reefs: (1) 1–10 m scale quadrats, (2) 10–100 m scale line transects, and (3) 100+ m scale manta-tows, which entail towing a diver on a sled behind a boat, with the diver pausing periodically to record estimates of reef benthic cover. Quadrats and transects resolve reef elements at the scale of 1s–10s of centimeters, providing detailed and statistically rigorous estimates of reef community structure (Bouchon 1981). Manta-tows are less rigorous because they are conducted at a much larger spatial scale without spatial reference cues, which has two main drawbacks: a decrease in resolving power and a lack of repeatability (Bainbridge and Reichelt 1988; Miller and Müller 1999). Regardless of the methodology, in situ surveys directly cover only a small fraction of the world's 250,000–600,000 km^2 total reef area (Smith 1978; Kleypas 1997; Spalding and Grenfell 1997). Thus, these syntheses provide current best-estimates of reef status, but vast reef areas remain unsurveyed.

Even when surveying a single reef, cost and logistical considerations dictate a statistical sampling approach. Observations are made at discrete (either random or nonrandom) locations on the reef; large areas of the reef remain unobserved. Such sampling provides estimates of various statistics, generally the mean and variance of the observed parameters. However, reef communities exhibit a great deal of patchiness, and even intensive surveys may not adequately

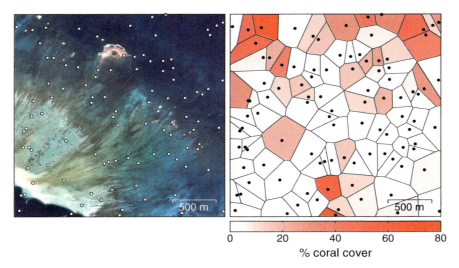

Fig. 1 Hypothetical field survey plan for a coral reef, with 100 randomly placed transect locations. (*Left*) Satellite image of reef, illustrating patchiness of communities; white dots are transect locations. (*Right*) Hypothetical survey results, with darker shades representing higher coral cover; black dots are transect locations. Based on this random sample, there is no information about boundaries between communities. Extrapolating transect results provides an inadequate picture of community distributions

capture community distributions (see example in Fig. 1). Extending surveys beyond a single reef, intensive in situ methods become intractable. Remote sensing is one tool that has potential for quantifying reef community structure and distribution at large scales (Mumby et al. 2001). This technology has been demonstrated to be the most cost-effective means for acquiring synoptic data on reef community structure (Mumby et al. 1999), and it is the only available tool that can acquire globally uniform data.

2 Brief History of Coral Reef Remote Sensing

Kuchler et al. (1988), Green et al. (1996), Mumby et al. (2004), and Andréfouët et al. (2005) provide thorough reviews of the history of coral reef remote sensing. Digital remote sensing of coral reefs began in the early 1970s with the launch of Landsat 1 (Smith et al. 1975). Initial investigations during the 1970s and 1980s relied on imagery from sensors onboard the Landsat and SPOT (*Satellite Pour l'Observation de la Terre*) satellites. This imagery was multispectral (two or three wavebands in the visible spectrum), broadband (each waveband 60–100 nm wide), and had moderate spatial resolution (ground sample distance, or pixel size, 20–80 m). The few, broad wavebands and moderate spatial resolution limited most studies to detection of reefs and delineation of reef geomorphology (e.g., Biña et al. 1978; Loubersac et al. 1988). Taking advantage of correlations between reef geomorphological and ecological zones, some investigations were able to delineate basic reef biotopes (e.g., Bour et al. 1986; Vercelli et al. 1988; Luczkovich et al. 1993; Ahmad and Neil 1994).

The 1990s saw the advent of hyperspectral imaging (or simply spectral imaging) from airborne platforms. Because airborne platforms were nearer to the reef, the resulting imagery had higher spatial resolution (0.5–20 m, depending on flight altitude). Also, in contrast to previous multispectral sensors, hyperspectral sensors possessed 10s–100s of narrow (5–10 nm) wavebands. The improved spatial and spectral resolutions revealed the potential for direct identification and mapping of reef communities (e.g., Hochberg and Atkinson 2000). However, processing these data into meaningful reef science products proved to be nontrivial. Especially problematic was the fact that no suitable algorithms existed to correct imagery for the confounding effects of the atmosphere, sea surface, and water column (Maritorena et al. 1994; Lee et al. 1998). Fundamental remote sensing research would be required to make coral reef applications operational.

A very important development in the 2000s was the widespread public availability of commercial satellite imagery from IKONOS and Quickbird. The spectral resolution of these data was roughly equivalent to that of Landsat (i.e., broadband multispectral), but the spatial resolution was much improved (2.4–4 m for color imagery, 0.6–1 m for panchromatic imagery). Investigation by remote sensing scientists into the utility of these data for mapping reef communities produced mixed results (Maeder et al. 2002; Mumby and Edwards 2002; Andréfouët et al. 2003). In general, commercial satellite imagery could distinguish between five and nine benthic classes with an accuracy of 50–80%, depending on the study. Typically, benthic classes were subjectively defined on a case-by-case basis, and they often did not align with fundamental community-types more familiar to reef scientists. However, for the first time, satellite observations provided detailed but relatively inexpensive views of coral reefs, views heretofore only available through costly aerial surveys. This spurred interest among the broader reef science and management communities, who began to utilize these images as contextual base maps for field studies, as well as generate their own remote sensing-derived products.

Remote sensing of coral reefs is an inherently interdisciplinary endeavor. The overarching scientific objective is to characterize some aspect of reef structure or function, and this requires background in reef ecology, geology, and/or biogeochemistry. However, by definition, observations are made at a distance, and information about the reef must be inferred indirectly. Linking remotely measured optical signals with reef structure, for example, requires knowledge of how light interacts with reef components, as well as how light is transmitted through the aquatic and atmospheric mediums. The actual light measurements require sophisticated sensor systems that integrate fore-optics (lenses), optical detectors, navigation units, data storage devices, motion-control units, and control electronics. If the sensor is on a satellite platform, then launch, orbital parameters, targeting, and data downlink become factors (not to mention heat dissipation, power considerations, etc.). Fortunately, no individual person is responsible for all of these aspects, but it is important to recognize that coral reef remote sensing is at the intersection of three disciplines: reef science, environmental optics, and engineering.

Coral reef remote sensing has diverged into paths of development and application. Basic research continues to advance remote sensing technologies, as well as methods for processing and analyzing imagery of shallow water systems in general. More targeted research focuses on development of the remote sensing tool to study specific coral reef processes. On the application side, reef scientists and managers now routinely incorporate remote sensing imagery and derived products into their ongoing projects.

3 Remote Sensing Basics

There are two methods of data acquisition in remote sensing, active and passive. A passive remote sensing system simply records the ambient energy, usually from the sun, reflected off a surface (equivalent to daylight photography). Multispectral and hyperspectral imagers are passive systems. In active remote sensing, a signal on a particular wavelength is transmitted, and the sensor records the reflection. An example of active remote sensing pertinent to coral reefs is LIDAR (LIght Detection And Ranging, hereafter referred to as lowercase "lidar"). With lidar, the sensor emits a very short laser pulse and very precisely measures the time it takes for the pulse to reflect from the reef surface and return to the sensor. The distance to the object is (essentially) half the travel time multiplied by the speed of light. Hydrographic lidars are flown on low-altitude aircraft and can achieve ground sample distances of 1–2 m; they have vertical resolutions on the order of 10 cm. These systems can generate very detailed bathymetric data over coral reefs to depths of ~40 m, depending on water clarity (Fig. 2). Such data can be incorporated into a seafloor classification

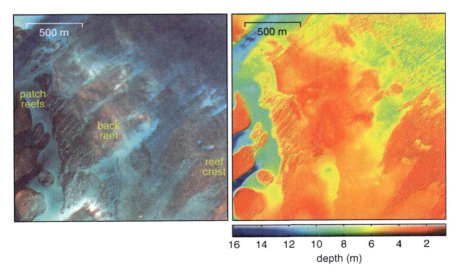

Fig. 2 Example lidar data set from Kaneohe Bay, Hawaii. (*Left*) Quickbird satellite scene for context. (*Right*) High-resolution lidar from SHOALS (Scanning Hydrographic Operational Airborne Lidar Survey). The lidar data highlight structural features that are not readily evident in the satellite image, such as spur and groove formation along right-hand side of the scene

procedure (Walker et al. 2008) or fused with passive data for further processing (Conger et al. 2006). As of early 2010, use of lidar data for reef mapping is increasing, but use of passive airborne and satellite multi- and hyperspectral imagery is far more prevalent.

Remote sensing imagery may be interpreted in many ways. Visual inspection is common for photographs, while computers are very useful for manipulations of digital data. A computer image processing system can enhance the image to improve its visual interpretability and perform perhaps the most common image analysis step: classification. Digital image data and classification products can serve as input to further analyses relevant to the reef scientist.

Classification of remote sensing imagery may either be supervised or unsupervised. In unsupervised classification, such as cluster analysis, a computer assigns pixels to classes based solely on statistical similarities; there are no predefined training classes. The computer calculates the similarity between each pair of pixels. Similar pixels are assigned to the same class, while dissimilar pixels are assigned to different classes. The result is a set of classes that are based on natural clustering within the entire data set. These classes have yet to be given appropriate ground cover labels. In fact, classes resulting from unsupervised classification do not necessarily correspond to ecologically meaningful classes.

Supervised classification utilizes training data to guide derivation of classification rules. Training data are comprised of pixels (or sometimes spectra) for which class membership is known. Classification rules assign unknown pixels to one of the predefined classes based on the statistical organization of the training data (as opposed to the entire data set). Importantly, the training data define the set of classes a priori. If training data have ecologically meaningful class labels, the classified map product shows the distributions of those classes. Visual interpretation of photographic imagery is a form of supervised classification. In this case, the classification rules are not defined statistically using a computer; rather, the classification rules are based on the photointerpreter's experience and/or knowledge of the reef in question.

4 Coral Reef Remote Sensing Considerations

The spectral and spatial requirements for coral reef remote sensing depend on the actual science question posed. To spectrally distinguish between basic reef bottom-types of coral, algae, and sand, a remote sensing system at a minimum must have several narrow wavebands (four or more, no more than 20-nm wide) that cover the range 500–580 nm (Hochberg and Atkinson 2003). To distinguish more bottom-types, such as various algal types, more narrow wavebands are required across the visible spectrum (Hochberg et al. 2003b). Atmospheric correction requires specific wavebands in the near infrared (NIR) and short-wave infrared (SWIR) region of the spectrum (700–1,500 nm, Gao et al. 2007), and water column correction benefits from more wavebands in the visible (VIS) region of the spectrum (400–700 nm, Lee and Carder 2002).

The literature is less clear on the issue of spatial resolution. Coral reef researchers may prefer high-resolution (1–4 m) commercial data, because they more closely match in situ observation scales. Modeling suggests that detection of bleached coral colonies requires spatial resolutions of 40–80 cm (Andréfouët et al. 2002). Forthcoming commercial satellite sensors approach this spatial resolution, but the issue remains whether an image can be acquired during the relatively short period of an ongoing bleaching event. Commercial high-resolution multispectral imagery has also been demonstrated to provide useful textural (as opposed to spectral) information that enhances detection of specific types of coral assemblages (Purkis et al. 2006). At the reef system scale, provided that spectral resolution is suitable for the desired application, spatial resolution requirements can probably be relaxed to several tens of meters. At a pixel size of 50×50 m^2, a single reef of 10 km^2 would comprise 4,000 pixels. Although this resolution may seem coarse to a scientist accustomed to views at the human scale, at the scale of the reef, 4,000 observations may constitute oversampling. A map product with this resolution would certainly convey the distribution of benthic communities across the reef.

An important consideration for any remote sensing study is calibration/validation, often referred to as groundtruth. Where possible, every remote sensing study should include a field campaign to measure both ecological and optical parameters. Data should be collected as near in time to the remote sensing acquisition as possible. With more stable ecological parameters, such as benthic cover on a relatively undisturbed reef, there is some discretion. Optical properties are much more temporally variable and are only valid if measured during or near in time to remote sensing data acquisition.

The rationale for optical measurements is that such ancillary data would greatly improve image processing, and they would be useful for validation of radiative transfer inversion results. The primary optical properties of interest to coral reef remote sensing are spectral reflectance of the seafloor, the spectral diffuse attenuation coefficient of the water, and the radiance leaving the water. Of these, seafloor spectral reflectance is the most readily measured; portable, diver-operated spectrometer systems are available for approximately US$5,000. Costs are much more prohibitive for the equipment necessary to measure water-leaving radiance and diffuse attenuation. Measuring optical properties is nontrivial, in terms of both logistics and cost; in many cases, it is simply not possible. In practice, most coral reef remote sensing applications forgo optical measurements entirely, relying instead on standard field-observation-based supervised classification.

Measuring reef ecological parameters is much more important to interpretation of remote sensing imagery, as well as to demonstration of product reliability. Field data guide the training process of supervised classification. Field data also provide a known point of reference on which to verify the accuracy of the remote sensing product (Congalton and Green 1999). It is crucial that the field sampling unit matches the sampling unit in the remote sensing data. For high-resolution commercial satellite imagery, a good sampling unit would be a series of quadrats or short transects covering an area equivalent to several image pixels. For moderate-resolution data, a sampling unit would necessarily consist of a series of longer transects or a manta-tow, again covering several image pixels. Quadrats and transects are preferred because they provide quantitative and objective measures of community structure.

One question that often arises is whether the same data can be used to both calibrate and validate remote sensing products. For small sample sizes, resubstitution of the training data biases the accuracy assessment, resulting in overly optimistic error rates (Rencher 2002). For large sample sizes, this bias is small and can generally be ignored. To remove the bias entirely, it is necessary to partition the training data and the accuracy assessment data. A simple technique is to divide the field observations into two groups, one for training and one for assessment. This technique has two disadvantages: it requires large sample sizes, and it does not evaluate the classification function that is used in practice (the error estimate may be very different from that estimated using the entire data set). In computer-based classification, full cross-validation is recommended. All observations but one are used to train the supervised classification, which is then applied to the withheld observation. The procedure is repeated for each observation. The advantages of cross-validation are that sample sizes are reduced and that it evaluates classification functions that are very similar to the one used in practice. The result is an unbiased estimate of classification error rates. In cases where manual digitization is used to generate map products, it is not possible to partition the training data, and it is necessary to acquire entirely independent validation data.

5 Remote Sensing of Optically Shallow Waters

Much of the research into remote sensing of optically shallow waters (i.e., where the seafloor is shallow enough to be detectable) centers on correcting remote sensing data for radiative transfer effects. The technical aim of remote sensing is to arrive at earth-surface spectral reflectances (proportion of light flux reflected from an object), which provide insight into the nature of the object that is remotely sensed. The problem is that, even in the absence of clouds, there are five different origins (fluxes) of light received by a remote sensor above a coral reef (Kirk 1994). These are (1) scattering of sunlight within the atmosphere, (2) reflection of the direct solar beam at the ocean surface, (3) reflection of skylight (previously scattered sunlight) at the ocean surface, (4) upward scattering of sunlight within the water, and (5) reflection of sunlight at the reef surface. In addition to scattering and reflection, absorption processes also affect the light fluxes in the water and atmosphere. Of the five fluxes, only light reflected off the reef surface can provide information about the reef's community structure, while the remaining fluxes obscure the desired signal. Deterministic remote sensing of coral reef communities requires accounting of all five light fluxes reaching the remote sensor (Fig. 3).

The signal received by a satellite sensor pointed at the ocean is far different from the original signal at the sea floor (Fig. 4). As much as 90% of the satellite signal arises from scattering within the atmosphere (Kirk 1994). Aerosols are particularly problematic because they are highly variable and their scattering effects are highly significant. The traditional solution relies on the fact that seawater absorbs light very strongly at NIR and SWIR wavelengths, essentially rendering the ocean black (Gordon 1997; Hooker and McClain 2000). Any light measured at these wavelengths arises either from the sea surface or the atmosphere. Aerosol values are derived from satellite measurements at two NIR/SWIR wavelengths. The ratio of these values is then used to select an aerosol model, which in turn is used to extrapolate aerosol values at VIS wavelengths. This atmospheric correction approach has been successfully applied to ocean color imagery as far back as the Coastal Zone Color Scanner in the early 1980s, and it continues to undergo refinement (e.g., Gao et al. 2007).

Glint is reflection of sunlight and skylight from the sea surface; glint can obscure subsurface features and seriously compromise seafloor detections. Correcting for glint is an area of active research. One promising approach (Hochberg et al. 2003a; Hedley et al. 2005) is to determine the spatial distribution of relative glint intensity using a waveband in the NIR (black ocean assumption). Glint at VIS wavelengths is determined via regression against the NIR relative glint values. Subtracting the regression values effectively removes glint effects (Fig. 3b).

Correcting for water column effects in shallow water remote sensing data has been an ongoing subject of research since the 1970s (Lyzenga 1978, 1985; Bierwirth et al. 1993). Water molecules, viruses, bacteria, and microplankton, among others, all absorb and scatter light, and the effects

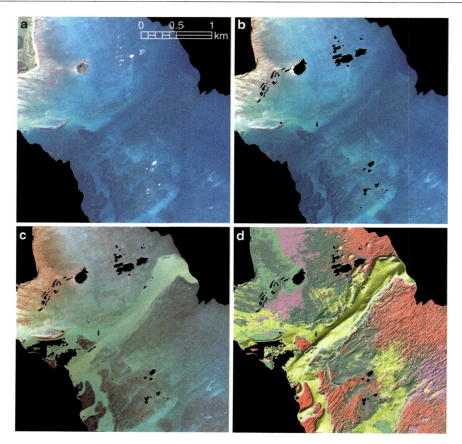

Fig. 3 Coral reef remote sensing analysis using AVIRIS (Airborne Visible InfraRed Imaging Spectrometer) scene of Kaneohe Bay, Hawaii. (**a**) Initial image. (**b**) Image after corrections for atmospheric and sea surface effects (clouds, breakers, and land surfaces are masked to highlight aquatic area). (**c**) Image after corrections for water column effects; this is essentially the reef without water. (**d**) Image classified to illustrate distributions of different reef community-types (reds – coral-dominated; greens/pink – algal-dominated; yellow – sand-dominated). Classification image is draped over reef bathymetric surface to highlight three-dimensional complexity

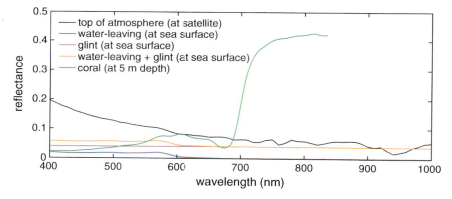

Fig. 4 Example of reflectances (upwelling light flux divided by downwelling light flux) relevant to coral reef remote sensing. Coral reflectance at the sea floor is modified greatly by absorption and scattering in the 5-m-deep water, resulting in water-leaving reflectance. Total sea surface reflectance is the sum of water-leaving and glint reflectances. This signal is further modified through absorption and scattering by atmospheric gases and aerosols. The remote sensing problem is to derive sea floor reflectance from the top-of-atmosphere signal. Values shown in this figure were derived using the Hydrolight and HydroMOD radiative transfer models (Data courtesy of Curtis D. Mobley)

are wavelength-dependent. The remote sensing problem is that a given pixel has unknown water column optical properties, unknown water depth, and unknown bottom reflectance. Further, reef water optical properties vary widely on spatial scales of 10s–100s of meters and temporal scales of hours (Fig. 5). As of early 2010, there are two approaches to deconvolving the water column and seafloor signals from hyperspectral data that have demonstrated favorable results

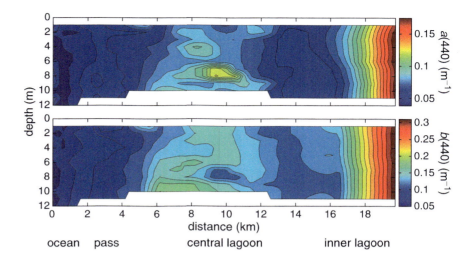

Fig. 5 Optical properties of New Caledonia's southeastern lagoon, as measured on an offshore-onshore transect of vertical profiles 11 March 2005. (*Top*) Absorption (**a**) at 440 nm. (*Bottom*) Scattering (**b**) at 440 nm

(Lee et al. 1998; Mobley et al. 2005). As in atmospheric correction, these approaches compare remotely sensed values with modeled values. The best-fit modeled values are taken to represent the water depth, water optical properties, and seafloor composition. In their current states, both approaches perform very well at retrieving water depth and reasonably well at retrieving water constituents, but seafloor composition retrievals appear to have mixed success. The effective maximum depth for such water column corrections is controlled largely by water clarity, but technological issues such as sensor noise can also impact retrievals. In general, given relatively clear and calm water, retrievals of seafloor depth are good to ~25 m.

Studies based on in situ spectral reflectances have fairly well established that basic reef bottom-types – algae, coral, and sand – are spectrally distinguishable from each other (Holden and LeDrew 1998; Holden and LeDrew 1999; Hochberg and Atkinson 2000, 2003; Kutser et al. 2003; Karpouzli et al. 2004). Sand is detectable because it is bright relative to other reef bottom-types. Corals are distinguished by a spectral feature near 570 nm that results from peridinin in their zooxanthellae (Hochberg et al. 2004). Chlorophytes, phaeophytes, and rhodophytes are also distinguishable from each other (Hochberg et al. 2003b), due to the spectral expressions of their own characteristic suites of photosynthetic accessory pigments (Kirk 1994). These studies using in situ spectral data are important, because if discrimination were not possible in situ, then it would likely not be possible using remote sensing. However, Purkis et al. (2006) demonstrated that textural information (spatial patterning as a function of color variation) can enhance detection of some bottom-types over purely spectral classifications. Individual case studies have demonstrated that classifications based on spectral reflectance can be scaled up to remote sensing imagery (Hochberg and Atkinson 2000; Andréfouët et al. 2004b), but the process has yet to be automated.

Development of accurate automated radiative transfer inversion algorithms is a primary focus of ongoing shallow water remote sensing research. These algorithms are intended for future airborne and satellite missions such as HyspIRI (Hyperspectral Infrared Imager) that focus on shallow water ecosystems, including coral reefs. Satellites acquire images globally, generating an enormous data volume. Automated processing is a basic requirement. Therefore, inversion techniques must be effective across the gamut of atmosphere and water column conditions.

6 Coral Reef Remote Sensing Applications

By far, the predominant applications of remote sensing to coral reefs are to delineate reef geomorphology and to determine distributions of benthic communities. There have been numerous demonstrations of the utility of remote sensing data to other areas of reef science. The following is a small sample to illustrate the breadth of the studies.

Andréfouët et al. (2004a) used IKONOS imagery to estimate percent cover of an invasive algae on Tahitian reefs, then used an empirical relationship to scale the percent cover estimates to biomass. Harborne et al. (2006) evaluated beta-diversity for nearly the entire reefscape of St. John, U.S.V.I., based on a map of benthic community structure derived from airborne multispectral imagery. Ortiz and Tissot (2008) manually digitized habitat maps of marine protected areas on Hawaii's Big Island using aerial photographs and lidar data. They found that in all areas surveyed, yellow tang recruits preferred coral-rich areas, while the distribution and abundance of adults varied greatly between sites. Brock et al. (2006a) used rugosity values derived from high-spatial-resolution lidar data to locate clusters of massive coral colonies growing among seagrass beds in the Florida Keys. Kuffner et al. (2007) found

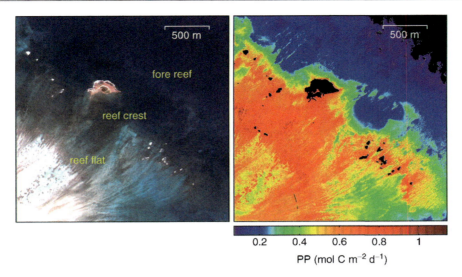

Fig. 6 Demonstration of gross primary production (PP) derived from remote sensing, as shown in Hochberg and Atkinson (2008). (*Left*) Quickbird scene showing Kaneohe Bay, Hawaii, fore reef, reef crest, and reef flat. (*Right*) Remote sensing estimate of PP based on optical absorbance and light-use efficiency have ranges and distributions comparable to those measured in situ, but further work is necessary to parameterize photosynthetic light-use efficiencies of reef communities

that fish species richness and abundance in Biscayne Bay, Florida, were correlated with reef rugosity values derived from airborne lidar data. Purkis et al. (2008) found a similar correlation of reef fish diversity and abundance with rugosity derived from IKONOS imagery for Diego Garcia (Chagos Archipelago). Palandro et al. (2008) processed a time-series of 28 Landsat scenes to detect changes in the coral reefs of the Florida Keys. Their results mirror the habitat decline observed by long-term in situ monitoring. Phinney et al. (2001) retraced the Caribbean-wide mortality of *Diadema antillarum* during 1983–1984 using satellite ocean color data and changes in reef habitats detected in Landsat imagery. Finally, Goreau and Hayes (1994) laid the groundwork for detection of coral mass bleaching events through tracking of sea surface temperature "hot spots" that indicate warming in the oceanic waters near reefs. Though this last application does not rely on direct sensing of reefs, it has proven remarkably accurate at predicting bleaching events, and it is an operational product of NOAA's Coral Reef Watch.

Remote sensing appears particularly suited to large-scale determination of reef biogeochemical rates. Atkinson and Grigg (1984) initially demonstrated the use of remote sensing imagery to scale modal rates of reef productivity and calcification (Kinsey 1985). This approach has since been repeated by other researchers using various image sources (Andréfouët and Payri 2000; Brock et al. 2006b; Moses et al. 2009). Hochberg and Atkinson (2008) proposed a remote sensing approach for estimating community- to reef-scale primary production based on optical absorbance and light-use efficiency. This type of model was first suggested by Monteith (1972) for crop plants and is now routinely used in terrestrial remote sensing studies (e.g., Running et al. 2004). Initial results are very encouraging (Fig. 6), but a good deal of fundamental research is required to further this topic, as there are no published values of photosynthetic light-use efficiency at the coral reef community scale.

Remote sensing has also entered the realm of purely applied reef science. The U.S. Coral Reef Task Force stated a national need for comprehensive coral reef maps that create accurate baselines for long-term monitoring; illustrate important community-scale trends in coral reef health over time; characterize habitats for place-based conservation measures such as MPAs; and enable scientific understanding of the large-scale oceanographic and ecological processes affecting reef health (U.S. Coral Reef Task Force 1999). To meet these needs, the National Oceanic and Atmospheric Administration initiated numerous regional mapping projects extensively utilizing IKONOS imagery. With funding from the National Aeronautics and Space Administration, the Millennium Coral Reef Mapping Project mapped the locations, extents, and geomorphologies of all the coral reefs in the world, based on Landsat imagery. These projects employed the most basic of image analysis methods: they relied almost exclusively on manual digitization to generate map products, essentially applying the same techniques that had been in use prior to the advent of digital remote sensing. However, these large-scale mapping efforts clearly demonstrate the utility of remote sensing data for coral reef study.

Commercial image data and geographic information systems are becoming ever more ubiquitous. Satellite and sensor systems continue their technological advances, and inversion algorithms continue to be refined and validated. It is certain that as more remote sensing data and products become available to the coral reef science community, many more applications will be identified and pursued.

7 Conclusion

Since its inception, coral reef remote sensing has shown promise as a tool to expand our understanding of reef system function. There is no doubt that remote sensing provides a unique, broad perspective of reefs. The potential power of remote sensing, especially from satellite platforms, lies in the ability to make routine observations of large and remote areas. The challenge remains to effectively extract from the remote sensing data information that is pertinent to reef science and management. As that challenge is met more and more, remote sensing can become a very strong complement to, and a link between, in situ observations.

Acknowledgments Curtis D. Mobley provided data and very useful comments. I am especially grateful to Ali L. Hochberg for her assistance in the preparation of this manuscript.

References

Ahmad W, Neil DT (1994) An evaluation of Landsat Thematic Mapper (TM) digital data for discriminating coral reef zonation: Heron Reef (GBR). Int J Remote Sens 15:2583–2597

Andréfouët S, Payri C (2000) Scaling-up carbon and carbonate metabolism of coral reefs using in-situ data and remote sensing. Coral Reefs 19:259–269

Andréfouët S, Berkelmans R, Odriozola L, Done T, Oliver J, Muller-Karger F (2002) Choosing the appropriate spatial resolution for monitoring coral bleaching events using remote sensing. Coral Reefs 21:147–154

Andréfouët S, Kramer P, Torres-Pulliza D, Joyce KE, Hochberg EJ, Garza-Perez R, Mumby PJ, Riegl B, Yamano H, White WH, Zubia M, Brock JC, Phinn SR, Naseer A, Hatcher BG, Muller-Karger FE (2003) Multi-site evaluation of IKONOS data for classification of tropical coral reef environments. Remote Sens Environ 88:128–143

Andréfouët S, Zubia M, Payri C (2004a) Mapping and biomass estimation of the invasive brown algae *Turbinaria ornata* (Turner) J. Agardh and *Sargassum mangarevense* (Grunow) Setchell on heterogeneous Tahitian coral reefs using 4-meter resolution IKONOS satellite data. Coral Reefs 23:26–38

Andréfouët S, Payri C, Hochberg EJ, Hu C, Atkinson MJ, Muller-Karger FE (2004b) Use of in situ and airborne reflectance for scaling-up spectral discrimination of coral reef macroalgae from species to communities. Mar Ecol Prog Ser 283:161–177

Andréfouët S, Hochberg EJ, Chevillon C, Muller-Karger FE, Brock JC, Hu C (2005) Multi-scale remote sensing of coral reefs. In: Miller RL, Castillo CED, McKee BA (eds) Remote sensing of coastal aquatic environments: technologies. Techniques and applications. Springer, Dordrecht, pp 299–317

Atkinson MJ, Grigg RW (1984) Model of coral reef ecosystem. II. Gross and net benthic primary production at French Frigate Shoals, Hawaii. Coral Reefs 3:13–22

Bainbridge SJ, Reichelt RE (1988) An assessment of ground truth methods for coral reef remote sensing data. Proc 6th Int Coral Reef Symp 2:439–444

Bierwirth PN, Lee TJ, Burne RV (1993) Shallow sea-floor reflectance and water depth derived by unmixing multispectral imagery. Photogr Eng Remote Sens 59:331–338

Biña RT, Carpenter K, Zacher W, Jara R, Lim JB (1978) Coral reef mapping using Landsat data: follow-up studies. Proc 12th Int Symp Remote Sens Environ 3:2051–2070

Bouchon C (1981) Comparison of two quantitative sampling methods used in coral reef studies: the line transect and the quadrat methods. Proc 4th Int Coral Reef Symp 2:375

Bour W, Loubersac L, Rual P (1986) Thematic mapping of reefs by processing of simulated SPOT satellite data: application to the *Trochus niloticus* biotope on Tetembia Reef (New Caledonia). Mar Ecol Prog Ser 34:243–249

Brock JC, Wright CW, Kuffner IB, Hernandez R, Thompson P (2006a) Airborne lidar sensing of massive stony coral colonies on patch reefs in the northern Florida reef tract. Remote Sens Environ 104:31–42

Brock JC, Yates KK, Halley RB, Kuffner IB, Wright CW, Hatcher BG (2006b) Northern Florida reef tract benthic metabolism scaled by remote sensing. Mar Ecol Prog Ser 312:123–139

Buddemeier R, Smith SV (1999) Coral adaptation and acclimatization: a most ingenious paradox. Am Zool 39:1–9

Chabanet P, Ralambondrainy H, Amanieu M, Faure G, Galzin R (1997) Relationships between coral reef substrata and fish. Coral Reefs 16:93–102

Congalton RG, Green K (1999) Assessing the accuracy of remotely sensed data: principles and practices. Lewis, Boca Raton

Conger CL, Hochberg EJ, Fletcher CH, Atkinson MJ (2006) Decorrelating remote sensing color bands from bathymetry in optically shallow waters. IEEE Trans Geosci Remote Sens 44:1655–1660

Connell JH (1997) Disturbance and recovery of coral assemblages. Coral Reefs 16:S101–S113

Gao BC, Montes MJ, Li RR, Dierssen HM, Davis CO (2007) An atmospheric correction algorithm for remote sensing of bright coastal waters using MODIS land and ocean channels in the solar spectral region. IEEE Trans Geosci Remote Sens 45:1835–1843

Gordon HR (1997) Atmospheric correction of ocean color imagery in the Earth observing system era. J Geophys Res Atmos 102:17081–17106

Goreau TJ, Hayes RL (1994) Coral bleaching and ocean "hot spots". Ambio 23:176–180

Green EP, Mumby PJ, Edwards AJ, Clark CD (1996) A review of remote sensing for the assessment and management of tropical coastal resources. Coastal Manage 24:1–40

Harborne AR, Mumby PJ, Zychaluk K, Hedley JD, Blackwell PG (2006) Modeling the beta diversity of coral reefs. Ecology 87:2871–2881

Hedley JD, Harborne AR, Mumby PJ (2005) Simple and robust removal of sun glint for mapping shallow-water benthos. Int J Remote Sens 26:2107–2112

Hochberg EJ, Atkinson MJ (2000) Spectral discrimination of coral reef benthic communities. Coral Reefs 19:164–171

Hochberg EJ, Atkinson MJ (2003) Capabilities of remote sensors to classify coral, algae and sand as pure and mixed spectra. Remote Sens Environ 85:174–189

Hochberg EJ, Atkinson MJ (2008) Coral reef benthic productivity based on optical absorptance and light-use efficiency. Coral Reefs 27:49–59

Hochberg EJ, Andréfouët S, Tyler MR (2003a) Sea surface correction of high spatial resolution Ikonos images to improve bottom mapping in near-shore environments. IEEE Trans Geosci Remote Sens 41:1724–1729

Hochberg EJ, Atkinson MJ, Andréfouët S (2003b) Spectral reflectance of coral reef bottom-types worldwide and implications for coral reef remote sensing. Remote Sens Environ 85:159–173

Hochberg EJ, Atkinson MJ, Apprill A, Andrefouet S (2004) Spectral reflectance of coral. Coral Reefs 23:84–95

Hoegh-Guldberg O, Mumby PJ, Hooten AJ, Steneck RS, Greenfield P, Gomez E, Harvell CD, Sale PF, Edwards AJ, Caldeira K, Knowlton N, Eakin CM, Iglesias-Prieto R, Muthiga N, Bradbury RH, Dubi A, Hatziolos ME (2007) Coral reefs under rapid climate change and ocean acidification. Science 318:1737–1742

Holden H, LeDrew E (1998) Spectral discrimination of healthy and non-healthy corals based on cluster analysis, principal components analysis, and derivative spectroscopy. Remote Sens Environ 65:217–224

Holden H, LeDrew E (1999) Hyperspectral identification of coral reef features. Int J Remote Sens 20:2545–2563

Hooker SB, McClain CR (2000) The calibration and validation of SeaWiFS data. Prog Oceanogr 45:427–465

Karpouzli E, Malthus TJ, Place CJ (2004) Hyperspectral discrimination of coral reef benthic communities in the western Caribbean. Coral Reefs 23:141–151

Kinsey DW (1985) Metabolism, calcification and carbon production I: systems level studies. Fifth Int Coral Reef Congr 4:505–526

Kirk JTO (1994) Light and photosynthesis in aquatic environments. Cambridge University Press, Cambridge

Kleypas JA (1997) Modeled estimates of global reef habitat and carbonate production since the last glacial maximum. Paleoceanography 12:533–545

Kleypas JA, Buddemeier RW, Archer D, Gattuso J-P, Langdon C, Opdyke BN (1999) Geochemical consequences of increased atmospheric carbon dioxide on coral reefs. Science 284:118–120

Kuchler DA, Biña RT, Claasen DvR (1988) Status of high-technology remote sensing for mapping and monitoring coral reef environments. Proc 6th Int Coral Reef Symp 1:97–101

Kuffner IB, Brock JC, Grober-Dunsmore R, Bonito VE, Hickey TD, Wright CW (2007) Relationships between reef fish communities and remotely sensed rugosity measurements in Biscayne National Park, Florida, USA. Environ Biol Fish 78:71–82

Kutser T, Dekker AG, Skirving W (2003) Modeling spectral discrimination of Great Barrier Reef benthic communities by remote sensing instruments. Limnol Oceanogr 48:497–510

Lee ZP, Carder KL (2002) Effect of spectral band numbers on the retrieval of water column and bottom properties from ocean color data. Appl Opt 41:2191–2201

Lee ZP, Carder KL, Mobley CD, Steward RG, Patch JS (1998) Hyperspectral remote sensing for shallow waters. 1. A semianalytical model. Appl Opt 37:6329–6338

Light PR, Jones GP (1997) Habitat preference in newly settled coral trout (*Plectropomus leopardus*, Serranidae). Coral Reefs 16:117–126

Loubersac L, Dahl AL, Collotte P, Lemaire O, D'Ozouville L, Grotte A (1988) Impact assessment of Cyclone Sally on the almost atoll of Aitutaki (Cook Islands) by remote sensing. Proc 6th Int Coral Reef Symp 2:455–462

Luczkovich JJ, Wagner TW, Michalek JL, Stoffle RW (1993) Discrimination of coral reefs, seagrass meadows, and sand bottom types from space: a Dominican Republic case study. Photogram Eng Remote Sens 59:385–389

Lyzenga DR (1978) Passive remote sensing techniques for mapping water depth and bottom features. Appl Opt 17:379–383

Lyzenga DR (1985) Shallow-water bathymetry using combined lidar and passive multispectral scanner data. Int J Remote Sens 6:115–125

Maeder J, Narumalani S, Rundquist DC, Perk RL, Schalles J, Hutchins K, Keck J (2002) Classifying and mapping general coral-reef structure using Ikonos data. Photogr Eng Remote Sens 68:1297–1305

Maritorena S, Morel A, Gentili B (1994) Diffuse reflectance of oceanic shallow waters: influence of water depth and bottom albedo. Limnol Oceanogr 39:1689–1703

Miller I, Müller R (1999) Validity and reproducibility of benthic cover estimates made during broadscale surveys of coral reefs by manta tow. Coral Reefs 18:353–356

Miller MW, Weil E, Szmant AM (2000) Coral recruitment and juvenile mortality as structuring factors for reef benthic communities in Biscayne National Park, USA. Coral Reefs 19:115–123

Mobley CD, Sundman LK, Davis CO, Bowles JH, Downes TV, Leathers RA, Montes MJ, Bissett WP, Kohler DDR, Reid RP, Louchard EM, Gleason A (2005) Interpretation of hyperspectral remote-sensing imagery by spectrum matching and look-up tables. Appl Opt 44:3576–3592

Monteith J (1972) Solar radiation and productivity in tropical ecosystems. J Appl Ecol 9:747–766

Moses CS, Andrefouet S, Kranenburg CJ, Muller-Karger FE (2009) Regional estimates of reef carbonate dynamics and productivity using Landsat 7 ETM+, and potential impacts from ocean acidification. Mar Ecol Prog Ser 380:103–115

Mumby PJ, Edwards AJ (2002) Mapping marine environments with IKONOS imagery: enhanced spatial resolution can deliver greater thematic accuracy. Remote Sens Environ 82:248–257

Mumby PJ, Green EP, Edwards AJ, Clark CD (1999) The cost-effectiveness of remote sensing for tropical coastal resources assessment and management. J Environ Manage 55:157–166

Mumby PJ, Chisolm JRM, Clark CD, Hedley JD, Jaubert J (2001) A bird's-eye view of the health of coral reefs. Nature 413:36

Mumby PJ, Skirving W, Strong AE, Hardy JT, LeDrew EF, Hochberg EJ, Stumpf RP, David LT (2004) Remote sensing of coral reefs and their physical environment. Mar Pollut Bull 48:219–228

Ortiz DM, Tissot BN (2008) Ontogenetic patterns of habitat use by reef-fish in a Marine Protected Area network: a multi-scaled remote sensing and in situ approach. Mar Ecol Prog Ser 365:217–232

Palandro DA, Andrefouet S, Hu C, Hallock P, Muller-Karger FE, Dustan P, Callahan MK, Kranenburg C, Beaver CR (2008) Quantification of two decades of shallow-water coral reef habitat decline in the Florida Keys National Marine Sanctuary using Landsat data (1984–2002). Remote Sens Environ 112: 3388–3399

Phinney JT, Muller-Karger F, Dustan P, Sobel J (2001) Using remote sensing to reassess the mass mortality of Diadema antillarum 1983–1984. Conserv Biol 15:885–891

Purkis SJ, Myint SW, Riegl BM (2006) Enhanced detection of the coral Acropora cervicornis from satellite imagery using a textural operator. Remote Sens Environ 101:82–94

Purkis SJ, Graham NAJ, Riegl BM (2008) Predictability of reef fish diversity and abundance using remote sensing data in Diego Garcia (Chagos Archipelago). Coral Reefs 27:167–178

Rencher AC (2002) Methods of multivariate analysis, 2nd edn. Wiley, New York

Running SW, Nemani RR, Heinsch FA, Zhao MS, Reeves M, Hashimoto H (2004) A continuous satellite-derived measure of global terrestrial primary production. Bioscience 54:547–560

Smith SV (1978) Coral-reef area and contributions of reefs to processes and resources of the world's oceans. Nature 273:225–226

Smith VE, Rogers RH, Reed LE (1975) Automated mapping and inventory of Great Barrier Reef zonation with LANDSAT. IEEE Ocean '75 1:775–780

Smith SV, Buddemeier RW (1992) Global change and coral reef ecosystems. Annu Rev Ecol Syst 23:89–118

Spalding MD, Grenfell AM (1997) New estimates of global and regional coral reef areas. Coral Reefs 16:225–230

U.S. Coral Reef Task Force (1999) The National Action Plan to conserve coral reefs. U.S. Department of the Interior and U.S. Department of Commerce

Vercelli C, Gabrie C, Ricard M (1988) Utilisation of SPOT-1 data in coral reef cartography Moorea Island and Takapoto Atoll, French Polynesia. Proc 6th Int Coral Reef Symp 2:463–468

Waddell J, Clarke A (eds) (2008) The state of coral reef ecosystems of the United States and Pacific freely associated states: 2008. NOAA Technical Memorandum NOS NCCOS 73. NOAA/NCCOS Center for Coastal Monitoring and Assessment's Biogeography Team, Silver Spring

Walker BK, Riegl B, Dodge RE (2008) Mapping coral reef habitats in southeast Florida using a combined technique approach. J Coast Res 24:1138–1150

Wilkinson CR (ed) (2008) Status of coral reefs of the world: 2008. Global Coral Reef Monitoring Network and Reef and Rainforest Research Center, Townsville

Coral Taxonomy and Evolution

John (Charlie) E.N. Veron

Abstract Taxonomy has changed from being a science in its own right to being a servant of other disciplines. It must now reflect natural order, incorporating changes in space (biogeography) as well as time (evolution). Environmental and geographic variations in corals form continuities that override concepts of taxonomically defined species being natural units. Natural order is seen in the existence of syngameons, a highly intuitive genetic concept that replaces traditional Darwinian order. Evolution of this order produces reticulate patterns in both geographic space and evolutionary time. These patterns are molded by environmental mechanisms, not natural selection. Relevant evolutionary theories are compared in historical context and the mechanisms that link reticulate evolution with Darwinian concepts are discussed.

Keywords Evolution • reticulate evolution • coral

1 Taxonomy

Humans need names and descriptions in order to convey information about the natural world. Taxonomy must meet this need of human communication, yet at the same time it must reflect the complexities of natural organization. This organization is one of endless complexity. It is one that is, and will remain, the Achilles' heel of taxonomy.

1.1 Traditional Concepts of Species

For over 100 years, species have been considered to be the building blocks of Nature, blocks that have a time and place of origin, which can evolve, and which maintain themselves as discrete entities by being reproductively isolated. This is

J.E.N. Veron (✉)
Coral Reef Research, 10 Benalla Road, Oak Valley,
Townsville, 4811, Australia
e-mail: j.veron@coralreefresearch.com

a logical concept, but one which breaks down, for most species, when confronted with the realities of taxonomy and geographic variation. The alternative, where most species have none of these attributes, is initially less intuitive but ultimately is highly explanatory of what can actually be observed in the real world.

Concepts of what species are have long been debated, a debate driven as much by misinterpretation as information. The debate, however, has mostly been ignored by taxonomists (who often believe they "know" what their species are for one reason or another) as well as most other biologists (who often consider it armchair philosophy and thus largely irrelevant to the needs of reality). Be that as it may, species have variously been considered to be (a) self-defining natural units, (b) human-defined units composed of assemblages united by common descent, or (c) genetically defined units resulting from Darwinian natural selection and/or reproductive isolation. The last involves a group of concepts about species, sometimes called the "neo-Darwinian synthesis" or the "biological species concept" (depending on a wealth of detail) that embodies the notions of the building blocks referred to above.

1.2 Classification

Museum displays of thirty ago provided much of the scientific basis for what was then known about coral species. The specimens came from voyages of discovery from the tropical world, and they mostly went to the museums of Europe, Japan, and America. At that time, each new form was described as a new species and given a Latin name. Where major collections were made in a single country, the corals were usually given a name that, in practice, had relevance to just that country. As a result, an account of the "Corals of Japan" was quite different, for example, from an account of the "Corals of the Marshall Islands," or wherever else.

These accounts were often very different, the main reason being that taxonomists of the time did not have the advantage of scuba diving; thus, they had little opportunity to study corals in any natural habitat other than reef flats. They could not, for example, see how a single species gradually changes its growth form down a reef slope in response to decreasing light and turbulence. Most coral taxonomists (and there were plenty of them) simply concluded that any distinctive growth form was a distinct species.

In all, about 400 taxonomic publications about reef corals, and descriptions of some 2,000 "nominal" species (species in name if nothing else) stem from the prescuba era. The issues this raises are many, for the rules of nomenclature state that in most cases the oldest name given to a species is the "correct" name. Unfortunately, that frequently means that the "correct" name is based on a poorly described specimen from an unknown environment and sometimes from an unknown country. The author of a new species in those days usually had no concept of what a species might actually be, and they usually had no idea of how their species might vary with environment.

Modern coral taxonomy appears to follow the path of its predecessors, but this appearance is superficial. Taxonomy has become the servant of other disciplines and must therefore incorporate the many forms of morphological and genetic variations within species that occurs in response to environment and geographic distance. In this chapter (as in taxonomy), species are defined as naturally occurring units that are visibly distinct, both in situ and in the laboratory. This distinction is best seen when two species occur together in situ – where neither environmental nor geographic variations affect the distinction. Ideally, taxonomy should also incorporate the existence of syngameons.

1.2.1 Syngameons

The concept of the syngameon, first recognized by botanists (Grant 1981) and introduced to the marine world through corals (Veron 1995), is important for the understanding of the evolutionary mechanisms of corals and the geographic patterns they form. A syngameon, by definition, is a reproductively isolated unit in time as well as geographic space. In concept, so are neo-Darwinian species. In reality, syngameons are nothing like any of the concepts of species referred to above as they incorporate geographic variation and the spectrum of genetic links geographic variation creates among different species (Fig. 1).

A syngameon may be a single species, or it may be a cluster of different species, which have variable genetic links with other members of the syngameon. Where a syngameon contains several species, a single component species may be distinct at a single geographic location, but because it intergrades with other species at other locations, it will become submerged in a mosaic of variation at other locations. The geographic range and morphological variation of the single "species" will depend on taxonomic decisions. These decisions

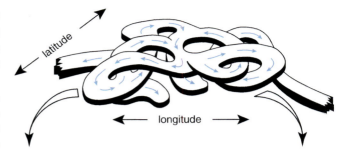

Fig. 1 Conceptual paths of gene flow of a single species. Some populations in a species' distribution range are downstream from all other populations. However, most populations import genetic information from upstream locations and export it to downstream locations. This process creates geographic variation within the range of the species and ultimately is responsible for the formation of syngameons (After Veron 2000)

will be arbitrary if they impose divisions in natural continua rather than reflect natural units. The syngameon as a whole is not morphologically visible unless its component species are determined genetically or experimentally (through cross-fertilization trials) in every part of the species' possible distribution range (Veron 1995).

2 Variation in Species

Like most plants, corals have morphological characteristics that vary according to (a) the environment in which they grow, (b) their geographic location, and (c) genetic links among component populations.

2.1 Environmental Variation

Most colonial corals show morphological variation within a single colony. Colonies of some species have variable growth forms, which are at least partly genetically regulated independently of environment. However, most species exhibit variations that are clearly associated with different parts of the colony having different micro-environments. Differences among micro-environments include growing space, light availability, sedimentation, and fish grazing. The morphological variations that result affect both growth form and corallite structure (Fig. 2).

Different colonies (or individuals) of all coral species vary along environmental gradients (Fig. 3). Divers can readily see this happening as they descend down a reef slope. This variation is also seen if a coral is transplanted from one habitat to another. It will usually change its appearance to reflect the conditions of the new habitat. Most plants do likewise; these responses are not under direct genetic

Coral Taxonomy and Evolution

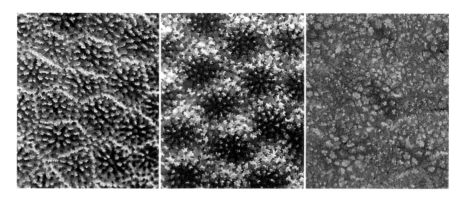

Fig. 2 Variation in the structure of corallites in a single specimen of *Porites lutea* (×10) (After Veron 1995)

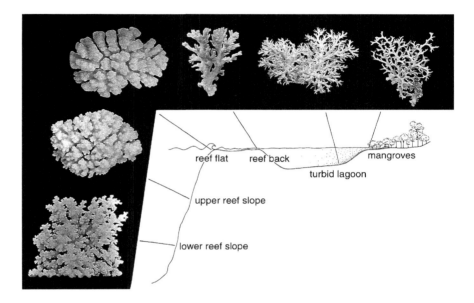

Fig. 3 Variation in the skeletal structure of colonies of *Pocillopora damicornis* from a wide range of habitats at a single geographic location (After Veron 1995)

control, although particular genotypes may be associated with particular morphologies.

The most important environmental factors controlling growth form are exposure to wave action, levels of illumination, sediment load, and exposure to currents. The different morphologies that result must be accommodated by taxonomic descriptions and understood by researchers in other fields when identifying corals.

2.2 Geographic Variation

There are two distinct categories of geographic variation in corals: those which are the outcome of environmental factors, and those which are genetic. Geographic variation that is the outcome of environment is best seen in high latitude coral communities where low temperature and nonreef habitats result in colonies being distinct from their tropical reef-dwelling counterparts. Likewise, there are whole geographic regions where the water is almost always clear (such as the Red Sea or Bahamas) and other regions where the water is usually turbid. Corals from these regions commonly have distinct points of morphological detail that are primarily due to environmental differences.

Geographic variation that is genetically based affects almost all species in some way or other. Again like most plants, the appearance of a single species changes (gradually or abruptly) from one region to the next. These changes may not concern the individual who is interested in the species of a single country, but they will definitely concern anybody who tries to answer the question "what are species?" (see below). Of more practical importance, it will also concern anybody who tries to identify a species in one country from information about that species from another country.

2.3 Genetic Links Among Populations

It is common for a series of adjacent colonies of the same species, in the same environment, to display a wide range of colors and to have a variety of morphological differences.

In such cases, the presence or absence of morphological continuities among colonies, or populations, can be used to distinguish groups of species (commonly called "sibling species") from a single variable species (commonly called "polymorphic species"). Although the differences seen are widespread, intermediate colonies can usually be found.

Genetic bridges and barriers generate reticulate patterns in time and space (see below), patterns that maintain the species' genetic heterogeneity. The taxonomic issues that arise are of endless complexity. They inevitably lead to the conclusion that there are no fundamental differences between species and subspecies taxonomic levels.

3 Taxonomic Issues

Geographic variation (whatever its cause) is far more difficult to study than environmental variation as knowledge of it must be accumulated from separate studies in different countries: it cannot be directly observed. The main taxonomic issues are: (a) species become progressively less recognizable as single units with increasing geographic range, (b) taxonomists are forced to make arbitrary decisions, (c) synonymies vary geographically. In each case, the more detailed a taxonomic study is, the greater the problem becomes.

3.1 Taxonomic Certainty and Geographic Range

With corals as with most plants, most species *do* exist as more-or-less definable units in single geographic regions, such as the Red Sea, the Indonesian/Philippines archipelago, the Great Barrier Reef, or the Caribbean. However, widespread species commonly show sufficient geographic variation that they could reasonably be divided into several separate "sibling species" were it not for the fact that these smaller units form continua. Thus, for example, the majority of species of the Red Sea also occur in Indonesia, but many are sufficiently different in the Red Sea that they would be considered to be distinct species if they were transplanted from the Red Sea to Indonesia (Fig. 4).

3.2 Arbitrary Decisions

In theory, taxonomy should accommodate biogeographic patterns. In practice, doing so creates issues like those created when a flat map is projected onto a sphere. The bigger the area of the map, the greater the distortion that results. It is possible to modify the flat map using different types of projections, but this only changes the nature of the distortion.

In traditional taxonomy, geographic variation within a species is accommodated by creating divisions within the species, such as varieties, races, or geographic subspecies. In reality, geographic variation repeatedly overrides the morphological boundaries of individual species. In other words, natural continua go beyond the taxonomic or morphological boundaries of a single species. This cannot be accommodated by creating divisions within species. The problem remains if the species unit is "split" into smaller units or "lumped" into larger units and it is not solvable by further or more detailed study. Ultimately, the only unit in Nature that is real is the syngameon referred to above.

The reasons why syngameons are not used in taxonomy are: (1) they can only be determined with any degree of certainty through exhaustive cross-fertilization studies in all geographic regions where their component species occur, (2) they are not likely to have distinguishing morphological characteristics, and (3) they would include so many morphological species that they would need to be redivided into subunits of some kind in order to be useful.

This issue will always force taxonomists to make an arbitrary decision as to what a particular species is. Some groups of species may be distinctive in some geographic regions and not in others. The outcome of detailed studies of these species may either be a single "species complex" or a group of similar species. In either case, species descriptions

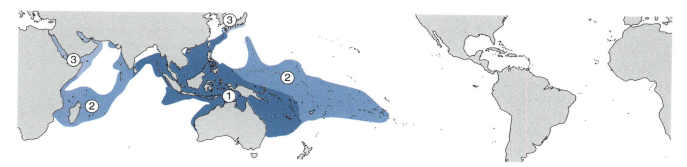

Fig. 4 Geographic variation from a taxonomic point of view. A species that is taxonomically straightforward in the central part of its range (area 1) may have morphological uncertainties in more peripheral parts (area 2) and form distinct subspecies at extremities (area 3) (After Veron 2000)

and distribution maps artificially simplify the reality of the complex.

3.3 Geographic Variation in Synonymies

Synonymies are intended to clarify the taxonomic history of an individual species and as such are commonly considered to be a formal statement about the status of that species. However, if species can vary geographically, it follows that detailed synonymies may also vary geographically. In practice, this consideration is seldom taken into account.

4 Natural Organization

The above considerations argue that the term "species" can legitimately have a wide range of meanings. Within a single region, species are usually morphologically distinguishable from other species and are genetically semi-isolated from other species. Over their full geographic range, species vary morphologically and genetically, and are not necessarily morphologically or genetically isolated from other species. Geographic patterns of species combine with distribution changes to produce networks of genetic links. These links are not observable in single geographic regions and for this reason are not generally recognized in taxonomic studies of corals or most other groups of organisms.

In theory at least, corals vary according to the spectrum of habitats they occupy, the size of their geographic range, and the extent of gene flow within that range. In practice, variation that is correlated with habitat (such as the variations of *Pocillopora damicornis* illustrated above) can be satisfactorily described and illustrated so that the species is a single identifiable unit. However, genetic variation defies the use of units if they merge with other units – if they create continuities with other units. When this happens, units (species) can only have arbitrary boundaries. This certainly happens with corals and probably does so with most other marine life.

General principles about species over large geographic ranges are: (1) their geographic boundaries merge with other species, (2) their morphological boundaries merge with other species, (3) there are no definable distinctions between species and subspecies taxonomic levels (Veron 1995).

Nature, therefore, is mostly composed of continua in space and time, continua that will always defy human attempts to make taxonomic units. If taxonomists create units (as they must if their taxonomy is to be meaningful), it must be remembered that these units are artificial: they lack discrete morphological boundaries and they also lack discrete geographic boundaries. Also, as our knowledge of coral biology increases, it is becoming clear that the ecological, physiological, and reproductive characters of individual species also vary geographically. It is a human dilemma that we need taxonomic units at all to communicate information about Nature's organization, when Nature is not necessarily divided into units of any kind.

5 Evolutionary Mechanisms

What happens to geographically variable species in evolutionary time? The issues that arise are not just theoretical. They make it necessary to consider evolutionary change in a way that is very different from that which is generally accepted in both popular and scientific literature. This introduces reticulate evolution, a concept fundamentally distinct from neo-Darwinian evolution that involves a different way of looking at what species are and how they change.

5.1 Reticulate Evolution

Evolutionary change can seldom be observed directly: it must be reconstructed, as is commonly done using phylogenetic trees. Trees are also one way of envisaging reticulate evolution. Such a tree might have a main trunk for the family, several large branches for genera, and many fine branches for the species. Species that are currently alive will be the tips of the uppermost branches and these will be trimmed to uniform height to represent present time. The branch tips, each representing a single species, can be converted into distribution maps, each map representing the present-day distribution of one species. If a single branch (or species) is then sliced into a sequence of horizontal layers, and each layer is turned into a distribution map, each map will indicate the distribution of the species at progressively distant points in time. If the maps are viewed like the pages of a book, the pattern will change sequentially back in time. These changes will not be just distribution changes, they will also be genetic changes occurring in response to changes in ocean currents. As a result, a species (or map) at one point in time is not the same as it is in another point in time: it has been genetically as well as geographically changed. These changes do not occur uniformly, they occur irregularly over the species' geographic range and evolutionary history. Geographic space and evolutionary time interact. The species may break apart, then reform into a slightly different species. This creates a "reticulate" pattern in both geographic space and evolutionary time (Veron 2002).

As Fig. 5 illustrates, basic issues that arise with reticulate evolution are: (a) no species has a clearly defined time of origin, (b) rates of evolution and extinction are similar over

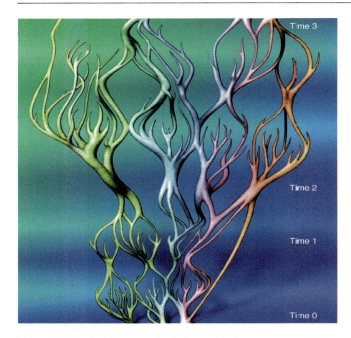

Fig. 5 A hypothetical view of reticulate evolutionary change within a group of species. At the bottom (Time 0), the group forms three distinct species each of which is widely dispersed by strong currents. At Time 1, the group forms many indistinct small species units that are geographically isolated because currents are weak. At Time 2, the group forms four species that are again widely dispersed by strong currents. Over the long geological interval to Time 3, the group has been repackaged several times (After Veron 2000)

the geological interval represented by the diagram, (c) there are no differences between mainstream species (represented by the thicker branches) and subspecies (represented by thinner branches), (d) there are no clearly defined differences between "species" and "hybrids," (e) the total amount of genetic information represented by the diagram has not greatly changed over time, but has been repackaged into different "species" units, (f) extinction occurs through repackaging as well as terminations of lineages.

5.2 Ocean Currents and Reticulate Patterns

Ocean currents are the vehicle of dispersion and the pathways of genetic connectivity. When a species breaks apart (as a result of weak ocean currents), it becomes many species. When it re-forms (as a result of strong ocean currents), it may again be a single species or it may be more than one species. It may also contain genes from other species. This is reticulate "repackaging," and it occurs constantly at all scales of space and time. Importantly, the repackaging is not confined to a single phylogeny, it involves many phylogenies simultaneously (Fig. 6).

Reticulate evolution is largely under physical environmental control rather than biological control. Although competition will always have a place, the physical control changes patterns

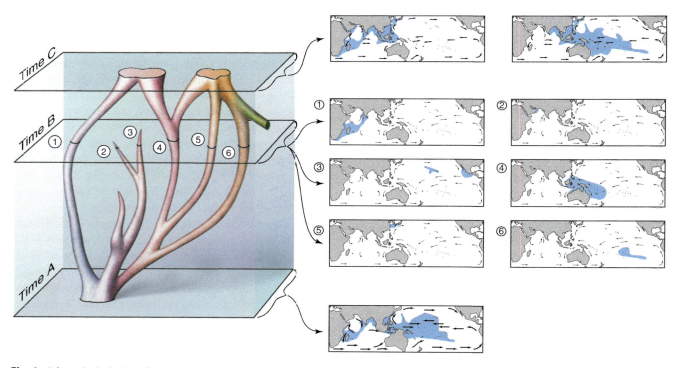

Fig. 6 A hypothetical view of evolutionary change in a small phylogeny (*left*) correlated with distribution patterns (maps, *right*) (see text) (After Veron 2000)

of genetic connections: it acts on genetic composition rather than morphology. This is again in sharp contrast with a major aspect of the neo-Darwinism where evolution is primarily controlled by competition between species, a process that creates morphological changes. Reticulate evolution is therefore a mechanism of slow arbitrary change rather than a mechanism for progressive improvement.

Reticulate evolution is primarily driven by changes in surface circulation patterns causing changes to the dispersal patterns of larvae. If currents remained constant throughout evolutionary time, the oceans would be divisible into source areas (where larvae come from) and destination areas (where larvae go to), and geographic gradations within species would be generally predictable. However, with the exception of the great ocean gyres, currents are not constant. Sea levels may fluctuate more than 100 m, oceanic passages are opened and closed by tectonic movements, and the Earth goes through cyclical climate changes due to variations in the Earth's orbit around the Sun. These, and probably several other types of geo-climatic cycles, cause variations in ocean currents. These changes open and close genetic contacts: they generate reticulate patterns.

The concept of reticulate evolution is highly explanatory of the observations about coral taxonomy and biogeography. It is also strongly supported by what we know of coral reproduction and is starting to be supported by genetic studies.

The main points illustrated in Fig. 6 are: (a) At Time A, currents are strong and the phylogeny exists as a single widespread species. (b) At Time B, currents are weak and the phylogeny has broken up into six geographically isolated indistinct or "sibling" species. (Two of these species (labeled 2 and 3) are doomed to subsequent extinction.) (c) At Time C, currents are strong again and the phylogeny has been repackaged into two widespread species, which are reproductively isolated. The species on the right is part of the phylogeny of another species (colored green). Importantly, there is no distinction between geographic (sympatric) and nongeographic (allopatric) origination because evolution is driven by environmental change, not biological competition.

5.3 Competing Hypotheses

Theories of biogeography and evolution have always been interrelated. Historically, there have been five main schools:

5.3.1 Darwin's Centers of Origin

Darwin (1859) proposed that dispersion was primarily driven by the evolution of new species through natural selection, where older and less competitive species are displaced away from centers of origin to more distant places by fitter and more competitive, descendant species. This is but one aspect of Darwin's theory of the origin of species, yet is one of his most thoughtful for it integrated the known distribution of many plant and animal groups with their fossil record.

Darwin's Centers of Origin theory, or refinements of it, dominated all coral biogeographical thinking until the theory of continental drift appeared to make a fundamental aspect of it – geographic centers – untenable (Briggs 1984). Interest in this theory, at least as far as corals are concerned, is historical for there are no identifiable centers of origin, only centers of diversity created by overlapping species ranges rather than endemism.

5.3.2 Croizat's Panbiogeography

Panbiogeography is very much the subject of Leon Croizat, biogeography's most prolific writer and an outspoken opponent of both Darwin's Centers of Origin and vicariance biogeography (see below) (Croizat 1981). Panbiogeography is primarily a method (rather than a theory) where "road maps" of taxa are superimposed to produce "dispersion highways" (Craw 1988). The intersection of many highways denotes the existence of "gates" or "nodes." These are places of high connectivity from which dispersion occurs.

Panbiogeography has more relevance to terrestrial life than marine where the biogeographic base map for marine life continually changes with the vagaries of paleoclimates, ocean currents, and long-distance dispersal (Craw and Weston 1984).

5.3.3 Vicariance Biogeography

Vicariance biogeography was a major advance on Darwin's Centers of Origin evolutionary theory (McCoy and Heck 1976). It is based on the notion that if one or more barriers form across a species' distribution range, the divided population will diverge in time, forming two or more distinct populations. If these become reproductively isolated, they would be two or more distinct species when the barriers are removed. This process on a large scale would lead to patterns of species that have no center of origin. Importantly, there is no requirement for dispersion to occur.

Vicariance speciation requires that an ancestral species occupies the geographic area of its descendants and that this area becomes divided up as those descendants progressively evolve. The process, therefore, produces ever-increasing numbers of species, each occupying an ever-decreasing geographic range. The result is ever-increasing endemicity; hence, a close association between the concepts of endemicity and vicariance.

There are significant arguments against vicariance as a general principle in coral biogeography: (a) dispersion is actually of overwhelming importance to corals, (b) the place of origin of coral species is not correlated with place of occurrence (Fig. 6), (c) the process of evolutionary change neither commences nor ceases in any particular point in time (Fig. 5), and (d) there has been no exponential increase in numbers of coral species in any geographic area. On these grounds, vicariance cannot be accepted as an evolutionary mechanism in its own right. More importantly, it follows that cladistics as an analytical method must lead to faulty conclusions, as originally stated by its founder (Hennig 1966). Both vicariance and cladistics (almost universally used in genetic analysis, also in many biogeographic analyses) are based on the same supposition that there is no lateral gene flow (or hybridization) among the species under question (Sober 1988). Put another way, both vicariance and cladistics require the absence of syngameon formations if analyses produced are numerically valid.

5.3.4 Dispersion and the Founder Principle

The concept of dispersion biogeography, proposed by Mayr (1942) and expanded by Carson (1971), is based on the notion that a species may come to occupy a previously unoccupied place through dispersion (a "founder" event). It may then be genetically isolated by the formation of a barrier and in time form two or more species. As with vicariance, speciation is allopatric and the outcome is seen in patterns of endemism. The main difference between this theory and vicariance is that this theory requires dispersion to occur and vicariance requires dispersion not to occur. If species co-occur after speciation, additional dispersion is equally necessary for both theories.

Dispersion biogeography is generally linked with Center of Origin biogeography as the alternative to vicariance (Simberloff 1981), but, as with so much biogeography, this highlights extremes rather than common ground. In reality, most evolutionary issues concerning spatial separation between populations of the same species are irrelevant to corals because of their capacity for long-distance dispersion.

5.3.5 Equilibrium Theory

The equilibrium theory of island biogeography (Macarthur and Wilson 1963, 1967), an essentially ecological theory, had a major impact on biogeographic concepts when it was first proposed. When rates of immigration to an island are balanced by local extinction, the number of species (but not necessarily the identity of those species) remains approximately constant. For any particular island, there is a dynamic balance between immigration, local extinction, proximity to a continental mainland, and the area of the island. Equilibrium theory has largely gone out of fashion although it is a useful concept for zooxanthellate corals on a regional scale.

5.3.6 Competing Hypotheses in Summary

Differences in principle between the five main schools of evolutionary theory considered here are as follows: (a) Center of Origin biogeography alone requires sympatric speciation. (b) Center of Origin biogeography requires that dispersion occurs after speciation, vicariance biogeography requires that it does not occur during speciation, and dispersion biogeography requires that it does occur before speciation. Panbiogeography has no geographic requirements. (c) Vicariance biogeography leads to increased biodiversity while the other concepts do not necessarily alter species diversity at the place of origin. (d) All concepts have a causal links between place (or pattern) of evolution and place (or pattern) of occurrence. The patterns are different in each case.

It is emphasized that boundaries between these theories are continually blurred, especially as the extreme complexity of the organization of life does not lend itself to reductionism, nor to rules of any kind. Nevertheless, reticulate and Darwinian evolution appear to be mutually exclusive theories.

5.4 Where Reticulate Evolution and Darwinian Evolution Meet

The theory of reticulate evolution, alone among its many rivals, requires no geographic restrictions, does not depend on barriers forming or going, and does not result in changes in species diversity. At least as far as corals are concerned, the theory provides a sound reason for observable taxonomic dilemmas, is seen in biogeographic patterns, is genetically sound, is reflected in the fossil record, and accords well with reproductive biology. It is for these reasons that reticulate evolution appears to be in direct conflict with Darwin's concept of evolution, which is now so widely accepted and which is so intuitive and logical.

However, these two theories are compatible. It can reasonably be supposed that most life, terrestrial as well as marine, does not exist as reproductively isolated species but rather as syngameons of some sort. (This is seen, for example, in domesticated plants and animals, almost all of which can be artificially hybridized.) These syngameons can undergo change through reticulate evolution, but not natural selection. However, depending on many circumstances, individual species (or genetic units of any kind) may become temporarily genetically isolated, perhaps through the formation of an environmental barrier. If these isolated units become genetically cohesive, they would then be able to respond to selective pressure and would be able to undergo Darwinian evolution. The level of isolation required would be unendingly variable among different taxa. If the unit continued to remain isolated, it may continue to evolve as a single species or develop into a syngameon. Alternatively, the newly formed unit may in time link back into the syngameon from which it came. In this case, it will cease to evolve through natural selection. There is, of course, an almost infinite array of such possibilities in the dynamics of life.

Acknowledgments The author thanks Dr. M.G. Stafford-Smith for critically reviewing the manuscript.

References

Briggs JC (1984) Centers of biogeography. In: Biography monograph 1. University of Leeds, Leeds
Carson HL (1971) Speciation and the founder principle. Stadler Genet Symp Univ Mo 3:51–70
Craw R (1988) Panbiogeography: methods and synthesis. In: Meyers AA, Giler PS (eds) Analytical biogeography. Chapman & Hall, London, pp 405–435
Craw R, Weston P (1984) Panbiogeography: a progressive research program? Syst Zool 33:1–13
Croizat L (1981) Biogeography: past, present and future. In: Nelson G, Rosen DE (eds) Vicariance biogeography. A critique. Columbia University Press, New York, pp 501–523
Darwin C (1859) On the origin of species my means of natural selection, or the preservation of favoured races in the struggle for life. Murray, London, pp 703
Grant V (1981) Plant speciation. Columbia University Press, New York
Hennig W (1966) Phylogenetic systematics. University of Illinois Press, Urbana
Macarthur RH, Wilson EO (1963) An equilibrium theory of insular biogeography. Evolution 17:373–387
Macarthur RH, Wilson EO (1967) The theory of island biogeography. Princeton University Press, Princeton
Mayr E (1942) Systematics and the origin of species. Columbia University Press, New York
McCoy ED, Heck KL (1976) Biogeography of corals, seagrasses and mangroves: an alternative to the centre of origin concept. Syst Zool 25:201–210
Simberloff D (1981) There have been no statistical tests of cladistic biogeographical hypotheses. In: Nelson G, Rosen DE (eds) Vicariance biogeography. A critique. Columbia University Press, New York, pp 40–63
Sober E (1988) The conceptual relationship of cladistic phylogenetics and vicariance biogeography. Syst Zool 37(2):45–53
Veron JEN (1995) Corals in space and time. Cornell University Press, New York
Veron JEN (2000) Corals of the world, vol 3. Australian Institute of Marine Science, Australia
Veron JEN (2002) Reticulate evolution in corals. Proc 9th Intern Coral Reef Symp 2000:43–48

The Coral Triangle

John (Charlie) E.N. Veron, Lyndon M. DeVantier, Emre Turak, Alison L. Green, Stuart Kininmonth, M. Stafford-Smith, and N. Peterson

Abstract Spatial analyses of coral distributions at species level delineate the Coral Triangle and provide new insights into patterns of diversity and endemism around the globe. This study shows that the Coral Triangle, an area extending from the Philippines to the Solomon Islands, has 605 zooxanthellate corals including 15 regional endemics. This amounts to 76% of the world's total species complement, giving this province the world's highest conservation priority. Within the Coral Triangle, highest richness resides in the Bird's Head Peninsula of Indonesian Papua, which hosts 574 species, with individual reefs supporting up to 280 species ha^{-1}. The Red Sea/Arabian region, with 364 species and 27 regional endemics, has the second highest conservation priority. Reasons for the exceptional richness of the Coral Triangle include the geological setting, physical environment, and an array of ecological and evolutionary processes. These findings, supported by parallel distributions of reef fishes and other taxa, provide a clear scientific justification for the Coral Triangle Initiative, arguably one of the world's most significant reef conservation undertakings.

Keywords Coral triangle • coral reefs • biogeography • conservation

J.E.N. Veron (✉), L.M. DeVantier, and E. Turak
Coral Reef Research, 10 Benalla Road,
Townsville, Oak Valley, 4811, Australia
and
Australian Institute of Marine Science,
MSO, Townsville, 4811, Australia
e-mail: j.veron@coralreefresearch.com

M. Stafford-Smith
Coral Reef Research, 10 Benalla Road,
Townsville, Oak Valley, 4811, Australia

S. Kininmonth
Australian Institute of Marine Science,
MSO, Townsville, 4811, Australia
e-mail: Sgampel@kenes.com

A.L. Green and N. Peterson
The Nature Conservancy, 51 Edmondstone Street, New york,
South Brisbane, 4101, Australia

1 Introduction

Over the past several decades, biogeographers have proposed centers of marine biodiversity of varying shapes, all centered on the Indonesian/Philippines Archipelago. Some stem from biogeographic theory or geological history, others from coral and reef fish distributions. These centers have been given a variety of names: Wallacea, East Indies Triangle, Indo-Malayan Triangle, Western Pacific Diversity Triangle, Indo-Australian Archipelago, Southeast Asian center of diversity, Central Indo-Pacific biodiversity hotspot, Marine East Indies, among others (reviewed by Hoeksema2007).

It was not until the postwar era that coral biogeography came to the forefront of marine biogeography, a position initiated by the American paleontologist John Wells (1954) when he published a table of coral genera plotted against locations. Many reiterations of this table formed the basis of sequence of published maps (Stehli and Wells 1971; Rosen 1971; Coudray and Montaggioni 1982; Veron 1993). These publications, all at generic level, highlighted the Indonesian/Philippines Archipelago as the center of coral diversity. Significantly, they also included the Great Barrier Reef of Australia as part of that center.

This view was fundamentally altered when global distributions were first compiled at species level, an undertaking that needed a computer-based spatial database. This compilation clearly indicated that the Indonesian/Philippines Archipelago, but not the Great Barrier Reef, was the real center of coral diversity (Veron 1995), a pattern now well established (see below).

The significance of this seemingly innocuous finding was not lost on conservationists. It meant that the international focus for coral and, by extrapolation, reef conservation shifted from the highly regulated World Heritage province of the Great Barrier Reef to the relatively understudied region to the north, where reefs were largely unprotected, and where human population densities and consequent environmental impacts were high by most world standards.

Political response to the delineation of the Coral Triangle (CT) was prompt. In August 2007, President Yudhoyono of

Indonesia proposed a new, six-nation Coral Triangle Initiative (CTI), as a mechanism to conserve key components of the global center of coral reef biodiversity. In September 2007, 21 world leaders attending the Asia Pacific Economic Cooperation (APEC) summit in Sydney formally endorsed CTI. At the CTI Summit in May 2009, involving leaders from all six CT countries, extraordinary political commitments to coral reef and marine conservation were made. The CTI has become one of the biggest conservation initiatives ever undertaken in the marine world.

2 Delineating the Coral Triangle

Delineation of the CT (Veron et al. 2009b) was established by the spatial database *Coral Geographic*, a major update of the original species maps of Veron (2000). This database contains comprehensive global species maps of zooxanthellate coral distributions in GIS format, allowing them to be interrogated to compare geographic regions or to elucidate patterns of diversity and endemism.

The 798 species maps in the *Coral Geographic* database are each divided into 141 ecoregions (Fig. 1), an approach increasingly used in biogeography (Spalding et al. 2007). These maps, which include verified published occurrences of each species in each ecoregion together with original data are from two sources: (1) revisions of the database used to generate the species distribution maps of Veron (2000) and (2) species complements derived from original fieldwork by the first three authors in 83 of the 141 ecoregions. Continually updated details of this dataset, currently including >2,500 georeferenced sites linked to habitat data, will be available online in 2010.

In summary, coral ecoregions provide a blueprint for establishing a globally representative network of coral reef MPAs. The world's highest diversity occurs in the CT, an area where more than 500 coral species are found in each ecoregion (Fig. 2). Species attenuate latitudinally according to ocean temperature (a) northward to mainland Japan, dispersed by the Kuroshio, (b) southward along the west

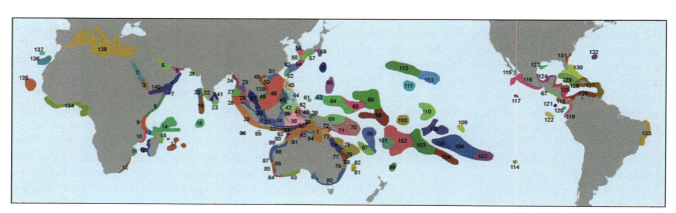

Fig. 1 Coral ecoregions of the world, delineated on the basis of known internal faunal and/or environmental uniformity and external distinctiveness from neighboring regions. The identity of the numbered ecoregions and the number of species in them is in Veron et al. (2009a) (From the spatial database *Coral Geographic*, see text)

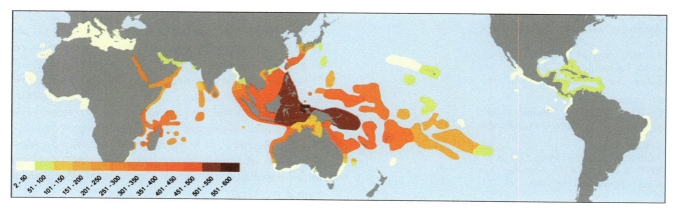

Fig. 2 Global biodiversity of zooxanthellate corals. Colors indicate total species richness of the world's 141 coral biogeographic "ecoregions" (From the spatial database Coral Geographic, see text)

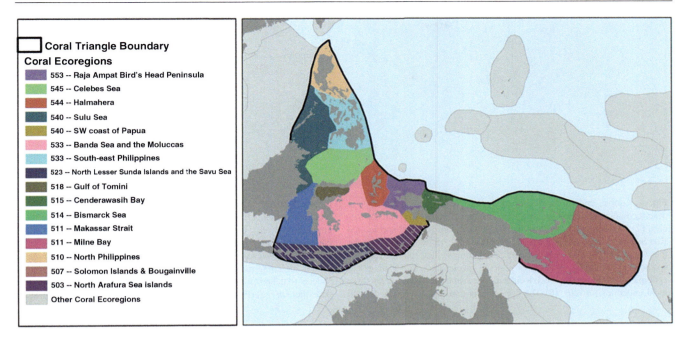

Fig. 3 Ecoregions and species richness of zooxanthellate corals of the CT (determined by the spatial database Coral Geographic, see text). A total of 1,118 sites were studied within the 16 ecoregions of this province (*left* panel, showing number of species per ecoregion); however, some islands of southern Indonesia (the two hatched ecoregions), especially their southern coastlines, remain data-deficient. Each ecoregion has >500 species (From the spatial database *Coral Geographic*, see text)

Australian coast, dispersed by the Indonesian Through-flow and the Leeuwin Current and (c) southward along the east Australian coast, dispersed by the East Australian Current. Species attenuate longitudinally eastward across the Pacific according to geographic distance and the concentration of reefs. The Indian Ocean has a more uniform longitudinal (east–west) distribution of diversity, although the eastern and western edges of this region are separated by a paucity of reefs in the equatorial center around Chagos (Sheppard 1999). All localities of the Atlantic are relatively depauperate and there are no obligate hermatypic species in common between the Atlantic and Pacific.

Sixteen ecoregions of the world have >500 species; these define the CT and reveal its internal components to the level or resolution of this study (Fig. 3).

3 Hotspots of Biodiversity and Endemism

"Hotspot" is a term frequently used by conservation biologists to denote a relatively restricted geographic area containing exceptionally high levels of biodiversity and/or endemism, a concept that has been effective in prioritizing conservation activities where resources are limited (Mittemeier et al. 1998). The term was introduced to the marine realm by Roberts et al. (2002) in a study based on fish, corals, molluscs, and lobsters, where 18 hotspots were identified in areas ranging in size from tiny Easter Island to the whole of the Great Barrier Reef. Methodologies and conclusions of that study were criticized as having questionable relevance to conservation (Hughes et al. 2002; Baird 2002). However, several conclusions of Hughes et al. (2002) with respect to levels of endemicity and the lack of concordance with richness are in turn not supported by the present study. Here, we demonstrate that certain reef areas do in fact rate highly in richness and endemicity, meeting both hotspot criteria.

Diversity may be the result of (1) a high level of endemism or (2) the overlap in the ranges of species with wide ranges (Veron 1995). The former category contributes 2.5% and 7.4% of the diversity of the CT and the Red/Arabian Sea region, respectively, and is a smaller component of all other diverse ecoregions. With 605 species, amounting to 83% of all the species of the Indo-Pacific or 76% of the species of the world, the coral diversity of the CT creates an overwhelming case for top conservation priority (see below). Within the Red Sea (with a total of 333 species), the Sinai Peninsula (205 species), northern Saudi Arabia (260 species), and Eritrea (219 species) have conservation merit based on diversity and endemism combined.

Patterns of endemism are created by isolation: either geographic distance (locations remote from centers of diversity) or geographic enclosure. It is a complex issue to address in corals and other taxa that are widely dispersed because it requires knowledge of where a species does not occur as well

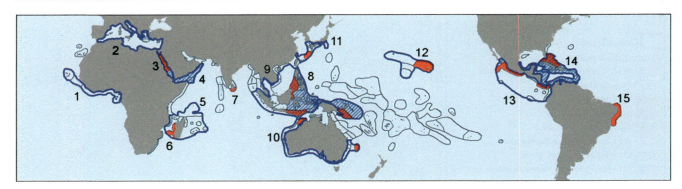

Fig. 4 Global endemism of zooxanthellate corals excluding species described after year 2000 (see text). (1) Eastern Atlantic: two regional species. (2) Mediterranean: three of the four species are regional species. (3) Red Sea: the north has six local endemics (*in red*); 15 regional species for the whole Red Sea. (4) Red Sea/Arabia combined: (hatched) 27 regional species. (5) SW Indian Ocean: four regional species including Madagascar endemics. (6) Madagascar: two local endemics (*in red*). (7) Sri Lanka: three local endemics (*in red*). (8) The CT: 15 regional species (hatched) with three centers of endemism (*in red*). (9) The CT and adjacent Asia: 41 regional species. (10) Australia: six regional species including two local endemics (*in red*). (11) Japan: six regional species including three local endemics. (12) Hawaii and Johnston Atoll: four regional species, of which three are endemic to Hawaii (*in red*). (13) Far Eastern Pacific: seven regional species including two local endemics (*in red*). (14) Caribbean: 24 regional species (hatched) including one endemic (*in red*). (15) Brazil: four local endemics (*in red*) (From the spatial database *Coral Geographic*, see text)

as where it does. Analyses thus require comprehensive data covering entire oceans. Levels of endemism also vary according to the size of selected areas and their geographic position relative to adjacent areas. Furthermore, new species are commonly thought to be endemic to their place of discovery, but are usually found elsewhere in subsequent studies. For these reasons, previous assessments of endemism are not adequately supported by relevant data. *Coral Geographic* reveals the best documented patterns of endemism to date, located in a complex of overlapping regions (Fig. 4).

4 Characteristics of the Coral Triangle

Indonesia and surrounding countries have attracted more attention from reef biogeographers than all other regions of the world combined (e.g. Briggs 2005). As noted above, the proposed centers of diversity derived from coral and reef fish distributions, biogeographic theory, and geological history have led to diverse opinions about where the global center of reef diversity actually is and what it should be called (reviewed by Hoeksema 2007). The present authors argue that the name "Coral Triangle" should be delineated by scleractinian coral diversity and that of coral-associated organisms, especially reef fishes for which comprehensive data are also available (Allen 2006, 2007). This name was first used in this context by Werner and Allen (1998) and has since gained wide acceptance in biogeographic, conservation, and faunistic studies.

The first comprehensive species-level diversity map (Veron 1995) showed that northern New Guinea was part of the center as earlier indicated by the distribution of mushroom corals (Hoeksema 1993). However, the Solomon Islands remained conspicuously data-deficient until a survey in 2004 (Green et al. 2006) revealed a high diversity of both corals (Veron and Turak 2006) and reef fish (Allen 2006), studies which formed the basis of a demarcation of the CT close to that which we now have (Green and Mous 2008).

The CT was delineated on the basis that it is an area that contains a high proportion of the species diversity of the Indo-Pacific and that this diversity occurs in an area small enough to permit meaningful conservation. With this delineation, the 16 ecoregions of the CT each host >500 reef coral species. If the CT were extended to include the northern Great Barrier Reef, Vanuatu, New Caledonia, and Fiji (an additional nine ecoregions), the total area would be approximately doubled for an addition of only 12 species including only three endemics. Conversely, if all conservation effort were centered in the Papuan Birds Head Peninsula of Western Papua diversity center (hosting 92% of the species of the CT), there would be little redundancy and extensive areas where human impact is low would be excluded (notably most of Papua New Guinea and the Solomon Islands).

The CT defined by *Coral Geographic* – that adopted by the CTI – is an area of 5.5×10^6 km^2 of ocean territory of Indonesia, the Philippines, Malaysia (Sabah), Timor Leste, Papua New Guinea, and the Solomon Islands – less than 1.6% of the world's total ocean area. Although the CT boundary was determined on the basis of coral diversity, this delineation does not provide any new biogeographic insights. The CT is not a distinct biogeographic unit, but comprises portions of two biogeographic regions (Indonesian-Philippines Region, and Far Southwestern Pacific Region, Veron 1995). Within the CT, highest richness resides in the Bird's Head Peninsula of Indonesian Papua, which hosts 574 species. Individual reefs there have up to 280 species ha^{-1}, over four times the total zooxanthellate scleractinian species richness of the entire Atlantic Ocean (Turak and DeVantier in press). Within the Bird's Head, The Raja Ampat Islands ecoregion has the world's coral biodiversity bullseye, with 553 species (Veron 2000; Turak and Souhoka 2003).

Importantly, boundaries of the Raja Ampat "bullseye" and the Birds Head diversity center are not highly distinctive. Indeed, more than 80% of all CT species are found in at least 12 of the 16 CT ecoregions. Nor is this region markedly distinct from neighboring ecoregions to the south and southeast. Ninety-five percent of CT species are found in one or more adjacent ecoregions (notably other parts of SE Asia including Malaysia, Thailand and Vietnam, Micronesia, the northern Great Barrier Reef, Vanuatu, New Caledonia, and Fiji), although all exhibit marked declines in species richness and ubiquitousness. It may therefore be asked: are the corals of the CT representative of larger areas; is their pattern of diversity seen in other taxa; and are the environmental and/or biological properties of the region unique? Many authors have commented on these questions in the context of evolutionary processes or conservation issues. They clearly have interlinked answers, involving the region's geological setting, its present physical environment, and an array of ecological issues that are not just relevant to corals.

Although the present chapter is about corals, the distribution of coral reef fishes, the only other major taxon for which comprehensive data are available (Allen 2007, who analyzed distribution patterns of 3,919 Indo-Pacific species) strongly supports the present delineation of the CT. In total, this area contains 52% of the reef fishes of the Indo-Pacific (37% of reef fishes of the world). Regional endemism is also significant in reef fish (100, 28, and 22 species for Indonesia, Philippines, and Papua New Guinea, respectively); however, the highest percentage of endemism, as for corals, is exhibited by remote island locations. Other major faunal groups, notably molluscs (Wells 2002) and crustaceans (De Grave 2001), have very high numbers of undescribed or cryptic species and thus are relatively little known at species level (e.g. Meyer et al. 2005). However, many biogeographic publications indicate that a wide variety of taxa reach maximum diversity in areas within the CT (reviewed by Hoeksema 2007). Although most of these taxa occupy shallow marine habitats, coral reefs are sometimes of secondary importance, as seen in the distributions of mangroves (Ricklefs and Latham 1993; Hogarth 1999; Groombridge and Jenkins 2002) and seagrass (Spalding et al. 2003), which also have the highest diversity within the CT. Even azooxanthellate corals, which have none of the physiological restrictions of zooxanthellate species, have a center of global diversity within the CT (Cairns 2007). These diversity maxima of fauna and flora, especially those not associated with reefs, are only seen in provinces the size of several ecoregions; areas large enough to contain an extreme diversity of habitats created by the complex coastlines of island archipelagos.

5 Reasons for Existence of the Coral Triangle

So much interest from so many points of view begs the question: why does the CT exist? There is no one simple answer, rather there are several interacting factors operating over different temporal and spatial scales.

5.1 Geological History

Two aspects of the geological history of the CT are relevant:

1. The southern half of the CT has been tectonically unstable as far back as the Eocene (38 million years ago), creating a constantly changing geography leading to repeated environmental perturbations, habitat complexity, and (it can only be presumed) evolutionary changes. The Philippines archipelago has had a different, perhaps a less dramatic, geological past, although the Miocene Ryukyu limestones, extending from Japan to Indonesia, show that reef distributions have changed beyond recognition over that time. The fossil record suggests that the corals of the CT are the world's youngest – less than half the mean age of their Caribbean counterparts. These relatively young genera either evolved in the region of the CT or have survived there since going extinct elsewhere (Stehli and Wells 1971; Veron 1995).

2. However important plate tectonic movements were to ocean circulation patterns of the distant past, they are small when compared with the impacts of sea-level changes during the Pleistocene. At least eight times during the last two million years, the shorelines of the CT region have alternated between present sea level and minus 130 m (approximately) (Siddall et al. (2003) (Fig. 5). All reefs were repeatedly aerially exposed, yet deep water remained in close proximity. The CT is thus characterized by complex island shorelines creating diverse shallow habitats adjacent to deep (>150 m) ocean. This created conditions for minimal broad-scale dislocation during times of rapid sea-level change, while also causing more localized changes in oceanographic patterns and isolation of populations to greater or lesser degree within marginal seas and large embayments, thus driving reticulation (see below).

5.2 Dispersion

1. The CT acts as a "catch-all" for larvae moving towards the region, entrained in both the South Equatorial Current and the North Equatorial Current (Jokiel and Martinelli 1992; Veron 1995)

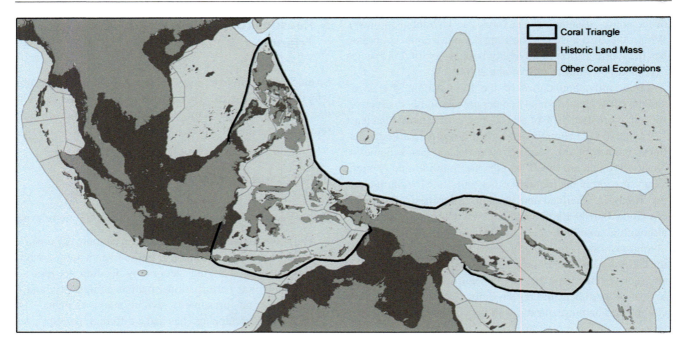

Fig. 5 At least eight times during the last 2 million years, the shorelines of the CT region alternated between those shown here. All reefs (which cannot be viewed at this scale) were repeatedly aerially exposed, yet deep water remained in close proximity. The CT is seen to be characterized by complex island shorelines creating diverse habitats and adjacent deep (>150 m) ocean providing minimal dislocation during times of sea-level change

2. Dispersion occurs away from the CT so that, at progressively increasing distance, species attenuate according to ocean temperature (a) northward to mainland Japan, dispersed by the Kuroshio, (b) southward along the west Australian coast, dispersed by the Indonesian Through-flow and the Leeuwin Current, and (c) southward along the east Australian coast, dispersed by the East Australian Current (Veron1995). This suggests that the CT is the most diverse part of the whole central Indo-Pacific simply because all other regions have attenuated species richness
3. Complex eddies created by the Indonesian Through-flow (Gordon and Fine 1996) drive genetic mixing, which constantly changes with wind, season, and (over geological time) sea level. Genetic mixing of this nature creates genetic heterogeneity through vicariance (see below); it also drives reticulate evolution (see below)

5.3 Biogeographic Patterns

Biodiversity is reflected in biogeographic patterns and the environments that created those patterns:

1. Diversity may be the result of (a) a high level of endemism or (b) the overlap in the ranges of species with wide ranges. Importantly, the first category contributes only 2.5% of the coral diversity of the CT. The biodiversity of corals is therefore due to the overlap of species ranges, ranges which extend eastwards into the Pacific and westwards into the Indian Ocean. Endemism becomes relatively more important with fish (Allen 2006) and other taxa in which species longevity is less than that of corals or which have a lesser capacity for long-distance dispersal or which are specialized for narrow niches.
2. Ocean temperatures of the CT are commonly near the thermal cap of 31°C (Kleypas et al. 2008). This temperature, or maxima close to it, is commonly maintained for months during the summer of much of the CT. It is a well-established maxim of biogeography ("Rapoport's Rule") that the mean latitudinal range of major taxa increases with increasing latitude (Stevens 1989). This is another way of saying that increasing latitude is correlated with increasing environmental tolerance. Perhaps, this is less well established for marine life (Clarke1992), but it does suggest that temperature tolerance is least limiting at equatorial latitudes.

5.4 Evolution

Because it links many scientific disciplines including taxonomy, biogeography, and genetics, the subject of evolution has led to several general theories, three of which (Darwin's Centres of Origin, Vicariance, and Reticulate Evolution) have special relevance to marine biodiversity, biogeography, and (therefore) the CT (see Veron, *Coral Taxonomy and Evolution*, this volume).

In summary, the diversity of the CT has no single explanation. Plate tectonics created the biogeographic template, one of the complex island coastlines and extreme habitat heterogeneity. Patterns of dispersion, mediated by ocean currents, formed sequences of attenuation away from the equator leaving the CT with the region's highest biodiversity. Many environmental parameters, especially ocean currents and temperature, underpin this pattern. Evolutionary patterns, the genetic outcomes of environmental drivers, show why the CT is a center of biodiversity.

6 Future Impacts in the Coral Triangle

Whatever the drivers of diversity once were, it is the future that matters as far as conservation is concerned. Anthropogenic increase in tropical ocean temperature has been less than 0.8°C to date but could be as much as 2–3°C this century (IPCC 2007a, b), further magnified 1–2°C above normal by El Niño events. The latter have already caused extensive mass bleaching of corals in most major regions of the world although not in the southern CT (in part due to upwelling), nor in the southern Red Sea (which has particularly temperature-tolerant corals). If temperature extremes of the western Pacific are capped at 31°C, the CT, particularly areas in the Pacific Warm Pool of eastern Indonesia to the Solomon Islands, will be relatively protected because corals there already tolerate that limit.

Ocean acidification is another matter. In a business-as-usual regime of anthropogenic carbon dioxide increase, calcification of equatorial corals will be marginal by the year 2030 (Guinotte et al. 2003; Veron et al. 2009a). Corals within the CT will be among the last to be affected (Hoegh-Guldberg et al. 2007, 2009) because carbon dioxide absorption will be slowest at the lowest (and therefore warmest) latitudes. From a conservation perspective, all corals are threatened by the dual process of mass bleaching and ocean acidification; however, those of the CT have a relatively good prognosis.

Human impacts are likely to be severe. Approximately 225 million people live within the CT, of which 95% live adjacent to the 131,254 km of coast. This population is largely dependent on the productivity of coral reefs and adjacent waters and a larger population is dependent on protein from these sources. There are no estimations of the value of the reefs of the CT; however, as Indonesia's reefs were valued at $1.6 billion annually in 2002 (Burke et al. 2002), the current value of the CT is at least $2.3 billion annually. Almost 90% of Southeast Asian fisheries are in shallow continental shelf waters including coral reefs, industries that provide an estimated 65–70% of the animal protein consumed in these countries (Valencia 1990; Chua and Garces 1994). Just what economic downturns are likely to follow in the wake of impacts from climate change are unknown, but they will not be counted as an annual cost for their effect will be permanent as far as humans are concerned. By one estimate, Indo-Pacific reefs are declining at >2% per year (Bruno and Selig 2007), a conclusion supported by others (Burke et al. 2002; Allen and Werner 2002; Bellwood et al. 2004; Hoeksema 2004; Briggs 2005; Mous et al. 2005; Carpenter et al. 2008). Furthermore, although correlations between ecological and economic impacts on reefs stemming from environmental decline cannot be estimated with certainty, it is likely that both are becoming increasingly difficult to reverse (Hoegh-Guldberg et al. 2009).

This decline is now a major focus of the CTI, which aims to bring together relevant governments in multilateral partnerships to safeguard the marine resources of the region. Local communities, governments, and nongovernment organizations are currently working together to establish networks of MPAs that are specifically designed to address climate change (West and Salm 2003), reinforced by other conservation strategies, notably ecosystem approaches to fisheries and integrated coastal management. The Indonesian Government has now pledged to conserve 10% of its marine environment in MPAs by 2010 and 20% by 2020 and has called on other governments of CT countries to do likewise. Furthermore, one of the world's first MPA networks specifically designed to address climate change is now being established in Papua New Guinea (Green et al. 2007), an approach that will now be applied to MPA design throughout the CT.

Despite a substantial commitment to coral reef conservation on the ground, drastic international action on greenhouse gas emissions is still required to ensure the long-term survival of coral reefs in the CT and elsewhere around the world. The future of these mega-diverse reefs, and the livelihoods of the people who depend on them, will be determined by the success of these undertakings.

Acknowledgments Dr. Gerry Allen (Western Australian Museum), Dr Peter Mous (Indonesian Ministry of Marine Affairs and Fisheries), and Sheldon Cohen (The Nature Conservancy) contributed to the preparation of this manuscript. Tim Simmonds (AIMS) assisted with map preparations. The authors particularly thank The Nature Conservancy for supporting all aspects of the work.

References

Allen GR (2006) Coral Reef Fish Diversity. In: Green A, Lokani P, Atu W, Ramohia P, Thomas P, Almany J (eds) Solomon Islands Marine Assessment, The Nature Conservancy Technical Report 1/06, 113–156

Allen GR (2007) Conservation hotspots of biodiversity and endemism for Indo-Pacific coral reef fishes. Aquat Conserv Mar Freshwater Ecosyst 18:541–556

Allen GR, Werner TB (2002) Coral reef fish assessment in the 'coral triangle' of southeastern Asia. Environ Biol Fishes 65:209–214

Baird AH et al (2002) Coral reef biodiversity and conservation. Science 296:1026–1027

Briggs JC (2005) Coral reefs: conserving the evolutionary sources. Biol Conserv 126:297–305

Bellwood DR, Hughes TP, Folke C, Nystrom M (2004) Confronting the coral reef crisis. Nature 429:827–833

Bruno JF, Selig ER (2007) Regional decline of coral cover in the Indo-Pacific: timing, extent, and subregional comparisons. Plos One 8:e711:1–8

Burke LE, Selig M, Spalding M (2002) Reefs at risk in Southeast Asia. World Resources Institute, Washington, DC

Cairns SD (2007) Deep-water corals: an overview with special reference to diversity and distribution of deep-water scleractinian corals. Bull Mar Sci 81:311–322

Carpenter KE et al (2008) One third of reef-building corals face elevated extinction risk from climate change and local impacts. Science 321:560–564

Chua TE, Garces LR (1994) Marine living resources management in the ASEAN region: lessons learned and the integrated management approach. In Ecology and conservation of Southeast Asia marine and freshwater environments including wetlands. Kluwer Academic Publishers, Belgium

Clarke A (1992) Is there a latitudinal species diversity cline in the sea? Trends Ecol Evol 7:286–287

Coudray J, Montaggioni L (1982) Coraux et recifs coralliens de la province Indo-Pacifique: repartitiongeographique et altitudinale en relation avec la tectonique globale. Bull Soc Géol Fr 24:981–993

De Grave S (2001) Biogeography of Indo-Pacific Pontoniinae shrimps (Crustacea: Decapoda): a PAE analysis. J Biogeog 28:1239–1253

Gordon A, Fine R (1996) Pathways of water between the Pacific and Indian Oceans in the Indonesian seas. Nature 379:146–149

Green AL, Mous P J (2008) Delineating the Coral Triangle. its ecoregions and functional seascapes,Version 5.0. TNC Coral Triangle Program Report 1/08.44 pp. Weblink http://conserveonline.org/workspaces/tnccoraltriangle/

Green AL, Lokani P, Atu W, Almany J (eds) (2006) Solomon Island Marine Assessment: Technical report of survey conducted May 13 to June 17, 2004. TNC Pacific Island Countries Report 1/06

Green AL et al (2007) Scientific design of a resilient network of marine protected areas, Kimbe Bay, West New Britain, Papua New Guinea. TNC Pacific Island Countries Report 2/07

Groombridge B, Jenkins MD (2002) World atlas of biodiversity: earth's living resources in the 21st century. University of California Press, Berkeley

Guinotte JM, Buddemeier RW, Kleypas JA (2003) Future coral reef habitat marginality: temporal and spatial effects of climate change in the Pacific basin. Coral Reefs 22:551–558

Hoegh-Guldberg O et al (2007). Coral reefs under rapid climate change and ocean acidification. Science 318:1737–1742

Hoegh-Guldberg O, Hoegh-Guldberg H, Veron JEN et al (2009) The Coral Triangle and climate change: ecosystems, people and societies at risk. World Wildlife Fund

Hoeksema BW (1993) Mushroom corals (Scleractinia: Fungiidae) of Madang Lagoon, northern Papua New Guinea: an annotated checklist with the description of *Cantharellus jebbi* spec. nov. Zoologische Mededelingen 67:1–19

Hoeksema BW (2004) Biodiversity and the natural resource management of coral reefs in Southeast Asia. In: Visser LE (ed) Challenging coasts. Transdisciplinary excursions into integrated coastal zone development. Amsterdam University Press, Amsterdam, pp 49–71

Hoeksema BW (2007) Delineation of the Indo-Malayan centre of maximum marine biodiversity: the Coral Triangle. In: Renema W (ed) Biogeography, time, and place: distributions, barriers, and islands. Springer, Dordrecht, pp 117–178

Hogarth PJ (1999) The biology of mangroves. Oxford University Press, Oxford

Hughes TP, Bellwood DR, Connolly SR (2002) Biodiversity hotspots, centers of endemism, and the conservation of coral reefs. Ecol Lett 5:775–784

IPPC (Intergovernmental Panel on Climate Change) (2007a) Summary for Policymakers. In Solomon S (ed) Climate change 2007: the physical science basis. Contribution of Working Group I to the Fourth Assessment Report of the Intergovernmental Panel on Climate change. Cambridge University Press, Cambridge, UK, New York

IPCC (HYPERLINK "http://www.ipcc-wg2.org/index.html" Intergovernmental Panel on Climate Change) (2007b) Climate change 2007: impacts, adaptation, and vulnerability. In Parry ML et al (eds) Contribution of Working Group II to the Third Assessment Report of the Intergovernmental Panel on Climate Change. Cambridge University Press, Cambridge, UK, New York

Jokiel P, Martinelli FJ (1992) The vortex model of coral reef biogeography. J Biogeog 19:449–458

Kleypas JA, Danabasoglu G, Lough JM (2008) Potential role of the ocean thermostat in determining regional differences in coral reef bleaching events. Geophys Res Lett 35:L03613. doi:10.1029/2007GL032257

Meyer CP, Geller JB, Paulay G (2005) Fine scale endemism on coral reefs: archipelagic differentiation in turbinid gastropods. Evolution 59:113–125

Mittemeier RAN, Myers JB, Thomsen GAB, da Fonseca, Olivieri S (1998) Biodiversity hotspots and major tropical wilderness areas: approaches to setting conservation priorities. Conserv Biol 12:516–520

Mous PJ, Muljadi A, Pet JS (2005) Status of coral reefs in and around Komodo National Park. Results of a bi-annual survey over the period 1996–2002. Pub. The Nature Conservancy Southeast Asia Center for Marine Protected Areas, Sanur, Bali

Ricklefs RE, Latham RE (1993) Global patterns of diversity in mangrove floras. In: Ricklefs RE, Schluter D (eds) Species diversity in ecological communities, historical and geographical perspectives. Chicago University Press, Chicago, pp 215–229

Roberts CM, McClean CJ, Veron JEN et al (2002) Marine biodiversity hotspots and conservation priorities for tropical reefs. Science 295:1280–1284

Rosen BR (1971) The distribution of reef coral genera in the Indian Ocean. In: DR Stoddart, Yonge CM (eds) Regional variation in Indian Ocean coral reefs (Symp Zool Soc Lond 28). Academic, London, pp 263–299

Sheppard CRC (1999) Corals of Chagos, and the biogeographical role of Chagos in the Indian Ocean. In: Sheppard CRC, Seaward MRD (eds) Ecology of the Chagos Archipelago. Linnean Society Occasional Publications 2. Westbury Academic and Scientific Publishing, Otley, pp 53–66

Siddall M, Rohling EJ, Almogi-Labin A et al (2003) Sea-level fluctuations during the last glacial cycle. Nature 423:853–858

Spalding MD, Fox HE, Allen GR et al (2007) Marine ecoregions of the world: a bioregionalisation of coastal and shelf areas. Bioscience 57:573–583

Spalding MD, Taylor M, Ravilious C et al (2003) Global overview. The distribution and status of seagrasses. In: Green EP, Short FT (eds) World atlas of seagrasses. University of California Press, Berkeley, pp 5–26

Stehli FG, Wells JW (1971) Diversity and age patterns in hermatypic corals. Syst Zool 20:115–126

Stevens GC (1989) The latitudinal gradient in geographical range: how so many species coexist in the tropics. Am Nat 133:240–256

Turak E, DeVantier L (2010) Biodiversity and conservation priorities of reef-building corals in the Papuan Bird's Head Seascape. In: Katz LS, Firman A, Erdmann MV (eds) A Rapid Marine Biodiversity Assessment of Teluk Cendrawasih and the FakFak-Kaimana Coastline of the Papuan Bird's Head Seascape, Indonesia. R.A.P.

Bulletin of Biological Assessment. Conservation International, Washington, DC

Turak E, Souhoka J (2003) Coral diversity and the status of coral reefs in the Raja Ampat Islands. In: Donnelly R, Neville D, Mous P (eds) Report on a rapid ecological assessment of the Raja Ampat Islands, Papua, Eastern Indonesia, held October 30–November 22, 2002. The Nature Conservancy Southeast Asia Center for Marine Protected Areas, Sanur, Bali, Indonesia

Valencia MJ (1990) International conflicts over marine resources in southeast Asia. Trends in Politization and Militarization. In Ghee LT Valencia MJ (eds) Conflicts over natural resources in southeast Asia and the Pacific. United Nations University Press, Tokyo, pp 94–144

Veron JEN (1993) A biogeographic database of hermatypic corals: species of the central Indo-Pacific, genera of the world. Aust Inst Mar Sci Monogr Ser 9

Veron JEN (1995) Corals in space and time: the biogeography and evolution of the scleractinia. Cornell University Press, Ithaca

Veron JEN (2000) Corals of the world (3 vols) Australian Institute of Marine Science, Townsville

Veron JEN et al (2009a) The coral reef crisis: the critical importance of <350 ppm CO_2. Mar Pollut Bull 58:1428–1437

Veron JEN et al (2009b) Delineating the Coral Triangle. Galaxea 19:91–100

Veron JEN, Turak E (2006) Coral diversity. In: Green AL, Lokani P, Atu W, Almany J (eds) Solomon Island Marine Assessment: Technical report of survey conducted May 13 to June 17 2004. TNC Pacific Island Countries Report 1/06: 37–63

Wells FE (2002) Centres of species richness and endemism of shallow-water marine molluscs in the tropical Indo-West Pacific. Proc Ninth Intern Coral Reef Symp Bali 2000(2):941–945

Wells JW (1954) Recent corals of the Marshall Islands. US Geol Surv Prof Pap 260:385–486

Werner TB, Allen GR (1998) A rapid biodiversity assessment of the coral reefs of Milne Bay Province, Papua New Guinea. RAP Bulletin of Biological Assessment 11. Conservation International, Washington, DC

West JM, Salm RV (2003) Resistance and resilience to coral bleaching: implications for coral reef conservation and management. Conserv Biol 17:956–967

Part III
Coral Biology: Symbiosis, Photosynthesis and Calcification

Sexual Reproduction of Scleractinian Corals

Peter L. Harrison

Abstract Sexual reproduction by scleractinian reef corals is important for maintaining coral populations and evolutionary processes. The ongoing global renaissance in coral reproduction research is providing a wealth of new information on this topic, and has almost doubled the global database on coral reproductive patterns during the past two decades. Information on sexual reproduction is now available for 444 scleractinian species, and confirms that hermaphroditic broadcast spawning is the dominant pattern among coral species studied to date. Relatively few hermaphroditic or gonochoric brooding species have been recorded. Multispecific coral spawning has been recorded on many reefs, but the degree of reproductive synchrony varies greatly within and among species at different geographic locations.

Keywords Sexual reproduction • asexual reproduction • scleractinia • sexuality • broadcast spawning • brooding • biogeographic patterns • proximate cues • mass spawning • multispecific spawning

1 Introduction

This review provides an overview of global knowledge and emerging patterns in the reproductive characteristics of the 444 scleractinian species for which information on sexual reproduction is available, and updates the previous review by Harrison and Wallace (1990), with particular emphasis on new discoveries and data from the past 2 decades. Sexual reproduction in scleractinian corals has been widely studied in many regions of the world, and substantial new information has become available since earlier detailed reviews on this subject by Fadlallah (1983), Harrison and Wallace (1990), and Richmond and Hunter (1990), with more recent topic reviews provided by Richmond (1997), Harrison and Jamieson (1999), Kolinski and Cox (2003), Guest et al. (2005a), Harrison and Booth (2007), and Baird et al. (2009a).

Scleractinian reef-building corals are foundation species on coral reefs because they provide the complex three-dimensional structure and primary framework of the reef, and essential habitats and other important resources for many thousands of associated species (reviewed by Harrison and Booth 2007). These corals are distinguished from other members of the Class Anthozoa (Phylum Cnidaria) such as soft corals, by their continuous hard calcium carbonate crystal exoskeleton, which form the essential building blocks of the reef ecosystem when cemented together by crustose coralline algae. The term "coral" is used in this review to refer to these hard corals from the Order Scleractinia.

Although corals are widely distributed throughout the world's seas and deeper ocean environments, they are particularly significant in shallow tropical and subtropical seas where the mutualistic symbiosis between the coral polyps and their endosymbiotic dinoflagellates (zooxanthellae) fuels light enhanced calcification and rapid growth of reef-building corals, resulting in coral reef development. Corals can be broadly divided ecologically, but not systematically, into reef-building (hermatypic) corals and non-reef-building (ahermatypic) corals. Hermatypic corals create the primary reef framework and most hermatypic species in shallow warm-water habitats normally contain millions of zooxanthellae (i.e., zooxanthellate); in contrast, although ahermatypic corals also secrete complex aragonite exoskeletons they do not usually contribute significantly to reef formation, and mostly lack zooxanthellae (e.g., Yonge 1973; Schuhmacher and Zibrowius 1985; Cairns 2007). However, some azooxanthellate corals do contribute to reef-building and are therefore hermatypic, including some deeper-water and deep-sea colonial cold-water corals that form reefs or bioherms, which provide important habitats for many other species (e.g., Brooke and Young 2003; Waller and Tyler 2005; Friewald and Roberts 2005).

The total number of extant scleractinian "species" is not known, and estimating global coral species richness is complicated by a number of issues (e.g., Veron 1995, 2000; Cairns 1999, 2007; Harrison and Booth 2007). These include the limited exploration of deeper reef, mesophotic, and

P.L. Harrison (✉)
School of Environmental Science and Management,
Southern Cross University, Lismore, NSW, 2480, Australia
e-mail: peter.harrison@scu.edu.au

deep-sea environments, as well as some shallow tropical reef regions where new species are likely to be found, and imperfect taxonomic resolution of highly variable species and potential cryptic species. Furthermore, the discovery of hybridization among some morphologically different coral morphospecies (e.g., Willis et al. 1992, 1997; Szmant et al. 1997; van Oppen et al. 2002; Vollmer and Palumbi 2002) challenges the application of the traditional biological-species concept based on reproductive isolation between different species, for some corals.

If we assume that the current primarily morphologically based taxonomy provides an appropriate indication of global coral species richness, there are probably at least 900 extant hermatypic scleractinian species (e.g., Wallace 1999; Veron 2000; J. Veron, personal communication). Of these, 827 zooxanthellate hermatypic coral species have been assessed for their conservation status (Carpenter et al. 2008). In addition, at least 706 azooxanthellate scleractinian species are known, including 187 colonial and 519 solitary coral species, with their most common depth range being 200–1,000 m (Cairns 2007). Of the more than 1,500 recognized coral species, aspects of sexual reproduction have now been recorded in at least 444 species, the vast majority being shallow-water zooxanthellate hermatypic coral species. This global knowledge base provides a wealth of information on the biology and ecology of coral reproduction that surpasses most other marine invertebrate groups, and corals therefore provide an important model for assessing life history and evolutionary theory.

This chapter focuses mainly on sexual reproductive characteristics in hermatypic species because these corals have been more extensively studied and are foundation species on coral reefs (Harrison and Booth 2007). Additional reference is made to sexual reproduction in some ahermatypic scleractinian species, and an overview of asexual reproduction in corals is provided below to highlight the diversity of reproductive processes exhibited by scleractinians.

2 Coral Life Cycle and Reproduction

Corals have a relatively simple life cycle involving a dominant benthic polyp phase and a shorter planula larval phase. The polyp phase is characterized by growth of tissues and skeleton that often includes one or more forms of asexual budding or reproduction; and repeated cycles of sexual reproduction (iteroparity) involving the production of gametes, fertilization, embryo development, and a larval phase that is usually planktonic and dispersive to some degree (Harrison and Wallace 1990). If the planula survives and successfully attaches and settles permanently on hard substratum (Fig. 1), it metamorphoses from the larval form into a juvenile polyp

Fig. 1 A brooded *Isopora cuneata* planula searching reef substratum covered with crustose coralline algae for a suitable site for initial attachment and metamorphosis into a juvenile polyp (Photo: author)

that initiates the formation of the calcium carbonate exoskeleton. Subsequent growth during an initial presexual juvenile phase leads to development of the adult form that becomes sexually reproductive, which completes the life cycle (Harrison and Wallace 1990).

Asexual reproduction produces genetically identical modules that may prolong the survival of the genotype, whereas sexual reproduction enables genetic recombination and production of new coral genotypes that may enhance fitness and survival of the species. Four basic patterns of sexual reproduction are evident among corals, which include: hermaphroditic broadcast spawners, hermaphroditic brooders, gonochoric broadcast spawners, or gonochoric brooders. Hermaphroditic corals have both sexes developed in their polyps and colonies, whereas gonochoric corals have separate sexes; and corals with these sexual patterns either broadcast spawn their gametes for external fertilization and subsequent embryo and larval development, or have internal fertilization and brood embryos and planula larvae within their polyps (reviewed by Harrison and Wallace 1990; Richmond and Hunter 1990). However, not all coral species are readily classified into these basic patterns, as mixed sexual patterns or both modes of development are known to occur in some species.

2.1 Asexual Budding and Reproduction

Different modes of asexual production can be distinguished; asexual budding of polyps leads to the formation of coral colonies, while various forms of asexual reproduction result in the production of new modules that form physically separate but genetically identical clones (ramets) (reviewed by Highsmith 1982; Cairns 1988; Harrison and Wallace 1990;

Richmond 1997). Most hermatypic coral species (Wallace 1999; Veron 2000) and about 26% of the known ahermatypic azooxanthellate coral species (Cairns 2007) form colonies via asexual budding of polyps and are therefore modular, iterative colonial organisms. Budded polyps usually form by growth and internal division of existing polyps (intratentacular budding) or development of new polyps from tissues adjacent to, or between, existing polyps (extra-tentacular budding) (reviewed by Vaughan and Wells 1943; Wells 1956; Cairns 1988; Veron 2000). In most colonies, these budded polyps remain interconnected (Fig. 2), and the colony is partly integrated via nerve and muscular networks within the thin veneer of tissues that overly the skeleton they secrete (e.g., Gladfelter 1983; Harrison and Booth 2007).

Asexual budding produces genetically identical polyps within each colony; however, DNA damage and somatic mutations can genetically alter cell lineages and may induce development of neoplasms (tumors) within colonies (e.g., Coles and Seapy 1998). In a few coral species, gregarious settlement of larvae and allogenic fusion of newly settled primary polyps produce chimeras that result in colonies composed of different genotypes (e.g., Hidaka et al. 1997; Hellberg and Taylor 2002; Puill-Stephan et al. 2009). This would confer some advantages of initial increased size and reduced mortality of chimeras during the vulnerable juvenile polyp phase of the life cycle. The increased genetic diversity may also enhance colony survival unless negative interactions and competition between cell lineages from different genotypes occur (Puill-Stephan et al. 2009).

The formation of coral colonies through modular iteration of budded polyps, and associated growth of their supportive and protective exoskeleton, provide important ecological and

Fig. 3 Large colonies have increased sexual reproductive output, such as the *Acropora tenuis* colony seen here spawning many thousands of buoyant egg–sperm bundles on the Great Barrier Reef (GBR) during the crepuscular period just after sunset (Photo: author)

evolutionary advantages for colonial species. Colonial growth enables corals to grow much larger than most single polyps; thus, colonies can occupy more space and more effectively compete for resources by growing above the reef substratum or over other benthos, and colonies can survive the death of individual polyps and partial colony mortality (e.g., Jackson and Coates 1986; Rosen 1990; Hughes et al. 1992). Increased size reduces the mortality risk in juvenile corals and increases colony biomass and resource acquisition, leading to increased reproductive output as the number of gravid polyps and fecundity increases with size and age (Fig. 3); hence, larger colonies can dominate gamete production within coral populations, unless reproductive senescence occurs (e.g., Kojis and Quinn 1981, 1985; Babcock 1984, 1991; Szmant-Froelich 1985; Rinkevich and Loya 1986; Harrison and Wallace 1990; Hall and Hughes 1996; Goffredo and Chadwick-Furman 2003; Zakai et al. 2006).

Corals exhibit a wide range of asexual reproductive processes that produce new clonal solitary corals or colonies (Highsmith 1982; Cairns 1988; Harrison and Wallace 1990). These processes include colony fragmentation resulting from storm and wave impacts or other damage, colony fission, longitudinal and transverse division, polyp expulsion or polyp "bail-out," growth and detachment of polyp balls in some *Goniopora* colonies, and budding of polyps from an anthocaulus or regenerating tissues in fungiids and some other corals (e.g., Wells 1966; Rosen and Taylor 1969; Sammarco 1982; Krupp et al. 1992; Kramarsky-Winter and Loya 1996; Kramarsky-Winter et al. 1997; Gilmour 2002; Borneman 2006). In addition, asexual production of brooded planulae occurs in populations of the common reef coral *Pocillopora damicornis* (Stoddart 1983; Ayre and Miller 2004; Sherman et al. 2006), and in *Tubastrea coccinea* and

Fig. 2 Integration of polyp tissues enables activities such as sexual reproduction to be coordinated within colonies that develop from asexual budding of coral polyps. Coordinated behavior is shown in this colony of *Acropora arabensis*, where bundles of eggs and sperm bulge under the oral disk of each polyp in the "setting" phase just prior to synchronized spawning (Photo: author)

Tubastrea diaphana (Ayre and Resing 1986). *Oulastrea crispata* may also brood asexually produced planulae during periods when sexual reproduction has ceased (Nakano and Yamazoto 1992; Lam 2000).

Asexual reproduction can therefore produce genetically identical ramets that may occupy substantial space on reefs and in some cases may disperse widely. The extent and importance of asexual versus sexual reproduction varies greatly among different populations of corals and among different coral species (e.g., Ayre et al. 1997; Ayre and Hughes 2000; Miller and Ayre 2004; Baums et al. 2006; Whitaker 2006; Sherman et al. 2006; Foster et al. 2007). The range of reproductive processes and modes in corals partly reflects the extraordinary ability of cnidarian cell lines to differentiate, dedifferentiate, and redifferentiate (e.g., Campbell 1974; Holstein et al. 2003), which provides their tissues with remarkable developmental plasticity and adaptability.

3 Historical Perspectives on Coral Reproduction

The current, extensive global knowledge of coral reproduction is a product of many decades of research; hence, it is useful to provide a brief overview of previous and recent research as an historical and geographic context for the more detailed summaries of sexual reproductive characteristics in the remaining sections of this chapter. Harrison and Wallace (1990) reviewed the history of research on coral reproduction and noted that sexual reproduction in scleractinians had been studied for 200 years, since Cavolini (c. 1790, cited in de Lacaze-Duthiers 1873) observed *Astroides* planulae in the Mediterranean region.

Most of the early reproduction studies focused on coral species that brooded planula larvae within their polyps, while early reports of broadcast spawning of gametes were either overlooked or dismissed as aberrant by subsequent researchers (see Harrison and Wallace 1990). This led to the dogma that corals were typically or uniformly viviparous brooders (e.g., Duerden 1902a; Hyman 1940; Vaughan and Wells 1943; Wells 1956). That misconception was rapidly overturned by research in the early 1980s, which demonstrated that broadcast spawning of gametes was the dominant mode of development in the majority of coral species studied worldwide (Fig. 4). Much of that new information resulted from the discovery of mass coral spawning on the Great Barrier Reef (GBR), which included records of broadcast spawning in more than 130 coral species (Harrison et al. 1984; Willis et al. 1985; Babcock et al. 1986). This clearly established broadcast spawning as the dominant mode of reproduction among reef corals studied in the Pacific region, and significantly changed our understanding of reef coral

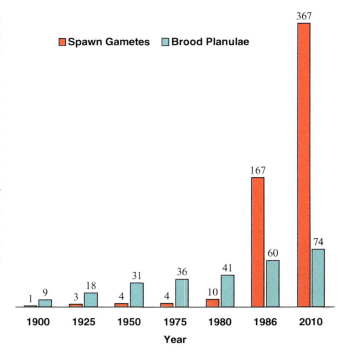

Fig. 4 Total number of coral species worldwide whose mode of development was recorded during different periods, from 1900 to the present. The trends highlight the historical bias towards brooding species prior to the 1980s, and the rapid increase in broadcast spawning species reported following the discovery of mass coral spawning on the GBR in the early 1980s. The species totals for broadcast spawning species (*red columns*) and brooding species (*pale blue columns*) include an additional ten species in 1986 (see Table 7.4 in Harrison and Wallace 1990), and an additional 13 species in 2010, in which both modes of development have been reported (see Table 4)

ecology. Somewhat different patterns of coral reproduction were evident in other reef regions in the Indo-Pacific (reviewed by Richmond and Hunter 1990; Harrison and Wallace 1990), and Szmant (1986) noted that brooding corals were relatively more abundant among coral species studied in the Atlantic region, and were characterized by small colony size.

By the late 1980s, at least 157 coral species were known to broadcast spawn gametes, while 50 species were recorded to brood their larvae, with another ten species recorded to have both modes of development (Fig. 4; see Harrison and Wallace 1990). Hermaphroditic broadcast spawners were by far the most common group recorded, gonochoric broadcast spawners were moderately common, while hermaphroditic brooders and gonochoric brooders were relatively uncommon (reviewed by Harrison and Wallace 1990; Richmond and Hunter 1990). Strong systematic trends were evident in sexual patterns, arrangement of gametes in polyps and sperm structure, whereas the mode of reproduction was more variable and a relatively plastic life history trait (Harrison 1985).

Table 1 Summary of global information on sexual patterns and mode of development for 444 scleractinian species

Sexual pattern Mode of development	Hermaphroditic	Mixed or contrasting sexuality reported	Gonochoric	Unknown sexual pattern	Total species
Spawn gametes	258	6	78	12	354
Spawn gametes and brood	8	1	4	–	13
Brood	25	5	15	16	61
Unknown mode of development	4	–	12	–	16
Total species	295	12	109	28	444

Information on reproductive characteristics was obtained primarily from analysis of all the available known publications on sexual reproduction in corals, with some additional data from summaries in other reviews (Fadlallah 1983; Harrison 1985; Harrison and Wallace 1990; Richmond and Hunter 1990; Richmond 1997; Harrison and Booth 2007; Baird et al. 2009a), conference abstracts, research reports and postgraduate theses, plus extensive information made available to the author from many colleagues around the world over the past 30 years, some of which is unpublished.

4 Recent Advances in Coral Reproduction Research

Over the last 20 years, research on coral reproduction has continued to grow substantially and has expanded into many reef regions that were not previously well represented, including equatorial and tropical regions of high coral species richness and biodiversity, and some subtropical reef regions (reviewed by Richmond 1997; Guest et al. 2005a; Harrison and Booth 2007; Baird et al. 2009a). This more recent research has resulted in substantial new information and has almost doubled the number of coral species for which sexual reproductive data are now available, from more than 230 species in the late 1980s (Harrison and Wallace 1990; Richmond and Hunter 1990) to at least 444 species by 2010 (Table 1). The current global data generally confirm and extend many of the trends and patterns highlighted in earlier studies, and some recent advances in our understanding of sexual reproduction in corals are summarized in Table 1.

4.1 Biogeographical Patterns of Coral Reproduction

As expected, multispecific coral spawning has now been recorded in many reef regions, but the scale of spawning and the degree of reproductive synchrony varies greatly within and among populations of different coral species in different regions, resulting in varied patterns of reproduction among coral assemblages (reviewed by Harrison and Wallace 1990; Richmond and Hunter 1990; Richmond 1997; Kolinski and Cox 2003; Guest et al. 2005a; Harrison and Booth 2007; Baird et al. 2009a). The global data show that reproductive patterns in different coral assemblages form a continuum, ranging from highly synchronized mass spawning events involving many colonies of many species from different families, through to a wide range of other synchronous multispecific spawning patterns involving fewer species and corals (summarized in Harrison and Wallace 1990; Richmond 1997; Harrison and Booth 2007; Mangubhai and Harrison 2009; Baird et al. 2009a). At the other extreme of this continuum, 24 coral species studied in the northern Gulf of Eilat in the Red Sea exhibit temporal reproductive isolation (Shlesinger and Loya 1985; Shlesinger et al. 1998). In contrast, more recent research has recorded highly synchronous maturation of gametes among many *Acropora* species on reefs 250 km south of Eilat in the Egyptian Red Sea, which is likely to result in large multispecific spawning events (Hanafy et al. 2010).

Coinciding with the increased range of reproductive patterns now evident among coral assemblages, some confusion has arisen over the use of the terms "mass spawning" and "multispecific spawning." For example, a few publications have described synchronous spawning by colonies of one or a few coral species in the Gulf of Mexico and Caribbean region as mass spawning (e.g., Gittings et al. 1992; Hagman et al. 1998a; Beaver et al. 2004; Bastidas et al. 2005), or have used the terms mass spawning and multispecific spawning interchangeably (e.g., Guest et al. 2005a), leading to different interpretations and application of these terms (e.g., Mangubhai and Harrison 2008a, 2009; cf. Guest et al. 2005a; Baird and Guest 2009). Harrison and Booth (2007) distinguished between the larger-scale synchronous mass spawning events currently known from some reefs on the GBR and Western Australia (WA), from other smaller-scale multispecific spawning events involving fewer species and corals. This distinction has the advantage of differentiating between reproductive patterns involving reproductive synchrony among fewer corals and species (i.e., multispecific spawning), from the largest synchronous mass spawning events recorded so far, and most recently published papers have tended to differentiate the terms "multispecific spawning" and "mass spawning."

By definition, multispecific spawning refers to synchronous spawning by two or more species, whereas the term mass spawning was initially used to describe the highly synchronous nocturnal spawning of many coral colonies of many species from a range of scleractinian families on the GBR (Harrison et al. 1983, 1984; Willis et al. 1985; Babcock et al. 1986; Wallace et al. 1986; Oliver et al. 1988; Harrison 1993). Synchronous spawning by more than 20–30 coral species involving many or most colonies of extensively sampled and monitored populations has been recorded on peak mass spawning nights after full moon periods in the austral late spring and early summer months on some GBR reefs, with split-spawning over consecutive lunar cycles occurring in some populations in some years (Willis et al. 1985; Babcock et al. 1986; Harrison and Wallace 1990). Mass spawning therefore represents an extreme form of multispecific spawning (Harrison and Booth 2007; Mangubhai and Harrison 2009). Willis et al. (1985) partly defined mass spawning as "the synchronous release of gametes by many species of corals, in one evening between dusk and midnight" to emphasize the large number and diversity of coral taxa participating in these events, but implicit in this definition was that it involves synchronous spawning by many individual corals in many of these species (e.g., Rosser and Gilmour 2008). Similar large-scale mass-spawning events have been reported among corals on some tropical and subtropical WA reefs in the austral autumn period involving synchronous spawning by up to 24 coral species from a wide range of genera and families (Simpson 1985, 1991; Simpson et al. 1993; Babcock et al. 1994). More recent research on some WA coastal and offshore reefs has confirmed a larger primary mass spawning period in autumn, with a smaller but ecologically significant multispecific spawning period involving fewer species and colonies occurring during late spring or early summer (Rosser and Gilmour 2008; Gilmour et al. 2009; Rosser and Baird 2009).

It is also important to note that some corals on GBR and WA reefs also, or only, spawn or release planulae during nonmass spawning periods. For example, some species reproduce on other lunar nights or during different lunar phases, or at other times of the year, and some corals exhibit biannual or multiple cycles of gametogenesis and breeding during the year (e.g., Marshall and Stevenson 1933; Kojis and Quinn 1982; Harriott 1983a, b; Harrison et al. 1984; Willis et al. 1985; Wallace 1985; Babcock et al. 1986; Kojis 1986a; Willis 1987; Harrison and Wallace 1990; Stobart et al. 1992; Tanner 1996; Baird et al. 2002, 2009a; Wolstenholme 2004; Rosser and Gilmour 2008; Gilmour et al. 2009; Rosser and Baird 2009; among others). Therefore, the mass coral spawning paradigm does not, and was never intended to, encompass all coral reproductive data from these regions. Future studies are likely to increase both the numbers of coral species known to have highly synchronous gametogenic and breeding patterns that participate in mass spawning events, as well as increase the numbers of species that do not participate or only partially spawn during mass spawning events. Broader scale surveys of *Acropora* assemblages at various sites along the GBR have also revealed a high degree of synchronous maturation of gametes in some *Acropora* populations, but less synchronous maturation and breeding periods in some other populations (e.g., Oliver et al. 1988; Hughes et al. 2000; Baird et al. 2002, 2009a, b).

Mass coral spawning is likely to occur in some other reef regions with diverse coral assemblages where gametogenic cycles and maturation are highly synchronized within and among populations of many species during brief seasonal breeding periods. Large multispecific coral spawning events have been increasingly recorded in some other Indo-Pacific regions, and some of these may prove to be of a similar scale compared to the GBR and WA mass spawning events, when further data on both the numbers of corals and the range of species that spawn synchronously become available.

Diverse reproductive patterns ranging from highly synchronized gamete maturation and multispecific spawning by more than ten coral species and many corals occurring during one or a few nights, through to less synchronous multispecific spawning by fewer corals and species and more extended breeding seasons have now been recorded from many Indo-Pacific regions, including: Japan (e.g., Heyward et al. 1987; Shimoike et al. 1992; Hayashibara et al. 1993; van Woesik 1995; Nozawa et al. 2006; Mezaki et al. 2007; Baird et al. 2009b), Taiwan (Dai et al. 1992; see also Kawaguti 1940), the Philippines (Bermas et al. 1992; Vicentuan et al. 2008), Palau (Kenyon 1995; Penland et al. 2004, and unpublished data; see also Kawaguti 1941a, b), Yap (Kenyon 1995), Guam (Heyward 1988a; Richmond and Hunter 1990; Richmond 1997), Singapore (Guest et al. 2002, 2005a, b), Thailand (Piromvaragorn et al. 2006; Kongjandtre et al. 2010), Indonesia (Edinger et al., unpublished data in Tomascik et al. 1997; Baird et al. 2009a), Papua New Guinea (Oliver et al. 1988; Baird et al. 2009a), Solomon Islands (Baird et al. 2001), Fiji and Western Samoa (Mildner 1991), Line Islands (Kenyon 2008), French Polynesia (Carroll et al. 2006), subtropical eastern Australia (Wilson and Harrison 1997, 2003; Harrison 2008), Egyptian Red Sea (Hanafy et al. 2010), and some other Indo-Pacific regions (e.g., Richmond 1997; Baird et al. 2009a). However, as noted for the GBR and WA reproductive patterns, other coral species are known to spawn or planulate at other times in these regions (reviewed by Harrison and Wallace 1990; Richmond and Hunter 1990; Richmond 1997; see also Kinzie 1993; Fan and Dai 1995, 1999; Dai et al. 2000; Bachtiar 2000; Diah Permata et al. 2000; Fan et al. 2002, 2006; Wallace et al. 2007; Villanueva et al. 2008; among others). In the South Pacific region, coincident spawning records after full moon periods in the austral spring and summer seasons occur on reefs throughout the GBR up to 1,200 km apart (e.g., Willis et al. 1985; Babcock et al. 1986; Oliver et al. 1988), and north to the Solomon Islands (Baird et al. 2001), and east to French Polynesia (Carroll et al. 2006).

Latitudinal trends in the seasonal patterns and timing of reproduction are evident in some Pacific regions. For example, peak reproduction tends to occur in late spring to early summer in lower latitude tropical reefs, with progressively later peaks in reproduction occurring in subtropical and higher latitude reefs (e.g., van Woesik 1995; Wilson and Harrison 1997, 2003; Nozawa et al. 2006; Mezaki et al. 2007; Harrison 2008; Baird et al. 2009b).

Less synchronous and more protracted spawning seasons have been recorded among hundreds of colonies of 20 *Acropora* species on equatorial reefs in Kenya where spawning occurred over 2–5 months within populations of different species, and gamete release in *Acropora* and some faviid species extended over a 9-month period from August to April (Mangubhai and Harrison 2006, 2008a, b, 2009; Mangubhai 2009). Although coincident spawning was recorded in colonies of two *Acropora* species, and some degree of multispecific spawning is likely to occur on these equatorial reefs, most species exhibited protracted and relatively asynchronous spawning with some level of temporal reproductive isolation between species (Mangubhai and Harrison 2006, 2008a, 2009). Similar patterns of asynchronous spawning over multiple lunar phases within populations of some coral species and some pulsed multispecific spawning events have recently been recorded from diverse coral assemblages in the Maldives (Harrison and Hakeem 2007, and unpublished data). These data support earlier hypotheses of protracted breeding seasons and less synchronous spawning nearer the equator (e.g., Orton 1920; Oliver et al. 1988), whereas more synchronous reproduction has been recorded among some species in other equatorial reefs (e.g., Baird et al. 2002; Guest et al. 2002, 2005a, b; Penland et al. 2004). Baird et al. (2009a) analyzed latitudinal trends in spawning synchrony using data from some Indo-Pacific *Acropora* assemblages and concluded that spawning synchrony peaked at midlatitudes on the central GBR and was lower near the equator (Singapore and Kenya), and lowest at subtropical eastern Australian reefs in the Solitary Islands. Hence, some reduction in spawning synchrony is apparent in at least some *Acropora* assemblages at high and low latitudes compared with midlatitude tropical GBR reefs.

Corals also successfully reproduce in more extreme temperate latitude environments such as in Kuwait (Harrison 1995; Harrison et al. 1997; Fig. 2), and at higher latitude regions where populations are at, or near, their latitudinal limits (e.g., van Woesik 1995; Wilson and Harrison 1997, 2003; Harii et al. 2001; Nozawa et al. 2006; Harrison 2008). Therefore, as noted by van Woesik (1995), there is no indication that these coral populations in marginal reef and cooler temperate environments are nonreproductive pseudopopulations.

Reproductive patterns documented in Hawaiian corals range from brooding throughout the year in a few species, to seasonal peak reproduction in spring and summer months for broadcast spawning and some brooding species, and although some species have overlapping spawning periods, there is no evidence of large multispecific or mass spawning events (e.g., Edmondson 1929, 1946; Harrigan 1972; Stimson 1978; Krupp 1983; Richmond and Jokiel 1984; Jokiel 1985; Jokiel et al. 1985; Heyward 1986; Hodgson 1988; Hunter 1988; Richmond and Hunter 1990; Harrison and Wallace 1990; Kenyon 1992; Schwartz et al. 1999; Neves 2000; reviewed by Kolinski and Cox 2003). Slightly different reproductive patterns are evident among the 14 coral species studied in the Equatorial Eastern Pacific region (EEP), where only two brooding species have been recorded, and extended breeding seasons occur in thermally stable environments but shorter seasonal breeding occurs during warm periods in seasonally varying upwelling environments (e.g., Glynn et al. 1991, 1994, 1996, 2000, 2008; Colley et al. 2002, 2006; reviewed by Glynn and Colley 2009). Some coral species in the EEP exhibit unusual or contrasting reproductive characteristics compared with other regions. For example, *Pavona varians* is a sequential hermaphrodite in the EEP but is recorded as gonochoric in other regions (see Table 3), and *P. damicornis* populations in the EEP do not brood planulae in contrast to most other regions (see Table 4).

Reproductive patterns have been widely studied in the Caribbean and other western Atlantic regions (reviewed by Szmant 1986; van Woesik et al. 2006). Multispecific spawning by a small number of coral species has been recorded on some reefs in the Caribbean (e.g., van Veghel 1993; Steiner 1995; de Graaf et al. 1999; Sanchez et al. 1999; Bastidas et al. 2005; Rodríguez et al. 2009), in the Gulf of Mexico (e.g., Gittings et al. 1992; Hagman et al. 1998a, b; Vize 2006), and subtropical Bermuda (Wyers et al. 1991). Long-term spawning records from the Gulf of Mexico show that the timing of spawning for species is highly consistent between years, but most species have unique spawning "windows" during the main spawning nights when no other coral species have been observed to spawn (Vize et al. 2005). In other western Atlantic corals, gametogenesis and spawning patterns are often synchronous within species or populations exhibit split-spawning over consecutive lunar cycles, but some species spawn at different lunar phases or different seasons and exhibit some degree of temporal reproductive isolation (e.g., Wyers 1985; Szmant 1986, 1991; Soong 1991; Steiner 1995; Acosta and Zea 1997; de Graaf et al. 1999; Mendes and Woodley 2002a; Vargas-Angel and Thomas 2002; Alvarado et al. 2004; Bastidas et al. 2005; Vargas-Angel et al. 2006; van Woesik et al. 2006; Rodriguez et al. 2009). An unusual feature of the western Atlantic coral fauna is that it contains a high proportion of brooding species (Szmant 1986) and reproductive patterns among Atlantic brooding species

Table 2 Summary of sexual patterns and mode of development for 444 scleractinian species, grouped according to families and molecular clades

Reproductive character	Sexual pattern		Mode of development				
Clade: Family "Complex" clade	H	Mixed or H/G	G	SG	SG+B	B	Total species sex/mode recorded
II. Dendrophylliidae	4	–	16	9	1	10	23
III. Poritidae	1	2	24	19	–	13	32
IV. Pachyseris	–	–	3	3	–	–	3
V. Euphyllidae etc.	2	2	3	6	–	2	8
VI. Acroporidae	155	–	–	150	–	6	157
VII. Agariciidae	3	3	5	9	–	4	13
VIII. Astrocoeniidae	–	–	2	2	–	–	2
IX. Siderastreidae	1	–	3	2	–	2	4
Flabellidae	–	1	7	2	–	4	8
Fungiacyathidae	–	–	1	1	–	–	1
Micrabaciidae	–	–	–	–	–	1	1
Totals for "complex"	166	8	64	203	1	42	252
"Robust" clade							
X. Pocilloporidae etc.	16	–	–	4	3	11	19
XI. Fungiidae etc.	5	3	22	22	5	–	31
XII. Meandrinidae	–	–	5	5	–	–	5
XIII. Oculinidae etc.	1	–	5	4	–	–	6
XIV.	–	–	3	4	–	–	4
XV. Diploastrea	–	1	–	1	–	–	1
XVI. Mont. cavernosa	–	–	1	1	–	–	1
XVII. "Faviidae" etc.	74	–	–	76	1	1	78
XVIII + XIX + XX.	16	–	–	17	–	–	17
XXI. Mussidae etc.	13	–	–	7	2	4	14
Caryophylliidae	4	–	7	8	1	2	13
Rhizangiidae	–	–	2	2	–	1	3
Totals for "Robust"	129	4	45	151	12	19	192
Total species	295	12	109	354	13	61	444

Systematic groupings of species and genera are aligned with clades identified by Fukami et al. (2008) and other recent phylogenetic analyses and groupings (e.g., Romano and Cairns 2000; Fukami et al. 2004b; Le Goff-Vitry et al. 2004; Kerr 2005; Benzoni et al. 2007; Nunes et al. 2008; Huang et al. 2009; Baird et al. 2009a) *H* hermaphroditic, *G* gonochoric; Mixed or contrasting sexual patterns reported (see Table 3); *SG* broadcast spawn gametes, *B* brood planulae, *SG+B* both modes of development reported (see Table 4)

range from synchronous seasonal planulation through to year-round planulae release (e.g., Wilson 1888; Duerden 1902a, b; Vaughan 1910; Lewis 1974; van Moorsel 1983; Szmant-Froelich et al. 1985; Tomascik and Sander 1987; Delvoye 1988; Soong 1991; Johnson 1992; McGuire 1998; Vermeij et al. 2003, 2004; Goodbody-Gringley and Putron 2009).

A wide range of reproductive patterns and differences in periods of spawning and planula release have also been recorded among South Atlantic corals from Brazil (e.g., Pires et al. 1999, 2002; Francini et al. 2002; Neves and Pires 2002; Lins de Barros et al. 2003; Neves and da Silveira 2003). The endemic gonochoric brooding coral *Siderastrea stellata* exhibits remarkably synchronous maturation of oocytes over a 20° range of latitude from equatorial to southern tropical locations, although planulation appears to begin slightly earlier at the southernmost site that is influenced by cold water upwelling (Lins de Barros and Pires 2007).

4.2 Environmental Influences on Coral Reproduction

Sexual reproductive processes in corals appear to be strongly influenced by various environmental factors that act as proximate factors regulating and synchronizing reproductive cycles and gamete maturation, and as ultimate causes that exert evolutionary selective pressure through enhanced reproductive success (reviewed by Harrison and Wallace 1990). For example, synchronized spawning within coral populations leads to higher concentrations of gametes that promotes enhanced fertilization success (e.g., Oliver and Babcock 1992; Willis et al. 1997; Levitan et al. 2004), which increases planula production and the probability of successful reproduction among corals that spawn together.

Earlier studies indicated that seasonal changes in sea temperature, day length, wind or current patterns, lunar cycles of

Table 3 Scleractinian species in which mixed or contrasting sexual patterns have been reported

Clade: Family species	Sexual pattern	Location and comments (References)
III. Poritidae		
Porites astreoides	H	Puerto Rico (e.g., Szmant-Froelich 1984; Szmant 1986)
	Mixed: H, f	Jamaica; 26H, 28f, 1m?; H colonies had f, m or H polyps (Chornesky and Peters 1987)
	Mixed: G, H	Panama; 79% of 168 colonies f, 4% m, 17% H (Soong 1991)
Porites brighami	G, H?	Hawaii (Kolinski and Cox 2003)
V. Euphyllidae etc.		
Galaxea fascicularis	Mixed: f and "H" colonies	GBR; pseudo-gynodioecious, have female, and H colonies with functional sperm+eggs that cannot be fertilized (Harrison 1988)
Galaxea astreata	Mixed: f and "H" colonies	GBR; pseudo-gynodioecious, have female, and H colonies with functional sperm+eggs that cannot be fertilized (Harrison 1988)
VII. Agariciidae		
Agaricia agaricites	G	Virgin Islands, Puerto Rico (Peters 1984)
	Mixed: G, H	Curaçao; 41m, 25f, 12H (Delvoye 1982, 1988)
	H	Curaçao (G.N.W.M. van Moorsel, personal communication in Fadlallah 1983)
Agaricia humilis	Mixed : H, G	Curaçao; 37H, 8m, 13f (Delvoye 1982, 1988)
	H	Curaçao (G.N.W.M. van Moorsel, personal communication in Fadlallah 1983)
	H	Jamaica (E. Chornesky, personal communication in Harrison and Wallace 1990)
Pavona varians	G	Eilat, Red Sea (Shlesinger et al. 1998)
	G	Hawaii (Mate 1998; Kolinski and Cox 2003)
	H sequential cosexual	Eastern Pacific; sequential cosexual H, some G colonies (Glynn et al. 2000; Glynn and Colley 2009)
Flabellidae		
Monomyces rubrum	H	South Africa, Maldives; polyps H, some have only m or f gametes, suggested protandrous hermaphrodite (Gardiner 1902a, b; 1904)
	G	New Zealand (Heltzel and Babcock 2002)
XI. Fungiidae		
Fungia scutaria	G (some H?)	Hawaii; likely G, >20 corals sampled with no evidence of H, but eggs from isolated corals fertilized, so some H? (Krupp 1983)
	G or H?	Eilat, Red Sea; small corals m, large corals f, so may exhibit protandrous sex change (Kramarsky-Winter and Loya 1998)
Heliofungia actiniformis	H sequential?	Palau; Abe (1937) concluded that this species is probably hermaphroditic with sperm developing before eggs in the same individual
	G	GBR (Willis et al. 1985; Babcock et al. 1986)
	G	GBR (Babcock et al. 1986)
Sandalolitha robusta	H	Hawaii; colonies change sex between years, indicating bidirectional sex change and sequential hermaphroditism (B. Carlson, personal communication)
XV.		
Diploastrea heliopora	G	GBR (Harrison 1985; Babcock and Harrison, unpubl. data in Harrison and Wallace 1990)
	H sequential	Singapore; 4 H colonies and 1 m colony recorded, and polyps in H colonies were either m or f with sequential development (Guest 2004)

H hermaphroditic, *G* gonochoric, *m* male, *f* female, numbers refer to numbers of colonies or solitary corals recorded with different sexes or sexual patterns, *?* uncertain, *GBR* Great Barrier Reef

night irradiance, and daily periods of light and dark may act as proximate cues operating on progressively finer time scales to synchronize reproductive cycles and breeding among many corals (e.g., Yonge 1940; Szmant-Froelich et al. 1980; Babcock 1984; Harrison et al. 1984; Jokiel et al. 1985; Simpson 1985; Fadlallah 1985; Willis et al. 1985; Babcock et al. 1986; Kojis 1986b; Hunter 1988; Oliver et al. 1988; Harrison and Wallace 1990; among others). More recent hypotheses have been proposed that correlate reproductive patterns with environmental factors that may act as additional or alternative proximate or evolutionary controls on sexual reproduction in corals. These include a combination of warm temperature and absence of heavy rainfall (e.g., Mendes and Woodley 2002a), solar insolation cycles (Penland et al. 2004; van Woesik et al. 2006), and the duration of regional calm periods that may enhance fertilization and larval retention (van Woesik 2009). However, as noted by Harrison and Wallace (1990), correlation does not prove causality; hence, rigorous manipulative experiments are required to unequivocally demonstrate the extent to

Table 4 Coral species in which both broadcast spawning (SG) and brooding (B) modes of development have been reported

Clade/Family: Species	Mode	Location and comments (References)
II. *Tubastrea coccinea*	B+SG	GBR (D. Ayre, personal communication in Harrison and Wallace 1990); planulae produced asexually (Ayre and Resing 1986)
	B	Hawaii (e.g., Edmondson 1929, 1946; Yonge 1932; Richmond 1997)
	B	Eastern Pacific; sperm release observed (Glynn et al. 2008)
X. *Pocillopora damicornis*	B	GBR (e.g., Stephenson 1931; Marshall and Stephenson 1933; Harriott 1983b; Muir 1984 – asexual planulae; Tanner 1996)
	B	Pacific region (e.g., Edmondson 1929, 1946; Kawaguti 1941a; Atoda 1947a; Harrigan 1972; Vandermeulen and Watabe 1973; Stimson 1978; Richmond 1982, 1987; Stoddart 1983; Richmond and Jokiel 1984; Jokiel 1985; Hodgson 1985; Dai et al. 1992; Diah Permata et al. 2000; Fan et al. 2002, 2006; Villanueva et al. 2008)
	SG	Eastern Pacific; no brooded planulae (e.g., Glynn et al. 1991)
	SG	Gulf of California; no B (Chavez-Romo and Reyes-Bonilla 2007)
	B+SG?	WA; broods asexual planulae, inferred to spawn gametes (Stoddart 1983; Stoddart and Black 1985; Ward 1992)
Pocillopora meandrina	B	Enewetak; recorded as *P. elegans* (Stimson 1978)
	SG	Hawaii (Stimson 1978; Fiene-Severns 1998; Kolinski and Cox 2003)
Pocillopora verrucosa	B	Enewetak (Stimson 1978); Philippines (Villanueva et al. 2008)
	SG	Hawaii (Kinzie 1993)
	SG	Red Sea (Shlesinger and Loya 1985; Fadlallah 1985; Shlesinger et al. 1998)
	SG	Indian Ocean (Sier and Olive 1994; Kruger and Schleyer 1998)
XI. *Fungia fungites*	SG	GBR (e.g., Willis et al. 1985; Babcock et al. 1986)
	B	Japan (Loya et al. 2009)
Heliofungia actiniformis	B	Palau (Hada 1932; Abe 1937)
	SG	GBR (e.g., Willis et al. 1985; Babcock et al. 1986)
Leptastrea purpurea	SG	Japan (Hayashibara et al. 1993); Palau (Penland et al., unpublished data)
	SG	Hawaii (G. Hodgson, unpublished data cited in Kolinski and Cox 2003)
	B	(P. Schupp, personal communication cited in Baird et al. 2009a)
Psammocora stellata	SG	Eastern Pacific (Glynn and Colley 2009)
	B?	Hawaii; planulae collected (Kolinski and Cox 2003)
Oulastrea crispata	SG+B	Japan and Hong Kong (Nakano and Yamazoto 1992; Lam 2000)
XVII. *Goniastrea aspera*	B	Palau (Abe 1937; Motoda 1939)
	SG+B	Japan (Sakai 1997; Nozawa and Harrison 2005)
	SG	Indo-Pacific: GBR (e.g., Babcock 1984; Harrison et al. 1984; Babcock et al. 1986); Japan (e.g., Heyward et al. 1987; Hayashibara et al. 1993; Mezaki et al. 2007); Taiwan (Dai et al. 1992); Palau (Penland et al. 2004); WA (Simpson 1985; Gilmour et al. 2009)
XXI. *Favia fragum*	B+SG?	Jamaica; mainly brooded planulae, occasionally spawned eggs and sperm – possibly abnormal (Duerden 1902a)
	B	Caribbean – Western Atlantic (e.g., Vaughan 1908, 1910; Matthai 1919, 1923; Lewis 1974; Szmant-Froelich et al. 1985; Szmant 1986; Soong 1991; Carlon and Olson 1993; Brazeau et al. 1998; Goodbody-Gringley and Putron 2009)
Manicina areolota	B+SG?	Bahamas; mainly brooded planulae, initially spawned abundant eggs and sperm – possibly abnormal (Wilson 1888)
	B	Caribbean – Western Atlantic (e.g., Duerden 1902a, c; Boschma 1929; Yonge 1935; Johnson 1992)
Caryophylliidae		
Caryophyllia smithii	B	Golfe du Lion; released advanced embryos (de Lacaze-Duthiers 1897)
	SG	England (e.g., Thynne 1859; Hiscock and Howlett 1977; Tranter et al. 1982)

Early reports of brooding in six other species are probably incorrect (see Table 7.4 in Harrison and Wallace 1990), hence are not included

GBR Great Barrier Reef, *WA* Western Australia, *?* uncertain

which different environmental factors regulate sexual reproductive cycles in corals.

Some corals have recently been shown to sense moonlight (Gorbunov and Falkowski 2002) via photosensitive cryptochromes (Levy et al. 2007), while other research has provided new insights into biochemical processes influencing gametogenesis and spawning (e.g., Atkinson and Atkinson 1992; Tarrant et al. 1999, 2004; Twan et al. 2006).

These biochemical and molecular approaches provide exciting prospects for further research to understand how corals perceive and integrate information on environmental cues (e.g., Vize 2009) to regulate their reproductive cycles.

Ongoing research is providing increasing evidence that sexual reproductive processes in corals are highly sensitive to a wide range of natural and anthropogenic stressors that cause sublethal stress, reduce fecundity, and impair reproductive success (reviewed by Loya and Rinkevich 1980; Harrison and Wallace 1990; Richmond 1993, 1997; Fabricius 2005). Environmental factors known to stress corals and negatively affect sexual reproduction include: thermal stress (e.g., Kojis and Quinn 1984; Edmunds et al. 2001; Nozawa and Harrison 2002, 2007; Bassim et al. 2002; Krupp et al. 2006; Negri et al. 2007; Meyer et al. 2008; Randall and Szmant 2009; Yakovleva et al. 2009), ultraviolet radiation (e.g., Gulko 1995; Wellington and Fitt 2003; Torres et al. 2008), coral bleaching (e.g., Szmant and Gassman 1990; Omori et al. 2001; Ward et al. 2002; Baird and Marshall 2002; Mendes and Woodley 2002b), lowered salinity (e.g., Edmondson 1946; Richmond 1993; Harrison 1995; Vermeij et al. 2006; Humphrey et al. 2008), and increased sedimentation and turbidity (e.g., Kojis and Quinn 1984; Jokiel 1985; Gilmour 1999; Fabricius et al. 2003; Humphrey et al. 2008).

Sublethal and toxic levels of pollutants also impair or prevent successful coral reproduction, including increased nutrients (e.g., Tomascik and Sander 1987; Ward and Harrison 1997, 2000; Harrison and Ward 2001; Bassim and Sammarco 2003), oil pollutants and dispersants (e.g., Loya 1976; Rinkevich and Loya 1977, 1979c; Loya and Rinkevich 1979; Peters et al. 1981; Guzmán and Holst 1993; Harrison 1993, 1995; Negri and Heyward 2000; Lane and Harrison 2002), trace metals (e.g., Heyward 1988b; Reichelt-Brushett and Harrison 1999, 2000, 2004, 2005; Negri and Heyward 2001), herbicides and insecticides (e.g., Negri et al. 2005; Markey et al. 2007), and mixtures of pollutants in storm water runoff (e.g., Richmond 1993). These stress effects are likely to be exacerbated by climate change impacts, including modified thermal environments that may disrupt reproductive cycles in corals and inhibit larval settlement and postsettlement survival. Furthermore, increased carbon dioxide absorption resulting in decreased seawater pH and aragonite saturation state ("ocean acidification") is likely to interfere with initiation of calcification in newly settled coral polyps and reef-building by hermatypic corals (e.g., Albright et al. 2008; Jokiel et al. 2008). As coral reproduction appears to have a narrower tolerance to stress than other life functions (Harrison and Wallace 1990), it is essential to maintain ecologically appropriate environmental conditions to enable successful reproduction by corals in future.

4.3 Molecular Perspectives on Coral Reproduction

Important research advances have also occurred in relation to other aspects of sexual reproduction in reef corals, and these coincide with the development of powerful molecular methods for analyzing genetic structure in populations, and phylogenetic and systematic relationships among taxa. Of particular significance is the discovery that hybridization and genomic introgression can occur among some morphologically different coral "species" (e.g., Willis et al. 1992, 1997; Miller and Babcock 1997; Szmant et al. 1997; Hatta et al. 1999; van Oppen et al. 2001, 2002; Vollmer and Palumbi 2002, 2004; Fukami et al. 2004a; Richards et al. 2008; among others), which has important implications for coral taxonomy and our understanding of scleractinian evolution (reviewed by Veron 1995, 2000; Willis et al. 2006; Harrison and Booth 2007). For example, the Caribbean coral *Acropora prolifera* is now recognized to be a first generation hybrid of *Acropora palmata* and *Acropora cervicornis* (van Oppen et al. 2001; Vollmer and Palumbi 2002, 2004). More complex patterns of hybridization versus reproductive isolation occur in some other corals, such as in the *Montastrea annularis* species complex in the Caribbean. Hybridization between *Montastrea* species in this complex has been demonstrated in cross-fertilization experiments in some regions, whereas subtle differences in the timing of gamete release and regional differences in gametic incompatibility between species provide mechanisms for reproductive isolation among these sympatrically spawning species in other Caribbean regions (e.g., Szmant et al. 1997; Knowlton et al. 1997; Levitan et al. 2004; Fukami et al. 2004a).

Advances in molecular research and oceanographic models have also enabled more detailed analyses of gene flow and connectivity among populations of reef corals in different regions. These studies have demonstrated a wide range of larval retention and dispersal patterns in both brooding and broadcast spawning corals, ranging from highly localized larval settlement and recruitment in some cases, through to ecologically significant gene flow and connectivity among coral populations on different reefs over distances of tens to many hundreds of kilometers (e.g., Ayre et al. 1997; Ayre and Hughes 2000, 2004; Nishikawa et al. 2003; Baums et al. 2006; van Oppen and Gates 2006; Underwood et al. 2007, 2009; Hellberg 2007; Miller and Ayre 2008; Noreen et al. 2009; reviewed by Jones et al. 2009).

This more recent genetic research coincides with substantially increased knowledge of coral larval settlement competency periods and settlement rates. Rapid settlement of brooded planulae after release from polyps has been well documented for many brooding species and is likely to enhance localized settlement (e.g., Duerden 1902a, b; Atoda 1947a, b; Harrigan 1972; Rinkevich and Loya 1979a, b; Gerrodette 1981; Richmond 1985, 1987, 1988; Harrison and Wallace 1990; Carlon and Olson 1993; Harii et al. 2001). Recent research has demonstrated that planktonic planulae of some broadcast spawning species can temporarily attach to hard substrata or benthic algae after a few days planktonic development before becoming fully competent to settle (e.g., Harrison 1997, 2006; Nozawa and Harrison 2002, 2005; Miller and Mundy 2003), and this precocious attachment is likely to enhance larval retention and settlement on, or near, their natal reef. Thus, planulae settlement competency varies within and among species such that although some planulae may settle locally, some planulae can be transported between reefs over larger geographic scales (e.g., Oliver and Willis 1987; Sammarco and Andrews 1988; Willis and Oliver 1990; Harrison and Wallace 1990; Sammarco 1994; Harrison and Booth 2007; Gilmour et al. 2009; Jones et al. 2009; among others). Furthermore, some brooded planulae and some planulae from broadcast spawning corals remain competent to settle for more than 2–3 months (e.g., Richmond 1987; Wilson and Harrison 1998; Harii et al. 2002; Nozawa and Harrison 2002, 2005), while a few planulae can survive for more than 200 days (Graham et al. 2008), which increases their potential for mesoscale and long-distance dispersal. Occasional long-distance larval dispersal is likely to contribute to the broad biogeographic ranges of some coral species (Harrison 2006).

Molecular research is also providing important new perspectives on the phylogenetic relationships of corals, which challenge the conventional systematic classification and arrangement of many scleractinian families and suborders that are based on skeletal morphology (e.g., Vaughan and Wells 1943; Wells 1956). Pioneering molecular studies identified two highly genetically divergent scleractinian clades – the "complex" coral clade and the "robust" coral clade that have different skeletal morphologies (Romano and Palumbi 1996, 1997). The "complex" clade corals tend to have relatively porous and more lightly calcified complex skeletons with a wide range of branching and other growth forms, whereas "robust" clade corals tend to have relatively robust and heavily calcified skeletons with mostly plate-like or massive growth forms (Romano and Palumbi 1996). Subsequent molecular research has provided further strong support for these divergent clades and has shown that many conventionally defined scleractinian suborders, families and even some genera are polyphyletic, and many groups contain representatives of both complex and robust clades, leading to a radical reappraisal of coral systematics and phylogeny (e.g., Romano and Cairns 2000; Chen et al. 2002; Fukami et al. 2004b, 2008; Le Goff-Vitry et al. 2004; Kerr 2005; Medina et al. 2006; van Oppen and Gates 2006; Benzoni et al. 2007; Nunes et al. 2008; Huang et al. 2009). These molecular phylogenetic hypotheses provide new opportunities for assessing the evolution of reproductive traits in corals (Baird et al. 2009a), which are described in detail in the remaining sections of this chapter.

5 Patterns of Sexual Reproduction

Information on sexual reproductive characteristics is available for at least 444 scleractinian species (Table 1). Among the 400 coral species in which both the sexual pattern and mode of development are known, hermaphroditic broadcast spawners remain the dominant category with 258 species (64.5%), gonochoric broadcast spawners are moderately common with 78 species (19.5%), while hermaphroditic brooders (25 species) and gonochoric brooders (15 species) are relatively uncommon (Table 1). In addition, 12 coral species exhibit mixed patterns of sexuality or have contrasting sexual patterns reported, both modes of development have been reported in 13 coral species, and a few species brood asexually derived planulae. Therefore, not all coral species can be strictly categorized in the four main patterns of sexual reproduction defined by the two main sexual patterns and modes of development (see Tables 1–4).

6 Sexual Patterns

Fadlallah (1983) and Harrison and Wallace (1990) reviewed the terminology associated with patterns of sexuality applicable to corals and noted the complexity of interpreting sexual patterns in some species. Most corals studied to date are either hermaphroditic or have separate sexes (gonochoric). Hermaphroditic coral species typically produce both male and female gametes in the polyps of colonies or solitary corals during their lifetime, and within this group, simultaneous and sequential hermaphroditism can be distinguished. A total of 295 solely hermaphroditic coral species have been reported to date (Tables 1 and 2). This represents 70.9% of the 416 species for which the sexual pattern is known; hence, hermaphroditism is clearly the dominant sexual pattern among coral species studied so far. In addition, some hermaphroditic colonies or solitary corals occur among the 13 species in which mixed or contrasting sexual patterns have been reported (Table 3).

Hermaphroditic coral species occur in 13 different families and clade groups identified by Fukami et al. (2008). Hermaphrodites dominate the families Acroporidae (155 species), faviids, and associated corals in clade group XVII (74 species), Pocilloporidae (16 species), and the 29 mussid and related coral species now grouped in clades XVIII–XXI (Fukami et al. 2008), where all species studied to date are hermaphroditic (Table 2).

Simultaneous hermaphrodites develop mature ova and mature sperm at the same time within the same individual (Policansky 1982), and the vast majority of known hermaphroditic coral species are simultaneous hermaphrodites that develop both ova and sperm within each polyp (Fig. 5). Interestingly, *Astroides calycularis* and *Cladopsammia rolandi* colonies were reported to be hermaphroditic, but different polyps were typically only female or male (monoecious pattern), with some *A. calycularis* polyps occasionally hermaphroditic (de Lacaze-Duthiers 1873; 1897).

Fig. 5 The great majority of corals studied to date are broadcast spawning simultaneous hermaphrodites, such as *Favites* brain corals that develop both eggs and sperm within each polyp, which are spawned in buoyant egg–sperm bundles (Photo: author)

Fertilization trials indicate that most simultaneous hermaphroditic corals are completely or partially self-sterile, although self-fertilization occurs in a few species (e.g., Heyward and Babcock 1986; Willis et al. 1997; Knowlton et al. 1997; Hagman et al. 1998b; Hatta et al. 1999; Carlon 1999; Miller and Mundy 2005).

Sequential hermaphrodites have ova and sperm that develop or mature at different times, and these may be used in the same breeding season (sequential cosexual), or individuals may exhibit true sex change over successive breeding seasons or over their lifetime (Ghiselin 1974; Policansky 1982). The four agariciid species *P. varians*, *Pavona gigantea*, *Pavona chiriquiensis*, and *Gardineroseris planulata* are sequential cosexual hermaphrodites (but with some gonochoric colonies) in the eastern Pacific region, and have multiple cycles of gamete development during breeding seasons whereby maturation of the sexes alternates, and early development of one sex coincides with maturation of the other sex (Glynn et al. 1996, 2000; Glynn and Colley 2009). The endemic Brazilian coral *Mussismilia hispida*, and three deep-sea *Caryophyllia* species also display cyclical sequential hermaphroditism, whereby gametes of both sexes occur on mesenteries but only gametes of one sex develop and mature at a time, followed sequentially by maturation and spawning of the other sex (Neves and Pires 2002; Waller et al. 2005). *Diploastrea heliopora* colonies in Singapore are hermaphroditic (with one male colony recorded), but individual polyps are either female or male with sequential patterns of egg and sperm development (Guest 2004).

Some corals produce only sperm during their initial growth and sexual development, but subsequently mature as simultaneous hermaphrodites and are therefore adolescent protandric hermaphrodites (e.g., Rinkevich and Loya 1979a; Kojis and Quinn 1985; Kojis 1986a; Harrison and Wallace 1990; Hall and Hughes 1996). This initial sex allocation to male function is in accordance with sex allocation theory (e.g., Charnov 1982), and indicates a lower investment of energy and other resources in producing only sperm during the early growth stages of colony growth, thereby enabling a larger investment in growth and survival until the mortality risk is lower and allocation to both female and male function is sustainable. Furthermore, variation in sex allocation can also occur in adult corals, as demonstrated by a long-term study of sexual reproduction in colonies of *Stylophora pistillata*, which showed that some hermaphroditic colonies with high fecundity became male or were sterile in subsequent breeding seasons, and vice versa (Rinkevich and Loya 1987).

True protandrous sex change from initial male function in small corals to female function in larger corals has recently been demonstrated for some fungiid mushroom corals, including *Fungia repanda*, *Fungia scruposa*, *Ctenactis echinata*, and *Ctenactis crassa*, and the two *Ctenactis* species exhibit bidirectional sex change (Loya and Sakai 2008; Loya et al. 2009). A similar observation of bidirectional sex change has been recorded in colonies of another fungiid *Sandalolitha robusta* (B. Carlson, personal communication). Another fungiid species, *F. scutaria*, is predominantly male at small sizes whereas large individuals are all females, which suggests that these fungiids are also protandrous hermaphrodites (Kramarsky-Winter and Loya 1998). Abe (1937) also suggested that sperm develop before eggs in individual *Heliofungia actiniformis* corals (Table 3), and the recent reports of protandrous sex change in other fungiids provide support for this conclusion. Protandry was previously reported in the flabellid coral *Monomyces rubrum* (Gardiner 1902a, b), although Matthai (1914) disputed this interpretation.

These observations indicate that sex change and labile sex allocation may be a feature of reproduction in some other fungiids and perhaps some other corals that are generally recorded as gonochoric (Table 3, Fig. 6). Sex allocation theory predicts that these patterns of sex change occur in response to differential male and female fitness as size and age increase (Charnov 1982). The labile sex allocation in fungiids resembles patterns known in dioecious plants that respond to changing environmental conditions, and different energy costs associated with male versus female function (Loya and Sakai 2008).

Gonochoric coral species are less commonly reported than hermaphroditic coral species and are characterized by colonies or solitary corals that have separate sexes and are therefore functionally only male or female during their lifetime (dioecious). A total of 109 coral species have been recorded as gonochoric (Tables 1 and 2), which represents 26.2% of the 416 scleractinian species for which the sexual pattern is known. Gonochoric coral species have been recorded in 16 families and clades (Table 2). Families and

Fig. 6 Fungiid mushroom corals have some individuals that spawn sperm (**a**), and some that spawn eggs (**b**), and have been generally classified as gonochoric species with separate sexes. However, some fungiids are now known to change sex during their lifetime and are therefore sequential hermaphrodites (Photos: author)

clades that have mainly gonochoric species recorded include the Dendrophylliidae (16 gonochoric species), Poritidae (24 gonochoric species), Flabellidae (seven gonochoric species), and Fungiidae (22 gonochoric species); but these families also have smaller numbers of hermaphroditic species recorded (Table 2). Families and clade groups in which all species studied so far are gonochoric include three *Pachyseris* species in clade IV, Astrocoeniidae (two species), Fungiacyathidae (only one species studied), Meandrinidae (five species), Rhizangiidae (two species), and clade XIV (three species).

Strong evidence for gonochorism occurs in some species of *Porites* (e.g., Kojis and Quinn 1982; Harriott 1983a; Soong 1991; Glynn et al. 1994; Shlesinger et al. 1998), *Turbinaria mesenterina* and some *Pavona* species (Willis 1987; Glynn and Colley 2009), and *Montastrea cavernosa* (e.g., Szmant 1991; Soong 1991), where male colonies and female colonies were repeatedly sampled with no evidence of sex change. In some other gonochoric species, large numbers of colonies or solitary corals have been sampled (e.g., Szmant-Froelich et al. 1980; Fadlallah 1982; Fadlallah and Pearse 1982a, b; Fine et al. 2001; Waller et al. 2002; Heltzel and Babcock 2002; Brooke and Young 2003; Goffredo et al. 2006; Lins de Barros and Pires 2007) and usually contain only male or female gametes, with no relationship between sex and coral size that would suggest protandry or protogynous hermaphroditism. However, in many gonochoric species, the evidence for separation of sexes is based primarily on limited sampling of a few corals or observations of spawning.

Some primarily gonochoric coral species also exhibit a low or variable incidence of hermaphroditism in some polyps or colonies within the population (see Table 1 in Fadlallah 1983; Table 7.3 in Harrison and Wallace 1990; Table 1 in Richmond and Hunter 1990). In cases where the hermaphrodites are relatively rare, the species can be considered to exhibit stable gonochorism (*sensu* Giese and Pearse 1974) and are classified as gonochoric (Tables 1 and 2).

In 12 other coral species, mixed or contrasting sexual patterns have been reported (Tables 1 – 3). In some of these species, different populations are either gonochoric or hermaphroditic, or populations contain mixtures of females, males, and hermaphrodites (Table 3). These species may have labile gonochorism (*sensu* Giese and Pearse 1974). Alternatively, some of these species may be sequential hermaphrodites with asynchronous or partly overlapping oogenic and spermatogenic cycles that were sampled when only one sex was evident in some corals in the population (Fadlallah 1983; Harrison and Wallace 1990). It is also possible that the different sexual patterns reported between populations of some species from different reef regions reflect taxonomic problems, or morphologically similar cryptic species that have evolved contrasting sexual patterns.

More unusual mixed sexual patterns have been recorded in a few other coral species. *Porites astreoides* colonies from Jamaica were either female or hermaphroditic, which is a gynodioecious sexual pattern (Chornesky and Peters 1987). In Panama, most colonies of this species were female, with a few males recorded, and 17% of 168 colonies were hermaphroditic, in which spermaries and eggs sometimes occurred in different polyps (Soong 1991). In contrast, *P. astreoides* from Puerto Rico were simultaneous hermaphrodites (Szmant-Froelich 1984; Szmant 1986). *Galaxea fascicularis* and *Galaxea astreata* were originally described as simultaneous hermaphrodites; however, subsequent research demonstrated that these species have populations composed of female colonies that spawn pinkish-red eggs, and hermaphroditic colonies that produce sperm and lipid-filled white eggs (Harrison 1988). Hermaphroditic *G. fascicularis* colonies produce functional sperm that can fertilize the spawned pigmented eggs from female colonies (Fig. 7). However, the white eggs contain unusually large lipid spheres and are not able to be fertilized, and function to lift the sperm bundles up to the water surface where the buoyant pigmented eggs accumulate, thereby potentially enhancing fertilization success (Harrison 1988). This pseudogynodioecious sexual pattern in at least some *Galaxea* species is therefore functionally gonochoric.

These more complex sexual patterns, documented sex change and the variable incidence of hermaphroditism in some gonochoric species highlight the need for detailed and careful analysis of reproductive patterns in corals, as the assumption that coral species are always simply hermaphroditic or gonochoric species is not supported by the range of sex allocation and sexual patterns evident. Unambiguous determination of sexual patterns requires long-term observations or repeated sampling of marked or mapped corals, and preferably large-scale sampling within different populations to characterize patterns at the species level.

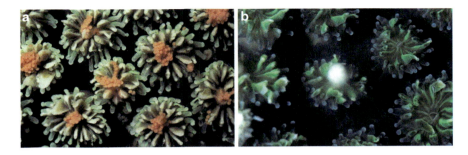

Fig. 7 *Galaxea fascicularis* populations have female colonies that spawn pigmented eggs (**a**), and other colonies that spawn functional sperm and white eggs that cannot be fertilized (**b**), and therefore have a pseudo-gynodioecious sexual pattern (Photos: author)

6.1 Systematic Trends in Sexual Patterns

Harrison (1985) noted that strong systematic trends were evident in sexual patterns and some other reproductive characteristics of corals. Subsequent reviews have expanded and provided further support for systematic trends in some reproductive characteristics based on conventional systematic classifications (Harrison and Wallace 1990; Richmond and Hunter 1990; Harrison and Jamieson 1999) and recent molecular phylogenies (Baird et al. 2009a). The available data indicate that sexual patterns are generally consistent within most coral species and genera studied to date, and in some families or clade groups. For example, sexual patterns are consistent in ten families and clades where at least two species have been studied (Table 2). However, as noted above, exceptions to these systematic trends are still evident within some species (Table 3) and genera, and an increasing number of families and clades now have both hermaphroditic and gonochoric species recorded (Table 2). Interestingly, similar proportions of species with hermaphroditic, gonochoric, and mixed or contrasting sexual patterns occur within both the "complex" and "robust" divergent coral clades (Table 2). Of the 105 coral genera for which sexual patterns have been recorded to date, 50 genera have only hermaphroditic species recorded, 38 genera have only gonochoric species recorded, while the remaining 17 genera have both sexual patterns or mixed sexuality recorded.

Baird et al. (2009a) concluded that the consistency of systematic trends in sexual patterns increased when global reproductive data are aligned with the revised molecular phylogeny of Fukami et al. (2008) and the "supertree" phylogeny of Kerr (2005), compared with the traditional morphological systematic organization. However, they noted that these molecular phylogenetic systematic arrangements do not resolve differences in sexuality within five of the families or clades defined by molecular groupings (Tables 1 and 2 in Baird et al. 2009a). In addition, when sequential hermaphrodites and species with mixed or contrasting sexuality (Table 3) are distinguished from gonochoric sexuality among species studied from the families Agariciidae (e.g., Delvoye 1982, 1988; Glynn et al. 1996, 2000; Glynn and Colley 2009) and Fungiidae (e.g., Loya and Sakai 2008), these families are not uniformly gonochoric (Tables 2 and 3). Furthermore, *Monomyces rubrum* may be a protrandrous hermaphrodite (Gardiner 1902a, b), and recent studies have indicated that *Pseudosiderastrea tayamai* (see Harii et al. 2009) and *Cladocora caespitosa* (see Kruzic et al. 2008) are hermaphroditic; hence, the sexual pattern in these species also contrasts with other members of these three families (Table 2). Overall, sexual patterns are consistent in 12 of the 21 families and clades listed by Baird et al. (2009a) that have had the sexual pattern of more than one species documented – the other nine families have species with hermaphroditic, gonochoric, and mixed or contrasting sexuality recorded. Therefore, systematic trends in sexuality are only consistent in some scleractinian families and clades currently defined by molecular phylogenies (Table 2), which implies that contrasting sexual patterns have evolved in a range of clades. The exceptions to stable systematic patterns in sexuality are interesting as they highlight taxa that should be targeted for further research to confirm their phylogenetic relationships and the evolution of hermaphroditism versus gonochorism in the Scleractinia (Harrison and Wallace 1990).

Molecular data indicate that *Alveopora* should be transferred from the Poritidae to the Acroporidae (Fukami et al. 2008), and this is in accordance with their sexual pattern; *Alveopora* species studied so far are hermaphroditic (e.g., Shlesinger and Loya 1985; Babcock et al. 1994; Harii et al. 2001; Nozawa et al. 2006), like all the Acroporidae, whereas most Poritidae are gonochoric (Table 2). Likewise, the transfer of the gonochoric species *Montastrea cavernosa* (e.g., Szmant-Froelich 1984; Szmant 1986), from the Faviidae to a separate clade (Fukami et al. 2008), results in a consistent hermaphroditic sexual pattern within all faviid species studied to date, including other *Montastrea* species (Table 2). Baird et al. (2009a) noted that the early record of *Isophyllia sinuosa* as being possibly gonochoric contrasted with other mussids that are hermaphroditic; however, Duerden (1902a) stated that only oocytes were seen in one colony, and hence, the sexual pattern of this mussid is not known and is not included in the summary tables in this chapter.

At present, the molecular relationships of many coral taxa have not yet been analyzed, and the phylogenetic relationships of some groups are uncertain or vary somewhat between studies and between molecular and morphological groupings (e.g., Kerr 2005; Fukami et al. 2008; Huang et al. 2009). Therefore, the recent molecular-based systematic groupings represent important new hypotheses that need to be further developed and tested using additional molecular and morphological data, to provide a comprehensive new systematic classification of the Scleractinia. Once this is achieved, it will provide a stronger basis for understanding systematic trends and evolution of sexual patterns in corals.

Szmant (1986) noted the dominance of hermaphroditic coral species and suggested that hermaphroditism might be the ancestral pattern in corals. However, analysis of scleractinian sperm morphology and ultrastructure has shown that the more primitive conical type of sperm is mostly associated with gonochoric coral species (e.g., Harrison 1985, 1990; Steiner 1991, 1993; Harrison and Jamieson 1999), which suggests that gonochorism is ancestral in the Scleractinia (Harrison 1985, 1990). Gonochorism is considered to be the ancestral condition of hexacorallian Anthozoa, including scleractinians (Daly et al. 2003), and Baird et al. (2009a) concluded that a phylogenetic analysis of reproductive traits also strongly supported gonochorism as the ancestral sexual state in corals.

Fig. 8 Most corals studied so far are broadcast spawning simultaneous hermaphrodites, such as *Favia* brain corals that forcibly eject egg–sperm bundles from the polyps during brief annual spawning events (Photo: author)

7 Mode of Development

Corals have two different primary modes of development: broadcast spawning of gametes from their polyps into the sea for external fertilization and development (spawners); or brooding of embryos and planula larvae within their polyps that are subsequently released, usually at an advanced stage of development (brooders). Brooding corals also spawn sperm, and among hermaphroditic brooders, the spawned sperm may enable self-fertilization of eggs within polyps of the same coral in some species (e.g., Brazeau et al. 1998; Gleason et al. 2001; Okubo et al. 2007), or fertilize eggs within polyps of other colonies of the same species leading to outcrossing (e.g., Ayre and Miller 2006).

The vast majority of coral species studied to date are broadcast spawning species (Fig. 8); hence, external fertilization leading to planktonic larval development is the dominant mode of development (Tables 1 and 2). Of the 428 coral species for which the mode of development has been recorded, 354 species (82.7%) are broadcast spawners, whereas 61 species (14.3%) are brooders (Tables 1 and 2). In at least another 13 coral species (3%), both modes of development have been recorded (Table 4), including at least two of the three species known to brood asexually generated planulae (e.g., Stoddart 1983; Ayre and Resing 1986). The number of broadcast spawning coral species has more than doubled since the late 1980s (Harrison and Wallace 1990; Richmond and Hunter 1990), whereas relatively few new records of brooding species have emerged in the last 20 years (Fig. 4).

Among the 13 coral species in which both modes of development have been reported (Table 4), some species clearly spawn gametes and brood planulae, while some species have different modes of development recorded in different regions. This may reflect differences between populations or may have arisen from unrecognized or unresolved taxonomic differences between the corals identified as the same species from different regions (see Stimson 1978). The early records of spawning in the Caribbean brooding species *Favia fragum* and *Manicina areolata* may be an artifact and were regarded as abnormal (see Wilson 1888; Duerden 1902a, b).

Initial spawning of gametes followed by brooding of some eggs and embryos that are retained in the polyps has been shown to occur in some *Goniastrea aspera* colonies in Japan (Sakai 1997; Nozawa and Harrison 2005). In contrast, colonies of this faviid species have only been observed to spawn gametes on the GBR and WA reefs, whereas this species was reported to brood and release planulae in Palau (Table 4).

Five fungiid coral species have both spawning and brooding modes of development recorded (Table 4), whereas other members of this family all spawn gametes (Table 2). *Heliofungia actiniformis* is of particular interest, as this is the only coral species known so far in which both modes of development have been recorded, and contrasting sexual patterns have been reported (Tables 1 to 4).

Both modes of development have been reported in three species of *Pocillopora* (Table 4), whereas three other *Pocillopora* species are reported to spawn gametes (Glynn et al. 1991; Kinzie 1993; Glynn and Colley 2009). The variable modes of development recorded in *P. damicornis* are particularly interesting, as this is one of the most widely studied brooding coral, and planulae in some populations are produced asexually (Table 4). Evidence for brooding of asexual planulae and inferred sexual reproduction via production and spawning of eggs and sperm in some populations of *P. damicornis*, is provided by histological and genetic studies (e.g., Stoddart 1983; Muir 1984; Stoddart and Black 1985; Ward 1992; Ayre and Miller 2004; Whitaker 2006; Sherman et al. 2006). In contrast, histological studies of *P. damicornis* colonies in Japan indicated that brooded planulae developed from eggs, and may be produced sexually (Diah Permata et al. 2000). A different reproductive pattern occurs in eastern Pacific and Gulf of California populations of *P. damicornis*, which are characterized by the production of eggs and sperm and inferred spawning of mature gametes, but there is no evidence of brooding or planulae production in these populations (Glynn et al. 1991; Colley et al. 2006; Chavez-Romo and Reyes-Bonilla 2007; Glynn and Colley 2009). The great variation in reproductive characteristics and life history traits recorded among populations identified as *P. damicornis* in different regions indicate that these characteristics are unusually variable in this "species"; alternatively, *P. damicornis* may be a species complex containing cryptic species with different reproductive patterns (e.g., Flot et al. 2008; Souter 2010).

At least one of the two *Tubastrea* species known to brood asexually generated planulae (Ayre and Resing 1986) is also thought to spawn gametes (Table 4); therefore the other *Tubastrea* species that produces asexual planulae may also exhibit both modes of development. *O. crispata* colonies in Japan and Hong Kong produce sperm and azooxanthellate eggs that are spawned over an extended period, and later develop brooded zooxanthellate planulae during the period when gametogenesis is not occurring, hence these planulae may be produced asexually (Nakano and Yamazoto 1992; Lam 2000). Production of brooded asexual planulae by locally adapted genotypes is likely to enhance local settlement and survival leading to higher recruitment success in some species, as predicted by the "strawberry coral model" (Williams 1975), whereas sexual reproduction in these species may enhance colonization of distant reefs. In contrast, Ayre and Miller (2004) concluded that although asexual production of brooded larvae was the primary mode of reproduction in a population of *P. damicornis* from the southern GBR, juveniles and adults displayed high genotypic diversity, with little evidence of asexually derived local recruitment.

The distinction between broadcast spawning and brooding modes of development becomes blurred in some species that exhibit intermediate modes of development, such as *Madracis* species, where internal fertilization leads to initial embryo and planula development during a short and variable "pseudobrooding" period of a few hours to days (Vermeij et al. 2003, 2004; de Putron 2004). *Eusmilia fastigiata* colonies spawn gametes through the polyp tentacles in Puerto Rico (Steiner 1995), whereas eggs or early stage embryos appear to be released through tentacles at Bonaire and Venezuela, which may indicate some degree of brooding or intermediate modes of development in these corals (de Graaf et al. 1999; Bastidas et al. 2005). Female *Stephanocoenia intersepta* colonies in Curaçao exhibit intratentacular fertilization, whereby the eggs are transferred into slits extending along the tentacles and held there for up to 10 min to increase exposure to spawned sperm and enhance fertilization success before the fertilized eggs are released (Vermeij et al. 2010); which is an intermediate mode of development. *Goniastrea favulus* spawns negatively buoyant eggs in a sticky mucus matrix that initially remains on or near the parent colony and may trap sperm, and this intermediate spawning strategy can lead to very high rates of fertilization (Kojis and Quinn 1981; Babcock 1984; Miller and Mundy 2005).

Both modes of development have been recorded within 13 families and clades, while six other families and clades in which at least two species have been studied have only broadcast spawners recorded (Table 2). Broadcast spawning dominates in the families Acroporidae (150 species), faviids and related clades (76 species), Fungiidae (22 species), and the mussid and related clades XVIII–XXI (24 species), while brooding is more common than broadcast spawning in the Family Pocilloporidae (Table 1). The proportion of broadcast spawning species is about 83% in both the "complex" and "robust" clades, whereas the proportion of brooding species is higher in the "complex" clade than in the "robust" clade (Table 2). All but one of the species reported to exhibit both modes of development occur in the "robust" clade (Table 2).

Overall, the occurrence of brooding corals among so many families and clades that are dominated by broadcast spawning species suggests that brooding has evolved in many coral taxa. Therefore, the mode of development appears to be a relatively plastic and variable life history trait compared with sexuality and other reproductive characteristics (Harrison 1985). Shlesinger et al. (1998) suggested that broadcast spawning was the ancestral mode of development in scleractinians and other Hexacorallia, and viviparous brooding is a derived reproductive characteristic.

8 Summary and Conclusions

The renaissance in coral reproduction studies that began in the 1980s has provided a wealth of data on many aspects of sexual reproduction in hermatypic zooxanthellate and some other corals. Information on sexual reproduction is available for at least 444 coral species, with hermaphroditic broadcast spawning the dominant pattern, and gonochoric broadcast spawning moderately common, whereas relatively few hermaphroditic brooding or gonochoric brooding species have been recorded. A small number of coral species have mixed or contrasting sexual patterns or both modes of development recorded, and a few species are known to produce asexual planulae. Multispecific spawning has been reported on many reefs where diverse coral assemblages have been studied; however, the degree of reproductive synchrony varies greatly within and among species at different locations. These differences in reproductive timing and synchrony result in a continuum of coral reproductive patterns ranging from temporal reproductive isolation, through various scales of multispecific spawning, to very large-scale synchronous mass spawning events involving many corals of many species from diverse families and clades.

The discovery of hybridization and genomic introgression between some morphologically different coral species that spawn synchronously, the existence of some morphologically similar but reproductively isolated cryptic species, and more recent molecular research showing that many conventionally defined scleractinian families, suborders and some genera are polyphyletic, have led to a radical reappraisal of coral taxonomy, systematics, and evolutionary processes in scleractinians. Molecular studies have also contributed to improved understanding of the extent to which coral larvae are retained on, or near, their natal reefs and contribute to localized recruitment, or are dispersed to other reefs providing connectivity over larger geographic scales, and these dispersal patterns vary among species, and over space and time. Increased knowledge of larval development and competency periods combined with data from population genetic studies have provided evidence for highly localized larval settlement and recruitment in some species, as well as varying scales of ecologically significant gene flow and connectivity among coral populations on different reefs over distances of tens to hundreds of kilometers.

Research has also clearly demonstrated that sexual reproductive processes are highly sensitive to a wide range of natural and anthropogenic stressors, which impair or block the critically important phases of reproduction and recruitment that are required to maintain and replenish coral populations. The magnitude of these stresses will be increasingly exacerbated by the rising tide of human populations using coral reef habitats, and by climate change impacts from modified thermal environments and altered seawater chemistry that can interfere with initiation of calcification in newly settled coral polyps and reef-building by hermatypic corals. Therefore, it is essential to manage reef environments to ensure that ecologically appropriate environmental conditions are maintained to enable successful reproduction by corals into the future.

Acknowledgments Thanks to Carden Wallace, Anna Scott, and Andrew Carroll for helpful comments on this chapter, and Peta Beeman and Barbara Harrison for assistance with its preparation. I also thank the many colleagues who have contributed unpublished data over many years and copies of their recent publications.

References

Abe N (1937) Post-larval development of the coral *Fungia actiniformis* var. *palawensis* Doderlein. Palao Trop Biol Stat Stud 1:73–93

Acosta A, Zea S (1997) Sexual reproduction of the reef coral *Montastrea cavernosa* (Scleractinia: Faviidae) in the Santa Marta area, Caribbean coast of Colombia. Mar Biol 128:141–148

Albright R, Mason B, Landon C (2008) Effect of aragonite saturation state on settlement and post-settlement growth of *Porites astreoides* larvae. Coral Reefs 27:485–490

Alvarado EM, Garcia RU, Acosta A (2004) Sexual reproduction of the reef building coral *Diploria labrynthiformis* (Scleractinia: Faviidae), in the Colombian Caribbean. Rev Biol Trop 52:859–868

Atkinson S, Atkinson MJ (1992) Detection of estradiol-17β during a mass coral spawn. Coral Reefs 11:33–35

Atoda K (1947a) The larva and postlarval development of some reef-building corals. I. *Pocillopora damicornis cespitosa* (Dana). Sci Rep Tohoku Univ (Ser 4) 18:24–47

Atoda K (1947b) The larva and postlarval development of some reef-building corals. II. *Stylophora pistillata* (Esper). Sci Rep Tohoku Univ (Ser 4) 18:48–64

Ayre DJ, Hughes TP (2000) Genotypic diversity and gene flow in brooding and spawning corals along the Great Barrier Reef, Australia. Evolution 54:1590–1605

Ayre DJ, Hughes TP (2004) Climate change, genotypic diversity and gene flow in reef-building corals. Ecol Lett 7:273–278

Ayre DJ, Miller KJ (2004) Where do clonal coral larval go? Adult genotypic diversity conflicts with reproductive effort in the brooding coral *Pocillopora damicornis*. Mar Ecol Prog Ser 277:95–105

Ayre DJ, Miller KJ (2006) Random mating in the brooding coral *Acropora palifera*. Mar Ecol Prog Ser 307:155–160

Ayre DJ, Resing JM (1986) Sexual and asexual production of planulae in reef corals. Mar Biol 90:187–190

Ayre DJ, Hughes TP, Standish RJ (1997) Genetic differentiation, reproductive mode, and gene flow in the brooding coral *Pocillopora damicornis* along the Great Barrier Reef. Aust Mar Ecol Prog Ser 159:175–187

Babcock RC (1984) Reproduction and distribution of two species of *Goniastrea* (Scleractinia) from the Great Barrier Reef Province. Coral Reefs 2:187–195

Babcock RC (1991) Comparative demography of three species of scleractinian corals using age and size dependent classifications. Ecol Monogr 61:225–244

Babcock RC, Bull GD, Harrison PL, Heyward AJ, Oliver JK, Wallace CC, Willis BL (1986) Synchronous spawnings of 105 scleractinian coral species on the Great Barrier Reef. Mar Biol 90:379–394

Babcock RC, Wills BL, Simpson CJ (1994) Mass spawning of corals on high latitude coral reef. Coral Reefs 13:161–169

Bachtiar I (2000) Reproduction of three scleractinian corals (*Acropora cytherea*, *A. nobilis*, *Hydnophora rigida*) in eastern Lombok Strait, Indonesia. Ilmu Kelautan 21:18–27

Baird AH, Guest JR (2009) Spawning synchrony in scleractinian corals: comment on Mangubhai & Harrison (2008). Mar Ecol Prog Ser 374:301–304

Baird AH, Marshall PA (2002) Mortality, growth and reproduction in scleractinian corals following bleaching on the Great Barrier Reef. Mar Ecol Prog Ser 237:133–141

Baird AH, Sadler C, Pitt M (2001) Synchronous spawning of *Acropora* in the Solomon islands. Coral Reefs 19:286

Baird AH, Marshall PA, Wolstenholme J (2002) Latitudinal variation in the reproduction of Acropora in the Coral sea. In: Proceedings of the 9th international coral reef symposium, vol 1, Bali, 2000, pp 385–389

Baird AH, Guest JR, Willis BL (2009a) Systematic and biogeographical patterns in the reproductive biology of scleractinian corals. Annu Rev Ecol Evol Syst 40:551–571

Baird AH, Birrel CL, Hughes TP, McDonald A, Nojima S, Page CA, Prachett MS, Yamasaki H (2009b) Latitudinal variation in reproductive synchrony in *Acropora* assemblages: Japan vs Australia. Galaxea 11:101–108

Bassim KM, Sammarco PW (2003) Effects of temperature and ammonium on larval development and survivorship in a scleractinian coral (*Diploria strigosa*). Mar Biol 140:479–488

Bassim KM, Sammarco PW, Snell TL (2002) Effects of temperature on success of (self and non-self) fertilization and embryogenesis in *Diploria strigosa* (Cnidaria, Scleractinia). Mar Biol 140:479–488

Bastidas C, Croquer A, Zubillaga AL, Ramos R, Kortnik V, Weinberger C, Marquez LM (2005) Coral mass- and split-spawning at a coastal and an offshore Venezuelan reefs, southern Caribbean. Hydrobiologia 541:101–106

Baums IB, Miller MW, Hellberg ME (2006) Geographic variation in clonal structure in a reef building Caribbean coral, *Acropora palmata*. Ecol Monogr 76:503–519

Beaver CR, Earle SA, Tunnell JW Jr, Evans EF, de la Cerda AV (2004) Mass spawning of reef corals within the Veracruz Reef System, Veracruz, Mexico. Coral Reefs 23:324

Benzoni F, Stefani F, Stolarski J, Pichon M, Mitta G, Galli P (2007) Debating phylogenetic relationships of the scleractinian *Psammocora*: molecular and morphological evidence. Contrib Zool 76:35–54

Bermas NA, Aliño PM, Atrigenio MP, Uychiaoco A (1992) Observations on the reproduction of scleractinian and soft corals in the Philippines. In: Proceedings of the 7th international coral reef symposium, vol 1, Gaum, 1992, pp 443–447

Borneman EH (2006) Reproduction in aquarium corals. In: Proceedings of the 10th international coral reef symposium, vol 1, Okinawa, 2004, pp 50–60

Boschma H (1929) On the postlarval development of the coral *Maeandra areolata* (L.). Pap. Tortugas Lab 26:129–147

Brazeau DA, Gleason DF, Morgan ME (1998) Self-fertilization in brooding hermaphroditic Caribbean corals: evidence from molecular markers. J Exp Mar Biol Ecol 231:225–238

Brooke S, Young CM (2003) Reproductive ecology of a deep-water scleractinian coral, *Oculina varicosa*, from the south-east Florida shelf. Cont Shelf Res 23:847–858

Cairns SD (1988) Asexual reproduction in solitary scleractinia. In: Proceedings of the 6th international coral reef symposium, vol 2, Townsville, 1988, pp 641–646

Cairns SD (1999) Species richness of recent scleractinia. Atoll Res Bull 459:1–46

Cairns SD (2007) Deep-water corals: an overview with special reference to diversity and distribution of deep-water scleractinian corals. Bull Mar Sci 81:311–322

Campbell RD (1974) Development. In: Muscatine L, Lenhoff HM (eds) Coelenterate biology. Reviews and new perspectives. Academic Press, New York, pp 179–210

Carlon DB (1999) The evolution of mating systems in tropical reef corals. Trends Ecol Evol 14:491–495

Carlon DB, Olson RR (1993) Larval dispersal distance as an explanation for adult spatial pattern in two Caribbean reef corals. J Exp Mar Biol Ecol 173:247–263

Carpenter KE, Abrar M, Aeby G, Aronson RB, Banks S et al (2008) One third of reef-building corals face elevated extinction risk from climate change and local impacts. Science 321:560–563

Carroll A, Harrison PL, Adjeroud M (2006) Sexual reproduction of *Acropora* reef corals at Moorea, French Polynesia. Coral Reefs 25:93–97

Charnov EL (1982) The theory of sex allocation, Princeton Monograph in Population Biology No. 18. Princeton University Press, Princeton, 355 pp

Chavez-Romo HE, Reyes-Bonilla H (2007) Sexual reproduction of the coral *Pocillopora damicornis* in the southern Gulf of California, Mexico. Ciencias Marinas 33:495–501

Chen CA, Wallace CC, Wolstenholme J (2002) Analysis of the mitochondrial 12S rRNA gene supports a two-clade hypothesis of the evolutionary history of scleractinian corals. Mol Phylogenet Evol 23:137–149

Chornesky EA, Peters EC (1987) Sexual reproduction and colony growth in the scleractinian coral *Porites astreoides*. Biol Bull 172:161–177

Coles SL, Seapy DG (1998) Ultra-violet absorbing compounds and tumorous growths on acroporid corals from Bandar Khayran, Gulf of Oman, Indian Ocean. Coral Reefs 17:195–198

Colley SB, Feingold JS, Pena J, Glynn PW (2002) Reproductive ecology of *Diaseris distorta* (Michelin) (Fungiidae) in the Galapagos Islands, Ecuador. In: Proceedings of the 9th international coral reef symposium, vol 1, Bali, 2000, pp 373–379

Colley SB, Glynn PW, Mayt AS, Mate JL (2006) Species-dependent reproductive responses of eastern Pacific corals to the 1997–1998 ENSO event. In: Proceedings of the 10th international coral reef symposium, vol 1, Okinawa, 2004, pp 61–70

Dai C-F, Soong K, Fan T-Y (1992) Sexual reproduction of corals in Northern and Southern Taiwan. In: Proceedings of the 7th international coral reef symposium, vol 1, Guam, 1992, pp 448–455

Dai C-F, Fan T-Y, Yu J-K (2000) Reproductive isolation and genetic differentiation of a scleractinian coral *Mycedium elephantotus*. Mar Ecol Prog Ser 201:179–187

Daly M, Fautin DG, Cappola VA (2003) Systematics of the Hexacorallia (Cnidaria: Anthozoa). Zool J Linnaean Soc 139:419–437

de Graaf M, Geertjes GJ, Videler JJ (1999) Observations on spawning of scleractinian corals and other invertebrates on the reefs of Bonaire (Netherlands Antilles, Caribbean). Bull Mar Sci 64:189–194

de Lacaze-Duthiers H (1873) Developpement des coralliaires. Actiniaires a Polypiers. Arch Zool Exp Gen 2:269–348

de Putron S (2004) Sexual reproduction of *Madracis mirabilis*: further evidence for a pseudo-brooding strategy. In: Proceedings of the 10th international coral reef symposium, Okinawa, 2004, p 190, Abstracts

Delvoye L (1982) Aspects of sexual reproduction in some Caribbean Scleractinia. Ph.D. thesis, Caribbean Marine Biological Institute Curacao, 93 pp

Delvoye L (1988) Gametogenesis and gametogenic cycles in *Agaricia agaricites* (L) and *Agarica humilis* verrill and notes on gametogenesis in *Madracis mirabilis* (Duchassiang & Michelotti) (Scleractinia). Uitgaven-Natuurwetenschappelijke S 123:101–134

Diah Permata W, Kinzie RA III, Hidaka M (2000) Histological studies on the origin of planulae of the coral *Pocillopora damicornis*. Mar Ecol Prog Ser 200:191–200

Duerden JE (1902a) West Indian madreporarian polyps. Mem Nat Acad Sci 8:402–597

Duerden JE (1902b) The morphology of the Madreporaria.-II. Increase of mesenteries in *Madrepora* beyond the protocnemic stage. Ann Mag Nat Hist (Ser 7) 10:96–115

Duerden JE (1902c) Aggregated colonies in madreporarian corals. Am Nat 36:461–472

Edmondson CH (1929) Growth of Hawaiian corals. Bernice P Bishop Mus Bull 58:1–38

Edmondson CH (1946) Behaviour of coral planulae under altered saline and thermal conditions. Bernice P Bishop Mus Occ Pap 18:283–304

Edmunds P, Gates R, Gleason D (2001) The biology of larvae from the reef coral *Porites astreoides*, and their response to temperature disturbances. Mar Biol 139:981–989

Fabricius KE (2005) Effects of terrestrial runoff on the ecology of corals and coral reefs: review and synthesis. Mar Poll Bull 50:125–146

Fabricius KE, Wild C, Wolanski E, Abele D (2003) Effects of transparent exopolymer particles and muddy terrigenous sediments on the survival of hard coral recruits. Estuar Coast Shelf Sci 57:613–621

Fadlallah YH (1982) Reproductive ecology of the coral *Astrangia lajollaensis*: sexual and asexual patterns in a kelp forest habitat. Oecologia 55:379–388

Fadlallah YH (1983) Sexual reproduction, development and larval biology in scleractinian corals. A review. Coral Reefs 2:129–150

Fadlallah YH (1985) Reproduction in the coral Pocillopora verrucosa on the reefs adjacent to the industrial city of Yanbu (Red Sea, Saudi Arabia). In: Proceedings of the 5th international coral reef congress, vol 4, Tahiti, 1985, pp 313–318

Fadlallah YH, Pearse JS (1982a) Sexual reproduction in solitary corals: overlapping oogenic and brooding cycles, and benthic planulas in *Balanophyllia elegans*. Mar Biol 71:223–231

Fadlallah YH, Pearse JH (1982b) Sexual reproduction in solitary corals: synchronous gametogenesis and broadcast spawning in *Paracyathus stearnsii*. Mar Biol 71:233–239

Fan T-Y, Dai C-F (1995) Reproductive ecology of the scleractinian coral *Echinopora lamellosa* in northern and southern Taiwan. Mar Biol 123:565–572

Fan T-Y, Dai C-F (1999) Reproductive plasticity in the reef coral *Echinopora lamellosa*. Mar Ecol Prog Ser 190:297–301

Fan T-Y, Li J-J, Ie S-X, Fang L-S (2002) Lunar periodicity of larval release by pocilloporid corals in southern Taiwan. Zool Stud 41:288–294

Fan T-Y, Lin K-H, Kuo F-W, Soong K, Liu L-L, Fang L-S (2006) Diel patterns of larval release by five brooding scleractinian corals. Mar Ecol Prog Ser 321:133–142

Fiene-Severns P (1998) A note on synchronous spawning in the reef coral *Pocillopora meandrina* at Molokini Islet, Hawai'i. In: Cox EF, Krupp DA, Jokiel PL (eds) Reproduction in reef corals – results of the 1997 Edwin W. Pauley Summer Program in marine biology. Hawai'i Institute of Marine Biology Technical Report 42, Kaneohe, pp 22–24

Fine M, Zibrowius H, Loya Y (2001) *Oculina patagonica*: a non-lessepsian scleractinian coral invading the Mediterranean Sea. Mar Biol 138:1195–1203

Flot J-F, Magalon H, Cruaud C, Couloux A, Tillier S (2008) Patterns of genetic structure among Hawaiian corals of the genus *Pocillopora* yield clusters of individuals that are compatible with morphology. Compt Rend Biol 331:239–247

Foster NL, Baums IL, Mumby PJ (2007) Sexual vs. asexual reproduction in an ecosystem engineer: the massive coral *Montastrea annularis*. J Anim Ecol 76:384–391

Francini CLB, Castro CB, Pires DO (2002) First record of a reef coral spawning event in the western South Atlantic. Invert Reprod Dev 42:17–19

Freiwald A, Roberts JM (eds) (2005) Cold-water corals and ecosystems. Springer, Berlin

Fukami H, Budd AF, Levitan DR, Kersanach R, Knowlton N (2004a) Geographic differences in species boundaries among members of the *Montastrea annularis* complex based on molecular and morphological markers. Evolution 58:324–337

Fukami H, Budd AF, Paulay G, Sole-Cava A, Chen CA et al (2004b) Conventional taxonomy obscures deep divergence between Pacific and Atlantic corals. Nature 427:832–835

Fukami H, Chen CA, Budd AF, Collins A, Wallace C, Chuang Y-Y, Chen C, Dai C-F, Iwao K, Sheppard C, Knowlton N (2008) Mitochondrial and nuclear genes suggest that stony corals are monophyletic but most families of stony corals are not (Order Scleractinia, Class Anthozoa, Phylum Cnidaria). PLoS ONE 3:1–9

Gardiner JS (1902a) South African corals of the genus *Flabellum*, with an account of their anatomy and development. Marine Investigations in South Africa, vol 2, Cape of Good Hope Department of Agriculture, Cape Town, pp 117–154

Gardiner JS (1902b) Some notes on variation and protandry in Flabellum rubrum, and senescence in the same and other corals. In: Proceedings of Cambridge philosophical society of biological science, vol 11, Cambridge, 1902, pp 463–471

Gardiner JS (1904) The Turbinolid corals of South Africa, with notes on their anatomy and variation. In: Marine Investigations in South Africa, vol 2, Cape of Good Hope Department of Agriculture, Cape Town, pp 95–129

Gerrodette T (1981) Dispersal of the solitary coral *Balanophyllia elegans* by demersal planular larvae. Ecology 62:611–619

Ghiselin MT (1974) The economy of nature and the evolution of sex. University of California Press, Berkeley, 346 pp

Giese AC, Pearse JS (1974) Introduction: general principles. In: Giese AC, Pearse JS (eds) Reproduction of marine invertebrates, vol I. Academic, New York, pp 1–49

Gilmour JP (1999) Experimental investigation into the effects of suspended sediment on fertilization, larval survival and settlement in a scleractinian coral. Mar Biol 135:451–462

Gilmour JP (2002) Substantial asexual recruitment of mushroom corals contributes little to population genetics of adults in conditions of chronic sedimentation. Mar Ecol Prog Ser 235:81–91

Gilmour JP, Smith LD, Brinkman RM (2009) Biannual spawning, rapid larval development and evidence of self seeding for scleractinian corals at an isolated system of reefs. Mar Biol 156:1297–1309

Gittings SR, Boland GS, Deslarzes KJP, Combs CL, Holland BS, Bright TJ (1992) Mass spawning and reproductive viability of reef corals at the east Flower Garden Bank, Northwest Gulf of Mexico. Bull Mar Sci 51:420–428

Gladfelter EH (1983) Circulation of fluids in the gastrovascular system of the reef coral *Acropora cervicornis*. Biol Bull 165:619–636

Gleason DF, Brazeau DA, Munfus D (2001) Can self-fertilizing coral species be used to enhance restoration of Caribbean reefs? Bull Mar Sci 69:933–943

Glynn PW, Colley SB (2009) Survival of brooding and broadcasting reef corals following large scale disturbances: is there any hope for broadcasting species during global warming? In: Proceedings of the 11th international coral reef symposium, vol 1, FL Landerdale, 2008, pp 361–365

Glynn PW, Gassman NJ, Eakin CM, Cortes J, Smith DB, Guzman HM (1991) Reef coral reproduction in the eastern Pacific: Costa Rica, Panama, and Galapagos Islands (Ecuador). I. Pocilloporidae. Mar Biol 109:355–368

Glynn PW, Colley SB, Eakin CM, Smith DB, Cortes J, Gassman NJ, Guzman HM, Del Rosario JB, Feingold JS (1994) Reef coral reproduction in the eastern Pacific: Costa Rica, Panama, and Galapagos Islands (Ecuador). II. Poritidae. Mar Biol 118:191–208

Glynn PW, Colley SB, Gassman N, Black K, Cortés J, Maté JL (1996) Reef coral reproduction in the eastern Pacific: Costa Rica, Panama, and Galapagos Islands (Ecuador). III. Agariciidae (*Pavona gigantea* and *Gardineroseris planulata*). Mar Biol 125:579–601

Glynn PW, Colley SB, Ting JH, Mate JL, Guzman HM (2000) Reef coral reproduction in the eastern Pacific: Costa Rica, Panama and Galapagos Islands (Ecuador). IV. Agariciidae, recruitment and recovery of *Pavona varians* and *Pavona* sp.a. Mar Biol 136:785–805

Glynn PW, Colley SB, Mate JL, Cortes J, Guzman HM, Bailey RL, Feingold JS, Enochs IC (2008) Reproductive ecology of the azooxanthellate coral *Tubastrea coccinea* in the Equatorial Eastern Pacific: Part V. Dendrophylliidae. Mar Biol 153:529–544

Goffredo S, Chadwick-Furman NE (2003) Comparative demography of mushroom corals (Scleractinia: Fungiidae) at Eilat, northern Red Sea. Mar Biol 142:411–418

Goffredo S, Airi V, Radetie J, Zaccanti F (2006) Sexual reproduction of the solitary sunset cup coral *Leptopsammia pruvoti* (Scleractinia, Dendrophylliidae) in the Mediterranean. 2. Quantitative aspects of the annual reproductive cycle. Mar Biol 148:923–931

Goodbody-Gringley G, Putron SJ (2009) Planulation patterns of the brooding coral *Favia fragum* (Esper) in Bermuda. Coral Reefs 289:59–963

Gorbunov MY, Falkowski PG (2002) Photoreceptors in the cnidarian hosts allow symbiotic corals to sense blue moonlight. Limnol Oceanogr 47:309–315

Graham EM, Baird AH, Connolly SR (2008) Survival dynamics of scleractinian coral larvae and implications for dispersal. Coral Reefs 27:529–539

Guest JR (2004) Reproduction of *Diploastrea heliopora* (Scleractinia: Faviidae) in Singapore. In: Proceedings of the 10th international coral reef symposium Abstracts, Okinawa, 2004, p 367

Guest JR, Chou LM, Baird AH, Goh BPL (2002) Multispecific, synchronous coral spawning in Singapore. Coral Reefs 21:422–423

Guest JR, Baird AH, Goh BPL, Chou LM (2005a) Seasonal reproduction in equatorial coral reefs. Invert Reprod Dev 48:207–218

Guest JR, Baird AH, Goh BPL, Chou LM (2005b) Reproductive seasonality in an equatorial assemblage of scleractinian corals. Coral Reefs 24:112–116

Gulko D (1995) Effects of ultraviolet radiation on fertilization and production of planula larvae in the Hawaiian coral *Fungia scutaria*. In Gulko D, Jokiel P (eds) Ultraviolet Radiation and Coral Reefs. HIMB Technical Report 41, Sea Grant, Honolulu, pp 135–147

Guzmán HM, Holst I (1993) The effects of chronic oil-sediment pollution on the reproduction of the Caribbean reef coral *Siderastrea siderea*. Mar Poll Bull 26:276–282

Hada Y (1932) A note of the earlier stage of colony formation with the coral, *Pocillopora cespitosa* Dana. Sci Rep Tohoku Univ (Ser. 4) 7:425–431

Hagman DK, Gittings SR, Deslarzes KJP (1998a) Timing, species participation and environmental factors influencing annual mass spawning at the Flower Garden Banks (Northwest Gulf of Mexico). Gulf Mexico Sci 16:170–179

Hagman DK, Gittings SR, Vize PD (1998b) Fertilization in broadcast-spawning corals of the Flower Garden Banks National Marine Sanctuary. Gulf Mexico Sci 16:180–187

Hall VR, Hughes TP (1996) Reproductive strategies of modular organisms: comparative studies of reef-building corals. Ecology 77:950–963

Hanafy MH, Aamer MA, Habib M, Rouphael AB, Baird AH (2010) Synchronous reproduction of corals in the Red sea. Coral Reefs 29:119–124

Harii S, Omori M, Yamakawa H, Koike Y (2001) Sexual reproduction and larval settlement of the zooxanthellate coral *Alveopora japonica* Eguchi at high altitudes. Coral Reefs 20:19–23

Harii S, Kayanne H, Takigawa H, Hayashibara T, Yamamoto M (2002) Larval survivorship, competency periods and settlement of two brooding corals, *Heliopora coerulea* and *Pocillopora damicornis*. Mar Biol 141:39–46

Harii S, Yasuda N, Rodriguez-Lanetty M, Irie T, Hidaka M (2009) Onset of symbiosis and distribution patterns of symbiotic dinoflagellates in the larvae of scleractinian corals. Mar Biol 156:1203–1212

Harrigan JF (1972) The planula larva of *Pocillopora damicornis*: lunar periodicity of swarming and substratum selection behaviour. Ph.D. thesis, University of Hawaii, Honolulu, 319 pp

Harriott VJ (1983a) Reproductive ecology of four scleractinian species at Lizard Island, Great Barrier Reef. Coral Reefs 2:9–18

Harriott VJ (1983b) Reproductive seasonality, settlement, and post-settlement mortality of *Pocillopora damicornis* (Linnaeus), at Lizard Island, Great Barrier Reef. Coral Reefs 2:151–157

Harrison PL (1985) Sexual characteristics of scleractinian corals: systematic and evolutionary implications. In: Proceedings of the 5th international coral reef congress, vol 4, Tahiti, 1985, pp 337–342

Harrison PL (1988) Pseudo-gynodioecy: an unusual breeding system in the scleractinian coral *Galaxea fascicularis*. In: Proceedings of the 6th international coral reef symposium, vol 2, Townsville, 1988, pp 699–705

Harrison PL (1990) Sperm morphology and fertilization strategies in scleractinian corals. Adv Invert Reprod 5:299–304

Harrison PL (1993) Coral spawning on the Great Barrier Reef. Search 24:45–48

Harrison PL (1995) Status of the coral reefs of Kuwait. Final report to the United Nations Industrial Development Organization and the United Nations Development Programme, UNIDO and UNDP, Vienna

Harrison PL (1997) Settlement competency periods and dispersal potential of planula larvae of the reef coral *Acropora longicyathus*. In: Proceedings of the Australian Coral Reef Society of the 75th Anniversary conference, Heron, 1997, p 263

Harrison PL (2006) Settlement competency periods and dispersal potential of scleractinian reef coral larvae. In: Proceedings of the 10th international coral reef symposium, vol 1, Okinawa, 2004, pp 78–82

Harrison PL (2008) Coral spawn slicks at Lord Howe Island, the world's most southerly coral reef. Coral Reefs 27:35

Harrison PL, Booth DJ (2007) Coral reefs: naturally dynamic and increasingly disturbed ecosystems. In: Connell SD, Gillanders BM (eds) Marine ecology. Oxford University Press, Melbourne, pp 316–377

Harrison PL, Hakeem A (2007) Asynchronous and pulsed multispecific reef coral spawning patterns on equatorial reefs in the Maldives Archipelago. In: Australian Coral Reef Society National Conference, Perth 2007

Harrison PL, Jamieson BGM (1999) Cnidaria and Ctenophora. In: Jamieson BGM (ed) Reproductive biology of invertebrates, Volume IX Part A, progress in male gamete ultrastructure and phylogeny. Oxford-IBH, New Delhi, pp 21–95

Harrison PL, Wallace CC (1990) Reproduction, dispersal and recruitment of scleractinian corals. In: Dubinsky Z (ed) Ecosystems of the world: coral reefs. Elsevier, Amsterdam, pp 133–207

Harrison PL, Ward S (2001) Elevated levels of nitrogen and phosphorus reduce fertilisation success of gametes from scleractinian reef corals. Mar Biol 139:1057–1068

Harrison PL, Babcock RC, Bull GD, Oliver JK, Wallace CC, Willis BL (1983) Recent developments in the study of sexual reproduction in tropical reef corals. In: Baker JT, Carter RM, Sammarco PW, Stark KP (eds) Proceedings of the inaugural Great Barrier Reef conference, JCU Press, Townsville, 1983, pp 217–219

Harrison PL, Babcock RC, Bull GD, Oliver JK, Wallace CC, Willis BL (1984) Mass spawning in tropical reef corals. Science 223:1186–1189

Harrison PL, Alhazeem SH, Alsaffar AH (1997) The ecology of coral reefs in Kuwait and the effects of stressors on corals. Kuwait Institute for Scientific Research Report No. KISR 4994, Kuwait

Hatta M, Fukami H, Wang W, Omori M, Shimoike K, Hayashibara YI, Sugiyama T (1999) Reproductive and genetic evidence for a reticulate evolutionary history of mass-spawning corals. Mol Biol Evol 16:1607–1613

Hayashibara T, Shimoike K, Kimura T, Hosaka S, Heyward A, Harrison P, Kudo K, Omori M (1993) Patterns of coral spawning at Akajima Island, Okinawa. Jpn Mar Ecol Prog Ser 101:253–262

Hellberg ME (2007) Footprints on water: the genetic wake of dispersal among reefs. Coral Reefs 26:463–473

Hellberg ME, Taylor MS (2002) Genetic analysis of sexual reproduction in the dendrophylliid coral *Balanophyllia elegans*. Mar Biol 141:629–637

Heltzel PS, Babcock RC (2002) Sexual reproduction, larval development and benthic planulae of the solitary coral *Monomyces rubrum* (Scleractinia: Anthozoa). Mar Biol 140:659–667

Heyward AJ (1986) Sexual reproduction in five species of the coral *Montipora*. In: Jokiel PL, Richmond RH, Rogers RA (eds) Coral reef population biology. Hawaii'i Institute of Marine Biology Technical Report No. 37, Hawaii, pp 170–178

Heyward AJ (1988a) Reproductive status of some Guam corals. Micronesica 21:272–274

Heyward AJ (1988b) Inhibitory effects of copper and zinc sulphates on fertilization in corals. In: Proceedings of the 6th international coral reef symposium, vol 2, Townsville, 1988, pp 299–303

Heyward AJ, Babcock RC (1986) Self- and cross-fertilization in scleractinian corals. Mar Biol 90:191–195

Heyward A, Yamazato K, Yeemin T, Minei M (1987) Sexual reproduction of corals in Okinawa. Galaxea 6:331–343

Hidaka M, Yurugi K, Sunagawa S, Kinzie RA III (1997) Contact reactions between young colonies of the coral *Pocillopora damicornis*. Coral Reefs 16:13–20

Highsmith RC (1982) Reproduction by fragmentation in corals. Mar Ecol Prog Ser 7:207–226

Hiscock K, Howlett RM (1977) The ecology of *Caryophyllia smithi* Strokes and Broderip on south-western coasts of the British Isles. In: Drew EA, Lythgoe JN, Woods JD (eds) Underwater research. Academic, New York, pp 319–334

Hodgson G (1985) Vertical distribution of planktonic larvae of the reef coral *Pocillopora damicornis* in Kaneohe Bay (Oahu, Hawaii). In: Proceedings of the 5th international coral reef congress, vol 4, Tahiti, 1985, pp 349–354

Hodgson G (1988) Potential gamete wastage in synchronously spawning corals due to hybrid inviability. In: Proceedings of the 6th international coral reef symposium, vol 2, Townsville, 1988, pp 707–714

Holstein TW, Hobmayer E, Technau U (2003) Cnidarians: an evolutionary conserved model system for regulation? Dev Dyn 226:257–267

Huang D, Meier R, Todd PA, Chou LM (2009) More evidence for pervasive paraphyly in scleractinian corals: systematic study of Southeast Asian Faviidae (Cnidaria: Scleractinia) based on molecular and morphological data. Mol Phylogenet Evol 50:102–116

Hughes TP, Ayre D, Connell JH (1992) The evolutionary ecology of corals. Trends Ecol Evol 9:292–295

Hughes TP, Baird AH, Dinsdale EA, Moltschaniwskyj NA, Pratchett MS, Tanner JE, Willis BL (2000) Supply-side ecology works both ways: the link between benthic adults, fecundity, and larval recruits. Ecology 81:2241–2249

Humphrey C, Weber M, Lott C, Cooper T, Fabricius K (2008) Effects of suspended sediments, dissolved inorganic nutrients and salinity on fertilisation and embryo development in the coral *Acropora millepora* (Ehrenberg, 1834). Coral Reefs 27:837–850

Hunter CL (1988) Environmental cues controlling spawning in two Hawaiian corals, *Montipora verrucosa* and *M. dilatata*. In: Proceedings of the 6th international coral reef symposium, vol 2, Townsville, 1988, pp 727–732

Hyman LH (1940) The invertebrates: Protozoa through Ctenophora, vol 1. McGraw-Hill, New York, 726 pp

Jackson JBC, Coates AG (1986) Life cycles and evolution of clonal (modular) animals. Philos Trans R Soc Lond B 313:7–22

Johnson KG (1992) Synchronous planulation of *Manicina areolata* (Scleractinia) with lunar periodicity. Mar Ecol Prog Ser 87:265–273

Jokiel PL (1985) Lunar periodicity of planula release in the reef coral *Pocillopora damicornis* in relation to various environmental factors. In: Proceedings of the 5th international coral reef congress, vol 4, Tahiti, 1985, pp 307–312

Jokiel PL, Ito RY, Liu PM (1985) Night irradiance and synchronization of lunar release of planula larvae in the reef coral *Pocillopora damicornis*. Mar Biol 88:167–174

Jokiel PL, Rogers KS, Kuffner IB, Andersson AJ, Cox EF, Mackenzie FT (2008) Ocean acidification and calcifying reef organisms: a mesocosm investigation. Coral Reefs 27:473–483

Jones GP, Almay GR, Russ GR, Sale PF, Steneck RS, van Oppen MJH, Willis BL (2009) Larval retention and connectivity among populations of corals and reef fishes: history, advances and challenges. Coral Reefs 28:307–325

Kawaguti S (1940) An abundance of reef coral planulae in plankton. Zool Mag (Tokyo) 52:31

Kawaguti S (1941a) On the physiology of reef corals. V. Tropisms of coral planulae, considered as a factor of distribution of the reefs. Palao Trop Biol Stat Stud 2:319–328

Kawaguti S (1941b) Materials for the study of reef corals. Kagaku Nanyo 3:171–176 (in Japanese)

Kenyon JC (1992) Sexual reproduction in Hawaiian *Acropora*. Coral Reefs 11:37–43

Kenyon JC (1995) Latitudinal differences between Palau and Yap in coral reproductive synchrony. Pac Sci 49:156–164

Kenyon JC (2008) *Acropora* (Anthozoa: Scleractinia) reproductive synchrony and spawning phenology in the Northern Line Islands, Central Pacific, as inferred from size classes of developing oocytes. Pac Sci 62:569–578

Kerr AM (2005) Molecular and morphological supertree of stony corals (Anthozoa: Scleractinia) using matrix representation parsimony. Biol Rev 80:543–558

Kinzie RA III (1993) Spawning in the reef corals *Pocillopora verrucosa* and *P. eydouxi* at Sesoko Island, Okinawa. Galaxea 11:93–105

Knowlton N, Maté JL, Guzmán HM, Rowan R, Jara J (1997) Direct evidence for reproductive isolation among three species of the *Montastrea annularis* complex in Central American (Panamá and Honduras). Mar Biol 127:705–711

Kojis BL (1986a) Sexual reproduction in *Acropora* (*Isopora*) species (Coelenterata: Scleractinia) I. *A. cuneata* and *A. palifera* on Heron Island reef, Great Barrier Reef. Mar Biol 91:291–309

Kojis BL (1986b) Sexual reproduction in *Acropora* (*Isopora*) (Coelenterata: Scleractinia) II. Latitudinal variation in A. palifera from the Great Barrier Reef and Papua New Guinea. Mar Biol 91:311–318

Kojis BL, Quinn NJ (1981) Aspects of sexual reproduction and larval development in the shallow water hermatypic coral, *Goniastrea australensis* (Edwards and Haime, 1857). Bull Mar Sci 31:558–573

Kojis BL, Quinn NJ (1982) Reproductive strategies in four species of *Porites* (Scleractinia). In: Proceedings of the 4th international coral reef symposium, vol 2, Manila, 1981, pp 145–151

Kojis BL, Quinn NJ (1984) Seasonal and depth variation in fecundity of *Acropora palifera* at two reefs in Papua New Guinea. Coral Reefs 3:165–172

Kojis BL, Quinn NJ (1985) Puberty in *Goniastrea favulus*. Age or size limited? In: Proceedings of the 5th international coral reef congress, vol 4, Tahiti, 1985, pp 289–293

Kolinski SP, Cox EF (2003) An update on modes and timing of gamete and planula release in Hawaiian scleractinian corals with implications for conservation and management. Pac Sci 57:17–27

Kongjandtre N, Ridgway T, Ward S, Hoegh-Guldberg O (2010) Broadcast spawning patterns of *Favia* species on the inshore reefs of Thailand. Coral Reefs 29:227–234

Kramarsky-Winter E, Loya Y (1996) Regeneration versus budding in fungiid corals: a trade off. Mar Ecol Prog Ser 134:179–185

Kramarsky-Winter E, Loya Y (1998) Reproductive strategies in two fungiid corals from the northern Red Sea: environmental constraints. Mar Ecol Prog Ser 174:175–182

Kramarsky-Winter E, Fine M, Loya Y (1997) Coral polyp expulsion. Nature 387:137

Kruger A, Schleyer MH (1998) Sexual reproduction in the coral *Pocillopora verrucosa* (Cnidaria: Scleractinia) in KwaZulu-Natal, South Africa. Mar Biol 132:703–710

Krupp DA (1983) Sexual reproduction and early development of the solitary coral *Fungia scutaria* (Anthozoa: Scleractinia). Coral Reefs 2:159–164

Krupp DA, Jokiel PL, Chartrand TS (1992) Asexual reproduction by the solitary scleractinian coral *Fungia scutaria* on dead parent coralla in Kaneohe Bay, Oahu, Hawaiian Islands. In: Proceedings of the 7th international coral reef symposium, vol 1, Guam, 1992, pp 527–533

Krupp DA, Hollingsworth LL, Peterka J (2006) Elevated temperature sensitivity of fertilization and early development in the mushroom coral *Fungia scutaria* Lamarck 1801. In: Proceedings of the 10th international coral reef symposium, vol 1, Okinawa, 2004, pp 71–77

Kruzic P, Zuljevic A, Nikolic V (2008) Spawning of the colonial coral *Cladocora caespitosa* (Anthozoa, Scleractinia) in the Southern Adriatic sea. Coral Reefs 27:337–341

Lacaze-Duthiers H de (1897) Faune du Golfe du Lion: Coralliaires zoanthaires sclerodermes. Arch Zool Exp Gen (Ser 3) 5:1–249

Lam KKY (2000) Sexual reproduction of a low-temperature tolerant coral *Oulastrea crispata* (Scleractinia: Faviidae) in Hong Kong, China. Mar Ecol Prog Ser 205:101–111

Lane A, Harrison PL (2002). Effects of oil contaminants on survivorship of larvae of the scleractinian reef corals *Acropora tenuis*, *Goniastrea aspera* and *Platygyra sinensis* from the Great Barrier Reef. In: Proceedings of the 9th international coral reef symposium, vol 1, Bali, 2000, pp 403–408

Le Goff-Vitry MC, Rogers AD, Bagalow D (2004) A deep-sea slant on the molecular phyologeny of the Scleractinia. Mol Phylogenet Evol 30:167–177

Levitan DR, Fukami H, Jara J, Kline D, McGovern TM, McGhee KE, Swanson CA, Knowlton N (2004) Mechanisms of reproductive isolation among sympatric broadcast-spawning corals of the *Montastraea annularis* species complex. Evolution 58:308–323

Levy O, Appelbaum L, Leggat W, Gothlif Y, Hayward DC, Miller DJ, Hoegh-Guldberg O (2007) Light-responsive cryptochromes from a simple multicellular animal, the coral *Acropora millepora*. Science 318:467–470

Lewis JB (1974) The settlement behaviour of planulae larvae of the hermatypic coral *Favia fragum* (Esper). J Exp Mar Biol Ecol 15:165–172

Lins de Barros M, Pires DO (2007) Comparison of the reproductive status of the scleractinian coral *Siderastrea stellata* throughout a gradient of 20° of latitude. Brazil J Oceanogr 55:67–69

Lins de Barros M, Pires DO, Castro CB (2003) Sexual reproduction of the Brazilian reef coral *Siderastrea stellata* Verrill, 1868 (Anthozoa, Scleractinia). Bull Mar Sci 73:713–724

Loya Y (1976) The Red Sea coral *Stylophora pistillata* is an r strategist. Nature 259:478–480

Loya Y, Rinkevich B (1979) Abortion effect in corals induced by oil pollution. Mar Ecol Prog Ser 1:77–80

Loya Y, Rinkevich B (1980) Effects of oil pollution on coral reef communities. Mar Ecol Prog Ser 3:167–180

Loya Y, Sakai K (2008) Bidirectional sex change in mushroom stony corals. Proc R Soc B 275:2335–2343

Loya Y, Sakai K, Heyward A (2009) Reproductive patterns of fungiid corals in Okinawa, Japan. Galaxea 11:119–129

Mangubhai S (2009) Reproductive ecology of the scleractinian corals *Echinopora gemmacea* and *Leptoria phrygia* (Faviidae) on equatorial reefs in Kenya. Invert Reprod Dev 22:213–228

Mangubhai S, Harrison PL (2006) Seasonal patterns of coral reproduction on equatorial reefs in Mombasa, Kenya. In: Proceedings of the 10th international coral reef symposium, vol 1, Okinawa, 2004, pp 106–114

Mangubhai S, Harrison PL (2008a) Asynchronous coral spawning patterns on equatorial reefs in Kenya. Mar Ecol Prog Ser 360:85–96

Mangubhai S, Harrison PL (2008b) Gametogenesis, spawning and fecundity of *Platygyra daedalea* (Scleractinia) on equatorial reefs in Kenya. Coral Reefs 27:117–122

Mangubhai S, Harrison PL (2009) Extended breeding seasons and asynchronous spawning among equatorial reef corals in Kenya. Mar Ecol Prog Ser 374:305–310

Markey KL, Baird AH, Humphrey C, Negri AP (2007) Insecticides and a fungicide affect multiple coral life stages. Mar Ecol Prog Ser 330:127–137

Marshall SM, Stephenson TA (1933) The breeding of reef animals. Part l. The corals. Sci Rep Great Barrier Reef Exped (1928–29) 3:219–245

Mate JL (1998) New reports on the timing and mode of reproduction of Hawaiian corals. In: Cox EF, Krupp DA, Jokiel PL (eds) Reproduction in reef corals – results of the 1997 Edwin W. Pauley Summer Program in marine biology. Hawai'i Institute of Marine Biology Technical Report 42, Kaneohe, p 7

Matthai G (1919) On *Favia conferta* Verrill, with notes on other Atlantic species of *Favia*. Br Mus (Nat Hist), Br Antarctic ("Terra Nova") Exped 1910, Zool 5:69–95

Matthai G (1914) A revision of the recent colonial Astraeidae possessing distinct corallites. Trans Linn Soc (Ser 2) Zool 17:1–140

Matthai G (1923) Histology of the soft parts of Astraeid corals. Q J Microsc Sci 67:101–122

McGuire MP (1998) Timing of larval release by *Porites astreoides* in the northern Florida Keys. Coral Reefs 17:369–375

Medina M, Collins AG, Takaoka TL, Kuehl JV, Boore JL (2006) Naked corals: skeleton loss in Scleractinia. Proc Natl Acad Sci U S A 103:9096–9100

Mendes JM, Woodley JD (2002a) Timing of reproduction in *Montastraea annularis*: relationship to environmental variables. Mar Ecol Prog Ser 227:241–251

Mendes JM, Woodley JD (2002b) Effect of the 1995–1996 bleaching event on polyp tissue depth, growth, reproduction and skeletal band formation in *Montastraea annularis*. Mar Ecol Prog Ser 235:93–102

Meyer E, Davies S, Wang S, Willis BL, Abrego D, Juenger TE, Matz MV (2008) Genetic variation in responses to a settlement cue and elevated temperature in the reef-building coral *Acropora millepora*. Mar Ecol Prog Ser 392:81–92

Mezaki T, Hayashi T, Iwase F, Nakachi S, Nozawa Y, Miyamoto M, Tominaga M (2007) Spawning patterns of high latitude scleractinian corals from 2002 to 2006 at Nishidomari, Otsuki, Kochi, Japan. Kurishio Biosphere 3:33–47

Mildner S (1991) Aspects of the reproductive biuology of selected scleractinian corals on Western Samoan and Fijian reefs. MSc thesis, James Cook University of North Queensland, Townsville, 81 pp

Miller KJ, Ayre DJ (2004) The role of sexual and asexual reproduction in structuring high latitude populations of the reef coral *Pocillopora damicornis*. Heredity 92:557–568

Miller KJ, Ayre DJ (2008) Population structure is not a simple function of reproductive mode and larval type: insights from tropical corals. J Animal Ecol 77:713–724

Miller K, Babcock R (1997) Conflicting morphological and reproductive species boundaries in the coral genus *Platygyra*. Biol Bull 192:98–110

Miller KJ, Mundy CN (2003) Rapid settlement in broadcast spawning corals: implications for larval dispersal. Coral Reefs 22:99–106

Miller KJ, Mundy CN (2005) In situ fertilization success in the scleractinian coral *Goniastrea favulus*. Coral Reefs 24:313–317

Motoda S (1939) Observation of period of extrusion of planula of *Goniastrea aspera* Verrill. Kagaku Nanyo 1:5–7 (in Japanese)

Muir PR (1984) Periodicity and asexual planulae production in Pocillopora damicornis (Linnaeus) at Magnetic Island. BSc Honours thesis, James Cook University of North Queensland, Townsville, 58 pp

Nakano Y, Yamazoto K (1992) Ecological study of reproduction of *Oulastrea crispata* in Okinawa. Behav Biol Ecol 1292

Negri AP, Heyward A (2000) Inhibition of fertilization and larval metamorphosis of the coral *Acropora millepora* (Ehrenberg, 1843) by petroleum products. Mar Poll Bull 41:420–427

Negri AP, Heyward A (2001) Inhibition of coral fertilization and larval metamorphosis by tributyltin and copper. Mar Poll Bull 41:420–427

Negri A, Vollhardt C, Humphrey C, Heyward A, Jones R, Eaglesham G, Fabricius K (2005) Effects of the herbicide diuron on the early life history stages of coral. Mar Poll Bull 51:370–383

Negri AP, Marshall PA, Heyward A (2007) Differing effects of thermal stress on coral fertilization and early embryogenesis in four Indo-Pacific species. Coral Reefs 26:759–763

Neves EG (2000) Histological analysis of reproductive trends of three *Porites* species from Kane'ohe Bay, Hawai'i. Pac Sci 54:195–200

Neves EG, da Silveira FL (2003) Release of planula larvae, settlement and development of *Siderastrea stellata* Verrill, 1868 (Anthozoa, Scleractinia). Hydrobiologia 501:139–147

Neves EG, Pires DO (2002) Sexual reproduction of Brazilian coral *Mussismilia hispida* (Verill, 1902). Coral Reefs 21:161–168

Nishikawa A, Katoh M, Sakai K (2003) Larval settlement rates and gene flow of broadcast-spawning (*Acropora tenuis*) and planula-brooding (*Stylophora pistillata*) corals. Mar Ecol Prog Ser 256:87–97

Noreen A, Harrison PL, van Oppen MJH (2009) Population structure and gene flow in the brooding reef coral *Seriatopora hystrix*. Proc R Soc B 276:3927–3935

Nozawa Y, Harrison PL (2002) Larval settlement patterns, dispersal potential, and the effect of temperature on settlement rates of larvae of the broadcast spawning reef coral, *Platygyra daedalea*, from the Great Barrier Reef. In: Proceedings of the 9th nternational coral reef symposium, vol 1, Bali, 2000, pp 409–416

Nozawa Y, Harrison PL (2005) Temporal settlement patterns of larvae of the broadcast spawning reef coral *Favites chinensis* and the broadcast spawning and brooding reef coral *Goniastrea aspera* from Okinawa, Japan. Coral Reefs 24:274–282

Nozawa Y, Harrison PL (2007) Effects of elevated temperature on larval settlement and post-settlement survival in scleractinian corals, *Acropora solitaryensis* and *Favites chinensis*. Mar Biol 152:1181–1185

Nozawa Y, Tokeshi M, Nojima S (2006) Reproduction and recruitment of scleractinian corals in a high-latitude coral community, Amakusa, southwestern Japan. Mar Biol 149:1047–1058

Nunes F, Fukami H, Vollmer SV, Norris RD, Knowlton N (2008) Re-evaluation of the systematics of the endemic corals of Brazil by molecular data. Coral Reefs 27:423–432

Okubo N, Isomura N, Motokawa T, Hidaka M (2007) Possible self-fertilization in the brooding coral *Acropora* (Isopora) *brueggemanni*. Zool Sci 24:277–280

Oliver JK, Babcock RC (1992) Aspects of the fertilization ecology of broadcast spawning corals: sperm dilution effects and in situ measurements of fertilization. Biol Bull 183:409–417

Oliver JK, Willis BL (1987) Coral-spawn slicks in the Great Barrier Reef: preliminary observations. Mar Biol 94:521–529

Oliver JK, Babcock RC, Harrison PL, Willis BL (1988) Geographic extent of mass coral spawning: clues to ultimate causal factors. In: Proceedings of 6th International Coral Reef Symposium, vol 2, Townsville, pp 803–810

Omori M, Fukami H, Kobinata H, Hatta M (2001) Significant drop of fertilization of *Acropora* corals in 1999: an after-effect of heavy coral bleaching? Limnol Oceanogr 46:704–706

Orton JH (1920) Sea-temperature, breeding and distribution in marine animals. J Mar Biol Ass UK 12:339–366

Penland L, Kloulechad J, Idip D, van Woesik R (2004) Coral spawning in the western Pacific Ocean is related to solar insolation: evidence of multiple spawning events in Palau. Coral Reefs 23:133–140

Peters EC (1984) A survey of cellular reactions to environmental stress and disease in Caribbean scleractinian corals. Helgolander Meeresunters 37:113–137

Peters EC, Meyers PA, Yevich PP, Blake NJ (1981) Bioaccumulation and histopathological effects of oil on a stony coral. Mar Poll Bull 12:333–339

Pires DO, Castro CB, Ratto CC (1999) Reef coral reproduction in the Abrolhos Reef Complex, Brazil: the endemic genus *Mussismilia*. Mar Biol 135:463–471

Pires DO, Castro CB, Ratto CC (2002) Reproduction of the solitary coral *Scolymia wellsi* Laboral (Cnidaria, Scleractinia) from the Abrolhos Reef Complex, Brazil. In: Proceedings of the 9th international coral reef symposium, vol 1, Bali, 2000, pp 381–384

Piromvaragorn S, Putchim L, Kongjantre N, Boonprakob R, Chankong A (2006) Spawning season of acroporid corals genus *Acropora* in the Gulf of Thailand. J Sci Res Chula Uni Section T 5:39–49

Policansky D (1982) Sex change in plants and animals. Ann Rev Ecol Syst 13:471–495

Puill-Stephan E, Willis BL, van Herwerden L, van Oppen MJH (2009) Chimerism in wild adult populations of the broadcast spawning coral *Acropora millepora* on the Great Barrier Reef. PLoS ONE 4:1–8

Randall CJ, Szmant AM (2009) Elevated temperature reduces survivorhsip and settlement of the larvae of the Caribbean scleractinian coral, *Favia fragum* (Esper). Coral Reefs 28:537–545

Reichelt-Brushett AJ, Harrison PL (1999) The effect of copper, zinc and cadmium on fertilization success of gametes from scleractinian reef corals. Mar Poll Bull 38:182–187

Reichelt-Brushett AJ, Harrison PL (2000) The effect of copper on the settlement success of larvae from the scleractinian coral *Acropora tenuis*. Mar Poll Bull 41:385–391

Reichelt-Brushett AJ, Harrison PL (2004) Development of a sublethal test to determine the effects of copper and lead on scleractinian coral larvae. Arch Environ Contam Toxicol 47:40–55

Reichelt-Brushett AJ, Harrison PL (2005) The effect of selected trace metals on the fertilization success of several scleractinian coral species. Coral Reefs 24:524–534

Richards Z, van Oppen MJH, Wallace CC, Willis BL, Miller DJ (2008) Some rare Indo-Pacific coral species are probable hybrids. PLoS ONE 3(9):1–8, e3240

Richmond RH (1982) Energetic considerations in the dispersal of *Pocillopora damicornis* (Linnaeus) planulae. In: Proceedings of the 4th international coral reef symposium, vol 2, Manila, 1981, pp 153–156

Richmond RH (1985) Reversible metamorphosis in coral planula larvae. Mar Ecol Prog Ser 22:181–185

Richmond RH (1987) Energetics, competency, and long-distance dispersal of planula larvae of the coral *Pocillopora damicornis*. Mar Biol 93:527–533

Richmond RH (1988) Competency and dispersal potential of planula larvae of a spawning versus a brooding coral. In: Proceedings of the 6th international coral reef symposium, vol 2, Townsville, 1988, pp 827–831

Richmond RH (1993) Coral reefs: present problems and future concerns resulting from anthropogenic disturbance. Am Zool 33:524–536

Richmond RH (1997) Reproduction and recruitment in corals: critical links in the persistence of reefs. In: Birkeland C (ed) Life and death of coral reefs. Chapman & Hall, New York, pp 175–197

Richmond RH, Hunter CL (1990) Reproduction and recruitment of corals: comparisons among the Caribbean, the Tropical Pacific, and the Red Sea. Mar Ecol Prog Ser 60:185–203

Richmond RH, Jokiel PL (1984) Lunar periodicity in larva release in the reef coral *Pocillopora damicornis* at Enewetak and Hawaii. Bull Mar Sci 34:280–287

Rinkevich B, Loya Y (1977) Harmful effects of chronic oil pollution on a Red sea scleractinian coral population. In: Proceedings of the 3rd international coral reef symposium, vol 2, Miami, 1977, pp 586–591

Rinkevich B, Loya Y (1979a) The reproduction of the Red Sea coral *Stylophora pistillata*. I. Gonads and planulae. Mar Ecol Prog Ser 1:133–144

Rinkevich B, Loya Y (1979b) The reproduction of the Red Sea coral *Stylophora pistillata*. II. Synchronisation in breeding and seasonality of planulae shedding. Mar Ecol Prog Ser 1:145–152

Rinkevich B, Loya Y (1979c) Laboratory experiments on the effects of crude oil on the Red Sea coral *Stylophora pistillata*. Mar Poll Bull 10:328–330

Rinkevich B, Loya Y (1986) Senescence and dying signals in a reef building coral. Experentia 42:320–322

Rinkevich B, Loya Y (1987) Variability in the pattern of reproduction of the coral *Stylophora pistillata* at Eilat, Red sea: a long-term study. Biol Bull 173:335–344

Rodríguez S, Alvizu A, Tagliafico A, Bastidas C (2009) Low lateral repopulation of marginal coral communities under the influence of upwelling. Hydrobiologia 624:1–11

Romano SL, Cairns SD (2000) Molecular phylogenetic hypothesis for the evolution of scleractinian corals. Bull Mar Sci 67:1043–1068

Romano SL, Palumbi SR (1996) Evolution of scleractinian corals inferred from molecular systematics. Science 271:640–642

Romano SL, Palumbi SR (1997) Molecular evolution of a portion of the mitochondrial 16S ribosomal gene region in Scleractinian corals. J Mol Evol 45:397–411

Rosen BR (1990) Coloniality. In: Briggs DR, Crowther PR (eds) Palaeobiology: a synthesis. Blackwell, Oxford, pp 330–335

Rosen BR, Taylor JD (1969) Reef coral from Aldabra: new mode of reproduction. Science 166:119–121

Rosser NL, Baird AH (2009) Multi-specific coral spawning in spring and autumn in far north-western Australia. In: Proceedings of the 11th international coral reef symposium, vol 1, Ft Lauderdale, 2008, pp 366–370

Rosser NL, Gilmour JP (2008) New insights into patterns of coral spawning on western Australian reefs. Coral Reefs 27:345–349

Sakai K (1997) Gametogenesis, spawning, and planula brooding by the reef coral *Goniastrea aspera* (Scleractinia) in Okinawa, Japan. Mar Ecol Prog Ser 151:67–72

Sammarco PW (1982) Polyp bail-out: an escape response to environmental stress and a new means of reproduction in corals. Mar Ecol Prog Ser 10:57–65

Sammarco PW (1994) Larval dispersal and recruitment processes in Great Barrier Reef corals: analysis and synthesis. In: Sammarco PW, Heron ML (eds) The bio-physics of marine larval dispersal. Coastal and estuarine studies, vol 45. American Geophysical Union, Washington, DC, pp 35–72

Sammarco PW, Andrews JC (1988) Localised dispersal and recruitment in Great Barrier Reef corals: the Helix experiment. Science 239:1422–1424

Sanchez JA, Alvarado EM, Gil MF, Charry H, Arenas OL, Chasqui LH, Garcia RP (1999) Synchronous mass spawning of *Montastraea annnularis* (Ellis & Solander) and *Montastraea faveolata* (Ellis & Solander) (Faviidae: Scleractinia) at Rosario Islands, Caribbean Coast of Colombia. Bull Mar Sci 65:873–879

Schuhmacher H, Zibrowius H (1985) What is hermatypic? A redefinition of ecological groups in corals and other organisms. Coral Reefs 4:1–9

Schwartz J, Krupp DA, Weis VM (1999) Late larval development and onset of symbiosis in the scleractinian coral *Fungia scutaria*. Biol Bull 196:70–79

Sherman CDH, Ayre DJ, Miller KJ (2006) Asexual reproduction does not produce clonal populations of the brooding coral *Pocillopora damicornis* on the Great Barrier Reef, Australia. Coral Reefs 25:7–18

Shimoike K, Hayashibara T, Kimura T, Omori M (1992) Observations of split spawning in *Acropora* spp. at Akajima Island, Okinawa. In: Proceedings of the 7th international coral reef symposium, vol 1, Guam, 1992, pp 484–488

Shlesinger Y, Loya Y (1985) Coral community reproductive patterns: Red Sea versus the Great Barrier Reef. Science 228:1333–1335

Shlesinger Y, Goulet TL, Loya Y (1998) Reproductive patterns of scleractinian corals in the northern Red Sea. Mar Biol 132:691–701

Sier CJS, Olive PJW (1994) Reproduction and reproductive variability in the coral *Pocillopora verrucosa* from the Republic of Maldives. Mar Biol 118:713–722

Simpson CJ (1985) Mass spawning of scleractinian corals in the Dampier archipelago and the implications for management of coral reefs in Western Australia. West Aust Dept Conserv Environ Bull 244:35

Simpson CJ (1991) Mass spawning of corals on Western Australian reefs and comparisons with the Great Barrier Reef. J R Soc West Aust 74:85–91

Simpson CJ, Cary JL, Masini RJ (1993) Destruction of corals and other reef animals by coral spawn slicks on Ningaloo Reef, western Australia. Coral Reefs 12:185–191

Soong K (1991) Sexual reproductive patterns of shallow-water reef corals in Panama. Bull Mar Sci 49:832–846

Souter P (2010) Hidden genetic diversity in a key model species of coral. Mar Biol 157:875–885

Steiner SCC (1991) Sperm morphology of scleractinians from the Caribbean. Hydrobiologia 216(217):131–135

Steiner SCC (1993) Comparative ultrastructural studies on scleractinian spermatozoa (Cnidaria: Anthozoa). Zoomorphology 113:129–136

Steiner SCC (1995) Spawning in scleractinian corals from SW Puerto Rico (West Indies). Bull Mar Sci 56:899–902

Stephenson TA (1931) Development and the formation of colonies in *Pocillopora* and *Porites*. Part I. Sci Rep Great Barrier Reef Exped (1928–29) 3:113–134

Stimson JS (1978) Mode and timing of reproduction in some common hermatypic corals of Hawaii and Enewetak. Mar Biol 48:173–184

Stobart B, Babcock RC, Willis BL (1992) Biannual spawning of three species of scleractinian coral from the Great Barrier Reef. In: Proceedings of the 7th international coral reef symposium, vol 1, Guam, 1992, pp 494–499

Stoddart JA (1983) Asexual production of planulae in the coral *Pocillopora damicornis*. Mar Biol 76:279–284

Stoddart JA, Black R (1985) Cycles of gametogenesis and planulation in the coral *Pocillopora damicornis*. Mar Ecol Prog Ser 23:153–164

Szmant AM (1986) Reproductive ecology of Caribbean reef corals. Coral Reefs 5:43–54

Szmant AM (1991) Sexual reproduction by the Caribbean reef corals *Montastrea annularis* and *M cavernosa*. Mar Ecol Prog Ser 74:13–25

Szmant AM, Gassman NJ (1990) The effects of prolonged "bleaching" on the tissue biomass and reproduction of the reef coral *Montastrea annularis*. Coral Reefs 8:217–224

Szmant AM, Weil E, Miller MW, Colon DE (1997) Hybridization within the species complex of the scleractinian coral *Montastrea annularis*. Mar Biol 129:561–573

Szmant-Froelich AM (1985) The effect of colony size on the reproductive ability of the Caribbean coral *Montastrea annularis* (Ellis and Solander). In: Proceedings of the 5th international coral reef congress, vol 4, Tahiti, 1985, pp 295–300

Szmant-Froelich AM (Oct 1984) Reef coral reproduction: diversity and community patterns. In: Advances in reef science. University of Miami, Coral gobles, pp 122–123

Szmant-Froelich AM, Yevich P, Pilson MEQ (1980) Gametogenesis and early development of the temperate coral *Astrangia danae* (Anthozoa: Scleractinia). Biol Bull 158:257–269

Szmant-Froelich AM, Reutter M, Riggs L (1985) Sexual reproduction of *Favia fragum* (Esper): lunar patterns of gametogenesis, embryogenesis and planulation in Puerto Rico. Bull Mar Sci 37:880–892

Tanner JE (1996) Seasonality and lunar periodicity in the reproduction of pocilloporid corals. Coral Reefs 15:59–66

Tarrant AM, Atkinson S, Atkinson MJ (1999) Estrone and estradiol-17β concentration in tissue of the scleractinian coral. Comp Biochem Physdiol A Montipora verrucosa 122:85–92

Tarrant AM, Atkinson MJ, Atkinson S (2004) Effects of steroidal estrogens on coral growth and reproduction. Mar Ecol Prog Ser 269:121–129

Thynne (1859) On the increase of madrepores. Ann Mag Nat Hist (Ser 3) 3:449–461

Tomascik T, Sander F (1987) Effects of eutrophication on reef-building corals III. Reproduction of the reef-building coral *Porites porites*. Mar Biol 94:77–94

Tomascik T, Mah AJ, Nontji A, Moosa MK (1997) The ecology of the Indonesian seas, Part 1. Periplus Editions, Hong Kong, 642 pp

Torres JL, Armstrong RA, Weil E (2008) Enhanced ultraviolet radiation can terminate sexual reproduction in the broadcasting coral species *Acropora cervicornis* Lamarck. J Exp Mar Biol Ecol 358:39–45

Tranter PRG, Nicholson DN, Kinchington D (1982) A description of the spawning and post-gastrula development of the cool temperate coral *Caryophyllia smithii* (Stokes and Broderip). J Mar Biol Assoc UK 62:845–854

Twan W-H, Hwang J-S, Lee Y-H, Wu H-F, Tung Y-H, Chang C-F (2006) Hormones and reproduction in scleractinian corals. Comp Biochem Physdiol A 144:247–253

Underwood JN, Smith LD, van Oppen MJH, Gilmour JP (2007) Multiple scales of genetic connectivity in a brooding coral on isolated reefs following catastrophic bleaching. Mol Ecol 16:771–784

Underwood JN, Smith LD, van Oppen MJH, Gilmour JP (2009) Ecologically relevant dispersal of corals on isolated reefs: implications for managing resilience. Ecol Appl 19:18–29

van Moorsel GWNM (1983) Reproductive strategies in two closely related stony corals (*Agaricia*, Scleractinia). Mar Ecol Prog Ser 13:273–283

van Oppen MJH, Gates RD (2006) Conservation genetics and the resilience of reef-building corals. Mol Ecol 15:3863–3883

van Oppen MJH, McDonald BJ, Willis BL, Miller DJ (2001) The evolutionary history of the coral genus *Acropora* (Scleractinia: Cnidaria) based on a mitochondrial and a nuclear marker: reticulation, incomplete lineage sorting or morphological convergence? Mol Biol Evol 18:1315–1329

van Oppen MJH, Willis BL, van Rheede T, Miller DJ (2002) Spawning times, reproductive compatibilities and genetic structuring in the *Acropora aspera* group: evidence for natural hybridization and semi-permeable species boundaries in corals. Mol Ecol 11: 1363–1376

van Veghel MLJ (1993) Multiple species spawning on Curacao reefs. Bull Mar Sci 52:1017–1021

van Woesik R (1995) Coral communities at high latitude are not pseudopopulations: evidence of spawning at 32°N, Japan. Coral Reefs 14:119–120

van Woesik R (2009) Calm before the spawn: global coral spawning patterns are explained by regional wind fields. Proc R Soc B 277: 715–722

van Woesik R, Lacharmoise F, Koksal S (2006) Annual cycles of solar insolation predict spawning times of Caribbean corals. Ecol Lett 9:390–398

Vandermeulen JH, Watabe N (1973) Studies on reef corals. I. Skeleton formation by newly settled planula larva of *Pocillopora damicornis*. Mar Biol 23:47–57

Vargas-Angel B, Thomas JD (2002) Sexual reproduction of *Acropora cervicornis* in nearshore waters off Fort Lauderdale, Florida, USA. Coral Reefs 21:25–26

Vargas-Angel B, Colley SB, Hoke SM, Thomas JD (2006) The reproductive seasonality and gametogenic cycle of *Acropora cervicornis* off Broward County, Florida, USA. Coral Reefs 25:110–122

Vaughan TW (1908) Geology of Florida keys and marine bottom deposits and recent corals of southern Florida. Carnegie Inst Wash Yearbook 7:131–136

Vaughan TW (1910) Geology of the keys, the marine bottom deposits and the recent corals of southern Florida. Carnegie Inst Wash Yearbook 8(1909):140–144

Vaughan TW, Wells JW (1943) Revision of the suborders, families and genera of the Scleractinia. Geol Soc Am Spec Pap 44:363 pp

Vermeij MJA, Sampayo E, Bröker K, Bak RPM (2003) Variation in planulae release of closely related coral species. Mar Ecol Prog Ser 247:75–84

Vermeij MJA, Sampayo E, Bröker K, Bak RPM (2004) The reproductive biology of closely related coral species: gametogenesis in *Madracis* from the southern Caribbean. Coral Reefs 23:206–214

Vermeij MJA, Fogarty ND, Miller MW (2006) Pelagic conditions affect larval behaviour, survival, and settlement patterns in the Caribbean coral *Montastrea faveolata*. Mar Ecol Prog Ser 310:119–128

Vermeij MJA, Barott KL, Johnson AE, Marhaver KL (2010) Release of eggs from tentacles in a Caribbean coral. Coral Reefs 29:411

Veron JEN (1995) Corals in space and time: the biogeography and evolution of the Scleractinia. UNSW Press, Sydney

Veron JEN (2000) Corals of the world. Australian Institute of Marine Science, Townsville

Vicentuan KC, Baria MV, Cabaitan PC, Dizon RM, Villanueva RD, Alino PM, Gomex ED, Guest JR, Edwards AJ, Heyward AJ (2008) Multi-species spawning of corals in north-western Philippines. Coral Reefs 27:83

Villanueva RD, Yap HT, Montaño MNE (2008) Timing of planulation by pocilloporid corals in the northwestern Philippines. Mar Ecol Prog Ser 370:111–119

Vize PD (2006) Deepwater broadcast spawning by *Montastrea cavernosa*, *Montastrea franksi*, and *Diploria strigosa* at the Flower Garden Banks, Gulf of Mexico. Gulf Mexico Sci 23:107–114

Vize PD (2009) Transcriptome analysis of the circadian regulatory network in the coral, *Acropora millepora*. Biol Bull 216:131–137

Vize PD, Embesi JA, Nickell M, Brown PD, Hagman DK (2005) Tight temporal consistency of coral mass spawning at the Flower Garden Banks, Gulf of Mexico, from 1997–2003. Gulf Mexico Sci 25(1): 107–114

Vollmer SV, Palumbi SR (2002) Hybridization and the evolution of reef coral diversity. Science 296:2023–2025

Vollmer SV, Palumbi SR (2004) Testing the utility of internally transcribed spacer sequences in coral phylogenetics. Mol Ecol 13: 2763–2772

Wallace CC (1985) Reproduction, recruitment and fragmentation in nine sympatric species of the coral genus *Acropora*. Mar Biol 88:217–233

Wallace CC (1999) Staghorn corals of the World: a revision of the coral genus *Acropora*. CSIRO, Collingwood

Wallace CC, Babcock RC, Harrison PL, Oliver JK, Willis BL (1986) Sex on the reef: mass spawning of corals. Oceanus 29:38–42

Wallace CC, Chen CA, Fukami H, Muir PR (2007) Recognition of separate genera within *Acropora* based on new morphological, reproductive and genetic evidence from *Acropora togianensis*, and the elevation of the subgenus *Isopora* Studer, 1878 to genus (Scleractinia: Astrocoeniidae; Acroporidae). Coral Reefs 26:231–239

Waller RG, Tyler PA (2005) The reproductive biology of two deep-water, reef-building scleractinians from the NE Atalntio Ocean. Coral Reefs 24:514–522

Waller RG, Tyler PA, Gage JD (2002) The reproductive ecology of the deep-sea scleractinian coral *Fungiacyathus marenzelleri* (Vaughan, 1906) in the Northeast Atlantic Ocean. Coral Reefs 21:325–331

Waller RG, Tyler PA, Gage JD (2005) Sexual reproduction in three hermaphroditic deep-sea *Caryophyllia* species (Anthozoa: Scleractinia) from the NE Atlantic Ocean. Coral Reefs 24:594–602

Ward S (1992) Evidence for broadcast spawning as well as brooding in the scleractinian coral *Pocillopora damicornis*. Mar Biol 112:641–646

Ward S, Harrison PL (1997) The effects of elevated nutrient levels on settlement of coral larvae during the ENCORE experiment, Great Barrier Reef, Australia. In: Proceedings of the 8th international coral reef symposium, vol 1, Panama, 1996, pp 891–896

Ward S, Harrison PL (2000) Changes in gametogenesis and fecundity of acroporid corals that were exposed to elevated nitrogen and phosphorus during the ENCORE experiment. J Exp Mar Biol Ecol 246:179–221

Ward S, Harrison P, Hoegh-Gulberg O (2002) Coral bleaching reduced reproduction of scleractinian corals and increases their susceptibility to future stress. In: Proceedings of the 9th international coral reef symposium, vol 2, Bali, 2000, pp 1123–1128

Wellington GM, Fitt WK (2003) Influence of UV radiation on the survival of larvae from broadcast-spawning reef corals. Mar Biol 143:1185–1192

Wells JW (1956) Scleractinia. In: Moore RC (ed) Treatise on invertebrate palaeontology. Pt. F. Coelenterata. Geological Society of America, University of Kansas Press, Lawrence, pp 328–444

Wells JW (1966) Evolutionary development in the scleractinian family Fungiidae. In: Rees WJ (ed) The Cnidaria and their evolution. Academic, London, pp 223–246

Whitaker K (2006) Genetic evidence for mixed modes of reproduction in the coral *Pocillopora damicornis* and its effect on population structure. Mar Ecol Prog Ser 306:115–124

Williams GC (1975) Sex and evolution. Princeton University Press, Princeton

Willis BL (1987) Morphological variation in the reef corals Turbinaria mesenterina and Pavona cactus: synthesis of transplant, histocompatibility, electrophoresis, growth, and reproduction studies. Ph.D. thesis, James Cook University of North Queensland, Townsville, 202 pp

Willis BL, Oliver JK (1990) Direct tracking of coral larvae: implications for dispersal studies of planktonic larvae in topographically complex environments. Ophelia 32:145–162

Willis BL, Babcock RC, Harrison PL, Oliver JK, Wallace CC (1985) Patterns in the mass spawning of corals on the Great Barrier Reef from 1981 to 1984. In: Proceedings of the 5th international coral reef congress, vol 4, Tahiti, 1985, pp 343–348

Willis BL, Babcock RC, Harrison PL, Wallace CC (1992) Experimental evidence of hybridisation in reef corals involved in mass spawning events. In: Proceedings of the 7th international coral reef symposium, vol 1, Guam, 1992, p 504

Willis BL, Babcock RC, Harrison PL, Wallace CC (1997) Experimental hybridization and breeding incompatibilities within the mating systems of mass spawning reef corals. Coral Reefs 16:S53–S65

Willis BL, van Oppen MJH, Miller DJ, Vollmer SV, Ayre DJ (2006) The role of hybridization in the evolution of reef corals. Annu Rev Ecol Evol Syst 37:489–517

Wilson HV (1888) On the development of *Manicina areolata*. J Morphol 2:191–252

Wilson JR, Harrison PL (1997) Sexual reproduction in high latitude coral communities at the Solitary Islands, Eastern Australia. In: Proceedings of 8th International Coral Reef Symposium, vol 1, Panama, pp 533–538

Wilson JR, Harrison PL (1998) Settlement-competency periods of larvae of three species of scleractinian corals. Mar Biol 131:339–345

Wilson JR, Harrison PL (2003) Spawning patterns of scleractinian corals at the Solitary Islands – a high latitudinal coral community in eastern Australia. Mar Ecol Prog Ser 260:115–123

Wolstenholme JK (2004) Temporal reproductive isolation and gametic compatibility are evolutionary mechanisms in the *Acropora humilis* species group (Cnidaria; Scleractinia). Mar Biol 144:567–582

Wyers SC (1985) Sexual reproduction of the coral *Diploria strigosa* (Scleractinia, Faviidae) in Bermuda: research in progress. In: Proceedings of the 5th international coral reef congress, vol 4, Tahiti, pp 1985, 301–306

Wyers SC, Barnes HS, Smith SR (1991) Spawning of hermatypic corals in Bermuda; a pilot study. Hydrobiologia 216(217):109–116

Yakovleva IM, Baird AH, Yamamoto HH, Bhagooli R, Nonaka M, Hidaka M (2009) Algal symbionts increase oxidative damage and death in coral larvae at high temperatures. Mar Ecol Prog Ser 378:105–112

Yonge CM (1932) A note of *Balanophyllia regia*, the only eupsammid coral in the British fauna. J Mar Biol Assoc UK 18:219–224

Yonge CM (1935) Studies on the biology of Tortugas corals I. Observations on *Maeandra areolata* Linn. Carnegie Inst Wash Pub 452:185–198

Yonge CM (1940) The biology of reef-building corals. Sci Rep Great Barrier Reef Exped (1928–29) 1:353–389

Yonge CM (1973) The nature of reef building (hermatypic) corals. Bull Mar Sci 23:1–5

Zakai D, Dubinsky Z, Avishai A, Caaras T, Chadwick NE (2006) Lunar periodicity of planula release in the reef-building coral *Stylophora pistillata*. Mar Ecol Prog Ser 311:93–102

Zooxanthellae: The Yellow Symbionts Inside Animals

Noga Stambler

Abstract Corals are associated with photosymbiotic unicellular algae and cyanobacteria. The unicellular algae are usually called zooxanthellae due to their yellow-brown color. The zooxanthellae are mainly classified as dinoflagellates to the genus Symbiodinium sp. The advantage of symbiosis is based on adaptations of transport and the exchange of nutritional resources, which allow it to be spread all over the tropical and some temperate oceans. Their existence over millions of years depends on the ability of the zooxanthellae, the host, and the holobiont as a whole unit to change, acclimate, and adapt in order to survive under developmental and stress.

Keywords Coral • *symbiodinium* • zooxanthellae • adaptation • carbon • host factor

1 Introduction

Corals are associated with photosymbiotic unicellular algae and cyanobacteria (e.g., as in the review of Venn et al. 2008 and Stambler 2010b). The unicellular algae are usually called zooxanthellae due to their yellow-brown color (Brandt 1883). The zooxanthellae are mainly classified as dinoflagellates to the genus *Symbiodinium* sp., to eight lineages (clades A–H, which are based on phylogenetic classification). Six of them (clades A–D, F, G) are found in scleractinian corals (LaJeunesse 2001; Coffroth and Santos 2005). This genetic diversity is part of the adaptation of the symbionts to the environment and, in many cases, they correlate with the diverse range of physiological properties in the host-symbiont assemblages (Stat et al. 2008b).

Photosynthetically fixed carbon is translocated from zooxanthellae to the host. Up to 95% of the fixed carbon may translocate under high-light conditions (e.g., Falkowski et al. 1984; Muscatine et al. 1984). The contribution of zooxanthellae to animal respiration (CZAR) is up to 100% of daily metabolic requirements and, in some cases, they provide even more than the total metabolic needs of the host animal (Muscatine et al. 1981, 1984; Davies 1984, 1991; Falkowski et al. 1984; Grottoli et al. 2006). However, it should be noticed that this contribution of the zooxanthellae to animal respiration decreased dramatically under bleaching in the case of *Porites compressa* (from 146% to 74%), *Montipora capitata* (from 132% to 41%), and *Porites lobata* (from 141% to 96%) (Palardy et al. 2008).

The advantage of symbiosis is based on adaptations of transport and the exchange of nutritional resources, which allow it to be spread all over the tropical and some temperate oceans. Carbon and nutrient fluxes between the host, the algae, and the environment are based on symbiotic relationships (Muscatine 1990; Yellowlees et al. 2008). These fluxes allow the corals and whole coral-reef communities to succeed at low concentrations of nitrogen (N) and phosphorus (P) in oligotrophic waters surrounding the reefs (Muscatine and Porter 1977). The assimilation of ammonium from the surrounding environment can be done by both the cnidarian host and the algae, as they both have the enzymes glutamine synthetase (GS) and glutamate dehydrogenase (Rahav et al. 1989). A variety of ammonium transporters, which are similar to bacterial transporters, exists in *Symbiodinium* (Leggat et al. 2007). Both symbiotic partners benefit from nitrogen recycling between animals and microorganisms. The host benefits from symbionts that act as a sink for potentially toxic nitrogenous waste compounds while the symbionts benefit from access to the N source for growth (e.g., Douglas 2008).

This symbiosis demands that the coral tolerates and recognizes the presence of the symbionts in its tissues, and that the algae will be able to survive in the tissue and develop some specific host–symbiont combinations (Weis 2008; Yellowlees et al. 2008). The algae and the host metabolism have to adjust to the conditions in order to increase their genotypic diversity even by lateral gene transfers between the endosymbionts and their cnidarian hosts (Furla et al. 2005). The success of the holobiont, the coral host, and the symbionts depends on the integrated physiological capacity of the symbiotic partners towards the environment (review: Trench 1993; Venn et al.

N. Stambler (✉)
The Mina & Everard Goodman Faculty of Life Sciences,
Bar-Ilan University, 52900, Ramat-Gan, Israel
e-mail: stambln@mail.biu.ac.il; drnogas@gmail.com

2008; Weis 2008; Yellowlees et al. 2008). Their existence over millions of years depends on the ability of the zooxanthellae, the host, and the holobiont as a whole unit to change, acclimate, and adapt in order to survive under developmental and stress conditions (review in Brown and Cossins 2011; Hoegh-Guldberg 2011; Lesser 2011; Stambler 2010a).

2 Geological History

The symbiotic scleractinian corals developed early in history, about 210 mya (Wood 1998, 1999). The coral taxa currently existing in symbiosis may or may not have been associated with dinoflagellates from that entire period (Trench 1993). Symbiosis with zooxanthellae evolved independently several times in the Triassic period; however, the first scleractinian corals did not form reefs and were solitary animals. The earliest true coral reefs date from the late Triassic period. At that time, the scleractinian corals that emerged in the warm waters of the Tethys Sea were predominantly zooxanthellate (Stanley and Swart 1995).

The *Symbiodinium* genus originated in the early Eocene period, 50 mya. The major diversification of extant *Symbiodinium* lineages started about 15 mya at the mid-Miocene, when Tethys Sea closure occurred and the ocean temperatures decreased (Pochon et al. 2006). *Symbiodinium* clades A and E were the first to initiate symbiosis (e.g., Karako-Lampert et al. 2004; Pochon et al. 2006; Stat et al. 2008b).

From an evolutionary point of view, in the later part of the Cenozoic era, the symbiotic algae in cnidarians were selected because of their reduced tolerance to elevated temperatures (Tchernov et al. 2004). However, the modern group of *Symbiodinium* sp. developed over millions of years and, as such, contains a broad diversity species that is differentiated physiologically (Tchernov et al. 2004).

3 Cellular Anatomy and the Symbiosome

Symbiodinium sp. size ranges from 6 μm to 15 μm diameter, and varies between genotype and host (see LaJeunesse 2001; LaJeunesse et al. 2005; Frade et al. 2008a). The morphological characteristics of *Symbiodinium* sp. were described by Freudenthal (1962) and appear in the outstanding works of Trench (and in review of Trench 1993). In *hospite* and in culture, the coccoid cells are limited by continuous cellulosic cell wall, which is external to the plasmalemma. There is a single chloroplast with thylakoids. The chloroplast thylakoids are stacked, and often arranged around the periphery of the cell. A pyrenoid body is connected to the chloroplast used for the storage of photosynthetic products, such as starch (Fig. 1). Specific crystalline material stores uric acid that can be mobilized rapidly and used as a nitrogen source, allowing the algal symbionts to grow in an N-poor environment (Clode et al. 2009). The nucleus has notable permanently condensed chromosomes, dinokaryon (the chromosome number varies among species 26–97). The cell contains 1.5–4.8 pg DNA (LaJeunesse et al. 2005). In *hospite*, the zooxanthallae are found only in the coccoid state and they divide mitotically. So far, coccoid cells from corals are found only in culture with flagellum, although the latter has been observed in algae of other hosts within a symbiosome (Trench 1993). The cells become motile either following karyokinesis and cytokinesis, or even without mitosis (Trench 1993). In addition to the lack of flagella, the zooxanthellae inside the coral tissue have slightly different cell-wall structure from the wall of the free-living *Symbiodinium* (Wakefield et al. 2000). In culture, the cells are motile swarmers, characteristic of gymnodinioid morphology (Figs. 2 and 3).

The zooxanthellae are located in vacuoles (symbiosomes) within the host's endoderm cells (Trench 1987). The symbiosome has been defined as the host-derived outer membrane together with the *Symbiodinium* cell and the space between the two. It also includes multilayered membranes derived from the *Symbiodinium* cell. Cultured zooxanthellae lack symbiosomes (Trautman et al. 2002). Following algal cytokinesis, each daughter cell is allocated to an individual symbiosome (Trench 1993). The symbiosome is composed of a zooxanthellae cell that rarely divides and is separated from the host gastrodermal cytoplasm by a symbiosome multimembrane complex (Wakefield et al. 2000; Kazandjian et al. 2008). In the sea anemone *Aiptasia pallida* and its endosymbiont *Symbiodinium bermudense*, the symbiosome membrane is a single, host-derived membrane, whereas the remaining membranes surrounding the algal cell are symbiont-derived

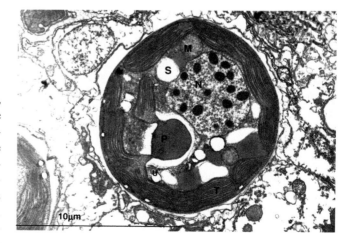

Fig. 1 Transmission electronic micrograph of zooxanthellae (*Symbiodinium* sp.). *T* thylakoid, *N* nucleus with condensed chromosomes, *M* mitochondria, *P* pyrenoid, *S* starch, *U* crystal of uric acid

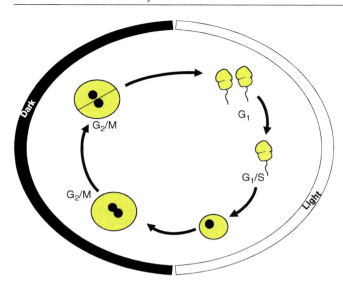

Fig. 2 The *Symbiodinium* cell cycle in culture during the light and dark cycle. G1, G1/S, G2/M phases (Based on Wang et al. 2008)

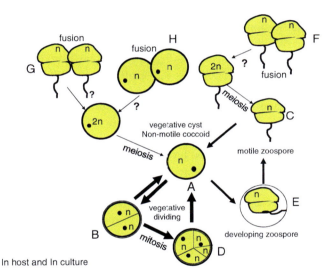

Fig. 3 Life cycle of *Symbiodinium* inside host tissue (*in hospite*) and in culture, including vegetative cyst, dividing vegetative cyst producing two daughter cells, dividing vegetative cyst producing three daughter cells, and zoospore. Different stages of cells with and/or without flagella. Fusion and meiosis of cells. n – haploid nuclear phase, 2n – diploid nuclear phase. (**a–c**) Observed in host and culture; (**c**) observed when zooxanthellae was released from the host; (**d**) is part of mitosis divisions; (**e**) observed in culture; and (**f, g**) are theoretical options (Adapted from Freudenthal 1962; Schoenberg and Trench 1980; Fitt and Trench (1983); Stat et al. (2006); and Lee John (personal comment))

translocation occur through and in conjunction with these membranes (Wakefield et al. 2000). The dinoflagellates occupy most of the interior of macerated host cells, leaving the host cytoplasm and cell membrane as a thin outer layer. As such, the symbiotic zooxanthellae in cnidarians live within an osmotically different environment from that of free-living dinoflagellates (Goiran et al. 1997; Mayfield and Gates 2007). This spatial arrangement may support the diffusion and transport of CO_2, bicarbonate ions, and nutrients from the environment to the algae (Muscatine et al. 1998). It could be that the symbiosome membranes also protect the symbionts from being digested (Chen et al. 2005). However, it should be noted that the symbiosome does not block the effects of the host-release factor (HRF), which stimulates photosynthate release, and the photosynthesis inhibiting factor (PIF) (Grant et al. 2003, see below).

Changes in the ultrastructure take place as part of photoacclimation, e.g., increase in thylakoid number under low light as well as under eutrophication (e.g., Falkowski and Dubinsky 1981; Stambler 1998).

Morphological changes occur in the symbionts under stress conditions. These changes include shrinkage of cell contents, increased vacuolization, and disorganized and loosened thylakoids (Franklin et al. 2004). At high temperature ($\geq 34°C$), cells that become apoptotic also become smaller in size and the cytoplasmic organelles within them fuse together, forming large organelle bodies. The last apoptotic stage occurs when the cell membrane ruptures (Strychar and Sammarco 2009).

The algal cells are usually arranged in a monolayer, resulting in millions of dinoflagellates per square centimeter of coral-colony surface (Drew 1972). Since symbiont types vary in cell size (LaJeunesse 2001; LaJeunesse et al. 2005) and cell density is preliminarily a consequence of space availability (Jones and Yellowlees 1997; Stambler and Dubinsky 2005), the density varies between holobionts. The average cell-specific density (CSD) ranged from 1.11 to 2.1. While in some species, e.g., *Stylophora pistillata*, the variation in distribution of the number of algae per host cell is minimal, in others, such as the sea anemone *Condylactis gigantea*, significant variation can be found (Muscatine et al. 1998). It is higher in fed compared to starved corals (e.g., Houlbreque et al. 2004).

The zooxanthellae population in a host can include one or more genotypes in different abundances that may change under different conditions (e.g., Goulet 2006, 2007; Baker and Romanski 2007, see below). Very rare abundance can be too low for detection even by molecular methods (Baker and Romanski 2007).

Cell density in bleaching corals can be reduced to a minimum and an undetected number, but in the case of host survival, this can be the inoculum for starting a new population in the host.

(Wakefield and Kempf 2001). The symbiosome membrane and the multilayer structure are intimately associated, and also have regular-interval interconnections. As such, they represent one functional unit (Kazandjian et al. 2008).

These membranes are part of the boundary between the host and the symbionts. The transport of gases and carbon

4 Division and Reproduction

It has been thought that the vegetative state of algae can be either haploid or diploid (Fitt and Trench 1983); however, based on molecular genetic evidence, it is now clear that algae from the genus *Symbiodinium*, both in culture and in *hospite*, are haploid (Santos and Coffroth 2003). Although sexual reproduction has not been observed, extensive recombination in *Symbiodinium* (LaJeunesse 2001) by fusion and meiosis (Fitt and Trench 1983; Fig. 3) is probably the reason for high allelic variability for allozymes, random-amplified polymorphic, and DNA fingerprints (LaJeunesse 2001; Santos and Coffroth 2003). However, there is little evidence of genetic recombination between *Symbiodinium* types (Santos et al. 2004; Sampayo et al. 2009). For example, in the tropical Atlantic, the frequency of multiple *Symbiodinium* alleles in a single coral colony of *Favia fragum* was found to be low (Carlon and Lippe 2008).

The cell-cycle process is regulated by light–dark stimulation. Some *Symbiodinium* algae lost motility when placed in either constant light or constant dark (Yacobovitch et al. 2004; Wang et al. 2008) while other *Symbiodinium* lost their motility pattern in constant light but kept their motility rhythms in constant dark (Fitt et al. 1981; Lerch and Cook 1984; Banaszak and Trench 1995). Sequential light followed by dark (12:12 h) entrained a single cell cycle, genotype B in culture, from the G1 to the S phase, and then to the G2/M phase, within these 24 h (Fig. 2). Blue light (450 nm) mimicked regular white light, while red and infrared light had little or no effect on entraining the cell cycle. Light treatment drove cells to enter the growing/DNA synthesis stage (i.e., G1 to S to G2/M), increasing motility and photosynthetic efficiency. Inhibition of photosynthesis stops the cell proliferation process. Darkness is required for the mitotic division stage, when cells return from G2/M to G1 (Wang et al. 2008).

The G1 phase of the *Symbiodinium* sp. cell cycle was extended dramatically in the symbiotic state. The slowing down in the cell cycle was also reflected in the usual low mitotic index (MI) and the low percentage of dividing cells (3–5%) (e.g., Hoegh-Guldberg et al. 1987; Wilkerson et al. 1988; Baghdasarian and Muscatine 2000). Daily and seasonal patterns can be observed in the MI (Wilkerson et al. 1988; Dimond and Carrington 2008), which is dependent on the photoacclimation of the algae (Titlyanov et al. 2001). As a result, the doubling times, which are a few days in culture (Fitt and Trench 1983), increase even to 29 days, i.e., several weeks, inside the host (Wilkerson et al. 1988). The doubling time of the zooxanthellae in the host is about 8 days when there is no limitation of nutrients (calculation based on data from Dubinsky et al. (1990)). Nutrition of the anemone *Aiptasia pulchella* caused variation in G1 phase-duration in the cell cycle of *Symbiodinium pulchrorum* (Smith and Muscatine 1999). The recovery and the re-established symbioses after bleaching depend on the doubling time of the algae, which can be less than 5 days (Toller et al. 2001).

5 Taxonomy from Morphology to Molecular Biology, Genus to Genotype

The dinoflagellate symbionts of coelenterates mainly belong to various species of the genus *Symbiodinium* (Freudenthal 1962) but in some cases they may belong to the genus *Amphidinium* (review in Trench 1997; Karako et al. 2002; Table 1). *Symbiodinium* species differ in cell morphology, ultrastructure, circadian rhythms, growth rates, host infectability, and photoacclimation (e.g., LaJeunesse 2001).

The genus *Symbiodinium* consists of eight lineages/subgeneric "clades" (A–H), each of which includes multiple types/genotypes (e.g., Baker 2003; Coffroth and Santos 2005; Pochon et al. 2006; LaJeunesse et al. 2008). Calcification of the different types of *Symbiodinium* is based on different nuclear, mitochondrial, and chloroplast genomes. Identification is based on: (a) nuclear ribosomal genes and spacer regions: the small subunit (SSU or 18S), large subunit (LSU or 28S), and internal spacer regions (ITS1 and 2); (b) mitochondrial cytochrome b (cytb); and (c) DNA chloroplast 23S rDNA) see in Coffroth and Santos 2005; Sampayo et al. 2009).

Clone and genotype level are separated based on microsatellite alleles) see in e.g., Coffroth and Santos 2005; Pettay and Lajeunesse 2007; Carlon and Lippe 2008).

Types are a much lower level of species and, in most cases, there is no attempt to arrange them according to species. However, while the genus *Symbiodinium* includes only a few species (i.e., *Symbiodinium bermudense*, *Symbiodinium cariborum*, *Symbiodinium corculorum*, *Symbiodinium goreauii*, *Symbiodinium kawagutii*, *Symbiodinium meandrinae*, *Symbiodinium muscatinei*, *Symbiodinium microadriaticum*, *Symbiodinium pilosum*, *Symbiodinium pulchroru*, *Symbiodinium trenchii*, and *Symbiodinium californium*), it includes many tens of types (LaJeunesse 2001; LaJeunesse et al. 2005, 2008; Table 1). Only a few studies correlated between species and type (LaJeunesse 2001; Table 1).

Corals mainly associate with *Symbiodinium* clade C and, in some cases, with clades A, B, D, F, and G. Some coral colonies harbor only a single symbiont type, while others harbor two or more types simultaneously (e.g., Rowan and Knowlton 1995; Baker 2003; Goulet 2006; Mieog et al. 2007; Abrego et al. 2008; LaJeunesse et al. 2008). Clade C is the most common and diverse clade in Indo-Pacific corals (LaJeunesse 2005). Clade C dominates the Indo-Pacific host fauna and shares dominance in the Atlantic-Caribbean with

Table 1 *Symbiodinium* species and genotype

Symbiodinium species	Genotype	Host	Reference
*Symbiodinium microadriaticum**	A, A1	*Cassiopeia xamachana* (Rhizostomeae)	LaJeunesse 2001; Stat et al. 2008b
*Symbiodinium microadriaticum** subsp. Condylactis	A1	*Condylactis gigantean* (Actiniaria)	LaJeunesse 2001
*Symbiodinium cariborum**	A1		LaJeunesse 2001
Symbiodinium microadriaticum subsp. Microadriaticum	A1		LaJeunesse 2001
*Symbiodinium corculorum***	A, A2	*Corculum cardissa*(Bivalvia)	LaJeunesse 2001; Stat et al. 2008b
*Symbiodinium meandrinae***	A2	*Meandrina meandrites* (Scleractinaria)	LaJeunesse 2001
*Symbiodinium pilosum***	A, A2	*Zoanthus sociatus* (Zoantharia)	LaJeunesse 2001; LaJeunesse et al. 2005; Stat et al. 2008b
Symbiodinium linucheae	A4	*Linuche unguiculata* (Coronatae)	LaJeunesse 2001
Symbiodinium microadriaticum	B	*Condylactis gigantea*	Karako-Lampert et al. 2005
*Symbiodinium pulchroru****	B1	*Aiptasia pulchella* (Actiniaria)	LaJeunesse 2001
*Symbiodinium bermudense****	B1	*Aiptasia tagetes* (Actiniaria)	LaJeunesse 2001
Symbiodinium muscatinei	B4		LaJeunesse 2001
Symbiodinium goreauii	C, C1	*Rhodactis lucida* (Corallimorph)	LaJeunesse 2001; LaJeunesse et al. 2005; Stat et al. 2008b
Symbiodinium trenchii	D		LaJeunesse et al. 2005
Symbiodinium californium	E	*Anthopleura elegantissima* (Actiniaria)	LaJeunesse 2001; LaJeunesse et al. 2005; Stat et al. 2008b
Symbiodinium varians	E		Stat et al. 2008b
Symbiodinium kawagutii	F1	*Montipora verrucosa* (Scleractinaria)	LaJeunesse 2001; LaJeunesse et al. 2005, 2008

Species marked with the same number of asterisks are probably the same species

clade B. C1 and C3, which are considered to be the ancestor core for separate types, are common in the Indo-Pacific and Atlantic-Caribbean (LaJeunesse 2005).

In general, but with some exceptions, clade C is the most widely distributed, and presumably has a wide temperature and salinity tolerance, dominating the tropical area (Baker 2003; Karako-Lampert et al. 2004). Some types of clade D are adapted to stress tolerance (Baker 2003). Clade B symbionts are specifically adapted to the lower light and cooler seas of higher-latitude environments in temperate areas (Rodriguez-Lanetty et al. 2001).

6 Inter- and Intrahost Transmission

The host can acquire its symbionts either from its parents or from the surrounding environment. In the case of asexual fragmentation reproduction, the symbionts are always a part of the new organisms. However, in the case of sexual reproduction, two options exist: (1) the symbionts are transferred directly from host to offspring in a process known as vertical, maternal, or closed-system transmission; (2) each generation must acquire new zooxanthellae from the surrounding seawater or, in rare cases, from a secondary host, in a process called open-system or horizontal transmission (see Karako et al. 2002; Coffroth and Santos 2005; Barneah et al. 2007a, b; Huang et al. 2008). This transmission strategy is common to most species, including coral colonies that are broadcast spawners.

Horizontal transmission is of symbionts released from different hosts that survive at the water column for short/long periods of time or via predation/infection of secondary hosts (Muller-Parker 1984; Barneah et al. 2007b). Even though *Symbiodinium* have been isolated from the water column as well as from coral sands (Loeblich and Sherley 1979; Gou et al. 2003; Coffroth et al. 2006; Hirose et al. 2008), only some of the *Symbiodinium,* specifically clade A, include free-living types, and only some of them can infect and become associated with the host (Coffroth et al. 2006). The free-living *Symbiodinium* have different characterization and their distribution changes around the world, e.g., only clade A exists in the Okinawa (Japan) sand (Hirose et al. 2008).

Generational shifts in symbiont type can occur in host broadcast spawners, whose larvae must acquire symbionts from environmental pools. However, hosts exhibiting vertical transmission (brooders) do not demonstrate this level of flexibility (LaJeunesse et al. 2004b; LaJeunesse 2005). Vertical transmission is an effective way to keep the symbiosis from one generation to another and guarantee maximum fit between the two components. However, vertical transmission may be disadvantageous: the symbionts might interfere with host developmental processes, consume limiting host nutrients, and it might be that the location of the larva development will not fit or be optimal for this symbiont genotype (Douglas 2008).

Horizontal strategy will be preferred when the vertical transmission is costly to the host. There are many types of symbionts in the environment, and control over creation of the holobiont is not by coral host alone (Genkai-Kato and Yamamura 1999).

Horizontal transmission maintains higher diversity and, by that, ensures the survival of the symbiosis, specifically under different environment conditions, similar to the advantage of sexual versus asexual reproduction (Stat et al. 2008b). In horizontal transmission, the invertebrates must be infected by *Symbiodinium* from environmental pools (Coffroth and Santos 2005), and this allows the host to become associated with *Symbiodinium* that are better adapted to local environmental conditions (van Oppen 2004; Coffroth and Santos 2005). However, a host with a horizontal acquisition system may fail to acquire symbionts (Genkai-Kato and Yamamura 1999). In the eastern Pacific, the corals *Pavona* and *Psammocora*, which rely on horizontal symbiont acquisition, harbored populations of *Symbiodinium* that were not found in the other host taxa (LaJeunesse et al. 2008).

Flexibility in the acquisition of symbionts should characterize the life history of coral species that must reacquire symbionts in each new generation (Baird et al. 2007). However, although corals obtaining their symbionts by horizontal transmission are expected to have more diverse symbionts associated with them compared to corals with vertical transmission, there are no clear statistics on that. Moreover, in acroporid corals, transmission mode does not affect symbiont diversity (van Oppen 2004). In the Great Barrier Reef (GBR), the majority of corals with a vertical strategy associated with genus-specific *Symbiodinium* type, although in some cases they associated with symbiont types similar to those found in hosts with a horizontal strategy (Stat et al. 2008a).

The different *Symbiodinium* types transfer through both horizontal and vertical transmission. *Symbiodinium* clade A transfers through a closed system in *Stylophora* in the Gulf of Eilat and in the Great Barrier Reef (Karako-Lampert et al. 2004), while *Acropora* acquires its clade A symbionts via horizontal transmission (Stat et al. 2008b). Homologous zooxanthellae of *Fungia scutaria* (Hawaii) are able to establish symbioses with larval hosts *Fungia scutaria* better than heterologous isolates, mixed with zooxanthellae from *Montipora verrucosa*, *Porites compressa,* and *Pocillopora damicornis* (Weis et al. 2001). This indicates a specific process occurring during infection and reorganization (Trench 1993).

The *Symbiodinium* population in coral larvae develops through vertical transmission partitioned according to coral species, while larvae develop through horizontal transmission strategy sharing a common symbiont type across the southern Great Barrier Reef environments (Stat et al. 2008a). Some vertical-transmission-strategy corals harbor one type almost exclusively: *Montipora digitata* and *Porites cylindrica* – Clsu10, and *Seriatopora hystrix* – Clsu 9. However, the pocillopordaii corals *Pocillopora damicornis* and *Stylophora pistillata* harbor different symbiont types in different colonies. In the horizontal-transmission corals *Acropora millepora, Acropora palifera, Favites abdita, Goniastrea favulus,* and *Lobophyllia corymbosa*, different types are found in colonies from the same species (Stat et al. 2008a).

7 Host Specificity

Symbiodinium types are found in diverse host taxon at different geographic locations, and/or under various environmental conditions. Symbiont types are not randomly distributed among cnidarians, mollusks, foraminifera, e.g., clade H is typical to foraminifera. There is no correlation between the scleractinian host taxa and phylogeny to symbiosis with microalgae (Trench 1987). Species of the same host generally harbor the same *Symbiodinium* clades but not always the same genotypes (Coffroth and Santos 2005, as review).

At the same location, different species will harbor different clades and genotypes of symbionts. For example, at the GBR, *Acropora tenuis* harbors types C1 and C2 while *Acropora millepora* harbors clade D; both coral species are broadcast spawning corals with horizontal transmission of symbionts (see Little et al. 2004). Most Hawaiian symbiont genotypes associate with a specific host genus/species and many *Symbiodinium* types from Hawaii differ from those identified in West and East Pacific hosts (LaJeunesse et al. 2004b).

The local environments influence, control, and determine specificity of the host and the *Symbiodinium* genotypes (LaJeunesse and Trench 2000; Rodriguez-Lanetty et al. 2001; Coffroth and Santos 2005, as review). Many host species are capable of symbiosis with more than one *Symbiodinium spp.*; these different symbiotic associations are typically partitioned by geographic location and physical conditions such as light and temperature (LaJeunesse 2002; LaJeunesse et al. 2004a, b). The combinations of coral species and *Symbiodinium* type change with regard to depth, irradiance, temperature gradients, latitude, and longitude.

In the eastern Pacific, colonies of *Pocillopora verrucosa, Pocillopora meandrina, Pocillopora capitata,* and *Pocillopora damicornis* host either *Symbiodinium* D1 or C1b-c. The partner combination of the holobiont appeared random and/or was patchy (LaJeunesse et al. 2008). A different pattern was found for *Porites panamensis* symbionts with clade C at the same location, where the C type was dependent on water temperature and/or depth (LaJeunesse et al. 2008). In the Gulf of Eilat, it seems that the story depends on coral species as well as location since, at the gulf, shallow-water colonies of *Stylophora pistillata* harbor clade A while deeper-water colonies harbor either clade A or C (Lampert-Karako et al. 2008). Different specificity was found for the genus *Madracis* from the southern Caribbean. This genus is dominantly associated with *Symbiodinium* clade B regardless of host species, depth, or within-colony position (Frade et al. 2008a, b). However, the specificity is of the zooxanthella genotype: type B15 occurred predominantly on the deeper reef in green and purple colonies, while type B7 was present in shallow environments in brown colonies (Frade et al. 2008a).

Changes in symbionts are more likely to occur between generations (Baird et al. 2007), under current environmental conditions. Even species that are flexible at the time of infection have strong fidelity as adults (Little et al. 2004). Juveniles of *Acropora tenuis* harbor mixed assemblages of symbionts, while adults usually host a single clade (Little et al. 2004). Larvae of *Fungia scutaria* can be infected by different symbionts from several hosts, while adult colonies usually host a single clade, although some horizontal strategies show evidence of specificity (Weis et al. 2001). In many cases, symbiont diversity is prevalent over time once established, even if environmental changes occur (Iglesias-Prieto et al. 2004; Goulet 2006, 2007), although this is not always the case (Baker and Romanski 2007; Abrego et al. 2008; Jones et al. 2008). For example, at Magnetic Island, *Acropora tenuis* juveniles initially establish symbiosis with a mix of genotype D and C1, and in less than 1 year they become dominated by genotype D; however, the adult colonies do not associate with type D (van Oppen et al. 2001). *Acropora millepora* shuffled their dominant symbiont population after bleaching from type C2 to D (Berkelmans and van Oppen 2006). The adult colonies do not harbor type D (van Oppen et al. 2001). However, the symbionts shuffled, i.e., a change occurred in the relative abundance of genetically different *Symbiodinium* types (*sensu* Baker 2003). This is one of the mechanisms that occurs in several coral species under stress conditions such as high temperature, causing bleaching (Baker 2001; Toller et al. 2001; Little et al. 2004; Baker and Romanski 2007; Jones et al. 2008; Thornhill et al. 2006). Symbiont shuffling is more likely a shifting, not a switching, in the symbiont community dominant in a colony (Jones et al. 2008).

Seriatopora hystrix (vertical transmission strategist) harbors unique symbiont types in different geographic locations within the Pacific, while *Acropora longicyathus* (horizontal transmission strategist) harbors the same types in reefs in Australia, Malaysia, and Japan (Loh et al. 2001). Horizontally transmitted associations are highly specific, despite the presence of a broad range of optional partners (Wood-Charlson et al. 2006). Only in some cases, the initial uptake of zooxanthellae by juvenile corals during natural infection is nonspecific; the association is flexible and depends on the dominant zooxanthellae (Little et al. 2004). With horizontal transmission, agglutination and phagocytosis assist in symbiont uptake by the animals (Rodriguez-Lanetty et al. 2006). Recognition between the symbionts and the coral occurs on a molecular level and depends on the identity of each partner (e.g., Belda-Baillie et al. 2002; Baker 2003; Rodriguez-Lanetty et al. 2006) to allow the phagocytosis. The specificity between *Fungia scutaria* and *Symbiodinium* sp. type C1f during the onset of symbiosis is mediated not only by recognition events before phagocytosis, but by subsequent cellular events occurring after the symbionts are incorporated into host cells (Rodriguez-Lanetty et al. 2006.(In the coral *Fungia scutaria*, the initial cell-surface cellular contact and recognition between the two partners evolve through a lectin/glycan (Wood-Charlson et al. 2006), while in the coral *Acropora millepora*, the protein Millectin, which is an ancient mannose-binding lectin, acts as a pattern recognition receptor (PRR). The Millectin can recognize carbohydrate structures on cells that are probably involved in recognition of the symbionts of the genus *Symbiodinium* (Kazandjian et al. 2008). It seems that an ancient role of C-type lectins in the innate immune response has been co-opted into the pathway that leads to the uptake of *Symbiodinium* by corals (Kazandjian et al. 2008).

Ecological dominance among clades differs between oceans (Baker 2003; LaJeunesse et al. 2003, LaJeunesse 2005). Although related as well as distant hosts harbor closely related symbionts of a similar type (Rowan and Powers 1991), high host specificity and coevolution occur. An example is the symbiotic type in *Porites*, *Montipora*, *Pocillopora*, and *Stylophora* from the Indo-Pacific and the Atlantic-Caribbean. Independent subclades of *Symbiodinium* spp. have evolved for *Porites*, a host genus common to both oceans, while each subclade has characteristic geographic distributions within each ocean (LaJeunesse 2005). Subclades within clade C that associated with different hosts indicated that host-symbiont specificity is part of the evolutionary process of development of new *Symbiodinium* species (LaJeunesse 2005).

The identification of the zooxanthellae, but not the host, on a molecular level ignores some of the specificity interaction between them. The change in the symbiont genotype associated with coral species is a long evolutionary train that differs from one location to another, from one host to another, and from one symbiont genotype to another, depending on the different environmental conditions leading to similar or different combinations.

8 The Host Factor and the Nature of Translocated Compounds

Glycerol, sugars, organic acids, amino acids, lipids, and polyunsaturated fatty acids are produced by the zooxanthellae and transferred to the host (review by Venn et al. 2008; Yellowlees et al. 2008; Stambler 2010b). The release of these photosynthetic products from isolated *Symbiodinium* cells is not triggered by changes in pH (Trench 1971).

In the host tissue, a compound described as host-release factor (HRF) stimulates the release of photosynthate from symbiotic algae (e.g., Muscatine 1967; Grant et al. 2006). The release of carbon by dinoflagellates incubated in HRF is always higher than by those incubated in seawater alone. Host-factor properties depend on the host and have different stabilities to heat (see in Biel et al. 2007). The HRF controls the amount of carbon translocated from the zooxanthellae to the host. Carbon is selectively released by the dinoflagellates to the incubation medium primarily as glycerol, with smaller amounts of glucose, organic acids, amino acids, and lipids (Fig. 4; Biel et al. 2007).

In the case of the temperate coral *Plesiastrea versipora*, the HRF, which has a low molecular weight (Mr < 1,000), stimulates the release of glycerol from its symbiotic dinoflagellate, which can then be utilized by the animal host for its own needs (Grant et al. 2006). The effect of HRF on algae is not related to changes in osmolarity (Grant et al. 2006). In this coral, HRF did not allow the glycerol to leak through the plasma membrane (Ritchie et al. 1993). However, the diversion of glycerol from the algae results in a partial decrease in the algal synthesis of triacylglycerol (TG) and starch (Grant et al. 2006).

HRF increased the ^{14}C-dihydroxyacetone phosphate pool, followed by a reduction by glycerol phosphate dehydrogenase to Gly-3-P. After glycerol-3-phosphate is saturated, Gly-3-P is shunted into a phosphatase reaction to remove the phosphate group and form glycerol (see Fig. 4; Biel et al. 2007). As a result, there is an increase in carbon fixation as glucose and starch released by dinoflagellates are incubated in HRF (Biel et al. 2007). According to Biel et al. (2007; Fig. 4), there is a connection between photosynthesis and respiration that can be seen at ultrastructural levels by the close location of chloroplasts and mitochondria (Biel et al. 2007; Fig. 1). The receptors and/or other transporters of the HRF signaling compounds are located on the algal cell membrane rather than on the host-derived symbiosome membrane (Grant et al. 2003).

The host factors are proteinaceous (Sutton and Hoeghguldberg 1990) and/or amino acids, including high concentrations of protein amino acids (Gates et al. 1995) and/or micromolar concentrations of the nonprotein amino acid, taurine (Wang and Douglas 1997). To better understand the host-factor mechanisms, synthetic host factor (SHF) was used. The synthetic host factor, consisting of aspartic acid, glutamic acid, M serine, histidine, glycine, arginine, taurine, alanine, tyrosine, methionine, valine, phenylalanine, isoleucine, leucine, and asparagine, at pH 8.3 and salinity of 33‰ (Gates et al. 1995; Biel et al. 2007; Stat et al. 2008b), has different effects on clades A and C from Hawaiian coral *Acrophora cytherea* (Stat et al. 2008b).

Photosynthesis inhibiting factor (PIF) is another host cell-signaling molecule, and is found in the coral *Plesiastrea versipora* (Grant et al. 2001). PIF partially inhibits photosynthetic carbon fixation in freshly isolated *Symbiodinium* from *Plesiastrea versipora* and the zoanthid *Zoanthus robustusas*, as well as algae from the coral *Montastraea annularis* in culture (Grant et al. 2006). The presence of symbiotic algae is not necessary for the production of the host signaling molecules HRF and PIF; they exist in the coral *Plesiastrea versipora* and are also expressed in naturally aposymbiotic colonies (Grant et al. 2004).

An advantage for both the algae and the corals is the removal by the algae of respiratory CO_2 and other metabolic breakdown products (NH_3, NO_3^-) from the host. The inorganic carbon (Ci) in the symbiosome is taken up by the *Symbiodinium* CO_2 concentrating mechanisms (CCMs) (Goiran et al. 1996; Leggat et al. 1999). In the chloroplast, the CCMs increase the availability of CO_2 to reach a high concentration that surrounds the enzyme ribulose-1,5-bisphosphate carboxylase/oxygenase (Rubisco), and, by that, enable carbon fixation (Rowan et al. 1996; review in Yellowlees et al. 2008). Carbonic anhydrase (CA), which is required for the supply of CO_2 for the activation of the Rubisco and essential in the acquisition of Ci, exists in both the host and the algae (review in Yellowlees et al. 2008). The corals maintain high CA activity and, in addition, have specific transporters for the delivery of bicarbonate ions to the symbiotic algae. These characteristics provide partial pressure of CO_2 in the immediate surroundings of the symbiont cells that is high enough to support photosynthetic carbon fixation (Allemand et al. 1998).

The symbiotic algae have a high-affinity phosphate transporter and can store polyphosphate when phosphate is replete. Phosphate uptake is dependent on light (Jackson et al. 1989; Jackson and Yellowlees 1990). Both the host and the alga are capable of ammonium assimilation, with both possessing the enzymes glutamine synthetase (GS) and glutamate dehydrogenase (Anderson and Burris 1987; Leggat et al. 2007; Yellowlees et al. 2008; Stambler 2010b; Fig. 5). *Symbiodinium*

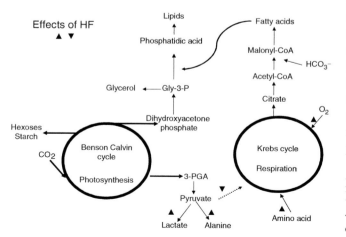

Fig. 4 Model of host-factor influence on photosynthesis and respiration of symbiotic algae. Arrow indicates increase or decrease of the biochemical reaction (Based on Biel et al. 2007)

Fig. 5 Model of the nitrogen cycle in the cnidarian association (After Stambler 2010b)

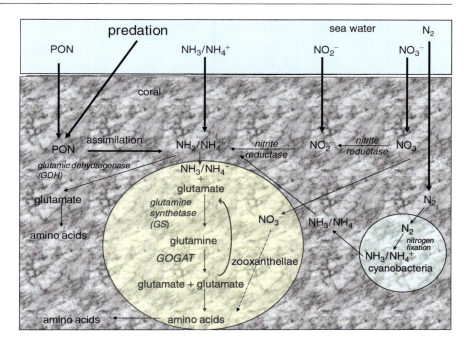

are also capable of utilizing nitrate as a nitrogen source (Fagoonee et al. 1999) and for translocation to the host (Tanaka et al. 2006), presumably as amino acids. Ammonium is the preferred nitrogen source. When it is available, nitrate uptake rates decrease significantly (Grover et al. 2003).

In the Gulf of Eilat, shallow-water colonies of *Stylophora pistillata* harbor clade A (Lampert-Karako et al. 2008) and can support more than 100% of the CZAR (Falkowski et al. 1984). This is not the case for the Hawaiian coral *Acrophora cytherea*, which harbors *Symbiodinium* clade A and which releases very little carbon that can be used for host nutrition (Stat et al. 2008b).

Elevated feeding rates of the coral enhanced rates of photosynthesis normalized per unit surface area. Although this feeding does not always correspond to a higher transfer of photosynthetic products, it might change the component transferred (review in Houlbreque and Ferrier-Pages 2009).

The rate and percentage of photosynthetic carbon translocated from the zooxanthellae to the host depend on the rate of their genotype, photosynthesis rate, acclimation and adaptation, and light and nutrient availability.

9 Population Dynamics and Controls

Mechanisms of zooxanthella regulation include pre- and postmitotic processes, before or after their division (e.g., Baghdasarian and Muscatine 2000; Dimond and Carrington 2008). The premitotic process includes: (a) density-dependent, negative feedback via space or nutrient limitation (e.g., Jones and Yellowlees 1997; see in Baghdasarian and Muscatine 2000); (b) host factor leading to release of photosynthate from the symbionts (Gates et al. 1995; Baghdasarian and Muscatine 2000); and (c) host effects, mainly inhibition of symbiont cell cycle (Smith and Muscatine 1999). Postmitotic regulation includes degradation, digestion, or expulsion of symbionts alone or accompanied by the division of host cells (Titlyanov et al. 1996; Hoegh-Guldberg et al. 1987; Gates et al. 1992; Baghdasarian and Muscatine 2000; Weis 2008; Fig. 6).

Under oligotrophic conditions, zooxanthellae are nitrogen- and phosphorus-limited. They cannot multiply due to this limitation and, as a result, much of the carbon fixed in photosynthesis is translocated to the host (Falkowski et al. 1993). The result is that the growth rate of the zooxanthellae is extremely slow, with doubling times as long as 70–100 days (growth rate, $\mu = 0.007$–0.001 day^{-1}) in the common Red Sea coral *Stylophora pistillala* (Falkowski et al. 1984) compared to the much higher growth rate of zooxanthellae cultured from the coral *Acropora* sp. (0.33–0.48 day^{-1}) (Taguchi and Kinzie 2001).

The growth rate does not significantly depend on the zooxanthella genotype; however, in culture, it can be more than twice as high under high light (HL) compared to low light (LL). In the case of genotype F2 from *Meandrina meandrites*, under HL it is 0.76 day^{-1} and under LL – 0.32 day^{-1}, while in genotype A2 *Montastrea* spp., under HL – 0.46 day^{-1} and under LL – 0.22 day^{-1} (Hennige et al. 2009). It should be noted that there is increasing evidence that the growth of the zooxanthellae both in culture and in the coral host is in association with bacteria. Recently,

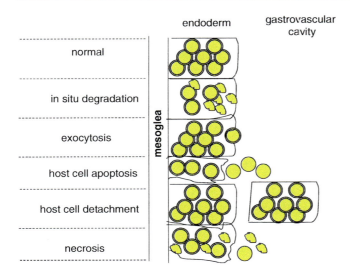

Fig. 6 Different types of cellular mechanisms of symbionts lost from host tissues. The zooxanthellae die or are killed by the host and are either expelled or digested; they undergo in situ degradation; symbiont exocytosis; host cells undergoing apoptosis or dying by necrosis release viable or degraded symbionts; symbionts lost together with host cell detachment (Adapted from Gates et al. 1992; Weis 2008)

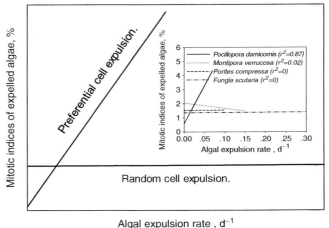

Fig. 7 Mitotic index as a function of algal expulsion rate: a. theoretical; b. in the field (Based on Baghdasarian and Muscatine 2000)

Agostinim et al. (2009) found that Vitamin B12 is produced by the bacteria that are translocated to the zooxanthellae.

The host regulates algal density by expulsion of dividing algal cells and not by digestion (Baghdasarian and Muscatine 2000). The rate of expulsion is an opposite function of the ability of host cells to accommodate new algal cells. However, this regulation is found in some corals, e.g., *Pocillopora damicornis*, but not in others, such as *Montipora verrucosa*, *Porites compressa*, and *Fungia scutaria*. Some species have very low daily rates of algal expulsion and, in such cases, this kind of regulation is not efficient (Baghdasarian and Muscatine 2000; Fig. 7).

In cases where regulation of the algae is based on expulsion of extra algae, when the algal division rate increases due to environmental changes such as increases in light, temperature, and nutrients, leading to increase in algal division rate and is expressed in MI, there is an increased rate of expulsion of algae (Baghdasarian and Muscatine 2000). Temperatures elevated by a few degrees resulted in higher MI and expulsion rates (Baghdasarian and Muscatine 2000). In the case of temperate coral *Astrangia poculata*, while MI was independent of symbiont density, the expulsion rates were dependent. For this coral, higher MI and expulsion rates were found in summer (Dimond and Carrington 2008). Expulsion rates change according to time of day and season (Hoegh-Guldberg et al. 1987; Dimond and Carrington 2008) because of variation in the algal cell cycle with and without dependence on host nutrition.

Once supplied with additional nutrients, either as inorganic compounds such as ammonium and phosphate or via zooplankton consumption by the host animal (e.g., Dubinsky et al. 1990; Falkowski et al. 1993; Dubinsky and Jokiel 1994), the zooxanthellae retain most of their photosynthetic products. The photosynthetic carbon is now utilized for the synthesis of zooxanthella biomass, accelerating their growth rates and increasing their densities up to fivefold (Dubinsky et al. 1990). Under eutrophication, there is also an increase in the number of zooxanthellae released to the surrounding water (Stimson and Kinzie 1991). Symbiont expulsion serves under eutrophication as the limiting overgrowth of the algae in the animal cells. The growth rate of the algal symbionts determines the upper limit of the standing stock and their rates of cell division (Smith and Muscatine 1999).

Under bleaching, the nitrogen uptake by corals supports zooxanthella recovery and increases their mitotic cell division (Rodrigues and Grottoli 2006).

After several days of starvation of the coral *Stylophora pistillata*, the zooxanthellae in its tissues exhibit degradation, possibly because the host used the algae as a food source (Titlyanov et al. 1996).

In the sea anemone *Anthopleura elegantissima*, no symbiosis-specific genes were involved in controlling and regulating symbiosis. Symbiosis is maintained by the varying expression of existing genes involved in vital cellular processes, including deregulation of the host cell cycle and suppression of apoptosis (Rodriguez-Lanetty et al. 2006).

There are several different types of cellular mechanisms by which zooxanthellae are lost from the host tissues: (1) in situ degradation: (a) programmed cell death (PCD) of the zooxanthellae (program always going to take place); (b) apoptosis and/or necrosis caused by biochemical and pathological processes under abnormal stimuli inducing cell death (see in Strychar et al. 2004a, b); (c) death and degradation because of stress effects, e.g., from reactive oxygen species (ROS); or (d) they are killed by the host cell (Gates et al. 1992; Weis 2008; Fig. 6). In the sea anemone *Aiptasia pulchella*, stress results in changes in lysosomal maturation and targeting of symbiosomes, resulting in symbiont digestion (Chen et al. 2005). After death, the symbionts will be digested in or expelled from the host. (2) Exocytosis, expelled freely. (3) Released inside detached host cell from the mesoglea. (4) This process releases viable or degraded symbionts with apoptotic host cells. (5) Necrosis of host causes the release of viable or degrading zooxanthellae. Loss of the zooxanthellae is into the gastrovascular cavity and, from there, to the surrounding water (Gates et al. 1992; Weis 2008; Fig. 6).

It is possible that in some cases, stress such as very high temperature causes the death of both the symbionts and the host simultaneously, while in other cases, death of the zooxanthellae or the coral will take place one after the other.

As a result of heat stress, apoptosis and necrosis occur simultaneously in both the sea anemone *Aiptasia sp.* tissues and its symbionts. Rate of apoptosis in the anemone endoderm increases within minutes of exposure to high temperature. Coincident with the timing of loss of zooxanthellae during bleaching, peak rates of apoptosis-like cell death is observed in the host. As exposure continues, apoptosis host cell number declines while necrosis cell number increases. Apoptosis and necrosis activity increases simultaneously in the zooxanthella cells dependent on temperature dose (Dunn et al. 2004). Apoptotic markers in response to heat stress in *Symbiodinium* spp. cells include: chromatin condensation, intact plasma membrane, vacuolization and vesicle formation, and cytoplasmic condensation (Dunn et al. 2004).

In the apoptosis of symbiont organisms in response to increased temperature and light, there is an increase in the activity of selective cysteine aspartate-specific proteases – caspase (Dunn et al. 2006; Richier et al. 2006; Weis 2008). During bleaching, apoptosis acts to maintain homeostasis, mitigate tissue damage, and remove dysfunctional symbionts (Dunn et al. 2004). Bleaching response might also represent a modified immune response that recognizes and removes dysfunctional symbionts (Dunn et al. 2007).

During stress leading to bleaching, although normal algae might be expelled under continuous and strong stress, usually the expulsion is of amorphous material and disorganized/digested cellular cells (Franklin et al. 2004). In the case of *Acropora hyacinthus* and *Porites solida* exposed to temperature stress as high as ≥34°C, it was not the host that was sensitive to temperature, but rather the symbionts, leading to expelled zooxanthellae characterized by irreversible ultrastructural and physiological changes symptomatic of cell degeneration and death (called apoptosis) or necrosis (Strychar and Sammarco 2009).

Holobiont corals with different clades may have developed different regulation mechanisms depending on the host, including its physiology and the environmental conditions (Rowan et al. 1997; Baghdasarian and Muscatine 2000), as well as mechanisms of the zooxanthellae themselves for expulsion from the host. Expulsion of zooxanthellae can occur under normal conditions, increased nutrition of the coral host, and under stress. Expulsion can be either during increasing symbiont density in the coral tissues (in the case of eutrophication) or during stress leading to bleaching (low density of zooxanthellae). Even though in recent years, study has focused on stress mechanisms causing bleaching and zooxanthellae loss from host tissues (the expulsion of zooxanthellae that in most cases are not viable and probably would not survive (Hill and Ralph 2007)), we should keep in mind that the release of zooxanthellae to the surrounding water is a natural control of their population number in the host, and is the only way that zooxanthellae are able to change their host. These released zooxanthellae are viable, and can infect other corals or larvae to create a new association by horizontal transmission.

10 Distribution Within Colony and Polyp

Corals harboring genetically mixed communities of *Symbiodinium* often show distribution patterns in accordance with differences in a light field across an individual colony. In *Acropora tenuis*, parts exposed to the Sun harbor type C2 while the shaded portions of the same colony harbor type C1 (van Oppen et al. 2001) *Symbiodinium* clade C is found predominately in the sides of *Montastraea* sp., while clade A is found predominately at the top of the same colonies (Rowan et al. 1997). In an individual colony of *Acropora valid*, Sun- and shade-adapted polyps were found to harbor either *Symbiodinium* clade C types alone or clades A and C simultaneously. Polyps harboring both clades A and C show higher metabolic activity of respiration and photosynthesis (Ulstrup et al. 2007).

Zooxanthellae are rare at the tip of stony corals (Fang et al. 1989). The tips that are exposed to high light have lower chlorophyll per coral unit area compared with the lower branches, which contain a higher concentration of photosynthetic pigments (Falkowski et al. 1984). In some cases, branches exposed to low light will have more zooxanthellae than those parts exposed to higher light, and/or more chlorophyll per algal cell (Titlyanov et al. 2001). Coral parts facing the dark will have no zooxanthellae (Dubinsky and Jokiel 1994; Titlyanov et al. 2001).

There is spatial heterogeneity in coral photosynthesis (Falkowski et al. 1984; Gladfelter et al. 1989; Kuhl et al. 1995; Ralph et al. 2002, 2005). Physiologically, zooxanthellae *in hospite* perform differently between Sun- and shade-adapted surfaces of individual colonies (Ralph et al. 2005), and between polyps and coenosarc tissue (Ralph et al. 2002). Polyps have lower photosynthesis available radiation (PAR) absorptivity than coenosarc tissue in *Acropora nobilis* (branching coral) and *Pavona decussata* (plate coral), whereas *Goniastrea australiensis* (massive coral) exhibits the opposite pattern. *Acropora nobilis* exhibits heterogeneity along the longitudinal axis of the branch; this can be differentiated from the effect of variations in illumination across the rugose and curved surfaces (Ralph et al. 2005). Differential bleaching responses between polyps and coenosarc tissue were found in *Pocillopora damicornis* but not in *Acropora nobilis* and *Cyphastrea serailia* (Hill et al. 2004).

11 Photosynthesis

Symbiodinium contain typical components of dinoflagellates: chlorophyll *a*, chlorophyll c_2, and carotenoids (peridinin, dinoxanthin, diadinoxanthin (DD, Dn,(diatoxanthin (Dt), and β-carotene) (e.g., Kleppel et al. 1989; Levy et al. 2006; Venn et al. 2006). The chlorophyll is part of the photosynthesis apparatus while Dn and Dt are part of the photoprotective xanthophyll (Brown et al. 2002). The zooxanthellae acclimate to photosynthesis under different light conditions. As light increases, the algal growth rate (μ), maximum photosynthesis, respiration, in vivo absorption (a*), and β carotene increase (Fig. 8). At the same time, chlorophyll *a* and *c*, peridinin concentrations, thylakoid area, the size of photosynthetic units (PSUs), and quantum yield (φ) decrease (Falkowski and Dubinsky 1981; Stambler and Dubinsky 2004; Fig. 8). The zooxanthellae photoacclimate, including, in addition to the adjustment of pigmentation, changes in the number of reaction centers in the light-harvesting photosystems (Falkowski and Dubinsky 1981; Dubinsky and Falkowski 2010). As a result of the photoacclimation of the zooxanthellae, corals can grow in shallow water exposed to sunlight and in deep water down to the photic zone (e.g., Mass et al. 2007; Stambler et al. 2008; Frade et al. 2008c). The photic zone is considered the depth from the surface to a depth of 1% of the sea subsurface light level. It should be noticed that photoinhibition is always observed in freshly isolated zooxanthellae (FIZ) (Fig. 9), but only occurs in shallow water at very high light intensity, inside the coral tissue (Hoegh-Guldberg and Jones 1999; review in Bhagooli and Hidaka 2004; Stambler and Dubinsky 2004; Levy et al. 2006; Frade et al. 2008c).

The light-saturated rate of photosynthesis (P_{max}), compensation light intensity (E_c), and light intensity of incipient saturation (E_k), all decrease with depth while the efficiency of photosynthesis (α) increases with depth (Mass et al. 2007; Hennige et al. 2008; Stambler et al. 2008). The response to light also depends on daily changes in light, for example, higher photosynthetic rates occur in the afternoon rather than the morning at the same PAR levels; however, this may vary significantly between species (Levy et al. 2004). Shallow-water coral reefs show a diurnal xanthophyll diadinoxanthin (Dn) and diatoxanthin (Dt) pattern, as well as changes in the photosynthesis parameter, such as quantum yield and photochemical efficiency (Fv/Fm). The potential of the maximum quantum yield of photochemistry in photosystem II (PSII) is determined in a dark-adapted state as the

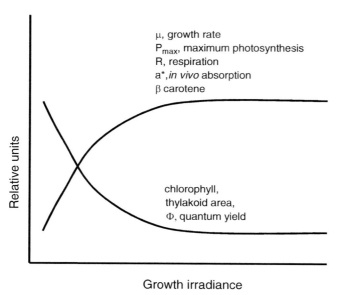

Fig. 8 Changes in chlorophyll concentration; P_{max}, maximum photosynthesis; R, respiration; a*, in vivo absorption; β carotene; thylakoid area; Φ, quantum yield; μ, growth rate as response to light intensity during growth (Based on Dubinsky et al. 1995)

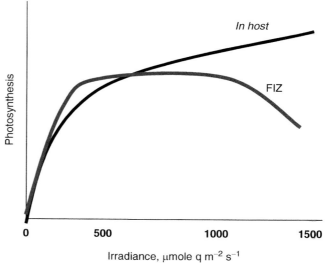

Fig. 9 Photosynthesis rate *in hospite* and in fresh isolated zooxanthellae (FIZ) as function of light intensity

ratio Fv/Fm = (Fm–F_0)/Fm, where F_0, Fm are the minimum and maximum yields of chlorophyll fluorescence, respectively, measured after a dark period (relative units). Fv, the variable fluorescent, is determined as Fv = (Fm–F_0) (Brown et al. 1999). The patterns of the xanthophyll cycling that exist in different *Symbiodinium* genotypes and the diel effective quantum yield of photosystem II, nonphotochemical quenching (NPQ) of the fluorescence, differ between corals species even when residing at the same depth (Warner and Berry-Lowe 2006).

The photosynthesis of freshly isolated zooxanthellae (FIZ) differs significantly from their photosynthesis in host tissue (Fig. 9) due to the different lights they are exposed to, their packaging in the tissue, and competition on CO_2 with the coral (Dubinsky et al. 1990; Stambler and Dubinsky 2005). Acclimation of the holobiont to different light levels involves the following coral host responses: (a) some host nonfluorescent pigments upregulate response to elevated irradiance. As a response, maximum photosynthesis per chlorophyll correlates with the concentration of an orange-absorbing nonfluorescent pigment (CP-580) in the coral *Montipora monasteriata* (Dove et al. 2008, Fig. 10); (b) under low light, changes in the shape of the colonies include flattening, and by that, they reduce the shading of the branches one of the other (e.g., Dustan 1975; Graus and Macintyre 1976; Stambler and Dubinsky 2005; Mass et al. 2007; Kaniewska et al. 2008); (c) increase in fluorescent host pigments acts as photoprotector under high light and UV; (d) changes in the skeleton of the coral host (Enriquez et al. 2005); and (e) changes in host-tissue thickness, e.g., tissue mass is smaller at the lower part of the colony (Anthony et al. 2002).

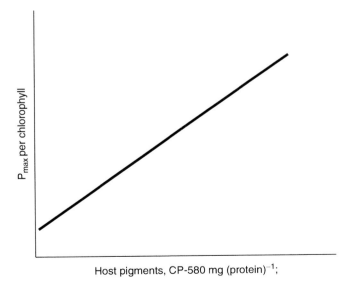

Fig. 10 Maximum photosynthesis per chlorophyll of the symbionts as a function of the host pigment concentrations (Based on Dove et al. 2008)

Stress that damages photosynthesis might lead to bleaching (Stambler and Dubinsky 2004; Stambler 2010a). Some of the damage is: (1) decrease in the efficiency of photosystem II (PSII) under high temperature, which causes a malfunction in the light reactions of photosynthesis; (2) degradation of the reaction center D1 protein, which occurs with temperature-dependent loss of PSII activity (Warner et al. 1999); (3) PSII damage that affects the impairment of the Calvin-Benson cycle and causes a decrease in carboxylation of ribulose 1,5 bisphosphate (RuBP) – Rubisco (Jones et al. 1998); (4) increase in reactive oxygen species (ROS) concentration, leading to cellular damage (Lesser 2006); and (5) damage of the thylakoid membranes, causing an increase in the rate of electron transport on the acceptor side of PSII with a simultaneous decrease in the maximum quantum yield of photochemistry in the reaction center (Tchernov et al. 2004). Combination of stress, specifically under high light and temperature, leads to chronic photoinhibition (Bhagooli and Hidaka 2004). pCO_2 enrichment of *Symbiodinium in hospite* of the coral *Acropora Formosa* caused an increase in chlorophyll *a* per cell under subsaturating light levels, thus supporting the idea that zooxanthellae are CO_2-limiting in the coral tissue. While light-enhanced dark respiration per cell increased due to an increase in the immediate products of the Calvin cycle, the dark respiration stayed the same; xanthophyll de-epoxidation increased; all of this leads to decreases in photosynthetic capacity per chlorophyll (Crawley et al. 2010). Expression of the first enzyme in the photorespiratory cycle, phosphoglycolate phosphatase (PGPase), was reduced by 50% under high CO_2 environment. This reduction in PGPase coincided with the decline in zooxanthella productivity (Crawley et al. 2010).

Physiological function of the symbionts is not always correlated with the clade level (Savage et al. 2002; LaJeunesse et al. 2003; Tchernov et al. 2004). Although photoacclimation of *Symbiodinium* genotype is variable, their light absorption per photosystem is similar (Hennige et al. 2009). In cultures and *in hospite,* most clade A *Symbiodinium* types, but not clades B, C, D, or F, show enhanced capabilities for alternative photosynthetic electron-transport pathways, including cyclic electron transport. Clade A undergoes pronounced light-induced dissociation of antenna complexes from photosystem II (PSII) reaction centers; this was not observed in other clades. As a result, clade A symbionts are resistant to high light intensities and high temperature, and, as such, survive bleaching (Lampert-Karako et al. 2008; Reynolds et al. 2008). Clades B, C, and D, found in symbioses in deeper waters than *Symbiodinium* clade A, benefit from enhanced light-harvesting capability. The *Symbiodinium* subclade B are found in shallow water, probably employing photoprotection mechanisms other than antenna translocation. *Symbiodinium* clades B and C from deeper-dwelling corals, susceptible to bleaching, can engage nonphotochemical quenching (NPQ) and varying degrees of chlororespiration (Reynolds et al. 2008).

12 Ecology: Geography, Temperature, and Host Effects

The geographical distribution of genotypically varying symbionts and their abundance is dependent on host specificity and tolerance to temperature and light variation.

Environmental gradients of light are one of the important controls of coral holobiont physiology, distribution, survival, and existence (Falkowski et al. 1990). At the GBR, in shallow water less than 3 m, *Stylophora pistillata* harbors C1, while 10 m is associated with C27 (LaJeunesse et al. 2003). The Caribbean *Montastraea spp.* hosts A and B clades in shallow waters (less than 6 m) and clade C symbionts at the deeper depths (Rowan and Knowlton 1995). The *Symbiodinium* types associated with *Montastraea* sp. and *Acropora* sp. depend on the irradiance that the colony is exposed to (Rowan et al. 1997; Ulstrup and van Oppen 2003). Different *Symbiodinium* types B show systematic patterns of distribution in different *Madracis* species over a depth and light gradient that the colony is exposed to. Brown colonies of *Madracis pharensis* from 10 m depth harbor *Symbiodinium* B7 while in deeper water (25 m), green and purple colonies of the same species are associated with type B15 (Frade et al. 2008a, b), whose larger cells are found at lower densities in the coral tissue compared to type B7. These two types show different adaptation and acclimation to light. Chlorophyll concentration per cell was higher in type B15. α, the initial slope of the photosynthesis versus irradiance curve (P versus E), i.e., the ratio of photosynthesis to light under light limitation, was higher for type B15 when normalized to algae cells (Frade et al. 2008b). In spite of this, the symbiont genotype in the *Madracis* colonies was dependent on the depths where they grow, and not on the different light microhabitats at each depth (Frade et al. 2008a).

Whether or not the distribution of coral species depends on its symbionts is not clear; however, in the eastern Pacific reefs, *Pocillopora verrucosa* with D1 type dominates shallow water while *Pavona gigantean* is associated with *Symbiodinium* C1c (Iglesias-Prieto et al. 2004). Nevertheless, juvenile corals harbored with clade C grow two to three times faster than those harbored with clade D (Little et al. 2004).

Only several host taxa are found in the western Pacific and Caribbean Oceans, which are dominated by a few prevalent generalist symbionts. In Hawaii, due to geographic isolation and low host diversity, a high proportion of coral species with vertical transmission have high symbiont diversity and specificity with no dominant generalist symbionts (LaJeunesse et al. 2004b; Fig. 11).

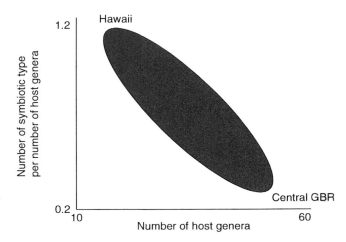

Fig. 11 The ratio between the number of symbiotic types per number of host genera and the number of host genera (Based on LaJeunesse et al. 2004b)

Corals associated with *Symbiodinium* live close to their upper thermal tolerance limits (Fitt et al. 2001). The coral *Acropora tenuis* response to bleaching is dependent on the *Symbiodinium* type with which the colony associates (Abrego et al. 2008). There are heat types within both clades C and D (Bhagooli and Hidaka 2004; Tchernov et al. 2004). For example, *Montipora digitat* with type C15 is more resistant to bleaching compared to other *Montipora* associated with other clade C types (LaJeunesse et al. 2003). *Symbiodinium* clade D (presumably D1) has been characterized as heat- or stress-tolerant based on increased frequency of this type within this clade in Caribbean and Indo-Pacific corals after bleaching events (Abrego et al. 2008; Jones et al. 2008). However, *Acropora tenuis* juveniles when hosting *Symbiodinium* type C1, demonstrate lower metabolic costs and higher physiological tolerance compared to juveniles with type D (Abrego et al. 2008). At higher eastern Pacific latitudes, *Pocillopora* spp naturally evolved and associated with type D, but not as a result of coral bleaching (Baker et al. 2004; LaJeunesse et al. 2008). It is possible that clade D includes algal types that differ in thermal tolerance (e.g., Tchernov et al. 2004). It should be noticed that phylotypes belonging to different genotypes can present similar patterns of sensitivity to elevated temperatures, but differ from their closely related sister phylotypes (Tchernov et al. 2004).

Since different *Symbiodinium* sp. have different temperature tolerances and, thus, different growth requirements (Fitt et al. 2000; Kinzie et al. 2001) over evolutionary timescales, colonies hosting thermally sensitive symbionts might become extinct by natural selection. Acquisition of less-sensitive symbiont populations (for example, clade D1 in the case of temperature increase) might result in colony survival (Buddemeier and Fautin 1993; Baker 2001; LaJeunesse et al. 2004a, b).

Zooxanthellae exposed to cold temperature stress decrease their maximum photochemical efficiency of PSII and may undergo chronic photoinhibition (Saxby et al. 2003; Thornhill et al. 2008). *Symbiodinium* type B2 associated with the temperate corals *Astrangia poculata* and *Oculina arbuscula* adapt and survive over extended periods of cold temperature stress and rapidly return to normal photosynthetic function when temperature increases (Thornhill et al. 2008).

Host-protective mechanisms against the stress of the holobiont include: the production of antioxidant enzymes (Lesser et al. 1990; Lesser 2006), mycosporine-like amino acids (MAAs) (Shick and Dunlap 2002), and fluorescent pigments (Salih et al. 2000). These protective mechanisms may be damaged under stress, for example, the fluorescent pigments are reduced at elevated temperatures (Dove 2004).

Under high light intensity, high rates of photosynthesis by the zooxanthellae generate high concentration of dissolved oxygen (Lesser 2006). These high concentrations can form reactive oxygen species (ROS). ROS causes major cellular damage, including oxidizing membranes, denaturing proteins, and damaged nucleic acids (Lesser 2006). ROS, especially the reactive nitrogen species nitric oxide (NO), may play a major role in bleaching (Bouchard and Yamasaki 2008; Weis 2008) even though both the host and the zooxanthellae have adaptations to prevent this damage. These adaptations include enzymes such as catalase, ascorbate peroxidase, and multiple isoforms of superoxide dismutase (SOD) (Richier et al. 2005, 2008; Lesser 2006; Weis 2008).

The mycosporine-like amino acids (MAAs) found in many coral dinoflagellate symbioses also originate in the endosymbionts and act as UV protectors (Shick and Dunlap 2002). For example, mycosporine-glycine, shinorine, porphyra-334, and palythine were detected in all *Symbiodinium* and their cnidarian hosts from the Mexican Caribbean (Banaszak et al. 2006). MAAs are present within the *Symbiodinium* and the host fractions (Banaszak et al. 2006). They are more concentrated in the tissues of the anthozoan host than in the zooxanthellae (Shick et al. 1995; Shick 2004; Furla et al. 2005). Under natural conditions, *Symbiodinium* clades do not influence the presence of MAAs in the symbionts from the Mexican Caribbean (Banaszak et al. 2006). This is in spite of the fact that in culture, only *Symbiodinium* clade A produces MAAs whereas other clades do not (Banaszak et al. 2006).

The thermal tolerance of *Symbiodinium* is dependent on genotype and adaptation: thermally tolerant type A1 increases light-driven O_2 consumption but not the amount of H_2O_2 produced, while sensitive type B1 increases the amount of H_2O_2 produced without an increase in light-driven O_2 consumption. In other words, the Mehler reaction, which elevates H_2O_2 production, is the response of clade B1 to temperature increase (Jones et al. 1998; Suggett et al. 2008).

Freshly isolated zooxanthellae respond to heat/light stress differently when they are *in hospite*, suggesting that the hosts play an important role in regulating the response of the holobiont (Bhagooli and Hidaka 2003). The advantage of the holobiont and its ability to survive depends on the host genotype (Baird et al. 2009) and on symbiont genotypes that change the mutualistic symbiosis interaction and efficiency under different environmental conditions. Hermatypic coral success and survival depend on the symbionts, host genomics, and the interaction between them in the local environment, with attention to the effect of the historical condition. Under continuous stress, evolutionary processes may shift the coral from a photoautotrophic to a heterotrophic situation, which already occurs under deep depths and bleaching conditions (Grottoli et al. 2006; Palardy et al. 2008; Houlbreque and Ferrier-Pages 2009). These processes may separate the symbiotic partners forever.

References

Abrego D, Ulstrup KE, Willis BL, van Oppen MJH (2008) Species-specific interactions between algal endosymbionts and coral hosts define their bleaching response to heat and light stress. Proc R Soc Lond Ser B-Biol Sci 275:2273–2282

Agostinim S, Suzuki Y, Casareto B, Nakano Y, Hidaka M, Badrun N (2009) Coral symbiotic complex: hypothesis through vitamin B12 for a new evaluation. Galaxea J Coral Reef Stud 11:1–11

Allemand D, Furla P, Benazet-Tambutte S (1998) Mechanisms of carbon acquisition for endosymbiont photosynthesis in Anthozoa. Can J Botany-Revue Canadienne de Botanique 76:925–941. In: 3rd international symposium on inorganic carbon acquisition by aquatic photosynthetic organisms Vancawer, British Columbia, 1997

Anderson SL, Burris JE (1987) Role of glutamine-synthetase in ammonia assimilation by symbiotic marine dinoflagellates (zooxanthellae). Mar Biol 94:451–458

Anthony KRN, Connolly SR, Willis BL (2002) Comparative analysis of energy allocation to tissue and skeletal growth in corals. Limnol Oceanogr 47:1417–1429

Baghdasarian G, Muscatine L (2000) Preferential expulsion of dividing algal cells as a mechanism for regulating algal-cnidarian symbiosis. Biol Bull 199:278–286

Baird AH, Cumbo VR, Leggat W, Rodriguez-Lanetty M (2007) Fidelity and flexibility in coral symbioses. Mar Ecol Prog Ser 347:307–309

Baird AH, Bhagooli R, Ralph PJ, Takahashi S (2009) Coral bleaching: the role of the host. Trends Ecol Evol 24:16–20

Baker AC (2001) Ecosystems - Reef corals bleach to survive change. Nature 411:765–766

Baker AC (2003) Flexibility and specificity in coral-algal symbiosis: diversity, ecology, and biogeography of *Symbiodinium*. Ann Rev Ecol Evol Syst 34:661–689

Baker AC, Romanski AM (2007) Multiple symbiotic partnerships are common in scleractinian corals, but not in octocorals: comment on Goulet (2006). Mar Ecol Prog Ser 335:237–242

Baker AC, Starger CJ, Mcclanahan TR, Glynn PW (2004) Corals' adaptive response to climate change. Nature 430:741

Banaszak AT, Trench RK (1995) Effects of ultraviolet (UV) radiation on marine microalgal-invertebrate symbioses: I. Response of the algal symbionts in culture and in hospite. J Exp Mar Biol Ecol 194:213–232

Banaszak AT, Santos MG, LaJeunesse TC, Lesser MP (2006) The distribution of mycosporine-like amino acids (MAAs) and the phylogenetic identity of symbiotic dinoflagellates in cnidarian hosts from the Mexican Caribbean. J Exp Mar Biol Ecol 337:131–146

Barneah O, Brickner I, Hooge M, Weis VM, Benayahu Y (2007a) First evidence of maternal transmission of algal endosymbionts at an oocyte stage in a triploblastic host, with observations on reproduction in *Waminoa brickneri* (Acoelomorpha). Invertebr Biol 126:113–119

Barneah O, Brickner I, Hooge M, Weis VM, LaJeunesse TC, Benayahu Y (2007b) Three party symbiosis: acoelomorph worms, corals and unicellular algal symbionts in Eilat (Red Sea). Mar Biol 151:1215–1223

Belda-Baillie CA, Baillie BK, Maruyama T (2002) Specificity of a model cnidarian-dinoflagellate symbiosis. Biol Bull 202:74–85

Berkelmans R, van Oppen MJH (2006) The role of zooxanthellae in the thermal tolerance of corals: a 'nugget of hope' for coral reefs in an era of climate change. Proc R Soc Lond Ser B-Biol Sci 273:2305–2312

Bhagooli R, Hidaka M (2003) Comparison of stress susceptibility of in hospite and isolated zooxanthellae among five coral species. J Exp Mar Biol Ecol 291:181–197

Bhagooli R, Hidaka M (2004) Photoinhibition, bleaching susceptibility and mortality in two scleractinian corals, *Platygyra ryukyuensis* and *Stylophora pistillata*, in response to thermal and light stresses. Comp Biochem Physiol A Mol Integr Physiol 137:547–555

Biel KY, Gates RD, Muscatine L (2007) Effects of free amino acids on the photosynthetic carbon metabolism of symbiotic dinoflagellates. Russ J Plant Physiol 54:171–183

Bouchard JN, Yamasaki H (2008) Heat stress stimulates nitric oxide production in Symbiodinium microadriaticum: a possible linkage between nitric oxide and the coral bleaching phenomenon. Plant Cell Physiol 49:641–652

Brandt K (1883) Über die morphologische und physiologische bedeutung des chlorophylls bei. Tieren Mitt Zool Sta Neapol 4:191–302

Brown B, Cossins A (2011) The potential for temperature acclimatisation of reef corals in the face of climate change. In: Dubinsky Z, Stambler N (eds) Coral reefs: an ecosystem in transition, Springer, Dordrecht

Brown BE, Ambarsari I, Warner ME, Fitt WK, Dunne RP, Gibb SW, Cummings DG (1999) Diurnal changes in photochemical efficiency and xanthophyll concentrations in shallow water reef corals: evidence for photoinhibition and photoprotection. Coral Reefs 18:99–105

Brown BE, Downs CA, Dunne RP, Gibb S (2002) Preliminary evidence for tissue retraction as a factor in photoprotection of corals incapable of xanthophyll cycling. J Exp Mar Biol Ecol 277:129–144

Buddemeier RW, Fautin DG (1993) Coral bleaching as an adaptive mechanism – a testable hypothesis. Bioscience 43:320–326

Carlon DB, Lippe C (2008) Fifteen new microsatellite markers for the reef coral *Favia fragum* and a new Symbiodinium microsatellite. Mol Ecol Resour 8:870–873

Chen MC, Hong MC, Huang YS, Liu MC, Cheng YM, Fang LS (2005) ApRab11, a cnidarian homologue of the recycling regulatory protein Rab11, is involved in the establishment and maintenance of the *Aiptasia-Symbiodinium* endosymbiosis. Biochem Biophys Res Commun 338:1607–1616

Clode PL, Saunders M, Maker G, Ludwig M, Atkins CA (2009) Uric acid deposits in symbiotic marine algae. Plant Cell Environ 32:170–177

Coffroth MA, Santos SR (2005) Genetic diversity of symbiotic dinoflagellates in the genus *Symbiodinium*. Protist 156:19–34

Coffroth MA, Lewis CF, Santos SR, Weaver JL (2006) Environmental populations of symbiotic dinoflagellates in the genus *Symbiodinium* can initiate symbioses with reef cnidarians. Curr Biol 16:985–987

Crawley A, Kline DI, Dunn S, Anthony K, Dove S (2010) The effect of ocean acidification on symbiont photorespiration and productivity in Acropora formosa. Glob Change Biol 16:851–863

Davies PS (1984) The role of zooxanthellae in the nutritional energy requirements of *Pocillopora eydouxi*. Coral Reefs 2:181–186

Davies PS (1991) Effect of daylight variations on the energy budgets of shallow-water corals. Mar Biol 108:137–144

Dimond J, Carrington E (2008) Symbiosis regulation in a facultatively symbiotic temperate coral: zooxanthellae division and expulsion. Coral Reefs 27:601–604

Douglas AE (2008) Conflict, cheats and the persistence of symbioses. New Phytol 177:849–858

Dove S (2004) Scleractinian corals with photoprotective host pigments are hypersensitive to thermal bleaching. Mar Ecol Prog Ser 272:99–116

Dove SG, Lovell C, Fine M, Deckenback J, Hoegh-Guldberg O, Iglesias-Prieto R, Anthony KRN (2008) Host pigments: potential facilitators of photosynthesis in coral symbioses. Plant Cell Environ 31:1523–1533

Drew EA (1972) The biology and physiology of alga-invertebrate symbioses. II. The density of symbiotic algal cells in a number of hermatypic hard corals and alcyonarians from various depths. J Exp Mar Biol Ecol 9:71–75

Dubinsky Z, Jokiel PL (1994) Ratio of energy and nutrient fluxes regulates symbiosis between zooxanthellae and corals. Pac Sci 48:313–324

Dubinsky Z, Falkowski P (2011) Light as a source of information and energy in zooxanthellate corals. In: Dubinsky Z and Stambler N (eds) Coral reefs: an ecosystem in transition, Springer, Dordrecht

Dubinsky Z, Matsukawa R, Karube I (1995) Photobiological aspects of algal mass culture. J Mar Biotech 2:61–65

Dubinsky Z, Stambler N, Benzion M, McCloskey LR, Muscatine L, Falkowski PG (1990) The effect of external nutrient resources on the optical-properties and photosynthetic efficiency of *Stylophora pistillata*. Proc R Soc Lond Ser B-Biol Sci 239:231–246

Dunn SR, Thomason JC, Le Tissier MDA, Bythell JC (2004) Heat stress induces different forms of cell death in sea anemones and their endosymbiotic algae depending on temperature and duration. Cell Death Differ 11:1213–1222

Dunn SR, Phillips WS, Spatafora JW, Green DR, Weis VM (2006) Highly conserved caspase and Bcl-2 homologues from the sea anemone *Aiptasia pallida*: lower metazoans as models for the study of apoptosis evolution. J Mol Evol 63:95–107

Dunn SR, Schnitzler CE, Weis VM (2007) Apoptosis and autophagy as mechanisms of dinoflagellate symbiont release during cnidarian bleaching: every which way you lose. Proc R Soc Lond Ser B-Biol Sci 274:3079–3085

Dustan P (1975) Growth and form in reef-building coral *Montastrea annularis*. Mar Biol 33:101–107

Enriquez S, Mendez ER, Iglesias-Prieto R (2005) Multiple scattering on coral skeletons enhances light absorption by symbiotic algae. Limnol Oceanogr 50:1025–1032

Fagoonee I, Wilson HB, Hassell MP, Turner JR (1999) The dynamics of zooxanthellae populations: a long-term study in the field. Science 283:843–845

Falkowski PG, Dubinsky Z (1981) Light-shade adaptation of *Stylophora-pistillata*, a hermatypic coral from the gulf of Eilat. Nature 289:172–174

Falkowski PG, Dubinsky Z, Muscatine L, Porter JW (1984) Light and the bioenergetics of a symbiotic coral. Bioscience 34:705–709

Falkowski PG, Jokiel P, Kinzie RI (1990) Irradiance and corals. In: Dubinsky Z (ed) Coral reefs. Ecosystems of the world, vol 25. Elsevier, Amsterdam, pp 89–107

Falkowski PG, Dubinsky Z, Muscatine L, McCloskey L (1993) Population-control in symbiotic corals. Bioscience 43:606–611

Fang LS, Chen YWJ, Chen CS (1989) Why does the white tip of stony coral grow so fast without zooxanthellae. Mar Biol 103:359–363

Fitt W, Trench R (1983) The relation of diel patterns of cell division to diel patterns of motility in the symbiotic. New Phytol 94:421–432

Fitt WK, Chang SS, Trench RK (1981) Motility pattern of different strains of the symbiotic dinoflagellate *Symbiodinium* (= *Gymnodinium*) *microadriaticum* Freudenthal in culture. Bull Mar Sci 31:436–443

Fitt WK, McFarland FK, Warner ME, Chilcoat GC (2000) Seasonal patterns of tissue biomass and densities of symbiotic dinoflagellates in reef corals and relation to coral bleaching. Limnol Oceanogr 45:677–685

Fitt WK, Brown BE, Warner ME, Dunne RP (2001) Coral bleaching: interpretation of thermal tolerance limits and thermal thresholds in tropical corals. Coral Reefs 20:51–65

Frade PR, Englebert N, Faria J, Visser PM, Bak RPM (2008a) Distribution and photobiology of *Symbiodinium* types in different light environments for three colour morphs of the coral *Madracis pharensis*: is there more to it than total irradiance? Coral Reefs 27:913–925

Frade PR, De Jongh F, Vermeulen F, Van Bleijswijk J, Bak RPM (2008b) Variation in symbiont distribution between closely related coral species over large depth ranges. Mol Ecol 17:691–703

Frade PR, Bongaerts P, Winkelhagen AJS, Tonk L, Bak RPM (2008c) In situ photobiology of corals over large depth ranges: a multivariate analysis on the roles of environment, host, and algal symbiont. Limnol Oceanogr 53:2711–2723

Franklin DJ, Hoegh-Guldberg P, Jones RJ, Berges JA (2004) Cell death and degeneration in the symbiotic dinoflagellates of the coral *Stylophora pistillata* during bleaching. Mar Ecol Prog Ser 272:117–130

Freudenthal HD (1962) *Symbiodinium* gen. nov. and *Symbiodinium microadriaticum* sp. nov., a zooxanthella, taxonomy, life cycle, and morphology. J Protozool 9:45–52

Furla P, Allemand D, Shick JM, Ferrier-Pages C, Richier S (2005) The symbiotic anthozoan: a physiological chimera between alga and animal. Integr Comp Biol 45:595–604

Gates RD, Baghdasarian G, Muscatine L (1992) Temperature stress causes host-cell detachment in symbiotic cnidarians – implications for coral bleaching. Biol Bull 182:324–332

Gates RD, Hoeghguldberg O, McFallngai MJ, Bil KY, Muscatine L (1995) Free amino-acids exhibit anthozoan host factor activity – they induce the release of photosynthate from symbiotic dinoflagellates in-vitro. Proc Natl Acad Sci U S A 92:7430–7434

Genkai-Kato M, Yamamura N (1999) Evolution of mutualistic symbiosis without vertical transmission. Theor Popul Biol 55:309–323

Gladfelter EH, Michel G, Sanfelici A (1989) Metabolic gradients along a branch of the reef coral *Acropora-palmata*. Bull Mar Sci 44:1166–1173

Goiran C, Allemand D, Galgani I (1997) Transient Na+ stress in symbiotic dinoflagellates after isolation from coral-host cells and subsequent immersion in seawater. Mar Biol 129:581–589

Goiran C, AlMoghrabi S, Allemand D, Jaubert J (1996) Inorganic carbon uptake for photosynthesis by the symbiotic coral/dinoflagellate association.1. Photosynthetic performances of symbionts and dependence on sea water bicarbonate. J Exp Mar Biol Ecol 199:207–225

Gou WL, Sun J, Li XQ, Zhen Y, Xin ZY, Yu ZG, Li RX (2003) Phylogenetic analysis of a free-living strain of *Symbiodinium* isolated from Jiaozhou Bay, PR China. J Exp Mar Biol Ecol 296:135–144

Goulet TL (2006) Most corals may not change their symbionts. Mar Ecol Prog Ser 321:1–7

Goulet TL (2007) Most scleractinian corals and octocorals host a single symbiotic zooxanthella clade. Mar Ecol Prog Ser 335:243–248

Grant AJ, Remond M, Withers KJT, Hinde R (2001) Inhibition of algal photosynthesis by a symbiotic coral. Hydrobiologia 461:63–69

Grant AJ, Trautman DA, Frankland S, Hinde R (2003) A symbiosome membrane is not required for the actions of two host signalling compounds regulating photosynthesis in symbiotic algae isolated from cnidarians. Comp Biochem Physiol A Mol Integr Physiol 135:337–345

Grant AJ, Starke-Peterkovic T, Withers KJT, Hinde R (2004) Aposymbiotic *Plesiastrea versipora* continues to produce cell-signalling molecules that regulate the carbon metabolism of symbiotic algae. Comp Biochem Physiol A Mol Integr Physiol 138:253–259

Grant AJ, Remond M, Starke-Peterkovic T, Hinde R (2006) A cell signal from the coral *Plesiastrea versipora* reduces starch synthesis in its symbiotic alga, *Symbiodinium* sp. Comp Biochem Physiol A Mol Integr Physiol 144:458–463

Graus RR, Macintyre IG (1976) Light control of growth form in colonial reef corals – computer-simulation. Science 193:895–897

Grottoli AG, Rodrigues LJ, Palardy JE (2006) Heterotrophic plasticity and resilience in bleached corals. Nature 440:1186–1189

Grover R, Maguer JF, Allemand D, Ferrier-Pages C (2003) Nitrate uptake in the scleractinian coral *Stylophora pistillata*. Limnol Oceanogr 48:2266–2274

Hennige SJ, Smith DJ, Perkins R, Consalvey M, Paterson DM, Suggett DJ (2008) Photoacclimation, growth and distribution of massive coral species in clear and turbid waters. Mar Ecol Prog Ser 369:77–88

Hennige S, Suggett DJ, Warner ME, McDougall KE, Smith DJ (2009) Photobiology of symbiodinium revisited: bio-physical and bio-optical signatures. Coral Reefs 28:179–195

Hill R, Ralph PJ (2007) Post-bleaching viability of expelled zooxanthellae from the scleractinian coral *Pocillopora damicornis*. Mar Ecol Prog Ser 352:137–144

Hill R, Schreiber U, Gademann R, Larkum AWD, Kuhl M, Ralph PJ (2004) Spatial heterogeneity of photosynthesis and the effect of temperature-induced bleaching conditions in three species of corals. Mar Biol 144:633–640

Hirose M, Reimer JD, Hidaka M, Suda S (2008) Phylogenetic analyses of potentially free-living *Symbiodinium* spp. isolated from coral reef sand in Okinawa, Japan. Mar Biol 155:105–112

Hoegh-Guldberg O (2011) The impact of climate change on coral reef ecosystems. In: Dubinsky Z and Stambler N (eds) Coral reefs: an ecosystem in transition, Springer, Dordrecht

Hoegh-Guldberg O, McCloskey LR, Muscatine L (1987) Expulsion of zooxanthellae by symbiotic cnidarians from the red-sea. Coral Reefs 5:201–204

Hoegh-Guldberg O, Jones RJ (1999) Photoinhibition and photoprotection in symbiotic dinoflagellates from reef-building corals. Mar Ecol Prog Ser 183:73–86

Houlbreque F, Ferrier-Pages C (2009) Heterotrophy in tropical scleractinian corals. Biol Rev Camb Philos Soc 84:1–17

Houlbreque F, Tambutte E, Allemand D, Ferrier-Pages C (2004) Interactions between zooplankton feeding, photosynthesis and skeletal growth in the scleractinian coral *Stylophora pistillata*. J Exp Biol 207:1461–1469

Huang HJ, Wang LH, Chen WNU, Fang LS, Chen CS (2008) Developmentally regulated localization of endosymbiotic dinoflagellates in different tissue layers of coral larvae. Coral Reefs 27:365–372

Iglesias-Prieto R, Beltran VH, LaJeunesse TC, Reyes-Bonilla H, Thome PE (2004) Different algal symbionts explain the vertical distribution of dominant reef corals in the eastern Pacific. Proc R Soc Lond Ser B-Biol Sci 271:1757–1763

Jackson AE, Miller DJ, Yellowlees D (1989) Phosphorus-metabolism in the coral zooxanthellae symbiosis – characterization and possible roles of 2 acid-phosphatases in the algal symbiont *Symbiodinium* sp. Proc R Soc Lond Ser B-Biol Sci 238:193–202

Jackson AE, Yellowlees D (1990) Phosphate-uptake by zooxanthellae isolated from corals. Proc R Soc Lond Ser B-Biol Sci 242:201–204

Jones RJ, Yellowlees D (1997) Regulation and control of intracellular algae (equals zooxanthellae) in hard corals. Philos Trans R Soc Lond B 352:457–468

Jones RJ, Hoegh-Guldberg O, Larkum AWD, Schreiber U (1998) Temperature-induced bleaching of corals begins with impairment of the CO_2 fixation mechanism in zooxanthellae. Plant Cell Environ 21:1219–1230

Jones AM, Berkelmans R, van Oppen MJH, Mieog JC, Sinclair W (2008) A community shift in the symbionts of a scleractinian coral following a natural bleaching event: field evidence of acclimatization. Proc R Soc Lond Ser B-Biol Sci 275:1359–1365

Kaniewska P, Anthony KRN, Hoegh-Guldberg O (2008) Variation in colony geometry modulates internal light levels in branching corals, *Acropora humilis* and *Stylophora pistillata*. Mar Biol 155:649–660

Karako S, Stambler N, Dubinsky Z, Seckbach J (2002) The taxonomy and evolution of the zooxanthellae-coral symbiosis. In: Symbiosis: mechanisms and model systems. Kluwer Academic Press, Dordrecht, pp 541–557

Karako-Lampert S, Katcoff DJ, Achituv Y, Dubinsky Z, Stambler N (2004) Do clades of symbiotic dinoflagellates in scleractinian corals of the Gulf of Eilat (Red Sea) differ from those of other coral reefs? J Exp Mar Biol Ecol 311:301–314

Karako-Lampert S, Katcoff DJ, Achituv Y, Dubinsky Z, Stambler N (2005) Physiology changes of Symbiodinium microadriaticum clade B as response to different environmental conditions. J Exp Mar Biol Ecol 318:11–20

Kazandjian A, Shepherd VA, Rodriguez-Lanetty M, Nordemeier W, Larkum AWD, Quinnell RG (2008) Isolation of symbiosomes and the symbiosome membrane complex from the zoanthid *Zoanthus robustus*. Phycologia 47:294–306

Kinzie RA, Takayama M, Santos SR, Coffroth MA (2001) The adaptive bleaching hypothesis: experimental tests of critical assumptions. Biol Bull 200:51–58

Kleppel GS, Dodge RE, Reese CJ (1989) Changes in pigmentation associated with the bleaching of stony corals. Limnol Oceanogr 34:1331–1335

Kuhl M, Cohen Y, Dalsgaard T, Jorgensen BB, Revsbech NP (1995) Microenvironment and photosynthesis of zooxanthellae in scleractinian corals studied with microsensors for O_2, pH and light. Mar Ecol Prog Ser 117:159–172

LaJeunesse TC (2001) Investigating the biodiversity, ecology and phylogeny of endosymbiotic dinoflagellates in the genus *Symbiodinium* using the ITS region: in search of a "species" level marker. J Phycol 37:866–880

LaJeunesse TC (2002) Diversity and community structure of symbiotic dinoflagellates from Caribbean coral reefs. Mar Biol 141:387–400

LaJeunesse TC (2005) "Species" radiations of symbiotic dinoflagellates in the Atlantic and Indo-Pacific since the miocene-pliocene transition. Mol Biol Evol 22:570–581

LaJeunesse TC, Trench RK (2000) Biogeography of two species of *Symbiodinium* (Freudenthal) inhabiting the intertidal sea anemone *Anthopleura elegantissima* (Brandt). Biol Bull 199:126–134

LaJeunesse TC, Loh WKW, van Woesik R, Hoegh-Guldberg O, Schmidt GW, Fitt WK (2003) Low symbiont diversity in southern Great Barrier Reef corals, relative to those of the Caribbean. Limnol Oceanogr 48:2046–2054

LaJeunesse TC, Thornhill DJ, Cox EF, Stanton FG, Fitt WK, Schmidt GW (2004a) High diversity and host specificity observed among symbiotic dinoflagellates in reef coral communities from Hawaii. Coral Reefs 23:596–603

LaJeunesse TC, Bhagooli R, Hidaka M, DeVantier L, Done T, Schmidt GW, Fitt WK, Hoegh-Guldberg O (2004b) Closely related *Symbiodinium* spp. differ in relative dominance in coral reef host communities across environmental, latitudinal and biogeographic gradients. Mar Ecol Prog Ser 284:147–161

LaJeunesse TC, Lambert G, Andersen RA, Coffroth MA, Galbraith DW (2005) *Symbiodinium* (Pyrrhophyta) genome sizes (DNA content) are smallest among dinoflagellates. J Phycol 41:880–886

LaJeunesse TC, Bonilla HR, Warner ME, Wills M, Schmidt GW, Fitt WK (2008) Specificity and stability in high latitude eastern Pacific coral-algal symbioses. Limnol Oceanogr 53:719–727

Lampert-Karako S, Stambler N, Katcoff DJ, Achituv Y, Dubinsky Z, Simon-Blecher N (2008) Effects of depth and eutrophication on the zooxanthellae clades of *Stylophora pistillata* from the Gulf of Eilat (Red Sea). Aquat Conserv Mar Freshwater Ecosyst 18: 1039–1045

Leggat W, Badger MR, Yellowlees D (1999) Evidence for an inorganic carbon-concentrating mechanism in the symbiotic dinoflagellate *Symbiodinium* sp. Plant Physiol 121:1247–1255

Leggat W, Hoegh-Guldberg O, Dove S, Yellowlees D (2007) Analysis of an EST library from the dinoflagellate (*Symbiodinium* sp.) symbiont of reef-building corals. J Phycol 43:1010–1021

Lerch AK, Cook CB (1984) Some effects of photoperiod on the motility rhythm of cultured zooxanthellae. Bull Mar Sci 34:477–483

Lesser MP, Stochaj WR, Tapley DW, Shick JM (1990) Bleaching in coral-reef anthozoans – effects of irradiance, ultraviolet-radiation, and temperature on the activities of protective enzymes against active oxygen. Coral Reefs 8:225–232

Lesser MP (2006) Oxidative stress in marine environments: biochemistry and physiological ecology. Annu Rev Physiol 68:253–278

Lesser M (2011) Coral bleaching: causes and mechanisms. Dubinsky Z and Stambler N (eds) Coral reefs: an ecosystem in transition, Springer, Dordrecht

Levy O, Dubinsky Z, Schneider K, Achituv Y, Zakai D, Gorbunov MY (2004) Diurnal hysteresis in coral photosynthesis. Mar Ecol Prog Ser 268:105–117

Levy O, Achituv Y, Yacobi YZ, Dubinsky Z, Stambler N (2006) Diel 'tuning' of coral metabolism: physiological responses to light cues. J Exp Biol 209:273–283

Little AF, van Oppen MJH, Willis BL (2004) Flexibility in algal endosymbioses shapes growth in reef corals. Science 304:1492–1494

Loeblich AR, Sherley JL (1979) Observations on the theca of the motile phase of free-living and symbiotic isolates of *Zooxanthella microadriatica* (Freudenthal) comb nov. J Mar Bio Assoc UK 59:195–205

Loh WKW, Loi T, Carter D, Hoegh-Guldberg O (2001) Genetic variability of the symbiotic dinoflagellates from the wide ranging coral species Seriatopora hystrix and *Acropora longicyathus* in the Indo-West Pacific. Mar Ecol Prog Ser 222:97–107

Mass T, Einbinder S, Brokovich E, Shashar N, Vago R, Erez J, Dubinsky Z (2007) Photoacclimation of *Stylophora pistillata* to light extremes: metabolism and calcification. Mar Ecol Prog Ser 334:93–102

Mayfield AB, Gates RD (2007) Osmoregulation in anthozoan-dinoflagellate symbiosis. Comp Biochem Physiol A Mol Integr Physiol 147:1–10

Mieog J, van Oppen MJH, Cantin N, Stam WT, Olsen JL (2007) Real-time PCR reveals a high incidence of Symbiodinium clade D at low levels in four scleractinian corals across the Great Barrier Reef: implications for symbiont shuffling. Coral Reefs 26:449–457

Muller-Parker G (1984) Dispersal of zooxanthellae on coral reefs by predators on cnidarians. Biol Bull 167:159–167

Muscatine L (1967) Glycerol excretion by symbiotic algae from corals and *Tridacna* and its control by the host. Science 156:516–519

Muscatine L, Porter JW (1977) Reef corals: mutualistic symbioses adapted to nutrient-poor environments. Bioscience 27:454–460

Muscatine L, McCloskey LR, Marian RE (1981) Estimating the daily contribution of carbon from zooxanthellae to coral animal respiration. Limnol Oceanogr 26:601–611

Muscatine L, Falkowski PG, Porter JW, Dubinsky Z (1984) Fate of photosynthetic fixed carbon in light and shade adapted colonies of the symbiotic coral *Stylophora pistillasta*. Proc R Soc Lond Ser B-Biol Sci 222:181–202

Muscatine L (1990) The role of symbiotic algal in carbon and energy flux in reef corals. In: Dubinsky Z (ed) Coral reefs. Elsevier, Dordrecht, pp 75–87

Muscatine L, Ferrier-Pages C, Blackburn A, Gates RD, Baghdasarian G, Allemand D (1998) Cell specific density of symbiotic dinoflagellates in tropical anthozoans. Coral Reefs 17:329–337

Palardy JE, Rodrigues LJ, Grottoli AG (2008) The importance of zooplankton to the daily metabolic carbon requirements of healthy and bleached corals at two depths. J Exp Mar Biol Ecol 367:180–188

Pettay DT, Lajeunesse TC (2007) Microsatellites from clade B *Symbiodinium* spp. specialized for Caribbean corals in the genus *Madracis*. Mol Ecol Notes 7:1271–1274

Pochon X, Montoya-Burgos JI, Stadelmann BJP (2006) Molecular phylogeny, evolutionary rates, and divergence timing of the symbiotic dinoflagellate genus *Symbiodinium*. Mol Phylogenet Evol 38:20–30

Rahav O, Dubinsky Z, Achituv Y, Falkowski PG (1989) Ammonium metabolism in the zooxanthellate coral, *Stylophora pistillata*. Proc R Soc Lond Ser B-Biol Sci 236:325–337

Ralph PJ, Gademann R, Larkum AWD, Kuhl M (2002) Spatial heterogeneity in active chlorophyll fluorescence and PSII activity of coral tissues. Mar Biol 141:639–646

Ralph PJ, Schreiber U, Gademann R, Kuhl M, Larkum AWD (2005) Coral photobiology studied with a new imaging pulse amplitude modulated fluorometer. J Phycol 41:335–342

Reynolds JM, Bruns BU, Fitt WK, Schmidt GW (2008) Enhanced photoprotection pathways in symbiotic dinoflagellates of shallow-water corals and other cnidarians (vol 105, pg 13674, 2008). Proc Natl Acad Sci U S A 105:17206–17206

Richier S, Furla P, Plantivaux A, Merle PL, Allemand D (2005) Symbiosis-induced adaptation to oxidative stress. J Exp Biol 208:277–285

Richier S, Sabourault C, Courtiade J, Zucchini N, Allemand D, Furla P (2006) Oxidative stress and apoptotic events during thermal stress in the symbiotic sea anemone, *Anemonia viridis*. FEBS J 273:4186–4198

Richier S, Cottalorda JM, Guillaume MMM, Fernandez C, Allemand D, Furla P (2008) Depth-dependant response to light of the reef building coral, *Pocillopora verrucosa*: implication of oxidative stress. J Exp Mar Biol Ecol 357:48–56

Ritchie RJ, Eltringham K, Hinde R (1993) Glycerol uptake by zooxanthellae of the temperate hard coral, *Plesiastrea-versipora* (lamarck). Proc R Soc Lond Ser B-Biol Sci 253:189–195

Rodrigues LJ, Grottoli AG (2006) Calcification rate and the stable carbon, oxygen, and nitrogen isotopes in the skeleton, host tissue, and zooxanthellae of bleached and recovering Hawaiian corals. Geochim Cosmochim Acta 70:2781–2789

Rodriguez-Lanetty M, Loh W, Carter D, Hoegh-Guldberg O (2001) Latitudinal variability in symbiont specificity within the widespread scleractinian coral *Plesiastrea versipora*. Mar Biol 138:1175–1181

Rodriguez-Lanetty M, Wood-Charlson EM, Hollingsworth LL, Krupp DA, Weis VM (2006) Temporal and spatial infection dynamics indicate recognition events in the early hours of a dinoflagellate/coral symbiosis. Mar Biol 149:713–719

Rowan R, Powers DA (1991) Molecular genetic identification of symbiotic dinoflagellates (zooxanthellae). Mar Ecol Prog Ser 71:65–73

Rowan R, Knowlton N (1995) Intraspecific diversity and ecological zonation in coral-algal symbiosis. Proc Natl Acad Sci U S A 92:2850–2853

Rowan R, Whitney SM, Fowler A, Yellowlees D (1996) Rubisco in marine symbiotic dinoflagellates: form II enzymes in eukaryotic oxygenic phototrophs encoded by a nuclear multigene family. Plant Cell 8:539–553

Rowan R, Knowlton N, Baker A, Jara J (1997) Landscape ecology of algal symbionts creates variation in episodes of coral bleaching. Nature 388:265–269

Salih A, Larkum A, Cox G, Kuhl M, Hoegh-Guldberg O (2000) Fluorescent pigments in corals are photoprotective. Nature 408:850–853

Sampayo EM, Dove S, LaJeunesse T (2009) Cohesive molecular genetic data delineate species diversity in the dinoflagellate genus *Symbiodinium*. Mol Ecol 18:500–519

Santos SR, Coffroth MA (2003) Molecular genetic evidence that dinoflagellates belonging to the genus *symbiodinium* freudenthal are haploid. Biol Bull 204:10–20

Santos SR, Shearer TL, Hannes AR, MA C (2004) Fine scale diversity and specificity in the most prevalent lineage of symbiotic dinoflagellates (*Symbiodinium*, Dinophyta) of the Caribbean. Mol Ecol 13:459–469

Savage AM, Trapido-Rosenthal H, Douglas AE (2002) On the functional significance of molecular variation in *Symbiodinium*, the symbiotic algae of Cnidaria: photosynthetic response to irradiance. Mar Ecol Prog Ser 244:27–37

Saxby T, Dennison WC, Hoegh-Guldberg O (2003) Photosynthetic responses of the coral *Montipora digitata* to cold temperature stress. Mar Ecol Prog Ser 248:85–97

Schoenberg DA, Trench RK (1980) Genetic-variation in *Symbiodinium* (=*gymnodinium*) microadriaticum freudenthal, and specificity in its symbiosis with marine-invertebrates.2. Morphological variation in *Symbiodinium microadriaticum*. Proc R Soc Lond Ser B-Biol Sci 207:429–444

Shick JM, Lesser MP, Dunlap WC, Stochaj WR, Chalker BE, Won JW (1995) Depth-dependent responses to solar ultraviolet-radiation and oxidative stress in the zooxanthellate coral *Acropora-microphthalma*. Mar Biol 122:41–51

Shick JM, Dunlap WC (2002) Mycosporine-like amino acids and related gadusols: biosynthesis, accumulation, and UV-protective functions in aquatic organisms. Annu Rev Physiol 64:223–262

Shick JM (2004) The continuity and intensity of ultraviolet irradiation affect the kinetics of biosynthesis, accumulation, and conversion of mycosporine-like amino acids (MAAS) in the coral *Stylophora pistillata*. Limnol Oceanogr 49:442–458

Smith GJ, Muscatine L (1999) Cell cycle of symbiotic dinoflagellates: variation in G(1) phase-duration with anemone nutritional status and macronutrient supply in the *Aiptasia pulchella-Symbiodinium pulchrorum* symbiosis. Mar Biol 134:405–418

Stambler N (1998) Effects of light intensity and ammonium enrichment on the hermatypic coral *Stylophora pistillata* and its zooxanthellae. Symbiosis 24:127–145

Stambler N, Dubinsky Z (2004) Stress effects on metabolism and photosynthesis of hermatypic corals. In: Rosenberg E, Loya Y (eds) Coral health and disease. Springer, Berlin, pp 195–215

Stambler N, Dubinsky Z (2005) Corals as light collectors: an integrating sphere approach. Coral Reefs 24:1–9

Stambler N, Levy O, Vaki L (2008) Physiological response of hermatypic Red Sea corals at distribution depth of 5–75 m. Isr J Plant Sci 56:45–53

Stambler N (2010) Coral symbiosis under stress. In: Seckbach J, Grube M (eds) Symbioses and stress. In cellular origin, life in extreme habitats and astrobiology, Vol 17, Part 3. 197–224, DOI: 10.1007/978-90-481-9449-0_10, Springer, Dordrecht

Stambler N (2011) Marine microralgae/cyanobacteria-invertebrate symbiosis, trading energy for strategic material. In: Dubinsky Z, Seckbach J (eds) All flesh is grass: plant-animal interactions. Cellular Origin, life in extreme habitats and astrobiology, Vol 17, Springer, Dordrecht

Stanley GD, Swart PK (1995) Evolution of the coral zooxanthellae symbiosis during the Ttriassic – a geochemical approach. Paleobiology 21:179–199

Stat M, Morris E, Gates RD (2008a) Functional diversity in coral-dinoflagellate symbiosis. Proc Natl Acad Sci U S A 105:9256–9261

Stat M, Loh WKW, Hoegh-Guldberg O, Carter DA (2008b) Symbiont acquisition strategy drives host-symbiont associations in the southern Great Barrier Reef. Coral Reefs 27:763–772

Stimson J, Kinzie RA (1991) The temporal pattern and rate of release of zooxanthellae from the reef coral *Pocillopora damicornis* (linnaeus) under nitrogen-enrichment and control conditions. J Exp Mar Biol Ecol 153:63–74

Strychar KB, Coates M, Sammarco PW, Piva TJ (2004a) Bleaching as a pathogenic response in scleractinian corals, evidenced by high concentrations of apoptotic and necrotic zooxanthellae. J Exp Mar Biol Ecol 304:99–121

Strychar KB, Sammarco PW, Piva TJ (2004b) Apoptotic and necrotic stages of *Symbiodinium* (Dinophyceae) cell death activity: bleaching of soft and scleractinian corals. Phycologia 43:768–777

Strychar KB, Sammarco PW (2009) Exaptation in corals to high seawater temperatures: low concentrations of apoptotic and necrotic cells in host coral tissue under bleaching conditions. J Exp Mar Biol Ecol 369:31–42

Suggett DJ, Warner ME, Smith DJ, Davey P, Hennige S, Baker NR (2008) Photosynthesis and production of hydrogen peroxide by *Symbiodinium* (Pyrrhophyta) phylotypes with different thermal tolerances. J Phycol 44:948–956

Sutton DC, Hoeghguldberg O (1990) Host-zooxanthella interactions in 4 temperate marine invertebrate symbioses - assessment of effect of host extracts on symbionts. Biol Bull 178:175–186

Taguchi S, Kinzie RA (2001) Growth of zooxanthellae in culture with two nitrogen sources. Mar Biol 138:149–155

Tanaka Y, Miyajima T, Koike I, Hayashibara T, Ogawa H (2006) Translocation and conservation of organic nitrogen within the coral-zooxanthella symbiotic system of *Acropora pulchra*, as demonstrated by dual isotope-labeling techniques. J Exp Mar Biol Ecol 336:110–119

Tchernov D, Gorbunov MY, de Vargas C, Narayan Yadav S, Milligan AJ, Haggblom M, Falkowski PG (2004) Membrane lipids of symbiotic algae are diagnostic of sensitivity to thermal bleaching in corals. Proc Natl Acad Sci U S A 101:13531–13535

Thornhill DJ, Fitt WK, Schmidt GW (2006) Highly stable symbioses among western Atlantic brooding corals. Coral Reefs 25:515–519

Thornhill DJ, Kemp DW, Bruns BU, Fitt WK, Schmidt GW (2008) Correspondence between cold tolerance and temperate biogeography in a western Atlantic *Symbiodinium* (Dinophyta) lineage. J Phycol 44:1126–1135

Titlyanov EA, Titlyanova TV, Leletkin VA, Tsukahara J, vanWoesik R, Yamazato K (1996) Degradation of zooxanthellae and regulation of their density in hermatypic corals. Mar Ecol Prog Ser 139:167–178

Titlyanov EA, Titlyanova TV, Yamazato K, van Woesik R (2001) Photoacclimation dynamics of the coral *Stylophora pistillata* to low and extremely low light. J Exp Mar Biol Ecol 263:211–225

Toller WW, Rowan R, Knowlton N (2001) Repopulation of zooxanthellae in the Caribbean corals *Montastraea annularis* and M-faveolata following experimental and disease-associated bleaching. Biol Bull 201:360–373

Trautman DA, Hinde R, Cole L, Grant A, Quinnell R (2002) Visualisation of the symbiosome membrane surrounding Cnidarian algal cells. Symbiosis 32:133–145

Trench RK (1971) Physiology and biochemistry of zooxanthellae symbiotic with marine coelenterates. 3. Effect of homogenates of host tissues on excretion f photosynthetic products in-vitro by zooxanthellae from two marine coelenterates. Proc R Soc Lond Ser B-Biol Sci 177:251–264

Trench RK (1987) Dinoflagellate in non-parasitic symbiosis. In: F.J.R. T (ed) The biology of dinoflagellate botanical monographs, vol 21, Blackwell Scientific, Oxford, pp 531–570

Trench RK (1993) Microalgal-invertebrate symbioses – a review. Endocytobiosis Cell Res 9:135–175

Trench RK (1997) Diversity of symbiotic dinoflagellate and the evolution of microalgal-invertebrate symbioses. Proc 8th Intl Coral Reef Symp 2:1275–1286

Ulstrup KE, Van Oppen MJH (2003) Geographic and habitat partitioning of genetically distinct zooxanthellae (*Symbiodinium*) in *Acropora* corals on the Great Barrier Reef. Mol Ecol 12:3477–3484

Ulstrup KE, van Oppen MJH, Kuhl M, Ralph PJ (2007) Inter-polyp genetic and physiological characterisation of *Symbiodinium* in an *Acropora valida* colony. Mar Biol 153:225–234

van Oppen MJH, Palstra FP, Piquet AM-T, Miller DJ (2001) Patterns of coral-dinoflagellate associations in *Acropora*: significance of local availability and physiology of *Symbiodinium* strains and host-symbiont selectivity. Proc R Soc Lond Ser B-Biol Sci 268:1759–1767

van Oppen MJH (2004) Mode of zooxanthellae transmission does not affect zooxanthellae diversity in acroporid corals. Mar Biol 144:1–7

Venn AA, Wilson MA, Trapido-Rosenthal HG, Keely BJ, Douglas AE (2006) The impact of coral bleaching on the pigment profile of the symbiotic alga, *Symbiodinium*. Plant Cell Environ 29:2133–2142

Venn AA, Loram JE, Douglas AE (2008) Photosynthetic symbioses in animals. J Exp Bot 59:1069–1080

Wakefield TS, Farmer MA, Kempf SC (2000) Revised description of the fine structure of in situ "Zooxanthellae" genus *Symbiodinium*. Biol Bull 199:76–84

Wakefield TS, Kempf SC (2001) Development of host- and symbiont-specific monoclonal antibodies and confirmation of the origin of the symbiosome membrane in a cnidarian-dinoflagellate symbiosis. Biol Bull 200:127–143

Wang JT, Douglas AE (1997) Nutrients, signals, and photosynthate release by symbiotic algae – The impact of taurine on the dinoflagellate alga *Symbiodinium* from the sea anemone *Aiptasia pulchella*. Plant Physiol 114:631–636

Wang LH, Liu YH, Ju YM, Hsiao YY, Fang LS, Chen CS (2008) Cell cycle propagation is driven by light-dark stimulation in a cultured symbiotic dinoflagellate isolated from corals. Coral Reefs 27:823–835

Warner ME, Berry-Lowe S (2006) Differential xanthophyll cycling and photochemical activity in symbiotic dinoflagellates in multiple locations of three species of Caribbean coral. J Exp Mar Biol Ecol 339:86–95

Warner ME, Fitt WK, Schmidt GW (1999) Damage to photosystem II in symbiotic dinoflagellates: a determinant of coral bleaching. Proc Natl Acad Sci U S A 96:8007–8012

Weis VM, Reynolds WS, deBoer MD, Krupp DA (2001) Host-symbiont specificity during onset of symbiosis between the dinoflagellates *Symbodinium* spp and planula larvae of the scleractinian coral *Fungia scutaria*. Coral Reefs 20:301–308

Weis VM (2008) Cellular mechanisms of Cnidarian bleaching: stress causes the collapse of symbiosis. J Exp Biol 211:3059–3066

Wilkerson F, Kobayashi D, Muscatine L (1988) Mitotic index and size of symbiotic algae in Caribbean reef corals. Coral Reefs 7:29–36

Wood R (1998) The ecological evolution of reefs. Annu Rev Ecol Evol Syst 29:179–206

Wood R (1999) Reef evolution. Oxford University Press, Oxford

Wood-Charlson EM, Hollingsworth LL, Krupp DA, Weis VM (2006) Lectin/glycan interactions play a role in recognition in a coral/dinoflagellate symbiosis. Cell Microbiol 8:1985–1993

Yacobovitch T, Benayahu Y, Weis VM (2004) Motility of zooxanthellae isolated from the Red Sea soft coral *Heteroxenia fuscescens* (Cnidaria). J Exp Mar Biol Ecol 298:35–48

Yellowlees D, Rees TAV, Leggat W (2008) Metabolic interactions between algal symbionts and invertebrate hosts. Plant Cell Environ 31:679–694

Light as a Source of Information and Energy in Zooxanthellate Corals

Zvy Dubinsky and Paul Falkowski

Keywords Underwater light • photoacclimation • spawning • lunar cycle • colony morphology • photoprotection

1 Introduction

Reef-building corals live in a mutualistic symbiosis with endocellular microalgae, the zooxanthellae that provide much of their animal host's metabolic energy needs through translocation of photosynthate. That dependence sets a limit on the depth distribution of zooxanthellate corals, commonly restricting it to depths where light exceeds 0.5% of its subsurface intensity. We review the various ways allowing corals to span over two orders of magnitude in the underwater light field over their bathymetric distribution range. These consist of adjustments by both components of the holobiont, the coral host, and its algal symbionts, and also affect the flows of energy and matter between them. Animal and algae optimize light harvesting and utilization on the molecular, biochemical, biophysical, metabolic, behavioral, and architectural levels. These photoacclimative responses result in corals exposed to the full blast of tropical sun being able to use such superabundant energy, while being spared from its destructive consequences. At the opposite extreme, corals under dim light make the most efficient and parsimonious use of it. We also bring up to date the ways in which moonlight triggers coral spawning cycles and also illustrate the synergy between visible light and UV in the photoacclimation process.

Z. Dubinsky(✉)
The Mina & Everard Goodman Faculty of Life Sciences,
Bar-Ilan University, Ramat-Gan, 52900, Israel
e-mail: dubinz@mail.biu.ac.il

P. Falkowski
Environmental Biophysics and Molecular Ecology Program,
Institute of Marine and Coastal Sciences
and Department of Earth and Planetary Sciences,
Rutgers University, New Brunswick, NJ 08901, USA
e-mail: falko@imcs.rutgers.edu

In addition, light sets internal clocks that control the synchronization of metabolism and tentacular activity with the diel cycles and provides cues for the timing of spawning activities.

2 The Underwater Light Field to Which Corals are Exposed

Paradoxically, light is one of the most predictable, yet stochastic, environmental variables. From knowledge of latitude alone, one can accurately predict the path of the Sun as it crosses the sky and thus number of hours of solar radiation on any day of the year at any place on this planet. This can be calculated over geological time from knowledge of the precession, obliquity, and eccentricity of the Earth's orbit. Indeed, day length is such a strong and predictable environmental signal that it has led to the evolution of endogenous "clock" genes across the tree of life – from cyanobacteria and eukaryotic algae to metazoans. Similarly, the predictability of the lunar cycle has been such a strong signal that it has become incorporated into several aspects of biological response functions, especially in reproduction; this queue is embedded in gamete release in cnidarians (especially corals) as well as humans. Thus, to a first order, light is a source of information for virtually all organisms on the planetary surface.

Solar radiation is also a source of energy. It is used by all photosynthetic organisms to drive redox reactions leading to the formation of organic matter (Falkowski and Raven 2007). This process, which is highly conserved in all oxygenic photosynthetic organisms, is critically dependent on spectral irradiance of light – that is the flux (intensity) and energy distribution ("color") of the photons incident on the surface of the organism. Spectral irradiance is far less predictable than day length. Solar radiation is attenuated in the atmosphere by aerosols and clouds, both of which vary in space and time. In the sea, the light field is further altered by water itself (which absorbs red and infrared radiation), dissolved and particulate organic matter, and local scattering conditions. Hence, it is not surprising that benthic organisms that depend on light for an energy source have developed a variety of strategies to optimize its acquisition and utilization.

In this chapter, we discuss light and its role in the photobiology of zooxanthellate corals in the context of the basic two functions: information and energy.

3 Light as an Informational Signal in Corals

Light perception by cnidarians is mediated by a set of chromophores, the cryptochromes, located in the ectoderm. These highly conserved proteins, found in bacteria and all eukaryotes, but not archea, bind FADH and absorb blue light. In corals, two genes, cry1 and cry2, encode for cryptochromes in the animal host, and these genes are used to sense lunar radiation. The lunar radiation effect is transduced throughout the animal and is manifested in enhanced feeding behavior (e.g., extended tentacles) (Gorbunov and Falkowski 2002), as well as a correlation with the reproductive cycle (Levy et al. 2007). The biophysical mechanism by which cryptochromes operate is not fully understood; however, the molecules undergo activation via phosphorylation (Lin and Shalitin 2003).

The perception of light is critical to spawning; indeed, one of the strongest cues for spawning in corals is the lunar cycle (reviewed in Harrison and Wallace (1990), Richmond and Hunter (1990), Richmond (1997)). Nevertheless, differences in environmental conditions between geographic regions result in variation in gamete release in the same species (Babcock et al. 1994; Harrison and Wallace 1990; Loya 2004; Oliver et al. 1988; Richmond 1997; Richmond and Hunter 1990). The degree of synchronization of spawning in pocilloporid corals varies considerably between regions and localities (reviewed in Tanner (1996)), the relative importance of lunar versus seasonal controls has not yet been clearly determined (Fautin 2002; Harrison and Wallace 1990; Shlesinger et al. 1998). The Red Sea common pocilloporid coral *Stylophora pistillata* has been the focus of extensive study, which also included its patterns of reproduction behavior (Rinkevich and Loya 1979). While several species of corals in the northern Red Sea spawn only in the course of a few summer nights, *S. pistillata* colonies do release planulae during most of the year (Shlesinger and Loya 1985; Zakai et al. 2006), on nights coinciding weakly with lunar phases (Figs. 1 and 2).

Although cryptochromes are clearly involved in linking coral reproductive cycles, spawning, and planulation to lunar phases, the signal transduction pathway is still not resolved. Of the environmental factors affecting coral reproductive patterns, the cycles of temperature, and day length (Harrison and Wallace 1990), are not related to that of the moon, and therefore are not involved in triggering the monthly larval release or spawning of corals. Hence, we

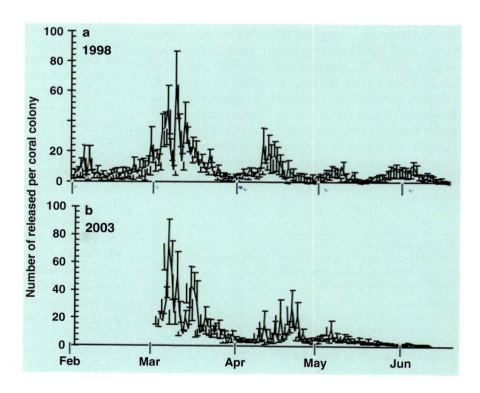

Fig. 1 Lunar periodicity of planulae release by 20 *Stylophora pistillata* colonies maintained in the laboratory at Eilat, northern Red Sea. Numbers in bold are axis labels representing the total number of planulae released each lunar night. Numbers around the circumference indicate lunar night (night one = full moon). Data were pooled from three lunar cycles during each of 2 years: (**a**) 1998 (**b**) 2003 (After Zakai et al. 2006)

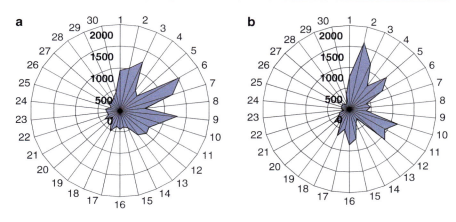

Fig. 2 *Stylophora pistillata*. Numbers of planulae released by coral colonies (*N*=20) under laboratory conditions at Eilat, northern Red Sea, over four complete lunar cycles in 1998 (February 9–June 24, 136 nights), and three complete lunar cycles in 2003 (March 17–June 22, 98 nights). Between laboratory periods, the corals were maintained on an underwater table at 10-m depth on the coral reef. FM=full moon. (**a**) 1998 (**b**) 2003 (After Zakai et al. 2006)

are left with those events whose patterns are controlled by – or covary with lunar phases, night irradiance, and tides. Night irradiance does control planulation cycles in *Pocillopora damicornis* (Jokiel 1985; Jokiel et al. 1985), and tidal cycles may also link lunar periodicity to planulae release in some cases (reviewed in Harrison and Wallace (1990)). It is plausible that the seeming lunar control of planulae release may actually be the phase shifted result of the lunar cycling of gametogenesis and fertilization and do not affect planulation (Fan et al. 2002; McGuire 1998; Szmant-Froelich et al. 1985). One is tempted to speculate that the more precise and narrow lunar timing of gamete release than that of planulation results from lunar triggering of gamete maturation, with planulae development and retention of mature planulae introducing some scatter in the timing of their release (Tanner 1996). The relation to maximal tidal range may have to do with these improving the broadcasting of propagules (Babcock et al. 1986, 1994; Oliver et al. 1988).

4 Fluorescent Proteins

One of the more curious sets of molecules in cnidarians are endogenous autofluorescent proteins, of which the green fluorescent protein (GFP) is the best known (Pieribone and Gruber 2005). These highly conserved molecules contain 238 amino acids that comprise 11 beta-sheets and fold to form a "can" with three amino acids, serine, glycine, and tyrosine, forming a posttranslationally modified fluorescent chromophore. The chromophore is excited in the near UV (395 nm) or blue (475 nm) and emits at 505–515 nm. Several natural variants of GFP are found, including proteins that fluoresce yellow and red, but not blue (Matz et al. 1999). The role of GFPs in corals is not clear. In other cnidarians, such as the jelly fish, *Aequoria victoria*, they are energetically coupled to the calcium-dependent luminescent protein aequorin, and are actively luminescent; that is

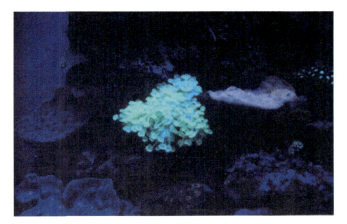

Fig. 3 Fluorescence in corals under UV light

the animal "glows." In corals however, the fluorescent proteins are not luminescent and their presence is only detectable by exposing the organism to an excitation light; the coloration is not often obvious under solar radiation (Fig. 3).

There is sometimes a general trend of decreasing fluorescent protein abundance with depth, leading to speculation that these molecules are maintained as a photoprotective agent or as a quencher of reactive oxygen species (Bou-Abdallah et al. 2006). The colored fluorescent proteins are not energetically coupled to the photosynthetic apparatus and play no role in excitation harvesting for energy; if such coupling were present, the excitation energy would be apparent in the red fluorescent bands of chlorophyll rather than the autofluorescent bands of the chromorphore.

5 Light as an Energy Source

Reef-building or hermatypic (sensu Schumacher and Zibrowius (1985)) corals, due to their dependence on the photosynthesis of the zooxanthellae (Brandt 1883), are mostly limited to the euphotic zone, which is operationally defined as the depth where light exceeds 1% of its subsurface intensity (Achituv and Dubinsky 1990; Stoddart 1969). Deeper than that, corals, including symbiotic species such as *Leptoseris fragilis* may still thrive (Fricke et al. 1987), however without developing into a reef framework.

Over that range (and below it), light decreases in near accordance with an exponential equation, while at the same time its spectral distribution also changes as the blue to red spectral domains ratio increases faster than the light's total (Bou-Abdallah et al. 2006) attenuation. Thus, the most vigorous development of coral reefs is limited to the upper 100 m, in clear transparent water, decreasing in turbid regions. Most coral reef-dominated shores are found in tropical oligotrophic "blue deserts" where the properties of the underwater light field are determined by the optics of water itself, with the blue (450 nm) domain of sunlight in the most penetrating one (Fig. 4).

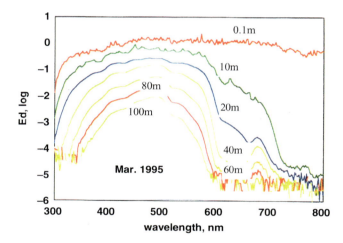

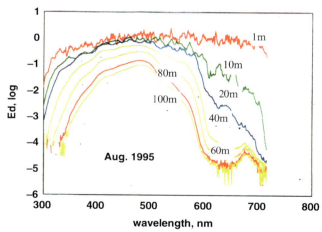

Fig. 4 Spectral distribution of underwater light in the Gulf of Elat (Aqaba) during high (March) and minimal chlorophyll concentrations (Iluz David, personal communication). The area is characterized by highly diverse coral reefs

Such waters differ significantly from phytoplankton-rich ones like high nutrient upwelling zones, where green (560 nm) rather than blue is the least absorbed light. In general, waters surrounding coral reefs fit the "type 1" category of Jerlov (1951, 1964, 1976). Due to the shortage of the main nutrients nitrogen and phosphorus in equatorial seas, these are termed "blue deserts," a name referring to the paucity of phytoplankton. Developed coral reefs are mostly limited to shallow coastal zones in areas above the 18° winter isotherm, between 30°N and 30°S. Due to their dependence on the photosynthesis of symbiotic algae, reefs are limited to the photic zone (Achituv and Dubinsky 1990; Falkowski et al. 1990; Stoddart 1969). UV is attenuated along with the visible spectrum with >1% of it being limited to the upper 30 m in clear waters, but significantly absorbed by the various dissolved organic materials constituting the "Gelbstoff" or gilvin (Kirk 1994).

6 The Zooxanthellae–Coral Association

Brandt (1883) baptized as "zooxanthellae" the endocellular, mostly dinoflagellate microalgae harbored in scores of aquatic invertebrates (Fig. 5).

It is the mutualistic symbiosis with zooxanthellae that forms the basis for the success of corals in space and time. The algal symbionts provide animal host with up to 95% of their energy requirements, as energy-rich products of their photosynthesis are "translocated" to the animal host (Falkowski et al. 1984; Muscatine and Porter 1977). The fact that *in hospice* algal densities reach $0.5-2.0 \times 10^6$ cells cm^{-2} (Fig. 6) while adjacent oligotrophic "blue desert" waters surrounding reefs are nearly devoid of phytoplankton indicates the advantages obtained by the zooxanthellae from their symbiosis associates. First among these are nitrogen and phosphorus, waste products of prey digestion by the animal host, and CO_2 from its respiration. The nutrients allow zooxanthellae to multiply at the rates at which they match Redfield ratios to the photosynthetic carbon flux (Falkowski et al. 1993). With the increasing availability of molecular biology's arsenal of tools, it has become evident that the traditionally adopted functional group, the zooxanthellae actually serves as an umbrella for some 70 diverse taxons and "clades" differing in their geographic, bathymetric, and thermal distributions, and in their host specificity. For a detailed discussion of the zooxanthellae, their biology, and ecology, see Stambler (this volume), (Karako et al. 2002; Baker 2003; Karako-Lampert et al. 2005; Lampert-Karako et al. 2008).

Fig. 5 Brandt's (1883) original drawings of zooxanthellae from different aquatic organisms

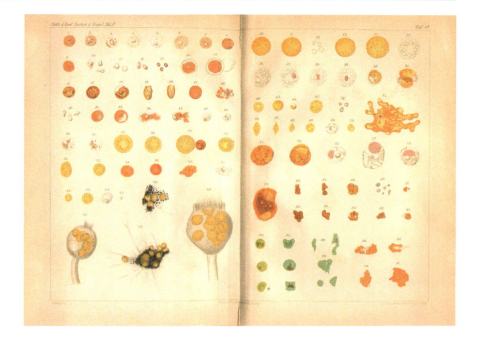

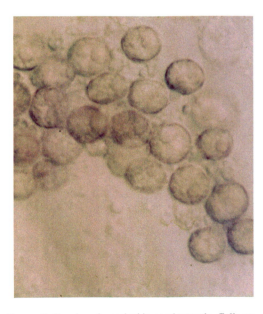

Fig. 6 Zooxanthellae densely packed in coral tentacle. Cells are ~10 μm diameter

7 Photoacclimation of the Zooxanthellae

Since zooxanthellate corals cover an irradiance range of over two orders of magnitude, over the typical bathymetric distribution of reefs, they provide a unique opportunity to examine the photoacclimative mechanisms of the algae and of their coral host. These also affect the energy and nutrient fluxes between the two partners optimizing the interactions among light harvesting, growth, and nutrition of the holobiont.

Fig. 7 Shade and high-light acclimated colonies of the Red Sea coral *Stylophora pistillata*

The zooxanthellae-harboring tissue of shallow-water corals is nearly transparent, with the white color of the skeleton showing through it. These differ dramatically in their pigmentation from coral colonies living under low light (Fig. 7). These may look as near perfect total light absorbers, harvesting ~100% of impinging light against <10% absorbed by their high-light, shallow-water conspecifics (Dubinsky et al. 1984; Stambler and Dubinsky 2005).

The difference in areal chlorophyll concentration between extremes of low-/high-light photoacclimated colonies may be as high as fivefold. Not only do chlorophyll a concentrations respond to ambient light, but so do the other main light-harvesting pigments, chlorophyll c and the

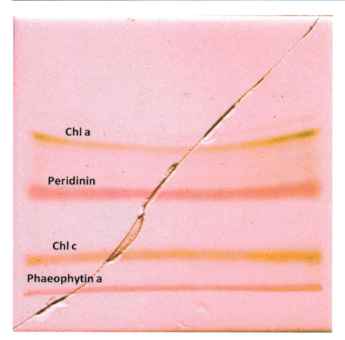

Fig. 8 Thin-layer chromatogram of zooxanthellae pigments. Note the bands of the light-harvesting carotenoid peridinin and of chlorophyll c, diagnostic of the affiliation of the zooxanthellae to the chromophyta (dinoflagellates, Pyrrhophyta)

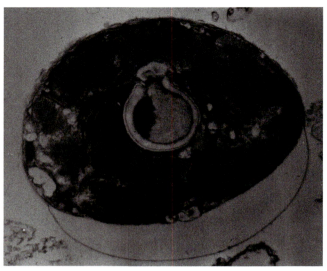

Fig. 9 Electron micrograph of a zooxanthella from a low-light colony of *Stylophora pistillata*. The cell is packed with thylakoids and no β carotene globules are seen. The central body is the pyrenoid

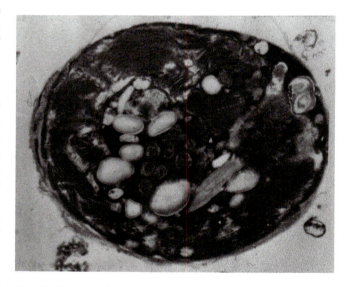

Fig. 10 Electron micrograph of a zooxanthella from a shallow-water, high-light colony of *Stylophora pistillata*. Note the few thylakoid cross sections and the globules of the photoprotective carotenoid β carotene

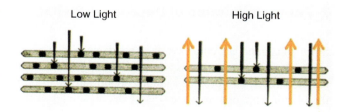

Fig. 11 Schematic representation of the photoacclimative modification of light harvesting. In the high-light acclimated coral, most light is reflected back from the skeleton

carotenoid peridinin, which allow the absorption of the green light not absorbed by the chlorophylls, and the subsequent energy transfer to the reaction centers of photosynthesis. The pigment assortment of zooxanthellae, the chlorophylls a and c, and the carotenoid peridinin reveals their taxonomic position as belonging to the Dinoflagellates in the Chromophyta (Fig. 8).

The up to fivefold increases in pigmentation of the zooxanthellae results in shade-adapted corals becoming near 100% absorbers. This increase in light harvesting is combined with a similar increase in the photosynthetic quantum yield of the zooxanthellae, reducing the ~200-fold in irradiance over the bathymetric range of corals such as *S. pistillata* to as little as four to fivefold difference in photosynthesis. Nevertheless, high-light corals of this species derive sufficient products of symbiont photosynthesis to provide up to 95% of their metabolic needs, whereas at the opposite end of their distribution, corals have to supplement the scant photosynthate produced under dim light, with energy obtained by predation on zooplankton (Dubinsky 2009; Dubinsky and Jokiel 1994; Falkowski et al. 1985).

Since the photosynthetic pigments are embedded in the thylakoid membranes, their area increases with low-light acclimation (Figs. 9–11). The changes in absorptivity result mainly from the photoacclimation of the zooxanthellae

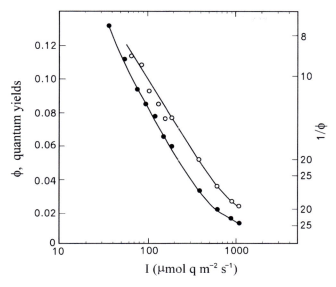

Fig. 12 The quantum yields of high- and low-light acclimated zooxanthellae exposed to different irradiances. Low-light full circles, high-light open circles, their ratio triangles. Hatched area is the ambient light from which the colonies were collected (After Dubinsky et al. 1984)

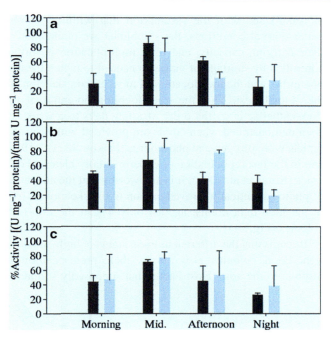

Fig. 13 Diel cycle of superoxide dismutase (SOD) enzyme activities from animal tissue and symbiotic zooxanthellae collected at the 5-m depth in Eilat, Red Sea. Measurements were performed between March 3 and 5, 2002. Activity is presented as % activity [(U mg^{-1} protein)/(max U mg^{-1} protein)] for (**a**) *Favia favus* ($N=3$), (**b**) *Goniopora lobata* ($N=2$), and (**c**) *Plerogyra sinuosa* ($N=3$). The *black bars* represent SOD activity of the animal tissue and the *gray bars* represent SOD activity of the zooxanthellae. Values are represented as means ± s.d. (night = 05.00 h; morning = 08.00 h; mid. = 10.00–14.00 h; afternoon = 16.00 h) (After Levy et al. 2006b)

manifested in their pigmentation, while their numbers remain in many cases nearly constant (Falkowski and Dubinsky 1981). High- to low-light transplantation experiments attempted during the 1981 study produced surprising results. While covering shallow-water colonies with neutral density plastic nets resulted in a fast increase in the pigmentation of the zooxanthellae and their photoacclimation was complete within less than 1 week, the transplantation of shade colonies to highlight failed and the colonies died. Only during the study of Cohen (2008), it was realized that the transferral of low-light colonies was possible only if undertaken in a stepwise fashion with 2–4 weeks at decreasing depths. This procedure succeeded with 100% survival.

Dubinsky et al. (1984), studying the photoacclimation of *Stylophora pistillata* reported that low-light corals absorb up to five times more of impinging light than their ivory-colored shallow-water conspecifics, findings also confirmed by Stambler and Dubinsky (2005). These low-light corals also utilize the absorbed light energy far more efficiently, as their photosynthetic quantum yields approach the theoretical limit of as little as only eight photons used in the evolution of one oxygen molecule (Fig. 12). The response of both zooxanthellae and host to the diel irradiance cycle include closely parallel fluctuations in the concentration of the enzymes superoxidismutase, catalase, and peroxidase (Fig. 13). All these constitute cellular defenses against high-light-generated free radicals stemming from the photosynthetic splitting of water (Levy et al. 2006a, b).

8 Energy and Nutrient Fluxes

The response of corals to light is also related to their nutrient status. In shallow waters and high light, the high-light-driven carbon influx cannot be matched by Redfield ratios of nitrogen and phosphorus in the oligotrophic waters surrounding reefs; thus, high-light corals depend on predation for additional nutrient supply. This situation is quite opposite to that in which deep water/low-light coral live. There, the photosynthate translocated from the zooxanthellae does not suffice to provide for the metabolic needs of the coral host, while the scant nutrients in the water keep the holobiont supplied with nutrient supply rates matching the limited carbon influx. Such "shade corals" depend on zooplankton capture for energy, rather than nutrients, unlike the shallow-water, high-light colonies (Dubinsky and Jokiel 1994; Falkowski et al. 1984; Titlyanov et al. 2000). A few salient points are worth pointing out. The in situ *in hospice* doubling rates of the zooxanthellae are very slow, ranging between 70 days for the high-light *S. pistillata* to 100 in the low-light ones. However, this slow rate suffices to maintain their densities in pace with the colony's growth. This growth rate seems to be in accordance with the assumption that within a colony under normal conditions, the symbionts are maintained at their carrying capacity, and their multiplication is further controlled by shortage of nutrients needed to match the carbon influx and the translocation of most carbon skeletons to the animal host. That the true doubling potential of the Zooxanthellae is on the order of a few days to a week has been demonstrated when corals are provided with seawater replete with nitrogen and phosphorus (Falkowski et al. 1993) and in the fast repopulation of experimentally bleached colonies (Koren et al. 2008). It is noteworthy that the allocation of photosynthetically derived carbon to the skeleton is 16% in the high-light corals, against only 4% in the dim-light colonies (Fig. 14).

It seems that the different nutrient status of high- and low-light coral colonies also affects their predisposition to respond to the hosts Host Factor that supposedly stimulates photosynthate leakage from the zooxanthellae (Eden et al. 1996). Freshly isolated zooxanthellae are exposed for a few hours in the light to ^{14}C-labeled sodium bicarbonate, and then, the fraction of label excreted in the dark is determined. The percentage of labeled compounds excreted when the algae are incubated in the dark in host tissue homogenate is compared to that excreted in the seawater control (Fig. 15). The nutrient-limited, high-light corals readily respond to host factor stimulation and excreted as much as 250% against seawater as 100%. Interestingly, no such response is seen in low-light corals (Fig. 16). We interpret these results to be due to the relatively useless carbon acquired under high light that under nitrogen and phosphorus limitation cannot be utilized by the zooxanthellae for cell doubling. In the shade, each carbon atom can be used with the corresponding nitrogen and phosphorus ratios for new cell building (Dubinsky and Berman-Frank 2001).

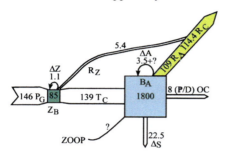 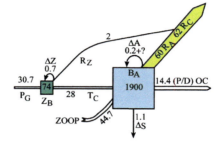

Fig. 14 Scaled models of carbon flow in mg C cm^{-2}day^{-1} of *S. pistillata* from high- (*left*) and low- (*right*) light acclimated colonies. In both cases, most of the photosynthetically assimilated carbon is translocated to the animal host; however, while it suffices to provide for all the metabolic needs in the high-light colonies, low-light ones have to supplement this by considerable predation (After Falkowski et al. 1984)

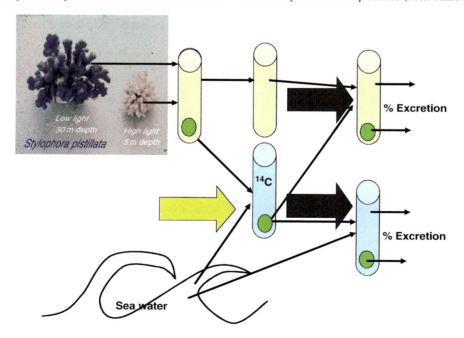

Fig. 15 Experimental procedure for the determination of Host Factor effect on photosynthate excretion by zooxanthellae. Zooxanthellae are separated from coral tissue homogenate, ^{14}C-labeled in the light and then transferred to host homogenate or seawater control. After dark incubation, the excreted fraction is determined using a liquid scintillation counter (After Eden et al. 1996)

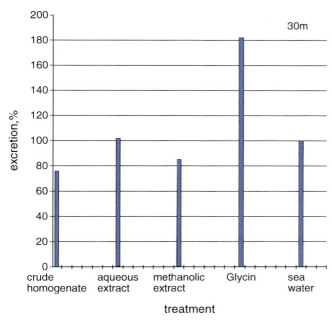

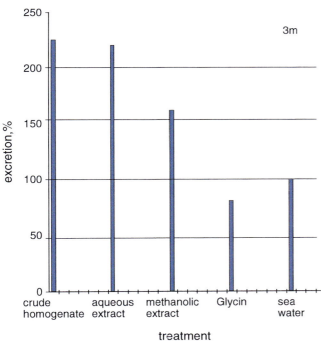

Fig. 16 Host factor activity on zooxanthellae from 3 m and 30 m deep *S. pistillata* colonies. The high-light, nutrient-limited algae incubated in host homogenate excreted 230% compared to 100% for seawater-incubated controls. The low-light, nutrient-replete ones did not respond at all (After Eden et al. 1996)

9 Colony Architecture

The optical, biophysical, and physiological acclimation processes of host and symbionts are accompanied by major changes in colony architecture, resulting in optimization of light

Fig. 17 Colony morphology of *Montastrea verrucosa* from Hawaii: *top*, high-light, shallow-water profusely branched, *bottom*, dim-light, horizontal, planar colony

harvesting and utilization over the two orders of magnitude range in irradiance levels over the common bathymetric range of coral reefs. In general, under high light, the tissue area over the projected "footprint" of the coral is maximized. The shadow cast by a colony is the amount of light available to the zooxanthellae for photosynthesis, thus, under high light that energy is far above optimal, and actually, might be lethal. Massive corals approach spherical or "bumpy" surfaces, increasing the surface exposed to sunlight by 4–5 times that of its projected area ($4\pi r^2/\pi r^2$). The same species, under low light exhibit a flat, surface-encrusting morphology, where the tissue area approaches the colonies "footprint." In branching species, two different strategies were observed. In species such as the Red Sea *Stylophora pistillata*, colonies growing on the reef table are profusely branched and of subspherical shape, with tissue areas up to ×20 the area of the circular shadow cast by them. In shaded crevices and at ~70 m, the depth limit of this species' distribution range, colonies branch in one plane, forming a horizontal fan-shaped structure with tissue area reduced to ×2 of its shadow. In the Hawaiian *Montipora verrucosa*, even in the same colony, one may see the upper parts profusely branched, while parts in the shade of the upper canopy spread out in a continuous planar sheet (Fig. 17). By these architectural adjustments, the zooxanthellae are exposed to the full amount of whatever little light reaching colonies under dim light, while "diluting" the intense solar light reaching near-surface reefs by "spreading" it over the largest tissue area that can be supported by the resulting products of symbiont photosynthesis (Stambler and Dubinsky 2005). The architectural changes in the branching *S. pistillata* have been quantified as the change in diameter versus height as colonies range between high-light subspherical to horizontally spread fans (Fig. 18). Concomitantly with the above changes in colony architecture and orientation, branches in the low-light colonies tend to be thinner and flattened rather than cylindrical when compared to those of high-light morphs (Fig. 19). The morphology changes are restricted by the slow growth rates of corals, but when corals are transplanted experimentally from high to low light or vice versa, the "new" growth is evident within weeks (Einbinder et al. 2009; Mass et al. 2007)

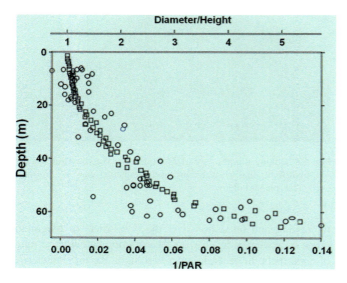

Fig. 18 Radius/height ratios in *S. pistillata* colonies over a 70-m depth light gradient. Colony changed progressively from subspherical to planar (After Mass et al. 2007)

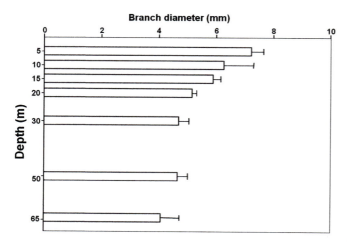

Fig. 19 Branch diameter in *S. pistillata* colonies is reduced with depth from ~8 to ~4 mm (After Mass et al. 2007)

It is noteworthy that UV acts synergistically with light by accelerating photoacclimation of corals to high irradiance (Cohen 2008). When corals were transferred in measured steps from 30 to 3 m, colonies exposed to both UV and light showed stronger and faster photoacclimation attributes than such that were exposed to attenuated UV. Interestingly, it was reported that sheltering from UV by compounds leached from seagrass, offered corals protection from bleaching in a shallow lagoon in the Seychelles (Iluz et al. 2008).

10 Conclusions

1. Zooxanthellate corals do thrive over bathymetric ranges where irradiance levels span as much as two orders of magnitude.
2. The response to the ambient light involves photoacclimative adjustments in the light harvesting and utilization of the symbionts, and metabolic changes in both host and symbiont.
3. Under high light, host and symbionts mobilize free-radical quenching mechanisms and compounds to mitigate photodynamic damage.
4. Major changes in colony architecture result in maximizing light available for symbiont photosynthesis under dim light and decreasing tissue area to light-harvesting ratios. The opposite takes place under high light as colony surface area is increased either by profuse branching or "bumpiness," thereby "diluting" the light reaching the zooxanthellae to optimal levels.
5. Low-light colonies have to predate on zooplankton as a source of energy to supplement that translocated from the light-limited algae. High-light growing corals, supplied with an abundant flux of carbon, have to rely on prey capture to obtain nitrogen and phosphorus.
6. The effect of Host Factor in the enhancement of translocation of photosynthate from symbionts to the coral host depends on the light intensity under which they grew. High-light, nutrient-limited shallow-water colonies readily respond to Host Factor stimulation while low-light, nutrient-sufficient ones do not respond to it.
7. Photoacclimation to high light of corals transferred from deep to shallow water is synergistically enhanced by UV radiation.

References

Achituv Y, Dubinsky Z (1990) Evolution and zoogeography of coral reefs ecosystems of the world. In: Dubinsky Z (ed) Coral reefs. Elsevier, Amsterdam

Babcock RC, Bull GD, Harrison PL, Heyward AJ, Oliver JK, Wallace CC, Willis BL (1986) Synchronous spawning of 105 sclereactinian coral species on the Great Barrier Reef. Mar Biol 90:379–394

Babcock RC, Wills BL, Simpson CJ (1994) Mass spawning of corals on a high latitude coral reef. Coral Reefs 13:161–169

Baker AC (2003) Flexibility and specificity in coral-algal symbiosis: diversity, ecology, and biogeography of Symbiodinium. Annu Rev Ecol Evol Syst 34:661–689

Bou-Abdallah F, Chasteen ND, Lesser MP (2006) Quenching of superoxide radicals by green fluorescent protein. Biochim Biophys Acta 1760:1690–1695

Brandt J (1883) Über die morphologische und physiologische bedeutung des chlorophylls bei tieren. Zool Stn Neapol 4:191–302

Cohen I (2008) Strategies of the coral Stylophora pistillata to acclimate to a "new depth". MSc thesis, Bar-Ilan University, Ramat-Gan, Israel

Dubinsky Z (2009) The light from the darkness, responses of zooxanthellate corals to the underwater light field. Galaxea 11:75–79

Dubinsky Z, Berman-Frank I (2001) Primary production and population growth in photosynthesizing organisms in aquatic ecosystems. Aquat Sci 63:4–17

Dubinsky Z, Jokiel P (1994) The ratio of energy and nutrient fluxes regulates the symbiosis between zooxanthellae and corals. Pac Sci 48:313–324

Dubinsky Z, Falkowski PG, Porter J, Muscatine L (1984) The absorption and utilization of radiant energy by light and shade-adapted colonies of the hermatypic coral, Stylophora pistillata. Proc R Soc Lond B 222:203–214

Eden N, Fomina I, Bil K, Titlyanov E, Dubinsky Z (1996) Photosynthetic capacity and composition of ^{14}C fixation products in symbiotic zooxanthellae of Stylophora pistillata in vivo under different light and nutrient conditions. In: Steinberger Y (ed) Preservation of our world in the wake of change, vol 6A/B. ISEEQS Publication, Jerusalem, pp 462–463

Einbinder S, Mass T, Brokovich E, Dubinsky Z, Erez Y, Tchernov D (2009) To eat or to sunbathe: morphological changes in the coral Stylophora pistillata along its whole bathymetrical distribution. Mar Ecol Prog Ser 381:167–174

Falkowski PG, Dubinsky Z (1981) Light-shade adaptation of Stylophora pistillata, a hermatypic coral from the Gulf of Eilat. Nature 289:172–174

Falkowski PG, Raven JA (2007) Aquatic photosynthesis. Princeton University Press, Princeton

Falkowski PG, Dubinsky Z, Muscatine L, Porter J (1984) Light and the bioenergetics of a symbiotic coral. Bioscience 34:705–709

Falkowski PG, Dubinsky Z, Wyman K (1985) Growth-irradiance relationships in phytoplankton. Limnol Oceanogr 30:311–321

Falkowski PG, Jokiel PL, Kinzie RA (1990) Irradiance and corals. In: Dubinsky Z (ed) Coral reefs. Ecosystems of the world. Elsevier Science, Amsterdam, pp 89–107

Falkowski PG, McClosky L, Muscatine L, Dubinsky Z (1993) Population control in symbiotic corals. Bioscience 43:606–611

Fan TY, Li JJ, Ie SX, Fang LS (2002) Lunar periodicity of larval release by pocilloporid corals in southern Taiwan. Zool Stud 41:288–294

Fautin DG (2002) Reproduction of Cnidaria. Can J Zool 80:1735–1754

Fricke HW, Vareschi E, Schlichter D (1987) Photoecology of symbiotic coral Leptoseris fragilis in the Red Sea twilight zone (an experimental study by submersible). Oceanologia 73:371–381

Gorbunov MY, Falkowski PG (2002) Photoreceptors in the cnidarian hosts allow symbiotic corals to sense blue moonlight. Limnol Oceanogr 47:309–315

Harrison PL, Wallace CC (1990) Reproduction, dispersal and recruitment of scleractinian corals. In: Dubinsky Z (ed) Ecosystems of the world. Coral reefs. Elsevier Science, Amsterdam, pp 133–207

Iluz D, Vago R, Chadwick NE, Hoffman R, Dubinsky Z (2008) Seychelles lagoon provides corals a refuge from bleaching. Res Lett Ecol. Article ID 281038

Jerlov NG (1951) Optical studies of ocean water. Rep Swedish Deep-Sea Exped 3:73–97

Jerlov NG (1964) Optical classification of ocean water. Physical aspects of light in the sea. University of Hawaii Press, Honolulu, pp 45–49

Jerlov NG (1976) Applied optics. Elsevier, Amsterdam

Jokiel PL (1985) Lunar periodicity of planula release in reef coral Pocillopora damicornis in relation to various environmental factors. In: Proceedings of the 5th international coral reef congress, vol 4, Tahiti, pp 307–312

Jokiel PL, Ito RY, Liu PM (1985) Night irradiance and synchronization of lunar spawning of larvae in the reef coral Pocillopora damicornis (Linnaeus). Mar Biol 88:167–174

Karako S, Stambler N, Dubinsky Z (2002) The taxonomy and evolution of the zooxanthellae-coral symbiosis. In: Seckbach J (ed) Symbiosis: mechanisms and model systems. Kluwer, Dordrecht, pp 539–557

Karako-Lampert S, Katcoff DJ, Achituv Y, Dubinsky Z, Stambler N (2005) Responses of Symbiodinium microadriaticum clade B to different environmental conditions. J Exp Mar Biol Ecol 318:11–20

Kirk JTO (1994) Light and photosynthesis in aquatic ecosystems, 2nd edn. Cambridge University Press, London/New York

Koren S, Dubinsky Z, Chomsky O (2008) Induced bleaching of Stylophora pistillata by darkness stress and its subsequent recovery. In: Proceedings of the 11th international coral reef symposium, Ft. Lauderdale, 7–11 July 2008

Lampert-Karako S, Stambler N, Katcoff DJ, Achituv Y, Dubinsky Z, Simon-Blecher N (2008) Effects of depth and eutrophication on the zooxanthella clades of Stylophora pistillata from the Gulf of Eilat (Red Sea). Aquat Conserv 18:1039–1045

Levy O, Achituv Y, Yacobi YZ, Stambler N, Dubinsky Z (2006a) The impact of spectral composition and light periodicity on the activity of two antioxidant enzymes (SOD and CAT) in the coral Favia favus. J Exp Mar Biol Ecol 328:35–46

Levy O, Achituv Y, Yacobi YZ, Dubinsky Z, Stambler N (2006b) Diel 'tuning' of coral metabolism: physiological responses to light cues. J Exp Biol 209:273–283

Levy O, Appelbaum L, Leggat W, Gothlif Y, Hayward DC, Miller DJ, Hoegh-Guldberg O (2007) Light-responsive cryptochromes from a simple multicellular animal, the coral Acropora millepora. Science 318:467–470

Lin C, Shalitin D (2003) Cryptochrome structure and signal transduction. Annu Rev Plant Biol 54:469–496

Loya Y (2004) The coral reefs of Eilat – past, present and future: three decades of coral community structure studies. In: Rosenberg E, Loya Y (eds) Coral health and disease. Springer, Berlin, pp 1–34

Mass T, Einbinder S, Brokovich E, Shashar N, Vago R, Erez J, Dubinsky Z (2007) Photoacclimation of Stylophora pistillata to light extremes: metabolism and calcification. Mar Ecol Prog Ser 334:93–102

Matz MV, Fradkov AF, Labas YA, Savitsky AP, Zaraisky AG, Markelov ML, Lukyanov SA (1999) Fluorescent proteins from nonbioluminescent Anthozoa species. Nat Biotechnol 17:969–973

McGuire MP (1998) Timing of larval release by Porites astreoides in the northern Florida Keys. Coral Reefs 17:369–375

Muscatine L, Porter JW (1977) Reef corals: mutualistic symbioses adapted to nutrient-poor environments. Bioscience 27:454–460

Oliver JK, Babcock RC, Harrison PL, Willis BL (1988) Geographic extent of mass coral spawning: clues to ultimate causal factors. In: Proceedings of the 6th international coral reef symposium, vol 2, Townsville, pp 853–859

Pieribone V, Gruber DF (2005) A glow in the dark. The Belknap Press/Harvard University Press, Cambridge, MA/London

Richmond RH (1997) Reproduction and recruitment in corals: critical links in the persistence of reefs. In: Birkeland C (ed) Life and death of coral reefs. Chapman and Hall, New York, pp 175–197

Richmond RH, Hunter CL (1990) Reproduction and recruitment of corals - comparisons among the Caribbean, the tropical Pacific, and the Red Sea. Mar Ecol Prog Ser 60:185–203

Rinkevich B, Loya Y (1979) The reproduction of the Red Sea coral Stylophora pistillata II. Synchronization in breeding and seasonality of planulae shedding. Mar Ecol Prog Ser 1:145–152

Schumacher H, Zibrowius H (1985) What is hermatypic? A redefinition of ecological groups in corals and other organisms. Coral Reefs 4:1–9

Shlesinger Y, Loya Y (1985) Coral community reproductive patterns: Red Sea versus the Great Barrier Reef. Science 228:1333–1335

Shlesinger Y, Goulet TL, Loya Y (1998) Reproductive patterns of scleractinian corals in the northern Red Sea. Mar Biol 132:691–701

Stambler N, Dubinsky Z (2005) Corals as light collectors: an integrating sphere approach. Coral Reefs 24:1–9

Stoddart DR (1969) Ecology and morphology of recent coral reefs. Biol Rev 44:433–498

Szmant-Froelich A, Reutter M, Riggs L (1985) Sexual reproduction of *Favia fragum* (Esper): lunar patterns of gametogenesis, embryogenesis and planulation in Puerto Rico. Bull Mar Sci 37:880–892

Tanner JE (1996) Seasonality and lunar periodicity in the reproduction of Pocilloporid corals. Coral Reefs 15:59–66

Titlyanov EA, Leletkin VA, Dubinsky Z (2000) Autotrophy and predation in the hermatypic coral *Stylophora pistillata* in different light habitats. Symbiosis 29:263–281

Zakai D, Dubinsky Z, Avishai A, Caaras T, Chadwick NE (2006) Lunar periodicity of planula release in the reef-building coral *Stylophora pistillata*. Mar Ecol Prog Ser 311:93–102

Coral Calcification, Cells to Reefs

Denis Allemand, Éric Tambutté, Didier Zoccola, and Sylvie Tambutté

Abstract In spite of more than one century and half of studies, mechanisms of coral biomineralization, leading to coral growth and reef formation, still remain poorly known, although major global threats to coral reefs, such as ocean acidification, primarily affect this process. Coral skeletons are used as environmental archives but the vital processes that govern incorporation of trace elements and stable isotope are still unknown.

Our knowledge on coral physiology is restricted to the organismal level due to the lack of appropriate cell model, however the advent of new approaches, such as coral genomic, is changing drastically our knowledge on these animals even if only a few data are available concerning the field of biomineralization.

This chapter reviews our present knowledge and discusses the different theories on coral calcification, from the molecular to the reef level. Conclusion is presented in a list of key issues to be resolved in order to understand the intimate mechanisms of calcification of corals, essential to determine the origin of the sensitivity of corals to ocean acidification, to improve paleoceanographic reconstructions or coral reef management, or "just" to understand how genes of a soft organism control the formation of an extracellular 3D-skeleton.

Keywords Calcification • biomineralization • physiology • calicodermis • ion transport • Ca^{2+}-ATPase • organic matrix • carbonic anhydrase • amorphous calcium carbonate • coral fibers • light-enhanced calcification (LEC) • galaxin

1 Introduction

Studies of coral calcification started with the pioneering descriptive work of Dana (1846), followed by other studies devoted to the morphology of coral exoskeleton for taxonomic purposes during the nineteenth century (Milne Edwards 1857; Ogilvie 1897). The first physiological approach was performed by Kawaguti and Sakumoto (1948), who demonstrated a relationship between light, zooxanthellae, and calcification, and shortly after Goreau (1959) laid the foundations of the physiology of coral calcification. However, after almost two centuries of studies, we must admit that our understanding of the intimate mechanisms of scleractinian coral calcification remains very weak as exemplified by the two following sentences concerning skeleton/tissue interactions taken one century apart from one another:

A question which has common interest both for zoologist and paleontologist is the relation of the soft parts of the polyp to the hard calcareous or horny skeleton produced in most corals.

Ogilvie 1897 (page 84)

The poor understanding of calcification mechanisms in corals results from a lack of information on tissue/skeleton interactions and temporal/spatial patterns in skeleton morphogenesis.

Le Tissier 1987 (abstract)

Coral skeletons constitute the basis of coral reefs, the world-largest bioconstruction, and are widely used for several purposes ranging from taxonomy (Wells 1956), recorders of environmental information (Barnes and Lough 1996) or as bioimplants for bone surgery (Demers et al. 2002). The phylogenetic position of corals as a sister group of bilateralia, the search for new biological models, the menace of seawater acidification that presently threatens coral reefs are giving a renewal of interest for coral biology, and particularly for coral calcification. However, our knowledge of calcification in corals is far behind our knowledge in other groups like molluscs, and, even in these groups, we are only at the dawn of the field of biomineralization.

Biomineralization is defined as both the study of biologically produced materials (or biominerals) and the processes that lead to their formation (Estroff 2008). Biomineralization is present in all kingdoms. A large part of our knowledge has been inferred from molluscs and coccolithophores; however, how a soft animal creates a 3D highly ordered mineral structure remains largely a mystery.

D. Allemand (✉), É. Tambutté, D. Zoccola, and S.Tambutté
Centre Scientifique de Monaco,
Avenue Saint-Martin, MC-98000, Monaco, Principality of Monaco
e-mail: allemand@centrescientifique.mc

Biomineralization is at the bridge between biology, structural biology, biochemistry, surgery, geology, mineralogy, paleontology, and engineering. It is also at the bridge between academic and applied research. Consequently, the study of biomineralization encompasses a large variety of methods and concepts.

The present review summarizes our knowledge on the various aspects of coral calcification, highlighting both solid knowledge and debated aspects with suggestions for future research directions. For further details, the reader is referred to previous general reviews (Milne Edwards 1857; Muscatine 1971; Buddemeier and Kinzie 1976; Johnston 1980; Constantz 1986; Barnes and Chalker 1990; Gattuso et al. 1999; Cohen and McConnaughey 2003; Allemand et al. 2004; Weis and Allemand 2009).

2 The Different Types of Biomineralization and the Coral Calcification

Lowenstam (1981) defined two types of biomineralization processes. The first one is called "biologically-induced mineralization." It corresponds to a process in which no specialized cellular or molecular machinery is set up to induce mineralization, with minerals showing crystal shapes similar to those formed by inorganic processes. In this type of biomineralization, minerals are precipitated along the surface of the cells, sometimes embedding them into the mineral. This process is common in prokaryotes, fungi, and some algae (Lowenstam and Weiner 1989; Mann 2001). The second process is the "organic matrix-mediated mineralization," further called by Mann (1983) "biologically controlled mineralization." It refers to a process restricted to a delineated environment and using an organic matrix framework within minerals (see Lowenstam and Weiner 1989 for a review). The vast majority of animal biomineralizations belongs to this type; however, that of coral mineralization still remains debated. For example, Barnes (1970) stated that the deposition of skeleton by corals is a physicochemical process and that only competitive crystal growth determines the morphology of the skeleton. Similarly, Constantz (1986) advocated a biologically induced process for corals and claimed that the "morphology [...] of the aragonite fibers [... is] entirely predictable by factors controlling abiotic, physio-chemical crystal growth." However, 6 years earlier, Johnston (1980), in a very thorough review on coral calcification, argued strongly for a biologically controlled process. Mann (1983), following Johnson's view, classified corals in biologically controlled mineralization. In their book, Lowenstam and Weiner (1989) nevertheless adopted an intermediary position by proclaiming that physicochemical processes predominate in corals. However, they also stated later in the chapter that, since aragonite crystal morphologies vary between taxa, this implies that the microenvironments must also vary between taxa suggesting a careful regulation of the ionic composition of the medium from which the crystals form. If today, data presented by Cuif and Dauphin (2005a) show that "coral fibres appear to be fully controlled structures," this idea is not totally accepted and, for example, Veis (2005) still quotes corals as an example of biologically induced mineralization.

In fact, two levels of control of coral calcification may exist: control by organic matrix and/or control of ion transport and delivery to the site of mineralization. From these, a combination of at least six hypotheses exist (Fig. 1). These different hypotheses are analyzed below.

3 The Site of Coral Calcification: The Subcalicoblastic Extracellular Calcifying Medium

The anatomy of corals is described in Chevalier (1987), Johnston (1980), and Fautin and Mariscal (1991). It is generally compared to a bag attached by its basis to the skeleton located outside the animal, with an oral tissue facing the seawater and an aboral tissue facing the skeleton. Cnidarians are the first animals showing real epithelia (i.e., exhibiting cell polarity, joined by cell junctions and associated with extracellular matrix, Tyler 2003). Each tissue is indeed composed of two epithelial cell layers, named epidermis (but generally referred as ectoderm) and gastrodermis (generally referred as endoderm), respectively, for the external and internal layers (Fautin and Mariscal 1991; Tyler 2003; Galloway

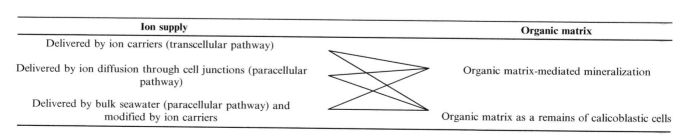

Fig. 1 Schema showing that the combination of different levels of control of coral calcification leads to six hypotheses

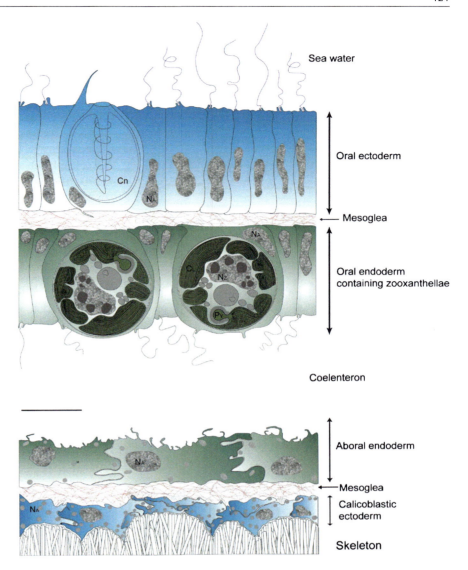

Fig. 2 Schematic section of the histology of a coral across its coenenchyme. Cn cnidocyte, N_A animal cell nucleus, N_Z zooxanthella cell nucleus, C_L chloroplast, P_Y pyrenoid. The small gray spots inside the cells are mitochondria. Scale bar: 5 μm

et al. 2007). These cell layers are separated by the mesoglea, a thin extracellular matrix of collagen (Schmid et al. 1999). The cavity delimited by the gastrodermis is called coelenteron or coelenteric cavity and corresponds to the gastrovascular cavity of the coral. Figure 2 shows a diagram of a histological section across coenenchyme (i.e., between two adjacent polyps).

3.1 The Skeletogenic Tissue: The Calicoblastic Epithelium

The tissue surrounding the skeleton was initially described by Von Heider (1881) who observed the line-forming cells in corals to which he gave the name of calycoblasts. He already suggested that these cells were at the origin of the skeleton. Hayasi (1937) further described the calicoblasts as a single layer of cells. Galloway et al. (2007) proposed to call it "calicodermis" as short-cut of "calicoblastic epidermis." The aboral tissue is thinner that the oral tissue and the calicoblastic cell layer is the thinnest (Johnston 1980; Tambutté et al. 2007a). While the oral ectoderm contains several specialized cells such as cnidocytes, mucocytes, and epitheliomuscular cells (Fautin and Mariscal 1991), the calicodermis, established during larval settlement (Vandermeulen 1975), consists entirely of calicoblastic cells and desmocytes (Muscatine et al. 1997; Goldberg 2001a; Tambutté et al. 2007a).

The shape of calicoblastic cells varies between long-thin-flat to thick and cup-like according to the calcification activity of the cells (Le Tissier 1990; Clode and Marshall 2002a; Tambutté et al. 2007a). Flat calicoblastic cells are associated with low calcification activity and cup-like cells

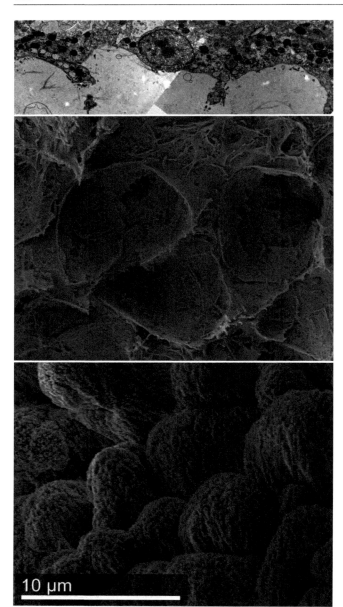

Fig. 3 Transmission (**a**) and scanning (**b**, **c**) electron microscopy of the calicodermis (**a**, **b**) and skeleton (**c**) showing the perfect correspondence between tissue and skeletal fibers. The scale is the same for the three pictures

are associated with high calcification activity (Tambutté et al. 2007a). In this case, there is a strict correspondence between tissue and skeletal fibers (Fig. 3), but a fiber is in fact produced by at least two contiguous calicoblastic cells (Tambutté et al. 2007a). These highly interdigitated cells always contains numerous mitochondria (Johnston 1980; Tambutté et al. 2007a) and, at least in some species, intracellular vesicles (Johnston 1980; Le Tissier 1990, 1991; Clode and Marshall 2002a, b).

3.2 The Subcalicoblastic Extracellular Calcifying Medium

The calicodermis is attached to the skeleton by desmocytes through a mortise and tenon system (Muscatine et al. 1997; Goldberg 2001a; Tambutté et al. 2007a). Whether this anchoring system allows a space between calicodermis and the skeleton is a matter of debate. Indeed, a large space is generally associated with a less direct control of crystal growth (Barnes 1970). Biologically-controlled biomineralization, as defined by Mann (1983), always occurs in a delineated compartment, but it is not clear how close the biomineral is to the cell membranes. For example, in molluscs, the nature of the interface between the mantle and the mineral is still debated, with some authors suggesting that an extrapallial fluid exists and plays a role as precursor medium (Moura et al. 2000), while others suggest that close contact is necessary to regulate the mineralization process by receiving feedback molecular signals from the newly formed biominerals (Marin et al. 2008). It should be stated however that biologically-controlled mineralization may occur far from the cells in a gel-like medium as exemplified in fish otoliths (see review by Allemand et al. 2007).

In corals, the presence of a space often greater than 1 μm in width has previously been described (Johnston 1980; Isa 1986; Le Tissier 1988a), with possible diurnal variations (Le Tissier 1988b). Barnes (1970, 1972) suggested that calcification primarily occurs within pockets created where calicodermis is lifted away from the skeletal surface. However, some authors suggested that this might be an artifact during sample preparation (Johnston 1980; Clode and Marshall 2002a, b; Tambutté et al. 2007a). On the other hand, the fact that a direct correspondence between calicodermis and crystal fibers was observed for different corals (*Pocillopra damicornis*: Johnston 1980, Brown et al. 1983; *Acropora cervicornis*: Gladfelter 1983; *Mycetophyllia reesi*: Goldberg 2001b; *Stylophora pistillata*: Tambutté et al. 2007a) leads to the hypothesis of a close contact. Tambutté et al. (2007a) demonstrated that a space may be artifactually produced in Field Emission Scanning Electron Microscopy (FESEM) during sample observation under vacuum. However, even a close contact does not preclude that a nanometric space may exist. It has been suggested, as early as the beginning of the twentieth century, that skeleton is deposited within a "colloidal matrix" (Duerden 1903). Hayasi (1937) observed an amorphous organic membrane between the calicodermis and the skeleton. Bryan and Hill (1941) postulated the site of mineralizaton to be a colloidal gel matrix secreted by the ectoderm. Clode and Marshall (2002a) observed a fibrillar mesh-like organic matrix between the calicodermis and the skeleton. This extracellular calcifying medium may be composed by hydrated proteins and glycosaminoglycans as a gel-like substance in which nucleation may occur (matrix compartment of Johnston 1980). Similarly, nacre formation occurs in molluscs in a hydrogel-like medium by the removal of part of the water leading to the mineral formation (Addadi et al. 2005). The presence of a hydrated gel as interface between the cells and the mineral as a precursor of the biomineral formation is supported by thermogravimetric analysis of skeletal powder that demonstrated a 1.5–2.2% of water inside the skeleton (Cuif et al. 2004a).

In conclusion, it appears that, at least in the few species studied, the interface between the cells and the skeleton is submicrometric and probably filled with an hydrogel. It is thus preferable to call this interface "extracellular calcifying medium" (ECM) rather than extracellular calcifying fluid (ECF).

3.3 ECM: Open or Closed Compartment?

The degree of isolation of the mineralizing medium from the environment gives information on the extent of biological control. Isolation may be achieved either by bordering the mineral within a membrane as is the case for spiculogenesis in gorgonians, or by forming an epithelium surrounding the mineral as in corals. In order to restrain the access of ions to the site of mineralization, this epithelium should possess junctions between calicoblastic cells allowing selective permeability. Two types of permeability junctions exist: tight junctions that are almost totally impermeable to ions and molecules and leaky junctions that allow some paracellular, sometimes selective, pathway (Frömter and Diamond 1972; Asano et al. 2003; Tyler 2003). Johnston (1980) was the first to describe junctions in the calicodermis, that appeared as "zonular and are presumed to completely encircle each ectodermal cells forming an occlusive band at least 0.25 μm deep." Junctions are present at the apical part (i.e., skeletal side) of all calicoblastic cells of the corals *Caryophyllia smithii* (Le Tissier 1990), *Galaxea fascicularis* (Clode and Marshall 2002a) and *Stylophora pistillata* (Tambutté et al. 2007a). Homologs of human contactin, a protein characteristic of septate junctions, are expressed in the coral *Acropora millepora*. Blast search for contactin on EST library (Meyer et al. 2009) gives a p = 3 e^{-15}. If the presence of junctions is firmly demonstrated, their nature and role are still debated (Magie and Martindale 2008). Clode and Marshall (2002a) called them "septate junctions" because of their ladder-like arrangement of transverse septa as in others anthozoans (Green and Flower 1980). Septate junctions are well described in many invertebrates where they are present in all ectodermally derived epithelia (Green and Bergquist 1982), but their permeability properties are a matter of debate. In *Drosophila* (Baumgartner et al. 1996),

in snails (Albrecht and Cavicchia 2001) as well as in *Bombyx mori* (Keil and Steinbrecht 1987), septate junctions are considered as forming "tight" epithelia and thus constitute an impermeable barrier to the diffusion of ions and molecules. On this basis, Clode and Marshall (2002a) conclude that calicodermis-bearing septate junctions also form a tight epithelia. However, in some invertebrates (insects, molluscs, and starfish), septate junctions present only selective permeabilities, giving the epithelium leaky permeability (Lane 1979; Dan-Sokhawa et al. 1995; Bleher and Machado 2004). In addition, Bénazet-Tambutté et al. (1996) have demonstrated that the oral epithelial layers are leaky but nevertheless display differential permeabilities depending on the size of the molecules and their charge. Even a leaky epithelium may restrict transport of ions and molecules.

The presence of septate junctions between calicoblastic cells strongly suggests that these cells are polarized, with apical and basolateral cell surfaces exhibiting different properties like distribution of ion carriers (Anderson et al. 2004). The fact that adjacent calicoblastic cells are often widely separated by large intercellular spaces, which may increase the absorption surface of ions and therefore the transepithelial flux (Berridge and Loschman 1972), may be an argument in favor of a transepithelial mode of transport according to a closed compartment. However, it is presently impossible to conclude on the permeability of the calicodermis without more information on the role of septate junctions in corals.

3.4 Physicochemical Characteristics of the Subcalicoblastic Extracellular Calcifying Medium

Determining ion concentrations and pH at the site of mineralization is essential to fully understand the intimate mechanisms of coral calcification. These characteristics are determined by the permeability of calicodermis. However, if a relatively large body of data exists concerning the physicochemical characteristics of coral surface or coelenteric cavity (Kühl et al. 1995; De Beer et al. 2000; Al-Horani et al. 2003a), there is only one paper that reports direct measurement of calcium concentration and pH in the ECM (Al-Horani et al. 2003b). This compartment appears to be alkaline and slightly more loaded with calcium compared to seawater, even at night (see Table 1). From these data, it therefore appears that metabolic energy is needed to transport calcium to the subcalicoblastic space. It should be noted however that results obtained by inserting a microsensor electrode with a tip diameter of 1–10 μm in a tight submicrometric scale space should be taken with caution since

Table 1 Ca^{2+} and pH values within coral tissues

Species	Localization	Ca^{2+} concentration (Δ mM compared to SW) Day	Night	pH Day	Night	Reference
Favia sp.	Coral tissue	–	–	8.6	7.3	Kühl et al. (1995)
Favia sp.	Polyp surface	−0.40	+0.40	8.0	7.0	De Beer et al. (2000)
Galaxea fascicularis	Polyp surface	−0.18 ± 0.06	+ 0.04 ± 0.06	8.49 ± 0.25	7.60 ± 0.22	Al-Horani et al. (2003b)
	Coelenteron	−0.24 ± 0.05	−0.06 ± 0.13	8.19 ± 0.20	7.61 ± 0.19	
	Calcifying medium (ECM)	+0.58 ± 0.29	+0.21 ± 0.25	9.2 ± 0.03	8.1 ± 0.08	

leakage and/or overlap with other compartments may happen. pH at the site of calcification may also be inferred from boron isotopic distribution, which is a marker of pH (Vengosh et al. 1991). Slightly alkaline values (from 0.04 to 0.3 pH unit) compared to the ambient seawater are obtained by these methods (Vengosh et al. 1991; Hemming and Hanson 1992; Gaillardet and Allègre 1995; Rollion-Bard et al. 2003; Hönisch et al. 2004; Reynaud et al. 2004). By comparing the morphology of synthetic and biogenically aragonite crystals, Holcomb et al. (2009) suggest that coral maintains a high saturation state of aragonite at the site of mineralization, which may be as much as seven times above that of the ambient seawater (Cohen et al. 2009).

Even if data are still scarce, they all suggest that the coral control the composition of the extracellular calcifying medium that is therefore physically isolated from the surrounding seawater, suggesting that the permeability of calicodermis is low. It is interesting to compare with other model organisms to determine if a general trend characterizes calcifying medium. Concerning calcium, it was shown that the pericrystalline fluid surrounding fish otoliths does not show higher calcium concentration compared to plasma but shows a very high turn-over rate (renewal rate of about ten times per day, see Allemand et al. 2007). Similar high turn-over rates were reported in the hen during eggshell formation, where a total renewal of the plasmatic pool of calcium occurs every 12 min (Nys 1990). In corals, the coelenteric cavity should be renewed 35 times per day in order to allow calcium supply for calcification (from data by Tambutté et al. 1996). In the crayfish *Cherax quadricarinatus*, calcium concentration also remains constant and similar to other tissues during the formation of the gastrolith (Shechter et al. 2008). On the other hand, it appears that calcifying fluids present generally high concentration of inorganic carbon. Total CO_2 is two to threefold more concentrated in trout endolymph compared to the plasma (Borelli et al. 2003a). Similar high total inorganic carbon concentration were found in the pericrystalline fluid of mollusc shell (Crenshaw 1972), and up to 130 mM of bicarbonate have been reported in the uterine fluid of hens during eggshell formation (Nys et al. 1991). These data strongly suggest that an active mechanism of transport is involved to supply carbon to the site of calcification;

however, data concerning DIC are lacking in corals. Concerning pH, values are always alkaline during $CaCO_3$ deposition. A pH between 8.5 and 9 was measured in mineralizing vacuoles of the perforate foraminifera, *Amphistegina lobifera* (Erez 2003). In the terrestrial crustacean *Porcellio scaber*, while hemolymph pH is about 7.6, pH of the calcifying fluid varies from 8.2, when $CaCO_3$ is produced, to 6.0 during resorption (Ziegler 2008). Similarly, high pH values (8.7) were reported in a crayfish by Shechter et al. (2008). However, the extrapallial fluid of three marine molluscs presents pH values below that of seawater, i.e., between 7.33 and 7.91 (Crenshaw 1972). Surprisingly, pH is not correlated with the rate of calcification in the fish endolymph as the highest pH (8.0) is found in the compartment that shows the lowest rate of calcification, whereas a pH of 7.4 is found in the compartment that shows the highest rate of calcification (see Allemand et al. 2007 for a review). All these results suggest that the relationship between alkaline fluid and $CaCO_3$ deposition is more complicated than is implied by the ions' simple chemical behavior.

This brief review shows that extracellular calcifying medium present common characteristics with generally a high pH and a high $CaCO_3$ saturation state. But due to the very low number of data concerning corals, it is urgent to develop tools to characterize ECM and particularly to measure ion concentration at the site of mineralization.

3.5 Site of the Initial Mineral Deposition

The initial site of mineral deposition has been debated for a long time. Von Heider (1881) claimed that the skeleton was initiated by intracellular precipitation of $CaCO_3$. Von Heider (1881) in Ogilvie 1897) postulated that ectodermal cells secreted calcareous elements extracellularly. Bourne (1887) and Fowler (1885) confirmed the conclusions of Koch. However, the debate springs up again with the observations of Kawaguti and Sato (1968) who claimed that calicoblastic cells accumulate "calcareous substances within them" before their deposition on the surface of the skeleton, resulting in its growth. Hayes and Goreau (1977) came to the same conclusions and stated that they bring "overwhelming evidence for an intracellular mode of calcification in Scleractinia." Finally, in his review, Johnston (1980) came to the conclusion that "there is no evidence for any preformed crystalline calcium carbonate in any [...] organelle within the calicoblastic ectoderm. At present, it can thus be concluded that calcification is initiated extracellularly."

4 Physiology of Coral Calcification

4.1 Coral Calcification: A Chemical Reaction with Four Molecules

Only a few models (for example, coccolithophores, Brownlee and Taylor 2002; foraminifera, Faber and Preisig 1994; Erez 2003; molluscs, Wheeler 1992) have been investigated concerning the physiology of the transport of ions for calcification, a process that remains poorly known. In corals, the first physiological data concerning ion supply for coral calcification were provided by Goreau (1959) and Goreau and Goreau (1959a, b).

The reaction of calcification in corals as in other animals making their skeleton with calcium carbonate is:

$$Ca^{2+} + CO_3^{2-} \Rightarrow CaCO_3$$

However, the $[CO_3^{2-}]/[HCO_3^-]$ ratio at physiological pH (between 7.5 and 9.0) is extremely low (i.e., 5×10^{-4} to 5×10^{-2}); consequently, this reaction may practically not contribute to the $CaCO_3$ formation (Ichikawa 2007), which therefore would result from:

$$Ca^{2+} + HCO_3^- \Rightarrow CaCO_3 + H^+$$

If the source of DIC is bicarbonate, there is a net production of proton in the ratio 1:1 per mole of calcium carbonate produced. This results in the release of one H^+ per Ca^{2+} ion. To maintain constant the pH within the calcifying fluid and facilitate mineral precipitation, this H^+ must be removed from the calcifying site and transported into the coelenteron.

If the source of DIC is CO_2, two H^+ per Ca^{++} ion are produced:

$$CO_2 + H_2O + Ca^{2+} \Rightarrow CaCO_3 + 2H^+$$

The source of DIC for calcification was investigated by using a double labeling technique with ^{45}Ca and $H^{14}CO_3$. It was suggested that a large part (40–60%) of the carbon used for $CaCO_3$ formation may originate from respired (metabolic) CO_2 (Erez 1978; Furla et al. 2000b). In the coral *S. pistillata*, the rate of respiration exceeds the rate of calcification by almost a factor of 4, suggesting that the production of metabolic CO_2 is indeed largely enough to supply calcification, particularly because calicoblastic cells contain numerous mitochondria. Furthermore, CO_2 concentration in the dark was found to be ninefold higher on the coral surface than in the surrounding seawater (de Beer et al. 2000).

These results confirm earlier results obtained using either ^{14}C-labeled food (Pearse 1970) or stable isotopes (Goreau 1977), but are in disagreement with the ones of Taylor (1983) who suggested that external HCO_3^- was the preferred substrate for skeletogenesis. Furla et al. (2000b) concluded that the pool of bicarbonate used for calcification within the animal tissue is small and presents a high turn-over rate. This assumption was based on the fact that there is no lag-phase in the incorporation of external DIC (see Fig. 3c of Furla et al. 2000b). However, a careful analysis of their data shows that the pool of tissular DIC was equilibrated only after 3 h. It can thus be suggested that the period used to monitor ^{14}C incorporation into skeleton (1 h) was not enough long to equilibrate pools and that the linear incorporation observed may just reflect a transient dilution of external labeled DIC into an internal pool before reaching its equilibrium after at least 3 h. Therefore, it can be suggested that external DIC is indeed diluted in a nonlabeled pool, but this pool is perhaps not metabolic CO_2 as suggested by Furla et al. (2000b) but rather intratissular bicarbonate. Such hypothesis may explain the conflicting observations mentioned above. It is strengthened by the observation by Furla et al. (2000b) who demonstrated the presence of a light-dependent DIC pool initially suggested to only play a role in carbon-concentrating mechanism (CCM) used for photosynthesis. This pool is increased 39-fold after 3 h of illumination. Also, high concentrations of carbonate ions have been visualized in the mesoglea layer and lateral intercellular spaces of *Porites porites* planulae using the von Kossa technique (Hayes and Goreau 1977).

However, whatever the carbon source for calcification, the ratio of H^+ produced per mole of calcium on the site of calcification cannot be determined since the carbon species transported from the calicoblastic cells to the calcifying medium (CO_2 and/or HCO_3^-) is unknown. Probably, the major part of DIC inside the cells is in the form of HCO_3^- after pH equilibration (given an intracellular pH of about 7.4 as in other marine invertebrates). This review shows the urgent need to reevaluate DIC metabolism for coral calcification.

4.2 Measurement of Coral Calcification

The rate of coral calcification can be measured by two kinds of methods (see Buddemeier and Kinzie 1976 for a review). The first ones are non-destructive. The most direct way, which can be used for mid- or long-term records, is to directly measure the size of the colony (linear growth or surface, which is in fact a measure of the growth rate) or the increase in weight by the buoyant weight technique (Jokiel et al. 1978). The alkalinity anomaly technique allows measurement of the rate of calcification under short-term conditions (≤1 h) (Smith and Kinsey 1978; Jacques and Pilson 1977; Chisholm and Gattuso 1991). The second kind of methods are destructive. Analysis of a slice of coral skeleton directly or under X-ray allows measuring long-term growth rates. The most sensitive method is however radioisotopic methods using ^{45}Ca as a tracer (Goreau 1959; Tambutté et al. 1995) that allows short incubation (1 min – 1 h). Using these different techniques, highly variable rates of light calcification have been recorded in Scleractinian corals, from about 6.4 nmol h^{-1} mg Prot^{-1} in *Oculina diffusa* to about 700 nmol h^{-1} mg Prot^{-1} in *Acropora formosa* (see review by Tentori and Allemand 2006).

The rate of calcification depends on the age of the coral. Young fragments of the coral *Madracis mirabilis* calcify faster than old fragments, the effect of age being absolute and independent of size (Elahi and Edmunds 2007). Similarly, spatial variations of calcification within a single colony were also observed (Marshall and Wright 1998; Al-Horani et al. 2005; Howe and Marshall 2002). On a given colony, the calcification rate may be 4–8 times faster at the apical parts of branching corals rather in the lateral and basal regions (Goreau and Goreau 1959a).

4.3 Ion Supply or Removal for Calcification: Paracellular and Passive or Transcellular and Active?

4.3.1 Ion Delivery to the Site of Mineralization: Three Possibilities

Figure 4 summarizes the different possible pathways of ion delivery to the ECM. Generally, two routes for ion and organic molecules to cross an epithelium are described: a paracellular and a transcellular route (Berridge and Loschman 1972). In the case of corals, two kinds of paracellular pathways can in fact be described, diffusion of ions or diffusion of bulk seawater. Use of each of these pathways depends on the permeability properties of the epithelium (i.e., on cell junctions).

For the first kind of paracellular route (Fig. 4a), ion or molecules diffuse between cells along their chemical (or electrochemical in the case of charged molecules) gradient. Permeation depends on the size of molecules (generally around 1 nm in diameter, Kottra and Frömter 1983) as well as on their charge. For example, the mammalian small intestinal paracellular pathway is at least four-times more permeable to

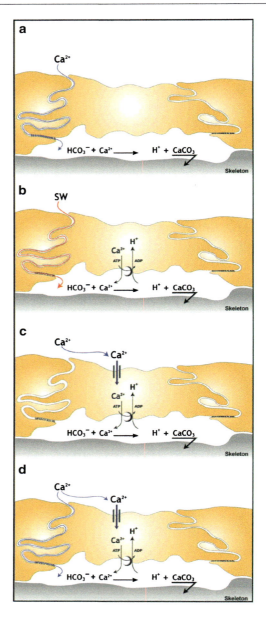

Fig. 4 Schematic view of the four hypotheses explaining transfer across the calicodermis to supply ions to the subcalicoblastic extracellular calcifying medium: (**a**) ions are provided by a passive paracellular pathway between calicoblastic cells, (**b**) bulk seawater provides the essential of ions by a paracellular pathway, (**c**) ions are supplied by an active transcellular pathway through calicoblastic cells, (**d**) combination of (**a**) and (**c**) where calcium ions are provided by both a transcellular and a paracellular pathway

K$^+$ than to Cl$^-$ ions. The overall permeability also varies more than 30-fold, depending on the tissue studied (Diamond and Wright 1969). Curiously, the oral layers of the coral *Heliofungia actiniformis* present a higher permeability to anions than to cations (Bénazet-Tambutté et al. 1996), whereas most other epithelia present a cation selectivity (Kottra and Frömter 1983). The second kind of paracellular route is non-selective and corresponds to a diffusion of bulk seawater between cells (Fig. 4b). The existence of such a pathway is suggested in the light of experiments using calcein. Calcein is a fluorescent, cell-impermeant dye known to react with calcium (Karoonuthaisiri et al. 2003). When added in seawater, this dye is rapidly incorporated in the newly formed skeleton (Erez and Tambutté, unpublished data). This observation suggest the presence of a largely nonselective paracellular pathway or even a water flow, allowing seawater to enter the subcalicoblastic extracellular calcifying medium without any physical barrier (Cohen and McConnaughey 2003; Erez and Braun 2007).

While the osmotic gradient is generally the motor of water transport, in the present cases, it is suggested that the seawater flow is mediated by polyp contraction in a way like ultrafiltration. The physicochemical characteristics (mainly pH) of this bulk seawater can be subsequently modified by the Ca^{2+}-ATPase in order to allow the formation of $CaCO_3$ (see Cohen and McConnaughey 2003). A similar mechanism has been observed in foraminifera (Erez 2003). However, it should be noted that calcein is known to fluoresce very weakly at high concentrations because of self-quenching, but its fluorescence increases at lower concentrations as quenching is reduced (Katsu et al. 2007). Therefore, it is possible that the amount of calcein incorporated within the skeleton may be very low, resulting from weak permeation. In that case, the importance of this passive uncontrolled supply to the calcification should be negligible. Moreover, Marschal et al. (2004) used calcein to label red coral spicules, which are intracellular biominerals, formed within a vacuole with no connection with the extracellular seawater. This raises the question of the mechanism of calcein transport from seawater to the biomineral.

In the transcellular route, molecules are absorbed at one side of the epithelial cell, transported within the cell cytoplasm, and from there, molecules are transported intracellularly toward the opposite plasma membrane on the other side of the epithelium (Fig. 4c). The first process is passive and only depends on the (electro)chemical gradient of the transported molecule. The second process is active and against a concentration gradient and therefore needs energy supply (Berridge and Loschman 1972).

The distinction between these paracellular and transcellular pathways is generally based upon pharmacological experiments using carrier inhibitors. Calcification was shown to be inhibited by verapamil, a well-known inhibitor of Ca^{2+}-channels (Marshall 1996; see Sect. 3.3.3). However, due to the fact that calcium ions are involved in many physiological processes, it can be argued that verapamil inhibits calcification indirectly via a Ca^{2+}-dependent process (for example, exocytosis of organic matrix vesicles). The following paragraphs will present in more details the present knowledge of ion transport, but as will be shown below, even if there are many arguments in favor of an active transport of ions, no definitive evidence allows us to really decide at present between an active or passive route. Perhaps, both paracellular and transcellular modes of transport coexist for the supply of ions (Fig. 4d).

4.3.2 Energetic Dependence of Ion Transport

Following our lack of knowledge on the permeability properties of the calicodermis (see Sect. 3.3), it is very difficult to give a definitive answer to the question "is ion supply/removal for calcification a passive or an active process?" First evidence for an active transport dependent on a source of energy came from the experiments of Chalker and Taylor (1975) and Chalker (1976) on *Acropora cervicornis* and *A. formosa*. They showed, by using several metabolic inhibitors (sodium azide, sodium cynanide, 2,4-dinitrophenol, CCCP, ethacrynic acid, and DCMU), that calcification in the light requires an energy-requiring transport. They concluded that energy is needed for the active transport of calcium and/or bicarbonate ions, or for organic matrix synthesis (see below). Curiously, dark calcification rates were not affected by these inhibitors, which led the authors to suggest that dark rates do not require metabolic energy from oxidative phosphorylation or rather that they are only due to isotopic exchange. This last hypothesis was supported by Marshall (1996) using CCCP, which found that $^{45}Ca^{2+}$ uptake in the dark was the same in dead or live colonies. However, he found at the same time that dark calcification was inhibited by the calcium channel inhibitor verapamil, suggesting yet a biologically controlled process, even in the dark. Using the alkalinity anomaly technique, which avoids isotopic exchange phenomena and possible artifacts, Barnes (1985) found that both CCCP and cyanide greatly inhibit (up to 80%) dark calcification rates suggesting that dark calcification is a real phenomenon as further supported by Howe and Marshall (2002). Interestingly, dark calcification rate was inhibited at lower CCCP concentrations than those inhibiting light-enhanced rates, suggesting that "photosynthesis can maintain the basic calcification mechanism when respiration is uncoupled" (Barnes 1985). This author also observed that cyanide failed to fully inhibit respiration even at 100 μM that may explain the above results on dark calcification. By comparing dose-response curves of the effect of cyanide on respiration and calcification, he concludes that ATP production may be sufficient to sustain dark rate and becomes limiting at high concentrations of the inhibitor. A recent paper comparing light-enhanced and dark calcification during the course of a bleaching event (Moya et al. 2008a) demonstrated that dark calcification was affected well after light calcification suggesting that dark calcification either needs less energy than light calcification or possesses mechanisms with higher affinity for ATP. From all these data, it can be concluded that dark calcification is dependent, as light calcification, on metabolic energy and thus results on an active process, but presumably it can still occur under low metabolic rates.

4.3.3 Calcium Transport

The initial rate of calcium deposition by corals always follows Michaelis-Menten kinetics (Chalker 1976; Krishnaveni et al. 1989; Tambutté et al. 1996). Such a function demonstrates that a carrier-dependent step is involved in the transcellular Ca^{2+} transport and would therefore suggest an active transepithelial Ca^{2+} pathway. However, Constantz (1986) suggested another interpretation of these data. He indeed suggested that "diffusion of Ca^{2+} to the subectodermal space was significantly reduced at low Ca^{2+} concentration and that the initial velocities of $CaCO_3$ precipitation thus followed abiotic, diffusion-controlled kinetics."

The affinity for calcium of the process of calcium deposition lies in the range of 3.5–9.09 mM for the four species studied (Table 2). It appears that the process is saturated at ambient seawater calcium concentration for two species (*Acropora formosa* and *Stylophora pistillata*) while it is not saturated for *Acropora cervicornis* (see Table 2). Curiously, conflicting data were obtained for *Galaxea fascicularis* by the same laboratory under similar experimental conditions: Krishnaveni et al. (1989) and Ip and Krishnaveni (1991) found respectively the process saturated around 25 and 10 mM. When calcification rate is plotted against aragonite saturation, a plateau is also obtained but at a level slightly lower that the control aragonite saturation (Gattuso et al. 1998). Surprisingly, Swart (1979) found that addition of calcium into seawater increased the rate of calcification of the coral *Acropora squamosa* until an inhibition was observed at higher concentrations (above 13 mM). A similar increase was observed by Marshall and Clode (2002) with *Galaxea fascicularis* without any inhibition even with addition of 5 mM calcium.

Table 2 Calcium affinity for calcification and calcium concentration at the plateau in different corals

Coral species	Km (mM)	Saturation (mM)	References
Acropora cervicornis	4.1	30	Chalker 1976
A. formosa	5.5	10	Chalker 1976
Galaxea fascicularis	9	25	Krishnaveni et al. 1989
G. fascicularis	3.5	10	Ip and Krishnaveni 1991
Stylophora pistillata	5.2	10	Calculated from Tambutté et al. 1996

No pool of calcium has been detected by isotopic methods within the coral tissue and the duration of transepithelial transport is about 1 min (Tambutté et al. 1996). Tambutté et al. (1996) and Marshall (1996) suggested that Ca^{2+} moves passively thanks to the large gradient between seawater and the cell cytoplasm (at least 100,000-fold) across verapamil-sensitive Ca^{2+} channels located at the basolateral side of the calicoblastic cells. From there, Ca^{2+} moves intracellularly by an unknown mechanism to the other side of the cell. This transcellular transport may be achieved by Ca^{2+}-binding proteins (CaBP). While their potential involvement in Ca^{2+} transport is presently unknown, several CaBP were identified in Cnidarians, such as cytosolic calmodulin (Yuasa et al. 2001), calbindin, or CaBP found in Ca-storage organelles, calsequestrin, or calreticulin. For example, blast on *Acropora millepora* EST library (Meyer et al. 2009) gives $p = e^{-142}$ and 71% homology for human calreticulin. The Ca^{2+}-buffering properties of these proteins enable to increase the total calcium concentration substantially while submicromolar free Ca^{2+} is preserved and therefore cytotoxicity avoided (Feher et al. 1992). Consequent amplification of the Ca^{2+} gradient thereby leads to an increased diffusion rate. Transcellular transport of Ca^{2+} may also be achieved via an organellar route such as vesicles in which X-Ray microanalysis evidenced calcium store (Clode and Marshall 2002b), endoplasmic reticulum, or mitochondria (Klein and Ahearn 1999). Finally, Ca^{2+} has to be transported against its electrochemical gradient across the apical plasma membrane to the other side of the epithelium. Apical Ca^{2+} export is active and probably carried out by a plasma membrane Ca^{2+}-ATPase (PMCA) since transport is inhibited by ethacrynic acid (Marshall 1996) or ruthenium red (Marshall 1996; Al-Horani et al. 2003b). The presence of a Na^+/Ca^{2+} exchanger (NCX) has also been hypothesized on the basis of the inhibitory effect of ouabain, an inhibitor of the Na^+/K^+-ATPase (Marshall 1996). Stable isotope fractionation of Ca^{2+} confirms the presence of a carrier-mediated step (Böhm et al. 2006). However, molecular characterization of these transport systems is poor, and only two genes coding for carrier proteins involved in calcium transport have been cloned in the coral *Stylophora pistillata*: the α1 subunit of an L-Type Ca^{2+}-channel (Zoccola et al. 1999) that was specifically immunolocalized within ectodermal layers and a PMCA (Zoccola et al. 2004) expressed in the calicodermis.

4.3.4 Dissolved Inorganic Carbon (DIC)

It was established by several methods that the major source of carbon (≈60–70%) for calcification is internal CO_2 (Goreau 1977; Erez 1978; Furla et al. 2000b; see Sect. 2.2); however, as discussed above (see Sect. 3.1), this assumption may be questioned. Concerning the mechanism of DIC transport across the different cell layers, this also remains an open debate. A Michaelis-Menten relationship between [HCO_3^-] and $^{45}Ca^{2+}$ incorporation into the skeleton was demonstrated for *Stylophora pistillata* (Furla et al. 2000b) and *Madracis mirabilis* (Jury et al. 2010), suggesting a carrier-mediated step. If for *Stylophora pistillata*, the process saturates at 1 mM HCO_3^-, i.e., below the normal seawater concentration (Furla et al. 2000b), the saturation was reached for around 3 mM in *Madracis mirabilis* (Jury et al. 2009), a result that is in agreement with the fact that addition of bicarbonate into seawater may enhance calcification (Marubini and Thake 1999; Herfort et al. 2008; Marubini et al. 2008), suggesting that in normal seawater, calcification may be DIC-limited.

Because of the high sensitivity to DIDS, a known inhibitor of anion carrier, of $^{45}Ca^{2+}$ incorporation into skeleton, both in the light and in the dark (respectively 85–100% inhibition in the coral *Stylophora pistillata*), and to iodide, a competitor for HCO_3^- anion exchangers (which induced a 60% inhibition in the night while no inhibition is recorded in the light), it was suggested that DIC supply to the site of calcification is mediated by a Cl^-/HCO_3^- exchanger (Tambutté et al. 1996; Furla et al. 2000b). Such a carrier-mediated transport may help to concentrate DIC into the extracellular calcifying medium (see above). However, it should be noted that DIDS is also able to inhibit in vitro coral H^+-ATPase (Furla et al. 2000a) and that CO_2 may directly diffuse from the calicoblastic cells to the ECM, supposed to be alkaline compared to the cells (see Sect. 3.4). Carbonic anhydrase, an enzyme that speeds up the equilibrium between CO_2 and HCO_3^-, is also involved in the process of calcification (see Sect. 4.5). It should be noted that, up to now, no molecular characterization of any HCO_3^- carrier has been published and only hypothesis based on pharmacological data are available. Interestingly, two genes coding for a Na^+-independent Cl^-/HCO_3^- exchanger (p = 3 e^{-38} with human erythrocyte band 3) and a Na^+-driven Cl^-/HCO_3^- exchanger (p = 1 e^{-137} with the mosquito exchanger) were found by blast search on the EST library of *Acropora millepora* (Meyer et al. 2009).

4.3.5 Removal of H⁺

In order to avoid decrease of pH at the site of mineralization due to the release of H^+ from the reaction $HCO_3^- \Leftrightarrow CO_3^{2-} + H^+$, it is necessary to titrate the H^+ produced. One hypothesis commonly accepted is based on the fact that the PMCA acts as a $Ca^{2+}/2H^+$ exchanger. In this way, the transport of calcium from the calicoblastic cells to the extracellular calcifying medium allows the removal of H^+ (McConnaughey and Falk 1991; see for a review Cohen and McConnaughey 2003). pH values have been determined in cnidarian cells to be about 7.0 (Venn et al. 2009), but no data are available concerning the mechanism of pH regulation in the calicoblastic cells and the mode of H^+ exit from these cells. It has been suggested (see below) that once released within the coelenteron, H^+ may be titrated by hydroxyl ions produced by the carbon-concentrating mechanism used for zooxanthellae photosynthesis (Furla et al. 1998, 2000b).

4.3.6 What About Other Mineralizing Organisms?

As stated above, the physiology of ion transport for calcification in mineralizing organisms is poorly known, but many authors suggest a passive supply of calcium. In the freshwater clam, *Unio complanatus*, Hudson (1992), using isolated mantle epithelium, did not found any evidence for net active transepithelial Ca^{2+} transport toward the shell. Similarly, Bleher and Machado (2004) concluded that Ca^{2+} may diffuse paracellularly from the hemolymph toward the extrapalial fluid across the septate junctions of the outer mantle epithelium of the freshwater mollusc *Anodonta cygnea*. These authors suggest however that paracellular diffusion coexists with transcellular transport with a contribution up to 50% of the total flux. In the fish endolymph, it is suggested that Ca^{2+} is supplied by simple diffusion, while active mechanisms are present for inorganic carbon and pH regulation, which play a major role in the regulation of otolith calcification (Allemand et al. 2007). In the foraminifera, Erez (2003) suggested that ions are supplied by vacuolization of seawater and subsequent modification of the content of these vacuoles (elevation of the pH, reduction of Mg/Ca ratio, possible CO_3^{2-} transport) allowing calcification. On the contrary, in the terrestrial crustacean, *Porcellio scaber*, no Ca^{2+} gradient was identified between the hemolymph and the ecdysial space, suggesting that a simple mechanism for paracellular Ca^{2+} transport cannot be sufficient, thus indicating the need for a transcellular mechanism (Ziegler 2008).

4.4 The First Mineral: Amorphous or Crystalline?

The exoskeleton of corals is made of aragonite. Aragonite is the orthorhombic polymorph of $CaCO_3$. It is less stable than calcite at standard temperature and pressure, and tends to alter to calcite on time scales of 10^7–10^8 years. The initial step of crystal nucleation is kinetically unfavorable, and for example experimentally spontaneous nucleation of calcite does not occur in seawater until $\Omega_{calcite} > 20$–25 (Ridgwell and Zeebe 2005).

It now becomes apparent that during the formation of calcite or aragonite in a wide range of calcifying organisms, such as plants, crustaceans, porifera, molluscs, echinoderms, or ascidians, a transient amorphous phase is first formed

(Cohen and McConnaughey 2003, 2003). It is believed that such amorphous phases are important for the subsequent development of crystalline structures (Aizenberg et al. 2003). Amorphous calcium carbonate (ACC) is unstable thermodynamically, and because of its instability and ease of dissolution, it may be used as a temporary storage readily available to a crystallization event. In cnidarians, ACC was found in spicules of the gorgonian, *Leptogorgia* (Weiner et al. 2003). Its presence in corals was suggested by Marshall and Clode (2002) and recently confirmed by Constantz (2008). Using Raman microscopy, he showed the presence of ACC within the scleractinian centers of calcification. Tambutté et al. (1996) found, by performing ^{45}Ca efflux in the coral *Stylophora pistillata*, the presence of a small calcium compartment included within the skeletal compartment. This skeletal compartment has a rapid half-time (12.9 min) compared to the bulk of skeleton (167 h). We can suggest that this compartment may correspond to a transient compartment composed of ACC before its transformation into aragonite.

4.5 Carbonic Anhydrase: A Key Enzyme

Carbonic anhydrase (CA) is a monomeric metalloenzyme that catalyzes the reversible hydration of carbon dioxide to form carbonic acid, which immediately ionizes to give hydrogen and bicarbonate ions following the reaction:

$$CO_2 + H_2O \Leftrightarrow HCO_3^- + H^+$$

The pK of this reaction is close to 6 in seawater, consequently at a normal pH of 8.2, the reaction is largely shifted to the right. According to the value reported by Al-Horani et al. (2003b), the ratio [HCO_3^-]/[CO_2] in the extracellular calcifying medium would vary between 134 and 1,900, respectively, in the night and during the day. It should be noted however that the pK greatly depends on the environment. For example, it varies from 6.75 to 7.41 when the enzyme is fixed respectively on porous silica or on graphite (Crumbliss et al. 1988).

There are at least five classes of CA with polyphyletic origin: α (vertebrates, invertebrates, bacteria and some chlorophytes), β (eubacteria and chlorophytes), γ (archea and some eubacteria), and δ and ζ (marine diatoms) (Supuran and Scozzafava 2007; Xu et al. 2008). Except δ and ζ class, which use cadmium, all other CA use zinc as cofactors. Among numerous roles in physiological process, carbonic anhydrases play an important role in biomineralization from invertebrates to vertebrates even if this role is not well understood (Mitsunaga et al. 1986; Kakei and Nakahara 1996; Miyamoto et al. 2005; Tohse et al. 2006).

In corals, the first mention of the presence of CA came from the inhibitory effect of 1 mM diamox (acetazolamide) on the rate of ^{45}Ca uptake (Goreau 1959). Hayes and Goreau (1977) localized by histochemistry CA in the membrane of the epidermal cells planulae of *Porites porites*. Similarly, Isa and Yamazato (1984) localized histochemically CA exclusively within the calicodermis of the coral *Acropora hebes*. Recently, CA was specifically immunolocalized within the calicodermis of the nonsymbiotic coral, *Tubastrea aurea* (Tambutté et al. 2007b). Curiously, the antibodies used were raised against cyanobacterial β-CA. Since no bacteria were colocalized with CA, this suggests that prokaryotic genes present within the coral genome (Technau et al. 2005) are expressed. Isa and Yamazato (1984) demonstrated that CA activity was correlated with the rate of calcification. All subsequent studies also found that sulfonamides (diamox, AZ, or ethoxyzolamide, EZ) inhibit calcification (Tambutté et al. 1996; Furla et al. 2000b; Al-Horani et al. 2003a, b; Marshall and Clode 2003; Tambutté et al. 2007b). The extent of inhibition goes up to 73% (using 300 μM of EZ on $^{45}Ca^{2+}$ incorporation in the skeleton of *Stylophora pistillata*, Tambutté et al. 1996), suggesting that CA, although important, is not essential to the calcification process. However, no dose-response curve of AZ or EZ on coral calcification was ever published. Inhibition also varies following the light condition. Goreau (1959) found that when incubated in the light with 1 mM AZ, $^{45}Ca^{2+}$ incorporation in the skeleton of *Porites divaritaca* decreased to a level similar to the dark rate, i.e., an inhibition of 51%. When incubated in the dark, the rate of calcification upon addition of AZ further decreases by 34% (Fig. 5). A similar trend was obtained with *Oculina diffusa*. This led Goreau (1959) to suggest that the action of CA and that of photosynthesizing zooxanthellae are "similar and probably synergistic." The subsequent inhibition observed at night however suggests an additional role of CA and photosynthetic zooxanthellae.

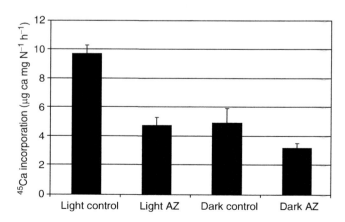

Fig. 5 Effect of 1 mM Diamox (acetazolamide, AZ) on the rate of calcification in the light and in the dark of apical polyps of the branching coral, *Porites divaritaca*. X-axis: treatment (Redrawn from Goreau 1959)

Recently, Moya et al. (2008b) cloned, sequenced, and localized an α-CA from the coral *Stylophora pistillata*, named STPCA. STPCA is a secreted isoform, and, owing to its specific secretion by the calicodermis, they proposed that this enzyme plays a direct role in biomineralization. Furthermore, the first inhibition studies on an invertebrate CA were performed on STPCA at the molecular level using simple anions or sulfonamides/sulfamates (Bertucci et al. 2009a, b). These authors also showed that STPCA can be activated by amino acids and amines (Bertucci et al. 2010). From these studies, it appears that STPCA has a catalytic activity for the CO_2 hydration reaction similar to that of the extracellular secreted human CA VI. By real-time PCR, Moya et al. (2008b) demonstrated that STPCA gene is twofold more expressed in the dark than in the light leading to the hypothesis that up-regulation of STPCA gene allows to cope with night acidosis. This result reinforces the pharmacological experiments of Goreau (1959) that suggested an additional role of CA at night. Furthermore, Grasso et al. (2008) have described, by using microarray analysis, two new CA genes (C007-E7 and A030-E11) in the coral *Acropora millepora*. Expression of these genes is developmentally regulated. While expression of the first CA suggests an involvement in the initiation of calcification at the metamorphosis, expression of the second CA is localized in septa in the adults.

Depending on the source of carbon delivered at the site of calcification, CA could help in supplying dissolved inorganic carbon (DIC) to the site of skeletogenesis or in removing carbonic acid from the site of skeletogenesis:

1. If CO_2 is the form of DIC delivered at the site of calcification (following a diffusion of metabolic CO_2 formed within the cytoplasm of calicoblastic cells to the calcifying fluid due to the probable high pH of this region, Furla et al. 2000b; Al-Horani et al. 2003b; Moya et al. 2008b), then extracellular CA may help in converting metabolic CO_2 into HCO_3^-, thus supplying calcification with inorganic carbon, following the equations (dark arrows: CA-mediated step):

$$CO_2 + H_2O \rightarrow H^+ + HCO_3^-$$

$$\frac{Ca^{2+} + HCO_3^- \rightarrow CaCO_3 + H^+}{Ca^{2+} + CO_2 + H_2O \rightarrow 2H^+ + CaCO_3}$$

The dissociation of the second reaction has a pK near 9. At pH below 8, only a minor quantity of CO_3^{2-} is present ($[CO_3^{2-}]/[HCO_3^-] < 0.01$). The two H^+ produced by these sets of reactions may then be removed from the site of calcification by the Ca^{2+}-ATPase present within the calicoblastic epithelium (Zoccola et al. 2004), which catalyzes the exchange $2H^+/Ca^{2+}$ and then helps in driving the reaction 2 to the right.

2. If HCO_3^- is the form of DIC delivered at the site of calcification, CA may then accelerate the dehydration of carbonic acid to buffer the acidity produced by the conversion of HCO_3^- into CO_3^{2-}. In order to proceed, this reaction needs a low $[CO_2]$ and a pH above 8 (Truchot 1987). Such mechanism was initially proposed by Goreau (1959) as a basis of light-enhanced calcification (see Sect. 5.1.4) following the equations:

$$Ca^{2+} + HCO_3^- \rightarrow CaCO_3 + H^+$$

$$\frac{H^+ + HCO_3^- \rightarrow CO_2 + H_2O}{Ca^{2+} + 2HCO_3^- \rightarrow CaCO_3 + CO_2 + H_2O}$$

In both cases, hydration/dehydration reactions may be mediated either by a CA anchored within the plasma membrane or present in the ECM as part of the organic matrix (Tambutté et al. 2007b).

From this short review, it appears that CA plays a major role in coral calcification as in other biomineralizing organisms. However, this role is not yet well fully understood and the number of isoforms is not known. It should be added that the role of CA is not restricted to the calcification process since CA is also involved in carbon-concentrating mechanisms for DIC supply to the symbiont photosynthesis (Weis et al. 1989; Furla et al. 2000a; see for a review Allemand et al. 1998a; Leggat et al. 2002).

4.6 The Key Role of Organic Matrix

The presence of organic matrix (OM) in a biomineral is the key feature indicating a "biological control" of the mineralization process. In corals, such a presence was debated until recently. In the biomedical field, where coral skeletons are used as bioimplants, it is often admitted that the OM composition is limited to the presence of single amino acids (Demers et al. 2002). In the geochemical field, it is generally believed that the OM only consists of the remains of animal tissue pinched between fully developed crystals (Constantz 1986). Compared to other calcifying invertebrates such as molluscs, arthropods, and echinoderms, very few data are available for coral OM (for review see Watanabe et al. 2003; Tambutté et al. 2007c). However, even in the other well studied model organisms, only putative functions are attributed to OM such as concentration of precursor ions, constitution of a tridimensional framework, template for nucleation of crystals, determination of calcium carbonate polymorph, control of crystal elongation, inhibition of crystal growth, determination of spatial arrangement of crystal units, involvement in enzymatic functions and cell signaling (Addadi and Weiner 1985, 1989; Marin et al. 2008). OM also leads to the strength of the coral skeleton. Indeed, while it ranks among the weaker skeletal materials (12–81 MN m^{-2}), excepted for cancelous bone (4 MN m^{-2}), it is higher than engineering materials like concrete (32 MN m^{-2}), probably because of its low OM content (Chamberlain 1978).

4.6.1 Content in OM

In corals, the presence of OM in the skeleton was first reported in the nineteenth century when Silliman (1846) claimed that 4–8% of skeletal weight was organic. About 100 years later, Wainwright (1963) estimated that OM only represents 0.01–1% of skeletal weight, and, later on, Johnston (1980) warned that when identifying OM it was important to distinguish between contaminants (soft tissues and endoliths) and real OM. The problem with endoliths (microalgae, fungi, and cyanobacteria) is that they can remove part of the matrix or modify its composition (Vertino et al. 2007). The same is true if tissues contaminate the preparation, and so precautions must be taken by carefully cleaning the powder of skeleton in a bleach solution. Once these precautions have been taken, the best estimate of OM content is obtained by thermogravimetric analysis on clean powder of skeleton (without demineralization) that gives a value of 1–2.5%, respectively, for the nonhydrated and the hydrated form of OM (Cuif et al. 2004a; Tambutté et al. unpublished results). However, biochemical studies mainly concern OM extracted after demineralization of powdered skeleton, which means that several steps are necessary before analysis. These steps consist in removal of the tissues, powdering of the skeleton, cleaning of the powder, dissolution of the powder, removal of salts, and separation of soluble (SOM) from insoluble OM (IOM). Consequently, some OM components can be lost during this long extraction process. For example, Puverel et al. (2007) have shown that low molecular weight components (<3.5 kDa), which are lost during removal of the salts, represent a large amount of organic matrix. One must also keep in mind that IOM is usually discarded for analysis due to the difficulty in resolubilizing this fraction and so that most values and biochemical characterization only concern the components of the SOM.

4.6.2 Synthesis of OM

As early as 1903, Duerden noted that there was "an organic secretion external to the coral skeletogenic tissues." Then, many observations in microscopy confirmed the presence of an organic sheet close to the calicoblastic cells, visualized after decalcification of the skeleton (Wainwright 1963; Johnston 1980; Le Tissier 1990, 1991; Yamashiro and Samata 1996; Goldberg 2001b; Clode and Marshall 2002b). Direct evidence of the synthesis of OM by calicoblastic cells was provided by immunohistochemistry. Puverel et al. (2005a), using antibodies raised against organic matrix, showed that only the calicodermis was specifically labeled. Moreover, they invalidated the hypothesis that OM consists of the remains of calicoblastic cells incorporated into the skeleton as proposed by some authors (Ogilvie 1897; Constantz 1986; Ramseyer et al. 1997), since the skeleton was not labeled with antibodies against proteins of the calicoblastic cell membranes. Concerning the mode of secretion of OM, intracellular vesicles within calicoblastic cells were observed (Johnston 1980; Le Tissier 1990, 1991; Clode and Marshall 2002b) and proposed to contain calcium and organic matrix (Johnston 1980; Marshall and Wright 1993; Clode and Marshall 2002b). Small vesicles were also sometimes observed between the calicoblastic cells and the skeleton (Le Tissier 1990). However, Clode and Marshall (2002a) suggested that these exocytosed vesicles were an artifact to fixation. Still, it cannot be completely excluded that exocytotic vesicles exist since experiments performed with agents inhibiting cytoskeleton activity prevent calcification, suggesting that exocytosis is involved in this process (Allemand et al. 1998b).

4.6.3 Biochemical Characterization of OM

Chitin was reported in some corals (Wainwright 1963), but it was later suggested that its presence was limited to some coral species (Young et al. 1971); however, no recent studies have looked for this polysaccharide. Hyaluronan-like substances were suggested in one coral species, *Mycetophyllia reesi* (Goldberg 2001b). The presence of lipids in the OM of several species was also reported (Bergmann and Lester 1940; Young et al. 1971; Isa and Okazaki 1987; Stern et al. 1999; Watanabe et al. 2003). Finally, most of the papers report the presence of proteins that can be sulfated, glycosylated or not (Wainwright 1963; Young 1971; Constantz and Weiner 1988; Dauphin 2001; Watanabe et al. 2003; Fukuda et al. 2003; Puverel et al. 2005b), and sugars sulfated or not (Goreau 1956; Constantz and Weiner 1988; Goldberg 2001b; Cuif et al. 2003; Puverel et al. 2005b). The protein to sugar ratios in the SOM vary among the species (Dauphin 2001; Puverel et al. 2005b). The content in proteins in the SOM is more constant among species than the sugar content (Puverel et al. 2005b). For example, in *Stylophora pistillata* and *Pavona cactus*, the protein content is not statistically different (respectively about 0.013% and 0.021% of the skeleton weight), whereas the content in total glycosaminoglycans and sulfated glycosaminoglycans is, respectively, 16- and 17-fold higher in *S. pistillata* (Puverel et al. 2005b). Finally, it is important to note that organic matrix contains water (see above and Cuif et al. 2004b), suggesting that matrices are present as hydrogels (Dauphin et al. 2008). The ratio of water in these hydrogels probably varies from species to species (Dauphin et al. 2008).

One characteristic of proteins extracted in the soluble organic matrix is their richness in acidic amino acids, and more specifically in aspartic acid (Young 1971; Mitterer 1978; Constantz and Weiner 1988; Cuif et al. 1999; Puverel et al. 2005b). The acidic feature of these proteins renders their biochemical characterization difficult (for review see Marin et al. 2008). The first organic matrix protein fully characterized in corals is galaxin, in *Galaxea fascicularis* (Fukuda et al. 2003). This protein is glycosylated, has an apparent molecular weight of 53 kDa, does not bind calcium, and its amino acid sequence is characterized by a tandem repeat structure. By EST analysis, three genes related to galaxin have been identified in the coral *Acropora millepora* and one of them is a galaxin ortholog named Amgalaxin (Reyes-Bermudez et al. 2009). All these proteins share the presence of repetitive motifs containing dicysteine residues. Other proteins have only been visualized on gels (Watanabe et al. 2003; Puverel et al. 2005b), and partial sequences have been determined for some of them (Puverel et al. 2005b). Among these partial sequences, one is noteworthy since it contains a series of 36 consecutive residues of Asp. Concerning sugars, it has been shown that they consist in monosaccharides (Gautret et al. 1997; Cuif et al. 1999) and in glycosaminoglycans (Puverel et al. 2005b).

One interesting point is to determine how many proteins are present in the OM of corals. However, the task is difficult for different reasons: (1) some proteins can be lost during the preparation of OM (see Sect. 4.6.1), (2) SOM is generally studied but IOM is usually discarded (see Sect. 4.6.1), (3) migration of acidic proteins in gels is difficult (Marin et al. 2008). If we refer to one-dimensional gel, it appears that there are less than ten proteins in the OM of corals (Puverel et al. 2005b; Fukuda et al. 2003; Watanabe et al. 2003). However, this number is certainly underestimated and linked to the difficulties in OM biochemical analysis. Indeed, whereas by this method the number of OM proteins in molluscs is estimated to be a few than ten of molecules, it has been suggested by working at the secretome level, that hundreds of proteins are likely to be contributing to shell fabrication and patterning (Jackson et al. 2006), and 520 proteins were identified in the acid-soluble organic matrix of the chicken calcified eggshell (Mann et al. 2006).

4.6.4 Role of OM in Calcification

Wainwright (1963) suggested that the limiting factor in the calcification process may well be the rate of synthesis of OM. Then, it has been shown by physiological studies using labeled bicarbonate or amino acids that organic matrix synthesis is a prerequisite step in the calcification process (Young 1973; Allemand et al. 1998b). Yamashiro and Samata (1996) suggested that OM observed in the initial settlement site of larvae may act as barriers for the calicoblastic cells rather than as sites for calcification, but this has never been confirmed. As they observed that the distribution pattern of organic matrix is not uniform across the skeletal surface, Clode and Marshall (2002a) suggested that the sheath of OM may act as an inhibitory structure, which may control and restrict $CaCO_3$ deposition or alternatively it may take on a structural rather than a regulatory role. Cuif and Dauphin (2005a, b) proposed a two-step mode of growth of coral skeleton, in which the biomineralization process starts with the secretion of a proteoglycan matrix where fibers and centers of calcification (COC) are simultaneously produced (see Sect. 6.1).

Due to the abundance of acidic amino acids in the OM, and especially because of long sequences of Asp (see Sect. 4.6.3), it has been suggested that OM can bind calcium with a high capacity and a low affinity (Puverel et al. 2005b) and thus can concentrate and deliver calcium when necessary for calcification since this binding is reversible. Since calcium binding capacity was also measured with skeletal phospholipids, it was proposed that they could serve as templates for $CaCO_3$ nucleation (Isa and Okazaki 1987; Fukuda et al. 2003). Aspartic acid residues have been also suggested to alter the equilibrium thermodynamics of the growth surface (Teng et al. 1998). The role of Galaxin still remains enigmatic even if the abundance of Cys residues may suggest that galaxin proteins are highly cross-linked to one another through intermolecular sulfide bonds to form a macromolecular network, thereby providing a structural framework to the matrix (Fukuda et al. 2003). OM may also play a catalytic role since it possesses a carbonic anhydrase (CA) activity (Watanabe et al. 2003; Tambutté et al. 2007b) that can eliminate the kinetic barrier to the interconversion of inorganic carbon at the calcification site. It has thus been suggested that CA may directly hydrate CO_2 and form CO_3^{2-} through nucleophilic reactions (Nakata et al. 2002; Ichikawa 2007). Finally, it can be assumed that OM plays a key role in the morphology of the skeletal elements since the quantity, the structure, and the composition of OM is taxonomy linked (Cuif et al. 1996; Dauphin et al. 2006, 2008).

4.6.5 Interaction of OM and Calcium Carbonate

At the macroscale level, it appears that OM is the ghost of the demineralized skeleton, suggesting that OM (which in this case consists in the insoluble fraction) plays a role of frame for calcification. At the microscale level, there is still a perfect correspondence between the skeletal units and the OM pattern as can be seen in Fig. 6.

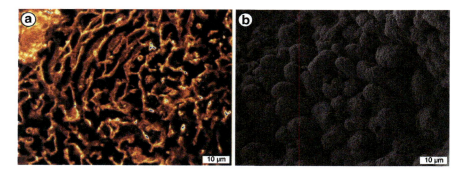

Fig. 6 Localization of organic matrix between skeletal fibers as shown by immunolabeling using antisoluble organic matrix antibodies (**a**) and correspondence with the skeletal fibers observed under scanning electron microscopy (**b**)

At the beginning on the twentieth century, it was hypothesized that organic material forms thin membranes surrounding groups of crystals (Krempf 1907; Hayasi 1937). Later it was reported that calcification centers are richer in organic compounds than the fibers (Alloiteau 1957). The organic envelopes surrounding the group of fibers were revealed by demineralization and either staining (Johnston 1980) or immunolabeling (Puverel et al. 2005a). Organic material in calcification centers and fibers was recorded by fluorescence microscopy (Gautret et al. 2000; Stolarski 2003), scanning electron microscopy (Perrin 2003), or immunolabeling (Puverel et al. 2005a). By Raman microspectroscopy, Perrin and Smith (2007) have confirmed the presence of OM in centers of calcification and fibers with a higher content in the former. The presence of sulfate in the skeleton at the microscale level has been evidenced by XANES spectroscopy (Cuif et al. 2003), but its origin either inorganic or organic has been a matter of debate (Sinclair 2004). By AFM phase images, it has been shown that OM is present at the nanoscale level since nodular nanograins of calcium carbonate are surrounded by an organic cortex (Cuif and Dauphin 2005a, b) of about 6 nm thickness (Dauphin et al. 2008).

Based on operational protocols, OM can be qualified as "intra or non-intracrystalline." "Non-intracrystalline" refers to the OM that is removed by bleach treatment and that can be degraded by endoliths (Ingalls et al. 2003). Intracrystalline organic matrix refers to OM that is occluded within the mineral structure and that is not removed from a finely powered skeleton by exposure to a strong oxidant, such as commercial bleach (Ingalls et al. 2003). Dauphin et al. (2006) have shown that organic matrices occluded in the aragonitic crystals of coral skeletons are not fully destroyed even when heated up to 300°C. Tambutté et al. (2007b) have shown that the enzymatic activity of OM is kept as long as OM is embedded in the mineral. All these observations indicate how intimate and protective is the relationship between mineral and organic components.

Perrin and Smith (2007) have observed that organic decay resulting from the hydrolysis of organic matrices occurs rapidly (in a few years) and is one of the key-mechanisms driving the first steps of diagenesis. Nyberg et al. (2001) have also observed rapid decrease in the total amino acid concentration of 33% over hundreds of years. However, other studies have shown preservation of OM through time (Ingalls et al. 2003; Gupta et al. 2006; Gautret 1993; Sorauf 1999; Muscatine et al. 2005). Indeed, Ingalls et al. (2003) have shown high degree of preservation of OM over several hundred of years and experiments performed on skeletons from the Triassic show that the matrix is preserved (Gautret 1993; Sorauf 1999; Muscatine et al. 2005). Ingalls et al. (2003) underline the fact that "intracrystalline OM may be better protected from diagenesis that nonintracrystalline OM." Here again, these observations suggest that OM is preserved in the mineral.

4.6.6 Comparison with OM from Other Invertebrates

It appears that secreted proteins of OM from different organisms generally show no similarity to other proteins in public databases (Jackson et al. 2006; Marin et al. 2008); however, some common features can be drawn. The abundance of Asp in OM proteins is a common feature with other invertebrate models such as molluscs (see Marin et al. 2008 for detailed review). The presence of consecutive Asp residues has also been reported for Aspein in the pearl oyster *Pinctada fucata*. The composition of aspein is remarkable since Asp residues represent the major part of the whole protein with the main body composed of many poly-Asp blocks (of 2–10 Asp residues long) interspersed by Ser-Gly dipeptides. However, even if the reason is not known, it is interesting to note that, in corals, acidic proteins are found in the aragonitic skeleton whereas in molluscs, they are mostly found in the calcitic prisms and not in the aragonitic tablets (Marin et al. 2008). Concerning Galaxin, this protein does not exhibit significant high sequence similarity to proteins in sequence databases. However, tandem repeat regions rich in cysteine have also been found in proteins isolated from the calcified molluscan exoskeletons (Watanabe et al. 2003).

One other characteristic of OM proteins from different invertebrate models is their modular organization, each module corresponding to a functional domain. Some domains are clearly identified, like the carbonic anhydrase domains found in the nacrein of Molluscs (Miyamoto et al. 1996, 2005) and in corals (Fukuda et al. 2003). Moreover, carbonic anhydrase activity is a common feature of OM from different models since this activity has been reported in a number of calcifying organisms (Borelli et al. 2003a, b). Another point to mention concerns cross-reactivity with antibodies against OM. Immunologic interphylum and interclass cross-reactivities have been observed in a number of cases: vertebrate to echinoderm (Veis et al. 1986), echinoderm to prochordates (Lambert and Lambert 1996), molluscs to brachiopods (Marin unpublished data in Marin et al. 2008), and within the phylum Mollusca (Marin et al. 1996). Experiments using antibodies raised against the purified molluscs OM proteins, mucoperlin, OMP1, and otoline do not show any positive reactivity with the coral *Stylophora pistillata* (Tambutté, unpublished results). Cross-reactions in corals with antibodies against the whole SOM show positive reactivity among families (Puverel et al. 2005a). Cross-reactivities between corals and fish otolith OM also show positive results, and, more specifically, antibodies against Starmaker (a protein from the OM of fish otolith Söllner et al. 2003) show positive reactivity in corals (L. Pereira-Mouriès, personal communication, 2003). Thus, even if cross-reactivities may be fortuitous, due to similar overall topographies of the epitopes, they may be also highly significant of the presence of short conserved epitopes across calcifying phyla as suggested by Marin et al. (2008).

Bone morphogenetic protein (BMP), a protein initially discovered in the bone organic matrix (Urist 1965) and shown to be able to induce bone formation when implanted ectopically, was also found to be expressed in the coral larvae (Lelong et al. 2001). BMP was recently demonstrated to be specifically expressed in the calicodermis of the adult stage of the coral *Stylophora pistillata* (Zoccola et al. 2009). This suggests that some basic mechanisms may have evolved in the common ancestor of metazoans, supporting the "ancient heritage" hypothesis of the origin of biomineralization (Marin et al. 2008).

4.6.7 Conclusions

In corals, to date, only one protein has been fully characterized in the organic matrix, whereas more than 40 proteins have been sequenced in molluscs and more than 500 were putatively present. However, even in molluscs, the function of these proteins is still enigmatic. Indeed, many of these proteins exhibit sites for posttranslational modifications, which are not biochemically characterized but which can change the function of the proteins. Moreover, most of the putative functions are deduced from in vitro experiments that do not mimic the right conditions in which mineralization occurs. The whole genome sequencing of corals will be helpful to determine proteins involved in calcification even if seeking out the gene of interest will be a long task. Microarray analysis or differential display between different stages of larva development will help in this task (Grasso et al. 2008). A proteomics approach will also remain a good tool even if OM proteins usually show low homologies in database. Indeed, the use of de novo sequencing will be helpful to overcome this obstacle. Even when many more proteins are identified in corals, we will be facing the same limitations as models that are more ahead in terms of progress. Indeed, enumerating precisely the proteins is a necessary required step for drawing the outline of the different functional domains and proteins families, for phylogenetic purposes, but an exhaustive catalog of OM proteins will not help so much in understanding the biomineralization process. Determining the expression of genes will, for example, be a necessary tool to better understand calcification (Moya et al. 2008c).

Finally, one of the most promising approaches to relate structure to function consists in reverse genetics. Indeed, the work performed by knocking down the gene of Starmaker in fish embryos has allowed the determination of the role of such a protein in the mineralization process (Söllner et al. 2003). In cnidarians, this approach has been successful in solitary sea anemones (Dunn et al. 2007), but until now, there are no data for corals in the literature, and no success was obtained in preliminary experiments performed on colonial adults (Zoccola, unpublished data). Taking into account that the number of proteins involved in biomineralization can be estimated as high as several hundreds (Mann et al. 2006) and that other components such as sugars and lipids are also present in OM, one can imagine that we are far from revealing the secrets of OM. The study of matrix proteins is still in its infancy, and a large number of issues remain to be studied.

4.7 The Cost of Calcification

No data are available concerning the energetic cost of coral calcification, and very few data are published on other calcifying organisms. It is known for example that ion transport activity is a costly process. For example, Na$^+$/K$^+$-ATPase uses 6–40% of the daily cell ATP (Hochachka et al. 1996). In coccolithophores, calcium transport represents a large part of the total cost of calcification (Anning et al. 1996). In molluscs, Palmer (1992) estimates that the cost of calcification would be near 6% of the total respiratory loss but would be equivalent to 75–140% of the energy invested in somatic growth and reproduction, respectively. This author estimates that the total cost of calcification would be 1–2 J mg^{-1} CaCO$_3$. Twenty-two percent of this cost is due to protein synthesis for shell material having 1.5% organic matrix. In the case of coral skeleton, using these values for respectively an organic matrix content of 0.1–1%, the total cost of coral calcification would be 0.80–1.85 J mg^{-1} CaCO$_3$. These values have to be compared to the net energy budget measured for the coral *Porites porites* by Edmunds and Davies (1986). Total energy budget of the animal host is 495.8 J g^{-1} dry tissue day^{-1}. As these authors reported a skeletal growth of 81.7 mg CaCO$_3$ g^{-1} dry tissue day^{-1}, and taking into account a cost varying from 0.80 to 1.85 J mg^{-1} CaCO$_3$ (see above), one may infer that the total cost of skeleton formation for *P. porites* varies from 65 to 151 J g^{-1} dry tissue day^{-1}, i.e., 13–30% of the total energetic cost of the coral. For comparison, tissue growth is assumed to represent about 40 J g^{-1} dry tissue day^{-1}, i.e., 8% of the total energetic cost of the coral. We should add to these data the results obtained by Fine and Tchernov (2007) who observed that coral subjected to acidic conditions (7.4) for 12 months totally dissolved their skeletons with a simultaneous three-times increase of tissue biomass, suggesting a reallocation process of energy after total arrest of calcification and thus confirming the high cost of calcification. This also confirms the assumption of Anthony et al. (2002) who state that energy allocation in corals is prioritized for skeletal rather tissue growth in normal conditions.

5 Environmental Control of Calcification

Even if, as described in the previous sections, the process of calcification still remains enigmatic on numerous aspects, the existing data allow us to conclude that calcification is a biologically-controlled process. However, although the microstructure of a coral is only determined by its genetic characteristics and so used as criteria in taxonomy, the macrostructure is under both genetic and environmental control (see Kaandorp in this book). Environmental parameters such as light, temperature, nutrition, pCO$_2$, salinity, and turbidity also control the rate of calcification. We will examine here the effects of light and temperature, other parameters being presented in other chapters of this book (see Erez et al., Chap. 10, this volume; Ferrier-Pagès et al., Chap. 15, this volume).

5.1 Light

Symbiotic corals contain intracellular zooxanthellae that perform photosynthesis under light conditions and coral morphology is adapted to optimize light collection for zooxanthellae photosynthesis. As an example, the coral *Stylophora pistillata* shows different ecomorphs varying from a branching to a flat form, respectively, in light and shade conditions (Gattuso 1987). Most of the symbiotic scleractinian corals are found in the euphotic zone, from the surface to 80 m even if the presence of *Leptoseris* at 165 m depth has been reported (Kahng and Maragos 2006). The rates of calcification of symbiotic corals are higher in the light than in the dark, a process described under the term of "light-enhanced calcification" (LEC).

5.1.1 Background History of LEC

The first in situ observations indicating a tight relationship between light and calcification date from the years 1930s. Yonge (1931) indicated that "the association between corals and zooxanthellae is… probably an indispensable factor in the necessarily exceptional powers of growth and repair possessed by the marine communities known as coral reefs." Yonge and Nicholls (1931) also suggested that the effect of light on coral growth was probably through the photosynthesis of zooxanthellae. Moreover, Kawaguti (1937) observed that corals could not develop in shaded areas and thus that light conditions had to be taken into account when performing growth experiments. A few years later, the first laboratory experiments were set by Kawaguti and Sakumoto (1948) who showed that corals "consume calcium from the surrounding seawater and accumulate it in the light." Finally, Goreau and Goreau (1959a) observed that corals deposit calcium fastest in sunlight, less during cloudy weather and least in darkness. Moreover, they observed that bleached colonies deposited calcium at lower rates in the light than normal colonies with zooxanthellae did in darkness.

5.1.2 LEC During a Daily Cycle

The compilation of 108 data from 26 papers of the literature shows that the median value of enhanced calcification by light is around 3 (Gattuso et al. 1999). The large range of ratios observed has been attributed to the methodology, to environmental parameters or to biological factors for in situ as well as for laboratory experiments. An explanation of this variability may come from the work of Schneider and Erez (2006). These authors showed that the magnitude of LEC is in fact not constant and varies according to pH and the carbonate chemistry, but the difference between light and dark rates of calcification remains constant. Concerning the biological parameters, the endogenous circadian rhythms have been implicated, and it was suggested to make experiments at fixed hours in order to compare the results (Buddemeier and Kinzie 1976; Tambutté et al. 1995, 1996; Houlbrèque et al. 2004). When light varies during a daily cycle, the calcification rates vary (Roth et al. 1982). When light is maintained constant, contradictory results have been obtained. Indeed, under these conditions, some studies show that rates vary (Clausen and Roth 1975a; Chalker 1977; Al-Horani et al. 2007) or not (Moya et al. 2006). Such differences can be attributed to a difference among species. Levy et al. (2007) have shown that corals possess light receptors (cryptochromes), whose expression varies during a daily cycle, but they did not measure the rate of calcification, so it is difficult to conclude on the existence of a rhythm for LEC. It is interesting to note that Schneider et al. (2009) observed in situ a phenomenon of hysteresis for both calcification rates and photosynthesis.

5.1.3 Controversy on LEC

As described above, as soon as the first experiments on LEC were performed, it appeared that zooxanthellae were at the origin of the process. Later on, the use of DCMU, an inhibitor of photosynthesis, allowed to conclude that the inhibition of photosynthesis was responsible for a decrease in calcification, and thus that LEC was directly linked to photosynthesis and not to light (Vandermeulen 1975). However Marshall (1996) have suggested that in fact photosynthesis does not enhance calcification but rather calcification is inhibited in the dark due to the presence of zooxanthellae, a process called dark-repressed calcification. In this case, the zooxanthellae in the dark would decrease calcification by storing the metabolites produced by the host. Since in the dark, bleached corals do not show higher rates of calcification than symbiotic corals (Moya et al. 2008a), this hypothesis can be challenged. Other authors have suggested than rather than light-enhanced calcification, the process should be described as "transcalcification" or stimulation of photosynthesis by calcification (McConnaughey and Whelan 1997). In this hypothesis, calcification would enhance photosynthesis by supplying in the coelenteron a source of protons that would help in converting bicarbonate into CO_2 thus available for photosynthesis. The fact that biphosphonate, a chelating agent of calcium that inhibits calcium carbonate deposition, has no effect on photosynthesis (Yamashiro 1995) suggests that calcification is not enhancing photosynthesis; however, since this inhibitor acts only on amorphous calcium carbonate (Marshall and Clode 2002), it is no possible to conclude. Marshall and Clode (2002) have also demonstrated that increasing calcium concentration in seawater increases calcification that in turn stimulates carbon incorporation into the tissues, which should indicate an increase in photosynthesis. However, since they did not observe increase in net photosynthesis (in term of net O_2 production), nor in respiration, it is impossible to conclude. Photosynthesis is not increased when increasing seawater calcium concentration (Gattuso et al. 2000), but it was shown, for the same coral species, that calcification is saturated at ambient seawater concentration (Tambutté et al. 1996). Curiously, Schneider and Erez (2006) observed that photosynthesis remains constant when calcification is increased by changing the concentration of carbonates although a correlation was found between these two parameters when all data are pooled. So, this controversy on LEC versus calcification enhanced photosynthesis is still under debate.

5.1.4 Hypothesis for LEC

For many years, different hypothesis have been proposed to explain LEC (for review see Gattuso et al. 1999; Allemand et al. 2004). They can be schematically divided into two groups, one based on inorganic chemistry and the other based on organic chemistry. It should be noted however that, among all the hypothesis proposed for LEC, none is exclusive and most can even be complementary. To decide among them, further experiments are necessary.

LEC and Inorganic Chemistry

To better understand these hypothesis, one must keep in mind the following equations:

$$\text{Photosynthesis}: CO_2 + H_2O \rightarrow CH_2O + O_2$$
$$\text{Respiration}: CH_2O + O_2 \rightarrow CO_2 + H_2O$$
$$\text{Calcification}: Ca^{2+} + 2HCO_3^- \rightarrow CaCO_3 + H_2CO_3$$
$$\text{or } Ca^{2+} + 2HCO_3^- \rightarrow CaCO_3 + CO_2 + H_2O$$
$$\text{or } Ca^{2+} + CO_2 + H_2O \rightarrow CaCO_3 + 2H^+$$

This group of hypothesis can be divided into subgroups.

LEC and Inorganic Carbon

Goreau (1959) proposed that the kinetics at which carbonic acid is removed from the calcification site during calcium carbonate precipitation depends on the fixation of CO_2 for zooxanthellae photosynthesis. However, it has been shown that the major source of inorganic carbon for photosynthesis is external bicarbonate (Furla et al. 1998; Allemand et al. 1998a). Furla et al. (2000a, b) suggested that the removal of protons from the calcification site would enhance calcification. Indeed, as stated in previous paragraphs, the only measurements performed in the calcifying medium (Al-Horani et al. 2003b; Rollion-Bard et al. 2003) show that pH is more basic than seawater, confirming that protons are removed from this medium. Measurements of pH in the coelenteron show that the values are higher in the light than in the dark due to the transformation of bicarbonate into CO_2 for photosynthesis (Furla et al. 2000a, b; Al-Horani et al. 2003a, b). This pH could create a favorable gradient of protons between the calcifying medium and the coelenteron and thus help in the removal of protons from the calcifying medium. The kinetics of pH transition between light and dark conditions is compatible with the kinetics of calcification rates during such a transition and could validate this hypothesis (Moya et al. 2006).

Goreau (1959) and Furla et al. (2000a, b) have shown that the effect of carbonic anhydrase inhibitors on calcification is higher in the dark than in the light, suggesting an increase of carbonic anhydrase activity in the dark. A similar light/dark regulation was previously observed in fish otoliths in which CA shows a marked diurnal variation of gene expression with ca. fourfold increase of expression at night compared to morning expression (Tohse et al. 2006). In corals, it has been suggested that the up-regulation of CA expression during the dark (Moya et al. 2008b) facilitates the continuous production of bicarbonate from metabolic CO_2 in spite of an increase in H^+ concentration into the subcalicoblastic environment (from 9.3 in the light to 8.2 in the dark, Al-Horani et al. 2003b). This result also shows that despite an enhancement of CA expression in the dark, a change in CA expression is likely not directly involved in light-enhanced calcification. Recent results, showing that STPCA can be activated by amino acids and amines (Bertucci et al. 2010), suggest that LEC can be explained by the activation of CA by photosynthates provided by the zooxanthellae. Indeed, it has been shown that zooxanthellae can deliver amino acids to the host (Wang and Douglas 1999), but so far, the nature and structure of these amino acids that could be involved, in vivo, in CA activation are still unknown.

LEC and Supply of Ions

Mueller (1984), by a pharmacological approach, suggested that during the day, the Ca^{2+} channel supposed to regulate the supply of Ca^{2+} for calcification, is activated by a phosphorylation following the action of a cyclic adenosine 3′,5′-monophosphate, and thus this mechanism would allow more calcium to enter the cells. Sanderman (2008) suggested that the entry of calcium into the calicodermis is stimulated in the light by lipid peroxidation induced by hydrogen peroxide (H_2O_2) generated during photosynthesis. Marshall (1996) has shown that the mechanism of ion transport is different between light and dark conditions. Ca^{2+}-ATPase pump was suggested to be inactive during the night (Cohen and McConnaughey 2003) and stimulated by light during the day (Al-Horani et al. 2003b). Moya et al. (2008c) have studied the expression of two genes coding for proteins involved in calcium transport, the calcium channel and the Ca^{2+}-ATPase and determined that they do not vary between light and dark. However, this does not preclude that these proteins could be regulated by posttranslational modifications.

LEC and Removal of Phosphates

Simkiss (1964) proposed that zooxanthellae may remove phosphates produced by the animal metabolism. No measurement of phosphates at the calcification site has been performed, but when phosphate concentration is increased, the skeleton shows morphological alterations (Rasmussen 1988) and decrease in calcification rates (Ferrier-Pagès et al. 2000). However, the absorption of phosphate has been shown in isolated zooxanthellae (Jackson and Yellowlees 1990) but not *in hospite*.

LEC and Organic Chemistry

This group of hypothesis can be divided into four subgroups:

LEC and Supply of Precursors for Organic Matrix Synthesis

Wainwright (1963) was the first to suggest that the formation of organic matrix is a limiting factor for calcification and that organic carbon transferred from zooxanthellae to the host could be used, directly or indirectly, for the synthesis of organic matrix. A few years later, Muscatine and Cernichiari (1969) have effectively shown that zooxanthellae are able to transfer molecules to the host and that these molecules are incorporated into the skeleton. Cuif and colleagues have then determined that the composition in amino acids and sugars is different between symbiotic and nonsymbiotic corals (Gautret et al. 1997; Cuif et al. 1999). Muscatine et al. (2005) have also shown that the stable isotope composition ($\delta^{13}C$ and $\delta^{15}N$) of OM is different between symbiotic and nonsymbiotic corals. Finally, Houlbrèque et al. (2004) have shown that the amount of organic matrix incorporated in dark conditions is lower than in light conditions. The fact that the timing of the transition of the calcification rate between light and dark conditions (Moya et al. 2006) and the time required for an amino acid to be absorbed and incorporated into OM (Allemand et al. 1998b) are similar, support the hypothesis of a light-dependent role of zooxanthellae in the synthesis of OM precursors. In any case, all these results show that zooxanthellae influence both the quantity and the quality of organic matrix. Zooxanthellae can either directly supply sugars, lipids or proteins or indirectly supply precursor molecules. By immunohistochemistry using antibodies against OM, it has been shown that only the calicoblastic cells are responsible for the synthesis and secretion of OM (Puverel et al. 2005a, b), suggesting an indirect role of zooxanthellae.

LEC and Oxygen

It has been proposed that oxygen is a limiting parameter for calcification and so in the night calcification would be inhibited by the lack of oxygen (Al-Horani et al. 2007) whereas production of oxygen by photosynthesis would increase coral metabolism and calcification (Rinkevich and Loya 1984). Recently, Colombo-Pallota et al. (2010) have shown that both oxygen and glycerol are necessary for LEC.

LEC and Supply of ATP

Crossland and Barnes (1974) have succeeded in stimulating the incorporation of calcium in the skeleton by adding ATP in the incubation medium. Then, Chalker and Taylor (1975) have shown that the inhibition of phosphorylative oxidation leads to a total inhibition of calcification in the light whereas it does not inhibit the dark calcification. They have thus suggested that photosynthesis produces molecules that can be used for respiration and production of ATP. They have suggested that this ATP could play a role in LEC by stimulating ion transport and/or organic matrix synthesis. This hypothesis was validated by Colombo-Pallota et al. (2010).

LEC and Nitrogen

Crossland and Barnes (1974) have suggested that the nitrogen metabolism of the animal combined to the production of glyoxilic acid by the zooxanthellae favor calcification by regulating the pH on the site of calcification. Such a hypothesis has never been further explored.

5.1.5 Paradox of LEC

Despite all the studies and hypothesis on LEC, a paradox rapidly appeared: in the 1950s, Goreau indicated "although the zooxanthellae seem to play an important role in determining calcification rates in reef-building corals, certain, as yet unknown, physiological factors operate to control the basic mineralization process in a manner which bears no obvious relationship to the number of algae present in a given species" (Goreau 1959a). If Goreau was a kind of visionary of the biological control of calcification, he was also "avant-gardiste" in sensing the difficulty to determine the exact role of zooxanthellae in LEC. In fact, his conclusions came from the observation, later on also reported by several authors, that the tips of coral branches always contain less zooxanthellae than the base whereas they calcify more (Goreau and Goreau and Goreau 1959a; Pearse and Muscatine 1971; Fang et al. 1989). This observation was then extended and generalized to septae (Marshall and Wright 1998) and to new mineralizing front of colonies growing on slides (Raz-Bahat et al. 2006; Tambutté et al. 2007a) where calcification rates are the highest whereas zooxanthellae concentration is the lowest. The transfer of molecules from the base to the apex has been shown and could explain the paradox of LEC (Pearse and Muscatine 1971; Taylor 1977), but other studies challenge this hypothesis (Rinkevich and Loya 1984). Even if not considering the long distance that molecules have to cover from base to apex, molecules have at least to be transported from the zooxanthellae, inside the oral endodermal cells, to the calicodermis located at 10–30 μm away in the coral *S. pistillata* (Tambutté et al. 2007a), and separated by coelenteron, cell membranes, and mesoglea. So what is transported and how it is transported is still a mystery. In fact, depending on the scale of observation, the role of zooxanthellae in LEC is more or less obvious: at the macroscale level, it is most probably that zooxanthellae enhance calcification under light conditions, at the microscale level many studies remain to be performed in order to understand how zooxanthellae play a role in LEC.

5.1.6 Conclusion on LEC

As described in the previous sections, calcification by itself is still a black box. If we add photosynthesis to calcification and their interactions, we increase the difficulty of understanding calcification. However, symbiotic corals are at the origin of the success of coral reefs and disrupting the association coral-zooxanthellae lead to bleaching which can be irreversible so the study of symbiosis in its aspect related to calcification is of major importance. The study of nonsymbiotic corals is also necessary especially to better understand the basic mechanisms of calcification, but we cannot preclude that symbiotic corals have evolved their mechanism of calcification compared to nonsymbiotic for example by transfer of genes. Once whole genome sequencing of the holobionte will be performed, the differential expression of genes of interest under light and dark conditions will help in revealing some aspects of LEC.

5.2 Temperature

Temperature is known to be a major determinant of habitat suitability. Its effect on coral distribution was recognized as early as 1843 by Dana (Dana 1843). Vaughan (1919) determined that the optimum temperature range for the formation of coral reefs is 25–29°C, but in some reef flats, higher (up to 38.9°C at Palao, Motoda 1940) or lower (12.5°C in the Arabian Gulf, Coles and Fadlallah 1991) temperatures have been recorded (see also for a review Kleypas et al. 1999). Due to the world climate change, appraising effect of temperature on coral reefs is particularly important.

Among other effects on coral physiology, temperature affects calcification following a bell-shaped relationship with an optimum around 25–27°C (Clausen 1971; Clausen and Roth 1975b; Jokiel and Coles 1977; Krishnaveni et al. 1989; Reynaud-Vaganay et al. 1999; Marshall and Clode 2004). This optimum depends on the temperature to which the coral was adapted (Clausen and Roth 1975b). Temperate corals show indeed lower optimum (around 18°C for *Plesiastrea versipora*, Howe and Marshall 2002). Adaptation occurs after a period of one to several weeks after the establishment of a new temperature regime (Clausen and Roth 1975b; Buddemeier and Kinzie 1976). In some cases, a linear relationship was found (Reynaud et al. 2004) probably because the optimum was not reached. Data collected from wild corals also show linear relationships (McNeil et al. 2004) probably due to local adaptation. The presence of a sharp temperature optimum, often with 1–2°C range (Clausen 1971; Reynaud-Vaganay et al. 1999; Marshall and Clode 2004), suggests that specific, but still unknown, physiological mechanism(s) is (are) directly dependent on temperature. By using Ca^{2+}-sensitive microsensors, Al-Horani (2005) showed that calculated Ca^{2+} fluxes was maximum at 26°C and then decreased to reach zero at 32°C. In addition to Ca^{2+} fluxes, other potential targets may include enzyme activity, protein (organic matrix) synthesis, CO_2 solubility (and therefore carbonate chemistry), symbiotic relationships… Temperature may also affect the skeletal microstructure as Howe and Marshall (2002) found that the temperate coral, *Plesiastrea versipora*, has a temperature-dependent diel pattern of crystal deposition with a shift from small spheroidal crystals at normal temperature to small needle-shaped crystals at higher temperature.

6 Unity and Diversity of Coral Skeletons

Modern coral skeletons are exclusively made of the orthorhombic metastable form of $CaCO_3$, aragonite. However, contrary to the common belief that scleractinian corals form only aragonitic skeletons, a Cretaceous scleractinian coral, *Coelosmilia* sp., with a calcitic skeleton was recently discovered (Stolarski et al. 2007). Interestingly, Ries et al. (2006) showed that, in laboratory conditions, three coral species (*Acropora cervicornis*, *Montipora digitata*, and *Porites cylindrica*) began producing calcite in "Cretaceous seawater" in which the Mg/Ca ratio was lower than that of modern seawater (respectively, 1.0 versus 5.2). Indeed, the corals produced progressively higher percentages of calcite and calcified at lower rates. This result may explain why coral reefs decreased from the fossil record during the beginning of the Cretaceous period (120 million years ago) and only reappeared 35 million years ago. It also shows that unlike previously believed, the concentration of magnesium in seawater may affect the $CaCO_3$ polymorph deposited. However, the fact that coral transferred to artificial "Cretaceous seawater" still precipitated two-thirds of their skeletal material as aragonite supports the hypothesis that corals exert significant control over their skeletal mineralogy (Ries et al. 2006).

6.1 The Basic Mechanism: Fibers and Centers of Mineralization

Since Bryan and Hill (1941), it has been generally assumed that corals initiate skeletal growth by forming the centers of calcification (COCs) initially observed by Ogilvie (1897). The skeletal unit, the sclerodermite or fiber, formed by a single orthorhombic crystal of aragonite, was supposed to develop from the COC that may act as a nucleation center. Although COCs were suggested to be under biological control, the successive growth of biomineral fibers was supposed to follow a purely physicochemical process (Constantz 1986). This interpretation was challenged by Cuif and Dauphin (1998) who suggested that these two structures are different mineralizing entities. COC are localized at the growing tips of structural units and are suggested to make the structural (and not functional, i.e., nucleating) framework of the future skeleton. Fibers use the COCs as a physical support to thicken the skeleton but these two structures are produced simultaneously under biological control in different regions. Consequently, Cuif and Dauphin (2005b) suggested that the term COC should be replaced by Early Mineralization Zones (EMZ). However, Holcomb et al. (2009), by comparing the morphology of synthetic aragonite grown at different aragonite saturation states and coral crystals, suggested that transition from centers of calcification to fibers may result from local change in saturation state of the coral's calcifying fluid. When saturation state is high, granular crystals precipitate at the tips of the existing skeletal elements forming COC, while as saturation decreases aragonite fibers are formed.

If EMZ and fibers are always observed in every type of skeletal architecture, their position and proportion greatly vary from one coral to another and is genetically determined (Cuif and Dauphin 1998). These two mineralizing entities have different chemical and biochemical properties (Allison 1996; Cuif and Dauphin 1998; Cohen et al. 2001; Adkins et al. 2003; Clode and Marshall 2003a; Cuif et al. 2003; Rollion-Bard et al. 2003; Meibom et al. 2003, 2004, 2006, 2007). The organic matrix content of EMZ appears different, both qualitatively and quantitatively, from adjacent fibers (Cuif et al. 2003), suggesting that these two structures are either deposited by different calicoblastic cells or are under different control. The fact that no major morphological differences are observed between calicoblastic cells above EMZ

Coral Calcification, Cells to Reefs

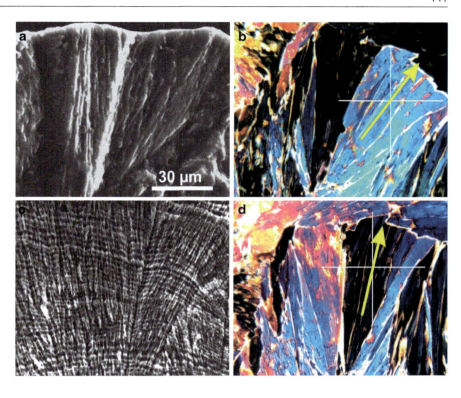

Fig. 7 Three different aspects of the same group of fibers from the coral *Favia stelligera*. (**a**) Scanning electron microscopy picture of fibers on a fracture surface of a corallite, (**b** and **c**) ultrathin section observed in polarized light. Fibers are groups of skeletal subunits with similar behavior with respect to polarization direction (orientation of the *yellow arrow*) (**d**) aspect of the same skeletal area after polishing and acidic etching. Organization of fibers is still recognizable but growth layers have been made visible due to differences in sensitivity to dissolution (From Cuif and Dauphin 2005a)

or fibers but that the calicodermis is thinner above EMZ from skeletal spines (Brown et al. 1983; Tambutté et al. 2007a) supports the second hypothesis. In any case, this suggests a high level of biological control of mineralization.

6.2 Concentric Layers: Annual, Diel Patterns

Presence of bands within scleractinian coral skeletons is well known (Ma 1934) and has been demonstrated to be annual (Knutson Knutson et al. 1972). Highsmith (1979) proposed that these bands result from a differential annual pattern of calcification rates: low-density bands correspond to a high organic matrix content (under moderate or low temperature) while high-density bands corresponds to a high mineral content. Similar pattern is observed in fish otoliths (Jolivet et al. 2008). However, the relative role of factors controlling banding is poorly understood: seasonal variations in seawater temperature (Dodge and Vaisnys 1975), in light intensity (Wellington and Glynn 1983), water turbidity and sedimentation (Dodge and Vaisnys 1975), reproduction and nutrient availability (Wellington and Glynn 1983) have been suggested to influence the formation of density bands. Lunar cycles of density banding have also been described (Buddemeier and Kinzie 1976). It should be noted however that density banding pattern is present in deep ahermatypic corals (Muscatine 1971; Adkins et al. 2004). We also recently observed annual density banding in *Porites* cultured in aquarium conditions for 15 years (unpublished data).

While fibers show a monocrystalline behavior under polarized light (which means for a crystallographer a single crystal), they appear constituted by a series of growth layers after etching with a mild acid or proteases demonstrating their polycrystalline nature (Fig. 7). Observed at the nanometer scale, coral skeleton shows growth layers of about 2 μm thick (see Fig. 6 of Cuif and Dauphin 2005a, b) that roughly correspond to a daily growth layer. These layers are not restricted to a single fiber but are present on the whole growth surface, suggesting that the calicodermis possesses a synchronous secretory activity. The mechanism underlying this cell coordination is presently totally unknown, but may involve gap junctions allowing direct transfer of signals between cells, although proteins involved in these junctions were not found in EST libraries. The formation of growth layers was recently attributed to a cyclic change of saturation state of the extracellular calcifying medium (Holcomb et al. 2009).

6.3 Nanograins as Units of Mineralization?

Use of atomic force microscopy has led to a new understanding of coral calcification. The daily 2 μm-thick growth layers show, when observed under atomic force microscopy (AFM), rounded granules of 40–80 nm in diameter (Stolarski 2003;

Cuif and Dauphin 2005a). Each of these nanograins appears surrounded by an organic cortex (Cuif and Dauphin 2005a). Similar small nodular structures of about 37 nm in diameter were previously observed by Clode and Marshall (2003b) upon organic matrix. These authors suggested that these granules correspond to Ca-rich regions detected by X-ray microanalysis. Nanograins were also observed in a wide range of biominerals (Oaki et al. 2006; Vielzeuf et al. 2008; see Alivisatos 2000 for a review). It now appears that these nanograins embedded within organic matrix may be the fundamental building block for the construction of biominerals. In this view, binding between OM and the mineral phase can be performed either with cations and anions separately, with the crystalline nuclei or even with ACC (Cölfen and Mann 2003). The OM is therefore no longer viewed as a static structural template but as comprising a dynamic interface that is inherently reactive and subject to mesoscale transformation (Cölfen and Mann 2003).

Following these observations, Cuif and Dauphin (2005b) updated by Cuif et al. (2008) suggested that a two-step cyclical process might describe the basic mechanism of calcification. In a first step, the calicodermis produces an organic cortex containing amorphous $CaCO_3$ stabilized by linkage to organic compounds. In a second step, an unknown mechanism leads to the crystallization process by itself producing the 2 μm-thick elemental growth layer. In this model, crystallization occurs within an organic hydrogel and OM components prevent formation of compact crystals with typical growth plans and surfaces, leading to a reticulate crystal. OM component are thus pushed out of the crystal lattice and appears as irregular envelopes as seen at the AFM. Whether the organic cortex is secreted by the calicodermis or formed outside remains an open question. In some species, intracellular vesicles were observed; however, no evidence was found for the presence of crystals in any organelles (see Sect. 3.5). However, Clode and Marshall (2003b) suggested that the extracellular granules observed in *Galaxea fascicularis* contained amorphous $CaCO_3$, a hypothesis strengthened by the recent observation made on the mollusc *Pinctada margaritifera* (Baronnet et al. 2008), showing that the content of nanogranules is not crystalline but amorphous (see Sect. 4.4).

In the light of all these data, we can suggest a mechanism of $CaCO_3$ deposition in coral (Fig. 8). Calicoblastic cells may secrete nanosized organic cortex. The high content of anionic residues of these "nanocortex" may help in sequestering

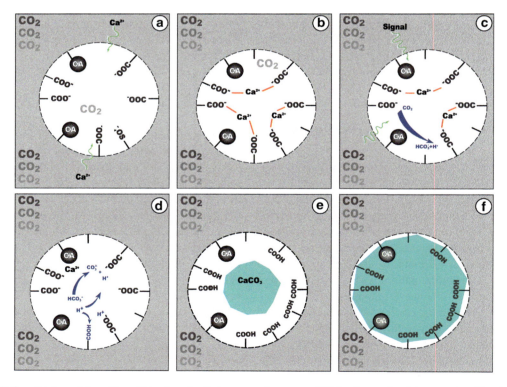

Fig. 8 Hypothetical mechanism of deposition of biomineral units (nanograins) in corals. (**a**) Secretion by the calicoblastic cells or auto-assemblage of a cortex ("nanocortex") of organic matrix (OM), (**b**) concentration of calcium within the nanocortex due to negative charge attraction, (**c**) activation of OM-linked carbonic anhydrase by an unknown signal leading to CO_2 hydration and H^+ formation, (**d**) buffering of H^+ by negative charges on OM leading to local increase in pH, release of free Ca^{2+} within the nanocortex and equilibration between HCO_3^- and CO_3^{2-} facilitated by the local increase in pH, (**e**) nucleation of $CaCO_3$ crystal and formation of the nanograin (in blue), (**f**) growth of the $CaCO_3$ crystal embedding part of OM, leaving a layer of organic cortex at the surface. In gray, organic hydrogel filling the extracellular calcifying medium

ionic calcium present within the ECM or amorphous $CaCO_3$. At this step, an unknown signal, such as hydrolysis of matrix proteins as in dentinogenesis (Fincham et al. 1999), will lead to an activation of organic matrix-linked CA leading to a rapid increase of CO_2 hydration according to the pH of the ECM and subsequent local release of H^+. Protons released can be buffered by carboxyl residues, therefore leading to the release of Ca^{2+}, which then crystallizes with CO_3^{2-} in aragonite. Later, nanograins will aggregate by coupled inorganic–organic interactions to produce, by mesoscale transformation, crystallographically oriented hybrid nanostructures (Cölfen and Mann 2003). Secretion of magnesium may form the transition between the different layers (Meibom et al. 2004).

7 Conclusions

Unraveling the mechanisms used by organisms to transform ions in solution into a highly ordered, genetically determined, three-dimensional mineral will be a central challenge for the present century. In the context of coral, while a large body of literature was published, we have to admit that the major steps of coral calcification remain to be discovered. This review shows that numerous questions were pending:

- What is the nature of the extracellular calcifying medium? Fluid or hydrogel?
- What are the physicochemical parameters of the extracellular calcifying medium?
- How ions or trace metals reach the site of mineralization? Active transepithelial transport, supply from bulk seawater, diffusion of some ions through cell junctions, or a combination of these different mechanisms?
- What is the limiting factor for calcification, pH, DIC, Ca^{2+} supply, and/or organic matrix synthesis?
- How many carbonic anhydrases are involved and what is(are) their exact role(s) in calcification?
- How secreted organic matrix may control the macro- and microstructure of the skeleton?
- How many organic macromolecules are required to form a coral skeleton?
- How does the calicodermis control the formation of the "unit layer"?
- How is the calcification process controlled by environmental factors?

In order to progress in the understanding of these processes, the contribution of new approaches, such as genomic, will be very important by revealing the presence of particular gene-encoded proteins, but it will not be sufficient without large progress in the physiology and cell biology of corals. Comparative physiological studies will be also important since most of our present knowledge comes from few species (mainly *Stylophora pistillata*, *Galaxea fascicularis*, and *Acropora* sp.).

Acknowledgments We would like to thank our colleagues for stimulating discussions and comments on part of our manuscript, particularly to Anne Cohen, Jean-Pierre Cuif, Jonathan Erez, Maoz Fine, Jean-Pierre Gattuso, Anne Juillet-Leclerc, Nicole Mayer-Gostan, Anders Meibom, Patrick Payan, Stéphanie Reynaud, and Alina Szmant. We are grateful to Alexander Venn for his comments on both the content and English form. We also thank Paola Furla, Aurélie Moya, Lucilia Pereira-Mouriès, and Sandrine Puverel, who shared our passion for coral physiology during their doctoral or postdoctoral research and Christian Söllner for the gift of antibodies. We also thank Séverine Lotto, Nathalie Techer and Natacha Segonds for their excellent technical work and the staff of the library of the Oceanographic museum as well as that of aquarium for their continuous help. Finally, we also remember all the stimulating discussions with Len Muscatine in the lab or in front of a glass of French swine during his visits in Monaco and we dedicate this review to him. Work performed at the Scientific Center of Monaco was funded by the Government of the Principality of Monaco.

References

Addadi L, Weiner S (1985) Interactions between acidic proteins and crystals: stereochemical requirements in biomineralization. Proc Natl Acad Sci USA 82:4110–4114

Addadi L, Berman A, Moradian Oldak J, Weiner S (1989) Structural and stereochemical relations between acidic macromolecules of organic matrices and crystals. Connect Tissue Res 21:127–135

Addadi L, Joester D, Nudelman F, Weiner S (2005) Mollusk shell formation: a source of new concepts for understanding biomineralization processes. Chem Eur J 12:980–987

Adkins JF, Boyle EA, Curry WB, Lutringer A (2003) Stable isotopes in deep-sea corals and a new mechanism for 'vital effects'. Geochim Cosmochim Acta 67:1129–1143

Adkins JF, Henderson GM, Wang S-L, O'Shea S, Mokadem F (2004) Growth rates of the deep-sea scleractinia *Desmophyllum cristagalli* and *Enallopsammia rostrata*. Earth Planet Sci Lett 227:481–490

Aizenberg J, Weiner S, Addadi L (2003) Coexistence of amorphous and crystalline calcium carbonate in skeletal tissues. Connect Tissue Res 44(suppl1):20–25

Albrecht EA, Cavicchia JC (2001) Permeability barrier in the mantle epithelium lining the testis in the apple snail *Pomacea canaliculata* (Gastropoda: Ampullariidae). Tissue Cell 33:148–153

Al-Horani FA (2005) Effects of changing seawater temperature on photosynthesis and calcification in the scleractinian coral *Galaxea fascicularis*, measured with O_2, Ca^{2+} and pH microsensors. Sci Mar 69:347–354

Al-Horani FA, Al-Moghrabi SM, de Beer D (2003a) Microsensor study of photosynthesis and calcification in the scleractinian coral, *Galaxea fascicularis*: active internal carbon cycle. J Exp Mar Biol Ecol 288:1–15

Al-Horani FA, Al-Moghrabi SM, de Beer D (2003b) The mechanism of calcification and its relation to photosynthesis and respiration in the scleractinian coral *Galaxea fascicularis*. Mar Biol 142:419–426

Al-Horani FA, Ferdelman T, Al-Moghrabi SM, de Beer D (2005) Spatial distribution of calcification and photosynthesis in the scleractinian coral *Galaxea fascicularis*. Coral Reefs 24:173–180

Al-Horani FA, Tambutté É, Allemand D (2007) Dark calcification and the daily rhythm of calcification in the scleractinian coral, *Galaxea fascicularis*. Coral Reefs 26:531–538

Alivisatos AP (2000) Naturally aligned nanocrystals. Science 289:736–737

Allemand D, Furla P, Bénazet-Tambutté S (1998a) Mechanisms of carbon acquisition for endosymbiont photosynthesis in Anthozoa. Can J Bot 76:925–941

Allemand D, Tambutté É, Girard J-P, Jaubert J (1998b) Organic matrix synthesis in the scleractinian coral *Stylophora pistillata*: Role in biomineralization and potential target of the organotin tributyltin. J Exp Biol 201:2001–2009

Allemand D, Ferrier-Pagès C, Furla P, Houlbrèque F, Puverel S, Reynaud S, Tambutté É, Tambutté S, Zoccola D (2004) Biomineralisation in reef-building corals: from molecular mechanisms to environmental control. C R Palévol 3:453–467

Allemand D, Mayer-Gostan N, De Pontual H, Bœuf G, Payan P (2007) Fish otolith calcification in relation to endolymph chemistry. In: Baeuerlein E (ed) Handbook of biomineralization. Wiley-VCH, Weinheim, pp 291–308

Allison N (1996) Comparative determination of trace and minor elements in coral aragonite by ion microprobe analysis with implications from Phuket, southern Thailand. Geochim Cosmochim Acta 60:3457–3470

Alloiteau J (ed) (1957) Contribution à la systématique des Madréporaires fossiles. CNRS, Paris

Anderson MJ, Van Itallie CM, Fanning AS (2004) Setting up a selective barrier at the apical junction complex. Curr Opin Cell Biol 16:140–145

Anning T, Nimer N, Merrett MJ, Brownlee C (1996) Costs and benefits of calcification in coccolithophorides. J Mar Syst 9:45–56

Anthony KRN, Connolly SR, Willis BL (2002) Comparative analysis of energy allocation to tissue and skeletal growth in corals. Limnol Oceanogr 47:1417–1429

Asano A, Asano K, Sasaki H, Furuse M, Tsukita S (2003) Claudins in *Caenorhabditis elegans*: their distribution and barrier function in the epithelium. Curr Biol 13:1042–1046

Barnes DJ (1970) Coral skeletons: an explanation of their growth and structure. Science 170:1305–1308

Barnes DJ (1972) The structure and formation of growth-ridges in scleractinian coral skeletons. Proc R Soc Lond B 182:331–350

Barnes DJ (1985) The effect of photosynthetic and respiratory inhibitors upon calcification in the staghorn coral, *Acropora formosa*. In: Delesalle B, Galzin R, Salvat B (eds) Proceeding of the fifth international coral reef congress. Museum National d'Histoire Naturelle (National Museum of Natural History) and the Ecole Pratique des Hautes Etudes (Practical School of Advanced Studies), Tahiti, pp 161–166

Barnes DJ, Chalker BE (1990) Calcification and photosynthesis in reef-building corals and algae. In: Dubinsky Z (ed) Coral reefs. Elsevier, Amsterdam, pp 109–131

Barnes DJ, Lough JM (1996) Coral skeletons: storage and recovery of environmental information. Global Change Biol 2:569–582

Baronnet A, Cuif J-P, Dauphin Y, Farre B, Nouet J (2008) Crystallization of biogenic Ca-carbonate within organo-mineral micro-domains. Structure of the calcite prisms of the Pelecypod *Pinctada margaritifera* (Mollusca) at the submicron to nanometre ranges. Mineralog Mag 72:539–548

Baumgartner S, Littleton JT, Broadie K, Bhat MA, Harbecke R, Lengyel JA, Chiquet-Ehrismann R, Prokop A, Bellen HJ (1996) A *Drosophila* neurexin is required for septate junction and blood-nerve barrier formation and function. Cell 87:1059–1068

Bénazet-Tambutté S, Allemand D, Jaubert J (1996) Permeability of the oral epithelial layers in cnidarians. Mar Biol 126:43–53

Bergmann W, Lester D (1940) Coral-reefs and the formation of petroleum. Science 92:452–453

Berridge MJ, Loschman J (1972) Transporting epithelia. Academic, New York

Bertucci A, Innocenti A, Zoccola D, Scozzafava A, Allemand D, Tambutté S, Supuran CT (2009a) Carbonic anhydrase inhibitors: inhibition studies of a coral secretory isoform with inorganic anions. Bioorg Med Chem Lett 19:650–653

Bertucci A, Innocenti A, Zoccola D, Scozzafava A, Tambutté S, Supuran CT (2009b) Carbonic anhydrase inhibitors. Inhibition studies of a coral secretory isoform by sulfonamides. Bioorg Med Chem 17:5054–5058

Bertucci A, Zoccola D, Tambutté S, Vullo D, Supuran CT (2010) Carbonic anhydrase activators. The first activation study of a coral secretory isoform with amino acids and amines. Bioorg Med Chem. doi:10.1016/j.bmc.2010.01.059

Bleher R, Machado J (2004) Paracellular pathway in the shell epithelium of *Anodonta cygnea*. J Exp Zool 301A:419–427

Böhm F, Gussone N, Eisenhauer A, Dullo W-C, Reynaud S, Paytan A (2006) Calcium isotope fractionation in modern scleractinian corals. Geochim Cosmochim Acta 70:4452–4462

Borelli G, Guibbollini ME, Mayer-Gostan N, Priouzeau F, De Pontual H, Allemand D, Puverel S, Tambutté É, Payan P (2003a) Daily variations of endolymph composition: relationship with the otolith calcification process in trout. J Exp Biol 206:2685–2692

Borelli G, Mayer-Gostan N, Merle P-L, De Pontual H, Boeuf G, Allemand D, Payan P (2003b) Composition of biomineral organic matrices with special emphasis on turbot (*Psetta maxima*) otolith and endolymph. Calcif Tissue Int 72:717–725

Bourne GC (1887) On the anatomy of *Mussa* and *Euphyllia* and the morphology of the Madreporian skeleton. Q J Micr Sci XXVIII:21–51, + Plate III/IV

Brown BE, Hewit R, Le Tissier MD (1983) The nature and construction of skeletal spines in *Pocillopora damicornis* (Linnaeus). Coral Reefs 2:81–89

Brownlee C, Taylor AR (2002) Algal calcification and silification. Encyclopedia of life sciences. MacMillan, London, pp 1–5

Bryan WH, Hill D (1941) Spherulitic crystallization as a mechanism of skeletal growth in the hexacorals. Proc R Soc Queensl 52:78–91

Buddemeier RW, Kinzie RA (1976) Coral growth. Oceanogr Mar Biol Annu Rev 14:183–225

Chalker BE (1976) Calcium transport during skeletogenesis in hermatypic corals. Comp Biochem Physiol 54A:455–459

Chalker BE (1977) Daily variation in the calcification capacity of *Acropora cervicornis*. In: Taylor DL (ed) Proc Third Int Coral Reef Symp Rosenstiel School of Marine and Atmospheric Science, Miami, Florida, pp 417–423

Chalker BE, Taylor DL (1975) Light-enhanced calcification, and the role of oxidative phosphorylation in calcification of the coral *Acropora cervicornis*. Proc R Soc Lond B 190:323–331

Chamberlain JA Jr (1978) Mechanical properties of coral skeleton: compressive strength and its adaptative significance. Paleobiology 4:419–435

Chevalier J-P (1987) Ordre des Scléractiniaires. In: Doumenc D (ed) Cnidaires Anthozoaires. Masson, Paris, pp 403–764

Chisholm J, Gattuso J-P (1991) Validation of the alkalinity anomaly technique for investigating calcification and photosynthesis in coral reef communities. Limnol Oceanogr 36:1232–1239

Clausen C (1971) Effects of temperature on the rate of ^{45}Calcium uptake by *Pocillopora damicornis*. In: Lenhoff HM, Muscatine L, Davis LV (eds) Experimental coelenterate biology. University of Hawaii Press, Honolulu, pp 246–260

Clausen CD, Roth AA (1975a) Estimation of coral growth rates from laboratory ^{45}Ca incorporation rates. Mar Biol 33:85–91

Clausen CD, Roth AA (1975b) Effect of temperature and temperature adaptation on calcification rate in the hermatypic coral *Pocillopora damicornis*. Mar Biol 33:93–100

Clode PL, Marshall AT (2002a) Low temperature FESEM of the calcifying interface of a scleractinian coral. Tissue Cell 34:187–198

Clode PL, Marshall AT (2002b) Low temperature X-ray microanalysis of calcium in a scleractinian coral: evidence of active transport mechanisms. J Exp Biol 205:3543–3552

Clode PL, Marshall AT (2003a) Skeletal microstructure of *Galaxea fascicularis* exsert septa: A high- resolution SEM study. Biol Bull 204:146–154

Clode PL, Marshall AT (2003b) Calcium associated with a fibrillar organic matrix in the scleractinian coral *Galaxea fascicularis*. Protoplasma 220:153–161

Cohen AL, McConnaughey TA (2003) Geochemical perspectives on coral mineralization. Rev Mineral Geochem 54:151–187

Cohen AL, Layne GD, Hart SR, Lobel SR (2001) Kinetic control of skeletal Sr/Ca in a symbiotic coral: implications for the paleotemperature proxy. Paleoceanography 16:20–26

Cohen AL, McCorkle DC, De Putron S, Gaetani GA, Rose KA (2009) Morphological and compositional changes in the skeletons of new coral recruits reared in acidified seawater: insights into the biomineralization response to ocean acidification. Geochem Geophys Geosyst 10:1–12

Coles SL, Fadlallah YH (1991) Reef coral survival an mortality at low temperatures in the Arabian Gulf: new species-specific lower temperature limits. Coral Reefs 9:231–237

Cölfen H, Mann S (2003) Higher-order organization by mesoscale self-assembly and transformation of hybrid nanostructures. Angew Chem Int Ed 42:2350–2365

Colombo-Pallotta MF, Rodríguez-Román A, Iglesias-Prieto R (2010) Calcification in bleached and unbleached *Montastraea faveolata*: evaluating the role of oxygen and glycerol. Coral Reefs. On line doi 10.1007/s00338-010-0638-x

Constantz BR (1986) Coral skeleton construction: a physiochemically dominated process. Palaios 1:152–157

Constantz B (2008) Raman spectroscopy of the initial mineral phase of coral skeleton. In: Abstract Book, Oral Mini-Symposium 3: calcification and coral reef – past and future, The 11th ICRS, Fort Lauderdale, FL, p 14

Constantz BR, Weiner S (1988) Acidic macromolecules associated with the mineral phase of scleractinian coral skeletons. J Exp Zool 248:253–258

Crenshaw MA (1972) The inorganic composition of molluscan extrapallial fluid. Biol Bull 143:506–512

Crossland CJ, Barnes DJ (1974) The role of metabolic nitrogen in coral calcification. Mar Biol 28:325–332

Crumbliss AL, Mc Lachlan KL, O'Daly JP, Henkens RW (1988) Preparation and activity of carbonic anhydrase immobilized on porous silica beads and graphite rods. Biotechnol Bioeng 31: 796–801

Cuif J-P, Dauphin Y (1998) Microstructural and physico-chemical characterization of 'centers of calcification' in septa of some recent scleractinian corals. Paläontologische Zeitschrift 72:257–270

Cuif J-P, Dauphin Y (2004) The environment recording unit in coral skeletons: structural and chemical evidences of a biochemically driven stepping-growth process in coral fibres. Biogeosci Discuss 1:625–658

Cuif J-P, Dauphin Y (2005a) The environment recording unit in corals skeletons – a synthesis of structural and chemical evidences for a biochemically driven, stepping-growth process in fibres. Biogeosciences 2:61–73

Cuif J-P, Dauphin Y (2005b) The two-step mode of growth in the scleractinian coral skeletons from the micrometre to the overall scale. J Struct Biol 150:319–331

Cuif J-P, Dauphin Y, Denis A, Gautret P, Marin F (1996) The organo-mineral structure of coral skeletons: a potential source of new criteria for Scleractinian taxonomy. Bull Inst Océanogr Monaco 14: 359–367

Cuif J-P, Dauphin Y, Freiwald A, Gautret P, Zibrowius H (1999) Biochemical markers of zooxanthellae symbiosis in soluble matrices of skeleton of 24 Scleractinia species. Comp Biochem Physiol 123A:269–278

Cuif J-P, Dauphin Y, Doucet J, Salome M, Susini J (2003) XANES mapping of organic sulfate in three scleractinian coral skeletons. Geochim Cosmochim Acta 67:75–83

Cuif J-P, Dauphin Y, Berthet P, Jegoudez J (2004) Associated water and organic compounds in coral skeletons: quantitative thermogravimetry coupled to infrared absorption spectrometry. Geochem Geophys Geosyst 5:1–9

Cuif J-P, Dauphin Y, Farre B, Nehrke G, Nouet J, Salomé M (2008) Distribution of sulphated polysaccharides within calcareous biominerals suggests a widely shared two-step crystallization process for the microstructural growth units. Mineralog Mag 72:233–237

Dana JD (1843) On the temperature limiting the distribution of corals. Am J Sci 45:130–131

Dana JD (1846) Structure and classification of zoophytes. United States Exploring Expedition during the years 1838, 1839, 1840, 1841, 1842, under the Command of Charles Wilkes, U.S.N. 7:1–740

Dan-Sokhawa M, Hiroyuki K, Koichi N (1995) Paracellular, transepithelial permeation of mcromolecules in the body wall epithelium of starfish embryo. J Exp Zool 271:264–272

Dauphin Y (2001) Comparative studies of skeletal soluble matrices from some Scleractinian corals and Molluscs. Int J Biol Macromol 28:293–304

Dauphin Y, Cuif J-P, Massard P (2006) Persistent organic components in heated coral aragonitic skeletons-Implications for palaeoenvironmental reconstructions. Chem Geol 231:26–37

Dauphin Y, Cuif J-P, Williams CT (2008) Soluble organic matrices of aragonitic skeletons of Merulinidae (Cnidaria, Anthozoa). Comp Biochem Physiol 150B:10–22

De Beer D, Kühl M, Stambler N, Vaki L (2000) A microsensor study of light enhanced Ca^{2+} uptake and photosynthesis in the reef-building hermatypic coral *Favia* sp. Mar Ecol Prog Ser 194:75–85

Demers C, Reggie Hamdy C, Corsi K, Chellat F, Tabrizian M, Yahia L (2002) Natural coral exoskeleton as a bone graft substitute: A review. Biomed Mater Eng 12:15–35

Diamond JM, Wright EM (1969) Biological membranes: physical basis of ion and nonelectrolyte selectivity. Annu Rev Physiol 31: 581–646

Dodge RE, Vaisnys JR (1975) Hermatypic coral growth-banding as environmental recorder. Nature 258:706–708

Duerden JE (1903) West Indian Madreporian polyps. Memoirs Natl Acad Sci 8:401–599, + 25 plates

Dunn SR, Phillips WS, Green DR, Weis VM (2007) Knockdown of Actin and caspase gene expression by RNA interference in the symbiotic anemone *Aiptasia pallida*. Biol Bull 212:250–258

Edmunds PJ, Davies PS (1986) An energy budget for *Porites porites* (Scleractinia). Mar Biol 92:339–347

Elahi R, Edmunds PJ (2007) Tissue age effects calcification in the Scleractinian coral *Madracis mirabilis*. Biol Bull 212:20–28

Erez J (1978) Vital effect on stable-isotope composition seen in foraminifera and coral skeletons. Nature 273:199–202

Erez J (2003) The source of ions for biomineralization in foraminifera and their implications for paleoceanographic proxies. Rev Mineral Geochem 54:115–149

Erez J, Braun A (2007) Calcification in hermatypic corals is based on direct seawater supply to the biomineralization site. In: Goldschmidt Conference Abstracts 2007 Cologne, Germany, pp A260.

Estroff LA (2008) Introduction: biomineralization. Chem Rev 108:4329–4331

Faber WW, Preisig HR (1994) Calcified structures and calcification in protists. Protoplasma 181:78–105

Fang LS, Chen YWJ, Chen CS (1989) Why does the white tip of stony coral grow so fast without zooxanthellae? Mar Biol 103:359–363

Fautin DG, Mariscal RN (1991) Cnidaria: Anthozoa. In: Harrison FW, Westfall JA (eds) Placozoa, Porifera, Cnidaria, and Ctenophora. Wiley-Liss, New York, pp 267–358

Feher JJ, Fullmer CS, Wasserman RH (1992) Role of facilitated diffusion of calcium by calbindin in intestinal calcium absorption. Am J Physiol 262:C517–C526

Ferrier-Pagès C, Gattuso J-G, Dallot S, Jaubert J (2000) Effect of nutrient enrichment on growth and photosynthesis of the zooxanthellate coral *Stylophora pistillata*. Coral Reefs 19:103–113

Fincham AG, Moradian-Oldak J, Simmer JP (1999) The structural biology of the developing dental enamel matrix. J Struct Biol 126: 270–299

Fine M, Tchernov D (2007) Scleractinian coral species survive and recover from decalcification. Science 315:1811

Fowler GH (1885) The anatomy of Madreporaria. Part I. Q J Micr Sci Lond 25:577–599

Frömter E, Diamond J (1972) Route of passive ion permeation in epithelia. Nat New Biol 235:9–13

Fukuda I, Ooki S, Fujita T, Murayama E, Nagasawa H, Isa Y, Watanabe T (2003) Molecular cloning of a cDNA encoding a soluble protein in the coral exoskeleton. Biochem Biophys Res Comm 304:11–17

Furla P, Bénazet-Tambutté S, Jaubert J, Allemand D (1998) Functional polarity of the tentacle of the sea anemone *Anemonia viridis*: role in inorganic carbon acquisition. Am J Physiol (Regul Integr Comp Physiol) 274:R303–R310

Furla P, Allemand D, Orsenigo MN (2000a) Involvement of H$^-$-ATPase and carbonic anhydrase in inorganic carbon uptake for endosymbiont photosynthesis. Am J Physiol (Regul Integr Comp) 278:R870–R881

Furla P, Galgani I, Durand I, Allemand D (2000b) Sources and mechanisms of inorganic carbon transport for coral calcification and photosynthesis. J Exp Biol 203:3445–3457

Gaillardet J, Allègre C (1995) Boron isotopic compositions of corals: seawater or diagenesis record? Earth Planet Sci Lett 136:665–676

Galloway SB, Work TM, Bochsler VS, Harley RA, Kramarsky-Winters E, Mc Laughlin SM, Meteyer CU, Morado JF, Nicholson JH, Parnell PG, Peters EC, Reynolds TL, Rotstein DS, Sileo L, Woodley CM (2007) Coral disease and health workshop: coral histopathology workshop II. National Oceanic and Atmospheric Administration, Silver Spring

Gattuso J-P (1987) Écomorphologie, métabolisme, croissance et calcification du scléractiniaire à zooxanthelles *Stylophora pistillata* (Golfe d'Aqaba, Mer rouge). Influence de l'éclairement. PhD thesis, Aix-Marseille II

Gattuso J-P, Frankignoulle M, Bourge I, Romaine S, Buddemeier RW (1998) Effect of calcium carbonate saturation of seawater on coral calcification. Glob Planet Change 18:37–46

Gattuso J-P, Allemand D, Frankignoulle M (1999) Photosynthesis and calcification at cellular, organismal and community levels in coral reefs: A review on interactions and control by carbonate chemistry. Am Zool 39:160–183

Gattuso J-P, Reynaud-Vaganay S, Furla P, Romaine-Lioud S, Jaubert J, Bourge I, Frankignoulle M (2000) Calcification does not stimulate photosynthesis in the zooxanthellate scleractinian coral *Stylophora pistillata*. Limnol Oceanogr 45:246–250

Gautret P, Marin F (1993) Evaluation of diagenesis in scleractinian corals and calcified demosponges by substitution index measurement and intraskeletal organic matrix analysis. Cour Forsch Senckenb 164:317–327

Gautret P, Cuif J-P, Freiwald A (1997) Composition of soluble mineralizing matrices in zooxnathellate and non-zooxanthellate scleractinian corals: biochemical assessment of photosynthetic metabolism through the study of a skeletal feature. Facies 36:189–194

Gautret P, Cuif J-P, Stolarski J (2000) Organic components of the skeleton of scleractinian corals. Evidence from *in situ* acridine orange staining. Acta Palaeontol Pol 45:107–118

Gladfelter EH (1983) Skeletal development in *Acropora cervicornis*: II. Diel patterns of calcium carbonate accretion. Coral Reefs 2:91–100

Goldberg WM (2001a) Desmocytes in the calicoblastic epithelium of the stony coral *Mycetophyllia reesi* and their attachment to the skeleton. Tissue Cell 33:388–394

Goldberg WM (2001b) Acid polysaccharides in the skeletal matrix and calicoblastic epithelium of the stony coral *Mycetophyllia reesi*. Tissue Cell 33:376–387

Goreau T (1956) Hystochemistry of mucopolysaccharide-like substances and alkaline phosphatase in madreporaria. Nature 177:1029–1030

Goreau TF (1959) The physiology of skeleton formation in corals. I. A method for measuring the rate of calcium deposition by corals under different conditions. Biol Bull 116:59–75

Goreau TJ (1977) Coral skeletal chemistry: physiological and environmental regulation of stable isotopes and trace metals in *Montastrea annularis*. Proc R Soc Lond B 196:291–315

Goreau TF, Goreau NI (1959a) The physiology of skeleton formation in corals. II. Calcium deposition by hermatypic corals under different conditions. Biol Bull 117:239–250

Goreau TF, Goreau NI (1959b) The physiology of skeleton formation in corals. III. Calcium rate as a function of colony weight and total nitrogen in the reef coral *Manicina areolota* (Lin.). Biol Bull 118:419–429

Grasso LC, Maindonald J, Rudd S, Hayward DC, Saint R, Miller DJ, Ball EE (2008) Microarray analysis identifies candidates genes for key roles in coral development. BMC Genom 9:540. doi:10.1186/1471-2164-9-540

Green C, Bergquist PR (1982) Phylogenetic relationships within the invertebrata in relation to the structure of septate junctions and the development of 'occluding' junctional types. J Cell Sci 53:279–305

Green CR, Flower NE (1980) Two new septate junctions in the phylum coelenterata. J Cell Sci 42:43–59

Gupta L, Suzuki A, Kawahata H (2006) Aspartic acid concentrations in coral skeletons as recorders of past disturbances of metabolic rates. Coral Reefs 25:599–606

Hayasi K (1937) On the detection of calcium in the calicoblasts of some reef corals. Palaeo Trop Biol Sta Stud 2:169–176

Hayes RL, Goreau NI (1977) Intracellular crystal-bearing vesicles in the epidermis of scleractinian corals, *Astrangia danae* (Agassiz) and *Porites porites* (Pallas). Biol Bull 152:26–40

Hemming NG, Hanson GN (1992) Boron isotopic composition and concentration in modern marine carbonates. Geochim Cosmochim Acta 56:537–543

Herfort L, Thake B, Taubner I (2008) Bicarbonate stimulation of calcification and photosynthesis in two hermatypic corals. J Phycol 44:91–98

Highsmith RC (1979) Coral growth rates and environmental controm of density banding. J Exp Mar Biol Ecol 37:105–125

Hochachka PW, Buck LT, Doll CJ, Land SC (1996) Unifying theory of hypoxia tolerance: molecular/metabolic defense and rescue mechanisms for surviving oxygen lack. Proc Nat Acad Sci USA 93:9493–9498

Holcomb M, Cohen AL, Gabitov RI, Hutter JL (2009) Compositional and morphological features of aragonite precipitated experimentally from seawater and biogenically by corals. Geochim Cosmochim Acta 73:4166–4179

Hönisch B, Hemming NG, Grottoli AG, Amat A, Hanson GN, Bijma J (2004) Assessing scleractinian corals as recorders for paleo-pH: Empirical calibration and vital effects. Geochim Cosmochim Acta 68:3675–3685

Houlbrèque F, Tambutté É, Richard C, Ferrier-Pagès C (2004) Importance of the micro-diet for scleractinian corals. Mar Ecol Prog Ser 282:151–160

Howe SA, Marshall AT (2002) Temperature effects on calcification rate and skeletal deposition in the temperate coral, *Plesiastrea versipora* (Lamarck). J Exp Mar Biol Ecol 275:63–81

Hudson RL (1992) Ion transport by the isolated mantle epithelium of the freshwater clam, *Unio complanatus*. Am J Physiol 263: R76–R83

Ichikawa K (2007) Buffering dissociation/formation reaction of biogenic calcium carbonate. Chem Eur J 13:10176–10181

Ingalls AE, Lee C, Druffel ERM (2003) Preservation of organic matter in mound-forming coral skeletons. Geochim Cosmochim Acta 67:2827–2841

Ip YK, Krishnaveni P (1991) Incorporation of strontium ($^{90}Sr^{++}$) into the skeleton of the hermatypic coral *Galaxea fascicularis*. J Exp Zool 258:273–276

Isa Y (1986) An electron microscope study on the mineralization of the skeleton of the staghorn coral *Acropora hebes*. Mar Biol 93:91–101

Isa Y, Okazaki M (1987) Some observations on the Ca^{2+}-binding phospholipids from scleractinian coral skeletons. Comp Biochem Physiol 87B:507–512

Isa Y, Yamazato K (1984) The distribution of carbonic anhydrase in a staghorn coral *Acropora hebes* (Dana). Galaxea 3:25–36

Jackson AE, Yellowlees D (1990) Phosphate uptake by zooxanthellae isolated from corals. Proc R Soc Lond B 242:201–204

Jackson DJ, Mcdougall C, Green K, Simpson F, Wörheide G, Degnan BM (2006) A rapidly evolving secretome builds and patterns a sea shell. BMC Biol 4:40. doi:10.1186/1741-7007-4-40

Jacques TG, Pilson MEQ (1977) Laboratory observations on respiration, photosynthesis and factors affecting calcification in the temperate coral *Astrangia danae*. In: Taylor DL (ed) Proceedings of third international coral reef symposium – Rosenstiel School of Marine and Atmospheric Science, Miami, FL, pp 455–461

Johnston IS (1980) The ultrastructure of skeletogenesis in zooxanthellate corals. Int Rev Cytol 67:171–214

Jokiel PL, Coles SL (1977) Effects of temperature on the mortality and growth of Hawaiian reef corals. Mar Biol 43:201–208

Jokiel PL, Maragos JE, Franzisket L (1978) Coral growth: buoyant weight technique. In: UNESCO (ed) Coral reefs: research methods, Paris, pp 529–541

Jolivet A, Bardeau J-F, Fablet R, Paulet Y-M, De Pontual H (2008) Understanding otolith biomineralization processes: new insights into microscale spatial distribution of organic and mineral fractions from Raman microspectrometry. Anal Bioanal Chem 392:551–560

Jury CP, Whitehead RF, Szmant AM (2010) Effects of variations in carbonate chemistry on the calcification rates of *Madracis mirabilis* (Duchassaing 1861): bicarbonate concentrations best predict calcification rates. Global Change Biol 16:1632–1644

Kahng SE, Maragos JE (2006) The deepest zooxanthellate scleractinian corals in the world? Coral Reefs 25:254

Kakei M, Nakahara H (1996) Aspects of carbonic anhydrase and carbonate content during mineralization of the rat enamel. Biochim Biophys Acta 1289:226–230

Karoonuthaisiri N, Titiyevskiy K, Thomas JL (2003) Destabilization of fatty acid-containing liposomes by polyamidoamine dendrimers. Colloids Surf B 27:365–375

Katsu T, Imamura T, Komagoe K, Masuda K, Mizushima T (2007) Simultaneous measurements of K^{+} and calcein release from liposomes and the determination of pore size formed in a membrane. Anal Sci 23:517–522

Kawaguti S (1937) On the physiology of reef corals II. The effect of light on colour and form of reef corals. Palao Trop Biol Sta Stud 177:177–186

Kawaguti S, Sakumoto D (1948) The effect of light on the calcium deposition of corals. Bull Oceanogr Inst Taïwan 4:65–70

Kawaguti S, Sato K (1968) Electron microscopy on the polyp of staghorn corals with special reference to its skeleton formation. Biol J Okayama Univ 14:87–98

Keil TA, Steinbrecht RA (1987) Diffusion barriers in silkmoth sensory epithelia: application of lanthanum tracer to olfactory sensilla of *Antheraea polyphemus* and *Bombyx mori*. Tissue Cell 19: 119–134

Klein MJ, Ahearn GA (1999) Calcium transport mechanisms of crustacean hepatopancreatic mitochondria. J Exp Zool 283: 147–159

Kleypas JA, McManus JW, Menez LAB (1999) Environmental limits to coral reef development: Where do we draw the line? Am Zool 39:146–159

Knutson DW, Buddemeier RW, Smith SV (1972) Coral chronometers: seasonal growth bands in reef corals. Science 177:270–272

Kottra G, Frömter E (1983) Functional properties of the paracellular pathway in some leaky epithelia. J Exp Biol 106:217–229

Krempf A (1907) Sur la formation du squelette chez les hexacoralliaires à polypier. CR Acad Sci Paris Ser D 144:157–159

Krishnaveni P, Chou LM, Ip YK (1989) Deposition of calcium ($^{45}Ca^{2+}$) in the coral *Galeaxea fascicularis*. Comp Biochem Physiol 94A:509–513

Kühl M, Cohen Y, Dalsgaard T, Jorgensen BB, Revsbech NP (1995) Microenvironment and photosynthesis of zooxanthellae in scleractinian corals studied with microsensors for O_2, pH and light. Mar Ecol Prog Ser 117:159–172

Lambert G, Lambert CC (1996) Spicule formation in the New Zealand ascidian *Pyura pachydermatina* (Chordata, Ascidiacea). Connect Tissue Res 34–5:25–31

Lane NJ (1979) Freeze fracture and tracer studies on the intercellular junctions in insect rectal tissues. Tissue Cell 11:481–506

Le Tissier MDAA (1987) The nature and construction of skeletal spines in *Pocillopora damicornis* (Linnaeus). PhD thesis, University of Newcastle upon Tyne, UK, 140 pp

Le Tissier MDAA (1988a) Patterns of formation and the ultrastructure of the larval skeleton of *Pocillopora damicornis*. Mar Biol 98:493–501

Le Tissier MDAA (1988b) Diurnal patterns of skeleton formation in *Pocillopora damicornis* (Linnaeus). Coral Reefs 7:81–88

Le Tissier MDAA (1990) The ultrastructure of the skeleton and skeletogenic tissues of the temperate coral *Caryophyllia smithii*. J Mar Biol Ass UK 70:295–310

Le Tissier MDAA (1991) The nature of the skeleton and skeletogenic tissues in the Cnidaria. Hydrobiologia 216/217:397–402

Leggat W, Marendy EM, Baillie B, Whitney SM, Ludwig M, Badgaer MR, Yellowlees D (2002) Dinoflagellate symbioses: strategies and adaptations for the acquisition and fixation of inorganic carbon. Funct Plant Biol 29:309–322

Lelong C, Mathieu M, Favrel P (2001) Identification of new bone morphogenetic protein-related members in invertebrates. Biochimie 83:423–426

Levy O, Appelbaum L, Gothlif Y, Hayward DC, Miller DJ, Hoegh-Guldberg O (2007) Light-responsive cryptochromes from a simple multicellular animal, the coral *Acropora millepora*. Science 318:467–469

Lowenstam HA (1981) Minerals formed by organisms. Science 211:1126–1131

Lowenstam HA, Weiner S (1989) On biomineralization. Oxford University Press, New York/Oxford

Ma TYH (1934) On the season change of growth in a reef coral, *Favia speciosa* (Dana) and the temperature of the japanese seas during the latest geological times. Proc Imp Acad (Tokyo) 10:353–356

Magie CR, Martindale MQ (2008) Cell-cell adhesion in the Cnidaria: insights into the evolution of tissue morphogenesis. Biol Bull 214:218–232

Mann S (1983) Mineralization in biological systems. Struct Bond 54:125–174

Mann S (2001) Biomineralization. Principles and concepts in bioinorganic materials chemistry. Oxford University Press, New York

Mann K, Macek B, Olsen JV (2006) Proteomic analysis of the acid-soluble organic matrix of the chicken calcified eggshell layer. Proteomics 6:3801–3810

Marin F, Smith M, Isa Y, Muyzer G, Westbroek P (1996) Skeletal matrices, muci, and the origin of invertebrate calcification. Proc Nat Acad Sci USA 93:1554–1559

Marin F, Luquet G, Marie B, Medakovic D (2008) Molluscan shell proteins: Primary structure, origin, and evolution. Curr Top Dev Biol 80:209–276

Marschal C, Garrabou J, Harmelin JG, Pichon M (2004) A new method for measuring growth and age in the precious red coral *Corallium rubrum* (L.). Coral Reefs 23:423–432

Marshall AT (1996) Calcification in hermatypic and ahermatypic corals. Science 271:637–639

Marshall AT, Clode PL (2002) Effect of increased calcium concentration in sea water on calcification and photosynthesis in the scleractinian coral *Galaxea fascicularis*. J Exp Biol 205:2107–2113

Marshall AT, Clode PL (2003) Light-regulated Ca^{2+} uptake and O_2 secretion at the surface of a scleractinian coral *Galaxea fascicularis*. Comp Biochem Physiol 136A:417–426

Marshall AT, Clode P (2004) Calcification rate and the effect of temperature in a zooxanthellate and an azooxanthellate scleractinian reef coral. Coral Reefs 23:218–224

Marshall AT, Wright OP (1993) Confocal laser scanning light microscopy of the extra-thecal epithelia of undecalcified scleractinian corals. Cell Tissue Res 272:533–543

Marshall AT, Wright A (1998) Coral calcification: autoradiography of a scleractinian coral *Galaxea fascicularis* after incubation in ^{45}Ca and ^{14}C. Coral Reefs 17:37–47

Marubini F, Thake B (1999) Bicarbonate addition promotes coral growth. Limnol Oceanogr 44:716–720

Marubini F, Ferrier-Pagès C, Furla P, Allemand D (2008) Coral calcification responds to seawater acidification: a working hypothesis towards a physiological mechanism. Coral Reefs 27: 491–499

McConnaughey TA, Falk RH (1991) Calcium-proton exchange during algal calcification. Biol Bull 180:185–195

McConnaughey TA, Whelan JF (1997) Calcification generates protons for nutrient and bicarbonate uptake. Earth Sci Rev 42:95–117

McNeil BI, Matear RJ, Barnes DJ (2004) Coral reef calcification and climate change: The effect of ocean warming. Geophys Res Lett 31:1–4

Meibom A, Stage M, Wooden J, Constantz BR, Dunbar RB, Owen A, Grumet N, Bacon CR, Chamberlain C (2003) Monthly Strontium/Calcium oscillations in symbiotic coral aragonite: Biological effects limiting the precision of the paleotemperature proxy. Geophys Res Lett 30:71–74

Meibom A, Cuif J-P, Hillion F, Constantz BR, Juillet-Leclerc A, Dauphin Y, Watanabe T, Dunbar RB (2004) Distribution of magnesium in coral skeleton. Geophys Res Lett 31:L23306. doi:10.1029/2004GL021313

Meibom A, Yurimoto H, Cuif J-P, Domart-Coulon I, Houlbrèque F, Constantz B, Dauphin Y, Tambutté É, Tambutté S, Allemand D, Wooden J, Dunbar R (2006) Vital effects in coral skeletal composition display strict three-dimensional control. Geophys Res Lett 33:L11608. doi:10.1029/2006GL025968

Meibom A, Mostefaoui S, Cuif J-P, Dauphin Y, Houlbrèque F, Dunbar R, Constantz B (2007) Biological forcing controls the chemistry of reef-building coral skeleton. Geophys Res Lett 34:L02601. doi:10.1029/2006GL028657

Meyer E, Aglyamova GV, Wang S, Buchanan-Carter J, Abrego D, Colbourne JK, Willis BL, Matz MV (2009) Sequencing and de novo analysis of a coral larval transcriptome using 454 GS-Flx. BMC Genom 10:219

Milne Edwards H (1857) Histoire naturelle des Coralliaires ou polypes proprement dits, Lib. Encyclopédique de Roret, Paris, France

Mitsunaga K, Akasaka K, Shimada H, Fujin Y, Yasumasu I, Numandi H (1986) Carbonic anhydrase activity in developing sea urchin embryos with special reference to calcification of spicules. Cell Differ 18:257–262

Mitterer RM (1978) Amino acid composition and metal binding capability of the skeleton protein of corals. Bull Mar Sci 28:173–180

Miyamoto H, Miyashita T, Okushima M, Nakano S, Morit T, Matsushiro A (1996) A carbonic anhydrase from the nacreous layer in oyster pearls. Proc Natl Acad Sci USA 93:9657–9660

Miyamoto H, Miyoshi F, Kohno J (2005) The Carbonic Anhydrase domain protein Nacrein is expressed in the epithelial cells of the mantle and acts as a negative regulator in calcification in the Mollusc *Pinctada fucata*. Zool Sci 22:311–315

Motoda S (1940) The environment and the life of masive reef coral, *Goniastrea aspera* Verrill inhabiting the reef flatin Palao. Palao Trop Biol Stn Stud 2:61–104

Moura G, Vilarinho L, Santos AC, Machado J (2000) Organic compounds in the extrapalial fluid and haemolymph of *Anodonta cygnea* (L.) with emphasis on the seasonal biomineralization process. Comp Biochem Physiol 125B:293–306

Moya A, Tambutté S, Tambutté É, Zoccola D, Caminiti N, Allemand D (2006) Study of calcification during a daily cycle of the coral *Stylophora pistillata*. Implications for "Light-Enhanced Calcification". J Exp Biol 209:3413–3419

Moya A, Ferrier-Pagès C, Furla P, Richier S, Tambutté É, Allemand D, Tambutté S (2008a) Calcification and associated physiological parameters during a stress event in the scleractinian coral *Stylophora pistillata*. Comp Biochem Physiol 151A:29–36

Moya A, Tambutté S, Lotto S, Allemand D, Zoccola D (2008b) Carbonic anhydrase in the scleractinian coral *Stylophora pistillata*: characterization, localization, and role in biomineralization. J Biol Chem 283:25475–25484

Moya A, Tambutté S, Béranger G, Gaume B, Scimeca J-C, Allemand D, Zoccola D (2008c) Cloning and use of a coral 36B4 gene to study differential expression of genes in "light-enhanced calcification" of corals. Mar Biotech 10:653–663

Mueller E (1984) Effects of a calcium channel blocker and an inhibitor of phosphodiesterase on calcification in *Acropora formosa*. In: Proceedings of Advances in Reef Science, Joint meeting of the Atlantic Reef Committee and the Internaional Society for Reef Studies, Miami, FL, pp 87–88

Muscatine L (1971) Calcification in corals. In: Lenhoff HM, Muscatine L, Davis LV (eds) Experimental coelenterate biology. University of Hawaii Press, Honolulu, Joint meeting of the Atlantic Reef Committee and the International Society for Reef Studies, Miami, FL, pp 227–237

Muscatine L, Cernichiari E (1969) Assimilation of photosynthetic products of zooxanthellae by a reef coral. Biol Bull 137:506–523

Muscatine L, Tambutté É, Allemand D (1997) Morphology of coral desmocytes, cells that anchor the calicoblastic epithelium to the skeleton. Coral Reefs 16:205–213

Muscatine L, Goiran C, Land L, Jaubert J, Cuif J-P, Allemand D (2005) Stable isotopes ($\delta C13$ and $\delta N15$) of organic matrix from coral skeleton. Proc Natl Acad Sci USA 102:1525–1530

Nakata K, Shimomura N, Shiina N, Izumi M, Ichikawa K, Shiro M (2002) Kinetic study of catalytic CO_2 hydration by water-soluble model compound of carbonic anhydrase and anion inhibition effect on CO_2 hydration. J Inorg Biochem 89:255–266

Nyberg J, Csapo J, Malmgren BA, Winter A (2001) Changes in the D- and L-content of aspartic acid, glutamic, acid, and alanine in a scleractinian coral over the last 300 years. Org Geochem 32:623–632

Nys Y (1990) Régulation endocrinienne du métabolisme calcique chez la poule et calcification de la coquille. Thèse de doctorat, Université de Paris, Paris, p 6

Nys Y, Zawadzki J, Gautron J, Mills AD (1991) Whitening of brown-shelled eggs: mineral composition of uterine fluid and rate of protoporphyrin deposition. Poult Sci 70:1236–1245

Oaki Y, Kotachi A, Miura T, Imai H (2006) Bridged nanocrystals in biominerals and their biomimetics: classical yet modern crystal growth on the nanoscale. Adv Func Mater 16:1633–1639

Ogilvie MM (1897) Microscopic and systematic study of madreporarian types of corals. Phil Trans R Soc Lond 187B:83–345

Palmer AR (1992) Calcification in marine molluscs. How costly is it? Proc Natl Acad Sci USA 89:1379–1387

Pearse VB (1970) Incorporation of metabolic CO_2 into coral skeleton. Nature 228:383

Pearse VB, Muscatine L (1971) Role of symbiotic algae (zooxanthellae) in coral calcification. Biol Bull 141:350–363

Perrin C (2003) Compositional heterogeneity and microstructural diversity of coral skeletons: implications for taxonomy and control on early diagenesis. Coral Reefs 22:109–120

Perrin C, Smith DC (2007) Decay of skeletal organic matrices and early diagenesis in coral skeletons. CR Palevol 6:253–260

Puverel S, Tambutté É, Zoccola D, Domart-Coulon I, Bouchot A, Lotto S, Allemand D, Tambutté S (2005a) Antibodies against the organic matrix in scleractinians: a new tool to study coral biomineralization. Coral Reefs 24:149–156

Puverel S, Tambutté É, Pereira-Mouries L, Zoccola D, Allemand D, Tambutté S (2005b) Soluble organic matrix of two Scleractinian corals: Partial and comparative analysis. Comp Biochem Physiol 141B:480–487

Puverel S, Houlbrèque F, Tambutté É, Zoccola D, Payan P, Caminiti N, Tambutté S, Allemand D (2007) Evidences of low molecular weight components in the organic matrix of the reef-building coral, *Stylophora pistillata*. Comp Biochem Physiol 147A:850–856

Ramseyer K, Miano TM, D'Orazio V, Wildberger A, Wagner T, Geister J (1997) Nature and origin of organic matter in carbonates from speleotherms, marine cements and coral skeletons. Org Geochem 26:361–378

Rasmussen CE (1988) The use of strontium as an indicator of anthropogenically altered environmental parameters. In: Proceedings of the 6th International Coral Reef Symposium, vol 2, Townsville, pp 325–330

Raz-Bahat M, Erez J, Rinkevich B (2006) In vivo light-microscopic documentation for primary calcification processes in the hermatypic coral *Stylophora pistillata*. Cell Tissue Res 325:361–368

Reyes-Bermudez A, Lin Z, Hayward DC, Miller DJ, Ball EE (2009) Differential expression of three galaxin-related genes during settlement and metamorphosis in the scleractinian coral *Acropora millepora*. BMC Evol Biol 9:1–29

Reynaud S, Hemming NG, Juillet-Leclerc A, Gattuso J-P (2004) Effect of pCO_2 and temperature on the boron isotopic composition of the zooxanthellate coral *Acropora* sp. Coral Reefs 23:539–546

Reynaud-Vaganay S, Gattuso J-P, Cuif J-P, Jaubert J, Juillet-Leclerc A (1999) A novel culture technique for scleractinian corals: application to investigate changes in skeletal $\partial^{18}O$ as a function of temperature. Mar Ecol Progr Ser 180:121–130

Ridgwell A, Zeebe RE (2005) The role of the global carbonate cycle in the regulation and evolution of the Earth system. Earth Planet Sci Lett 234:299–315

Ries JB, Stanley SM, Hardie LA (2006) Scleractinian corals produce calcite, and grow more slowly, in artificial Cretaceous seawater. Geol Soc Am 34:525–528

Rinkevich B, Loya Y (1984) Does light enhance calcification in hermatypic corals? Mar Biol 80:1–6

Rollion-Bard C, Chaussidon M, France-Lanord C (2003) pH control on oxygen isotopic composition of symbiotic corals. Earth Planet Sci Lett 215:275–288

Roth AA, Clausen CD, Yahiku PY, Clausen VE, Cox WW (1982) Some effects of light on coral growth. Pac Sci 36:65–81

Sanderman IM (2008) Light driven lipid peroxidation of coral membranes and a suggested role in calcification. Rev Biol Trop 56:1–9

Schmid V, Ono S, Reber-Muller S (1999) Cell-substrate interactions in cnidaria. Microsc Res Tech 44:254–268

Schneider K, Erez J (2006) The effect of carbonate chemistry on calcification and photosynthesis in the hermatypic coral *Acropora eurystoma*. Limnol Oceanogr 51:1284–1293

Schneider K, Levy O, Dubinsky Z, Erez J (2009) In situ diel cycles of photosynthesis and calcification in hermatypic corals. Limmol Oceanogr 54:1995–2002

Shechter A, Berman A, Singer A, Freiman A, Grinstein M, Erez J, Aflalo ED, Sagi A (2008) Reciprocal changes in calcification of the gastrolith and cuticle during the molt cycle of the red claw crayfish *Chera quadricarinatus*. Biol Bull 214:122–134

Silliman B (1846) On the chemical composition of the calcareous corals. Am J Sci Arts 51:189–199

Simkiss K (1964) Phosphates as crystal poisons of calcification. Biol Rev 39:487–505

Sinclair D (2004) Interactive comment on "The environment recording unit in coral skeletons: structural and chemical evidences of a biochemically driven stepping-growth process in coral fibres" by J.P. Cuif and Y. Dauphin. Biogeosci Disc 1(2004):265–272, Biogeosci Disc 1: 265–272

Smith SV, Kinsey W (1978) Calcification and organic carbon metabolism as indicated by carbon dioxide. UNESCO, Paris

Söllner C, Burghammer M, Busch-Nentwich E, Berger J, Schwarz H, Riekel C, Nicolson T (2003) Control of crystal size and lattice formation by Starmaker in otolith biomineralization. Science 302:282–286

Sorauf JE (1999) Skeletal microstructure, geochemistry, and organic remnants in Cretaceous scleractinian corals: Santonian Gosau beds of Gosau. Austria J Paleontol 73:1029–1041

Stern B, Abbott GD, Collins MJ, Armstrong HA (1999) Development and comparison of different methos for the extraction of biomineral associated lipids. Anc Biomol 2:321–324

Stolarski J (2003) Three-dimensional micro- and nanostructural characteristics of the scleractinian coral skeleton: a biocalcification proxy. Acta Palaeontol Pol 4:497–530

Stolarski J, Meibom A, Przeniosto R, Mazur M (2007) A cretaceous scleractinian coral with a calcitic skeleton. Science 318:92–94

Supuran CT, Scozzafava A (2007) Carbonic anhydrases as targets for medicinal chemistry. Bioorg Med Chem 15:4336–4350

Swart PK (1979) The effect of seawater calcium concentrations on the growth and skeletal composition of a scleractinian coral, *Acropora squamosa*. J Sediment Pet 49:951–954

Tambutté É, Allemand D, Bourge I, Gattuso J-P, Jaubert J (1995) An improved ^{45}Ca protocol for investigating physiological mechanisms in coral calcification. Mar Biol 122:453–459

Tambutté É, Allemand D, Mueller E, Jaubert J (1996) A compartmental approach to the mechanism of calcification in hermatypic corals. J Exp Biol 199:1029–1041

Tambutté É, Allemand D, Zoccola D, Meibom A, Lotto S, Caminiti N, Tambutté S (2007a) Observations of the tissue-skeleton interface in the scleractinian coral *Stylophora pistillata*. Coral Reefs 26:517–529

Tambutté S, Tambutté É, Zoccola D, Caminiti N, Lotto S, Moya A, Allemand D, Adkins J (2007b) Characterization and role of carbonic anhydrase in the calcification process of the azooxanthellate coral *Tubastrea aurea*. Mar Biol 151:71–83

Tambutté S, Tambutté É, Zoccola D, Allemand D (2007c) Organic matrix and Biomineralization of scleractinian corals. In: Baeuerlein E (ed) Handbook of biomineralization: biology aspects and structure formation. Wiley-VCH, Weinheim, pp 243–259

Taylor DL (1977) Intra-colonial transport of organic compund and calcium in some atlantic reef corals. In: Taylor DL (ed) Proc Third Int Coral Reef Symp. Rosenstiel School of Marine and Atmospheric Science, Miami, FL, pp 431–436

Taylor DL (1983) Mineralization in symbiotic systems. Endocytobiology 2:689–697

Technau U, Rudd S, Maxwell P, Gordon PMK, Saina M, Grasso LC, Hayward DC, Sensen CW, Saint R, Holstein TW, Ball EE, Miller DJ (2005) Maintenance of ancestral complexity and non-metazoan genes in two basal cnidarians. Trends Genet 21:633–639

Teng HH, Dove PM, Orme CA, De Yoreo JJ (1998) Thermodynamics of calcite growth: baseline for understanding biomineral formation. Science 282:724–727

Tentori E, Allemand D (2006) Light-enhanced calcification and dark decalcification in isolates of the soft coral *Cladiella* sp. during tissue recovery. Biol Bull 211:193–202

Tohse H, Murayama E, Ohira T, Takagi Y, Nagasawa H (2006) Localization and diurnal variations of carbonic anhydrase mRNA expression in the inner ear of the rainbow trout *Oncorhynchus mykiss*. Comp Biochem Physiol 145C:257–264

Truchot J-P (1987) Comparative aspects of extracellular acid-base balance. Springer, Berlin/Heidelberg

Tyler S (2003) Epithelium - The primary building block for Metazoan complexity. Integr Comp Biol 43:55–63

Urist MR (1965) Bone: formation by autoinduction. Science 150:893–899

Vandermeulen JH (1975) Studies on reef corals. III. Fine structural changes of calicoblast cells in *Pocillopora damicornis* during settling and calcification. Mar Biol 31:69–77

Vaughan TW (1919) Corals and the formation of coral reefs. Ann Rep Smithson Inst 17:189–238

Veis A (2005) A window on biomineralization. Science 307:1419–1420

Veis DJ, Albinger TM, Clohisy J, Rahima M, Sabsay B, Veis A (1986) Matrix proteins of the teeth of the sea urchin *Lytechinus variegatus*. J Exp Zool 240:35–46

Vengosh A, Kolodny Y, Starinsky A, Chivas AR, McCulloch MT (1991) Coprecipitation and isotopic fractionation of boron in modern biogenic carbonates. Geochim Cosmochim Acta 55:2901–2910

Venn AA, Tambutté É, Lotto S, Zoccola D, Allemand D, Tambutté S (2009) Imaging intracellular pH in a reef coral and symbiotic anemone. Proc Natl Acad Sci USA 106:16574–16579

Vertino A, Stolarski J, Beuck L (2007) Organo-mineral skeleton of deep-water scleractinia: shelter and "snack" for bioeroding organisms. In: 10th International symposium on fossil Cnidaria and Porifera, St Petersburg, Russia, p 98

Vielzeuf D, Garrabou J, Baronnet A, Grauby O, Marschal C (2008) Nano to macroscale biomineral architecture of red coral (*Corallium rubrum*). Am Mineral 93:1799–1815

Von Heider A (1881) Die Gattung *Cladocora* Ehrenb. Sber Akad Wiss Wien 84:634–637

Wainwright SA (1963) Skeletal organization in the coral, *Pocillopora damicornis*. Q J Micr Sci 104:169–183

Wang JT, Douglas AE (1999) Essential amino acid synthesis and nitrogen recycling in an alga-invertebrate symbiosis. Mar Biol 135:219–222

Watanabe T, Fukuda I, China K, Isa Y (2003) Molecular analyses of protein components of the organic matrix in the exoskeleton of two scleractinian coral species. Comp Biochem Physiol 136B:767–774

Weiner S, Levi-Kalisman Y, Raz S, Addadi L (2003) Biologically formed amorphous calcium carbonate. Connect Tissue Res 44(suppl1):214–218

Weis V, Allemand D (2009) What determines coral health? Science 324:1153–1155

Weis VM, Smith GJ, Muscatine L (1989) A 'CO_2 supply' mechanism in zooxanthellate cnidarians: role of carbonic anhydrase. Mar Biol 100:195–202

Wellington GM, Glynn PW (1983) Environmental influences on skeletal banding in Eastern Pacific (Panama) Corals. Coral Reefs 1:215–222

Wells JW (1956) Scleractinia. In: Moore RC (ed) Treatise of invertebrate paleontology. Geological Society of America, Lawrence, pp F328–F444

Wheeler AP (1992) Mechanisms of molluscan shell formation. In: Bonucci E (ed) Calcification in biological systems. CRC, Boca Raton, pp 179–216

Xu Y, Feng L, Jeffrey PD, Shi Y, Morel FM (2008) Structure and metal exchange in the cadmium carbonic anhydrase of marine diatoms. Nature 452:56–61

Yamashiro H (1995) The effects of HEBP, an inhibitor of mineral deposition, upon photosynthesis and calcification in the scleractinian coral, *Stylophora pistillata*. J Exp Mar Biol Ecol 191:57–63

Yamashiro H, Samata T (1996) New type of organic matrix in corals formed at the decalcified site: structure and composition. Comp Biochem Physiol 113A:297–300

Yonge CM, Nicholls AG (1931) Studies on the physiology of corals. V. On the relationship between corals and zooxanthellae. Sci Rep Gt Barrier Reef Exped 1:177–211

Young SD (1971) Organic material from scleractinian coral skeletons. I. Variation in composition between several species. Comp Biochem Physiol 40B:113–120

Young SD (1973) Calcification and synthesis of skeletal organic material in the coral, *Pocillopora damicornis* (L.) (Astrocoeniidae, Scleractinia). Comp Biochem Physiol 44A:669–672

Young SD, O'Connor JD, Muscatine L (1971) Organic material from scleractinian coral skeletons. II. Incorporation of ^{14}C into protein, chitin and lipid. Comp Biochem Physiol 40B:945–958

Yuasa HJ, Suzuki T, Yazawa M (2001) Structural organization of lower marine nonvertebrate calmodulin genes. Gene 279:205–212

Ziegler A (2008) The cationic composition and pH in the moulting fluid of *Porcellio scaber* (Crustacea, Isopoda) during calcium carbonate deposit formation and resorption. J Comp Physiol B 178:67–76

Zoccola D, Tambutté É, Sénegas-Balas F, Michiels J-F, Failla J-P, Jaubert J, Allemand D (1999) Cloning of a calcium channel α1 subunit from the reef-building coral, *Stylophora pistillata*. Gene 227:157–167

Zoccola D, Tambutté É, Kulhanek E, Puverel S, Scimeca J-C, Allemand D, Tambutté S (2004) Molecular cloning and localization of a PMCA P-type calcium ATPase from the coral *Stylophora pistillata*. Biochim Biophys Acta 1663:117–126

Zoccola D, Moya A, Béranger GE, Tambutté É, Allemand D, Carle GF, Tambutté S (2009) Specific expression of BMP2/4 ortholog in biomineralizing tissues of corals and action on mouse BMP receptor. Mar Biotech 11:260–269

Coral Calcification Under Ocean Acidification and Global Change

Jonathan Erez, Stéphanie Reynaud, Jacob Silverman, Kenneth Schneider, and Denis Allemand

Abstract Coral reefs are unique marine ecosystems that form huge morphological structures (frameworks) in today's oceans. These include coral islands (atolls), barrier reefs, and fringing reefs that form the most impressive products of $CaCO_3$ biomineralization. The framework builders are mainly hermatypic corals, calcareous algae, foraminifera, and mollusks that together are responsible for almost 50% of the net annual $CaCO_3$ precipitation in the oceans. The reef ecosystem acts as a huge filtration system that extracts plankton from the vast fluxes of ocean water that flow through the framework. The existence of these wave resistant structures in spite of chemical, biological, and physical erosion depends on their exceedingly high rates of calcification. Coral mortality due to bleaching (caused by global warming) and ocean acidification caused by atmospheric CO_2 increase are now the major threats to the existence of these unique ecosystems. When the rates of dissolution and erosion become higher than the rates of precipitation, the entire coral ecosystem starts to collapse and will eventually be reduced to piles of rubble while its magnificent and high diversity fauna will vanish. The loss to nature and to humanity would be unprecedented and it may occur within the next 50 years. In this chapter, we discuss the issue of ocean acidification and its major effects of corals from the cell level to the reef communities. Based on the recently published literature, it can be generalized that calcification in corals is strongly reduced when seawater become slightly acidified. Ocean acidification lowers both the pH and the CO_3^{2-} ion concentration in the surface ocean, but calcification at the organism level responds mainly to CO_3^{2-} and not to pH. Most reports show that the symbiotic algae are not sensitive to changes in the carbonate chemistry. The potential mechanisms responsible for coral sensitivity to acidification are either direct input of seawater to the biomineralization site or high sensitivity of the enzymes involved in calcification to pH and/or CO_2 concentrations. Increase in pH at the biomineralization site is most probably the most energy demanding process that is influenced by ocean acidification. While hermatypic corals and other calcifiers reduce their rates of calcification, chemical and biological dissolution increase and hence net calcification of the entire coral reef is decreasing dramatically. Community metabolism in several sites and in field enclosures show in some cases net dissolution. Using the relations between aragonite saturation (Ω_{arag}) and community calcification, it is possible to predict that coral reefs globally may start to dissolve when atmospheric CO_2 doubles.

Keywords Barrier reefs • fringing reefs • ocean acidification • enzymes • bio-fieters

1 Introduction

1.1 The Ecological Importance of Coral Calcification

Coral colonies with their myriad of calcifying polyps are responsible for producing the reef bioherm, that is, the large-scale physical structure that is the essence of coral reef functionality. Coral reefs function essentially as huge bio-filters, which filter out plankton from the vast amounts of seawater that flow through the reef systems. In doing so, coral reefs are obtaining nutrients and carbon from the particulate phase that supports their high gross production, their ecological

---

J. Erez (✉), J. Silverman, and K. Schneider
Institute of Earth Sciences, The Hebrew University
of Jerusalem, Jerusalem 91904, Israel
e-mail: erez@vms.huji.ac.il

S. Reynaud and D. Allemand
Centre Scientifique de Monaco, Avenue Saint-Martin,
MC-98000, Monaco, Principality of Monaco
e-mail: allemand@centrescientifique.mc

J. Silverman and K. Schneider
Carnegie Institute, Stanford, USA

diversity, and their evolutionary success in the oligotrophic ocean (e.g., Erez 1990, and references therein). While many of the reef-dwelling planktonivorous organisms and filter feeders are not necessarily corals (e.g., fish, tube worms, sponges, bryozoans, crinoids, mollusks, and many others), all these organisms, both sessile and vagile, require the physical structure of the coral reef for their existence, proper function, and shelter. The essential basis for these unique ecosystems is therefore the ability of corals to precipitate $CaCO_3$ at a rate that exceeds the fast erosion rates of biological, chemical, and physical nature to which coral reefs are exposed (e.g., Goreau 1959; Hutchings 1986; Lazar and Loya 1991; Hutchings et al. 2005). The net $CaCO_3$ deposition of an entire coral reef ecosystem is a delicate balance between $CaCO_3$ precipitation, mechanical erosion, and dissolution processes. Net precipitation of $CaCO_3$ allows coral reefs to exist, expand, and cope with sea-level rise or basement subsidence. Net dissolution and physical erosion on the other hand will shrink the volume of a coral reef ecosystem to the point that it will stop to function as a wave resistant structure (Glynn and Colgan 1992). Intensive activity of boring organisms as well as coral respiration and microbial heterotrophic activity within corals and the reef framework are responsible for localized low pH values (e.g., Enmar et al. 2000; Tribble et al. 1989, 1990; Yates and Halley 2006). These metabolic processes may cause skeleton dissolution beneath the coral tissue even in live corals (see below). It is well documented that in coral reefs where the corals have died (e.g., because of thermal bleaching), the entire framework decomposed to rubble within a few years (Glynn 1993, 1996). Noting the delicate balance between growth and erosion in coral reefs, it is not surprising that the saturation with respect to $CaCO_3$ (in particular the mineral aragonite, which is precipitated by corals – Ω_{arag}) plays an important role in the existence and proliferation of coral reefs. In fact, it has been suggested (Buddemeier and Fautin 1996a, b; Kleypas et al. 1999) that the global distribution of coral reefs in the warm parts of the surface ocean (above 18°C and 30° latitude north and south of the equator) could be related to the carbonate chemistry of these waters. This is caused by a reduction in CO_2 solubility with increasing temperature, so that at higher temperatures less CO_2 dissolves in the surface ocean, the concentration of total dissolved inorganic carbon (C_T) decreases, and the CO_3^{2-} ion concentration becomes lower, that is, Ω_{arag} decreases. Applying the same logic, it is possible to understand how atmospheric CO_2 increase is affecting and will affect the $CaCO_3$ budget of these sensitive ecosystems. Since alkalinity is a conservative property of seawater (see below), the increase in atmospheric CO_2 causes an increase in $CO_{2(aq)}$ and in C_T, thus lowering the pH and increasing HCO_3^- at the expense of CO_3^{2-}. This is lowering Ω_{arag} and with it the rate of coral calcification.

1.2 Global and Local Environmental Changes and Their Effects on Coral Reef Calcification

1.2.1 Global Warming and Bleaching

Global warming is one of the major anthropogenic environmental consequences of fossil fuel burning and atmospheric CO_2 increase (IPCC 2007). While this may have a positive effect on high-latitude coral reefs, which are located at the boarder of their low thermal tolerance, the main influence of oceanic warming is massive mortality of corals all over the world as a result of thermal bleaching (Hoegh-Guldberg 1999, chapters by Lesser and Hoegh-Guldberg in this book). It is now clear that excess warming of roughly 1–2°C above the recorded maximum long-term average, for a period of 1–2 weeks, will cause massive coral bleaching involving release or death of algal symbionts and often massive mortality of the coral host (Goreau 1964, 1990, 1992; Glynn 1993; Hoegh-Guldberg 1999). This phenomenon became apparent in the early 1980s and is now occurring at a frequency of almost every year in different oceanic regions. A massive bleaching event occurred in the Indian Ocean in the year 1997–1998 associated with an unusual El Niño year. While this is not an issue directly relevant to coral calcification, it results in dramatic decrease in live coral cover and hence affects the ratio between gross calcification and dissolution (see below). While calcification decreases in direct proportion with coral mortality, dissolution and erosion apparently increase with coral mortality, to the point where net dissolution in the reef may be observed.

1.2.2 Eutrophication

By their very nature, coral reefs thrive in the shallow (0–60 m) tropical and subtropical benthic environments that are well illuminated in order to allow the symbiotic algae to carry out their part in the symbiosis. These regions have been increasingly populated by humans, resulting in profoundly destructive influences on nearby coral reefs (e.g., Indonesia, Hawaii, Caribbean, Fiji, Maldives, and Australia). In many of these locations, raw or treated sewage and agricultural effluents rich in nutrients eventually reach coral reef habitats. This nutrient input lowers the calcification rates and enhances growth of macro-algae, which disrupts the balance between benthic autotrophs and heterotrophs to the point that it becomes a real threat for the existence of these coral reefs. It has recently been demonstrated in the Northern Gulf of Eilat (Lazar et al. 2008; Silverman et al. 2007b) that eutrophication due to mariculture activity (intensive fish cage farming) lowers the calcification of the entire coral reef community. While the exact mechanisms that are responsible for this effect are still not fully clear, it is well documented (Marubini and Davies 1996; Ferrier-Pagès et al. 2001) that higher NO_3^- concentrations lower calcification rates. In addition, intensive growth of macroalgae over live corals due to eutrophication smothers the coral tissues and leads to their mortality (see also chapter by Fabricius in this book).

1.2.3 Coral Breakage by Tourism, Boating, and Fishing

Coral reef ecosystems are perhaps the most important marine attractions for the general public. The beauty and the colorful tropical fish, corals, mollusks, sponges, and other associated fauna, together with their warm, calm, and clear blue water make these ecosystem frequently visited by tourists and SCUBA divers. Recent studies have shown that over the years, boating, diving, and snorkeling activities around coral reefs have caused severe breakage and deterioration of the coral reefs (Wilkinson and Buddemeier 1994; Wilkinson 2004).

1.2.4 Ocean Acidification

Atmospheric CO_2 increase mainly due to fossil fuels combustion, cement production, or land-use change is a well-documented process in the past several decades (IPCC 2007). Equilibration of atmospheric CO_2 with the surface waters of the oceans leads to pH lowering and this process, described as ocean acidification, has already decreased the oceanic surface pH by roughly 0.1 units since the preindustrial age (Brewer 1978; Chen and Millero 1979; Feely et al. 2004; Orr et al. 2005; Kleypas et al. 2006). It is expected that, by the year 2100, the oceanic pH will be lowered by at least an additional 0.2–0.3 pH units (Zeebe and Wolf-Gladrow 2001; Caldeira and Wickett 2003; Sabine et al. 2000; Feely et al. 2004). As will be described below, this process decreases the rates of calcification of corals and many other $CaCO_3$-precipitating organisms in the oceans (see reviews by Raven et al. 2005; Kleypas et al. 2006; Hoegh-Guldberg et al. 2007; Veron et al. 2009). Thus, ocean acidification, combined with coral mortality due to thermal bleaching (Silverman et al. 2009), may cause most coral reefs to start dissolving when atmospheric CO_2 levels will reach values as low as 450 ppm (Hoegh-Guldberg et al. 2007). This level of atmospheric CO_2 may be reached between the years 2030 and 2050 (IPCC 2007). This acute environmental issue has focused renewed interest into the biomineralization mechanisms of hermatypic corals as a function of the pH, pCO_2, CO_3^{2-}, and Ω_{arag} of the water in which they grow (see also chapter by Allemand et al. in this book). Finally, it must be stated that the combined effect of all these stressors on reef calcification may become a real threat for the existence of coral reefs in the next few decades.

2 Basics of Coral Calcification Relevant to Ocean Acidification

Coral calcification is addressed in detail in the chapter by Allemand et al. (this volume). Here, we will address only two issues that have direct implications to ocean acidification. One is the phenomenon of light-enhanced calcification because of its potential connection to the carbonate chemistry both in individual corals and on an entire reef ecosystem. The second is the hypothesis that seawater may be the mother liquor from which $CaCO_3$ is precipitated during the biomineralization process in corals (Erez and Braun 2007). In Sect. 3.2.1, we will address the direct effects of seawater acidification on coral calcification.

2.1 Light and Dark Calcification, the Effect of the Symbiotic Algae, and the Classical Calcification Hypothesis of Goreau

Hermatypic coral calcification is strongly connected to light intensity in their environment. The well-known phenomenon of light-enhanced calcification (LEC) has intrigued coral reef researchers since it was discovered in the late 1930s (Yonge and Nicholls 1931). A major contribution to this field were the seminal studies of Goreau (1959), who, using radioactive ^{45}Ca, was able to show that LEC is common to many hermatypic corals as well as to calcareous algae. Later studies of Goreau and Goreau (1959) proposed that LEC is strongly connected to the photosynthetic activity of their symbionts. The main effect of the symbionts on host calcification is supposed to be mediated by the removal of CO_2 for their photosynthesis, as demonstrated by these widely used set of stoichiometric equations:

$$\underset{\leftarrow \text{ dissolution}}{\overset{\text{calcification} \rightarrow}{Ca^{2+} + 2HCO_3^- \leftrightarrow CaCO_3 + H_2O + CO_2}} \underset{\leftarrow \text{ oxidation}}{\overset{\text{photosynthesis} \rightarrow}{\leftrightarrow CaCO_3 + CH_2O + O_2}} \quad (1)$$

Reaction (1) shows that calcification and photosynthesis enhance each other. Indeed, the major calcifiers in the ocean are photosynthetic calcareous algae (mainly coccolithophores), photosynthetic symbiont-bearing foraminifera, and hermatypic corals, which all show LEC. In the open ocean, there is a clear decoupling between the forward reactions that are limited to the photic zone and the backward reactions that involve organic matter oxidation, formation of CO_2 (causing acidic conditions), and then dissolution of $CaCO_3$. These backward reactions occur mainly on the ocean floor within the sediments. But in corals and coral reefs the back reactions (organic matter oxidation and $CaCO_3$ dissolution) may become important if calcification is lowered and oxidation of organic matter (i.e., respiration) and dissolution are increasing.

While LEC in hermatypic corals has been demonstrated beyond any doubt in many independent studies, the exact mechanism for this phenomenon is not clear yet. In addition to CO_2 removal, it has been speculated that the symbionts contribute to the host calcification by providing energy, by removing inhibitors (like phosphate), by supplying precursors for the synthesis of organic matrix, or by buffering the H^+ produced during calcification (see Allemand et al. this book for a review). The symbionts, concentrated mainly in the oral epithelium and the tentacles, are physically far away from the calcifying tissue (the aboral calicoblastic epithelium). Hence statements like that of Langdon and Atkinson (2005) or McConnaughey (2003) that "calcification and photosynthesis are thought to share a common dissolved inorganic carbon (DIC) pool which is in the cytosol of the host cell" are incorrect. Similarly, the idea of competition for DIC between photosynthesis and calcification based on nutrient enrichment experiments (Langdon and Atkinson 2005 and references therein) is doubtful, because of the bare fact that LEC exists, and that attempts to separate the two processes by DCMU or other inhibitors failed (see below). On the other hand, in several species of corals, the fastest calcifying parts, for example, the apical polyps in *Acropora* (Pearse and Muscatine 1971; Fang et al. 1989; Gladfelter 2007) or the branch tips in *Stylophora* are white and without symbionts. Decoupling between photosynthesis and LEC has been reported to a certain extent by Ohde and Hossain (2004) and in the mesocosm experiments of Leclercq et al. (2000, 2002). Similar observations were also made by Schneider and Erez (2006), under variations in carbonate chemistry of the water in which *Acropora eurystoma* was growing. It is clearly seen that when carbonate ion, pH, or DIC are changing, calcification is changing both in the light and in the dark, while photosynthesis of the symbionts remains constant. As mentioned above, experiments using DCMU and other pharmacological agents in light experiments (Vandermeulen et al. 1972; Chalker and Taylor 1975; Barnes 1985) showed that lower photosynthetic rates were associated with lower calcification rates and led to the conclusion that the two processes are coupled. Similar experiments with inhibition of the enzyme carbonic anhydrase (CA) activity yielded reduction in both photosynthesis and LEC (see Allemand et al. this book). Others suggested that dark calcification is causing reduction in calcification (Rinkevich and Loya 1984; Marshall 1996) perhaps due to respiration. Using Ca^{2+} and pH microelectrodes and enzyme inhibitors, Al-Horani et al. (2003) suggested that the activity of Ca^{2+}-ATPase that elevates the pH at the site of biomineralization is light dependent and this may be independent of the symbiont activity. Finally, it should be mentioned that McConnaughey (2003) and Schneider and Erez (2006) proposed that host calcification is actually helping symbionts photosynthesis by generation of protons, which may be transported to the microenvironment where the symbionts photosynthesize and help to acidify it. This has been challenged by Gattuso et al. (2000), but their experiments were carried out at normal (and not elevated) pH values and hence did not really test the hypothesis. On the other hand, Furla et al. (1998) suggested that host carbon-concentrating mechanisms may buffer H^+ generated during the calcification process.

2.2 Direct Supply of Seawater to the Biomineralization Site

As discussed already by Allemand et al. (this book), the possibility of direct seawater supply into the calicoblastic space (as narrow as it may be) has been suggested based on observations that the cell impermeable fluorescent dyes (both Calcein and FITC-Dextran) were incorporated into the skeleton of microcolonies completely covered by tissue (Erez and Braun 2007). This was observed in three species (*Acropora* spp., *Stylophora pistillata*, and *Pocillopora damicornis*). The simplest interpretation of these observations is that seawater with the fluorescent dye arrives to the site of biomineralization, thus providing high concentration of Ca^{2+} (around 10 mM) and at least 2 mM of DIC. The microelectrode study of Al-Horani et al. (2003) may support this possibility because the Ca^{2+} concentrations measured in the sub-calicoblastic extracellular calcifying medium were just slightly above that of seawater (+0.58 ± 0.29 mM compared to ambient seawater). The Ca^{2+} concentrations showed light and dark fluctuations that were simultaneous with pH fluctuations between 9.3 and 8.5 with the higher values in the light. This correlation, and the observation that Ruthenium Red (a Ca^{2+}-ATPase inhibitor) blocks the Ca^{2+} and pH dynamics, suggest that Ca^{2+}-ATPase, in addition to its role in supplying calcium to the calcification site, would play a major role in increasing the pH of the calcifying fluid. If this is an important route of the ion supply for biomineralization in corals, then their sensitivity to ocean acidification can readily be explained (see below).

2.3 Information from Shell Chemistry and Isotopes

There is an increasing body of evidence that during their calcification process, corals show tight response to the chemical and physical parameters in their surrounding water (e.g., Cohen and McConnaughey 2003). This is the reason why their skeletons are widely used to reconstruct paleotemperatures and other paleoceanographic conditions particularly during the Holocene and the Pleistocene when good preservation of their aragonite is evident. Many coral skeletons have the advantage of displaying clear annual banding that are wide enough to record seasonal information. This has been used as a very powerful tool to test the skeletal response in terms of stable isotopes and trace elements ($\delta^{18}O$, $\delta^{13}C$, Sr, Mg, Ba, and others) of naturally growing corals to the seasonal changes in their environmental water. Using this technique, the so-called field calibrations have clearly demonstrated, despite appreciable species and individual "vital effects," that the coral skeletons are highly valuable recorders of past oceanic conditions. There is much to learn about biomineralization in corals based on the way in which they record the changes in their water chemistry and isotopes. This information helps to develop deeper understanding of the mechanisms of biomineralization and their environmental sensitivity particularly to the ongoing process of bleaching and ocean acidification. In the context of this chapter, there is one very important conclusion that comes from many of the detailed studies that have looked at coral skeletal trace metal and isotopic chemistry. Although not always completely spelled out, many of the authors who worked on this subject imply that seawater is the mother liquor from which the coral skeleton is precipitated (Cohen and McConnaughey 2003; Erez and Braun 2007). The deviations from expected chemical or isotopic equilibrium with seawater are often explained by one of the following three mechanisms: (1) kinetic effects, that is, the fast rate of aragonite precipitation (e.g., McConnaughey 1989); (2) precipitation from a semi-closed seawater reservoir showing Rayleigh fractionation processes (mainly for trace elements, Gaetani and Cohen 2006; Gagnon et al. 2007); (3) increase of the pH and/or the Ca^{2+} concentration of the original seawater that enter the biomineralization site. This is usually attributed to the activity of Ca^{2+}-ATPase (e.g., Allison et al. 2005; Allison and Finch 2007; Rollion-Bard et al. 2003; Zoccola et al. 2004). But in all these studies, except the biochemical one of Zoccola et al. (2004), it is implied that the organism brings seawater to the site of biomineralization. If indeed ambient seawater is the starting solution from which the coral skeletons is precipitated, then its initial pH and carbonate ion concentration will influence the rate of precipitation and this may be the prime reason for their high sensitivity to ocean acidification as discussed below.

3 Sensitivity of Corals and Coral Reefs to Changes in the Carbonate Chemistry of the Water

3.1 Carbonate Chemistry of Seawater and the Use of Variable Experimental Techniques

The carbonate chemistry of seawater is rather complex because $CO_{2(g)}$ reacts with the seawater to produce dissolved CO_2 ($CO_{2(aq)}$), which is hydrated to form carbonic acid (H_2CO_3) which dissociates into bicarbonate (HCO_3^-) and carbonate (CO_3^{2-}) ions, and their associated protons (reactions 3–6). The sum of all these species is defined as "dissolved inorganic carbon" (DIC), which is often named "total inorganic carbon" or C_T or ΣCO_2 (Eq. 9).

$$CO_{2(g)} \overset{K_H}{\leftrightarrows} CO_{2(aq)2} \quad (3)$$

$$CO_2 + H_2O \rightleftarrows H_2CO_3 \quad (4)$$

$$H_2CO_3 \overset{K_1}{\leftrightarrows} HCO_3^- + H^+ \quad (5)$$

$$HCO_3^- \overset{K_2}{\leftrightarrows} CO_3^{2-} + H^+ \quad (6)$$

Also note that carbonic acid is only a small fraction of the $CO_{2(aq)}$ (Eq. 7) and therefore it is often common to combine reactions 4 and 5 to give Eq. 8:

$$\frac{CO_{2(aq)}}{H_2CO_3} \cong 650 \quad (7)$$

$$CO_{2(aq)} + H_2O \rightleftarrows HCO_3^- + H^+ \quad (8)$$

$$DIC = C_T = CO_{2(aq)} + H_2CO_3 + HCO_3^- + CO_3^{2-} \quad (9)$$

A second parameter relevant to the carbonate system is the total alkalinity, which stems from the charge balance in seawater (Eq. 10). Rearranging Eq. 10, Total Alkalinity (ALK) is defined as the difference between the cations of the strong bases and the anions of the strong acids present in seawater equaling the sum of the anions of the week acids in the seawater (Eq. 11):

$$Na^+ + K^+ + 2Ca^{2+} + 2Mg^{2+} - Cl^- - 2SO_4^{2-} - HCO_3^- - 2CO_3^{2-} - B(OH)_4^- - OH^- + H^+ = 0 \quad (10)$$

$$Na^+ + K^+ + 2Ca^{2+} + 2Mg^{2+} - Cl^- - 2SO_4^{2-} = HCO_3^- + 2CO_3^{2-} + B(OH)_4^- + OH^- - H^+ \equiv ALK \quad (11)$$

The concept of alkalinity is not trivial (see Zeebe and Wolf-Gladrow 2001), but its measurement is relatively simple and it is a very useful parameter to study the role of $CaCO_3$ dynamics in the oceanic carbon cycle. This is because alkalinity is conservative with salinity and its value changes mainly by precipitation and dissolution of $CaCO_3$, which cause alkalinity to decrease and increase, respectively.

Including CO_2 exchange with the atmosphere, there are four major reactions (reactions (3)–(6)) that are involved in the oceanic carbonate system. The natural processes, such as gas exchange, photosynthesis, respiration, $CaCO_3$ precipitation, and dissolution (Eq. 1), are all influencing the series of reactions shown below. It should also be remembered that the equilibrium constants (K_H, K_0, K_1, and K_2) are all empirical constants that are strongly influenced by temperature, salinity, and pressure. An excellent summary of the carbonate system in seawater and its interaction with the atmosphere is given in Zeebe and Wolf-Gladrow (2001). Briefly, it is possible to consider this system for first approximation as being in thermodynamic equilibrium between the various carbonate species and also with the atmosphere. The distribution of chemical species of the carbonate system is a function of the pH and it is governed by the empirical equilibrium constants mentioned above (Fig. 1). There are several programs available on the web that can be used to calculate accurately the carbonate chemistry of seawater under different conditions (e.g., CO2sys – Lewis and Wallace 1998; Seacarb program – Lavigne et al. 2008). As an example of the variability in the carbonate system at various atmospheric CO_2 (pCO_2) levels and different temperatures, see Table 1 below.

Solubility of CO_2 in surface ocean seawater is a function of atmospheric CO_2 partial pressure (pCO_2) with a strong temperature effect (according to Henry's law). Because CO_2 is a reactive gas, it equilibrates with the oceanic carbonate system (according to Eqs. 3–6). For a given alkalinity, this increases the $CO_{2(aq)}$ and the HCO_3^- concentrations and decreases the CO_3^{2-} concentration while at the same time DIC is increasing (Fig. 2).

A most useful graphic presentation of the variability in the carbonate system is the famous Deffeyes diagrams (Deffeyes 1965), showing pH, pCO_2, or Ω_{arag} isolines as a function of alkalinity and DIC (Figs. 3–6). Different processes

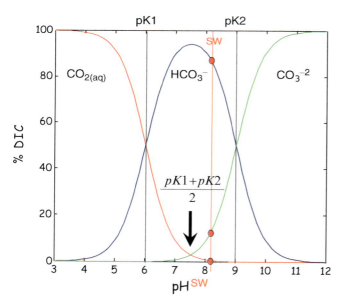

Fig. 1 The distribution of inorganic carbon species as a function of pH at equilibrium in a closed system. The values plotted are based on the equilibrium constants shown in Eqs. 2–8. Present-day seawater values are marked by the *red line*. At the relevant pH range for ocean acidification, the dominant species is HCO_3^- but the changes in it are small. The changes in $CO_{2(aq)}$ and in CO_3^{2-} on the other hand are large and they are relevant both for photosynthesis and for calcification

Table 1 Future carbonate chemistry of surface ocean water for different pCO_2 scenarios, and alkalinity of 2,300 μeq kg^{-1} at temperatures 24°C and 28°C (the *top* 2 blocks) and salinity 35. Such values can ideally be obtained experimentally by bubbling air with different pCO_2 values. In the bottom two blocks, similar experimental simulations, which are based on constant DIC but variable alkalinity, are shown for the same temperatures. As can be seen, the pCO_2 and pH are very similar, the CO_3^{2-} and Ω_{arag} are slightly higher, and the HCO_3^- ion is slightly higher at the low pCO_2 values and slightly lower at the high pCO_2 values. These differences are within the experimental errors and diurnal changes of most of the published studies

Temp	Atm pCO_2	pH	DIC	A_T	$CO_{2(aq)}$	HCO_3^{-1}	CO_3^{-2}	Ω_{arag}
24	180	8.30	1,809	2,306	5	1,461	342	5.4
	280	8.15	1,913	2,306	8	1,630	274	4.3
	380	8.05	1,980	2,306	11	1,739	230	3.6
	480	7.96	2,029	2,306	14	1,817	198	3.1
	560	7.91	2,059	2,306	16	1,864	179	2.8
	720	7.82	2,106	2,306	21	1,935	150	2.4
	1,120	7.65	2,179	2,306	33	2,040	107	1.7
28	180	8.28	1,769	2,306	5	1,395	369	6.0
	280	8.14	1,876	2,306	7	1,568	300	4.8
	380	8.04	1,946	2,306	10	1,681	254	4.1
	480	7.96	1,997	2,306	13	1,763	221	3.6
	560	7.90	2,029	2,306	15	1,813	201	3.2
	720	7.81	2,078	2,306	19	1,889	169	2.7
	1,120	7.65	2,156	2,306	29	2,004	122	2.0
24	180	8.33	2,000	2,564	5	1,590	405	6.4
	280	8.17	2,000	2,413	8	1,696	296	4.7
	380	8.05	2,000	2,325	11	1,755	234	3.7
	480	7.96	2,000	2,265	14	1,793	193	3.1
	560	7.90	2,000	2,229	16	1,814	170	2.7
	720	7.79	2,000	2,177	21	1,843	136	2.2
	1,120	7.61	2,000	2,101	33	1,877	91	1.4
28	180	8.33	2,000	2,625	5	1,543	452	7.3
	280	8.17	2,000	2,465	7	1,657	335	5.4
	380	8.05	2,000	2,369	10	1,723	267	4.3
	480	7.96	2,000	2,304	13	1,766	222	3.6
	560	7.90	2,000	2,265	15	1,790	195	3.2
	720	7.80	2,000	2,208	19	1,823	158	2.5
	1,120	7.61	2,000	2,125	29	1,865	106	1.7

can be represented as vectors on these diagrams, including CO_2 increase and removal, photosynthesis and respiration, and calcification and $CaCO_3$ dissolution. The expected changes in pH, $CO_{2(aq)}$, CO_3^{2-}, or Ω_{arag} can easily be obtained from such diagrams as a function of alkalinity and/or DIC.

As can be seen in Fig. 3, CO_2 invasion and release are horizontal vectors on this diagram and cause a decrease and increase in pH, respectively, without any effect on alkalinity. The slight slope off the horizontal in the photosynthesis–respiration arrows is caused by the changes in NO_3^- that are

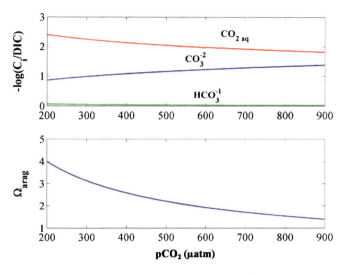

Fig. 2 Changes in DIC speciation (*top* panel) and Ω_{arag} of ocean surface water as a function of atmospheric pCO_2 in equilibrium. Note that while $CO_{2(aq)}$ and HCO_3^- are increasing CO_3^{2-} is decreasing and accordingly, Ω_{arag} is also decreasing

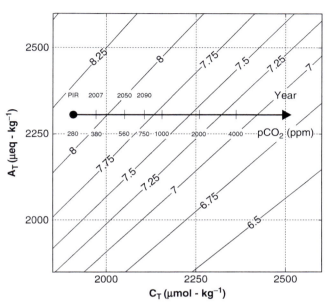

Fig. 4 The trend of ocean acidification shown on a Deffeyes diagram (as in Fig. 3) showing iso-contours of pH are plotted as a function of total alkalinity (A_T) and total dissolved inorganic carbon in seawater (C_T). Atmospheric pCO_2 values and their expected timing according to IPCC (2007) are shown as a horizontal line increasing C_T without changes in A_T. *PIR* preindustrial rate

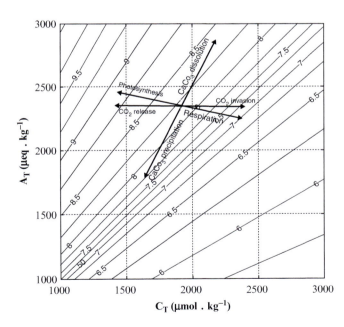

Fig. 3 Deffeyes diagram showing iso-contours of pH are plotted as a function of total alkalinity (A_T) and total dissolved inorganic carbon in seawater (C_T). Thermodynamic dissociation constants of the carbonate system were calculated as a function salinity (35), temperature (25°C) for surface water according to the relations presented in Roy et al. (1993) for the total hydrogen pH scale

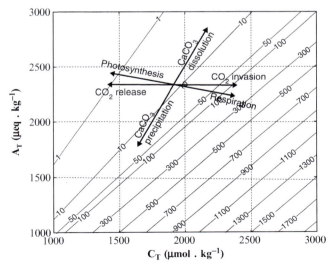

Fig. 5 Iso-contours of $CO_{2(aq)}$ (in units of $\mu mol.kg^{-1}$) are plotted as a function of A_T and C_T calculated as in Fig. 3. Note that $CaCO_3$ precipitation will increase $CO_{2(aq)}$ and hence will cause a net flux of CO_2 to the atmosphere while $CaCO_3$ dissolution will cause the opposite trends

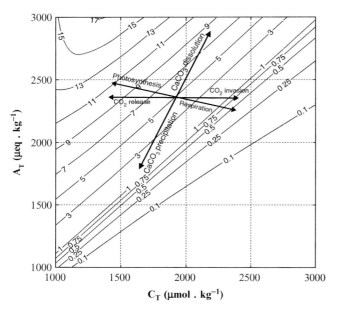

Fig. 6 Iso-contours of Ω_{arag} are plotted as a function of A_T and C_T calculated as in Fig. 3. The solubility product of aragonite in seawater was calculated as a function of salinity and temperature for surface water following Mucci (1983)

slightly the HCO_3^- concentration (Table 1, Fig. 2). Decreased concentration of CO_3^{2-} lowers the saturation level (Ω) for the carbonate minerals that are widely used by marine organisms to build the $CaCO_3$ shells, mainly calcite, aragonite, and high-Mg calcite (Kleypas et al. 1999; Feely et al. 2004; Orr et al. 2005). Corals are particularly sensitive to this effect because they are precipitating the less stable mineral, aragonite (Table 1).

During experiments with individual corals or with coral mesocosms, there are several ways to manipulate the carbonate system (Table 2) and, except for minor differences, they should all be similar. The most commonly used technique is to bubble air with a known concentration of CO_2 in seawater prior or during their introduction to the tank experiment. While this technique mimics well the processes involved, there are however some pitfalls in this approach: the exchange needs to be performed at constant temperature (the experimental temperature) and the seawater should be in equilibrium with the gas mixtures. Frequent temperature changes may prevent equilibration and this may cause a severe problem if large amounts of seawater are pumped directly from a nearby fluctuating natural reef environment which may also fluctuate in its alkalinity. Intensive bubbling may cause evaporation with significant increase in salinity, which elevates the alkalinity and this is often neglected in the calculations. In addition, the organisms themselves may change the water carbonate chemistry especially between light and dark cycles. An alternative approach is to increase or decrease the alkalinity under constant DIC conditions by adding HCl or freshly prepared NaOH and then doing the entire experiment in sealed containers without contact with the atmosphere. These experiments are almost identical to CO_2 bubbling with the exception that the bicarbonate concentration is slightly lower (Table 1). For example, for atmospheric CO_2 doubling to 560 ppm using the gas exchange method, the following values will be obtained (24°C): pH=7.91, $CO_{2(aq)}$=16 μM, HCO_3^- = 1,864 μM, CO_3^{2-} = 179 μM, and Ω_{arag} =2.8. By adding HCl, these values will be pH =7.90 $CO_{2(aq)}$=16 μM, HCO_3^- = 1,814 μM, CO_3^{2-} = 170 μM, and Ω_{arag}=2.7. These differences are much lower than the experimental fluctuations described above. Other experiments may be performed by keeping the pH constant and changing all the other parameters, or by keeping the $CO_{2(aq)}$ constant and changing all the other parameters (e.g., Schneider and Erez 2006). Adding $NaHCO_3$ is often used to increase the DIC but this should be done with a control on the pH (by alkalinity manipulation). In all these experiments, it is important to make sure that the residence time of water in the experiment is short enough so that the physiology of the corals is not influencing the carbonate chemistry beyond certain limits that are acceptable experimentally. A second consideration is how the physiological rates are measured. In most of the experiments, the initial and final buoyant weights are used to estimate the

associated with photosynthesis and respiration according to the modified Redfield equation with a slope of 16:106 (Broecker and Peng 1982). Dissolution and precipitation of $CaCO_3$ appear on this diagram as vectors with slope of 2:1 for alkalinity to DIC ratio. This is because one mole of $CaCO_3$ represents two equivalents of alkalinity. An important use of this diagram is that it allows an easy prediction of the pH changes that will occur as a result of several simultaneous processes (the vectors showed in the diagram), which may be additive in vector form. As can be seen in Fig. 4, the direct effect of atmospheric CO_2 invasion is to increase DIC along a constant alkalinity value and hence there is a significant decrease in the pH. Accordingly, future scenarios of CO_2 levels are depicted in Fig. 4. For pCO$_2$ of 1,000 ppm expected at the end of the twenty-first century, the surface ocean pH will be lower than 7.75. This is a value that many researchers have used in their experiments with corals (see below). It is also possible to see that the decrease in pH as a function of DIC is faster below 7.75 than it is between 8.25 and 8.0, which would bear more dramatic consequences if atmospheric CO_2 will increase between 1,000 and 4,000 ppm.

The Revelle factor (or the buffer factor) gives the ratio between the change in pCO$_2$ and the change in DIC in an open system (Broecker and Peng 1982) and its values for surface tropical ocean water is roughly 10–12. DIC increase without changes in alkalinity decreases the pH of the surface ocean and the carbonate system responds by significant lowering of the concentration of the CO_3^{2-} while increasing

Table 2 Summary of most of the available experimental and field data on coral physiology in response to ocean acidification. For comparison between experiments, the reduction in the rate for pCO₂ doubling is calculated

Coral species	References	Methods	Duration of incubation	Experimental range	Relation	Calcification at 2×atm CO₂ (%) or equivalent	Comments
Acropora cervicornis	Chalker (1976)	Ca	2 h	0–40 mM Ca	Michaelis–Menten	−22	
Acropora formosa	Chalker (1976)	Ca	2 h	0–16 mM Ca	Michaelis–Menten	−20	
Galaxea fascicularis	Krishnaveni et al. (1989)	Ca	2 h	0–30 mM Ca	Michaelis–Menten	−19	
Stylophora pistillata	Tambutté et al. (1996)	Ca	1.25 h	0–20 mM Ca	Michaelis–Menten	−22	
Stylophora pistillata	Gattuso et al. (1998)	Ca	2.5 h	Ω_{arag}: 1–5.8	Michaelis–Menten	−12	
Okinawa coral reef	Ohde and Van Woesik (1999)	In situ	–	Ω_{arag}: 2.2–6.3	Exponential	−75	
Porites porites	Marubini and Thake (1999)	NaHCO₃	32 days	+2 mM HCO₃⁻	–		Addition in SW of 2 mM HCO₃⁻ causes a doubling of coral growth
Biosphere	Langdon et al. (2000)	CaCl₂, Na₂CO₃, NaHCO₃	3.8 years	Ω_{arag}: 1.5–5	Linear	−40	Long-term experiment, indicates that corals do not seem to be able to acclimate to changing Ω_{arag}
Mesocosm	Leclercq et al. (2000)	CO₂ injection	24 h	Ω_{arag}: 1.3–5.4	Linear	−21	
Stylophora pistillata	Furla et al. (2000)	HCl/NaOH/ NaHCO₃	3 h	0–3 mM HCO₃⁻	Michaelis–Menten	No effect on ⁴⁵Ca but a 40% reduction in ¹⁴C-HCO₃⁻ incorporation	Saturation at 1 mM HCO₃⁻
Porites compressa	Marubini et al. (2001)	HCl/NaOH	6 weeks	Ω_{arag}: 2.4–5	–	−11	Rising CO₂ will impact corals living at all depths
Mesocosm	Leclercq et al. (2002)	CO₂ injection	9–30 days	Ω_{arag}: 1.5–4	Linear	−14	
Galaxea fascicularis	Marshall and Clode (2002)	Ca	2–4 h	+ 2.5 and 5 mM Ca	–	–	Addition in SW of Ca²⁺ causes an increase of coral calcification
Biosphere	Langdon et al. (2003)	HCl/Na₂CO₃/ NaHCO₃	5 weeks	Ω_{arag}: 2.8–4	–	−85	
Stylophora pistillata	Reynaud et al. (2003)	CO₂ injection	5 weeks	460–760 µatm CO₂	–	From 0% (at 25°C) to 50% (at 28°C)	Interaction between CO₂ and temperature (25 and 28°C): the effect of CO₂ is enhanced by increased temperature
Acropora verweyi	Marubini et al. (2003)	HCl/NaOH	8 days	Ω_{arag}: 2.3–4.4	–	−18	Intraspecific variation from −5 to −32%
Pavona cactus	Marubini et al. (2003)	HCl/NaOH	8 days	Ω_{arag}: 2.2–4.4	–	−18	

(continued)

Table 2 (continued)

Coral species	References	Methods	Duration of incubation	Experimental range	Relation	Calcification at 2×atm CO_2 (%) or equivalent	Comments
Galaxea fascicularis	Marubini et al. (2003)	HCl/NaOH	8 days	Ω_{arag}: 2.2–4.4	–	–16	
Turbinaria reniformis	Marubini et al. (2003)	HCl/NaOH	8 days	Ω_{arag}: 2.2–4.4	–	–13	
Porites lutea	Ohde and Hossain (2004)	HCl/NaOH	1 week	Ω_{arag}: 1.1–7.7	Linear	–45	
Madracis mirabilis	Horst (2004)	HCl/NaOH	6 days	395–692 µatm CO_2	–	–35	Interaction between CO_2 and temperature (26, 28 and 30°C); the effect of CO_2 is independent on temperature
Porites rus	Horst (2004)	HCl/NaOH	8 days	440–635 µatm CO_2	–	–50	Interaction between CO_2 and temperature (25, 26 and 28°C); the effect of CO_2 is independent on temperature
Acropora cervicornis	Renegar and Riegl (2005)	CO_2 injection	16 weeks	360–750 µatm CO_2	–	–70	
Porites compressa and Montipora capitata	Langdon and Atkinson (2005)	HCl/NaOH	2 weeks	Ω_{arag}: 1.6–3	Linear	44 (summer) to –80 (winter)	
Acropora eurystoma	Schneider and Erez (2006)	HCl/NaOH/ NaHCO$_3$	2 h	Ω_{arag}: 1.5–6.4	Linear	–55	
Porites lutea	Hossain and Ohde (2004)	HCl/NaOH	3–6 h	Ω_{arag}: 1.5–5.2	Linear	–42	
Fungia sp.	Hossain and Ohde (2004)	HCl/NaOH	3–6 h	Ω_{arag}: 1.5–5.2	Linear	–42	
Red Sea	Silverman et al. (2007a)	In situ	–	Ω_{arag}: 3.7–4.5	Curvilinear	–55	
Stylophora pistillata	Marubini et al. (2008)	HCl/NaOH/ NaHCO$_3$	8 days	Ω_{arag}: 1.5–9.4	Curvilinear	–15	
Acropora sp.	Herfort et al. (2008)	NaHCO$_3$	8 h	0.5–8 mM HCO$_3$	–	–	Calcification not saturated at 8 mM HCO$_3^-$
Porites porites	Herfort et al. (2008)	NaHCO$_3$	8 h	0.5–8 mM HCO$_3$	–	–	Calcification saturated at 8 mM HCO$_3^-$
Acropora intermedia	Anthony et al. (2008)	CO_2 injection	8 days	Ω_{arag}: 1.5–7.1	–	–17 at 26°C, 0 at 29°C	
Porites lobata	Anthony et al. (2008)	CO_2 injection	8 days	Ω_{arag}: 1.5–7.1	–	–12 at 26°C, +23 at 29°C	
Porites astreoides	Albright et al. (2008)	HCl	1 month	Ω_{arag}: 2.2–3.2	Linear	–78	Effect on skeletal growth rate of juveniles
Mesocosm	Jokiel et al. (2008)	HCl	10 months	+365 µatm	–	–20	
Oculina arbuscula	Ries et al. (2009)	CO_2 injection	60 days	Ω_{arag}: 0.8–3.2	–	No effect up to Ω_{arag} = 1.8, drastic decrease at 0.8	

Species	Reference	Method	Duration	Range	Relationship	Effect	Notes
Acropora formosa	Crawley et al. (2009)	CO_2 injection	4 days	Present to 1,500 µatm	—	—	Effect on photorespiration, decrease of P_{max} per unit of chlorophyll
Favia fragum	Cohen et al. (2009)	HCl	8 days	Ω_{arag}: 0.22–3.71	—	−25%	Calcification still detected at Ω_{arag} = 0.22
Cladocora caespitosa	Rodolfo-Metalpa et al. (2009)	CO_2 injection	1 month	Ω_{arag}: 1.95–3.86	—	No effect	Whatever the temperature (ambient or +3°C)
Cladocora caespitosa	Rodolfo-Metalpa et al. (2009)	CO_2 injection	1 year	Ω_{arag}: 1.95–3.86	—	No effect	Whatever the temperature (ambient or +3°C), no relationship between calcification rates and Ω_{arag}
Madracis mirabilis	Jury et al. (2009)	CO_2 injection/ HCl-NaOH	2 h	Ω_{arag}: 1.74–4.30	Michaelis–Menten with $[CO_3^{2-}]$, no relationship with pH	Effects dependant on SW chemistry	Calcification rate not clearly attributable to changes in $[CO_3^{2-}]$ or Ω_{arag}; corals responded to variation in $[HCO_3^-]$
Mesocosms (*Montipora capitata*)	Andersson et al. (2009)	HCl	1 day	Ω_{arag}: 1.4–2.8	Linear in function to $[CO_3^{2-}]/\Omega_{arag}$	−101%	Reef decalcification under SW at 2 ambient CO_2

calcification. This requires a relatively long experimental period of several weeks, and does not allow an estimate of the photosynthetic or respiratory rates. In other experiments, alkalinity depletion and oxygen dynamics are used to estimate calcification and photosynthesis. One additional factor that varies between experiments is the ranges of CO_2, pH, and/or DIC. When predictions related to global CO_2 increase are made, it is important to use only data that change these parameters within realistic ranges (i.e., pCO_2 up to 1,200 ppm and/or pH of 7.6). The DIC levels should also be manipulated accordingly. For example, doubling of HCO_3^- concentration in seawater is equivalent to increasing pCO_2 to ~2,800 ppm in addition to almost doubling the alkalinity, which is not a realistic scenario.

3.2 The Effects of Ocean Acidification on Individual Corals and the Connection to Cell- and Tissue-Level Processes

As explained above (see also Table 1), global increase of atmospheric CO_2 will induce a decrease of seawater pH, $[CO_3^{2-}]$, Ω_{arag}, and an increase in dissolved CO_2 and $[HCO_3^-]$. In addition, such changes will also affect the buffering capacity of the seawater. All these changes may directly affect calcification in hermatypic corals by acting on the animal host, but may also indirectly affect calcification by their influence on the symbionts and hence on the light-enhanced calcification (LEC). We will review here physiological parameters involved in coral calcification likely to be affected by rising CO_2 and its correlated parameters.

3.2.1 Direct Effects on Coral Calcification

The influence of all the above processes on individual corals and on coral-dominated mesocosms is well documented and almost always shows a similar response. When CO_2 level in seawater is elevated or when seawater is acidified, corals slow down their calcification rates (Gattuso et al. 1999; Marubini and Atkinson 1999; Langdon et al. 2000, 2003; Leclercq et al. 2000, 2002; Marubini et al. 2001, 2003, 2008; Langdon and Atkinson 2005; Schneider and Erez 2006; Ohde and Hossain 2004; Andersson et al. 2009; Cohen et al. 2009). A summary of most of the existing data is given in Table 2. The response may change between species and also within species. In some of these experiments, in addition to CO_2, nutrients (mainly NO_3^-, NH_3, and PO_4^{3-}) and temperature were also manipulated. In most of these experiments, the rate of calcification was the only parameter measured, but in some experiments photosynthesis of the symbiotic algae, photorespiration, and dark respiration were also measured (e.g., Reynaud et al. 2003; Ohde and Hossain 2004; Schneider and Erez 2006; Marubini et al. 2008; Crawley et al. 2009). In most of these experiments, several of the carbonate species were changing simultaneously and there was no attempt to test or separate which parameter is actually controlling the calcification of the corals. Such a systematic study was carried out by Schneider and Erez (2006), showing that the main controlling factor is the CO_3^{2-} ion concentration. This was done by keeping the pH or the pCO_2 constant and changing the CO_3^{2-} (Figs. 7 and 8) by changing DIC. The conclusion of this study was that the CO_3^{2-} is most probably the main factor, controlling the rate of calcification of corals.

The study of Marubini et al. (2008) also supports the same idea, although the relation between CO_3^{2-} and calcification was slightly exponential, but in addition it suggests that other parameters such as a change in intracellular pH may play a role. However, Jury et al. (2009) challenged this view and claimed that corals responded strongly to variation in bicarbonate concentration rather than in carbonate or pH. Their conclusion however ignores the obvious fact that CO_3^{2-} was also changing under their experimental conditions. In general, almost all studies, which addressed the ocean acidification issue, presented their data as a function of CO_3^{2-} or Ω_{arag}. If indeed the CO_3^{2-} ion is the main controlling factor of coral calcification, then the future response of corals can be predicted according to various scenarios of ocean acidification (see below). The main question now is why corals are so sensitive to the CO_3^{2-} ion concentration or to Ω_{arag}? This may be surprising if one considers that calcification in corals as in other eukaryotic organisms is fully controlled genetically and physiologically (see Allemand et al., this volume). One answer may be that direct supply of seawater to the site of biomineralization is the main source of ions for the calcification process (see above), but transepithelial transport processes were also suggested to be involved in the supply of ions (see Allemand et al. this volume). Furthermore, bringing seawater to the site of calcification is not enough in order to precipitate $CaCO_3$ at the rate that hermatypic coral grows and most probably the pH in the extracellular calcifying medium (ECM) must be elevated by the coral in order to convert most of the DIC into CO_3^{2-}.

To do this, the pH has to be well above 9 (see Fig. 9). Indeed the microelectrode study of Al-Horani et al. (2003) shows that the pH in the subcalicoblastic space may be 9.3 under light conditions. Most probably, the elevation of pH at the calcification site is mediated by the enzyme Ca^{2+}-ATPase, which has been localized on the lower membrane of the calicoblastic cells (Zoccola et al. 2004). Elevating the pH at the calcification site may also serve for another purpose that is to enhance the flux of $CO_{2(aq)}$ from the cytosol across the membranes of the calicoblastic epithelium. The source of

Coral Calcification Under Ocean Acidification and Global Change

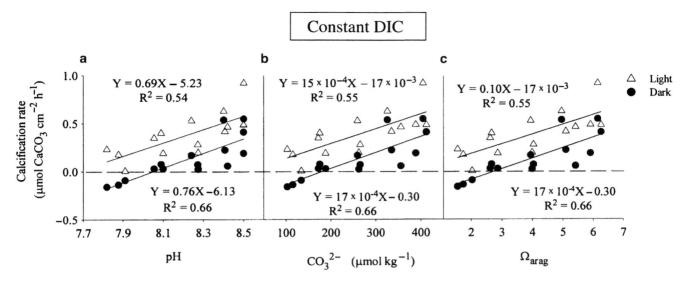

Fig. 7 The response of the coral *Acropora eurystoma* to changes in carbonate chemistry (modified from Schneider and Erez 2006). Light (*open triangles*) and dark (*black circles*) calcification rates are shown as a function of pH, CO_3^{2-}, and Ω_{arag} at constant DIC. Note that the slopes of light and dark calcification rates are parallel suggesting a constant difference between the two at different carbonate chemistry of the seawater

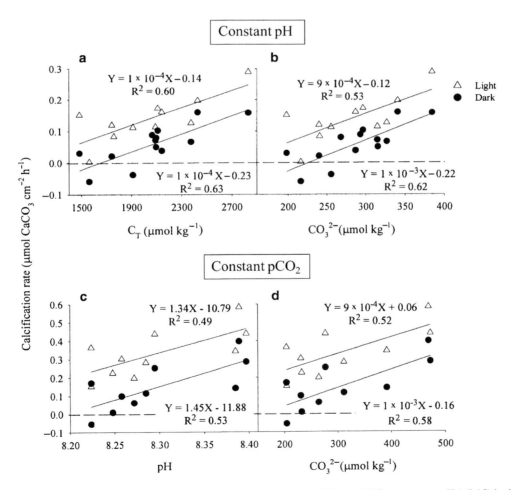

Fig. 8 The response of the coral *Acropora eurystoma* to changes in carbonate chemistry (modified from Schneider and Erez 2006). Light (*open triangles*) and dark (*black circles*) calcification rates are shown as a function of C_T and CO_3^{2-} at constant pH (~8.15) in the top part and at constant pCO_2 (~380) in the bottom panel. As can be seen, the CO_3^{2-} ion is the controlling factor of calcification rates

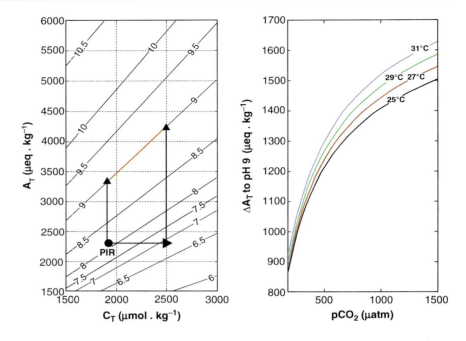

Fig. 9 The left panel shows the influence of ocean acidification on calcification in corals and foraminifera. Assuming that the organism has to elevate the pH of the calcifying fluid to 9 it has to increase the alkalinity from 2.300 to 3,400 μeq kg^{-1} for PIR conditions and to values that are much higher than that due to ocean acidification. The right panel shows that this increase (ΔA_T) will be higher at higher temperatures due to global warming

this $CO_{2(aq)}$ may be respiration but also seawater DIC by a concentrating mechanism that the corals may have. A strong evidence for such a mechanism is the large internal carbon pool for calcification inferred from ^{14}C incorporation kinetics, which lag behind the ^{45}Ca incorporation (Erez 1978; Furla et al. 2000, Allemand et al. this book). If this is indeed the pathway of DIC for coral calcification, then it may be possible to explain the sensitivity of corals to ocean acidification as follows: the CO_3^{2-} concentration at the ECM is the boundary condition that the coral has to obtain. Under lower concentration of CO_3^{2-} ion in the ambient seawater, the corals will have to spend more energy to elevate its concentration at the calcification site (Fig. 9). It should be mentioned here that a very similar scenario was described in foraminifera (Erez 2003), which are also very sensitive to ocean acidification (e.g., Bijma et al. 1999; Barker and Elderfield 2002; Erez 2003; Russell et al. 2004). Bringing seawater to the calcification site in foraminifera is mediated by large seawater vacuoles, and similar to corals, they need also to elevate the pH and accumulate inorganic carbon for calcification. It has been argued before (Erez 2003) that this is indeed necessary because in seawater the ratio of Ca^{2+} to CO_3^{2-} is well below 10 (even at pH 9). Recently, this carbon-concentrating mechanism in foraminifera has been demonstrated using pH imaging and confocal microscopy (Bentov et al. 2009). This mechanism is utilizing $CO_{2(aq)}$ obtained from acidic seawater vacuoles, which diffuses into large alkaline vacuoles through the cytosol. The alkaline vacuoles are then transported to the site of biomineralization where they are exocytosed. A similar rationale can be applied for coral calcification but apparently this happens without the involvement of vacuoles.

It should also be mentioned that CO_3^{2-} (or Ω_{arag}) is not always the controlling parameters and the correlation between ambient Ω and calcification rates is not always obvious (McConnaughey et al. 2000; Ries et al. 2009). In some cases, increasing CO_2 level (decreasing $[CO_3^{2-}]$ or Ω) did not induce a decrease in calcification both for tropical (Reynaud et al. 2003) or temperate corals (Rodolfo-Metalpa et al. 2009). Furthermore, the effect on calcification of increasing CO_2 level is highly variable with values ranging from 0 to −56% of inhibition (reviewed by Kleypas et al. 2006, see Table 2) and even in strong undersaturated conditions (Ω_{arag} =0.22), calcification may still occur (Cohen et al. 2009). Also, it may be argued that although coral calcification is extracellular, the coral skeleton is isolated from seawater by two to four layers of tissue (see Allemand et al., this volume) and thus may not be directly exposed to seawater Ω (McConnaughey et al. 2000).

One additional source of information on coral calcification in view of ocean acidification is the information coming from analysis of growth and calcification rates that are obtained from skeletal records of massive corals and in particular the genus *Porites*. Several papers published on this subject (Lough and Barnes 2000; McNeil et al. 2004) have all claimed that in the GBR (Australia), the record shows an increase in the rate of calcification during the twentieth century and they attributed this to the increased temperatures that the corals were exposed to. However, two more recent reports, Cooper et al. (2008) and more importantly De'ath et al. (2009), looked at similar records in more detail and extended their observations into later years (up to the year 2005). These records (of 328 corals from 69 reefs in the GBR) clearly show (Fig. 10) that until 1970 there was indeed

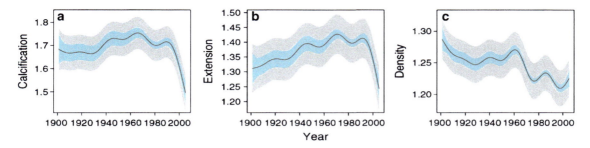

Fig. 10 Natural growth records as a function of time of the coral *Porites* sp. from the Great Barrier Reef (Australia) presented by De'ath et al. (2009). Calcification seems to slightly increase between 1920 and 1970

a slow but consistent increase in the rate of calcification which could have been attributed to temperature increase (McNeil et al. 2004). However, after 1970 calcification rates slowed down and after 1990 there is a sharp decrease (of more then 14%) in the calcification rates until 2005. This record is of prime importance because it represents a large number of measurements representing *in situ* growth records. But the caveat is that this record cannot be directly attributed to ocean acidification because the carbonate chemistry in each location has not been monitored. Still, this record is in agreement with almost all laboratory and field measurements (see below), and hence the most probable cause for this decline is ocean acidification. It seems that the decline is slightly lower than expected (only 14%) relative to the expected decline based on other field data (e.g., Silverman et al. 2007a) but this may be explained by the positive effect of temperature on these corals during most of the twentieth century.

3.2.2 Potential Effect on Symbiotic Algae

Because of LEC, we may assume that any effect on symbionts may also affect host calcification. The most obvious effect of elevated CO_2 is a fertilization of symbiont photosynthesis as observed for land plants and therefore a consecutive increase in light-enhanced calcification. However, a review of the literature concerning effects of CO_2 increase on photosynthesis leads in most cases to the conclusion that there is no effect (Goiran et al. 1996; Schneider and Erez 2006; Marubini et al. 2008). In one case, elevated pCO_2 even lead to a slight decrease of symbiont photosynthesis (Reynaud et al. 2003). Langdon et al. (2003) also showed that net community production (in the Biosphere II system) did not change in response to elevated pCO_2 and the same was observed by Leclercq et al. (2002) on small mesocosms. This lack of effect should not be surprising since it is has been suggested that corals rely on HCO_3^- for photosynthesis (Burris et al. 1983; Goiran et al. 1996). Increasing pCO_2 has a stimulatory effect on CO_2 users, such as sea grasses, which rely mainly on dissolved CO_2 for photosynthesis (Zimmerman et al. 1997). Indeed, several studies showed that HCO_3^- is not the limiting factor for zooxanthellae photosynthesis that appears to be saturated at present ambient concentration (K_m for HCO_3^- = 408 µM for the coral *Galaxea fascicularis*, Goiran et al. 1996). This however cannot be generalized since other authors reported that addition of HCO_3^- to seawater results in increased photosynthesis (Weis 1993, Herfort et al. 2008; Marubini et al. 2008). In these studies, however, the increase in HCO_3^- was completely outside of any future CO_2 increase scenario. Recently, a second effect of elevated CO_2 was reported by Anthony et al. (2008): they showed that increasing CO_2 concentration may lead to coral bleaching which in turn can affect coral calcification. Unexpectedly, CO_2 appears more effective in triggering coral bleaching than in decreasing coral calcification. These experiments however were performed under highly variable environmental conditions, and their definition of bleaching is ambiguous. Additional observations are needed to confirm these results. CO_2 enrichment was also shown to affect symbiont photorespiration and productivity (Crawley et al. 2009), which may have some implication for calcification.

Although warming and acidification of the ocean occur simultaneously with increasing atmospheric pCO_2 levels, there are only few experimental studies published to date on their combined effect on calcification and photosynthesis in corals. Reynaud et al. (2003) were the first to show that warming may aggravate the effect of acidification. These authors showed that for the coral *Stylophora pistillata*, net photosynthesis is lower at 25°C than at 28°C, and calcification was lower at high pCO_2 only at the higher temperature, while at 25°C calcification did not change at 734 ppm compared to 450 ppm. The calcification results at 25°C were surprising and in disagreement with other data published on corals (Marubini et al. 2003; Schneider and Erez, 2006). Anthony et al. (2008) showed that an intermediate-CO_2 treatment (520–705 ppm) had no impact on *Acropora intermedia* productivity at low temperature, but increased by 40% in the warm treatment (28–29°C). However, at the highest CO_2 treatment (1,010–1,360 ppm), productivity dropped to near zero for both temperature groups, which does not agree with all other papers. Calcification decreases with increasing pCO$_2$ at the lower temperature (25–26°C), while in the higher temperature (28–29°C) the decrease is significant only in the highest pCO$_2$ (1,000–1,300 ppm). In *Porites lobata*, productivity was marginally enhanced by warming at the control CO$_2$, but fell by 80% in the warm, intermediate-CO$_2$ group (opposite to the pattern for *Acropora*). High CO$_2$ led to a 30% drop in productivity in *Porites* (relative to the control) in cool conditions, and dropped to near zero at the highest CO$_2$ dosing, analogous to the pattern for *Acropora*. These results are in agreement with those discussed below for *A. eurystoma* (assuming that 25–26°C are the optimal temperatures for calcification). However, for *P. lutea* the difference between temperatures did not show clear difference in calcification. It should be noted that in this experiment there was a high variability within each pCO$_2$ level (380, 520–700, and 1,000–1,300 ppm) and accordingly in the pH and Ω_{arag}. Similarly, Rodolfo-Metalpa et al. (2009) did not show any effect of temperature (ambient vs. +3°C) on the CO$_2$ enrichment (400 vs. 700 ppm) on the temperate coral, *Cladocora caespitosa*. In an experiment carried out in the laboratory of J. Erez, calcification, net photosynthesis and respiration in *Acropora eurystoma* were studied in response to combination of three pH levels (7.9, 8.3, and 8.5) and three temperatures (21°C, 24°C, and 29°C) in short (1.5 h) incubations (Schneider 2006). Net photosynthesis decreased and respiration increased with temperature increase, whereas the pH had no effect on net photosynthesis and respiration. Calcification *versus* temperature expressed an optimum at 24°C within each pH as expected from previous experiments dealing solely with the temperature (Jokiel and Coles 1977; Coles and Jokiel 1978), while calcification *versus* pH was positively correlated, as expected from previous publications, on the carbonate system effect in corals (Ohde and Hossain 2004; Schneider and Erez 2006). Interestingly, when calcification data was sorted according to temperatures and plotted versus CO_3^{2-} concentration, the highest slope was at 24°C, the optimum temperature for calcification (Fig. 11). These results suggest that the temperature affects the sensitivity of calcification to changes in the carbonate chemistry/aragonite saturation state. This also suggests that increase in temperature may affect the sensitivity to ocean acidification in high-latitude reefs as they approach warmer (and supposedly more optimal) temperatures.

From the available data it is clear that there is a species specific response to the combined effect, but there is still a need for further experiments to test more than two temperatures. Finally, we cannot ignore the lack of a standard experimental setup and calculation methods for the metabolic rates (calcification, net photosynthesis, and respiration), which may introduce significant variation between the different studies.

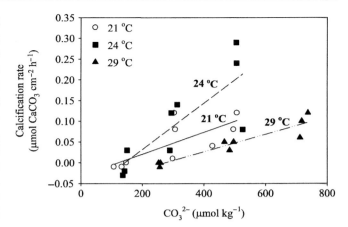

Fig. 11 The combined effect of CO_3^{2-} and temperature on calcification for the coral *Acropora eurystoma* (see experimental details in the text above). This data set suggests that the response of the coral to CO_3^{2-} is more sensitive (higher slope) near the optimum temperature for its calcification

3.2.3 Direct Effect of CO$_2$

It is well known that animals are directly affected by CO$_2$ under a process called respiratory acidosis but effective concentrations of CO$_2$ are generally higher than 1,000 ppm, that is, far above the expected global changes (Melzner et al. 2009). However, some invertebrates are affected by values in the range of that expected by climatic models. Indeed, hematophagous parasites and phytophagous insects use CO$_2$ to locate their food by following the pCO$_2$ gradients produced by its metabolism (respiration and/or photosynthesis). For example, in blood parasites, small variations of CO$_2$ from 10 to 300 ppm are enough to activate flight or searching behavior (Takken and Knols 1999). In ticks, perception of CO$_2$ changes as small as 10 ppm initiate questing behavior (Sage 2002). The sensitivity to CO$_2$ is even higher in lepidopteran herbivores. For example, the response threshold of the CO$_2$ receptor of labial organ of the moth *Heliothis armigera* is estimated to be just 0.5 ppm CO$_2$ over ambient levels brought about by photosynthetic activity of their plant foods (Stange 1992). These effects are all with respect to CO$_2$ gas partial pressure.

CO$_2$ changes are detected by CO$_2$-sensitive neurons identified in many insect species (Kwon et al. 2007). The mechanisms of CO$_2$ sensing is not well known and may involve either direct binding of CO$_2$ to a receptor, or binding of HCO_3^- or perhaps an effect mediated by a change in intracellular pH similar to acid-sensing taste cells in vertebrates (Putnam et al. 2004). In any case, this undoubtedly showed that cell physiology might be affected by very small changes in CO$_2$ such as expected to occur in the next century. A direct mediation of CO$_2$, as well as intracellular pH (see below), rather than a role of CO_3^{2-} ion (or Ω_{arag}) in the mechanism of action of ocean acidification may also explain why calcifying and non-calcifying marine invertebrates are both affected by ocean acidification (Pörtner 2008). It was also shown that doubling of atmospheric CO_2 might cause plant respiration to vary between a 20% increase and a 60% decrease with a 15–20% reduction being the average (Sage 2002). In corals, it has been shown that the rate of dark respiration remained constant (Reynaud et al. 2003; Langdon et al. 2003; Schneider and Erez 2006) when pCO_2 increased. These results suggest that changes in coral respiration cannot explain the observed changes in calcification but we probably need more data to conclude this definitely.

3.2.4 pH-Mediated Effects

This is the most obvious potential mechanism of coral sensitivity to CO_2. Cells regulate their intracellular pH (pHi) by maintaining a difference with the extracellular medium (Roos and Boron 1981). Consequently, when extracellular pH (pHe) varies, pHi also varies with an intensity depending on the intracellular buffering capacity. In general, intracellular variations are a roughly linear function of pHe (Johnson and Epel 1981), and the concentrations are lower by approximately a factor of 2–4 compared to the external variations (Tufts and Boutilier 1989; Wahl et al. 1996). Upon hydration, CO_2 leads to acidification of the medium. Doubling of CO_2 will induce a pH decrease of seawater by about 0.2 pH units since the industrial revolution, which may reduce pHi within ranges that may affect cell physiology (see below). Recently, the first measurement of pHi in cnidarian cells (the coral, *Stylophora pistillata* and the sea anemone, *Anemonia viridis*) was determined using fluorescent dyes and confocal microscopy (Venn et al. 2009). pHi was found lower of about 1 pH unit than the surrounding seawater. Light induced a slight alkalinization in symbiont-bearing cells for both coral and sea anemone. Unfortunately, the effects of changes in seawater pH on pHi in corals remain unknown.

A decrease of pH may affect calcification in two ways: directly, by decreasing the concentration and the rate of CO_3^{2-} formation within the ECM, and indirectly, by altering coral metabolism. While expected pH changes are very small, they may affect biological processes by altering enzyme activity through protonation of amino acid residues such as histidine, leading to conformational changes in proteins coupled to a change in their activity (Lehninger et al. 2004). For example, a decrease of extracellular pH of about 0.2 pH units induces a decrease of the activity of human metalloproteases by 10% (Fukasawa et al. 1998). Such proteases may be involved in organic matrix maturation (see Allemand et al. this book). A similar pH decrease induces a decrease of the carrier-mediated Na^+-dependent uptake of inorganic phosphate in pylori caeca of rainbow trout by 15% (Sugiura and Ferraris 2004), or a decrease of the activity of alkaline phosphatase, an enzyme shown to be involved in coral calcification (Domart-Coulon et al. 2001) by 50% (Xu et al. 2006). Carbonic anhydrase, an enzyme playing a major role in both calcification and carbon supply to photosynthesis (see chapters by Allemand et al. and by Furla et al. in this book) is highly sensitive to pH and a decrease of 0.2 pH units (from 8.2 to 8.0) induces a 50% change in the K_{cat} of a model enzyme (Nakata et al. 2002). Key enzymes of glycogenolysis and glycolysis, for example, the phosphorylase kinase, the phosphofructokinase, and the pyruvate dehydrogenase multienzyme complex, have also been demonstrated to be highly pH sensitive, being activated by intracellular alkalinization and suppressed by acidification in the physiological pH range (Deitmer and Rose 1996). Respiratory rate of vertebrates may increase by a factor of 5–6 upon a change of only 0.1 pH units (Fencl et al. 1966). Conversely, in invertebrates a decrease of seawater pH by 0.2 pH units leads to a decrease of about 10% of the rate of O_2 consumption by the marine worm *Sipunculus nudus* (Langenbuch and Pörtner 2002). Protein synthesis is also a process very sensitive to pH (Hofmann and Hand 1994; Kwast and Hand 1996; Langenbuch and Pörtner 2003), the effect being even more drastic upon hypoxia (Hofmann and Hand 1994). Metal speciation is also greatly affected by pH (Knutzen 1981), leading to either higher metal toxicity or to alteration in metal delivery to metalloproteins such as SODs, peroxydases, catalases, carbonic anhydrases, cytochromes, ferritin, nitrogenase, urease, and arginase. It should also be noted that the effect of pH decrease might be enhanced by the increase in temperature since there is an inverse relation between these two parameters (Roos and Boron 1981).

These data undoubtedly show that expected changes in CO_2 concentration might affect coral physiology through alteration in pH. However presently, no evidence shows pH to be directly involved as a mediator for reduced of calcification and recently Jury et al. (2009) found no correlation between the rate of calcification and the seawater pH, a result in contrast with other published data (Schneider and Erez 2006; Marubini et al. 2008). Furthermore, the couple pHe–pHi may be a unifying parameter, which is operative both in calcifiers and non-calcifiers to set CO_2 sensitivity (Pörtner 2008). It is important to design experiments that test the pH sensitivity of the major enzymes or metabolic pathways of corals. It should be noted however that Schneider and Erez (2006) showed that under constant pHe, calcification was still affected by the $[CO_3^{2-}]$, suggesting that this parameter is the main regulator of coral calcification.

3.2.5 Change in the Buffering Capacity

The buffering capacity (β, expressed as mM. pH units^{-1}) is defined as:

$$\beta = dB/dpH$$

where dB is the amount of base (or H$^+$) added to the solution, and dpH is the change in pH of the solution due to that base (or acid) addition. When β increases, a pH change induced by a given amount of base (or acid) is decreasing. Marubini et al. (2008) found an inverse relationship between coral calcification and β. Their data suggest that low buffering capacity of the media is associated with high rates of coral calcification. Allemand et al. (2007) reported the same observation in fish otoliths during metabolic acidosis, where a low buffering capacity was reported close to a site of high calcification rate. We may hypothesize that a high β may prevent the organism to control and probably elevate the pH in the ECM, as indeed shown by Al-Horani et al. (2003), thus reducing the carbonate supply to calcification site.

3.2.6 Conclusion: Origins of the Sensitivity of Corals to Ocean Acidification

Despite the central role of CO_2 in animal or plant physiology, some of the most fundamental questions about how it interacts with biological system and how organisms detect it remain unanswered (Kwon et al. 2007). This is true of course also in corals. This brief review of current literature concerning the effects of pCO_2 increase and seawater acidification on corals and their proposed mechanisms, shows that targets are numerous and that it is impossible today to give a final conclusion. If a decrease of seawater CO_3^{2-} concentration is the most common explanation, it is based on the fact that coral skeleton is directly exposed to seawater which is still debated (see chapter by Allemand et al. in this book). However, physiological evidences, at least in insects, undoubtedly showed that changes as small as 0.5 ppm in CO_2 levels, that is, about one third of the measured present annual increase in atmospheric CO_2, may elicit physiological responses. It should also be noted that Todgham and Hofmann (2009) showed that the transcriptome of sea urchin larvae were highly modified after 40 h in seawater at pH 7.96 (pCO_2 = 540 ppm). Particularly, expression of genes coding for acid–base and ion carriers as well as organic matrix proteins was decreased. Such an effect strongly suggests that ocean acidification may directly alter the basic physiology of coral in many different ways.

One of the questions that remains to be answered is how can a relatively small decrease in Ω or small changes in pH lead to a decrease in calcification whereas corals are daily submitted to large changes in physicochemical parameters and are known to be tolerant to anoxia (and therefore to respiratory acidosis)! For example, changes in pHe during a daily cycle in the coral tissue (Kühl et al. 1995; Furla et al. 1998; Al-Horani et al. 2003) or within the ecosystem (Suzuki et al. 1995) are up to 1.5 pH units, that is, far higher than expected changes due to global change, suggesting that coral cells are already adapted to large variations in external pH. A second major question lies in the problem of adaptation. It was known that in fishes, while acid-tolerant species exist, species sensitive to acid exposure lack the ability to develop any tolerance to chronic low pH exposure (Reid 1995). The fact that some coral species are insensitive (*Cladocora caespitosa*, e.g., Rodolfo-Metalpa et al. 2009) while some others are greatly sensitive (*Fungia* sp., Hossain and Ohde 2004) greatly suggests that different patterns are prevalent in corals, but may also suggest regional differences in adaptation (temperate vs. tropical). Such a diversity may also be present at the species level as coral population may possess a far higher diversity than initially thought (Császár et al. 2009), a characteristic that may facilitate their adaptation. Finally, it was shown that coral may adapt by becoming sea anemone-like animals (Fine and Tchernov 2007), which is both a good way to save the species but a very bad solution for coral reefs! However, this was shown on one species of Mediterranean corals and this observation has not been extended yet to other coral species.

This review on mechanisms of impact of seawater acidification on coral calcification emphasizes the urgent need to develop models to study the effect of increase of CO_2 at the cellular level. Identification of the mechanisms underlying the sensitivity of coral calcification to CO_2 will hopefully allow better understanding of their potential capacity for adaptation.

3.3 Ocean Acidification and Coral Reefs at the Community Level

While many studies have measured the community calcification of coral reefs in situ (mainly during the 1970s–1990s, e.g., Barnes and Taylor 1973; Barnes and Crossland 1977, 1980; Barnes and Chalker 1988; Kinsey 1978; Kinsey and Davies 1979; Gattuso et al. 1993; Pichon 1995; Gattuso et al. 1996, 1997, 1999), there was no attempt to link the changes in these rates with changes in the carbonate chemistry of the water. This may in part be because the seasonal variations in tropical and equatorial reefs were small. Two recent studies attempted to do just this: Ohde and Van Woesik (1999) in Okinawa and Silverman et al. (2007a) in the Red Sea. These studies were carried out in higher latitude reefs where seasonal changes were large enough to demonstrate such relations.

3.3.1 Community Calcification as a Function of Ω_{arag}

In both Okinawa and the Red Sea, strong decline in calcification was observed as a function of decrease in Ω_{arag} (or in the concentration of the CO_3^{2-} ion). These studies are in good agreement with the laboratory and mesocosms studies (Langdon et al. 2003). The most detailed study is that of Silverman et al. (2007a) who made 22 repeated studies on one coral reef over the course of 2 years in the Gulf of Aqaba (Eilat, Red Sea). This study demonstrated that relatively small changes in Ω_{arag} cause a large change in community calcification (Fig. 12).

The relations developed during this study were then used to predict the future effect of atmospheric CO_2 increase on coral reef calcification globally (Silverman et al. 2009; Fig. 13). In this and in other studies, net dissolution rates have been reported (Yates and Halley 2006; Silverman et al. 2007a) for the first time in coral reefs. This phenomenon is surprising because Ω values are not lower than 1 (see below). These phenomena may increase if atmospheric CO_2 continues to build up and eventually many reefs may start to dissolve.

3.3.2 CaCO$_3$ Dissolution in Coral Reefs

It is difficult to estimate the rates of dissolution in coral reefs and therefore it is also difficult to estimate the gross production (this is similar to the problem of gross photosynthesis and light time respiration). The main reason is that at least in the past, most reefs showed both in the daytime and at night, net accumulation of $CaCO_3$ (i.e., decrease in the alkalinity relative to the open ocean). There are several approaches toward this cardinal question of dissolution which may become the dominant process in coral reefs in view of ocean acidification. On the individual colony level, it is possible to estimate bioerosion from direct calcification measurements (e.g., Lazar and Loya 1991; Ohde and Hossain 2004; Schneider and Erez 2006; Tentori and Allemand 2006). The latter three studies showed alkalinity increase (i.e., dissolution) in dark and low carbonate ion concentrations. Lazar and Loya (1991) showed significant reduction in calcification rate in colonies that were infected with boring bivalves compared to their controls (without the bivalves), and obtained the rates of dissolution from the difference between the two. Alternatively, the metabolism (i.e., respiration) of the bioeroders can be determined and their potential rate of bioerosion can be estimated. At the community levels, estimates of dissolution are obtained either from confined volume experiments (Yates and Halley 2003, 2006) or from nighttime alkalinity increase relative to the open sea or from other community metabolism estimates. Several of these reports on dissolution at the community level have noticed that dissolution occurred in the dark despite the fact that Ω_{arag} is above 1 (Kinsey 1978; Barnes and Devereux 1984; Gattuso et al. 1993, 1997; Conand et al. 1997; Boucher et al. 1998; Yates and Halley 2003, 2006; Silverman et al. 2007a). A more elegant way to achieve dissolution estimate was utilized by Silverman et al. (2007a) by calculating the addition of alkalinity to the reef water based on the exchange of reef water with open seawater (with higher alkalinity). The actual observed alkalinity within the reef was higher than expected (from the exchange), indicating that there is some alkalinity added within the reef system which can be calculated as a dissolution rate. As summarized by Yates and Halley (2006), with time there are increasing reports of net dissolution in coral reef environments over a 24-h integrals (Conand et al. 1997; Gattuso et al. 1997; Boucher et al. 1998; Yates and Halley 2003, 2006; Silverman et al. 2007a). Again the Ω_{arag} values in these reports were above 1.

The calcification-dissolution process should therefore be described as follows:

Gross calcification (G) = Net calcification (N) − Dissolution (D)
Hence D = N-G and when N = 0, D = − G

If inorganic chemistry is controlling the system then D should be noticeable only when Ω_{arag} of the ambient water is below 1. For some individual corals (Schneider and Erez 2006) or mesocosm experiments (Leclercq et al. 2000, 2002), as well as in various community metabolism studies

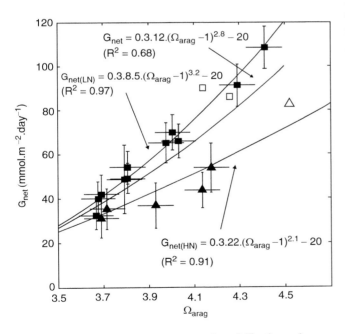

Fig. 12 The relations between net community calcification and aragonite saturation (Ω_{arag}) for the Nature Reserve reef in the Northern Gulf of Eilat (Silverman et al. 2007a). The variability in Ω_{arag} was seasonal and the measurements were carried out during 2000–2002. Empty markers indicate early 1990s measurements, normalized for live coral coverage, which was twice that of the current study

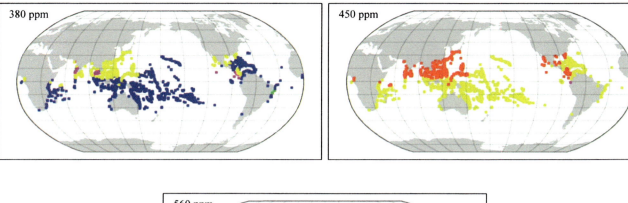

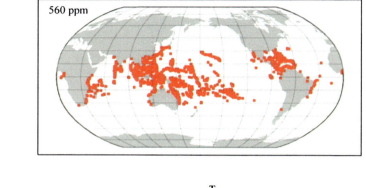

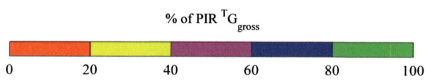

Fig. 13 Reduction in gross community calcification of coral reefs in percentage relative to preindustrial rates (PIR, 280 ppm) for different atmospheric pCO$_2$ scenarios. The relations between community calcification and Ω_{arag} are based on the relations shown in Fig. 12

(Silverman et al. 2007a) or incubations over variable tropical substrates (Yates and Halley 2006), the net rates of calcification become 0 at values of Ω_{arag} which are significantly higher than 1. Using the relations between G and Ω_{arag}, we can calculate G for values of Ω_{arag} above 1 for the situation of $N=0$ and this should be equal to the rate of dissolution. In principle, we can also derive the relations between D and Ω_{arag}. As an example we can use the equation given by Silverman et al. (2007a) for low nitrate conditions to calculate D values for $N = 0$ as follows:

$$D = G = A \cdot 8.5 \cdot (\Omega_{arag} - 1)^{3.2}$$

(where A is the proportion of live coral cover)

For $A=1$ when $\Omega_{arag} = 1.5, 2, 2.5$, D will be 1, 10, and 30 mmol CaCO$_3$ m^{-2} d^{-1}, respectively. If A is lower, it will decrease these values accordingly. This rough calculation does not take into account the dependence of D on Ω_{arag} which may also be important, but there is no information on it that we are aware of. It should be noted that most of this dissolution is probably occurring in sediment or rock interstitial water within live or dead coral reef framework substrate as a result of microbial respiration and/or the metabolic activity of the corals or other boring organisms. In this context, it is important to note that in the Moloki reef flats $N=0$ was achieved in all the sub-environments sampled (Yates and Halley 2006). The average Ω_{arag} value for these sites was ~2.5 suggesting considerably high dissolution rates at the range of 3–6 mmol CaCO$_3$ m^{-2} d^{-1} with $A = 0.1$ and $A = 0.2$, respectively.

Another approach to the dissolution of CaCO$_3$ in coral reefs (which may yield an upper limit to the dissolution process) is to make the extreme assumption that the difference between light and dark calcification (both for individual corals and on the community level) is caused by differences in Ω_{arag} in the calcifying fluid (or seawater) at the biomineralization site. This fluid becomes more basic due to the photosynthetic activity of the symbionts (and the rest of the primary producers on the community level), and more acidic due to respiration of the corals and their associated boring organisms and microbial communities. As suggested by Schneider and Erez (2006) this difference may be in the range of 0.3–0.4 pH units, most of it due to respiration.

It is possible that gross calcification remains uniform but dissolution increases in the dark to make the light dark difference in the net calcification. If one accepts this possibility, then it might explain the dependence of net calcification on Ω_{arag} due to the dependence on dissolution on Ω_{arag}. It may also explain why there is such a good agreement between this "biological" process and the inorganic precipitation experiments of Burton and Walter (1987) as noted by Silverman et al. (2007a) on the community level. The idea here is that gross calcification (G) may remain constant (and perhaps independent of Ω_{arag}), but dissolution (D) is controlled directly by Ω_{arag} (or the CO_3^{2-} ion concentration). The standard field and laboratory measurements are net calcification (N) which is the sum of G and D, is directly sensitive to Ω_{arag}. This hypothesis may also explain the sensitivity of coral calcification to ocean acidification despite the very low changes in Ω_{arag} that are produced by atmospheric CO_2 increase.

3.4 Implications for the Future Existence of Coral Reefs

Perhaps the most severe consequence of ocean acidification is its harmful effect on coral reef ecosystems oceanwide. If our (rather conservative) estimates for coral response are correct and the IPCC (2007) estimates of atmospheric CO_2 (business as usual) will materialize, it is expected that within a few tens of years, most coral reefs in the world may reach net dissolution (Silverman et al. 2009). This is not a reversible process because once the framework starts to dissolve and the physical forces (waves and storms) and the biological (bioerosion and respiration) continue their erosive action, the whole framework may disintegrate into a pile of rubble within a few years as was documented by Glynn (1993, 1996) in the Galapagos reefs. While some coral species as well as other organisms in this community may continue to exist, the ecosystem as a whole will cease to function. The filtration of plankton without the framework will become less efficient and the shelter of the framework used by so many fish and other organisms will disappear. The most diverse oceanic ecosystem will cease to function and the losses to mankind will be beyond repair. In this context, it has recently been shown (Fine and Tchernov 2007) that under very high CO_2 levels the aposymbiotic coral *Oculina patagonica* may lose their skeleton and still survive, suggesting that the effect of ocean acidification on coral reefs may not be so severe. However, one should not confuse the performance of a single coral species, which may continue to live without skeleton, with the endurance of the reef framework serving as a wave resistant structure. Corals without skeleton or with a lightly calcified one will not be able to construct and maintain a reef structure, which is the key element of the whole reef ecosystem.

4 General Conclusions

While it is now firmly established that ocean acidification greatly impact the marine environment, and particularly coral reefs, the parameters underlying this CO_2 sensitivity remain to be fully investigated. Most of the published data demonstrated that a decrease in the CO_3^{2-} of seawater reduces coral growth, but there are some exceptions that need to be explained. Most of the experimental data were acquired using addition of strong acid rather than CO_2 bubbling that is probably a better way to simulate the future global changes. However, the difference between the two methods is negligible for realistic future CO_2 scenarios. Different species may respond differently and the length of the experiment may also be an important factor that introduces variability between different studies. However, it is interesting to note that short experiments (of a few hours) often give similar results to those that lasted weeks, months, and even years. It should also be mentioned that long-term growth records of coral collected in the natural environment, show reduction in calcification during the last 20 years, and ocean acidification seems to be the most plausible explanation for this observation. Community metabolism studies of coral reef ecosystems also show strong dependence of calcification on the CO_3^{2-} ion concentration. Using the relations between these parameters and future CO_2 and temperature increase scenarios it is possible to predict that coral reefs globally will start to dissolve when atmospheric CO_2 will double (560 ppm).

Future research will have to answer important questions such as the capacity of corals to adapt to ocean acidification or what physiological mechanisms define the limits of tolerance to elevated CO_2 levels. These are major questions for the existence of coral reef in the future. Another important question is the actual mechanism that causes coral sensitivity to ocean acidification. For this purpose, cell models of coral calcification are urgently needed. They will also allow studying the combined effect of other parameters such as light, temperature, nutrients, and salinity in combination with atmospheric pCO_2. Our knowledge on the effect of seawater acidification on corals is very recent, dating from the end of 1990s and is limited to less than 21 species (Table 2) out of more than 1,300 existing ones (Cairns 1999). Obviously, a survey of sensitivity of more reefs forming species is also needed. Finally, although it is out of the scope of this review, we also have to value the cost of loosing coral reef ecosystems and the cost of preserving them in order to provide policy recommendations for reef management.

Acknowledgments We would like to thank our colleagues for stimulating discussions during the writing of this review. Financial support for the Israeli and Monegasc teams came from Israel Science Foundation (grant 206/01–13.0 to J.E.) and from the government of Monaco, respectively. The authors want to thank their institutes (Institute of Earth Sciences at the Hebrew University of Jerusalem, the IUI Eilat and the Centre Scientifique de Monaco) for their logistic and financial support as well as colleagues for fruitful discussions. Thanks are due to Jean-Pierre Gattuso for his comments on this manuscript.

References

Al-Horani F, Al-Moghrabi SM, de Beer D (2003) Microsensor study of photosynthesis and calcification in the scleractinian coral, *Galaxea fascicularis*: active internal carbon cycle. J Exp Mar Biol Ecol 288:1–15

Albright R, Mason B, Langdon C (2008) Effect of aragonite saturation state on settlement and post-settlement growth of *Porites astreoides* larvae. Coral Reefs 27:485–490

Allemand D, Mayer-Gostan N, de Pontual H, Bœuf G, Payan P (2007) Fish otolith calcification in relation to endolymph chemistry. In: Bauerlein E (ed) Handbook of biomineralization, vol 1. Wiley-VCH, Weinheim, pp 291–308

Allison N, Finch AA (2007) High temporal resolution Mg/Ca and Ba/Ca records in modern *Porites lobata* corals. Geochem Geophys Geosyst 8:Q05001. doi:10.1029/2006GC001477

Allison N, Finch AA, Newville M, Sutton SR (2005) Strontium in coral aragonite. 3. Sr coordination and geochemistry in relation to skeletal architecture. Geochim Cosmochim Acta 15:3801–3811

Andersson AJ, Kuffner IB, Mackenzie FT, Jokiel PL, Rodgers KS, Tan A (2009) Net loss of $CaCO_3$ from a subtropical calcifying community due to seawater acidificaton: mesocosm-scale experimental evidence. Biogeosciences 6:1811–1823

Anthony K, Kline DI, Diaz-Pulido G, Dove S, Hoegh-Guldberg O (2008) Ocean acidification causes bleaching and productivity loss in coral reef builders. Proc Natl Acad Sci USA 105:17442–17446

Barker S, Elderfield H (2002) Foraminiferal to glacial-interglacial changes in atmospheric CO_2. Science 297:833–836

Barnes DJ (1985) The effect of photosynthetic and respiratory inhibitors upon calcification in the staghorn coral, *Acropora formosa*. In: Delesalle B, Galzin R, Salvat B (eds) Proceeding of the fifth International Coral Reef Congress, Tahiti, pp 161–166

Barnes DJ, Chalker BE (1988) Calcification and photosynthesis in reef-building corals and algae. In: Dubinsky Z (ed) Coral reefs. Elsevier, Amsterdam, pp 110–131

Barnes DJ, Crossland CJ (1977) Coral calcification: sources of error in radioisotope techniques. Mar Biol 42:119–129

Barnes DJ, Crossland CJ (1980) Diurnal and seasonal variations in the growth of a staghorn coral measured by time-lapse photography. Limnol Oceanogr 25:1113–1117

Barnes DJ, Devereux MJ (1984) Productivity and calcification on a coral reef - A survey using pH and oxygen-electrode techniques. J Exp Mar Biol Ecol 79:213–231

Barnes DJ, Taylor DL (1973) *In situ* studies of calcification and photosynthetic carbon fixation in the coral *Montastrea annularis*. Helgoländer Wiss Meeresunters 24:284–291

Bentov S, Erez J, Brownlee C (2009) The role of seawater endocytosis in the biomineralization process in calcareous foraminifera. Proc Natl Acad Sci USA 106(51):21500–21504

Bijma J, Spero HJ, Lea DW (1999) Reassessing foraminiferal stable isotope geochemistry: Impact of the oceanic carbonate system (experimental results). In: Fischer G, Wefer G (eds) Uses of proxies in paleoceanography: examples from the South Atlantic. Springer Verlag, Berlin/Heidelberg, pp 489–512

Boucher G, Clavier J, Hily C, Gattuso J-P (1998) Contribution of soft-bottoms to the community metabolism (primary production and calcification) of a barrier reef flat (Moorea, French Polynesia). J Exp Mar Biol Ecol 225:269–283

Brewer PG (1978) Direct observation of the oceanic CO_2 increase. Geophys Res Lett 5(12):997–1000

Broecker WS, Peng TH (1982) Tracers in the sea. Lamont-Doherty Geological Observatory. Columbia University, Palisades

Buddemeier RW, Fautin DG (1996a) Saturation state and the evolution and biogeography of symbiotic calcification. Bull Inst Océanogr Monaco 14:23–32

Buddemeier RW, Fautin DG (1996b) Global CO_2 and evolution among the Scleractinia. Bull Inst Océanogr Monaco 14:33–38

Burris JE, Porter JW, Laing WA (1983) Effects of carbon dioxide concentration on coral photosynthesis. Mar Biol 75:113–116

Burton EA, Walter LM (1987) Relative precipitation rates of aragonite and Mg calcite from seawater: temperature or carbonate ion control. Geology 15:111–114

Cairns SD (1999) Species richness of recent scleractinia. Atoll Res Bull 459:1–46

Caldeira K, Wickett ME (2003) Anthropogenic carbon and ocean pH. Nature 425(6956):365

Chalker BE (1976) Calcium transport during skeletogenesis in hermatypic corals. Comp Biochem Physiol 54A:455–459

Chalker BE, Taylor DL (1975) Light-enhanced calcification, and the role of oxidative phosphorylation in calcification of the coral *Acropora cervicornis*. Proc Roy Soc Lond B 190:323–331

Chen CTA, Millero FJ (1979) Gradual increase of oceanic CO_2. Nature 277:205–206

Cohen AL, McConnaughey TA (2003) Geochemical Perspectives on Coral Mineralization. In: Dove PM, Weiner S, Yoreo JJ (eds) Biomineralization, vol 54, Reviews in Mineralogy and Geochemistry. Mineralogical Society of America and Geochemical Society, Washington, DC, pp 151–187

Cohen AL, McCorkle DC, De Putron S, Gaetani GA, Rose KA (2009) Morphological and compositional changes in the skeletons of new coral recruits reared in acidified seawater: insights into the biomineralization response to ocean acidification. Geochem Geophys Geosyst 10:1–12

Coles SL, Jokiel PL (1978) Synergistic effects of temperature, salinity and light on the hermatypic coral *Montipora verrucosa*. Mar Biol 49:187–195

Conand C, Chabanet P, Cuet P, Letourneur Y (1997) The carbonate budget of a fringing reef in La Reunion Island (Indian Ocean): sea urchin and fish bioerosion and net calcification. In: Lessios HA, MacIntyre IG (eds) Proc 8th Int Coral Reef Symp, vol 1. Smithsonian Tropical Research Institute, Balboa, pp 953–958

Cooper TF, De'Ath G, Fabricius KE, Lough JM (2008) Declining coral calcification in massive *Porites* in two nearshore regions of the northern Great Barrier Reef. Glob Change Biol 14:529–538

Crawley A, Kline DI, Dunn S, Dove S (2009) The effects of ocean acidification on symbiont photorespiration and productivity in *Acroporaformosa*.GlobChangeBiol.doi:10.1111/j.1365-2486.2009.01943.x

Czászár NBM, Seneca FO, van Oppen MJH (2009) Variation in antioxidant gene expression in the scleractinian coral *Acropora millepora* under laboratory thermal stress. Mar Ecol Prog Ser 392:93–102

De'ath G, Lough JM, Fabricius KE (2009) Declining Coral Calcification on the Great Barrier Reef. Science 323:116–119

Deffeyes KS (1965) Carbonate equilibria: a graphic and algebric approach. Limnol Oceanog 10:412–426

Deitmer JW, Rose CR (1996) pH regulation and proton signalling by glial cells. Prog Neurobiol 48:73–103

Domart-Coulon IJ, Elbert DC, Scully EP, Calimlim PS, Ostrander GK (2001) Aragonite crystallization in primary cell cultures of multicellular isolates from a hard coral *Pocillopora damicornis*. Proc Natl Acad Sci USA 98:11885–11890

Enmar R, Stein M, Bar-Matthews M, Sass E, Katz A, Lazar B (2000) Diagenesis in live corals from the Gulf of Aqaba I: the effect on paleo-oceanography tracers. Geochim Cosmochim Acta 64:3123–3132

Erez J (1978) Vital effect on stable-isotope composition seen in foraminifera and coral skeletons. Nature 273:199–202

Erez J (1990) On the importance of food sources in coral-reef ecosystems. In: Dubinsky Z (ed) Coral reefs. Elsevier, Amsterdam, pp 411–441

Erez J (2003) The Source of ions for Biomineralization in Foraminifera and their implications for Paleoceanographic Proxies. In: Dove PM, Yoreo JJD, Weiner S (eds) Reviews in mineralogy and geochemistry, vol 54. Mineralogical Society of America, Washington, DC, pp 115–149

Erez J, Braun A (2007) Calcification in hermatypic corals is based on direct seawater supply to the biomineralization site, 17th Annual V M Goldschmidt, Cologne, Germany. Geochimi Cosmochim Acta 71(15):A260–A260

Fang LS, Chen YWJ, Chen CS (1989) Why does the white tip of stony coral grow so fast without zooxanthellae? Mar Biol 103:359–363

Feely RA, Sabine CL, Lee K, Berelson W, Kleypas J, Fabry VJ, Millero FJ (2004) Impact of anthropogenic CO_2 on the $CaCO_3$ system in the oceans. Science 305:362–366

Fencl V, Miller TB, Pappenheimer JR (1966) Studies on the respiratory response to disturbances of acid-base balance, with deductions concerning the ionic composition of cerebral interstitial fluid. Am J Physiol 210:459–472

Ferrier-Pagès C, Schoelzke V, Jaubert J, Muscatine L, Hoegh-Guldberg (2001) Response of a scleractinian coral, *Stylophora pistillata*, to iron and nitrate enrichment. J Exp Mar Biol Ecol 259:249–261

Fine M, Tchernov D (2007) Scleractinian coral species survive and recover from decalcification. Science 315:1811

Fukasawa K, Fukasawa KM, Kanai M, Fujii S, Hirose J, Harada M (1998) Dipeptidyl peptidase III is a zinc metallo-exopeptidase. Biochem J 329:275–282

Furla P, Bénazet-Tambutté S, Jaubert J, Allemand D (1998) Functional polarity of the tentacle of the sea anemone *Anemonia viridis*: role in inorganic carbon acquisition. Am J Physiol (Regul Integr Comp Physiol) 274:R303–R310

Furla P, Galgani I, Durand I, Allemand D (2000) Sources and mechanisms of inorganic carbon transport for coral calcification and photosynthesis. J Exp Biol 203:3445–3457

Gaetani GA, Cohen AL (2006) Element partitioning during precipitation of aragonite from seawater: a framework for understanding paleoproxies. Geochim Cosmochim Acta 70:4617–4634

Gagnon AC, Adkins JF, Fernandez DP, Robinson LF (2007) Sr/Ca and Mg/Ca vital effects correlated with skeletal architecture in a scleractinian deep-sea coral and the role of Rayleigh fractionation. Earth Planet Sci Lett 261:280–295

Gattuso J-P, Pichon M, Delesalle B, Frankignoulle M (1993) Community metabolism and air-sea CO_2 fluxes in a coral reef ecosystem (Moorea, French Polynesia). Mar Ecol Prog Ser 96:259–267

Gattuso J-P, Pichon M, Delesalle B, Canon C, Frankignoulle M (1996) Carbon fluxes in coral reefs. 1. Lagrangian measurement of community metabolism and resulting air-sea CO_2 disequilibrium. Mar Ecol Prog Ser 145:109–121

Gattuso J-P, Payri CE, Pichon M, Delesalle B, Frankignoulle M (1997) Primary production, calcification, and air-sea CO_2 fluxes of a macroalgal-dominated coral reef community (Moorea, French Polynesia). J Phycol 33:729–738

Gattuso J-P, Frankignoulle M, Bourge I, Romaine S, Buddemeier RW (1998) Effect of calcium carbonate saturation of seawater on coral calcification. Global Planet Change 18:37–46

Gattuso J-P, Allemand D, Frankignoulle M (1999) Photosynthesis and calcification at cellular, organismal and community levels in coral reefs: A review on interactions and control by carbonate chemistry. Amer Zool 39:160–183

Gattuso J-P, Reynaud-Vaganay S, Furla P, Romaine-Lioud S, Jaubert J, Bourge I, Frankignoulle M (2000) Calcification does not stimulate photosynthesis in the zooxanthellate scleractinian coral *Stylophora pistillata*. Limnol Oceanogr 45:246–250

Gladfelter EH (2007) Skeletal development in *Acropora palmata* (Lamark 1816): a scanning electron microscope (SEM) comparison demonstrating similar mechanisms of skeletal extension in axial versus encrusting growth. Coral Reefs 28:883–892

Glynn PW (1993) Coral reef bleaching: ecological perspectives. Coral Reefs 12:1–17

Glynn PW (1996) Coral reef bleaching: facts, hypotheses and implications. Glob Change Biol 2:495–509

Glynn PW, Colgan MW (1992) Sporadic disturbances in fluctuating coral reef environments: El Nino and coral reef development in the eastern Pacific. Am Zool 32:707–718

Goiran C, Al-Moghrabi S, Allemand D, Jaubert J (1996) Inorganic carbon uptake for photosynthesis by the symbiotic coral/dinoflagellate association. I. Photosynthetic performances of symbionts and dependence on sea water bicarbonate. J Exp Mar Biol Ecol 199:207–225

Goreau TF (1959) The physiology of skeleton formation in corals. I. A method for measuring the rate of calcium deposition by corals under different conditions. Biol Bull 116:59–75

Goreau TF (1964) Mass expulsion of zooxanthellae from Jamaican Reed communities after Hurricane Flora. Science 145:383–386

Goreau TJ (1990) Coral bleaching in Jamaica. Nature 343:417

Goreau TJ (1992) Bleaching and reef community change in Jamaica: 1951-1991. Am Zool 32:683–695

Goreau TF, Goreau NI (1959) The physiology of skeleton formation in corals. II. Calcium deposition by hermatypic corals under different conditions. Biol Bull 117:239–250

Herfort L, Thake B, Taubner I (2008) Bicarbonate stimulation of calcification and photosynthesis in two hermatypic corals. J Phycol 44:91–98

Hoegh-Guldberg O (1999) Climate change, coral bleaching and the future of the world's coral reefs. Mar Freshwater Res 50:839–866

Hoegh-Guldberg O, Mumby P-J, Hooten AJ, Steneck RS, Greenfield P, Gomez E, Harvell CD, Sale PF, Edwards AJ, Caldeira K, Knowlton N, Eakin CM, Iglesias-Prieto R, Muthiga N, Bradbury RH, BDubi A, Hatziolos ME (2007) Coral reefs under rapid climate change and ocean acidification. Science 318:1737–1742

Hofmann GE, Hand SC (1994) Global arrest of translation during invertebrate quiescence. Proc Natl Acad Sci USA 91:8492–8496

Horst GP (2004) Effects of temperature and CO_2 variation on calcification and photosynthesis of two branching reef corals. Master of Science in Biology, California State University, Northridge

Hossain MMM, Ohde S (2004) Calcification of cultured *Porites* and *Fungia* under different aragonite saturation state of seawater at 25°C. Proceedings of the 10th International Coral Reef Symposium. Japanese Coral Reef Society, Okinawa, Japan, pp 597–606

Hutchings PA (1986) Biological destruction of the reef. Coral Reefs 4:239–252

Hutchings PA, Peyrot-Clausade M, Osnorno A (2005) Influence of land runoff on rates and agents of bioerosion of coral substrates. Mar Pollut Bull 51:438–447

IPCC (2007) Climate change 2007: the Physical science basis. Contribution of the working group I to the 4th assessment report of the Intergovernmental panel on climate change. Cambridge University, Cambridge, UK and New York

Johnson CH, Epel D (1981) Intracellular pH of sea urchin eggs measured by the dimethyloxazolidinedione (DMO) method. J Cell Biol 89:284–291

Jokiel PL, Coles SL (1977) Effects of temperature on the mortality and growth of Hawaiian reef corals. Mar Biol 43:201–208

Jokiel PL, Rodgers KS, Kuffner IB, Andersson AJ, Cox EF, MacKenzie FT (2008) Ocean acidification and calcifying reed organisms: a mesocosm investigation. Coral Reefs 27:473–483

Jury CP, Whitehead RF, Szmant AM (2009) Effects of variations in carbonate chemistry on the calcification rates of *Madracis mirabilis* (Duchassaing 1861): bicarbonate concentrations best predic calcification rates. Glob Change Biol. doi:10.1111/j.1365-2486.2009.02057.x

Kinsey DW (1978) Alcalinity changes and coral reef calcification. Limnol Oceanogr 23:989–991

Kinsey DW, Davies PJ (1979) Effects of elevated nitrogen and phosphorus on coral reef growth. Limnol Oceanogr 24:935–940

Kleypas JA, Buddemeier RW, Archer D, Gattuso JP, Langdon C, Opdyke BN (1999) Geochemical consequences of increased atmospheric carbon dioxide on coral reefs. Science 284:118–120

Kleypas JA, Feely RA, Fabry VJ, Langdon C, Sabine CL, Robbins LL (2006) Impacts of ocean acidification on coral reefgs and other marine calcifiers: a guide for future research. NSF – NOAA – USGS

Knutzen J (1981) Effect of decreased pH on marine organisms. Mar Pollut Bull 12:25–29

Krishnaveni P, Chou LM, Ip YK (1989) Deposition of calcium ($^{45}Ca^{2+}$) in the coral *Galeaxea fascicularis*. Comp Biochem Physiol 94A:509–513

Kühl M, Cohen Y, Dalsgaard T, Jorgensen BB, Revsbech NP (1995) Microenvironment and photosynthesis of zooxanthellae in scleractinian corals studied with microsensors for O_2, pH and light. Mar Ecol Prog Ser 117:159–172

Kwast KE, Hand SC (1996) Oxygen and pH regulation of protein synthesis in mitochondria from *Artemia franciscana* embryos. Biochem J 313:207–213

Kwon JY, Dahanukar A, Weiss LA, Carlson JR (2007) The molecular basis of CO_2 reception in *Drosophila*. Proc Natl Acad Sci USA 104:3574–3578

Langdon C, Atkinson MJ (2005) Effect of elevated pCO_2 on photosynthesis and calcification of corals and interactions with seasonal change in temperature/irradiance and nutrient enrichment. J Geophys Res 110:C09S07

Langdon C, Takahashi T, Sweeney C, Chipman D, Goddard J, Marubini F, Aceves H, Barnett H, Atkinson MJ (2000) Effect of calcium carbonate saturation state on the calcification rate of an experimental coral reef. Glob Biogeochem Cycles 14:639–654

Langdon C, Broecker W, Hammond D, Glenn E, Fitzsimmns K, Nelson SG, Peng T-H, Hajdas I, Bonani G (2003) Effect of elevated CO_2 on the community metabolism of an experimental coral reef. Global Biogeochem Cycles 11:1–14

Langenbuch M, Pörtner HO (2002) Changes in metabolic rate and N excretion in the marine invertebrate *Sipunculus nudus* under conditions of environmental hypercapnia: identifying effective acid-base variable. J Exp Biol 205:1153–1160

Langenbuch M, Pörtner HO (2003) Energy budget of hepatocytes from Antarctic fish (*Pachycara brachycephalum* and *Lepidonotothen kempi*) as a function of ambient CO_2: pH-dependent limitations of cellular protein synthesis. J Exp Biol 206:3895–3903

Lavigne H, Proye A, Gattuso J-P (2008) Portions of code and/or corrections were contributed by Epitalon JM, Hofmann A, Gentili B, Orr J and Soetaert K. seacarb: calculates parameters of the seawater carbonate system. R package version 2.2. http://cran.at.r-project.org/web/packages/seacarb/index.html

Lazar B, Loya Y (1991) Bioerosion of coral reefs - A chemical approach. Am Soc Limnol Oceanogr 36(2):377–383

Lazar B, Erez J, Silverman J, Rivlin T, Rivlin A, Dray M, Meeder E, Iluz D (2008) Recent environmental changes in the chemical-biological oceanography of the Gulf of Aqaba (Eilat). In: Por FD (ed) Aqaba-Eilat, The improbable Gulf. Environment, biodiversity and preservation. Magnes, Jerusalem

Leclercq N, Gattuso J-P, Jaubert J (2000) CO_2 partial pressure controls the calcification rate of a coral community. Glob Change Biol 6:329–334

Leclercq N, Gattuso J-P, Jaubert J (2002) Primary production, respiration, and calcification of a coral reef mesocosm under increased CO_2 partial pressure. Limnol Oceanogr 47:558–564

Lehninger AL, Cox MM, Nelson DL (2004) Principles of biochemistry. W.H. Freeman, New York

Lewis E, Wallace DWR (1998) Program developed for CO_2 system calculations. ORNL/CDIAC-105. Carbon Dioxide Information Analysis Center, Oak Ridge National Laboratory, US Department of Energy, Oak Ridge

Lough JM, Barnes DJ (2000) Environmental controls on growth of the massive coral *Porites*. J Exp Mar Biol Ecol 245:225–243

Marshall AT (1996) Calcification in hermatypic and ahermatypic corals. Science 271:637–639

Marshall AT, Clode PL (2002) Effect of increased calcium concentration in sea water on calcification and photosynthesis in the scleractinian coral *Galaxea fascicularis*. J Exp Biol 205:2107–2113

Marubini F, Atkinson MJ (1999) Effects of lowered pH and elevated nitrate on coral calcification. Mar Ecol Prog Ser 188:117–121

Marubini F, Davies PS (1996) Nitrate increases zooxanthellae population density and reduces skeletogenesis in corals. Mar Biol 127:319–328

Marubini F, Thake B (1999) Bicarbonate addition promotes coral growth. Limnol Oceanogr 44:716–720

Marubini F, Barnett H, Langdon C, Atkinson MJ (2001) Dependence of calcification on light and carbonate ion concentration for the hermatypic coral *Porites compressa*. Mar Ecol Prog Ser 220:153–162

Marubini F, Ferrier-Pages C, Cuif J-P (2003) Suppression of growth scleractinian corals by decreasing ambient carbonate ion concentration: a cross-family comparison. Proc Royal Soc B 270:1–23

Marubini F, Ferrier-Pagès C, Furla P, Allemand D (2008) Coral calcification responds to seawater acidification: a working hypothesis towards a physiological mechanism. Coral Reefs 27:491–499

McConnaughey T (1989) ^{13}C and ^{18}O isotopic disequilibrium in biological carbonates. I. Patterns. Geochim Cosmochim Acta 53:151–162

McConnaughey TA (2003) Sub-equilibrium oxygen-18 and carbon-13 levels in biological carbonates: carbonate and kinetic models. Coral Reefs 22:316–327

McConnaughey TA, Adey WH, Small AM (2000) Community and environmental influences on reef coral calcification. Limnol Oceanogr 45:1667–1671

McNeil BI, Matear RJ, Barnes DJ (2004) Coral reef calcification and climate change: The effect of ocean warming. Geophys Res Lett 31:L22309

Melzner F, Gutowska MA, Langenbuch M, Dupont S, Lucassen M, Thorndyke M, Bleich M, Pörtner H-O (2009) Physiological basis for high CO_2 tolerance in marine ectothermic animals: pre-adaptation through lifestyle and ontogeny? Biogeosciences 6:4693–4738

Mucci A (1983) The solubility of calcite and aragonite in seawater at various salinities, temperature, and one atmosphere total pressure. Am J Sci 283:780–799

Nakata K, Shimomura N, Shiina N, Izumi M, Ichikawa K, Shiro M (2002) Kinetic study of catalytic CO_2 hydration by water-soluble model compound of carbonic anhydrase and anion inhibition effect on CO_2 hydration. J Inorg Biochem 89:255–266

Ohde S, Hossain MMM (2004) Effect of $CaCO_3$ (aragonite) saturation state of seawater on calcification of *Porites* coral. Geochem J 38:613–621

Ohde S, Van Woesik R (1999) Carbon dioxide flux and metabolic processes of a coral reef. Okinawa. Bull Mar Sci 65:559–576

Orr JC, Fabry VJ, Aumont O, Bopp L, Doney SC, Feely RA, Gnanadesikan A, Gruber N, Ishida A, Joos F, Key RM, Lindsay K, Maier-Reimer E, Matear R, Monfray P, Mouchet A, Najjar RG,

Plattner G-K, Rodgers KB, Sabine CL, Sarmiento JL, Schlitzer R, Slater RD, Totterdell IJ, Weirig M-F, Yamanaka Y, Yool A (2005) Anthropogenic ocean acidification over the twenty-first century and its impact on calcifying organisms. Nature 437:681–686

Pearse VB, Muscatine L (1971) Role of symbiotic algae (zooxanthellae) in coral calcification. Biol Bull (Woods Hole) 141: 350–363

Pichon M (1995) Coral Reef Ecosystems. In: Nierenberg WA (ed) Encyclopedia of environmental biology, vol 1. Academic, San Diego, pp 425–443

Pörtner H-O (2008) Ecosystem effects of ocean acidification in times of ocean warming: a physiologist's view. Mar Ecol Prog Ser 373: 203–217

Putnam RW, Filosa JA, Ritucci NA (2004) Cellular mechanisms involved in CO_2 and acid signaling in chemosensitive neurons. Am J Physiol Cell Physiol 287:C1493–C1526

Raven J, Caldeira K, Elderfield H, Hoegh-Guldberg O, Liss P, Riebesell U, Shepherd J, Turley C, Watson A (2005) Ocean acidification due to increasing atmospheric carbon dioxide. The Royal Society, Cardiff, p 60

Reid SD (1995) Adaptation to and effects of acid water on the fish gill. In: Hochachka PW, Mommsen TP (eds) Biochemistry and molecular biology of fishes, vol 5. Elsevier Science, New York, pp 213–217

Renegar DA, Riegl BM (2005) Effect of nutrient enrichment and elevated CO_2 partial pressure on growth rate of Atlantic scleractinian coral *Acropora cervicornis*. Mar Ecol Prog Ser 293:69–76

Reynaud S, Leclercq N, Romaine-Lioud S, Ferrier-Pagès C, Jaubert J, Gattuso J-P (2003) Interacting effects of CO_2 partial pressure and temperature on photosynthesis and calcification in a scleractinian coral. Glob Change Biol 9:1660–1668

Ries JB, Cohen AL, McCorkle DC (2009) Marine calcifiers exhibit mixed responses to CO_2-induced ocean acidification. Geology 37: 1131–1134

Rinkevich B, Loya Y (1984) Does light enhance calcification in hermatypic corals? Mar Biol 80:1–6

Rodolfo-Metalpa R, Martin S, Ferrier-Pagès C, Gattuso J-P (2009) Response of the temperate coral *Cladocora caespitosa* to mid- and long-term exposure to pCO_2 and temperature levels projected in 2100. Biogeosciences 6:7103–7131

Rollion-Bard C, Chaussidon M, France-Lanord C (2003) pH control on oxygen isotopic composition of symbiotic corals. Earth Planet Sci Lett 215:275–288

Roos A, Boron WF (1981) Intracellular pH. Physiol Rev 61:296–434

Roy RN, Roy LN, Vogel KM, Porter-Moore C, Pearson T, Good CE, Millero FJ, Campbell DM (1993) The dissociation-constants of carbonic-acid in seawater at salinities 5 to 45 and temperatures 0°C to 45°C. Mar Chem 44:249–267

Russell AD, Honisch B, Spero HJ et al (2004) Effects of seawater carbonate ion concentration and temperature on shell U, Mg, and Sr in cultured planktonic foraminifera. Geochim Cosmochim Acta 68(21):4347–4361

Sabine CL, Wanninkhof R, Key RM, Goyet C, Millero FJ (2000) Seasonal CO_2 fluxes in the tropical and subtropical Indian Ocean. Mar Chem 72:33–53

Sage RF (2002) How terrestrial organisms sense, signal, and respond to carbon dioxide. Integr Comp Biol 42:469–480

Schneider K (2006) The effects of anthropogenic global changes on the calcification and photosynthesis of hermatypic corals. Ph.D. thesis, The Hebrew University of Jerusalem

Schneider K, Erez J (2006) The effect of carbonate chemistry on calcification and photosynthesis in the hermatypic coral *Acropora eurystoma*. Limnol Oceanogr 51:1284–1293

Silverman J, Lazar B, Erez J (2007a) The effect of aragonite saturation, temperature and nutrients on the community calcification rate of a coral reef. J Gephys Res 112. doi:10.1029/2006JC003770

Silverman J, Lazar B, Erez J (2007b) Community metabolism of a coral reef exposed to naturally varying dissolved inorganic nutrient loads. Biogeochemistry 84(1):67–82

Silverman J, Lazar B, Cao L, Caldeira K, Erez J (2009) Coral reefs may start dissolving when atmospheric pCO_2 doubles. Geophys Res Lett 36:659–662

Stange G (1992) High resolution measurement of atmospheric carbon dioxide concentration changes by the labial palp organ of the moth *Heliothis armigera* (Lepidoptera: Noctuidae). J Com Physiol A 171:317–324

Sugiura SH, Ferraris RP (2004) Contributions of different NaPi cotransporter isoforms to dietary regulation of P transport in the pyloric caeca and intestine of rainbow trout. J Exp Biol 207: 2055–2064

Suzuki A, Nakamori T, Kayanne H (1995) The mechanism of production enhancement in coral reef carbonate systems: model empirical results. Sediment Geol 99:259–280

Takken W, Knols BGJ (1999) Odor-mediated behavior of afro-tropical malaria mosquitoes. Annu Rev Entomol 44:131–157

Tambutté É, Allemand D, Mueller E, Jaubert J (1996) A compartmental approach to the mechanism of calcification in hermatypic corals. J Exp Biol 199:1029–1041

Tentori E, Allemand D (2006) Light-Enhanced Calcification and dark decalcification in isolates of the soft coral *Cladiella* sp. during tissue recovery. Biol Bull 211:193–202

Todgham AE, Hofmann GE (2009) Transcriptomic response of sea urchin larvae *Strongylocentrotus purpuratus* to CO_2-driven seawater acidification. J Exp Biol 212:2579–2594

Tribble, GW, Sansone FJ, Li Y-H, Smith SV, Buddemeier RW (1989) Material fluxes from a reef framework. Proc Sixth Int Coral Reef Symp 2, Townsville, pp 577–582

Tribble GW, Sansone FJ, Smith SV (1990) Stoichiometric modeling of carbon diagenesis within a coral reef framework. Geochim Cosmochim Acta 54:2439–2449

Tufts BL, Boutilier RG (1989) The absence of rapid chloride/bicarbonate exchange in lamprey erythrocytes: implications for CO_2 transport and ion distributions between plasma and erythrocytes in the blood of *Petromyzon marinus*. J Exp Biol 144:565–576

Vandermeulen JH, Davis ND, Muscatine L (1972) The effect of inhibitors of photosynthesis on zooxanthellae in corals and other marine invertebrates. Mar Biol 16:185–191

Venn AA, Tambutté É, Lotto S, Zoccola D, Allemand D, Tambutté S (2009) Intracellular pH in Symbiotic Cnidarians. Proc Natl Acad Sci USA 39:16574–16579

Veron JEN, Hoegh-Guldberg O, Lenton TM, Lough JM, Obura DO, Pearce-Kelly P, Sheppard CRC, Spalding M, Stafford-Smith MG, Rogers AD (2009) The Coral reef crisis: the critical importance of <350 ppm CO_2. Mar Pollut Bull 58:1428–1436

Wahl ML, Coss RA, Bobyock SB, Leeper DB, Owen CS (1996) Thermotolerance and intracellular pH on two chinese hamster cell lines adapted to growth at low pH. J Cell Physiol 166:438–445

Weis VM (1993) Effect of dissolved inorganic carbon concentration on the photosynthesis of the symbiotic sea anemone *Aiptasia pulchella* (Carlgren): role of carbonic anhydrase. J Exp Mar Biol Ecol 174:209–225

Wilkinson C (2004) Status of coral reefs of the world: Global Coral Reef Monitoring Network and Australian Institute of Marine Science, Townsville, Queensland, Australia, p 301

Wilkinson C, Buddemeier (1994) Global climate change and coral reefs: implications for people and reefs. Report of the UNEP-IOC-ASPEI-IUCN Global Task Team on the implications of climate change on coral reefs

Xu Y, Wahlund TM, Feng L, Shaked Y, Morel FMM (2006) A novel alkaline phosphatase in the coccolithopore *Emiliana huxleyi* (Prymnesiophyceae) and its regulation by phosphorus. J Phycol 42:835–844

Yates KK, Halley RB (2003) Measuring coral reef community metabolism using new benthic chamber technology. Coral Reefs 22:247–255

Yates KK, Halley RB (2006) CO_3^{2-} concentration and pCO_2 thresholds for calcification and dissolution on the Molokai reef flat, Hawaii. Biogeosciences 3:357–369

Yonge CM, Nicholls AG (1931) Studies on the physiology of corals. V. On the relationship between corals and zooxanthellae. Scient Rep Gt Barrier Reef Exped 1:177–211

Zeebe RE, Wolf-Gladrow D (2001) CO_2 in seawater: equilibrium, kinetics, isotopes. Elsevier Oceanography Series, 65, Amsterdam, p 346

Zimmerman RC, Kohrs DG, Steller DL, Alberte RS (1997) Impacts of CO_2 enrichment on productivity and light requirements of eelgrass. Plant Physiol 115:599–607

Zoccola D, Tambutté É, Kulhanek E, Puverel S, Scimeca J-C, Allemand D, Tambutté S (2004) Molecular cloning and localization of a PMCA P-type calcium ATPase from the coral *Stylophora pistillata*. Biochim Biophys Acta 1663:117–126

Simulating and Quantifying the Environmental Influence on Coral Colony Growth and Form

Jaap A. Kaandorp, Maxim Filatov, and Nol Chindapol

Abstract Understanding the growth process of scleractinian corals is crucial to study their role in the marine ecosystem and to obtain insight into their susceptibility to changes in the external physical environment. In this chapter, we describe a method for obtaining three-dimensional images of coral colonies and quantifying morphological properties of complex-shaped colonies. We introduce a method to simulate the accretive growth process in corals and models for simulating the influence of light on the local growth process and the influence of advection-diffusion on the local absorption of nutrients (e.g., inorganic carbon) at the surface of the coral. The morphometric analysis can be used to do a quantitative comparison of real and simulated forms and to identify missing parameters in the growth model. The model of the physical environment can be used to study the hydrodynamics and local distribution of nutrients and light in coral morphologies

Keywords Morphogenesis • scleractinian corals • simulation models • computed tomography scanning • morphometrics • hydrodynamics • advection-diffusion • iinorganic carbon • skeletonization algorithms • medial axis • accretive growth model

1 Introduction

Many scleractinian corals show a high degree of morphological plasticity due to variations in the physical environment. A detailed review on the morphological plasticity in corals and the influence of the environment can be found in a recent paper by (Todd 2008). Two dominant parameters influencing morphological plasticity are water movement and the availability of light in photosynthetic organisms. Environmental parameters, closely linked to water movement, are the supply of suspended material in filter-feeding organisms and sedimentation. Sedimentation may also strongly influence the availability of light. Furthermore, there exists evidence about the effect of gravity on the colony morphology (Meroz et al. 2002).

There are a large number of studies (Todd 2008) where colonies have been transplanted from one environment to a different one to identify the impact of the physical environment on the growth form of the colony. There are a number of problematic issues in these experiments. The first issue is that many transplantation experiments were done with colonies that were not necessarily of the same genotype and in this non-clonal approach it cannot be excluded that morphological changes are caused by the genotype and not by the influence of the environment. By using clone-mates (fragments of the same coral colony) it is basically possible to determine the morphological response to the environment. The second issue is that the interaction between the physical environment (flux of nutrients over boundary layers, local flow velocities, and local light intensities) and local growth velocities is very difficult (if not impossible) to assess in detail in experiments. A possible solution is to use simulation models where these quantities can be estimated in great detail. By using simulation models, it also possible to obtain a deeper insight into the morphogenetic process itself and to find mathematical rules capturing the biomechanics of the growth process of a coral colony and the impact of the physical environment on morphogenesis. The third issue is that the morphological changes at the colony level due to transplantation experiments, especially in three-dimensional complex-shaped branching forms, are difficult to be interpreted and to be quantified. A solution here is to use detailed morphometric methods to detect local changes in the growth form.

In Graus and Macintyre (1982), transplantation experiments with non-clonal colonies of *Montastrea annularis* and morphological simulation models were used to investigate

J.A. Kaandorp (✉), M. Filatov, and N. Chindapol
Section Computational Science,
University of Amsterdam, Kruislaan 403,
1098 SJ, Amsterdam, The Netherlands
e-mail: J.A.Kaandorp@uva.nl; M.V.Filatov@uva.nl;
N.Chindapol@uva.nl;

the influence of local light intensities on the growth process and the overall colony morphology. *M. annularis* shows a hemispherical colony form under circumstances with a maximum light intensity, when the colony grows close to the water surface. The colony gradually transforms from hemispherical through column-shaped and tapered forms to a substrate covering plate when the light intensity decreases. In Muko et al. (2000), a combination of transplantation experiments (using clonal transplants) and simulation models was used to demonstrate that the morphological plasticity in *Porites sillimaniani* is related to the availability of light. In the transplantation experiments it was shown that colonies under high-light conditions developed branches, while under low-light conditions remained flat.

Water flow has a strong influence on the growth process of scleractinian corals. Veron and Pichon (1976) present several series of growth forms of scleractinians (e.g., *Pocillopora damicornis* and *Seriatopora hysterix*), which are arranged along a gradient of increasing water movement. Both species show a gradual transformation from a compact shape with a relatively low branch spacing, under exposed conditions, to a thin-branching shape with a relatively larger branch spacing under sheltered conditions. In the Caribbean coral *Madracis mirabilis* a similar range, compact growth forms with a low branch spacing gradually changing into more open thin-branching forms with a larger branch spacing, is found. Two examples of growth forms of the two extremes, the thin-branching low-flow morph collected from a depth of 20 m and the compact hemispherical high-flow morph collected from a depth of 6 m, are shown in Fig. 1. At the site where the samples were collected water movement at 6 m is on average 2.3 times higher compared to the site at 20 m (Kaandorp et al. 2003). Earlier experiments with branching corals in flume studies (Chamberlain and Graus 1977) showed that densely packed branching colonies, comparable to *M. mirabilis* colonies, act as a solid body. Water flow (even for relatively high flow velocities such as 20 cm s^{-1}) starts to circumvent the colony and a stagnant region develops inside the colony. The authors suggest that the water velocity inside colonies reaches an upper limit, a relatively low (compared with the water velocity externally of the colony) saturation velocity. (Lesser et al. 1994) provide data about flow velocities around *Pocillopora damicornis* colonies. For high-flow morphs, an external flow velocity of about 8 cm s^{-1} and inside colony flow of around 1 cm s^{-1} were found. For the low-flow morphs, an external flow velocity of about 3 cm s^{-1} and an inside colony flow of 0.8 cm s^{-1} were reported.

Hydrodynamics affects the distribution of food particles (Sebens et al. 1997; Anthony 1999) and dissolved material (Sorokin 1993; Lesser et al. 1994; Marubini and Thake 1999; Marubini et al. 2002; Reidenbach et al. 2006) in the immediate environment of the coral. For zooxanthellate corals, calcification depends on a phototrophic component related to local availability of light and local gradients of dissolved inorganic carbon (Lesser et al. 1994; Marubini and Thake 1999; Marubini et al. 2002; Allemand et al. this book), and a heterotrophic component related to the uptake of nutrients from the environment. The relative contribution of the phototrophic and heterotrophic components can be estimated from skeletal $\delta^{13}C$ and $\delta^{18}O$ isotopes (McConnaughey et al. 1997; Heikoop et al. 2000; Maier et al. 2003). In Maier et al. (2003), it was demonstrated that the calcification of *M. mirabilis* is mainly supported by photosynthesis based on the analysis of the $\delta^{13}C$ and $\delta^{18}O$ isotopes. If photosynthesis is the main source of energy then local gradients of inorganic carbon will play a crucial role in the morphogenesis of *M. mirabilis* as they represent a limiting resource to skeleton formation. In recent field experiments by Mass and Genin (2008) it was found that *Pocillopora verrucosa* develop asymmetrical colonies due to the influence of an asymmetric directed flow. The increase in fluid motion around organism is believed to increase nutrient transport and uptake and ultimately enhances the organism's growth rate (Webster and Weissburg 2009).

In a previous study (Kaandorp et al. 2003) we carried out simulation experiments using actual morphologies of *M. mirabilis* in a laminar flow regime. The actual morphologies *M. mirabilis* were obtained using Computed Tomography

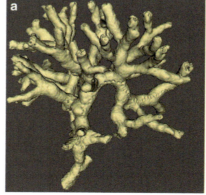

Fig. 1 Surface renderings of CT scans of *Madracis mirabilis*: (**a**) lateral view of a thin-branching low-flow morph, (**b**) top view of a compact high-flow morph. Both colonies are visualized on the same scale. The dimensions of object a and b are, respectively, 13 × 8 × 11 cm and 11 × 10 × 6 cm

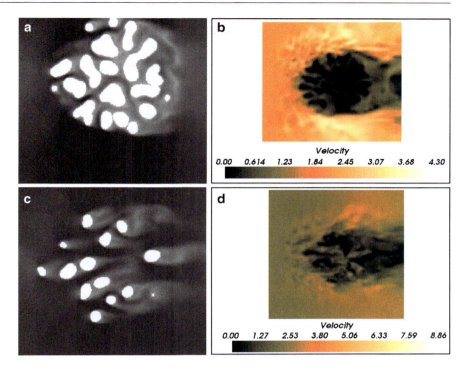

Fig. 2 (**a** and **b**) Sections through the simulated flow velocity pattern around a surface rendering of the high-flow morph of the *M. mirabilis* colony shown in Fig. 1b. (**c** and **d**) Low-flow morphology in Fig 1a. The flow was directed from the left to the right. The absolute size of the flow vector is visualized in the pictures. Black indicates a maximum size of the flow vector and white refers to a value of near zero in (**a** and **c**). In (**b** and **d**) the color represents the absolute size of the flow vector measured in cm s^{-1}

(CT) scanning techniques. These simulations showed that around the branching morphologies relatively large diffusive boundary layers are formed, which act as stagnant regions, and therefore limiting the transport of inorganic carbon from the water column to the coral colony. In Fig. 2a, a section parallel to the substratum plane, through a simulated flow pattern formed around a three-dimensional image of *M. mirabilis* (in this example the same colony as shown in Fig. 1b was used), is shown. In this picture the flow was directed from the left to the right. The absolute size of the flow vector is visualized in the picture. Black indicates in Fig. 2a and c a maximum size of the flow vector and white a value of near zero. This picture shows that between the branches stagnant areas develop with flow velocities near zero. In these diffusion-dominated, stagnant areas relatively large boundary layers are formed, so photosynthetic bicarbonate assimilation will result in a local depletion and gradients of inorganic carbon around the colony.

Morphological plasticity is directly related to various biologically relevant parameters, for example, the diameter of the branches, branching rate, branching angles, and branch spacing. In studies on particle capture in the branching scleractinian coral *M. Mirabilis* and the influence of hydrodynamics (Sebens et al. 1997), it was demonstrated that branch diameters and branch spacing are crucial morphological properties. The diameter of the branches and spacing between branches is variable and may be controlled by a combination of hydrodynamics and genetics. Sebens et al. (1997) argue that through modifications of its branch structure and branch spacing, *M. mirabilis* can function efficiently as a passive suspension feeder over a wide range of exposure to water movement. In a study by Bruno and Edmunds (1998) on *M. mirabilis*, it is demonstrated that by increasing the branch spacing the thickness of the diffusive boundary layer is effectively decreased and mass transport and high respiration rates even under low-flow conditions can be maintained.

In this chapter, we describe a method for obtaining three-dimensional images of coral colonies and quantifying morphological properties of the three-dimensional complex-shaped colonies (Kruszynski et al. 2007). We introduce in this chapter a method to simulate the accretive growth process in corals and models for simulating the influence of light on the local growth process and the influence advection-diffusion on the local absorption of nutrients (e.g., inorganic carbon) at the surface of the coral. This introduction is based on a number of earlier publications (Kaandorp and Kuebler 2001; Kaandorp and Sloot 2001; Merks et al. 2003; Kaandorp et al. 2005). The simulation models of growth and form of the coral can be combined with the three-dimensional images of the actual corals in different ways: the morphometric method can be used to do a quantitative comparison of real and simulated forms. The model of the physical environment can be used to study the hydrodynamics and local distribution of nutrients and light in coral morphologies. The morphological simulation model, the accretive growth model in combination with an advection-diffusion model, can be used to study the hypothesis that external gradients of inorganic carbon in the boundary layers of the colony and local light intensities are shaping the coral.

2 Three-Dimensional Images of Coral Colonies Obtained Using Computer Tomography Scanning

Three-dimensional images of the colonies were obtained using X-ray Computed Tomography (CT) scanning techniques (Kaandorp and Kuebler 2001). The CT scan data was stored in DICOM format (a general data format used for medical images). In the case of the objects shown in Fig. 1, the CT scan data consists of $512 \times 512 \times z$ ($20 \leq z \leq 50$) three-dimensional pixels, the so-called voxels. The slice thickness of the CT scan data is 2.5 mm in the *xy* direction. Each voxel represents a density value between 0 and 2^{12}, where 0 is the lowest density (the air around the coral skeleton), while high values indicate the calcium carbonate of the coral skeleton. In Fig. 1a and b, two examples are shown of reconstructed three-dimensional images of the *M. mirabilis* colonies. The three-dimensional reconstruction was done using a surface rendering technique based on the marching cube technique (Lorensen and Cline 1987). The surface is constructed, approximately, at the boundary between air and the calcium carbonate skeleton of the coral. With this technique, using the original data set of $512 \times 512 \times z$ voxels, an image is reconstructed with an equal resolution in *x*, *y*, and *z* directions (see for details Schroeder et al. 1997). Only the surface of the coral is visualized, without any surface structures such as corallites. On the voxels representing the surface of the corals a triangulated mesh was constructed using the surface rendering technique.

3 Morphometrics of Three-Dimensional Complex-Shaped Branching Colonies

An important complication in the morphological analysis of growth forms of scleractinian corals and many other marine sessile organisms is that the growth forms are usually indeterminate and complex. Most methods for the analysis of growth and form use landmark-based geometric morphometrics (Bookstein 1991). These methods are more suitable for unitary organisms and less applicable for the analysis of indeterminate growth forms of modular organisms (Harper et al. 1986). In a number of cases, the organism is built from well-defined modules (e.g., the corallites and polyps in scleractinian corals), in other cases the module itself has no well-defined shape, but an irregular and indeterminate form (e.g., an osculum and its corresponding aquiferous system in sponges). For organisms with well-defined modules it is possible to apply landmark-based methods for the morphometrics of individual modules. For example, Budd et al. (1994) and Budd and Guzman (1994) used landmark-based methods to measure the corallites in a scleractinian coral, Todd et al. (2001) measured the morphological variation of the polyps in a scleractinian coral.

In previous studies (Kaandorp 1999; Abraham 2001; Sanchez and Lasker 2003; Shaish et al. 2007), methods have been developed for the morphological analysis of branching marine sessile organisms. In these studies, the analysis is based on two-dimensional images of the branching object. An important limitation in the two-dimensional studies is that this method only works well for growth forms that tend to form a branching pattern in one plane, for example, the sponge *Raspailia inaequalis* analyzed by Abraham (2001).

Three-dimensional images of branching objects can be analyzed based on morphological skeletons (an overview of these methods is provided by Jonker and Vossepoel 1995). The morphological skeleton can be obtained with an image processing algorithm. Skeletonization algorithms reduce the object to a network of thin lines, one pixel or voxel thick, running through the centers of the object (Fig. 3). A morphological skeleton or medial axis has the same topology (a similar branching structure) as the original object, and occupies the same spatial extent in the image. It is possible to detect the root of the morphological skeleton and to find the successive branching points. The hierarchical ordering of the branches in Fig. 3 can be used to measure various biologically relevant parameters such as the diameter of the branches, branching rate, branching angles, and branch spacing.

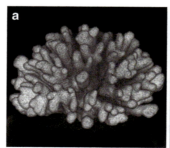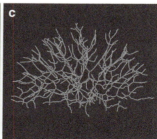

Fig. 3 Skeletonization of the coral shown in Fig 1b. (**a**) original object (**b**) surface reconstruction of the object (**c**) medial axis of the object with the same branching structure as the original object

4 The Accretive Growth Model

In the simulations, we used an accretive growth model (Kaandorp and Kuebler 2001; Kaandorp and Sloot 2001; Merks et al. 2003; Kaandorp et al. 2005). In the simulation the process is modeled as a surface-normal deposition process, where new material is deposited along the normal vectors that are constructed on the surface of the previous growth stage. In X-ray studies, it was demonstrated that the corallite tends to be set normal with respect to the previous growth layers (Darke and Barnes 1993; Le Tissier et al. 1994). This normal deposition process can especially very well be visualized in *M. annularis*. In Fig. 4, a volume rendered (Schroeder et al. 1997) slice of this coral is shown. In this picture the slice is slightly rotated. Part of the original surface of the colony is shown and a length section is made through the colony. At the surface the position of the corallites is visible as dark-colored pores, while the annual growth of the colony is visible as density bands in the section of the colony. From the density bands it is, at least in theory, possible to reconstruct the surface of the colony in earlier growth stages. This section shows that the corallites are set perpendicular with respect to the previous growth layer. In the study by Le Tissier et al. (1994), examples are shown of X-ray pictures of a branching coral (*Porites porites*) in which annual growth is visible in X-ray pictures of sections through the colony and where it can also be observed that the corallites are set perpendicular with respect to previous growth layers. The corallite moves outward with the (living) peripherical tissue and leaves a growth trajectory that can be reconstructed by connecting the positions of the center of a corallite in the successive growth layers.

In the simulation model it is assumed that the living tissue deposits new layers of material (aragonite) on top of the previous layers, which remain unchanged. The growth layers are represented by layers of triangles, which are again organized in a polygonal pattern. This is demonstrated in Fig. 5, where two initial simulated growth stages are shown. In Fig. 5a, the initial object used in the simulations is shown: a sphere tessellated with triangles, where the triangles are organized in a pattern of pentagons and hexagons. In Fig. 5b, the next growth stage is shown in which a layer is constructed on top (in the vertical direction) of the previous growth stage. The polygons in Fig. 5 represent the corallites in the simulations. In the growth model, we tested the hypothesis that dissolved inorganic carbon used by photosynthesis is the main limiting factor in the calcification process, and growth is limited by the local amount of nutrient available to the simulated corallite. Furthermore, we tested the hypothesis local light intensities on the surface of the coral are limiting photosynthesis and coral growth. We assume that inorganic carbon in the immediate environment is exclusively distributed by diffusion. The basic idea of the simulation model is shown in Fig. 6: nutrients are released from the top

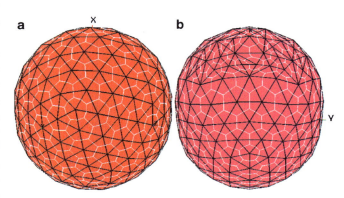

Fig. 5 Two successive objects generated with the radiate accretive growth model. The object shown in (**a**) is the initial object used in the simulations. The polygons shown in both object represent the corallites in the simulations

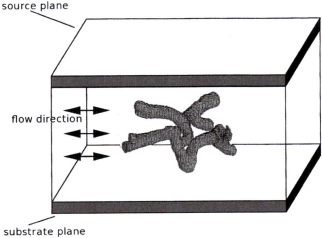

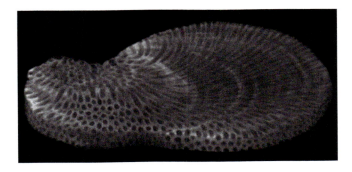

Fig. 4 Volume rendered slice of *Montastrea annularis*. The slice is slightly rotated showing part of the surface and a length section through the colony. The position of the corallites is visible as dark-colored pores at the surface, while the annual growth of the colony is visible as density bands in the section of the colony

Fig. 6 Simulation box. In the advection-diffusion simulations it is assumed that the top plane is the source of nutrients, while the nutrients are absorbed at the surface of the simulated growth form. In the light simulations, it is assumed that the light direction corresponds to the vertical and that all light is absorbed at the surface of the object

plane in the simulation box and a simple parallel light source is assumed, where the light direction corresponds to the vertical.

We used a cellular-automata-based particle model, the "moment propagation" method (Lowe and Frenkel 1995; Merks et al. 2002) in conjunction with the lattice Boltzmann method (Chopard and Droz 1998; Succi 2001) to model the dispersion of nutrient in the external environment by advection-diffusion. The lattice Boltzmann method is especially suitable for developing scalable simulations (using distributed computation) of advection-diffusion processes and modeling boundary layers in complex three-dimensional geometries (Kaandorp et al. 1996; Koponen et al. 1998; Kandhai et al. 2002). In the advection-diffusion simulations, it was assumed that nutrient is absorbed at the centers of simulated corallites (the polygons shown in Fig. 5). In the simulations it was assumed that the source of nutrient is located at the top plane of the cube enclosing the simulated object (the simulation box in Fig. 6). With this advection-diffusion simulation the gradients around the object are computed for every growth stage. An example of a section through a simulated growth form is shown in Fig. 7a. In this simulation nutrient was distributed by diffusion and exclusively driven by the absorption of simulated nutrient. The section through the simulated growth form is shown in white, while the concentration of nutrient around the object is displayed using a color-gradient red-yellow-green-blue. Red indicates the highest concentration (located at the top plane of simulation box). The solid lines in the picture depict isosurfaces with equal concentrations around the object. In the simulations nutrient is all the time absorbed at the surface of the object and at the substrate plane, resulting in nutrient depleted region close to the object (blue) where the nutrient concentration is zero.

The linear extension rate of the simulated skeleton in Fig. 7b is driven by the amount of absorbed simulated nutrient. Quantitatively local growth at surface of object can be directly related to the amount of absorbed simulated nutrient. The addition of layers on top of previous layers is visualized in Fig. 7b. In the simulations there are only local interactions between the corallites, which are closely packed on a growth layer. The size of the simulated corallite varies around a certain (species-specific) mean size of the corallite. Simulated corallites that become too large split up into new ones, while small ones are deleted. For further details about the splitting up and deletion of triangles, we refer to Merks et al. (2003). After each growth step, where a new layer of triangles is constructed on top of the previous one, the gradients computation is repeated again. In this chapter, the simulated growth forms were obtained within 80 growth steps. In Fig. 8, a series of

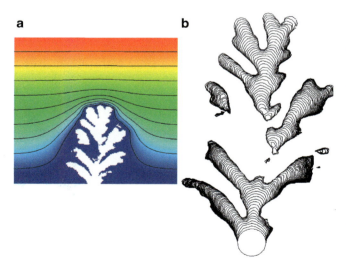

Fig. 7 Sections through a simulated growth form. (**a**) The section through the simulated growth form is shown in white, while the concentration of nutrient around the object is displayed using a color-gradient red-yellow-green-blue. *Red* indicates the highest concentration, while *blue* indicates a depletion of nutrient and visualizes the region with a concentration of near zero. In this example, nutrient was exclusively distributed by diffusion. The solid lines in the picture indicate isosurfaces with equal concentrations. (**b**) Section showing the addition of layers on top of previous layers in the simulated accretive growth process

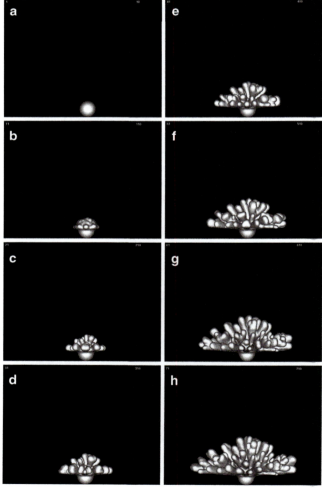

Fig. 8 Series of successive simulated growth stages using the accretive growth model under diffusion limited conditions. The shape of the object is displayed after every tenth growth stage

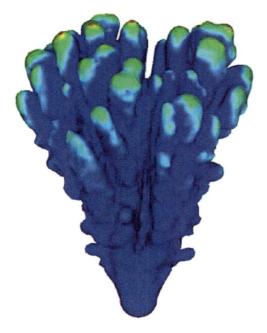

Fig. 9 Example of a simulated growth form where the accretive growth process is driven by a combination of local absorbed simulated nutrient and local light intensities

successive simulated growth stages are shown of the nutrient-driven growth model. In Fig. 8, the shape of the object is displayed after every tenth growth stage. The growth process starts with an initial spherical object. In Fig. 9, an example is shown of object generated by the accretive growth model where the linear extension rate of the simulated skeleton is driven by local absorbed simulated nutrient and local absorbed simulated light intensities (both nutrient and light are assumed to contribute equally to the linear extension rate).

5 Comparison Between Coral Colony Morphologies and Simulated Forms

The only species-specific information included in our model is the average size of a corallite: in our model a parameter controlling the average size of the polygons representing the simulated corallites. With the incorporation of simple local rules, controlling the size of individual simulated corallites, and local gradients it is possible to generate branching morphologies which approximate the morphologies of surface renderings of CT scans of *M. mirabilis* shown in Fig. 1b. A diffusion-limited environment is sufficient to get the correct gradients in the simulated morphogenesis. When comparing simulated and actual objects in Fig. 1 we see that the branch spacing, which is quite regular and characteristic in *M. mirabilis* (Sebens et al. 1997), is also very regular in the simulated objects. In Fig. 8, it can be seen that the initial symmetry of the spherical object disappears after a few growth steps;

the basic symmetry breaking mechanism is the fact that the surface is tessellated with discrete polyps. During the formation of branches easily configurations will occur where it is not possible to tessellate the branches in a symmetrical way with discrete polyps and the symmetry will disappear. The results in Fig. 8 indicate that the model can be used to approximate a branching coral with relatively large, closely packed, and undifferentiated corallites (e.g., *M. mirabilis*). Figure 9 demonstrates the impact of local light intensities on the growth process, in this example the simulated growth form tends to form a more upward-branching pattern, which is in a number of cases also observed in some corals (e.g., *Acropora palmata*). By using a detailed morphometric comparison between actual growth forms and the simulated forms it is possible to distinguish which type of growth model approximates the coral colonies the best (Filatov et al. in prep.).

Intricately branched corals appear to be more susceptible to mortality than massive corals when bleaching occurs during periods of high sea-surface temperatures (Davies et al. 1997; Nakamura and van Woesik 2001; Nakamura et al. 2003). One hypothesis is that in intricately branched corals mass transfer through the boundary layer is strongly diffusion-dominated and can disperse fewer of the metabolic by-products that arise during warmer periods (Nakamura and van Woesik 2001; Nakamura et al. 2003).

Our results confirm the central role of diffusion-dominated boundary layers. The experiments in Fig. 2 show that diffusion-dominated regions tend to develop within a laminar flow. In these cases, the distribution of nutrients may be approximated by a diffusion-limited model. Although we have set our advection-diffusion model exclusively to the diffusion-dominated regime, it can be expected that advection together with turbulent diffusion and mixing still plays an important role in the distribution of nutrients. For this reason we think it is important to use an advection-diffusion model in combination with a morphogenetic model of the coral.

A major challenge is to understand the coupling between gene regulation of growth of the corallites, colony, and the impact of the physical environment. Although the developmental biology of individual polyp in scleractinians (Ball et al 2002, 2004) is relatively simple compared with some other metazoans, there is very little understanding about the role of gene regulation at the colony level and signaling processes between polyps. Many details are still missing in the understanding of the regulation of the calcification process (Allemand et al. 2004, 2010), gene regulation, and influence of the environment. A major challenge in computational biology is to develop simulation models here where gene regulation, calcification, growth by accretion, and the influence light and hydrodynamics are coupled. This is a typical multiscale modeling problem where processes on very different spatiotemporal scales need to be coupled.

Acknowledgments This work is supported by grants from the Netherlands Organization for Scientific Research (VEARD project, 643.100.601) and the EC (MORPHEX, NEST contract no. 043322).

References

Abraham ER (2001) The fractal branching of an arborescent sponge, Mar Biol 138: 503–510

Allemand D, Ferrier-Pagès, Furla C, Houlbrèque P, Puverel FS, Reynaud S, Tambutté É, Tambutté S, Zoccola D (2004) Biomineralisation in reef-building corals: from molecular mechanisms to environmental control. C R Palevol 3:453–467

Anthony KRN (1999) Coral suspension feeding on fine particulate matter. J Exp Mar Biol Ecol 232:85–106

Ball EE, Hayward DC, Reece-Hoyes JS, Hislop NR, Samuel G, Saint R, Harrison PL, Miller DJ (2002) Coral development: from classical embryology to molecular control. Int J Biol 46:671–678

Ball EE, Hayward DC, Saint R, Miller DJ (2004) A simple plan – cnidarians and the origins of developmental mechanisms. Nat Genet 5:567–577

Bookstein FL (1991) Morphometric tools for landmark data: geometry and biology. Cambridge University Press, New York

Bruno JF, Edmunds PJ (1998) Metabolic consequences of phenotypic plasticity in the coral *Madracis mirabilis* (Duchassaing and Michelotti): the effect of morphology and water flow on aggregate respiration. J Exp Mar Biol Ecol 229:187–195

Budd AF, Johnson KG, Potts DC (1994) Recognizing morphospecies in colonial reef corals: I. Landmark-based methods. Paleobiology 20: 484–505

Budd AF, Guzman HM (1994) Siderastrea glynni, a new species of scleractinian coral (cnidaria: anthozoa) from the eastern pacific, Proc Biol Soc Wash 107:591–599

Chamberlain JA, Graus RR (1977) Water flow and hydromechanical adaptations of branched reef corals. Bull Mar Sci 25:112–125

Chopard B, Droz M (1998) Cellular automata modelling of physical systems. Cambridge Univerity Press, Cambridge

Darke WM, Barnes DJ (1993) Growth trajectories of corallites and ages of polyps in massive colonies of reef-building corals of the genus *Porites*. Mar Biol 117:321–326

Davies JM, Dunne RP, Brown BE (1997) Coral bleaching and elevated sea-water temperature in Milne Bay Province Papua New Guinea 1996. Mar Fresh W Res 48:513–518

Filatov M, Kruszynski KJ, Kaandorp JA, van Liera R, Vermeij M, Bak R Morpho-space exploration of simulated coral morphologies and three-dimensional images of scleractinian corals (in preparation)

Graus RR, Macintyre IG (1982) Variation in growth forms of the reef coral *Montastrea annularis* (Ellis and Solander): a quantitative evaluation of growth response to light distribution using computer simulation. Smithson Contr Mar Sci 12:441–464

Harper JL, Rosen BR, White J (1986) The growth and form of modular organisms. The Royal Society London

Heikoop JM, Dunn JJ, Risk MJ, Schwarcz HP, McConnaughey TA, Sandeman IM (2000) Separation of kinetic and metabolic isotopic effects in carbon-13 records preserved in reef coral skeletons. Geochim Cosmochim Acta 64:975–987

Jonker PP, Vossepoel AM (1995) Mathematical morphology in 3D images: comparing 2D & 3D skeletonization algorithms, pp 83–108, In: BENEFIT Summer School on Morphological Image and Signal Processing, Zakopane, K. Wojciechowski (ed.), Silesian Technical University, ACECS, Gliwice, Poland

Kaandorp JA (1999) Morphological analysis of growth forms of branching marine sessile organisms along environmental gradients. Mar Biol 134:295–306

Kaandorp JA, Kuebler JE (2001) The algorithmic beauty of seaweeds, sponges and corals. Springer, Heidelberg

Kaandorp JA, Sloot PMA (2001) Morphological models of radiate accretive growth and the influence of hydrodynamics. J Theor Biol 209:257–274

Kaandorp JA, Lowe CP, Frenkel D, Sloot PMA (1996) The effect of nutrient diffusion and flow on coral morphology. Phys Rev Let 77: 2328–2331

Kaandorp JA, Koopman EA, Sloot PMA, Bak RPM, Vermeij MJA, Lampmann LEH (2003) Simulation and analysis of flow patterns around the scleractinian coral *Madracis mirabilis* (Duchassaing and Michelotti). Philos Trans R Soc Lond B 358:1551–1557

Kaandorp JA, Sloot PMA, Merks RMH, Bak RPM, M.J.A. Vermeji MJA, Maier C (2005) Morphogenesis of the branching reef coral *Madracis mirabilis*, Proc Roy Soc B 272:127–133

Kandhai D, Hlushkou D, Hoekstra AG, Sloot PMA, Van As H, Tallarek U (2002) Influence of stagnant zones on transient and asymptotic dispersion in macroscopically homogeneous porous media. Phys Rev Lett 88:234501–234504

Koponen A, Kandhai D, Hellen E, Alava M, Hoekstra A, Kataja M, Niskanen K, Sloot PMA, Timmonen J (1998) Permeability of three-dimensional random fiber webs. Phys Rev Lett 80:716–719

Kruszynski K, Kaandorp JA, van Liere R (2007) A computational method for quantifying morphological variation in scleractinian corals. Coral Reefs 26:831–840

Lesser MP, Weis VM, Patterson MR, Jokiel PL (1994) Effects of morphology and water motion on carbon delivery and productivity in the reef coral, *Pocillopora damicornis* (Linnaeus): diffusion barriers, inorganic carbon limitation, and biochemical plasticity. J Exp Mar Biol Ecol 178:153–179

Le Tissier MD'AA, Clayton B, Brown BE, Spencer Davies P (1994) Skeletal correlates of coral density banding and an evaluation of radiography as used in scelerochronology. Mar Ecol Prog Ser 110: 29–44

Lorensen WE, Cline HE (1987) Marching cubes: a high resolution 3D surface construction algorithm. ACM Comput Graph 21:163–169

Lowe CP, Frenkel D (1995) The super long-time decay of velocity fluctuations in a two-dimensional fluid. Physica 220:251–260

Maier C, Patzold J, Bak RPM (2003) The skeletal isotopic composition as an indicator of ecological plasticity in the coral genus *Madracis*. Coral Reefs 22:370–380

Marubini F, Thake B (1999) Bicarbonate addition promotes coral growth. Limnol Oceanogr 44:716–720

Marubini F, Ferrier-Pages C, Cuif J (2002) Suppression of skeletal growth in scleractinian corals by decreasing ambient carbonate-ion concentration: a cross-family comparison. Proc R Soc Lond B 270:179–184

Mass T, Genin A (2008) Environmental versus intrinsic determination of colony symmetry in the coral *Pocillopora verrucosa*. Mar Ecol Prog Ser 2008 369:131–137

McConnaughey TA, Burdett J, Whelan JF, Paul CK (1997) Carbon isotopes in biological carbonates: respiration and photosynthesis. Geochim Cosmochim Acta 61:611–622

Merks RMH, Hoekstra AG, Sloot PMA (2002) The moment propagation method for advection-diffusion in the lattice Boltzmann method: validation and Peclet number limits. J Comput Phys 183:563–576

Merks RMH, Hoekstra AG, Kaandorp JA, Sloot PMA (2003) Models of coral growth: Spontaneous branching, compactification and the Laplacian growth assumption. J Theor Biol 224:153–166

Meroz E, Brickner I, Loya Y, Retzman-Shemer A, Ilan M (2002) The effect of gravity on coral morphology. Proc R Soc B 269:717–720

Muko S, Kawasaki K, Sakai K, Takasu F, Shigesada N (2000) Morphological plasticity in the coral *Porites sillimaniani* and its adaptive significance. Bull Mar Sci 66:225–239

Nakamura T, van Woesik R (2001) Water-flow rates and passive diffusion partially explain differential survival of corals during the 1998 bleaching event. Mar Ecol Prog Ser 212:301–304

Nakamura T, Yamasaki H, van Woesik R (2003) Water flow facilitates recovery from bleaching in the coral *Stylophora pistillata*. Mar Ecol Prog Ser 256:287–291

Reidenbach MA, Koseff JR, Monismith SG, Steinbuck JV, Genin A (2006) Effects of waves, unidirectional currents, and morphology on mass transfer in branched reef corals. Limnol Oceanogr 51:1134–1141

Sanchez JA, Lasker HR (2003) Patterns of morphological integration in marine modular organisms: supra-module organization in branching octocoral colonies. Proc Roy Soc B 270:2039–2044

Schroeder W, Martin K, Lorensen B (1997) The visualization toolkit: an object-oriented approach to 3D graphics, 2nd edn. Prentice-Hall, Upper Saddle River

Sebens KP, Witting J, Helmuth B (1997) Effects of water flow and branch spacing on particle capture by the reef coral *Madracis mirabilis* (Duchassaing and Michelotti). J Exp Mar Biol Ecol 211: 1–28

Shaish L, Abelson A, Rinkevich B (2007) How plastic can phenotypic plasticity be? The branching coral *Stylophora pistillata* as a model system. PLoS ONE 2007 2(7):e644

Sorokin YI (1993) Coral reef ecology. Springer, Heidelberg

Succi S (2001) The lattice Boltzmann equation: for fluid dynamics and beyond. Oxford University Press, Oxford

Todd PA, Sanderson PG, Chou LM (2001) Morphological variation in the polyps of the scleractinian coral Favia speciosa (Dana) around Singapore, Hydrobiologia 444:227–235

Todd PA (2008) Morphological plasticity in scleractinian corals. Biol Rev 83(3):315–337

Veron JEN, Pichon M (1976) Scleractinia of eastern Australia part Families Thamnasteriidae, Astrocoeniidae, Pocilloporidae. Australian Government Publishing Service, Canberra

Webster DR, Weissburg MJ (2009) The hydrodynamics of chemical cues among aquatic organisms. Annu Rev Fluid Mech 41:73–90

Physiological Adaptation to Symbiosis in Cnidarians

Paola Furla, Sophie Richier, and Denis Allemand

Abstract Up to the nineteenth century, cnidarians, among other organisms such as echinoderms and sponges, were classified as zoophytes, or animal – plant. This term, initially used by Wotton in 1552 and later by Linné and Cuvier (Daudin 1926), was only referring at this time to the external shape of the organisms, fixed branched one. This term was abandoned in the twentieth century; it is however curious to note that biological and physiological reasons might support this term. Indeed, Brandt at the end of the nineteenth century showed the presence of photosynthetic algae inside the tissues of these animals. He suggested that these algae were symbiotic and called them zooxanthellae (Brandt 1881; see Perru 2003, for a review). Zooxanthellae belong to the dinoflagellate phylum and were initially considered as a single species, called Symbiodinium microadriaticum (Freudenthal 1962). It was presently shown to be highly diverse and subdivided in Symbiodinium clades (Pochon et al. 2006). Associations of these different clades with their host did not evolve randomly and members of the same cnidarian species generally harbor the same Symbiodinium clade(s) (see review by Coffroth and Santos 2005, Stambler 2010 this book).

Keywords Endosymbiosis • mutualism • symbiodinium • carbon concentrating mechanism • carbonic anhydrase • ROS • superoxide dismutase • catalase • MAAs • pocilloporins

P. Furla (✉)
UMR 7138 SAE Systématique Adaptation Evolution,
Université de Nice-Sophia Antipolis, F-06108, Nice, France
e-mail: Paola.Furla@unice.fr

S. Richier
Laboratoire d'Océanographie de Villefranche,
UMR 7093, CNRS, Université Pierre et Marie Curie,
F-06234, Villefranche-sur-Mer Cedex, France
e-mail: Sophie.Richier@obs-vlfr.fr

D. Allemand
Centre Scientifique de Monaco,
Avenue Saint Martin, MC-98000, Monaco, France
e-mail: allemand@centrescientifique.mc

1 The Coral/Zooxanthella Holobiont: A Chimera?

Up to the nineteenth century, cnidarians, among other organisms such as echinoderms and sponges, were classified as zoophytes, or animal – plant. This term, initially used by Wotton in 1552 and later by Linné and Cuvier (Daudin 1926), was only referring at this time to the external shape of the organisms, fixed branched one. This term was abandoned in the twentieth century; it is however curious to note that biological and physiological reasons might support this term. Indeed, Brandt at the end of the nineteenth century showed the presence of photosynthetic algae inside the tissues of these animals. He suggested that these algae were symbiotic and called them zooxanthellae (Brandt 1881; see Perru 2003, for a review). Zooxanthellae belong to the dinoflagellate phylum and were initially considered as a single species, called *Symbiodinium microadriaticum* (Freudenthal 1962). It was presently shown to be highly diverse and subdivided in *Symbiodinium* clades (Pochon et al. 2006). Associations of these different clades with their host did not evolve randomly and members of the same cnidarian species generally harbor the same *Symbiodinium* clade(s) (see review by Coffroth and Santos 2005, Stambler this book, Stambler and Dubinsky this book and Allemand et al. this book).

Almost half of the cnidarians (Douglas et al. 1993) harbor in their tissue such photosynthetic symbionts. Symbionts may be transmitted either vertically, directly from parents to progeny ("closed" system), or horizontally ("open" system) through the environment. This last mode of transmission seems to be used by about 85% of zooxanthellate cnidarians (Schwarz et al. 2002) and could offer the opportunity for the host to become associated with zooxanthellae that are better adapted to a particular environment (Baker et al. 2004). However, a controversial theory on virulence predicts that horizontal transmission of symbionts may promote the evolution of parasitism, altering the cost/benefit ratio of the symbiosis (Fine 1975; Sachs and Wilcox 2006).

This mutual association probably appeared about 200 million years ago. It has totally modified the life of some cnidarians, the scleractinian corals, let them to develop in oligotrophic waters and build the biggest bioconstruction of the world, the coral reefs. If this symbiosis is at the origin of important mutual benefits, particularly trophic benefits (Trench 1987; Goodson et al. 2001; Stambler and Dubinsky this book), it will also require costly adaptations (Bronstein 2001). The presence of the photosynthetic symbiont inside the cnidarians constraints the host to behave as a plant: living in shallow waters highly exposed to light, absorbing CO_2 and rejecting O_2. Moreover, except in some scyphozoans, localization of symbionts is restricted to endodermal cells, where they settle in membrane-bounded vacuoles derived from the animal cell wall (Wakefield and Kempf 2001), known as the symbiosome. This intracellular location could allow the host to control its symbionts by regulating the transfer of organic or inorganic compounds right through this symbiosome membrane (Roth et al. 1988) but also involve for the nutrients to cross multitude of membrane barriers.

What are the major adaptations used to allow this mutualistic association? First, the intracellular localization of zooxanthellae let them experiencing ion concentration very different from the external seawater. Second, zooxanthellae need to derive its CO_2 supply for photosynthesis from the host. Third, as a consequence of photosynthesis, the host has to be resistant to the permanent hyperoxia present during the day. Finally, in order to optimize symbiont photosynthesis, the host needs to be distributed in the well-illuminated euphotic shallow waters.

The purpose of this chapter is to review the chimeric nature of this mutualistic association and how both partners adapted to this unusual situation by evolving characteristics usually associated with phototrophic organisms.

2 First Adaptation: A Marine Microalgae Living in an Intracellular Medium

The first consequence of the intracellular location of the zooxanthella is the loss of the two flagella characteristic of the dinoflagellata phylum and a modification of cell membranes (Trench 1987, Stambler this book).

The second consequence concerns the zooxanthellae environment. Due to their intracellular location, they experience ionic conditions probably very different from those present in seawater. Indeed, marine invertebrates cells usually have a lower intracellular Na^+ and Ca^{2+} levels, a lower pH, and a higher K^+ compared with seawater (Kirschner 1991). For Na^{++}, seawater concentration is about 500 mM and intracellular concentrations are usually in the 20–50 mM range. However, in symbiotic cnidarians, higher concentrations, from 60 mM to even 140 mM, were reported for the coral *Galaxea fascicularis* (Goiran et al. 1997) and the tropical symbiotic sea anemone *Condylactis gigantea*, respectively (Lopez et al. 1991). Freshly isolated zooxanthellae have a Na^+ concentration of 30–35 mM (Goiran et al. 1997). Free intracellular Ca^{2+} concentration in cnidarians, as in other cells, is about 100,000-fold lower than in seawater, that is, about 70 nM versus 10 mM (Sawyer and Muscatine 2001). Animal intracellular pH is more than a pH unit lower than surrounding seawater (Venn et al. 2009), whereas zooxanthella intracellular pH is similar to the intracellular pH of marine microalgae (Priouzeau, unpublished). However, ionic conditions within the perisymbiotic space are totally unknown. Very few studies are devoted to ionic regulation in corals (see Seibt and Schlichter 2001; Mayfield and Gates 2007). Goiran et al. (1997) showed that regulation of Na^+ concentration differed in free-living conditions compared to *in hospite* ones. In order to maintain constant the Na^+ concentration when zooxanthellae are transferred from their host into seawater, an ouabain-sensitive, K^+-independent Na^+-ATPase is stimulated within the first 10 min following the transfer. This Na^+-ATPase is probably repressed *in hospite*, suggesting that the intracellular localization of the symbiont save ionic regulation costs. Similarly, one should expect that regulation of Ca^{2+} concentration of zooxanthellae *in hospite* is easier than in seawater; however, no data are presently available. The ionic regulation in both host and symbiont cells should then be an important topic to future work. It could help to understand the physiological adaptations involved in this symbiosis and to interpret the recent rushing data coming from genomic analyses.

3 Second Adaptation: The Need of a Permanent Supply of CO_2 for Symbiont Photosynthesis

Photosynthetic organisms usually absorb CO_2 directly from their environment. In contrast, the photosynthetic partner *in hospite* due to its intracellular location cannot directly absorb CO_2, which therefore needs to be supplied by the animal host. Unfortunately, metabolic CO_2 from both partners is not enough to fulfill the needs of photosynthesis, since a net photosynthetic rate is measured, forcing the host to absorb itself the CO_2 from seawater. Although the total concentration of dissolved inorganic carbon (DIC) in seawater is about 2.4 mM, dissolved CO_2 represents only about 0.5% of this (i.e., 12 μM), far too low to allow optimal carbon fixation by the symbiont's RubisCo enzyme (Ribulose-1, 5-biphosphate carboxylase-oxygenase). Indeed, dinoflagellates have the form II Rubisco (Rowan et al. 1996), whose oxygenase activity has a higher affinity for O_2 compared to form I Rubisco (Jordan and Ogren 1981). Consequently, in normal condition, the zooxanthella Rubisco would not yield net carbon fixation if the oxygenase activity is not limited. In the zooxanthella, the first strategy to reduce the Rubisco oxygenase activity is to isolate the enzyme in the chloroplastidian pyrenoid, lacking in photosystems and consequently in oxygen production site (Leggat et al. 2002). Moreover, *in hospite,* carboxylation activities were favored by efficient external DIC transports and by CO_2-concentrating mechanism (CCM). CCM are well described in many phototrophs to increase the concentration of CO_2 in the proximity of Rubisco (Raven 1990, 2003; Giordano et al. 2005). All CCM have similar physiological outcomes, that is, negligible inhibition of CO_2 fixation by O_2, low CO_2 compensation points, high affinity for external CO_2.

The absorption of DIC from seawater implies therefore a transport across the animal cell membranes, at least the symbiosome and the endodermal cell membranes. In fact, experiments using perfused tentacles and pieces of tentacles inserted into Ussing chambers showed that while a passive diffusional, paracellular pathway may exist across ectodermal cells, the major part of DIC (85%) enters via an active transcellular pathway across the ectodermal cell layers and the endodermal membranes (Bénazet-Tambutté et al. 1996; Furla et al. 1998a, b; see review by Allemand et al. 1998). This means that DIC has to cross four animal membranes (apical and basolateral ectodermal plasma membranes, endodermal plasma membrane, symbiosome membrane) before being absorbed by the symbiont. Therefore, the host plays a major role in the uptake and supply of DIC for symbiont photosynthesis, controlling by this way the photosynthesis of its symbionts.

In seawater, both CO_2 and HCO_3^- are potential sources of DIC. While the first is freely diffusible into lipid bilayer membranes, the second is a charged ion, unable to diffuse through this barrier and needs a carrier protein. By using a wide range of methods (see review by Allemand et al. 1998), it has been shown that the host uses the seawater HCO_3^- and absorbs it in a manner similar to that demonstrated in vertebrate renal proximal tubules (Gluck and Nelson 1992). Uptake of DIC is mediated at the ectodermal plasma membrane level by a H^+-ATPase (Furla et al. 2000a), leading to the protonation of HCO_3^- to carbonic acid, following the reaction:

$$HCO_3^- + H^+ \rightarrow H_2CO_3$$

A membrane-bound carbonic anhydrase (CA) then dehydrates carbonic acid into CO_2:

$$H_2CO_3 \rightarrow CO_2 + H_2O$$

The uncharged CO_2 molecule then diffuses into the ectodermal cell, following the concentration gradient created by the extrusion of H^+ in the external medium. Once in the animal cytoplasm, CO_2 is equilibrated with HCO_3^- according to the intracellular pH (Venn et al. 2009) by another CA isoform (Furla et al. 2000a), which prevents back-diffusion of CO_2. The mechanism of transfer of HCO_3^- from ectodermal to endodermal cells and across the symbiosome membrane is presently totally unknown but may involve Cl^-/HCO_3^- antiport since the whole process is highly sensitive to inhibitors of anion transport such as DIDS (Al-Moghrabi et al. 1996; Furla et al. 1998a). CA is a very important enzyme in corals involved both in the regulation of calcification process (see Allemand et al. this book) and in DIC supply to photosynthesis. Weis's group (Weis 1991; Weis and Reynolds 1999) showed that its expression in sea anemones is enhanced by the presence of symbionts. By a genomic approach, Ganot et al. (2008) confirmed that CA expression is under the control of the symbiotic state.

This system therefore acts like plant CCMs. Indeed, it has been observed that the DIC pool in the tissues of the coral *Stylophora pistillata* increases 39-fold upon illumination (Furla et al. 2000b), leading to a ratio DICint/DICext of about 61, a value close to that reported for various micro- and macroalgae (Aizawa and Miyachi 1986; Bowes and Salvuci 1989). Therefore, the holobiont possesses a CCM, including inducible host CA in proportion to the density of zooxanthellae (see reviews Allemand et al. 1998; Leggat et al. 2002; Furla et al. 2005). This allows the intracellular symbionts to fix inorganic carbon actively, paralleling the example of micro- and macroalgae.

Thus, although the metabolic activities of all animal cells involve elimination of inorganic carbon, the coral host constantly absorbs inorganic carbon to supply its symbionts with enough CO_2 for fixation by Rubisco to support the high level of net primary productivity.

In parallel, zooxanthellae adapt themselves to the symbiotic way of life. Indeed, the ionic environmental conditions are very different in seawater (free-living life style) and in the symbiosome (symbiotic life style). Goiran et al. (1996) and Al-Moghrabi et al. (1996) demonstrated that while in seawater the DIC uptake was mainly mediated by a Na^+ antiport with HCO_3^- using the huge Na^+ gradient present in this medium, in the host, uptake is mediated by a H^+-ATPase, possibly leading to HCO_3^- protonation and CO_2 formation. The presence of this H^+-ATPase in the zooxanthellae genome was recently demonstrated (Bertucci et al. 2010). This enzyme belongs to the type IIIa of plasma membrane ATPases (P-type ATPase). Its level of expression is totally dependent on the symbiotic state since there is no detectable expression in cultured zooxanthellae. Similar adaptations of zooxanthellae to symbiosis were found in the giant clam. Upon isolation from the clam zooxanthellae changes within 2 days their source of DIC from CO_2 in symbiosis to HCO_3^- in free-living state

(Leggat et al. 1999). These data demonstrate that the environmental conditions are capable of regulating expression of specific DIC carriers. It remains to be determined whether it is the interaction between the host and the symbionts or the ionic environmental conditions that control in fine the expression.

4 Third Adaptation: Withstand Hyperoxia

Zooxanthella photosynthesis produces classically oxygen, leading to a local hyperoxia as in plants (D'Aoust et al. 1976; Dykens and Shick 1982). O_2 partial pressure increases indeed in the light at the surface of the coral (Shashar et al. 1993) or in the coelenteric cavity of a sea anemone (Richier et al. 2003) by about 3–3.5-fold normoxia. Such high concentrations of O_2 lead to the overproduction of reactive oxygen species (ROS) (Dykens et al. 1992). In hyperoxia conditions, several ROS can be produced: radical forms as O_2^- (superoxide ion) or OH^- (hydroxyl radical) or nonradical forms as H_2O_2 (hydrogen peroxide).

Usually, ROS cause protein oxidation, lipid peroxidation, and DNA degradation (Halliwell and Guteridge 1999); however, no increase of damages has been recorded in symbiotic cnidarians tissues either in terms of lipid peroxidation (measured by MDA assay) or protein carbonylation during natural light-induced hyperoxia in both host and symbionts (Richier et al. 2005). This suggests that, as in plants, symbiotic cnidarians possess efficient antioxidant defenses against oxidative stress.

Antioxidative defense involved both nonenzymatic and enzymatic mechanisms (Halliwell and Guteridge 1999). Nonenzymatic mechanisms include low-molecular weight compounds such as bilirubin, melatonin, melanin, pocilloporins, or mycosporine-like amino acids (see below).

4.1 High Diversity of Enzymatic Antioxidative Defense is a Consequence of Symbiosis

Enzymatic defenses are composed of a set of enzymes following the following general pathway (Halliwell and Guteridge 1999):

$O_2^- + O_2^- + 2H^+ \rightarrow H_2O_2 + O_2$ Superoxide dismutases

$H_2O_2 + H_2O_2 \rightarrow 2H_2O + O_2$ Catalases

$H_2O_2 + RH_2 \rightarrow 2H_2O + R$ Peroxydases

Dykens and Shick (1982) found that the activity of superoxide dismutases (SODs), the first enzyme involved in ROS scavenging was higher in light compared to dark conditions. By using electrophoresis gels in non-denaturating conditions, Richier et al. (2003) found up to seven SOD isoforms within the symbiotic host cells of the temperate sea anemone *Anemonia viridis*, while the nonsymbiotic sea anemone, *Actinia schmidti*, harbors only three SOD isoforms (Richier et al. 2005), as classically observed in animal cells (Halliwell and Guteridge 1999). This result is surprising since such a high diversity in SOD isoforms is generally a plant feature. A similar diversity (5–6 SOD isoforms) was observed in zooxanthellae of different clades that inhabit coral and sea anemone tissues (Richier et al. 2003, 2005). Using specific inhibitors, Richier et al. (2003) described the type of SOD present in the symbiotic association. Surprisingly, the animal host expresses the three known SOD families, Fe-, CuZn-, and MnSOD (Table 1), while usually animals only expressed CuZn- and MnSOD. An extracellular CuZn-SOD was also identified by a molecular approach (Plantivaux et al. 2004).

These results show that the high resistance of symbiotic cnidarians to hyperoxia is due to enhanced oxidative defense

Table 1 Structural and biochemical characteristics of SOD isoforms expressed by the *A. viridis/Symbiodinium* sp holobiont (Modified from Richier 2004 and Plantivaux 2006). nd : not determined

Family	Isoform	PM (kDa)	pI	Intracellular localization	Conformation	Host/symbiont location	Identification methods
FeSOD	FeI	88	nd	nd	Tetramer	Host endoderm symbiont	Native gel Molecular cloning
	FeII	72	nd	nd	nd	Host endoderm symbiont	Native gel
MnSOD	MnI	100	5.1	Mitochondria	Tetramer	Host (ecto-/endoderm) symbiont	Native gel Molecular cloning
	MnII	31.4	4.1	Cytosol	Dimer	Host endoderm symbiont	Native gel
	MnIII	31.8	4.09	Cytosol	Dimer	Host endoderm symbiont	Native gel
	MnIV	31.3	4.02	Cytosol	Dimer	Host endoderm symbiont	Native gel
CuZnSOD	CuZna	80	nd	Extracellular SOD (EC-SOD)	Tetramer	Host (ecto-/endoderm)	Molecular cloning
	CuZnb	60	4.8	Cytosol	Tetramer	Host (ecto-/endoderm)	Native gel Molecular cloning

both in terms of intensity and diversity. This unusual pattern is not restricted to temperate sea anemones. Shick and Dykens (1985) found a good correlation between activities of both SOD and catalase and chlorophyll concentration within 34 species of symbiotic invertebrates. A high diversity in SOD isoforms was demonstrated in three scleractinan corals (*Stylophora pistillata*, *Plerogyra sinuosa*, and *Pocillopora verrucosa*) and in two sea anemones (*Aiptasia pulchella* and *Entacmea quadricolor*) (Richier et al. 2005, 2008; Furla et al. 2005). Therefore, the unusual pattern of SODs, either in terms of activity or isoform diversity between symbiotic and nonsymbiotic hexacorallians, suggests that the diversification of SOD in animal host cell was a consequence of symbiotic relationships. The origin of the diversification is actually investigated and it could in part be due to horizontal gene transfer between host and symbiont (Furla et al. 2008).

The other antioxdant enzymes having a key role in the symbiosis relationship are the catalases. These enzymes, present in host and symbiont cells, catalyze the decomposition of hydrogen peroxide to water and oxygen. Merle et al. (2007) demonstrated the importance of catalase in symbiotic cnidarian by inhibiting them with the drug. The dysfunctioning of catalases were not lethal but induced symbiosis breakdown (bleaching). Previous works by Lesser (1997) had also demonstrated the importance of catalase in preventing symbiont photosystem disruption during heat stress. Therefore, the enzymatic antioxidant defenses, as SOD or catalases, are of primary importance in the adaptation of cnidarians to symbiosis-induced hyperoxic environment.

4.2 Nonenzymatic Antioxidative Mechanisms

Together with antioxidant enzymes, cnidarians have also water-soluble nonenzymatic antioxidants, such as ascorbic acid or glutathione (Lesser 2005). Carotenoids and tocopherols are the major lipid-soluble antioxidants that scavenge ROS (Edge et al. 1997). The carotenoids, peridinin, dinoxanthin, diadinoxanthin, and diatoxanthin are found in symbionts, while β-carotene is present in both animal host and symbionts (Mobley and Gleason 2003). It has been shown that a large part of host carotenoids are derived from zooxanthellae metabolism (Mobley and Gleason 2003). However, carotenoids are also found in nonsymbiotic cnidarians, as the animal host may also obtain them by feeding on carotenoid-rich zooplankton (Sebens et al. 1996). In all cases, the production of carotenoids depends on environmental conditions and increases in corals subjected to high levels of solar radiations (Ambarsari et al. 1997).

Other compounds may act as nonenzymatic antioxidants. Coral fluorescent proteins, homologous to the green-fluorescent protein (GFP), are animal-encoded proteins that give pink, purple, or blue colors to the organism (Dove et al. 2001; see reviews by Tsien 1998; Alieva et al. 2008). Bou-Abdallah et al. (2006) described that GFP from the hydromedusa *Aequorea victoria* could quench superoxide anion radicals (O_2^-) and exhibit SOD-like activity by competing with cytochrome c for reactions with O_2^-.

Usually considered as a UV screen, mycosporine-like amino acids (MAAs) are low-molecular weight molecules that absorb UV radiations and thus play a major role as UV protectants (Shick and Dunlap 2002). However, some MAAs may also protect organisms by scavenging ROS such as the oxo-MAA mycosporine-glycine (Dunlap and Yamamoto 1995). This MAA is predominant in phototrophic symbioses (Dunlap et al. 2000).

5 Fourth Adaptation: Withstand Solar Radiations

In order to allow optimal photosynthesis of their symbionts, host distribution is generally restricted to shallow waters and therefore symbiotic cnidarians had to tolerate high solar irradiation and protect themselves against sunburns. Such protection is mediated by two types of adaptations: ultraviolet screens and pigments.

5.1 Ultraviolet Screens

Major ultraviolet (UV)-absorbing compounds are mycosporine-like amino acids (MAAs) which include 13 different molecules identified in reef-building corals (Shick and Dunlap 2002). MAAs mainly act as natural sunscreens that absorb and dissipate solar ultraviolet radiations (UVR). Their concentrations in animal and algal tissues increase when UVR intensities are experimentally enhanced (Shick et al. 1999), or when corals are transplanted from greater to shallower depths (Shick et al. 1996).

Since glyphosate, an inhibitor of several enzymes of the shikimate pathway, decreased MAA production in the coral *Stylophora pistillata* exposed to UVR, it was concluded that MAAs in symbiotic anthozoans are synthesized via the shikimate pathway (Shick et al. 1999). This pathway being supposed to be absent in animals (Bentley 1990), zooxanthellae have been assumed to be the main MAAs-synthesizing partner. However, zooxanthellae in culture often produce a more limited number of MAAs than that found in the holobiont from which they were isolated (Shick and Dunlap 2002). In the coral *Stylophora pistillata*, up to seven different MAAs were found in host tissues while zooxanthellae did not produce in vitro MAAs (Shick et al. 1999). All these results suggest that the host may play a role in MAA biosynthesis. Shick et al. (1999) suggested that the host may metabolize the primary MAAs to produce secondary MAAs. A bioinformatic study of the genome of the nonsymbiotic sea anemone, *Nematostella vectensis*, suggested another alternative (Starcevic et al. 2008). This study demonstrated the unexpected presence of genes encoding enzymes for the shikimate pathway within the genome of *N. vectensis*. Molecular evidence led the authors to suggest horizontal transfer of ancestral genes of the shikimic acid pathway into the genome of the sea anemone. The donors were identified as unknown bacteria and two dinoflagellates (*Oxyrrhis marina* and *Heterocapsa triquetra*). It remains to be determined if these genes are expressed and active in *N. vectensis*.

5.2 Host and Symbiont Pigments

Apart from the brownish pigmentation caused by the photoactive compounds of the dinoflagellate symbionts, most of the reef anthozoan colors come for the animal host tissues (Oswald et al. 2007). Referred to as pocilloporins or fluorescent protein (FPs), these pigments have been identified in both symbiotic and nonsymbiotic cnidarians (Matz et al. 1999; Dove et al. 2001; Wiedenmann et al. 2004; Alieva et al. 2008). Due to their large diversity, they cover a great range of absorption and emission spectra (Prescott et al. 2003). However, their biological roles remain unclear and several hypotheses have been proposed.

The first role attributed to FPs is to act as photoprotectants that shield the zooxanthellae photosynthetic machinery of corals settled in high light conditions (Salih et al. 1998; Dove 2004) by scattering the light reaching the coral. However, their concentration did not vary along a depth gradient (Mazel et al. 2003). They are nevertheless predominant in symbiotic species living in photic environments (Mazel et al. 2003). The second role suggested for FP is to enhance photosynthesis in low-light conditions by amplifying the levels of photosynthetic active radiation (PAR) available for zooxanthellae harbored by colonies growing under low light habitats (Schlichter and Fricke 1990). This mechanism involves capture by FPs of short-wavelength light (including UV) and reemission into longer-wavelength in the PAR region of the spectrum, susceptible to be absorbed by the primary photosynthetic algal pigments. However, this hypothesis has been challenged by Gilmore et al. (2003) who showed that there is inefficient energy transfer from the FP to the zooxanthellae chlorophyll, at least in shallow corals. Another role, suggested by Bou-Abdallah et al. (2006), is to act as ROS scavengers. They have demonstrated that FP from the hydromedusa *Aequorea victoria* quenches O_2^- radicals as an SOD-like activity and then contributes to the total antioxidant capacities. Other proposed roles include aposematic coloration (Matz et al. 2006) and immune defense of the holobiont facing virulent aggressors (Palmer et al. 2008).

Whatever the role of FPs, they appear to play a major role in the functioning of the symbiotic association, representing an example of symbiosis partnership.

6 Conclusion

Deep physiological adaptations are supported by the host cnidarian and their zooxanthellae in order to live in symbiosis with success. Optimization of not only CO_2 absorption but also nitrogen uptake (Grover et al. 2002, 2003, 2006; see for review Furla et al. 2005) is crucial for the trophic benefits obtained by the symbiosis. Resistance to hyperoxia is also of primary importance in order to allow a large biomass of zooxanthella inside the host tissue (around 3.10^6 zoox/mg protein) and high primary production. Protection against solar radiations by MAAs or Pocilloporins is also requested for a large seawater surface distribution.

Additional molecular and physiological adaptations have to be discovered and the new data coming from genomic analysis will help to diversify the field. Recently, Rodriguez-Lanetty et al. (2006) have demonstrated, in the symbiotic sea anemone *Anthopleura elegantissima*, an upregulation of the scavenger receptor SR-B1 expression, a member of the CD36 family which is known to control the immune response of the host by the symbiont (Stafford et al. 2002). A modification of the immunorecognition by the host itself or by the symbionts could then been a novel aspect of adaptation for the establishment of the symbiosis.

Despite the original and deep adaptations involved in maintaining the symbiosis, external perturbations could disrupt them. This is the case during bleaching episodes where the expulsion of the zooxanthellae induces the loss of cnidarian coloring. Several studies have been suggested that the cellular causes of bleaching are associated to an overproduction of ROS. During hyperthermia or UV exposures, the additional ROS are not neutralised by the antioxidant defenses of the holobiont and lead to cell damages and cell death.

References

Aizawa K, Miyachi S (1986) Carbonic anhydrase and CO_2-concentrating mechanisms in microalgae and cyanobacteria. FEMS Microbiol Rev 39:215–233

Alieva NO, Konzen KA, Field SF, Meleshkevitch EA, Hunt ME, Beltran-Ramirez V, Miler DJ, Wiedenmann J, Salih A, Matz MV (2008) Diversity and evolution of coral fluorescent proteins. PLoS ONE 3:e2680

Allemand D, Furla P, Bénazet-Tambutté S (1998) Mechanisms of carbon acquisition for endosymbiont photosynthesis in Anthozoa. Can J Bot 76:925–941

Al-Moghrabi S, Goiran C, Allemand D, Speziale N, Jaubert J (1996) Inorganic carbon uptake for photosynthesis by the symbiotic coral/dinoflagellate association. II. Mechanisms for bicarbonate uptake. J Exp Mar Biol Ecol 199:227–248

Ambarsari I, Brown BE, Barlow RG, Britton G, Cummings D (1997) Fluctuations in algal chlorophyll and carotenoid pigments during solar bleaching in the coral *Goniastrea aspera* at Phuket Thailand. Mar Ecol Prog Ser 159:303–307

Baker AC, Starger CJ, McClanahan TR, Glynn PW (2004) Shifting to new algal symbionts may safeguard devastated reefs from extinction. Nature 430:CBl153

Bénazet-Tambutté S, Allemand D, Jaubert J (1996) Inorganic carbon supply to symbiont photosynthesis of the sea anemone, *Anemonia viridis*: role of the oral epithelial layers. Symbiosis 20:199–217

Bentley R (1990) The shikimate pathway – a metabolic tree with many branches. Crit Rev Biochem Mol Biol 25:307–384

Bertucci A, Tambutté É, Tambutté S. Allemand D, Zoccola D (2010) Symbiosis-dependant gene expression in coral-dinoflagellate association: cloning and characterization of a P-type H^+-ATPase gene. Proc R Soc B 277:87–95

Bou-Abdallah F, Chasteen ND, Lesser MP (2006) Quenching of superoxide radicals by green fluorescent protein. Biochim Biophys Acta 1760:1690–1695

Bowes G, Salvuci ME (1989) Plasticity in the photosynthetic carbon metabolism of submersed aquatic macrophytes. Aquat Bot 34:233–266

Brandt K (1881) Uber das Zusammenleben von Algen und Tieren. Biologisches Centralblatt 1:524–527

Bronstein JL (2001) The costs of mutualism. Amer Zool 41:825–839

Coffroth MA, Santos SR (2005) Genetic diversity of symbiotic dinoflagellates in the genus *Symbiodinium*. Protist 156:19–34

D'Aoust BG, White R, Wells JM, Olsen DA (1976) Coral-algal associations: Capacity for producing and sustaining elevated oxygen tensions in situ. Undersea Biomed Res 3:35–40

Daudin H (1926) Cuvier et Lamarck. Les classes zoologiques et l'idée de série animale. Félix Alcan, Paris, pp 1790–1830

Douglas AE, McAuley PJ, Davies PS (1993) Algal symbiosis in cnidarian. J Zool 231:175–178

Dove S (2004) Scleractinian corals with photoprotective host pigments are hypersensitive to thermal bleaching. Mar Ecol Prog Ser 272:99–116

Dove SG, Hoegh-Guldberg O, Ranganathan S (2001) Major colour patterns of reef-building corals are due to a family of GFP-like proteins. Coral Reefs 19:197–204

Dunlap WC, Yamamoto Y (1995) Small-molecule antioxidants in marine organisms: antioxidant activity of mycosporine-glycine. Comp Biochem Physiol 112B:105–114

Dunlap WC, Shick JM, Yamamoto Y (2000) UV protection in marine organisms. I. Sunscreens, oxidative stress and antioxidants. In: Yoshikama Y, Toyokuni S, Yamamoto Y, Naito Y (eds) Free radicals in chemistry, biology and medicine. OICA International, London

Dykens JA, Shick JM (1982) Oxygen production by endosymbiotic algae controls superoxyde dismutase activity in their animal host. Nature 297:579–580

Dykens JA, Shick JM, Benoit C, Buettner GR, Winston GW (1992) Oxygen radical production in the sea anemone *Anthopleura elegantissima* and its endosymbiotic algae. J Exp Biol 168:219–241

Edge R, McGarvey DJ, Truscott TG (1997) The carotenoids as antioxidants - a review. J Photochem Photobiol B 41:189–200

Fine PEM (1975) Vectors and vertical transmission: an epidemiological perspective. Ann NY Acad Sci 266:173–194

Freudenthal HD (1962) *Symbiodinium* gen. nov. *Symbiodinium microadriaticum* sp. nov., a zooxanthella: Taxonomy, life cycle, and morphology. J Protozool 9:45–52

Furla P, Bénazet-Tambutté S, Jaubert J, Allemand D (1998a) Functional polarity of the tentacle of the sea anemone *Anemonia viridis*: Role in inorganic carbon acquisition. Amer J Physiol (Regul Integr Comp Physiol) 274:R303–R310

Furla P, Bénazet-Tambutté S, Jaubert J, Allemand D (1998b) Diffusional permeability of dissolved inorganic carbon through the isolated oral epithelial layers of the sea anemone, *Anemonia viridis*. J Exp Mar Biol Ecol 221:71–88

Furla P, Allemand D, Orsenigo MN (2000a) Involvement of H^+-ATPase and carbonic anhydrase in inorganic carbon uptake for endosymbiont photosynthesis. Amer J Physiol (Regul Integr Comp) 278:R870–R881

Furla P, Galgani I, Durand I, Allemand D (2000b) Sources and mechanisms of inorganic carbon transport for coral calcification and photosynthesis. J Exp Biol 203:3445–3457

Furla P, Allemand D, Ferrier-Pages C, Shick M (2005) The symbiotic anthozoan: physiological chimera between alga and animal. Integr Comp Biol (formerly Am Zool) 45:595–604

Furla P, Richier S, Merle P-L, Garello G, Plantivaux A, Forcioli D, Allemand D (2008) Roles and origins of superoxide dismutases in a symbiotic cnidarian. In: Organisms SEaEGoCR (ed) 11th international coral reef symposium, Fort Lauderdale, 2008

Ganot P, Moya A, Deleury E, Allemand D, Furla P, Sabourault C (2008) Exploring symbiotic interactions in the sea anemone-zooxanthellae model by large-scale ests analysis. In: Organisms SEaEGoCR (ed) 11th International coral reef symposium, Fort Lauderdale, 2008

Gilmore AM, Larkum AWD, Salih A, Itoh S, Shibata Y, Bena C, Yamasaki H, Papina M, Van Woesik R (2003) Simultaneous time resolution of the emission spectra of fluorescent proteins and zooxanthellar chlorophyll in reef-building corals. Photochem Photobiol 77:515–523

Giordano M, Beardall J, Raven JA (2005) CO_2 concentrating mechanisms in algae: Mechanisms, environmental modulation, and evolution. Annu Rev Plant Biol 56:99–131

Gluck S, Nelson R (1992) The role of the V-ATPase in renal epithelium H^+ transport. J Exp Biol 172:205–218

Goiran C, Al-Moghrabi S, Allemand D, Jaubert J (1996) Inorganic carbon uptake for photosynthesis by the symbiotic coral/dinoflagellate association I. Photosynthetic performances of symbionts and dependence on sea water bicarbonate. J Exp Mar Biol Ecol 199:207–225

Goiran C, Allemand D, Galgani I (1997) Transient Na^+ stress in symbiotic dinoflagellates after isolation from coral-host cells and subsequent immersion in seawater. Mar Biol 129:581–589

Goodson MS, Whitehead F, Douglas AE (2001) Symbiotic dinoflagellates in marine cnidaria: diversity and function. Hydrobiologia 461:79–82

Grover R, Maguer JF, Reynaud-Vaganay S, F-P C (2002) Uptake of ammonium by the scleractinian coral *Stylophora pistillata*: Effect of feeding, light, and ammonium concentrations. Limnol Oceanogr 47:782–790

Grover R, Maguer JF, Allemand D, Ferrier-Pages C (2003) Nitrate uptake in the scleractinian coral *Stylophora pistillata*. Limnol Oceanogr 48:2266–2274

Grover R, Maguer J-F, Allemand D, Ferrier-Pagès C (2006) Urea uptake by the scleractinian coral *Stylophora pistillata*. J Exp Mar Biol Ecol 332:216–225

Halliwell B, Guteridge JMC (1999) Free radicals in biology and medicine, 3rd edn. Oxford University Press, New York

Jordan DB, Ogren WL (1981) Species variation in the specificity of ribulose biphosphate carboxylase/oxygenase. Nature 291:513–515

Kirschner LB (1991) Water and ions. In: Prosser CL (ed) Environmental and metabolic animal physiology. Wiley-Liss, New York, pp 13–107

Leggat W, Badger MR, Yellowlees D (1999) Evidence for an inorganic carbon-concentrating mechanism in the symbiotic dinoflagellate *Symbiodinium* sp. Plant Physiol 121:1247–1255

Leggat W, Marendy EM, Baillie B, Whitney SM, Ludwig M. Badgaer MR, Yellowlees D (2002) Dinoflagellate symbioses: strategies and adaptations for the acquisition and fixation of inorganic carbon. Funct Plant Biol 29:309–322

Lesser MP (1997) Oxidative stress causes coral bleaching during exposure to elevated temperature. Coral Reefs 16:187–192

Lesser MP (2005) Oxidative stress in marine environments: biochemistry and physiological ecology. Annu Rev Physiol 68:253–257

Lopez I, Egea R, Herrera FC (1991) Are cœlenterate cells permeable to large anions? Comp Biochem Physiol 100A:193–198

Matz MV, Fradkov AF, Labas YA, Savitsky AP, Zaraisky AG, Markelov ML, Lukyanov SA (1999) Fluorescent proteins from non bioluminescent Anthozoa species. Nat Biotechnol 17:969–973

Matz MV, Marshall NJ, Vorobyev M (2006) Are corals colorful? Photochem Photobiol 82:345–350

Mayfield AB, Gates RD (2007) Osmoregulation in anthozoan-dinoflagellate symbiosis. Comp Biochem Physiol 147A:1–10

Mazel CH, Lesser MP, Gorbunov MY, Barry TM, Farrel JH. Wyman KD, Falkowski PG (2003) Green-fluorescent proteins in Caribbean corals. Limnol Oceanogr 48:402–411

Merle PL, Sabourault C, Richier S, Allemand D, Furla P (2007) Catalase characterization and implication in bleaching of a symbiotic sea anemone. Free Radic Biol Med 42(2):236–246

Mobley KB, Gleason DF (2003) The effect of light and heterotrophy on carotenoid concentrations in the Caribbean anemone *Aiptasia pallida* (Verrill). Mar Biol 143:629–637

Oswald F, Schmitt F, Leutenegger A, Ivanchenko S, D'Angelo C, Salih A, Maslakova S, Bulina M, Schirmbeck R, Nienhaus GU, Matz MV, Wiedenmann J (2007) Contributions of host and symbiont pigments to the coloration of reef corals. FEBS J 274:1102–1109

Palmer CV, Mydlarz LD, Willis BL (2008) Evidence of an inflammatory-like response in non-normally pigmented tissues of two scleractinian corals. Proc R Soc Ser B Biol Sci 275:2687–2693

Perru O (2003) De la Société à la Symbiose: Une histoire des découvertes sur les associations chez les êtres vivants. Librairie philosophique J. Vrin. Institut de l'Institut Interdisciplinaire d'Etudes Epistémologiques, Paris, Lyon

Plantivaux A (2006) Adaptation aux stress oxydants chez un Cnidaire symbiotique : approche biochimique et génomique: rôle de la Cu/Zn-SOD. Ph.D. thesis. Nice-Sophia Antipolis University, France

Plantivaux A, Furla P, Zoccola D, Garello G, Forcioli D, Richier S, Merle P-L, Tambutté É, Tambutté S, Allemand D (2004) Molecular characterization of two CuZn-superoxide dismutases in a sea anemone. Free Radic Biol Med 37:1170–1181

Pochon X, Montoya-Burgos JI, Stadelmann B, Pawlowski J (2006) Molecular phylogeny, evolutionary rates, and divergence timing of the symbiotic dinoflagellate genus *Symbiodinium*. Mol Phylogenet Evol 38:20–30

Prescott M, Ling M, Beddoe T (2003) The 2.2 Å crystal structure of a pocilloporin pigment reveals a nonplanar chromophore conformation. Structure 11:275–284

Raven JA (1990) Sensing pH? Plant Cell Environ 13:721–729

Raven JA (2003) Inorganic carbon concentrating mechanisms in relation to the biology of algae. Photosynth Res 77:155–171

Richier S (2004) Mécanismes de résistance d'une endosymbiose marine méditerranénne aux stress oxydatifs. Ph.D. thesis. Nice-Sophia Antipolis University, France

Richier S, Merle P-L, Furla P, Pigozzi D, Sola F, Allemand D (2003) Characterization of superoxide dismutases in anoxia- and hyperoxia-tolerant symbiotic cnidarians. Biochim Biophys Acta 1621:84–91

Richier S, Furla P, Plantivaux A, Merle P-L, Allemand D (2005) Symbiosis-induced adaptation to oxidative stress. J Exp Biol 208:277–285

Richier S, Cottalorda J-M, Guillaume M, Fernandez C, Allemand D, Furla P (2008) Depth-dependant response to light of the reef building coral, *Pocillopora verrucosa*: implication of oxidative stress. J Exp Mar Biol Ecol 357:48–56

Rodriguez-Lanetty M, Phillips WS, Weis VM (2006) Transcriptome analysis of a cnidarian-dinoflagellate mutualism reveals complex modulation of host gene expression. BMC Genomics 7:1–11

Roth E, Jeon K, Stacey G (1988) Homology in endosymbiotic systems: The term symbiosome. In: Palacios RV D (ed) Molecular genetics of plant-microbe interactions. The American Phytopatological Society, St Paul, pp 220–225

Rowan R, Whitney SM, Fowler A, Yellowlees D (1996) Rubisco in marine symbiotic dinoflagellates: Form II enzymes in eukaryotic oxygenic phototrophs encoded by a nuclear multigene family. Plant Cell 8:539–553

Sachs JL, Wilcox TP (2006) A shift to parasitism in the jellyfish symbiont *Symbiodinium microadriaticum*. Proc R Soc B 273:425–429

Salih A, Hoegh-Guldberg O, Cox G (1998) Photoprotection of symbiotic dinoflagellates by fluorescent pigments in reef. In: Greenwood JG, Hall NJ (eds) Proceedings of the Australian coral ref society 75th anniversary conference, Heron Island, 1997, pp 218–230

Sawyer SJ, Muscatine L (2001) Cellular mechanisms underlying temperature-induced bleaching in the tropical sea anemone *Aiptasia pulchella*. J Exp Biol 204:3443–3456

Schlichter D, Fricke HW (1990) Coral host improves photosynthesis of endosymbiotic algae. Naturwissenschaften 77:447–450

Schwarz JA, Weis VM, Potts DC (2002) Feeding behavior and acquisition of zooxanthellae by planula larvae of the sea anemone *Anthopleura elegantissima*. Mar Biol 140:471–478

Sebens KP, Vandersall KS, Savina LA, Graham KR (1996) Zooplankton capture by two scleractinian corals, *Madracis mirabilis* and *Montastrea cavernosa*, in a field enclosure. Mar Biol 127:303–317

Seibt C, Schlichter D (2001) Compatible intracellular ion composition of the host improves carbon assimilation by Zooxanthellae in mutualistic symbioses. Naturwissenschaften 88:382–386

Shashar N, Cohen Y, Loya Y (1993) Extreme diel fluctuations of oxygen in diffusive boundary layers surrounding stony corals. Biol Bull 185:455–461

Shick JM, Dunlap WC (2002) Mycosporine-like amino acids and related gadusols: Biosynthesis, accumulation, and UV-protective functions in aquatic organisms. Annu Rev Physiol 64:223–262

Shick JM, Dykens JA (1985) Oxygen detoxification in algal-invertebrate symbioses from great barrier reef. Oecologia 66:33–41

Shick JM, Lesser MP, Jokiel PL (1996) Effects of ultraviolet radiation on corals and other coral reef organisms. Glob Change Biol 2:527–545

Shick JM, Romaine-Lioud S, Ferrier-Pagès C, Gattuso J-P (1999) Ultraviolet-B radiation stimulates shikimate pathway-dependent accumulation of mycosporine-like amino acids in the coral *Stylophora pistillata* despite decreases in its population of symbiotic dinoflagellates. Limnol Oceanogr 44:1667–1682

Starcevic A, Akthar S, Dunlap WC, Shick JC, Hranueli D, Cullum J, Long PF (2008) Enzymes of the shikimic acid pathway encoded in the genome of a basal metazoan, *Nematostella vectensis*, have microbial origins. Proc Natl Acad Sci 105:2533–2537

Trench RK (1987) Dinoflagellates in non-parasitic symbioses. In: Taylor FJR (ed) The biology of dinoflagellates. Blackwell Scientific Publications, Oxford, pp 530–570

Tsien RY (1998) The green fluorescent protein. Annu Rev Biochem 67:509–544

Venn AA, Tambutté É, Lotto S, Zoccola D, Allemand D, Tambutté S (2009) Intracellular pH in symbiotic cnidarians. Proc Natl Acad Sci 106(39):16574–17579

Wakefield TS, Kempf SC (2001) Development of host- and symbiont-specific monoclonal antibodies and confirmation of the origin of the symbiosome membrane in a cnidarian-dinoflagellate symbiosis. Biol Bull 200:127–143

Weis VM (1991) The induction of carbonic anhydrase in the symbiotic sea anemone *Aiptasia pulchella*. Biol Bull 180:496–504

Weis VM, Reynolds WS (1999) Carbonic anhydrase expression and synthesis in the sea anemone *Anthopleura elegantissima* are enhanced by the presence of dinoflagellate symbionts. Physiol Biochem Zool 72:307–316

Wiedenmann J, Ivanchenko S, Oswald F, Nienhaus GU (2004) Identification of GFP-like proteins in non-bioluminescent, azooxanthellate Anthozoa opens new perspectives for bioprospecting. Mar Biotechnol 6:270–277

Part IV
The Coral Reef Ecosystem: Bacteria, Zooplankton, Algae, Invertebrates, Fishes and Model

Biogeochemistry of Nutrients

Marlin J. Atkinson

Abstract Previous reviews of coral reef biogeochemistry are briefly summarized, as well as biogeochemical reactions and C:N:P pool sizes of water, biota, and sediments. Ranges of gross and net productivity are given for characteristic reef habitats. Rates of nutrient uptake and release are also presented, and a generalized kinetic approach to reef biogeochemistry using mass transfer theory is considered. The implication of mass transfer limitation of biogeochemical reactions is discussed in the wider context of how reefs function. It is important to take a unified approach in normalizing biogeochemical rates so that the quantification of all biogeochemical reactions can be achieved, and a more global understanding of the limits and constraints of coral reef processes and metabolism.

Keywords Nutrients • nitrogen • phosphorus • productivity • metabolism • calcification • mass transfer • chemistry

1 Introduction

Biogeochemistry is the study of biologically mediated chemical compounds that influence geological processes. Coral reefs are vibrant, living structures that maintain themselves at sea level through the combined biogenic calcification of a variety of organisms. Thus, the study of biogeochemistry is central to understanding material and energy fluxes in coral reefs, which ultimately set limits to metabolic performance of the ecosystem. Biogeochemistry involves the study of relationships between carbon and nutrient fluxes, where nutrients are broadly defined as all compounds necessary to support the turnover of carbon in living organisms, and generally include the compounds of the elements (N, P, Si, S, I, Fe, etc.). There are four major reviews of biogeochemical compounds and/or nutrients in the literature: Crossland (1983); D'Elia and Wiebe (1990); Szmant (2002); Atkinson and Falter (2003).

In this chapter, I provide a historical view of coral reef biogeochemistry by summarizing these reviews, which span 25 years and reflect the ideas of their time. The high biomass communities of coral reefs can achieve high areal productivity. Much of the biogeochemical study of organisms and communities thus far has been to understand how high areal productivity is maintained in low-nutrient water, and how this, in turn, provides and sustains a diverse food web. Anyone with an interest in these topics should read these reviews, and explore the cited literature. I would point out, however, that these reviews also pose fundamental questions about reef function, and the implicit role of biogeochemistry in reef community structure.

Enough is now known to delineate answers to most of those questions. I do not want to repeat the referencing of the literature in these reviews, so only the most recent literature is cited here. Current literature on nutrients and coral reefs is disparate, primarily because reef productivity and nutrient cycles have been studied from many different disciplines, including geochemistry, marine biology, zoogeography, oceanography, and ecology. For this reason, I offer here some of my own insight into the underlying thinking of the literature – a perspective not necessarily revealed in the papers themselves. I also will elucidate the recent viewpoint on the dynamic nature of biogeochemical fluxes on reefs.

2 Summary of Reviews

The Crossland review (1983) raised four fundamental questions regarding productivity and nutrient relationships: (1) Given the low nutrient concentrations of most coral reefs, how is high primary productivity and biomass sustained? (2) Is there seasonality of stored nutrient pools, and subsequent release or recycling? (3) Is "tight recycling" between organisms and communities more important than continual

M.J. Atkinson (✉)
Hawaii Institute of Marine Biology,
P.O. Box 1346, Kaneohe, HI, 96744, USA
e-mail: mja@hawaii.edu

nutrient input? (4) Is the knowledge of inorganic versus organic nutrients sufficient for interpretation of the nutrients required for growth?

The first question was already 30 years old when Crossland wrote, having been originally posed by Odum and Odum (1955), who were interested in how ecosystems process energy. The most revealing aspect of these questions is the attention to "tight recycling" and the search to find other sources of nutrients. The implicit assumption was that nutrients in the water column could not support this high level of productivity, so many biogeochemical studies sought to explain high productivity in low-nutrient water. These studies indentified other sources of nutrients, including rain, groundwater, recycling through sediments, sporadic mixing events through internal waves, and endolithic upwelling, to name the most popular. Crossland (1983) pointed out that nutrient concentrations in ocean water change very little as water moves across shallow reefs, even though one could find apparently differing levels of nutrient production and consumption across different habitats and zones. The explanation for this was that if productivity was low overall, then one would expect to find little change in nutrients. This view was actually a departure from the decade earlier, when the implicit assumptions were that instantaneous photosynthesis should be matched by simultaneous nutrient uptake, thus stimulating changes in uptake rates during day–night periods. Crossland concluded that: (1) nutrient concentration revealed minimal information and it was thus necessary to understand fluxes (rates per unit area), (2) concentrations are low and reflect the nearby ocean, (3) both dissolved and particulate nutrient fluxes should be studied in multi-disciplinarily teams, (4) sources and sinks of nutrients must be identified, and (5) global comparative studies should be conducted to characterize coral reefs.

The D'Elia and Wiebe review (1990) focused on the role of nitrogen, phosphorus, and silica compounds in the context of the relationship between productivity and nutrient cycles, and nutrient input. The review is comprehensive, covering most of the literature and summarizing what was known about the various biogeochemical reactions and pathways on coral reefs. They discussed all the known biochemical reactions and introduced Michaelis–Menton kinetics (better termed Monad kinetics) to describe kinetics of nutrient reactions with the benthos. The authors, however, provided no summary of areal rates of biogeochemical reactions, nor did they explain how these rates might compare with productivity. The reader is left to conclude several points from that review: (1) nutrient concentrations are low and reflect the neighboring ocean; (2) every biogeochemical pathway found in plankton systems can be found in a coral reef ecosystem; (3) biogeochemical rates are faster than plankton systems, but actually are unknown; (3) coral reefs show substantial rates of nitrogen fixation, nitrification, and denitrification, with a tendency to export nitrogen compounds; (4) reefs show a small uptake of phosphate, which appears to predict net production with a very high CNP stoichiometery, compared to the Redfield ratio. D'Elia and Wiebe also concluded that: (1) nutrient recycling on reefs must occur, but the process is poorly understood and (2) productivity is probably nutrient-limited and, thus, changes in nutrient input may change community structure, but not overall production. The authors also touched on some of my earlier works, which indicated that phosphate uptake is slow and appears to be related to net primary production of plants with high C:N:P ratios. Even in this 1990 review, there are no summaries of actual uptake rates, mostly because the studies had different ways of normalizing the rates, making it difficult to make inter-comparisons.

By the mid-1980s, it had become abundantly clear that rates of nutrient uptake into benthos were relatively slow, compared to the rate of advection of water. This was a surprise to many of us, and we had a very difficult time understanding this result. In the early 1980s, I asked a simple question of six coral reef scientists: if ^{32}P is put into the water on the upstream reef crest, how far before half is removed? All of these researchers answered less than 10 m. In a later publication (Atkinson and Smith 1987) we showed, using ^{32}P and ^{3}H$_2$O, that water traveled 450 m across a reef flat with only 11% of ^{32}P removed. We realized that we were all missing some very basic information. Nevertheless, the central idea was that the relatively closed, efficient nutrient cycles on reefs created a well-balanced system, capable of high carbon production and consumption. The notion prevailed that external sources of nutrients would affect productivity and alter food-chain dynamics. Simple food-chain models of the time showed that little of the gross primary production was moving up the food web, and that very high ecological efficiency was necessary to support biomass of higher trophic levels.

Based on the D'Elia and Wiebe review and predominant ideas of nutrient recycling, external nutrient input became one of the major explanations for shifts in community structure from coral to algae. Szmant (2002) took this issue on directly in her review, which is a critique of a large amount of literature. The article primarily compares the impacts of nutrients with overfishing, sedimentation and storms on phase shifts from coral to algae. Szmant argues there is little evidence of direct effects of nutrient loading on the phase shifts, except in restricted embayments, where nutrients can build up in the sediments and be released to sustain high plankton populations. She concludes that nutrients are a major factor in the decline of a few reefs, but appears to play a lesser role than sedimentation, overfishing, and the effects of global warming (e.g., bleaching). Szmant also points out that there is little solid evidence that reefs respond to nutrient input, or even that corals require low-nutrient water. The paper does not summarize the biogeochemistry; instead it discusses the effects of nutrient perturbation on biogeochem-

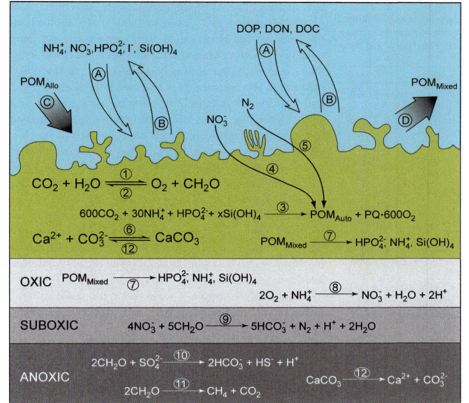

Fig. 1 Biogeochemical reactions and pathways occurring in coral reefs: the top layer is the benthos, the light gray layer represents surface oxic sediments, medium gray layer is the suboxic sediments, and the dark gray layer is the anoxic sediments. POM is particulate organic matter with different types of POM denoted by the subscript: auto=autotrophic; mixed=autochthonous; all=autochthonous plus allochthonous; allo=allochthonous

istry. Her paper is a must-read. Like the other reviews, however, she presents no rates and no clarification about how nutrients interact with the carbon cycle. None of these three reviews report nutrient fluxes, making it difficult to calculate rates of nutrient inputs, recycling, and exports. The last review paper (my own, with Jim Falter) is a general review of coral reef biogeochemistry. It summarizes basic chemical reactions, concentrations, and pool sizes for water, biota and sediment, and introduces the mass transfer approach to replace Michaelis–Menton kinetics. Some of that information is summarized here. Biogeochemical reactions that occur on reefs are tabulated and illustrated in Fig 1. I would note that all biogeochemical reactions that occur in other aquatic environments also occur on reefs. The biogeochemical reactions on reefs tend to be faster and occur in closer proximity compared to other aquatic environments.

3 Nutrient Pool Sizes

Table 1 provides the range of the C:N:P pool sizes in water, benthos, and sediments in mmol m^{-2}. The major point here is that the water column is about three orders of magnitude lower in mass per area than the benthos or the sediments. The sediment pool looks particularly large because it is summed over the depth of 1 m. It is probably better to think of available nutrients over a depth of about 10 cm in such an environment; thus, the values for the sediment pools can be divided by 10. Turnover times can be calculated by dividing the pool size by the rate of exchange.

4 Productivity

Gross primary production varies 100–2,000 mmol C m^{-2} day^{-1}, depending on habitat and light (Table 2). Low-relief, sand communities are the lowest (100–300 mmol C m^{-2} day^{-1}), with reef flats being moderate (350–500 mmol C m^{-2} day^{-1}) and high-relief communities of coral and algae showing maximal values (1,000–2,000 mmol C m^{-2} day^{-1}). These overall metabolic rates are consistent between reefs, suggesting they are independent of species composition. These basic communities can be identified easily from air photography and satellite imagery, making ecosystem-wide

Table 1 Mass of phosphorus (P), fixed nitrogen (N), and organic carbon (C), in mmol m^{-2} in 1 m of the water column above the benthos, in the living benthos (autotrophic and heterotrophic) and in the topmost 1 m of sediments (solid and dissolved phases). Phosphorus and fixed nitrogen include all inorganic and organic forms

Pool	P	N	C
1-m water column	0.16–0.90[a]	<7[a]	10–250[a]
Living benthos			
Autotrophic	50[b]	1,350[b]	22,400[b]
Heterotrophic	50[c]	~2,000[i]	~5,000[d]
1-m sediments			
Solid phase	1,400[h] (10,000)[e,f,g,h]	34,000[i]	300,000[h]
Dissolved phase	<2[j]	<50[j]	5–125[k]

[a] Values for the water column are based on the data presented in Table 3

[b] Values for benthic autotrophs calculated from dry-weight biomass estimates (Odum and Odum 1955) and assuming a C:N:P ratio of 550:30:1 (Atkinson and Smith 1983)

[c] Benthic heterotroph-phosphorus content calculated using dry biomass estimates (Odum and Odum 1955) and assuming a phosphorus content of 1% by weight (Pomeroy and Kuenzler 1969)

[d] Benthic heterotrophic biomass calculated assuming nearly all of the biomass is composed of CH$_2$O. Organic phosphorus and total phosphorus (in parentheses), and organic carbon calculated based on the data of

[e] Entsch et al. (1983)

[f] Atkinson (1987a)

[g] Szmant and Forrester (1996)

[h] Suzumura et al. (2002), assuming a porosity of 0.5 (Buddemeier and Oberdorfer 1988) and a sediment density of 2.7 g cm^{-3}

[i] Solid phase nitrogen content measured by Szmant and Forrester (1996)

[j] Estimates for sediment dissolved pool sizes assume pore water DIP concentrations <3 μM and DIN concentrations <80 μM, a sediment porosity of 0.5, and that dissolved inorganic pools are much greater than dissolved organic pools

[k] This estimate assumes pore water dissolved organic carbon concentrations are equal to ambient water dissolved organic carbon concentrations (Tribble et al. 1990)

Table 2 Gross primary production (P), community respiration (R), net community production (NCP), and net community calcification (G) in mmol C m^{-2} day^{-1} for various communities as originally tabulated by Kinsey (1985), with additional data from Gattuso et al. (1993, 1996); Kraines et al. (1996, 1997); Boucher et al. (1998); and Andrefouet and Payri (2000)

Habitat	P	R	NCP	G
Reef flat	640	600	−220–	130
	(330–1,580)	(290–1,250)	310	(20–250)
Algal	460	300	0–130	90
pavement	(170–580)	(40–560)		(70–110)
High	1,200	1,300	−830–	240
coverage	(660–1,920)	(500–2,000)	250	(110–320)
Sandy	130	130	−40–30	35
areas	(80–230)	(90–200)		(10–70)
Shallow	450	430	−200–	40
lagoon	(210–1,080)	(180–790)	280	(20–55)
Entire reef	390	370	0–70	45
system	(190–640)	(190–570)		(3–135)

Means are in bold; ranges provided in parentheses

calculations of gross production possible. In a more recent paper (Hochberg and Atkinson 2008), we showed that productivity can be modeled from remote-sensing image data by calculating the quanta absorbed (*not* incident irradiance), and multiplying by a constant value photosynthetic efficiency. The assumption is that on a large scale, the variability in photosynthetic efficiency is averaged and tends toward a constant value. This assumption needs to be tested more thoroughly; it is likely that the constant will vary with depth and season. Nevertheless, using this approach effectively predicts the general observations that high biomass, three-dimensional communities have the highest productivity, and that productivity is higher in summer than in winter. In the near future, I expect more accurate estimates of productivity from this light absorption approach than those derived from identifying habitat and applying an average metabolic value based on standard metabolism (Andrefouet and Payri 2000).

Community respiration varies over the same range as gross production, and the two are positively correlated (Table 2). Communities or bottom types with high production tend to exhibit high respiration, indicating that much of the respiration of organic material occurs within the habitat, or more probably, within the organism that produces the carbon Falter et al. (2001, 2009). In plant-dominated communities, dark respiration is highly correlated to the previous day's gross productivity, indicating a large labile pool of carbon and increased light respiration.

There are, however, patterns of net production and consumption on reefs. Reef crests of hard substratum with high water velocity and dominated by coralline algae and turfs have net autotrophy, whereas back-reef areas tend to be dominated by heterotrophic feeders. Nevertheless, there are a few studies identifying or combining net carbon production or consumption with changes in nutrient flux (Miyajima et al. 2007). Needless to say, these patterns have proven challenging to measure in the field.

There have been enough nutrient uptake rates calculated to make some generalizations. The first is that uptake rates for phosphate vary from 0.1 to 1 mmol m^{-2} day^{-1}, while those of ammonia and nitrate vary from 0.1 to 10 mmol m^{-2} day^{-1} (Table 3). These rates of uptake are actually very slow compared to the advective flux of nutrients and the nutrients required for gross primary productivity. These rates are slow because of mass transfer limitation, which I explain later in this chapter. Nutrient flux can be estimated by multiplying the nutrient concentration in the water by a rate constant, here denoted by the symbol S. The rate constant, S, is between about 1 and 15 m day^{-1}. We might measure rate constants near 20, but this will be a very high biomass community under strong currents and wave motion (Table 3).

Table 3 Uptake (**A**) and release (**B**) rates of dissolved inorganic phosphorus (DIP), ammonium, and nitrate in mmol m^{-2} day^{-1}; and (**C**) uptake rate coefficients (S, where rate = S × concentration) for each nutrient in m day^{-1}. Values are single values, ranges, or means with errors, depending on the source of data. See details of publications for determination of errors. Publications reporting significant uptake or release, but no rates are not listed in this table; citations to those publications are in cited literature

(A) Uptake			
DIP	NH$_4^+$	NO$_3^-$	Reference
0.9 ±0.2	–	–	Pilson and Betzer (1973)
0.2–1.1	–	–	Atkinson (1987b)
1.1 ±0.2	–	–	Bilger and Atkinson (1992)
1.2 ±0.6	7.8 ±4.5	–	Steven and Atkinson (2003)
0–0.8	0.1–4.8	0.2–5	Atkinson et al. (2001)
0.7–1.0	0.8–2.3	0.8–2.3	Falter (2002)
(B) Release			
0.9	4.3	–	Steven and Atkinson (2003)
0.5 ±0.3	2.1 ±1.4	ns	Atkinson et al. (2001)
0.25	2	4	Tribble et al. (1990)[a]
(C) S			
4.5	–	–	Pilson and Betzer (1973)[a]
9	–	–	Atkinson (1987)
1.2–15	–	–	Atkinson and Bilger (1992)
0.4–5.2	2.2–12.7	–	Bilger and Atkinson (1995)
–	1.4–12.4	–	Thomas and Atkinson (1997)
2.9–4.1	6.1–12.7	–	Larned and Atkinson (1997)
5.8 ±3.4	11.1 ±6.4	–	Steven and Atkinson (2003)
7.4 ±2.0	7.5 ±2.1	4.4 ±0.4	Atkinson et al. (2001)
9.3 ±1.3	15.5 ±2.1	15.5 ±2.1	Falter (2002)

[a] Represents data that has been reinterpreted

5 Mass Transfer

There is now strong evidence that mass transfer theory can be applied to the understanding of biogeochemistry of coral reefs. In this section, I explain how this approach can be used to answer some of the fundamental questions and apparent paradoxes posed in the reviews discussed above.

Mass transfer describes the mass exchange of a dissolved compound between a fluid and a reactive surface, and thus it is useful in describing the exchange of biogeochemical compounds between water and coral reef communities. D'Elia and Wiebe presented the use of Michaelis–Menton kinetics (Monad) to describe the relationship between nutrient concentration and a metabolic response of the benthos. Michaelis–Menton kinetics were originally developed to explain enzyme kinetics in homogeneous solutions and were later applied to plankton and benthos feeding to fit data to a hyperbolic function. The Michaelis–Menton hyperbolic function usually has a region of low nutrient concentration where the metabolic rate is directly proportional to concentration. It is in this region that the rate-limiting step for nutrient exchange is diffusion through a diffusive boundary layer. In the case of nutrient uptake into macroalgae or a hard bottom covered with a variety of autotrophs, the demand for nutrients is so great at the surface that nutrient uptake rates do not become maximal, or "saturated", until nutrient concentrations of tens or hundreds of mmol m^{-3}. In the region of low concentration, a rate can become mass transfer limited and show a dependence on water velocity. The water velocity dependence arises from the fact that the thickness of the diffusive boundary layer is related to water velocity. The value of a first-order rate constant for uptake (Table 3C) is proportional to the friction of the bottom, the water velocity, and the diffusion constant of the compound (Falter et al. 2004).

This approach gives us some means to calculate a maximum rate of uptake or release of a dissolved compound between benthos and water. So far, we have no indication that any measurements of nutrient uptake or release in the field are greater than the reported or calculated mass transfer rates. Mass transfer rates are affected by waves and turbulence, although there has been some confusion in the literature. Waves enhance mass transfer especially at lower frequencies in more branched or porous beds. The reversing acceleration of the wave penetrates into the branched structure. Enhancements over the equation presented above are about 30% (Lowe et al. 2007). While it is widely stated that mass transfer is related to turbulence in the water, this is not true (Falter et al. 2007). Large-scale turbulence in water generally cascades to smaller scale turbulence until the energy is "dissipated"; the idea being that this turbulence controls the thickness of the boundary layer. In fact, external turbulence has only a small effect, which has been quantified for coral reef surfaces and is proportional to the turbulent intensity to the 0.1 root. This statement translates into only a 5–10% effect on the first-order rate constants (Table 3C) in water that is highly turbulent with 25% turbulent intensity (Falter et al. 2007).

We have parameterized the rate kinetics as a turbulent boundary, where the top or outer transition region of the boundary layer is exposed to turbulence. This turbulence is generated locally by the friction of the water flow against a stationary surface. Thus, these are inherently turbulent boundary layers that are not greatly affected by what we termed "external turbulence" or turbulence introduced into the mean current. So the most important parameters for mass exchange rates are nutrient concentrations, water velocity, and friction, not turbulence. This approach confines the uptake and release of dissolved nutrients into fairly narrow ranges and gives us insight into the limits of coral reef function. The most recent paper that summarizes these equations is Falter et al. 2004. So, what does mass transfer really tell us about reef biogeochemical cycles?

6 Implications of Mass Transfer and Questions Revisited

Mass transfer limits on biogeochemical rates dictate that the exchange of a dissolved compound will be relatively slow compared to the advection of water past the surface. If the nutrient uptake rate = St U_b (C), where St is a dimensionless constant termed the Stanton number, U_b is the velocity of the bulk water above the reef and C is the concentration of the nutrient. St is the ratio of uptake into the surface divided by the advection past the surface (St = uptake rate to surface/U_b C). St is of order 10^{-3}–10^{-4}, meaning the uptake velocity is slow – only 10 m day^{-1} compared to the velocity of flowing water, which is usually 10^4 m day^{-1}. So the relatively constant concentrations of nutrients across reef transects of 10–100 m exist because uptake or release is barely detectable. One must expand water sampling to several 100 m and at least 2-h water residence time to detect natural changes in concentrations. This is not rapid uptake and release, so readily assumed decades ago, nor rapid recycling through the water column.

The slow uptake of nutrients, controlled by mass transfer, relative to the fixation of carbon gives high C/N/P ratios of most biomass. Thus, most plants will show nutrient-limited responses of photosynthesis and growth. Given mass transfer limitation and the low N:P ratio of dissolved compounds in seawater, plants should be, in general, N-limited for photosynthesis and growth, which is consistent with nutrient limitation growth experiments on reef macrophytes. Mass transfer also predicts that nutrient availability will be highest where there is high nutrient concentration, high water velocities, and high drag. Thus, the highest nutrient availability will be at the shallow fore-reefs, and if nutrient delivery controls net community production, then one would predict reef communities would grow best at the margins. We would also expect to find plants with highest nutrient content in higher water velocities. There is recent evidence that fish reduce energy by swimming in large turbulent flows, so we might even predict higher trophic transfer and efficiency in the fore-reef, reef crest regions. Mass transfer also predicts a relatively low net primary production and sets limits for ecosystem performance. We think this is about 150–200 mmol C m^{-2} day^{-1} (Falter et al. 2009), based on detailed studies linking carbon to nutrient cycles in Kaneohe Bay, Hawaii. Mass transfer also tells us that there is usually not enough N as ammonia and nitrate to support net primary production, so the reef must accommodate with either N-fixation or particle removal and N-remineralization. Mass transfer tells us that, in most cases, nutrient input is relatively small, but this is probably highly variable, both seasonally and across oceanic regions. A complete model has not yet been developed.

Most importantly, the mass transfer approach tells us that the way we have been studying biogeochemistry of reefs lacks control of key parameters. If a biogeochemical rate is a function of nutrient concentration, water velocity, and friction, then we must design facilities to simulate these properties. It turns out this is difficult to achieve. Just spinning water in a small beaker and producing water velocity does not mean that water is forced across the surfaces of the test organisms. Using the mass transfer approach mandates that experimental facilities be created in which the dissipation of energy by friction on the bottom and the water velocity are appropriately simulated. Mass transfer sets limits to biogeochemical rates; a few insights can be made. The following are examples: mass transfer tells us that to get a significant change in the benthic nutrient pools, a large change in nutrient concentration of the water is required for days to weeks to get sufficient nutrients into the benthos. The lack of response of the coral microatolls to nutrient loading in the ENCORE experiment is one example (Koop et al. 2001). Mass transfer also tells us that we do not need to look for novel sources of nutrients to explain high primary productivity. Mass transfer tells us that some organic compounds may be an important source of nutrients, but there are few measurements of the kinetics of organic matter in the literature, and the most recent ones show uptake is not near mass transfer (Yahel et al. 2003). Mass transfer tells us that we must first understand both nitrogen fixation and particulate and dissolved organic nitrogen, to further understand how the reef provides for sufficient nitrogen. Mass transfer does not explain how particles and organisms interact; however, it can be used to compare potential mass transfer rates with particle removal and recycling of C:N:P from particles.

7 Summary: A Mass Transfer Interpretation of Coral Reef Biogeochemistry

In summary, I would like to present my thoughts on how these systems function and answer some of the questions raised in previous reviews, particularly answering the four questions raised by Crossland (1983). Reefs biogeochemistry is controlled by a few simple parameters that continually adjust to the changing conditions of the ocean. Ocean conditions, such as nutrient and particle concentrations and waves, strongly influence the relative rates of biogeochemical processes. Phosphate is taken up at mass transfer limits and controls the net primary production of the system through the stoichiometry of the organic matter being produced. This C:N:P ratio is highly variable but much higher than the Redfield ratio: 400–800: 20–40:1. Both ammonia and nitrate are also taken up at near mass transfer rates; however, there is not sufficient nitrogen to satisfy the N:P ratio of uptake. Nitrogen fixers and combinations of filter-feeding organisms near fore-reefs and reef flats produce nitrogen, exporting to the water column and even near surface sediments. This nitrogen is available, but usually concentrations of nitrogen increase across reef flats and lagoons, simply because the system has a tendency to increase the N:P portion of the water column. We witness this on most high residence time reef environments. Because mass transfer kinetics are relatively slow compared to advection, concentrations of ammonia, nitrate, and organic nitrogen increase. Particles are the most variable input; in some reefs, particle input is negligible, while in others, it is significant. Productivity is strongly a function of light, as discussed above, so we can get varying relationships between carbon fixation and nutrient input, remineralization and recycling. Nitrification is probably only 10% of nitrogen uptake, and denitrification varies, depending on hydrodynamics in the sediments. Rates are strongly a function of hydrodynamics, so well-flushed systems with higher nutrients provide communities with a large potential to have high net primary production. Much of the carbon produced each day is consumed during the course of that day, mostly through autotrophs. Thus, the actual net growth of the community is controlled by the mass transfer of phosphate. In an algal-dominated system, which is not heavily grazed, the uptake of phosphate can be measured at the mass transfer limit. In communities with high topography and lots of grazers, phosphate is remineralized and net changes in P across a reef are less than mass transfer rates. These generalization require further measurement in the field and testing. Thus, specific answers to Crossland's questions are : (1) high C:N:P ratios of organic matter and mass transfer limitation of nutrient uptake dictates the relationship between high primary productivity and low-nutrient water; (2) yes, there is extremely seasonality in primary productivity and net primary productivity based on changes in light and nutrient uptake; (3) No, tight recycling is less important than continual nutrient input; (4) No, we still need more knowledge on how organic forms of nutrients contribute to reef function.

For reef flats with nutrient subsidies, such as groundwater, the elevated N from nitrate can actually match in the right N:P ratio for macroalgae and the mass transfer limited rate for net community production will generally tend toward production of algae. These algae will increase in biomass, and in terms of respiration rates, can increase until reef flats become anoxic, leading to further eutrophy. We can now quantify this effect for any given reef. Nutrients in the ranges generally observed in the environment do not kill coral directly. However, increasing oxygen demand in the elevated biomass, and the resulting lack of oxygen, certainly can kill coral quickly.

High gross primary production in low-nutrient water arises from the high C:N:P ratio of organic matter. Rapid growth occurs where there are higher currents, in which nutrient boundary layers are thinned by the action of the waves, and the transport of nutrients to and from surfaces is enhanced. These regions have higher net growth, and also with waves, aids in the export of detritus.

If we are going to increase our understanding of the fundamental dynamics of coral reefs, we need to build new facilities, and normalize rates to area. Only in this way will communities under different hydrodynamic conditions, roughness, and morphologies be studied and compared.

References

Andrefouet S, Payri C (2000) Scaling-up carbon and carbonate metabolism of coral reefs using in-situ data and remote sensing. Coral Reefs 19:259–269

Atkinson MJ (1987a) Low phosphorus sediments in a hypersaline marine bay. Estuar Coast Shelf Sci 24:335–348

Atkinson MJ (1987b) Rates of phosphate uptake by coral reef flat communities. Limnol Oceanogr 32:426–435

Atkinson MJ, Bilger RW (1992) Effect of water velocity on phosphate uptake in coral reef-flat communities. Limnol Oceanogr 37: 273–279

Atkinson MJ, Falter JL (2003) Coral Reefs. In: Black K, Shimmield G (eds) Biogeochemistry of marine systems. CRC Press, Boca Raton pp 40–64

Atkinson MJ, Smith SV (1983) C:N:P ratios of benthic marine plants. Limnol Oceanogr 28:568–574

Atkinson MJ, Smith DF (1987) Slow uptake of ^{32}P over a barrier reef flat. Limnol Oceanogr 32:436–441

Atkinson MJ, Falter JL, Hearn CJ (2001) Nutrient dynamics in the biosphere 2 coral reef mesocosm: water velocity controls NH_4 and PO_4 uptake. Coral Reefs 20:341–346

Bilger RW, Atkinson MJ (1992) Anomalous mass transfer of phosphate on coral reef flats. Limnol Oceanogr 37:261–272

Bilger RW, Atkinson MJ (1995) Effects of nutrient loading on mass-transfer rates to a coralreef community. Limnol Oceanogr 40:279–289

Boucher GJ, Clavier CH, Gattuso JP (1998) Contributions of soft-bottoms to the community metabolism (primary production and calcification) of a barrier reef flat (Moorea, French Polynesia). J Exp Mar Biol Ecol 225:269–283

Buddemeier RW, Oberdorfer JA (1988) Hydrogeology and hydrodynamics of coral reef pore waters. Proc Sixth Int Coral Reef Symp 2:485–490

Crossland CJ (1983) Dissolved nutrients in coral reef waters. In: Barnes DJ (ed) Perspectives in coral reefs. Australian institute of marine science, Brian Couston Publishers, Manuka, pp 56–68

D'Elia C, Wiebe W (1990) Biogeochemical nutrient cycles in coral reef ecosystems. In: Dubinsky Z (ed) Coral reefs: ecosystems of the world series, vol 25. Elsevier Science, Amsterdam, pp 49–74

Entsch B, Boto KG, Sim RG, Wellington JT (1983) Phosphorus and nitrogen in coral reef sediments. Limnol Oceanogr 28:465–476

Falter JL (2002) Mass transfer limits to nutrient uptake by shallow coral reef communities. Ph.D. dissertation, University of Hawaii, Hondulu, 126 pp

Falter JL, Atkinson MJ, Langdon C (2001) Production-respiration relationships at different time- scales within the Biosphere 2 coral reef biome. Limnol Oceanogr 46:1653–1660

Falter JL, Atkinson MJ, Merrifield MA (2004) Mass-transfer limitation of nutrient uptake by a wave-dominated reef flat community. Limnol Oceanogr 49:1820–1831

Falter JL, Atkinson MJ, Lowe RJ, Monismith SG, Koseff JR (2007) Effects of non-local turbulence on the mass transfer of dissolved species to model coral reefs. Limnol Oceanogr 52:274–285

Falter, Atkinson, Schar DW, Lowe, Monismith (2010) Short-term coherency between gross primary production and community respiration in an algal-dominated reef flat. Coral Reefs. doi 10.1007/s0038-010-0671-9

Gattuso JP, Pinchon M, Delasalle B, Frankignoulle M (1993) Community metabolism and air- sea CO_2 fluxes in a coral reef ecosystem (Moorea, French Polynesia). Mar Ecol Prog Ser 96:259–267

Gattuso JP, Pinchon M, Delesalle B, Canon C, Frankignoulle M (1996) Carbon fluxes in coral reefs. I. Lagrangian measurement of community metabolism and resulting air-sea CO_2 disequilibrium. Mar Ecol Prog Ser 145:109–121

Hochberg EJ, Atkinson MJ (2008) Coral reef benthic productivity based on optical absorptance and light-use efficiency. Coral Reefs 27:49–59

Kinsey DW (1985) Metabolism, calcification, and carbon production: I. Systems level studies. Proc 5th Int Coral Reef Congr 4:505–526

Koop K, Booth D, Broadbents A, Brodie J et al (2001) ENCORE: the effect of nutrient enrichment on coral reefs. Synthesis of results and conclusions. Mar Pollut Bull 42:91–120

Kraines S, Suzuki Y, Yamada K, Komiyama H (1996) Separating biological and physical changes in dissolved oxygen concentration in a coral reef. Limnol Oceanogr 41:1790–1799

Kraines S, Suzuki Y, Omori T, Shitashima K, Kanahara S, Komiyama H (1997) Carbonate dynamics of the coral reef system at Bora Bay, Miyako Island. Mar Ecol Prog Ser 156:1–16

Larned ST, Atkinson MJ (1997) Effects of water velocity on NH_4 and PO_4 uptake and nutrient-limited growth in the macroalgae *Dictyosphaeria cavernosa*. Mar Ecol Prog Ser 157:295–302

Lowe RJ, Falter JL, Koseff JR, Monismith SG, Atkinson MJ (2007) Spectral wave flow attenuation within submerged canopies: implications for wave energy dissipation. J Geophys Res 112:C05018. doi:10.1029/2006JC003605

Miyajima T, Tanaka Y, Koike I, Yamano H, Kayanne H (2007) Evaluation of spatial correlation between nutrient exchange rates and benthic biota in a reef-flat ecosystem by GIS-assisted flow-tracking. J Oceanogr 63(4):643–659

Odum HT, Odum EP (1955) Trophic structure and productivity of a windward coral reef community on Eniwetok Atoll. Ecol Monogr 25:1415–1444

Pilson MEQ, Betzer SB (1973) Phosphorus flux across a coral reef. Ecology 54:1459–1466

Pomeroy LR, Kuenzler EJ (1969) Phosphorus turnover by coral reef animals. Symp Radio-ecol, AEC TID 4500(3):474–482

Steven ADL, Atkinson MJ (2003) Nutrient uptake by coral-reef micro-atolls. Coral Reefs 22:197–204

Suzumura M, Miyajima T, Hata H, Umezuwa Y, Kayanne H, Koike I (2002) Cycling of phosphorus maintains the production of microphytobenthic communities in carbonate sediments of a coral reef. Limnol Oceanogr 47:771–781

Szmant A (2002) Nutrient Enrichment on coral reefs: is it a major cause of coral reef decline? Estuaries 25(4):743–766

Szmant A, Forrester A (1996) Water column and sediment nitrogen and phosphorus distribution patterns in the Florida Keys, USA. Coral Reefs 15:21–41

Thomas FIM, Atkinson MJ (1997) Ammonium uptake by coral reefs: effects of water velocity and surface roughness on mass transfer. Limnol Oceanogr 42:81–88

Tribble GW, Sansone FJ, Smith SV (1990) Stoichiometric modeling of carbon diagenesis within a coral reef framework. Geochim Cosmochim Acta 54:2439–2449

Yahel G, Sharp JH, Marie D, Hase C, Genin A (2003) In-situ feeding and element removal in the symbiotic-bearing sponge *Theonella swinhoei*: Bulk DOC is the major source for carbon. Limnol Oceanogr 48(1):141–149

The Role of Dissolved Organic Nitrogen (DON) in Coral Biology and Reef Ecology

Yoshimi Suzuki and Beatriz E. Casareto

Abstract The recent studies on the behavior, role, and characteristics of DON (dissolved organic nitrogen), including nitrogen fixation and organic nitrogen compounds in corals and coral reefs were reviewed. DON is one of the important components as well as DOC for understanding biogeochemical cycling and ecosystem function in coral reefs. Recent results and views have been integrated into a new concept for the nitrogen cycle in coral reefs.

Keywords DON • nitrogen organic compounds • chemical symbiosis • internal cycling • nitrogen fixation

1 Significance and Newly Raised Questions Regarding DON

Along with carbon and other nutrients, nitrogen is an essential element in coral reef ecosystems, and is present at very low concentrations in the waters surrounding coral reefs. Therefore, nitrogen is one of the key limiting elements in these oligotrophic areas. However, both primary production and biodiversity are high in coral reefs. This constitutes a "paradox," particularly with respect to the imbalance of consumption and production between inorganic and organic nitrogen (Suzuki et al. 2000), because in coral reefs a large part of the organic matter is produced in dissolved forms. This means that dissolved organic nitrogen (DON) is an important component for determining coral reef metabolism and biogeochemical cycles. Shiroma et al. (2008) tried to measure nitrate and ammonium concentrations in the gastrovascular cavity, and found about 100 times higher concentrations of these nutrients compared with those in seawater. This suggests that internal recycling of organic and inorganic nitrogen compounds is possibly an important source stemming from the degradation of organic nitrogen.

Measurements of DON using a high-temperature catalytic oxidation and persulfate oxidation (Sharp 1983) method or ultraviolet oxidation (Armstrong and Tibbitts 1968) were confirmed in a recent publication (Bronk et al. 2000). To calculate DON concentrations, one must first obtain the total dissolved nitrogen (TDN) concentration, and the inorganic nitrogen (ammonium + nitrate + nitrite) concentrations, which are subtracted from TDN. The residual is defined as DON.

However, compared to the open ocean, few studies have been conducted on DON in coral reefs. In open ocean water, DON concentrations range from 0.8 to 13 µM N (mean of 5.8 ± 2.0 µM N in surface water, decreasing with depth to a mean of 3.9 ± 1.8 µM N (Abell et al. 2000). Nevertheless, studies of the chemical composition of the DON pool, such as urea, amino acids, humic acid, and fulvic substances, have been conducted (Capone 2000; Metzler et al. 2000; Billen 1991; Keil and Kirchman 1999; Harvey and Boran 1985). Various sources of DON in the open ocean have been reported, including phytoplankton (Antia et al. 1991; Bronk and Ward 1999), N_2 fixers (Karl et al. 1992; Hansell and Feely 2000), bacteria (Jorgensen et al. 1999), and micro- and macro-zooplankton (Miller and Glibert 1998).

There are many unresolved questions about DON in coral reefs, including its sources, distribution, seasonal changes, chemical characteristics, degradation, ecological role, turnover time, etc. What are the fundamental problems? (1) Does the presence of DON in coral reef waters force us to reevaluate whether coral reefs are truly oligotrophic? (2) How much do internal sources of nitrogen contribute to the nitrogen cycle, compared with external sources? (3) Do DOC (dissolved organic carbon) and DON behave differently in chemical and biological processes in coral reef water? (4) How much do the characteristics of DON at the molecular level identify with biological processes such as photosynthesis, degradation, and excretion? (5) How do we understand the nitrogen cycle including DON in coral reefs? This chapter summarizes recent results regarding DON and other nutrients in coral reefs.

Y. Suzuki (✉)
Graduate School of Science and Technology, Shizuoka University, 836 Oya Suruga shizuoka 422-8529, Japan
e-mail: seysuzu@ipc.shizuoka.ac.jp

B.E. Casareto
Graduate School of Science and Technology, Shizuoka University, 836 Oya Suruga shizuoka 422-8529, Japan

2 Concentration and Distribution of DON, DIN, and PON in Coral Reef Water

In four sites within Kaneohe Bay, Oahu, Hawaii, DON and DIN (dissolved inorganic nitrogen) concentrations were measured. Mean DON concentrations in the water column at the Moku o Loe reef slope were 6.73 μM N, ~10 times higher than DIN concentrations (Larned 1998). As to the temporal and special scaling of planktonic responses, in Kaneohe Bay, Oahu, Hawaii, DON concentrations ranged from 5.7 to 8.3 μM N, DIN concentrations were 0.2 to 0.5 μM N, and PON (particulate organic nitrogen) were 5.1 to 46 mg m^{-3} (Cox et al. 2006). Both sets of results from these same sites confirm that DON concentrations are greater by a factor of 10 than DIN concentrations. Glibert and O'Neil (1999) had reported DON release by *Trichodesmium* spp. They measured DON concentrations from 3.5 to 15 μM N in Bora Bay, Miyako, Okinawa Japan. Import and export fluxes of HMW-DON (high molecular weight: greater than 1,000 daltons) and LMW-DON (low molecular weight: less than 1,000 Daltons) on a coral reef were studied (Suzuki et al. 2000) (Table 1).

DON concentrations ranged from 5.5 to 11.2 μM N in the water column, and were slightly higher at low tide. These authors reported that in reef water 30–40% of DON is in the low-molecular-weight fraction (less than 1,000 Daltons), and HMW-DON is being exported to the ocean. Miyajima et al. (2005) studied the distribution and partitioning of nitrogen and phosphorus in a fringing lagoon of Ishigaki Island, Okinawa, Japan. They found DON concentrations ranging from 4 to 8 μM N for offshore water and from 3.8 to 7.8 μM N for stagnant lagoon water, with no significant difference in DON concentrations between night and day in the lagoon water. They reported that DIN concentrations were relatively constant at 0.2 μM N for offshore water and ranged from 0.7 to 1.7 μM N for lagoon water. They also reported that PON concentrations were 0.4–1.0 μM N for both offshore and lagoon waters. They showed that DON and PON increased when gross primary production (GPP) was high (Miyajima et al. 2007). Fairoz et al. (2008) showed that DON concentrations of 3.3 μM N in the water column in Sesoko, Okinawa, Japan were related to the thermal stress of corals, and observed an increase of DON concentrations to 29.9 μM N at 35 °C. This higher value is due to the release of coral mucus. However, there are very few reports about temporal and special changes of DON, PON, and DIN, and their relationships among biological and physical parameters in coral reef waters.

There are even fewer reports about DON and PON concentrations in the pore waters of coral reefs. Higher concentrations of DON ranging from 24 to 38 μM N in pore waters of sediments in Sesoko, Okinawa, Japan were observed, which are higher than those in the water column (Suzuki unpublished data). In general, DIN, especially ammonium concentrations in pore water are higher than those in the water column around coral reefs (Suzuki et al. 2000), suggesting that sediments are one of the important sources of organic matter. This means that for understanding the fate of DON and PON, the sources and sinks (i.e., production and consumption) of DON and PON need to be studied further.

3 Behavior of DON in Coral Reef Water

Dissolved organic nitrogen (DON) is produced largely through its direct excretion by plants and by corals, which comprise up to 20–30% of their primary production (Sorokin 1995). The particulate organic matter is formed from the disintegrated feces of phytophages, from the crushed thalli of dead macrophytes, as well as from the microbial and algal

Table 1 Concentrations of DOC and DON in different molecular size in coral reef water at Miyako Island, Okinawa (μM) (based on Suzuki et al. 2000)

	Station-O5	Station-M2	Station-L2
Molecular weight	DOC-ocean	DOC-reef	DOC-lagoon
>10,000	18.2	20.1	27.4
10,000–1,000	13.0	11.1	18.8
<1,000	52.1	40.1	42.0
Total	83.3	71.3	88.2
Molecular weight	DON-ocean	DON-reef	DON-lagoon
>10,000	2.31	2.91	2.83
10,000–1,000	1.46	1.11	1.71
<1,000	3.51	2.49	2.49
Total	7.29	6.51	7.03
Molecular weight	C/N-ocean	C/N-reef	C/N-lagoon
>10,000	7.9	6.9	9.7
10,000–1,000	8.9	10.0	11.0
<1,000	14.8	16.1	16.9
Total	11.4	10.9	12.5

biomass washed out from sediments and periphyton by surf (Bronk 2002). This cycling of organic matter and nutrients among the different organisms maintains the coral reef ecosystem (Odum and Heald 1955). Yet little is known about the sources and sinks of DON in coral reefs. We need to know the fate of DON within coral reefs in order to understand the interactions between coral ecosystems and material cycling.

DON is a heterogeneous mixture of biologically labile compounds with likely turnover times on the order of days to weeks as well as persistent refractory components. A large number of compounds has been identified within the DON pool, including urea, dissolved amino acids, dissolved free amino acids, humic acids and fulvic substances, and nucleic acids.

DOC, DON, and nutrients were measured in coral reefs at Miyako Island, Okinawa, Japan for understanding the difference in behavior between DOC and DON (Suzuki et al. 2000). Organic compounds are produced by primary production in biologically active waters and released by metabolism or breakdown of living organisms. According to a degradation experiment, they had divided these compounds into two categories: labile and refractory fractions. Their results showed that fresh and labile organic compounds in coral reefs are greater than those in open ocean. Dissolved organic matter was divided into three different molecular weight fractions using ultrafiltration: below 1,000, 1,000–10,000 Dalton, and above 10,000. The difference of concentrations between day and night was 15–22 µM C for DOC and 3–4 µM N for DON, respectively. High-molecular-weight DOC and DON (above 1,000 Dalton) had increased from 22 µM (DOC in daytime: 96) to 37 µM C (DOC in nighttime: µM C) for DOC and 2.8 µM N (DON in daytime: 6.0) to 6.1 µM N (DON in night time: 9.4) for DON.

These results indicate that high-molecular-weight dissolved organic matter (HMW-DOM) had been produced during nighttime through metabolism by the coral reef while low-molecular-weight dissolved organic matter (LMW-DOM: below 1,000 Dalton) was almost constant over the day. The C/N ratio of HMW-DOM increased from daytime to nighttime to about 4.5, suggesting that the contribution of coral metabolism to the increase of DOM in the water during nighttime is important. Stepanauskas et al. (1999) reported the molecular size of DON in Swedish wetlands. DON is present as HMW with 23% as an average value. It is generally accepted that the behavior of DON is mostly the same as that of DOC, and that therefore it is possible to calculate the DON fluxes and biomass simply by using the C:N ratio. However in coral reefs, this has not been verified.

Little is known about the sources and sinks of DON in coral reefs. Most studies estimate that important sources of DON include mucus, bacteria, nitrogen fixers, and other organisms (Glibert and O'Neil 1999; Suzuki et al. 2000; Grover et al. 2008). Yet it is not known how much DON is produced by each organism group, respectively. Also, few reports exist on the degradation of DOM. According to degradation experiments of mucus and DOM (Fairotz et al. unpublished data), the degradation rate of DON is slower than that of DOC, and therefore the DOC:DON ratio decreases with time. Hence, we need to study further the sources and sinks of DOC and DON, respectively, in coral reef water.

4 Nitrogen Compounds in DON

4.1 Urea

Urea is a low-molecular-weight (LMW) organic compound, which is produced by the decomposition and metabolism of nitrogenous organic compounds. Urea is occasionally treated as inorganic nitrogen as in Capone (2000). Urea concentrations range from 0 to 13 µM in coastal areas, with higher concentrations in more productive areas (Metzler et al. 2000), but no reports exist of its concentrations in coral reef water.

4.2 Dissolved Free Amino Acids (DFAA)

Dissolved free amino acids (DFFA), with concentrations ranging between 1 and 700 nM (Bronk 2002), is an important component of DON, constituting approximately 10% of the DON pool. There are several studies of DFAA uptake by corals (Ferrier 1991; Hoegh-Guldberg and Williamson 1999; Grover et al. 2008). Grover et al. (2008) studied the importance of DFAA as a nitrogen source for the scleractinian coral *Stylophora pistillata*. Experiments were performed using ^{15}N-enriched DFAA, and percentage ^{15}N enrichment was measured both in animal tissue and in the zooxanthellae at different DFFA concentrations. They reported that DFAA uptake was correlated with light, and that DFFA constituted 24% of the organic matter that was taken up by *S. pistillata*. This study was the first to use the ^{15}N technique to measure DFAA uptake rates in scleractinian corals. Its most important finding was that ^{15}N enrichment occurred rapidly in the zooxanthellae, in which ^{15}N was detected within less than 2 h. This report reveals that DFAA can represent an important source of nitrogen for corals at in situ concentrations (200–500 nM), with uptake rates as high as those measured for DIN at the same concentrations. Ambariyanto and Hoegh-Guldberg (1999) had also studied the net uptake of DFAA by the giant clam, *Tridacna maxima* at One Tree Island, near the southern end of the Great Barrier Reef, Australia. But they found that DFAA did not supply significant amounts of energy (only 0.1%) and nitrogen (only 1%) for giant clams.

4.3 Dissolved Combined Amino Acids (DCAA)

Fairoz et al. (2008) studied the behavior of DCAA and DFAA in corals under thermal stress. Experiments were done at Sesoko, Okinawa, Japan, using 10 branches of *Montipora digitata*. They reported that organic matter concentrations increased to 29.9 µM N from 3.3 µM N for DON, and to 0.22 µM P from 0.03 µM P for DOP (dissolved organic phosphate), and that DCAA concentrations also increased from 25 to 414 µg l^{-1}. DCAA consisted of peptides and proteins, suggesting that organic compounds like proteinaceous matter were released from coral at the higher temperature. They also reported that the DON released from coral is mainly high molecular weight with an increase in proline concentration. Amounts and compositions of amino acids are useful indexes for understanding coral stress and symbiosis.

4.4 Humic Acid and Fulvic Substances

Humic substances are the most hydrophobic component of the DON pool, composed of organic acids (500–10,000 MW) and operationally defined according to their retention on hydrophobic resins (Thurman 1985; Aiken 1988). In seawater, humic substances typically make up 10–20% of the DON pool, while hydrophilic fulvic acids can contribute 50% or more (Thurman 1985). In general, humic substances arise from microbial degradation of organic matter produced by marine organisms, but details remain unknown (Hatcher and Spiker 1988). For coral reef water particularly, no reports exist, but it is generally known that humic and fulvic substances are refractory with a long turnover time. Therefore, we need to study this subject further to understand the cycles of nutrients and organic matter in coral reefs.

4.5 Nitrogen Fixation

Nitrogen fixation by *Trichodesmium* spp. has been shown to be a significant source of new nitrogen to tropical and subtropical marine ecosystems in which it occurs (Karl et al. 1992; Capone 2000; Kayane et al. 2005). Recently using flow cytometry (Olson et al. 1993), picophytoplankton can be accurately enumerated and discriminated into two main groups: picoeukaryotic algae and cyanobacteria (*Tricodesmium, Prochlorococcus, Synechococcus*) (e.g. Waterbury et al. 1979; Chisholm et al. 1988; Casareto et al. 2000, 2003). *Trichodesmium* spp. has been studied extensively in the open ocean (e.g. Campbell et al. 1997) and in coral reefs (Charpy 2005). Glibert and O'Neil (1999), in their study of DON release and amino acid oxidase activity by *Trichodesmium* spp., reported that DON concentrations are often in the range of 5–10 µM N, and that *Trichodesmium* spp. may release a significant fraction of newly fixed nitrogen in the form of DON. They studied concentrations of DON and NH_4^+ within a *Trichodesmium* spp. bloom in the Great Barrier Reef, and found that concentrations inside the blooms were 3.8 µM N higher than those outside the blooms. This means that nitrogen fixers are important producers of DON in coral reef water.

For cyanobacteria, there are several reports that biological nitrogen fixation appears to make a major contribution to N supply in coral reef ecosystems, as has been shown for the Eniwetok Atoll (Webb et al. 1975), the Great Barrier Reef (Larkum et al. 1988), and for the lagoons of Tikehau Atoll (Charpy-Roubaud et al. 1990) and New Caledonia (Charpy et al. 2007). Many studies on N_2 fixation have dealt with shallow coral reefs (Larkum et al. 1988; Shashar et al. 1994; Charpy-Roubaud and Larkum 2001); however, they did not identify N_2 fixers in coral rubble. Casareto et al. (2008) found that coral reefs are sites of high nitrogen fixation activity where marine cyanobacteria are the major contributors. In coral reefs, cyanobacteria can be found in very diverse environments: in the water column, on sandy bottoms, on coral rubble as endolithic and epilithic forms, or forming microbial mats. The purpose of the present review is to evaluate N_2 fixation rates in different environments (sandy bottom, coral rubble, and cyanobacteria mats) and their contribution to net primary production. Two different fringing coral reef sites were studied by Casareto et al. (2008): Sesoko at Okinawa, Japan, and La Reunion in the Indian Ocean. N_2 fixation and primary production rates were measured using ^{13}C and ^{15}N fixation techniques. Daily N_2 fixation in the sandy bottom was about 3 nM N (µg Chl-a)$^{-1}$, with nearly 6% contribution to the primary production. In coral rubble it ranged from 0.5 to 7 nM N (µg Chl-a)$^{-1}$ with 2.4 to 28% contribution to primary production, while in cyanobacteria mats it varied between 0.2 and 242 nM N (µg Chl-a)$^{-1}$ with contributions of 2.8 to 42% of primary production. Differences in N_2 fixation rates between daytime and nighttime indicated the presence of both heterocystous and non-heterocystous cyanobacteria.

Casareto et al. (2008) also evaluated the contribution of nitrogen fixation to primary production, and for coral rubble from La Reunion this contribution varied from 24% to 28% for endolithic algae and from 2.4% to 17% for epilithic plus endolithic algae. At Sesoko, the contribution of N_2 fixation to primary production of coral rubbles was lower and varied

Table 2 N$_2$ fixation rates during night and 24 h of coral rubble, cyanobacteria mats, and sandy bottom in comparable units of mg N m^{-2} time^{-1} (Casareto et al. 2005)

Type of sediment	Location	N$_2$ fixation 12 h(dark) mg N m^{-2} time^{-1}	N$_2$ fixation 24 h
Coral gravel	Sesoko	1.45 ± 0.84	2.37 ± 1.93
	Le Reunion	0.57 ± 0.37	2.07 ± 1.2
Cyanobacteria mat	Sesoko	64.14 ± 3.05	94.81 ± 7.42
	Le Reunion	27.12 ± 7.32	96.98 ± 2.28
Sand	Sesoko	0.20 ± 0.17	3.08 ± 1.79

between 3.5% and 17% for endolithic algae and 4% and 10% in the case of epilithic plus endolithic algae. In the case of cyanobacteria mats, the contribution of N$_2$ fixation to the total primary production varied from 2.8% to 42% at La Reunion and from 12% to 28.5% at Sesoko. On the sandy bottom, the contribution of new production to the total primary production was 5.7% (Table 2).

Shashar et al. (1994) studied nitrogen fixation in stony corals, especially to seek evidence for coral–bacteria interactions. They showed that nitrogen-fixing bacteria found in the skeleton of corals benefit from organic carbon excreted by the coral tissue, and that those interactions between the nitrogen-fixing organisms and the coral may be of major importance for the nitrogen budget of the corals. Charpy et al. (2007) studied the benthic nitrogen fixation using the acetylene reduction technique, and estimated that nitrogen fixation by benthic cyanobacteria was 16.4 ± 5.4 mg N$_2$ m^{-2} d^{-1} at 21 m depth, representing 19% of the nitrogen requirement for benthic primary production. Charpy et al. (2010) reported that dinitrogen-fixing organisms in cyanobacterial mats were studied in two shallow coral reef ecosystems: La Reunion Island, southwestern Indian Ocean, and Sesoko (Okinawa) Island, and western north Pacific Ocean. Rapidly expanding benthic miniblooms, frequently dominated by a single cyanobacterial taxon, were identified by microscopy and molecular tools. In addition, nitrogenase activity by these blooms was measured in situ. Dinitrogen fixation and its contribution to mat primary production were calculated using ^{15}N$_2$ and ^{13}C methods. Dinitrogen-fixing cyanobacteria from mats in La Reunion and Sesoko showed few differences in taxonomic composition. The 24-h nitrogenase activity, as measured by acetylene reduction, varied between 11 and 324 nmoles C$_2$H$_2$ reduced μg^{-1} Chl-a. The highest values were achieved by the heterocystous *Anabaena* sp., performed mostly during the day. Highest values for nonheterocystous cyanobacteria were achieved by *Hydrocoleum coccineum* mostly during the night. Daily nitrogen fixation varied from nine (*Leptolyngbya* sp.) to 238 nM N$_2$μg^{-1} Chl day^{-1} (*H. coccineum*). Primary production rates ranged from 1,321 (*Symploca hydnoides*) to 9,933 nM C μg^{-1} Chl day^{-1} (*H. coccineum*). Dinitrogen fixation satisfied between 5% and 21% of the nitrogen required for primary production. Figure 1

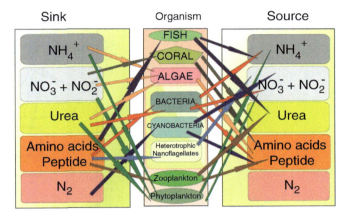

Fig. 1 Relationships among nitrogen compounds (ammonium, nitrate + nitrite, urea, amino acids + peptide, and N$_2$) and different biology in coral reefs

summarizes the correlation among the different nitrogen compounds and different coral reef organisms.

5 Role of DON in the Coral Reef Ecosystem

We need to fundamentally reconsider the meaning of "oligotrophic area" with respect to coral reefs. Although this classification was based on their low nitrate concentrations, the turnover times and recycling of nutrients are very rapid within coral reefs, which means that degradation rate of organic matters in dissolved forms and particulate forms is very rapid (Suzuki et al. 2000). Because this was not known until recently, the role of DON in the nitrogen cycle of coral reefs has not yet been considered correctly. When we consider the fate of the DON, which is present at higher concentrations than nitrate or ammonium, its relatively slow degradation compared with that of DOC and the high production of mucus by coral reef organisms, the contribution of DON to the nitrogen cycle within coral ecosystems is greater than previously thought. It was previously thought that the bulk of nutrients were supplied from outside the reef by some combination of water exchange with the open ocean (Charpy-Roubaud et al. 1999), nitrogen fixation (Charpy 2005), atmospheric input, and riverine input (Johannes et al. 1983).

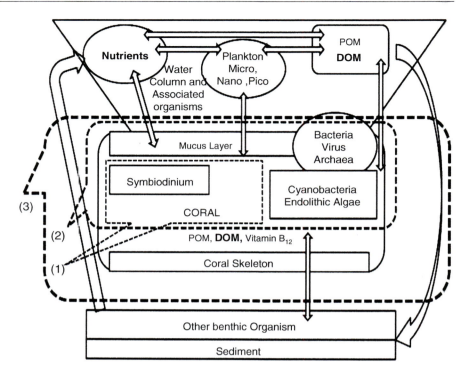

Fig. 2 Proposed conceptual illustration for the chemical symbiosis with connection to detection of environment stress and physiological change in coral biological system from a chemical prospective. Highlighted words reflect results from this paper. The dashed cages represent concepts of (a) biological symbiosis, (b) holobiont and (c) chemical symbiosis

Hence, all sources of nutrients were considered as external inputs to coral reef.

However, the recent results concerning DON in coral reefs lead us to new concepts. We propose to include more chemical perspectives on the coral biological system and to investigate the dynamics of chemical cycling by introducing the concept of chemical symbiosis within the coral holobiont. The schematic diagram in Fig. 2 explains the basic concept of chemical symbiosis.

Coral symbiosis is well explained in coral biology and widely used in literature to show the coral animal and zooxanthellae relationship in hermatypic corals (Muscatine et al. 1985) and in microbial perspective coral systems are also explained as coral holobionts (Rohwer et al. 2002). However, there are similarities and some differences from our proposed chemical symbiosis concept, which is based on a different perspective. Our chemical symbiosis describes the exchange of chemical substances such as organic material, nutrients, and vitamin B_{12} (Agostini et al. 2008; Agostini et al. 2009), which are essential for the living coral biological system and for the production of the nonliving organic matrix (coral skeleton). This concept of chemical symbiosis also attempts to describe the physiological response to environmental stress, from the viewpoint of exchange of chemicals, by which chemical signals can serve as physiological indicators either in the environment or within the coral organism. This is needed to generate new insights of coral reef biogeochemical cycles through the dynamic cycling of DON. Recycling of

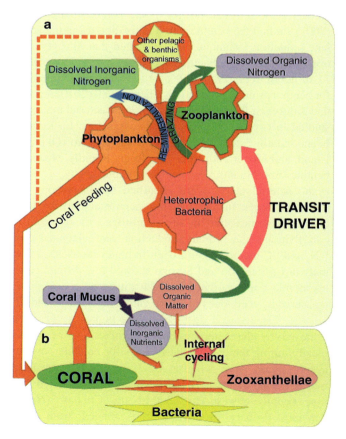

Fig. 3 Nitrogen cycle in coral reef: DON is an important internal source of inorganic nitrogen

DON may determine internal cycling of nitrogen in coral reefs, suggesting that we must reconsider the definition of oligotrophic area for coral reefs (Fig. 3).

References

Abell J, Emerson S, Renaud P (2000) Distributions of TOP, TON and TOC in thye North Pacific subtropical gyer: Implications for nutrient supply in the surface ocean and remineralization in the upper thermocline. J Mar Res 58:203–222

Agostini S, Suzuki Y, Casareto BE, Nakano Y, Fairoz MFM, Shiroma K, Irikawa A, Daigo K (2008) New approach to study the coral symbiotic complex: Application to vitaminB$_{12}$ 917921

Agostini S, Suzuki Y, Casareto BE, Nakano Y, Hidaka M, Badrun N (2009) Coral symbiotic complex: hypothesis through vitamin B$_{12}$ for a new evaluation. Galaxea J Coral Reef Studies 11:1–11

Aiken GR (1988) A critical evaluation of the use of macroporus resins for the isolation of aquatic humic substances. In: Frimmel FH, Christman RF (eds) Humic substances and their role in the environment. Wiley, New York, pp 15–28

Ambariyanto, Hoegh-Guldberg O (1999) Net uptake of dissolved amino acids by the giant clam, *Tridacna maxima*: Alternative sources of energy and nitrogen? Coral Reefs 17:1–7

Armstrong FAJ, Tibbitts S (1968) Photochemical combustion of organic matter in sea water for nitrogen, phosphorus and carbon determination. J Marbiol Assoc UK 48:143–152

Antia NJ, Harrrsion PJ, Oliveira L (1991) Phycological reviews: the role of dissolved organic nitrogen in phytoplankton nutrition, cellbiology and ecology. Phycologia 30:1–89

Billen G (1991) Protein degradation in aquatic environments. In: Chrost RJ (ed) Microbial enzymes in aquatic environments. Springer, New York, pp 123–143

Bronk DA, Ward BB (1999) Gross and net nitrogen uptake and DON release in the euphotic zone of Monterey Bay, California. Limonol Oceanogr 44:573–585

Bronk DA, Lomas M, Gilbert PM, Schukert KJ, Sandeerson MP (2000) Total dissolved nitrogen analysis: comparison between the persulfate, UV and high temperature oxidation method. Mar Chem 69:163–178

Bronk DA (2002) Dynamics of dissolved organic nitrogen. In: Hansell DA, Carlson CA (eds) Biogeochemistry of marine dissolved organic matter. Academic, London/San Diego, pp 153–247

Campbell L, Liu HB, Nolla HA, Vaulot D (1997) Annual variability of phytoplankton and bacteria in the subtropical North Pacific Ocean at station ALOHA during the 1991–1994 ENSO event. Deep-Sea Res 44:167–192

Capone DG (2000) The marine nitrogen cycle. In: Kirchman DL (ed) Microbial ecology of the oceans. Wiley, New York, pp 455–494

Casareto BE, Suzuki Y, Yoshida K, Ishikawa Y, Kurosawa K (2000) Benthic primary production in a coral reef at Bora Bay of Miyako Island, Okinawa. Proc 9th ICRS, Bali, 1:95–100

Casareto BE, Charpy L, Blancholt J, Suzuki Y, Kurosawa K, Ishikawa Y (2005) Phototrophic prokaryotes in Bora Bay, Miyako Island, Okinawa, Japan. Proc. 10th ICRS, Okinawa, pp 851–860

Casareto BE. Charpy L, Langlade MJ, Suzuki T, Ohba H, Niraula M, Suzuki Y (2008) Nitrogen fixation in coral reef environments. Proc. 11th ICRS, Ft. Lauderale, pp 890–894

Charpy-Roubaud CDJ, Charpy L, Cremoux JL (1990) Nutrient budget of the Lagoonal Waters in an Open Central South Pacific Atool Tikechau Tuamou French Polynesia. Mar Biol 107:67–74

Charpy-Roubaud CDJ, Charpy L, Larkum AWD (2001) Atmospheric dinitrogen fixation by benthic communities of Tikehau Lagoon (Tuamotu Archipelago French Polynesia) and its contribution to benthic primary production. Mar Biol 139:991–997

Charpy L (2005) Importance of photosynthetic picoplankton in coral reef ecosystems. Vie Et Milieu 55:217–223

Charpy L, Alliod R, Rodier M, Golubic S (2007) Benthic nitrogen fixation in the SW New Caledonia lagoon. Aquat Microb Ecol 47:73–81

Charpy L, Palinska KA, Casareto BE, Langlade MJ, Suzuki Y, Abrd RMN, Golubic S (2010) Dinitrogen-fixing cyanobacteria in microbial mats of two shallow coral reef ecosystems. Microb Ecol 59:174–186

Chisholm SW, Olson RJ, Zettler ER, Waterbury J, Goerick R, Welschmyer N (1988) A novel free-living prochlorophyte occurs at high cell concentrations in the oceanic euphotic zone. Nature 334:340–343

Cox EF, Ribes M, Kinzie RA (2006) Temporal and special scalling of planktonic responses to nutrient into a subtropical embayment. Mar Ecol Prog Ser 324:19–35

Fairoz MFM, Suzuki Y, Casareto BE, Agostini S, Shiroma K, Charpy L (2008) Role of organic matter in chemical symbiosis at coral reefs: release of organic nitrogen and amino acids under heat stress, Proc. 11th ICRS, July, Ft. Lauderdale, pp 895–899

Ferrier MD (1991) Net uptake of dissolved free amino acids by four scleractinian corals. Coral Reefs 10:183–187

GLibert P, O'Neil JM (1999) Dissolved organic nitrogen release and amino acid oxidase activity by *Trichodesmium spp*. Bull de l' Inst Oceanogr Monaco special 19:265–270

Grover R, Maguer JF, Allemand D, Ferrier-Pages C (2008) Uptake of dissolved free amino acids by the scleractinian coral *Stylophora pistillata*. J Exp Biol 211:860–865

Hatcher PG, Spiker EC (1988) Selective degradation of plant biomolecules. In: Frimmel FH, Christman RF (eds) Humic substances and their role in the environment. Wiley, New York, pp 15–28

Hansell DA, Feely RA (2000) Atmospheric intertropical convergence impacts surface ocean carbon and nitrogen biogeochemistry in the western tropical Pacific. Geophys Res Lett 27:1013–1016

Harvey GR, Boran DA (1985) Geochemistry of humic substances in seawater. In: Aiken GR, McKnight DM, Wershaw RL, MacCarthy P (eds) Humic substances in soil sediment and water. Wiley, New York, pp 223–247

Hoegh-Guldberg O, Williamson J (1999) Availability of two forms of dissolved nitrogen to the coral *Pocillopora damicornis* and its symbiotic zooxanthellae. Mar Biol 133:561–570

Johannes RE, Wiebe WJ, Crossland CJ (1983) Three patterns of nutrient flux in a coral reef community. Mar Ecol Prog Ser 12:131–136

Jorgensen NOG, Tranvik LJ, Berg GM (1999) Occurrence and bacterial cycling of dissolved nitrogen in the Gulf of Riga, the Baltic Sea. Mar Ecol Prog Ser 191:1–18

Karl D, Letelier R, Hebel DV, Bird DF, Winn CD (1992) The role of nitrogen fixation in biogeochemical cycling in the subtropical North Pacific Ocean. Nature 388:533–538

Kayane H, Hirota M, Yamamuro M, Koike I (2005) Nitrogen fixation of filamentous cyanobacteria in a coral reef measured using three different methods. Coral Reefs 24:197–200

Keil RG, Kirchman DL (1999) Utilization of dissolved protein and amino acids in the northern Sargasso Sea. Aquat Microb Ecol 18:293–300

Larkum AWD, Kennedy IR, Muller WJ (1988) Nitrogen fixation on a coral reef. Mar Biol 98:143–155

Larned ST (1998) Nitrogen-versus phosphorus-limited growth and sources of nutrients for coral reef macroalgae. Mar Biol 132:409–421

Metzler P, Glibert P, Gaeta S, Ludlam J (2000) Contrasting effects of substrate and grazer manipulations on picoplankton in oceanic and coastal waters off Brazil. J Plankton Res 22:77–99

Miller CAQ, Glibert PM (1998) Nitrogen excretion by the calanoid copepod *Acartia tonsa*: Results from mesocosmos experiments. J Planton Res 20:1767–1780

MIyajima T, Tanaka Y, Koike I (2005) Determining [15] N enrichment of dissolved organic nitrogen in environmental waters by gas chromatography/negative–ion chemical ionization mass spectrometry. Limonol Oceanogr Methods 3:164–173

Muscatine L, McCloskey LR, Loya L (1985) A comparison of the growth rates of zooxanthellae by a reef coral. Biol Bull 137:506–523

Odumn H.T. and E.P. Odumn (1955) Trophic structure and productivity of windward coral reef community on Eniwetok Atoll. Ecol Monogr 25:291–320

Olson RJ, Zettler ER, Durand MD (1993) Phytoplankton analysis using flow cytometory. In: Kemp PF, Sherr BF, Sherr EB, Cole JJ (eds) Handbook of methods in aquatic micro ecology. Lewis Publication, Boca Raton, pp 175–186

Rohwer F, Seguritan V, Azam F, Knowlton N (2002) Diversity and distribution of coral-associated bacteria. Mar Ecol Prog Ser 243:1–10

Shashar N, Cohen Y, Loya Y, Sar N (1994) Nitrogen fixation (acetylene reduction) in stony corals: evidence for coral-bacteria interactions. Mar Ecol Prog Ser 111:259–264

Sharp JH (1983) The distribution of inorganic nitrogen and dissolved and particulate organic nitrogen in the sea. In: Capone DG, Carpenter EJ (eds) Nitrogen in the marine environment. Plenum, New York, pp 101–120

Sorokin (1995) In: Sorokin (ed) Coral reef ecology. Springer, New York

Shiroma K, Suzuki Y, Daigo K, Agostini S, Fairoz MFM, Casareto BE (2008) Nitrogen dynamics in symbiotic relationships in corals, Proc. 11th ICRS, Ft. Lauderale, pp 931–934

Stepanauskas R, Edling H, Travik LJ (1999) Differential dissolved organic nitrogen availability and bacterial aminopeptidase activity in limnic and marine waters. Microb Ecol 38:264–272

Suzuki Y, Casareto BE, Yoshida K, Kurosawa K (2000) Import and export fluxes of HMW-DOC and LMW-DOC in coral reef at Miyako Island, Okinawa, Proceeding of the 9th ICRS, Bali, 555–559

Thurman EM (1985) Organic geochemistry of natural waters. M. Nijhoff & W. Junk Publication, Boston, p 497

Waterbury JB, Watson SW, Brand LE (1979) Widespread occurrence of a unicellular marine plankton, cyanobacterium. Marine Biological Association, vol 277. Citadel Hill, Plymouth, pp 293–294

Webb KL, DuPaul WD, Wiebe W, Sottile W, Johannes RE (1975) Enewetak Atoll: Aspects of the nitrogen cycle on a coral reef. Limnol Oceanogr 20:198–210

The Role of Plankton in Coral Trophodynamics

Christine Ferrier-Pagès, Mia Hoogenboom, and Fanny Houlbrèque

Abstract Historically, reef-building corals have been considered to be photoautotrophs due to their symbiosis with dinoflagellates that transfer photosynthetically fixed carbon to the animal tissue. Nevertheless, corals also obtain carbon heterotrophically through capture of plankton, ingestion of suspended particulate matter, and uptake of dissolved organic compounds. This review assesses the effects of heterotrophy on coral physiology, and how strongly feeding on all of these food sources contributes to coral energy budgets. Evidence in the literature demonstrates that feeding has a positive effect on coral tissue, enhancing the growth of both partners of the symbiosis. Nevertheless, the effects of feeding are light dependent: in general, tissue quality (lipid and protein composition) is enhanced in the presence of an adequate food source only under low-light conditions or in bleached corals. On the other hand, growth rates are typically highest under conditions of high light and food availability. However, under low-light conditions, feeding can provide a mechanism to maintain skeletal growth rates even though photosynthesis is reduced. Overall, a strong interaction between autotrophy and heterotrophy is apparent for scleractinian corals. Feeding can play a central role in maintaining physiological function when autotrophy is reduced. Moreover, taking all food sources into account, heterotrophy contributes more strongly to coral energy budgets than was previously thought. Nevertheless, not all symbiotic corals can sufficiently upregulate heterotrophic feeding to compensate for reduced photosynthesis, and identifying which coral species are facultative heterotrophs should be a focus of future research.

Keywords Feeding • heterotrophy • photosynthesis • coral physiology

C. Ferrier-Pagès (✉) and M. Hoogenboom
Centre Scientifique de Monaco, c/o Musée Océanographique,
Avenue Saint-Martin, Monaco, MC, 98000
e-mail: ferrier@centrescientifique.mc

F. Houlbrèque
International Atomic Energy Agency, Marine Environment Laboratories, 4 Quai Antoine 1er, Monaco, MC 98000

1 Introduction

Reef-building corals have been considered to be mainly photo-autotrophs, because they live in symbiosis with unicellular dinoflagellates (zooxanthellae) that transfer large amounts of photosynthetically fixed carbon to their host (Muscatine and Porter 1977). These photosynthates, often deficient in nitrogen and phosphorus, are thought to be exuded as mucus (Crossland 1987; Wild et al. 2004) or used as fuel for respiration, rather than assimilated into biomass (Falkowski et al. 1984; Davies 1991). Essential nutrients for growth and reproduction must therefore be acquired through heterotrophic feeding (Sebens et al. 1996; Anthony and Fabricius 2000; Ferrier-Pagès et al. 2003). The Scientific Reports of the Great Barrier Reef Expedition (1928–1929) of C.M. Yonge were among the first investigations into heterotrophic behavior of corals (Yonge 1930a,b; Yonge and Nicholls 1931). Since these famous works, numerous studies have confirmed that most coral species can in fact be active heterotrophs (Goreau and Goreau 1960; Goreau et al. 1971; Muscatine 1973; Wellington 1982; Sebens et al. 1996; Grottoli 2002; Houlbrèque et al. 2004a,b; Palardy et al. 2005, 2006).

Heterotrophic feeding by corals takes many forms, ranging from capture of live organic matter (LOM), uptake of dissolved organic material (DOM), and/or ingestion of suspended particulate matter (SPM, Fig. 1). LOM is considered to be the most important of these food sources, and corals are able to capture particles of a wide size range (from 0.4 μm to 2 mm) through nematocyst discharges, tentacle grabbing, or by mucus adhesion (reviewed by Muscatine 1973). Picoplankton are the smallest organisms that corals commonly ingest, both taxa that are free-living in the water column (Sorokin 1973; Farrant et al. 1987; Bak et al. 1998; Ferrier-Pagès et al. 1998; Houlbrèque et al. 2004b), and taxa directly associated with the coral mucus layer (Rohwer et al. 2001). Indeed, it has been suggested that corals develop a bacterial farm around them in order to be continuously fed. Evidence for this phenomenon comes from Herndl and Velimirov (1985) who found a large bacterial population within the coelenteron of four Anthozoan species.

Other forms of LOM that provide a food source for corals include nanoplankton (Ferrier-Pagès et al. 1998; Houlbrèque

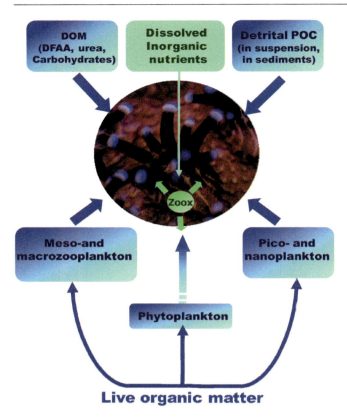

Fig. 1 Corals acquire nutrients through the animal feeding (heterotrophic feeding is represented by blue arrows): on dissolved organic matter (DOM), detrital particulate organic matter (POM), and live organic matter (LOM) (pico- and nanoplankton and meso- and macrozooplankton). The ingestion of phytoplankton has only been shown for soft corals. Corals can acquire nutrients via autotrophy (autotrophic nutrition is represented by green arrows), by transfer of photosynthates produced by the symbiotic dinoflagellates, which pump dissolved inorganic nutrients from seawater

et al. 2004b; Kramarsky-Winter et al. 2006) and meso-/macrozooplankton (Coles 1969; Johannes et al. 1970; Johannes and Tepley 1974; Porter 1974; Sebens et al. 1996; Palardy et al. 2006) (Fig. 1). Most studies on grazing rates have been performed using zooplankton, including copepods, eggs, larvae, and demersal zooplankton (i.e., Titlyanov et al. 2000a; Grottoli 2002; Ferrier-Pagès et al. 2003; Fabricius and Metzner 2004; Palardy et al. 2005, 2006; Grottoli et al. 2006). In general, corals can ingest from 0.5 to 2 prey items per polyp (Sebens et al. 1996) with ingestion rates depending on plankton density (Ferrier-Pagès et al. 2003; Palardy et al. 2005) or species (Palardy et al. 2005) as well as on water flow rates around colonies (Sebens and Johnson 1991; Sebens et al. 1998). The type of zooplankton found in the gut contents of corals is diverse (Sebens et al. 1996; Palardy et al. 2005; 2006) and is more strongly influenced by the feeding effort of coral colonies than by prey availability or polyp size (Palardy et al. 2005, 2006). Finally, although ingestion of phytoplankton has been demonstrated for soft corals (Fabricius et al. 1995), this has not yet been observed among the scleractinia.

Uptake of DOM mainly concerns carbohydrates, dissolved free amino acids and urea in nanomolar concentrations (Ferrier 1991; Al-Moghrabi et al. 1993; Grover et al. 2006, 2008) (Fig. 1). Uptake rates depend on DOM external concentration as well as on light intensity since photosynthesis enhances DOM uptake (Grover et al. 2006, 2008). Uptake of dissolved organic compounds may occur via diffusion or more certainly via active transport (Grover et al. 2006, 2008). Finally, corals can ingest detrital organic matter either in suspension (SPM, suspended particulate matter), trapped in the sediment (Anthony 1999; Rosenfeld et al. 1999; Anthony and Fabricius 2000; Mills et al. 2004) or in the form of mucus (Wild et al. 2004; Huettel et al. 2006). Generally, massive species with large polyps tend to have higher SPM feeding rates than branching ones with small polyps (Anthony and Fabricius 2000). Conversely to DOM, SPM uptake increases when symbiont photosynthesis decreases (Anthony and Fabricius 2000; Grottoli et al. 2006).

Such a wide diversity of food sources for coral heterotrophy indicates that this feeding mode may account for a large part of the energetic budget of corals, bringing carbon, nitrogen, phosphorus, and other nutrients not supplied by the photosynthesis of the symbionts (Muscatine and Porter 1977; Fitt and Cook 2001; Titlyanov et al. 2000a). Previous papers have reviewed the different methods by which corals can catch their food (Muscatine 1973), as well as the type of prey ingested (i.e., Anthony 1999; DiSalvo 1972; Sorokin 1973; Sebens et al. 1996; Ferrier-Pagès et al. 1998; Palardy et al. 2005). The aim of this review is to assess the effects of heterotrophy on coral physiology and its importance in coral trophodynamics. Indeed, while the importance of autotrophy for the nutritional energy of symbiotic corals has been widely assessed throughout the past 30 years (Muscatine and Porter 1977; Muscatine 1980; Falkowski et al. 1984; Muscatine et al. 1984; Cook et al. 1988; Davies 1991; Muller-Parker et al. 1994a,b; Swanson and Hoegh-Guldberg 1998; Wang and Douglas 1999; Cook and Davy 2001; LaJeunesse 2001), the impact of heterotrophy on coral metabolism has attracted far less attention.

2 Effect of Heterotrophy on Coral Physiology

2.1 Effect of Heterotrophy on Tissue Growth

2.1.1 Animal Tissue Fraction

A common method for identifying food sources (Peterson and Fry 1987) and quantifying carbon and nitrogen fluxes between trophic levels (Rau et al. 1992) is analysis of the isotopic composition of animal tissue (particularly ^{13}C and ^{15}N). In general, the $\delta^{13}C$ signature of a consumer is similar to that of its diet, while $\delta^{15}N$ is enriched by 3–4‰ with each successive trophic level (Rau et al. 1983; Owens 1987). The first general evidence that feeding affects coral tissue comes from changes observed in the $\delta^{13}C$ and $\delta^{15}N$ isotopic signatures. Muscatine et al. (1989) were the first to measure the isotopic signature of coral tissues and found a significant enrichment in both ^{13}C and ^{15}N with depth. This result was explained by a lower photosynthesis to respiration ratio for corals from deep water together with an increase in heterotrophic nutrition. More recently, Reynaud et al. (2002) confirmed these initial findings by experimentally measuring, both for animal tissue and zooxanthellae, a 1.5‰ difference in the $\delta^{13}C$ signature of fed and starved *Stylophora pistillata* colonies. Nevertheless, such isotopic enrichment was not observed for shallow-water corals (Yamamuro et al. 1995), or for corals living in inshore waters and receiving large amounts of $\delta^{15}N$-depleted terrestrial particulate and dissolved organic matter (Sammarco et al. 1999).

In most species, the effect of heterotrophy on animal tissue growth is mainly represented by an increase in protein and/or lipid concentrations per unit skeletal surface area (Anthony and Fabricius 2000; Anthony et al. 2002; Ferrier-Pagès et al. 2003; Houlbrèque et al. 2003). Generally, feeding causes a strong increase in tissue growth compared to skeletal growth. Anthony et al. (2002) suggested that either tissue may react more rapidly than the skeleton to availability of resources, or that the energy content of the tissue may represent the major component of total energy investment in coral growth. Lipids, which represent a major energy reserve for corals (Edmunds and Davies 1986; Harland et al. 1993), are highly influenced by feeding in many coral species. Overall, lipid concentrations are increased in fed corals, both for healthy (Al-Moghrabi et al. 1995; Treignier et al. 2008) and bleached colonies (Grottoli et al. 2006; Rodrigues and Grottoli 2007).

There is increasing evidence that light/photosynthesis and feeding interact to determine tissue properties (Figs. 2 and 3). Firstly, healthy colonies of *Galaxea fascicularis* showed an increase in saturated and mono-unsaturated fatty acids when experimentally maintained under low light and fed *Artemia salina* (Al-Moghrabi et al. 1995). When kept in the dark for 20 days, poly-unsaturated fatty acids also significantly increased in fed colonies but decreased in unfed colonies. Similarly, colonies of *Turbinaria reniformis* doubled all classes of lipids when maintained under low light (100 μmole photons $m^{-2}s^{-1}$) and fed with natural zooplankton (Fig. 2, Treignier et al. 2008). In such fed colonies, fatty acids, sterols, and alcohols increased from 100 to 250, 40 to 120, and 10 to 15 $\mu g\,cm^{-2}$, respectively. However, an increase in lipid stocks was not observed in *T. reniformis* maintained under high light (300 μmoles photons $m^{-2}s^{-1}$), because energy gained by feeding was directed into skeletal growth (Fig. 3, Treignier et al. 2008). Additional evidence for an interaction between photosynthesis and feeding comes from observations made on bleached corals. Feeding has been shown to be very

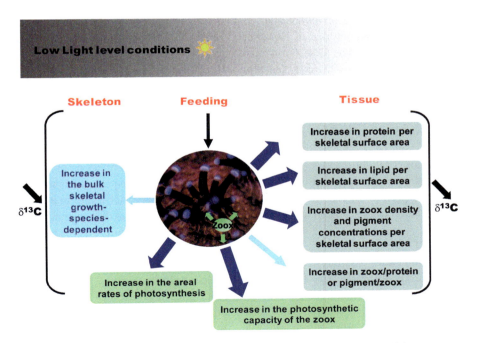

Fig. 2 Effects of feeding on corals maintained under low-light levels. Information was obtained with experiments performed either on *Stylophora pistillata* for most parameters, or *Turbinaria reniformis* for lipids. Zoox = zooxanthellae. Thick arrows represent a large effect of feeding, while small arrows represent a small effect of feeding

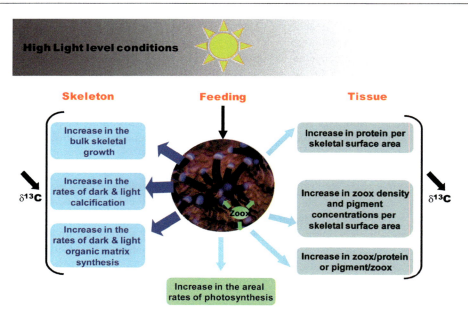

Fig. 3 Effects of feeding on corals maintained under high-light levels. Information was obtained with experiments performed either on *Stylophora pistillata* for most parameters, or *Turbinaria reniformis* for lipids. Zoox = zooxanthellae. Thick arrows represent a large effect of feeding, while small arrows represent a small effect of feeding

important for lipid stocks when corals bleach and translocation of algal photosynthates is greatly reduced. Although it is not the rule for all coral species (Grottoli et al. 2004), a decrease in storage lipids (i.e., wax esters, triacylglycerols, and polyunsaturated fatty acids) has been measured in thermally bleached corals (Grottoli et al. 2004; Yamashiro et al. 2005; Bachok et al. 2006; Papina et al. 2007). This suggests that the animal draws from its energy reserves to compensate for the decrease in the photosynthetic lipid production. However, corals able to catch zooplankton, and thus able to replenish their energy reserves, are less likely to die from bleaching than poor plankton consumers (Grottoli et al. 2006). This is the case for *Montipora capitata*, which increased its grazing rates more than fivefold when bleached and was thus able to acquire sufficient carbon from heterotrophy to meet its metabolic energy requirements, and to restore its lipid reserves (Grottoli et al. 2006). For this coral, the average percent contribution of heterotrophically acquired carbon to daily animal respiration (CHAR) therefore increased from 20 to 100%, demonstrating that heterotrophy played a central role in resilience to bleaching.

Feeding also results in higher protein concentration per unit surface area, both for healthy (Szmant-Froelich and Pilson 1980; Kim and Lasker 1998; Ferrier-Pagès et al. 2003; Houlbrèque et al. 2003, 2004a), and bleached corals (Rodrigues and Grottoli 2007). Indeed, it has been suggested that the major role of zooplankton capture could be to provide the symbiosis with essential amino acids, (Rahav et al. 1989), since animals were thought to be incapable of synthesizing them de novo. In the coral *Oculina arbuscula*, the ingested ^{15}N-labeled brine shrimp was indeed recovered in the protein fraction after 4 h for the zooxanthellae, and in the amino-acid pool that was then converted into protein for the animal fraction (Piniak et al. 2003). In the branching tropical coral *Stylophora pistillata*, a twofold increase in protein per surface area (from 0.42 to 0.73 mg cm^{-2}) was observed after 4 weeks of experimental feeding with *Artemia salina* prey (Houlbrèque et al. 2003, 2004a) or with natural zooplankton (Ferrier-Pagès et al. 2003). For this species, an interaction between light and feeding on the tissue growth rate was again observed. Estimates of tissue growth rates, based on the protein and weight values, ranged from 0.1 to 0.3% day^{-1} in starved and fed corals maintained under low light, respectively, and from 0.4 to 0.6% day^{-1} for the same corals maintained under high light. For healthy corals, feeding had a stronger impact on protein under low light, when carbon from photosynthesis is not sufficient for metabolic requirements (Ferrier-Pagès et al. 2003). For bleached corals, protein contents decrease in parallel with lipids during the stress (e.g., *Montipora capitata*, Rodrigues and Grottoli 2007). As was the case for lipids, particulate feeding by bleached corals of this species lead to increased protein concentrations within 2 months (from 0.2 to 0.3 g DW^{-1}, Rodrigues and Grottoli 2007). Moreover, stable isotope analyses (^{13}C) of host tissue and zooxanthellae indicated that fixed carbon was heterotrophically acquired during the first month of recovery from bleaching, before photoautotrophic acquisition resumed after 4–8 months (Rodrigues and Grottoli 2007).

2.1.2 Algal Fraction

In addition to its effects on coral tissue, heterotrophic feeding influences the symbiont population. Indeed, since nutrients are continuously exchanged between the host and its symbionts, feeding affects zooxanthellae metabolism. Several authors have observed translocation of nutrients from the coral animal to the symbionts (D'Elia and Cook 1988; Dubinsky et al. 1990; Piniak et al. 2003). A depression of N uptake by symbionts was observed in colonies of the hermatypic coral *Madracis mirabilis* when fed zooplankton to repletion (D'Elia and Cook 1988). Transfer of ^{15}N-labeled prey from the animal to the symbionts was also shown to occur in less than 10 min in the coral *Oculina arbuscula* (Piniak et al. 2003). Such transfer of nutrients explains the general increase in zooxanthellae densities per skeletal surface area that has been observed in healthy fed colonies (Muscatine et al. 1989; Dubinsky et al. 1990; Titlyanov et al. 2000a,b, 2001; Ferrier-Pagès et al. 2003; Houlbrèque et al. 2003, 2004a). Similarly, during a bleaching event, colonies of *Montipora capitata* that presented a high feeding rate were able to maintain symbionts at the same density as unbleached colonies, whereas zooxanthellae densities for a species with a lower feeding capacity (*Porites compressa*) rapidly decreased. The increase in zooxanthellae density in response to feeding lends support to the hypothesis that zooxanthellae are nitrogen limited (Dubinsky et al. 1990). A comparable increase in density is also observed when dissolved inorganic nitrogen is supplied to the corals (Dubinsky et al. 1990). Overall, whenever corals are enriched in inorganic or organic nutrients, there is an increase in the nitrogen content of the zooxanthellae and a corresponding decrease in the C/N ratio (Snidvongs and Kinzie III, 1994; Grover et al. 2002).

Due to the general increase in zooxanthellae densities per skeletal surface area in fed corals, chlorophyll concentrations per square centimeter are often higher in fed *versus* starved corals (Dubinsky et al. 1990; Stambler et al. 1991; Titlyanov et al. 1999) maintaining chlorophyll per algal cell constant. However, a feeding-related increase in chlorophyll per zooxanthellae has also been observed (Titlyanov et al. 2000a, 2001; Ferrier-Pagès et al. 2003; Houlbrèque et al. 2003). It must be noted that the zooxanthellae increase per skeletal surface area after feeding is partially due to the general thickening of the tissue above the skeleton. When algal densities are expressed per amount of animal tissue protein, data show that the animal protein/algal density ratio is either maintained constant (Fitt et al. 1982; Houlbrèque et al. 2003), decreases (Muller-Parker 1985; Al-Moghrabi et al. 1995), or increases in favor of the algal component (Clayton and Lasker 1984; Ferrier-Pagès et al. 2003; Houlbrèque et al. 2004a). In the latter situation, growth of the symbionts can be higher than the growth of the animal cells, leading to the occurrence of multiple symbionts within the same animal cell. This number of symbionts per host cell has been called the cell-specific density or CSD (Muscatine et al. 1998). Most corals collected in the field are characterized by a predominance of host cells containing a single dinoflagellate (singlets, 62.3–80.4% of cells) followed in decreasing frequency by those containing two (doublets, 28–34%), and three (triplets, 3.0–0.7%) dinoflagellates. However, some species, such as *Acropora palmata* and *Madracis mirabilis* present 20–50% of doublets, respectively, suggesting a higher capacity for heterotrophy (Sebens et al. 1996; Muscatine et al. 1998). Several authors (Titlyanov et al. 2000a, 2001; Houlbrèque et al. 2003) have also noted that the influence of heterotrophy on algal growth is light dependent, with the biggest effect of feeding observed under low light (Fig. 2). In such cases, the positive effect of feeding on algal density may be a strategy to increase rates of photosynthesis and energy production. In temperate corals, the few studies performed have also shown a temperature-feeding interaction on zooxanthellae density but the direction of this effect requires further investigation (Howe and Marshall 2001; Miller 1995; Rodolfo-Metalpa et al. 2008).

In conclusion, feeding has a positive effect on coral tissue, enhancing the growth of both partners of the symbiosis. This means that nutrients ingested by the coral animal also benefit the algal partner (e.g., Piniak et al. 2003). Nutrient exchanges between both partners are also observed with inorganic nutrients, which are first taken up by the zooxanthellae and then transferred to the coral host (Hoegh-Guldberg and Smith 1989; Dubinsky and Stambler 1996; Marubini and Davies 1996; Grover et al. 2002, 2003). Finally, the effect of feeding on coral tissue is light dependent and affected by zooxanthellae densities: feeding has the greatest impact on symbiont dynamics either under low-light conditions or in bleached corals.

2.2 Effect of Heterotrophy on Rates of Photosynthesis

The effect of heterotrophy on rates of photosynthesis has not been well investigated and further research is needed to understand all aspects of this relationship. There is some experimental evidence that indicates an increase in areal rates of photosynthesis in fed corals, due to increased zooxanthellae density and chlorophyll content per skeletal surface area (Dubinsky et al. 1990; Titlyanov et al. 2000a,b, 2001; Houlbrèque et al. 2003, 2004a). Houlbrèque et al. (2004a) measured both a change in the maximum net photosynthetic rate and in the light intensity at which photosynthesis approaches saturation. Nevertheless, the literature gives contradictory results regarding the effects of feeding on the photosynthetic capacity of zooxanthellae (i.e., photosynthesis per cell or per chlorophyll). While some studies showed no change in rates of photosynthesis per cell with feeding (Houlbrèque et al. 2003, 2004a) or even a decrease (Dubinsky et al. 1990), others demonstrate the opposite effect (Titlyanov et al. 2001; Davy et al. 2006). Titlyanov et al. (2001) showed that zooxanthellae photosynthetic capacity was enhanced by feeding under low light due to increased photoacclimation potential compared to that of starved corals. A very recent study (Griffin et al. sbm) performed on *Pocillopora damicornis* confirms that feeding increases zooxanthellae viability and improves their photosynthetic efficiency (ΦPSII), indicating that photosynthetic activity is constrained in the absence of a heterotrophic supplement to nutrition (Houlbrèque et al. 2003, 2004a). It is generally thought that nitrogen supply through heterotrophy drives the enhancement of symbiont photosynthetic capacities. Nitrogen is required for photo-adaptation or photoacclimation (Dubinsky et al. 1990; Titlyanov et al. 2001), and starvation induces an increase in the ratio of glutamine/glutamate suggesting a lack of nitrogen (for *Plesiastrea versipora*, Davy et al. 2006). Differences in the effects of feeding on the rates of photosynthesis of different coral species might therefore originate from species-specific differences in internal stores of nitrogen, either from the host or the zooxanthellae themselves. An alternative explanation is that the photobiological response to heterotrophy is mainly due to improved host–symbiont coupling (Furla et al. 2005), since pigment content and the ratio of chlorophyll-a to chlorophyll-c_2 did not change.

Recent studies (Griffin et al. sbm) also indicate that heterotrophic feeding increases bleaching resilience: colonies of *P. damicornis* fed brine shrimp and experiencing a heat shock did not show a decline in zooxanthellae photosynthetic function. However, such a decline was observed in starved corals and was consistent with previous studies showing photosynthetic impairment at temperatures above 31 °C (Hill and Ralph 2006). This suggests that either the fed host provided nutritional support to prevent damage to the photosynthetic apparatus of the zooxanthellae, or the host's demand for photosynthate was reduced allowing the symbiont to use these energy sources for their own survival.

Increased areal rates of photosynthesis do not always result in higher transfer of photosynthates from the zooxanthellae to the coral host. The first studies of this phenomenon were performed with inorganic nitrogen supply, and demonstrated an inverse relationship between nitrogen enrichment (which enhanced zooxanthellae growth) and carbon excretion in the coral *Porites astreoides* (McGuire and Szmant 1997) and in another anthozoan (green hydra, McAuley 1992). Davy and Cook (2001) also demonstrated lower carbon translocation in *Artemia salina* fed sea anemones compared to starved ones for *Aiptasia pallida*. The lower transfer of zooxanthellate photosynthates in fed animals has been explained by retention of photosynthates for the symbiont's own requirements (Davy and Cook 2001). Another factor to take into account is the quality of the photosynthates transferred. Nutrient-replete zooxanthellae mainly transfer amino acids to the host in addition to glucose and glycerol (Swanson and Hoegh-Guldberg 1998; Wang and Douglas 1999). Nutrient limitation (i.e., reduced feeding) might reduce amino-acid synthesis and induce a shift toward translocation of carbon-enriched compounds. Clearly, the effects of feeding on photosynthetic efficiency and carbon translocation require further research.

2.3 Effect of Heterotrophy on Skeletal Growth

In addition to its effects on coral tissue, heterotrophic feeding influences skeleton formation. General evidence for this comes firstly from the observed correlation between the $\delta^{13}C$ isotopic signature of tissue and skeleton (Heikoop et al. 2000). Usually, corals deposit a calcium carbonate skeleton that is depleted in ^{13}C relative to ambient seawater, as a result of kinetic and metabolic fractionation (McConnaughey 1989). However, physiological processes can alter the skeletal $\delta^{13}C$ signature. Elevated photosynthesis generally results in $\delta^{13}C$ depletion (Swart et al. 1996; Juillet-Leclerc et al. 1997) whereas respiration, as well as coral spawning, causes an enrichment (Swart et al. 1996; Kramer et al. 1993; Gagan et al. 1996). Theoretically, increased heterotrophic feeding by corals should lead to a decrease in skeletal $\delta^{13}C$ because zooplankton is depleted in ^{13}C relative to seawater (Rau et al. 1992). However, studies of this effect have produced conflicting results. Using a 19-year seasonal skeletal record of *Porites*, Felis et al. (1998) measured ^{13}C depletions that coincided with large, interannual plankton blooms, and suggested that corals have increased heterotrophy during these events. Reynaud et al. (2002) found no effect of feeding on the skeletal $\delta^{13}C$ signature of *Stylophora pistillata* potentially due to the fact that the *Artemia salina* prey used during the experiment were not as depleted in ^{13}C as natural zooplankton. Conversely, Muscatine et al. (2005) showed higher $\delta^{13}C$ of the skeletal organic matrix of non-symbiotic corals, which rely on heterotrophy, compared to symbiotic ones. Grottoli (2002) also observed an increase in skeletal $\delta^{13}C$ for colonies of *Porites compressa* fed with brine shrimps in high concentrations. In the latter study, it was hypothesized that an increase in zooxanthellae and rates of photosynthesis following input of nitrogen from feeding drove the increase in skeletal $\delta^{13}C$ (Grottoli 2002). Overall, although it is clear that feeding influences skeleton formation, how the interaction between photosynthesis and feeding moderates this effect warrants further investigation.

Different terms can be used to describe skeletal growth in corals. The first term is linear extension rate (LER), which is most often expressed in millimeters of skeleton accreted. LER can be measured from skeletal banding seen on X-radiographs of thin slices of coral skeleton cut along the growth axis (Lough and Barnes 1997), or by staining the skeleton with a dye (usually Sodium Alizarin Sulfonate) and measuring the amount of calcium carbonate deposited above the stain line (Barnes and Crossland 1980). The buoyant weight technique (Jokiel et al. 1978; Spencer-Davies 1989) is another method that measures bulk skeletal growth rate (most often expressed in% day^{-1} or in mg g^{-1}). This growth rate is the product of skeletal density and extension rate and is obtained by weighing the coral in seawater where the skeletal and seawater densities are known. Finally, calcification (most often expressed as nmoles Ca^{2+} mg protein^{-1} d^{-1}) is the term employed for skeletal growth when the incorporation of the radiotracer ^{45}Ca is measured in the skeleton (Tambutté et al. 1995). All these different techniques for measurements of skeletal growth rates have been employed to assess the effect of heterotrophy on coral calcification.

Of the above techniques, bulk skeletal weight increases have most often been used to investigate the effects of feeding on calcification. Johannes (1974) was one of the first to work on this subject and found that corals grew equally fast in 1 μm-filtered seawater as in unfiltered seawater. Although the amount of food in the two water types was not assessed, this result suggests that food availability had a negligible effect on skeletal growth. Later Wellington (1982) used field manipulations of light and zooplankton concentrations to show that reduced feeding decreased skeletal growth for only one of three study species (*Pavona clavus*) but had no effect on two other coral species. In agreement with this result, a study of the effects of light intensity and suspended particulate matter (SPM) concentrations showed that a coral with a high capacity to utilize SPM as a food source (*Goniastrea retiformis*) had slightly (10%) higher growth rates when grown under high SPM concentrations and high light (Anthony and Fabricius 2000). Conversely to the above observations, bulk skeletal growth of the coral *Stylophora pistillata* was highly enhanced (30%) when colonies were experimentally fed during 8 weeks with natural zooplankton, although the effect of feeding was light dependent (Ferrier-Pagès et al. 2003; Houlbrèque et al. 2003, 2004a). In the latter studies, fed corals kept at low light maintained a constant growth rate over time, growth was strongly suppressed in starved corals and the highest growth rates were observed for fed corals maintained under high light (as previously noticed for *G. retiformis*, Anthony and Fabricius 2000). Collectively, these studies indicate that feeding has a positive effect on growth rates for certain coral species, but that light intensity is also an important factor (Figs. 2 and 3). Nevertheless, this effect is by no means apparent for all species: feeding may have no effect on skeletal growth (e.g., Wellington 1982; Anthony and Fabricius 2000), or may even reduce growth rates (Grottoli 2002). In the latter example, linear extension rates of the coral *Porites compressa* decreased when colonies were exposed to very high plankton concentrations (5–60 times greater than those measured on the reef). Grottoli (2002) hypothesized that very high feeding rates overstimulate zooxanthellae growth and decouple the coral–algal symbiosis.

More recently, an interaction between light and feeding was confirmed using experiments on ^{45}Ca incorporation into the skeleton of the coral *Stylophora pistillata* (Houlbrèque et al. 2003, 2004a) (Figs. 2 and 3). The use of this radioisotope allows short-term measurements of dark and light calcification rates, which were both two to three times higher in corals fed during 5 weeks with natural zooplankton and *Artemia salina* nauplii. Light calcification rates ranged from 100 to 250 nmoles Ca^{2+} cm^{-2} h^{-1} in starved and fed corals, respectively, and dark calcification rates ranged from 40 to 80 nmoles Ca^{2+} cm^{-2} h^{-1} for the same corals. The increase in calcium carbonate deposition was linked to an increase in organic matrix synthesis (Houlbrèque et al. 2004a). Calcification indeed consists of two processes: deposition of an organic matrix layer followed by deposition of a calcium carbonate (CaCO$_3$) layer (Allemand et al. 1998). This organic matrix potentially plays a key role in processes such as crystal size, growth and orientation, and regulation of skeletal formation (Weiner and Addadi 1991; Falini et al. 1996; Belcher et al. 1996), and is composed of various amino acids with a composition that differs between symbiotic and non-symbiotic species (Cuif and Gautret 1995). Houlbrèque et al. (2004a) demonstrated that dark calcification rates were more strongly enhanced by feeding than were light calcification rates. This is due to the fact that, under illumination, there is a close coupling between deposition of the organic matrix and the CaCO$_3$ layers, whereas in darkness organic matrix deposition is usually depressed compared to the deposition of calcium carbonate.

In summary, feeding can enhance skeletal growth through three mechanisms:

1. Heterotrophy can stimulate calcification through tissue growth and enhanced supply of dissolved inorganic carbon (DIC). DIC necessary for calcification can be acquired from seawater bicarbonate (Gattuso et al. 1999; Marubini and Thake 1999) or from respired CO$_2$ (Erez 1978; Furla et al. 2000). Since feeding clearly enhances tissue growth and biomass, calcification can be stimulated by an increased supply of external DIC, via additional transporting molecules or of internal DIC, via enhanced respiration rates (Houlbrèque et al. 2003). Such tissue thickening might serve as a storage strategy when prey is available, allowing a subsequent skeletal growth followed

by thinning of the tissue (Barnes and Lough 1993). High tissue biomass can also supply additional energy, especially for the dark processes such as for the calcium/proton pump (McConnaughey 1989; McConnaughey and Whelan 1997; Anthony et al. 2002).
2. Feeding can indirectly enhance calcification by increasing the photosynthetic process. Photosynthesis supplies ATP for the proton pump, which in turn facilitates transport of carbon for calcification (McConnaughey 1989).
3. Feeding can enhance the construction of the organic matrix by providing some necessary external amino acids. As seen earlier, there is a tight coupling between organic matrix synthesis and calcification, the enhancement of the first process leading to a parallel enhancement of the second one.

In conclusion, the effect of feeding on skeletal growth is species dependent with some species having higher heterotrophic capacities than others. The effect is also light dependent since the highest skeletal growth rates are obtained for fed corals incubated under high light (Fig. 3). Under low light, feeding can maintain, or even enhance, skeletal growth rates that would otherwise be reduced due to lower photosynthetic energy acquisition (Fig. 2). It therefore appears that skeletal growth has a high-energy demand: growth is enhanced when both autotrophy and heterotrophy supply energy to the symbiosis.

3 Energetic Inputs from Heterotrophy

The contribution of heterotrophic feeding to the energy budgets of corals in their natural habitat is not well understood. Due to the difficulty of monitoring *in situ* rates of predation, most studies of coral feeding are experimental and field-based estimates of the energetic input from feeding are therefore rare. Moreover, to date no model has taken into account the potential energy acquisition summed over all types of food available to corals: typically, studies of coral feeding have only considered one type of prey at a time. Finally, nutrient assimilation efficiencies for the different types of food that corals can ingest are not well known because they are mainly deduced from the "egesta" method (Conover 1966; Anthony 1999), which assumes that only the organic component of the food is significantly affected by digestion. Only one study (Piniak et al. 2003) has used the more precise ^{15}N method. All of these factors mean that the relative contributions of autotrophy and heterotrophy to carbon budgets of corals are unknown. In this section, we draw together data from the literature to quantify the magnitude of carbon acquisition from different food sources for several coral species, and from all food sources for a single coral species for which data is available (*Stylophora pistillata*).

Based on experimental work using *Artemia salina* (at a feeding density of 100 Artemia l^{-1}) or natural zooplankton (at 1,500 cells l^{-1}), estimates of carbon acquisition from plankton feeding range from 24 to 600 µg C cm^{-2} d^{-1} (Fig. 4, based on a carbon content of 0.15 µg C per zooplankton prey, Ribes et al. 1998). This broad range of values indicates that feeding capacity is highly species specific, with individual polyps of different coral species capturing between 2 and 50 prey items per hour (see Clayton and Lasker 1982; Sebens and Johnson 1991; Johnson and Sebens 1993; Sebens et al. 1998; Ferrier-Pagès et al. 2003). Although experimental work indicates that zooplankton feeding can make a substantial contribution to daily carbon input, estimates based on field measurements yield much lower values. Indeed, the only study performed on corals maintained under natural conditions has estimated that zooplankton generates approximately 5 µg C cm^{-2} d^{-1} for colonies of *Pavona cactus*, *Pavona gigantea*, and *Pocillopora damicornis* (Palardy et al. 2005). Although experimental measurements of feeding in *Pavona* sp. are not available, these field estimates of feeding for *P. damicornis* are 50-fold lower than experimental estimates (Fig. 4). This inconsistency is most likely due to the fact that the field study did not include predation on the demersal zooplankton community, which migrates near corals during the night and is generally present at a much greater density than planktonic zooplankton (more than 3,000 cells l^{-1}, Heidelberg et al. 2004; Holzman et al. 2005). In fact, a 40% depletion (or 2.60 mg l^{-1}) of demersal zooplankton by reef organisms has been observed during the night (Yahel et al. 2005; Heidelberg et al. 2004). Although no studies have assessed carbon gain by corals at such plankton densities, natural rates of plankton feeding are likely to be higher than previously observed.

In addition to zooplankton, corals also prey on pico- and nanoplankton. Although studies of this feeding mode are rare, pico- and nanoplankton feeding is estimated to yield carbon uptake of approximately 3 µg C cm^{-2} d^{-1} for *S. pistillata* and 30 µg C cm^{-2} d^{-1} for *G. fascicularis* (Houlbrèque et al. 2004b, based on a polyp density of 360 and 1.2 polyps cm^{-2}, respectively). Finally, ingestion of dissolved organic carbon (DOC, Houlbrèque et al. 2004b) and suspended particulate matter (SPM, Anthony 1999) yields from 3 to 580 µg C cm^{-2} d^{-1} (Mills et al. 2004; Anthony 1999; Anthony and Fabricius 2000; Anthony and Connolly 2004). As is the case for zooplankton feeding, SPM ingestion rates are highly species specific (Fig. 4). Very high ingestion rates have been measured for the species *Montastrea franski*, and *Siderastrea radians* (from 474 to 584 µg C cm^{-2} d^{-1}), whereas values for a range of other species are in the vicinity of 10–100 µg C

Fig. 4 Carbon gain through various feeding modes for different coral species

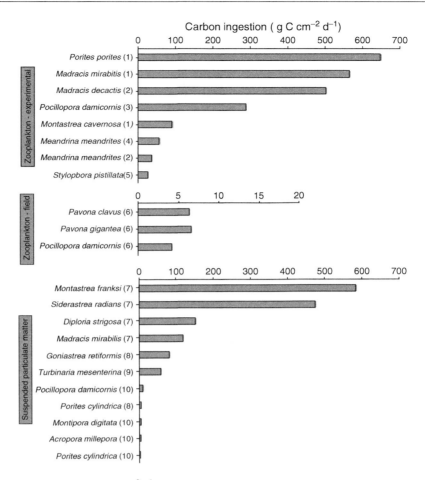

cm^{-2} d^{-1}. Based on these ingestion rates, studies of SPM feeding have therefore concluded that heterotrophic carbon supply varies from 15 to 35% of the daily metabolic demand in healthy corals (Porter 1976; Sorokin 1993; Grottoli et al. 2006) and may reach 100% in bleached corals (Grottoli et al. 2006). Clearly, a considerable body of evidence now disputes the early view that heterotrophic feeding makes only a minor contribution to the carbon budgets of scleractinian corals (e.g., Muscatine and Porter 1977; Davies 1991).

In fact, relative to carbon acquisition via photosynthesis, it can be demonstrated that heterotrophic feeding contributes significantly to coral energy budgets: even under conditions that have traditionally been considered autotrophic. For the species *Stylophora pistillata*, which has been well studied by many authors, estimates of the daily net carbon fixed by zooxanthellae range from 25 to 123 μg C cm^{-2} d^{-1} in shade- and light-adapted colonies, respectively (Muscatine et al. 1984).

Taking all forms of feeding into account, daily carbon acquisition via heterotrophy reaches 18 μg C cm^{-2} d^{-1} at the minimum (Fig. 5). This value is based on the lowest observed measurements of carbon acquired from zooplankton feeding (5 μg C cm^{-2} d^{-1} for zooplankton, Palardy et al. 2005), 8 μg C cm^{-2} d^{-1} for pico- and nanoplankton (Houlbrèque et al. 2004b) and 5 μg C cm^{-2} d^{-1} for DOC/SPM (Anthony 1999; Houlbrèque et al. 2004b). Therefore, the lower bound of estimates of heterotrophically acquired carbon is in fact more than 70% of the value for carbon acquired through symbiont photosynthesis for shade-adapted corals. If predation on demersal zooplankton is included into this estimate, an additional gain of 24 μg C cm^{-2} d^{-1} (Ferrier-Pagès et al. 2003), total daily heterotrophically acquired carbon reaches a maximum estimate of 42 μg C cm^{-2} d^{-1}. This represents more than one-third of the total carbon brought by photosynthesis in light-adapted colonies.

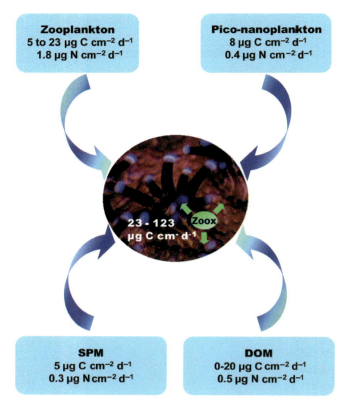

Fig. 5 Auto- and heterotrophic acquisition of carbon and nitrogen in the species *Stylophora pistillata*. *DOC*: dissolved organic carbon; *SPM*: suspended particulate matter. See "Energetic inputs from heterotrophy" for calculations

and 48 µg N cm^{-2} d^{-1} (Mills 2000; Mills et al. 2004; Anthony 1999; Anthony and Fabricius 2000; Anthony and Connolly 2004), based on sediment nitrogen content of 0.41% by weight (Anthony and Fabricius 2000). Pico- and nanoplankton ingestion can yield 0.8–6 µg N cm^{-2} d^{-1} (Houlbrèque et al. 2004b), while dissolved organic matter can contribute between 0.1 and 16 µg N cm^{-2} d^{-1} (Ferrier 1991; Badgley et al. 2006; Hoegh-Guldberg and Williamson 1999; Grover et al. 2006, 2008). Indeed, for the species *Stylophora pistillata* it has been estimated that even at the lower range of the concentrations commonly found in seawater (approximately 0.2–0.3 µM), dissolved organic nitrogen can contribute at least 11% of the total daily nitrogen required for tissue growth (0.5 µg N cm^{-2} d^{-1}, Grover et al. 2008). Drawing together all of these sources of nutrient acquisition, it is evident that nitrogen uptake can exceed 3 µg N cm^{-2} d^{-1} (based on values for *Stylophora pistillata*, Fig. 5). Unfortunately, there is insufficient data available to document the uptake of other major nutrients (e.g., phosphorus) contributed by the different feeding modes. Although some studies agree that corals need to take up organic phosphorus from an external source (D'Elia 1977), studies of nutrient uptake and utilization are lacking. What limited evidence there is indicates that feeding on bacteria would yield approximately 3 µg P d^{-1} (Sorokin 1973), a value comparable to uptake of nitrogen.

4 Perspectives and Directions for Future Research

Although investigation of the importance of heterotrophic feeding for coral metabolism has a long history (Yonge 1930a,b; Goreau et al. 1971), interest in this subject has only recently been regained. It is now evident that, taking into account feeding on all possible sources, heterotrophy contributes more to the carbon budget of corals than previously expected. However, many questions regarding the interactions between heterotrophy, autotrophy, energy allocation, and environmental conditions remain unanswered. To resolve these questions, we need first to have a better quantification of the amount of carbon translocated by the zooxanthellae to the host in different environmental conditions. Most estimates of carbon translocation are based on the "contribution of zooxanthellae to animal respiration" or "CZAR" equation presented by Muscatine et al. (1981). However, this equation is based on several assumptions, in particular, that the respiration of the coral host compared to symbionts is based upon the relative biomass of the two partners (Muscatine et al. 1981; Smith and Muscatine 1986; Verde and McCloskey 1996). Moreover, estimates of translocation from this method tend to be higher than those based on direct measurements (using ^{14}C labeling techniques, Trench 1979). Therefore, new

In addition to providing a supplementary source of carbon, heterotrophy is a vital source of nitrogen, phosphorus, and other limiting nutrients for the coral symbiosis (Houlbrèque et al. 2004b; Grover et al. 2006, 2008). This is evidenced by the fact that efficiency with which heterotrophically acquired nutrients are assimilated into tissue varies between 33% and 100% for suspended particulate matter (Anthony 1999; Mills 2000; Mills et al. 2004), and reaches 70–100% for zooplankton (Bythell 1988; Piniak et al. 2003). Among the total amount of nutrient acquired, the proportion of ingested prey materials utilized by the symbiotic algae is fairly consistent, ranging from 15 to 25% (Cook 1972; Szmant-Froelich 1981; Piniak et al. 2003). Based on these values, the coral *Stylophora pistillata* fed with natural zooplankton (ca. 1,500 prey l^{-1}) can therefore gain more than 1.8 µg N cm^{-2} d^{-1} (Ferrier-Pagès et al. 2003), representing approximately one-third of the nitrogen required for tissue growth. In addition to zooplankton feeding, ingestion of pico- and nanoplankton together with dissolved and particulate organic matter can be more than sufficient to sustain tissue growth in several coral species (Hoegh-Guldberg and Williamson 1999; Ferrier-Pagès et al. 2003). Depending on species-specific feeding rates, SPM can deliver between 0.3

techniques need to be developed to improve our understanding of CZAR.

Secondly, we need to better define the importance of the different food sources for corals, how the dependence on particular sources may vary across different environmental conditions, and which specific nutrient (carbon, nitrogen, phosphorus) is mainly derived from feeding. Even though some studies have individually assessed the grazing rates on zooplankton (Sebens et al. 1996), pico- and nanoplankton (Houlbrèque et al. 2004b), dissolved/particulate organic matter (Anthony 1999) or sediment (Anthony 1999), none of them have measured, or even estimated, the total amount of energy gained by feeding on all potential sources for a given coral species in a given environment. Therefore, it is not known on which food source corals are most reliant, let alone the capacity of coral species to switch between nutrition modes depending on their habitat. Once feeding rates on the different prey have been accurately measured, both under laboratory and field conditions, it will be possible to estimate how strongly heterotrophic feeding varies seasonally due to changes in plankton concentration, water flow, and turbidity.

Another key question that is poorly understood is how energy acquired through photosynthesis compared to heterotrophy is allocated between symbiont population growth, coral tissue growth, skeletal growth, and reproduction. Indeed, some studies have shown that corals used heterotrophic energy differently depending on the light level under which they were grown (i.e., depending on energy gain through photosynthesis). For example, in *T. reniformis* feeding increased lipid stocks under low light, whereas it enhanced growth under high light (Treignier et al. 2008). Similarly, for *S. pistillata* feeding enhanced growth more strongly under high light compared with low light (Houlbrèque et al. 2003). These few observations provide compelling evidence that corals adopt different energy allocation strategies depending on light and food availability. However, more research is needed to understand how corals use their energy sources to cope with environmental constraints, and what mechanisms are involved in the enhancement of growth by feeding. Furthermore, it remains unclear precisely how feeding enhances photosynthesis, and under which environmental conditions such enhancement occurs. In other words, we need to know when corals allocate food to the zooxanthellae to enhance their photosynthetic capacities, and when they sequester nutrients for use by the host tissue. All these questions can first be investigated under experimental laboratory conditions, but such studies must also be extended to natural conditions.

Recent work has highlighted the importance of heterotrophic feeding as a source of carbon for corals during bleaching events (Grottoli et al. 2006). There is now clear evidence that feeding rates on zooplankton can increase dramatically in bleached corals and provide them with up to 100% of their daily metabolic demand (Grottoli et al. 2006). Nevertheless, not all species are capable of upregulating heterotrophy sufficiently to compensate for reduced photosynthesis. In the Grottoli et al. (2006) study only one of three study species was able to do so (*Montipora capitata* compared with *Porites compressa* and *Porites lobata*). Similarly, Anthony and Fabricius (2000) found that where *Goniastrea retiformis* was able to increase sediment feeding sufficiently to compensate for lower photosynthesis in shaded conditions, the same was not observed for another species (*Porites cylindrica*). Clearly, more research needs to be done to determine which corals are more "heterotrophic" as they are probably the species most resistant to bleaching.

Finally, from a broader perspective, further research must be conducted into the trophic links between plankton, corals, and other organisms. It is well known that there is tight recycling of nutrients within reef ecosystems, for example, coral-dwelling fishes excrete waste nutrients that are subsequently taken up by corals (e.g., Meyer and Schultz 1985). Moreover, mucus released by corals functions as a trap for LOM and SPM (Wild et al. 2004), and can form an important food source for other reef-dwelling organisms (Richman et al. 1975). Numerous species of reef fish also rely on coral tissue and/or coral larvae as a food source (Pratchett 1995; Pratchett et al. 2001). Few of these trophic interactions have been quantified and therefore little is known about the importance of coral heterotrophy for the overall health of reef ecosystems.

5 Conclusions

A strong interaction between autotrophy and heterotrophy is apparent for scleractinian corals. Feeding plays a central role in maintaining coral physiological functioning whenever autotrophy is insufficient, such as for corals living in shaded conditions or experiencing a bleaching event (Anthony and Fabricius 2000; Grottoli et al. 2006). Moreover, the available literature indicates that heterotrophy contributes a larger proportion of total carbon acquisition than was previously expected. In light of the predicted increase in the frequency and severity of bleaching events (Hoegh-Guldberg 1999), this indicates that corals able to increase their feeding effort when necessary will be more resilient to stresses and may come to dominate the reef community. Nevertheless, the available experimental evidence indicates that most symbiotic corals cannot rely solely heterotrophic nutrition (Clayton and Lasker 1982; Grottoli et al. 2006). Therefore, predictive models of climate impacts on reef-building corals should take into account the potential for heterotrophic feeding to mitigate environmental stressors. Identifying which coral species are facultative heterotrophs should be a focus of future research.

References

Al-Moghrabi S, Allemand D, Jaubert J (1993) Valine uptake by the scleractinian coral *Galaxea fascicularis* characterization and effect of light and nutritional status. J Comp Physiol B 163:355–362

Al-Moghrabi S, Allemand D, Couret JM (1995) Fatty acid of the scleractinian coral *Galaxea fascicularis*: effect of light and feeding. J Comp Physiol B 165:183–192

Allemand D, Tambutté E, Girard JP, Jaubert J (1998) Organic matrix synthesis in the scleractinian coral *Stylophora pistillata*: role in biomineralization and potential target of the organotin tribulyltin. J Exp Biol 201:2001–2009

Anthony KRN (1999) Coral suspension feeding on fine particulate matter. J Exp Mar Biol Ecol 232:85–106

Anthony KRN, Connolly SR (2004) Environmental limits to growth: physiological niche boundaries of corals along turbidity-light gradients. Oecologia 141(3):373–384

Anthony KRN, Fabricius KE (2000) Shifting roles of heterotrophy and autotrophy in coral energetics under varying turbidity. J Exp Mar Biol Ecol 252:221–253

Anthony KRN, Connolly SR, Willis BL (2002) Comparative analysis of energy allocation to tissue and skeletal growth in corals. Limnol Oceanogr 47:1417–1429

Bachok Z, Mfilinge P, Tsuchiya M (2006) Characterization of fatty acid composition in healthy and bleached corals from Okinawa, Japan. Coral Reefs 25:545–554

Bagdley BD, Lipschultz F, Sebens K (2006) Nitrate uptake by the reef coral Diploria strigosa: effects of concentration, water flow and irradiance. Mar Biol 149(2):327–338

Bak RPM, Joenje M, DeJong I, Lambrechts DYM, Newland G (1998) Bacterial suspension feeding by coral reef benthic organisms. Mar Ecol Prog Ser 175:285–288

Barnes DJ, Crossland CJ (1980) Diurnal and seasonal variations in the growth of a staghorn coral measured by time-lapse photography. Limnol Oceanogr 25:1113–1117

Barnes DJ, Lough JM (1993) On the nature and causes of density banding in massive coral skeletons. J Exp Mar Biol Ecol 167:91–108

Belcher AM, Wux XH, Christensen RJ, Hansma PK, Stucky GD, Morse DE (1996) Control of crystal phase switching and orientation by soluble mollusc shell proteins. Nature 381:56–58

Bythell JC (1988) A total nitrogen and carbon budget for the elkhorn coral *Acropora palmata* (Lamarck). Proc 6th Int Coral Reef Symp 2:535–540

Clayton WS Jr, Lasker KE (1982) Effects of light and dark treatments of feeding by the reef coral *Pocillopora damicornis*. J Exp Mar Biol Ecol 63:269–280

Clayton WS Jr, Lasker KE (1984) Host feeding regime and zooxanthellae photosynthesis in the anemone *Aiptasia pallida*. Biol Bull 167:590–600

Cook CB (1972) Benefit for symbiotic zoochlorella from feeding by green hydra. Biol Bull 142:236–242

Cook CB, Davy SK (2001) Are free amino acids responsible for the "host factor" effects on symbiotic zooxanthellae in extracts of host tissue? Hydrobiologia 461(1–3):71–78

Cook CB, D'Elia CF, Muller-Parker G (1988) Host feeding and nutrient sufficiency for zooxanthellae in the sea anemone *Aiptasia pallida*. Mar Biol 98:253–262

Coles SL (1969) Quantitative estimate of feeding and respiration of three corals. Limnol Oceanogr 14:949–953

Conover RJ (1966) Assimilation of organic matter by zooplankton. Limnol Oceanogr 11:338–345

Crossland CJ (1987) In situ release of mucus and DOC-lipid from the corals *Acropora variabilis* and *Stylophora pistillata* in different light regimes. Coral Reefs 6:35–43

Cuif JP, Gautret P (1995) Glucides et protéines de la matrice soluble des biocristaux de scléractiniaires Acroporidés. C R A S Paris II 320:273–278

D'Elia CF (1977) The uptake and release of dissolved phosphorus by reef corals. Limnol Oceanogr 22:301–315

D'Elia CF, Cook CB (1988) Methylamine uptake by zooxanthellae-invertebrate symbioses: insights into host ammonium environment and nutrition. Limnol Oceanogr 33:1153–1165

Davies PS (1991) Effect of daylight variations on the energy budgets of shallow water corals. Mar Biol 108:137–144

Davy SK, Cook CB (2001) The relationship between nutritional status and carbon flux in the zooxanthellate sea anemones *Aiptasia pallida*. Mar Biol 119:999–105

Davy SK, Withers KJT, Hinde R (2006) Effects of host nutritional status and seasonality on the nitrogen status of zooxanthellae in the temperate coral *Plesiastrea versipora* (Lamarck). J Exp Mar Biol Ecol 335:256–265

DiSalvo LH (1972) Bacterial counts in surface open waters of Eniwetok Atoll, Marshall Islands. Atoll Res Bull 151:1–5

Dubinsky Z, Stambler N, Ben-Zion M, McCloskey LR, Muscatine L, Falkowski PG (1990) The effect of external nutrient resources on the optical properties and photosynthetic efficiency of *Stylophora pistillata*. Proc R Soc Lond B 239:231–246

Edmunds PJ, Davies PS (1986) An energy budget for *Porites porites* (Scleractinian). Mar Biol 92:339–347

Erez J (1978) Vital effect on stable-isotope composition seen foraminifera and coral skeletons. Nature 273:199–202

Fabricius KE, Metzner J (2004) Scleractinian walls of mouths: predation on coral larvae by corals. Coral Reefs 23:245–248

Fabricius KE, Yahel G, Genin A (1995) Herbivory in asymbiotic soft corals. Science 268:90–93

Falini G, Albeck S, Weiner S, Addadi L (1996) Control of aragonite or calcite polymorphism by mollusk shell macromolecules. Science 271:67–69

Falkowski PG, Dubinsky Z, Muscatine L, Porter JW (1984) Light and bioenergetics of a symbiotic coral. Bioscience 11:705–709

Farrant PA, Borowitzka MA, Hinde R, King RJ (1987) Nutrition of the temperate Australian coral *Capnella gaboensis*. Mar Biol 95:575–581

Felis T, Pätzold J, Loya Y, Wefer G (1998) Vertical water mass mixing and plankton blooms recorded in skeletal stable carbon isotopes of a Red Sea coral. J Geophys Res 103:731–739

Ferrier MD (1991) Net uptake of dissolved free amino acids by four scleractinian corals. Coral Reefs 10:183–187

Ferrier-Pagès C, Gattuso JP, Cawet G, Jaubert J, Allemand D (1998) Release of dissolved organic carbon and nitrogen by the zooxanthellate coral *Galaxea fascicularis*. Mar Ecol Prog Ser 172:265–274

Ferrier-Pagès C, Witting J, Tambutté E, Sebens KP (2003) Effect of natural zooplankton feeding on the tissue and skeletal growth of the scleractinian coral *Stylophora pistillata*. Coral Reefs 22:229–240

Fitt WK, Cook CB (2001) The effects of feeding or addition of dissolved inorganic nutrients in maintaining the symbiosis between dinoflagellates and a tropical marine cnidarian. Mar Biol 139:507–517

Fitt WK, Pardy RL, Littler MM (1982) Photosynthesis, respiration, and contribution to community productivity of the symbiotic sea anemone *Anthopleura elegantissima* (Brandt 1835). J Exp Mar Biol Ecol 61:213–232

Furla P, Allemand D, Orsenigo MN (2000) Involvement of H+-ATPase and carbonic anhydrase in inorganic carbon uptake for endosymbiont photosynthesis. Am J Physiol 278:870–881

Furla P, Allemand D, Shick JM, Ferrier-Pagès C, Richier S, Plantivaux A, Merle PL, Tambutté S (2005) The symbiotic anthozoan: a physiological chimera between alga and animal. Int Comp Biol 45(4):595–604

Gagan MK, Chivas AR, Isdale PJ (1996) Timing coral based climatic histories using ^{13}C enrichments driven by synchronized spawning. Geology 24:1009–1012

Gattuso JP, Allemand D, Frankignoulle M (1999) Photosynthesis and calcification at cellular, organismal and community levels in coral reefs: a review of interactions and control by carbonate chemistry. Integr Comp Biol 39(1):160–183

Goreau TF, Goreau NI (1960) Distribution of labeled carbon in reef-building corals with and without zooxanthellae. Science 131:668–669

Goreau TF, Goreau NI, Yonge CM (1971) Reef corals: autotrophs or heterotrophs? Biol Bull 141:247–260

Grottoli AG (2002) Effect of light and brine shrimp on skeletal delta C-13 in the Hawaiian coral *Porites compressa*: a tank experiment. Geochim Cosmochim Acta 66:1955–1967

Grottoli AG, Rodrigues LJ, Juarez C (2004) Lipids and stable carbon isotopes in two species of Hawaiian corals, *Porites compressa* and *Montipora verrucosa*, following a bleaching event. Mar Biol 145:621–631

Grottoli A, Rodrigues L, Palardy J (2006) Heterotrophic plasticity and resilience in bleached corals. Nature 440:1186–1189

Grover R, Maguer JF, Reynaud-Vaganay S, Ferrier-Pagès C (2002) Uptake of ammonium by the scleractinian coral *Stylophora pistillata*: effect of feeding, light and ammonium concentrations. Limnol Oceanogr 47:782–790

Grover R, Maguer JF, Allemand D, Ferrier-Pagès C (2003) Nitrate uptake by the scleractinian coral *Stylophora pistillata*. Limnol Oceanogr 48(6):2266–2274

Grover R, Maguer JF, Allemand D, Ferrier-Pagès C (2006) Urea uptake by the scleractinian coral *Stylophora pistillata*. J Exp Mar Biol 332:216–225

Grover R, Maguer J-F, Allemand D, Ferrier-Pagès C (2008) Uptake of dissolved free amino acids (DFAA) by the scleractinian coral *Stylophora pistillata*. J Exp Biol 211:860–865

Harland AD, Navarro JC, Davies PS, Fixter LM (1993) Lipids of some Carribbean and Red Sea corals: total lipid, wax esters, triglycerides and fatty acids. Mar Biol 117:113–117

Heidelberg KB, Sebens KP, Purcell JE (2004) Composition and sources of near reef zooplankton on a Jamaican forereef along with implications for coral feeding. Coral Reefs 23:263–276

Heikoop JM, Dunn JJ, Risk MJ, McConnaughey TA, Sandman IM (2000) Separation of kinetic and metabolic effect in carbon-13 records preserved in reef coral skeletons. Geochim Cosmochim Acta 64:975–987

Herndl GJ, Velimirov B (1985) Bacteria in the coelenteron of Anthozoa: control of coelenteric bacterial density by the coelenteric fluid. J Exp Mar Biol Ecol 93:115–130

Hill R, Ralph PJ (2006) Photosystem II heterogeneity of *in hospite* zooxanthellae in scleractinian corals exposed to bleaching conditions. Photochem Photobiol 82:1577–1585

Hoegh-Guldberg O (1999) Coral bleaching, climate change, and the future of the world's coral reefs. Mar Freshwater Res 50:839–866

Hoegh-Guldberg O, Smith GJ (1989) Influence of the population density of zooxanthellae and supply of ammonium on the biomass and metabolic characteristics of the reef corals *Seriatopora hystrix* and *Stylophora pistillata*. Mar Ecol Prog Ser 57:173–186

Hoegh-Guldberg O, Williamson J (1999) Availability of two forms of dissolved nitrogen to the coral *Pocillopora damicornis* and its symbiotic zooxanthellae. Mar Biol 133:561–570

Holzman R, Reidenbach MA, Monismith SG, Koseff JR, Genin A (2005) Near-bottom depletion of zooplankton over a coral reef. II. Relationships with zooplankton swimming ability. Coral Reefs 24:87–94

Houlbrèque F, Tambutté E, Ferrier-Pagès C (2003) Effects of zooplankton availability on the rates of photosynthesis, and tissue and skeletal growth in the scleractinian coral *Stylophora pistillata*. J Exp Mar Biol Ecol 296:145–166

Houlbrèque F, Tambutté E, Allemand D, Ferrier-Pagès C (2004a) Interactions between zooplankton feeding, photosynthesis and skeletal growth in the scleractinian coral *Stylophora pistillata*. J Exp Biol 207:1461–1469

Houlbrèque F, Tambutté E, Richard C, Ferrier-Pagès C (2004b) Importance of a micro-diet for scleractinian corals. Mar Ecol Prog Ser 282:151–160

Howe SA, Marshall AT (2001) Thermal compensation of metabolism in the temperate coral, *Plesiastrea versipora* (Lamarck, 1816). J Exp Mar Biol Ecol 259:231–248

Huettel M, Wild C, Gonelli S (2006) Mucus trap in coral reefs: formation and temporal evolution of particle aggregates caused by coral mucus. Mar Ecol Prog Ser 307:69–84

Johannes RE (1974) Sources of nutritional energy for reef corals. Proc 2nd Int Coral Reef Symp 1:133–137, Brisbane

Johannes RE, Tepley L (1974) Examination of feeding on the reef coral *Porites lobata* in situ using time lapse photography. Proc 2nd Int Coral Reef Symp 1:127–131, Brisbane

Johannes RE, Cole S, Kuenzel NT (1970) The role of zooplankton in the nutrition of some scleractinian corals. Limnol Oceanogr 15:579–586

Johnson AS, Sebens KP (1993) Consequences of flattened morphology: Effects of flow on feeding rates of the scleractinian coral *Meandrina meandrites*. Mar Ecol Prog Ser 1–2:99–104

Juillet-Leclerc A, Gattuso JP, Montaggioni LF, Pichon M (1997) Seasonal variation of primary productivity and skeletal δ^{13}C and δ^{18}O in the zooxanthellate scleractinian coral *Acropora formosa*. Mar Ecol Prog Ser 157:109–117

Jokiel PL, Maragos JE, Franzisket L (1978) Coral growth: buoyant weight technique. In: Stoddart DR, Johannes RE (eds) Coral reefs: research methods. UNESCO monographs on oceanographic methodology, Paris, pp 529–542

Kim K, Lasker HR (1998) Allometry of resource capture in colonial cnidarians and constraints on modular growth. Funct Ecol 12:646–654

Kramarsky-Winter E, Harel M, Siboni N, Ben Dov E, Brickner I, Loya Y, Kushmaro A (2006) Identification of a protist-coral association and its possible ecological role. Mar Ecol Prog Ser 317:67–73

Kramer PA, Swart PK, Szmant AM (1993) The influence of different sexual reproductive patterns on density banding and stable isotopic compositions of corals. Proc 7th Int Coral Reef Symp 1:222

LaJeunesse TC (2001) Investigating the biodiversity, ecology and phylogeny of endosymbiotic dinoflagellates in the genus *Symbiodinium* using the ITS region: in search of a "species" level marker. J Phycol 37:866–880

Lough JM, Barnes DJ (1997) Several centuries of variation of skeletal extension, density and calcification in massive Porites colonies from the Great Barrier Reef: a proxy for seawater temperature and a background of variability against which to identify unnatural change. J Exp Mar Biol Ecol 211:29–67

Marubini F, Davies PS (1996) Nitrate increases zooxanthellae population density and reduces skeletogenesis in corals. Mar Biol 127:319–328

Marubini F, Thake B (1999) Bicarbonate addition promotes coral growth. Limnol Oceanogr 44(3):716–720

McAuley PJ (1992) The effect of maltose release on growth and nitrogen metabolism of symbiotic Chlorella. Br Phycol J 27:417–422

McConnaughey TA (1989) ^{13}C and ^{18}O isotopic disequilibrium in biological carbonates: I Patterns. Geochim Cosmochim Acta 53:151–162

McConnaughey TA, Whelan FF (1997) Calcification generates protons for nutrient an bicarbonate uptake. Earth Sci Rev 42:92–117

McGuire MP, Szmant AM (1997) Time course of physiological responses to NH4 enrichment by a coral-zooxanthellae symbiosis. Proc 8th Coral Reef Symp 1:909–914

Meyer JL, Schultz ET (1985) Tissue condition and growth rate of corals associated with schooling fish. Limnol Oceanogr 30:157–166

Miller MW (1995) Growth of a temperate coral: effects of temperature, light, depth and heterotrophy. Mar Ecol Prog Ser 122:217–225

Mills MM (2000) Corals feeding on sediments? Ingestion, assimilation, and contributions to coral nutrition. PhD thesis, University of Maryland, College Park

Mills MM, Lipschultz F, Sebens KP (2004) Particulate matter ingestion and associated nitrogen uptake by four species of scleractinian corals. Coral Reefs 23:311–323

Muller-Parker G (1985) Effect of feeding regime and irradiance on the photophysiology of the symbiotic sea anemone Aiptasia pulchella. Mar Biol 90:65–74

Muller-Parker G, Cook CB, D'elia CF (1994a) Elemental composition of the coral Pocillopora damicornis exposed to elevated seawater ammonium. Pac Sci 48:234–246

Muller-Parker G, McCloskey LR, Hoegh-Guldberg O, McAuley PJ (1994b) Effect of ammonium enrichment on animal and algal biomass of the coral Pocillopora damicornis. Pac Sci 48:273–282

Muscatine L (1973) Nutrition of corals. In: Jones OA, Endean R (eds) Biology and geology of coral reefs. Academic, New York, pp 77–115

Muscatine L (1980) Productivity of zooxanthellae. In: Falkowski PG (ed) Primary productivity in the Sea. Plenum Publishing Corporation, New York, pp 381–402

Muscatine L, Porter JW (1977) Reef corals: mutualistic symbioses adapted to nutrient-poor environments. Bioscience 27:454–460

Muscatine L, McCloskey LR, Marian RE (1981) Estimating the daily contribution of carbon from zooxanthellae to coral animal respiration. Limnol Oceanogr 26:601–611

Muscatine L, Falkowski PG, Porter JW, Dubinsky Z (1984) Fate of photosynthetic fixed carbon in light- and shade-adapted colonies of the symbiotic coral Stylophora pistillata. Proc R Soc Lond B 222:181–202

Muscatine L, Fallowski PG, Dubinsky Z, Cook PA, McCloskey LR (1989) The effect of external nutrient resources on the population dynamics of zooxanthellae in a reef coral. Proc R Soc Lond B 236(1284):311–324

Muscatine L, Ferrier-Pagès C, Blackburn A, Gates RD, Baghdasarian G, Allemand D (1998) Cell-specific density of symbiotic dinoflagellates in tropical anthozoaires. Coral Reefs 17:329–337

Muscatine L, Goiran C, Land L, Jaubert J, Cuif J-P, Allemand D (2005) Stable isotopes ($\delta^{15}N$ and $\delta^{13}C$) of organic matrix from coral skeleton. Proc Natl Acad Sci 102(5):1525–1530

Owens NJP (1987) Natural variations in ^{15}N in the marine environment. Adv Mar Biol 24:389–451

Palardy JE, Grottoli AG, Matthews KA (2005) Effects of upwelling, depth, morphology and polyp size on feeding in three species of Panamian corals. Mar Ecol Prog Ser 300:79–89

Palardy JE, Grottoli AG, Matthews KA (2006) Effect of naturally changing zooplankton concentrations on feeding rates of two coral species in the Eastern Pacific. J Exp Mar Biol Ecol 331:99–107

Papina M, Meziane T, Van Woesik R (2007) Acclimation effect on fatty acids of the coral Montipora digitata and its symbiotic algae. Comp Biochem Physiol B 147:583–589

Peterson BJ, Fry B (1987) Stable isotopes in ecosystem studies. Annu Rev Ecol Syst 18:293–320

Piniak G, Lipschultz F, McClelland J (2003) Assimilation and partitioning of prey nitrogen within two anthozoans and their endosymbiotic zooxanthellae. Mar Ecol Prog Ser 262:125–136

Porter JW (1974) Zooplankton feeding by the Caribbean reef-building coral Montastrea cavernosa. Proc 2nd Int Coral Reef Symp 1:111–125

Porter JW (1976) Autotrophy, heterotrophy and resource partitioning in Caribbean reef-building corals. Am Nat 110(975):731–742

Pratchett MS (1995) Dietary overlap among coral-feeding butterflyfishes (Chaetodontidae) at Lizard Island, northern Great Barrier Reef. Mar Biol 148:373–38

Pratchett MS, Gust N, Goby G, Klanten SO (2001) Consumption of coral propagules represents a significant trophic link between corals and reef fish. Coral Reefs 20:13–17

Rahav O, Dubinsky Z, Achituv Y, Falkowski PG (1989) Ammonium metabolism in the zooxanthellate coral Stylophora pistillata. Proc R Soc Lond B 236:325–337

Rau GH, Mearns AJ, Young DR, Olson RJ, Schafer HA, Kaplan IR (1983) Animal $^{13}C/^{12}C$ correlates with trophic level in pelagic food webs. Ecology 64:1314–1318

Rau GH, Ainley DG, Bengtson JL, Torres JJ, Hopkins TL (1992) $^{15}N/^{14}N$ and $^{13}C/^{12}C$ in Weddell sea birds, seals, and fish: implications for diet and trophic structure. Mar Ecol Prog Ser 84:1–8

Reynaud S, Ferrier-Pagès C, Sambrotto R, Juillet-Leclerc A, Jaubert J, Gattuso J-P (2002) Effect of feeding on the carbon and oxygen isotopic composition in the tissue and skeleton of the scleractinian coral Stylophora pistillata. Mar Ecol Prog Ser 238:81–89

Ribes M, Coma R, Gili J-M (1998) Heterotrophic feeding by gorgonian corals with symbiotic zooxanthella. Limnol Oceanogr 43:1170–1179

Richman S, Loya Y, Slobodkin LB (1975) The rate of mucus production by corals and its assimilation by the coral reef copepod Acartia negligens. Limnol Oceanogr 20:918–923

Rodolfo-Metalpa R, Peirano A, Houlbrèque F, Abbate M, Ferrier-Pagès C (2008) Effects of temperature, light and heterotrophy on the growth rate and budding of the temperate coral Cladocora caespitosa. Coral Reefs 27:17–25

Rodrigues L, Grottoli AG (2007) Energy reserves and metabolism as indicators of coral recovery from bleaching. Limnol Oceanogr 52:1874–1882

Rohwer F, Breitbart M, Jara J, Azam F, Knowlton N (2001) Diversity of bacteria associated with the Carribean coral Montastrea franksii. Coral Reefs 20:85–91

Rosenfeld M, Bresler V, Abelson A (1999) Sediment as a possible source of food for corals. Ecol Lett 2:345–348

Sammarco PW, Risk MJ, Schwarcz HP, Heikoop JM (1999) Cross-continental shelf trends in coral delta N-15 on the Great Barrier Reef: further consideration of the reef nutrient paradox. Mar Ecol Prog Ser 180:131–138

Sebens KP, Johnson AS (1991) Effects of water movement on prey capture and distribution of reef corals. Hydrobiologia 226:91–101

Sebens KP, Vandersall KS, Savina LA, Graham KR (1996) Zooplankton capture by two scleractinian corals Madracis mirabilis and Montastrea cavernosa in a field enclosure. Mar Biol 127:303–317

Sebens KP, Grace S, Helmuth B, Maney E, Miles J (1998) Water flow and prey capture by three scleractinian corals, Madracis mirabilis, Montastrea cavernosa, and Porites porites in a field enclosure. Mar Biol 131:347–360

Smith GJ, Muscatine L (1986) Carbon budgets and regulation of the population density of symbiotic algae. Endocyt C Res 3:212–238

Snidvongs A, Kinzie RA (1994) Effects of nitrogen and phosphorus enrichment on in vivo symbiotic zooxanthellae of Pocillopora damicornis. Mar Biol 118:705–711

Sorokin YI (1973) Role of microflora in metabolism and productivity of Hawaiian Reefs. Oceanology-USSR 13:262–267

Sorokin YI (1993) Coral reef ecology. Ecological studies. Springer, Berlin, p 465

Spencer-Davies P (1989) Short-term growth measurements of corals using an accurate buoyant weighing technique. Mar Biol 101(3): 389–395

Stambler Z, Dubinsky N (1996) Marine pollution and coral reefs. Glob Change Biol 2:511–526

Stambler N, Popper N, Dubinsky Z, Stimson J (1991) Effects of nutrient enrichment and water motion on the coral Pocillopora damicornis. Pac Sci 45:299–307

Swanson R, Hoegh-Guldberg O (1998) Amino acid synthesis in the symbiotic sea anemone *Aiptasia pulchella*. Mar Biol 131:83–93

Swart PK, Leder JJ, Szmant AM, Dodge RE (1996) The origin of variations in the isotopic record of scleractinian corals. II. Carbon. Geochism Cosmochim Acta 60:2871–2886

Szmant-Froelich A (1981) Coral nutrition: comparison of the fate of 14C from ingested labelled brine shrimp and from the uptake of NaH14CO3 by its zooxanthellae. J Exp Mar Biol Ecol 55:133–144

Szmant-Froelich A, Pilson MEQ (1980) The effects of feeding frequency and symbiosis with zooxanthellae on the biochemical composition of *Astrangia danae* Milne Edwards and Haime 1849. J Exp Mar Biol Ecol 48:85–97

Tambutté E, Allemand D, Bourge I, Gattuso JP, Jaubert J (1995) An improved ^{45}Ca protocol for investigating physiological mechanisms in coral calcification. Mar Biol 122:453–459

Titlyanov EA, Titlyanova TV, Tsukahara J, Van Woesik R, Yamazato K (1999) Experimental increases of zooxanthellae density in the coral *Stylophora pistillata* elucidate adaptive mechanisms for zooxanthellae regulation. Symbiosis 26:347–362

Titlyanov EA, Bil' K, Fomina L, Titlyanova T, Leletkin V, Eden N, Malkin A, Dubinsky Z (2000a) Effects of dissolved ammonium addition and host feeding with *Artemia salina* on photoacclimation of the hermatypic coral *Stylophora pistillata*. Mar Biol 137:463–472

Titlyanov EA, Tsukahara J, Titlyanova TV, Leletkin VA, Van Woesik R, Yamazato K (2000b) Zooxanthellae population density and physiological state of the coral *Stylophora pistillata* during starvation and osmotic shock. Symbiosis 28:303–322

Titlyanov EA, Titlyanova TV, Yamazato K, Van Woesik R (2001) Photo-acclimation of the hermatypic coral *Stylophora pistillata* while subjected to either starvation or food provisioning. J Exp Mar Biol Ecol 257:163–181

Treignier C, Grover R, Tolosa I, Ferrier-Pagès C (2008) Effect of light and feeding on the Fatty acid and sterol composition of zooxanthellae and host tissue isolated from the scleractinian coral *Turbinaria reniformis*. Limnol Oceanogr 53(6):2702–2710

Trench RK (1979) The cell biology of plant-animal symbiosis. Ann Rev Plant Physiol 30:485–531

Verde EA, McCloskey LR (1996) Photosynthesis and respiration of two species of algal symbionts in the anemone *Anthopleura elegantissima* (Brandt)(Cnidaria; Anthozoa). J Exp Mar Biol Ecol 195:187–202

Wang JT, Douglas AE (1999) Nitrogen recycling or nitrogen conservation in an alga-invertebrate symbiosis? J Exp Biol 201:2445–2453

Weiner S, Addadi L (1991) Acidic macromolecules of mineralized tissues. The controllers of crystal formation. Trends Biochem Sci 16:252–256

Wellington GM (1982) An experimental analysis of the effects of light and zooplankton on coral zonation. Oecologia 52:311–320

Wild C, Huettel M, Klueter A, Kremb SG, Rasheed MYD, Jørgensen BB (2004) Coral mucus as an energy carrier and particle trap in the reef ecosystem. Nature 428:66–70

Yahel R, Yahel G, Berman T (2005) Diel pattern with abrupt crepuscular changes of zooplankton over a coral reef. Limnol Oceanogr 50:930–944

Yamamuro M, Kayanne H, Minagawa M (1995) Carbon and nitrogen stable isotopes of primary producers in coral reef ecosystems. Limnol Oceanogr 40:617–621

Yamashiro H, Oku H, Onaga K (2005) Effect of bleaching on lipid content and composition of Okinawan corals. Fish Sci 71:448–453

Yonge CM (1930a) Studies on the physiology of corals. I. Feeding mechanisms and food. Sci Rep Great Barrier Reef Exped 1:13–57

Yonge CM (1930b) Studies on the physiology of corals. III. Assimilation and excretion. Sci Rep Great Barrier Reef Exped 1:83–91

Yonge CM, Nicholls AG (1931) Studies on the physiology of corals. IV. The structure, distribution and physiology of the zooxanthellae. Sci Rep Great Barrier Reef Exped 1:135–176

Fish or Germs? Microbial Dynamics Associated with Changing Trophic Structures on Coral Reefs

Elizabeth A. Dinsdale and Forest Rohwer

1 Introduction

Overfishing major predators has dramatically changed the trophic structures of coral reefs. Here, we argue that the photosynthate, which would normally support the large predators via trophic transfers, is being used by microbes (Bacteria and Archaea) on degraded reefs. The supply of higher concentrations of photosynthate to the microbes increases their population size and enables heterotrophic microbes to dominate the community. In turn, the heterotrophic microbes detrimentally affect the corals causing disease outbreaks and death, which causes the phase shift from coral to fleshy algae (including macroalgae and turf algae) dominated reefs. To succeed, conservation and restoration efforts need to understand and consider the influence of microbes.

2 Trophic Structure on Coral Reefs

Most coral reefs exist today with few top-level predators: a trophic structure that is very different from historical coral reefs (Jackson et al. 2001; Pandolfi et al. 2005; Essington et al. 2006; Hughes et al. 2007; Jackson 2008). Friedlander and DeMartini (2002) showed that the fish biomass on nearly pristine coral reefs in the Northwest Hawaiian Islands have a trophic structure resembling an inverted pyramid, where more than 54% of the fish biomass is apex predators. In comparison, in the Main Hawaiian Islands apex predators make up less than 3% of the trophic structure (Friedlander and DeMartini 2002). Similarly, on the remote and completely protected Kingman Reef in the Northern Line Islands (Halpern et al. 2008; Knowlton and Jackson. 2008), an inverted trophic biomass pyramid also exists, where 85% of the total fish biomass are sharks and large piscivores. On the most heavily fished reefs of Kiritimati, which has the highest levels of human activity in the same island chain, top predators are essentially nonexistent (DeMartini et al. 2008; Sandin et al. 2008). Calculations based on the number of monk seals, a major predator in the Caribbean, prior to fishing also support the idea that the trophic structure in the Caribbean would have been an inverted biomass pyramid (McClenachan and Cooper 2008).

Coupled with the removal of apex predators on coral reefs is an increase in fleshy algae cover (Done 1992; Hughes 1994; Mumby et al. 2006; Hughes et al. 2007). These phase shifts on coral reefs have occurred from the Caribbean to the Pacific (Done 1992; Hughes 1994). The phase shifts often coincide with an increase in nutrient inputs from industrialized agriculture, including dissolved forms of inorganic nitrogen and phosphate. Whether the removal of fish or increase in dissolved inorganic nitrogen and phosphate compounds is the major factor leading to the death of corals has received much debate in the literature (Littler et al. 1991; Hughes 1994; Lapointe 1997; McCook et al. 2001; Smith et al. 2001; Thacker et al. 2001; Boyer et al. 2004; Albert et al. 2008).

3 Herbivores

On most occasions, phase shifts are associated with overfishing (Munro 1983; Hughes 1994). However, the trajectory of a coral reef from high-coral-covered to fleshy-algae-dominated does not always occur immediately after the removal of herbivorous fish (Bellwood et al. 2004; Ledlie et al. 2007). The time lag can occur because of the presence of other herbivorous groups, such as sea urchins, on the reef. It is likely that the large numbers of sea urchins help maintain the high cover of corals in the Caribbean until a disease killed most of them (Lessios 1988; Forocucci 1994; Lessios 2005). Sea urchins do not appear to be able to protect Pacific coral reefs,

---

E.A. Dinsdale (✉) and F. Rohwer
Biology Department, San Diego State University, 5500 Campanile Drive, San Diego, CA 92182, USA
e-mail: Elizabeth_dinsdale@hotmail.com; forest@sunstroke.sdsu.com

possibly because these sea urchins do not feed on fleshy algae with high tannin and phenol content (Coppard and Campbell 2007; Nordemar et al. 2007). The types and size of herbivores present affects their efficiency in removing fleshy algae and therefore the rate of the phase shift (Bellwood et al. 2006; Fox and Bellwood 2008; Lokrantz et al. 2008).

Predators exert various pressures on the trophic structure of the ecosystem (Connell 1998; Carpenter et al. 2008). One expected outcome of removing predators is an increase in prey biomass, which are mostly herbivores. On coral reefs with high and low predator biomass, the biomass of the prey species remains relatively constant (DeMartini et al. 2008; Sandin et al. 2008) and in protected coral reefs where there are higher predator biomass there are more herbivores (Mumby et al. 2006; Kramer and Heck 2007; Mumby et al. 2007). Therefore, even when the herbivores are not removed by fishing, without predators they are not effective at removing fleshy algae.

4 Dissolved Inorganic Nitrogen and Soluble Reactive Phosphorus

Increases in dissolved inorganic nitrogen (DIN) and soluble reactive phosphorus (SRP) are implicated in the death of corals and have received much research attention (Littler et al. 1991; Lapointe 1997; Boyer et al. 1999; Haynes 2001; Koop et al. 2001; McClanahan et al. 2003; D'Croz et al. 2005; Schaffelke et al. 2005; Littler et al. 2006; Bell et al. 2007). A threshold value for both of these nutrients has been proposed, above which a reef is considered eutrophic and the coral cover is expected to decline (Lapointe 1997). Proving the effect of increased concentrations of DIN and SRP on coral health has been difficult. A large field experiment that increased concentration of these nutrients by approximately 10 times found that DIN and SRP did not kill corals (Koop et al. 2001). In the highest concentration phase of the same experiment, where concentrations of DIN and SRP were increased by 30 times, one coral species showed increased mortality levels compared to the control sites (Koop et al. 2001). Similarly, large-scale treatments of nubbins from several Caribbean corals did not find any adverse effects of either DIN or SRP (acting singly or in combination) (Kuntz et al. 2005; Kline et al. 2006), suggesting that nutrients on coral reefs are not the direct cause of coral mortality.

Many nutrient studies have been conducted on the Great Barrier Reef, because it is situated adjacent to an area of high levels of industrial agriculture runoff. In a review of these studies, Furnas et al. (2005) showed that nutrient concentrations were low given the high levels of inputs (except during floods). It was proposed that the low levels were maintained by pelagic phytoplankton communities. Most of the nutrients delivered to the benthic community are delivered as organic matter from the phytoplankton (Furnas et al. 2005; Atkinson 2010).

5 Dissolved Organic Carbon and Coral Reef Microbes

In contrast to DIN and SRP, an increased organic carbon load (by approximately twofold to threefold) causes coral bleaching and mortality (Hodgson 1990; Fabricius et al. 2003; Kuntz et al. 2005; Kline et al. 2006). These deleterious effects increase exponentially with time. Some species-level differences have been observed: *Montastraea annularis* (Kline et al. 2006) and *Agaricia tenuifolia* (Kuntz et al. 2005) showed high levels of mortality associated with organic carbon load, while *Porites furcata* was less susceptible (Kuntz et al. 2005). Similarly, *Oxypora glabra*, *Porites lobata*, and *Pocillopora meandrina* experienced high levels of mortality when they were exposed to organic loaded sediment, whereas *Montipora verrucosa* was less susceptible (Hodgson 1990). Organic-carbon-killing of corals, whether in the form of simple sugars, polysaccharides, or complex compounds associated with sediment, is negated by antibiotics (Hodgson 1990; Kuntz et al. 2005; Kline et al. 2006), which suggests microbes are playing a role.

The standing stock of dissolved organic carbon (DOC) is determined by a combination of carbon fixation via photosynthesis, consumption via heterotrophic bacterial growth, and import/export from exogenous sources (Carlson 2002). In general, the standing stock of DOC in the ocean is 70–80 μM and grossly consists of two pools. Labile DOC (mostly fresh photosynthate) (Carlson 2002; Carlson et al. 2002), which is rapidly turned over within minutes to days, and represents a major dynamical component of the marine carbon cycle (Opsahl and Benner 1997). Most of the measured standing stock is actually refractory DOC, which can exist for thousands of years (Carlson 2002). The DOC pool consists of thousands, and possibly millions, of different compounds. Together, these attributes (i.e., the rapid turnover of some DOC components and the complex structure) make it difficult to measure DOC and many of the earlier measurements on coral reefs were unreliable (UNESCO 1994).

Much of the DOC on coral reefs is produced by *in situ* photosynthesis (i.e., is not transported onto the reef) (Sakka et al. 2002). Different forms of DOC are produced by the actions of microbes, plankton (both phytoplankton and zooplankton), and other larger organisms, including corals and sponges (Ducklow and Mitchell 1979b; Crossland et al. 1980; Crossland 1987). Mucus and coral spawn have been shown to be broken down such that they contribute to the

DOC pool (Ducklow and Mitchell 1979b; Crossland 1987; van Duyl and Gast 2001; Wild et al. 2004a, b, 2005; Allers et al. 2008; Patten et al. 2008b; Wild et al. 2008). Mean DOC measurements (including both the liable and refractory fractions) associated with coral reefs range from 34 to 160 μM (Pages et al. 1997; Torreton et al. 1997; van Duyl and Gast 2001; Dinsdale et al. 2008), suggesting there is variable production and use of DOC on each individual reef. DOC flux is higher above the coral reef structure than surrounding "oceanic" water (Hata et al. 2002). DOC levels are highest in the water above the surface of the corals and decrease within the crevices between coral colonies and the reef structure (van Duyl and Gast 2001; van Duyl et al. 2006). The labile component of the DOC is difficult to measure, but has been estimated to be approximately 20% in some coral reef waters (Sakka et al. 2002).

Whether the water column-associated microbes or benthic community are driving variations in the DOC pool is a major outstanding question. In the open ocean (where more research has been conducted), microbes consume most of the liable DOC and have been shown to be carbon limited (Carlson et al. 1996; Carlson et al. 2002; Carlson et al. 2004). On coral reefs, complex interactions between DOC, microbial growth, and filter feeders (e.g., sponges) exist (de Goeij and van Duyl 2007; de Goeij et al. 2008a,b). Using labeled carbon, it was shown that both the sponge cells and the associated microbes assimilated DOC (de Goeij et al. 2008a), which may account for much of the consumption of the DOC pool on coral reefs. DOC can also be assimilated into the sediments (Wild et al. 2008) but this has been estimated to be low (Charpy and Charpy-Roubaud 1990). Despite this complexity, one take-home message from the numerous studies conducted in other marine environments is that higher production of labile DOC will support more heterotrophic microbes (Fuhrman et al. 1989; Carlson et al. 1996; Carlson et al. 2002; Carlson et al. 2004) and this may occur on coral reef ecosystems (Charpy and Charpy-Roubaud 1990).

6 Microbes and the Coral Holobiont

Microbes are an integral part of the coral holobiont (Ducklow and Mitchell 1979a; Rohwer et al. 2002; Rohwer and Kelley 2004; Rosenberg and Loya 2004) and provide important nutrients and other resources to the coral host (Reshef et al. 2006). In addition to zooxanthellae, Bacteria, Archaea, Fungi, and viruses are all found in association with corals and many form species-specific associations (Rohwer et al. 2001; Bourne and Munn 2005; Beman et al. 2007; Marhaver et al. 2008; Patten et al. 2008a; Littman et al. 2009). Functional roles of coral-associated microbes include ammonia, nitrate, sulfur, and carbon metabolism (Lessar et al. 2004; Beman et al. 2007; Lesser et al. 2007; Wegley et al. 2007; Siboni et al. 2008). Symbiotic cyanobacteria produce nitrogen that is taken up by the zooxanthellae (Lesser et al. 2004; Lesser et al. 2007). Together, these studies demonstrate the fundamental roles that the microbes play in the growth and survival of the coral holobiont.

Much of the energy in the coral holobiont is provided via photosynthesis by the zooxanthellae (Falkowski et al. 1993; Yellowlees et al. 2008). The amount of DIN and SRP in the water column affects the rate of production and translocation of products by the zooxanthellae (Muscatine and D'elia 1978; Tanaka et al. 2006; Yellowlees et al. 2008). Simple sugars are converted by the animal into complex carbohydrates, lipids, and released as coral mucus or DOC (Wild et al. 2004b). The exact composition of coral mucus is unknown and varies from species to species (Ducklow and Mitchell 1979b). The mucus forms a layer between the animal and surrounding seawater and is colonized by hundreds of millions of microbes per square centimeter (Wegley et al. 2004; Koren and Rosenberg 2006; Johnston and Rohwer 2007; Klaus et al. 2007; Lampert et al. 2008). Metagenomic analysis of these coral-associated microbes shows that they encode enzymes to metabolize the coral mucus and transport the resulting sugars into their cells for energy (Wegley et al. 2007). The species-specific nature of microbial association with corals (Rohwer et al. 2002; Bourne and Munn 2005; Littman et al. 2009) may be based on a mutualistic relationship of mucus composition and enzymes to metabolize the mucus, as first suggested by Ducklow and Mitchell (1979a). Disruption of this relationship by addition of exogenous DOC leads to uncontrolled microbial growth and coral death (Hodgson 1990; Kuntz et al. 2005; Kline et al. 2006).

7 Local Connections Between Coral Disease, Fishing, and Fleshy Algae

There has been an increase in the types and incidence of coral disease over the last 30 years (Green and Bruckner 2000; Aronson and Precht 2001; Harvell et al. 2002; Harvell et al. 2007). While the number of described disease morphologies has increased, specific pathogen for each morphology has not been identified to date (Harvell et al. 1999; Rosenberg and Loya 2004; Willis et al. 2004; Rosenberg et al. 2007; Toledo-Hernandez et al. 2008; Rosenberg and Kurshmaro 2010). This suggests that the increased level of disease may be associated with a change in the relationship between microbes and corals, rather than the introduction of new specific pathogenic microbial strains. The relationship between specific and opportunistic pathogens and the influence of host response, infective dose, and environmental conditions is visualized in Fig. 1. Specific pathogens, such as those associated with black band disease, white band disease, etc., described in Rosenberg and Kurshmaro (2010), would cause disease when the host response is strong, even in pristine environments. Diseases, such as bacterial bleaching, would occur when the microbes are present and there is an environmental trigger, such as, increased temperature, which reduces host response or increasing microbial virulence allowing the microbe to cause disease. If part of the normal microbial flora increases in abundance it may lead to disease, as was shown in the dosing experiment by Kline et al. (2006). Where environmental influences increase, such as higher temperatures, more dissolved organic carbon, etc., the microbial dose rate increases and the host response is weakened, microbes that are normally present within the environment may opportunistically cause disease. We believe that most of the world's coral reefs are approaching the right-hand side of the continuum and overfishing and/or nutrient additions are the drivers of increased coral diseases leading to increased mortality and the mechanism of phase shifts.

One of the most straightforward ways that fishing changes coral reefs is by increasing the number and length of coral – fleshy algal interaction zones. Smith et al. (2006) showed that when corals are placed immediately next to fleshy algae, the corals die. Death occurs even when there is an intervening filter that prevents the passage of microbes and viruses to the corals but allows the flow of dissolved compounds like DOC. Addition of antibiotics into the experimental setup prevented coral death, implying the involvement of microbes. The pattern occurred across a range of corals and fleshy algae, with some species-level variation. Oxygen microprobe readings showed that the coral – fleshy algal interface was hypoxic. The hypoxia was relieved by antibiotic addition. Together with the previously mentioned DOC experiments, these results suggest that DOC released by the fleshy algae enhances growth of microbes on the coral surface. These microbes then grow so fast that they use up all the oxygen and smother the coral.

In the Smith et al. (2006) experiment, coral death was associated with change in the microbial community already present within the water column or associated with the corals. Nugues et al. (2004) showed that fleshy algae harbor coral pathogens, which would enhance the negative effects of fleshy algae on corals via microbial activity. Vermeij et al. (2008) showed that algal exudates kill coral recruits and the addition of antibiotics alleviates this mortality.

8 Large-Scale Connections Between Coral Disease, Fishing, Fleshy Algae, and Eutrophication

In addition to the microbial activity associated with fleshy algae – coral interaction zones, the microbe–coral–fleshy algae DOC model could also work at larger scales. Figure 2 outlines how DOC loading might work at a reef scale. On a "Healthy Coral Reef," primary production is nutrient limited (DIN and SRP are represented by *green* arrows). The limited nutrients are used by the fleshy algae and the zooxanthallae for photosynthesis. Most of the organic carbon produced from photosynthesis is assimilated (via trophic transfer) by the complex macroorganism-dominated food web, which includes the fish and corals. As organic carbon (represented by the *black* arrows) moves through the food web, it is respired and lost as carbon dioxide (~90% from trophic level to trophic level). The inverted trophic biomass pyramid described for reefs with minimal fishing (Friedlander and DeMartini 2002; Mora et al. 2006; DeMartini et al. 2008; Sandin et al. 2008) suggests there is a movement of carbon through the food web until it is stored in the apex predators. There are relatively low numbers of sponges on the healthy coral reef (Aerts 1998), because DOC and the associated microbial food web is small. Coral growth is successful in the minimal nutrient conditions, because the coral animal is able catch plankton and their associated microbes fix nitrogen (Lesser et al. 2007; Wegley et al. 2007). DOC released into the water column is used by the water column microbes, directly by some filter feeders (link not shown in diagram),

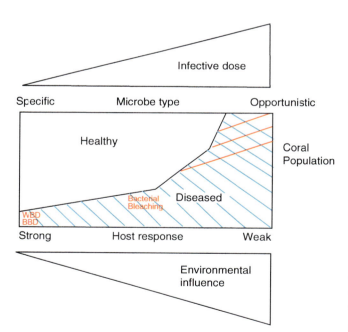

Fig. 1 The proposed interactions between host response, microbe type, infective dose, and environmental conditions, and how they lead to an increase in the prevalence of disease in the coral populations are provided in a stylized diagram. The red hatched area is the condition that most of the world's coral reefs are approaching. BBD=black band disease, WBD=white band disease, and bacterial bleaching are defined in Rosenberg and Kurshmaro (2010)

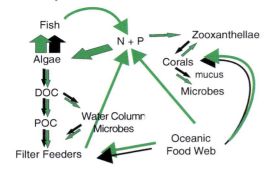

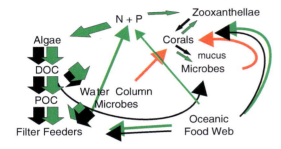

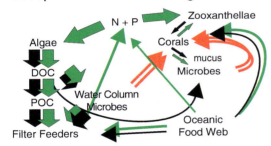

Fig. 2 A model of the proposed response of the microbes within a coral reef ecosystem under three different conditions: (**a**) healthy, (**b**) overfished, and (**c**) overfished with eutrophication. The model is stylized and only shows major interactions. The green arrows represent the flow of nutrients (dissolved inorganic nitrogen and soluble reactive phosphate), the black arrows represent the flow of organic carbon and the red arrows suggest a negative impact of one organism on the next. Note: healthy is being used to denote a coral reef prior to industrialized fishing

and can spontaneously form into particulate organic matter (Chin et al. 1998; Verdugo et al. 2004). The particulate matter is used by benthic filter feeders and pelagic planktivores. The carbon assimilated by the microbes passes to the macroorganism food web via the microbial loop (Azam et al. 1983; Azam and Smith 1991; Azam 1998) (but for simplification this link is not shown either).

Overfishing disrupts the system by removing the macroorganisms, such as sharks, that are main consumers of primary production via trophic transfers (Hairston and Hairston 1993; Hairston and Hairston 1997). The removal of the apex predators does not appear to cause a predation release that allows lower trophic levels to increase and consume more fleshy algae (Mumby et al. 2006; Newman et al. 2006; Mumby et al. 2007; Sandin et al. 2008). In fact, it appears that the remaining herbivores are less effective in consuming fleshy algae and there is a reduction in the amount of carbon that is drawn up the macroorganismal food web. This excess carbon (released by the fleshy algae and not drawn up the macrobial food web) is available for the microbial food web, causing them to become more abundant. Increased microbial activity directly threatens corals (*red* arrows), increasing the level of coral disease. Both the water column microbes and those associated with the coral may be involved in the death of the corals. The extra microbes support more filter feeders (Aerts 1998; Bak et al. 1998; Yahel et al. 2003) and their bioeroding activity may help explain the rapid loss of rugosity on declining reefs.

Eutrophication, by providing greater amounts of DIN and SRP to the fleshy algae, increases fleshy algae growth and thus increases DOC production further (Schaffelke and Klumpp 1998). Zooxanthellae are also released from their nutrient limitation (Hoegh-Guldberg and Smith 1989; Falkowski et al. 1993; Dubinsky and Stambler 1996; Grover et al. 2003; Hoegh-Guldberg et al. 2004; Grover et al. 2006). Increased zooxanthellae production may increase the number of coral-associated microbes (Ducklow and Mitchell 1979c; Pascal and Vacelet 1981) and encourages microbes that have more pathogenic characteristics (i.e., ones that grow more rapidly and consume oxygen). Nutrient additions may also increase the severity of an existing coral disease lesion (Voss and Richardson 2006).

In this model, overfishing and eutrophication are synergistic. Note that eutrophication by itself can also lead to phytoplankton blooms that kill coral reefs (Abram et al. 2003) and fleshy algae that directly overgrows the coral (McCook et al. 2001). In the model, a positive feedback occurs because coral death, due to microbial activity, creates more free space for fleshy algal growth, which in turn leads to more DOC release into the water. The proposed positive feedback is not at odds with the alternative stable state hypotheses (Done 1992; Knowlton 1992; Hughes 1994; Tanner et al. 1994; Bellwood et al. 2004; Pandolfi et al. 2005; Hughes et al. 2007; Knowlton and Jackson. 2008), rather it is a mechanism for the system bifurcation. As stated by Mora (2006; 2008), "the drivers of coral degradation have been challenging" to identify – we suggest that microbialization of the reef may be the driver.

To directly investigate this model in the field, the microbial communities on coral reefs with varying fish and benthic communities were studied in the Northern Line Islands. As mentioned above, this archipelago lies in a remote area of the Pacific Ocean, with a gradient of human and fishing activities, from preindustrial to those experienced on most coral reefs today (DeMartini et al. 2008; Dinsdale et al. 2008; Halpern et al. 2008; Knowlton and Jackson. 2008; Sandin et al. 2008). Consistent with the proposed model, there was a positive correlation between microbial number and fishing activity across the four islands. There were 10 times more microbes per ml of seawater on Kiritimati (which had essentially no apex fish predators, extremely low coral cover, and high fishing levels) than there were on Kingman (the relatively pristine coral reef) (Dinsdale et al. 2008). A more extensive survey conducted on the coral reefs, the entire way around the island of Kiritimati, which provides a fishing gradient without the latitudinal gradient of the Northern Line Islands expedition, showed that microbial number were positively correlated with the area of the reef that received the most fishing pressure (McDole et al. 2008).

There were also dramatic changes in microbial trophic interactions in the Northern Line Islands. On Kingman, there were approximately equal numbers of microbial heterotrophs and autotrophs. As the amount of human activity increased and the fish biomass declined, the microbes were dominated by heterotrophs. On Kiritimati approximately 30% of the microbes were classified as "super-heterotrophs," Bacteria that live in extremely energy-rich environments. These microbes are classically thought of as opportunistic pathogens, and include representatives that most people have heard of including, *Escherichia coli*, *Staphylococcus*, *Streptococcus*, *Enterobacteria*, etc. (Dinsdale et al. 2008).

These super-heterotrophs would be expected to: (a) act as opportunistic pathogens, and (b) consume large amounts of DOC. The benthic observations from the Northern Line Islands support both of these points. First, the presence of the super-heterotrophic Bacteria was directly correlated with a known historical decline in coral cover, as well as an increase in the prevalence of unhealthy surviving corals (Dinsdale et al. 2008; Knowlton and Jackson. 2008; Sandin et al. 2008). Second, the standing stock of DOC was much lower on the severely degraded coral reef because the super-heterotrophic microbes were eating it (Dinsdale et al. 2008).

9 The DDAMed Model

The ways DOC, Disease, fleshy Algae, and Microbes (DDAM) work together to kill corals is summarized in Fig. 3. The feedback loop is driven by overfishing and/or eutrophication, which encourage algal growth. It can also be driven by other factors like bleaching or heavy metal poisoning, which will compromise the coral's ability to mount an immune response to the super-heterotrophic microbes. Increasing sea surface temperatures may be particularly problematic because it will: (a) cause bleaching, (b) increase

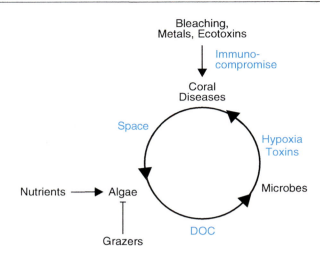

Fig. 3 The proposed feedback loop created by the elements dissolved organic carbon (DOC), Disease, fleshy Algae, and Microbes (DDAMed), which work together to undermine coral health. The removal of grazers and/or increase in nutrients allows higher cover of fleshy algae, releasing more DOC, which enhances microbial growth leading to low oxygen levels that cause disease in corals. Coral disease may be exacerbated by other environmental variables that compromise the immune response by the corals, all of which leads creates more bare space for fleshy algae recruitment

microbial activity, and (c) enhance photosynthesis and DOC release. Recent studies suggest that increased temperature is also related to reduction in herbivores, which will further increase fleshy algae (Smith 2008).

10 The Future

To further understand the relationship between microbes, DOC, and coral reef decline, there are many research questions that need to be answered. The production rate of microbes on coral reefs associated with varying levels of human activity needs to be estimated. The consumption rate of DOC and production by fleshy algae needs to be measured. In particular, the amount of labile DOC produced by different types of fleshy algae needs to be quantified (e.g., do turf algae produce more labile DOC than *Halimedia* spp.?). A major unanswered question is why removal of predators does not lead to a dramatic increase in the biomass of lower trophic levels (Mumby et al. 2006; Newman et al. 2006; Sandin et al. 2008). It is this inability of the macroorganisms to assimilate the organic carbon on disturbed reefs that appears to favor the microbes. Corals can adapt and respond to microbial changes, but this area of research is in its infancy (Mullen et al. 2004; Rosenberg et al. 2007). The movement of microbes onto and off the reef also needs to be considered in the future. The viral shunt, which cycles DOC within the microbial loop (Suttle 2007), will play a role in the proposed

DDAMed model, but to date there is not enough research to include these organisms in the model. And for restoration, experiments that increase fish on overfished reefs and assess the response of the microbial community need to be performed.

It currently appears that human activities on coral reefs are doing everything to give microbes the upper hand. To conserve these remarkable ecosystems understanding how to prevent microbial heterotrophic dominance or how to reestablish a balanced microbial community on coral reefs is required.

Acknowledgments The authors would like to thank the participants of the Line Islands Expedition and the San Diego Coral Club for their helpful discussions, criticisms, and insights.

References

Abram N, Gagan M, McCulloch M, Chappell J, Hantoro W (2003) Coral reef death during the 1997 Indian Ocean dipole linked to Indonesian wildfires. Science 301:952–955

Aerts LAM (1998) Sponge/coral interactions in Caribbean reefs: analysis of overgrowth patterns in relation to species identity and cover. Mar Ecol Prog Ser 175:241–249

Albert S, Udy J, Tibbetts IR (2008) Responses of algal communities to gradients in herbivore biomass and water quality in Marovo Lagoon, Solomon Islands. Coral Reefs 27:73–82

Allers E, Niesner C, Wild C, Pernthaler J (2008) Microbes enriched in seawater after addition of coral mucus. App Environ Microbiol 74:3274–3278

Aronson RB, Precht WF (2001) White-band disease and the changing face of Caribbean reefs. Hydrobiologia 460:25–38

Atkinson JM (2010) Biogeochemical cycles on coral reefs. In: Dubinsky Z, Stambler N (eds) Coral reefs and ecosystem in transition. Springer, Berlin

Azam F (1998) Microbial control of oceanic carbon flux: the plot thickens. Science 280:694–696

Azam F, Smith DC (1991) Bacterial influence on the variability in the ocean's biogeochemical state: a mechanistic view. In: Demers S (ed) Particle analysis in oceanography. Springer, Berlin, pp 213–235

Azam F, Fenchel T, Field JG, Gray JS, Reil LAM, Thingstad F (1983) The ecological role of water-column microbes in the sea. Mar Ecol Prog Ser 10:257–263

Bak RPM, Joenje M, de Jong I, Lambrechts DYM, Nieuwand G (1998) Bacterial suspension feeding by coral benthic organisms. Mar Ecol Prog Ser 175:285–288

Bell PRF, Lapointe BE, Elmetri I (2007) Reevaluation of ENCORE: support for the eutrophication threshold model for coral reefs. Ambio 36:416–424

Bellwood DR, Hughes TP, Folke C, Nystrom M (2004) Confronting the coral reef crisis. Nature 429:827–833

Bellwood DR, Hughes TP, Hoey AS (2006) Sleeping functional group drives coral-reef recovery. Curr Biol 16:2434–2439

Beman JM, Roberts KJ, Wegley L, Rohwer F, Francis CA (2007) Distribution and diversity of archaeal ammonia monooxygenase genes associated with corals. App Environ Microbiol 73:5642–5647

Bourne DG, Munn CB (2005) Diversity of bacteria associated with the coral *Pocillopora damicornis* from the Great Barrier Reef. Environ Microbiol 7:1162–1174

Boyer JN, Fourqurean JW, Jones RD (1999) Seasonal and long-term trends in the water quality of Florida Bay (1989-1997). Estuaries 22:417–430

Boyer KE, Fong P, Armitage AR, Cohen RA (2004) Elevated nutrient content of tropical fleshy algae increases rates of herbivory in coral, seagrass, and mangrove habitats. Coral Reefs 23:530–538

Carlson CA (2002) Production and removal processes. In: Hansell DA, Carlson CA (eds) Biogeochemistry of marine dissolved organic matter. Academic, Amsterdam, pp 91–153

Carlson CA, Ducklow HW, Sleeters TD (1996) Stocks and dynamics of bacterioplankton in the northwestern Sargasso Sea. Deep Sea Res II 43:491–515

Carlson CA, Giovannoni SJ, Hansell DA, Goldberg SJ, Parsons R, Otero MP, Vergin K, Wheeler BR (2002) Effect of nutrient amendments on bacterioplankton production, community structure, and DOC utilization in the northwestern Sargasso Sea. Aquat Microb Ecol 30:19–36

Carlson CA, Giovannoni SJ, Hansell DA, Goldberg SJ, Parsons R, Vergin K (2004) Interactions between DOC, microbial processes, and community structure in the mesopelagic zone of the northwestern Sargasso Sea. Limnol Oceanogr 49:1073–1083

Carpenter SR, Brock WA, Cole JJ, Kitchell JF, Pace ML (2008) Leading indicators of trophic cascades. Ecol Lett 11:128–138

Charpy L, Charpy-Roubaud CJ (1990) Trophic structure and productivity of the lagoonal communities of Tikehau Atoll (Tuamotu Archipelago, French Polynesia). Hydrobiologia 207:43–52

Chin W-C, Orellana MV, Verdugo V (1998) Spontaneous assembly of marine dissolved organic matter into polymer gels. Nature 391:568–572

Connell SD (1998) Effects of predators on growth, mortality and abundance of a juvenile reef-fish: evidence from manipulations of predator and prey abundance. Mar Ecol Prog Ser 169:251–261

Coppard SE, Campbell AC (2007) Grazing preferences of diadematid echinoids in Fiji. Aquat Bot 86:204–212

Crossland CJ (1987) *In situ* release of mucus and DOC-lipid from the corals *Acropora variabilis* and *Stylophora pistillata* in different light regimes. Coral Reefs 6:35–42

Crossland CJ, Barnes DJ, Borowitzka MA (1980) Diurnal lipid and mucus production in the Staghorn coral *Acropora acuminata*. Mar Biol 60:81–90

D'Croz L, Del Rosario JB, Gondola P (2005) The effect of fresh water runoff on the distribution of dissolved inorganic nutrients and plankton in the Bocas del Toro archipelago, Caribbean Panama. Caribbean J Sci 41:414–429

de Goeij JM, van Duyl FC (2007) Coral cavities are sinks of dissolved organic carbon (DOC). Limnol Oceanogr 52:2608–2617

de Goeij JM, Moodley L, Houtekamer M, Carballeira NM, van Duyl FC (2008a) Tracing C^{-13}-enriched dissolved and particulate organic carbon in the bacteria-containing coral reef sponge *Halisarca caerulea*: evidence for DOM feeding. Limnol Oceanogr 53:1376–1386

de Goeij JM, van den Berg H, van Oostveen MM, Epping EHG, Van Duyl FC (2008b) Major bulk dissolved organic carbon (DOC) removal by encrusting coral reef cavity sponges. Mar Ecol Prog Ser 357:139–151

DeMartini EE, Friedlander AM, Sandin SA, Sala E (2008) Differences in fish-assemblage structure between fished and unfished atolls in the northern Line Islands, central Pacific. Mar Ecol Prog Ser 365:199–215

Dinsdale EA, Pantos O, Smriga S, Edwards RA, Wegley L, Angly F, Brown E, Haynes M, Krause L, Sala E, Sandin SA, Vega TR, Willis BL, Knowlton N, Azam F, Rohwer F (2008) Microbial ecology of four coral atolls in the Northern Line Islands. PLoS ONE 3:e1584

Done TJ (1992) Phase shifts in coral reef communities and their ecological significance. Hydrobiologia 247:121–132

Dubinsky Z, Stambler N (1996) Marine pollution and coral reefs. Glob Change Biol 2:511–526

Ducklow HW, Mitchell R (1979a) Bacterial populations and adaptations in the mucus layers on living corals. Limnol Oceanogr 24:715–725

Ducklow HW, Mitchell R (1979b) Composition of mucus released by coral reef coelenterates. Limnol Oceanogr 24:706–714

Ducklow HW, Mitchell R (1979c) Observations on naturally and artificially diseased tropical corals: a scanning electron microscope study. Microb Ecol 5:215–223

Essington TE, Beaudreau AH, Wiedenmann J (2006) Fishing through marine food webs. Proc Natl Acad Sci 103:3171–3175

Fabricius K, Wild C, Wolanski E, Abele D (2003) Effects of transparent exopolymer particles (TEP) and muddy terrigenous sediments on the survival of hard coral recruits. Coast Shelf Sci 57:613–621

Falkowski PG, Dubinsky Z, Muscatine L, McCloskey L (1993) Population control in symbiotic corals. Bioscience 43:606–611

Forcucci D (1994) Population density, recruitment and 1991 mortality event of *Diadema antillarum* in the Florida Keys. Bull Mar Sci 54:917–928

Fox RJ, Bellwood DR (2008) Remote video bioassays reveal the potential feeding impact of the rabbitfish *Siganus canaliculatus* (f: Siganidae) on an inner-shelf reef of the Great Barrier Reef. Coral Reefs 27:605–615

Friedlander AM, DeMartini EE (2002) Contrasts in density, size, and biomass of reef fishes between the northwestern and the Main Hawaiian Islands: the effects of fishing down apex predators. Mar Ecol Prog Ser 230:253–264

Fuhrman JA, Sleeter TD, Carlson CA, Proctor LM (1989) Dominance of bacterial biomass in the Sargasso Sea and its ecological implications. Mar Ecol Prog Ser 57:207–217

Furnas M, Mitchell A, Skuza M, Brodie J (2005) In the other 90%: phytoplankton responses to enhanced nutrient availability in the Great Barrierr Reef Lagoon. Mar Pollut Bull 51:253–265

Green EP, Bruckner AW (2000) The significance of coral disease epizootiology for coral reef conservation. Biol Conserv 96:347–361

Grover R, Maguer JF, Allemand D, Ferrier-Pages C (2003) Nitrate uptake in the scleractinian coral *Stylophora pistillata*. Limnol Oceanogr 48:2266–2274

Grover R, Maguer JF, Allemand D, Ferrier-Pages C (2006) Urea uptake by the scleractinian coral *Stylophora pistillata*. J Exp Mar Biol Ecol 332:216–225

Hairston NG Jr, Hairston NG Sr (1993) Cause-effect relationships in energy flow, trophic structure, and interspecific interactions. Am Natl 142:379–411

Hairston NGJ, Hairston NGS (1997) Does food web complexity eliminate trophic-level dynamics? Am Nat 149:1001–1007

Halpern BS, Walbridge S, Selkoe KA, Kappel CV, Micheli F, D'Agrosa C, Bruno JF, Casey KS, Ebert C, Fox HE, Fujita R, Heinemann D, Lenihan HS, Madin EMP, Perry MT, Selig ER, Spalding M, Steneck R, Watson R (2008) A global map of human impact on marine ecosystems. Science 319:948–952

Harvell CD, Kim K, Burkholder J, Colwell RR, Epstein PR, Grimes J, Hofmann EE, Lipp EK, Osterhaus ADME, Overstreet R, Porter JW, Smith GW, Vasta GR (1999) Emerging marine diseases – climate links and anthropogenic factors. Science 285:1505–1510

Harvell CD, Mitchell CE, Ward JR, Altizer S, Dobson AP, Ostfeld RS, Samuel MD (2002) Climate warming and disease risk for terrestrial and marine biota. Science 296:2158–2162

Harvell D, Jordan-Dahlgren E, Merkel S, Rosenberg E, Raymundo L, Smith GJ, Weil E, Willis BL (2007) Coral disease, environmental drivers and the balance between coral and microbial associates. Oceanography 20:172–195

Hata H, Kudo S, Yamano H, Kurano N, Kayanne H (2002) Organic carbon flux in Shiraho coral reef (Ishigaki Island, Japan). Mar Ecol Prog Ser 232:129–140

Haynes D (2001) Great Barrier Reef water quality: current issues. Great Barrier Reef Marine Park Authority, Townsville, p 90

Hodgson G (1990) Tetracycline reduces sedimentation damage to corals. Mar Biol 104:493–496

Hoegh-Guldberg O, Smith GJ (1989) Influence of the population-density of zooxanthellae and supply of ammonium on the biomass and metabolic characteristics of the reef corals *Seriatopora hystrix* and *Stylophora pistillata*. Mar Ecol Prog Ser 57:173–186

Hoegh-Guldberg O, Muscatine L, Goiran C, Siggaard D, Marion G (2004) Nutrient-induced perturbations to delta C^{-13} and delta N^{-15} in symbiotic dinoflagellates and their coral hosts. Mar Ecol Prog Ser 280:105–114

Hughes TP (1994) Catastrophes, phase shift, and large-scale degradation of a Caribbean reef. Science 265:1547–1551

Hughes TP, Rodrigues MJ, Bellwood DR, Ceccarelli D, Hoegh-Guldberg O, McCook L, Moltschaniwskyj N, Pratchett MS, Steneck RS, Willis BL (2007) Phase shifts, herbivory, and resilience of coral reefs to climate change. Curr Biol 17:1–6

Jackson JBC (2008) Ecological extinction and evolution in the brave new ocean. Proc Natl Acad Sci 105:11458–11465

Jackson JBC, Kirby MX, Berger WH, Bjorndal KA, Botsford LW, Bourque BJ, Bradbury RH, Cooke R, Erlandson J, Estes JA, Hughes TP, Kidwell S, Lange CB, Lenihan HS, Pandolfi JM, Peterson CH, Steneck RS, Tegner MJ, Warner RR (2001) Historical overfishing and the recent collapse of coastal ecosystems. Science 293:629–638

Johnston IS, Rohwer F (2007) Microbial landscapes on the outer tissue surfaces of the reef-building coral *Porites compressa*. Coral Reefs 26:375–383

Klaus JS, Janse I, Heikoop JM, Sanford RA, Fouke BW (2007) Coral microbial communities, zooxanthellae and mucus along gradients of seawater depth and coastal pollution. Environ Microbiol 9:1291–1305

Kline DL, Kuntz NM, Breitbart M, Knowlton N, Rohwer F (2006) Role of elevated organic carbon levels and microbial activity in coral mortality. Mar Ecol Prog Ser 314:119–125

Knowlton N (1992) Thresholds and multiple stable states in coral reef community dynamics. Am Zoo 32:674–682

Knowlton NK, Jackson JBC (2008) Shifting baselines, local impacts, and global change on coral reefs. PLoS Biol 6:215–220

Koop K, Booth D, Broadbent A, Brodie J, Bucher D, Capone D, Coll J, Dennison W, Erdmann M, Harrison P, Hoegh-Guldberg O, Hutchings P, Jones GB, Larkum AWD, O'Neil J, Steven A, Tentori E, Ward S, Williamson J, Yellowlees D (2001) ENCORE: the effect of nutrient enrichment on coral reefs. Synthesis of results and conclusions. Mar Pollut Bull 42:91–120

Koren O, Rosenberg E (2006) Bacteria associated with mucus and tissues of the coral *Oculina patagonica* in summer and winter. Appl Environ Microbiol 72:5254–5259

Kramer KL, Heck KL (2007) Top-down trophic shifts in Florida keys patch reef marine protected areas. Mar Ecol Prog Ser 349:111–123

Kuntz NM, Kline DL, Sandin SA, Rohwer F (2005) Pathologies and mortality rates caused by organic carbon and nutrient stressors in three Caribbean coral species. Mar Ecol Prog Ser 294:173–180

Lampert Y, Kelman D, Nitzan Y, Dubinsky Z, Behar A, Hill RT (2008) Phylogenetic diversity of bacteria associated with the mucus of Red Sea corals. FEMS Microbiol Ecol 64:187–198

Lapointe BE (1997) Nutrient thresholds for bottom-up control of macroalgal blooms on coral reefs in Jamaica and southeast Florida. Limnol Oceanogr 42:1119–1131

Ledlie MH, Graham NAJ, Bythell JC, Wilson SK, Jennings S, Polunin NVC, Hardcastle J (2007) Phase shifts and the role of herbivory in the resilience of coral reefs. Coral Reefs 26:641–653

Lesser MP, Mazel CH, Gorbunov MY, Falkowski PG (2004) Discovery of symbiotic nitrogen-fixing cyanobacteria in corals. Science 305:997–1000

Lesser MP, Falcón LI, Rodríguez-Román A, Enríquez S, Hoegh-Guldberg O, Roberto Iglesias-Prieto R (2007) Nitrogen fixation by symbiotic cyanobacteria provides a source of nitrogen for the scleractinian coral *Montastraea cavernosa*. Mar Ecol Prog Ser 346:143–152

Lessios HA (1988) Mass mortality of *Diadema antillarum* in the Caribbean: what have we learned? Ann Rev Ecol Syst 19:371–393

Lessios HA (2005) *Diadema antillarum* populations in Panama twenty years following mass mortality. Coral Reefs 24:125–127

Littler MM, Littler DS, Titlyanov EA (1991) Comparisons of N- and P-limited productivity between high granitic islands versus low carbonate atolls in the Seychelles Archipelago: a test of the relative-dominance paradigm. Coral Reefs 10:199–209

Littler MM, Littler DS, Brooks BL, Lapointe BE (2006) Nutrient manipulation methods for coral reef studies: a critical review and experimental field data. J Exp Mar Biol Ecol 336:242–253

Littman RA, Willis BL, Pfeffer C, Bourne DG (2009) Diversity of coral-associated bacteria differ with location, but not species for three Acroporids on the Great Barrier Reef. FEMS Microbiol Ecol 68:152–156

Lokrantz J, Nystrom M, Thyresson M, Johansson C (2008) The non-linear relationship between body size and function in parrotfishes. Coral Reefs 27:967–974

Marhaver KL, Edwards RA, Rohwer F (2008) Viral communities associated with healthy and bleaching corals. Environ Microbiol 10:2277–2286

McClanahan T, Sala E, Stickels P, Cokos B, Baker A, Starger C (2003) Interaction between nutrients and herbivory in controlling algal communities and coral condition on Glover's Reef, Belize. Mar Ecol Prog Ser 261:135–147

McClenachan L, Cooper AB (2008) Extinction rate, historical population structure and ecological role of the Caribbean monk seal. Proc R Soc Lond B 275:1351–1358

McCook LJ, Jompa J, Diaz-Pulido G (2001) Competition between corals and algae on coral reefs: a review of evidence and mechanisms. Coral Reefs 19:400–417

McDole T, Edwards RA, Dinsdale EA, Walsh S, Donovan M, Rohwer F (2008) Microbial dynamics of Kiritimati Atoll. 11th Int. Coral Reef Symp. Fort Lauderdale

Mora C, Andrefoutet S, Costello MJ, Kranenburg C, Rollo A, Veron J, Gaston KJ, Myers RA (2006) Coral reefs and the global network of marine protected areas. Science 312:1750–1751

Mora C (2008) A clear human footprint in the coral reefs of the Caribbean. Proc R Soc B 275:767–773

Mullen K, Peters E, Harvell CD (2004) Coral resistance to disease. In: Rosenberg E, Loya Y (eds) Coral health and disease. Springer, Berlin, pp 377–399

Mumby PJ, Dahlgren CP, Harborne AR, Kappel CV, Micheli F, Brumbaugh DR, Holmes KE, Mendes JM, Broad K, Sanchirico JN, Buch K, Box S, Stoffle RW, Gill AB (2006) Fishing, trophic cascades, and the process of grazing on coral reefs. Science 311:98–101

Mumby PJ, Harborne AR, Williams J, Kappel CV, Brumbaugh DR, Micheli F, Holmes KE, Dahlgren CP, Paris CB, Blackwell PG (2007) Trophic cascade facilitates coral recruitment in a marine reserve. Proc Natl Acad Sci 104:8362–8367

Munro JL (1983) Caribbean coral reef fisheries, 2nd edn. ICLARM Stud. Rev. ICLARM, Makati, pp 1–276

Muscatine L, D'elia CF (1978) Uptake, retention, and release of ammonium by reef corals. Limnol Oceanogr 23:725–734

Newman MJH, Paredes GA, Sala E, Jackson JBC (2006) Structure of Caribbean coral reef communities across a large gradient of fish biomass. Ecol Letters 9:1216–1227

Nordemar I, Sjoo GL, Mork E, McClanahan TR (2007) Effects of estimated herbivory on the reproductive potential of four East African algal species – a mechanism behind ecosystem shifts on coral reefs? Hydrobiologia 575:57–68

Nugues MM, Smith GW, van Hooidonk RJ, Seabra MI, Bak RPM (2004) Algal contact as a trigger for coral disease. Ecol Lett 7:919–923

Opsahl S, Benner R (1997) Distribution and cycling of terrigenous dissolved organic matter in the ocean. Nature 386:480–482

Pages J, Torreton J, Sempere R (1997) Dissolved organic carbon in coral-reef lagoons by high temperature catalytic oxidation and UV spectrometry. C R Acad Sci Ser II A Sci Terre Planetes 324:915–922

Pandolfi JM, Jackson JBC, Baron N, Bradbury RH, Guzman HM, Hughes TP, Kappel CV, Micheli F, Ogden CJ, Possingham HP, Sala E (2005) Are U. S. coral reefs on the slippery slope to slime? Science 307:1725–1726

Pascal H, Vacelet V (1981) Bacterial utilization of mucus on the coral reef of Aquaba (Red Sea). Proc 4th Int Coral Reef Symp 1:669–677

Patten NL, Harrison PL, Mitchell JG (2008a) Prevalence of virus-like particles within a staghorn scleractinian coral (*Acropora muricata*) from the Great Barrier Reef. Coral Reefs 27:569–580

Patten NL, Mitchell JG, Middelboe M, Eyre BD, Seuront L, Harrison PL, Glud RN (2008b) Bacterial and viral dynamics during a mass coral spawning period on the Great Barrier Reef. Aquat Microb Ecol 50:209–220

Reshef L, Koren O, Loya Y, Zilber-Rosenberg I, Rosenberg E (2006) The coral probiotic hypothesis. Environ Microbiol 8:2068–2073

Rohwer F, Kelley S (2004) Culture-independent analyses of coral-associated microbes. In: Rosenberg E, Loya Y (eds) Coral health and disease. Springer, Berlin, pp 265–277

Rohwer F, Breitbart M, Jara J, Azam F, Knowlton N (2001) Diversity of Bacteria associated with the Caribbean coral *Montastraea franksi*. Coral Reefs 20:85–91

Rohwer F, Seguritan V, Azam F, Knowlton N (2002) Diversity and distribution of coral associated bacteria. Mar Ecol Prog Ser 243:1–10

Rosenberg, Kurshmaro EA (2010) Microbial diseases of corals: pathology and ecology. In: Dubinsky Z, Stambler N (eds) Coral reefs and ecosystem in transition. Springer, Berlin

Rosenberg E, Loya Y (2004) Coral health and disease. Springer, Berlin

Rosenberg E, Koren O, Reshef L, Efrony R, Zilber-Rosenberg I (2007) The role of microorganisms in coral health, disease and evolution. Nat Rev Microbiol 5:355–362

Sakka A, Legendre L, Gosselin M, Niquil N, Delesalle B (2002) Carbon budget of the planktonic food web in an atoll lagoon (Takapoto, French Polynesia). J Plankton Res 24:301–320

Sandin S, Smith JE, DeMartini E, Dinsdale EA, Donner SD, Friedlander AM, Konotchick T, Malay M, Maragos J, Obura D, Pantos O, Paulay G, Richie M, Rohwer F, Schroeder RE, Walsh S, Jackson JBC, Knowlton N, Sala E (2008) Baselines and degradation of coral reefs in the northern Line Islands. PLoS ONE 3:e1548

Schaffelke B, Klumpp DW (1998) Short-term nutrient pulses enhance growth and photosynthesis of the coral reef macroalga *Sargassum baccularia*. Mar Ecol Prog Ser 170:95–105

Schaffelke B, Mellors J, Duke NC (2005) Water quality in the Great Barrier Reef region: responses of mangrove, seagrass and macroalgal communities. Mar Pollut Bull 51:279–296

Siboni N, Ben-Dov E, Sivan A, Kushmaro A (2008) Global distribution and diversity of coral-associated Archaea and their possible role in the coral holobiont nitrogen cycle. Environ Microbiol 10:2979–2990

Smith TB (2008) Temperature effects on herbivory for an Indo-Pacific parrotfish in Panama: implications for coral-algal competition. Coral Reefs 27:397–405

Smith JE, D SC, Hunter CL (2001) An experimental analysis of the effects of herbivory and nutrient enrichment on benthic community dynamics on a Hawaiian reef. Coral Reefs 19:332–342

Smith JE, Shaw M, Edwards RA, Obura DO, Pantos O, Sala E, Sandin SA, Rohwer F (2006) Indirect effects of algae on coral:algae-mediated, microbe-induced coral mortality. Ecol Letters 9:835–345

Suttle CA (2007) Marine viruses – major players in the global ecosystem. Nat Rev Microbiol 5:801–812

Tanaka Y, Miyajima T, Koike I, Hayashibara T, Ogawa H (2006) Translocation and conservation of organic nitrogen within the coral-zooxanthellae symbiotic system of *Acropora pulchra*, as demonstrated by dual isotope-labeling techniques. J Exp Mar Biol Ecol 336:110–119

Tanner J, Hughes T, Connell J (1994) Species coexistence, keystone species, and succession – a sensitivity analysis. Ecology 75: 2204–2219

Thacker R, Ginsburg D, Paul V (2001) Effects of herbivore exclusion and nutrient enrichment on coral reef fleshy algae and cyanobacteria. Coral Reefs 19:318–329

Toledo-Hernandez C, Zuluaga-Montero A, Bones-Gonzalez A, Rodriguez JA, Sabat AM, Bayman P (2008) Fungi in healthy and diseased sea fans (*Gorgonia ventalina*): is *Aspergillus sydowii* always the pathogen? Coral Reefs 27:707–714

Torreton J-P, Pages J, Dufour P, Cauwet G (1997) Bacterioplankton carbon growth yield and DOC turnover in some coral reef lagoons. Proc 8th Int Coral Reef Symp 1:947–952

UNESCO (1994) Protocols for the Joint Global Ocean Flux study (JGOFS) Core Measurements. IOC report Manuals and Guidelines 170

van Duyl FC, Gast GJ (2001) Linkage of small-scale spatial variations in DOC, inorganic nutrients and bacterioplankton growth with different coral reef water types. Aquat Microb Ecol 24:17–26

van Duyl FC, Scheffers SR, Thomas FIM, Driscoll M (2006) The effect of water exchange on bacterioplankton depletion and inorganic nutrient dynamics in coral reef cavities. Coral Reefs 25:23–36

Verdugo V, Alldredge AL, Azam F, Kirchman DI, Passow U, Santschi PH (2004) The oceanic gel phase: a bridge in the DOM-POM continuum. Mar Chem 92:67–85

Vermeij MJA, Smith JE, Smith CM, Vega TR, Sandin SA (2008) Survival and settlement success of coral planulae independent and synergistic effects of fleshy algae and microbes. Oecologia 159:325–336

Voss JD, Richardson LL (2006) Nutrient enrichment enhances black band disease progression in corals. Coral Reefs 25:569–576

Wegley L, Yu Y, Breitbart M, Casas V, Kline DL, Rohwer F (2004) Coral-associated Archaea. Mar Ecol Prog Ser 273:89–96

Wegley L, Edwards RA, Rodriguez-Brito B, Liu H, Rohwer F (2007) Metagenomic analysis of the microbial community associated with the coral *Porites astreoides*. Environ Microbiol 9:2707–2719

Wild C, Huettel M, Klueter A, Kremb SG, Rasheed MY, Jorgensen BB (2004a) Coral mucus functions as an energy carrier and particle trap in the reef ecosystem. Nature 428:66–70

Wild C, Rasheed MY, Werner U, Franke U, Johnson R, Huettel M (2004b) Degradation and mineralization of coral mucus in reef environments. Mar Ecol Prog Ser 267:169–171

Wild C, Woyt H, Huettel M (2005) Influence of coral mucus on nutrient fluxes in carbonate sands. Mar Ecol Prog Ser 287:87–98

Wild C, Jantzen C, Struck U, Hoegh-Guldberg O, Huettel M (2008) Biogeochemical responses following coral mass spawning on the Great Barrier Reef: pelagic-benthic coupling. Coral Reefs 27:123–132

Willis BL, Page C, Dinsdale EA (2004) Coral disease on the Great Barrier Reef. In: Rosenberg E, Loya Y (eds) Coral health and disease. Springer, Berlin, pp 69–104

Yahel G, Sharp JH, Marie D, Hase C, Genin A (2003) *In situ* feeding and element removal in the symbiont-bearing sponge *Theonella swinhoei*: bulk DOC is the major source for carbon. Limnol Oceanogr 48:141–149

Yellowlees D, Rees TAV, Leggat W (2008) Metabolic interactions between algal symbionts and invertebrate hosts. Plant Cell Environ 31:679–694

Coral Reef Algae

Peggy Fong and Valerie J. Paul

Abstract Benthic macroalgae, or "seaweeds," are key members of coral reef communities that provide vital ecological functions such as stabilization of reef structure, production of tropical sands, nutrient retention and recycling, primary production, and trophic support. Macroalgae of an astonishing range of diversity, abundance, and morphological form provide these equally diverse ecological functions. Marine macroalgae are a functional rather than phylogenetic group comprised of members from two Kingdoms and at least four major Phyla. Structurally, coral reef macroalgae range from simple chains of prokaryotic cells to upright vine-like rockweeds with complex internal structures analogous to vascular plants. There is abundant evidence that the historical state of coral reef algal communities was dominance by encrusting and turf-forming macroalgae, yet over the last few decades upright and more fleshy macroalgae have proliferated across all areas and zones of reefs with increasing frequency and abundance. Ecological processes that sustain these shifts from coral- to algal-dominated tropical reefs include increases in open suitable substrate due to coral mortality, anthropogenic increases in nutrient supply, reductions in herbivory due to disease and overfishing, and the proliferation of algae with chemical defenses against herbivory. These shifts are likely to be accelerated and the algal state stabilized by the impacts of invasive species and climate change. Thus, algal-dominated tropical reefs may represent alternative stable states that are resistant to shifts back to coral domination due to the strength and persistence of ecological processes that stabilize the algal state.

Keywords Macroalgae • cyanobacteria • chemical defenses • nutrients • herbivory • climate change • invasive species • diversity • coral/algal competition

P. Fong (✉)
Department of Ecology and Evolutionary Biology, University of California Los Angeles, 621 Young Drive South, 90095-1606, Los Angeles, CA, USA
e-mail: pfong@biology.ucla.edu

V.J. Paul
Smithsonian Marine Station at Fort Pierce, 701 Seaway Drive, 34949, Fort Pierce, FL, USA
e-mail: Paul@si.edu

1 Importance of Coral Reef Algae

Coral reefs are one of the most diverse and productive ecosystems on the planet, forming heterogeneous habitats that serve as important sources of primary production within tropical marine environments (Odum and Odum 1955; Connell 1978). Coral reefs are located along the coastlines of over 100 countries and provide a variety of ecosystem goods and services. Reefs serve as a major food source for many developing nations, provide barriers to high wave action that buffer coastlines and beaches from erosion, and supply an important revenue base for local economies through fishing and recreational activities (Odgen 1997).

Benthic algae are key members of coral reef communities (Fig. 1) that provide vital ecological functions such as stabilization of reef structure, production of tropical sands, nutrient retention and recycling, primary productivity, and trophic support. Macroalgae of an astonishing range of diversity, abundance, and morphological form provide these equally diverse ecological functions. For example, crustose coralline red algae which resemble "pink paint on a rock" are known to make essential contributions of calcium carbonate in the form of calcite "cement" that consolidates the larger volume of less-dense calcium carbonate produced by corals and other animals (Littler and Littler 1988; Littler and Littler 1995). However, the importance of this biotic process relative to submarine lithification has been questioned, especially in deeper reef zones (Steneck and Testa 1997; Macintyre 1997). These same crustose coralline red algae form intertidal ridges at the crest of reefs that protect more delicate forms of coral and invertebrates of the backreef zone from the full force of oceanic waves (Littler and Littler 1988). Calcifying green algae from an entirely different algal division are also important sources of the carbonate that forms key reef habitats. Some genera of siphonaceous (multinucleate but single-celled) green algae form stony plates strengthened by aragonite; this form of calcium carbonate ultimately comprises a large portion of the sandy sediments in backreefs and lagoons. One study found that green algae of the genus *Halimeda* contributed an average of 40% of the volume of sand that comprised

Fig. 1 Algae are key members of coral reef communities: (**a**) the brown tube sponge *Agelas wiedenmayeri* being overgrown by *Palisada* (*Laurencia*) *poiteaui* and *Dictyota* spp. at a depth of 7 m in the Florida Keys, (**b**) *Dictyota pulchella* and other macroalgae growing on the blue sponge (*Aiolochroia crassa*) at a depth of 10 m in the Florida Keys, (**c**) the zoanthid *Palythoa caribbea* being overgrown by *Dictyota pulchella* and other macroalgae at a depth of 7 m in the Florida Keys, and (**d**) the Christmas tree worm *Spirobranchus giganteus* next to the brown alga *Dictyota* sp. at a depth of 10 m in the Florida Keys (Photographs by Raphael Ritson-Williams)

the barrier reef sediment leeward of a 9-km long emergent reef crest in Belize (Macintyre et al. 1987). When considered globally, carbonate from *Halimeda* spp. accounts for an estimated 8% of total global production (Hillis 1997).

A highly diverse assemblage of tropical reef algae is responsible for a large amount of the nutrient retention and recycling that contributes to the high level of primary productivity typical of coral reefs and provides trophic support to the incredible diversity of consumers. Although tropical reefs generally occur in highly oligotrophic oceanic water, areal rates of primary productivity are comparable to some of the most productive terrestrial ecosystems such as tropical forests (Mann 1982). Shallow reef flats sustain rates of primary productivity over an order of magnitude higher than the surrounding oceanic water and the bulk of this productivity is due to algae (Littler and Littler 1988). These highly productive reef flats are typically covered by crustose coralline red algae and algal turfs comprised of a diverse assemblage of filamentous algae and cropped bases of larger forms. Areas of highest wave energy are often covered by crustose coralline algae, thick and leathery brown algae such as *Turbinaria* and *Sargassum*, and calcified green algae such as *Halimeda* that can withstand wave impact (Littler and Littler 1988). Sand plains surrounding reefs also contain a high diversity of algae as the relatively flat topography of these regions provides a refuge from herbivores. Tropical reefs have been called "oases of productivity in a nutrient desert," and this productivity has been attributed to many processes including: (1) high advection rates supplying low concentrations yet high volumes of oceanic water, (2) efficient nutrient uptake by algae of low concentrations of nutrients, (3) extensive nitrogen fixation by cyanobacteria and bacteria, and (4) a highly diverse and spatially rugose set of habitats that dramatically increase the retention and recycling capability of the ecosystem (Littler and Littler 1988).

2 Diversity

Marine macroalgae or "seaweeds" are a functional rather than phylogenetic group comprised of members from two Kingdoms and at least four major Phyla including the Cyanobacteria (prokaryotic blue-green algae, sometimes termed Cyanophyta), Chlorophyta (green algae), Heterokontophyta (including brown algae, Class Phaeophyceae), and Rhodophyta (red algae) (Lee 2008). Coral reef algae have representatives across this wide range of taxonomic diversity. Structurally, coral reef macroalgae encompass a diversity of forms ranging from simple chains of prokaryotic cells to single-celled yet multinucleate thalli up to a meter in length and upright vine-like rockweeds with complex internal structures analogous to vascular plants. All seaweeds at some stage of their life cycles are unicellular (usually as reproductive stages such as spores or zygotes), and they are viewed as "primitive" photosynthetic organisms because of their relatively simple construction and their long evolutionary history.

Prokaryotic blue-green algae, or Cyanobacteria, are the oldest group with fossils dating back over three billion years (Schopf 2000). Well-preserved fossil cyanobacteria are nearly indistinguishable in morphology from their extant relatives and can be found in intertidal and shallow marine environmental settings like those inhabited by cyanobacteria today. Many of these first algal fossil remains are stromatolites, structures formed in shallow tropical waters when cyanobacterial mats accrete layers by trapping, binding, and cementing sedimentary grains (Lee 2008). Photosynthesis by these early primary producers was responsible for much of the oxygen that eventually built up to the levels that exist today (~20%) (Canfield 1999; Kasting and Siefert 2002; Kerr 2005). Evolution of eukaryotic algae occurred much later, about 700–800 million years ago, though this date is difficult to accurately pinpoint as most groups were composed of soft tissue that would not have been preserved reliably in the fossil record (Lee 2008).

Cyanobacteria are ubiquitous worldwide. On tropical reefs, they are often found forming mats along reef margins or on coral (Smith et al. 2009), may be epiphytic on other algae and other reef organisms (Paul et al. 2005; Ritson-Williams et al. 2005; Fong et al. 2006), rapidly colonize open space opportunistically after disturbances (Littler and Littler 1997), and may bloom in response to nutrient enrichment (Littler et al. 2006; Ahern et al. 2007, 2008; Paerl et al. 2008; Arthur et al. 2009).

Although the marine green algae (Chlorophyta) range from cold temperate to tropical waters, green algae reach their highest diversity and natural abundance in tropical and subtropical regions, with several families such as the Caulerpaceae and Udoteaceae very abundant in coral reef and associated seagrass habitats (Dawes 1998). Often overlooked, but very abundant are filamentous green algae that bore into coral skeleton and proliferate widely, with high rates of productivity (Littler and Littler 1988).

The brown algae (Class Phaeophyceae) are almost exclusively marine and primarily dominant in temperate waters. However, some genera of complex and structurally robust forms such as *Turbinaria* and *Sargassum* dominate in high-energy reef zones (e.g., Stewart 2006). Other groups of fast-growing and more opportunistic genera, such as *Dictyota*, may form seasonal blooms on reefs, covering up to 40% of the benthos in some areas of the Florida Keys (Lirman and Biber 2000; Kuffner et al. 2006).

The Rhodophyta (red algae) are the most diverse group of macroalgae. At present, the approximately 4,000 named species of red algae exceed the number of species in all other groups combined (Lee 2008). The most common forms of red algae on coral reefs include crustose members of the family Corallinaceae as well as a high diversity of small, less-obvious filamentous species that comprise algal turfs. Both groups are ubiquitous across reef zones. However, there are some genera of upright and branching calcifying forms such as *Galaxaura* and branching or flattened foliose red algae in the genera *Gracilaria*, *Laurencia*, *Asparagopsis*, and *Halymenia* that can be quite conspicuous and abundant on reefs under certain conditions because their structural and chemical defenses make them resistant to herbivores.

Communities of tropical macroalgae are often extremely speciose (Littler and Littler 2000, http://www.algaebase.org) and field identification of many species proves challenging. This is especially true for members of the Rhodophyta, as they are often differentiated by microscopic reproductive structures (Abbot 1999; Lee 2008). This was the impetus for field ecologists to develop a functional grouping system using morphological forms. The underlying theory was that algae of very similar morphological form may function in a community and ecosystem more similarly than those that are morphologically diverse yet more closely related phylogenetically. Steneck and Watling (1982) first classified macroalgae into seven functional groups based on susceptibility to grazing by gastropods: groups were filamentous, crustose, foliose, corticated foliose, corticated macrophyte, leathery macrophyte, and articulated calcareous algae.

Littler and Littler (1984) proposed a somewhat different set of functional form groups based on a broader set of characteristics such as nutrient uptake rates, productivity, turnover rates, and resistance to herbivory. They argued that algae that share these functional characteristics would perform similarly in response to variation in environmental conditions, regardless of differences in taxonomy. The functional form groups proposed by Littler and Littler (1984) include sheetlike, filamentous, coarsely branched, thick-leathery, jointed-calcareous, and crustose forms (Table 17.1). These groups are arranged in a spectrum from fast-growing opportunistic species that are most susceptible to herbivory (sheetlike) to the slowest growing persisters that are not readily consumed (crustose). While functional form groupings have been used extensively over the last 2 decades, there is a burgeoning recognition that macroalgae do not always fit neatly into discreet categories. Using the example of flattened sheetlike algae mentioned above, members of the genus *Dictyota* in the Phaeophyceae should also be included in this category. However, while they are fast-growing nutrient specialists (Fong et al. 2003), and therefore share this characteristic of the functional form group, they are also chemically defended, making some species resistant to even relatively high levels of herbivory.

Table 1 Functional-form groups of predominant macroalgae: their characteristics and representative taxa

	Functional-form group	External morphology	Comparative anatomy	Thallus size/texture	Example Genera
1	Sheet-like Algae	Flattened or thin tubular (folisoe)	1- several cell layers thick	Soft, flexible	*Ulva* *Halymenia*
2	Filamentous Algae	Delicately branched	Uniseriate, multiseriate, or lightly corticated	Soft, flexible	*Chaetomorpha* *Cladophora* *Gelidium* *Caulerpa*
3	Coarsely Branched Algae	Terete, upright, thicker branches	Corticated	Wiry to fleshy	*Acanthophora* *Laurencia*
4	Thick Leathery Macrophytes	Thick blades and branches	differentiated, heavily corticated, thick walled	Leathery-rubbery	*Sargassum* *Turbinaria*
5	Jointed Calcareous Algae	Articulated, calcareous, upright	Calcified, genicula, flexible intergenicula	Stony	*Galaxaura* *Amphiroa*
6	Crustose Algae	Epilithic, prostrate, encrusting	Calcified, heterotrichous	Stony and tough	*Porolithon* *Hydrolithon*

Source: adapted from Littler and Littler (1984)

3 Distribution and Abundance

There is abundant evidence that the historical state of coral reef algal communities was dominance by encrusting and turf-forming macroalgae (e.g., Odum and Odum 1955; Littler and Littler 1984, 1988). Crustose coralline members of the Rhodophyta grow ubiquitously on solid substrates of coral reefs intertidally down to at least 260 m (Littler et al. 1986). Thus, some members of this group are able to tolerate the most extreme low-light conditions found on coral reefs, while others of this same form can tolerate exposure to both extremely high irradiance and desiccation. In general, maximum abundance of crustose coralline algae occurs in shallow turbulent areas. Although crustose coralline algae are both abundant and widely distributed, productivity levels are low relative to other algal groups, suggesting a "persister" life-history strategy. Algal turfs are also ubiquitous on hard substrates throughout tropical reef ecosystems. While turfs are often dominated by filamentous members of the Rhodophyta, they also can include filamentous green algae and cyanobacteria, and cropped bases of larger algae. In contrast to crusts, turfs are characterized by extremely high rates of primary productivity, though biomass is usually very low (<0.27 kg m^{-2}), suggesting an opportunistic life-history strategy where success is a result of growing slightly faster than herbivores can consume them (Carpenter 1986; Duffy and Hay 1990, 2001).

Historically, other types of macroalgae with a more upright, foliose morphology were restricted to backreefs, lagoons, or deeper reef areas. Typical standing stocks were 3.0–3.5 kg m^{-2}, though rarely could be as high as 10 kg m^{-2} (Littler and Littler 1988). Calcareous and siphonaceous Chlorophyta usually dominated on rubble or soft-sediment areas of the backreef and lagoon, and often were found in association with sea grasses or mangroves. These are areas unsuited to most other forms of macroalgae due to the lack of hard substrate, but many calcareous green algae have rhizome-like structures that act as anchors in soft substrates. One genus of calcareous chlorophyte, *Halimeda*, can also be found across most zones of the reef as well as forming extensive meadows in deeper water off the forereef (Littler and Littler 1988; Fukunaga 2008). Natural *Halimeda* populations are commonly 100 plants per square meter, but may reach densities up to 500 plants per square meter. With the exception of a few genera adapted to high wave action, most members of the Phaeophyceae historically reached the highest abundance in backreef and lagoon habitats, where there was sufficient hard-bottomed or rubble habitat with relatively low topographic relief. Historical studies of the distribution and abundance of Cyanobacteria are relatively rare, because cyanobacteria were often considered a component of the turf algae in ecological studies (Littler and Littler 1988). It was believed that, with the exception of cryptic boring algae and members of microalgal filamentous communities, independent, macroscopic Cyanobacteria were largely limited to intertidal or very shallow water habitats.

Over the last few decades, macroalgae of all Divisions have been documented to proliferate across all areas and zones of reefs with increasing frequency and abundance. Green algae such as *Dicytospheria cavernosa* have dominated reefs of Kaneohe Bay (Stimson et al 2001). Brown algae in the family Dictyotaceae, including *Dictyota* spp. and sometimes *Padina* spp., *Stypopodium zonale*, and *Lobophora variegata*, have become increasingly dominant on the reefs, not just the sand and rubble plains of the backreef (Rogers et al. 1997; Lirman and Biber 2000; Kuffner et al. 2006). Blooms of upright and branching red algae such as *Acanthophora spicifera* and *Gracilaria* spp. are dominating reefs, sometimes lasting years (Fong et al. 2006). Clearly, the distribution and abundance of coral reef macroalgae has undergone rapid change over the last few decades, and those changes appear to be accelerating (although a meta-analysis by Bruno et al. 2009 showed little change in upright fleshy and calcareous forms of macroalgae over the past decade). Although many investigations into the mechanisms of change and the impacts on coral reef structure and functioning have been conducted and are discussed below, there are still considerable knowledge gaps that must be addressed.

4 Ecological Processes Controlling Algal Populations and Communities

Mechanisms that control distribution and abundance of coral reef algae are the same as for other primary producers: geographic limits for growth are set by temperature and light and for removal by grazing and physical disturbance. Within these geographical limits, biomass accumulation is controlled by many interacting biotic and abiotic factors including availability of suitable substrate, light quantity and quality, nutrients, intra- and interspecific competition, and herbivory.

4.1 Factors Limiting Settlement and Growth: Suitable Substrate

The overwhelming majority of marine algae need little more than "hard" substrata to settle, and thus are able to recruit to a variety of available surfaces throughout depths where enough light penetrates. These surfaces can be abiotic or biotic and include rock, coral rubble, shells of animals live or dead, sea grasses, mangrove roots, and other algae. Microbial biofilms on settlement surfaces may be an important cue for settlement of some algal spores (Amsler 2008b). Some benthic organisms, like many crustose coralline algae, have "antifouling" mechanisms that entail sloughing of outer layers where epiphytes (algae that live on other primary producers) recruit. Others, like corals, slough mucus off their surfaces to rid themselves of algal settlers. Chemical defenses are also employed by marine algae and invertebrates to deter algal settlers and other fouling organisms (Lane and Kubanek 2008; Chadwick and Morrow 2010). Thus, not all hard substrate is equally "available" for algal recruitment.

Not all algae require hard substrata throughout their entire life cycle. As stated earlier, most seaweeds have some form of single-celled swimming or floating stage as part of their complex life cycles, usually a reproductive structure such as a zoospore or gamete. But a pelagic stage is not always limited to single cells. Some macrophytic forms of algae detach and form floating rafts (e.g., Stewart 2006; Bittick et al. 2010) or mats that may drift along the bottom or settle onto benthic communities (Holmquist 1994). These may be very important to dispersal of both the alga and its associated community. They may also have negative effects on the community upon which they land. For example, species of *Dictyota* in the Florida Keys undergo frequent fragmentation, often as a result of herbivory by fishes (Herren et al. 2006). However, fragments quickly entangle, form holdfasts, and become epiphytic on other organisms including other algae, corals, and sponges (Beach et al. 2003). Negative effects of epiphytic *Dictyota* on other algae include reduced growth due to shading and chemically mediated elevation of respiration.

There are some forms of algae, predominantly siphonaceous green algae, which do not require hard substrata at any stage of their life cycle. They are able to recruit to and establish in soft sediment areas with rhizomes adapted for attachment in soft substrata. These algae include members of the genera *Halimeda, Caulerpa, Penicillus,* and *Udotea,* which are very important primary producers in backreef and lagoon habitats, and often are associated with seagrass beds.

Algae are rapid and efficient initial colonizers of space on almost any area of coral reefs that has been opened by disturbances. Several examples show how hurricane damage on reefs can lead to rapid colonization by algae (Hughes 1994; Fong and Lirman 1996; Rogers et al. 1997). Within days, coral skeleton bared by hurricane damage was colonized by filamentous green algae that succeeded to turf dominated by filamentous red algae within a month (Fong and Lirman 1996). When openings in otherwise healthy coral colonies were small, areas were recovered by coral in a matter of months; however, larger openings were only partially recovered by coral and the rest remained algal turf. Algae also rapidly colonize dead coral following episodes of coral bleaching and mortality (Littler and Littler 1997; Baker et al. 2008). Of course, algae can also be rapidly removed from reefs when storms or high waves impact reef habitats, which can open up hard substrata for recruitment of other types of algae or corals and other benthic invertebrates (Becerro et al. 2006).

4.2 Factors Limiting Settlement and Growth: Light

Coral reef algae depend on light for use in photosynthesis. Although tropical waters are clear with high-light penetration compared to temperate zones, there are still patterns of light attenuation with depth that change both the quality and the quantity of available light for photosynthesis. Light is reduced exponentially with depth following the Beer–Lambert law, $I_z = I_0 e^{K_{dz}z}$, where I_z is irradiance at depth z, I_0 is surface irradiance, and K_{dz} is the attenuation coefficient for downwelling irradiance. The surface irradiance reaching a reef at a given depth is influenced by properties of the water that affect the attenuation coefficient. Light is attenuated by both absorption and scattering. Scattering of light by water molecules and particulate matter is greatest for the shorter high-energy wavelengths, while absorption is greatest in longer, lower-energy red wavelengths. Attenuation rate is increased by suspended sediment, detrital particles, dissolved organic matter, and biota. The net result is that most red light is absorbed in the first few meters of depth, while blue and green light penetrate the deepest (Fig. 2).

Coral reef algae have adapted to life at depth, though adaptations to the changes in quality and quantity in light vary across algal divisions. Like all primary producers, algae absorb light for photosynthesis in the visible wavelengths between 400 and 700 nm. This part of the spectrum is called photosynthetically active radiation or PAR. However, not all wavelengths within this active range are equally useful across all algal groups. Each algal pigment has a different action spectrum: an action spectrum is the rate of a physiological activity, in this case, absorption of light, plotted against wavelength (Fig. 2). Peaks in this spectrum show which wavelengths of light are most effective in fueling photosynthesis for each algal pigment.

All algae (as well as all terrestrial plants) contain chlorophyll a, a pigment that absorbs light maximally in the red and blue wavelengths (Fig. 2) and reflects green light, resulting in a green appearance. Accessory pigments in algae absorb light in different wavelengths than chlorophyll a and funnel this energy for use in photosynthesis. The major algal divisions contain a variety of accessory pigments that enable capture of different portions of PAR. Green algae contain chlorophyll b as their major accessory pigment. This pigment absorbs maximally in only a slightly different range of wavelengths than chlorophyll a, and therefore does not greatly extend the ability to absorb different wavelengths. Thus, based on pigment content alone, one would predict the depth distribution of green algae should be shallower than other divisions. Brown algae have auxiliary pigments such as fucoxanthins that absorb a broader spectrum of light of blue and green wavelengths than chlorophyll alone; these pigments should enable them to proliferate to deeper depths relative to green algae. Red algae contain phycobilin pigments such as phycoerthyrin, which absorb maximally in the blue and green wavelengths that penetrate deepest in clear waters. The red algae therefore have the potential to grow at the deepest depth compared to the other divisions.

It is important to note that the presence of various algal pigments only sets the potential depth distribution of the macroalgae that contain them. Although auxiliary pigments enable algae to extend their ranges deeper, this does not mean that they are limited to those depths. For example, while crustose coralline red algae are the deepest-living marine macroalgae and have been found as deep as 268 m in the very clear waters of the Bahamas (Littler et al. 1986), they are also common across all reefs zones including the intertidal. In addition, while tropical green algae do proliferate in shallow water, some genera, such as *Halimeda*, have been found to form expansive meadows quite deep, down to 75 m in clear water off Malta (Larkum et al. 1967) and to a maximum depth of 118 m in Hawaii (Runcie et al. 2008). Another species of green algae, *Johnson-sealinkia profunda*, is the deepest growing frondose macroalga, recorded at 200 m in the Bahamas (Littler and Littler 1988). In comparison, *Lobophora variegata* only grew down to 100 m in this same area, and was found to proliferate in the 20–30 m range on reefs of Curaçao (Nugues and Bak 2008). Clearly, while pigments control the potential for growth and proliferation of algae at certain depths, other factors such as grazing, wave energy, disturbance, and other adaptations to low light all contribute to algal depth zonation.

Algae can adapt to changing light conditions across a variety of temporal scales, from within minutes to over a season.

Fig. 2 Different divisions of algae have adapted to the varied light regimes that occur along depth gradients in oceanic waters. In clear water, chlorophyll suffices for all groups, as it absorbs blue light that penetrates in all depths. However, in more turbid water, light attenuates more quickly and accessory pigments fill in the "green window," expanding the potential depth distribution of different divisions of algae (Artwork by Kendal Fong, photographs by Raphael Ritson-Williams and V. Paul)

Light availability can vary over short timescales in response to changes in cloud cover, variation in shading by the macroalgal community during wind or wave events, and changes in light attenuation with sediment resuspension. Although seasonal variability in light regime in most coral reef systems is lower than in temperate zones, irradiance levels do vary across this scale. Algae can adapt very quickly to reduced light levels by increasing pigment content, which increases light utilization efficiency and allows macroalgae to photosynthesize more effectively at low-light levels. There are respiratory costs to the production and maintenance of higher pigment content in low-light plants, so at high irradiances pigment concentrations are typically lower. Algae also react to high-light intensities. Photoinhibition may occur at high irradiances or under high UV stress, and damage to algal photosystems during this process may not be reversible.

4.3 Factors Limiting Settlement and Growth: Nutrients

Coral reefs have long been known to be "oases of productivity in a nutrient desert." Over 4 decades ago in a seminal paper, Odum and Odum (1955) estimated that supply of allochthonous nutrients to coral reefs of Eniwetok Atoll via advection of oceanic water was not sufficient to sustain the high levels of productivity characteristic of these systems. This finding motivated a plethora of research into how high levels of productivity are supported on coral reefs that was focused on other allochthonous sources of nutrients (e.g., organic and particulate nutrients, plankton, nitrogen fixation, symbiosis), the relative importance of N and P limitation, and efficient uptake, storage, retention, and recycling of water column nutrients by primary producers. Many studies have also tried to quantify the relative importance of nutrient limitation compared to other ecological process in determining the distribution and abundance of coral reef algal communities.

4.3.1 Allochthonous Versus Autochthonous Sources of Nutrients

In theory, the primary source of "new" or allochthonous nutrients to coral reefs should largely depend on the geographic location of the reef system, the type of coral reef (fringing, barrier, or atoll), as well as the type of terrestrial system to which it is coupled. The geographic location of a reef affects many global factors that may influence nutrient supplies, such as currents (Garzon-Ferreira et al. 2004), atmospheric deposition (Muhs et al. 2007), and degree of isolation from major continental sources. Both the type of reef and the type of associated terrestrial system (if any) will greatly influence the connectivity of the reef system to oceanic versus terrestrial sources of nutrients and therefore their relative importance. For example, nutrient supply to the fringing reef along the shore of Moorea, French Polynesia, a volcanic island with high topographic relief, is likely to be more influenced by watershed characteristics such as rainfall, groundwater, and soil stability than the barrier reef of this same island. And the nutrient supply to both the fringing and barrier reefs of Moorea should be influenced more by the presence of the island than nearby Tetiaroa Atoll, with its low relief set of sandy islands.

Coral reef systems occur across such a large diversity of geographic locations, associated landmasses, and reef types, as well as many other factors, that it is difficult to make generalizations about overall effects on allochthonous nutrient supply. However, recent work across marine systems has suggested that sources of new nitrogen to all marine systems are rapidly increasing globally as a result of anthropogenic alterations of the global nitrogen cycle (for reviews, see Vitousek et al. 1997; Downing et al. 1999), though the importance of these increases in coral reefs is highly debated (for a review, see Fong 2008). Overall, it is likely that coral reef systems more closely associated with terrestrial watersheds will be affected the most by increasing anthropogenic nitrogen supplies compared to open oceanic reef systems.

In situ or autochthonous sources of nutrients to coral reef algae include nitrogen fixation, recycling from other biota, regeneration from deposition of organic matter in any associated sediment areas, and regeneration of nutrients from primary producers during decomposition (Fong 2008). Cyanobacteria abound on coral reefs, and many may fix nitrogen. They are important components of the ubiquitous algal turf communities and live within coral skeletons (Littler and Littler 1988) and epiphytically on other algae (Fong et al. 2006). Recently, blooms were found on coral communities of the reefs themselves (Paul et al. 2005; Smith et al. 2009). Nitrogen fixation on coral reefs may be very high, especially following disturbances that open new substrate for colonization by filamentous cyanobacteria (Larkum 1988). Some measures are equal to the highest rates found in terrestrial systems (Capone 1983). Animals in coral reefs that live in close association with algae may increase local nutrient supplies by releasing nutrient-rich waste products (e.g., Williams and Carpenter 1988). Flocculent material, most likely at least partially of biotic origin, settling on the surfaces of algal thalli has been identified as a source of nutrients to algae in coral reefs (Schaffelke 1999). In tropical reef systems, recycling from sediments may only be a significant contribution in areas subject to long-term nutrient enrichment like Kanoehe Bay, Hawaii (Stimson and Larned 2000). However, in other types of soft sediment and enriched systems, macroalgal interception of nutrient fluxes is ecologically important as it may uncouple sedimentary and water column N linkages

(Valiela et al. 1992, 1997), reducing supplies to other producer groups (Fong et al. 1998). Thus, more research on this source of nutrients is warranted on coral reefs.

4.3.2 N Versus P Limitation of Coral Reef Algae

For tropical marine algae, quantifying the relative roles of nitrogen (N) and phosphorus (P) limitation has been an important focus of research in recent years. In contrast to temperate systems where N limitation is the paradigm (Fujita et al. 1989; Thybo-Christesen et al. 1993; Rivers and Peckol 1995; Taylor et al. 1995; Sfriso and Marcomini 1997; Gallegos and Jordan 1997), many tropical studies have found P to limit productivity and growth more frequently than N (e.g., Lapointe 1987, 1989; Lapointe et al. 1992). Others found stimulation by both N and P (Schaffelke and Klumpp 1998; Lapointe 1987; Paerl et al. 2008), or dual limitation (Fong et al. 2003). There are several possible explanations why there is high spatial and temporal variability in the relative importance of N and P limitation in coral reef algae. The strength of P limitation has been related to the amount of P-adsorbing carbonate in sediments (Delgado and Lapointe 1994; Lapointe et al. 1992; McGlathery et al. 1994), the habitat or substrate type the algae occupied (Lapointe 1989; Littler and Littler 1988), and the type of island adjacent to the coral reef (Littler et al. 1991). Others have hypothesized that the relative importance of N and P limitation should vary across a nutrient supply gradient, with N increasing in importance in more eutrophic systems due to saturation of the P-adsorption capacity of sediments (Delgado and Lapointe 1994; Downing et al. 1999). Studies in Kaneohe Bay, Hawaii, an area with a history of nutrient enrichment, supported this hypothesis as N limited nine of ten species tested from a broad range of functional forms (Larned and Stimson 1997; Larned 1998). However, Fong et al. (2003) found that several species of macroalgae in Puerto Rico differed in their response to N and P additions depending on their nutrient status; algae from nutrient-replete environments did not respond strongly to either nutrient. Clearly, there is high spatial and temporal variability in the relative importance of N and P limitation in coral reef algae. More research is needed to further our understanding of these patterns of variability.

Three approaches have been used extensively to determine if N and/or P limits productivity of algae, including those on coral reefs. In the first approach, N:P ratios of dissolved inorganic nutrients in the water column have been used as a measure of nutrient availability; water column N:P ratios are compared to nutrient requirements of algae to determine limitation (e.g., Redfield et al. 1963; Lapointe 1989; Duarte 1992). However, different species or functional forms of algae may require nutrients in differing proportions. One study found that one alga was limited by N and another by P when grown together in seawater of the same N:P ratio (Fong et al. 1993), although this study was conducted with algae from warm temperate, not tropical areas, leaving this an open question for tropical algae. In addition, water column measures provide only a snapshot in time and may not adequately characterize availability in areas where nutrients are supplied in pulses as in the tropics (McCook 1999) or in many estuaries (Fry et al. 2003; Boyle et al. 2004). In the second approach, N:P ratios in algal tissue have been used to predict nutrient limitation (e.g., Lapointe et al. 1992). However, this method also has limitations, because differing uptake and storage capacities of algae may confound the relationship. For example, if both N and P are abundant in the water, and an alga has a greater uptake ability and storage capacity for N than P, then the resultant high tissue N:P ratio would indicate P limitation when limitation by nutrients was not occurring.

The third approach used to determine N or P limitation is factorial enrichment experiments adding N and P alone and in combination, and quantifying responses such as photosynthesis, growth, and changes in tissue N and P content (e.g., Lapointe 1987, 1989; Fong et al. 1993; Larned 1998; Fong et al. 2006). When addition of a nutrient increased any of these response variables, it was considered to be in limiting supply. This approach has been used across a range of scales, from laboratory or field microcosm dose-response experiments (e.g., Fong et al. 2003; Paerl et al. 2008) to large-scale and longer-term field experiments (e.g., Koop et al. 2001). The challenges when using this approach include choice of realistic experimental enrichment levels, effectively scaling up the relatively short-term and small-scale responses from microcosm experiments, and effectively enriching in situ experiments in high-energy environments. The advantages, however, are that the experimental approach provides direct rather than indirect evidence of limitation. Although at present the disadvantages of this approach require careful interpretation and application of experimental results, the strength provided by direct evidence warrants further research and methodological development.

4.3.3 Efficient Nutrient Uptake by Coral Reef Algae

Predicting nutrient uptake rates of coral reef algae from water column sources is complex. For years, nutrient uptake was thought to be a simple function of water column inorganic nutrient concentration that could be described by Michaelis-Menten uptake kinetics (e.g., Fong et al. 1994b), an approach proven successful for phytoplankton (Droop 1983; Sommer 1991). More recently, the recognition that macroalgae are often both relatively stationary in a dynamic flow environment and have significant nutrient storage capacities that reflect past nutrient supplies has focused current research into the physics of dissolved solute transport across boundary layers as well as the influence of the biological condition of algal thalli on uptake (for a review, see Hurd 2000). Both of these must be incorporated into our understanding of nutrient availability to and uptake by coral reef algae.

Uptake of nutrients from the water column depends on many factors, including the nature and velocity of water flow, water nutrient concentrations of biologically useable substrates, and algal metabolic demand (for a review, see Hurd 2000). For small filamentous or crustose algae completely within the benthic boundary layer, current speeds may be much reduced compared to those that extend into the overlying faster water flow (Carpenter and Williams 1993; Nepf and Koch 1999). Attention to near-bottom flow speeds is essential on coral reefs, as these forms of algae are usually spatially dominant.

Algal morphology also affects uptake of nutrients. Morphology controls the algal height with respect to the benthic boundary layer, and thallus structure may enhance uptake by flexibility in flowing water (Hurd 2000). In addition, some algal growth forms, such as densely packed mats, may reduce uptake by reducing current speeds within the mat (Fong et al. 2001). Morphology also determines the physical surface area of an algal thallus containing uptake sites. Several studies have identified a positive relationship between an alga's surface area to volume ratio and uptake of nutrients (e.g., Hein et al. 1995).

Once nutrients cross the boundary layer and contact the surface of the algal thallus, they must be transported across the cell membrane and then assimilated into organic compounds followed by incorporation into proteins and macromolecules for growth (McGlathery et al. 1996; Cohen and Fong 2004). This process has been best studied for nitrogen (N). Uptake rate varies among the commonly co-occurring forms of N available in coastal marine waters, NH_4^+ and NO_3^-. For example, some algae have strong preferences for NH_4^+, while others take up either inorganic N source (e.g., Hanisak 1983; Lotze and Schramm 2000; Naldi and Wheeler 2002). Uptake of NH_4^+ is less energetically costly, because NO_3^- must first be reduced to NH_4^+ by nitrate reductase before assimilation (Hurd et al. 1995); thus, energetics may explain preference for NH_4^+ by some macroalgae. However, NH_4^+ storage capacity may be limited due to toxicity (Waite and Mitchell 1972; Haines and Wheeler 1978; Lotze and Schramm 2000), and therefore assimilation rate into inorganic molecules may limit maximum uptake rate of this form of N. The ability of opportunistic macroalgae to take up both forms of N simultaneously may be one mechanism that results in algal blooms (Thomas and Harrison 1987; Cohen and Fong 2004).

Algal demand, a function of algal tissue nutrient status, also influences nutrient uptake rates of macroalgae. Tissue nutrient status reflects the history of nutrient supply to an alga, as algae subject to excessive or pulses of nutrients may store nutrients for future growth (Wheeler and North 1980; Lapointe and Duke 1984; Fong et al. 1994a, b). Several investigators (e.g., Fujita 1985; McGlathery et al. 1996; Fong et al. 2003; Kennison 2008) found that algae with nutrient-enriched tissues always took up N more slowly than nutrient-depleted algae. These studies demonstrate that algae with higher internal nutrient content will have lower metabolic demand and therefore slower N uptake rates.

Pedersen (1994) separated uptake of N into three phases; although this work was on temperate opportunistic forms of algae, there is reason to believe it is applicable to many coral reef algae with this same strategy. The first phase, surge uptake, is transiently enhanced nutrient uptake by nutrient-limited algae that may last only minutes to hours. Surge uptake has been documented in tropical algae from nutrient-poor sites in Puerto Rico (Fong et al. 2003). Because water column nutrients in tropical systems are characteristically low, experimental nutrient pulses to macroalgae in these experiments were also relatively low (~20 μM N), yet the algal response was rapid and of equal magnitude to bloom-forming species from temperate estuaries subjected to nutrients an order of magnitude higher in concentration (Kennison 2008). This suggests that tropical macroalgae may be especially well adapted to take advantage of pulses of nutrients through surge uptake. During the second phase, internally controlled phase of uptake, the rate-limiting step is assimilation of N into organic compounds (Fujita et al. 1988; Rees et al. 1998). This occurs when external N is maintained at a relatively high concentration for enough time that storage pools within algal tissue begin to fill (Fujita et al. 1988; McGlathery et al. 1996; Lotze and Schramm 2000; Cohen and Fong 2004). The third phase of uptake, externally controlled uptake, occurs at low substrate concentrations and is regulated by the rate of nutrient transport across the alga's surface (Pedersen 1994). This is a function of mass transport to the thallus surface as well as diminishing water column nutrient concentration.

The complex and interacting factors and processes affecting nutrient uptake rates of marine macroalgae may mask increasing nutrient supplies to coral reefs as these systems undergo phase shifts from coral- to macroalgal-dominated reefs. As opportunistic algae have rapid nutrient uptake rates, phase shifts to these functional forms may change nutrient dynamics by increasing algal uptake, enhancing sequestering of nutrients within algal tissues, and accelerating recycling of pulses of nutrients within these systems (Valiela et al. 1997). In coral reefs, algae respond quickly to even low levels of nutrient enrichment by enhanced uptake rates (Fong et al. 2001, 2003). Thus, high nutrient uptake in macroalgal-dominated systems may mask increasing supplies by maintaining low water column concentrations, the usual metric to assess increasing nutrient supplies. This suggests we must rethink our current method for managing water quality in coral reef ecosystems.

4.3.4 Nutrient Storage and Retention by Coral Reef Algae

The capacity to store nutrients, represented by standing stock or biomass, varies tremendously across coral reef algal communities. As an example, consider the tremendous differences in biomass between one study of coral reef turf algae (e.g., 0.03–0.6 kg wet wt m^{-2} based on a wet:dry weight ratio of 10:1; Foster 1987) and another of blooms of macroalgae (up to 10 kg m^{-2}; Littler and Littler 1988). Although both algal-dominated communities may be highly productive in terms of gross primary productivity, they clearly occupy different ends of a spectrum in terms of biomass accumulation and therefore N storage capacity. Although there are exceptions, in general opportunistic species with simple thallus forms such as those that dominate coral reef turfs often have low levels of biomass despite relatively rapid growth rates due to short life spans, susceptibility to removal by physical disturbance, and grazing by herbivores. In contrast, persisters such as crustose or upright calcified forms often have high standing stocks due to longevity of individual thalli, investment in structure to withstand physical disturbances, and chemical or structural defenses as protection from herbivores. Although more typical of temperate zones, reservoirs of nutrients stored in coral reef algae may also undergo cyclical and/or seasonal patterns (e.g., Lirman and Biber 2000).

4.3.5 Recycling of Nutrients by Coral Reef Algae: Turnover Rates

Algal turnover rates can affect different processes and storage compartments of the nutrient cycle on coral reefs, including rates of microbial transformation, supply of nutrients to other primary producers, and sediment and water nutrient pools. Turnover rates of nutrients stored in algal biomass depend on rates of consumption by grazers, recycling due to death and decomposition, and export. Overall, coral reef algal communities dominated by turfs turnover more rapidly compared to systems dominated by persisters like calcified or fleshy algae (e.g., Fong 2008).

On a global scale, direct consumption of algae by grazers was estimated as 33.6% of total macroalgal net primary productivity, demonstrating the general importance of macroalgae as the base of grazing food webs in all coastal ecosystems (Duarte and Cebrian 1996). However, on pristine, low-nutrient coral reefs this value may be much higher (see below), as there are a multitude of large herbivorous fishes and sea urchins. Experimental and historical evidence abounds to show that herbivores consume a large proportion of macroalgal productivity (for a review, see Jackson et al. 2001), functioning to accelerate turnover of nutrients stored in tissue.

Death and subsequent decomposition of macroalgal detritus result in release and recycling of stored nutrients. Processing of N through detrital pathways comprises about a third of macroalgal net primary productivity globally (Duarte and Cebrian 1996), while estimates for recycling within coral reef algae are much higher (Duarte and Cebrian 1996). When algae decompose, they release organic N to the water. In addition, recent studies demonstrated that substantial dissolved organic N also "leaks" from healthy thalli (Tyler et al 2003; Fong et al. 2003); healthy macroalgae in growth phase may release 39% of gross primary productivity which is processed via detrital and microbial pathways. Some organic compounds in the water can be taken up directly or are quickly remineralized to inorganic forms that may fuel productivity of other primary producers. If burial in the sediment occurs, remineralization may be slower; thus, sediments may act as a slow release fertilizer and enhance algal productivity when external supplies are low (Stimson and Larned 2000).

Duarte and Cebrian (1996) calculated that a global average of 43.5% of macroalgal net primary production is exported. Export is a function of standing stock, water motion, and algal morphology. For example, *Turbinaria*- and *Sargassum*-dominated reef crests most likely export far more nutrients than those dominated by crustose corallines, despite equally vigorous wave action in both communities, due to their vastly larger standing stock. In contrast, lagoonal systems may export fewer nutrients as there is both less physical disturbance and lower water exchange with oceanic waters to detach and then remove biomass. Thus, algae in lagoons may represent a larger and relatively more stable reservoir of nutrients than those on the reefs themselves.

4.4 Factors Causing Removal: The Importance of Herbivory in Limiting Algal Proliferation

Herbivory on coral reefs can be intense; however, because of overfishing on modern reefs areas of highest herbivory may be limited to remote reefs or those in well-enforced marine protected areas (Jackson 2008). Coral reef herbivores can remove almost 100% of the biomass produced daily by marine algae in certain reef habitats (Carpenter 1986; Hay 1991; Hay and Steinberg 1992; Choat and Clements 1998), and the feeding activities of marine herbivores are an important ecological force controlling the structure and dynamics of algal communities (Ogden and Lobel 1978; Hay 1991; Hay and Steinberg 1992; Paul et al. 2001; Hughes et al. 2007; Amsler 2008a; Bellwood and Fulton 2008). Almost all algal biomass in the ocean is exposed to consumers because most marine algae do not produce underground parts equivalent to the roots of terrestrial plants, although at least one species of *Caulerpa* is known to absorb nutrients through its underground rhizomes (Williams 1984; Ceccherelli and Cinelli 1997). Experimental studies conducted in the Caribbean in the 1980s showed the importance of fish and sea urchins in influencing the abundance and distribution of algae in different reef habitats (Sammarco 1983; Hay 1984; Carpenter 1986; Lewis 1986; Taylor et al. 1986; Littler et al. 1989).

Herbivores consume food that is relatively low in nutritional value and high in indigestible structural material (Mattson 1980) strategies that enhance nutrient uptake (Cruz-Rivera and Hay 2000). Coral reef fishes may eat many times their required energetic needs in order to gain enough nitrogen from seaweeds (Hatcher 1981). Seaweeds are eaten by diverse vertebrates, especially fishes and turtles in tropical waters, and invertebrate consumers that vary in their selectivity and impact on different algae (Bjorndal 1980; Horn 1989; Choat 1991; Hay 1991; Hixon 1997). Invertebrate herbivores include gastropods (snails, limpets, sacoglossans, sea hares, cephalaspideans, and chitons), sea urchins, crabs, amphipods, isopods, shrimps, polychaetes, and copepods (Hay et al. 1987; Hay and Fenical 1988; John et al. 1992; Hixon 1997). The importance of different herbivore groups varies geographically, herbivore species diversity increases toward the tropics (Gaines and Lubchenco 1982; Horn 1989; Choat 1991; Hillebrand 2004; Floeter et al. 2005). Herbivore diversity in the tropics has for the evolution of seaweeds, as evidenced by the increased diversity of defenses and higher chemical defenses in tropical algae when compared to temperate macroalgae (Vermeij 1987; Steinberg and van Altena 1992; Bolser and Hay 1996; Cronin et al. 1997).

4.5 Factors Causing Removal: Chemical Defenses and Interactions

Seaweeds have several mechanisms for tolerating or resisting herbivory, and these defensive strategies have been discussed previously (Duffy and Hay 1990; Hay 1991). Many seaweeds can deter herbivores by morphological, structural, and chemical defenses or by associating with deterrent seaweeds or other benthic organisms that reduce herbivore foraging (Duffy and Hay 1990; Wahl and Hay 1995). Structural defenses such as calcification and toughness are common in certain groups of green and red seaweeds and have been previously discussed (Paul and Hay 1986; Steneck 1988; Duffy and Hay 1990). Chemical defenses of seaweeds have been reviewed, and it is not our intent to comprehensively review this topic (Paul et al. 2001, 2007; Paul and Ritson-Williams 2008; Amsler 2008a). Often, several defensive mechanisms may be functioning simultaneously (Hay et al. 1987; Paul et al. 2001) and the importance of multiple defenses may be very significant in herbivore-rich tropical waters. The common co-occurrence of $CaCO_3$ and chemical defenses in tropical seaweeds has been suggested to be adaptive, because the high diversity of tropical herbivores limits the effectiveness of any single defense (Hay 1997; Paul 1997; Paul et al 2001). For example, combinations of $CaCO_3$ and seaweed extracts have been tested as feeding deterrents and both additive (Schupp and Paul 1994) and synergistic (Hay et al. 1994) effects of these combined defenses have been observed.

Many possible defensive functions for algal natural products (often called secondary metabolites) have been proposed including antimicrobial, antifouling, and antifeedant activities (Paul et al. 2001; Paul and Ritson-Williams 2008; Lane and Kubanek 2008). To date, the role of these compounds as defenses toward herbivores has been best studied. Recent studies have clearly shown that many seaweed natural products function as feeding deterrents toward herbivores (e.g., Hay 1997; Paul et al. 2001; Amsler 2008a). However, many compounds may also have other roles or may function simultaneously as defenses against pathogens, fouling organisms, and herbivores, thereby increasing the adaptive value of these metabolites (Paul and Fenical 1987; Schmitt et al. 1995). Some algal secondary metabolites do show antimicrobial or antifouling effects (Paul and Ritson-Williams 2008; Lane and Kubanek 2008).

Thousands of natural products have been isolated from marine red, brown, and green algae, and the majority of these have come from tropical algae (Maschek and Baker 2008; MarinLit 2009). In general, these compounds occur in relatively low concentrations, ranging from 0.2% to 2% of algal dry mass, although compounds such as the polyphenolics in brown algae can occur at concentrations as high as 15% of algal dry mass (Hay and Fenical 1988; Steinberg 1992). Except for metabolites from phytoplankton and cyanobacteria, very few nitrogenous compounds have been isolated from macroalgae (Ireland et al. 1988; MarinLit 2009). Some cyanobacteria, red algae, and a few green algae incorporate halides from seawater into the organic compounds they produce (Fenical 1975, 1982; Ireland et al. 1988; Hay and Fenical 1988). Bromine and chlorine are the most common halides found in marine algae. Halogenating enzymes such as bromoperoxidases and chloroperoxidases function in the biosynthesis of these halogenated compounds (Butler and Walker 1993). The majority of macroalgal compounds are terpenoids, especially sesqui- and diterpenoid. Acetogenins (acetate-derived metabolites), including unusual fatty acids, constitute another common class of algal secondary metabolites (Ireland et al. 1988; Maschek and Baker 2008). Most of the remaining metabolites result from mixed biosynthesis and are often composed of terpenoid and aromatic portions.

Cyanobacteria are often heavily chemically defended. High abundances of *Lyngbya* spp. and *Oscillatoria* spp. have been observed on coral reefs, where benthic mats of cyanobacteria can cover thousands of square meters (Paul et al. 2005; Paul et al. 2007). These benthic, filamentous cyanobacteria produce a wide variety of secondary metabolites, many of which are toxic or pharmacologically active (Moore 1981, 1996; Gerwick et al. 1994; Nagle and Paul 1999; Tan 2007). Their ability to fix nitrogen may explain their production of many nitrogen-containing secondary metabolites including peptides and lipopeptides. Cyclic peptides and depsipeptides are the major types of compounds isolated from marine cyanobacteria (Moore 1996), and some of these compounds have been shown to deter herbivores (Thacker et al. 1997; Nagle and Paul 1999; Cruz-Rivera and Paul 2007; Paul et al. 2007). In contrast, the opisthobranch sea hare *Stylocheilus striatus* specializes on *Lyngbya majuscula* and prefers artificial diets containing compounds produced by cyanobacteria; however, high concentrations of some of these metabolites can still deter feeding by *Stylocheilus* (Paul and Pennings 1991; Nagle et al. 1998; Capper and Paul 2008). The sea hares *S. longicauda* and *Dolabella auricularia* sequester cyanobacterial compounds from their diets, gaining protection from fish and invertebrate predators (Paul and Pennings 1991; Pennings and Paul 1993; Pennings et al. 1996, 1999).

The marine green algae contain a suite of compounds that provide chemical defenses. Most of the compounds isolated from the green algae are terpenes; sesquiterpenes and diterpenes are particularly common (Hay and Fenical 1988). Tropical green algae of the order Caulerpales have been especially well studied; members of this group, including species of *Caulerpa* and *Halimeda*, contain acyclic or monocyclic sesqui- and diterpenoids (Paul and Fenical 1986, 1987), which are known to defend against herbivores (Paul et al. 2001; Erickson et al. 2006). Chemical defenses in calcified green seaweeds may be particularly important against herbivores such as parrot fishes and sea urchins that can readily consume calcified foods (Pennings and Svedberg 1993; Schupp and Paul 1994; Pitlik and Paul 1997). In a series of laboratory and field experiments designed to examine differences among fish species in their responses to both chemical and structural defenses in *Halimeda*, Schupp and Paul (1994) found that *Halimeda* diterpenes can limit feeding by the parrot fish *Scarus sordidus*, which is not affected by the levels of $CaCO_3$ found in *Halimeda* spp. In contrast, the rabbitfish *Siganus spinus* and the surgeonfish *Naso lituratus* were deterred by $CaCO_3$ in their diets but were unaffected by *Halimeda* diterpenes. In general, combined defenses ($CaCO_3$ and terpenes) increased the number of fish species that were deterred relative to either single defense, which may explain the abundance of *Halimeda* spp. and other calcified green algae in many reef habitats.

Brown algae are the only seaweeds that produce polyphenolic compounds. Although these compounds may function like terrestrial tannins by binding proteins or other macromolecules, they are structurally different compounds that are complex polymers derived from a simple aromatic precursor, phloroglucinol (1, 3, 5-trihydroxybenzene) (Fenical 1975; Ragan and Glombitza 1986; Targett and Arnold 1998). These metabolites are often termed "phlorotannins" to distinguish them from the terrestrial tannins. Polyphenolics in brown algae may function as defenses against herbivores (Steinberg 1992; Targett and Arnold 1998), as antifoulants (Sieburth and Conover 1965; Lau and Qian 1997, but see Jennings and Steinberg 1997), as chelators of metal ions (Ragan and Glombitza 1986), and in UV absorption (Pavia et al. 1997).

In addition to polyphenolics, brown algae in the order Dictyotales produce nonpolar metabolites such as terpenes, acetogenins, and compounds of mixed terpenoid-aromatic biosynthesis (Maschek and Baker 2008). *Sargassum* species also produce acetogenins and compounds of mixed terpenoid-aromatic biosynthesis (Maschek and Baker 2008). Brown algal compounds, especially compounds from *Dictyota* spp., have been shown to deter a variety of herbivores in temperate and tropical waters (Hay et al. 1987, 1998; Paul et al. 1988; Hay 1991; Pereira and da Gama 2008).

The greatest variety of secondary metabolites is probably found among the red algae where all classes of compounds except phlorotannins are represented and many metabolites are halogenated (Fenical 1975; Faulkner 1984; Maschek and Baker 2008). Red seaweeds from the families Bonnemaisoniaceae, Rhizophyllidaceae, and Rhodomelaceae are rich in halogenated compounds that range from halogenated methanes, haloketones, and phenolics to more complex terpenes (Fenical 1975, 1982; Faulkner 1984; Marshall et al. 1999). The red algal genus *Laurencia*, the subject of extensive investigations, produces over 500 compounds (MarinLit 2009), many of which are halogenated and of unique structural types (Erickson 1983; Faulkner 1984).

Usually, the presence or absence of deterrent secondary metabolites in seaweeds correlates well with the susceptibility of seaweeds toward herbivores. Seaweeds that are least palatable to grazing fishes often employ chemical and structural defenses (Hay 1984, 1997; Paul and Hay 1986; Hay et al. 1994; Schupp and Paul 1994; Meyer and Paul 1995; Paul et al. 2001). The common method of testing for feeding deterrent effects against herbivores has been to incorporate seaweed extracts or isolated metabolites at natural concentrations into a palatable diet, either a preferred seaweed or an artificial diet, and then to compare feeding rates of the grazers on treated foods with those on appropriate controls (Hay et al. 1998). Deterrent effects observed in these assays appear to be based primarily on the taste of the treated food. If a compound is deterrent toward an herbivore, the degree of avoidance is often related to the concentration of the extract or metabolite in the diet. These methods do not assess toxicity or other physiological effects on the consumers or possible detoxification methods by herbivores (Sotka and Whalen 2008). Predictions about the toxic or deterrent effects of particular secondary metabolites toward natural predators may be difficult to make based upon chemical structures or results of pharmacological assays. Field and laboratory assays with natural herbivores are important for examining these ecological interactions (Hay et al. 1998).

A variety of compounds from all classes of marine algae have now been tested for their effects on feeding by many different temperate and tropical herbivores. Many of these compounds effectively deter feeding by herbivores. However, there is considerable variance in the responses of different types of herbivores to even very similar compounds. There is also considerable variation among different herbivores, even closely related species, in their responses to secondary metabolites from seaweeds. Thus, as diversity of herbivores increases, the probability of having herbivores that are not affected by any particular type of algal defense undoubtedly increases, and in these cases, complex mixtures of secondary metabolites (Biggs 2000) or multiple types of defenses may be particularly important (Paul and Hay 1986; Schupp and Paul 1994; Paul 1997).

Benthic community structure on coral reefs is strongly influenced by the chemical defenses of seaweeds as well as benthic invertebrates. Many macroalgae living on coral reef slopes, where herbivory is most intense, contain structural or chemical defenses that allow them to establish populations in the presence of abundant and diverse herbivores (Hay 1991; Paul 1992; Paul et al. 2001). In studies of algal succession on artificial reefs on Guam, Tsuda and Kami (1973) suggested that selective browsing by herbivorous fishes on macroalgae removed potential competitors and favored the establishment of unpalatable cyanobacteria. It is likely that this same model functions during phase shifts on coral reefs, resulting in establishment of primarily unpalatable macroalgae and cyanobacteria in coral reef habitats experiencing reduced herbivory. The result is the predominance of chemically defended seaweeds, including species of *Halimeda, Dictyota,* and *Lobophora,* which are the seaweeds most often implicated in phase shifts on Caribbean reefs (Rogers et al. 1997; McClanahan et al. 1999, 2000).

4.6 Benthic-Community-Level Interactions: Nutrient Supply Shapes Community Structure

The extent to which nutrient supply and thus limitation of coral reef algae shapes the benthic structure of coral reefs is an extremely controversial issue. While some believe nutrients are of key importance, others question whether algae in tropical systems are usually, or ever, limited by nutrients. To support the latter belief, some reason that if tropical algal turfs are extremely productive even in areas characterized by low water column nutrients (Hatcher 1988; McCook 1999) they cannot be nutrient limited. To support this, Williams and Carpenter (1998) provided evidence that supply of solutes to algal turfs may be more limited by boundary layers than concentration. A second rationale focuses on field studies where water nutrient concentrations do not correlate with algal growth or abundance in the field (e.g., McCook 1999; Thacker and Paul 2001).

Microcosm and in situ field experimental studies of nutrient limitation of tropical algae have had conflicting results, adding fuel to this controversy. Across many systems and scales, effects of nutrient additions have varied from no effects (Delgado et al. 1996; Larkum and Koop 1997; Miller et al. 1999; Koop et al. 2001) to orders of magnitude differences in effects on photosynthesis, growth, and biomass accumulation (Schaffelke and Klumpp 1997, 1998; Lapointe 1987, 1989; Smith et al. 2001). Interpretation of experimental results is limited, in part, by the difficulty of relating results of laboratory or microcosm studies of the effects of nutrient addition to natural growth in high-energy, high-flow environments with variable nutrient supply typical of coral reefs (Fong et al. 2006), and the related methodological challenge of effectively conducting in situ experiments in these same environments (reviewed in McCook 1999). Factors potentially confounding experimental microcosm approaches include, among others, the lack of recognition of the role of nutrient history and status of the experimental macroalgae (Fong et al. 2003). Many experiments are conducted without setting them in the context of the natural spatial and temporal variability inherent in the environment. On the other hand, several processes that are difficult to control in the field may confound in situ enrichment experiments. On the local scale, preferential selection for enriched, fast-growing algae by herbivores may effectively mask the effects of enrichment in small-scale in situ experiments (Boyer et al. 2004; Fong et al. 2006). In addition, sources of nutrients in tropical systems may not be solely from the water column but may flux from sediments (e.g., Stimson and Larned 2000) and therefore not be included in traditional supply estimates. Because of these complex processes and many possible confounding factors, the importance of nutrient limitation for tropical algae is not well understood.

Although we are far from resolving this controversy, it is of paramount importance to continue to our efforts to unravel the complexity. Nutrient supplies to marine systems are certain to continue to increase with coastal development, and interactions with other anthropogenic stressors are equally certain to have complex and often unexpected effects.

4.7 Benthic-Community-Level Interactions: Positive Algal Cues for Coral Larvae

Many marine invertebrate larvae use chemical cues to determine the appropriate habitat for settlement (Pawlik 1992; Hadfield and Paul 2001; Ritson-Williams et al. 2009), and crustose coralline algae are known to induce settlement and metamorphosis for a variety of marine invertebrate larvae (Hadfield and Paul 2001). Settlement is defined as the larval behavioral response that occurs when a pelagic larva descends to the bottom and moves over a substrate with or without attaching to it; it is often considered a reversible process. Metamorphosis includes the subsequent morphological and physiological changes that pelagic larvae undergo to become benthic juveniles. Chemical cues are implicated for settlement and metamorphosis of corals and other coral reef invertebrates; thus, changes in benthic communities can affect the settlement and metamorphosis of many invertebrate larvae (Ritson-Williams et al. 2009; Chadwick and Morrow 2010).

Crustose coralline algae serve as cues for the settlement of coral larvae, although different species of corals display different degrees of specificity in their requirements for crustose coralline algae (CCA) to induce metamorphosis (Morse et al. 1994; Morse and Morse 1991; Harrington et al. 2004; Ritson-Williams et al. 2010). An insoluble, cell-wall polysaccharide (a type of sulfated lipoglycosaminoglycan) is one type of compound that induces the settlement of many species of coral larvae including *Agaricia* spp. in the Caribbean and *Acropora* spp. in the Pacific (Morse and Morse 1991; Morse et al. 1994, 1996). It has been suggested that many different corals require the same type of algal cue for the induction of settlement and metamorphosis (Morse et al. 1996).

Acropora spp. settle and metamorphose in the presence of *Titanoderma prototypum* and *Hydrolithon* spp., but do not require CCA for settlement and metamorphosis (Harrington et al. 2004; Ritson-Williams et al. 2010). Some species of CCA induce very low rates of settlement and metamorphosis of coral larvae, indicating the species-specific nature of these coral–algal interactions (Ritson-Williams et al. 2010). Studies with larvae of *Acropora millepora*, a common Indo-Pacific coral species, and coral larvae collected from natural slicks after mass spawning events also demonstrated the role that coralline algae play in inducing settlement and metamorphosis of acroporid larvae (Heyward and Negri 1999). Chemical extracts of the algae and the coral skeleton were also active with up to 80% of larvae metamorphosing in 24 h. Larvae of the corals *Acropora tenuis* and *A. millepora* in Australia had the highest rates of settlement in response to the coralline alga *Titanoderma prototypum*, which also caused the lowest coral post-settlement mortality of the algae tested. Methanol extracts of *T. prototypum* and *Hydrolithon reinboldii* both induced high rates of metamorphosis at natural concentrations (Harrington et al. 2004).

Coralline algae have been identified as a positive settlement cue for some corals, but it is unclear if the algae themselves or biofilms present on these algae are responsible for the observed settlement behavior (Johnson et al. 1991; Webster et al. 2004). A recent study on Guam found that larvae of the spawning species *Goniastrea retiformis* preferred substrate covered with crustose coralline algae, but the reef-flat brooding coral *Stylaraea punctata* preferred biofilmed rubble (Golbuu and Richmond 2007). Similarly, larvae of the pocilloporid *Stylophora pistillata* did not require coralline algae for metamorphosis (Baird and Morse 2004). Johnson et al. (1991) noted that unique bacteria occur on the surfaces of crustose coralline algae and that they could serve as the sources of inducers for settlement of corals and other invertebrates. Biofilms were isolated from the coralline alga *Hydrolithon onkodes*, and one bacterium alone, *Pseudoalteromonas* sp., was enough to induce settlement and metamorphosis of *Acropora millepora* larvae (Negri et al. 2001). When *H. onkodes* was sterilized in an autoclave, and treated with antibiotics it still induced significantly more settlement and metamorphosis than seawater or terracotta tiles. It is likely that the compounds that stimulate coral larval settlement may be more diverse in reef habitats than previously recognized, and they appear to be associated with only certain species of crustose coralline algae.

4.8 Benthic-Community-Level Interactions: Negative Algal Cues for Coral Larvae

Macroalgae and benthic cyanobacteria can negatively impact the settlement of coral larvae (Kuffner and Paul 2004; Kuffner et al. 2006). The cyanobacterium *Lyngbya majuscula* reduced the survivorship of *Acropora surculosa* larvae and settlement and metamorphosis of *Pocillopora damicornis* in studies conducted on Guam (Kuffner and Paul 2004). In the Florida Keys, USA, two brown algae, *Dictyota pulchella* and *Lobophora variegata*, reduced the total number of settlers of the brooding coral *Porites astreoides*, while the cyanobacterium *Lyngbya polychroa* caused avoidance behavior such that more larvae settled away from the settlement tile (Kuffner et al. 2006). The mechanisms that caused larval avoidance are unclear, but these macrophytes are known to be chemically rich and defended from some herbivores. Baird and Morse (2004) showed similar effects of *Lobophora* sp. on larvae of the corals *Acropora palifera* and *Stylophora pistilata* and suggested that the alga contained compounds that inhibited larval metamorphosis. In contrast, *Favia fragum* larvae had high rates of settlement and metamorphosis onto live *Halimeda opuntia* when offered with coral rubble (Nugues and Szmant 2006).

Filamentous algal turfs, which can trap sediment, were tested alone and in combination with sediments to determine their effect on the settlement of larvae of the spawning Pacific coral *Acropora millepora* (Birrell et al. 2005). There was reduced settlement in response to one of the algal turfs regardless of the presence of sediment. The other algal turf did not reduce *A. millepora* settlement unless sediment was added (Birrell et al. 2005).

Waterborne compounds from macroalgae have been demonstrated to influence settlement and metamorphosis of larvae of corals and other invertebrates (Walters et al. 1996; Birrell et al. 2008a; Miller et al. 2009). Both positive and negative effects were observed for seawater collected from aquaria that had contained macroalgae for the settlement of larvae of the spawning coral *Acropora millepora* onto fragments of the crustose coralline alga *Hydrolithon reinboldii* (Birrell et al. 2008a). Miller et al. (2009) also saw variability in the responses of larvae of three coral species to macroalgal exudates from different benthic macroalgal assemblages in the Florida Keys. These results underscore the complexity of the effects of macroalgae on coral larval settlement behavior.

4.9 Benthic-Community-Level Interactions: Algal/Coral Competition

Ritson-Williams et al. (2009) and Birrell et al. (2008b) recently reviewed ecological mechanisms affecting coral recruitment and concluded that competition between macroalgae and larval and juvenile coral is a very important ecological force shaping coral reef community structure. Further, Chadwick and Morrow (2010) determined competition between corals and macroalgae occurs through a wide variety of mechanisms, including both physical and chemical processes that can impact all stages of the coral life cycle. Seven mechanisms of competition were identified including preemption of space, shading, allelopathy, attraction of settling larvae to ephemeral algal surfaces, abrasion, basal encroachment, and increased sedimentation. However, a recurring theme of these recent reviews is that the nature and importance of competitive interactions between corals and macroalgae varies greatly across different species of macroalgae (for reviews, see McCook et al. 2001; Birrell et al. 2008b; Ritson-Williams et al. 2009; Chadwick and Morrow 2010). For example, in Roatan, *Lobophora variegata* shading increased mortality of juvenile *Agaricia agaricites*, while the mere presence of the alga reduced coral growth (Box and Mumby 2007). In contrast, shading by *Dictyota pulchella* only affected coral growth. Thus, the diversity, abundance, and spatial placement of the macroalgal community with respect to the coral community must be considered when assessing the importance of competition between these groups (Fig. 3).

Reductions in recruitment in areas that have shifted from coral- to algal-dominated reefs (Edmunds and Carpenter 2001; Birrell et al. 2005) are thought to be due in part to chemically induced mortality or the increased biomass of fleshy algae

Fig. 3 Examples of algal/coral competition: (**a**) the algae *Dictyota pulchella*, *Lobophora variegata*, and *Halimeda* sp. growing over the corals *Montastraea* sp. and *Porites astreoides* at a depth of 5 m in Belize, (**b**) the alga *Laurencia obtusa* growing on top of the coral *Acropora palmata* at a depth of 1 m on the reef crest in Belize, (**c**) the alga *Caulerpa macrophysa* growing next to the coral *Porites astreoides* on a reef flat in Belize, and (**d**) the alga *Halimeda copiosa* growing on *Leptoseris cucullata* at a depth of 30 m in Belize (Photographs by Raphael Ritson-Williams)

functioning as a reservoir for coral pathogens (Nugues et al. 2004; Ritson-Williams et al. 2009; Chadwick and Morrow 2010; Rasher and Hay 2010). Bak and Borsboom (1984) proposed that reduction in water flow adjacent to macroalgae could cause increased coral mortality through changes in the flow regime and increased allelochemical concentrations. Most recently, enhanced microbial activity caused by algal exudates has been identified (Smith et al. 2006; Vermeij et al. 2009), and Kline et al. (2006) determined elevated levels of dissolved organic carbon, which can occur in areas of high algal biomass, increased the growth rate of microbes living in the mucopolysaccharide layer of corals. These studies all suggest that the detrimental effect of algae on corals could be mediated by stimulation of microbial concentrations in the vicinity of a coral colony or recruit, but it is not clear at present whether such stimulation occurs by stimulating the microbial community directly through the release of dissolved organic carbon or by lowering the coral's resistance to microbial infections through allelopathy or other mechanisms.

Several studies have identified the role of macroalgal community structure in mediating the outcome of competition between macroalgae and larval and juvenile corals. For example, Vermeij (2006) attributed large reductions in coral recruitment in Curacao over the last 20 years to shifts in macroalgal dominance from CCA to fleshy macroalgae creating a less-suitable habitat for successful coral recruitment. Other recent studies in the Caribbean showed a clear pattern of increased coral recruitment in places where *Diadema* urchin recovery and grazing had reduced fleshy algal abundance and simultaneously increased the population density of juvenile corals (Edmunds and Carpenter 2001; Aronson et al. 2004; Macintyre et al. 2005).

4.10 Benthic-Community-Level Interactions: Invasive Species

Invasions of nonindigenous organisms into marine habitats have been ranked among the most serious sources of stress to marine ecosystems (Carlton and Geller 1993). During the last century, and especially over the last 3 decades, the frequency of invasions of exotic marine species has increased in temperate coastal regions around the world (for reviews, see Carlton and Geller 1993; Ruiz et al. 1997). Invasions of marine algae are also increasing in frequency, with the total number of introduced seaweeds documented at 277 as of 2007 (Williams and Smith 2007). Invasions of marine algae, especially the invasion of the "killer alga" *Caulerpa taxifolia* into the Mediterranean, represent some of the most dramatic and well-known examples of the strong adverse effects invasive species may have on native populations and communities (Bourdouresque et al. 1992). Although studies of the frequency and impacts of invasions of marine algae in tropical regions are relatively rare (Coles and Eldredge 2002), global patterns of invasion suggest that the situation will only worsen with time (Williams and Smith 2007).

A recent review identified patterns of invasion of marine algae that varied across algal taxonomic group, morphology, and functional form (Williams and Smith 2007). Although most of the studies summarized were in temperate regions, these patterns can be used to identify which groups warrant future attention by coral reef management groups. Overall, several larger algal families have disproportionately more numbers of invasive species than would be expected by chance. For the green algae (Chlorophyta), the incidence of successful invasive species is high in several families, including the Caulerpaceae, Ulvaceae, and Codiaceae. Each of these families is important in tropical habitats, especially on coral reefs. *Caulerpa* (Bourdouresque et al. 1992) and *Codium* (Lapointe et al. 2005) are two of the most famous invasive genera on coral reefs due to their wide proliferation and strong negative community impacts. Families of red algae (Rhodophyta) that include disproportionately more species of successful invaders include, among others, the Rhodomelaceae, Cystocloniaceae, and Gracilariaceae; these families contain some of the invaders with the greatest known ecological impact on tropical reefs. For example, *Acanthophora spicifera* in the Rhodomelaceae has invaded coral reefs worldwide (http://www.issg.org/database/species/distribution), forming stable blooms on coral reefs in Pacific Panama via protection by an associational defense with epiphytic cyanobacteria (Fong et al. 2006). Among the Cystocloniaceae, all invasive species belong to the genus *Hypnea*, identifying the unique invasive abilities of this genus. One of these species, *Hypnea musciformis*, forms widespread and destructive blooms on reefs off the coast of Maui (Smith et al. 2002). The Gracilariaceae contain the highly successful invader *Gracilaria salicornia*, also common in Hawaii (Fig. 4). Although the Areschougiaceae does not contain a disproportionately large number of invasive species, it does contain some of the most successful genera, including *Kappaphycus* (Hawaii, Smith et al. 2002; India, Chandrasekaran et al. 2008) and *Eucheuma*. One dominant red algal family on coral reefs, the Corallinaceae (calcifying reds), though speciose, has fewer invasive species than would be predicted by chance, suggesting that it may not have the characteristics that enhance invasiveness.

The success of invasions among algal groups is likely determined by characteristics of both the invader and the invaded habitat. One caveat, however, is that most invasions are only noted when there is proliferation of the large macroscopic form of the alga. Since most algae have bi- or triphasic life cycles, one must keep in mind that the success of some invasions may be due to some unknown characteristics of the diminutive or microscopic phase. In a

Fig. 4 Many species of invasive red algae proliferate on Hawaiian coral reefs including (**a**) *Kappaphycus alvarezii*, (**b**) *Gracilaria salicornia*, and (**c**) *Acanthophora spicifera*. (**d**) In some areas, such as the coast of Maui, blooms of *Hypnea musciformis* become so prolific that they detach, form floating rafts, and deposit on the beach (Photographs by Jennifer Smith)

global assessment of successful invasions across functional forms (Williams and Smith 2007), invasion success ranked as corticated macroalgae = filamentous > corticated foliose > leathery > siphonaceous > crustose. However, these ranks changed greatly among habitat types. Areas characterized by frequent physical disturbance were successfully invaded more often by filamentous or coarsely branched algae such as those in the genera *Acanthophora*, *Cladophora*, and *Polysiphonia*. In contrast, areas subject to lower disturbance and perhaps higher productivity were more likely to be successfully invaded by corticated or leathery macroalgae such as *Kappaphycus*, *Eucheuma*, and some of the more robust forms of *Gracilaria*. Different levels of herbivory may also play a role in invasion success, because many of the filamentous or coarsely branched forms are not grazer resistant.

A comparison of the most and least successful invasive algal forms may provide some insight as to overall characteristics that enhance invasion success. Some of the most successful algal invaders are siphonaceous green macroalgae in the Order Bryopsidales. Common genera include *Codium*, *Caulerpa*, and *Bryopsis*, all of which are important members of tropical algal communities with broad habitat adaptability allowing them to colonize both hard and soft substrata. Morphologically, these algae are single celled yet multinucleate and range from simple to quite complex pseudoparenchymatous thalli. This thallus construction allows for rapid growth, efficient wound healing, and prolific asexual reproduction via fragmentation. In contrast, articulated and crustose calcareous algae are the least successful invaders; crustose calcareous algae have only experienced a single invasion, while articulated calcified forms have never been documented to successfully invade a new habitat. Explanations may include relatively specific habitat requirements, narrow physiological tolerances, and a lack of appropriate propagules to disperse via anthropogenic means.

Differences in success of invasions among algal groups may also be strongly influenced by the strength and mode of the invasion vector. However, in 40% of algal invasions globally, the vector is unknown (Williams and Smith 2007). Of those that have been identified, hull fouling and aquaculture account for the majority of successful invasions. Of invaders carried to new habitats on the hulls of ships, over 50% are filamentous or sheetlike. Dispersal via aquaculture can be direct, through escape of cultured algae. In this case, the characteristics of the invaders are determined by the traits that also make them commercially valuable. These are often corticated or leathery macroalgae with copious amounts of agar, carrageenan, or other valuable products. They are also often fast growing and tolerant of a wide range of environmental conditions including herbivores because of their tough or leathery thalli. Dispersal through agriculture may also be indirect, for example, when an invasive alga is brought in accidentally with another aquacultured organism such as an epiphyte attached to shellfish. These are often red corticated algae, though there have been several documented cases of larger leathery forms transported this way (Williams and Smith 2007). In contrast to invasive animals, ballast water of ships account for a relatively small proportion of algal invasions, ~10% globally. It is likely that lack of light in ballast tanks may account for this difference. The aquarium industry accounts for <1% of algal invasions globally, though it accounts for perhaps the most noteworthy, the invasion of *Caulerpa taxifolia* into the Mediterranean (Meinesz 1999).

The effects of invasive algae on native marine populations, communities, and ecosystems have only been studied in 6% of the cases reviewed by Williams and Smith (2007). Most commonly, negative effects on the abundance of native seaweeds or epiphytes were detected, though this result is hardly universal. Some studies tested the effects on animals, usually whether native herbivores avoided invasive algae. Overall, the findings did not support the "enemy release hypothesis" – invasive species may not be preferred, but were consumed, even the heavily chemically defended *Caulerpa taxifolia*. Yet, herbivores do not seem to effectively control the spread of invasive algae, contrasting with plant invasions on land, in freshwater, or in salt marshes. One explanation is that herbivores may enhance spread of invasive algae through increasing fragmentation. The effect of invasive algae on native algal community structure or ecosystem processes is virtually unstudied. The little evidence that exists suggests that few communities are resistant to invasions, and that both disturbance and nutrient enrichment may enhance invasion success.

Very little is known about introduced algal species in the tropics. This may be due to the lack of historical records and the relative paucity of tropical phycologists. Algal invasions, however, may be of special concern in tropical reefs systems because of the growing aquaculture industry in many of these regions. Many of the most invasive and destructive tropical red algae such as *Eucheuma* and *Kappaphycus* are often cultured right next to coral reefs, and significant negative impacts have begun to be recorded (Smith et al. 2002; Chandrasekaran et al. 2008). Aquaculture near coral reefs is predicted to have a major economic impact in the future (Williams and Smith 2007). Recommendations to limit these impacts include prevention, early detection and rapid responses to eradicate new invaders, and control of present invaders. All of these will require a much more thorough understanding of the processes that control invasion success on coral reefs.

4.11 Climate Change

Climate change is producing a suite of changes in global environmental drivers, including sea-level rise, increased temperature, increased CO_2 in the air and water, ocean acidification, and changes in weather patterns (Parmesan and Yohe 2003). Although perhaps the least studied of all with the highest uncertainties, there are indications that ocean circulation may change (Diaz-Pulido et al. 2007). The IPCC (2007) predicts that CO_2 and temperature will continue to rise with greatest warming at the highest latitudes resulting in further ice melt and rising sea level, putting large deltas and island nations at greater risk of flooding and coastal erosion. In wetter tropical areas, more heat extremes, heavy precipitation, and higher river run off are predicted, while in drier subtropical areas there will be a decrease in precipitation with more heat waves and drought. Despite this variance in total precipitation across the tropics, the frequency of heavy precipitation events will continue to increase greatly over most areas of the globe, with increased flooding, erosion of terrestrial sediments, physical disturbances, and potentially enhanced nutrient supply to coastal ecosystems. There will be increased tropical cyclone intensity with more poleward shifts in their paths causing major physical disturbances. Tropical regions recognized as particularly vulnerable to this suite of changes include small islands due to their close association with the oceanic environment. Tropical ecosystems noted to be especially vulnerable are mangroves and coral reefs. These global changes will directly affect overall productivity of coral reef algae as well as species distributions, abundances, and diversity. There will also be indirect effects on algae through negative effects on corals. Both direct and indirect effects will result in changes in reef community structure and ecosystem functioning.

Sea-level rise, which is expected to continue for centuries even if green house gas concentrations stabilize today (IPCC 2007), may have relatively minor direct effects on coral reef algae. Even the highest range of predicted rates of rise, 59 cm over the next 100 years (IPCC 2007), are generally not thought to be rapid enough to exceed the ability of coral reef accretion processes to "keep up" (Smith and Buddemeier 1992). One caveat is that these predictions by IPCC do not include rapid dynamical changes in ice cover, which may result in a more rapid and higher rise of up to 7 m. In addition, predictions of reef accretion are based on current or past growth rates (e.g., Smith and Buddemeier 1992; Kan and Kawana 2006), which may be drastically reduced due to other effects of climate change such as acidification (Hoegh-Guldberg et al. 2007). Given these qualifiers, however, for most reefs effects on algae most likely will be limited to minor shifts in patterns of zonation due to increased water depth, with possible losses of the deepest communities. To compensate, rising seas will flood intertidal and shallow subtidal areas and coastal low lands, creating new shallow water habitat that is available for algal colonization and growth (Erez et al. 2010, this volume). In addition, some lagoons that at present lack a good ocean connection may become well flushed and more productive (Smith and Buddemeier 1992). This type of compensation may not always be possible, however, especially for island nations, where low lands are often tightly constrained between the ocean and mountains or are filled with high-density development that may be actively protected (seawalls, filling), limiting the expansion of shallow habitat.

Sea-level rise may have far more significant indirect than direct effects on coral reef algae through interactions with associated terrestrial systems. When reefs are associated with continents or islands, key risks include flooding and coastal erosion (Heberger et al. 2009), which may reduce photic zone depth and reduce suitable habitat for algae, especially on fringing reefs. In addition, erosion of fringing reef flats due to increased wave action as water deepens may reduce the ability of these systems to provide protection from storms (Sheppard et al. 2005), thus forming a positive feedback accelerating erosion and flooding. Because of these indirect effects, the net effect of sea-level rise on coral reef algal communities is difficult to predict, but most likely will be minor relative to the responses to other factors associated with climate change.

The IPCC (2007) predicts sea surface temperatures (SST) will continue to rise in the range of 1.1–6.4°C over the next 100 years, with the greatest uncertainty in predictive ability for the tropics. Paleoclimate records suggest that the tropics are more buffered to temperature changes than other regions (e.g., Smith and Buddemeier 1992), though most agree that the past does not provide an adequate model for anthropogenic climate change. Even if water temperatures remain within the lowest range of predictions, there will be strong effects on coral reef algae, both directly and indirectly through the negative effects on coral. There is a plethora of evidence that corals bleach with high temperature anomalies of as little as 1–2°C over a period of a few weeks and suffer mortality at higher intensities or longer durations (e.g., Hoegh-Guldberg 1999; Hoegh-Guldberg et al. 2007; Hueerkamp et al. 2001; D'Croz et al. 2001). Recent evidence also suggests corals may suffer more outbreaks of disease with thermal stress (Bruno et al. 2007). There is a preponderance of evidence that when corals die they are quickly replaced by algal communities (e.g., Littler and Littler 1997; Hughes et al. 2007). Some evidence suggests that coral reefs subject to higher natural variability in temperature may be more susceptible to climate change (McClanahan et al. 2009), while other studies found variability in response was dependent on the rate and duration of temperature rise (Fong and Glynn 2000; Glynn and Fong 2006). There is little doubt, however, that continued sea surface warming will open a significant amount of space presently occupied by corals for colonization by algae (Hoegh-Guldberg et al. 2007), transforming many coral reefs to algal reefs.

Direct effects of rising SST are also predicted to have strong direct effects on coral reef algal distribution and abundance. Overall, the largest global effect will be poleward shifts in geographic ranges of species that will ultimately alter the composition of all marine communities (for a review, see Hawkins et al. 2008), including coral reef algae. Tropical and subtropical species of algae are predicted to expand their ranges both north and south, while temperate and polar species undergo a latitudinal retreat. Small and fragmented edge populations may be more susceptible to changes in thermal stress because of inherently lower fitness and lower adaptive capacity due to random genetic drift (Pearson et al. 2009); this may effectively enhance the invasive ability of tropical algae. In addition, in temperate latitudes warming may result in deepening thermoclines and relaxation of cold water upwelling, thus enabling warm water species to more easily jump gaps in distribution, especially along the western margins of major continents. Overall, rising SST is expected to expand the distribution and enhance the abundance of tropical algal communities at the expense of both coral and temperate algal communities. There is some experimental evidence that warming may favor algal turf over crustose coralline algae (Diaz-Pulido et al. 2007). Certain groups of especially thermal tolerant or thermophilic algae such as the Cyanobacteria can be expected to thrive, possibly leading to more frequent harmful cyanobacterial blooms (Hallock 2005; Paul 2008; Paerl and Huisman 2008).

Experimental evidence is beginning to accrue that increased CO_2 concentrations and the resultant acidification of seawater may have strong effects on all primary producers, including coral reef algae. At present, CO_2 concentrations in the atmosphere are ~380 ppm, a level unprecedented in at least the last 650,000 years (IPCC 2007), with predictions of ~500 ppm by 2100 (Hoegh-Guldberg et al. 2007). Experiments with elevated CO_2 and lowered pH revealed strong negative effects on tropical crustose calcareous algae (CCA). In Hawaii, decreased pH treatments decreased growth and benthic cover of CCA by 85% compared to algae in experimental microcosms subject to ambient seawater; rhodoliths actually decreased in weight by 250%, showing a net dissolution of algal-derived $CaCO_2$ over time (Jokiel et al. 2008; Kuffner et al. 2008). Another study on Australia's Great Barrier Reef demonstrated that CCA is even more sensitive than coral to elevated CO_2; while CO_2 bleaches both coral and CCA, it has more extreme effects on CCA including net negative primary productivity and dissolution of algal carbonate (Anthony et al. 2008). In a review, Diaz-Pulido et al. (2007) suggested upright calcifying forms of macroalgae may also be affected by acidification and that this may lead to reduction in production of sand and loss of habitat.

Results of aquaculture optimization studies also identified the role of rising CO_2 in changing the growth rates of algae. Overall, increased supplies of CO_2 greatly accelerated growth of algae in the genus *Gracilaria* that cannot utilize HCO_3^- as a carbon source (Friedlander and Levy 1995; Israel et al. 2005). CO_2 effects, however, are not uniform across groups. Aquacultured *Porphyra* dramatically decreased growth with increased CO_2, most likely related to increases in dark respiration rates (Israel et al. 1999). Positive effects of increased CO_2 were especially important in low turnover aquaculture systems (Friedlander and Levy 1995), suggesting that changes in CO_2 may be especially important in highly productive systems with low turnover such as tropical lagoons. Blooms of native and invasive species of *Gracilaria* are already prolific on some coral reefs, and these studies suggest they will only intensify. Thus, rising CO_2 and acidification will change the structure of tropical algal communities by shifting dominance from CCA toward fleshy and invasive species.

How climate change will affect the supply of nutrients to coastal ecosystems with resultant effects on coral reef algae is difficult to predict (Diaz-Pulido et al. 2007). Although complex, the supply of nutrients to surface waters is, at least in part, a function of temperature, and has shown a decreasing trend with increasing temperature over the last 3 decades in the Northern Hemisphere (Kamykowski and Zentara 2005). A strengthening and deepening of the thermocline may cause severe nutrient limitation, especially in coral reef regions where thermocline shoaling may provide a significant source of nutrients such as to the Great Barrier Reef and the Florida reef tract (Leichter et al. 1996; Wolanski and Pickard 1983). Upwelling may be suppressed, limiting oceanic nutrient supplies to some reef systems that are subject to seasonal upwelling such as those in the Gulf of Panamá and the Galapagos Islands (Glynn and Maté 1997). Whether this reduction in supply will be matched or greatly exceeded by changes in terrestrial supplies is highly debated (Jickells 1998; Steneck et al. 2002), but most likely depends on many local factors including accelerating variance in precipitation and the level of watershed development. If supplies are enhanced, then there may be a net increase in benthic algal productivity as well as increases in phytoplankton. However, the net effects of changes in nutrient supply on coral reef algae are likely to vary greatly across tropical regions and to be driven by interactions with local and global changes in other anthropogenically influenced environmental factors.

Current and projected rising sea levels with increased coastal erosion as well as more frequent and intense storms (IPCC 2007) will act to increase the supply of terrestrial sediments to coral reefs; this will only be accelerated by local anthropogenic changes due to coastal development. Sediment particles smother reef organisms and reduce light for photosynthesis (for a review, see Rogers 1990), effectively reducing habitat area by decreasing the depth of the photic zone. Increased sedimentation has been documented to have many negative effects on corals that reduce reef accretion rates and may make them more susceptible to rising sea level. Although studies on the direct effects of sediments on coral reef algae are limited, evidence from the Mediterranean demonstrated that sediment effects vary across algal functional forms. High levels of sedimentation decreased the growth of the dominant algal turfs, facilitating coexistence with upright macroalgae such as *Halimeda, Dictyota,* and *Padina* (Airoldi and Cinelli (1997); Airoldi 1998). However, coral reef turfs are more tolerant of sedimentation than corals as increased sedimentation favored growth of turfs that then reduced recruitment of coral and CCA (Nugues and Roberts 2003). Another study also suggested coral reef turfs may be more tolerant of high levels of sediment than would be predicted from studies in the Mediterranean. Bellwood and Fulton (2008) suggest that sediments provide a refuge from herbivory for coral reef algal turfs, thus facilitating dominance of this functional form as an alternative stable state. As sedimentation is predicted to increase with climate and other anthropogenic factors, it is essential that we further our understanding of the effect of enhanced sediment supply on coral reef algae.

While each of the individual factors that are predicted to change with climate will have significant effects on coral reef algae, evidence is beginning to accumulate that interactions among factors may be even more important. However, interactive effects may be quite complex, limiting our predictive abilities without detailed experiments. For example, while increased temperature and nutrients may increase algal recruitment and growth overall, they should also increase the number of grazers, driving algal communities toward opportunistic species (Lotze and Worm 2002). O'Connor (2009) also found that warming strengthens herbivore–algal interactions, shifting important trophic pathways. Recent multifactorial experiments showed that some stages in the complex life cycles typical of algae may be more sensitive to interactions among climate-related factors, making certain stages important bottlenecks limiting the ability of algal dominants to survive climate change over the long term (Fredersdorf et al. 2009). Aquaculture studies identified an important interaction between rising CO_2 and nutrients, with positive CO_2 effects on growth of *Gracilaria* being accelerated with pulsed nutrient supplies (Friedlander and Levy 1995). Thus, increased storms combined with rising CO_2 will act to facilitate the already prolific blooms of these opportunists on some reefs. Overall, it appears that interacting factors associated with climate change will synergistically enhance algal blooms, and may shift communities beyond a state from which they can return.

In summary, our changing climate may result in irreversible changes in coral reef benthic producer communities. Widespread coral mortality will open space for colonization by algae, producing a shift from coral- to algal-dominated reefs. Sea-level rise may increase algal-dominated areas as well, though these areas may be severely stressed by erosion and sedimentation. Tropical algal distributions will expand poleward, resulting in shifts in community composition due to differential responses of algal species to individual and interactive factors driven by climate change. Upward negative cascades may occur in these newly colonized latitudes as trophic pathways are altered. On tropical reefs there will be a shift from CCA (reef building) to other types of algae, especially chemically defended species where herbivores are protected, or fast-growing opportunists in areas of reduced herbivory. Shifts toward invasive and/or opportunistic species of algae may also occur as these species have appropriate physiologies to take advantage of changing environmental conditions. If nutrient supplies increase, further shifts toward dominance by nutrient specialists that are disturbance tolerant may occur. These shifts will be most dramatic in lagoons with low turnover or flushing rates and that are associated with developed or developing watersheds. Overall, climate change, with other interacting anthropogenic changes, will synergistically enhance algal blooms on coral reefs worldwide.

5 Phase Shifts, Alternative Stable States, and the Stability of Algal-Dominated Tropical Reefs

There is a preponderance of evidence that reductions in live coral cover have been followed by phase shifts to algal dominance (Hughes 1994; McCook 1999; Diaz-Pulida and McCook 2002; McClanahan and Muthiga 2002; Rogers and Miller 2006), although one meta-analysis suggests that recent increases in dominance by upright algae may have been overestimated (Bruno et al. 2009). Coral reef phase shifts are defined as a transition from a community dominated by reef-building organisms to one dominated by non-reef-building organisms, most often macroalgae (Done 1992). Phase shifts have been attributed to reduced herbivory (Hughes 1994; Hughes and Connell 1999), increased nutrients (Banner 1974; Smith et al. 1981; Lapointe et al. 2005; Fabricius 2005), and reductions in live coral cover due to environmental stress and disturbance (McClanahan et al. 1999; Diaz-Pulida and McCook 2002). Regardless of the cause, reductions in coral dominance can result in increased rates of bioerosion that exceed accretion, leading to gradual destruction of the reef framework and habitat loss (Glynn and Maté 1997).

Though all agree that coral reefs are experiencing high levels of degradation, the relative importance of herbivory and nutrients in regulating phase shifts to algal dominance has been hotly debated (e.g., Aronson et al. 2003; Hughes et al. 2003; Pandolfi et al. 2003; Burkepile and Hay 2006). Though increased nutrients and reductions in herbivory may initiate and commonly contribute to the persistence of phase shifts (Fig. 5), there appears to be increasing evidence that elevated SST due to global warming may be the greatest threat to coral reefs globally (Hoegh-Guldberg 1999; Knowlton 2001; Walther et al. 2005; Hughes et al. 2003;

Hoegh-Guldberg 2004; Pandolfi et al. 2005). Following coral bleaching and mortality events, broad areas of reefs are often overgrown by macroalgae, which can competitively exclude corals once established (Coyer et al. 1993; McCook 1999, 2001; Jompa and McCook 2003; Rogers and Miller 2006; Kuffner et al. 2006). Therefore, the capacity of reefs to weather bleaching events may be determined by the presence of intact herbivore populations and low levels of nutrients, which may help limit the proliferation of macroalgae on reefs. The recent proliferation and persistence of algal dominance on coral reefs requires further study, given the potential interactions between temperature stress and human-initiated changes in top-down and bottom-up forces.

Once established, these algal-dominated phase shifts can be difficult to reverse for many reasons already discussed. Algae that dominate communities following the reduction of herbivory are often chemically defended species such as *Halimeda* and *Dictyota* spp., as well as various filamentous cyanobacteria, which are relatively unpalatable to herbivores (Fig. 6). Studies of phase shifts from coral- to macroalgal-dominated communities have often overlooked filamentous cyanobacteria, which generally have been grouped with turf algae in ecological research (e.g., Steneck and Dethier 1994). Benthic cyanobacteria may play an important role in these phase shifts, as they can be early colonizers of dead coral and disturbed substrates (Tsuda and Kami 1973; Larkum 1988), and in some cases have been observed to overgrow coral branches and kill coral polyps (Littler and Littler 1997). In addition, many types of macroalgae can inhibit the recruitment of corals and other invertebrates through a variety of mechanisms (Birrell et al. 2008b; Chadwick and Morrow 2010), which can lead to the persistence of algal-dominated reef communities.

5.1 The Nature of Transitions to Algal Domination of Tropical Reefs: Phase Shifts Versus Alternative Stable States

Dramatic shifts in populations, communities, and ecosystems worldwide in response to natural and anthropogenic alterations in environmental conditions have focused much attention on the nature of change and the processes that drive them (Scheffer et al. 2001; Beisner et al. 2003; Didham et al. 2005; Jackson 2008). Ecological theory predicts that there are many possible patterns of change for populations and communities. Some studies found ecological communities shifted in a relatively simple, linear, or linearizable manner, where a given change in a controlling environmental variable produced a predictable community response (Fig. 7a). Much recent attention, however, has focused on nonlinear shifts, with extreme nonlinearities occurring in response to both

Fig. 5 An example of multiple stresses causing a partial shift away from a coral-dominated state. This *Pocillopora* reef was treated with nutrient addition and herbivory reduction with cages. The time series is: (**a**) before manipulation, (**b**) 6 weeks, (**c**) 10 weeks, (**d**) 16 weeks, (**e**) 22 weeks, and (**f**) 2 days after end of manipulations. Algal cover developed quickly, but was rapidly grazed down after cages were removed, suggesting that herbivory is a stabilizing force on coral reefs (Photographs by Ranjan Muthukrisnan)

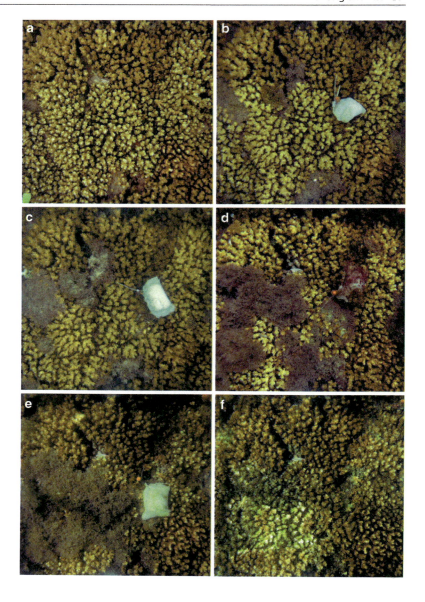

Fig. 6 Blooms of *Acanthophora spicifera* on some reefs of Pacific Panamá were stabilized for at least 5 years by epiphytic cyanobacteria that defended the alga from herbivory: (**a**) Algal bloom on Uva Island reef. (**b**) Cyanobacteria coating the algal thalli (Photographs by Tyler Smith)

natural and anthropogenic changes in environmental forcing functions. Examples span a diversity of ecosystems including the collapse of most major marine fisheries (Roughgarden and Smith 1996; Myers and Worm 2003), desertification of the sub-Sahara (Xue and Shukla 1993), and shifts from coral- to algal-dominated tropical reefs (Hughes et al. 2007).

The rising frequency of rapid collapses from one community state to another, often less desirable, state has renewed interest in whether this process is a simple and symmetrically reversible phase shift (Fig. 7b), or if these shifts represent alternative stable states (Scheffer et al. 2001; Beisner et al. 2003; Didham et al. 2005). It is important to distinguish between these two methods of community change because management strategies must be very different for each if the goal is to promote and stabilize the initial, often "desirable" state. For phase shifts, incremental changes of environmental stressors may cause little change until a critical threshold is reached; at that point, shifts from one community dominant to another are rapid and catastrophic over a very short range of environmental conditions. Thus, in the forward direction, the past history of environmental change will not aid prediction of future change. However, backward shifts are equally rapid, and occur predictably at the same level of environmental stress as the forward shift, making management strategies of these systems conceptually straightforward – one must reverse the environmental condition back to the point that caused the forward shift (e.g., Suding et al. 2004).

The clearest, strongest evidence that communities can exist as alternative stable states is to reverse the environmental condition and observe a backward transition at a different point than caused the forward transition, proving hysteresis (e.g., Knowlton 2001). However, these data are often difficult, expensive, or logistically impossible to obtain. Two key aspects of alternative stable state theory that distinguish alternative stable state transitions in communities from phase shifts and can be measured on the spatial and temporal scales appropriate to reefs are the processes that cause transitions among states and those mechanisms that stabilize them (Scheffer et al. 2001). Transitions among community states can be due to either change in the environment or disturbance. Forward shifts from one state to another (e.g., F_1 in Fig. 7c, d) occur at very different points or conditions of the environmental driver than backward shifts (e.g., F_2) that restore the initial state. Thus, there is a range of environmental conditions, between points F_1 and F_2, where either of two stable states can occur. As a result, there is limited ability to predict the community state based on the current condition of the environment. For alternative stable states, managers would have to reverse the environmental condition far beyond the point of forward shift, to F_2, to restore the initial community, often a very difficult and expensive management option. The only other way to shift communities between states is to subject them to large disturbances that push the state beyond the unstable

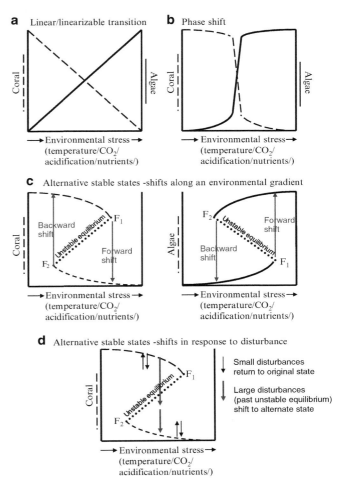

Fig. 7 Conceptual diagram showing the difference between community changes with predictable Y values (e.g., coral or algal cover) for each X value (environmental stresses) such as (**a**) linear and (**b**) phase shift relationship and (**c** and **d**) alternative stable states where there can be two Y values for a range of X values. One key difference that characterizes alternative stable states is that forward and backward transitions occur at different environmental conditions, thus requiring a greater "correction" of an environmental condition to reestablish the prior, often more desirable state, than it took to degrade it. A second difference that characterizes alternative stable states is that they are maintained by negative, stabilizing feedbacks, which make them resilient to small disturbances, and are only shifted to the other state by large disturbances. Disturbances of varying magnitudes are shown by arrows pushing the community away from the stable equilibrium (Artwork by Kendal Fong)

equilibrium (represented by the dotted lines in Fig. 7c, d). Once past the unstable equilibrium, negative feedbacks will function to stabilize the community in the alternate state.

Another key prediction from theory is that, in order for alternative stable states to exist, negative feedback mechanisms must exist to stabilize each state within the range of environmental conditions, F_1 to F_2. Stabilizing mechanisms can include abiotic or biotic processes, but must be strong enough to buffer these states across a range of environmental conditions (between F_1 and F_2). The existence of stabilizing mechanisms provides resilience to small disturbances that only push the community within the basin of attraction of the existing state (does not cross the line of unstable equilibrium in Fig. 7c, d). Thus, at any environmental condition between F_1 and F_2, theory predicts that small disturbances will result in a return to the stable state. To manage any tropical reef system to maintain the coral state it is key to identify and protect the positive feedback mechanisms that support that state.

Field surveys and experiments that demonstrate rapid transitions among community states are often unable to distinguish between phase shifts and alternative stable states as limited temporal and/or spatial scales usually restrict observations of transitions to single events. The types of evidence that can distinguish between these patterns of change and support that tropical reefs exist as alternative stable states include: (1) persistent dominance by one state or the other across a wide range of environmental stress, and (2) the existence of strong and consistent negative feedbacks that stabilize each state.

There are many examples of rapid, catastrophic collapses from coral to algal dominance, and evidence is accumulating that these algal states may be stable or resistant to reversal. Coral to algal transitions have been documented in response to stressors such as nutrients, overfishing, diseases, and increased sedimentation, as well as large disturbances such as hurricanes and ENSO-associated increases in sea surface temperature (e.g., Hughes 1994; Fong and Lirman 1996; Glynn and Maté 1997; Nugues and Roberts 2003; Rogers and Miller 2006; Hughes et al. 2007; Baker et al. 2008, and many others). A few studies have shown that prior to these transitions, coral states were relatively stable. For example, push cores demonstrated that coral-dominated states were relatively stable over at least 3,000 years on two Caribbean coral reefs (Aronson et al. 2004). However, there is a paucity of evidence for recovery from environmental stressors, which could be due to continued stress (but see Idjadi et al. 2006; Diaz-Pulido et al. 2009). To our knowledge, Kaneohe Bay represents the only recorded case of a backward transition from algal to coral dominance due to a reversal in the environmental stress of eutrophication (Smith et al. 1981). Even in this system, however, more recent phase shifts toward other invasive algae have occurred (Stimson and Conklin 2008). In addition, evidence for recovery (transition back to coral state) from disturbances such as ENSO is very mixed, with areas in the Indian Ocean recovering rapidly, areas of the Pacific with patchy recovery (some reefs recovered rapidly, others did not recover), and a general lack of recovery in the Western Atlantic (Baker et al. 2008). In the Caribbean, this lack of recovery may be due to a recruitment bottleneck when coral abundance is extremely low (Mumby 2009). These patterns are consistent with alternative stable state theory, which predicts both stability across a range of environmental stresses and that recovery from disturbance will either be rapid or the algal state will become stable depending on the magnitude of the disturbance (Fig. 7c, d).

There is mounting evidence that numerous negative feedbacks stabilize the algal state as predicted by alternative stable state theory. As discussed above, some of the ecological processes that stabilize the algal state include algal defenses, nitrogen fixation by cyanobacteria ameliorating nutrient limitation, and inhibition of coral recruitment by some forms of algae (Ritson-Williams et al. 2009; Birrell et al. 2008b). In addition, algal states often lack the topographic complexity of coral states, thereby reducing herbivory. Herbivory can also be suppressed by sediment deposition that is enhanced by algal turfs, thereby stabilizing this algal state (Bellwood and Fulton 2008). Clearly, it is vital to continue research into the stabilizing mechanisms for both coral- and algal-dominated states, as these may provide some realistic management strategies for future reef conservation and restoration.

Acknowledgments We thank Zvy Dubinsky for the opportunity to write this review. The manuscript was improved by comments from R. Muthukrisnan, T. Kane, M. Littler, D. Littler, V. Pittman, R. Ritson-Williams, and C. R. Fong. Kendal Fong created artwork for the figures and photographs were provided by Raphael Ritson-Williams, Tyler Smith, Jennifer Smith, and Ranjan Muthukrisnan. Peggy Fong's research has been supported by NSF (with Peter Glynn), EPA, UCLA's Academic Senate, and the Department of Ecology and Evolutionary Biology at the University of California Los Angeles. Valerie J. Paul acknowledges the Smithsonian Hunterdon Oceanographic Endowment and the Marine Science Network for research support. Funding received from NOAA's ECOHAB program (the Ecology and Oceanography of Harmful Algae Blooms) Project NA05NOS4781194, the Florida Sea Grant College Program Grant No. NA06OAR4170014, Mote Marine Laboratory's Protect our Reefs Grants Program, and the U. S. Geological Survey Eastern Region State Partnership Program has supported Valerie J Paul's work on marine algal chemical defenses, coral–algal interactions, and marine plant–herbivore interactions. This is contribution number 816 from the Smithsonian Marine Station at Fort Pierce.

References

Abbot IA (1999) Marine red Algae of the Hawaiian Islands. Bishop Museum Press, Honolulu

Ahern KS, Udy JW, Ahern CR (2007) Nutrient additions generate prolific growth of Lyngbya majuscula (cyanobacteria) in field and bioassay experiments. Harmful Algae 6:134–151

Ahern KS, Ahern CR, Udy JW (2008) In situ field experiment shows Lyngbya majuscula (cyanobacterium) growth stimulated by added iron, phosphorus and nitrogen. Harmful Algae 7:389–404

Airoldi L (1998) Roles of disturbance, sediment stress, and substratum retention on spatial dominance in algal turf. Ecology 79:2759–2770

Airoldi L, Cinelli F (1997) Effects of sedimentation on subtidal macroalgal assemblages: an experimental study from a Mediterranean rocky shore. J Exp Mar Biol Ecol 215:269–288

Amsler CD (ed) (2008a) Algal chemical ecology. Springer, Berlin

Amsler CD (2008b) Algal sensory chemical ecology. In: Amsler CD (ed) Algal chemical ecology. Springer, Berlin, pp 297–309

Anthony KRN, Kline DI, Diaz-Pulida G, Hoegh-Guldberg O (2008) Ocean acidification causes bleaching and productivity loss in coral reef builders. Proc Natl Acad Sci U S A 105:17442–17446

Aronson RB, Bruno JF, Precht WF, Glynn PW, Harvel CD, Kaufman L, Rogers CS, Shinn EA, Valentine JF (2003) Causes of coral reef degradation. Science 302:1502

Aronson RB, Macintyre IG, Wapnick CM, O'Neill MW (2004) Phase shifts, alternative state states, and the unprecedented convergence of two reef systems. Ecology 85:1876–1891

Arthur KE, Paul VJ, Paerl HW, O'Neil JM, Joyner J, Meickle T (2009) Effects of nutrient enrichment of the cyanobacterium *Lyngbya* sp. on growth, secondary metabolite concentration and feeding by the specialist grazer *Stylocheilus striatus*. Mar Ecol Prog Ser 394:101–110

Baird AH, Morse ANC (2004) Induction of metamorphosis in larvae of the brooding corals *Acropora palifera* and *Stylophora pistilata*. Mar Freshwater Res 55:469–472

Bak RPM, Borsboom JLA (1984) Allelopathic interaction between a reef coelenterate and benthic algae. Oecologia 63:194–198

Baker AC, Glynn PW, Riegl B (2008) Climate change and coral reef bleaching: an ecological assessment of long-term impacts, recovery trends, and future outlook. Estuar Coast Shelf Sci 80:435–471

Banner AH (1974) Kaneohe Bay, Hawaii, urban pollution and a coral reef ecosystem. Proc 2nd Int Symp on Coral Reefs 2:685–702

Beach K, Walters L, Borgeas H, Smith C, Coyer J, Vroom P (2003) The impact of *Dictyota* spp. on *Halimeda* populations of Conch Reef, Florida Keys. J Exp Mar Biol Ecol 297:141–159

Becerro MA, Bonito V, Paul VJ (2006) Effects of monsoon-driven wave action on coral reefs on Guam and implications for coral recruitment. Coral Reefs 25:193–199

Beisner BE, Haydon DT, Cuddington K (2003) Alternative stable states in ecology. Front Ecol Environ 1:376–382

Bellwood DR, Fulton CJ (2008) Sediment-mediated suppression of herbivory on coral reefs: decreasing resilience to rising sea levels and climate change? Limnol Oceanogr 53:2695–2701

Biggs JS (2000) The role of secondary metabolite complexity in the red alga Laurencia palisada as a defense against diverse consumers. M.S. thesis, University of Guam, Mangilao, Guam

Birrell CL, McCook LJ, Willis BL (2005) Effects of algal turfs and sediment on coral settlement. Mar Pollut Bull 51:408–414

Birrell CL, McCook LJ, Willis BL, Harrington L (2008a) Chemical effects of macroalgae on larval settlement of the broadcast spawning coral *Acropora millepora*. Mar Ecol Prog Ser 362:129–137

Birrell CL, McCook LJ, Willis BL, Diaz-Pulido GA (2008b) Effects of benthic algae on the replenishment of corals and the implication for the resilience of coral reefs. Oceanogr Mar Bio Annu Rev 46:25–63

Bittick SJ, Bilotti ND, Peterson HA, Stewart HL (2010) *Turbinaria ornata* as an herbivory refuge for associate algae. Mar Biol 157:317–323

Bjorndal KA (1980) Nutrition and grazing behavior of the green turtle *Chelonia mydas*. Mar Biol 56:147–154

Bolser RC, Hay ME (1996) Are tropical plants better defended? Palatability and defenses of temperate vs. tropical seaweeds. Ecology 77:2269–2286

Bourdouresque CF, Meinesz A, Verlaque A, Knoepffler-Peguy M (1992) The expansion of the tropical algae *Caulerpa taxifolia* in the Mediterranean. Cryptogam Algal 13:144–145

Box SJ, Mumby PJ (2007) Effects of macroalgal competition on growth and survival of juvenile Caribbean corals. Mar Ecol Prog Ser 342:139–149

Boyer KE, Fong P, Armitage AR, Cohen RA (2004) Elevated nutrient content of macroalgae increases rates of herbivory in coral, seagrass, and mangrove habitats. Coral Reefs 23:530–538

Boyle KA, Fong P, Kamer K (2004) Spatial and temporal patterns in sediment and water column nutrients in an eutrophic southern California estuary. Estuaries 27:254–267

Bruno JF, Selig ER, Casey KS, Page CA, Willis BL, Harvell CD, Sweatman H, Melendy AM (2007) Thermal stress and coral cover as drivers of coral disease outbreaks. PLoS Biol 5:e124. doi:10.1371/journal.pbio.0050124

Bruno JF, Sweatman H, Precht WF, Selig ER, Schute VG (2009) Assessing evidence of phase shifts from coral to macroalgal dominance on coral reefs. Ecology 90:1478–1484

Burkepile DE, Hay ME (2006) Herbivore vs. nutrient control of marine primary producers: context-dependent effects. Ecology 87:3128–3139

Butler A, Walker JV (1993) Marine haloperoxidases. Chem Rev 93:1937–1944

Canfield DE (1999) A breath of fresh air. Nature 400:503–504

Capone DG (1983) Benthic nitrogen fixation. In: Carpenter EJ, Capone DG (eds) Nitrogen in the marine environment. Springer, New York, pp 105–137

Capper A, Paul VJ (2008) Grazer interactions with four species of *Lyngbya* in southeast Florida. Harmful Algae 7:717–728

Carlton JT, Geller JB (1993) Ecological roulette, the global transport of nonindigenous marine organisms. Science 261:78–82

Carpenter RC (1986) Partitioning herbivory and its effects on coral reef algae communities. Ecol Monogr 56:345–363

Carpenter RC, Williams SL (1993) Effects of algal turf canopy height and microscale substratum topography on profiles of flow speed in a coral forereef environment. Limnol Oceanogr 38:687–694

Ceccherelli D, Cinelli F (1997) Short-term effects of nutrient enrichment of the sediment and interactions between the seagrass *Cymodoceae nodosa* and the introduced green alga *Caulerpa taxifolia* in a Mediterranean bay. J Exp Mar Biol Ecol 217:165–177

Chadwick NE, Morrow KM (2010) Competition among sessile organisms on coral reefs. In: Dubinsky Z, Stambler N (eds.) Coral reefs: an ecosystem in transition. Springer, Dordrecht

Chandrasekaran S, Nagendran NA, Pandiaraja D, Krishnankutty N, Kamalakannan B (2008) Bioinvasion of *Kappaphycus alvarezii* on corals in the Gulf of Mannar. India Curr Sci 94:1167–1172

Choat JH (1991) The biology of herbivorous fishes on coral reefs. In: Sale PF (ed) The ecology of fishes on coral reefs. Academic, San Diego, pp 120–155

Choat JH, Clements KD (1998) Vertebrate herbivores in marine and terrestrial environments: a nutritional ecology perspective. Annu Rev Ecol Syst 29:375–403

Cohen RA, Fong P (2004) Nitrogen uptake and assimilation in *Enteromorpha intestinalis* (L.) Link (Chlorophyta): using ^{15}N to determine preference during simultaneous pulses of nitrate and ammonium. J Exp Mar Biol Ecol 309:67–77

Coles SL, Eldredge LG (2002) Nonindigenous species introduction on coral reefs: a need for information. Pac Sci 56:191–209

Connell JH (1978) Diversity in tropical rain forests and coral reefs. Science 199:1302–1310

Coyer JA, Ambrose RF, Engle JM, Carroll JC (1993) Interactions between corals and algae on a temperate zone rocky reef: mediation by sea urchins. J Exp Mar Biol Ecol 167:21–37

Cronin G, Paul VJ, Hay ME, Fenical W (1997) Are tropical herbivores more resistant than temperate herbivores to seaweed chemical defenses? Diterpenoid metaboites from *Dictyota acutiloba* as feeding deterrents for tropical versus temperate fishes and urchins. J Chem Ecol 23:289–302

Cruz-Rivera E, Hay ME (2000) Can quantity replace quality? Food choice, compensatory feeding, and fitness of marine mesograzers. Ecology 81:201–219

Cruz-Rivera E, Paul VJ (2007) Chemical deterrence of a cyanobacterial metabolite against generalized and specialized grazers. J Chem Ecol 33:213–217

D'Croz L, Maté JL, Oke JE (2001) Responses to elevated sea water temperature and UV radiation in the coral *Porites lobata* from upwelling and non-upwelling environments on the Pacific coast of Panamá. Bull Mar Sci 69:203–214

Dawes CJ (1998) Marine botany. Wiley, New York

Delgado O, Lapointe BE (1994) Nutrient-limited productivity of calcareous versus fleshy macroalgae in a eutrophic, carbonate-rich tropical marine environment. Coral Reefs 13:151–159

Delgado O, Rodriguez-Prieto C, Gacia E, Ballesteros E (1996) Lack of severe nutrient limitation in *Caulerpa taxifolia* (Vahl) C. Agardh, an introduced seaweed spreading over the oligotrophic Northwestern Mediterranean. Bot Mar 39:61–67

Diaz-Pulida G, McCook LJ (2002) The fate of bleached corals: patterns and dynamics of algal recruitment. Mar Ecol Prog Ser 232:115–126

Diaz-Pulido G, McCook LJ, Larkum AWD, Lotze HK, Raven JA, Schaffelke B, Smith JE, Steneck RS (2007) Vulnerability of macroalgae of the Great Barrier Reef to climate change. In: Marshall PA, Johnson J (eds) Climate change and the great barrier reef. Great Barrier Reef Marine Park Authority, Townsville, pp 153–192

Diaz-Pulido G, McCook LJ, Dove S, Berkelamns R, Roff G, Kline DI, Weeks S, Evans RD, Williamson DH, Hoegh-Guldberg O (2009) Doom and boom on a resilient reef: climate change, algal overgrowth and coral recovery. PLoS ONE 4:e5239. doi:10.1371/journal.pone.0005239

Didham RK, Watts CH, Norton DA (2005) Are systems with strong underlying abiotic regimes more likely to exhibit alternative stable states? Oikos 110:409–416

Done TJ (1992) Phase shifts in coral reef communities and their ecological significance. Hydrobiologia 247:121–132

Downing JA, McClain M, Twilley RW, Melack JM, Elser J, Rabalais NN, Lewis WM, Turner RE, Corredor J, Soto D, Yanez-Arancibia A, Kopaska JA, Howarth RW (1999) The impact of accelerating land-use change on the N-cycle of tropical aquatic ecosystems: current conditions and projected changes. Biogeochemistry 46:109–148

Droop MR (1983) 25 years of algal growth kinetics: a personal view. Bot Mar 26:99–112

Duarte CM (1992) Nutrient concentration of aquatic plants: patterns across species. Limnol Oceanogr 37:882–889

Duarte CM, Cebrian J (1996) The fate of marine autotrophic production. Limnol Oceanogr 41:1758–1766

Duffy JE, Hay ME (1990) Seaweed adaptations to herbivory. Bioscience 40:368–375

Duffy JE, Hay ME (2001) The ecology and evolution of marine consumer-prey interactions. In: Bertness MD, Gaines SD, Hay ME (eds) Marine community ecology. Sinauer Associates, Inc, Sunderland, pp 131–157

Edmunds PJ, Carpenter RC (2001) Recovery of *Diadema Antillarum* reduces macroalgal cover and increases abundance of juvenile corals on a Caribbean reef. Proc Natl Acad Sci 98:5067–5071

Erickson KL (1983) Constituents of *Laurencia*. In: Scheuer PJ (ed) Marine natural products: chemical and biological perspectives, vol 5. Academic, New York, p 131

Erickson AA, Paul VJ, Van Alstyne KL, Kwiatkowski LM (2006) Palatability of macroalgae that employ different types of chemical defenses. J Chem Ecol 32:1883–1895

Fabricius KE (2005) Effects of terrestrial runoff on the ecology of corals and coral reefs: review and synthesis. Mar Poll Bull 50:125–146

Faulkner DJ (1984) Marine natural products: metabolites of marine algae and herbivorous marine mollusks. Nat Prod Rep 1:251–280

Fenical W (1975) Halogenation in the Rhodophyta: a review. J Phycol 11:245–259

Fenical W (1982) Natural products chemistry in the marine environment. Science 215:923–928

Floeter SR, Behrens MD, Ferreora CEL, Paddack MJ, Horn MH (2005) Geographical gradients of marine herbivorous fishes: patterns and processes. Mar Biol 147:1435–1447

Fong P (2008) Macroalgal dominated ecosystems. In: Capone DG, Carpenter EJ (eds) Nitrogen in the marine environment. Springer, New York

Fong P, Glynn PW (2000) A regional model to predict coral population dynamics in response to El Niño-Southern Oscillation. Ecol Appl 10:842–854

Fong P, Lirman D (1996) Hurricanes cause population expansion of the branching coral, *Acropora palmata*. Mar Ecol 16:317–335

Fong P, Boyer KE, Zedler JB. (1998) Developing an indicator of nutrient enrichment in coastal estuaries and lagoons using tissue nitrogen content of the opportunistic alga, Enteromorpha intestinalis (L. Link). J Exp Mar Biol Ecol 231:63–79

Fong P, Zedler JB, Donohoe RM (1993) Nitrogen versus phosphorus limitation of algal biomass in shallow coastal lagoons. Limnol Oceanogr 38:906–923

Fong P, Donohoe RM, Zedler JB (1994a) Nutrient concentration in tissue of the macroalga *Enteromorpha* spp. as an indicator of nutrient history: an experimental evaluation using field microcosms. Mar Ecol Prog Ser 106:273–281

Fong P, Foin TC, Zedler JB (1994b) A simulation model of lagoon algae based on nitrogen competition and internal storage. Ecol Monogr 64:225–247

Fong P, Kamer K, Boyer KE, Boyle KA (2001) Nutrient content of macroalgae with differing morphologies may indicate sources of nutrients to tropical marine systems. Mar Ecol Prog Ser 220:137–152

Fong P, Boyer KE, Kamer K, Boyle KA (2003) Influence of initial tissue nutrient status of tropical marine algae on response to nitrogen and phosphorus additions. Mar Ecol Prog Ser 262:111–123

Fong P, Smith TB, Wartian MJ (2006) Epiphytic cyanobacteria maintain shifts to macroalgal dominance on coral reefs following enso disturbance. Ecology 87:1162–1168

Foster SA (1987) The relative impacts of grazing by Caribbean coral reef fishes and *Diadema*: effects of habitat and surge. J Exp Mar Biol Ecol 105:1–20

Fredersdorf J, Müller R, Becker S, Wienke C, Bischof K (2009) Interactive effects of radiation, temperature, and salinity on different life history stages of the Arctic kelp *Alaria esculenta* (Phaeophyceae). Oecologia 160:483–492

Friedlander M, Levy I (1995) Cultivation of *Gracilaria* in outdoor tanks and ponds. J Appl Phycol 7:315–324

Fry B, Gace A, McClelland JW (2003) Chemical indicators of anthropogenic nitrogen loading in four Pacific estuaries. Pac Sci 57:77–101

Fujita RM (1985) The role of nitrogen status in regulating transient ammonium uptake and nitrogen storage by macroalgae. J Exp Mar Biol Ecol 92:283–301

Fujita RM, Wheeler PA, Edwards RL (1988) Metabolic regulation of ammonium uptake by *Ulva rigida* (Chlorophyta): a compartmental analysis of the rate-limiting step for uptake. J Phycol 24:560–566

Fujita RM, Wheeler PA, Edwards RL (1989) Assessment of macroalgal nitrogen limitation in a seasonal upwelling region. Mar Ecol Prog Ser 53:293–303

Fukunaga A (2008) Invertebrate community associated with the macroalga *Halimeda kanaloana* meadow in Maui, Hawaii. Int Rev Hydobiol 93:328–334

Gaines SD, Lubchenco J (1982) A unified approach to marine plant-herbivore interactions. II. Biogeography. Ann Rev Ecol Syst 13:111–138

Gallegos CL, Jordan TE (1997) Seasonal progression of factors limiting phytoplankton pigment biomass in the Rhode River estuary, Maryland (USA): II. Modeling N versus P limitation. Mar Ecol Prog Ser 161:199–212

Garzon-Ferreira J, Cortes J, Croquer A (2004) Southern tropical America: coral reef status and consolidation as GCRMN regional node. Status Coral Reefs World 2:509–522

Gerwick WH, Roberts MA, Proteau PJ, Chen JL (1994) Screening cultured marine microalgae for anticancer-type activity. J Appl Phycol 6:143–149

Glynn PW, Fong P (2006) Patterns of reef coral recovery by the regrowth of surviving tissues following the 1997–1998 El Niño warming and 2000, 2001 upwelling events in Panamá, eastern Pacific. In: Proceedings of the 10th international coral reef symposium, Okinawa, 2004, pp 624–630

Glynn PW, Maté JL (1997) Field guide to the Pacific coral reefs of Panama. Proc 8th Intl Coral Reef Smp 1:145–166

Golbuu Y, Richmond RH (2007) Substratum preferences in planula larvae of two species of scleractinian corals, *Goniastrea retiformis* and *Stylaraea punctata*. Mar Biol 152:639–644

Hadfield MG, Paul VJ (2001) Natural chemical cues for settlement and metamorphosis of marine-invertebrate larvae. In: McClintock J, Baker B (eds) Marine chemical ecology. CRC Press, Boca Raton, pp 431–462

Haines KC, Wheeler PA (1978) Ammonium and nitrate uptake by the marine macrophytes *Hypnea musciformis* (Rhodophyta) and *Macrocystis pyrifera* (Phaeophyta). J Phycol 14:319–324

Hallock P (2005) Global change and modern coral reefs: new opportunities to understand shallow-water carbonate depositional processes. Sed Geol 175:19–33

Hanisak MD (1983) The nitrogen relationships of marine macroalgae. In: Carpenter EJ, Capone DG (eds) Nitrogen in the marine environment. Academic, New York, pp 699–730

Harrington L, Fabricius K, De'ath G, Negri A (2004) Recognition and selection of settlement substrata determine post-settlement survival in corals. Ecology 85:3428–3437

Hatcher BG (1981) The interaction between grazing organisms and the epilithic algal community of a coral reef: a quantitative assessment. Proc 4th Int Coral Reef Symp 2:515–524

Hatcher BG (1988) Coral reef primary productivity: a beggar's banquet. Trends Ecol Evol 3:106–111

Hawkins SJ, Moore PJ, Burrows MT, Poloczanska E, Mieszkowska N, Herbert RJH, Jenkins SR, Thompson RC, Genner MJ, Southward AJ (2008) Complex interactions in a rapidly changing world: responses of rocky shore communities to recent climate change. Clim Res 37:123–133

Hay ME (1984) Pattern of fish and urchin grazing on Caribbean coral reefs: are previous results typical? Ecology 65:446–454

Hay ME (1991) Fish-seaweed interactions on coral reefs: effects of herbivorous fishes and adaptations of their prey. In: Sale PF (ed) The ecology of fishes on coral reefs. Academic, New York, pp 96–119

Hay ME (1997) The ecology and evolution of seaweed-herbivore interactions on coral reefs. Coral Reefs 16:67–76

Hay ME, Fenical W (1988) Marine plant-herbivore interactions: the ecology of chemical defense. Ann Rev Ecol Syst 19:111–145

Hay ME, Steinberg PD (1992) The chemical ecology of plant-herbivore interactions in marine versus terrestrial communities. In: Rosenthal GA, Berenbaum MR (eds) Herbivores: their interactions with secondary plant metabolites, vol. II., Ecological and evolutionary processes. Academic, San Diego

Hay ME, Duffy JE, Pfister CA, Fenical W (1987) Chemical defense against different marine herbivores: are amphipods insect equivalents? Ecology 68:1567–1580

Hay ME, Kappel QE, Fenical W (1994) Synergisms in plant defenses against herbivores: interaction of chemistry, calcification, and plant quality. Ecology 75:1714–1726

Hay ME, Stachowicz JJ, Cruz-Rivera E, Bullard S, Deal MS, Lindquist N (1998) Bioassays with marine and freshwater macroorganisms. In: Haynes KF, Millar JG (eds) Methods in chemical ecology, vol. 2, bioassay methods. Chapman and Hall, Norwell, pp 39–141

Hein M, Pedersen MF, Sand-Jensen K (1995) Size-dependent nitrogen uptake in micro- and macroalgae. Mar Ecol Prog Ser 118:247–253

Heberger MH, Cooley P, Herrera P, Gleick H, Moore E (2009) The impacts of sea-level rise on the California coast. A paper from: California climate change center. http://www.pacinst.org/reports/sea_level_rise/

Herren LW, Walters LJ, Beach KS (2006) Fragmentation generation, survival, and attachment of *Dictyota* spp. at Conch Reef in the Florida Keys, USA. Coral Reefs 25:287–295

Heyward AJ, Negri AP (1999) Natural inducers for coral larval metamorphosis. Coral Reefs 18:273–279

Hillebrand H (2004) On the generality of the latitudinal diversity gradient. Am Nat 163:192–211

Hillis L (1997) Coralgal reefs from a calcareous green alga perspective, and a first carbonate budget. Proc 8th Int Coral Reef Symp 1:761–766

Hixon MA (1997) Effects of reef fishes on coral and algae. In: Birkeland C (ed) Life and death of coral reefs. Chapman and Hall, New York, pp 230–248

Hoegh-Guldberg O (1999) Climate change, coral bleaching, and the future of the world's coral reefs. Mar Freshwater Res 50:839–866

Hoegh-Guldberg O (2004) Coral reefs in a century of rapid environmental change. Symbiosis 37:1–7

Hoegh-Guldberg O, Mumby PJ, Hooten AJ, Steneck RS, Greenfield P, Gomez E, Harvell CD, Sale PF, Edwards AJ, Caldeira K, Knowlton N, Eakin CM, Iglesias-Priet R, Muthiga N, Bradbury RH, Dubi A, Hatziolos ME (2007) Coral reefs under rapid climate change and ocean acidification. Science 318:1737–1742

Holmquist JG (1994) Benthic macroalgae as a dispersal mechanism for fauna: influence of a marine tumbleweed. J Exp Mar Biol Ecol 180:235–251

Horn MH (1989) Biology of marine herbivorous fishes. Oceanogr Mar Biol Annu Rev 27:167–272

Hueerkamp C, Glynn PW, D'Croz L, Maté JL, Colley SB (2001) Bleaching and recovery of five eastern Pacific corals in an El Niño-related temperature experiment. Bull Mar Sci 69:215–236

Hughes TP (1994) Catastrophes, phase shifts, and large-scale degradation of a Caribbean coral reef. Science 265:1547–1551

Hughes TP, Connell JH (1999) Multiple stressors on coral reefs: a long term perspective. Limnol Oceanogr 44:932–940

Hughes TP, Baird AH, Bellwood DR, Card M, Connolly SR, Folke C, Grosberg R, Hoegh-Guldberg O, Jackson JBC, Kleypas J, Lough JM, Marshall P, Nyström M, Palumbi SR, Pandolfi JM, Rosen B, Roughgarden J (2003) Climate change, human impacts, and the resilience of coral reefs. Science 301:929–933

Hughes TP, Rodrigues MJ, Bellwood DR, Ceccarelli D, Hoegh-Guldberg O, McCook LJ, Moltschaniwskyj N, Pratchett MS, Steneck RS, Willis B (2007) Phase shifts, herbivory, and the resilience of coral reefs to climate change. Curr Biol 17:360–365

Hurd CL (2000) Water motion, marine macroalgal physiology, and production. J Phycol 36:453–472

Hurd CL, Berges JA, Osborne J, Harrison PJ (1995) An in vitro nitrate reductase assay for marine macroalgae: optimization and characterization of the enzyme for *Fucus gardneri* (Phaeophyta). J Phycol 31:835–843

Idjadi JA, Lee SC, Bruno JF, Precht WF, Allen-Requa L, Edmunds PJ (2006) Rapid phase-shift reversal on a Jamaican reef. Coral Reefs 25:209–211

IPCC (Intergovernmental Panel on Climate Change) (2007) Summary for policymakers. In: Solomon S, Qin D, Manning M, Chen Z, Marquis M, Avery KB, Tignor M, Miller HL (eds) Climate change 2007: the physical science basis. Contribution of working group I to the fourth assessment report of the intergovernmental panel on climate change, Cambridge University Press, Cambridge

Ireland CM, Roll DM, Molinski TF, Mckee TC, Zabriskie TM, Swersey JC (1988) Uniqueness of the marine chemical environment: categories of marine natural products from invertebrates. In: Fautin D (ed) Biomedical importance of marine organisms. California Academy of Sciences, San Francisco, pp 41–57

Israel A, Katz S, Dubinski Z, Merrill JE, Friedlander M (1999) Photosynthetic inorganic carbon utilization and growth of *Porphyra linearis* (Rhodophyta). J Appl Phycol 11:447–453

Israel A, Gavriele J, Glazer A, Friedlander M (2005) Utilization of flue gas from a power plant for tank cultivation of the red seaweed *Gracilaria cornea*. Aquaculture 249:311–316

Jackson JBC (2008) Ecological extinction and evolution in the brave new ocean. Proc Natl Acad Sci U S A 105:11458–11465

Jackson JBC, Kirby MX, Berger WH, Bjorndal KA, Botsford LW, Bourque BJ, Bradbury RH, Cooke R, Erlandson J, Estes JA, Hughes TP, Kidwell S, Lange CB, Lenihan HS, Pandolfi JM, Peterson CH

(2001) Historical overfishing and the recent collapse of coastal ecosystems. Science 293:629–638

Jennings JG, Steinberg PD (1997) Phlorotannins versus other factors affecting epiphyte abundance on the kelp *Ecklonia radiata*. Oecologia 109:461–473

Jickells TD (1998) Nutrient biogeochemistry of the coastal zone. Science 281:2127–2129

John DM, Hawkins SJ, Price J (eds) (1992) Plant-animal interactions in the marine benthos. Clarendon Press, Oxford

Johnson CR, Muir DG, Reysenbach AL (1991) Characteristic bacteria associated with surfaces of coralline algae: a hypothesis for bacterial induction of marine invertebrate larvae. Mar Ecol Prog Ser 74:281–294

Jokiel PL, Rodgers KS, Kuffner IB, Andersson AJ, Cox EF, Mackenzie FT (2008) Ocean acidification and calcifying reef organisms: a mesocosm investigation. Coral Reefs 27:473–483

Jompa J, McCook LJ (2003) Coral-algal competition: macroalgae with different properties have different effects on corals. Mar Ecol Prog Ser 258:87–95

Kamykowski D, Zentara S (2005) Changes in world ocean nitrate availability through the 20th century. Deep-Sea Res 52:1719–1744

Kan H, Kawana T (2006) 'Catch-up' of a high-latitude barrier reef by back-reef growth during post-glacial sea-level rise, Southern Ryukus, Japan. Proceedings of the 10th international coral reef symposium, Okinawa, 2004, pp 494–503

Kasting JF, Siefert JL (2002) Life and the evolution of Earth's atmosphere. Science 296:1066–1068

Kennison RL (2008) Evaluating ecosystem function of nutrient retention and recycling in excessively eutrophic estuaries. Ph.D. dissertation. University of California, Los Angeles

Kerr RA (2005) The story of O_2. Science 308:1730–1732

Kline DI, Kuntz NM, Breitbart M, Knowlton N, Rohwer F (2006) Role of elevated organic carbon levels and microbial activity in coral mortality. Mar Ecol Prog Ser 314:119–125

Knowlton N (2001) The future of coral reefs. Proc Nat Acad Sci U S A 98:5419–5425

Koop K, Booth D, Broadbent A, Brodie J, Bucher D, Capone D, Coll J, Dennison W, Erdmann M, Harrison P, Hoegh-Guldberg O, Hutchings P, Jones GB, Larkum AWD, O'Neil J, Steven A, Tentori E, Ward S, Williamson J, Yellowlees D (2001) ENCORE: the effect of nutrient enrichment on coral reefs. Synthesis of results and conclusions. Mar Poll Bull 41:91–120

Kuffner IB, Paul VJ (2004) Effects of the benthic cyanobacterium *Lyngbya majuscula* on the larval settlement of the reef corals *Acropora surculosa* and *Pocillopora damicornis*. Coral Reefs 23:455–458

Kuffner IB, Walters LJ, Becerro MA, Paul VJ, Ritson-Williams R, Beach K (2006) Inhibition of coral recruitment by macroalgae and cyanobacteria. Mar Ecol Prog Ser 323:107–117

Kuffner IB, Anderson AJ, Jokiel PL, Rodgers KS, Mackenzie FT (2008) Decreased abundance of crustose coralline algae due to ocean acidification. Nat Geosci 1:114–117

Lane AL, Kubanek J (2008) Secondary metabolite defense against pathogens and biofoulers. In: Amsler CD (ed) Algal chemical ecology. Springer, Berlin, pp 229–243

Lapointe BE (1987) Phosphorus- and nitrogen-limited photosynthesis and growth of *Gracilaria tikvahiae* (Rhodophyceae) in the Florida Keys: an experimental field study. Mar Biol 93:561–568

Lapointe BE (1989) Macroalgal production and nutrient relations in oligotrophic areas of Florida Bay. Bull Mar Sci 44:312–323

Lapointe BE, Duke CS (1984) Biochemical strategies for growth of *Gracilaria tikvahiae* (Rhodophyta) in relation to light intensity and nitrogen availability. J Phycol 20:488–495

Lapointe BE, Littler MM, Littler DS (1992) Nutrient availability to marine macroalgae in siliciclastic versus carbonate-rich coastal waters. Estuaries 15:75–82

Lapointe BE, Barile PJ, Littler MM, Littler DS (2005) Macroalgal blooms on southeast Florida coral reefs II. Cross-shelf discrimination of nitrogen sources indicates widespread assimilation of sewage nitrogen. Harmful Algae 4:1106–1122

Larkum AWD (1988) High rates of nitrogen fixation on coral skeletons after predation by the crown of thorns starfish *Acanthaster planci*. Mar Biol 97:503–506

Larkum AWD, Koop K (1997) ENCORE, algal productivity and possible paradigm shifts. Proc 8th Int Coral Reef Symp 2:881–884

Larkum AWD, Drew EA, Crossett RN (1967) The vertical distribution of attached marine algae in Malta. J Ecol 55:361–371

Larned ST (1998) Nitrogen- versus phosphorus-limited growth and sources of nutrients for coral reef macroalgae. Mar Biol 132:409–421

Larned ST, Stimson J (1997) Nitrogen-limited growth in the coral reef chlorophyte *Dictyosphaeria cavernosa*, and the effect of exposure to sediment-derived nitrogen on growth. Mar Ecol Prog Ser 145:95–108

Lau SCK, Qian PY (1997) Phlorotannins and related compounds as larval settlement inhibitors of the tube-building polychaete *Hydroides elegans*. Mar Ecol Prog Ser 159:219–227

Lee RE (2008) Phycology. Cambridge University Press, Cambridge

Leichter JJ, Wing SR, Miller SL, Denny MW (1996) Pulsed delivery of subthermocline water to Conch Reef (Florida Keys) by internal tidal bores. Limnol Oceanogr 41:1490–1501

Lewis SM (1986) The role of herbivorous fishes in the organization of a Caribbean reef community. Ecol Monogr 56:184–200

Lirman D, Biber P (2000) Seasonal dynamics of macroalgal communities of the Northern Florida Reef tract. Bot Mar 43:305–314

Littler MM, Littler DS (1984) A relative-dominance model for biotic reefs. Proceedings of the joint meeting of the atlantic reef committee society of reef studies, Miami, 1984

Littler MM, Littler DS (1988) Structure and role of algae in tropical reef communities. In: Lembi CA, Waaland JR (eds) Algae and human affairs. Cambridge University Press, Cambridge, pp 29–56

Littler MM, Littler DS (1995) Selective herbivore increases biomass of its prey: a chiton-coralline reef-building association. Ecology 76:1666–1681

Littler MM, Littler DS (1997) Disease-induced mass mortality of crustose coralline algae on coral reefs provides rationale for the conservation of herbivorous fish stocks. Proc 8th Int Coral Reef Symp 1:719–723

Littler DS, Littler MM (2000) Caribbean reef plants. Offshore Graphics, Washington, DC

Littler MM, Littler DS, Blair SM, Norris JN (1986) Deep-water plant communities from and uncharted seamount off San Salvador Island, Bahamas: distribution, abundance, and primary Productivity. Deep-Sea Res 33:881–892

Littler MM, Taylor PR, Littler DS (1989) Complex interactions in the control of coral zonation in a Caribbean reef flat. Oecologia 80:331–340

Littler MM, Littler DS, Titlyanov EA (1991) Comparisons of N- and P-limited productivity between high granitic islands vs. low carbonate atolls in the Seychelles Archipelago: a test of the relative-dominance paradigm. Coral Reefs 10:199–209

Littler MM, Littler DS, Lapointe BE, Bariles PJ (2006) Toxic cyanobacterial associated with groundwater conduits in the Bahamas. Coral Reefs 25:186

Lotze HK, Schramm W (2000) Ecophysiological traits explain species dominance patterns in macroalgal blooms. J Phycol 36:287–295

Lotze HK, Worm B (2002) Complex interactions of climate change and ecological controls on macroalgal recruitment. Limnol Oceanogr 47:1734–1741

Macintyre IG (1997) Reevaluating the role of crustose coralline algae in the construction of coral reefs. Proc 8th Int Coral Reef Symp 1:725–730

Macintyre IG, Glynn PW, Hinds F (2005) Evidence of the role of *Diadema antillarum* in the promotion of coral settlement and survivorship. Coral Reefs 24:273

Macintyre IG, Graus RR, Reinthal PN, Littler MM, Littler DS (1987) The barrier reef sediment apron: tobacco reef, Belize. Coral Reefs 6:1–12

Mann KH (1982) Ecology of coastal waters: a systems approach. Blackwell, Oxford

MarinLit: A Marine Literature Database (2009) MarinLit database, Department of Chemistry, University of Canterbury, Christ church. http://www.chem.canterbury.ac.nz/marinlit/marinlit.shtml

Marshall RA, Harper DB, McRoberts WC, Dring MJ (1999) Volatile bromocarbons produced by *Falkenbergia* stages of *Asparagopsis* sp. (Rhodophyta). Limnol Oceanogr 44:1348–1352

Maschek JA, Baker BJ (2008) The chemistry of algal secondary metabolites. In: Amsler CD (ed) Algal chemical ecology. Springer, Berlin, pp 1–24

Mattson WJ Jr (1980) Herbivory in relation to plant nitrogen content. Ann Rev Ecol Syst 11:119–161

McCalanahan TR, Muthiga NA (2002) An ecological shift in a remote coral atoll of Belize over 25 years. Environ Conserv 25:122–130

McClanahan TR, Aronson RB, Precht WF, Muthiga NA (1999) Fleshy algae dominate remote coral reefs of Belize. Coral Reefs 18:61–62

McClanahan TR, Ateweberhan M, Omukoto J, Pearson L (2009) Recent seawater temperature histories, status, and predictions for Madagascar's coral reefs. Mar Ecol Prog Ser 380:117–128

McClanahan TR, Berbman K, Huitric M, McField M, Elfwing T, Nystrom N, Nordemar I (2000) Response of fishes to algae reduction on Glovers Reef. Belize Mar Ecol Prog Ser 206:273–282

McCook LJ (1999) Macroalgae, nutrients and phase shifts on coral reefs: scientific issues and management consequences for the Great Barrier Reef. Coral Reefs 18:357–367

McCook LJ (2001) Competition between corals and algal turfs along a gradient of terrestrial influence in the nearshore central Great Barrier Reef. Coral Reefs 19:419–425

McCook L, Jompa J, Diaz-Palido G (2001) Competition between corals and algae on coral reefs: a review of evidence and mechanisms. Coral Reefs 19:400–417

McGlathery KJ, Marino R, Howarth RW (1994) Variable rates of phosphorus uptake by shallow marine carbonate sediments: mechanisms and ecological significance. Biogeochemistry 25:127–146

McGlathery KJ, Pedersen MF, Borum J (1996) Changes in intracellular nitrogen pools and feedback controls on nitrogen uptake in *Chaetomorpha linum* (Chlorophyta). J Phycol 32:393–401

Meinesz A (1999) Killer algae. University of Chicago Press, Chicago

Meyer KD, Paul VJ (1995) Variation in aragonite and secondary metabolite concentrations in the tropical green seaweed *Neomeris annulata*: effects on herbivory by fishes. Mar Biol 122:537–545

Miller MW, Hay ME, Miller SL, Malone D, Sotka EE, Szmant AM (1999) Effects of nutrients versus herbivores on reef algae: a new method for manipulating nutrients on coral reefs. Limnol Oceanogr 44:1847–1861

Miller MW, Valdivia A, Kramer KL, Mason B, Williams DE, Johnston L (2009) Alternate benthic assemblages on reef restoration structures and cascading effects on coral settlement. Mar Ecol Prog Ser 387:147–156

Moore RE (1996) Cyclic peptides and depsipeptides from cyanobacteria: a review. J Ind Microbiol 16:134–143

Moore RE (1981) Constituents of blue-green algae. In: Scheuer PJ (ed) Marine natural products, vol 3. Academic, New York, pp 1–52

Morse ANC, Iwao K, Baba M, Shimoike K, Hayashibara T, Omori M (1996) An ancient chemosensory mechanism brings new life to coral reefs. Biol Bull 191:149–154

Morse DE, Morse ANC (1991) Enzymatic characterization of the morphogen recognized by *Agaricia humilis* (Scleractinian Coral) larvae. Biol Bull 181:104–122

Morse DE, Morse ANC, Raimondi PT, Hooker N (1994) Morphogen-based chemical flypaper for *Agaricia humilis* coral larvae. Biol Bull 186:172–181

Muhs DR, Budahn JR, Prospero JM, Carey SN (2007) Geochemical evidence of African dust inputs to soils of western Atlantic islands: Barbados, the Bahamas, and Florida. J Geophys Res Part F Earth Surf 112:1–26

Mumby PJ (2009) Phase shifts and the stability of macroalgal communities on Caribbean coral reefs. Coral Reefs 28:761–773

Myers RA, Worm B (2003) Rapid worldwide depletion of predatory fish communities. Nature 423:280–283

Nagle DG, Paul VJ (1999) Production of secondary metabolites by filamentous tropical marine cyanobacteria: ecological functions of the compounds. J Phycol 35:1412–1421

Nagle DG, Camacho FT, Paul VJ (1998) Dietary preferences of the opisthobranch mollusc *Stylocheilus longicauda* for secondary metabolites produced by the tropical cyanobacterium *Lyngyba majuscula*. Mar Biol 132:267–273

Naldi M, Wheeler PA (2002) ^{15}N measurements of ammonium and nitrate uptake by *Ulva fenestrata* (Chlorophyta) and *Gracilaria pacifica* (Rhodophyta): comparison of net nutrient disappearance, release of ammonium and nitrate, and ^{15}N accumulation in algal tissue. J Phycol 38:135–144

Negri AP, Webster NS, Hill RT, Heyward AJ (2001) Metamorphosis of broadcast spawning corals in response to bacteria isolated from crustose algae. Mar Ecol Prog Ser 223:121–131

Nepf HM, Koch EW (1999) Vertical secondary flows in submersed plant-like arrays. Limnol Oceanogr 44:1072, 1080

Nugues MM, Bak RPM (2008) Long-term dynamics of the brown macroalga *Lobophora variegata* in Curaçao. Coral Reefs 27:389–393

Nugues MM, Roberts CM (2003) Coral mortality and interaction with algae in relation to sedimentation. Coral Reefs 22:507–516

Nugues MM, Szmant AM (2006) Coral settlement onto *Halimeda opuntia*: a fatal attraction to an ephemeral substrate? Coral Reefs 25:585–591

Nugues MM, Smith GW, Hooidonk RJ (2004) Algal contact as a trigger for coral disease. Ecol Lett 7:919–923

O'Connor MI (2009) Warming strengthens an herbivore-plant interaction. Ecology 90:388–398

Odgen JC (1997) Ecosystem interactions in the tropical coastal seascape. In: Birkeland C (ed) Life and death of coral reefs. Chapman and Hall, New York, pp 288–295

Ogden JC, Lobel PS (1978) The role of herbivorous fishes and urchins in coral reef communities. Environ Biol Fish 3:49–63

Odum HT, Odum EP (1955) Trophic structure and productivity of a windward coral reef community on Eniwetok Atoll. Ecol Monogr 25:291–320

Pavia H, Cervin G, Lindgren A, Aberg P (1997) Effects of UV-B radiation and simulated herbivory on phlorotannins in the brown alga *Ascophyllum nodosum*. Mar Ecol Prog Ser 157:139–146

Paerl HW, Huisman J (2008) Blooms like it hot. Science 320:57–58

Paerl HW, Joyner JA, Joyner AR, Arthur KE, Paul VJ, O'Neil JM, Heil CA (2008) Co-occurrence of dinoflagellate and cyanobacterial harmful algal blooms in southwest Florida coastal waters: dual nutrient (N and P) input controls. Mar Ecol Prog Ser 371:143–153

Pandolfi JM, Bradbury RH, Sala E, Hughes TP, Bjorndal KA, Cooke RG, McArdle D, McClenachan L, Newman MJH, Paredes G, Warner RR, Jackson JBC (2003) Global trajectories of the long-term decline of coral reef ecosystems. Science 301:955–958

Pandolfi JM, Jackson JBC, Baron N, Bradbury RH, Guzman HM, Hughes TP, Kappel CV, Micheli F, Ogden JC, Possingham HP, Sala E (2005) Are U.S. coral reefs on the slippery slope to slime? Science 307:1725–1726

Parmesan C, Yohe G (2003) A globally coherent fingerprint of climate change impacts across natural systems. Nature 421:37–42

Paul VJ (1992) Ecological roles of marine natural products. Cornell University Press, Ithaca

Paul VJ (1997) Secondary metabolites and calcium carbonate as defenses of calcareous algae on coral reefs. Proc 8th Int Coral Reef Symp 1:707–712

Paul VJ (2008) Global warming and cyanobacterial harmful algal blooms. In: Hudnell HK (ed) Cyanobacterial harmful algal blooms: state of the science and research needs, advances in experimental medicine and biology. Springer, New York, pp 239–257

Paul VJ, Fenical W (1986) Chemical defense in tropical green algae, order Caulerpales. Mar Ecol Prog Ser 34:157–169

Paul VJ, Fenical W (1987) Natural products chemistry and chemical defense in tropical marine algae of the phylum Chlorophyta. In: Scheuer PJ (ed) Bioorganic marine chemistry. Springer, Berlin, pp 1–29

Paul VJ, Hay ME (1986) Seaweed susceptibility to herbivory: chemical and morphological correlates. Mar Ecol Prog Ser 33:255–264

Paul VJ, Pennings SC (1991) Diet-derived chemical defenses in the sea hare *Stylocheilus longicauda* (Quoy and Gaimard 1824). J Exp Mar Biol Ecol 151:227–243

Paul VJ, Ritson-Williams R (2008) Marine chemical ecology. Nat Prod Rep 25:662–695

Paul VJ, Cruz-Rivera E, Thacker RW (2001) Chemical mediation of macroalgal-herbivore interactions: ecological and evolutionary perspectives. In: McClintock J, Baker B (eds) Marine chemical ecology. CRC Press, LLC, Boca Raton, pp 227–265

Paul VJ, Arthur KE, Ritson-Williams R, Ross C, Sharp K (2007) Chemical defenses: from compounds to communities. Biol Bull 213:226–251

Paul VJ, Thacker R, Banks K, Golubic S (2005) Benthic cyanobacterial bloom impacts on the reefs of South Florida (Broward County, USA). Coral Reefs 24:693–697

Paul VJ, Wylie C, Sanger H (1988) Effects of algal chemical defenses toward different coral-reef herbivorous fishes: a preliminary study. Proc 6th Int Coral Reef Symp 3:73–78

Pawlik JR (1992) Chemical ecology of the settlement of benthic marine invertebrates. Oceanogr Mar Biol Annu Rev 30:273–335

Pearson GA, Lago-Leston A, Mota C (2009) Frayed at the edges: selective pressure and adaptive response to abiotic stressors are mismatched in low diversity edge populations. J Ecol 97:450–462

Pedersen MF (1994) Transient ammonium uptake in the macroalga *Ulva lactuca* (Chlorophyta): nature, regulation, and the consequences for choice of measuring technique. J Phycol 30:980–986

Pennings SC, Paul VJ (1993) Sequestration of dietary metabolites by three species of sea hares: location, specificity, and dynamics. Mar Biol 117:535–546

Pennings SC, Svedberg JM (1993) Does $CaCO_3$ in food deter feeding by sea urchins? Mar Ecol Prog Ser 101:163–167

Pennings SC, Paul VJ, Dunbar DC, Hamann MT, Lumbang WA, Novack B, Jacobs RS (1999) Unpalatable compounds in the marine gastropod *Dolabella auricularia*: distribution and effect of diet. J Chem Ecol 25:735–755

Pennings SC, Weiss AM, Paul VJ (1996) Secondary metabolites of the cyanobacterium *Microcoleus lyngbyaceus* and the sea hare *Stylocheilus longicauda*: palatability and toxicity. Mar Biol 123:735–743

Pereira RC, da Gama BAP (2008) Macroalgal chemical defenses and their role in structuring tropical marine communities. In: Amsler CD (ed) Algal chemical ecology. Springer, Berlin, pp 25–55

Pitlik TJ, Paul VJ (1997) Effects of toughness, calcite level, and chemistry of crustose coralline algae (Rhodophyta, Corallinales) on grazing by the parrotfish *Chlorurus sordidus*. Proc 8th Int Coral Reef Symp 1:701–706

Ragan MA, Glombitza K (1986) Phlorotannins, brown algal polyphenols. In: Round FE, Chapman DJ (eds) Progress in phycological research, vol 4. Biopress Limited, Bristol, pp 129–241

Rasher, DB, Hay ME (2010) Chemically rich seaweeds poison corals when not controlled by herbivores. Proc Natl Acad Sci 107:9683–9688

Redfield AC, Ketchum BA, Richards FA (1963) The influence of organisms on the chemical composition of sea-water. In: Hill MN (ed) The sea, vol 2. Wiley, New York

Rees TAV, Grant CM, Harmens HE, Taylor RB (1998) Measuring rates of ammonium assimilation in marine algae: use of the protonophore carbonyl cyanide m- chlorophenylhydrazone to distinguish between uptake and assimilation. J Phycol 34:264–272

Ritson-Williams R, Arnold S, Fogarty N, Steneck R, Vermeij M, Paul VJ (2009) New perspectives on ecological mechanisms affecting coral recruitment on reefs. Smithson Contrib Mar Sci 38:437–457

Ritson-Williams R, Paul VJ, Arnold SN, Steneck R (2010) Larval settlement preferences and post-settlement survival of the threatened Caribbean corals *Acropora palmata* and *A. cervicornis*. Coral Reefs 29:71–81

Ritson-Williams R, Paul VJ, Bonito V (2005) Marine benthic cyanobacteria overgrow coral reef organisms. Coral Reefs 24:629

Rivers JS, Peckol P (1995) Interaction effects of nitrogen and dissolved inorganic carbon on photosynthesis, growth, and ammonium uptake of the macroalgae *Cladophora vagabunda* and *Gracilaria tikvahiae*. Mar Biol 121:747–753

Rogers CS (1990) Responses of coral reefs and reef organisms to sedimentation. Mar Ecol Prog Ser 62:185–202

Rogers CS, Miller J (2006) Permanent 'phase shift' or reversible declines in coral cover? Lack of recovery of two coral reefs in St John, U.S. Virgin Islands. Mar Ecol Prog Ser 306:103–114

Rogers CS, Garrison V, Grober-Dunsmore R (1997) A fishy story about hurricanes and herbivory, seven years of research on a reef in St. John, U.S. Virgin Islands. Proc 8th Int Coral Reef Symp 1:555–560

Roughgarden J, Smith F (1996) Why fisheries collapse and what to do about it. Proc Nat Acad Sci U S A 93:5078–5083

Ruiz GM, Carlton JT, Grosholz D, Hines AH (1997) Global invasions of marine and estuarine habitats by nonindigenous species: mechanisms, extent, and consequences. Am Zool 37:621–632

Runcie JW, Gurgel CFD, Mcdermid KJ (2008) *In situ* photosynthetic rates of tropical marine macroalgae at their lower depth limit. Eur J Phycol 43:377–388

Sammarco PW (1983) Effects of fish grazing and damselfish territoriality on coral reef algae. I. Algal community structure. Mar Ecol Prog Ser 13:1–14

Schaffelke B (1999) Particulate organic matter as an alternative nutrient source for tropical *Sargassum* species (Fucales, Phaeophyceae). J Phycol 35:1150–1157

Schaffelke B, Klumpp DW (1997) Growth of germlings of the macroalga *Sargassum baccularia* (Phaeophyta) is stimulated by enhanced nutrients. Proc 8th Int Coral Reef Symp 2:1839–1842

Schaffelke B, Klumpp DW (1998) Nutrient-limited growth of the coral reef macroalga *Sargassum baccularia* and experimental growth enhancement by nutrient addition in continuous flow culture. Mar Ecol Prog Ser 164:199–211

Scheffer M, Carpenter S, Foley JA, Folkes C, Walker B (2001) Catastrophic shifts in ecosystems. Nature 413:591–596

Schopf JW (2000) The fossil record: tracing the roots of the Cyanobacterial lineage. In: Whitton BA, Potts M (eds) The ecology of cyanobacteria: their diversity in time and space. Kluwer, Dordrecht, pp 13–35

Schupp PJ, Paul VJ (1994) Calcification and secondary metabolites in tropical seaweeds: variable effects on herbivorous fishes. Ecology 75:1172–1185

Schmitt TM, Hay ME, Lindquist N (1995) Antifouling and herbivore deterrent roles of seaweed secondary metabolites: constraints on chemically-mediated coevolution. Ecology 76:107–123

Sfriso A, Marcomini A (1997) Macrophyte production in a shallow coastal lagoon. Part I: Coupling with chemico-physical parameters and nutrient concentrations in waters. Mar Environ Res 44:351–375

Sheppard C, Dixon DJ, Gourlay M, Sheppard A, Payet R (2005) Coral mortality increases wave energy reaching shores protected by reef flats: examples from the Seychelles. East Coast Shelf Sci 64:223–234

Sieburth JM, Conover JT (1965) *Sargassum* tannin, an antibiotic which retards fouling. Nature 208:52–53

Smith JE, Hunter CL, Smith CM (2002) Distribution and reproductive characteristics of nonindigenous and invasive marine algae in the Hawaiian Islands. Pac Sci 56:299–315

Smith JE, Kuwabara J, Coney J, Flanagan K, Beets J, Brown D, Stanton F, Takabayashi M, duPlesses S, Griesemer BK, Barnes S, Turner J (2009) An unusual cyanobacterial bloom in Hawaii. Coral Reefs 27:851

Smith E, Shaw M, Edwards A, Obura D, Pantos O, Sala E, Sandin SA, Smriga S, Hatay M, Rohwer FL (2006) Indirect effects of algae on coral: algae-mediated, microbe induced coral mortality. Ecol Lett 9:835–845

Smith JE, Smith CM, Hunter CL (2001) An experimental analysis of the effects of herbivory and nutrient enrichment on benthic community dynamics on a Hawaiian reef. Coral Reefs 19:332–342

Smith SV, Buddemeier RW (1992) Global change and coral reef ecosystems. Ann Rev Ecol Syst 23:89–118

Smith SV, Kimmerer WJ, Laws EA (1981) Kaneohe Bay Hawaii sewage diversion experiment: perspective on ecosystem responses to nutrient perturbation. Pac Sci 35:279–395

Sommer U (1991) A comparison of the Droop and Monod models of nutrient limited growth applied to natural populations of phytoplankton. Funct Ecol 5:535–544

Sotka EE, Whalen KE (2008) Herbivore offense in the sea: the detoxification and transport of secondary metabolites. In: Amsler CD (ed) Algal chemical ecology. Springer, Berlin, pp 203–228

Steinberg PD (1992) Geographical variation in the interaction between marine herbivores and brown algal secondary metabolites. In: Paul VJ (ed) Ecological roles of marine natural products. Comstock Publishing Associates, Ithaca, pp 51–92

Steinberg PD, van Altena I (1992) Tolerance of marine invertebrate herbivores to brown algal phlorotannins in temperate Australasia. Ecol Monogr 62:189–222

Steneck RS (1988) Herbivory on coral reefs: a synthesis. Proc 6th Int Coral Reef Symp 1:37–49

Steneck RS, Dethier MN (1994) A functional group approach to the structure of algal-dominated communities. Oikos 69:476–498

Steneck RS, Testa V (1997) Are calcareous algae important to reefs today or in the past? Symposium summary. Proc 8th Int Coral Reef Symp 1:685–688

Steneck RS, Watling L (1982) Feeding capabilities and limitation of herbivorous mollusks: a functional group approach. Mar Biol 68:299–319

Steneck RS, Graham MH, Bourque BJ, Corbett D, Erlandson JM, Estes JA, Tegner MJ (2002) Kelp forest ecosystems: biodiversity, resilience and future. Environ Conserv 29:436–459

Stewart HL (2006) Ontogenetic changes in buoyancy, breaking strength, extensibility and reproductive investment in a drifting macroalga *Turbinaria ornata* (Phaeophyta). J Phycol 42:43–50

Stimson J, Conklin E (2008) Potential reversal of a phase shift: the rapid decrease in the cover of the invasive green macroalga *Dictyosphaeria cavernosa* Forsskal on coral reefs in Kaneohe Bay, Oahu, Hawaii. Coral Reefs 27:717–726

Stimson J, Larned ST (2000) Nitrogen efflux from the sediments of a subtropical bay and the potential contribution to macroalgal nutrient requirements. J Exp Mar Biol Ecol 252:159–180

Stimson J, Larned ST, Conklin E (2001) Effects of herbivory, nutrients, and introduced algae on the distribution and abundance of the invasive macroalga *Dictyosphaeria* cavernosa in Kaneohe Bay, Hawaii. Coral Reefs 19:343–357

Suding KN, Gross KL, Houseman GR (2004) Alternative states and positive feedbacks in restoration ecology. Trends Ecol Evol 19:46–53

Tan LT (2007) Bioactive natural products from marine cyanobacteria for drug discovery. Phytochemistry 68:954–979

Targett NM, Arnold TM (1998) Predicting the effects of brown algal phlorotannins on marine herbivores in tropical and temperate oceans. J Phycol 34:195–205

Taylor DI, Nixon SW, Granger SL, Buckley BA, McMahon JP, Lin HJ (1995) Responses of coastal lagoon plant communities to different forms of nutrient enrichment: a mesocosm experiment. Aquat Bot 52:19–34

Taylor PR, Littler MM, Littler DS (1986) Escapes from herbivory in relation to the structure of mangrove island macroalgal communities. Oecologia 69:481–490

Thacker RW, Paul VJ (2001) Are benthic cyanobacteria indicators of nutrient enrichment? Relationships between cyanobacterial abundance and environmental factors on the reef flats of Guam. Bull Mar Sci 69:497–508

Thacker RW, Nagle DG, Paul VJ (1997) Effects of repeated exposures to marine cyanobacterial secondary metabolites on feeding by juvenile rabbitfish and parrotfish. Mar Ecol Prog Ser 167:21–29

Thomas TE, Harrison PJ (1987) Rapid ammonium uptake and nitrogen interactions in five intertidal seaweeds grown under field conditions. J Exp Mar Biol Ecol 107:1–8

Thybo-Christesen M, Rasmussen MB, Blackburn TH (1993) Nutrient fluxes and growth of *Cladophora sericea* in a shallow Danish bay. Mar Ecol Prog Ser 100:273–281

Tsuda RT, Kami HT (1973) Algal succession on artificial reefs in marine lagoon environment in Guam. J Phycol 9:260–264

Tyler AC, McGlathery KJ, Andersen IC (2003) Benthic algae control sediment-water column fluxes of nitrogen in a temperate lagoon. Limnol Oceanogr 48:2125–2137

Valiela I, Foreman K, LaMontagne M, Hersh D, Costa J, Peckol P, DeMeo-Andreson B, D'Avanzo C, Babione M, Sham CH, Brawley J, Lajtha K (1992) Couplings of watersheds and coastal waters sources and consequences of nutrient enrichment in Waquoit Bay Massachusetts. Estuaries 15:443–457

Valiela I, McClelland J, Hauxwell J, Behr PJ, Hersh D, Foreman K (1997) Macroalgal blooms in shallow estuaries: controls and ecophysiological and ecosystem consequences. Limnol Oceanogr 42:1105–1118

Vermeij GJ (1987) Evolution and escalation: an ecological history of life. Princeton University Press, Princeton

Vermeij MJA (2006) Early life-history dynamics of Caribbean coral species on artificial substratum: the importance of competition, growth, and variation in life-history strategy. Coral Reefs 25:59–71

Vermeij MJA, Smith JE, Smith CM, Thurber RV, Sandin SA (2009) Survival and settlement success of coral planulae: independent and synergistic effects of macroalgae and microbes. Oecologia 159:325–336

Vitousek PM, Aber JD, Howarth RW, Likens GE, Matson PA, Schindler DW, Schlesinger WH, Tilman DG (1997) Human alteration of the global nitrogen cycle: sources and consequences. Ecol Appl 7:737–750

Wahl M, Hay ME (1995) Associational resistance and shared doom: effects of epibiosis on herbivory. Oecologia 102:329–340

Waite T, Mitchell R (1972) The effect of nutrient fertilization on the benthic alga *Ulva lactuca*. Bot Mar 25:151–156

Walters LJ, Hadfield MG, Smith CM (1996) Waterborne chemical compounds in tropical macroalgae: positive and negative cues for larval settlement. Mar Biol 126:383–393

Walther G, Post E, Convey P, Menzei A, Parmesan C, Beebee TJC, Fromentins J, Hoegh-Guldberg O, Bairlein F (2005) Ecological responses to recent climate change. Nature 416:389–395

Webster NS, Smith LD, Heyward AJ, Watts JEM, Webb RI, Blackall LL, Negri AP (2004) Metamorphosis of a scleractinian coral in response to microbial biofilms. Appl Environ Microbiol 70:1213–1221

Wheeler PA, North WJ (1980) Effect of nitrogen supply on nitrogen content and growth rate of juvenile *Macrocyctis pyrifera* (Phaeophyta) sporophytes. J Phycol 16:577–582

Williams SL (1984) Uptake of sediment ammonium and translocation in a marine green macroalga *Caulerpa cupressoides*. Limnol Oceanogr 29:374–379

Williams SL, Carpenter RC (1988) Nitrogen-limited primary productivity of coral reef algal turfs: potential contribution of ammonium excreted *by Diadema antillarum*. Mar Ecol Prog Ser 47:145–152

Williams SL, Carpenter RC (1998) Effects of unidirectional and oscillatory water flow nitrogen fixation (actylene reduction) in coral reef algal turfs, Kaneohe Bay, Hawaii. J Exp Mar Biol Ecol 226:293–316

Williams SL, Smith JE (2007) A global review of the distribution, taxonomy, and impacts of introduced seaweeds. Ann Rev Ecol Syst 38:327–359

Wolanski E, Pickard GL (1983) Upwelling by internal tides and Kelvin waves at the continental shelf break on the Great Barrier Reef. Aust J Mar Freshwater Res 34:65–80

Xue Y, Shukla J (1993) The influence of land surface properties on sahel climate. Part I: desertification. J Climate 6:2232–2245

Invertebrates and Their Roles in Coral Reef Ecosystems

Peter W. Glynn and Ian C. Enochs

1 Introduction

There are some fundamental generalizations that can be made about the biology and ecology of invertebrates associated with coral reefs. For example, it is widely accepted that coral reefs support the highest biodiversity of all marine ecosystems, and that invertebrates contribute dominantly to this condition. It is also acknowledged that numerous invertebrate taxa are involved in highly complex and coevolved relationships with metazoans, unicellular protists, and multicellular algae. Further, during the past few decades it has been demonstrated that certain invertebrate consumers can have strong and widespread effects on coral abundances, community structure, and the integrity of reef formations.

There are many things we do not know about invertebrates inhabiting coral reef environments. Some invertebrate consumers undergo surges or drastic declines in population abundances, with far-reaching effects on reef communities. A notable example is the sea star corallivore *Acanthaster planci*, which has experienced population outbreaks on numerous Indo-Pacific coral reefs during the past several decades and probably earlier (Plate 1a). In outbreak densities, this sea star has consumed large quantities of reef corals, eliminating virtually all live coral cover over reef tracts of tens of hundreds of kilometers in extent. Since the early 1980s, the abrupt decline of the herbivorous sea urchin *Diadema antillarum* in the Caribbean has allowed the proliferation of algae to the detriment of coral cover on many reefs. While various hypotheses have been proposed to explain these population changes, no uncontested mechanisms of causation have been forthcoming. Nonetheless, important advances have been made since the 1990s and our understanding of invertebrate interactions within reef ecosystems demands a fresh reexamination in light of new developments. Some of the most recent, in-depth treatments of these subject areas, for example those in Dubinsky (1990), Birkeland (1997), Reaka-Kudla (1997), and Karlson (1999), are now over 10 years old.

In this review, we concentrate on invertebrates that directly or indirectly affect corals, coral community composition, and the structure/integrity of reef frameworks. The various topics covered are focused dominantly around trophic and habitat relationships, which represent essential resources that directly or indirectly link all reef organisms. Particular subject areas that will be addressed include: (1) a brief history of studies that have contributed to our understanding of the functional roles of reef invertebrates, (2) an overview of the abundance and diversity of reef invertebrate taxa, (3) reef habitats, (4) biotic interactions, (5) trophodynamic considerations, and (6) the outlook and implications of research into invertebrate abundances and interactions in coral reef ecosystems. Some topics are introduced only briefly because they are covered in depth elsewhere in this volume. Reference will be made to relevant chapters that more fully address particular subjects.

2 Historical Overview

From the late 1800s to the first half of the twentieth century, there has been sporadic progress in our understanding of the roles of invertebrates influencing coral reef community structure and dynamics. Early efforts were largely directed toward taxonomic and systematic studies, with the naming of new species and development of classification schemes. A few examples of these pioneering contributors are Wimmer (1879) treating the molluscan fauna of the Galápagos Islands, Gravier (1901) on the polychaetous annelids of the Red Sea, and Rathbun (1933) who documented occurrences and named new species of brachyuran crabs in Puerto Rico and the Virgin Islands. Such studies remain fundamentally important in ongoing efforts of biodiversity research, biogeographic analyses, and assessing changes (e.g., range reductions, extinctions) in modern invertebrate taxa. Our brief historical overview focuses on relatively recent advances in invertebrate studies; a broader historical perspective of coral reef research is presented by Richmond in this book.

P.W. Glynn (✉) and I.C. Enochs
Division of Marine Biology and Fisheries, Rosenstiel School of Marine and Atmospheric Science, University of Miami, Miami, FL
e-mail: pglynn@rsmas.miami.edu; ienochs@rsmas.miami.edu

Plate 1 (a) *Acanthaster planci* feeding on acroporid corals, Palau (Photo by C. Birkeland); (b) *Terpios hoshinota* overgrowing the coral *Porites rus*, Gun Beach, Guam; (c) *Ceratoporella nicholsoni*, Discovery Bay, Jamaica, in a cavern on the forereef slope (Photo by P. Willenz); (d) *Lysmata* sp. cleaning *Gymnothorax fimbriatus*, Milne Bay, Papua New Guinea; (e) Didymozoid digenetic flukes (*Didymocystis* sp.) attached to the gill filaments of the yellowfin grouper, *Mycteroperca venenosa*, Puerto Rico (Photo by L. Wiliams); (f) A basket star (*Astrophyton muricatum*) adhering to a gorgonian (*Pseudopterogorgia* sp.), Key Largo Dry Rocks, Florida Keys; (g) *Amphiprion percula* associated with their anemone host (*Heteractis magnifica*), Milne Bay, Papua New Guinea; (h) *Dermatolepis dermatolepis* seeking refuge within the spines of *Centrostephanus coronatus*, Clipperton Atoll (Photos (d, f, g) are courtesy of and copyrighted by Michael C. Schmale)

With extended expeditionary activities, such as the Great Barrier Reef Expedition (1928–1929) in Australia, and establishment of the first laboratories on coral reefs, for example, the Tortugas Laboratory of the Carnegie Institution of Washington (1903–1939) in the Florida Keys, and the Palau Tropical Biological Station (1934–1943) in the Caroline Islands, coral reef studies experienced significant and rapid advancements. Early expeditions resulted in the publication of inventories and relative abundances of metazoans, as well as the recognition of zonation patterns across reefs on the Great Barrier Reef (Stephenson et al. 1931; Otter 1937), in New Caledonia (Catala 1950), and in French Polynesia (Morrison 1954). From detailed underwater studies of the micro- and macrofauna of coral reefs in the Red Sea and Maldive Islands, a chief objective of the Xarifa Expedition (1957–1958), Gerlach (1959) was one of the first workers to articulate the importance of habitat niches and their influence on biotic diversity (Fig. 1). A brief history of this pioneering effort, including the participants and their contributions, was published recently in the journal *Coral Reefs* (Wallace and Zahir 2007). Beginning in the late 1960s and early 1970s, interest in reef invertebrates accelerated, due in part to the devastating effects of *Acanthaster planci* predation on reef corals in the Indo-Pacific region (e.g., Chesher 1969; Endean 1973). Before these severe impacts, it was widely held that reef corals were largely immune to predation (Wells 1957). To understand how massive sea star outbreaks occurred, a major shift in research focus was initiated, from descriptive to community-oriented functional approaches.

The 1970s and early 1980s also experienced a convergence of events that further stimulated research on reef-associated invertebrates. The first International Coral Reef Symposium

Fig. 1 Early iconography of macrofaunal associates on a Maldivian branching coral (Gerlach 1959). A variety of sessile, sedentary, and bioeroding taxa of sponges, cnidarians, polyclad flatworms, polychaetous annelids, molluscs, crustaceans, echinoderms, ascidians, and fishes are illustrated

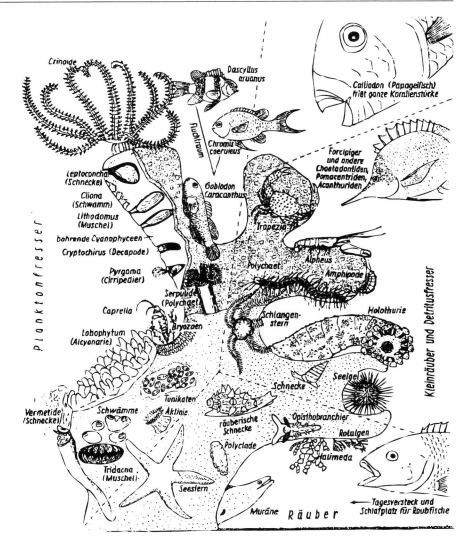

at Mandapam Camp, India, held under the auspices of the Marine Biological Association of India, was convened in January 1969. Of the seven themes presented at this symposium, the section offering invertebrate studies, "Coral Reefs as Biotopes: Invertebrates", was the largest with 11 papers read. The subjects ranged from the depletion of zooplankton over reefs to ecological studies on sponges, cnidarians, molluscs, and crustaceans. David Stoddart's landmark review, *Ecology and morphology of Recent coral reefs,* also was published in 1969. In this paper, he made the important connection between coral mortality by *Acanthaster*, and the subsequent attrition of dead colonies from bioerosion, thus leading inevitably to the breakdown and disintegration of reef structures. Additional invertebrate studies were published in 1971, in *Regional Variation in Indian Ocean Coral Reefs*, edited by David Stoddart and Sir Maurice Yonge. These studies included quantitative documentation of invertebrate diversity in sedimentary and hard surface reef habitats, and the roles of echinoids and *Acanthaster* in effecting coral community structure.

The *Biology and Geology of Coral Reefs*, a four-volume series (1973–1977) edited by O.A. Jones and R. Endean, introduced several topics related to reef invertebrates, including species diversity, ecological interactions (predation, competition), bioerosion, and reviews of coral crustacean symbionts, and reef-associated echinoderm taxa. A special session was devoted to research on *Acanthaster* in 1973 at the Second International Coral Reef Symposium (Australia) to address large-scale coral mortality observed in several areas of the Indo-Pacific region. Animated discussion centered on the causes of outbreaks, whether these resulted from an absence of predator control or occurred naturally when conditions favored the formation of feeding or spawning aggregations. Advances in the functional roles of reef invertebrates have continued at later symposia, for example, in 1977 (Miami), 1981 (Manila), and at subsequent venues convened approximately every 4 years. Establishment of the International Society for Reef Studies in 1981, and publication of the society's journal *Coral Reefs* in 1982, have encouraged international collaboration and quickened the pace of invertebrate reef studies.

3 Overview of Major Invertebrate Taxa

Over a decade ago, the total number of all described species occurring on global coral reefs was estimated at ~93,000 (Reaka-Kudla 1997). Assuming that coral reefs and rainforests are equally diverse, and adjusting for the area differences of each ecosystem – global coral reefs ~5% of the area of rainforests – Reaka-Kudla (1997) further estimated that the expected species richness of global coral reefs would approach 950,000 species. There is no firm consensus about this value. Some authors predict coral reef species richness to be up to three times higher (Knowlton 1993; Small et al. 1998), while others offer sharp criticisms of the assumptions inherent in these estimates and predictions of much lower values (May 1992, 1994). In addition to the basic assumptions underlying these estimates, several other potential sources of error contribute to the discrepancies. For example (a) numerous reefs in species-rich areas are still undersampled and unstudied; (b) many, perhaps the majority of reef species, are of small size and members of the cryptos; (c) some unknown number of existing synonymies and morphologically cryptic sibling species will reduce or elevate counts, respectively; and (d) presently there is only a modest effort devoted to systematic studies. In addition, with the recent large numbers of coral reefs that have been severely impacted, an unknown number of species are likely going extinct daily and will continue to disappear in the near future. According to Wilkinson (2008), an estimated 19% of global coral reefs have been "effectively lost," and 35% will be in critical and threatened stages of decline within the next 10–40 years. Carpenter et al. (2008) have predicted that one third of all reef-building corals now face an elevated extinction risk. The obligate metazoan symbionts of corals would be especially vulnerable if this prediction is realized (Glynn, 2011).

Of 35 recognized metazoan phyla, all but three are known with species that inhabit coral reefs (Paulay 1997). Pogonophorans were formerly recognized as a distinct phylum (Pogonophora), but are now classified as a family (Siboglinidae) in the phylum Annelida (Rouse 2001). They have not been found on coral reefs. Although a detailed inventory of reef taxa is unavailable for any reef ecosystem or region, it is probable that invertebrate species outnumber all other presently known biota, including protistans, algae, fungi, and fishes (Reaka-Kudla 1997; Bouchet 2006). Invertebrate metazoan taxa that are especially diverse or involved in key ecological interactions on reefs belong to eight phyla: Porifera, Cnidaria, Annelida, Sipuncula, Arthropoda, Mollusca, Echinodermata, and Chordata. These taxa are introduced briefly here with general information on their species richness, abundance, geographic occurrences, habitat niches, ecological functions, and key references to the literature. Estimates of known species numbers for phyla are mostly from Bouchet (2006). It is cautioned that these can vary widely, in excess of 100% for some phyla. Sources are noted for species richness estimates for lower ranking taxa. Examples of the other 24 less well-known phyla, categorized according to their modes of feeding, conclude this section.

3.1 Major Taxa

To provide a frame of reference for the range of invertebrates associated with coral communities, a brief overview of selected taxa and where they occur on reefs follows. Eight major reef phyla and 24 minor phyla, all with reef-associated species, are briefly featured to provide an inventory of the metazoan taxa discussed later in this chapter. Classification schemes are presented for the eight major phyla, along with representative genera, and the minor phyla are listed in Table 1. Because of the broad scope, complexity and rich literature characterizing this subject, landmark studies and reviews will be referenced to help the interested reader gain access to specific topics. The distinctions "major" and "minor" are clearly arbitrary. For example, some species groups that are designated as major may occur only rarely or be absent altogether from some reefs and reef regions. And minor taxa may occur in high abundances on reefs, but are not readily visible due to their cryptic habits. Additionally, an uncommon or rare species may become locally abundant if experiencing rapid population growth.

Phylum Porifera*
    Class Calcispongiae (=Calcarea)**
        Subclass Calcinea
            Order Murrayonida (*Murrayona, Paramurrayona, Lelapiella*)
            Order Clathrinida (*Leucetta*)
        Subclass Calcaronea
            Order Lithonida (*Tulearinia*)
            Order Baeriida (*Lepidoleucon*)
    Class Demospongiae
        Order Homosclerophorida
            Family Plakinidae (*Plakortis, Plakinastrella*)
        Order Hadromerida
            Family Clionaidae (*Cliona, Spheciospongia*)
            Family Suberitidae (*Terpios*)
            Family Acanthochaetetidae (*Acanthochaetetes, Willardia*)
            Family Alectonidae (*Alectona*)

Table 1 Minor invertebrate taxa associated with coral reefs with information on habitat niches, dominant biotic interactions (where known), and relevant literature citations. Taxa are listed in the approximate phylogenetic order followed in Brusca and Brusca (2003). Cryptos refers to organisms that are largely hidden and protected from full or direct exposure to major environmental factors (Kobluk 1988). Glynn and Enochs (personal observations) denote information from original field observations in the eastern tropical Pacific (Panamá). Other sources consulted were Parker (1982), Paulay (1997), and Brusca and Brusca (2003)

Taxon	Habitat niche	Biotic interactions	Authority
Placozoa	Cryptos, water column	Graze on algal films, scavengers	Pearse and Voigt (2007)
Rhombozoa	Cephalopod nephridia	Parasites	Hochberg (1982, 1983); Castellanos-Martínez et al. (2007)
Ctenophora	Epizoic on octocorals, water column	Micropredators; *Coeloplana* sp. consumes *Artemia* nauplii	Fricke (1970); Rudman (1999); F.M. Bayer (unpublished manuscript)
Acoelomorpha	Epizoic on corals	Possibly remove coral mucus, may interfere with coral photosynthesis	Barneah et al. (2007)
Platyhelminthes	Cryptos, micro-/epibenthos, parasitic	Trematodes and cestodes: ecto-/endoparasites of fishes, turbellarians: carnivores, scavengers	Plaisance and Kritsky (2004); Glynn and Enochs (personal observations); Newman et al. (2003)
Nemertea	Cryptos, epibenthic, demersal reef plankton	Predators, scavengers	Porter (1974); Devaney and Eldredge (1987); Thiel and Kruse (2001)[a]; Glynn and Enochs (personal observations)
Rotifera	Cryptos, epibenthic, meiofauna	Suspension feeders, predators, parasites	Gourbault and Renaud-Mornant (1990)
Gastrotricha	Meiofauna (coarse sediments), inner reef flat	Feeding mode unknown	Thomassin et al. (1976); Hochberg (2008)
Kinorhyncha	Meiofauna	Deposit feeders and grazers	Higgins (1983); Netto et al. (1999); Sørensen (2006)
Nemata (=Nematoda)	Meiofauna, cryptos, epibenthic, epizoic, parasitic	Diverse feeding types: deposit feeding, predation, scavengers, ecto-/endoparasites	Porter (1974); Thomassin et al. (1976); Alongi (1986, 1989); Gourbault and Renaud-Mornant (1990); Netto et al. (1999)
Nematomorpha	Planktonic, parasitic	Parasitize decapod crustaceans, slight effects on host	Dorier (1965)[a]
Acanthocephala	Parasitic	Intestinal parasites in teleosts, intermediate hosts are crustaceans, uncommon	Amin (1982)[a]
Entoprocta	Epibenthic, cryptos, epizoic on sponges, ectoprocts, ascidians and other invertebrates	Suspension feeding	Brusca and Brusca (2003)[a]
Gnathostomulida	Meiofauna; coarse reef sediments, near seagrasses and mangroves	Graze bacteria and fungal microflora	Sterrer (2000)
Priapula	Infauna, meiofauna	Predation, detritivores	Brusca and Brusca (2003)[a]
Loricifera	Meiofauna	May ingest microbes	Brusca and Brusca (2003)[a]
Echiura	Cryptos, reef sediments, cavities between corals	Suspension and deposit feeding	Stephen and Edmonds (1972)[a]; Edmonds (1987), Glynn and Enochs (personal observations)
Tardigrada	Meiofauna, inner reef flat	Diverse feeding types, unknown on reefs	Thomassin et al. (1976), Netto et al. (1999)
Pycnogonida	Cryptos, epizoic, epiphytic	Suctorial predators on algae and microinvertebrates, scavengers, predation (coral polyps)	Chevalier (1987), Paulay et al. (2003)
Phoronida	Coral rock/rubble/sand	Suspension feeding, sediment stabilization	Bailey-Brock and Emig (2000)
Ectoprocta	Cryptos, epibenthic	Suspension feeding, calcifiers, framework stabilization: encrusting, infilling	Cuffey (1972); Buss (1979); Jackson and Winston (1982); Jackson (1984)
Brachiopoda	Cryptos	Suspension feeding, calcifiers, secondary bioconstructors	Jackson et al. (1971)
Chaetognatha	Reef holoplankton	Predators	Glynn (1973); Renon and Lefevre (1985); Sorokin (1990)
Hemichordata (Enteropneusta)	Reef sediments	Deposit feeders	Kicklighter et al. (2003)
Hemichordata (Pterobranchia)	Epibenthic	Suspension feeding	Dilly (1985); Dilly and Ryland (1985)

[a] General references, or with emphasis on extratropical or non-reefal species

Order Chondrosida
　　Family Chondrillidae (*Chondrilla*)
Order Poecilosclerida***
　　Family Iotrochotidae (*Iotrochida*)
　　Family Tedaniidae (*Tedania*)
　　Family Mycalidae (*Mycale*)
　　Family Merliidae (*Merlia*)
Order Halichondrida
　　Family Halichondriidae (*Halichondria*)
Order Agelasida
　　Family Astroscleridae (*Ceratoporella, Goreauiella, Stromatospongia*)
Order Haplosclerida
　　Family Callyspongiidae (*Callyspongia*)
　　Family Chalinidae (*Chalinula, Haliclona*)
　　Family Niphatidae (*Niphates, Gelliodes*)
　　Family Petrosiidae (*Petrosia*)
　　Family Calcifibrospongiidae (*Calcifibrospongia*)
　　Family Phloeodictyidae (*Siphonodictyon*)
Order Dictyoceratida
　　Family Irciniidae (*Ircinia*)
Order Dendroceratida
Order Verongida
　　Family Aplysinidae (*Aplysina, Verongula*)
Order Verticillitida**** (*Vaceletia*)
Class Hexactinellida

*Classification after Hooper and van Soest (2002), which is based on morphological characters. With the recent application of molecular methodologies, the phylogenetic and evolutionary relationships within this phylum are undergoing rapid and significant changes (e.g., Erpenbeck and Wörheide 2007).

**Philippe et al. (2009) have provided strong evidence that the Class Calcispongiae shares basic synapomorphies with other poriferans and should not be treated as a separate taxonomic entity removed from Porifera as suggested by Manuel et al. (2003). It should be noted that Sperling et al. (2009) have suggested that Porifera is paraphyletic, consisting of three separate lineages (Demospongiae + Hexactinellida, Calcarea (Calcispongiae) and Homoscleromorpha).

***Only representative families are listed, that is those with species commonly found in coral reef communities.

****ptVaceletia* is an extant representative of the extinct Sphinctozoa. Its placement here is provisional since this taxon bears no precise affinity with any extant demosponge orders (Vacelet 2002).

Sponges are diverse and abundant on all coral reefs with hundreds and probably more than 1,000 species present on large reef tracts such as the Great Barrier Reef, Australia (van Soest 1990; Hooper and Lévi 1994). Global marine sponge richness according to the World Porifera Database is 8,318 (as of 10/9/2009) though this number is likely inflated due to the inclusion of subspecies and varieties. Hooper and van Soest (2002) suggest the existence of ~15,000 species, an estimate based on the rate of discovery of new species and cryptic species complexes. Another indication of the richness of Porifera is the large number of taxa not listed above, for example, Hadromerida consisting of 13 families and Poecilosclerida with 25 families. In addition, the described species of *Mycale* include 236 species, *Callyspongia* 183 species, and *Haliclona* 429 species. A few representative reef genera are noted under each higher taxon. Calcareous sponges (Calcispongiae) secrete spicules of calcium carbonate (as the mineral calcite) are all marine, and contribute relatively few species to the sponge faunas of coral reefs. Members of the Class Demospongiae are the most prevalent on coral reefs in terms of species and biomass. Their skeleton is composed of single-rayed (monaxonic) or four-rayed (tetraxonic) siliceous spicules, often supplemented or replaced by spongin, an organic collagenous fiber forming a network. Glass sponges (Hexactinellida), which possess a syncytium of somatic cells and triaxonic silica spicules, occur at depths below coral reefs and in polar (at least Antarctic) waters.

Diaz and Rützler (2001) made a strong case for sponge species playing an essential functional role in coral reef ecosystems. Based on Caribbean studies, they noted six biological and ecological properties that support this assertion, namely sponges (1) demonstrate one of the highest diversities of all reef macrobiota, (2) contribute high abundance/area coverage and biomass, (3) exhibit numerous symbiotic associations with microorganisms, (4) are highly competitive for space, (5) exercise various levels of impact on reef frameworks (both positive and negative), and (6) perform high volume exchanges with the water column. Regarding the latter, reef sponges are suspension feeders, capturing mainly bacteria, POM and absorbing DOM (de Goeij et al. 2008a, b).

Several of the relatively uncommon calcareous sponge species associated with reefs occur in caves, tunnels, and deep forereef habitats, mostly in the Indo-Pacific region. *Paramurrayona*, however, is found in cryptic habitats on reefs in both the Indo-Pacific and Caribbean (Jamaica) regions. *Leucetta* is particularly well adapted to shallow areas and is perhaps the most frequently encountered calcareous sponge genus inhabiting coral reefs.

Demosponges, presently numbering 83 families in 14 orders, contribute conspicuously to the reef macrobiota (Plate 1b), but also live cryptically on calcareous substrates within reef structures and endolithically in coral skeletons where they excavate complex chamber networks. This class contains about 85% of all described recent sponge species (Hooper and van Soest 2002).

Hypercalcified sponges, formerly referred to as sclerosponges or coralline sponges, are characterized by a compound basal skeleton of calcium carbonate (as the minerals aragonite or calcite, depending on the species), siliceous spicules and organic fibrous elements (Plate 1c). The massive calcareous skeleton contributes to the construction of reef frameworks. Eight named species occur in three orders of the Class Calcispongiae, and 15 species belong to five orders in the Class Demospongiae, underlining their polyphyletic status (Vacelet et al. in press). These sponges occur on Indo-Pacific reefs, including those in the Red Sea, as well as in the Caribbean Sea where extant populations were first discovered (Hartman and Goreau 1966, 1972). Hypercalcified sponges at shallow depths (~5–20 m) are usually cryptic in framework cavities and caves, while individuals at greater depths (~30–200 m) frequently occupy forereef wall sites (Jackson et al. 1971; Basile et al. 1984). Members of the family Astroscleridae, for example, *Ceratoporella nicholsoni* and *Goreauiella auriculata*, share affinities with extinct reef-building chaetetids and stromatoporoids.

For comprehensive reviews on the history of sponge research and classification, see Rützler (2004) and Hooper and van Soest (2002), respectively.

*Phylum Cnidaria**
   Class Hydrozoa
      Order Hydroida
         Suborder Anthomedusae
            Family Milleporidae (*Millepora*)
            Family Stylasteridae (*Distichopora, Stylaster*)
         Suborder Leptomedusae (*Antennella, Hydrodendron, Plumularia*)
   Class Anthozoa
      Subclass Octocorallia (=Alcyonaria)**
         Order Alcyonacea***
            Stolonifera group (*Clavularia, Coelogorgia, Sarcodictyon, Tubipora*)
            Alcyoniina group (*Cladiella, Sarcophyton, Sinularia, Xenia*)
            Scleraxonia group (*Briareum, Iciligorgia*)
            Suborder Holaxonia (*Euplexaura, Pseudopterogorgia*)
            Suborder Calcaxonia (*Ellisella, Plumarella*)
         Order Helioporacea (*Heliopora*)
         Order Telestacea (*Telesto*)
         Order Pennatulacea (*Cavernulina, Pteroeides*)
      Subclass Hexacorallia (=Zoantharia)
         Order Actiniaria (*Aiptasia, Bartholomea*)
         Order Scleractinia (*Acropora, Pocillopora, Porites*)
         Order Zoanthiniaria (*Palythoa, Zoanthus*)
         Order Corallimorpharia (*Actinodiscus, Rhodactis, Ricordia*)
      Subclass Ceriantipatharia
         Order Antipatharia (*Antipathes, Cirrhipathes*)
         Order Ceriantharia (*Arachnanthus, Cerianthus*)

*Members of two classes (Scyphozoa and Cubozoa) not considered here occasionally enter reef environments in the plankton and may be important in the polypoid stage. In addition, sessile stauromedusae may occur on reefs.

**Brusca and Brusca (2003) list eight orders belonging to this subclass, five with species commonly encountered on coral reefs. We group the orders Gorgonacea and Stolonifera within the order Alcyonacea, following Bayer (1981) and Fabricius and Alderslade (2001). Based on the discovery of new species and continuing morphological study, it is now clear that octocoral species in the former two orders form a complete series of growth forms, from simple soft corals to complex gorgonians, hence the single order Alcyonacea.

***Due to a lack of consistent differences in growth form and colony morphology, the stoloniferans, alcyoniinans, and scleraxonians are treated as separate groups within the Alcyonacea rather than formal orders or suborders.

In addition to zooxanthellate scleractinian corals, several other cnidarian taxa are often abundant, and occasionally predominant, in reef communities. Among cnidarian hydrozoans, zooxanthellate species of fire corals in the genus *Millepora* (Milleporidae) construct calcareous skeletons, some of which contribute significantly to reef framework structures. Informative sources on the taxonomy, distribution, and biology of *Millepora* can be found in Boschma (1948), Lewis (1989), and Razak and Hoeksema (2003). *Millepora* spp. occur on coral reefs worldwide with at least 13 valid species: seven in the Indo-Pacific and six in the western Atlantic. The identity of three to four species, based on morphometrics, is in need of further study. Milleporid hydrocorals provide habitats for numerous associates, for example, polychaete worms, crustaceans, and echinoderms (see Section Habitat Providers).

Another group of hydrozoans that construct sturdy, calcareous skeletons are the stylasterines or lace corals (Stylasteridae). These hydrocorals are azooxanthellate, form relatively small colonies, and occur mostly in deep water. A few species are occasionally present on shallow exposed substrates, but more commonly occur in dimly lit reef habitats. *Distichopora* and *Stylaster* are widespread genera on coral reefs (Cairns 1983, 1991; Veron 2000).

Benthic hydrozoans (Anthomedusae and Leptomedusae) are usually the least conspicuous of this class on coral reefs, due to their generally small size and dull-colored exoskeleton. Several species are members of the cryptos, but some form large colonies with ~30-cm long quill-like stems, for example, *Pennaria disticha* and *Macrorhynchia phoenicea* in waters surrounding Guam (Kirkendale and Calder 2003). In terms of species richness and abundance, hydroids often contribute importantly to reef biotas. For example, Kirkendale and Calder (2003) predicted that the hydroid fauna of Guam, when adequately studied, will likely include 100 species or more. A good number of these species rarely inhabit reefs, but occur on artificial structures in harbors.

A diverse assemblage of anthozoan taxa (Anthozoa) occurs on coral reefs and often contributes importantly to the structure of reef communities. The three subclasses adopted by most present-day taxonomists are the Octocorallia (=Alcyonaria), Hexacorallia, and Ceriantipatharia. The octocoral orders noted here are the

Alcyonacea (soft corals, sea fans, sea whips), Helioporacea (blue coral), and Telestacea.

The large size and complex colony morphology of many octocorals provide shelter for a rich assemblage of invertebrate associates (and fishes). Additionally, octocorals contribute importantly to reef community biomass with some species facilitating reef consolidation and framework construction. In terms of taxonomic diversity, octocorals rival the zooxanthellate scleractinian corals. In the tropical Indo-Pacific, the Alcyonacea contain around 90 genera and 23 families compared with zooxanthellate scleractinian corals with 89 genera in 18 families (Fabricius and Alderslade 2001). Although an exact species number is unknown, Fabricius and De'ath (2001) noted the occurrence of several hundred soft coral species on the Great Barrier Reef. The *Sinularia* genus alone boasts more than 120 species.

Tubipora musica, the organ-pipe coral, is the most widely recognized species in the Stolonifera group. It forms calcareous, multitiered colonies, is zooxanthellate, and confined to the Indo-Pacific region. The dark-red skeleton is formed of vertically oriented tubes consisting of fused sclerites. The tubes are partitioned horizontally by thin stolon plates, but the elegant skeleton is nearly invisible when the gray to light yellow polyps are extended.

The dominant reef-dwelling octocorals in the Indo-West Pacific are soft corals, members of the Alcyoniina group. Some of these species deposit large solid bases on the reef flat and thus contribute to reef building (Schuhmacher 1997). Soft corals and gorgonians (formerly Gorgonacea) occur on coral reefs in all major biogeographic regions, but the former dominate coral communities in the Indo-West Pacific and the latter in the western Atlantic. Two soft coral species (Scleraxonia group) that are often important space occupiers in the western Atlantic are *Briareum asbestinum* and *Erythropodium caribaeorum*. Gorgonian (Holaxonia and Calcaxonia) faunal richness in the western Atlantic is typically high, as exemplified for Cuba with 55 (Alcolado et al. 2003) and Panamá with 38 reef-dwelling species (Guzmán 2003).

Heliopora coerulea, a member of the Helioporacea, is the only extant zooxanthellate reef species in this order. This is the only octocoral that forms a massive aragonite skeleton. It is locally abundant on many western Pacific and Indian Ocean reefs. *Heliopora* contributes importantly to the shallow lagoon community of the Shiraho reef on Ishigaki Island, southern Ryukyus Islands (Harii and Kayanne 2003). A third order of octocorals is the telestaceans, which form flexible and branching colonies often associated with fouling communities. Finally, the order Pennatulacea, comprising the sea pens and sea pansies, live on soft-bottom habitats in the Indo-Pacific region. Sea pens possess a large primary axial polyp, supported by an internal calcareous axis, which extends the length of the colony. Sea pen species may attain up to 1 m in length, but generally are not visible because they remain retracted in the sediment during the day.

Three of the four orders of hexacorallians are soft bodied and common members of coral communities. The true sea anemones (Actiniaria) are solitary or occur as clonal aggregations. They typically possess two siphonoglyphs (ciliated grooves extending down the pharynx) and can reach a large size. Some species of clownfish anemones (Stichodactylidae) can attain nearly 1 m in diameter and an age exceeding 100 years. Zoanthids (Zoanthiniaria) often form extensive patches in shallow reef zones. The polyps arise from a basal mat or stolon, commonly forming sheets or carpets on reef substrates. Some species are zooxanthellate and others epizootic, often found associated with sponges, black corals, or gorgonians. Corallimorpharians (Corallimorpharia) are anemone-like, commonly occurring on reefs. Like scleractinian corals, they do not possess a siphonoglyph, and their nematocysts and other anatomical structures suggest a close phylogenetic relationship with the stony corals (Fautin and Lowenstein 1992). Recent studies, however, are not in agreement with the validity of this proposed affinity (Medina et al. 2006; Fukami et al. 2008).

The black or thorny corals (Antipatharia) and tube anemones (Ceriantharia) make up the subclass Ceriantipatharia. Black corals form fan or whiplike colonies that occasionally occur in shallow water, but are generally more abundant below 30–50 m depth. They are named for their brown to black axial skeleton, which is composed of chitinous fibrils and protein. Tube anemones are large, solitary polyps that live in vertical tubes in soft sediments. They display two whorls of elongate tentacles arising from the oral disk. These anemones are capable of rapid contraction into sturdy tubes made of unique specialized cnidae (ptychocysts) and mucus.

*Phylum Annelida**
  Class Polychaeta****
    Scolecida******
      Family Opheliidae (*Armandia, Polyophthalmus*)
    Palpata
      Aciculata
        Phyllodocida
          Family Chrysopetalidae (*Chrysopetalum*)
          Family Hesionidae (*Hesione*)
          Family Nereididae (*Ceratonereis, Nereis, Perinereis*)
          Family Phyllodocidae (*Phyllodoce*)
          Family Polynoidae (*Allmaniella, Perolepis, Thormora*)
          Family Sigalionidae (*Sthenelais, Thalenessa*)
          Family Syllidae (*Exogone, Odontosyllis, Typosyllis*)
        Eunicida

Family Amphinomidae (*Hermodice, Notopygos, Pherecardia*)
Family Dorvilleidae**** (*Dorvillea*)
Family Eunicidae**** (*Eunice, Nematonereis, Palola*)
Family Lumbrineridae**** (*Lumbrineris*)
Family Oenonidae (*Oenone*)
Canalipalpata
Spionida
Family Spionidae**** (*Dipolydora, Scolelepis*)
Terebellida
Family Cirratulidae**** (*Dodecaceria, Timarete*)
Family Terebellidae (*Pista, Terebella, Thelepus*)
Sabellida
Family Sabellariidae (*Idanthyrsus, Lygdamis*)
Family Sabellidae**** (*Megalomma, Sabellastarte*)
Family Serpulidae (*Salmacina, Serpula, Spirobranchus*)
Class Myzostomida (*Notopharyngoides, Myzostoma*)
Class Clitellata
Order Hirudinida (*Myzobdella*)

*Here, we follow the classical treatment of Echiura as a phylum but it is noted that recent molecular evidence supports its inclusion in Annelida.

**This class includes 83 families, about 1,000 genera, and over 13,000 known species (Wilson et al. 2003). Only some of the more prominent families occurring on coral reefs are listed herein.

***This classification follows Rouse and Fauchald (1997), and is also adopted by Glasby et al. (2000). These authors omitted the recognition of Linnaean categories (orders or other higher-level taxa) due to the difficulty of establishing monophyly.

****Families with species that bore into coral and other calcareous skeletons (Hutchings 1986).

Nearly 30 families of polychaetous annelid worms occur on coral reefs, and are often abundant members of the epibenthos and cryptic fauna. Two thirds of the invertebrates found in one small coral colony consisted of 1,441 polychaetes (Grassle 1973). Eye-catching epibenthic species include tubicolous suspension-feeding species in the families Serpulidae and Sabellidae. Cryptic polychaetes include numerous predatory and scavenging species that reside in cavities as well as detritivores, endolithic borers, and micro- and meiofaunal deposit feeders in reef sediments. Commensal species from various families associate with sponges, cnidarians, other polychaetes, echinoderms, molluscs, and crustaceans. Relatively few polychaetes are ecto- or endoparasitic (Pettibone 1982). The bioerosive destruction of reef carbonates is probably the most important overall effect of polychaete worms (Hutchings 1986). Some other ecological interactions, however, can have considerable local importance. These involve, for example, corallivory, the protection of coral hosts from crown-of-thorns sea star attack, and scavenging.

Myzostomes are minute, highly modified annelid worms that occur as commensals or echinoderm parasites. Most species are associated with reef crinoids and feed on particles in their host's ambulacral grooves (Grygier 2000). Their diversity corresponds to gradients in their host's diversity, which is highest on reefs in the Indo-West Pacific region.

Marine leeches, members of the order Hirudinea, occur as ectoparasites on many species of bony and cartilaginous reef fishes in the western Pacific and Caribbean Sea. These blood-sucking parasites are most commonly attached to fish gills, but there is a record of their occurrence on the bodies of portunid swimming crabs (Williams et al. 1994).

Phylum Sipuncula
Class Phascolosomida
Order Aspidosiphoniformes (*Aspidosiphon, Lithacrosiphon*)
Order Phascolosomiformes (*Phascolosoma*)
Class Sipunculida
Order Golfingiaformes
Family Themistidae (*Themiste*)
Family Phascolionidae (*Phascolion*)
Family Golfingiidae (*Golfingia, Nephasoma*)
Order Sipunculiformes (*Sipunculus, Siphonosoma*)

Sipunculans or peanut worms possess a body plan similar to that of annelid worms, but without segmentation. The body is peanut shaped and turgid when contracted. The body is divisible into a retractable introvert and thicker trunk. A mouth surrounded by feeding tentacles is located at the tip of the introvert. While Sipuncula contains only about 320 named species worldwide, it is here considered an important group in coral reefs because there are several species that actively erode coral skeletons. Several phascolosomid sipunculans are known to bore into corals. Scoffin et al. (1980) reported that species in the genera *Lithacrosiphon*, *Paraspidosiphon* (Aspidosiphoniformes), and *Phascolosoma* (Phascolosomiformes) were among the most important bioeroding taxa on a fringing reef in Barbados. Evidence from a study of sipunculid burrow linings suggests that the penetration of limestone rock is accomplished by both chemical and mechanical abrasion (Williams and Margolis 1974). Several species in the Class Sipunculida also nestle under coral rubble (e.g., *Golfingia* in the family Golfingiidae, and *Siphonosoma* and *Sipunculus* in the order Sipunculiformes) or live inside vacated gastropod shells (*Phascolion*, in the family Phascolionidae). Cutler (1995), Rice (1976), Rice and Macintyre (1982), and Schulze (2005) are good sources for further information on sipunculid biology and ecology.

Phylum Arthropoda
    Subphylum Crustacea
        Class Malacostraca
            Order Stomatopoda (*Gonodactylus, Haplosquilla, Lysiosquillina*)
            Order Decapoda
                Infraorder Caridea (*Alpheus, Synalpheus, Hippolyte, Periclimenes*)
                Infraorder Brachyura (*Domecia, Portunus, Tetralia, Trapezia*)
                Infraorder Anomura (*Aniculus, Calcinus, Petrolisthes*)
                Infraorder Palinura (*Biarctus, Panulirus, Scyllarus*)
            Order Mysida (*Anisomysis, Doxomysis, Mysidium*)
            Order Tanaidacea (*Apseudes, Leptochelia*)
            Order Isopoda (*Cymothoa, Elthusa*)
            Order Amphipoda (*Ampithoe, Gammaropsis, Hyale*)
        Class Maxillopoda
    Subclass Thecostraca
        Infraclass Cirripedia (*Balanus, Chthamalus, Lithotrya, Savignium*)
    Subclass Copepoda (*Acartia, Oithona, Paeonodes, Xarifia*)
    Class Ostracoda (*Bishopina, Caudites, Loxoconcha, Propontocypris*)

Nine major taxa of crustaceans (Crustacea) occur in coral communities. Stomatopods or mantis shrimps (Stomatopoda) are notable predators that live in burrows in soft sediments or cavities in loose and solid calcareous substrates. Stomatopods are raptorial carnivores, employing their subchelae to "spear" or "club" such prey as crustaceans and fishes. About 350 named species in 12 families exist globally with most occurring in tropical seas. Recently, the named stomatopods of Guam increased from 7 to 34 species (Ahyong and Erdmann 2003), suggesting that much basic systematic work remains in coral reef areas.

The decapods (Decapoda) include shrimps (Caridea), true crabs (Brachyura), hermit and porcelain crabs (Anomura), and spiny and Spanish lobsters (Palinura). The majority of species in these groups are vagile, but some are sedentary or restricted to particular coral colonies or sites within a colony. Some shrimps and swimming crabs spend part of their time in the water column. Lobsters often undergo extensive nocturnal foraging excursions, from sheltered reefs to grass beds and sand plains, and some hermit crabs (and land crabs) live along the shore and return to the sea only to release their hatching larvae.

Of the 13 families of tropical shallow-water shrimps listed by Bruce (1976), the Palaemonidae, Gnathophyllidae, Alpheidae, and Hippolytidae are particularly species rich, and occur as commensals on a great variety of invertebrate hosts. Hosts are usually large in relation to the size of the shrimp associates, and include species in the following phyla: Porifera, Cnidaria, Annelida, Mollusca, Arthropoda (crustaceans), Echinodermata, and Chordata (ascidians). Scleractinian corals host no fewer than 13 genera of shrimps in the subfamily Pontoniinae (Palaemonidae) alone. In addition, *Periclimenes* (Pontoniinae) contains 44 species that are commensal associates with sponges, cnidarians, a nudibranch, and all reef echinoderms. A more recent taxonomic treatment of the pontoniine shrimps, including observations on their ecology, is available in Bruce (1998). This study lists 48 species in 16 genera that are obligate symbionts of 23 coral genera, most commonly encountered on branching species of *Acropora* and genera in the family Pocilloporidae.

Bruce (1976) also noted that some shrimps (e.g., species of *Periclimenes*) form mutualistic relationships with reef fishes. Cleaner shrimps associate temporarily with a variety of fishes that visit specific reef sites or "cleaning stations" (Plate 1d). While initially regarded as a strong mutualism, with the cleaners benefiting from a food supply of ectoparasites and the fish hosts from the removal of parasites, some workers now regard this relationship as a one-sided commensalism (Poulin and Grutter 1996). Losey (1987) in particular concluded that cleaner shrimps (and cleaner fishes) are more accurately classified as "behavioral parasites." The cleaners benefit from feeding, exploiting the propensity of fish hosts to engage in tactile stimulation.

Another well-studied shrimp-fish relationship ubiquitous on coral reefs involves gobies and alpheid shrimps. Karplus (1987) recognized 70 goby species that formed mutualistic relationships with 13 species of *Alpheus*. Shrimps can be considered the host partners because they construct burrows in which both species take shelter. When outside of burrows, shrimps are typically in contact with gobies, which alert the shrimp of predators through tactile communication (e.g., tail flicks that are detected by the shrimp's antennae). A shrimp species in French Polynesia has been shown to prefer burrowing in a mixed sand-rubble substrate, thus reducing the chances of burrow collapse. This habitat choice in turn affects the distribution and abundance of its goby partner (Thompson 2004). The stability of this mutualism has been demonstrated recently by field studies involving the manipulation of goby immigration and predation on the population dynamics of this partnership (Thompson 2005).

Brachyuran crabs are abundant and diverse in coral communities, with hundreds of species belonging to numerous families. Some well-represented families include the spider crabs (Majidae), cancer crabs (Cancridae), parthenopod crabs (Parthenopidae), swimming crabs (Portunidae), grapsid crabs (Grapsidae), mud crabs (Xanthidae), trapezioid crabs (Domeciidae, Tetraliidae, and Trapeziidae), pea crabs (Pinnotheridae), and coral gall crabs (Cryptochiridae). [Cryptochiridae Paulson, 1875, has been shown to be a senior synonym of Hapalocarcinidae Calman, 1900 (Kropp and Manning 1985)]. Although crabs are generally free ranging, the majority are secretive and leave their refuges only when attracted by potential prey (e.g., dead, moribund, or live animals) or at night when foraging. Most portunids are aggressive carnivores and are likely to be seen on open reef surfaces.

Many crabs live symbiotically with a variety of metazoan hosts. The trapezioid crabs in the family Trapeziidae, represented by five genera (*Hexagonalia*, *Quadrella*, *Tetralia*, *Tetraloides* and *Trapezia*) and about 40 species, are obligate symbionts of cnidarians. They move among the branches of scleractinian corals and occur in association with their hosts across the Pacific and Indian Oceans (Castro 2000). *Trapezia* is the most speciose genus with 23 species (Castro 2003), all restricted to pocilloporid coral hosts (*Pocillopora*, *Seriatopora* and *Stylophora*). The eight presently known species of *Tetralia* and *Tetraloides* are restricted to Indo-West Pacific species of *Acropora*. These crabs feed on coral mucus, detritus and zooplankton captured by coral polyps, and fat bodies produced by the coral hosts (*Trapezia* mutualists of *Pocillopora*) (Patton 1976; Stimson 1990). *Quadrella* spp. and *Hexagonalia* spp. are associated with soft corals. Pea crabs and coral gall crabs are sedentary, the former parasitic or commensal with polychaetes, molluscs, or echinoderms, and the latter forming chambers in coral skeletons where the females are permanently imprisoned. There are 18 genera and about 30 named species of obligate coral gall crabs, with most present in the Indo-Pacific region (Kropp and Manning 1987; Kropp 1990). Gall crabs have been found on numerous species in at least ten scleractinian families, with most inhabiting zooxanthellate corals. These crabs feed on coral mucus, debris, and some species consume coral tissues. The latter are best regarded as parasites. Additional information on the systematics of all extant 6,793 valid species and subspecies of brachyurans can be found in Ng et al. (2008).

Hermit (Paguroidea) and porcelain (Porcellanidae) crabs are among the more common groups of anomurans found in coral communities. Most species of hermit crabs are omnivorous detritivores (Hazlett 1981), but two are known to scrape and ingest coral tissue (Glynn et al. 1972) and one species feeds on plankton by filtering water with densely setose second antennae (Schuhmacher 1977). Some hermit crabs are involved in symbiotic associations with sea anemones that exhibit complex behavior (Ross 1970). Porcellanid crabs are filter feeders and are commonly associated with sponges, sea anemones, and tube-dwelling polychaetes. Coral-associated porcellanids in the Moluccas and eastern Australia are noted in Haig (1979, 1987). Usually secretive during daylight hours, spiny and slipper lobsters (Palinura) typically forage widely over reefs at night. Palinuran lobsters are omnivorous scavengers, but also feed on live prey such as sea urchins (Sonnenholzner et al. 2009).

While the remaining seven higher-order taxa of crustaceans are small and usually inconspicuous on reefs, they are diverse, numerous, and often involved in a variety of trophic roles, for example, predation, parasitism, herbivory, suspension feeding, scavenging, detritivory, grazing on microorganism films, and serving as prey for larger consumers. Tanaids (Tanaidacea) are minute members of the benthos, commonly living in burrows or tubes associated with reef sediments and algae (Peyrot-Clausade 1977; Robichaux et al. 1981; Klumpp et al. 1988; Cruz-Rivera and Paul 2006). Feeding habits include scavenging, detritivory, suspension feeding, and predation. Mysids (Mysida) enter reef waters in the plankton and exist as resident reef populations. These mysids often form large swarms in reef lagoons, attaining densities of 0.5–1.5 million indiv. m^{-3} (Carleton and Hamner 1989). Mysids are mostly suspension feeders and some species are predators. Isopods (Isopoda), amphipods (Amphipoda), and ostracods (Ostracoda) engage in nearly all types of feeding, ranging from predation, parasitism, and suspension feeding to scavenging and detritivory. While most barnacles (Cirripedia) and copepods (Copepoda) are suspension feeders, some obligate symbionts of live corals engage in other types of feeding activities. No fewer than nine pyrgomatine barnacle genera totaling 45 species occur as symbionts in seven scleractinian suborders and in the hydrocoral genus *Millepora* (Fig. 2). Slightly more than one quarter ($n = 57$ species) of all known faviinad corals host a variety of barnacle species. The majority of species in the Dendrophylliina and Caryophilliina, 92.9% (13 of 14 species) and 62.5% (five of eight species), respectively, host barnacle symbionts.

Compared with barnacles, copepod symbioses involve several ordinal-level taxa, numerous families and species, and occur over a far wider array of invertebrate hosts. Approximately one third of the ~11,500 recognized species in the early 1990s are parasites or associates of fishes and invertebrates (nearly equally divided), especially in tropical regions (Humes 1994). Among four copepod orders, frequent invertebrate hosts include sponges, cnidarians, polychaete worms, sipunculans, molluscs, echinoderms, and ascidians (Gotto 1979). More copepod symbionts are found on cnidarians than any other invertebrate phylum. Over 2 decades ago, Humes (1985a) noted that the greatest number of copepod associates, 172 species in the order Poecilostomatoida, occurred on scleractinian corals (Humes 1985b). About one half of these belong to the family Xarifiidae. *Xarifa*, the most speciose genus, contains 75 species living on 93 species of corals present from the Red Sea eastward to New Caledonia. These taxa are named after the Xarifa Expedition, in recognition of S.A. Gerlach's discovery of this unique

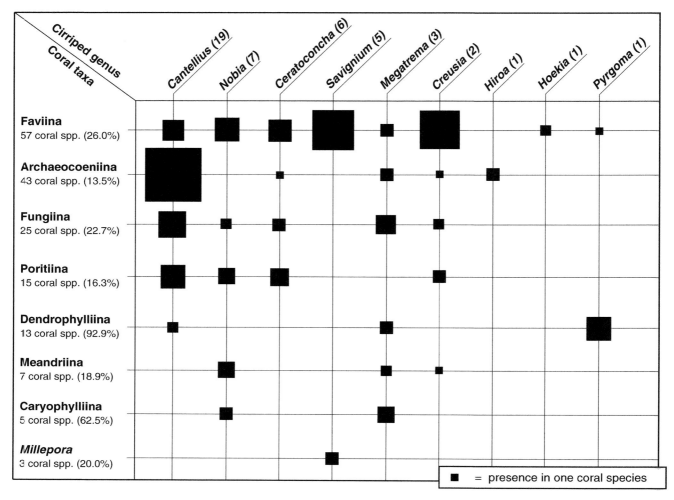

Fig. 2 Symbiotic occurrence of pyrgomatine barnacle genera among dominantly zooxanthellate scleractinian coral suborders and the hydrocoral genus *Millepora* (Entries are modified and updated from Ross and Newman (1973), from data in Ogawa and Matsuzuki (1992)). The number of species in each barnacle genus is denoted along the top row. Square block areas are proportional to the numbers of coral species harboring barnacles. The number of coral species hosting barnacles is denoted for each coral taxon. The percentage of the total number of coral species within each taxon with barnacle associates is also shown. Scleractinian nomenclature and species numbers follow Veron (2000). Number of *Millepora* species after Razak and Hoeksema (2003) and Lewis (1989)

symbiotic relationship. In the West Indies, numerous copepod associates belonging to the Asterocheridae and Corallovexiidae also occur in reef-building corals (Stock 1988). These were undiscovered until the 1970s and 1980s because of their endoparasitic lifestyle and difficulty of removal from their hosts. In general, the copepod associates of corals are minute (<3 mm in length), and the xarifiids typically have an elongate body, weakly defined segmentation, and reduced appendages (Fig. 3).

Phylum Mollusca[*]
    Subphylum Aculifera
        Class Aplacophora (*Epimenia*)
        Class Polyplacophora (*Acanthopleura, Chiton*)
    Subphylum Conchifera
        Class Bivalvia
            Subclass Pteriomorphia
                Order Mytiloida
                    Family Mytilidae (*Botula, Brachidontes, Lithophaga, Modiolus*)
                Order Pectinoida
                    Family Pectinidae (*Chlamys, Pecten, Pedum*)
            Subclass Heterodonta
                Order Veneroida
                    Family Tridacnidae (*Hippopus, Tridacna*)
                    Family Trapeziidae (*Coralliophaga, Trapezium*)
                    Family Petricolidae (*Petricola, Rupellaria*)
                Order Myoida
                    Superfamily Gastrochaenoidea
                        Family Gastrochaenidae (*Gastrochaena, Spengleria*)

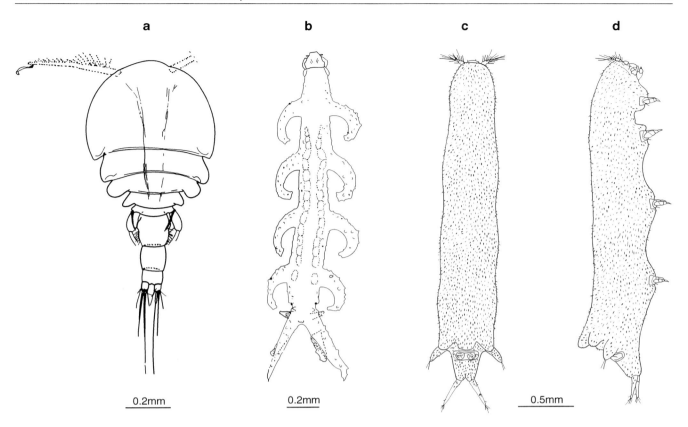

Fig. 3 Copepod associates of zooxanthellate scleractinian corals: (**a**) *Asteropontopsis faviae* (Asterocheridae), dorsal view of female from *Favia fragum*, Curaçao (From Stock 1987); (**b**) *Corallonoxia longicauda* (Corallovexiidae), dorsal view of female from *Meandrina meandrites*, Curaçao (From Stock 1975); (**c, d**) *Xarifia exserens* (Xarifiidae), dorsal (**c**) and lateral (**d**) views of female from *Galaxea fascicularis* (Humes 1985b)

    Superfamily Pholadomyoidea
      Family Pholadidae (*Aspidopholas, Jouannetia, Parapholis*)
Class Scaphopoda (*Cadulus, Dentalium*)
Class Cephalopoda
    Subclass Nautiloidea (*Allonautilus, Nautilus*)
    Subclass Coleoidea
      Order Sepioidea (*Sepia*)
      Order Teuthoidea (*Loligo, Sepioteuthis*)
      Order Octopoda (*Hapalochlaena, Octopus*)
Class Gastropoda
    Patellogastropoda
      Family Acmaeidae (*Acmaea*)
    Vetigastropoda
      Family Turbinidae (*Astraea*)
    Neritimorpha
      Family Neritidae (*Nerita, Neritina*)
    Caenogastropoda
      Family Vermetidae (*Dendropoma, Petaloconchus*)
      Family Cypraeidae (*Cypraea*)
      Family Ovulidae (*Cyphoma, Simnia*)
      Family Cassidae (*Cassis, Cypraecassis, Phalium*)
      Family Ranellidae (*Charonia, Cymatium*)
      Family Epitoniidae (*Epitonium*)
      Family Eulimidae (*Melanella*)
      Family Muricidae (*Coralliophila, Drupella, Magilus, Quoyula*)
      Family Conidae (*Conus*)
    Heterobranchia
    Opisthobranchia
      Order Sacoglossa
        Family Elysiidae (*Elysia*)
      Order Anaspidea
        Family Aplysiidae (*Dolabella*)
      Order Nudibranchia
        Family Doridomorphidae (*Doridomorpha*)
        Family Pinufiidae (*Pinufius*)
        Family Aeolidiidae (*Aeolidia, Spurilla*)
        Family Tergipedidae (*Cuthona, Phestilla*)
    Pulmonata (*Onchidium, Siphonaria*)

*The higher classification of molluscs is presently in a state of flux. We follow the classification scheme of Bouchet and Rocroi (2005) that updates traditional categories with recent cladistic analyses in an attempt to represent monophyletic relationships.

Molluscs are the most species rich of all marine invertebrate phyla with over 52,000 described species worldwide (Bouchet 2006). It is estimated that this number represents only about half of the living molluscan fauna. Nearly all molluscs associated with coral reefs share several synapomorphies, for example, a mantle with shell glands that secrete calcareous spicules, shell plates or shells, and a muscular foot that in cephalopods is modified into tentacles surrounding the mouth. A radula is present in herbivorous and carnivorous molluscs. Two subphyla, representing distinct grades of organization, are generally recognized, namely the Aculifera and Conchifera. The auculiferans comprise two classes, the shell-less Aplacophora with aragonitic spicules in the cuticle, and the polyplacophorans that possess multiple (usually eight) shell plates. Members of the Conchifera possess a single (primary) shell as adults or at some earlier stage of development.

Solenogasters are generally small vermiform molluscs that belong to the Aplacophora. About 200 species have been described, most living below the depths of tropical coral reefs. They are free living or epizoic, and are often found in association with various cnidarians. Some feeding observations were possible on a shallow-occurring solenogaster (*Epimenia arabica*) described from reefs in the Red Sea (Salvini-Plawen and Benayahu 1991). This species, which can attain 21 cm in body length, was found living among and feeding on an alcyonacean between 2 and 5 m depth.

Chitons (Polyplacophora), with eight shell plates, are represented by several species that are often abundant on reef flats and in other shallow reef zones. Their intense grazing on micro-filamentous algae leads to significant bioerosion of carbonate substrates.

Pelecypods (Bivalvia), often cryptic, are primarily suspension feeders with numerous species associated with coral reefs (Morton 1983). Bivalves that bore and nestle in corals are common in the Pteriomorphia and Heterodonta. Nestling bivalves, for example species of *Barbatia*, *Pedum*, and *Tridacna*, contribute in a small way to the material addition of reef frameworks and the binding of reef structures. Coral borers can exhibit a high diversity, as exemplified by the families and species involved in this activity in some regions. In one reef area sampled in southeastern Thailand, Valentich-Scott and Tongkerd (2008) documented the presence of 21 coral-boring species in five families. Representative genera included *Aspidopholas*, *Botula*, *Coralliophaga*, *Gastrochaena*, *Leiosolenus*, *Lithophaga*, *Petricola*, and *Spengleria*. Some reef bivalves engage in phototrophy, for example tridacnid clams host zooxanthellae that are harvested and utilized by the clams. Additionally, a nestling species of oyster (*Pedum spondyloideum*) has been shown to defend surrounding coral tissue from attack by *Acanthaster* (De Vantier and Endean 1988).

Tusk shells (Scaphopoda), with a tubular shell open at both ends, are micro-carnivores that burrow in sediment. Although we could not find a specific example of a scaphopod occurring in reef sediments, it is likely that the numerous records from shallow (<10 m) soft bottoms in coral reef regions would provide appropriate habitat conditions for members of this taxon (Steiner and Kabat 2004).

Cephalopods possess a sophisticated nervous system that is manifested in complex behaviors. Reef-associated species occur in all of the major taxa – chambered nautiluses (Nautiloidea), cuttlefishes (Sepioidea), squids (Teuthoidea), and octopuses (Octopoda). Only nautiluses have an externally chambered shell, but all cephalopods possess large, complex eyes, circumoral appendages (tentacles), a radula and beak, and a muscular funnel (siphon). Most non-nautiloid cephalopods possess internal shells.

The shelled nautiloids are a fossil-rich group that underwent a series of radiations during the Paleozoic era. While over 17,000 fossil species have been described, only two genera survive. There are four well-established species of *Nautilus* and possibly two species of *Allonautilus*, all present in the Indo-Pacific region (Jereb and Roper 2005). *Nautilus*, and likely *Allonautilus*, occur on steep forereef slopes, moving from deeper water (~200–500 m) after the sunset to 60–70 m and shallower where they prey opportunistically on crustaceans, polychaete worms, molluscs, and probably scavenge dead animals (Hanlon and Messenger 1996; Ward and Saunders 1997).

Cuttlefishes and squids are voracious carnivores that actively swim over reefs. Cuttlefishes occur on coral reefs in the Indo-Pacific, and squids are found in all major reef regions. They visually locate and feed on a variety of metazoan prey, including other cephalopods and fishes. Octopods are mostly benthic predators, often territorial and cryptic. Many new species, more than 150 from the Indo-Pacific alone, have been discovered but not yet described during the past 17 years (Norman and Hochberg 2005).

Gastropoda is the largest of all classes, with more than half of all living molluscan species. Some families and genera are especially diverse, for example 270 valid species of *Conus* and 141 species of *Cypraea* are presently recognized (SMEBD 2009). Gastropod species occupy all reef habitats and engage in a variety of feeding interactions: herbivory, predation (including corallivory), suspension feeding, deposit feeding, and parasitism (Kohn 1983). Herbivore grazing by species of *Acmaea*, *Astraea*, and *Nerita* promote reef bioerosion, and muricid corallivores such as *Coralliophila*, *Drupella* and *Magilus* consume coral tissues, sometimes in significant amounts. All 20 of the molluscan corallivores listed by Rotjan and Lewis (2008) are gastropods: 16 prosobranch species and four opisthobranchs. Some gastropod predators possess unique adaptations for subduing relatively large and agile prey species. For example, *Cassis tuberosa* can grip *Diadema* with its foot and then drill a hole through the test with its proboscis. Drilling is accomplished by the secretion of sulfuric acid and rasping by the radula. *Charonia tritonis*, the triton trumpet, pursues, captures, and consumes adult *Acanthaster*, and cone snails prey on a variety of invertebrates and fishes (Kohn and Nybakken 1975). The unique toxoglossate radula in cones is harpoon-like, situated at the end of a long proboscis. Cones can deliver a rapid strike, injecting a potent venom (neurotoxin) that quickly immobilizes prey (Shimek and Kohn 1981). The immobilized prey is then swallowed whole. Some muricid gastropods lack a radula but are, nonetheless, capable of feeding on prey such as scleractinian corals. *Coralliophila* accomplishes this by employing its muscular proboscis as a pump, probably with the aid of enzymatic secretions that loosen coral tissues (Ward 1965).

Phylum Echinodermata[*]
   Class Crinoidea
      Order Comatulida (*Comactinia, Comanthus, Nemaster, Tropiometra*)
   Class Asteroidea
      Order Valvatida (*Culcita, Linckia, Nidorellia, Oreaster, Pentaceraster*)
      Order Spinulosida (*Acanthaster*)
   Class Ophiuroidea
      Order Phrynophiurida (*Astroboa, Astrophyton, Ophiomyxa, Schizostella*)
      Order Ophiurida (*Ophiactis, Ophiocoma, Ophiolepis, Ophiothrix*)
   Class Echinoidea
      Order Cidaroida (*Eucidaris*)
      Order Diadematoida (*Astropyga, Diadema, Echinothrix*)
      Order Temnopleuroida (*Toxopneustes, Tripneustes*)
      Order Echinoida (*Echinometra, Echinostrephus, Heterocentrotus*)
      Order Holectypoida (*Echinoneus*)
      Order Clypeasteroida (*Clypeaster*)
      Order Spatangoida (*Metalia, Brissus, Meoma*)
   Class Holothuroidea
      Order Dendrochirotida (*Afrocucumis, Pentacta*)
      Order Aspidochirotida (*Actinopyga, Holothura, Stichopus, Thelenota*)
      Order Apodida (*Euapta, Synapta*)

[*]Classification after Hendler et al. (1995) and sources cited therein.

Members of the Echinodermata are exclusively marine and number about 7,000 described extant species in five classes. Nearly twice as many fossil species are known, distributed among about 16 classes. The number of species inhabiting coral reefs can be approximated from Hendler et al. (1995) and Pawson (1995) who estimated that ~180 and ~1,500 echinoderm species potentially occur on reefs in the Greater Caribbean and Indo-Pacific regions, respectively. This amounts to a conservative estimate of 1,680 species or about one quarter (24%) of the total echinoderm fauna. Four distinct synapomorphies neatly distinguish the members of this phylum and are largely responsible for the functional biology of echinoderms: (1) radially symmetrical five-part body organization of adults, (2) calcitic skeleton composed of numerous ossicles, (3) water vascular system, and (4) mutable collagenous tissue. All five classes of echinoderms are abundant on coral reefs and often contribute importantly to overall ecosystem biomass, with some species playing pivotal keystone roles in controlling coral community structure. Numerous studies have documented the strong effects of corallivore and herbivore consumers in this phylum.

About 600 species of living comatulid (non-stalked) crinoids are known (Hendler et al. 1995). Comatulid or feather star crinoids, the usual taxon inhabiting coral reefs, grasp the substrate with flexible jointed appendages (cirri) when resting or feeding. They are motile, usually cryptic during the day, and emerge on open surfaces at night. While generally sedentary, moving only short distances, some species are capable of swimming. *Oligometra serripinna*, a widespread Indo-West Pacific species, swims by flailing its arms and thus may quickly change its position (Clark 1976). Crinoids typically spread their arms perpendicular to ambient currents to entrap suspended particulate organic matter and plankton, which is transported to the mouth along ciliary mucoid tracks in ambulacral grooves. Local species richness and population densities can reach high levels on reefs bordering continental margins in the Indo-West Pacific and Caribbean regions (Birkeland 1989a). In waters off Eilat, *Lamprometra klunzingeri* was found at densities up to 70 indiv. m^{-2} (Fishelson 1968). Crinoids are absent from several outlying central Pacific reefs (e.g., Hawaiian, Society, Line Islands) and eastern Pacific reefs. In the western Atlantic, the richest areas are the Straits of Florida, Bahamas, and Caribbean Sea with about 12 species potentially inhabiting reefs in this region (Meyer et al. 1978). Comatulid crinoids host numerous symbiotic taxa, including protists (dinoflagellates, foraminifers, ciliates), sponges, cnidarians (hydroids, hydrocorals), platyhelminthes, polychaete annelids (four families, including myzostomids that are specialized crinoid parasites), gastropods, crustaceans (carideans, anomurans, brachyurans, isopods, amphipods, copepods, ostracods, cirripedes), bryozoans, echinoderms, and chordates (tunicates and gobiesocid fishes) (e.g., Meyer and Ausich 1983; Zmarzly 1984; Fabricius and Dale 1993; Huang et al. 2005).

Approximately, 1,800 species of living asteroids or sea stars are known, present at all ocean depths and latitudes. The highest diversity on coral reefs occurs in the Indo-West Pacific and Red Sea regions with relatively few species on Caribbean reefs (Hendler et al. 1995). Due in large measure to the calcareous skeletal plates and mutable connective tissue, sea stars can quickly transform their bodies from a rigid to a flexible state, which is of benefit in sheltering and maneuvering to feed (Birkeland 1989b). Juvenile sea stars and the smaller species tend to be cryptic. Large foraging species (e.g., *Acanthaster*, *Culcita*) are usually present on reef surfaces where potential prey are located. Sea stars possess a diversity of feeding habits, involving carnivory, cannibalism, herbivory, scavenging, deposit feeding, and suspension feeding (Sloan 1980; Jangoux and Lawrence 1982). Those feeding on sessile animals, for example solitary molluscs or colonial cnidarians, typically extrude their stomach from the disk and envelope the prey. The nine valid species of asteroid corallivores listed in Rotjan and Lewis (2008) are characteristically extra-oral feeders. The most significant corallivore species, members of the genera *Acanthaster* and *Culcita*, occur on most Indo-Pacific coral reefs. None of these species or other asteroid corallivores are known in the wider Caribbean region. Sea star population densities are generally not high on reefs, but some species (members of *Pentaceraster* and *Oreaster*) sometimes form large aggregations, and *Acanthaster planci* outbreaks may number into the tens of thousands of individuals per hectare. Relative abundances of asteroids (plus echinoids and holothuroids) are noted for six west Pacific coral reef sites in Pearse (2009).

Living brittle stars and basket stars (Ophiuroidea) number about 2,000 known species, and are well represented on coral reefs. Most species are relatively small and secretive, living on various invertebrate hosts, in live and dead coral, reef rubble and sedimentary deposits. The maximum dimensions of small species that occur as epizoic symbionts are in the millimeter size range. Free-living cryptic species are frequently centimeters in size, and some basket stars can attain over a meter in diameter when fully expanded and feeding. While generally inconspicuous, brittle star population densities on reefs are commonly 20–40 indiv. m^{-2}, reaching abundances 10–100 times higher in soft sediments (Hendler et al. 1995). In the *Heliopora* zone at Enewetak, *Ophiocoma erinaceus* occurred at estimated densities of 150–15,000 indiv. m^{-2} (Chartock 1983). Brittle stars are frequently attacked by fishes and various invertebrates. Several species exhibit behaviors and defensive properties that minimize predation events: (a) self-mutilation, that is, the autotomy of arms and other body parts; (b) distastefulness and luminescence; (c) rapid escape response; and (d) "stop-motion reflex" that is triggered by approaching predators (Hendler et al. 1995). Ophiuroid feeding behaviors are highly diverse, involving the uptake of dissolved and particulate organic matter, the capture of small metazoans, and particle uptake from sediments. On eastern Pacific reefs in Panamá, brittle stars emerge from pocilloporid coral frameworks at night and extend their arms in the water column to capture food. Basket stars with numerous arm bifurcations (Gorgonocephalidae) are nocturnal suspension feeders (Clark 1976). They form tight interwoven knots during daylight hours and unfurl at night when feeding. When assuming a feeding posture, the arm branches are unfolded and form a parabolic filtration fan oriented to intercept moderately strong currents. The concave side of the fan faces into the current in both basket stars and crinoids. Plankters are caught and held by the ultimate fine arm branches and later ingested when the basket star knots up during the day (Meyer and Lane 1976).

Echinoids (sea urchins and sand dollars) number about 900 described species although less than one tenth of these are present in any given reef region. An indication of the importance of echinoids on coral reefs can be judged from Birkeland's (1989a) treatment, in which he devoted about 50% of his general echinoderm review to the influence of echinoids on coral communities. The ubiquity, generally high abundances, destructive feeding activities, and limestone excavating (bioerosive) effects are in large measure responsible for this. Most "regular" echinoids are slow moving or sedentary, largely confined to particular reef sites (zones) or limestone depressions and burrows. "Irregular" echinoids, with elongate tests (e.g., sea biscuits and heart urchins), usually bury themselves in reef sediments. Some sea urchin species exhibit high local abundances, for example *Diadema mexicanum* with maximum densities of 50–150 indiv. m^{-2} and *Echinometra lucunter* with densities of 100 indiv. m^{-2} (Table 2). Sea urchins can exact enormous impacts on coral community structure at such elevated abundances, a combination of high recruitment, survivorship, and availability of trophic resources. In most sea urchins, feeding depends on Aristotle's lantern, a complex masticatory apparatus of muscles and five calcareous, protractible teeth. This is a versatile feeding organ that effectively removes fleshy and skeleton-bearing benthic prey and many species can penetrate and erode limestone substrates.

Hendler et al. (1995) noted that there are about 1,250 known sea cucumber (Holothuria) species, and the majority of these occur in shallow tropical waters. The predominant holothurians on coral reefs belong to the orders Aspidochirotida and Apodida. Dendrochirotids are more typical of temperate waters, perhaps because of a greater supply of their plankton food source at higher latitudes (Birkeland 1989a). Reef species generally occur in low-energy microhabitats in sediments of high organic content. An exception is *Actinopyga mauritiana*, an Indo-West Pacific species that is found in high energy surge channels (Birkeland 1989a). *Holothuria atra*, one of the more common large central Pacific holothurians, attains densities of 5–35 indiv. m^{-2} on the reef flat at Enewetak Atoll, Marshall Islands. Many holothurians are toxic, most likely an evolutionary response to intense predation on coral reefs (Bakus et al. 1986). Holothurians are often exposed on shallow sand patches; however, many species also are concealed during daylight hours, emerging to forage only at night. The majority of reef species are deposit feeders, employing 10–30 retractile feeding tentacles that surround the mouth. Some species are highly selective with ingested food including organic detritus, microorganisms, and meiofauna. Like many other echinoderm taxa, sea cucumbers host

Table 2 The abundance, biomass, and diversity of selected invertebrate taxa

Taxa	Location	Substrate	Abundance[a]	Biomass	Reference
Porifera					
Multiple species	U.S. Virgin Iss.	Reef	$\bar{x} = 0.89$ m^{-2}	na	Beets and Lewand (1986)
Actiniaria					
Multiple species	U.S. Virgin Iss.	Reef	$\bar{x} = 0.88$ m^{-2}	na	Beets and Lewand (1986)
Gorgonacea					
Gorgonia spp.	U.S. Virgin Iss.	Reef	$\bar{x} = 0.25$ m^{-2}	na	Beets and Lewand (1986)
Polychaeta					
Multiple species	Kaneohe Bay, Hawaii	Endolithic coral rock	$\bar{x} = 81{,}200$ m^{-2} SD = 29,200	$\bar{x} = 52.9$ g m^{-2} [c] SD = 35	Brock and Brock (1977)
Multiple species	Indian Ocean	Endolithic coral rock	$\bar{x} = 58{,}200$ m^{-2} SD = 34,300	na	Kohn and Lloyd (1973)
Multiple species	Aldabra Atoll	Coral rock	$\bar{x} = 30.23$ l^{-1} SE = 3.02	na	Brander et al. (1971)
Multiple species	Watamu, Kenya	Coral rock	$\bar{x} = 9.49$ l^{-1} SE = 2.31	na	Brander et al. (1971)
Multiple species	Heron Island, GBR	*Pocillopora damicornis*	1,441 in 4.7 kg coral colony	na	Grassle (1973)
Multiple species	Enewetak Atoll	Coral rock (endolithic)	$\bar{x} = 87{,}925$ m^{-2}	$\bar{x} = 8.3$ g m^{-2} [c]	Bailey-Brock et al. (1980)
Multiple species	Enewetak Atoll	Coral rock (epilithic)	$\bar{x} = 61{,}213$ m^{-2}	$\bar{x} = 21.0$ g m^{-2} [c]	Bailey-Brock et al. (1980)
Polychaeta					
Multiple species	Davies Reef, GBR	Inside damselfish territories	7,041 m^{-2}	0.58 g m^{-2} [c]	Klumpp et al. (1988)
Multiple species	Davies Reef, GBR	Outside damselfish territories	2,809 m^{-2}	0.24 g m^{-2} [c]	Klumpp et al. (1988)
Pherecardia striata	Panamá, eastern Pacific	Carbonate framework	Median per site = 0–380 m^{-2}	na	Glynn (1984)
Sipuncula					
Multiple species	Enewetak Atoll	Coral rock	58 in 7 coral colonies	na	Highsmith (1981)
Multiple species	Carrie Bow Cay, Belize	*Porites astreoides* (reef crest)	300–1,200 m^{-2}	na	Rice and Macintyre (1982)
Crustacea					
Stomatopoda	Aldabra Atoll	Coral rock	$\bar{x} = 4.77$ l^{-1} SE = 1.12	na	Brander et al. (1971)
Stomatopoda	Watamu, Kenya	Coral rock	$\bar{x} = 5.92$ l^{-1} SE = 1.46	na	Brander et al. (1971)
Trizopagurus magnificus	Señora Islet, Pearl Iss., Panamá	Peripheral branches of *Pocillopora* spp.	13–54 m^{-2}	na	Glynn et al. (1972)
Amphipoda	Davies Reef, GBR	Inside damselfish territories	8,390 m^{-2}	0.37 g m^{-2} [c]	Klumpp et al. (1988)
Amphipoda	Davies Reef, GBR	Outside damselfish territories	1,137 m^{-2}	0.08 g m^{-2} [c]	Klumpp et al. (1988)
Mollusca					
Multiple species	Davies Reef, GBR	Inside damselfish territories	2,738 m^{-2}	0.27 g m^{-2} [c]	Klumpp et al. (1988)
Mollusca					
Multiple species	Davies Reef, GBR	Outside damselfish territories	908 m^{-2}	0.30 g m^{-2} [c]	Klumpp et al. (1988)
Cerithiidae	Enewetak Atoll	na	Up to 800 m^{-2}	na	Kohn (1987)
Predatory gastropods	Enewetak Atoll,	Windward platform	$\bar{x} = 5.2$ m^{-2}	na	Kohn (1987)
Octopus briareus	U.S. Virgin Iss.	Reef	$\bar{x} = 0.01$ m^{-2}		Beets and Lewand (1986)
Echinodermata					
Multiple species	Aldabra Atoll	Coral rock	0.4 l^{-1}	na	Brander et al. (1971)
Multiple species	U.S. Virgin Iss.	Reef	$\bar{x} = 5.99$ m^{-2}	na	Beets and Lewand (1986)
Diadema antillarum	Barbados (pre '83 mortality)	Reef (fringing)	$\bar{x} = 9.26$ m^{-2}	na	Hunte et al. (1986)

(continued)

Table 2 (continued)

Taxa	Location	Substrate	Abundance[a]	Biomass	Reference
D. antillarum	Barbados (post '83 mortality)	Reef (fringing)	$\bar{x} = 0.72$ m^{-2}	na	Hunte et al. (1986)
Echinometra lucunter	Virgin Iss.	Reef (algal ridge)	100 m^{-2}	na	Ogden (1977)
Primarily *D. mexicanum*	Uva Island, Panamá (lower seaward slope)	Inside damselfish territory	10–40 m^{-2}	na	Glynn (1988a)
Primarily *D. mexicanum*	Uva Island, Panamá (lower seaward slope)	Outside damselfish territory	50–150 m^{-2}	na	Glynn (1988a)
Holothuria atra	Enewetak Atoll	Reef (reef flat)	5–35 m^{-2}	na	Bakus (1973)
Tunicata					
Multiple species	U.S. Virgin Iss.	Reef	$\bar{x} = 0.74$ m^{-2}	na	Beets and Lewand (1986)

Key: SD, standard deviation; na, not available; GBR, Great Barrier Reef, Australia; SE, standard error of the mean

[a] Abundances are indicated as numbers of individuals per area or volume unless otherwise noted

[b] Ash free dry weight

[c] Dry weight

diverse symbionts, namely protistans, flatworms, polychaetes, crustaceans (copepods, decapods), pycnogonids, and molluscs (gastropods, bivalves).

Relevant sources treating coral reef echinoderms are Bakus (1973), Clark (1976), and Birkeland (1989a). Additional echinoderm studies relating to coral reefs can be found in the series of papers edited by Jangoux and Lawrence (1983–2001), and the proceedings of various colloquia and conferences (e.g., Jangoux 1980; De Ridder et al. 1990; Féral and David 2003; Heinzeller and Nebelsick 2004). Hyman's (1955) classic treatise offers a wealth of ecological information, especially regarding the older literature. Several studies on Atlantic and eastern Pacific reef echinoderms were published recently under the editorship of Alvarado and Cortés (2005, 2008).

Phylum Chordata
    Subphylum Urochordata
        Class Ascidiacea (*Clavelina, Diplosoma, Ecteinascidia, Trididemnum*)
        Class Thaliacea (*Doliolum, Salpa*)
        Class Appendicularia (*Fritillaria, Oikopleura*)
    Subphylum Cephalochordata (*Asymmetron, Branchiostoma, Epigonichthys*)

The ascidians or sea squirts (Ascidiacea), members of the Subphylum Urochordata, are the only invertebrate chordates that are generally diverse and abundant on global coral reefs. The estimated number of described ascidian species is 4,900 (Bouchet 2006); the species richness of the remaining two urochordate classes (Thaliacea and Appendicularia) has not been evaluated, but Pechenik (2005) estimates that at least 90% of all described urochordates are ascidian species. Most ascidians occur at higher latitudes and at greater depths than on coral reefs. Numerous tropical species, however, are associated with coral reef communities worldwide, on firm substrates in both cryptic and epibenthic habitats (Kott 1980).

Ascidians occur as solitary individuals, as social aggregations, or as colonies, the latter two a result of asexual budding. Colonies can reach up to 50 cm in greatest dimension. Most species are suspension feeders. Some tunicates, especially several didemnid species, form apparently obligate symbiotic associations with the prokaryotic alga genus *Prochloron* (Kott 1982; Kott et al. 1984; Lewin and Cheng 1989). The algae are usually embedded in the test or occur in the cloaca of colonial species. Hirose et al. (1996) found that *Prochloron* can also occur intracellularly in the mesenchymal cells of the test in *Lissoclinum*, a western Pacific ascidian. In some species, photosynthetically fixed carbon is translocated to the tunicate host (Pardy and Lewin 1981; Griffiths and Thinh 1983). Factors contributing to the success of *Trididemnum solidum* on Caribbean reefs (Bak et al. 1981, 1996), and probably to other ascidian species in the Indo-Pacific, are: (1) epibenthic dominance of colonies due to asexual division, budding, fusion, and rapid growth; (2) effective competition through overgrowth of non-ascidian benthos and possible allelochemical effects; (3) low predation pressure and high regeneration capacity; (4) prolonged life span and iteroparous larval production. The translocation of photosynthate probably contributes to accelerated growth (Sybesma et al. 1981) and antipredatory chemical defenses limit mortality (Stoecker 1980).

Thaliaceans (salps) and appendicularians (larvaceans) are often present in plankton entering reefs (Sorokin 1990). Appendicularians, especially the genus *Oikopleura*, occur consistently in reef waters (Glynn 1973; Porter et al. 1977).

3.2 Minor Taxa

Due to the diversity of lifestyles, the reef-associated species of 24 minor invertebrate phyla cannot be categorized neatly in terms of their habitat niches or trophic interactions. Cognizant of this disparity many, if not the majority of species, can be assigned to four broad trophic groups, namely, (1) epibenthic consumers, (2) infaunal consumers, (3) parasites, and (4) microphagous meiofauna.

3.2.1 Epibenthic Consumers

Nine minor phyla contain consumer species that feed epibenthically on reefs: Placozoa, Ctenophora, Acoelomorpha, Nemertea, Entoprocta (=Kamptozoa), Pycnogonida, Ectoprocta (=Bryozoa), Brachiopoda, Hemichordata (Pterobranchia), and Chaetognatha. The feeding habits in these groups are highly varied, ranging from (a) grazing on algal films; (b) suspension feeding on plankton and particulate organic matter; (c) feeding on coral mucus and polyps; (d) feeding on invertebrates, including moribund and dead tissues; (e) ingesting detritus; to (f) suctorial feeding on algae and microinvertebrates. Information as to the main diets of species of Placozoa, Ctenophora, and Acoelomorpha is limited.

3.2.2 Infaunal Consumers

The phyla Priapula, Echiura, Phoronida, and Hemichordata (Enteropneusta) are typically associated with reef sediments. Fewer than 20 species of priapulids are known. Since they range in size from 0.5 mm to 20 cm, the smaller species would best be classified with the meiofauna. Priapulids are predominantly predatory, feeding on invertebrates, but one meiofaunal species that inhabits reef sediments (*Tubiluchus corallicola*) consumes organic detritus (van der Land 1970). Most known echiurans, phoronids, and hemichordates are suspension feeders, but an abundant echiuran (*Anelassorhynchus* sp.) in reef sediments in the eastern Pacific is a deposit feeder. Phoronids occupy permanent chitinous tubes and tend to stabilize soft sediments where they are abundant. Several of the lesser-known crustaceans, for example cephalocarids (López-Cánovas and Lalana 2001), thalassinideans (Suchanek 1983), cumaceans (Emery 1968; Alldredge and King 1977), and nebaliaceans (Modlin 1996) also occupy reef sediments where they feed on particulate organic matter within sediments or by filter feeding at the sediment–water interface. Often present in meiofauna collections, the adult stages of the majority of these taxa generally exceed 0.5 mm in greatest body dimension. The larger species, especially the mud shrimp (Thalassinidea) that construct extensive burrow holes, are active in bioturbation (Suchanek 1983; Rowden and Jones 1993).

3.2.3 Parasites

Four phyla are notable for their parasitic members, which often infect coral reef species. Rombozoans number only about 65 species and all parasitize the nephridia and associated structures of cephalopods, particularly cuttlefishes and octopuses (Hochberg 1982, 1983). All four classes of Platyhelminthes – Turbellaria (turbellarians), Monogenea (gill worms), Trematoda (flukes) (Plate 1e), and Cestoda (tapeworms) – possess parasites that infect reef invertebrates and fishes. The literature on fish parasites is especially rich because studies have mainly focused on this commercially important group (e.g., Dyer et al. 1985; Williams et al. 1996). Larvae of the Nematophora, horse hair or Gordian worms, parasitize decapod crustaceans. Both larval and adult stages of Acanthocephala, thorny-headed worms, occur as intestinal parasites in reef fishes.

3.2.4 Microphagous Microbenthos and Meiofauna

Minute metazoans, generally in the size range 0.1–0.5 mm (greatest body dimension), inhabit reef sediments and firm surfaces associated with turf algae and other benthic macrobiota. These small organisms belong in large part to the phyla Rotifera, Gastrotricha, Kinorhyncha, Nemata (=Nematoda), Gnathostomulida, Loricifera, and Tardigrada. Some of the larger protistans, such as ciliates, are also members of the meiofauna. Polychaete worms, micromolluscs, and copepods often contribute importantly to the meiofauna of reef sediments (Fenchel 1978; Coull 1990). It is necessary to add to this list as well the larvae and recently settled early life history stages of numerous macrobiota, which temporarily share these fine-scale microhabitats with the permanent meiofauna. Suspension feeders are relatively uncommon, particularly where sediment movement and abrasion occur. The feeding behaviors of the majority of the meiofauna can be broadly categorized as microphagous, which includes feeding on bacteria, diatoms, other benthic microalgae, protistans such as foraminifers and dinoflagellates, and particulate organic matter. Although virtually invisible, the meiofauna is not only important in nutrient regeneration, but also as a food source for larger metazoans (Fenchel 1978; Coull 1990).

4 Invertebrate Reef Habitats

Reef invertebrates can be broadly divided into those that are exposed, and those that remain hidden, the cryptic fauna. Many metazoans, however, are visible or invisible depending on the tidal phase, time of day, or developmental stage. For example, on shallow reef zones at low tide, suspension feeders will retract into their tubes and swimmers will retreat into reef recesses. During nocturnal hours, many species that live hidden in the reef during the day will emerge to feed on open surfaces or in the water column. And over the longer term, as early developmental stages mature, species will migrate from frameworks and epibenthic refugia to more open habitats.

Non-hidden, conspicuous taxa can be further divided into benthic and pelagic lifestyles. Pelagic reef invertebrates include nektonic species that actively control their movements. While relatively uncommon in reef ecosystems, examples of invertebrate taxa that utilize this lifestyle are primarily cephalopods, including squids (Teuthoidea), cuttlefishes (Sepioidea), and nautiluses (Nautilidae). More common are invertebrate reef plankton species, which live primarily in the water column but do not exhibit pronounced control over their movement. Examples of reef holoplankton (planktonic throughout their life cycle) include all manner of crustaceans (Copepoda, Mysida, Euphausiacea), chaetognaths (Chaetognatha), cnidarians (Scyphozoa, Hydrozoa), pteropod molluscs (Thecosomata, Gymnosomata), and chordate salps (Thaliacea). It should be noted that many of these planktonic taxa may exhibit degrees of directed movement, including vertical migration as well as bursts of swimming over short distances.

Exposed benthic fauna (epibenthic species) can be further divided into motile and sessile taxa. Motile invertebrate epifauna on coral reefs may often be conspicuous and have consequently evolved complex defensive mechanisms for dealing with high predation pressure. These taxa include echinoids with long venomous spines, gastropod molluscs with hard calcitic shells, as well as a variety of holothurian, opisthobranch, and polyclad flatworm species that are rendered unpalatable by the production or incorporation of various toxic compounds. Sessile epibenthic fauna utilize similar mechanisms to avoid predation. They exhibit a variety of growth forms ranging from encrusting and laminar to complex branching and mounding morphologies. Examples of these taxa include hydrozoan and all anthozoan cnidarians, poriferans, bryozoans, foraminiferans, tunicates, as well as sedentary polychaetes (Serpulidae, Sabellidae).

Cryptofauna (sensu Bakus 1966) otherwise known as coelobites (Ginsburg and Schroeder 1973), or sciaphiles (Laborel 1960) remain hidden throughout the majority of their lives. Composed primarily of invertebrate taxa, this community is thought to have a biomass greater than or equal to epibenthic communities (Ginsburg 1983). Kobluk (1988) divides this community into the sessile, vagrant, and endolithic cryptos. Hutchings (1983) adopts the terms "true borers" and "opportunistic" species to distinguish between bioeroding and epilithic forms. Another detailed classification scheme is presented by Ginsburg (1983) who divides the cryptic community into six categories based on what he terms "mode of life." His categories, with selected invertebrate reef fauna, are:

1. Encrusting – bryozoans, foraminiferans, poriferans, scleractinians, tunicates.
2. Attached – bivalves, crinoids, gorgonians, poriferans, scleractinians, tunicates.
3. Boring – annelids, bivalves, carideans, cirripedes, poriferans, sipunculans.
4. Burrowing – annelids, brachyurans, carideans.
5. Vagile – decapods, gastropods, ophiuroids.
6. Nektonic, planktonic – brachyurans, carideans, mysids, cephalopods.

4.1 The Cryptic Reef Habitat

The complex, anastomosing cryptic habitat is extensive. In many coral reefs, the surface area of habitable cryptic spaces is greater than that of the surface (Jackson et al. 1971; Buss and Jackson 1979; Logan et al. 1984; Richter and Wunsch 1999). The size of the cryptic habitat is poorly understood and highly variable, dependent on the locality, hydrographic conditions, and species composition of framework structures. Estimates of the proportion of cryptic reef volume are 30–50% (Bermuda, patch reefs; Garrett et al. 1971), >50% (Bonaire, leeward reefs; Kobluk and van Soest 1989), and 75–90% (general estimates; Ginsburg 1983).

4.1.1 Classification

The cryptic reef habitat is diverse, dependent on scale and the varying morphology of those structural species responsible for its construction (Ginsburg 1983). Various authors have used different conventions to subdivide the cryptic reef habitat based upon a specific crypt's size and structure (Garrett et al. 1971; Ginsburg 1983; Fagerstrom 1987; Kobluk 1988). According to Ginsburg (1983), the cryptic habitat can be divided into intraskeletal and interskeletal cavities, borings, and interparticle cavities.

Intraskeletal cavities are those recesses and voids, naturally occurring within the skeleton of various framework taxa. The size and shape of intraskeletal cavities may range from tiny empty corallites in the skeleton of a scleractinian coral to the relatively cavernous volume of an abandoned mollusc shell. Organisms occupying these voids include most coral reef taxa ranging from cryptic polychaetes to large octopods.

Interskeletal cavities (referred to by Ginsburg as growth, framework, and shelter cavities) consist of empty spaces between dead or living structural taxa. The shape of these crypts is dependent on the taxa that compose them and upon the various taphonomic processes that have subsequently altered their form. Metazoan communities occupying interskeletal crypts are often highly diverse. Enochs and Hockensmith (2009) observed over 117 taxa (crustaceans, molluscs, echinoderms, polychaetes, sipunculans) occupying the sheltering branches of living *Pocillopora damicornis*.

Cavities created by endolithic bioeroders are often occupied by a suite of nonboring taxa. In the eastern Pacific, xanthid crabs, ophiuroids, as well as numerous species of polychaetes can be found within the bore holes left behind by dead lithophage bivalves. Similar observations by Hutchings (1986) and McCloskey (1970) led them to conclude that boring species are of great importance in determining cryptofaunal community structure. Boring species sculpt solid blocks of carbonate into highly complex substrates suitable for occupation by opportunistic species.

Interparticle cavities consist of voids between sediment grains. A diverse assemblage of invertebrates of varying size are present in the sediments on and surrounding reefs as well as sediments within cavities and underlying coral frameworks. Depending on body dimensions, these organisms comprise the meiofauna (62 μm–0.5 mm) or macrofauna (≥1 mm). It should be noted that reef sediments are often created due to biological erosive and taphonomic processes and therefore these communities may also be indirectly dependent on boring/eroding cryptofauna.

4.1.2 Cryptic Environment

The cryptic environment is different than that of the exposed reef in that its three-dimensional framework structure provides shelter to the organisms living there. This shelter may offer a greater degree of environmental stability in contrast to the ever-changing physical and chemical conditions on the reef surface.

Shelter from intense wave action has been found to have widespread effects on cryptic organism abundances. Reduced water movement allows greater access to suspended organic matter. This in turn provides an ideal habitat for filter feeding organisms (Wood 1999). Suspended matter eventually settles out in calm sheltered environments and leaves nutrient-rich deposits that are utilized by deposit feeders. This process has been used to explain reduced biomass and biodiversity of communities living within crevices too sheltered to experience the deposition of suspended organic matter (Fagerstrom 1987).

Cryptic environments receive significantly less visible light (0.0004–0.06% of surface illumination; Ginsburg 1983; Logan 1981), experience spectral filtering of incident light (Kobluk 1988), and receive less ultraviolet radiation than exposed reef surfaces (Jokiel 1980). This affects the associated faunal community in a variety of ways. The absence of light restricts algae and other phototrophs, yet it is inviting to heterotrophic invertebrates that are accustomed to lower light conditions (Garrett et al. 1971; Ginsburg 1983; Logan et al. 1984; Fagerstrom 1987; Kobluk 1988).

Another important characteristic of the cryptic environment is the availability of shelter from surface-living predators. Some studies have suggested that predation pressure is lower in the cryptic environment (Jackson and Buss 1975). The concept of a direct relationship between shelter and prey abundance is not completely clear. Certain predators, such as gonodactylid stomatopods, have also shown higher abundances correlated with increasing availability of cryptic shelter (Steger 1987). Analysis of the cryptic fish population of Uva Reef, Panamá revealed that all species collected were carnivores (Glynn 2006). The presence of these cryptic predators may influence the recruitment of invertebrate prey species into the same environment.

4.2 Habitat Providers

For the purposes of this review, the term "habitat provider" refers to a species offering structure or substrate that another species lives in or on respectively. The habitat provider may confer a variety of benefits to its associates including physical shelter from predation, camouflage, stability, range expansion, and/or advantageous positioning. Habitat providers are not utilized exclusively as food sources by their associates. Admittedly, reef metazoan interactions are not always "clear-cut" and ambiguities may arise. Obligate symbionts may obtain shelter inside their host or utilize their host's exteriors as substrates. Their hosts are herein considered as habitats. Non-obligate symbionts may utilize multiple microhabitats, including those present on a single host species. It may be argued that habitat interactions involving two or more living individuals are always symbiotic.

4.2.1 Porifera

The number of studies concerning invertebrates associated with sponge taxa is most likely second only to that of scleractinians. Sponges provide a complex habitat and shelter from predation for organisms living within their extensive canal networks. Pearse (1934) recorded as many 17,128 animals (22 species) living within a single loggerhead sponge (*Spheciospongia vesparia*). Sponge specimens collected from the Dry Tortugas averaged 0.16 associates cm^{-3} and those from Bimini contained an average of 0.035 indiv. cm^{-3} (Pearse 1950). Associates included carideans, brachyurans, a porcelanid crab, stomatopods, amphipods, fishes, bivalves, gastropods, an annelid, an anemone, holothurians, and ophiuroids. Greater numbers of taxa were observed in other sponge species in this same study. Closer examination of sponge tissues may yield abundances of invertebrate metazoans far in excess of those recorded by Pearse. The minute polychaete worm *Haplosyllis spongicola* has been found inhabiting the reef sponge *Theonella swinhoei* at mean densities of 73 indiv. cm^{-3} (Magnino et al. 1999).

Some interactions between sponges and their associates may be termed mutualistic. The brittle star *Ophiothrix lineata* feeds on detritus on the outer surface of *Callyspongia vaginalis*, effectively cleaning the inhalant canals of its host (Hendler 1984). Feeding behavior was only observed during the night; *O. lineata* remained hidden within the sponges' interstices during the day, presumably to avoid diurnal invertivores. Shelter within sponge tissues may also provide the, albeit counterintuitive, benefit of facilitating dispersal. Richards et al. (2007) have postulated that sponges sheltering amphipod associates may frequently become dislodged and swept across large distances by currents. This would effectively aid in the dispersal of the amphipods and increase genetic connectivity between distant, otherwise separated populations of sponge associates. Associations with sponges may have potentially adverse effects as well. Duffy (1992), for example, has shown that alpheid shrimps associated with sponges are more frequently parasitized by epicaridean isopods. For a comprehensive review of inquiline associates of poriferans in all reef habitats, see Wulff (2006).

4.2.2 Scleractinia

Scleractinian corals provide habitat for most reef-associated biota. They may provide permanent shelter to excavating and boring endolithic (e.g., sponges, lithophage bivalves, polychaetes, sipunculans) or sessile cryptic species (e.g., sponges, bryozoans, tunicates). Other cryptic motile taxa (e.g., annelids, crustaceans, molluscs, echinoderms, fishes) may intermittently occupy reef coral interstitial spaces for both shelter and feeding purposes. Planktonic invertebrates have been observed to undergo diel vertical migration into and out of coral frameworks (Porter and Porter 1977). A variety of primarily nektonic vertebrates (fishes and turtles) may forage within reef coral interstices, feeding on the rich supply of invertebrates available there.

The morphology of some coral taxa strongly affects the abundance and number of crustacean species associates. Vytopil and Willis (2001) found that tightly branching acroporid colonies supported higher abundances and greater species richness of *Tetralia* crabs and palaemonid shrimps than open-branched host colonies (Fig. 4). Within coral species, interbranch spacing demonstrated a higher correlation with crustacean abundances than interbranch volume or live surface area, suggesting that protection from predators may be responsible for increased abundances of metazoan associates. Shirayama and Horikoshi (1982) observed that invertebrate communities associated with branching corals were composed primarily of epilithic species (motile and sessile), while massive corals were predominantly occupied by boring species. Again, this suggests that habitat architecture and shelter from predation is an important factor in structuring cryptic invertebrate communities associated with scleractinian corals.

4.2.3 Alcyoniina (Gorgonacea)

Goh et al. (1999) observed that of the 31 known gorgonian species from the reefs of Singapore, no fewer than 16 had associated fauna. While these workers included predators and organisms involved in brief undetermined interactions as associates, 29 invertebrate species (seven unidentified) and one fish species were listed. Patton (1972) pointed out that gorgonian tissues are often unpalatable to fishes and therefore may reduce predation pressure for closely associated species. A conspicuous example of gorgonians providing habitat structure to reef invertebrates is their close association with ophiuroid basket stars (Plate 1f). Basket stars are nocturnal planktivores that orient themselves in strong water currents to maximize prey capture (Hendler et al. 1995). Gorgonian corals orient themselves at angles maximizing water flow incident on their surface (Grigg 1972), which may also prove beneficial to their basket star associates.

4.2.4 Actiniaria

A total of 26 species of fishes in the genera *Amphiprion* (Plate 1g) and *Premnas* are known obligate associates of ten species of anemones (Fautin 1991). Among five species of anemones collected by Chadwick et al. (2008), they recorded three species of symbiotic fishes, one species of mysid shrimp and nine species of caridean shrimps. Numerous nematocyst-armed tentacles provide shelter from predation for those species that are adapted to live among them, though some species of caridean shrimps may also feed on the anemone's tissues (Chadwick et al. 2008).

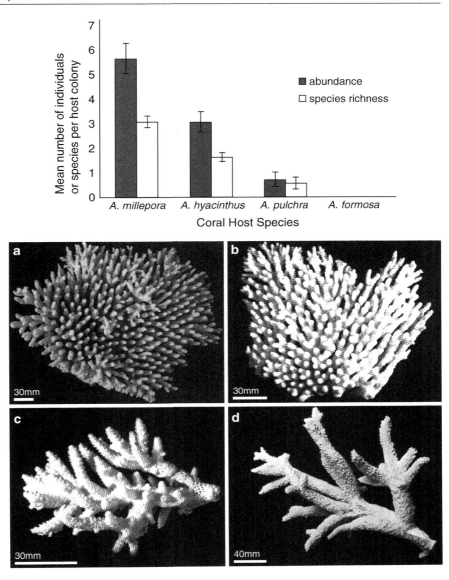

Fig. 4 Relationship between host coral colony (*Acropora* spp.) branch morphology and species richness and abundance of obligate crustacean symbionts. (a) *A. millepora*; (b) *A. hyacinthus*; (c) *A. pulchra*; (d) *A. formosa* (modified alter Vytopil and Willis 2001; a–c, from Wallace 1999; d, from Veron and Wallace 1984)

Removal of anemonefish symbionts can have detrimental effects on their anemone hosts. In many cases, soon after removal of the anemonefish, anemone hosts were preyed upon by butterflyfish (*Chaetodon fasciatus*) (Porat and Chadwick-Furman 2004). Additionally, long-term study (4 years) showed that anemone growth was positively correlated with the presence of fish symbionts and those anemones that hosted few or no anemonefish died. It has been postulated that anemonefish provide nutritional benefits to their hosts in the form of waste products (Fautin 1991). Recent laboratory experiments, however, have shown that anemonefish do not excrete phosphates in quantities sufficient to meet their hosts' requirements (Godinot and Chadwick 2009). Instead, it is likely that ammonia-rich waste products from anemonefish benefit the host's zooxanthellae and thereby reduce the anemone's reliance on heterotrophy (Roopin and Chadwick 2009).

4.2.5 Antipatharia

Black corals (Antipatharia) usually do not occur in high abundances on coral reefs, however, in deeper reef zones they may coexist with hermatypic scleractinians. For this reason, literature concerning antipatharian associates within coral reef ecosystems is rare. Pearse et al. (1998) describe an ophiuroid, *Ophiocanops fugiens*, that associates with three species of black corals in waters off Indonesia. Boland and Parrish (2005) have documented fishes sheltering within black coral branches off Maui, Hawaii. There are numerous examples of invertebrate taxa associating with antipatharians in non-reef environments and it is likely that many similar associations also exist on coral reefs.

4.2.6 *Millepora*

Lewis (2006) reviewed commensal and symbiotic associations with milleporid corals. He described *Millepora* as an ideal habitat as it excludes sediments and is protected by abundant and potent cnidae. Associates include barnacles, alpheid shrimps, tanaids, as well as serpulid and spionid polychaetes. Interactions of metazoan associates are often complex. In the case of the commensal alpheid associates, the shrimps were found to live within the tubes formed by milleporid tissues growing around polychaete burrows, an interaction involving three disparate phyla.

4.2.7 Polychaeta

Polychaete worms belonging to the families Chaetopteridae, Sabellidae, Sabellariidae, Serpulidae, and Terebellidae construct conspicuous tubelike structures in reef formations (Bailey-Brock 1976). These tubes may be inhabited by a variety of small motile cryptofauna once the worms die. Metazoans known to utilize worm tubes for shelter include crustaceans (e.g., alpheid shrimps; Lewis 2006) as well as blennies (Chaenopsidae) (Luckhurst and Luckhurst 1978) and presumably other non-tube-forming annelids.

4.2.8 Crinoidea

Some of the earliest records of diverse fauna associated with crinoids are those of Alcock (1902), who described the cryptically colored crustacean associates of crinoids collected from coral communities off the coast of India. Potts (1915) reviewed metazoan associates of coral reef crinoids from the tropical western Pacific. He listed more than 20 species of crustaceans, echinoderms, annelids, and molluscs, all of which exhibited degrees of host mimicry. He noted that many of the 11 species he collected exhibited morphs matching the natural color variation of their hosts. More recently, Morton and Mladenov (1992) collected two species of polychaetes, two myzostomid worms, seven species of crustaceans, four species of ophiuroids, and a gastropod, living within the microhabitat provided by a single species of crinoid (*Tropiometra afra-macrodiscus*) in waters off Hong Kong. As mentioned earlier, myzostomid worms are often associated with crinoids (Grygier 2000).

4.2.9 Ascidiacea

A review of endobionts inhabiting coral reef ascidians in New Caledonia can be found in Monniot et al. (1991). Bivalve molluscs in the genus *Musculus* attach themselves to the exterior of ascidians and slowly work their way into the tunic of their hosts until only their siphon is visible. Similarly, amphipods are known to envelope themselves in the tunic of ascidians by using their pereiopods to lever the cavity open. With their bodies completely sheltered, the amphipods extend their feeding appendages and antennae into the water to feed on passing food particles (Monniot et al. 1991). Additionally, Chavanich et al. (2007) have found amphipods inhabiting the branchial chamber of reef-dwelling colonial ascidians in Thailand. Other crustacean associates include caridean shrimps that mimic the color patterns of their host as well as a diverse array of commensal copepods (+200 species, >80 genera, four families) (Monniot et al. 1991).

4.2.10 Motile Taxa

Dead and discarded gastropod shells are common throughout most marine ecosystems, where they are frequently used as shelter for a variety of metazoan taxa (McLean 1983; Gutiérrez et al. 2003). Coral reefs are no exception, as abundant and diverse gastropod populations produce shell shelters utilized by other invertebrates to cope with high predation pressure. Hermit crabs (Paguroidea) are a conspicuous example of an invertebrate taxon occupying molluscan shells. Pagurids, in turn, may host a diverse array (>500 species worldwide) of epi- and endobiotic invertebrate associates (Williams and McDermott 2004). Other taxa found within gastropod shells include sipunculid worms (Alcock 1902; Schulze 2005), tanaids (McSweeny 1982), fish (McLean 1983), octopuses (Mather 1982), amphipods (Carter 1982), and some polychaetes. Taxa found to settle on the hard substrate provided by the external surface of mollusc shells (often while colonized by pagurids or living molluscs) include hydroids (Rees 1967), sponges (Sandford and Brown 1997), bryozoans (Taylor et al. 1989), anemones (Cowles 1920), as well as foraminiferans, brachiopods, polychaetes, barnacles, and corals (Lescinsky et al. 2002). These primarily filter feeding taxa may benefit from the raised and mobile nature of their substrate.

A variety of taxa associate with echinoids, utilizing the protection conferred by the sea urchin's sharp spines (Plate 1h). One particularly interesting interaction involves the sea urchin *Astropyga radiata* and the cardinal fish *Siphamia argentea*. These fish normally hide among the sea urchin's spines but when their densities become too high, they school in a closely packed ball immediately above the sea urchin, collectively mimicking the color and form of their protective echinoid associates (Fricke 1970). Coppard and Campbell (2004) collected associates of the sea urchin species belonging to the genera *Diadema* and *Echinothrix* in Fiji. They recorded 13 species living on the tests and among the spines of the sea urchins. These included eight species of fishes, two caridean shrimps (striped for camouflage among the spines), a novel copepod species, as well as two species of benthic ctenophores. In the case of the ctenophores, the authors postulated that sea urchins provide a more ideal feeding substrate (elevated position for filter feeding) in addition to protection from predators. Other diadematid echinoid associates are noted in Gooding (1974), Grygier and Newman (1991), and Hayes (2007).

Sea stars (Asteroidea) are known to provide a habitat for a variety of species. Associates of the corallivorous asteroid *Acanthaster planci* have been extensively reviewed (Canon 1972; Eldredge 1972; Moran 1986; Birkeland and Lucas 1990). Many of these species (including pontoniid shrimps, an apogonid fish, three species of brittle star, one species of ctenophore) utilize the spines of *A. planci* as shelter from potential predators (Birkeland and Lucas 1990). Other metazoans are presumed to benefit from the shelter of *A. planci*'s ambulacral grooves (>3 polychaete species), gut lumen (two carapid fishes, one turbellarian), and coelomic cavity (one carapid fish) (Birkeland and Lucas 1990).

As previously mentioned, the basket star *Astrophyton muricatum* is often observed in close association with various octocoral species. Additionally, a variety of taxa are known to associate with *A. muricatum* (reviewed by Hendler et al. 1995). The caridean shrimp *Periclimenes perryae* is an obligate associate of *A. muricatum*, while other taxa associate facultatively with either the inside (polychaete worms, copepods) or exterior of the basket star (alpheid shrimps, amphipods, copepods, isopods, molluscs, ophiuroids).

In addition to the aforementioned associations, endobionts are found within numerous other motile reef invertebrates. A total of 984 holothurian species collected from southern Vietnam contained 1,111 metazoan associates. These belonged to seven species, including one epibiotic polychaete parasite, two species of inquiline fishes, two species of externally associated brachyuran crabs, a shrimp, as well as a parasitic gastropod. In addition to holothurians, opisthobranch molluscs have recently been observed to host endosymbiotic pearlfish (Carapidae) (Glynn et al. 2008). As previously mentioned, many symbiotic interactions including parasitisms (e.g., saculinid barnacles on decapods) and mutualisms (e.g., photosynthetic dinoflagellates) exist where the symbiont utilizes the host as a habitat. Finally, hard-bodied motile invertebrates (e.g., slipper lobsters; Bailey-Brock 1976) invariably experience fouling (considered by some to be commensalism or parasitism) by epibiotic organisms that utilize this motile substrate as a microhabitat.

4.3 Bioerosion

Numerous species in several invertebrate phyla are known to erode coral skeletons and reef carbonate frameworks. These species, collectively termed bioeroders, are largely responsible for the coral rubble and carbonate sediments that are ubiquitous features of reef habitats. Additionally, many bioeroding taxa are known to weaken the skeletons of living scleractinians. The affected colonies are often more susceptible to fragmentation, which may ultimately lead to the asexual reproduction of the eroded species. Invertebrate phyla with bioeroding species include Mollusca (lithophage and tridacnid bivalves, gastropods, chitons), Echinodermata (sea urchins), Porifera (notably clionaid sponges), Sipuncula, Polychaeta (cirratulid, eunicid, sabellid, spionid worms), and Crustacea (hermit crabs, barnacles, pistol shrimps). For an in-depth review of bioerosion and its ecological impacts on coral reefs, refer to Chapter 25 by Tribollet and Golubic. Comprehensive reviews of invertebrate metazoans involved in coral reef bioerosion may also be found in Hutchings (1986), Glynn (1997), and Perry and Hepburn (2008).

4.4 Framework Consolidation

In addition to eroding and weakening carbonates, some poriferans play an important role in framework consolidation and substrate stabilization in the reef environment. The simple binding of carbonate fragments by sponges may promote reef accretion and increase the survivorship of adult and juvenile corals.

Enhancement of reef accretion by Porifera may occur via a variety of ways. In its simplest form, large sponges may support or consolidate otherwise bioeroded coral colonies, thereby aiding in their persistence and growth (Hartman 1977). Additionally, sponges can act as temporary stabilizers, quickly attaching to loose carbonates and restricting mobility so that slower growing, more permanent framework stabilizers may take hold (Wulff and Buss 1979; Wulff 1984; Rasser and Riegl 2002; Rützler 2004). Wulff (1984) observed that preliminary sponge binding and consolidation can occur within 1 month of rubble formation and that the subsequent, more permanent, cementation by crustose coralline algae, bryozoans, and other invertebrates may take up to 7 months. Additionally, cryptic sponges may occupy cavities within reef frameworks and effectively increase sedimentation in these microenvironments (Rützler 2004). Compaction and cementation of these sediments result in a stronger, more stable reef structure.

In the reef environment, high turbulence caused by dynamic flow regimes and storm events as well as bioturbation can

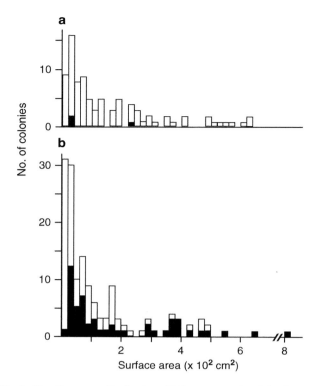

Fig. 5 Size-frequency distribution of foliaceous corals originally present on (**a**), an unmanipulated patch reef and (**b**), an equivalent reef with sponges removed. Occluded bars represent number of coral colonies that became dislodged from the reefs during the 6 months of the experiment (Wulff and Buss 1979)

lead to the creation and frequent movement of loose carbonate sediments. Mobile framework rubble can abrade or overtop living coral tissues and may even lead to destabilization and upturning of entire coral colonies. Often, the binding processes detailed above will reduce the detrimental impacts of substrate mobility.

The importance of some of these processes was experimentally validated by Wulff and Buss (1979) who manipulated patch reefs in San Blas, Panamá by removing binding sponges. They observed that patch reefs lacking sponges had higher mortality of small corals and lost, on average, more coral surface area than unmanipulated reefs (Fig. 5). This increased mortality was attributed to the need for corals to have a stable substrate on which to settle and grow. Areas without sponges had more mobile benthic topography and therefore increased juvenile coral mortality.

4.5 Growth Form Modification

It is well established that coral morphology is highly dependent on various environmental conditions (Done 1983; Todd 2008) including, but not limited to wave action and flow (Veron and Pichon 1976), depth and light (Barnes 1973; Jaubert 1977), sedimentation (Stafford-Smith and Ormond 1992), aerial exposure (Scoffin and Stoddart 1978), and competition (Lang and Chornesky 1990). In addition, a variety of invertebrate phyla are known to modify the skeletal architecture of various reef structural taxa, including scleractinian and milleporid corals, as well as sclerosponges and octocorals (Puce et al. 2008). While the intricacies of these cryptic interactions are poorly understood, it is believed they are more complex than direct interspecific competition for space. The ecological impact of invertebrate-mediated growth-form modification is most likely minimal on large spatial scales. Alteration of a coral's morphology, however, may have profound impacts on small-scale hydrodynamics, symbiont nutrition, and the microhabitat of coral associates.

The sponge *Mycale laevis* has been found to encrust the undersides of platy corals in the Caribbean and to subsequently modify the periphery of the coral substrate (Goreau and Hartman 1966). Edges of the coral colony fold around the sponge osculae, where effluent water may provide nutritional benefits to the host coral. The sponge in turn benefits from the expanding available habitat created by the outward growth of the coral colony.

Perhaps the most widely studied example of coral skeletal modification by an invertebrate is that of the coral gall crab (Cryptochiridae). Gall crabs cause the formation of two distinct features in their coral host. In massive corals (e.g., faviids), they form pits or depressions in the surface of the coral colony (e.g., Abelson et al. 1991; Simon-Blecher et al. 1999). In branching corals (e.g., pocilloporids), a bivalved shelter may form around the symbiont crab (e.g., Abelson et al. 1991). The mechanism of formation and purpose of these structures have been debated in the literature.

Gall crabs have classically been considered to be suspension feeders, capitalizing on nanoplankton incident on the gall shelter (Potts 1915). Kropp (1986) refuted this claim and argued that a lack of setae sufficient for filter feeding as well as direct observations indicated that cryptochirids were feeding on coral mucus. Simon-Blecher et al. (1999) have since reported uptake of carbon from carbon-labeled corals by the pit-forming crab *Cryptochirus coralliodytes*. Additionally, they found that the gut of the same crab species contained mucopolysaccharides, further supporting the hypothesis that the crabs were feeding on coral mucus (Simon-Blecher et al. 1999). Abelson et al. (1991) observed cryptochirids of both the pit and gall types, and used flow tanks to determine the effects of these skeletal deformities on surrounding waters. They concluded that pit crabs were feeding on organic materials that accumulated in their shelters. Gall crabs, however, were found to use their maxillipeds to capture particles from surrounding waters. The gall shape was observed to allow flow through the characteristically shaped shelter, thereby increasing particle deposition (Abelson et al. 1991).

Other crustacean taxa induce the formation of galls in corals (reviewed in Dojiri 1988). Several species of copepods in the order Poecilostomatoida form galls in various scleractinian species (e.g., Dojiri 1988; Kim and Yamashiro 2007) and species in the order Siponostomatoida were observed to create galls in hydrocorals (Stock 1981). Male–female pairs of the pontoniid shrimp *Paratypton siebenrocki* live in galls in acroporid corals (Bruce and Trautwein 2007). Still other decapod crustaceans, such as *Domecia acanthophora*, create "resting places" or depressions in coral skeletons, where their sedentary behavior presumably induces the coral to grow around their abodes (Patton 1967).

Sessile invertebrate metazoans, epilithic on a coral surface, may induce changes in the skeletal architecture of structural reef taxa. Barnacles associated with *Millepora* spp. influence the development of more robust and bladed colony morphologies (Vago et al. 1998). Polychaetes affect skeletal morphology in scleractinian corals (Randall and Eldredge 1976; Liu and Hsieh 2000; Wielgus et al. 2002), milleporid corals (Lewis 1998, 2006), and hypercalcified sponges (personal observations). Polychaete-mediated form modifications are diverse and taxa specific. Examples include raised ridges on a colony surface (e.g., Lewis 2006), grooves and tubes between coral calices (syllid worms with faviid and musid corals, Randall and Eldredge 1976), discrete elevations on the surface of massive corals (spionid polychaetes and poritid corals; Liu and Hsieh 2000), branch formation (spionid polychaetes and montiporid corals; Liu and Hsieh 2000), and the development of conical growths (spionid polychaetes and various coral species; Wielgus et al. 2002).

5 Biotic Interactions

Several reviews and general surveys addressing invertebrate biotic interactions in coral communities were published in the 1980s and 1990s. These papers addressed predation, herbivory, symbiosis, competition, and indirect effects, and are considered herein. Jackson (1983) and Karlson (1999) offered all-inclusive treatments, discussing topics critical to understanding coral population dynamics. We touch only briefly on herbivory and competition since these biotic interactions are addressed in depth by Fong and Paul and Chadwick and Morrow in this book, respectively.

5.1 Predation

While predation studies on reefs are often directed toward metazoans that feed on corals, one must recognize that many scleractinian coral species are effective predators, capturing and ingesting zooplankton and particulate organic matter of animal origin. This feeding mode was well established in numerous early studies (e.g., Abe 1938; Yonge 1930, 1973; Porter 1974). A detailed treatment of this subject can be found in Glynn (1988b, 1990), and Carpenter (1997) as well as in Ferrier-Pagés et al. in Chapter 15 of this volume.

5.1.1 Corallivores

Invertebrate corallivores consume coral tissues with sometimes dramatic effects. Small coral colonies may be consumed completely by relatively large corallivores, such as sea stars and gastropods, or incompletely, resulting in partial predation with patches of surviving tissues. If predation is intermittent, surviving tissues may heal with regeneration and re-sheeting of damaged areas. Micropredators that live and feed on corals during their entire adult lives, for example some acoelomorph worms, flatworms, prosobranch and nudibranch gastropods, copepods, and decapod crustaceans (crabs and shrimps) consume small amounts of tissue, secretory products (e.g., mucus and fat bodies), and exogenous food captured by the host. Many of these kinds of corallivores probably have a modest negative effect on coral hosts and can just as easily be classified as parasites (Castro 1988; Glynn 1988b).

In a current inventory of mostly larger corallivores (Rotjan and Lewis 2008), fishes greatly outnumber invertebrates by over two to one, that is,. 114:51 species, respectively. With respect to the invertebrates, 21 species of echinoderm corallivores were listed, 20 species of molluscs, nine species of crustaceans, and a single annelid worm species. This review offers information on geographic occurrence, feeding mode, inflicted damage, and prey species. Rates of consumption are known for only about one third of the species listed. Noted below are the feeding effects of some corallivore taxa, selected on the basis of their having major impacts on particular coral taxa or coral community structure (Carpenter 1997).

Acanthaster planci, dubbed "an extraordinary species" by Birkeland and Lucas (1990), is the most notorious of all corallivores. It has devastated tens of thousands of hectares of reef-building corals in recent decades (Moran 1986; Birkeland and Lucas 1990). Due to some confusion surrounding the taxonomic status of *Acanthaster planci*, a brief digression on this problem is in order. Rotjan and Lewis (2008) noted a second species of *Acanthaster*, namely *Acanthaster ellisii*, but this taxon was listed in synonymy with *A. planci* by Birkeland and Lucas (1990). The latter authors did recognize a second species, namely *Acanthaster brevispinus*, but this sea star is an omnivore inhabiting soft-bottom off-reef areas. Based on molecular data across the entire distribution of *Acanthaster*, Vogler et al. (2008) found that *A. planci* consists of four deeply diverged clades that form a pan-Indo-Pacific species complex. These four newly recognized species are distributed in the (1) eastern and western Pacific Ocean, (2) northern and (3) southern Indian Ocean, and (4) Red Sea. The following information on the feeding biology and ecology of *Acanthaster* refers to corallivores, but does not take into account the above-noted species and regional differences.

Widely distributed throughout the Indo-Pacific region, adult *Acanthaster* feed predominantly on reef corals under natural field conditions. Where present on reefs, population densities can range from low/moderate (1–10^2 indiv. ha^{-1}) to extremely high (>10^3–10^5 indiv. ha^{-1}). In outbreak aggregations on Australian and Samoan reefs, >40,000 to ~500,000 sea stars were removed from small areas over 18-month periods. The pliable nature of its body, extrudable stomach, and prehensile arms allows *Acanthaster* to feed on corals exhibiting a variety of colony morphologies (Birkeland 1989b). The stomach is extruded and positioned to envelope the coral prey (an entire small colony or portion of a large colony), and then digestive enzymes are secreted, liquefying coral tissues that are absorbed externally. Although only coral tissues are removed, leaving the skeleton intact, there is often collateral mortality of sessile or slow-moving coral symbionts. Virtually all zooxanthellate coral taxa are consumed, but with definite species preferences depending on sea star abundances, availability of prey, and symbiotic crustacean agonistic behaviors (Glynn 1982b; 1983b; Pratchett 2001). Daily consumption ranges from 100 to 200 cm^2 of soft tissue. Large feeding aggregations can remove from 0.5 to 5–6 km^2 of live coral cover in a single year. Considerable controversy surrounding the causes of population outbreaks and fluctuations in abundances in the past can be found in Endean and Cameron (1990), and a collection of papers in Wilkinson (1990) and Wilkinson and Macintyre (1992). For evidence relating elevated phytoplankton production to *Acanthaster* reproductive success and larval survival, see Birkeland (1982), Brodie et al. (2005), and Houk et al. (2007). It is possible that these divergent hypotheses are due to species differences in life histories and/or ecological and behavioral traits. Other asteroid corallivores include two species of *Culcita*, which prey heavily on juvenile corals in some areas of the Indo-Pacific (Thomassin 1976; Glynn and Krupp 1986), and at least six other species (in six genera) that are facultative corallivores.

Several species of echinoids are present globally on all reefs; two genera, namely *Diadema* and *Eucidaris*, can cause considerable damage to corals. Species in these genera are primarily herbivores, but often prey on live corals when abundant. Population densities can range from 20 to 50 indiv. m^{-2}, and for species of *Diadema* on some dead and eroding reefs as high as 100 indiv. m^{-2} (Plate 2a). Sea urchin abundances have been shown to be controlled by the abundances of their fish predators on east African reefs (McClanahan and Mutere 1994; McClanahan et al. 2009). An assemblage of sea urchins, including species of *Echinometra*, *Echinostrephus*, and *Echinothrix*, showed mean densities of 0.1–0.2 indiv. m^{-2} in areas protected from fishing (more abundant fish predators of sea urchins) and 5.3–7.6 indiv. m^{-2} on unprotected reefs. Mean sea urchin densities were as high as 32.7 indiv. m^{-2} on one unprotected reef. Aristotle's lantern, a complex masticatory apparatus, consists of five mineralized calcareous teeth that can excavate both hard and soft substrates. When feeding on corals, soft tissues as well as skeleton are fragmented and removed. An eastern tropical Pacific endemic sea urchin, *Eucidaris galapagensis*, is perhaps one of the most destructive echinoids, likely limiting reef growth and recovery in the Galápagos Islands (Glynn et al. 1979; Glynn 1994) (Plate 2b). Another cidaroid, *Prionocidaris baculosa*, was reported to injest both live and dead coral in the Gulf of Suez (Pearse 1969).

No fewer than five families of prosobranch gastropods contain species that consume coral tissues (Robertson 1970; Hadfield 1976). *Drupella*, a member of the family Muricidae, is an Indo-Pacific genus that has had devastating impacts on corals in the Red Sea (Schuhmacher 1992; Antonius and Riegl 1997), Japan and the Philippines (Moyer et al. 1982), Hong Kong (Morton et al. 2002), on the Great Barrier Reef (Baird 1999), and in northwestern Australia (Turner 1992, 1994). Like *Acanthaster*, thousands of individuals can form feeding aggregations that result in large-scale coral mortality. Feeding is most active at night, with individuals especially targeting injured corals or corals already being eaten. Four species of *Drupella* are obligate corallivores and share several feeding characteristics with *Acanthaster*. *Drupella* has a highly specialized radula that effectively removes mostly retracted tissues from corallites without leaving any radular rasping marks. Since *Drupella* is secretive and usually forms feeding aggregations, these behaviors complicate attempts to determine "normal" and "outbreak" population densities (Turner 1994). Reported densities range from 0 to 0.5 indiv. m^{-2} to 1,500 indiv. 0.5 m^{-2} (Moyer et al. 1982). Sampling in back reef areas on the Ningaloo Reef (western Australia) in the late 1980s disclosed densities of 1–2 million indiv. km^{-1} (Stoddard 1989). Mean per capita feeding rates at Ningaloo were 2.6 cm^2 d^{-1}, and ranged from ~0.5 to 10 cm^2 d^{-1}. *Drupella* shows a strong preference for acroporid (*Acropora*, *Montipora*) and pocilloporid (*Stylophora*, *Pocillopora*, *Seriatopora*) corals, but with declining abundances of these taxa the snails may become conditioned to feed on less-abundant species in other families. *Quoyula*, another muricid corallivore, is sedentary and lives attached to pocilloporid corals, causing only local tissue damage. It lowers its proboscis into the mouths of surrounding polyps and removes their gut contents by a suction force (Schuhmacher 1992).

Plate 2 (**a**) *Diadema antillarum* grazing on a reef limestone substrate, Makerel Reef, San Blas Islands, Panamá (Photo by H. Lessios); (**b**) *Eucidaris galapagensis* grazing on a dead coral framework at Onslow Island, Galápagos Islands, Ecuador; (**c**) *Coralliophila abbreviata* feeding on *Acropora palmata*, Key Largo, Florida Keys (Photo by L. Johnston); (**d**) Atlantic *Hermodice carunculata* consuming the branch tips of a Pacific *Pocillopora damicornis* colony in a captive prey choice experiment (Glynn 1982b); (**e**) *Pherecardia striata* feeding on an arm of the sea star *Leiaster analogus*, Uva Island, Gulf of Chiriquí, Panamá; (**f**) *Hymenocera picta* on an arm of *Acanthaster planci*, Uva Island; (**g**) Barnacle symbionts (*Nobia* sp.) associated with the coral *Galaxea* sp., Sunrise Atoll, north of New Caledonia (Photo by P. Laboute); (**h**) An infestation of acoelomorph worms (*Waminoa brickneri*) covering a live colony of *Stylophora pistillata*, Aqaba, Jordan (Photo by R. Sulam)

Corallivore species are also known in three nudibranch (Opisthobranchia) genera, namely *Phestilla*, *Pinufius*, and *Cuthona* (Rudman 1981). Rudman (1981) noted that *Aeolidia edmondsoni*, which is listed by Rotjan and Lewis (2008), is a junior synonym of *Phestilla lugubris*. Feeding mostly on *Porites*, these nudibranchs ingest coral tissue whole, but utilize different components. For example, *P. lugubris* does not digest spirocysts, *Phestilla minor* selectively digests spirocysts, *Cuthona poritophages* selectively digests zooxanthellae, and *Pinufius rebus* sequesters zooxanthellae and possibly satisfies part of its nutrient requirements from a stable symbiosis with these algae. Rudman (1982) has also surmised that *Doridomorpha gardineri* feeds on the zooxanthellate octocorallian *Heliopora coerulea*. Hadfield (1976), however, has warned that a close association with potential prey or even the occurrence of zooxanthellae and other cnidarian products in gut contents are no guarantee that corallivory is the principal feeding mode. Wägele and Johnsen (2001) have offered convincing evidence that several species of the nudibranch clade Cladobranchia contain functional photosynthetic zooxanthellae that may contribute to the host's metabolic needs. From starvation experiments under light and dark treatments, Hoegh-Guldberg and Hinde (1986) and Hoegh-Guldberg et al. (1986) demonstrated an important role of zooxanthellae for the survival of a non-corallivore nudibranch.

The large majority of invertebrate molluscan corallivores occur in the Indo- and eastern Pacific, 17 of 20 species considered in Rotjan and Lewis (2008). Widespread Indo-Pacific taxa include the nudibranch *Phestilla*, and the prosobranch gastropods *Drupella*, *Coralliophila*, and *Quoyula*. *Coralliophila abbreviata* has had a large impact on acroporid corals in the western Atlantic (Knowlton et al. 1990;

Miller 2001; Baums et al. 2003) (Plate 2c). All but one of the molluscan species listed by Rotjan and Lewis (2008) are obligate corallivores.

To the nine arthropod corallivore species listed in Rotjan and Lewis (2008) that occur on Indo-Pacific and eastern Pacific reefs, it is necessary to add two facultative hermit crab corallivores present in the Caribbean region (Gilchrist 1985). These are *Petrochirus diogenes* and *Paguristes* sp., which were observed to severely damage small colonies of *Acropora* and *Agaricia*. An obligate shrimp (*Alpheus lottini*) and several crab species (*Trapezia* spp., *Tetralia* spp.) were classified as corallivores by Rotjan and Lewis (2008), but these are probably best regarded as symbiotic mutualists and are considered below. Additionally, the single obligate cirriped species (*Pyrgoma*) is a parasite and is discussed later in this context.

Many hermit crab species feed facultatively on coral tissues within reef ecosystems: *Aniculus elegans*, *Calcinus obscurus*, and *Trizopagurus magnificus* on *Pocillopora* spp. in the eastern Pacific and *Petrochirus diogenes* and *Paguristes* sp. on *Acropora* spp. and *Agaricia* spp. in the Caribbean (Glynn et al. 1972; Gilchrist 1985). The chelipeds are employed in picking and scraping tissues, which are then transferred to the maxillipeds and eventually to the mouth. Estimates of the population densities on a study reef in the Pearl Islands (Pacific Panamá) were 27.5 indiv. m^{-2} or 275,000 indiv. ha^{-1} for *T. magnificus*, and 0.02 indiv. m^{-2} or 200 indiv. ha^{-1} for *A. elegans*. Hermit crabs are often abundant on reefs, but have never been reported at outbreak densities.

Only one polychaete worm species commonly feeds on corals, namely *Hermodice carunculata*, a member of the fireworm family Amphinomidae (Fauchald and Jumars 1979) (Plate 2d). The feeding habits of this worm, a temperate to tropical Atlantic/Mediterranean species, were first described by Marsden (1962, 1963) from reefs in Barbados and Jamaica. It feeds on corals, especially branching species (*Porites*, *Acropora*), sea anemones (Lizama and Blanquet 1975), and probably other cnidarians. Another amphinomid polychaete, the cosmopolite *Eurythoe complanata*, was briefly noted by Hartman (1954) to feed on corallines, coral colonies, or microorganisms associated with corals in the Marshall Islands. Her observations are in need of verification. When *H. carunculata* feeds on a branching coral the buccal mass is folded over the projecting branch tip and the tissues appear to be sucked up and passed into the anterior region of the digestive tract. Tissue consumption amounts to ~13 cm^2 d^{-1} (Witman 1988). Due to the motility and cryptic habits of *H. carunculata*, reliable measures of population densities are difficult to obtain. Most workers, however, report abundance estimates of <1 indiv. m^{-2}. Although *Hermodice* has not been reported to have a strong impact on undisturbed coral populations, its feeding activities (along with other corallivores) can quickly cause the collapse of corals already decimated by disease, storms, or bleaching disturbances (Knowlton et al. 1990).

5.1.2 Other Kinds of Predators

Probably no reef metazoan is immune from predation. Even *Acanthaster*, abundantly armed with venomous spines, is attacked and eaten by an array of piscine and invertebrate predators. Nine invertebrate predators of juvenile and adult *Acanthaster*, belonging to anemone, coral, crab, shrimp, and annelid worm taxa, are known (Moran 1986). The conspicuous nature of zooxanthellate corals has greatly facilitated observations on corallivores. This is not the case with many other reef predators that are motile and cryptic. Here, we briefly note a few non-coral predators that likely play an important role in reef trophic relationships.

Nemerteans occur on coral reefs, but remain essentially unstudied in reef habitats. Nemerteans are efficient predators in high latitude intertidal areas because of a unique prey-capturing proboscis with potent toxins that can be rapidly everted. They also possess a well-developed chemosensory ability to follow prey trails (Thiel and Kruse 2001).

Pherecardia striata, an amphinomid polychaete worm that is abundant on eastern Pacific coral reefs, attacks wounded or moribund sea stars and various live crustaceans (Glynn 1984) (Plate 2e). Glynn (1982a) argued that *P. striata* and *Hymenocera picta*, the harlequin shrimp, can together control *Acanthaster* population abundances on reefs in Panamá (Plate 2f). Other crustaceans that are mostly generalist predators include stomatopods, portunid crabs, and isopods, which prey on a variety of metazoans, for example polychaete worms, crustaceans, molluscs, and fishes. From a study conducted in the U.S. Virgin Islands, Reaka-Kudla (1987) found that several stomatopod species consumed high proportions of brachyuran and hermit crabs, as well as gastropods and bivalve molluscs. She also demonstrated size-dependent intraspecific predation with large individuals preying on smaller stomatopods, that is, cannibalism. When stomatopods and octopods compete for lairs, generally large individuals of the former will consume smaller individuals of the latter and vice versa (Cronin et al. 2006). In an attempt to quantify the predation pressure of *Gonodactylus bredini* on a Caribbean reef in Panamá, Caldwell et al. (1989) estimated population densities and prey consumption of this stomatopod. Field sampling revealed densities of 8–20 indiv. m^{-2}, and predation rates on their most frequent prey (gastropods in the genus *Cerithium* and hermit crabs in vacated shells) of ~24 indiv. d^{-1}.

All cephalopod taxa are macrophagous carnivores with high feeding rates, especially young stages and active swimming species. For example, a laboratory-reared juvenile squid was found to consume daily 30–60% of its own body weight and octopods 20–40% (Boucher-Rodoni et al. 1987). Due to sampling difficulties (octopuses are cryptic and frequently change dens) only rough estimates are available on their abundances. An in-depth study of the foraging behavior of *Octopus cyanea* reported ten occupied dens along a 300-m transect of reef habitat in French Polynesia (Forsythe and Hanlon 1997). This is equivalent to a population density of about 7 indiv. ha^{-1}. The gastropod genus *Conus*, especially prominent on Indo-Pacific reefs, contains numerous specialist species vis-à-vis habitat niches and prey preferences (Kohn 1968; Kohn and Nybakken 1975). Most species are primary carnivores, specializing on polychaete annelids, gastropods, and fishes. Population densities on Indian Ocean atolls ranged from 0.03 to 0.15 indiv. m^{-2} on topographically diverse substrates to 1 indiv. m^{-2} on more uniform limestone benches. Feeding rates, with maximum estimates of 10% of body weight consumed per day, are appreciably lower than for cephalopods.

Crustaceans inhabiting *Porites* colonies in Hawaii, including a large xanthid crab, prey on the nudibranch corallivore *Phestilla sibogae*, thus limiting coral tissue mortality (Gochfeld and Aeby 1997). This represents an indirect trophic interaction, a topic that is considered in more detail below. Portunid or swimming crabs in the genera *Charybdis*, *Portunus* and *Thalamita* are aggressive predators that frequent coral reefs. Several of the 54 species of swimming crabs listed for the Marianas Islands occur in reef habitats (Paulay et al. 2003). Crustacean micropredators, such as cirolanid isopods, were observed to prey on field-caged juvenile pomacentrids (Jones and Grutter 2008). Whether this is commonplace under natural conditions awaits further study.

5.2 Herbivory

Several invertebrate herbivore species are abundantly represented in the phyla Arthropoda (e.g., amphipods, brachyurans, isopods, tanaids), Mollusca (e.g., limpets, gastropods, chitons, opisthobranchs), Echinodermata (e.g., echinoids, asteroids), and Annelida (e.g., eunicid, syllid polychaetes). Of all of these taxa, echinoids (primarily of the genus *Diadema*) have received the greatest attention in coral reef ecosystems. For reviews of herbivory and herbivores, see Steneck (1988), Hay (1991, 1997), Carpenter (1997), and Hixon (1997) as well as Fong and Paul's Chapter 17 in this book.

5.3 Other Consumers

Numerous feeding behaviors other than predation and herbivory have been described for a variety of reef taxa, but their effects on other biota have been little studied. With the changing condition and degradation of many global reefs, it is likely that some of these unexplored trophic interactions may exacerbate or accelerate the declines of calcifying organisms.

Suspension feeding metazoans are abundantly represented on reefs by species in the following taxa: Porifera, Cnidaria (Scleractinia, Hydroida, Octocorallia, Actiniaria, Zoanthidea, Corallimorpharia), Rotifera, Entoprocta, Annelida, Crustacea (Decapoda, Mysida, Cirripedia, Copepoda, Ostracoda), Mollusca (Gastropoda, Bivalvia), Phoronida, Ectoprocta, Brachiopoda, Echinodermata (Crinoidea, Ophiuroidea), and Urochordata. These organisms are efficient at food capture; capable of filtering an array of suspended particle sizes ranging from ≤2 μm particulate organic matter and bacterioplankton to phytoplankton and zooplankton (see Sorokin 1990, 1993 for summaries). Where abundant, suspension feeders can cause high mortality of bacteria, phytoplankton, and meroplankton. The high depletion rates of meroplankton significantly impact the recruitment success of numerous reef species. Field experiments in the Caribbean have demonstrated high rates of bacterivory by sessile suspension feeding cryptofauna with depletion of 60–100% of bacteria under high benthic cover (Buss and Jackson 1981; Scheffers et al. 2004). Some suspension feeders that utilize mucus nets can even influence the growth form, growth rate, and survival of coral host substrates (Smalley 1984; Colgan 1985; Kappner et al. 2000; Zvuloni et al. 2008).

Deposit feeders are represented by numerous species inhabiting reef sediments. Many of these organisms are minute, members of the meiofauna (e.g., Gastrotricha, Kinorhyncha, Nemata, Annelida, Tardigrada), and some are relatively large (e.g., Priapula, Sipuncula, Echiura, Annelida, Echinodermata). In terms of ecological services, the majority of species utilizing this feeding mechanism facilitate the decomposition and re-mineralization of organic matter (carcasses, feces, detritus) in concert with microbial communities. Reef community production is dependent on the recycling of organic matter back to inorganic forms (e.g., Hatcher 1997).

It is important to recognize that there exist a multitude of variations on the principal feeding mechanisms and behaviors thus far considered, with a wide range of effects on prey species. For example, some potential prey that are injured after attack by a predator may eventually die due to the debilitating effects of scavengers (Glynn 1982b). In addition, there is ample evidence that many reef invertebrates, both adults and larval stages, are also capable of absorbing dissolved organic matter, consisting dominantly of free amino acids and carbohydrates (Ferrier 1991; Hoegh-Guldberg 1994; Ambariyanto and Hoegh-Guldberg 1999). The extent to which such uptake satisfies nutritional needs is still largely unresolved. Next, we consider the biotic interactions of species living in intimate association, that is, symbiosis (mutualism, commensalism, parasitism).

5.4 Symbiosis

The varied relationships and adaptations of invertebrate symbioses far exceed the imagined possibilities and are one of the great marvels, and in many instances conundrums, in coral community species interactions. Bouchet (2006) alludes to the symbiosis guild as the "black box" of coral reef biodiversity because so little is known of the species involved in such interrelationships. Symbiosis is here defined as an intimate and protracted heterospecific association between two or more organisms, including mutualism, commensalism, and parasitism. In mutualisms, both partners derive a benefit, whereas commensal relationships benefit one partner with no discernible effects on the other. The special case of parasitism involves one species that derives a nutritional benefit from another (host) species. In addition, the parasite is invariably smaller than its host.

5.4.1 Mutualisms and commensalisms

It is often difficult to distinguish between mutualisms and commensalisms. In several instances, symbiotic associates that were formerly regarded as commensals have been shown to be mutualists. As noted earlier, there are myriads of species that live attached to other animals (epizoites), and probably the majority of these are commensals. Concentrated field studies, however, have revealed a plethora of subtle interspecific interactions that are best classified as mutualisms. Selected examples of these are examined here because of their fascinating nature and in many instances likely coevolved adaptations.

From carefully designed field experiments, Wulff (1997) demonstrated an unexpected mutualism in reef sponges. She found that growth rates and survival of Caribbean sponges are enhanced when heterospecific sponges adhere to each other. Although the mechanism is unknown, the sponges can better withstand environmental hazards, such as predation, sedimentation, mechanical damage, and diseases, when intertwined in multispecies associations.

Sessile polychaete worms embedded in coral skeletons have long been assumed to be commensals. The activities of *Spirobranchus giganteus* in Australia, however, protect the surrounding tissues of its coral host (*Porites* spp.) from attack by *Acanthaster* (De Vantier et al. 1986). In the Red Sea, coral tissues adjacent to this same polychaete species have been shown to have higher survival potential from bleaching and predator disturbances than distant tissues (Ben-Tzvi et al. 2006). In both cases, the surviving tissue patches aide in coral host recovery. In the Red Sea example, it was hypothesized that coral tissues surrounding worm tubes benefited from (1) improved water circulation, (2) improved dispersal of waste products, and (3) increased availability of nutrients.

Two examples of bivalve molluscs embedded in coral skeletons are also suggestive of mutualistic associations. *Pedum spondyloideum*, a scallop embedded in *Porites* spp., was found to impede foraging *Acanthaster* as noted above for the polychaete worm *Spirobranchus* (De Vantier and Endean 1988). Also, Mokady et al. (1998) presented evidence suggesting that boring bivalves may interact in a mutualistic fashion with corals by providing essential nutrients (e.g., ammonia).

A large number of crustacean symbionts formerly regarded as commensals or parasites have been shown to interact advantageously with their hosts. Cleaning symbioses, involving diverse species of shrimps and juvenile fishes that remove necrotic tissues and parasites from a variety of fish hosts, are observed on virtually all coral reefs (Losey 1987, 1993; Poulin 1993). Since cleaning behavior and parasite removal have not always shown a strong or beneficial effect for client fishes, Losey (1977, 1979) proposed that tactile stimuli rather than fish survivorship was the prime driver of this interaction. More recent studies on the Great Barrier Reef, however, have shown that the removal of parasitic gnathiid isopod larvae from host fishes has a significant influence on host cleaning (Grutter 1995; Grutter and Poulin 1998). Moreover, Becker and Grutter (2004) have demonstrated conclusively that cleaner shrimp do remove and consume large numbers of isopod and copepod ectoparasites from coral reef fishes in the wild. These workers recorded 43 shrimp species involved in cleaning, which is nearly double the number noted in a recent review of cleaning symbiosis (Côté 2000). Also, in an aquarium study in the Caribbean, Bunkley-Williams and Williams (1998) demonstrated that a cleaner shrimp (*Periclimenes pedersoni*) could effectively remove juvenile cymothoid isopod ectoparasites from grunts. Grutter's (2002) review supports the hypothesis that cleaning behavior is mutualistic and probably a co-evolved interaction. As Côté (2000) noted, the nature of cleaning interactions may be site and time dependent, greatly influenced by local environmental conditions such as parasite infestation rates.

Shrimps associated with giant sea anemones display interactions ranging from obligate mutualism to occasional parasitism (Chadwick et al. 2008). Red Sea obligate anemone/shrimp associations revealed reciprocal benefits under some conditions (mutualism), and occasional parasitism (feeding on anemone host tentacles) under conditions of limited food supply.

Probably the best evidence supporting crustacean/coral mutualisms derives from the various activities of crab and shrimp symbionts promoting the survivorship of their hosts. Decapod crustaceans associated with corals were classified as parasites by several workers before beneficial cleaning effects and defensive behaviors were demonstrated. Trapeziid crabs and alpheid shrimps obtain shelter and nutriments from their coral hosts and in turn promote the survivorship of their hosts by improved cleansing activities. Coral host health and survivorship have been demonstrated for pocilloporid species in the laboratory (Glynn 1983a) and field (Stewart et al. 2006). The movements and feeding activities of the symbionts promote circulation and prevent the accumulation of sediment and fouling organisms. The defensive behavior of crustacean symbionts in preventing corallivore attack is more immediate and dramatic. Several species of *Trapezia* and the pistol shrimp *Alpheus lottini* effectively defend their pocilloporid coral hosts from *Acanthaster* attack (Glynn 1983b, 1985, 1987; Pratchett et al. 2000; Pratchett 2001). This protection is effective in coral communities with low to moderate abundances of *Acanthaster*. At *Acanthaster* outbreak densities, however, the crustacean guards are usually overwhelmed and their coral hosts consumed.

Numerous barnacle species in nine genera of Pyrgomatinae occur exclusively embedded in the skeletons of scleractinian and hydrocoral cnidarians (Plate 2g). Most barnacle symbionts are suspension feeders (commensals), species in at least four genera are parasites, consuming host tissues (Ross and Newman 2000), and studies of a barnacle associate of the hydrocoral genus *Millepora* support a mutually beneficial nutritional role (Cook et al. 1991; Achituv and Mizrahi 1996). Carbon, phosphorous, and nitrogen excreted by barnacle symbionts are absorbed by zooxanthellae and then incorporated into the tissues of the hydrocoral. Organic matter and zooxanthellae expelled by the coral host also probably contribute carbon to the barnacle (Achituv et al. 1997).

5.4.2 Parasitism

While all coral reef higher taxa have parasites or act as hosts, some of these interactions (e.g., invertebrate parasites of fishes) are more widely known (Kinne 1980). Examples of metazoan parasites associated with coral hosts can be found in Cheng (1986), Castro (1988), Peters (1997), and Kaiser and Bryce (2001). For a novel example of brood parasitism in an echinoderm, see Hendler et al. (1999). Kinne's (1980–1990) edited four-volume treatise of marine diseases, with numerous examples of metazoan parasites as agents of disease, is the most comprehensive treatment available, but examples of reef-associated parasites are inconveniently dispersed throughout the volumes. For an appreciation of the diversity of metazoan parasites observed on coral reef fishes, the reader is referred to Williams et al. (1996) and Grutter (1994), who listed 18 and 20 invertebrate taxa, respectively. Surveys conducted at two localities on the Great Barrier Reef found 36 fish species in 12 families infested by gnathiid isopod parasites (Grutter and Poulin 1998).

Parasites result in negative effects to the host, which can lower a host's fitness through energy and/or material losses, reproductive impairment, and changes in behavior. In extreme cases of high parasite loading or virulence, infestation can result in death. The causative agent of some diseases on reefs can be initiated by or due entirely to an invertebrate parasite. From an ecological perspective, an important challenge in parasitic interactions is to quantify changes in host fitness and determine their effects at population and community levels. Kinne (1980) offered the following caveat, slightly paraphrased here, which continues to confound workers in this field. In most instances, it is not possible to differentiate between parasitism and other symbiotic associations. It is certain that upon further study some of the guild assignments offered here (i.e., in Kinne's treatise) will change. We briefly examine a few of the more thoroughly studied examples of flatworms and crustaceans, chiefly from the perspective of damage to the hosts.

Flatworms infest corals as both external and internal parasites. In Hawaii, *Montipora verrucosa* hosts the obligate ectoparasite *Prosthiostomum* sp., a polyclad flatworm that feeds on the coral's coenosarc and polyps (Jokiel and Townsley 1974). When abundant these worms can kill large patches of tissue, especially on the lower parts of colonies where they shelter. Heavy infestations affected nearly half of the *M. verrucosa* population on the Coconut Island reef in 1972. At that time, the part of Kaneohe Bay studied suffered from eutrophic pollution. It was concluded that such stressful conditions could be an important factor in limiting the abundance and distribution of this coral. Acoelomorph flatworms belonging to the genus *Waminoa* have been shown to infest at least 14 zooxanthellate coral species at Eilat, Red Sea (Barneah et al. 2007) (Plate 2h). While these worms may remove surface mucus layers or interfere with their host's photosynthetic efficiency when present in high numbers, further study is necessary to determine their particular symbiotic status. A final and undisputed example of a flatworm parasite involves the digenetic trematode *Podocotyloides stenometra* that infests the intermediate host coral *Porites compressa* in Hawaii (Aeby 1992, 1998). Encysted metacercaria in the tentacles of *Porites* polyps stimulate the production of pink or purple nodules in adjacent tissues. The infected swollen polyps are unable to retract normally and are more susceptible to predation by a butterflyfish corallivore, which is the definitive host. Aeby (1992) has suggested that benefits may accrue to (a) the butterflyfish with an increased feeding efficiency on infected polyps, (b) the trematode by enhanced transmission and completion of its life cycle, and (c) the coral by ridding itself of infected polyps that can be regenerated. Pink nodules have been observed on poritid corals on the Great Barrier Reef (Willis et al. 2004) and in the Galápagos Islands (Vera and Banks 2009), but this abnormality is not necessarily a response to parasitic infection. Corallivores, bioeroders, competitors, and epizoites can also stimulate the production of pink growth irregularities (e.g., Benzoni et al. 2009).

Parasitism has evolved among several crustacean reef taxa, including brachyuran crabs, barnacles, copepods, and isopods. Two families of crabs, the Eumedonidae and

Pinnotheridae (pea crabs), contain species that are obligate associates of a variety of invertebrates. Eumedonids are found on cnidarians and more frequently on echinoderm hosts, often cryptically colored on external body surfaces (Stevcic et al. 1988). Since the trophic status of most eumedonids is unknown, they have only tentatively been classified as parasites. As an example, *Echinoecus pentagonus*, extensively studied by Castro (1971), can be regarded as a parasite since it inhabits the peristome and test of a sea urchin where males and immature females feed on epithelial tissues and tube feet. Their effects are minimal because of the host's ability to quickly regenerate damaged tissues. Pinnotherid crabs live within the mantle cavities of bivalves, and are also associated with various species of polychaetous annelids, echinoderms, and tunicates (Cheng 1967; Kinne 1983). Some species cause damage to the ctenidial tissues of bivalves, but this may be an indirect effect due to crabs feeding on food strings collected by the hosts. Because they cause pathological changes and are metabolically dependent on their host's food collection, they may be considered as parasites.

Among the numerous pyrgomatine barnacles that utilize live coral skeletons as substrate, four genera belonging to the tribe Hoekiini parasitize species of the zooxanthellate coral genus *Hydnophora* (Ross and Newman 1995, 2000). These barnacles are able to prevent the overgrowth of their hosts, and unlike setose-feeding pyrgomatids, subsist on coral tissues and possibly organic matter obtained by absorption. An isotopic study demonstrated that the coral host was the source of carbon for coral-inhabiting barnacles in the Red Sea (Achituv et al. 1997).

The diversity of copepod associates inhabiting marine invertebrates is vast, with a minimum of 1,727 species recognized in the early 1990s (Humes, 1994). Cnidarians, molluscs, and echinoderms harbor the highest numbers of copepod species, and many of these are assumed to be parasitic (e.g., Stock 1975; Humes 1985b). Within these host phyla, the highest copepod diversities occur in scleractinian corals (Humes 1985a), pelecypod, and opisthobranch gastropods (Cheng 1967), as well as holothurians and ophiuroids (Jangoux 1990). Copepods occur on corals both externally and within their polyps, and in molluscs, within the mantle cavity and attached to ctenidia. In holothurians, copepods occur primarily in the coelomic cavity and digestive tract, and in ophiuroids on outer surfaces, in cysts, bursae, and internally in the stomach. Most copepods located externally possess relatively untransformed bodies. In contrast, the bodies of endoparasites are typically small (<3 mm in length), elongate, with weakly defined segmentation and reduced appendages. Scleractinian corals host the greatest number of copepod associates with about one half in the Indo-Pacific family Xarifiidae and with 75 species of *Xarifia* associated with 93 species of corals (Humes 1985a). Caribbean corals are parasitized by highly transformed poecilostomatoid copepods in the family Corallovexiidae, which were assumed to be parasitic (Stock 1975). *Xarifia* was observed tearing coral polyp tissue with its claws (Gerlach quoted in Humes 1960). Stock also observed corallovexids eating coral tissue but did not offer any additional information. Whether the majority of these copepods are parasites or commensals is largely unknown because of the absence of rigorous experimental documentation.

Several recent laboratory and field studies have provided strong evidence linking early developmental stages of isopods to the debilitation and mortality of early life history stages of reef fishes. An experimental study conducted at Lizard Island, Great Barrier Reef, demonstrated high mortality rates of newly settled apogonid fishes by mancae, a pre-juvenile stage of a cymothoid isopod (Fogelman and Grutter 2008). The mancae of *Anilocra apogonae*, whose adult phase parasitizes an adult apogonid species, attach temporarily to the anterior body of several species of young apogonids where they feed on the host's blood. Young fishes are particularly vulnerable, experiencing reduced growth and high mortality. Only about one half of two experimentally infested fishes survived to day 5 (Fig. 6). The young stages of the isopod's definitive fish host had a higher survival rate, with about one half of the experimentally infested fish surviving to day 18.

Because crustacean parasite larvae are mobile and can attack multiple juvenile fishes, they are often classified as micropredators (Kuris and Lafferty 2000). Other examples of crustacean parasites considered as micropredators are the larval stages of gnathiid isopods (e.g., Grutter and Poulin 1998; Finley and Forrester 2003; Grutter et al. 2008; Jones and Grutter 2008; Penfold et al. 2008). Like cymothoid mancae, these reef-based larval isopods can have strong negative effects on juvenile fishes recruiting to reefs. Finley and

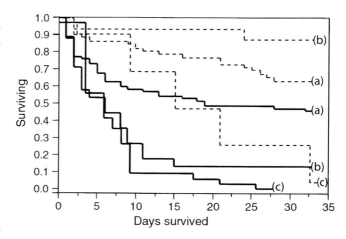

Fig. 6 Proportion of infested (*solid bars*) and uninfested (*dashed bars*) young apogonid fishes surviving parasitic mancae attack by the cymothoid isopod *Anilocra apogonae*. (**a**) *Apogon trimaculatus*, (**b**) *Apogon nigrofasciatus*, and (**c**) *Cheilodipterus quinquelineatus* (Modified from Fogelman and Grutter 2008)

Forrester (2003) highlighted the importance of ectoparasites on small reef fish survivorship, calling attention to the possible role of indirect effects (e.g., parasite impacts increasing the vulnerability of hosts), and recommended greater attention to studies of parasite/host interactions. Strathmann et al. (2002) and others have suggested that the vulnerability of early fish developmental stages to high abundances of reef parasites and micropredators may be an important selective force in the evolution of the pelagic phase in the large majority of reef fishes.

5.5 Competition

Competition is a ubiquitous feature of reef ecosystems, manifested in the daily pursuit of food, space, and reproductive success. Mechanisms of competition in the reef ecosystem are wide ranging. In corals alone, competition may occur through simple overgrowth and direct resource consumption or from more complex allelopathy and development of specialized competitive structures (e.g., sweeper tentacles and polyps, mesenterial filaments). Effects range from cryptic interaction occurring at the organismal level to large-scale, ecosystem-altering processes. For a comprehensive review of competitive interactions involving coral reef organisms, see Buss (1986), Lang and Chornesky (1990) as well as Chadwick and Morrow's in this book.

5.6 Indirect Effects

Numerous examples of indirect species interactions in coral communities have been proposed. The influence of one species on another can be significantly affected, either positively or negatively, by a third (or more) species in an interaction web. In an attempt to clarify terminology and identify the mechanisms involved in multispecies interactions, Miller and Kerfoot (1987) offered a useful distinction between three classes of indirect effects, namely trophic linkage, as well as both behavioral and chemical responses. Studies in rocky intertidal communities have revealed several variations on these basic themes, their frequent occurrence and dependence on the abundance, competitive abilities, behavior, and other traits of the interactors (Menge 1995; Abrams et al. 1996). Some examples of indirect effects in coral reef communities are noted in Glynn (1988b), Carpenter (1997), Hixon (1997), and Pennings (1997). The reader is referred especially to Pennings (1997) for a review of the variety of indirect interactions observed on coral reefs. The latter author underlined the difficulties in documenting the full effects of these interactions on coral reefs.

Indirect trophic interactions may involve polychaete worms, sea urchins, sea stars, molluscs, and crustaceans. For example, intense sea urchin (*Diadema*) grazing in the Caribbean can limit algal populations, which would benefit slower growing zooxanthellate corals (Hughes 1994). The direct feeding effects of Indo-Pacific corallivores (*Acanthaster* and *Drupella*) on reef corals could indirectly provide free substrate space for the recruitment and growth of non-coral benthos such as algae, sponges, alcyonarians, and ascidians. While such effects can be pronounced they are often localized because corallivore feeding aggregations are notoriously patchy due to chance recruitment events. The likelihood that certain invertebrate corallivores, for example the polychaete worm *Hermodice carunculata* (Sussman et al. 2003) and the gastropod molluscs *Coralliophila abbreviata* (Williams and Miller 2005) and *Cyphoma gibbosum* (Burkepile and Hay 2007), serve as vectors of coral disease transmission is another step removed from a direct predatory effect.

Crabs can prevent the overgrowth of algae on corals (Coen 1988a, b), and obligate crab and shrimp symbionts can clean (remove sediment and potentially fouling species) and protect their coral hosts from predation (Glynn 1983b, 1987; Pratchett et al. 2000). Field experiments testing coral host defense by crustacean guards have shown this to be an effective antipredatory behavior throughout the Indo-Pacific region (Glynn 1976, 1983b, 1987; Pratchett 2001). On Indo-Pacific reefs, *Acanthaster* preys on generally abundant and poorly defended acroporid species and avoids pocilloporid corals with aggressive *Trapezia* guards, thus increasing the prevalence of the latter. In the eastern Pacific, abundant and defended pocilloporid corals usually predominate with *Acanthaster* preying on uncommon and unprotected non-pocilloporid species (Glynn 1987). If pocilloporid corals with crustacean guards form a barrier around preferred non-pocilloporid corals, the latter can avoid attack (Glynn 1985), an associational form of defense. The influence of these interactions, of course, depends on *Acanthaster* abundance, feeding pressure, and persistence in local coral patches.

Indirect cascading effects have been proposed from correlation analyses relating predator and prey abundances. Comparisons of fishing intensity on protected and unprotected reefs in Kenya have suggested consumer interactions resulting from overfishing of predators, corresponding increases in sea urchin abundances and declines in algal biomass and coral cover (e.g., McClanahan 1995). Another fisheries exploitation study has offered evidence for a causal relationship between reef fish invertivore densities and *Acanthaster* corallivory at 13 island sites in Fiji (Dulvy et al. 2004). With increasing fishing pressure sea star predator abundances declined, allowing increases in *Acanthaster* densities resulting in the decline of their reef coral prey and increases in filamentous algae. Thus, the overexploitation of predatory fishes had an indirect negative effect on reef-building corals and a positive effect on algal abundance.

In early studies of fish-host cleaning behavior by shrimp and fish cleaners, convincing evidence that parasite loads could influence this behavior was not forthcoming. Thus, Losey and Margules (1974) and Losey (1977, 1979) hypothesized that fish posturing behavior was more a response to tactile stimuli than to parasite infections. More recently, field and laboratory studies of fish isopod ectoparasite loads offer evidence that cleaning behavior interactions may well result in an indirect benefit to fish cleanees vis-à-vis fecundity and mortality. For example, Grutter (1995) found that fishes with more isopod parasites spent more time at cleaning stations than less-parasitized fishes. Juvenile fishes can be killed outright by parasites, and likely indirectly by parasites altering their host's behavior making them more susceptible to predation (Barber et al. 2000).

In spite of the difficulties in establishing the true nature of cleaning symbiosis, whether it is a form of mutualism or parasitism, indirect interactions are involved in this symbiosis. When first documented in the 1960s, cleaner crustaceans and fishes were assumed to indirectly benefit fish clients by reducing ectoparasite loads and to help in wound healing (Limbaugh et al. 1961; Feder 1966). Cleaners benefit from food intake, but also risk increased predation and parasitism. Potential indirect costs to clients are increased predation, en route to or at cleaning stations, and risk of injury by cleaners or cleaner mimics (Côté 2000).

It has been argued that chemical defenses have evolved on coral reefs due to intense biotic pressures from predators, competitors, pathogens, and fouling organisms (Hay 1992). Most studies in this field have focused on the secondary metabolites of algae involved in direct defenses and numerous indirect effects that benefit various invertebrates (Hay and Fenical 1988; Hay 1992; Paul 1992a). An example of the latter is the ascoglossan *Elysia halimedae* that occurs exclusively on the green alga *Halimeda*. When feeding, *E. halimedae* concentrates on tissues with a high concentration of a diterpenoid, which is converted to a closely related diterpene alcohol and sequestered by the ascoglossan (Paul and Van Alstyne 1988). This compound effectively deters fish predators. Examples of invertebrate herbivores (e.g., crabs and an amphipod) that associate with noxious host plants have also been described (reviewed in Hay 1992). Such associates do not sequester defensive metabolites, but experience reduced exposure to potential predators by association with unpalatable plants. This sort of deception has been observed in specialist mesograzers that can tolerate their host plant's chemical defenses. An intriguing case of an amphipod on Caribbean reefs, *Pseudamphithoides incurvaria*, limits fish predation by constructing and living in a portable domicile in the chemically defended algal genus *Dictyota* (Lewis and Kensley 1982; Hay and Fenical 1988). The amphipod's protective pod is carried around and consumed when in need of nutriment. Reef invertebrates feeding on sessile metazoan taxa, such as cnidarians, sponges, and tunicates, can also sequester secondary metabolites that may serve to deter predators (Paul 1992b). Finally, some species of ctenophores, flatworms, and nudibranchs that feed on cnidarians retain their prey's unfired nematocysts for defensive purposes, a phenomenon known as kleptocnidae (Greenwood and Mariscal 1984; Greenwood 2009). The ingested nematocysts in some aeolid nudibranchs are actively selected and deposited in cerata, dorsally located and conspicuously colored fleshy outgrowths that likely serve an aposomatic function (Todd 1981).

In concluding this brief survey, it can be seen that the variety and complexity of invertebrate indirect interactions on coral reefs are truly impressive. Because these types of interactions often demonstrate strong geographic and temporal variability, it is difficult to generalize and predict their effects.

6 Trophic Interactions

A coral reef ecosystem is a complex web of trophic interactions, with food and energy moving through complex pathways among countless species. Invertebrate metazoans play a critical role in these pathways and consequently in the overall functioning of the reef ecosystem. Invertebrates can be found in all feeding guilds in reef ecosystems, acting as ecologically significant predators, parasites, herbivores, scavengers, detritivores, as well as suspension and deposit feeders. The sheer magnitude of their biomass demands inclusion in trophic modeling and their role as a food source for fishes makes them important for coral reef fisheries and various ecological services.

6.1 Guilds and Reef Invertebrates

A "guild" in the ecological literature is used to describe a group of organisms that utilize the same resources in a similar way (Simberloff and Dayan 1991). This definition is admittedly vague, open to wide interpretation, and highly scale dependent, but of great importance to ecological theory and trophodynamics. The term "functional group" is often used interchangeably with "guild" and is often employed in the aquatic invertebrate literature. However, according to its original use (Cummins 1974), "functional groups" may refer simply to the resource being used, and not to the methods by which it is obtained.

The guild concept is important when considering reef invertebrates because of the wide variety of food sources and feeding methods employed. On many reefs, invertebrates are ecologically dominant herbivores (e.g., echinoids, opisthobranch molluscs), carnivores (e.g., stomatopodcrustaceans, cephalopod molluscs), and detritivores (e.g., decapod crustaceans, annelids). They utilize diverse feeding mechanisms including, but not limited to, deposit (e.g., holothurians, echiurans) and suspension feeding (e.g., sponges, bryozoans), scavenging (e.g., amphinomid polychaetes, pagurids), and predation (e.g., cirolanid isopods, portunid crabs). Consideration of these categories has probably alerted the reader to specific examples of overlap. As previously mentioned, these concepts are scale specific, dependent on the level of detail applied to the categories of resources considered and how those resources are being consumed. For instance, invertebrate carnivores may include predators, grazers, parasites, parasitoids, or scavengers. Predators alone may include planktivores, invertivores, piscivores, and corallivores. These groups may utilize foraging, ambushing, suspension feeding, or adaptations and combinations thereof. For a detailed review of the abundances, composition, and role of different invertebrate trophic groups of Enewetak Atoll, see Kohn (1987). For a review of the various types of feeding interactions that occur on coral reefs as well as how these interactions affect reef systems, see Glynn (1990).

Suspension feeders belong to one of the most studied feeding guilds in coral reef ecosystems. Members of this group are especially important as they provide a means of translating the often abundant reef plankton into benthic biomass (Sorokin 1993). In areas with high planktonic throughput, filter feeders are vital to energy/nutrient capture and assimilation. Recently, some researchers have proposed that reef cavities and high abundances of suspension feeding cryptic fauna act as a biocatalytic filter (Richter and Wunsch 1999; Richter et al. 2001). Water flowing through the high surface area of interconnecting crypts is filtered by suspension feeders that remove large amounts of suspended materials. Buss and Jackson (1981) have postulated that the efficiency of particle depletion may be high enough to affect the zonation of organisms within reef caves.

6.2 Food Webs

The next level of complexity, building upon the guild concept, seeks to draw relationships between groups of trophically related species. Within the coral reef ecosystem, the diverse variety of feeding interactions cannot be overemphasized. Simplified interpretations of these interactions may be portrayed as food webs or networks. Food-web theory (reviewed in Pimm et al. 1991) may help in the understanding of ecosystem dynamics and increase our ability to predict an ecosystem's response to a variety of perturbations. Web complexity may have profound implications for the resilience of an ecosystem or the response of multispecies assemblages to the addition or removal of a component species. Glynn (2004, 2008) has found that even in the depauperate reef systems of the eastern Pacific, food-web complexity is high. It remains to be seen whether this is true for all reef systems. It is likely, however, that high-biodiversity Indo-Pacific reef systems display levels of food-web complexity orders of magnitude higher than that of their eastern Pacific counterparts.

Despite the aforementioned information obtained from this type of analysis, coral reef ecosystem food webs remain poorly studied and uncommon in the literature. The reasons for the paucity of this type of information are many. Abundances for all but the most conspicuous taxa are lacking and most organisms remain hidden within frameworks and other reef habitat niches. Perhaps just as importantly, interspecific interactions are often temporally and spatially cryptic. These problems are compounded in the marine realm, where field conditions make frequent and prolonged observation difficult. Nocturnal feeding behaviors or infrequent and brief diurnal predation events may go unnoticed despite daily observation. Invertebrate metazoans that remain poorly enumerated due to cryptic lifestyles are even more poorly understood when it comes to behavior and diet. Despite these issues, subtle interactions such as harlequin shrimp (*Hymenocera picta*) and a polychaete worm (*Pherecardia striata*) feeding on the crown of thorns sea star (*Acanthaster planci*) may have profound effects on coral mortality and distribution (Glynn 1977, 1984). This particular interaction remained unnoticed until 1970, when it was first observed in captivity (Wickler and Seibt 1970).

In addition to functioning as herbivores and carnivores, perhaps one of the most ubiquitous features of invertebrates in reef food webs is their prevalence in lower trophic levels, where they elevate energy from less-desirable food sources to higher trophic levels (e.g., Glynn 2008). As detritivores, invertebrates utilize biotic by-products and decomposed organic matter. Feces, among other things, are consumed by a diverse assemblage of deposit and filter feeding invertebrates. While coprophagy may be common among reef fishes (Robertson 1982), feces quickly fall to the benthos and accumulate in reef cavities, thereby favoring consumption by smaller, more cryptic invertebrates.

Perhaps the most studied pathways of by-product consumption are those that originate from scleractinian corals. Coral mucus, naturally secreted by the coral animal as a cleaning, feeding, and defensive mechanism, may be an important source of nutrients for a variety of reef organisms. Some invertebrates have been observed to feed directly upon mucus (e.g., *Mithrax* spp., Coffroth 1990), however it is likely that much of the mucus is consumed by benthic microorganisms (Wild et al. 2004), which may in turn be consumed by metazoan meiofauna and detritivores. Similarly, mass coral spawning events produce large amounts of unfertilized gametes that are fed on by prokaryotes, which in turn act as food for abundant invertebrate meiofauna and suspension feeders (Guest 2008).

6.3 Quantitative Modeling

A pioneering attempt to quantify the biomass and trophic groups of organisms occupying a reef ecosystem was that of Odum and Odum (1955). In their calculations, the biomass of lower trophic level invertebrates greatly exceeded that of reef fishes. Across four independently sampled zones, invertebrates accounted for 96.9% of the herbivore biomass and 87.2% of the carnivore biomass. Across all sites and between both trophic groups, invertebrate biomass exceeded piscine biomass by at least 170% and in areas where fishes were present, invertebrate biomass was up to 1,489 times that of fishes. Invertebrate biomass was especially prevalent on the algal coral ridge at 162 g m^{-2} dry weight, followed closely by zones of large and small coral colonies with 128 and 123 g m^{-2} dry invertebrate biomass, respectively.

More recently, workers have begun quantitatively modeling reef trophic interactions. This work has primarily developed from the pioneering approach of Polovina (1984) and the ECOPATH model. At its heart, the ECOPATH approach is a system of mass-balanced equations for a steady state system. A variety of equations feed into the central concept that production of biomass for a given species, minus predation, minus other biomass loss (natural mortality) is equal to zero. Solving these systems of linear equations allows estimation of population parameters for otherwise poorly studied taxa such as coral reef invertebrates.

Arguably, the most complex and thorough model of a reef ecosystem to date is Opitz's (1996) ECOPATH model of a Caribbean coral reef. Her fully parameterized 50-box model (modified in Fig. 7) estimates invertebrate biomass (wet weight) as being roughly 1.6 kg m^{-2} of reef habitat. This value is 15.3 times higher than vertebrate biomass including fishes, seabirds, and turtles. An ECOPATH model constructed for a reef system in Moorea (Arias-González et al. 1997) resulted in a more conservative ratio of invertebrate to vertebrate biomass of 2.6. In this study, invertebrates comprised 0.34 kg m^{-2} (wet weight). These differences may be ecosystem specific or simply artifacts of how the models were parameterized. Regardless, relative to fish communities invertebrate biomass is numerically dominant.

The magnitude of the invertebrate community and high degree of trophic connectivity within a coral reef ecosystem is highlighted in Fig. 7. In this box model, the size of each rectangle is proportional to the logarithm of the biomass of that functional group. Invertebrate biomass has been highlighted in gray and clearly shows the prevalence of these taxa. Green lines denote trophic pathways involving the direct consumption of invertebrate prey, whereas red represents other trophic interactions and blue involves vertebrate and invertebrate consumption. These color codes, along with the central location of invertebrate taxa, highlight the integral nature of invertebrate communities underpinning the trophic dependencies of a coral reef ecosystem.

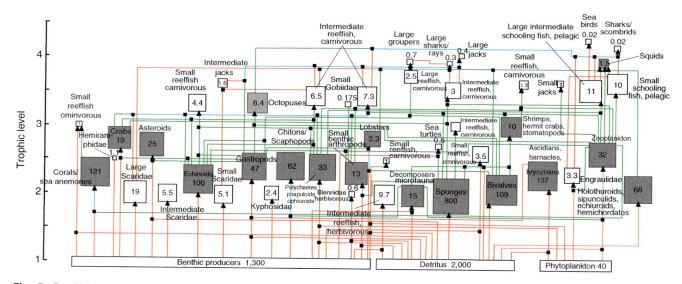

Fig. 7 Graphical representation of the trophic interactions of a Caribbean coral reef from the 50-box ECOPATH model of Opitz (1996). The size of each box is proportional to the logarithm of the biomass (indicated in g m^{-2} wet weight) of the corresponding groups. The graphic has been modified to highlight the high proportion of invertebrate biomass (*gray boxes*) and trophic pathways involving the direct consumption of invertebrate taxa (*green lines*). Pathways involving the direct consumption of non-invertebrate (primary producers and vertebrates) are indicated in *red*. Pathways involving the nonexclusive consumption of invertebrate taxa are indicated in *blue*

Despite the marked prevalence of invertebrates in Opitz's model, she observes that the data used to parameterize this component are lacking. This is primarily due to insufficient data concerning abundances and biomass of cryptic invertebrates. Information on species-specific feeding preferences, metabolic requirements, interspecific interactions, and reproductive biology of these hidden taxa is largely absent from the literature. This information is of paramount importance for understanding the ecology of these organisms, and by extension, their contribution to the ecology of coral reef ecosystems.

6.4 Prey for Fishes

To develop an understanding of the role of invertebrate taxa as food sources for fishes, one need only consider the diverse array of feeding behaviors and adaptive morphologies employed by invertivore fishes on coral reefs (reviewed in Hiatt and Strasburg 1960; Hobson 1974, 1991; Wainwright and Bellwood 2002). Squirrelfishes (Holocentridae) emerge from shelters at night to feed on benthic crustaceans, polychaetes, and molluscs. Moray eels (Muraenidae) utilize their elongate, slender bodies and strong olfactory senses to locate and prey on otherwise concealed invertebrates. Some triggerfishes (Balistidae) employ their robust jaws to break apart or perturb reef carbonates, thereby exposing previously sheltered invertebrates. Nocturnal planktivorous fishes (e.g., Apogonidae) often have larger mouths than diurnal planktivores (e.g., Pomacentridae), presumably an adaptation reflecting the relatively larger size of nocturnal invertebrate zooplankton (Hobson 1974). Many corallivore fishes (e.g., Chaetodontidae, Monacanthidae) have elongate mouths that allow them to bite off coral polyps or colony branch tips. Goat fishes (Mullidae) employ chemosensory barbels to locate infaunal invertebrates, concealed in reef sediments. Additionally, many transient fishes, not considered to be obligate reef associates (e.g., Carangidae), may regularly enter reef ecosystems and exploit abundant invertebrate food sources.

In a review of studies from multiple localities, Bakus (1966) noted that carnivores comprised approximately 65% of reef fishes, while herbivores accounted for only 25% (both species richness and biomass were considered). In a study of 5,526 West Indian reef fishes, Randall (1967) found that arthropods were the dominant food source of 90 of the 212 sampled fish species. In contrast, only 48 species of fishes had stomach contents consisting of more than 50% fish prey. It is a testament to the diversity and importance of invertebrate reef taxa that the number of species of invertivore fishes feeding primarily on a single taxon (Crustacea) was approximately twice the number of species of piscivorous fishes. It should be noted that other studies (e.g., Odum and Odum 1955) have found the biomass of herbivorous fishes to be higher than that of carnivorous fishes in reef ecosystems.

There is often a relationship between the abundances as well as species richness of fishes and invertebrates in reef ecosystems. Wolf et al. (1983) observed that invertebrates may impact the distribution and abundance of invertivore fishes occupying artificial reef frameworks in St. Croix. These authors postulated that the settlement of many fish species to reefs was positively influenced by the previous colonization of invertebrate food sources. Klumpp et al. (1988) have found that invertebrate microherbivores can be important consumers of benthic reef algae, especially in damselfish territories. In unsheltered reef areas, not guarded by damselfishes, these authors postulated that reduced cryptic invertebrate abundances were due in part to intense predation by fishes. In fact, high invertivory in the reef environment is considered by some (e.g., Bakus 1966, 1981; Faulkner 1992; Paul 1992b) to be a probable driver behind the evolution of the cryptic behaviors, antipredatory morphologies, and chemical defenses, so widespread among reef invertebrates.

Often fish–invertivore interactions are most apparent in reef environments that experience human exploitation. Ashworth et al. (2004) have observed higher abundances of invertebrates (holothurians, diadematid echinoids, conchs) in heavily fished areas compared with no-take marine reserves, suggesting that predation pressure by fishes may be exerting top-down control on invertebrate populations. Similarly, Dulvy et al. (2002) found the richness of vagile epifaunal invertebrates to be negatively correlated with fishing pressure, again suggesting a close trophic connection between pelagic fishes and benthic invertebrate populations. In extreme situations, overfishing of invertivore fishes may result in grossly elevated abundances of sea urchins (e.g., *Echinometra mathaei*) that may in turn competitively exclude less-abundant species as well as contribute to decreases in coral cover, benthic complexity, and ultimately lead to phase shifts (McClanahan and Muthiga 1988).

7 Outlook and Implications

7.1 Major Recent Developments

Global climate change, related species population responses, and the predictions of major perturbations to marine ecosystems in the near future have dominated coral reef studies during the past few decades. Already, widespread disturbances to coral reefs resulting from climate change stressors and various other anthropogenic influences have been abundantly documented. Besides obvious effects caused by increases in ocean acidification, such as depressed coral calcification and colony growth (e.g., Gattuso et al. 1998; Kleypas and

Langdon 2006; Hoegh-Guldberg et al. 2007), several recent studies are demonstrating negative effects in invertebrate reproduction and early life history stages. Demonstrated effects from elevated acidification experiments, to projected levels in the near future, include (1) reduced fertilization in a sea urchin (Havenhand et al. 2008), (2) reduced viability in early molluscan and echinoid development (Kurihara and Shirayama 2004; Kurihara et al. 2007; Ellis et al. 2009), and (3) impairment of calcification of the exoskeleton in a lobster (Arnold et al. 2009). For reviews of the myriad impacts that ocean acidification can have on diverse invertebrate and vertebrate taxa, see Shirayama and Thornton (2005) and Fabry et al. (2008). See Byrne et al. (2009), however, who showed that sea urchin fertilization was compromised by elevated temperature and not increased acidification. These results suggest that numerous invertebrate taxa with calcareous skeletons, such as sponges, annelid worms, bryozoans, brachiopods, molluscs, crustaceans, and tunicates, would likely be affected by ocean acidification at early life history stages.

Corallivory is now recognized as a significant determinant of coral community structure, with diverse taxa involved in this mode of feeding. Considering the exceptionally high mortality rates of early life history stages of corals, methods should be developed to determine the causative agents. From laboratory observations in the western Atlantic, Cooper et al. (2007) found that swarms of the ciliate protistan *Helicostoma nonatum* could consume *Porites* spat within 6 h. The same or a closely related ciliate possibly causes brown band syndrome in corals on the Great Barrier Reef, Australia (Willis et al. 2004). Could similar pathogens or micropredators contribute toward high rates of early post-settlement mortality in natural reef settings? It has also been recognized during the past few decades of elevated coral mortality that weakened corals (from bleaching, disease, interalia) become more susceptible to predator attack. Our understanding of the contributions of non-coral invertebrate predators and parasites, particularly cryptic and motile taxa, to reef energetics and community structure are woefully inadequate. Their effects must be considerable in light of the high numbers and diversity of taxa so far known.

The rapidly advancing discipline of phylogenetic analysis, utilizing molecular, biochemical and new morphological approaches, is allowing for a better understanding of invertebrate evolutionary relationships (e.g., Giribet et al. 2007; Dunn et al. 2008; Philippe et al. 2009). Molecular sequence data, permitting the examination of genotypic diversity directly, are offering unprecedented insights of deep phylogenetic relationships as well as discrimination of taxonomic boundaries among living Metazoa. Several of the classification schemes introduced earlier in this review are a result of this research. Like phylogenies constructed from classical morphological traits, however, these molecular-based classifications are not without considerable controversy.

7.2 Likely Future Research Foci

Compared with nontropical regions, there are relatively few documented cases of nonindigenous invertebrate species introductions into coral reef areas (Coles and Eldredge 2002; Hutchings et al. 2002). One highly contested example involves the probable invasion of a Red Sea stomatopod crustacean (*Gonodactylus falcatus*), resulting in the virtual displacement of a native Hawaiian species (*Pseudosquilla ciliata*). This invasion has not led to the extinction of *P. ciliata*, but to a greatly diminished abundance in its dead coral habitat. The most recent reported surveys indicate that *G. falcatus* now outnumbers *P. ciliata* by 100 to 1 (Coles and Eldredge 2002). A second example involves two azooxanthellate corals that have invaded coral reef areas in the southwestern Atlantic (de Paula and Creed 2004). These two species, *Tubastraea coccinea* and *Tubastraea tagusensis*, now occur in shallow coral communities where they compete with the zooxanthellate Brazilian endemic *Mussismilia hispida*. Field evidence suggests that the two alien species are superior competitors, causing necrosis, tissue death, and skeletal deformation at and near contact zones in *M. hispida* (Creed 2006).

Are these examples unusual or an indication of the lack of studies on coral reefs addressing such problems? It has been hypothesized that the dense species packing and high connectivity of western Pacific coral reefs may make it more difficult for invasive species to become established (Coles and Eldredge 2002; Hutchings et al. 2002). This possibility must be balanced with likely changes in reef species community composition, such as declines in population abundances, displacements, and extinctions vis-à-vis global climate change. Thus, expected declines in coral community biodiversity may render these assemblages more vulnerable to nonindigenous species disturbances. Finally, measures must be taken to prevent the introductions of coral reef species known or suspected of causing major disruptions. Two examples of concern are the Indo-Pacific and tropical Atlantic corallivores *Acanthaster planci* and *Hermodice carunculata*, respectively, which occur on opposite sides of the Isthmus of Panamá. These species have been shown to feed on exotic coral prey (Glynn 1982b), and *H. carunculata* is a potential reservoir and vector of a coral-bleaching pathogen (Sussman et al. 2003).

Recent years have seen a rise in collaborative research projects with goals of cataloguing and describing coral reef invertebrate biodiversity. Large-scale biodiversity censuses have been conducted in New Caledonia (Mollusca; Bouchet et al. 2002), Guam and the Marianas Islands (Paulay 2003), Moorea (Biocode project, http://bscit.berkeley.edu/biocode), as well as the French Frigate Shoals and Australia (CReefs project, http://www.creefs.org). These projects are illuminating a level of biodiversity that dwarfs our current records of reef invertebrate biota.

To deal with this growing wealth of information, a variety of databases are currently under construction containing systematic, morphological, genetic, and ecological data for a variety of invertebrate reef taxa. These databases are often limited to a single high-level taxon (e.g., Hexacorallia, Fautin and Buddemeier 2006; Indo-Pacific Mollusca, OBIS Indo-Pacific Molluscan Database, http://clade.ansp.org/obis/find_mollusk.html; Central/southeastern Pacific Crustacea, Poupin 2003). Several institutions, however, are constructing comprehensive multi-taxa biodiversity databases for specific geographic locations (e.g., Australian faunal directory, ABRS 2009; Cook Island Biodiversity Database, McCormack 2007).

There is a growing trend toward the digitization and free dissemination of taxonomic literature (e.g., http://www.biodiversitylibrary.org/, http://www.decapoda.nhm.org/). Interactive keys to marine phyla found on reefs are being developed and made publically available (e.g., www.crustacea.net). Several international agencies are collaborating to create the Integrated Taxonomic Information System (www.itis.gov), which provides researchers with information on the validity of species names and their taxonomic classification. Additionally, museums and research universities (e.g., Florida Museum of Natural History, Harvard University Museum of Comparative Zoology, U.S. National Museum of Natural History), which in the past have been poorly accessible depositories of taxonomic information, are in the process of digitizing their collection manifests, allowing researchers across the globe to quickly find and access sample material.

Genetic barcoding, which has proven to be a controversial but viable technique among some marine taxa (e.g., Crustacea, Costa et al. 2007), will most likely be employed on a greater scale and reliable techniques will be developed for more phyla. These molecular techniques will doubtless become more widespread but they will not be able to completely replace morphological and developmental-based taxonomy.

In the future, these institutions, databases, and web tools will allow for a greater degree of collaboration and user accessibility, thereby providing scientists with the information and instruments necessary to conduct increasingly complex, ecosystem-based coral reef research. Already, web portals such as World Register of Marine Species (WoRMS, SMEBD 2009, http://www.marinespecies.org) are facilitating the communication and collaboration of multiple institutions, databases, and researchers. As these data portals mature, they will most likely lead to greater ease of invertebrate identifications, which may in turn direct more research toward these poorly studied taxa.

We are presently at a critical juncture in light of recent disturbances and consequent degradation affecting global coral reefs. How might invertebrates exacerbate or diminish coral reef decline, and what will be the fate of invertebrate taxa in a world of waning coral reef development? Is it possible that some non-symbiotic invertebrate taxa may assume more prominent roles in the development or degradation of tropical marine communities? These questions will likely drive reef invertebrate research in the coming decades.

Acknowledgments Several persons helped with various aspects of this review for which we are grateful: Literature – Viktor W. Brandtneris, William E. Browne, Anne C. Campbell, Angela C. Clark, Daniel J. Diresta, Elizabeth A. Fish, Lyza Johnston, William A. Newman, and Peter K. Swart; Photographs – Orit Barneah, Charles Birkeland, Lyza Johnston, P. Laboute, Haris A. Lessios, and Michael C. Schmale; Taxonomic advice – Leslie Harris, José H. Leal, Charles Messing, David L. Pawson, Rob W.M. van Soest, Nancy Voss, and Philippe Willenz; Overall review – Richard C. Brusca, Daniel J. Diresta, John S. Pearse, and Bernhard Riegl. Our contributions to this review have been supported by the U.S. National Science Foundation (Biological Oceanography Program), grant OCE-0526361 and earlier awards.

References

Abe N (1938) Feeding behaviour and the nematocysts of *Fungia* and 15 other species of coral. Palau Trop Biol Stn Stud 1:469–521

Abelson A, Galil BS, Loya Y (1991) Skeletal modifications in stony corals caused by indwelling crabs: hydrodynamical advantages for crab feeding. Symbiosis 10:233–248

Abrams P, Menge BA, Mittelbach GG, Spiller D, Yodzis P (1996) The role of indirect effects in food webs. In: Polis GA, Winemiller KO (eds) Food webs: integration of pattern and dynamics. Chapman and Hall, New York, pp 371–395

ABRS (2009) Australian Faunal Directory. Australian Biological Resources Study, Canberra. http://www.environment.gov.au/biodiversity/abrs/onlineresources/fauna/afd/index.html

Achituv Y, Mizrahi L (1996) Recycling of ammonium within a hydrocoral (*Millepora dichotoma*)-zooxanthellae-cirripede (*Savignium milleporum*) symbiotic association. Bull Mar Sci 58:856–860

Achituv Y, Brickner I, Erez J (1997) Stable carbon isotope ratios in Red Sea barnacles (Cirripedia) as an indicator of their food source. Mar Biol 130:243–247

Aeby GS (1992) The potential effect the ability of a coral intermediate host to regenerate has had on the evolution of its association with a marine parasite. In: Proceedings of 7th international coral reef symposium, vol 2, Guam, 1992, pp 809–815

Aeby GS (1998) A digenean metacercaria from the reef coral, *Porites compressa*, experimentally identified as *Podocotyloides stenometra*. J Parasitol 84:1259–1261

Ahyong ST, Erdmann MV (2003) The stomatopod Crustacea of Guam. Micronesica 35–36:315–352

Alcock A (1902) A naturalist in Indian seas. John Murray, London, 328 pp

Alcolado PM, Claro-Madruga R, Menéndez-Macias G, García-Parrado P, Martínez-Daranas B, Sosa M (2003) The Cuban coral reefs. In: Cortés J (ed) Latin American coral reefs. Elsevier, Amsterdam, pp 53–75

Alldredge AL, King JM (1977) Distribution, abundance, and substrate preferences of demersal reef zooplankton at Lizard Island Lagoon, Great Barrier Reef. Mar Biol 41:317–333

Alongi DM (1986) Population structure and trophic composition of the free-living nematodes inhabiting carbonate sands of Davis Reef, Great Barrier Reef Australia. Aust J Mar Freshw Res 37:609–619

Alongi DM (1989) The role of soft-bottom benthic communities in tropical mangrove and coral reef ecosystems. CRC Crit Rev Aquat Sci 1:243–280

Alvarado JJ, Cortés J (eds) (2005) Research on echinoderms in Latin America. Rev Biol Trop (Int J Trop Biol) 53(suppl 3):1–387

Alvarado JJ, Cortés J (eds) (2008) Research on echinoderms in Latin America II. Rev Biol Trop (Int J Trop Biol) 56:1–364

Ambariyanto A, Hoegh-Guldberg O (1999) Net uptake of dissolved free amino acids by the giant clam, *Tridacna maxima*: alternative sources of energy and nitrogen? Coral Reefs 18:91–96

Amin OM (1982) Acanthocephala. In: Parker SP (ed) Synopsis and classification of living organisms. McGraw-Hill Book Co, New York, pp 467–474

Antonius A, Riegl B (1997) A possible link between coral diseases and a corallivorous snail (*Drupella cornus*) outbreak in the Red Sea. Atoll Res Bull 447:1–9

Arias-González JE, Delesalle B, Salvat B, Galzin R (1997) Trophic functioning of the Tiahura reef sector, Moorea Island, French Polynesia. Coral Reefs 16:231–246

Arnold KE, Findlay HS, Spicer JI, Daniels CL, Boothroyd D (2009) Effect of CO_2-related acidification on aspects of the larval development of the European lobster, *Homarus gammarus* (L.). Biogeoscience 6:1747–1754

Ashworth JS, Ormond RFG, Sturrock HT (2004) Effects of reef-top gathering and fishing on invertebrate abundance across take and no-take zones. J Exp Mar Biol Ecol 303:221–242

Bailey-Brock JH (1976) Habitats of tubicolous polychaetes from the Hawaiian Islands and Johnston Atoll. Pac Sci 30:69–81

Bailey-Brock JH, Emig CC (2000) Hawaiian Phoronida (Lophophorata) and their distribution in the Pacific region. Pac Sci 54:119–126

Bailey-Brock JH, White JK, Ward LA (1980) Effects of algal turf and depressions as refuges on polychaete assemblages of a windward reef bench at Enewetak Atoll. Micronesica 16:43–58

Baird A (1999) A large aggregation of *Drupella rugosa* following the mass bleaching of corals on the Great Barrier Reef. Reef Res 9:6–7

Bak RPM, Sybesma J, van Duyl FC (1981) The ecology of the tropical compound ascidian *Trididemnum solidum*. II. Abundance, growth and survival. Mar Ecol Prog Ser 6:43–52

Bak RPM, Lambrechts DYM, Joenje M, Nieuwland G, Van Veghel MLJ (1996) Long-term changes on coral reefs in booming populations of a competitive colonial ascidian. Mar Ecol Prog Ser 133:303–306

Bakus GJ (1966) Some relationships of fishes to benthic organisms on coral reefs. Nature 210:280–284

Bakus GJ (1973) The biology and ecology of tropical holothurians. In: Jones OA, Endean R (eds) Biology and geology of coral reefs 1, Biol 1. Academic, New York, pp 325–367

Bakus GJ (1981) Chemical defense mechanisms and fish feeding behavior on the Great Barrier Reef, Australia. Science 211:497–499

Bakus GJ, Targett NM, Schulte B (1986) Chemical ecology of marine organisms: an overview. J Chem Ecol 12:951–987

Barber I, Hoare D, Krause J (2000) Effects of parasites on fish behaviour: a review and evolutionary perspective. Rev Fish Biol Fish 10:131–165

Barneah O, Brickner I, Hooge M, Weis VM, LaJeunesse TC, Benayahu Y (2007) Three party symbiosis: acoelomorph worms, corals and unicellular algal symbionts in Eilat (Red Sea). Mar Biol 151:1215–1223

Barnes DJ (1973) Growth in colonial scleractinians. Bull Mar Sci 23:280–298

Basile LL, Cuffey RJ, Kosich DF (1984) Sclerosponges, pharetronids, and sphinctozoans (relict cryptic hard-bodied Porifera) in the modern reefs of Enewetak Atoll. J Paleontol 58:636–650

Baums IB, Miller MW, Szmant AM (2003) Ecology of a corallivorous gastropod, *Coralliophila abbreviata*, on two scleractinian hosts. II. Feeding, respiration and growth. Mar Biol 142:1093–1101

Bayer FM (1981) Key to the genera of Octocorallia exclusive of Pennatulacea (Coelenterata: Anthozoa), with diagnoses of new taxa. Proc Biol Soc Wash 94:901–947

Becker JH, Grutter AS (2004) Cleaner shrimp do clean. Coral Reefs 23:515–520

Beets J, Lewand L (1986) Collection of common organisms within the Virgin Islands National Park/Biosphere Reserve. Biosphere Reserve Research Report 3, 45 p. Virgin Islands Resource Management Cooperative, Virgin Islands National Park

Ben-Tzvi O, Einbinder S, Brokovish E (2006) A beneficial association between a polychaete worm and a scleractinian coral? Coral Reefs 25:98

Benzoni F, Galli P, Pichon M (2009) Pink spots on *Porites*: not always a coral disease. Coral Reefs. doi:10.1007/s00338-009-0571-z

Birkeland C (1982) Terrestrial runoff as a cause of outbreaks of *Acanthaster planci* (Echinodermata: Asteroidea). Mar Biol 69:175–185

Birkeland C (1989a) The influence of echinoderms on coral-reef communities. In: Jangoux M, Lawrence JM (eds) Echinoderm studies 3. Balkema, Rotterdam, pp 1–79

Birkeland C (1989b) The Faustian traits of the crown-of-thorns starfish. Am Sci 77:154–163

Birkeland C (1997) Life and death of coral reefs. Chapman and Hall, New York

Birkeland C, Lucas JS (1990) *Acanthaster planci*: major management problem of coral reefs. CRC Press, Boca Raton, pp 257

Boschma H (1948) The species problem in *Millepora*. Zool Verh Leiden 1:1–115

Boucher-Rodoni R, Boucaud-Camou E, Mangold K (1987) Feeding and digestion. In: Boyle PR (ed) Cephalopod life cycles 2. Comparative reviews. Academic, London, pp 85–108

Bouchet P (2006) The magnitude of marine biodiversity. In: Duarte CM (ed) The exploration of marine biodiversity: scientific and technological challenges. Fundación BBVA, Bilbao, pp 33–64

Bouchet P, Lozouet P, Maestrati P, Héros V (2002) Assessing the magnitude of species richness in tropical marine environments: exceptionally high numbers of molluscs at a New Caledonia site. Biol J Linn Soc 75:421–436

Bouchet P, Rocroi J-P (2005) Classification and nomenclature of gastropod families. Malacologia 47:1–397

Brander KM, McLeod A, Humphreys WF (1971) Comparison of species diversity and ecology of reef-living invertebrates on Aldabra Atoll and at Watamu, Kenya. Symp Zool Soc Lond 28:397–431

Brodie J, Fabricius K, De'ath G, Okaji K (2005) Are increased nutrient inputs responsible for more outbreaks of crown-of-thorns starfish? An appraisal of the evidence. Mar Poll Bull 51:266–278

Brock RE, Brock JH (1977) A method for quantitatively assessing the infaunal community in coral rock. Limnol Oceanogr 22:948–951

Bruce AJ (1976) Shrimps and prawns of coral reefs, with special reference to commensalisms. In: Jones OA, Endean R (eds) Biology and geology of coral reefs 3, biol. 2. Academic, New York, pp 37–94

Bruce AJ (1998) New keys for the identification of Indo-West Pacific coral associated pantoniine shrimps, with observations on their ecology. Ophelia 49:29–46

Bruce AJ, Trautwein SE (2007) The coral gall shrimp, *Paratypton siebenrocki* Balss, 1914 (Crustacea: Decapoda: Pontoniinae), occurrence in French Polynesia, with possible abbreviated larval development. Cah Biol Mar 48:225–228

Brusca RC, Brusca GJ (2003) Invertebrates, 2nd edn. Sinauer Association, Inc, Sunderland

Bunkley-Williams L, Williams EH (1998) Ability of pederson cleaner shrimp to remove juveniles of the parasitic cymothoid isopod, *Anilocra haemuli*, from the host. Crustaceana 71:862–869

Burkepile DE, Hay ME (2007) Predator release of the gastropod *Cyphoma gibbosum* increases predation on gorgonian corals. Oecologia 154:167–173

Buss LW (1979) Bryozoan overgrowth interactions: the interdependence of competition for space and food. Nature 281:475–477

Buss LW (1986) Competition and community organization on hard surfaces in the sea. In: Diamond J, Case TJ (eds) Community ecology. Harper & Row, New York, pp 517–536

Buss LW, Jackson JBC (1979) Competitive networks: nontransitive competitive relationships in cryptic coral reef environments. Am Nat 113:223–234

Buss LW, Jackson JBC (1981) Planktonic food availability and suspension-feeder abundance: evidence of in situ depletion. J Exp Mar Biol Ecol 49:151–161

Byrne M, Ho M, Selvakumaraswamy P (2009) Temperature, but not pH, compromises sea urchin fertilization and early development under near-future climate change scenarios. Proc R Soc B Biol Sci 276:1883–1888

Cairns SD (1983) A generic revision of the Stylasteridae (Coelenterata: Hydrozoa). Part 1. Description of the genera. Bull Mar Sci 33: 427–508

Cairns SD (1991) A generic revision of the Stylasteridae (Coelenterata: Hydrozoa). Part 3. Keys to the genera. Bull Mar Sci 49:538–545

Caldwell RL, Roderick GK, Shuster SM (1989) Studies of predation by *Gonodactylus bredini*. In: Ferrero EA (ed) Biology of stomatopods. Selected symposia and monographs U.Z.I., 3 Mucchi, Modena, pp 117–131

Canon LRG (1972) Biological associates of *Acanthaster planci*. Crown-of-Thorns Seminar. University of Queensland, Brisbane, pp 9–18

Carleton JH, Hamner WM (1989) Resident mysids: community structure, abundance and small-scale distributions in a coral reef lagoon. Mar Biol 102:461–472

Carpenter RC (1997) Invertebrate predators and grazers. In: Birkeland C (ed) Life and death of coral reefs. Chapman & Hall, New York, pp 198–229

Carpenter KE, Abrar M, Aeby G, Aronson RB, plus 35 other authors (2008) One-third of reef-building corals face elevated extinction risk from climate change and local impacts. Science 321:560–563

Carter JW (1982) Natural history observations on the gastropod shell-using amphipod *Photis conchicola* Alderman, 1936. J Crust Biol 2: 328–341

Castellanos-Martinez S, Gómez-del Prado Rosas C, Hochberg FG (2007) Dicyemid parasites in the kidneys of *Octopus hubbsorum* Berry, 1953 in Bahia de La Paz, BCS, Mexico. WSM Annual Report 40, 39

Castro P (1971) Nutritional aspects of the symbiosis between *Echinoecus pentagonus* and its host in Hawaii, *Echinothrix calamaris*. In: Cheng TC (ed) Aspects of the biology of symbiosis. University Park Press, Baltimore, pp 229–247

Castro P (1988) Animal symbioses in coral reef communities: a review. Symbiosis 5:161–184

Castro P (2000) Biogeography of trapezoid crabs (Brachyura, Trapeziidae) symbiotic with reef corals and other cnidarians. In: von Vaupel Klein JC, Schram FR (eds) The biodiversity crisis and crustacea. Proceedings 4th International Crustacean Congress, Amsterdam, vol 2. Crustacean Issues 12:65–75

Castro P (2003) The trapezoid crabs (Brachyura) of Guam and Northern Mariana Islands, with the description of a new species *Trapezia latreille*, 1828. Micronesica 35–36:440–455

Catala R (1950) Contribution à l'étude écologique des îlots coralliens du Pacifique sud. Bull Biol France Belgique (Paris) 84:234–310

Chadwick NE, Ďuriš Z, Horká I (2008) Biodiversity and behavior of shrimps and fishes symbiotic with sea anemones in the Gulf of Aqaba, northern Red Sea. In: Por FD (ed) Aqaba-Eilat, the improbable Gulf: environment, biodiversity and preservation. The Hebrew University Magnes Press, Jerusalem, pp 209–239

Chartock MA (1983) Habitat and feeding observations on species of *Ophiocoma* (Ophiocomidae) at Enewetak. Micronesica 19:131–149

Chavanich S, Ketdecha N, Viyakarn V, Bussarawit S (2007) Preliminary surveys of the commensal amphipod, *Leucothoe spinicarpa* (Abildgaard, 1789), in the colonial tunicate, *Ecteinascidia thurstoni* Herdman, 1891, in the Andaman Sea, Thailand. Pub Seto Mar Biol Lab Spec Pub Ser 8:97–101

Cheng TC (1967) Marine molluscs as hosts for symbioses, with a review of known parasites of commercially important species. Adv Mar Biol 5:1–424

Cheng TC (1986) General parasitology. Academic, New York

Chesher RH (1969) Destruction of Pacific coral reefs by the sea star *Acanthaster planci*. Science 165:280–283

Chevalier J-P (1987) Ordre des Scléractiniaires (Scleractinia Bourne, 1900; Madreporaria (Pars) Milne Edwards et Haime, 1857; Hexacorallia (Pars) Haeckel, 1866). In: Grassé PP (ed) Traité de zoologie: anatomie, systématique, biologie, III. Masson et Cie Éditeurs, Paris, pp 679–753

Clark AM (1976) Echinoderms of coral reefs. In: Jones OA, Endean R (eds) Biology and geology of coral reefs, biol 2, vol 3. Academic, New York, pp 95–123

Coen LD (1988a) Herbivory by crabs and the control of algal epibionts on Caribbean host corals. Oecologia 75:198–203

Coen LD (1988b) Herbivory by Caribbean majid crabs: feeding ecology and plant susceptibility. J Exp Mar Biol Ecol 122:257–276

Coffroth MA (1990) Mucous sheet formation on poritid corals: An evaluation of coral mucus as a nutrient source on reefs. Mar Biol 105:39–49

Coles SL, Eldredge LG (2002) Nonindigenous species introductions on coral reefs: a need for information. Pac Sci 56:191–209

Colgan MW (1985) Growth rate reduction and modification of a coral colony by a vermetid mollusk, *Dendropoma maxima*. In: Proceedings of 5th International Coral Reef Congress, vol 6, Tahiti, 1985, pp 205–210

Cook PA, Stewart BA, Achituv Y (1991) The symbiotic relationship between the hydrocoral *Millepora dichotoma* and the barnacle *Savignium milleporum*. Hydrobiologia 216(217):285–290

Cooper W, Lirman D, Schmale M, Lipscomb D (2007) Consumption of coral spat by histophagic ciliates. Coral Reefs 26:249–250

Coppard SE, Campbell AC (2004) Organisms associated with diadematid echinoids in Fiji. In: Heinzeller T, Nebelsick JH (eds) Echinoderms: München. Taylor & Francis, London, pp 171–176

Costa FO, deWaard JR, Boutillier J, Ratnasingham S, Dooh RT, Hajibabaei M, Hebert PDN (2007) Biological identifications through DNA barcodes: the case of the Crustacea. Can J Fish Aquat Sci 64:272–295

Côté IM (2000) Evolution and ecology of cleaning symbioses in the sea. Oceanogr Mar Biol Ann Rev 38:311–355

Coull BC (1990) Are members of the meiofauna food for higher trophic levels? Trans Am Microsc Soc 109:233–246

Cowles RP (1920) Habits of tropical crustacea: III Habits and reactions of hermit crabs associated with sea anemones. Philipp J Sci 15:81–90

Creed JC (2006) Two invasive alien azooxanthellate corals, *Tubastraea coccinea* and *Tubastraea tagusensis*, dominate the native zooxanthellate *Mussismilia hispida* in Brazil. Coral Reefs 25:350

Cronin TW, Caldwell RL, Marshall J (2006) Learning in stomatopod crustaceans. Int J Comp Psychol 19:297–317

Cruz-Rivera E, Paul VJ (2006) Feeding by coral reef mesograzers: algae or cyanobacteria. Coral Reefs 25:617–627

Cuffey RJ (1972) The roles of bryozoans in modern coral reefs. Int J Earth Sci 61:542–550

Cummins KW (1974) Structure and function of stream ecosystems. BioScience 24:631–641

Cutler EB (1995) The Sipuncula: their systematics, biology, and evolution. Cornell University Press, New York, pp 453

De Paula AF, Creed JC (2004) Two species of the coral *Tubastraea* (Cnidaria, Scleractinia) in Brazil: a case of accidental introduction. Bull Mar Sci 74:175–183

De Ridder C, Dubois P, Lahaye MC, Jangoux M (eds) (1990) Echinoderm research. In: Proceedings of 2nd European conference, echinoderms, Brussels, AA Balkema, Rotterdam

Devaney DM, Eldredge LG (1987) Phylum Nemertea (Rhynchocoela). In: Devaney DM, Eldredge LG (eds) Reef and shore fauna of Hawaii Section 2: Platyhelminthes through Phoronida and Section 3: Sipuncula through Annelida. Bishop Museum Special Publication 64 (2 and 3), Bishop Museum Press, Honolulu, pp 59–69

De Vantier LM, Endean R (1988) The scallop *Pedum spondyloideum* mitigates the effects of *Acanthaster planci* on the host coral *Porites*: host defense facilitated by exaptation? Mar Ecol Prog Ser 47:293–301

De Vantier LM, Reichelt RE, Bradbury RH (1986) Does *Spirobranchus giganteus* protect host *Porites* from predation by *Acanthaster planci*: predation pressure as a mechanism of coevolution? Mar Ecol Prog Ser 32:307–310

Diaz MC, Rützler K (2001) Sponges: an essential component of Caribbean coral reefs. Bull Mar Sci 69:535–546

Dilly PN (1985) The habitat and behavior of *Cephalodiscus gracilis* (Pterobranchia, Hemichordata) from Bermuda. J Zool Lond (A) 207:223–239

Dilly PN, Ryland JS (1985) An intertidal *Rhabdopleura* (Hemichordata, Pterobranchia) from Fiji. J Zool Lond (A) 205:611–623

Done TJ (1983) Coral zonation: its nature and significance. In: Barnes DJ (ed) Perspectives on coral reefs. Australian Institute of Marine Science, Townsville, pp 107–147

Dorier A (1965) Classe des Gordiacés, V Siebold 1843 (= Nematomorpha Vejdovsky 1886). In: Grassé PP (ed) Traité de zoologie: anatomie, systématique, biologie, IV. Masson et Cie Éditeurs, Paris, pp 1201–1222

Dojiri M (1988) *Isomolgus desmotes*, new genus, new species (Lichomolgidae), a gallicolous poecilostome copepod from the scleractinian coral *Seriatopora hystrix* Dana in Indonesia, with a review of gall-inhabiting crustaceans of anthozoans. J Crust Biol 8:99–109

Dubinsky Z (ed) (1990) Ecosystems of the world 25: coral reefs. Elsevier, Amsterdam

Duffy JE (1992) Host use patterns and demography in a guild of tropical sponge-dwelling shrimps. Mar Ecol Prog Ser 90:127–138

Dulvy NK, Freckleton RP, Polunin NVC (2004) Coral reef cascades and the indirect effects of predator removal by exploitation. Ecol Lett 7:410–416

Dulvy NK, Mitchell RE, Watson D, Sweeting CJ, Polunin NVC (2002) Scale-dependant control of motile epifaunal community structure along a coral reef fishing gradient. J Exp Mar Biol Ecol 278:1–29

Dunn CW, Hejnol A, Matus DQ, Pang K, Browne WE, Smith SA, Seaver E, Rouse GW, Obst M, Edgecombe GD, Sørensen MV, Haddock SHD, Schmidt-Rhaesa A, Okusu A, Kristensen RM, Wheeler WC, Martindale MQ, Giribet G (2008) Broad phylogenomic sampling improves resolution of the animal tree of life. Nature 452:745–749

Dyer WG, Williams EH Jr, Williams LB (1985) Digenetic trematodes of marine fishes of the western and southwestern coasts of Puerto Rico. Proc Helminthol Soc Wash 52:85–94

Edmonds SJ (1987) Phyla Sipuncula and Echiura. In: Devaney DM, Eldredge LM (eds) Reef and shore fauna of Hawaii. Section 2: Platyhelminthes through Phoronida and Section 3: Sipuncula through Annelida. Bishop Museum Special Publication 64 (2 and 3), Bishop Museum Press, Honolulu, pp 185–212

Eldredge LG (1972) Associates of *Acanthaster planci*. In: Tsuda RT (ed) Proceedings of the University of Guam-Trust Territory *Acanthaster planci* (Crown of Thorns Starfish) Workshop. University Guam Marine Laboratory, Mangilao, Guam. Technical Report 3

Ellis RP, Bersey J, Rundle SD, Hall-Spencer JM, Spicer JI (2009) Subtle but significant effects of CO_2 acidified seawater on embryos of the intertidal snail, *Littorina obtusata*. Aquat Biol 5:41–48

Emery AR (1968) Preliminary observations on coral reef plankton. Limnol Oceanogr 13:293–303

Endean R (1973) Population explosions of *Acanthaster planci* and associated destruction of hermatypic corals in the Indo-west Pacific region. In: Jones OA, Endean R (eds) Biology and geology of coral reefs, 2, biology 1. Academic, New York, pp 389–438

Endean R, Cameron AM (1990) *Acanthaster planci* population outbreaks. In: Dubinsky Z (ed) Coral reefs, ecosystems of the world 25. Elsevier, Amsterdam, pp 469–492

Enochs IC, Hockensmith G (2009) The effects of coral mortality on the community composition of cryptic metazoans associated with *Pocillopora damicornis*. In: Proceedings of 11th international coral reef symposium, vol 26, Ft. Lauderdale, 2008, pp 7–11

Erpenbeck D, Wörheide G (2007) On the molecular phylogeny of sponges (Porifera). Zootaxa 1668:107–126

Fabricius K, Alderslade P (2001) Soft corals and sea fans. A comprehensive guide to the tropical shallow-water genera of the Central-West Pacific, the Indian Ocean and the Red Sea. Australian Institute of Marine Science, Townsville

Fabricius KE, Dale MB (1993) Multispecies associations of symbionts on shallow water crinoids of the central Great Barrier Reef. Coenoses 8:41–52

Fabricius KE, De'ath G (2001) Biodiversity on the Great Barrier Reef: large-scale patterns and turbidity-related local loss of soft coral taxa. In: Wolanski E (ed) Oceanographic processes of coral reefs: physical and biological links in the Great Barrier Reef. CRC Press, London, pp 127–144

Fabry VJ, Seibel BA, Feely RA, Orr JC (2008) Impacts of ocean acidification on marine fauna and ecosystem processes. ICES J Mar Sci 65:414–432

Fagerstrom JA (1987) The evolution of reef communities. Wiley, New York

Fauchald K, Jumars PA (1979) The diet of worms: a study of polychaete feeding guilds. Oceanogr Mar Biol Annu Rev 17:193–284

Faulkner DJ (1992) Chemical defenses of marine mollusks. In: Paul VJ (ed) Ecological roles of marine natural products. Comstock Publication Association, Ithaca, pp 119–163

Fautin DG (1991) The anemonefish symbiosis: What is known and what is not. Symbiosis 10:23–46

Fautin DG, Lowenstein JM (1992) Phylogenetic relationships among scleractinians, actinians, and corallimorpharians (Coelenterata: Anthozoa). In: Proceedings of 7th international coral reef symposium, vol 2, Guam, 1992, pp 665–670

Fautin DG, Buddemeier RW (2006) Biogeoinformatics of the hexacorals. http://www.kgs.ku.edu/Hexacoral/

Feder HM (1966) Cleaning symbiosis in the marine environment. In: Henry SM (ed) Symbiosis, vol I. Academic, New York, pp 327–380

Fenchel TM (1978) The ecology of micro- and meiobenthos. Ann Rev Ecol Syst 9:99–121

Féral JP, David B (2003) Echinoderm research 2001: proceedings of the sixth European conference on echinoderm research. AA Balkema, Rotterdam

Ferrier MD (1991) Net uptake of dissolved free amino acids by four scleractinian corals. Coral Reefs 10:183–187

Finley RJ, Forrester GE (2003) Impact of ectoparasites on the demography of a small reef fish. Mar Ecol Prog Ser 248:305–309

Fishelson L (1968) Gamete shedding behaviour of the feather-star *Lamprometra klunzingeri* in its natural habitat. Nature 219:1063

Fogelman RM, Grutter AS (2008) Mancae of the parasitic cymothoid isopod, *Anilocra apogonae*: early life history, host-specificity, and effect on growth and survival of preferred young cardinal fishes. Coral Reefs 27:685–693

Forsythe JW, Hanlon RT (1997) Foraging and associated behavior by *Octopus cyanea* Gray, 1849 on a coral atoll, French Polynesia. J Exp Mar Biol Ecol 209:15–31

Fricke HW (1970) Ein mimetisches Kollektiv-Beobachtungen an Fischschwarmen, die Seeigel nachahmen. Mar Biol 5:307–314

Fukami H, Chen CA, Budd AF, Collins A, Wallace C, Chuang YY, Chen C, Dai C-F, Iwao K, Sheppard C, Knowlton N (2008) Mitochondrial and nuclear genes suggest that stony corals are monophyletic but most families of stony corals are not (Order Scleractinia, Class Anthozoa, Phylum Cnidaria). PLoS ONE 3:1–9

Garrett P, Smith DL, Wilson AO, Patriquin D (1971) Physiography, ecology, and sediments of two Bermuda patch reefs. J Geol 79:647–668

Gattuso JP, Frankignoulle M, Bourge L, Romaine S, Buddemeier RW (1998) Effect of calcium carbonate saturation of seawater on coral calcification. Glob Planet Change 18:37–46

Gerlach SA (1959) Über das tropische Korallenriff als Lebensraum. Verh Dt Zool Ges 39S:356–363

Gilchrist SL (1985) Hermit crab corallivore activity. In: Proceedings of 5th international coral reef congress, vol 5, Tahiti, 1985, pp 211–214

Ginsburg RN (1983) Geological and biological roles of cavities in coral reefs. In: Barnes DJ (ed) Perspectives on coral reefs. Clouston, Australia, pp 148–153

Ginsburg RN, Schroeder JH (1973) Growth and submarine fossilization of algal cup reefs, Bermuda. Sedimentology 20:575–614

Giribet G, Dunn CW, Edgecombe GD, Rouse GW (2007) A modern look at the Animal Tree of Life. Zootaxa 1668:61–79

Glasby CJ, Hutchings PA, Fauchald K, Paxton H, Rouse GW, Russell CW, Wilson RS (2000) Class Polychaeta. In: Barnes PL, Ross GJB, Glasby CJ (eds) Polychaetes & allies: the southern synthesis. Fauna of Australia. Vol. 4A Polychaeta, Myzostomida, Pogonophora, Echiura, Sipuncula. CSIRO, Melbourne, pp 1–296

Glynn PW (1973) Ecology of a Caribbean coral reef. The *Porites* reef-flat biotope: Part II. Plankton community with evidence for depletion. Mar Biol 22:1–21

Glynn PW (1976) Some physical and biological determinants of coral community structure in the eastern Pacific. Ecol Monogr 46:431–456

Glynn PW (1977) Interactions between *Acanthaster* and *Hymenocera* in the field and laboratory. In: Proceedings of 3rd international coral reef symposium, vol 1, Miami, 1977, pp 209–215

Glynn PW (1982a) *Acanthaster* population regulation by a shrimp and a worm. In: Proceedings of 4th international coral reef symposium, vol 2, Manila, 1981, pp 607–612

Glynn PW (1982b) Coral communities and their modifications relative to past and prospective Central American seaways. Adv Mar Biol 19:91–132

Glynn PW (1983a) Increased survivorship in corals harbouring crustacean symbionts. Mar Biol Lett 4:105–111

Glynn PW (1983b) Crustacean symbionts and the defense of corals: coevolution on the reef? In: Nitecki MH (ed) Coevolution. University of Chicago Press, Chicago, pp 111–178

Glynn PW (1984) An amphinomid worm predator of the crown-of-thorns sea star and general predation on asteroids in eastern and western Pacific coral reefs. Bull Mar Sci 35:54–71

Glynn PW (1985) El Niño-associated disturbance to coral reefs and post disturbance mortality by *Acanthaster*. Mar Ecol Prog Ser 26:295–300

Glynn PW (1987) Some ecological consequences of coral-crustacean guard mutualisms in the Indian and Pacific Oceans. Symbiosis 4:303–324

Glynn PW (1988a) El Niño warming, coral mortality and reef framework destruction by echinoid bioerosion in the eastern Pacific. Galaxea 7:129–160

Glynn PW (1988b) Predation on coral reefs: some key processes, concepts and research directions. In: Proceedings of 6th international coral reef symposium, vol 1, Townsville, 1988, pp 51–62

Glynn PW (1990) Feeding ecology of selected coral-reef macroconsumers: patterns and effects on coral community structure. In: Dubinsky Z (ed) Coral reefs, ecosystems of the world 25. Elsevier, Amsterdam, pp 365–391

Glynn PW (1994) State of coral reefs in the Galapagos Islands: natural vs. anthropogenic impacts. Mar Pollut Bull 29:131–140

Glynn PW (1997) Bioerosion and coral growth: a dynamic balance. In: Birkeland C (ed) Life and death of coral reefs. Chapman & Hall, New York, pp 68–94

Glynn PW (2004) High complexity food webs in low-diversity eastern Pacific reef-coral communities. Ecosystems 7:358–367

Glynn PW (2006) Fish utilization of simulated coral reef frameworks versus eroded rubble substrates off Panamá, eastern Pacific. In: Proceedings of 10th international coral reef symposium, Okinawa, 2004, pp 250–256

Glynn PW (2008) Food-web structure and dynamics of eastern tropical Pacific coral reefs: Panama and Galapagos Islands. In: McClanahan T, Branch GM (eds) Food webs and the dynamics of marine reefs. Oxford University Press, pp 185–209

Glynn PW (2011) In tandem reef coral and cryptic metazoan declines and extinctions. Bull Mar Sci 87

Glynn PW, Krupp DA (1986) Feeding biology of a Hawaiian sea star corallivore, *Culcita novaeguineae* Miller and Troschel. J Exp Mar Biol Ecol 96:75–96

Glynn PW, Stewart RH, McCosker JE (1972) Pacific coral reefs of Panamá: structure, distribution, and predators. Geol Rundsch 62:483–519

Glynn PW, Wellington GM, Birkeland C (1979) Coral reef growth in the Galapagos: limitation by sea urchins. Science 203:47–49

Glynn PW, Enochs IC, McCosker JE, Graefe AN (2008) First record of a pearlfish, *Carapus mourlani*, inhabiting the aplysiid opisthobranch mollusc *Dolabella auricularia*. Pac Sci 62:593–601

Gochfeld DJ, Aeby GS (1997) Control of populations of the coral-feeding nudibranch *Phestilla sibogae* by fish and crustacean predators. Mar Biol 130:63–69

Godinot C, Chadwick NE (2009) Phosphate excretion by anemonefish and uptake by giant sea anemones: demand outstrips supply. Bull Mar Sci 85:1–9

Goh NKC, Ng PKL, Chou LM (1999) Notes on the shallow water gorgonian-associated fauna on coral reefs in Singapore. Bull Mar Sci 65:259–282

de Goeij JM, Moodley L, Houtekamer M, Carballeira NM, Van Duyl FC (2008a) Tracing C-13-enriched dissolved and particulate organic carbon in the bacteria-containing coral reef sponge *Halisarca caerulea*: evidence for DOM feeding. Limnol Oceanogr 53:1376–1386

de Goeij JM, van den Berg H, van Oostveen MM, Epping EHG, Van Duyl FC (2008b) Major bulk dissolved organic carbon (DOC) removal by encrusting coral reef cavity sponges. Mar Ecol Prog Ser 357:139–151

Gooding RU (1974) Animals associated with the sea urchin, *Diadema antillarum*. In: Bright TJ, Pequegnat LH (eds) Biota of the west flower bank. Gulf Publishing Co., Houston, pp 333–336

Goreau TF, Hartman WD (1966) Sponge: effect on the form of reef corals. Science 151:343–344

Gotto RV (1979) The association of copepods with marine invertebrates. In: Russell FS, Yonge CM (eds) Advances in marine biology, vol 16. Academic, London, pp 1–109

Gourbault N, Renaud-Mornant J (1990) Micro-meiofaunal community structure and nematode diversity in a lagoonal ecosystem (Fangataufa, Eastern Tuamotu Archipelago). PSZNI Mar Ecol 11:173–189

Grassle JF (1973) Variety in coral reef communities. In: Jones OA, Endean R (eds) Biology and geology of coral reefs, biology 1, vol 2. Academic, New York, pp 247–270

Gravier C (1901) Contribution à l'étude des annelids polychètes de la Mer Rouge. Nouv Arch Mus Paris Ser 4:147–268

Greenwood PG (2009) Acquisition and use of nematocysts by cnidarian predators. Toxicon 54:1065–1070

Greenwood PG, Mariscal RN (1984) Immature nematocyst incorporation by the aeolid nudibranch *Spurilla neapolitana*. Mar Biol 80:35–38

Griffiths DJ, Thinh LV (1983) Transfer of photosynthetically fixed carbon between the prokaryotic green alga *Prochloron* and its ascidian host. Aust J Mar Freshw Res 34:431–440

Grigg R (1972) Orientation and growth form of sea fans. Limnol Oceanogr 17:185–192

Guest J (2008) How reefs respond to mass coral spawning. Science 320:621–623

Grutter AS (1994) Spatial and temporal variations of the ectoparasites of seven reef fish species from Lizard Island and Heron Island, Australia. Mar Ecol Prog Ser 115:21–30

Grutter AS (1995) Relationship between cleaning rates and ectoparasite loads in coral reef fishes. Mar Ecol Prog Ser 118:51–58

Grutter AS (2002) Cleaning behaviour from the parasite's perspective. Parasitol Suppl 124:565–581

Grutter AS, Poulin R (1998) Intraspecific and interspecific relationships between host size and the abundance of parasitic larval gnathiid isopods on coral reef fishes. Mar Ecol Prog Ser 164:263–271

Grutter AS, Pickering JL, McCallum H, McCormick MI (2008) Impact of micropredatory gnathiid isopods on young coral reef fishes. Coral Reefs 27:655–661

Grygier MJ (2000) Class Myzostomida. In: Beesley PL, Ross GJB, Glasby CJ (eds) Polychaetes & allies: the southern synthesis. Fauna of Australia. Vol. 4A polychaeta, myzostomida, pogonophora, echiura, sipuncula. CSIRO Publishing, Melbourne, pp 297–329

Grygier MJ, Newman WA (1991) A new genus and two new species of Microlepadidae (Cirripedia: Pedunculata) found on western Pacific diadematid echinoids. Galaxea 10:1–22

Gutiérrez JL, Jones CG, Strayer DI, Iribarne OO (2003) Molluscs as ecosystem engineers: the role of shell production in aquatic habitats. Oikos 101:79–90

Guzmán HM (2003) Caribbean coral reefs of Panamá: present status and future perspectives. In: Cortés J (ed) Latin American coral reefs. Elsevier, Amsterdam, pp 241–274

Hadfield MG (1976) Molluscs associated with living tropical corals. Micronesica 12:133–148

Haig J (1979) Expédition Rumphius II (1975) Crustacés parasites, commensaux, etc. (Th. Monod et R. Serène, eds) V. Porcellanidae (Crustacea, Decapoda, Anomura). Bull Mus Natn Hist Nat, Paris, 4e sér, 1:119–136

Haig J (1987) Porcellanid crabs from the Coral Sea. Beagle Rec NT Museum Arts Sci 4:11–14

Hanlon RT, Messenger JB (1996) Cephalopod behaviour. Cambridge University Press, Cambridge

Harii S, Kayanne H (2003) Larval dispersal, recruitment, and adult distribution of the brooding stony octocoral *Heliopora coerulea* on Ishigaki Island, southwest Japan. Coral Reefs 22:188–196

Hartman O (1954) Marine annelids from the northern Marshall Islands. Prof Pap US Geol Surv 260Q:615–644

Hartman WD (1977) Sponges as reef builders and shapers. Am Assoc Petrol Geol Stud Geol 4:127–134

Hartman WD, Goreau TF (1966) *Ceratoporella*, a living sponge with stromatoporoid affinities. Am Zool 6:563–564

Hartman WD, Goreau TF (1972) *Ceratoporella* (Porifera: Sclerospongiae) and the chaetetid "corals". Trans Connecticut Acad Arts Sci 44:133–148

Hatcher BG (1997) Coral reef ecosystems: how much greater is the whole than the sum of the parts? Coral Reefs 16:S77–S91

Havenhand JN, Buttler FR, Thorndyke MC, Williamson JE (2008) Near-future levels of ocean acidification reduce fertilization success in a sea urchin. Curr Biol 18:651–652

Hay ME (1991) Fish-seaweed interactions on coral reefs: effects of herbivorous fishes and adaptations of their prey. In: Sale PF (ed) The ecology of fishes on coral reefs. Academic, New York, pp 96–119

Hay ME (1992) The role of seaweed chemical defenses in the evolution of feeding specialization and in the mediation of complex interactions. In: Paul VJ (ed) Ecological roles of marine natural products. Comstock Pub. Assoc, Ithaca, pp 93–118

Hay ME, Fenical W (1988) Marine plant-herbivore interactions: the ecology of chemical defense. Ann Rev Ecol Syst 19:111–145

Hayes FE (2007) Decapod crustaceans associating with the sea urchin *Diadema antillarum* in the Virgin Islands. Nauplius 15:81–85

Hazlett BA (1981) The behavioral ecology of hermit crabs. Ann Rev Ecol Syst 12:1–22

Heinzeller T, Nebelsick JH (eds) (2004) Echinoderms: München. Balkema, Leiden

Hendler G (1984) The association of *Ophiothrix lineata* and *Callyspongia vaginalis*: a brittlestar-sponge cleaning symbiosis? PSZNI Mar Ecol 5:9–27

Hendler G, Miller JE, Pawson DL, Kier PM (1995) Sea stars, sea urchins, and allies: echinoderms of Florida and the Caribbean. Smithsonian Institution Press, Washington, DC

Hendler G, Grygier MJ, Maldonado E, Denton J (1999) Babysitting brittle stars: heterospecific symbionts between ophiuroids (Echinodermata). Invert Biol 118:190–201

Hiatt RW, Strasburg DW (1960) Ecological relationships of the fish fauna on coral reefs of the Marshall Islands. Ecol Monogr 30:65–127

Higgins RP (1983) The Atlantic barrier reef ecosystem at Carrie Bow Cay, Belize, II: Kinorhyncha. Smith Contrib Mar Sci 18:1–131

Highsmith RC (1981) Coral bioerosion at Enewetak: agents and dynamics. Int Revue ges Hydrobiologia 66:335–375

Hirose E, Maruyama T, Cheng L, Lewin RA (1996) Intracellular symbionts of a photosynthetic prokaryote, *Prochloron* sp., in a colonial ascidian. Invert Biol 115:343–348

Hixon MA (1997) Effects of reef fishes on corals and algae. In: Birkeland C (ed) Life and death of coral reefs. Chapman & Hall, New York, pp 230–248

Hobson ES (1974) Feeding relationships of teleostean fishes on coral reefs in Kona, Hawaii. Fish Bull 72:915–1031

Hobson ES (1991) Trophic relationships of fishes specialized to feed on zooplankters above coral reefs. In: Sale PF (ed) The ecology of fishes on coral reefs. Academic, San Diego, pp 69–95

Hochberg FG (1982) The "kidneys" of cephalopods: a unique habitat for parasites. Malacologia 23:121–134

Hochberg FG (1983) The parasites of cephalopods: a review. Mem Natl Mus Vic 44:109–145

Hochberg R (2008) Gastrotricha of Bocas del Toro, Panama: a preliminary report. Meiofauna Mar 16:101–107

Hoegh-Guldberg O (1994) Uptake of dissolved organic matter by larval stage of the crown-of-thorns starfish *Acanthaster planci*. Mar Biol 120:55–63

Hoegh-Guldberg O, Hinde R (1986) Studies on a nudibranch that contains zooxanthellae I. Photosynthesis, respiration and the translocation of newly fixed carbon by zooxanthellae in *Pteraeolidia ianthina*. Proc R Soc Lond B 228:493–509

Hoegh-Guldberg O, Hinde R, Muscatine L (1986) Studies on a nudibranch that contains zooxanthellae II. Contribution of zooxanthellae to animal respiration (CZAR) in *Pteraeolidia ianthina* with high and low densities of zooxanthellae. Proc R Soc Lond B 228:511–521

Hoegh-Guldberg O, Mumby PJ, Hooten AJ, Steneck RS, Greenfield P, Gomez E, Harvell CD, Sale PF, Edwards AJ, Caldeira K, Knowlton N, Eakin CM, Iglesias-Prieto R, Muthiga N, Bradbury RH, Dubi A, Hatziolos ME (2007) Coral reefs under rapid climate change and ocean acidification. Science 318:1737–1742

Hooper JNA, Lévi C (1994) Biogeography of Indo-west Pacific sponges: Microcionidae, Raspailiidae, Axinellidae. In: van Soest RWM, Kempenvan TMG, Brakeman JC (eds) Sponges in time and space: biology, chemistry, paleontology. Proceedings of 4th international porifera congress, Amsterdam, 1993, AA Balkema, Rotterdam, pp 191–212

Hooper JNA, Van Soest RWM (eds) (2002) Systema Porifera: a guide to the classification of sponges, vols 1 & 2. Kluwer/Plenum, New York

Houk P, Bograd S, Van Woesik R (2007) The transition zone chlorophyll front can trigger *Acanthaster planci* outbreaks in the Pacific Ocean: historical confirmation. J Oceanogr 63:149–154

Huang HD, Rittschof D, Jeng MS (2005) Multispecies associations of macrosymbionts on the comatulid crinoid *Comanthina schlegeli* (Carpenter) in southern Taiwan. Symbiosis 39:47–51

Hughes TP (1994) Catastrophes, phase shifts, and large-scale degradation of a Caribbean coral reef. Science 265:1547–1551

Humes AG (1960) New copepods from madreporarian corals. Kiel Meeresfosch 16:229–235

Humes AG (1985a) Cnidarians and copepods: a success story. Trans Am Microsc Soc 104:313–320

Humes AG (1985b) A review of the Xarifiidae (Copepoda, Poecilostomatoida), parasites of scleractinian corals in the Indo-Pacific. Bull Mar Sci 36:467–632

Humes AG (1994) How many copepods? Hydrobiologia 292(293):1–7

Hunte W, Côté I, Tomascik T (1986) On the dynamics of the mass mortality of *Diadema antillarum* in Barbados. Coral Reefs 4:135–139

Hutchings PA (1983) Cryptofaunal communities of coral reefs. In: Barnes DJ (ed) Perspectives on coral reefs. Brian Clouston, Manuka, pp 200–208

Hutchings PA (1986) Biological destruction of coral reefs. Coral Reefs 4:239–252

Hutchings PA, Hilliard RW, Coles SL (2002) Species introductions and potential for marine pest invasions into tropical marine communities, with special reference to the Indo-Pacific. Pac Sci 56:223–233

Hyman LH (1955) The invertebrates: Echinodermata, the coelomate bilateria, vol 4. McGraw-Hill, New York

Jackson JBC (1983) Biological determinants of present and past sessile animal distributions. In: Tevesz MJS, McCall PL (eds) Biotic interactions in recent and fossil benthic communities. Plenum, New York, pp 39–120

Jackson JBC (1984) Ecology of cryptic coral reef communities: III Abundance and aggregation of encrusting organisms with particular reference to cheilostome Bryozoa. J Exp Mar Biol Ecol 75:37–57

Jackson JBC, Buss LW (1975) Allelopathy and spatial competition among coral reef invertebrates. Proc Natl Acad Sci U S A 72:5160–5163

Jackson JBC, Winston JE (1982) Ecology of cryptic coral reef communities: I Distribution and abundance of major groups of encrusting organisms. J Exp Mar Biol Ecol 57:135–147

Jackson JBC, Goreau TF, Hartman WD (1971) Recent brachiopod-coralline sponge communities and their paleoecological significance. Science 173:623–625

Jangoux M (ed) (1980) Echinoderms: present and past. In: Proceedings of European colloquim on echinoderms, Brussels, AA Balkema, Rotterdam

Jangoux M (1990) Diseases of Echinodermata. In: Kinne O (ed) Diseases of marine animals III, introduction, Cephalopoda, Annelida, Crustacea, Chaetognatha, Echinodermata, Urochordata. Biologische Anstalt Helgoland, Hamburg, pp 439–567

Jangoux M, Lawrence JM (eds) (1982) Echinoderm nutrition. AA Balkema, Rotterdam

Jangoux M, Lawrence JM (eds) (1983) Echinoderm studies, vol 1. AA Balkema, Rotterdam

Jangoux M, Lawrence JM (eds) (2001) Echinoderm studies, vol 6. AA Balkema, Rotterdam

Jaubert J (1977) Light, metabolism and growth forms of the hermatypic scleractinian coral *Synaraea convexa* Verrill in the lagoon of Moorea (French Polynesia). In: Proceedings of 3rd international coral reef symposium, vol 1, Miami, 1977, pp 483–488

Jereb P, Roper CFE (eds) (2005) Cephalopods of the world, an annotated and illustrated catalogue of cephalopod species known to date. Vol 1: Chambered nautiluses and sepioids (Nautilidae, Sepiidae, Sepiolidae, Sepiadariidae, Idiosepiidae and Spirulidae). FAO species catalogue for fishery purposes, no 4, vol 1. FAO United Nations, Rome

Jokiel PL (1980) Solar ultraviolet radiation and coral reef epifauna. Science 207:1069–1071

Jokiel PL, Townsley SJ (1974) Biology of the polyclad *Prosthiostomum* (*Prosthiostomum*) sp, a new coral parasite from Hawaii. Pac Sci 28:361–373

Jones CM, Grutter AS (2008) Reef-based micropredators reduce the growth of post-settlement damselfish in captivity. Coral Reefs 27:677–684

Jones OA, Endean R (1973–1977) Biology and geology of coral reefs vols. I–IV, Biol. 1 & 2, Geol. 1 & 2. Academic, New York

Kaiser KL, Bryce C.W. (2001) The recent molluscan marine fauna of Isla de Malpelo, Colombia. The Festivus Occas Pap 1 33:1–149

Kappner I, Al-Moghrabi SM, Richter C (2000) Mucus-net feeding by the vermetid gastropod *Dendropoma maxima* in coral reefs. Mar Ecol Prog Ser 204:309–313

Karlson RH (1999) Dynamics of coral communities. Kluwer, Dordrecht

Karplus I (1987) The association between gobiid fishes and burrowing alpheid shrimps. Oceanogr Mar Biol Ann Rev 25:507–562

Kicklighter CE, Kubanek J, Barsby T, Hay ME (2003) Palatability and defense of some tropical infaunal worms: alkylpyrrole sulfamates as deterrents to fish feeding. Mar Ecol Prog Ser 263:299–306

Kim IH, Yamashiro H (2007) Two species of poecilostomatoid copepods inhabiting galls on scleractinian corals in Okinawa, Japan. J Crust Biol 27:319–326

Kinne O (ed) (1980) Diseases of marine animals I: general aspects, Protozoa to Gastropoda. Wiley, Chichester

Kinne O (ed) (1983) Diseases of marine animals II: introduction, Bivalvia to Scaphopoda. Biologische Anstalt Helgoland, Hamburg, pp 467–1038

Kinne O (ed) (1984) Diseases of marine animals IV, part 1, introduction, Pisces. Biologische Anstalt Helgoland, Hamburg

Kinne O (ed) (1985) Diseases of marine animals IV, part 2, introduction, Reptilia, Aves, Mammalia. Biologische Anstalt Helgoland, Hamburg

Kinne O (ed) (1990) Diseases of marine animals III, introduction, Cephalopoda, Annelida, Crustacea, Chaetognatha, Echinodermata, Urochordata. Biologische Anstalt Helgoland, Hamburg

Kirkendale L, Calder DR (2003) Hydroids (Cnidaria: Hydrozoa) from Guam and the Commonwealth of the Northern Marianas Islands (CNMI). Micronesica 35–36:159–188

Kleypas JA, Langdon C (2006) Coral reefs and changing seawater carbonate chemistry. In: Phinney JT, Hoegh-Guldberg O, Kleypas J, Skirving W, Strong A (eds) Coral reefs and climate change: science and management, Coastal and Estuarine Studies 61. American Geophysical Union, Washington, DC, pp 73–110

Klumpp DW, McKinnon AD, Mundy CN (1988) Motile cryptofauna of a coral reef: abundance, distribution and trophic potential. Mar Ecol Prog Ser 45:95–108

Knowlton N (1993) Sibling species in the sea. Annu Rev Ecol Syst 24:189–216

Knowlton N, Lang JC, Keller BD (1990) Case study of natural population collapse: post-hurricane predation on Jamaican staghorn corals. Smith Contrib Mar Sci 31:1–25

Kobluk DR (1988) Cryptic faunas in reefs: ecology and geologic importance. Palaios 3:379–390

Kobluk DR, Van Soest RWM (1989) Cavity-dwelling sponges in a southern Caribbean coral reef and their paleontological implications. Bull Mar Sci 44:1207–1235

Kohn AJ (1968) Microhabitats, abundance and food of *Conus* on atoll reefs in the Maldive and Chagos Islands. Ecology 49:1046–1062

Kohn AJ (1983) Feeding biology of gastropods. In: Saleuddin ASM, Wilbur KM (eds) The Mollusca, vol 5, physiology, part 2. Academic, New York, pp 1–63

Kohn AJ (1987) Intertidal ecology of Enewetak Atoll. In: Devaney DN, Reese ES, Burch BL, Helfrich P (eds) The natural history of Enewetak Atoll volume I, the ecosystem: environments, biotas, and processes. U.S. Department of Energy, Office of Scientific and Technical Information, Oak Ridge, pp 139–157

Kohn AJ, Lloyd MC (1973) Polychaetes of truncated reef limestone substrates on eastern Indian Ocean coral reefs: diversity, abundance, and taxonomy. Int Rev Gesamten Hydrobiol 58:369–399

Kohn AJ, Nybakken JW (1975) Ecology of Conus on eastern Indian Ocean fringing reefs: diversity of species and resource utilization. Mar Biol 29:211–234

Kott P (1980) The ascidians of the reef flats of Fiji. Proc Linn Soc NSW 105:147–212

Kott P (1982) Didemnid-algal symbioses: host species in the western Pacific with notes on the symbiosis. Micronesica 18:95–127

Kott P, Parry DL, Cox GC (1984) Prokaryotic symbionts with a range of ascidian hosts. Bull Mar Sci 34:308–312

Kropp RK (1986) Feeding biology and mouthpart morphology of three species of coral gall crabs (Decapoda: Cryptochiridae). J Crust Biol 6:377–384

Kropp RK (1990) Revision of the genera of gall crabs (Crustacea: Cryptochiridae) occurring in the Pacific Ocean. Pac Sci 44:417–448

Kropp RK, Manning RB (1985) Cryptochiridae, the correct name for the family containing the gall crabs (Crustacea: Decapoda: Brachyura). Proc Biol Soc Wash 98:954–955

Kropp RK, Manning RB (1987) The Atlantic gall crabs, family Cryptochiridae (Crustacea: Decapoda: Brachyura). Smith Contrib Zool 462:1–21

Kurihara H, Shirayama Y (2004) Effects of increased atmospheric CO_2 on sea urchin early development. Mar Ecol Prog Ser 274:161–169

Kurihara H, Kato S, Ishimatsu A (2007) Effects of increased seawater pCO_2 on early development of the oyster Crassostrea gigas. Aquat Biol 1:91–98

Kuris AM, Lafferty KD (2000) Parasite-host modeling meets reality: adaptive peaks and their ecological attributes. In: Poulin R, Morand S, Skorping A (eds) Evolutionary biology of host-parasite relationships: theory meets reality. Elsevier, New York, pp 9–26

Laborel J (1960) Contribution à l'étude directe des peuplements benthiques sciaphiles sur substrat rocheux en Méditerranée. Rec Trav St Mar Endoume 33:117–173

Lang JC, Chornesky EA (1990) Competition between scleractinian reef corals: a review of mechanisms and effects. In: Dubinsky Z (ed) Coral reefs, ecosystems of the world 25. Elsevier, Amsterdam, pp 209–252

Lescinsky HL, Edinger E, Risk MJ (2002) Mollusc shell encrustation and bioerosion rates in a modern epeiric sea: taphonomy experiments in the Java Sea, Indonesia. Palaios 17:171–191

Lewin RA, Cheng L (1989) Collection and handling of Prochloron and its ascidian hosts. In: Lewin RA, Cheng L (eds) Prochloron, a microbial enigma. Chapman and Hall, New York, pp 9–18

Lewis JB (1989) The ecology of Millepora. A review. Coral Reefs 8:99–107

Lewis JB (1998) Reproduction, larval development and functional relationships of the burrowing, spionid polychaete Dipolydora armata with the calcareous hydrozoan Millepora complanata. Mar Biol 130:651–662

Lewis JB (2006) Biology and ecology of the hydrocoral Millepora on coral reefs. Adv Mar Biol 50:1–55

Lewis SM, Kensley B (1982) Notes on the ecology and behavior of Pseudamphithoides incurvaria (Crustacea, Amphipoda, Amphithoidae). J Nat Hist 16:267–274

Limbaugh C, Pederson H, Chace FA Jr (1961) Shrimps that clean fishes. Bull Mar Sci Gulf Caribb 11:237–257

Liu PJ, Hsieh HL (2000) Burrow architecture of the spionid polychaete Polydora villosa in the corals Montipora and Porites. Zool Stud 39:47–54

Lizama J, Blanquet RS (1975) Predation on sea anemones by the amphinomid polychaete, Hermodice carunculata. Bull Mar Sci 25:442–443

Logan A (1981) Sessile invertebrate coelobite communities from shallow reef tunnels, Grand Cayman, B.W.I. In: Proceedings of 4th international coral reef symposium, vol 2, Manila, 1981, pp 735–744

Logan A, Mathers SM, Thomas MLH (1984) Sessile invertebrate coelobite communities from reefs of Bermuda: species composition and distribution. Coral Reefs 2:205–213

López-Cánovas CI, Lalana R (2001) Benthic meiofauna distribution at three coral reefs from SW of Cuba. Rev Invest Mar 22:199–204

Losey GS (1977) The validity of animal models: a test for cleaning symbiosis. Biol Behav 2:223–238

Losey GS (1979) Fish cleaning symbiosis: proximate causes of host behavior. Anim Behav 27:669–685

Losey GS (1987) Cleaning symbiosis. Symbiosis 4:229–258

Losey GS (1993) Knowledge of proximate causes aids our understanding of function and evolutionary history. Mar Behav Physiol 23:175–186

Losey GS, Margules L (1974) Cleaning symbiosis provides a positive reinforcer for fish. Science 184:179–180

Luckhurst BE, Luckhurst K (1978) Diurnal space utilization in coral reef fish communities. Mar Biol 49:325–332

Magnino G, Lancioni T, Gaino E (1999) Endobionts of the coral reef sponge Theonella swinhoei (Porifera, Demospongiae). Invert Biol 118:213–220

Manuel M, Borchiellini C, Alivon E, Le Parco Y, Vacelet J, Boury-Esnault N (2003) Phylogeny and evolution of calcareous sponges: monophyly of Calcinea and Calcaronea, high level of morphological homoplasy, and the primitive nature of axial symmetry. Syst Biol 52:311–333

Marsden JR (1962) A coral-eating polychaete. Nature 193:598

Marsden JR (1963) The digestive tract of Hermodice carunculata (Pallas), Polychaeta: Amphinomidae. Can J Zool 41:165–184

Mather JA (1982) Choice and competition: their effects on occupancy of shell homes by Octopus joubini. Mar Behav Physiol 8:285–293

May RM (1992) Bottoms up for the oceans. Nature 357:278–279

May RM (1994) Biological diversity: differences between land and sea. Philos Trans R Soc Lond B 343:105–111

McClanahan TR (1995) A coral reef ecosystem-fisheries model: impacts of fishing intensity and catch selection on reef structure and processes. Ecol Model 80:1–19

McClanahan TR, Mutere JC (1994) Coral and sea urchin assemblage structure and interrelationships in Kenyan reef lagoons. Hydrobiologia 286:109–124

McClanahan TR, Muthiga NA (1988) Changes in Kenyan coral reef community structure and function due to exploitation. Hydrobiologia 166:267–276

McClanahan TR, Muthiga NA, Maina J, Kamukuru AT, Yahya SAS (2009) Changes in northern Tanzania coral reefs during a period of increased fisheries management and climatic disturbance. Aquat Conserv Mar Freshw Ecosyst 19:758–771

McCloskey LR (1970) The dynamics of the community associated with a marine scleractinian coral. Hydrobiologia 55:13–81

McCormack G (2007) Cook Islands Biodiversity Database, Version 2007.2. Cook Islands Natural Heritage Trust, Rarotonga. http://cookislands.bishopmuseum.org

McLean R (1983) Gastropod shells: a dynamic resource that helps shape benthic community structure. J Exp Mar Biol Ecol 69:151–174

McSweeny ES (1982) A new Pagurapseudes (Crustacea: Tanaidacea) from southern Florida. Bull Mar Sci 32:455–466

Medina M, Collins AG, Takaoka TL, Kuehl JV, Boore JL (2006) Naked corals: skeleton loss in Scleractinia. Proc Natl Acad Sci USA 103:9096–9100

Meesters E, Knijn R, Willemsen P, Pennartz R, Roebers G, Van Soest RWM (1991) Sub-rubble communities of Curaçao and Bonaire coral reefs. Coral Reefs 10:189–197

Menge BA (1995) Indirect effects in marine rocky intertidal interaction webs: patterns and importance. Ecol Monogr 65:21–74

Meyer DL, Ausich WI (1983) Biotic interactions among Recent and among fossil crinoids. In: Tevesz MJS, McCall PL (eds) Biotic interactions in recent and fossil benthic communities. Plenum, New York, pp 377–427

Meyer DL, Lane NG (1976) The feeding behavior of some Paleozoic crinoids and recent basketstars. J Paleontol 50:472–480

Meyer DL, Messing CG, Macurda DB Jr (1978) Biological results of the University of Miami Deep-Sea Expeditions. 129. Zoogeography of tropical Western Atlantic Crinoidea (Echinodermata). Bull Mar Sci 28:412–441

Miller MW (2001) Corallivorous snail removal: evaluation of impact on *Acropora palmata*. Coral Reefs 19:293–295

Miller TE, Kerfoot WC (1987) Redefining indirect effects. In: Kerfoot WC, Sih A (eds) Predation: direct and indirect impacts on aquatic communities. University Press of New England, Hanover, New Hampshire, pp 33–37

Mokady O, Loya Y, Lazar B (1998) Ammonium contribution from boring bivalves to their coral host – a mutualistic symbiosis? Mar Ecol Prog Ser 169:295–301

Modlin RF (1996) Contributions to the ecology of *Paranebalia belizensis* from the waters off central Belize, Central America. J Crust Biol 16:529–534

Monniot C, Monniot F, Laboute P (1991) Coral reef ascidians of New Caledonia. ORSTOM, Paris

Moran PJ (1986) The *Acanthaster* phenomenon. Oceanogr Mar Biol Annu Rev 24:379–480

Morrison JPE (1954) Ecological notes on the mollusks and other animals of Raroia. Atoll Res Bull 34:1–18

Morton B (1983) Coral-associated bivalves of the Indo-Pacific. In: Russell-Hunter WD (ed) The Mollusca, vol 6 Ecology. Academic, Orlando, pp 139–224

Morton B, Mladenov P (1992) The associates of *Tropiometra afra-macrodiscus* (Echinodermata: Crinoida) in Hong Kong. In: Morton B (ed) The marine flora and fauna of Hong Kong and Southern China III. Hong Kong University Press, Hong Kong, pp 431–438

Morton B, Blackmore G, Kwok CT (2002) Corallivory and prey choice by *Drupella rugosa* (Gastropoda: Muricidae) in Hong Kong. J Molluscan Stud 68:217–223

Moyer JT, Emerson WK, Ross M (1982) Massive destruction of scleractinian corals by the muricid gastropod, *Drupella*, in Japan and the Philippines. Nautilus 96:69–82

Netto SA, Waarwick RM, Attrill MJ (1999) Meiobenthic and macrobenthic community structure in carbonate sediments of Rocas Atoll (North-east, Brazil). East Coast Shelf Sci 48:39–50

Newman LJ, Paulay G, Ritson-Williams R (2003) Checklist of polyclad flatworms (Platyhelminthes) from Micronesian coral reefs. Micronesica 35–36:189–199

Ng PKL, Guinot D, Davie PJF (2008) Systema Brachyurorum: Part I. An annotated checklist of extant brachyuran crabs of the world. Raff Bull Zool 17(Suppl):1–286

Norman MD, Hochberg FG (2005) The current state of octopus taxonomy. Phuket Mar Biol Cent Res Bull 66:127–154

Odum HT, Odum EP (1955) Trophic structure and productivity of a windward coral reef community on Eniwetok Atoll. Ecol Monogr 25:291–320

Ogawa K, Matsuzaki K (1992) An essay on host specificity, systematic taxonomy, and evolution of the coral-barnacles. Bull Biogeogr Soc Jpn 47:1–16

Ogden JC (1977) Carbonate-sediment production by parrotfish and sea urchins on Caribbean reefs. In: Frost SH, Weiss MP, Saunders JB (eds) Reefs and related carbonates – ecology and sedimentology. Stud Geol 4. American Association of Petroleum Geologists, Tulsa, pp 281–288

Opitz S (1996) Trophic interactions in Caribbean coral reefs. ICLARM Technical Report 43

Otter GW (1937) Rock-destroying organisms in relation to coral reefs. Br Museum (Nat Hist) Gr Barrier Reef Exped 1928–29 Sci Rep 1:323–352

Pardy RL, Lewin RA (1981) Colonial ascidians with prochlorophyte symbionts: evidence for translocation of metabolites from alga to host. Bull Mar Sci 31:817–823

Parker SP (1982) Synopsis and classification of living organisms. McGraw-Hill, New York

Patton WK (1967) Studies on *Domecia acanthophora*, a commensal crab from Puerto Rico, with particular reference to modifications of the coral host and feeding habits. Biol Bull 132:56–67

Patton WK (1972) Studies on the animal symbionts of the gorgonian coral, *Leptogorgia virgulata* (Lamarck). Bull Mar Sci 22:419–431

Patton WK (1976) Animal associates of living reef corals. In: Jones OA, Endean R (eds) Biology and geology of coral reefs, 3. Biology 2. Academic, New York, pp 1–36

Paul VJ (1992a) Seaweed chemical defenses on coral reefs. In: Paul VJ (ed) Ecological roles of marine natural products. Comstock Publication Association, Ithaca, pp 24–50

Paul VJ (1992b) Chemical defenses of benthic marine invertebrates. In: Paul VJ (ed) Ecological roles of marine natural products. Comstock Publication Association, Ithaca, pp 164–188

Paul VJ, Van Alstyne KL (1988) Use of ingested algal diterpenoids by *Elysia halimedae* Macnae (Opisthobranchia: Ascoglossa) as antipredator defenses. J Exp Mar Biol Ecol 119:15–29

Paulay G (1997) Diversity and distribution of reef organisms. In: Birkeland C (ed) Life and death of coral reefs. Chapman & Hall, New York, pp 298–353

Paulay G (ed) (2003) Marine biodiversity of Guam and the Marianas: overview. Micronesica 35–36:3–682

Paulay G, Kropp R, Ng PKL, Eldredge LG (2003) The crustaceans and pycnogonids of the Mariana islands. In: Paulay G (ed) Marine biodiversity of Guam and the Marianas. Micronesica 35–36:456–513

Pawson DL (1995) Echinoderms of the tropical island Pacific: status of their systematics and notes on their ecology and biogeography. In: Maragos JE, Peterson MNA, Eldredge LG, Bardach JE, Takeuchi HE (eds) Marine and coastal biodiversity in the tropical island Pacific region. Vol 1, species systematics and information management priorities. East-West Center, Honolulu, pp 171–192

Pearse AS (1934) Inhabitants of certain sponges at Dry Tortugas: Pap. Tortugas Lab. Carnegie Inst Wash 28:117–124

Pearse AS (1950) Notes on the inhabitants of certain sponges at Bimini. Ecology 31:149–151

Pearse JS (1969) Reproductive periodicities of Indo-Pacific invertebrates in the Gulf of Suez. I. The echinoids *Prionocidaris baculosa* and *Lovenia elongata* (Gray). Bull Mar Sci 19:323–350

Pearse JS (2009) Shallow-water asteroids, echinoids and holothuroids at 6 sites across the tropical west Pacific, 1988–1989. Galaxea, J. Coral Reef Stud 11:1–9

Pearse VB, Voigt O (2007) Field biology of placozoans (*Trichoplax*): distribution, diversity, biotic interactions. Integr Comp Biol 47:677–692

Pearse VB, Pearse JS, Hendler G, Byrne M (1998) An accessible population of *Ophiocanops*, off NE Sulawesi, Indonesia. In: Mooi R, Telford M (eds) Echinoderms: San Francisco. AA Balkema, Rotterdam, pp 413–418

Pechenik JA (2005) Biology of the invertebrates, 5th edn. McGraw-Hill, Boston

Penfold R, Grutter AS, Kuris AM, McCormick MI, Jones CM (2008) Interactions between juvenile marine fish and gnathiid isopods: predation versus micropredation. Mar Ecol Prog Ser 357:111–119

Pennings SC (1997) Indirect interactions on coral reefs. In: Birkeland C (ed) Life and death of coral reefs. Chapman & Hall, New York, pp 249–272

Perry CT, Hepburn LJ (2008) Syn-depositional alteration of coral reef framework through bioerosion, encrustation and cementation: Taphonomic signatures of reef accretion and reef depositional events. Earth Sci Rev 86:106–144

Peters EC (1997) Diseases of coral reef organisms. In: Birkeland C (ed) Life and death of coral reefs. Chapman & Hall, New York, pp 114–139

Pettibone MH (1982) Annelida. In: Parker SP (ed) Synopsis and classification of living organisms, vol 2. McGraw-Hill, New York, pp 1–43

Peyrot-Clausade M (1977) Settlement of an artificial biota by a coral reef cryptofauna. In: Proceedings of 3rd international coral reef symposium, vol 1, Miami, 1977, pp 101–103

Philippe H, Derelle R, Lopez P, Pick K, Borchiellini C, Boury-Esnault N, Vacelet J, Renard E, Houliston E, Quéinnec E, Da Silva C, Wincker P, Le Guyader H, Leys S, Jackson DJ, Schreiber F, Erpenbeck D, Morgenstern B, Wörheide G, Manuel M (2009) Phylogenomics revives traditional views on deep animal relationships. Curr Biol 19:706–712

Pimm SL, Lawton JH, Cohen JE (1991) Food web patterns and their consequences. Nature 350:669–674

Plaisance L, Kritsky DC (2004) Dactylogyrids (Platyhelminthes: Monogenoidea) parasitizing butterfly fishes (Teleostei: Chaetodontidae) from the coral reefs of Palau, Moorea, Wallis, New Caledonia, and Australia: species of *Euryhaliotrematoides* n. gen. and *Aliatrema* n. gen. J Parasitol 90:328–341

Polovina JJ (1984) Model of a coral reef ecosystem: the ECOPATH model and its application to French Frigate Shoals. Coral Reefs 3:1–11

Porat D, Chadwick-Furman NE (2004) Effects of anemonefish on giant sea anemones: expansion behavior, growth, and survival. Hydrobiologia 530–531:513–520

Porter JW (1974) Zooplankton feeding by the Caribbean reef building coral *Montastrea cavernosa*. In: Proceedings of 2nd international coral reef symposium, vol 1, Brisbane, 1974, pp 111–125

Porter JW, Porter KG (1977) Quantitative sampling of demersal plankton migrating from different coral reef substrates. Limnol Oceanogr 22:553–556

Porter JW, Porter KG, Batac-Catalan Z (1977) Quantitative sampling of Indo-Pacific demersal reef plankton. In: Proceedings of 3rd international coral reef symposium, vol 1, Miami, 1977, pp 105–112

Potts F (1915) The fauna associated with the crinoids of a tropical coral reef: with especial reference to its colour variations. Mar Biol Carnegie Inst Wash 8:73–96

Poulin R (1993) A cleaner perspective on cleaning symbiosis. Rev Fish Biol Fish 3:75–79

Poulin R, Grutter AS (1996) Cleaning symbioses: proximate and adaptive explanations. BioScience 46:512–517

Poupin J (2003) Crustacean Decapoda and Stomatopoda of Easter Island and surrounding areas. A documented checklist with historical overview and biogeographic comments. Atoll Res Bull 500:1–50

Pratchett MS (2001) Influence of coral symbionts on feeding preferences of crown-of-thorns starfish *Acanthaster planci* in the western Pacific. Mar Ecol Prog Ser 214:111–119

Pratchett MS, Vytopil E, Parks P (2000) Coral crabs influence the feeding patterns of crown-of-thorns starfish. Coral Reefs 19:36

Puce S, Di Camillo CG, Bavestrello G (2008) Hydroids symbiotic with octocorals from the Sulawesi Sea, Indonesia. J Mar Biol Assoc UK 88:1643–1654

Randall JE (1967) Food habits of reef fishes of the West Indies. Stud Trop Oceanogr 5:665–847

Randall RH, Eldredge LG (1976) Skeletal modification by a polychaete annelid in some scleractinian corals. In: Mackie GO (ed) Coelenterate ecology and behavior. Plenum, New York, pp 453–465

Rasser MW, Riegl B (2002) Holocene coral reef rubble and its binding agents. Coral Reefs 21:57–72

Rathbun MJ (1933) Brachyuran crabs of Porto Rico and the Virgin Islands Sci Surv of Porto Rico and the Virgin Islands, vol 15, part 1. New York Academy of Sciences, New York

Rayand RC, Parrish FA (2005) A description of fish assemblages in the black coral beds off Lahaina, Maui, Hawai'i. Pac Sci 59:411–420

Razak TB, Hoeksema BW (2003) The hydrocoral genus *Millepora* (Hydrozoa: Capitata: Milleporidae) in Indonesia. Zool Verh Leiden 345:313–336

Reaka-Kudla ML (1987) Adult-juvenile interactions in benthic reef crustaceans. Bull Mar Sci 41:108–134

Reaka-Kudla ML (1997) The global biodiversity of coral reefs: a comparison with rain forests. In: Reaka-Kudla ML, Wilson DE, Wilson EO (eds) Biodiversity II: understanding and protecting our biological resources. Joseph Henry Press, Washington, DC, pp 83–108

Rees HJ (1967) A brief survey of the symbiotic associations of Cnidaria with Mollusca. Proc Malacol Soc Lond 37:213–231

Renon JP, Lefevre M (1985) Zooplankton. In: Proceedings of 5th international coral reef congress, vol 1, Tahiti, 1985, pp 387–392

Rice ME (1976) Sipunculans associated with coral communities. Micronesica 12:119–132

Rice ME, Macintyre IG (1982) Distribution of Sipuncula in the coral reef community, Carrie Bow Cay, Belize. In: Rützler K, Macintyre IG (eds) The Atlantic barrier reef ecosystems at Carrie Bow Cay, Belize. I Structure and communities. Smithsonian Institution Press, Washington, DC, pp 311–320

Richards VP, Thomas JD, Stanhope MJ, Shivji MS (2007) Genetic connectivity in the Florida reef system: comparative phylogeography of commensal invertebrates with contrasting reproductive strategies. Mol Ecol 16:139–157

Richter C, Wunsch M (1999) Cavity-dwelling suspension feeders in coral reefs – a new link in reef trophodynamics. Mar Ecol Prog Ser 188:105–116

Richter C, Wunsch M, Rasheed M, Kötter I, Badran MI (2001) Endoscopic exploration of Red Sea coral reefs reveals dense populations of cavity-dwelling sponges. Nature 413:726–730

Robertson DR (1982) Fish feces as fish food on a Pacific coral reef. Mar Ecol Prog Ser 7:253–265

Robertson R (1970) Review of the predators and parasites of stony corals, with special reference to symbiotic prosobranch gastropods. Pac Sci 24:43–54

Robichaux DM, Cohen AC, Reaka ML, Allen D (1981) Experiments with zooplankton on coral reefs, or, will the real demersal plankton please come up? PSZNI Mar Ecol 2:77–94

Roopin M, Chadwick NE (2009) Benefits to host sea anemones from ammonia contributions of resident anemonefish. J Exp Mar Biol Ecol 370:27–34

Ross DM (1970) The commensal association of *Calliactis polypus* and the hermit crab *Dardanus gemmatus* in Hawaii. Can J Zool 48:351–357

Ross A, Newman WA (1973) Revision of the coral-inhabiting barnacles (Cirripedia: Balanidae). San Diego Soc Nat Hist Trans 17:137–174

Ross A, Newman WA (1995) A coral-eating barnacle, revisited (Cirripedia, Pyrgomatidae). Contrib Zool 65:129–175

Ross A, Newman WA (2000) A new coral-eating barnacle: the first record from the Great Barrier Reef, Australia. Mem Queensland Museum 45:585–591

Rotjan RD, Lewis SM (2008) Impact of coral predators on tropical reefs. Mar Ecol Prog Ser 367:73–91

Rouse GW (2001) A cladistic analysis of Siboglinidae Caullery, 1914 (Polychaeta, Annelida): formerly the phyla Pogonophora and Vestimentifera. Zool J Linn Soc 132:55–80

Rouse GW, Fauchald K (1997) Cladistics and polychaetes. Zool Scripta 26:139–204

Rowden AA, Jones MB (1993) Critical evaluation of sediment turnover estimates for Callianassidae (Decapoda: Thalassinidae). J Exp Mar Biol Ecol 173:265–272

Rudman WB (1981) Further studies on the anatomy and ecology of opisthobranch molluscs feeding on the scleractinian coral *Porites*. Zool J Linn Soc 71:373–412

Rudman WB (1982) The taxonomy and biology of further aeolidacean and arminacean nudibranch molluscs with symbiotic zooxanthellae. Zool J Linn Soc 74:147–196

Rudman WB (1999) Benthic ctenophores. In: Sea slug forum. Australian Museum, Sydney. http://www.seaslugforum.net/factsheet.cfm?base=ctenopho

Rützler K (2004) Sponges on coral reefs: a community shaped by competitive cooperation. Boll Mus Ist Biol Univ Genova 68:85–148

Salvini-Plawen LV, Benayahu Y (1991) *Epimenia arabica* spec. nov., a solenogaster (Mollusca) feeding on the alcyonacean *Scleronephthya corymbosa* (Cnidaria) from shallow waters of the Red Sea. PSZNI Mar Ecol 12:139–152

Sandford F, Brown C (1997) Gastropod shell substrates of the Florida hermit-crab sponge, *Spongosorites suberitoides*, from the Gulf of Mexico. Bull Mar Sci 61:215–223

Schulze A (2005) Sipuncula (peanut worms) from Bocas del Toro, Panama. Caribb J Sci 41:523–527

Scoffin TP, Stoddart DR (1978) The nature and significance of micro-atolls. Philos Trans R Soc Lond Biol 284:99–122

Scoffin TP, Stearn CW, Boucher D, Frydl P, Hawkins CM, Hunter IG, MacGeachy JK (1980) Calcium carbonate budget of a fringing reef on the west coast of Barbados. Bull Mar Sci 30:475–508

Scheffers SR, Nieuwland G, Bak RPM, Van Duyl FC (2004) Removal of bacteria and nutrient dynamics within the coral reef framework of Curaçao (Netherlands Antilles). Coral Reefs 23:413–422

Schuhmacher H (1977) A hermit crab, sessile on corals, exclusively feeds by feathered antennae. Oecologia (Berl) 27:371–374

Schuhmacher H (1992) Impact of some corallivorous snails on stony corals in the Red Sea. Proc. 7th Int. Coral Reef Symp. Guam 2:840–846

Schuhmacher H (1997) Soft corals as reef builders. Proc 8th Int Coral Reef Symp Panama 2:499–502

Shimek R, Kohn AJ (1981) Functional morphology and evolution of the toxoglossan radula. Malacologia 20:423–438

Shirayama Y, Horikoshi M (1982) A new method of classifying the growth form of corals and its application to a field survey of coral-associated animals in Kabira Cove, Ishigaki Island. J Oceanogr Soc Jpn 38:193–207

Shirayama Y, Thornton H (2005) Effects of increased atmospheric CO_2 on shallow water marine benthos. J Geophys Res 110:C09S08. doi:10.1029/2004JC002618

Schulze A (2005) Sipuncula (peanut worms) from Bocas del Toro, Panama. Caribb J Sci 41:523–527

Simberloff D, Dayan T (1991) The guild concept and the structure of ecological communities. Ann Rev Ecol Syst 22:115–143

Simon-Blecher N, Chemedanov A, Eden N, Achituv Y (1999) Pit structure and trophic relationship of the coral pit crab *Cryptochirus coralliodytes*. Mar Biol 134:711–717

Sloan NA (1980) Aspects of the feeding biology of asteroids. Oceanogr Mar Biol Annu Rev 18:57–124

Small AM, Adey WH, Spoon D (1998) Are current estimates of coral reef biodiversity too low? The view through the window of a microcosm. Atoll Res Bull 458:1–20

Smalley TL (1984) Possible effects of intraspecific competition on the population structure of a solitary vermetid mollusk. Mar Ecol Prog Ser 14:139–144

SMEBD (2009) World register of marine species. http://www.marinespecies.org

Sonnenholzner JI, Ladah LB, Lafferty KD (2009) Cascading effects of fishing on Galápagos rocky reef communities: reanalysis using corrected data. Mar Ecol Prog Ser 375:209–218

Sørensen MV (2006) New kinorhynchs from Panama, with a discussion of some phylogenetically significant cuticular structures. Meiofauna Marine 15:51–77

Sorokin YI (1990) Plankton in the reef ecosystems. In: Dubinsky Z (ed) Coral reefs, ecosystems of the world 25. Elsevier, Amsterdam, pp 291–327

Sorokin YI (1993) Coral reef ecology. Ecological studies 102. Springer, Berlin

Sperling EA, Peterson KJ, Pisani D (2009) Phylogenetic-signal dissection of nuclear housekeeping genes supports the paraphyly of sponges and the monophyly of Eumetazoa. Mol Biol Evol 26:2261–2274

Stafford-Smith MG, Ormond RFG (1992) Sediment rejection mechanisms of 42 species of Australian scleractinian corals. Aust J Mar Freshw Res 43:683–705

Steger R (1987) Effects of refuges and recruitment on gonodactylid stomatopods, a guild of mobile prey. Ecology 68:1520–1533

Steiner G, Kabat AR (2004) Catalog of species-group names of Recent and fossil Scaphopoda (Mollusca). Zoosystema 26:549–726

Steneck RS (1988) Herbivory on coral reefs: a synthesis. Proc. 6th Int. Coral Reef Symp. Townsville 1:37–49

Stephen AC, Edmonds SJ (1972) The phyla Sipuncula and Echiura. Trustees British Museum (Natural History), London

Stephenson TA, Stephenson A, Tandy G, Spender M (1931) The structure and ecology of Low Isles and other reefs. Sci Rep Great Barrier Reef Exped 1928-29 Br Museum (Nat Hist) 3:1–112

Sterrer W (2000) Gnathostomulida in the Pelican Cays, Belize. Atoll Res Bull 478:265–271

Stevcic Z, Castro P, Gore RH (1988) Re-establishment of the family Eumedonidae Dana, 1853 (Crustacea: Brachyura). J Nat Hist 22:1,301–1324

Stewart HL, Holbrook SJ, Schmitt RJ, Brooks AJ (2006) Symbiotic crabs maintain coral health by clearing sediments. Coral Reefs 25:609–615

Stimson JS (1990) Stimulation of fat body production in the polyps of the coral *Pocillopora damicornis* by the presence of mutualistic crabs of the genus *Trapezia*. Mar Biol 106:211–218

Stock JH (1975) Corallovexiidae, a new family of transformed copepods endoparasitic in reef corals. Stud Fauna Curaçao other Caribb Isl 47:1–45

Stock JH (1981) Associations of Hydrocorallia Stylasterina with gall-inhabiting Copepoda Siphonostomatoidea from the South-West Pacific, 2. On six species belonging to four new genera of the copepod family Asterocheridae. Bijdr Dierk Amsterdam 51:287–312

Stock JH (1987) Copepoda Siphonostomatoida associated with West Indian hermatypic corals 1: associates of Scleractinia: Faviinae. Bull Mar Sci 40:464–483

Stock JH (1988) Copepods associated with reef corals: a comparison between the Atlantic and the Pacific. Hydrobiologia 167(168):545–547

Stoddart DR (1969) Ecology and morphology of Recent coral reefs. Biol Rev 44:433–498

Stoddart DR, Yonge M (eds) (1971) Regional variation in Indian Ocean coral reefs. Symposium Zoological Society of London, no 28. Academic, London

Stoddard J (1989) Fatal attraction. Landscope – W.A.'s Conservation, Forests and Wildlife Magazine, Winter 1989 ed. pp 14–20

Stoecker D (1980) Chemical defenses of ascidians against predators. Ecology 61:1,327–1334

Strathmann RR, Hughes TP, Kuris AM, Lindeman KC, Morgan SG, Pandolfi JM, Warner RR (2002) Evolution of local recruitment and its consequences for marine populations. Bull Mar Sci 70:377–396

Suchanek TH (1983) Control of seagrass communities and sediment distribution by *Callianassa* (Crustacea, Thalassinidae) bioturbation. J Mar Res 41:281–298

Sussman M, Loya Y, Fine M, Rosenberg E (2003) The marine fireworm *Hermodice carunculata* is a winter reservoir and spring-summer vector for the coral-bleaching pathogen *Vibrio shiloi*. Environ Microbiol 5:250–255

Sybesma J, van Duyl FC, Bak RPM (1981) The ecology of the tropical compound ascidian *Trididemnum solidum* III. Symbiotic association with unicellular algae. Mar Ecol Prog Ser 6:53–59

Taylor PD, Schembri PJ, Cook PL (1989) Symbiotic associations between hermit crabs and bryozoans from the Otago region, southeastern New Zealand. J Nat Hist 23:1059–1085

Thiel M, Kruse I (2001) Status of the Nemertea as predators in marine ecosystems. Hydrobiologia 456:21–32

Thomassin BA (1976) Feeding behavior of the felt-, sponge-, and coral-feeder sea stars, mainly *Culcita schmideliana*. Helgol Mar Res 28:51–65

Thomassin BA, Vivier MH, Vitiello P (1976) Distribution de la méiofaune et de la macrofaune des sables coralliens de la retenue d'eau épirécifale du Grand Récif de Tuléar (Madagascar). J Exp Mar Biol Ecol 22:31–53

Thompson AR (2004) Habitat and mutualism affect the distribution and abundance of a shrimp-associated goby. Mar Freshw Res 55:105–113

Thompson AR (2005) Dynamics of demographically open mutualists: immigration, intraspecific competition, and predation impact goby populations. Oecologia 143:61–69

Todd CD (1981) The ecology of nudibranch molluscs. In: Barnes M (ed) Oceanography and marine biology, an annual review, vol 19. Aberdeen University Press, Aberdeen, pp 141–234

Todd P (2008) Morphological plasticity in scleractinian corals. Biol Rev Camb Philos Soc 83:315–337

Turner SJ (1992) The egg capsule and early life history of the corallivorous gastropod *Drupella cornus* (Roeding, 1978). Veliger 35:16–25

Turner SJ (1994) The biology and population outbreaks of the corallivorous gastropod *Drupella* on Indo-Pacific reefs. Oceanogr Mar Biol Ann Rev 32:461–530

Vacelet J (2002) Recent 'Sphinctozoa', Order Verticillitida, Family Verticillitidae Steinmann, 1882. In: Hooper JNA, Van Soest RWM (eds) Systema Porifera: a guide to the classification of sponges. Kluwer/Plenum, New York, pp 1097–1098

Vacelet J, Willenz P, Hartman WD (in press) Hypercalcified Porifera: systematic descriptions of extant hypercalcified Demospongiae & Calcispongiae. Treat Invert Paleontol, Part E, vol 4

Vago R, Achituv Y, Vaky L, Dubinsky Z, Kizner Z (1998) Colony architecture of *Millepora dichotoma* Forskal. J Exp Mar Biol Ecol 224:225–235

Valentich-Scott P, Tongkerd P (2008) Coral-boring bivalve molluscs of southeastern Thailand, with the description of a new species. Raffles Bull Zool Suppl 18:191–216

Van der Land J (1970) Systematics, zoogeography, and ecology of the Priapulida. Zool Verh Leiden 112:1–118

Van Soest RWM (1990) Shallow-water reef sponges of eastern Indonesia. In: Rützler K (ed) New perspectives in sponge biology. Smithsonian, Washington, DC, pp 302–308

Van Soest RWM, Boury-Esnault N, Hooper JNA, Rützler K, de Voogd NJ, Alvarez B, Hajdu E, Pisera AB, Vacelet J, Manconi R, Schoenberg C, Janussen D, Tabachnick KR, Klautau M (2008) World Porifera database. http://www.marinespecies.org/porifera. Consulted on 2009-03-19 & 2009-08-08

Vera M, Banks S (2009) Health status of the coral communities of the northern Galapagos Islands Darwin, Wolf and Marchena. Galapagos Res 66:65–74

Vogler C, Benzie J, Lessios H, Barber P, Worheide G (2008) A threat to coral reefs multiplied? Four species of crown-of-thorns starfish. Biol Lett 4:696–699

Veron JEN, Pichon M (1976) Scleractinia of Eastern Australia, Part 1: Families Thamnasteriidae, Astrocoeniidae, Pocilloporidae. AIMS Monogr Ser 1:1–86

Veron JEN (2000) Corals of the world, vols 1–3. Australian Institute of Marine Science, Townsville

Vytopil E, Willis BL (2001) Epifaunal community structure in *Acropora* spp. (Scleractinia) on the Great Barrier Reef: implications of coral morphology and habitat complexity. Coral Reefs 20:281–288

Wägele H, Johnsen G (2001) Observations on the histology and photosynthetic performance of "solar powered" opisthobranchs (Mollusca, Gastropoda, Opisthobranchia) containing symbiotic chloroplasts or zooxanthellae. Org Divers Evol 1:193–210

Wainwright PC, Bellwood DR (2002) Ecomorphology of feeding in coral reef fishes. In: Sale PF (ed) Coral reef fishes. Dynamics and diversity in a complex ecosystem. Academic, Orlando, pp 33–55

Wallace CC, Zahir H (2007) The "*Xarifa*" expedition and the atolls of the Maldives, 50 years on. Coral Reefs 26:3–5

Ward J (1965) The digestive tract and its relation to feeding habits in the stenoglossan prosobranch *Coralliophila abbreviata* (Lamarck). Can J Zool 43:447–464

Ward PD, Saunders WB (1997) *Allonautilus*: a new genus of living nautiloid cephalopod and its bearing on phylogeny of the Nautilida. J Paleontol 71:1054–1064

Wells JW (1957) Corals. In: J.W. Hedgpeth (ed.) Treatise on marine ecology and paleoecology Vol 1. Ecology. Geol Soc Amer Mem 67. Waverly Press, Balitmore, pp 1087–1104

Wickler W, Seibt U (1970) Das Verhalten von *Hymenocera picta* Dana, einer Seesterne fressenden Garnele (Decapoda, Natantia, Gnathophyllidae). Z Tierpsychol 27:352–368

Wielgus J, Glassom D, Ben-Shaprut O, Chadwick-Furman N (2002) An aberrant growth form of Red Sea corals caused by polychaete infestations. Coral Reefs 21:315–316

Wild C, Huettel M, Klueter A, Kremb SG, Rasheed MYM, Jørgensen BB (2004) Coral mucus functions as an energy carrier and particle trap in the reef ecosystem. Nature 428:66–70

Wilkinson CR (1990) Acanthaster planci. Coral Reefs (special issue) 9:93–172

Wilkinson CR (2008) Status of coral reefs of the world: 2008. Global Coral Reef Monitoring Network and Reef and Rainforest Research Centre, Townsville, pp 298

Wilkinson CR, Macintyre IG (1992) The *Acanthaster* debate. Coral Reefs (special issue) 11:51–122

Williams JA, Margolis SV (1974) Sipunculid burrows in coral reefs: evidence for chemical and mechanical excavation. Pac Sci 28:357–359

Williams JD, McDermott JJ (2004) Hermit crab biocoenoses: a worldwide review of the diversity and natural history of hermit crab associates. J Exp Mar Biol Ecol 305:1–128

Williams EH Jr, Bunkley-Williams L, Burreson EM (1994) Some new records of marine and Freshwater leeches from Caribbean, southeastern U.S.A., eastern Pacific, and Okinawan animals. J Helminthol Soc Wash 61:133–138

Williams EH Jr, Bunkley-Williams L, Dyer WG (1996) Metazoan parasites of some Okinawan coral reef fishes with a general comparison to the parasites of Caribbean coral reef fishes. Galaxea 13:1–13

Williams DE, Miller MW (2005) Coral disease outbreak: pattern, prevalence and transmission in *Acropora cervicornis*. Mar Ecol Prog Ser 301:119–128

Willis BL, Page CA, Dinsdale EA (2004) Coral disease on the Great Barrier Reef. In: Rosenberg E, Loya Y (eds) Coral health and disease. Springer, Berlin, pp 69–104

Wilson RS, Hutchings PA, Glasby CJ (2003) Polychaetes: an interactive identification guide. CSIRO, Melbourne (CD-Rom)

Wimmer A (1879) Zur Conchylien-Fauna der Galápagos-Inseln. Sber. Akad. Wiss. Wien Math-Nat Classe (Abt I) 80:465–514

Witman JD (1988) Effects of predation by the fireworm *Hermodice carunculata* on milleporid hydrocorals. Bull Mar Sci 42:446–458

Wolf NG, Bermingham EB, Reaka-Kudla ML (1983) Relationships between fishes and mobile benthic invertebrates on coral reefs. In: Reaka ML (ed) The ecology of deep and shallow coral reefs. Symposia series for undersea research, vol 1. Office of Undersea Res., NOAA, Rockville, pp 69–78

Wood R (1999) Reef evolution. Oxford University Press, Oxford, p 414

Wulff JL (1984) Sponge-mediated coral reef growth and rejuvenation. Coral Reefs 3:157–163

Wulff JL (1997) Mutualisms among species of coral reef sponges. Ecology 78:146–159

Wulff JL (2006) Ecological interactions of marine sponges. Can J Zool 84:146–166

Wulff JL, Buss LW (1979) Do sponges help hold coral reefs together? Nature 281:374–475

Yonge CM (1930) Studies on the physiology of corals I. Feeding mechanisms and food. Sci Rep Gt Barrier Reef Exped 1:13–57

Yonge CM (1973) The nature of reef-building (hermatypic) corals. Bull Mar Sci 23:1–15

Zmarzly DL (1984) Distributions and ecology of shallow-water crinoids at Enewetak Atoll, Marshall Islands, with an annotated checklist of their symbionts. Pac Sci 38:105–122

Coral Reef Fishes: Opportunities, Challenges and Concerns

W. Linn Montgomery

Abstract Coral reef fishes represent superb models for test of biological theory in field or laboratory. Nonetheless, our knowledge comes from few locations, few biological disciplines, and few of the more than 70 families of fishes occupying coral reefs. Most reef fishes exhibit complex life histories involving distinctive pelagic larval stages, ecological and structural changes associated with settling on reefs, and a quest by growing juveniles for high quality adult habitat. Remarkable adaptations in development, physiology and behavior characterize these stages, yet opportunities to understand such adaptations decline with the destruction of reef habitat and fish populations by rapidly expanding human populations. Recent advances reflect the power of interdisciplinary collaboration, but the future of reefs and their fishes will likely depend on the ability of reef biologists to collaborate with caring non-scientists and provide knowledge, understanding and direction to those who govern.

Keywords Adults • Behavior • Conservation • Coral reef fishes • Ecology • Eggs • Exploitation • Feeding • Fishes • Larvae • Life history • Management • Physiology • Symbiosis

1 Introduction

During the last half-century, marine reef fish biologists made significant strides in testing biological theory developed primarily in terrestrial or freshwater systems as well as that with its origins in marine environments (see chapters in Sale 1991, 2002). Research continues to richen under the influence of new methods, concepts, and challenges that alter the face of coral reef fish studies. This dynamism demands much of both established workers and others new to the field who will shoulder the unanswered questions and challenges of the future.

W.L. Montgomery (✉)
Department of Biological Sciences, Northern Arizona University,
Flagstaff, AZ 86011-5640, USA
e-mail: Linn.Montgomery@nau.edu

Those who seek serious understanding of the present state of coral reef fish biology must begin with two recent and tremendously significant references: Peter Sale's (2002) edited volume on coral reef fish biology and Gene Helfman's (2007) recent tome on fish conservation. Conceptual foundations from these works should also be supplemented from a recent book on temperate North American systems and fishes (Allen LG et al. 2006). Collectively, these provide highly integrative summaries of a wealth of literature and identify directions for future research as well as conservation and management needs. Nonetheless, long-standing biases toward whole-organism and field studies continue to characterize coral reef fish research, perhaps consistent with a relative dearth of laboratories close to reefs that possess resources to support equipment-intensive research or controlled environments. As Choat and Robertson (2002, p. 58) note, "Much coral reef research is still expeditionary in nature."

That said, the use of interdisciplinary approaches to understanding the world of reef fishes continues to expand. As a result, content of this chapter was driven primarily by two questions. First, how should students considering careers in marine life sciences prepare to make truly significant contributions? Recent literature reflects tremendous growth in interdisciplinary research, application of new methods and technologies, and an array of anthropogenic impacts on reef fishes, as well as growing recognition that responses to these impacts demands consideration of the human condition from local to global scales. Second, what might be useful and interesting in recent literature for potential collaborators whose expertise spans organisms other than fishes, reductionist organizational levels below that of whole organisms, or ecosystems other than coral reefs? Our non-reef brethren may provide complementary approaches and powerful evolving methods while we provide expertise about and access to perhaps the globe's most glorious ecosystem and its inhabitants.

The chapter begins with a look at recent studies of coral reef fishes with respect to geographic, topical, and taxonomic emphases, driven essentially by a single question: What are we missing as we seek an integrative overview of world coral reef ecosystems? As I later return in more detail to anthropogenic threats to reefs and their fishes, a short section follows on aspects of human populations living near and exploiting reef systems. Subsequent sections focus on areas of active, integrative, and often groundbreaking research, with emphasis on recent studies integrating new approaches or technologies. Finally, I turn to issues surrounding direct and indirect threats to coral reef fishes and their home ecosystems, as well as to conservation efforts, concerns, and needs.

2 Biases in Knowledge of Coral Reef Fishes

Where are our strengths and gaps in terms of knowledge about coral reef fishes? After reading extensive and authoritative reviews such as those in Sale (1991, 2002), identifying significant gaps in our knowledge of coral reef fishes appears a daunting task. Nonetheless, it is an important step for those seeking to identify topics for research or even potentially productive career tracks. A crude overview may derive from electronic searches of major citation databases. For example, an online search on June 1, 2009, of the Cambridge Scientific Abstracts Illumina Biological Sciences database combining the keywords "coral reef(s)" and "fish(es)" anywhere in the record for the years 1999–2009 yielded 2,011 citations; the same search restricting output to journal articles only reduced that number to 1,846 citations. I am well aware that such information is imprecise and incomplete, yet broad patterns within these citations tell much about the state of the discipline. Searches for additional keywords within those 1,846 citations reflect strong research emphasis on certain regions, topics, and taxa.

We appear to know little about large fractions of the world's coral reefs and their fishes. For example, the *World Atlas of Coral Reefs* (Spalding et al. 2007) lists 105 nations with coral reefs (*n.b.* – use of "nation" here follows the New American Oxford Dictionary, McKean 2005: "a body of people who share a real or imagined common history, culture, language or ethnic origin, who typically inhabit a particular country or territory," thus skirting issues of political jurisdiction), and provides an estimate of the area of coral reefs for each. Of these nations, six lacked precise information about citations generated (France, UK, USA, US Minor Outlying Islands, New Zealand, and the Netherlands), although data were available for most individual islands or island groups under their jurisdictions. Of the 99 for which data were available, 70 were named specifically in keywords of fewer than 10 journal articles (Table 1; the same pattern appears if the search is expanded to terms found "anywhere" in the citation), yet these 70 account for >73,000 km² of reef, roughly 26% of the world total.

Asymmetric treatment of general topical areas is also evident (Table 2). As suggested above, whole organism research swamps more reductionist approaches. Ecological and

Table 1 A decade of publications on coral reef fishes: citations by nation

Number of citations	Nation or other political entity
316	Australia
50–69	Japan, Mexico, French Polynesia
30–49	Indonesia, Bahamas, New Caledonia, Papua New Guinea, Brazil, the Philippines, US Virgin Islands, Kenya, Puerto Rico, Barbados, Belize, the Netherlands Antilles
10–29	Israel, Panama, Fiji, Egypt, Tanzania, Palau, Jamaica, Solomon Islands, Columbia, Reunion, Seychelles, India, Taiwan
1–9	Cuba, Maldives, Micronesia, Mauritius, Wallis and Futuna Islands, Bermuda, American Samoa, Malaysia, Tonga, Ecuador, Turks and Caicos Islands, British Virgin Islands, Vanuatu, Kiribati, Viet Nam, Costa Rica, Venezuela, Chagos Archipelago, Cayman Islands, Guam, Madagascar, Sri Lanka, Jordan, Saudi Arabia, Eritrea, Thailand, Mozambique, Cook Islands, Honduras, Haiti, Comoros, Dominica, Clipperton, Qatar, Oman, Aruba, Grenada, Saint Vincent and the Grenadines, Mayotte, Guadeloupe, Martinique, Anguilla, Pitcairn, Sudan, United Arab Emirates, Somalia, Tuvalu, Iran, Yemen, Djibouti, Singapore, Trinidad and Tobago, Bangladesh, Nauru, Niue
0	Marshall Islands, Myanmar, China, Nicaragua, Dominican Republic, Bahrain, Western Samoa, Antigua and Barbuda, Brunei, Saint Kitts and Nevis, Saint Lucia, Kuwait, Cambodia, Tokelau, Spratly Islands

From a search of Cambridge Scientific Abstracts, 1 June 2009. A search for journal articles published 1999–2009 with "coral reef(s)" and "fish(es)" in keywords yielded 1,846 citations. Names of nations or other political entities tabulated here were then used within the basic search. Within each category, nations are listed in decreasing order of publications

behavioral studies prevail, but disciplinary descriptors tend to be used broadly. For example, 695 studies include "population" in keywords, but relatively few of those appear to address topics like "demography" (54 citations), "metapopulation" (7), or "population dynamics" (146) commonly associated with detailed studies of populations. Concepts like energy flow (3) and nutrient cycling (4) were rarely specified, although both are central to ecosystem (369) ecology. Detailed analysis of courtship behavior (6), critical to successful spawning, has received little recent attention with coral reef fishes, although behavior (482) and reproduction/ spawning (255) were common in keywords. Physiology (66) and cell biology (15) have received little attention, a conclusion supported by little activity on more specific topics in physiology and cell biology (e.g., endocrine* – 6, neurophys* – 5, renal – 0, membrane* – 2, endoplasmic – 1; the frequent reference to mitochondria is due primarily to the use of mtDNA in genetic, taxonomic, and biogeographic

Table 2 A decade of publications on coral reef fishes: conceptual areas

Topical categories		Life history stage	
Ecology*	515	Larva*	329
Population*	695	Juvenile*	285
Demograph*	54	Adult*	230
Population dynamics	146		
Communit*	564	**Interactions**	
Ecosystem*	369	Predat*	311
Energy flow	3	Compet*	101
Nutrient cycling	4	Parasit*	58
Biogeography*/zoogeograph*	79	Symbio*	51
Behavio(u)r*	482	Mutualis*	15
Reproduc*/spawn*	255	Commensal*	9
Courtship	6		
Aggression/fight*	22	**Feeding**	
Communicat*	17	Feed*/food*	443
Development*/embryolog*	208	Herbiv*	198
Evolution*	78	Carniv*	48
Physiolog*	66	Graz*/brows*	108
Endocrine*	6		
Neurophys*	5	**Fishing and fisheries**	
Osmoregulat*	0	Fisher*	411
Renal	0	Fishing	313
Genetic*	97		
Molecular	44	**Other**	
Cell(ular)	15	Aggregation*	66
Mitochondria*	57	Threatened/endangered	40
Membran*	2	Systematic*/taxonom*	100
Endoplasmic	1	New species (fishes)	22

Search as described in Table 1. Keywords tabulated here were used to refine the basic search. "*" indicates search for any word including the basal form of the term (e.g., ecolog* would retrieve citations including ecology, ecological, ecologist, etc., in keywords); "/" separating terms indicates simultaneous search for either term

work). Continued focus on larval and juvenile fishes recognizes the distinctive demands of these life history stages. Feeding and other interactions are common topics, although it is likely that future study of fishes in important commensal or mutualist relationships will likely increase our understanding of complexity in reef communities. It is heartening that fishing and fishers received considerable attention, for this is an area of growing concern for managers and conservationists. Related to this area is a growing but still sparse literature on aggregations in reef fishes, for spawning aggregations of commercially important species have become targets for highly destructive fisheries.

Two families of widespread commercial importance (Serranidae – groupers, sea basses; Lutjanidae – snappers) constitute two of the six most commonly named families of coral reef fishes (Table 3), flanked by four families that personify obvious, abundant, frequently colorful, and behaviorally and ecologically diverse coral reef fishes: Pomacentridae – damselfishes, Labridae – wrasses, Scaridae – parrotfishes (more properly included within the Labridae, but retained here as recognition of their distinctive characteristics and long historic use of the family name), Acanthuridae – surgeonfishes. Of the 77 families listed in Table 3, 24 yielded no citations, while another 33 families appeared in only 1–10 citations. While representatives of many of these families are small, cryptic, or rarely encountered, some are common and obvious. Examples of the latter include scorpionfishes (including brilliantly patterned lionfishes), pipefishes, porgies, barracudas, rudderfishes, triggerfishes, spadefishes, hawkfishes, trumpetfishes, cornetfishes, spiny puffers, mullets, sweepers, and the Moorish idol.

What can these listings tell us? Predictably, abundant citations derive from nations or regions near extensive coral reefs and with well-equipped laboratory facilities at field stations, universities/colleges, or research institutions. For example, a search like those above using only "Caribbean" or "Gulf of Mexico," waters whose combined states and islands harbor multiple laboratories and provide easy access to reefs, produced 355 citations. A scan of nations yielding no or few citations, yet encompassing a large fraction of the world's coral reefs, produce a more complex picture of factors that inhibit a stronger record of research. Many locations are, of course, isolated by distance, and thus expense, from research centers. Many lack research facilities, and those with laboratories or educational institutions may be constrained by funding to equip facilities, few faculty, and competing

Table 3 A decade of publications on coral reef fishes: citations per family in keywords

Family	Common name	Cited	Family	Common name	Cited
Pomacentridae	Damselfish	142	Priacanthidae	Bigeyes	2
Labridae	Wrasses	110	Carapidae	Pearlfish	2
Scaridae	Parrotfish	78	Cirrhitidae	Hawkfish	2
Serranidae	Sea basses	76	Terapontidae	Tigerperch	2
Acanthuridae	Surgeonfish	66	Bythitidae	Vivip. Brotulas	2
Lutjanidae	Snappers	65	Aulostomidae	Trumpetfish	1
Gobiidae	Gobies	65	Caracanthidae	Orb. Velvetfish	1
Chaetodontidae	Butterflyfish	62	Centriscidae	Shrimpfish	1
Apogonidae	Cardinalfish	49	Fistulariidae	Cornetfish	1
Lethrinidae	Emperors	32	Diodontidae	Spiny puffers	1
Haemulidae	Grunts	30	Microdesmidae	Wormfish	1
Blenniidae	Blennies	28	Malacanthidae	Tilefish	1
Carangidae	Jacks	24	Plesiopidae	Spiny basslets	1
Siganidae	Rabbitfish	21	Mugilidae	Mullets	1
Mullidae	Goatfish	18	Albulidae	Bonefish	0
Balistidae	Triggerfish	14	Ambassidae	Glassfish	0
Pomacanthidae	Angelfish	14	Aploactinidae	Velvetfish	0
Holocentridae	Soldierfish	11	Aplodactylidae	Marblefish	0
Monacanthidae	Filefish	11	Batrachoididae	Toadfish	0
Pseudochromidae	Dottiebacks	11	Berycidae	Berycids	0
Scorpaenidae	Scorpionfish	10	Cheilodactylidae	Morwongs	0
Synodontidae	Lizardfish	8	Clinidae	Clinids	0
Syngnathidae	Pipefish	8	Creediidae	Sandburrowers	0
Nemipteridae	Threadfin bream	7	Cynoglossidae	Tongue soles	0
Sparidae	Porgies	7	Dactylopteridae	Flying gurnards	0
Tripterygiidae	Triplefins	7	Echeneidae	Remoras	0
Sphyraenidae	Barracudas	5	Eleotridae	Sleepers	0
Kyphosidae	Rudderfish	5	Odacidae	Butterfish	0
Labrisomidae	Labrisomids	5	Ogcocephalidae	Frogfish	0
Tetraodontidae	Pufferfish	5	Opisthognathidae	Jawfish	0
Callionymidae	Dragonets	5	Pegasidae	Sea moths	0
Caesionidae	Fusiliers	4	Pempheridae	Sweepers	0
Ostraciidae	Boxfish	4	Pentacerotidae	Armorheads	0
Gobiesocidae	Clingfish	4	Plotosidae	Eeltail catfish	0
Centropomidae	Snooks	3	Solenostomidae	Ghost pipefish	0
Antennariidae	Anglerfish	3	Trichinotidae	Sand divers	0
Pinguipedidae	Sandperch	3	Xenisthmidae	Wrigglers	0
Chaenopsidae	Tube blennies	3	Zanclidae	Moorish idol	0
Ephippidae	Spadefish	2			

Search as described in Table 2. Families from Bellwood and Wainwright (2002, Figure 1) supplemented from personal experience

educational and professional demands on faculty, students, and others. For example, the University of the South Pacific operates 14 campuses on 11 islands across >30 million square kilometers (http://www.usp.ac.fj/) with approximately 13.4 million square kilometers of Exclusive Economic Zones (EEZ), yet biological and marine science programs and facilities are centered in Fiji. Finally, unstable political conditions and isolationist policies affect access to reefs controlled by others on the list. At a time when multiple damaging or destructive pressures influence coral reefs and their inhabitants, we cannot provide a broad-scale assessment of conditions in dispersed but significant areas.

Many of the same factors (e.g., distance, expense, and facilities) that inhibit research on reefs in general also inhibit the expansion of research into underrepresented disciplines. In addition, biologists are drawn to reefs by their beauty, biological diversity, complex interactions, and comfortable field working conditions – generally strong contrasts to laboratory settings. Nonetheless, powerful modern techniques often require sophisticated – and expensive – equipment, stable power supplies, controlled lab environments, and specialized storage of chemicals and samples. Reef biologists have been very successful in addressing reductionist research where samples from remote locations can be taken and fixed, frozen, or otherwise prepared for return to a better-equipped facility, but there are relatively few locations where one can collect a specimen on a reef and return it alive and healthy to a well-equipped facility in a short time.

Rearing of coral reef fishes in captivity eases this constraint for some taxa. Successful rearing by aquarists and researchers of >100 species in ≥24 families had occurred by publication of Leis and McCormick's (2002) review, most based on eggs from captive brood stocks and most from species with benthic eggs.

Despite such advances, few taxa of coral reef fishes have attracted extensive study. In some cases, the reasons are clear. For example, damselfishes, parrotfishes, and surgeonfishes number among the dominant herbivores, while wrasses are probably the most ecologically and anatomically diverse group of reef carnivores. These families, along with the butterflyfishes, also offer researchers a remarkable range of social and reproductive systems for development and test of ecological and behavioral theory. Concerns about excessive artisanal and commercial fisheries draw attention to groupers and snappers, although these families also exhibit considerable interspecific variation in ecological and behavioral traits. Although gobies are almost universally small fishes, they are common, readily maintained in experimental aquaria, and often involved in complex interactions such as cleaning symbiosis or as associates with burrowing shrimps. Despite the demonstrated utility of these various groups as models for addressing important biological concepts, many families remain rich resources for study, particularly as reef biologists attempt to assemble an integrated view of workings of coral reef ecosystems.

3 The Human Element in Study and Exploitation of Coral Reef Fishes

Before proceeding to descriptions of recent work on coral reef fishes, it may be wise to consider the human context within which reefs and fishes function. As noted above, many reef areas are poorly known, in part due to their isolation from strong research centers and management programs. With rare exception, however, inhabitants of islands and coastlines rely to varying degrees on reef resources, small fractions of world reefs are enclosed in protected areas (Mora et al. 2006), and few of these are managed or enforced effectively. Fortunately, many involved in reef and reef fish conservation and management at local, regional, and global levels exert truly heroic efforts for these ecosystems. Nonetheless, clouds persist on the horizon of the future, due at least in part to continued human population growth and frequent poverty in many areas. Doubling rates for nations with coral reefs cluster near or below 50 years (Fig. 1; the world mean is ca. 61 years), and the fraction of the population living below the national poverty line exceeds 20% in the majority of areas for which data were available (Fig. 1). National poverty lines are calculated on the basis of a nation's estimates of funds required to cover basic human necessities in that country, so are not easily comparable among nations. Nonetheless, the implications for reefs, ready sources of edible biomass, and supplemental incomes remain clear.

Conservationists, managers, and scientists have aligned to deal with existing near-emergency situations such as overfishing and other destructive practices on coral reefs. However, threats from human population dynamics require a constant view to the future and well-informed governmental action. Those with detailed knowledge of reef organisms and function must champion reef management and conservation in a political arena. Reef biologists must expand collaboration with political scientists and activists, convince administrators to place value on a scientist's provision of information and insights to government, and train students in avenues used to influence policy.

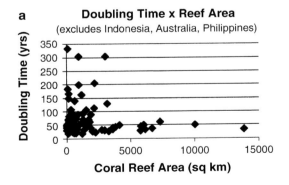
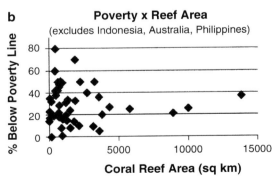

Fig. 1 Human population doubling times (**a**) and fraction of population below the poverty line (**b**) for countries with coral reefs. Reef areas, doubling times, and percent below poverty line for Indonesia, Australia, and the Philippines, respectively: reef area – 51,020, 48,960, and 25,060 km^2; doubling time – 61, 58, and 36 years; percent below poverty line – 17.8%, ND, and 30%. Doubling times (calculated from reports of annual population growth rates) and poverty estimates from World Bank (2008) and the US Central Intelligence Agency Factbook (https://www.cia.gov/library/publications/the-world-factbook/index.html). Reef areas from Spalding et al. (2001), supplemented with data from the United Nations Environment Programme's World Conservation Monitoring Centre and US National Oceanic and Atmospheric Administration

4 Complexity in the Early Life History of Coral Reef Fishes

A general view of the life history of a coral reef fish begins with a fertilized egg released into the water column above a reef or attached to a substrate on the reef. Hatching generally releases a pelagic larva that remains in the water column for days to months, depending on species and environmental circumstances, and then moves to and settles on a reef, transforming into a small juvenile. As they grow, juveniles may continue to occupy the settlement site or move into areas more consistent with adult habitat. During their adult lives, fishes practice a remarkable array of feeding, sheltering, reproductive and other behaviors.

Three major points become apparent as one delves into the specifics of any stage in this life history. First, there exist exceptions to any generality, opening the door to development and test of theory at almost any organizational level using animals amenable to field and often laboratory study. Second, coral reef fish research has gone beyond a time when it might have been characterized as almost exclusively both "expeditionary" and observational, and now represents a field driven by creative questions, rigorous experimental design, and focus on processes underlying success of any life history stage. Third, each life phase must be viewed as somewhat distinct in terms of ecological demands and threats, behavioral and physiological responses to these demands and threats, and the implications of each for evolution of adaptations spanning from cellular biology to those driving ecosystem function and biogeography. This view leads me to structure this chapter, at least initially, on the basis of life history stages rather than topics or subdisciplines (e.g., genetics, physiology, populations, communities, and ecosystems).

4.1 Eggs

Most coral reef fishes fertilize eggs externally (oviparity; for general information, see Thresher 1984; Paxton and Eschmeyer 1994). Some species release eggs into the water column (pelagic spawners such as wrasses, surgeonfishes, and groupers), while others attach eggs to reef substrata that have usually been prepared by the male parent who will subsequently defend the eggs (e.g., damselfishes, blennies). Members of some families, however, deviate from these general patterns by fertilizing eggs internally and bearing live young, nourished at least in part by maternal tissues (e.g., livebearing brotulas, Bythitidae) or having males brood fertilized eggs and hatchlings in their mouths (e.g., cardinalfishes, jawfishes). It is no surprise that adaptations exist to enhance reproductive success regardless of the general pattern. For example, buccal cavities of seven species of cardinalfishes from two genera (*Apogon*, *Cheilodipterus*) were injected with silicon rubber that, when cured, allowed measurement of buccal volume and shape. Males of five of the seven species exhibited a greater buccal volume compared to females, an apparent response to male-only mouthbrooding (Barnett and Bellwood 2005), with most of the difference due to greater height and perhaps somewhat greater expandability of the cavity in males.

Even within single lineages, variation within a family and among genera and species suggests diverse evolved responses among different life history stages to differing ecological conditions. Kavanagh and Alford (2003) summarized early life history data from ca. 170 species of benthic-spawning damselfishes, 85 of which included information on egg size and/or days to hatch (Table 4). Descriptions of reproductive biology of a variety of fishes frequently contain information about egg and clutch size (e.g., sea horses – Foster and Vincent 2004; halfbeaks [Hemiramphidae] – McBride and Thurman 2003; gobies – Privatera 2001, 2002; damselfish – Kokita 2003), but detailed descriptions of development of embryos appear rare (e.g., Kavanagh 2000; Yasir and Qin 2007; Olivotto et al. 2003). Furthermore, applications of molecular tools to study of reef fish development lag descriptive work on development. I was unable to find any reference to study of highly conserved HOX genes in coral reef fishes, despite the explosion of interest in and understanding of their central roles in establishing body organization and structure in vertebrates and invertebrates (see Carroll 2005 for an accessible introduction to such genes and the field of evolutionary development).

Eggs of coral reef fishes are much more than a packet that partially separates some invariant early developmental steps from their surroundings. Egg quality affects larval quality and depends to some degree on female condition. Females of a damselfish (*Pomacentrus amboinensis*) receiving supplemental foods produced eggs with larger oil globules and roughly 10% larger embryonic yolk reserves than controls, although egg size was not influenced (Gagliano and McCormick 2007), indicating that egg size may not always be the best indicator of either maternal reproductive investment or egg quality. In contrast, egg size increased in response to supplemental feeding in a damselfish exhibiting extended, biparental post-hatch brood care, *Acanthochromis polyacanthus* (5.8 mm^3 experimental vs. 4.4 mm^3 control; Donelson et al. 2008). Even transient food supplementation due to important ecosystem events may exert positive influence on egg and larval quality. Newly hatched larvae of *P. amboinensis* possessed larger yolk sacs when produced by mothers that fed heavily on coral eggs following mass, synchronous coral spawning (McCormick 2003).

Eggs and egg care reflect responses to both physicochemical and biological threats. Benthic eggs occur in condensed

Table 4 Egg and larval characteristics of genera of damselfishes (Pomacentridae)

Genus	Egg size (mm)	Days to hatch	Size (mm) at hatch	Pelagic larva (days)	Size (mm) at settling
Abudefduf (15)	1.2	5	3.1	23	12
Acanthochromis (1)	4	16	ND	0	5.5
Amblyglyphidodon (4)	1.4	4.5	3.4	16	10
Amblypomacentrus (1)	ND	4	2.8	17.5	8.5
Amphiprion (14)	2.6	9.5	4.3	12.5	6.5
Cheiloprion (1)	ND	ND	ND	17	8
Chromis (34)	0.75	3	2.3	31.5	11
Chrysiptera (13)	1.25	4.5	3.2	18.5	10
Dascyllus (5)	0.7	2.5	2.2	26.5	9
Dischistodus (5)	1.25	4.5	3.3	17.5	10
Hemiglyphidodon (1)	ND	ND	ND	18	10
Hypsipops (1)	2	17.5	ND	20	ND
Lepidozygus (1)	ND	ND	ND	22.5	ND
Microspathodon (3)	ND	4.5	ND	27	12
Neopomacentrus (5)	1.2	3.5	2.6	20	11
Nexilosus (1)	ND	ND	ND	30.5	ND
Paraglyphidodon (3+)	1.55	6	3.3	18.5	9
Parma (1)	1.1	9	ND	ND	ND
Plectroglyphidodon (6+)	0.9	3	2.5	30	ND
Pomacentrus (24)	1.65	4	3.2	24	12
Pomachromis (1)	ND	ND	2.5	ND	ND
Premnas (1)	1.95	6.5	3.9	10.5	7.5
Pristotus (1)	ND	ND	ND	25.5	ND
Stegastes (19+)	0.9	4	ND	32	12

Summarized from Appendix in Kavanagh and Alford (2003). Number of species considered for each genus in parentheses. Values are means of maximum and minimum sizes and days, excluding data for temperate species for genera with both tropical and temperate representatives

patches, usually in caves, crevices, or other enclosed spaces with potentially reduced water flow and thus potential for hypoxic conditions. Filial cannibalism in male beau gregory damselfish, *Stegastes leucostictus*, appears related to differences in oxygen levels. They regularly ingest some of their own eggs, but in a pattern spread throughout a clutch rather than in focused areas characteristic of feeding by egg predators (Payne et al. 2002). Rates of egg removal were higher at lower (3.74–3.95 mg l^{-1}) than higher oxygen levels (4.09–4.31 mg l^{-1}), and when egg densities were experimentally reduced ca. 40%, embryos developed more rapidly with greater hatching success compared to controls. Pelagic eggs likely face other environmental threats; the wrasse *Thalassoma duperrey* in Hawaii deposits UV-absorbing mycosporine-like amino acids (MAAs) derived from dietary sources in ovaries (eggs), but not testes (unpublished data cited in Zamzow 2004).

Predators on eggs may exert strong effects. Planktivorous triggerfishes (*Melichthys niger* and *M. vidua*) at a spawning aggregation site at Johnston Atoll focused attacks on three species of pair-spawning fishes (*Aulostomus chinensis, Bothus mancus, Ostracion meleagris*) whose egg volumes were much greater (ca. 15–34 mm^3 vs. 1.2–1.3 mm^3) than those of group or pair-spawning species that were also occasionally attacked (*Acanthurus nigroris, Chlorurus sordidus, Parupeneus multifasciatus;* Sancho et al. 2000). The behavior of the triggerfishes was very similar to that of fusiliers (*Caesio* sp.) that preyed on eggs of brown surgeonfish, *Acanthurus nigrofuscus* in the Red Sea (Myrberg et al. 1988 and unpublished; Mazeroll and Montgomery 1995). Fusiliers converge on the point of gamete release for small-group spawns (presumably a female and several males), bite actively for a few seconds, then move away, apparently having destroyed the entire clutch from that female's spawning. During mass-spawning events, when scores of small groups spawn within a few seconds, fusiliers disperse within the gamete cloud and feed actively for a much longer period than with single-group spawns, but the feeding is not focused on a single point for any individual fish and thus is distributed over eggs from many females.

Benthic eggs face threats from many reef predators applying various behavioral tactics. On the Great Barrier Reef, the damselfish *Pomacentrus amboinensis* suffers attacks from individual fishes or multispecies aggregations (Emslie and Jones 2001). Single predators, such as the common wrasse *Thalassoma lunare*, were usually repulsed by the defending male, so probably exert little impact on a per attack basis. Multispecies groups composed of *T. lunare*, bird-nose wrasse (*Gomphosis varius*), angelfish (*Centropyge bicolor*), sandperch (*Parapercis hexopthalma*), dottyback (*Pseudochromis fuscus*), and other *P. amboinensis*, however, generally eliminated all eggs from a nest. Some egg predators change tactics against different target species. A small (≤30 mm total length [TL]) Japanese clingfish, *Pherallodichthys meshimaensis* (Gobiesocidae), appears to be an obligate fish-egg-eater at larger sizes (>17 mm TL; Hirayama et al. 2005), using two tactics to take advantage of loss of attentiveness in nesting fishes. When associating with spawning triplefins (*Helcogramma obtusirostris*), 1–2 clingfish approach spawners as they lay eggs on benthic macroalgae, are rarely attacked (15% of pre-spawning approaches), and once egg-laying begins are completely ignored and pluck just-spawned eggs from immediately below the female. In contrast, clingfish in groups as large as 65 individuals clustered around holes occupied by a nest-defending male blenny, *Istiblennius edentulous* (Blenniidae), and entered to eat individual eggs. Seventy-seven percent of intrusions by clingfish occurred while the defending male was absent courting females or chasing conspecifics, and 88% of these were successful; only 10% of intrusions were successful with the male blenny present. Perhaps the most unusual form of threat to eggs deposited on reef surfaces comes from the turtle-headed sea snake, *Emydocephalus annulatus* (Hydrophiidae), which feeds only on fish eggs (primarily those of damselfish, blennies, and gobies), scraping them from surfaces with enlarged scales above the lip (Shine et al. 2003, 2004).

For embryos in eggs that survive, developmental patterns reflect the molding hand of evolutionary pressures. For example, larvae seeking a reef on which to settle approach playbacks of reef sounds (Leis et al. 2002, 2003; Simpson et al. 2004), but when might they acquire the ability to recognize such sounds? Heart rates of 3–5-day postfertilization embryos of two anemonefishes (genus *Amphiprion*) increased in response to playback of pure sounds at frequencies and sound pressure levels similar to reef noise, and became increasingly sensitive and broadened their frequency response range prior to hatching (Simpson et al. 2005; hatching in *Amphiprion* spp. generally falls at 6–9 days; Kavanagh and Alford 2003). Settling anemonefishes identify anemones by smell, and some learn the scent of the correct host anemone during egg or early larval stages (Kavanagh and Alford 2003). In a comparative study of development in four damselfishes, Kavanagh and Alford (2003) found the olfactory system developed earlier and more rapidly in the anemonefish, *Premnas biaculeatus*, than in three other damselfishes. Juveniles of *Acanthochromis polyacanthus* lack a larval phase, settle as juveniles after hatching, and so must be equipped for visual performance reached by other species after passing both egg and larval stages (Kavanagh 2000; Pankhurst et al. 2002; Kavanagh and Alford 2003). Embryonic development of their visual system is very rapid compared to other damselfishes, yet development of other systems proceeds slowly, with less complete development of skeletal and other systems at hatching. Compensation for smaller, less behaviorally competent hatchlings in this species may come from extended parental care of juveniles for up to 4 months after hatching (Kavanagh 2000).

4.2 Pelagic Larvae

In stark contrast to the growing yet sparse literature on eggs and pre-hatch embryos of coral reef fishes, truly astounding work focuses on larval stages. Early work on fish larvae focused on freshwater or temperate marine species, and often on those of commercial or sport fishing value. Work on coral reef fish larvae began to appear with some regularity in the 1970s, and led to the seminal book by Leis and Rennis (1980), followed by others of a similar nature (Leis 1984; Leis and Trnski 1988) and capped by Leis and Carson-Ewart (2004), all of which provided a foundation for more expansive and detailed studies. Leis' (1991) early review of larval fish biology dealt with horizontal and vertical distribution at different scales, feeding (primarily from gut content studies), and sparse information on growth and mortality (otolith-based studies were left to Victor 1991, in the same volume), but sections on behavior and sensory capabilities were admittedly based almost entirely on fishes from other systems.

A decade later, Leis and McCormick (2002) documented tremendous growth in knowledge about larval biology, ecology, and behavior, Cowen (2002) integrated information about larval distribution with oceanographic processes, Planes (2002) addressed applications of genetic analyses to understanding larval distributions, and Myrberg and Fuiman (2002) reviewed the primary sensory systems and capabilities of reef and other fishes, with particular attention to larval forms. Once perceived as developmentally and behaviorally simplistic passive drifters over long distances, larvae are now recognized as highly capable athletes, sensitive to diverse environmental cues, and at times able to function independently of the track of transporting water masses. Furthermore, their often striking structural characteristics (Fig. 2) likely reflect responses to largely unappreciated conditions and pressures in pelagic environs.

Much of what we know about events in the lives of larvae and postsettlement juveniles derives from studies of otoliths ("ear stones" that rest on beds of sensory hair cells in the inner ear), which grow incrementally in proportion to body growth (see Thorrold and Hare 2002). Daily, seasonal, or annular growth rings may be read at high magnification much like growth rings of trees, providing age, age-specific growth, and age of settling (Fig. 3). Chemical analyses of rings with electron microscope and other techniques also provide information on environmental chemistry at the time of deposition, reflecting such events as shifts among currents and other water masses. Many recent papers and other sources provide

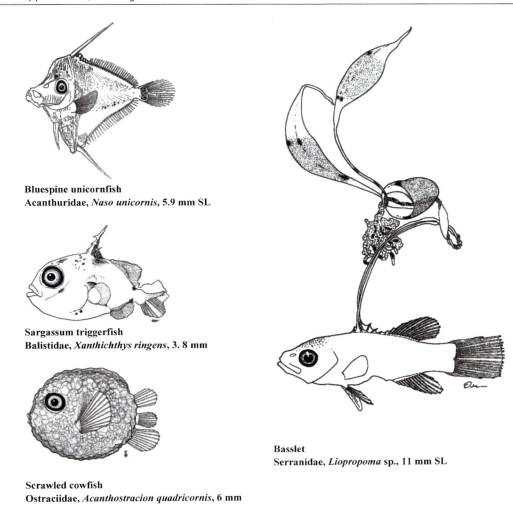

Fig. 2 Examples of pelagic larval stages of coral reef fishes, from Moser et al. (1984). Used with permission

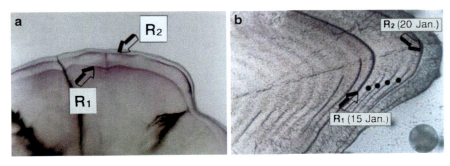

Fig. 3 Light microscope photographs of otoliths from a South Sulawezi cardinalfish, *Sphaeramia ordicularis*, showing the whole otolith (**a**) and ground otolith (**b**). Arrows indicate increments stained by immersion of the fish in a solution of the calcium stain, Alizarin complexone on January 15 (R1) and January 20 (R2) 1998. Dark circles on Fig.3b indicate daily growth increments between the days of staining. Used with permission from Dr. Satoshi Suyama, Tohoku National Fisheries Research Institute, Japan

descriptions of methods for preparing, reading, and validating age and growth estimates from otoliths (e.g., Green et al. 2009); studies cited below should provide an impression of the breadth and power of otolith analyses for teasing apart complexities in the early life history of fishes.

Many factors influence the high rates of larval mortality (Leis and McCormick 2002; Cowen 2002), including predation, maternal circumstances and condition, food supplies, larval growth rates, and a variety of physical and chemical factors that regulate to some degree the distributions of larvae.

Information about impacts of specific predators on reef fish larvae in pelagic waters appears sparse, although various authors have demonstrated effective escape responses of larvae to predator threat at several meters distance (see Leis and McCormick 2002). Valuable information may reside in plankton tow data or gut content studies of important pelagic predators. Coral reef fish larvae in plankton captured in a very large larval trawl (5 mm mesh, with mouth 18 m high; Bertrand et al. 2002) declined rapidly with distance from islands in the Society, Tuamotu and Marquesas Archipelagos (French Polynesia; Lo-Yat et al. 2006). Although many such larvae were taken >100 km from a source reef, few occurred beyond 300 km and pelagic fishes dominated beyond 150 km, suggesting potentially heavy loss to predators, which are certainly not in short supply. Predators taken in the same day and night trawls or from stomachs of three species of tuna captured by longline during the same cruises included representatives of five groups of holoplanktonic crustaceans, two types of meroplanktonic crustacean larvae, at least five families of cephalopods, four phyla of other invertebrate animals, ca. 19 families of vertically migrating fishes, and at least another six families of non-reef shallow water fishes.

In contrast to overt attacks from predators, physicochemical influences on mortality may be quite subtle. For example, egg quality in terms of rich yolk provisioning appears to increase early larval survival but, contrary to expectations, smaller larvae exhibit higher early survival than larger individuals (Gagliano et al. 2007). Maternal stress hormones influence development, but in complex ways. In *Pomacentrus amboinensis*, heightened cortisol levels in eggs increased egg mortality and anatomical asymmetry in larvae, extended hatching time slightly in high-cortisol treatments, but increased post-hatch survival (Gagliano and McCormick 2009). Almost 60% of hatchling *P. amboinensis* launched into a pelagic larval stage suffered otolith (ear stones associated with hearing) asymmetry, yet on average 63% of larvae returning to reefs possessed symmetrical otoliths (Gagliano et al. 2008). This difference suggests asymmetric individuals experience higher mortality or other cause of failure to return to reefs. In experiments playing reef sounds near larval light traps, asymmetric fish returned in greater frequency to low frequency (<573 Hz) than high frequency (>573 Hz) sounds.

Larvae that survive the pelagic phase of their life cycle possess remarkable abilities in terms of their ability to detect and move to reefs on which they may settle, but to what are they drawn? Invertebrate noises generally present throughout day and night (particularly from snapping shrimp) prevail at higher frequencies and might reflect potential prey for juvenile and adult fishes, while sounds produced by adult fishes at lower frequencies may mislead larvae to habitat suboptimal for larval settlement, predatory threats, or transient choruses unavailable as larvae approach the reef. In playback experiments with low-frequency (<570 HZ) and high-frequency (570–2,000 Hz) coral reef sounds played adjacent to nocturnal light traps, damselfishes, cardinalfishes, emperors, blennies, and gobies (Pomacentridae, Apogonidae, Lethrinidae, Blenniidae, and Gobiidae, respectively), as well as syngnathids (sea horses and pipefishes), responded most positively to high-frequency sounds (Simpson et al. 2008). Neural recording of brainstem activity in settlement-stage larvae of coral trout *Plectropomus leopardus* (Family Serranidae, sea basses and groupers) detected ready responses to frequencies of 100–800 Hz, but less consistent response to 1,200 or 2,000 Hz (Wright et al. 2008).

Some species respond vigorously to the smell of home reefs or conspecifics, a trait that may increase their likelihood of remaining in local waters until they settle. Settlement-stage larvae of six damselfishes and 10 cardinalfishes tested in a two-channel flume all preferred water from their home reef versus water from a foreign reef (Gerlach et al. 2007), with one exception: the normally nondispersing *Acanthochromis polyacanthus* that lacks a larval stage and remains with parents for several weeks after hatching. In similar choice experiments in Papua New Guinea (Dixson et al. 2008), an anemonefish, *Amphiprion percula*, preferred beach water to offshore or reef crest water, water from reefs with vegetated islands to water from emergent reefs without islands, and water with the scent of a host anemone or of rainforest vegetation to untreated offshore water. The host anemone in these tests, *Stichodactyla haddoni*, almost always occurred within 10 m of the shoreline, commonly beneath overhanging vegetation and in close proximity to leaf litter. Electrical activity in olfactory epithelium of *P. leopardus* demonstrated response to odors of alanine, proline, and water in which other *P. leopardus* larvae had been swimming (Wright et al. 2008).

As they near a reef, advanced stage larvae must first orient relative to the reef and thereafter can depend on their ability to swim rapidly for extended periods to arrive at a reef. In some cases, sensory modalities other than hearing and smell appear to underlie orientation. During in situ experiments with larvae captured in light traps and then released and followed morning or afternoon on windward or leeward sides of Lizard Island (Leis and Carson-Ewart 2003), 80–100% of released individuals exhibited directional movements from the point of release. Two species of *Chaetodon* moved offshore. In morning tests, the damselfish *Chromis atripectoralis* swam SE, onshore on the leeward side, but offshore on the windward side of the island. In the afternoon on the leeward side, *C. atripectoralis* swam SSW during sunny periods but lacked directionality during cloudy periods. Some species moved toward or away from shore depending on the distance of the release point from the island. *Pomacentrus lepidogenys* moved offshore when released 100 or 500 m from the reef, but toward shore at 1,000 m, suggesting assessment of both direction and distance to the reef.

Larvae swim astoundingly well. Leis and McCormick (2002) note that late-stage coral reef fish larvae swam at an average of 13.7 body lengths per second in flume experiments and field observations, then put this value in a more human perspective: "…a freestyle swimmer capable of 13.7 BL s^{-1} would swim the 100-m race in 3.6 s; the Olympic record is 48 s." Actual speeds averaged 20.6 cm s^{-1} with maxima of 65 cm s^{-1}, faster than most ambient currents near reefs. Larval endurance is also astounding, averaging 84 h and 41 km to exhaustion (unfed and in lab; maximum 289 h, 140 km), roughly equivalent to a human swimming 4,000 km. When Leis and Carson-Ewart (2003) captured larvae in light traps, then released and followed them during the day, mean swimming speeds of different species ranged 12–34 cm s^{-1}, generally well within the range of mean (windward 15 cm s^{-1}, leeward 12 cm s^{-1}) or maximum (windward 41 cm s^{-1}, leeward 31 cm s^{-1}) current speeds measured at the times of the experiments. More recently, Fisher et al. (2005) applied two slightly different protocols in experimental flumes to measure critical swimming speeds (approximate speeds at which fish fail to maintain position in a measured current for a set time period [2 or 5 min in the cited study]) for larvae of 89 species in 28 families from either Caribbean or Great Barrier Reef waters. Excluding a single unidentified holocentrid larva that swam 101 cm s^{-1}, average critical speeds for families ranged from 18 to 21 cm s^{-1} for reef-hugging blennies, nocturnal planktivorous cardinalfishes, schooling clupeids (herring and sardine relatives), and angelfishes to 67–75 cm s^{-1} for herbivorous rabbitfishes, nocturnal holocentrid squirrel-fishes, and pelagic predatory jacks, so no particular ecological pattern emerged relative to swimming speeds. The vast majority, however, attain speeds that allow them to move with considerable independence relative to common current speeds around reefs. Mean critical speeds of 82 of the 89 species exceeded the 13.5 cm s^{-1} average current speed around Lizard Island, site of an Australian Museum's Lizard Island Research Station, and most families averaging 2–3 times that value. Late-stage larvae of surgeonfishes, rabbitfishes, snappers, emperors, damselfishes, butterflyfishes, and threadfin breams swam more rapidly than mean transport current speeds for the Florida Keys, Bahamas, Barbados, and three Great Barrier Reef sites (Fisher 2005). When changes of swimming speeds with growth were taken into account, these data indicate that larval fishes can likely move somewhat independently of average current speeds for >50% of their larval life. One could question whether flume-based swimming speeds have ecological relevance. Larvae of three anemonefishes, a benthic damselfish and a cardinalfish reared and filmed undisturbed in captivity (Fisher and Bellwood 2003) swam consistently at relatively high speeds (3.7 to 5.7 BL s^{-1} for pomacentrids, ~1.5 BL s^{-1} for the apogonid) day and night.

4.3 Post-Settlement Larvae and Juveniles

Much as the release of eggs or newly hatched larvae into pelagic environs constitutes a major transition from reefs, return and settlement of pelagic larvae on reefs represents another series of new ecological hurdles for individuals and a time when settling larvae may express diverse behaviors. In Moorea, French Polynesia, members of a diverse array of larvae (14 families, 20 genera, 27 species) observed at night moving (or being swept) over the reef crest and into or through the lagoon either swam or floated passively, exhibited species-specific tendencies to travel in one of three vertical zones (<30 cm from surface, <30 cm from substrate, or between these zones), and settled near the crest if travelling near the bottom or farther into the lagoon if in surface or middle portions of the water column (Irisson and Lecchini 2008).

The settlement transition is predictably dangerous. Planes and Lecaillon (2001) caged patch reefs before and after removal of all other fishes, introduced presettlement larvae into the cages, and measured mortality after 36 h. Survival on removal reefs was much higher (mean ~83%, species-specific means ~ 34–100%) than on reefs where resident fishes remained (mean 47%, species-specific means ~ 24–100%).

This return and settlement of larvae is also critical for maintenance of populations and communities on reefs, attracting attention to factors operating on larval or settlement stages that might influence establishment of young fishes on reefs. Physical oceanographic processes exert some influence. For example, multivariate regression methods applied to water column and meteorological characteristics (e.g., seasonal variation in day length, wind speed/direction, water temperatures, chlorophyll a levels) explained >60% of variability in presettlement larval fish assemblages at a low-variability site in the New Caledonia lagoon (Carassou et al. 2008b). Once settled, complex behavioral responses to environmental cues exert strong influence, many of which may be suitable for experimentation in field or lab. For example, Lecchini et al. (2007) examined survival of a recently settled damselfish, *Chromis viridis* (~10 or 20 mm TL, ~3 or 16 days postsettlement, respectively), caged with their usual habitat (*Porites rus* coral colonies), artificial habitat (dense branched sticks on a concrete plate) with similar structure to *P. rus*, or completely dissimilar habitat (concrete block with 15 cm × 15 cm holes) in the presence of several types of predators. Predictably, mortality rose from *Porites* to sticks to block for both size classes and was greater for small than large fish in all cases. The authors then tested the effects on settling of the presence of larger conspecifics (~20 mm) or another small damselfish (*Dascyllus aruanus*) that often occupies *Porites*. When given the choice of an empty coral head or one occupied by larger conspecifics, small *Chromis* joined the larger conspecifics; when the choice was empty coral head versus one occupied by *Dascyllus*, 95% moved to the empty coral head. Finally, mortality of small *Chromis* was essentially equivalent when they were caged with coral heads alone, with larger *Chromis*, or with *Dascyllus*.

Clearly, local phenomena exert influence on the structure and function of reef fish communities. However, except perhaps for lakes, springs, and mountaintops, few major ecosystems are as spatially discontinuous as coral reefs on a broad geological scale. Early models and studies of such island – or island-like – systems focused naturally on the balance between rates of colonization by organisms from elsewhere (connectivity), the availability of suitable resources on an island (establishment), and the rates of extinction of often small island populations (persistence; e.g., MacArthur and Wilson 1967). As reef fish biologists move ever closer to an integrative view of egg-to-juvenile life history, growing attention has focused on connectivity among sources of larvae and settlement sites and the implications of connectivity for population and community persistence, diversity, and evolution. Leis (2002) put to rest the view that connectivity could be modeled from knowledge of duration of a passive larval life (see previous section) and "far-field" currents outside the immediate influence of reefs. He recognized the frequency and significance of self-recruitment, where individuals produced on a reef became entrained locally and settled on their natal reef, and that genetic studies suggesting widespread panmixis could be explained by infrequent colonization from afar rather than the proposed swamping of genetic differentiation by regular or frequent colonization.

Modern technology can provide striking insights into the sources and histories of colonists. Hamilton et al. (2008) applied elemental analysis of lead (Pb) levels in otoliths of the Caribbean bluehead wrasse, *Thalassoma bifasciatum*, using laser ablation inductively coupled plasma mass spectrometry (LA-ICPMS). This technique uses lasers to punch ~30×10 µm (diameter×depth; see Figure 7 in *Supporting* Information for Hamilton et al. 2008, http://www.pnas.org/content/105/5/1561/suppl/DC1) pits along a line that radiates from the central core of the otolith to its edge, then analyzes elemental composition of the vaporized material. Pb profiles distinguished recently settled juveniles from offshore and inshore waters and demonstrated that while more settlers were from inshore waters, late larval growth, condition at settlement and postsettlement survival were highest in specimens derived from offshore waters. Differences could not be ascribed to lead toxicity, and were more likely due to physiological and ecological factors. Nonetheless, the impact of larval life on postsettlement success may be significant and have implications for assessments of connectivity among isolated populations. Similar conclusions about strong influence of larval life on postsettlement survival were reached based on standard otolith methods by Gagliano et al. (2007).

In some cases, settlement involves a distinct metamorphosis from the larval form. For example, the larval serranid *Diploprion bifasciatum* sports long, fleshy extensions of its second and third dorsal fin spines, but these are lost within 24 h of settling (Fig. 2; McCormick et al. 2002). In contrast, the remaining 33 species in 13 families treated in that study exhibited changes in various measurements (standardized to standard length) of <5%, and all but one species exhibited pigment changes. Structural changes affected body depth, head shape, and lengths of fin supports (spines, soft rays) of dorsal and pectoral fins. Larvae of 28 species possessed a transparent body, snout or fins, but transparency was lost in all but five of these after only 4 days on the reef. Costs of such remodeling are not well understood, although postsettlement growth was influenced by length of larval life (and degree of differentiation), some species apparently cease feeding during a transition that may span several days, and metabolic costs must be associated with processes of elongation and absorption of tissues and production of new and relatively abundant pigments.

Costs of metamorphosis and a more integrated view of the nature of demands and changes will likely emerge from more reductionist studies. For example, larvae of 12 species in seven genera of wrasses and parrotfishes possess only cones (used for acuity and, potentially, color vision) in somewhat disorganized retinas, developing rods (low-light detectors), highly organized spatial arrangements of receptors, and retinomotor functions (movements of detectors under varying light conditions) only after settling (Lara 2001). This reconstruction of the visual system at and shortly after settling also coincides with increased visual acuity.

Larvae may settle in reef habitats they will continue to occupy throughout their lives or only transiently, and when shifts occur they may be over either short (meters) or long distances (kilometers). Fifteen abundant fishes at two sites in the lagoon at Moorea exhibited one of four patterns when habitat of recently settled juveniles was compared to that of their respective adults (Lecchini and Poignonec 2009). Habitat zones were categorized based on distance between reef crest and island, wave exposure, depth, and substrate (sand, coral rubble, coral slab, macroalgae, algal turf, dead coral, live coral). In 16 of the 30 comparisons of juvenile with adult habitat, larvae remained at the site where they settled. Of the 14 remaining comparisons, five species made long-range movements from nursery sites to a specific adult habitat, six moved to an area where adults used more habitats than settlers, and only three moved to an area where adults used fewer habitats than the settlers. A similar study tracked early juvenile to late juvenile or adult habitats for 39 species in 10 families in New Caledonia (Mellin et al. 2007). Juvenile assemblages differed between soft- and hard-bottom environs, and there were seasonal differences in habitat associations for some species. Most species exhibited no change in habitat with increasing size over either local (10–100 m) or long distances (1–10 km), although 23% did show a radical change in associated habitat over long distances.

5 Topics in the Lives of Adult Coral Reef Fishes

The emphasis on early life history stages in previous sections grew from the breadth and depth of conceptually and technically integrative studies directed at these stages, and from the remarkable discoveries about these heretofore poorly understood stages. Nonetheless, adult fishes form the foundation of the history and evolution of coral reef fish research, beginning with formal taxonomic descriptions in the early 1600s (see Pietsch 1995), and organized and revitalized through Linnaeus' (1758) inclusion of Artedi's 1738 ichthyological treatise in *Systema Naturae* (see Jordan 1917). Among the fishes listed in the 10th edition (1758) of *Systema Naturae*, the foundation from which modern taxonomy has grown, are several genera common to coral reefs, including *Muraena* (moray), *Scorpaena* (scorpionfish), *Chaetodon* (butterflyfish), *Fistularia* (cornetfish), *Balistes* (triggerfish), *Ostracion* (boxfish), *Tetraodon* (puffer), *Diodon* (spiny puffer), *Syngnathus* (sea horse), and *Pegasus* (sea moth).

As indicated previously, those desiring a broad yet detailed overview of reef fish biology should turn to chapters in Sale (2002). Topics addressed below represent those with which I have personal experience.

5.1 Trophic Links on Coral Reefs

Feeding ecology of reef fishes attracts considerable attention, in part because of the obvious links between feeding and various determinants of individual fitness (e.g., habitat selection and use, survival, and reproductive output) and because feeding activities are often easily observed and quantified and reflect subtle adaptations for overcoming prey defenses or other ecological constraints. Examples include the behavior of obligate egg-eating clingfish (Hirayama et al. 2005) described previously (see section on eggs), the tendency for many fishes to join multispecies aggregations to overcome defenses of territorial fishes protecting eggs or rich food resources (Montgomery 1981; Wolf 1985), cleaning symbioses where small fishes and invertebrates clean parasites and damaged tissues from larger fishes (Cote 2000; Grutter 2008), and the annoying (for both target fish and divers with exposed legs …) practice of a group of blennies to mimic cleaner fishes and attack other fishes in order to feed on healthy fins and scales (or occasional diver leg hairs; Cote and Cheney 2004; Cheney and Cote 2005).

For fishes, descriptions of gut contents have served historically as the primary indicators of diet, prey selection, role in resource partitioning, and trophic position in a community. High diversity in reef fish communities generated expectations of specialization in use of food resources, but these expectations were not met in a number of cases (Longenecker 2007). In a study of eight small, cryptic Hawaiian reef fishes, Longenecker (2007) demonstrated that inability to detect food resource partitioning may relate to a methodological weakness: imprecise identification of food items. He identified invertebrate prey to very fine levels – species, genus, or similar level of precision with morphotypes not attributable to described taxa or with fragmented prey items. Such precision is uncommon in dietary studies, with prey commonly identified to Family or higher-level taxon. Although six of Longenecker's species included a wide array of items in their diets (mean 43 items, range 21–75; the remaining two species specialized entirely on either coral or detritus), the minimum number of items required to constitute ≥50% of the diet averaged only 3.25 (range 3–5) and dietary overlap among species was at a level indicative of dietary specialization and, possibly, fine-scale food partitioning. Longenecker's data and analysis also reflect a long-proposed geographic pattern of increasing food specialization in the tropics compared to temperate areas, as well as that increased diversity among prey taxa may facilitate high species diversity among small predators.

Technical developments join conceptual advances in strengthening food web research. For example, application of stable isotope analyses now allow rapid definition of the trophic position of fishes in communities and can be achieved without lethal sampling. Metabolic processing of nutrients alters $^{13}C:^{12}C$ and $^{15}N:^{14}N$ stable isotope ratios, with enrichment of the heavier isotope in sample ratios relative to a known standard presented as $\delta^{13}C$ and $\delta^{15}N$ values. These values reflect both the likely origin of foods, reflected particularly in $\delta^{13}C$ values, and the position of an organism in a food web, indicated most clearly by $\delta^{15}N$ values. For example, Carassou et al. (2008a) supplemented information on diet with $\delta^{13}C$ and $\delta^{15}N$ analyses of 34 species in New Caledonia. Herbivores, defined on the basis of gut contents, exhibited low $\delta^{15}N$ consistent with their low trophic level, although parrotfishes, generally considered as herbivores tended to have slightly higher $\delta^{15}N$ than other herbivores, perhaps due to inclusion of detritus in their diets. The trophic levels of dietary reef carnivores and piscivores as defined by isotope signatures were lower than expected, suggesting greater omnivory and a broader food base than previously recognized. Frederich et al. (2009) related isotopic composition of 13 species of damselfishes (Pomacentridae) to gut contents, habitat, and social behavior. These species fell into three dietary and isotopic groups: zooplanktivores that fed in aggregations away from the reef surface, benthic primarily herbivores that held territories or small home ranges on the reef, and an intermediate group of species with a diversified diet that moved widely on or aggregated near the reef. Although all fishes exhibited higher $\delta^{15}N$ values than their foods, in this case $\delta^{15}N$ could not separate the three groups by trophic level.

One area of particularly integrative research deals with herbivorous fishes. Various primarily herbivorous parrotfishes, surgeonfishes, rudderfishes, rabbitfishes, angelfishes, and damselfishes rank among the most abundant larger fishes on coral reefs and constitute large fractions of reef fish biomass. Several factors drive the integrative interests in herbivorous fishes: ecological, behavioral, and morphological diversification among herbivorous fishes; potential difficulties for fishes in extracting sufficient energy and nutrients from algae laden with complex carbohydrates and various defensive compounds; the apparent ecological conundrum of low algal biomass supporting high herbivore biomass; and long-standing recognition that as one moves away from the tropics one encounters fewer fish herbivores but much higher algal standing crops (see reviews by Choat and Clements 1998; Harmelin-Vivian 2002; Wainwright and Bellwood 2002). Consistent with the multiple variables influencing herbivorous fish feeding, attempts to identify broad patterns in feeding selectivity of herbivores or their impacts on algal taxa often produce complex outcomes. For example, in feeding choice experiments with transplanted large-bodied algae, herbivores demonstrated no clear differences in use of algal taxa that differed in structure (thallus shape, fine vs. coarse branching, toughness, degree of calcification) or reported defenses (Mantyka and Bellwood 2007a). However, same experiments (Mantyka and Bellwood 2007b) detected considerable variation among herbivores in their individual degrees of selectivity for these algal taxa. Algal taxa large enough to collect and transplant for such experiments are often those passed over by herbivorous fishes that feed on multispecies mats or monocultures of small, delicate algae (Montgomery et al. 1980). Unfortunately, while gut content studies appear to reflect consistent feeding by many reef herbivores on small, delicate algae (e.g., Montgomery 1980; Hata et al. 2002; Jones et al. 2007), controlled feeding choice experiments with them probably await preparation of single-species cultures that can be offered to fishes in the field.

The integrative nature of research on reef herbivores received an important boost during the last decade or so from renewed focus on the biochemical composition of algal foods and the digestive processes of herbivorous fishes, due in particular to work by Kendall Clements, Michael Horn, and their students and colleagues in both tropical and temperate areas of the Pacific (e.g., Clements et al. 2009; Skea et al. 2007; Raubenheimer et al. 2005; Crossman et al. 2005; Horn et al. 2006; German and Horn 2006; Gawlicka and Horn 2006). Recently, Clements et al. (2009) took an important step toward bringing herbivorous fishes into step with the highly interdisciplinary form of nutritional ecology developed in various terrestrial systems. They recognize that an historic focus on factors (e.g., presumed defensive chemicals) leading to rejection of items from diets of herbivores must be complemented with greater focus on nutrients in the diet and nutrient demands of fishes, and they identify several issues with methods used in early attempts to quantify nutrient levels and quality in marine algae. It is striking, for example, that despite the wealth of information on the ecology and feeding biology of herbivorous coral reef fishes, the only detailed studies of activity of endogenous digestive enzymes or of digestive functions of gut microfloras deal with north and south Pacific temperate fishes (*loc. cit.*, and Mountfort et al. 2002).

5.2 Coral Reef Fishes as Habitats

Much has been made here and elsewhere about the diversity of fishes on coral reefs and their various behavioral and ecological roles in reef functions. Somewhat in contrast to the work described elsewhere in this volume that recognizes the multitude of symbioses (here used to encompass parasitism, commensalism, and mutualism) within coral reefs, relatively little work focuses on fishes as habitats for other contributors to the rich biotic diversity on reefs.

Descriptions of new parasites and quantification of parasite diversity on hosts suggests a wealth of unrecognized relationships. For example, Justine (2007) reported 16 species of monogenean, copepod, and isopod parasites on the gills of a single species of sea bass (*Epinephelus maculates*) from New Caledonia, some of which are probably specific to that host. Munoz et al. (2007) sought pattern in parasite communities of 14 wrasses but, except for a tendency for higher ectoparasite loads on larger hosts, found little evidence of effects of phylogeny or host characteristics.

Parasites can, of course, exert negative effects on host fish survival or performance. Adult or juvenile parasitic isopods can injure or kill fishes, leading some to view them as micropredators (Jones and Grutter 2008; Grutter et al. 2008; Fogelman and Grutter 2008), as well as exert more subtle physiological effects. For example, a large isopod attached externally to cardinalfish affects the host's behavior, metabolic rate, and rate of weight loss, possibly by inducing increased drag on an otherwise streamlined fish (Östlund-Nilsson et al. 2005). While description of heretofore undetected species and relationships among coral reef fishes and their occupants is important to understanding reef ecology, such close relationships may also be indicators of anthropogenic impacts on reefs. The nature of parasites from seven cardinalfishes in New Caledonia reflected environmental conditions influenced by community and industrial wastewater inputs (Sasal et al. 2007).

An intriguing symbiosis between a large unicellular organism and a Red Sea surgeonfish (Fishelson et al. 1985) hints at unappreciated biodiversity in relationships between fishes and heretofore unknown associates. Initially described as *Epulopiscium fishelsoni*, a likely protist on the basis of large size (to >600 μm), mobility, complex ultrastructure, and complex circadian growth and reproductive cycles (Montgomery and Pollak 1988), molecular studies subsequently identified these organisms as the largest Eubacteria (Angert et al. 1993). Apparently restricted to surgeonfishes, ~10 structural "morphs" of these symbionts (differing in maximum size, shape, and reproductive traits) occur in an array of host species from Hawaii, Guam, Japan, French Polynesia, the Great Barrier Reef, Tuvalu, South Africa, and the Caribbean (Clements et al. 1989; Montgomery, Clements, N. Grim, S. Treutner, H. Nakagawa, T. Umino, T. Shibuno unpublished). Phylogenetic analyses place *Epulopiscium* close to *Clostridium*, a genus of small bacteria well known from vertebrate intestines, and particularly *Metabacterium*, an occupant of guinea pig caecae (Angert et al. 1993, 1996). Recent studies describe the complex reproductive cycle of these microbes (Flint et al. 2005; Ward et al. 2009) and demonstrate that total DNA quantity can far exceed that of diploid eukaryote cells, and individual cells may possess tens of thousands of copies of the bacterial genome (Bresler et al. 1998; Mendell et al. 2008). Restriction of these bacteria to a single fish family, their complex life cycles that correlate to fish feeding and activity cycles, a variety of structurally distinct forms, and their widespread distribution through oceans and host species suggest they possess great potential for interdisciplinary studies of coevolution, host–microbe relationships and biogeography.

6 Crises in Management and Conservation of Coral Reef Fishes

Coral reef fishes face a multitude of threats to their health, population, and community structure, and even persistence (Helfman 2007). They are not alone in this regard, and coral reef conservationists and managers may benefit from lessons learned in freshwater and temperate marine systems that have been under intense and diverse pressures for perhaps longer periods than most reef systems. In North America, for example, inland and temperate marine fishes are imperiled by combinations of habitat loss; sedimentation; chemical pollution; dewatering; anthropogenic modifications to natural channels or flow regimes; overexploitation for subsistence, commercial, recreational, scientific, or educational purposes; intentional eradication with ichthyocides; indirect impacts of fishing pressure such as reduction or loss of host fish populations required by parasitic lampreys; disease and parasitism; impacts of nonindigenous organisms such as hybridization, competition, predation, and restriction of range; and predicted effects of global warming (Allen 2006; Horn and Stephens 2006; Jelks et al. 2008; Love 2006; Schroeter and Moyle 2006).

Coral reefs and their fishes face at least a subset of these same pressures, as well as others more specific to coral reef systems (see Helfman 2007). Threats to reefs and their fishes include physical habitat loss from "blast fishing" with dynamite and other explosives that destroys the biota and pulverizes and destabilizes the substrate to a degree that inhibits reestablishment of the original biota (Raymundo et al. 2007); various effects of urbanization (sedimentation, development, etc.; Wolanski et al. 2009); release of chemical pollutants (Anderson et al. 2003) with widespread obvious or subtle physiological effects (e.g., acidification; Guinotte and Fabry 2008); subsistence or commercial overfishing (Sala et al. 2001; Munro and Blok 2005; McClanahan et al. 2008); use of diver-delivered poisons, like cyanide, to capture fishes for the aquarium and live food fish trade (Maka et al. 2005); trap fishing (Hawkins et al. 2007); impacts of nonindigenous organisms (e.g., establishment of Pacific lionfish, *Pterois* spp., in the Western Atlantic; Freshwater et al. 2009; Barbour et al. 2010); and the potential effects of global climate change (Munday et al. 2008; Wilson et al. 2010).

Where knowledge remains insufficient to predict future impacts of a complex threat, even identifying the most productive research questions may prove difficult. Wilson et al. (2010) provide a guiding light in this regard for global climate change. They present 53 questions or sets of questions based on input from authorities in coral reef systems, each considered tractable for research, widely applicable across reef systems, and of high priority if we are to develop understanding of potential effects of climate change. Questions are placed in nine categories spanning from fundamental ecological to physiological, fish health, and management issues, many with citations to recent key references that serve as entries into relevant literature. As suggested earlier in this chapter, they also recognize the relative dearth of information about physiological responses of reef fishes to problems like increasing temperature and acidification and the historic tendency to focus on relatively few groups of fishes.

While many threats relate to ecosystem damage, some relate especially to remarkable aspects of the biology of the fishes themselves. Population and even species devastation of large groupers and other types of fishes traces to their formation of large spawning aggregations to which they travel from great distances (Claydon and Gordon 2005). Nassau grouper, *Epinephelus striatus*, once common and commercially important in the Caribbean and tropical western Atlantic, is considered endangered by the International

Union for Conservation of Nature, threatened by the American Fisheries Society (Sadovy and Domeier 2005), and of Special Concern by the US NOAA National Marine Fisheries Service (NOAA Fisheries), depressing descriptors for a species with at least 100 historically known spawning sites and aggregations involving as many as 100,000 or more individuals (Whaylen et al. 2004; Sala et al. 2001). This fish may travel at least 220 km to historic spawning sites (Bolden 2000), yet there appears to be little information available about mechanisms used in navigation or possibly learning of migratory paths for this or other exploited species. The potential loss of such detailed behavioral information may be appreciated more fully when compared with studies of aggregating wrasses or surgeonfishes that move long distances relative to their body size, display complex orientation, movement, and aggregating behaviors, and yet are not subjected to overfishing (e.g., Kiflawi and Mazeroll 2006).

In some cases, damage to fish communities may result from motivations or methods rarely encountered beyond coral reef systems. Despite its inherent appeal to those interested in reefs, diving-based ecotourism can lead to inadvertent damage to habitats, particularly those important for smaller species (Uyarra and Cote 2007). Methods that exert wide-ranging effects include blast fishing, cyanide application, and even spearfishing. Spearfishing in some developed nations evolved to resemble sport hunting, with either explicit regulations (e.g., limits on species and sizes) or understood guidelines on personal conduct (e.g., no use of SCUBA). However, subsistence and commercial spearfishing on a variety of Pacific islands contributes measurably to overfishing, focusing on surgeonfishes and parrotfishes (often 20–30+% of total spearfishing catches), followed by groupers, rabbitfishes, rudderfishes, and a variety of other species representing at least 13 different families (Gillett and Moy 2006). Controls among island nations are complicated by inconsistent regulations among islands and even villages, insufficient enforcement, spearfishing at night and with SCUBA, focus on particularly large unicornfishes and wrasses, and short-term but heavy localized impact of multiple commercial divers from large foreign vessels that appear, sweep a reef clean of desired fishes, and move on. In a more tightly regulated and sport-motivated environment (Great Barrier Reef, Australia), focus shifted for both spear and line fishers to preferred sea basses, with some purposeful take of targeted emperors, snappers, sweetlips, surgeonfish, parrotfish, wrasses, and mackerel (Frisch et al. 2008).

Any litany of threats to coral reef fishes and their habitats underscores that successful conservation or effective management of species and ecosystems will proceed most effectively through integration of detailed understanding of life histories, organismal physiology, and dynamics of populations, communities, and ecosystems (Helfman 2007). However, previous sections of this chapter suggest that we may be poorly armed to deal with threats to specific reefs and their fishes, for our knowledge of coral reef condition, threats and dynamics remains geographically spotty, as does the understanding of the roles and biology of many families of fishes. Fortunately, well-understood and broadly applicable physiological, behavioral, ecological, and oceanographic principles should continue to support development of robust conservation and management plans that can be implemented somewhat independent of geographic or community specifics. Such implementation likely faces constraints relating to governance and human population condition beyond the normal purview and experience of research biologists and educators.

Fortunately, scientists and educators are not alone in their concern for coral reefs and their inhabitants. For example, the Coral Reef Alliance (www.coral.org), a nongovernmental organization that describes itself as the "only international nonprofit organization that works exclusively to protect our planet's coral reefs," maintains a searchable database of groups and organizations involved in coral reef issues. A recent count of entries in their various categories of organizations included 461 nongovernmental/nonprofit organizations, 71 government agencies/programs, 50 scientific research institutes/universities, 49 marine protected areas/parks/reserves, 40 aquariums/zoos, 13 businesses, 10 international agencies/programs, 9 community groups, and 6 dive operations, clubs, or associations. There exists a rich potential for collaboration.

7 Perspectives

Can a focused review such as this contribute to welfare of a globally significant environment and resource? The most likely avenue may relate to a question posed at the beginning of the chapter: how should students considering careers in marine life sciences prepare to make truly significant contributions? First, recent literature exemplifies the power of robust, interdisciplinary, observational and experimental research. An implicit message weaves through such work: field observations generate questions about how nature works, but answers derive from rigorously designed field or lab studies likely to carry an individual beyond individual strengths and into the realm of collaboration. Insofar as studies of coral reef fishes are concerned, important field observations will come most frequently and rapidly to those who combine broad foundations in the concepts and methods of natural sciences with depth of understanding of the biology of organisms common to reefs.

Given looming human socioeconomic as well biological conservation and management issues affecting coral reefs, students should also prepare to exert influence over reef conservation and management decisions through educational or political activism. Such preparation need not require extensive formal education. Instead, one must attend to how decisions are made locally, regionally, or nationally (political approach), develop trusted relationships with those who make or directly influence decisions (personal approach), and provide decision-makers with scientifically robust information, intellectually sound commentary on proposed actions, and alternative recommendations for management and conservation that include predictions about the human and biological outcomes of these alternatives (scientific approach). These represent only minor redirection of the skills we all use to (a) understand and adhere to research guidelines imposed by governments where we work; (b) develop positive working relationships with collaborators, students, and employees; and (c) generate alternative hypotheses to explain biological phenomena and then test them based on predictions from these alternatives. Those willing and able to apply these skills to furthering conservation and management efforts will enter a new learning environment, develop connections and even friendships with an array of people fascinating for their diverse experiences and motivations, and bring honor to our profession through service to one of our planet's greatest ecosystem treasures.

References

Allen MJ (2006) Pollution. In: Allen LG, Pondella DJ II, Horn MJ (eds) The ecology of marine fishes: California and adjacent waters. University of California Press, Berkeley, pp 595–610

Allen LG, Pondella DJ II, Horn MJ (eds) (2006) The ecology of marine fishes: California and adjacent waters. University of California Press, Berkeley

Anderson E, JudsonB, Fotu ST, Thaman B (2003) Marine pollution risk assessment for the Pacific Islands region (PACPOL Project RA1). South Pacific regional environment programme, Apia, Samoa. Volume 1: Main report, 158 pp. Volume 2: Appendices, 85 pp

Angert ER, Clements KD, Pace NR (1993) The largest bacterium. Nature 362:239–241

Angert ER, Brooks AE, Pace NR (1996) Phylogenetic analysis of *Metabacterium polyspora*: clues to the evolutionary origin of daughter cell production in *Epulopiscium* species, the largest bacteria. J Bacteriol 178:1451–1456

Artedi P (1738) Ichthyologia sive opera omnia de piscibus scilicet: Bibliotheca ichthyologica. Philosophia ichthyologica. Genera piscium. Synonymia specierum. Descriptiones specierum. Omnia in hoc genere perfectiora quam antea ulla posthuma vindicavit, recognovit, coaptavit et editit Carolus Linnaeus. Wishoff, Lugduni Batavorum [Leiden]

Barbour AB, Montgomery ML, Adamson AA, Díaz-Ferguson E, Silliman BR (2010) Mangrove use by the invasive lionfish *Pterois volitans*. Mar Ecol Prog Ser 401:291–294

Barnett A, Bellwood DR (2005) Sexual dimorphism in the buccal cavity of paternal mouthbrooding cardinalfishes (Pisces: Apogonidae). Mar Biol 148:205–212

Bellwood DR, Wainwright PC (2002) The history and biogeography of fishes on coral reefs. In: Sale PF (ed) Coral reef fishes: Dynamics and diversity in a complex ecosystem. Academic, New York, pp 5–32

Bertrand A, Bard F-X, Josse E (2002) Tuna food habits related to the micronekton distribution in French Polynesia. Mar Biol 140:1023–1037

Bolden SK (2000) Long-distance movement of a Nassau grouper (*Epinephelus striatus*) to a spawning aggregation in the central Bahamas. Fish Bull 98:642–645

Bresler V, Montgomery WL, Fishelson L, Pollak P (1998) Gigantism in a bacterium, *Epulopiscium fishelsoni*, correlates with complex patterns in arrangement, quantity and segregation of DNA. J Bacteriol 180:5601–5611

Carassou L, Kulbicki M, Nicola TJR, Polunin NVC (2008a) Assessment of fish trophic status and relationships by stable isotope data in the coral reef lagoon of New Caledonia, southwest Pacific. Aquat Living Resour 21:1–12

Carassou L, Ponton D, Mellin C, Galzin R (2008b) Predicting the structure of larval fish assemblages by a hierarchical classification of meteorological and water column forcing factors. Coral Reefs 27:867–880

Carroll SB (2005) Endless forms most beautiful: The new science of EvoDevo and the making of the animal kingdom. W. W. Norton & Co., New York

Cheney KL, Cote IM (2005) Frequency-dependent success of aggressive mimics in a cleaning symbiosis. Proc R Soc Lond B Biol Sci 272:2635–2639

Choat JH, Clements KD (1998) Vertebrate herbivores in marine and terrestrial systems: a nutritional ecology perspective. Ann Rev Ecol Syst 29:375–403

Choat JH, Robertson DR (2002) Age-based studies. In: Sale PF (ed) Coral reef fishes: Dynamics and diversity in a complex ecosystem. Academic, New York, pp 57–80

Claydon J, Gordon JDM (2005) Spawning aggregations of coral reef fishes: characteristics, hypotheses, threats and management. Oceanogr Mar Biol Annu Rev 42:265–302

Clements KD, Sutton DC, Choat JH (1989) Occurrence and characteristics of unusual protistan symbionts from surgeonfishes Acanthuridae of the Great Barrier Reef Australia. Mar Biol 102:403–412

Clements KD, Raubenheimer D, Choat J (2009) Nutritional ecology of marine herbivorous fishes: ten years on. Funct Ecol 23:79–92

Cote IM (2000) Evolution and ecology of cleaning symbiosis in the sea. Oceanogr Mar Biol Annu Rev 38:311–355

Cote IM, Cheney KL (2004) Distance-dependent costs and benefits of aggressive mimicry in a cleaning symbiosis. Proc R Soc Lond Ser B Biol Sci 271:2627–2630

Cowen RK (2002) Larval dispersal and retention and consequences for population connectivity. In: Sale PF (ed) Coral reef fishes: Dynamics and diversity in a complex ecosystem. Academic, New York, pp 149–170

Crossman DJ, Choat JH, Clements KD (2005) Nutritional ecology of nominally herbivorous fishes on coral reefs. Mar Ecol Prog Ser 296:129–142

Dixson DL, Jones GP, Munday PL, Planes S, Pratchett MS, Srinivasan M, Syms C, Thorrold SR (2008) Coral reef fish smell leaves to find island homes. Proc Biol Sci 275:2831–2839

Donelson JM, McCormick MI, Munday PL (2008) Parental condition affects early life-history of a coral reef fish. J Exp Mar Biol Ecol 360:109–116

Emslie MJ, Jones GP (2001) Patterns of embryo mortality in a demersally spawning coral reef fish and the role of predatory fishes. Environ Bio Fishes 60:363–373

Fishelson L, Montgomery WL, Myrberg AA Jr (1985) A unique symbiosis in the gut of tropical herbivorous surgeonfish (Acanthuridae: Teleostei) from the Red Sea. Science 229:49–51

Fisher R (2005) Swimming speeds of larval coral reef fishes: impacts on self-recruitment and dispersal. Mar Ecol Prog Ser 285:223–232

Fisher R, Bellwood DR (2003) Undisturbed swimming behaviour and nocturnal activity of coral reef fish larvae. Mar Ecol Prog Ser 263:177–188

Fisher R, Leis JM, Clark DL, Wilson SK (2005) Critical swimming speeds of late-stage coral reef fish larvae: variation within species, among species and between locations. Mar Biol 147:1201–1212

Flint JF, Drzymalski D, Montgomery WL, Southam G, Angert ER (2005) Nocturnal production of endospores in natural populations of *Epulopiscium*-like surgeonfish symbionts. J Bacteriol 187:7460–7470

Fogelman RM, Grutter AS (2008) Mancae of the parasitic cymothoid isopod, *Anilocra apogonae*: early life history, host-specificity, and effect on growth and survival of preferred young cardinal fishes. Coral Reefs 27:685–693

Foster SJ, Vincent ACJ (2004) Life history and ecology of seahorses: implications for conservation and management. J Fish Biol 64:1–61

Frederich B, Fabri G, Lepoint G, Vandewalle P, Parmentier E (2009) Trophic niches of thirteen damselfishes (Pomacentridae) at the Grand Recif of Toliara, Madagascar. Ichthyol Res 56:10–17

Freshwater DW, Hines A, Parham S, Wilbur A, Sabaoun M, Woodhead J, Akins L, Purdy B, Whitfield PE, Paris CB (2009) Mitochondrial control region sequence analyses indicate dispersal from the US East Coast as the source of the invasive Indo-Pacific lionfish *Pterois volitans* in the Bahamas. Mar Biol 156:1213–1221

Frisch AJ, Baker R, Hobbs J-PA, Nankervis L (2008) A quantitative comparison of recreational spearfishing and linefishing on the Great Barrier Reef: implications for management of multi-sector coral reef fisheries. Coral Reefs 27:85–95

Gagliano M, McCormick MI (2007) Maternal condition influences phenotypic selection on offspring. J Anim Ecol 76:174–182

Gagliano M, McCormick MI (2009) Hormonally mediated maternal effects shape offspring survival potential in stressful environments. Oecologia 160:657–665

Gagliano M, McCormick MI, Meekan MG (2007) Survival against the odds: ontogenetic changes in selective pressure mediate growth-mortality trade-offs in a marine fish. Proc R Soc B Biol Sci 274:1575–1582

Gagliano M, Depczynski M, Simpson SD, Moore JAY (2008) Dispersal without errors: symmetrical ears tune into the right frequency for survival. Proc R Soc B Biol Sci 275:527–534

Gawlicka AK, Horn MH (2006) Trypsin gene expression by quantitative *in situ* hybridization in carnivorous and herbivorous prickleback fishes (Teleostei: Stichaeidae): ontogenetic, dietary, and phylogenetic effects. Physiol Biochem Zool 79:120–132

Gerlach G, Atema J, Kingsford MJ, Black KP, Miller-Sims V (2007) Smelling home can prevent dispersal of reef fish larvae. Proc Natl Acad Sci U S A 104:858–863

German DP, Horn MH (2006) Gut length and mass in herbivorous and carnivorous prickleback fishes (Teleostei: Stichaeidae): ontogenetic, dietary, and phylogenetic effects. Mar Bio 148:1123–1134

Gillett R, Moy W (2006) Spearfishing in the Pacific Islands. Current status and management issues, vol 19, FAO/FishCode Review. FAO, Rome, p 72

Green BS, Mapstone BD, Carlos G, Begg GA (eds) (2009) Tropical fish otoliths: information for assessment, management and ecology. Springer, Dordrecht

Grutter AS (2008) Interactions between gnathiid isopods, cleaner fish and other fishes on Lizard Island, Great Barrier Reef. J Fish Biol 73:2094–2109

Grutter AS, Pickering JL, McCallum H, McCormick MI (2008) Impact of micropredatory gnathiid isopods on young coral reef fishes. Coral Reefs 27:655–661

Guinotte JM, Fabry VJ (2008) Ocean acidification and its potential effects on marine ecosystems. Ann NY Acad Sci 1134:320–342

Hamilton SL, Regetz J, Warner RR (2008) Postsettlement survival linked to larval life in a marine fish. Proc Natl Acad Sci U S A 105:1561–1566

Harmelin-Vivian ML (2002) Energetics and fish diversity on coral reefs. In: Sale PF (ed) Coral reef fishes: Dynamics and diversity in a complex ecosystem. Academic, Amsterdam, pp 265–274

Hata H, Nishihira M, Kamura S (2002) Effects of habitat-conditioning by the damselfish *Stegastes nigricans* (Lacepede) on the community structure of benthic algae. J Exp Mar Biol Ecol 280:95–116

Hawkins JP, Roberts CM, Gell FR, Dytham C (2007) Effects of trap fishing on reef fish communities. Aquat Conserv Mar Freshw Ecosyst 17:111–132

Helfman GS (2007) Fish conservation: A guide to understanding and restoring global aquatic biodiversity and fishery resources. Island Press, Washington, DC

Hirayama S, Shiiba T, Sakai Y, Hashimoto H, Gushima K (2005) Fish-egg predation by the small clingfish *Pherallodichthys meshimaensis* (Gobiesocidae) on the shallow reefs of Kuchierabu-jima Island, southern Japan. Environ Bio Fishes 73:237–242

Horn MJ, Stephens JS Jr (2006) Climate change and overexploitation. In: Allen LG, Pondella DJ II, Horn MJ (eds) The ecology of marine fishes: California and adjacent waters. University of California Press, Berkeley, pp 621–635

Horn MH, Gawlicka AK, German DP, Logothetis EA, Cavanagh JW, Boyle KS (2006) Structure and function of the stomachless digestive system in three related species of New World silverside fishes (Atherinopsidae) representing herbivory, omnivory, and carnivory. Mar Biol 149:1237–1245

Irisson J-O, Lecchini D (2008) *In situ* observation of settlement behaviour in larvae of coral reef fishes at night. J Fish Biol 72:2707–2713

Jelks HL, Walsh SJ, Burkhead NM, Contreras-Balderas S, Díaz-Pardo E, Hendrickson DA, Lyons J, Mandrak NE, McCormick F, Nelson JS, Platania SP, Porter BA, Renaud CB, Schmitter-Soto JJ, Taylor EB, Warren ML Jr (2008) Conservation status of imperiled North American freshwater and diadromous fishes. Fisheries 33:372–407

Jones CM, Grutter AS (2008) Reef-based micropredators reduce the growth of post-settlement damselfish in captivity. Coral Reefs 27:677–684

Jones GP, Santana L, McCook LJ, McCormick MI (2007) Resource use and impact of three herbivorous damselfishes on coral reef communities. Mar Ecol Prog Ser 328:215–224

Jordan DS, Evermann BW (1917) The genera of fishes, from Linnaeus to Cuvier, 1758–1833, seventy-five years, with the accepted type of each. Part I. Leland Stanford Junior University Publications, University series, no 27: pp 161

Justine J-L (2007) Parasite biodiversity in a coral reef fish: twelve species of monogeneans on the gills of the grouper *Epinephelus maculatus* (Perciformes: Serranidae) off New Caledonia, with a description of eight new species of *Pseudorhabdosynochus* (Monogenea: Diplectanidae). Syst Parasitol 66:81–129

Kavanagh KD (2000) Larval brooding in the marine damselfish *Acanthochromis polyacanthus* (Pomacentridae) is correlated with highly divergent morphology, ontogeny and life-history traits. Bull Mar Sci 66:321–337

Kavanagh KD, Alford RA (2003) Sensory and skeletal development and growth in relation to the duration of the embryonic and larval stages in damselfishes (Pomacentridae). Biol J Linn Soc 80:187–206

Kiflawi M, Mazeroll AI (2006) Female leadership during migration and the potential for sex-specific benefits of mass spawning in the brown surgeonfish (*Acanthurus nigrofuscus*). Environ Bio Fishes 76:19–23

Kokita T (2003) Potential latitudinal variation in egg size and number of a geographically widespread reef fish, revealed by common-environment experiments. Mar Biol 143:593–601

Lara MR (2001) Morphology of the eye and visual acuities in the settlement-intervals of some coral reef fishes (Labridae, Scaridae). Environ Bio Fishes 62:365–378

Lecchini D, Poignonec D (2009) Spatial variability of ontogenetic patterns in habitat associations by coral reef fishes (Moorea lagoon – French Polynesia). Estuar Coast Shelf Sci 82:553–556

Lecchini D, Planes S, Galzin R (2007) The influence of habitat characteristics and conspecificson attraction and survival of coral reef fish juveniles. J Exp Mar Biol Ecol 341:85–90

Leis JM (1984) The larvae of Indo-Pacific coral reef fishes: a guide to identification. University of Hawaii Press, Honolulu

Leis JM (1991) The pelagic stage of reef fishes: the larval biology of coral reef fishes. In: Sale PF (ed) The ecology of fishes on coral reefs. Academic, New York, pp 183–230

Leis J (2002) Pacific coral-reef fishes: the implications of behaviour and ecology of larvae for biodiversity and conservation, and a reassessment of the open population paradigm. Environ Bio Fishes 65:199–208

Leis JM, Carson-Ewart BM (2003) Orientation of pelagic larvae of coral-reef fishes in the ocean. Mar Ecol Prog Ser 252:239–253

Leis JM, Carson-Ewart BM (2004) The larvae of Indo-Pacific coastal fishes: an identification guide to marine fish larvae. Fauna malesiana handbook 2. Brill Academic Publishers, Leiden

Leis JM, McCormick MI (2002) The biology, behavior and ecology of the pelagic, larval stage of coral reef fishes. In: Sale PF (ed) Coral reef fishes: dynamics and diversity in a complex ecosystem. Academic, New York, pp 171–199

Leis JM, Rennis DS (1980) The larvae of Indo-Pacific coral reef fishes. New South Wales University Press, Sydney

Leis JM, Trnski T (1988) Larvae of Indo Pacific Shorefishes. New South Wales University Press, Sydney

Leis JM, Carson-Ewart BM, Cato DH (2002) Sound detection in situ by the larvae of a coral-reef damselfish (Pomacentridae). Mar Ecol Prog Ser 232:259–268

Leis JM, Carson-Ewart BM, Hay AC, Cato DH (2003) Coral reef sounds enable nocturnal navigation by some reef-fish larvae in some places at some times. J Fish Biol 63:724–737

Linnaeus C (1758) Tomus I. Systema naturae per regna tria naturae, secundum classes, ordines, genera, species, cum characteribus, differentiis, synonymis, locis. Editio decima, reformata. Laurentii Salvii, Holmiae [Stockholm]

Longenecker K (2007) Devil in the details: high-resolution dietary analysis contradicts a basic assumption of reef-fish diversity models. Copeia 2007:543–555

Love MS (2006) Subsistence, commercial, and recreational fisheries. In: Allen LG, Pondella DJ II, Horn MJ (eds) The ecology of marine fishes: California and adjacent waters. University of California Press, Berkeley, pp 567–594

Lo-Yat A, Meekan M, Carleton J, Galzin R (2006) Large-scale dispersal of the larvae of nearshore and pelagic fishes in the tropical oceanic waters of French Polynesia. Mar Ecol Prog Ser 325:195–203

MacArthur RH, Wilson EO (1967) The theory of island biogeography. Princeton University Press, Princeton

Maka KWK, Yanaseb H, Renneberga R (2005) Cyanide fishing and cyanide detection in coral reef fish using chemical tests and biosensors. Biosens Bioelectron 20:2581–2593

Mantyka CS, Bellwood DR (2007a) Direct evaluation of macroalgal removal by herbivorous coral reef fishes. Coral Reefs 26:435–442

Mantyka CS, Bellwood DR (2007b) Macroalgal grazing selectivity among herbivorous coral reef fishes. Mar Ecol Prog Ser 352:177–185

Mazeroll AI, Montgomery WL (1995) Structure and organization of local migrations in brown surgeonfish (*Acanthurus nigrofuscus*). Ethology 99:89–106

McBride RS, Thurman PE (2003) Reproductive biology of *Hemiramphus brasiliensis* and *H. balao* (Hemiramphidae): maturation, spawning frequency, and fecundity. Biol Bull Mar Bio Lab Woods Hole 204:57–67

McClanahan TR, Hicks CC, Darling ES (2008) Malthusian overfishing and efforts to overcome it on Kenyan coral reefs. Ecol Appl 18:1516–1529

McCormick MI (2003) Consumption of coral propagules after mass spawning enhances larval quality of damselfish through maternal effects. Oecologia 136:37–45

McCormick MI, Makey L, Dufour V (2002) Comparative study of metamorphosis in tropical reef fishes. Mar Biol 141:841–853

McKean E (ed) (2005) The new Oxford American dictionary, 2nd edn. Oxford University Press, Oxford

Mellin C, Kulbicki M, Ponton D (2007) Seasonal and ontogenetic patterns of habitat use in coral reef fish juveniles. Estuar Coast Shelf Sci 75:481–491

Mendell JE, Clements KD, Choat JH, Angert ER (2008) Extreme polyploidy in a large bacterium. Proc Natl Acad Sci U S A 105:6730–6734

Montgomery WL (1980) Comparative feeding ecology of two herbivorous damselfishes (Pomacentridae: Teleostei) from the Gulf of California, Mexico. J Exp Mar Biol Ecol 47:9–24

Montgomery WL (1981) Mixed-species schools and the significance of vertical territories of damselfishes. Copeia 1981:477–481

Montgomery WL, Pollak PE (1988) *Epulopiscium fishelsoni* n.gen, n.sp., a protist of uncertain affinities from the gut of an herbivorous reef fish. J Protozool 35:565–569

Montgomery WL, Gerrodette T, Marshall LD (1980) Impact of grazing by the yellowtail surgeonfish, *Prionurus punctatus*, on algal communities in the lower Gulf of California, Mexico. Bull Mar Sci 30:901–908

Mora C, Andrèfouët S, Costello MJ, Kranenburg C, Rollo A, Veron J, Gaston KJ, Myers RA (2006) Coral reefs and the global network of marine protected areas. Science 312:1750–1751

Moser HG, Richards WJ, Cohen DM, Fahay MP, Kendall AW Jr, Richardson SL (eds) (1984) Ontogeny and systematics of fishes.. American Society of Ichthyologists and Herpetologists, Lawrence, Kansas, Spec. Publ. No. 1

Mountfort DO, Campbell J, Clements KD (2002) Hindgut fermentation in three species of New Zealand marine herbivorous fish. Appl Environ Microbiol 68:1374–1380

Munday PL, Jones GP, Pratchett MS, Williams AJ (2008) Climate change and the future for coral reef fishes. Fish Fish 9:261–285

Munoz G, Grutter AS, Cribb TH (2007) Structure of the parasite communities of a coral reef fish assemblage (Labridae): testing ecological and phylogenetic host factors. J Parasitol 93:17–30

Munro JL, Blok L (2005) The status of stocks of groupers and hinds in the northeastern Caribbean. Proc Gulf Caribb Fish Inst 56:283–294

Myrberg AA Jr, Fuiman LA (2002) The sensory world of coral reef fishes. In: Sale PF (ed) Coral reef fishes: Dynamics and diversity in a complex ecosystem. Academic, New York, pp 123–148

Myrberg AA Jr, Montgomery WL, Fishelson L (1988) The reproductive behavior of Acanthurus nigrofuscus (Forskal) and other surgeonfishes (Fam. Acanthuridae) off Eilat, Israel (Gulf of Aqaba, Red Sea). Ethology 79:31–61

Olivotto I, Cardinali M, Barbaresi L, Maradonna F, Carnevali O (2003) Coral reef fish breeding: the secrets of each species. Aquaculture 224:69–78

Östlund-Nilsson S, Curtis L, Nilsson GE, Grutter AS (2005) (2005) Parasitic isopod *Anilocra apogonae*, a drag for the cardinal fish *Cheilodipterus quinquelineatus*. Mar Ecol Prog Ser 287:209–216

Pankhurst PM, Pankhurst NW, Parks MC (2002) Direct development of the visual system of the coral reef teleost, the spiny damsel, *Acanthochromis polyacanthus*. Environ Bio Fishes 65:431–440

Paxton JR, Eschmeyer WN (eds) (1994) Encyclopedia of fishes. Academic, San Diego

Payne AG, Smith C, Campbell AC (2002) Filial cannibalism improves survival and development of beaugregory damselfish embryos. Proc R Soc Lond Ser B Biol Sci 269:2095–2102

Pietsch TW (ed) (1995) Historical portrait of the progress of ichthyology, from Its origins to our own time, by Georges Cuvier. Johns Hopkins University Press, Baltimore

Planes S (2002) Biogeography and larval dispersal inferred from population genetic analysis. In: Sale PF (ed) Coral reef fishes: dynamics and diversity in a complex ecosystem. Academic, New York, pp 201–220

Planes S, Lecaillon G (2001) Caging experiment to examine mortality during metamorphosis of coral reef fish larvae. Coral Reefs 20:211–218

Privatera LA (2001) Characteristics of egg and larval production in captive bluespotted gobies. J Fish Biol 58:1211–1220

Privatera LA (2002) Reproductive biology of the coral-reef goby, *Asterropteryx semipunctata*, in Kaneohe Bay, Hawaii. Environ Bio Fishes 65:289–310

Raubenheimer D, Zemke-White WL, Phillips RJ, Clements KD (2005) Algal macronutrients and food selection by the omnivorous marine fish *Girella tricuspidata*. Ecology 86:2601–2610

Raymundo LJ, Maypa AP, Gomez ED, Cadiz P (2007) Can dynamite-blasted reefs recover? A novel, low-tech approach to stimulating natural recovery in fish and coral populations. Mar Pollut Bull 54:1009–1019

Sadovy Y, Domeier M (2005) Are aggregation-fisheries sustainable? Reef fish fisheries as a case study. Coral Reefs 24:254–262

Sala E, Ballesteros E, Starr RM (2001) Rapid decline of *Nassau Grouper* spawning aggregations in Belize: fishery management and conservation needs. Fisheries 26:23–29

Sale PF (ed) (1991) The ecology of fishes on coral reefs. Academic, New York

Sale PF (ed) (2002) Coral reef fishes: dynamics and diversity in a complex ecosystem. Academic, New York

Sancho G, Petersen CW, Lobel PS (2000) Predator-prey relations at a spawning aggregation site of coral reef fishes. Mar Ecol Prog Ser 203:275–288

Sasal P, Mouillot D, Fichez R, Chifflet S, Kulbicki M (2007) The use of fish parasites as biological indicators of anthropogenic influences in coral-reef lagoons: a case study of Apogonidae parasites in New-Caledonia. Mar Pollut Bull 54:1697–1706

Schroeter RE, Moyle PB (2006) Alien fishes. In: Allen LG, Pondella DJ II, Horn MJ (eds) The ecology of marine fishes: California and adjacent waters. University of California Press, Berkeley, pp 611–620

Shine R, Shine T, Shine B (2003) Intraspecific habitat partitioning by the sea snake *Emydocephalus annulatus* (Serpentes, Hydrophiidae): the effects of sex, body size, and color pattern. Biol J Linn Soc 80:1–10

Shine R, Bonnet X, Elphick MJ, Barrott EG (2004) A novel foraging mode in snakes: browsing by the sea snake *Emydocephalus annulatus* (Serpentes, Hydrophiidae). Funct Ecol 18:16–24

Simpson SD, Meekan MG, McCauley RD, Jeffs A (2004) Attraction of settlement-stage coral reef fishes to reef noise. Mar Ecol Prog Ser 276:263–268

Simpson SD, Yan HY, Wittenrich ML, Meekan MG (2005) Response of embryonic coral reef fishes (Pomacentridae: *Amphiprion* spp.) to noise. Mar Ecol Prog Ser 287:201–208

Simpson SD, Meekan MG, Jeffs A, Montgomery JC, McCauley RD (2008) Settlement-stage coral reef fish prefer the higher-frequency invertebrate-generated audible component of reef noise. Anim Behav 75:1861–1868

Skea GL, Mountfort DO, Clements KD (2007) Contrasting digestive strategies in four New Zealand herbivorous fishes as reflected by carbohydrase activity profiles. Comp Biochem Physiol A Mol Integr Physiol 146:63–70

Spalding MD, Ravilious C, Green EP (2001) World atlas of coral reefs. University of California Press, Berkeley

Spalding MD, Ravilious C, Green EP (2007) World atlas of coral reefs, revised 5 September 2007. United Nations Environment Programme, World Conservation Monitoring Centre, Cambridge

Thorrold SR, Hare JA (2002) Otolith applications in reef fish ecology. In: Sale PF (ed) Coral reef fishes: Dynamics and diversity in a complex ecosystem. Academic, New York, pp 243–264

Thresher RE (1984) Reproduction in reef fishes. T.F.H. Publications, Inc. Ltd, Neptune City

Uyarra MC, Cote IM (2007) The quest for cryptic creatures: impacts of species-focused recreational diving on corals. Biol Conserv 136:77–84

Victor BC (1991) Settlement strategies and biogeography of reef fishes. In: Sale P (ed) The ecology of fishes on coral reefs. Academic, San Diego, pp 231–260

Wainwright PC, Bellwood DR (2002) Ecomorphology of feeding in coral reef fishes. In: Sale PF (ed) Coral reef fishes: Dynamics and diversity in a complex ecosystem. Academic, New York, pp 33–55

Ward RJ, Clements KD, Choat JH, Angert ER (2009) Cytology of terminally differentiated *Epulopiscium* mother cells. DNA Cell Biol 28:57–64

Whaylen L, Pattengill-Semmens CV, Semmens BX, Bush PG, Boardman MR (2004) Observation of the Nassau grouper, *Epinephelus striatus*, spawning aggregation site in Little Cayman, Cayman Islands, including multi-species spawning information. Environ Bio Fishes 70:305–313

Wilson SK, Adjeroud M, Bellwood DR, Berumen ML, Booth D, Bozec Y-M, Chabanet P, Cheal A, Cinner J, Depczynski M, Feary DA, Gagliano M, Graham NAJ, Halford AR, Halpern BS, Harborne AR, Hoey AS, Holbrook SJ, Jones GP, Kulbiki M, Letourneur Y, De Loma TL, McClanahan T, McCormick MI, Meekan MG, Mumby PJ, Munday PL, Öhman MC, Pratchett MS, Riegl B, Sano M, Schmitt RJ, Syms C (2010) Crucial knowledge gaps in current understanding of climate change impacts on coral reef fishes. J Exp Biol 213:894–900

Wolanski E, Martinez JA, Richmond RH (2009) Quantifying the impact of watershed urbanization on a coral reef: Maunalua Bay, Hawaii. Estuar Coast Shelf Sci 84:259–268

Wolf NG (1985) Food selection and resources partitioning by herbivorous fishes in mixed species groups. Proc 5th Int Coral Reef Congress Symp Semin (A) 4:23–28

World Bank (2008) Poverty data: A supplement to World Development Indicators 2008. International Bank for Reconstruction and Development/the World Bank, Washington, DC

Wright KJ, Higgs DM, Belanger AJ, Leis JM (2008) Auditory and olfactory abilities of larvae of the Indo-Pacific coral trout *Plectropomus leopardus* (Lacepede) at settlement. J Fish Biol 72:2543–2556

Yasir I, Qin JG (2007) Embryology and early ontogeny of an anemonefish, *Amphiprion ocellaris*. J Mar Biol Ass UK 87:1025–1033

Zamzow JP (2004) Effects of diet, ultraviolet exposure, and gender on the ultraviolet absorbance of fish mucus and ocular structures. Mar Biol 144:1057–1064

Competition Among Sessile Organisms on Coral Reefs

Nanette E. Chadwick and Kathleen M. Morrow

Abstract Competition among sessile organisms is a major process on coral reefs, and is becoming more important as anthropogenic disturbances cause shifts in dominance to non-reef builders such as macroalgae, soft corals, ascidians, and corallimorpharians. Long-term monitoring and field experiments have demonstrated that competition for limited space can exert major impacts on reef biodiversity and community composition across habitats and regions. Recent experiments also reveal increasingly important roles of allelopathic chemicals and the alteration of associated microbes in shaping competitive outcomes among benthic space occupiers. Competition impacts the recruitment, growth, and mortality of sessile reef organisms and alters their population dynamics. Co-settlement and aggregation of conspecific coral colonies may lead to intense intraspecific competition, including chimera formation and potential somatic and germ cell parasitism. The complexity of competitive outcomes and their alteration by a wide variety of factors, including irradiance, water motion, and nutrient levels, results in mostly circular networks of interaction, often enhancing species diversity on coral reefs. Competition is a model process for revealing impacts of human activities on coral reefs, and will become increasingly important as alternate dominants gain space at the expense of reef-building corals.

Keywords Interference competition • exploitation competition • competition • cnidarian • macroalgae • cyanobacteria • scleractinian • corallimorpharian • actinarian • sea anemone • ascidian • zoanthid • fungiid • hydrocoral • octocoral • soft coral • stony coral • coral • sponge • climate change • chimera • growth • mortality • reproduction • competitive network • coral–algal interaction • phase shift • feedback loop • model • allelopathy • herbivory • recruitment • antibiotic • microorganism • bacteria • abrasion • palytoxin • bleaching • disease • natural products • nematocyst • mucus • diversity • community structure • aggression • population • alternate dominant

1 Introduction

Scleractinian reef-building corals require space on hard substratum for the settlement and metamorphosis of their larvae into primary coral polyps. Due to their dependence on endo-symbiotic algae (zooxanthellae) for energy (Stambler in this book) and on the consumption of zooplankton for essential nutrients (Ferrier-Pages et al. 2010), the space they occupy must be exposed to adequate irradiance and to water currents carrying food. Thus, suitable space on shallow marine substratum often is a limiting resource for the settlement, growth, and reproduction of tropical reef corals (Connell et al. 2004, Birrell et al. 2008a, Foster et al. 2008). Competition among corals for substratum space is a major process on tropical reefs, and in some locations and time periods can control their patterns of diversity and abundance (Connell et al. 2004). Lang and Chornesky (1990) published the only major review of coral competition 2 decades ago. Since then, many advances have been made in our understanding of important aspects of competition among reef corals, especially in terms of associated fitness costs, long-term impacts on community structure, and effects of within-species interactions among coral colonies. Here we review the recent literature on coral competition, following a similar organizational format to that of Lang and Chornesky (1990).

Global climate change and increasing human impacts on coral reefs during the past 20 years have caused substantial decreases in the abundance of reef-building corals worldwide (Pandolfi et al. 2003, Bruno and Selig 2007). Widespread declines in coral cover, due in part to mass bleaching (Hoegh-Guldberg 2004), disease (Harvell et al. 1999), hurricanes (Holland and Webster 2007), and loss of grazers (Carpenter 1990), have exposed large areas of reef substratum and contributed to colonization by other benthic organisms (e.g., macroalgae, sponges, ascidians, and corallimorpharians). These trends

N.E. Chadwick (✉) and K.M. Morrow
Department of Biological Sciences, Auburn University, 101 Rouse Life Sciences Building, Auburn, AL 36849, USA
e-mail: chadwick@auburn.edu

have precipitated an increase in competitive interactions between corals and other sessile reef organisms (Done 1992, Bak et al. 1996, Griffith 1997, Hughes et al. 2007). As such, we broaden our review here to include competition among all major sessile organisms that occupy exposed space on coral reefs: cnidarians (stony corals, soft corals, hydrocorals, zoanthids, actiniarian sea anemones, corallimorpharian sea anemones), other sessile invertebrates (sponges, ascidians, see Glynn and Enochs 2010), and algae (red, green, and brown macroalgae, cyanobacterial algal mats, see Fong and Paul 2010). We do not discuss competition among organisms in cryptic habitats on reefs, such as in caves or crevices (reviewed in Knowlton and Jackson 2001).

This review includes five major topics on competition among sessile organisms on coral reefs, focusing on recent advances within each area: (1) methods of study; (2) mechanisms; (3) altering factors; (4) impacts on individuals, populations, and communities; and (5) conclusions and directions for future research.

2 Methods of Studying Competition Among Sessile Organisms on Reefs

2.1 Field Surveys at a Single Point in Time

Several field surveys of competitive interactions among reef cnidarians have been conducted over the past 20 years. They have revealed complex, circular networks of competitive dominance among species, and have supported ideas advanced in the 1980s (Paine 1984, Buss 1986) that the lack of a linear, transitive competitive hierarchy allows the coexistence of diverse cnidarians on coral reefs (Lang and Chornesky 1990, Knowlton and Jackson 2001). The space occupied by each sessile organism is likely to change at least slightly over ecological timescales (years to decades), as individuals and colonies are competitively damaged by some neighbors, while overgrowing others. This complexity of outcomes creates an ever-changing mosaic of space occupation on coral reefs, which is altered sporadically by disturbances such as storms, diseases, or predator outbreaks that remove individuals and open space for colonization (Connell et al. 2004).

A few field surveys have reported outcomes of within-genus or within-species competition in corals. Examination of competition among five species of *Porites* in Japan revealed a linear hierarchy of dominance (Rinkevich and Sakai 2001), but inclusion of more species may reveal circular networks within this genus. Chornesky (1991) surveyed contacts among colonies of *Agaricia tenuifolia* on Belize reefs, and concluded that they may benefit from intraspecific contact, due to interdigitation and mutual strengthening of their skeletons on turbulent reefs. Thus, some contact interactions among sessile organisms on reefs are standoffs with no clear effects, and some may be mutually beneficial interactions rather than competition.

A handful of studies during the 1990s described competitive networks for all coral species within a reef area. An intransitive, circular network occurs among major stony corals on reefs in Taiwan, with stony corals in general dominating soft corals (Dai 1990). A stony coral competitive network in the Red Sea is similar to that in Taiwan, in that it contains circular intransitive outcomes with no clear correlation between aggressive ranking and the relative abundance of species (Abelson and Loya 1999). The aggressive rank of some corals is similar between the Pacific Ocean and Red Sea, but varies widely for some species among regions, indicating that more information is needed on the consistency of competitive network outcomes among coral reef regions. In contrast, on temperate rocky reefs that contain a much simpler assemblage of only three species of stony corals and one corallimorpharian, the competitive hierarchy is linear in both field surveys and laboratory observations (Chadwick 1991).

Other studies have surveyed competitive outcomes among a subset of the cnidarians that occupy a given reef system. Three morphotypes of the Caribbean boulder coral *Montastraea annularis* were observed to compete with a wide variety of sessile organisms on reefs in Curacao (Van Veghel et al. 1996). A survey of brain corals *Platygyra daedalea* in the Red Sea revealed that they damage other corals and defend their space on reefs in about half of all observed interactions (Lapid et al. 2004). Also in the Red Sea, the clonal corallimorpharian *Rhodactis rhodostoma* is dominant over most contacted stony and soft corals, but some massive faviid corals cause unilateral damage to the corallimorpharian polyps (Langmead and Chadwick-Furman 1999a). In the Caribbean Sea, the congener *R. (Discosoma) sanctithomae* also damages most contacted corals (Miles 1991), and so do other corallimorpharian and actiniarian sea anemones (Sebens 1976). In a field survey in Jamaica, competitive contacts with an encrusting octocoral damage stony corals but not sea anemones and algae (Sebens and Miles 1988).

The recent proliferation of macroalgae on some coral reefs has prompted field surveys of the frequency of coral–algal contacts. However, in contrast to many reef cnidarians, algae may be seasonally ephemeral and often do not cause clear zones of tissue mortality on competitors. The more subtle impacts of competition with macroalgae have been revealed mostly through time-series surveys (see Sect. 2.2) and manipulative experiments (see Sect. 2.3). A photographic survey of coral reefs in Florida during the seasonal peak of macroalgal abundance (>50% cover) documented high frequencies of coral–algal interactions, in which the basal perimeter of coral colonies contacted mixed algal turfs and the macroalgae *Halimeda* and *Dictyota*

(Lirman 2001). A onetime survey in Hawaii used photo quadrats to demonstrate that the abundance of recent recruits of the coral *Montipora capitata* decreases significantly with the cover of fleshy algae (Vermeij et al. 2009).

Field surveys of competitive contacts among sessile reef organisms other than stony corals and algae are rare. In a cross-taxonomic study on competition among a Red Sea soft coral, scleractinian coral, hydrocoral, and sponge, Rinkevich et al. (1992) described a nontransitive circular network among the four examined species. Also in the Red Sea, a survey of contacts between scleractinian corals and the mucus nets of sessile vermetid gastropods revealed that corals are deformed under the mucus nets (Zvuloni et al. 2008). On southern Caribbean reefs in Colombia, field observations indicated that sponges overgrow corals in only 2.5% of interactions, and most contacts are standoffs with possibly mutual inhibition of growth (Aerts and van Soest 1997). A field survey in the northeastern Pacific documented that cold-water solitary corals frequently compete for space with colonial ascidians (Bruno and Witman 1996).

Interpretation of the results of onetime surveys is problematic, because the outcomes of some competitive interactions reverse after several weeks to months, and thus depend on the duration of previous contact (reviewed in Lang and Chornesky 1990, Chadwick-Furman and Rinkevich 1994, Langmead and Chadwick-Furman 1999b). Networks of dominance soon after ecological disturbances, when organisms have grown into contact only recently, may differ from those that develop after several months to years of competition. However, the onetime surveys that recently have been conducted in several geographical regions have confirmed the initial conclusions of Lang and Chornesky (1990) that circular, nontransitive competitive networks among reef organisms often prevent domination of the reef benthos by a few species, and thus enhance species diversity on coral reefs.

2.2 Long-Term Field Monitoring

Norström et al. (2009) recently reviewed case studies of long-term field observations (20–50 years) on the outcomes of competition among sessile reef organisms, and documented phase shifts in all major reef regions between stony corals and non-algal alternate dominants (corallimorpharians, actiniarian sea anemones, soft corals, ascidians, sponges, and sea urchins). Some long-term shifts appear to be caused directly by human interference. Positive feedback may occur between the high abundance of alternate dominants and the continued mortality of reef-building corals, thus reinforcing the alternate state (Fig. 1a). In one of the longest field observational studies to date, Connell et al. (2004) reported on 38 years of monitoring on the Great Barrier Reef in Australia. They observed intense competition among corals in some reef crest and exposed pool habitats, and frequent disturbances that interrupted competitive processes on the reef flat. They also found that the percent cover of macroalgae fluctuated widely from 15% to 85% in their monitoring plots. Coyer et al. (1993) recorded variation in coral–algal relative abundances over 10 years on temperate reefs in California, and concluded that the two varied inversely, implying space competition. Field monitoring in Taiwan revealed that actiniarian sea anemones replaced stony corals on anthropogenically impacted reefs over 20 years (Chen and Dai 2004).

Few other monitoring studies of competition on reefs have lasted longer than 1–3 years, constrained by the timescales of research grants and graduate student theses. An 18-month study on Australian reefs showed that macroalgae fluctuated from 41% to 56% cover, and contacted a majority of corals (Tanner 1995). The role of epiphytic cyanobacteria (putative genus *Lyngbya*) in maintaining the dominance of macroalgae on Pacific coral reefs in Panama was examined during 2 years of field surveys (Fong et al. 2006). Cyanobacterial epiphytes appeared to protect some red macroalgae from herbivory, allowing macroalgal-dominated communities to persist on reefs for several years following the pulse disturbance of ENSO in 1997–1998 that killed the previously dominant corals. This phase shift developed due to the stochastic availability of a large supply of algal and cyanobacterial recruits, followed by their mutually positive association and rapid growth that prevented coral reestablishment. Littler and Littler (2006) surveyed coral–algal interactions over 2 years in Belize and developed a Relative Dominance Model (RDM) of space occupation on coral reefs, in which the dominance of four main functional groups varies with levels of eutrophication and herbivory, often anthropogenically driven (Fig. 1b). Each factor can mediate the others; for example, high rates of herbivory can delay the impact of elevated nutrients, and low levels of nutrients can offset the impact of reduced herbivory. The RDM provides a simple and easily understood illustrative aid to the development of monitoring methods for coral reefs. In a recent review based on the RDM, Littler et al. (2009) suggest that coral recovery from phase shifts can be inhibited further by large-scale stochastic disturbances such as tropical storms, cold fronts, warming events, diseases, and predator outbreaks (Littler et al. 2009). Other models have emerged recently from survey data collected on coral reefs. A complex framework of feedback loops has been developed to describe the long-term outcomes of competition among algae and corals (Mumby and Steneck 2008). Mumby (2009) developed a simulation model in which he proposed that three possible equilibria exist on coral reefs; a stable coral-depauperate state, an unstable equilibrium at intermediate coral cover, and a stable coral-rich state. This model emphasizes the high temporal variability of macroalgal cover in comparison to the relatively slow changes that occur in coral cover over time,

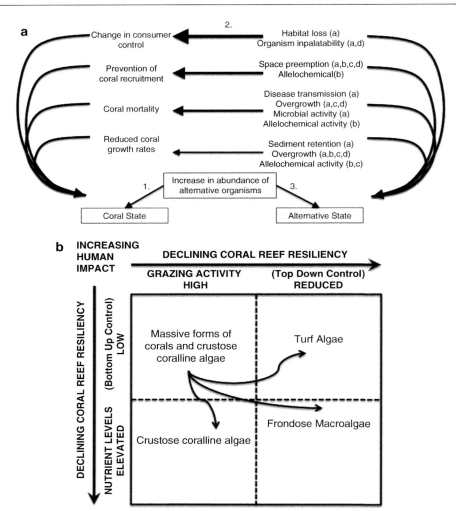

Fig. 1 Models of phase shifts on coral reefs developed from long-term surveys of competitive interactions. (**a**) Potential feedback mechanisms supporting competitive phase shifts on coral reefs to dominance by non-reef builders: (1) opening of space due to coral mortality leads to an increase in the abundance of alternative benthic organisms, then (2) various mechanisms of positive feedback (3) directly or indirectly reinforce the alternative states (reproduced with permission from Norström et al. 2009). (**b**) Relative Dominance Model, in which the abundance of four major functional groups of space occupiers varies with levels of human impact. This impact destabilizes the competitive equilibrium between corals and algae, with a high probability of shifting from dominance by corals to alternative dominance by algae (reproduced with permission from Littler and Littler 2006)

and the author suggests avoiding confusion by adopting the term "coral-depauperate state" rather than "macroalgal-dominated state."

The only long-term field monitoring study on coral–sponge interactions was conducted by López-Victoria et al. (2006) over 13 months on Colombian reefs in the Caribbean Sea, and showed that competitive outcomes varied among species pairs and with the angle of contact between opponents. Coral–ascidian interactions were examined during a 1.5-year study in the northern Red Sea, revealing that colonial ascidians seasonally overgrew some stony corals (Shenkar et al. 2008). A 5-year dataset (1996–2000) encompassing 20 reef sites in Florida documented a reduction in the percent cover of stony corals (from 8.1% to 4.6%), followed by an equivalent increase in the percent cover of macroalgae (from 5.7% to 9.6%, Maliao et al. 2008). Although these changes appear slight, according to Mumby's (2009) simulation model, reefs under chronic disturbance are expected to experience only very small random changes over this timescale.

A recent study disputes the claim that coral–algal phase shifts are increasing on present-day reefs, and contends that the replacement of corals by macroalgal canopies is less common than generally assumed (Bruno et al. 2009). The authors compiled data from multiyear reef surveys between 1996 and 2006, and concluded that most reefs worldwide lie between the extremes of coral- versus algal-dominated, not concordant with phase shifts between two stable states as defined in the literature (Done 1992, Knowlton 1992, Bellwood et al. 2004). Rather, most of the cover on tropical reefs in both the Caribbean and Indo-Pacific consists of organisms other than hard corals and macroalgae, such as sponges, soft corals, and corallimorpharians (Bruno et al. 2009), and the abundance of these replacement organisms appears to be increasing (Norström et al. 2009).

A limitation of long-term observational studies is that often they cannot elucidate the underlying mechanisms that drive observed changes. They have revealed that the intensity and outcome of some competitive interactions on reefs correlate with observed levels of natural and anthropogenic disturbances. However, to establish that competition is a primary driver of change in the relative abundances of reef organisms, and to reveal the mechanisms involved, manipulative experiments are needed to separate the impacts of various factors.

2.3 Field Experiments

During the past 20 years, field manipulations on coral reefs have served as a powerful tool to elucidate both the mechanisms and effects of competition among sessile organisms. As with observational studies, most have focused on competition among cnidarians, or between cnidarians and algae. All field experiments that extended >1 month have demonstrated impacts on at least one of the competitors. These recent field manipulations, plus some conducted prior to the first major review on coral competition (Lang and Chornesky 1990), have established that interference competition influences all examined types of benthic space occupiers on coral reefs.

Field experiments on within-species competition in sessile reef organisms have been conducted only for corals, mostly at Eilat in the Red Sea by Baruch Rinkevich and colleagues. Field experiments lasting up to 2 years have documented linear competitive hierarchies with clear dominants and subordinates, but some circular outcomes among intermediate genotypes within two species of common scleractinian corals, *Acropora hemprichi* (Rinkevich et al. 1994) and *Stylophora pistillata* (Chadwick-Furman and Rinkevich 1994). Similar experimental field contacts among ten genotypes of the hydrocoral *Millepora dichotoma* resulted in a more complex network of nontransitive outcomes (Frank and Rinkevich 1994). Within-species competition among colonies of the encrusting soft coral *Parerythropodium fulvum* also was investigated experimentally by Frank et al. (1996) on Eilat reefs for 2.5 months. A field experiment in Panama on competition within aggregations of the branching Caribbean soft coral *Plexaura homomalla* revealed a reduction in the growth of inner-aggregation colonies after 1 year (Kim and Lasker 1997). A 1-year field experiment in the Philippines showed that colonies of *Porites attenuata* spaced 5–10 cm apart inhibit branch production in neighbors but enhance initial upward extension, possibly via release of dissolved chemical signals (Raymundo 2001). Overall, these field manipulations of within-species competition reveal impacts in terms of overgrowth, intercolony damage, inhibition of lateral branching, and possibly interference with food intake among colonies of the same coral species. They have established that both stony and soft corals exhibit complex mechanisms of intraspecific competition, and interact in both linear hierarchies and circular networks of dominance among genotypes.

Several recent field experiments have examined interspecific competition among corals. Romano (1990) set up competitive contacts between two species of common reef corals in Hawaii, and observed a reversal in outcome after a few weeks, with significant negative effects on the long-term subordinate after 11 months. A freshwater flood ended her experiment on the reef flat following heavy rains on the adjacent island of Oahu, illustrating the unpredictable duration of field experiments on highly dynamic coral reefs. Elahi (2008) placed free-living fungiid corals in aggregations on reefs in Japan and observed negative impacts on their growth after only 1.5 months of both intra- and interspecific contact with other fungiids. A 1-year field experiment in a Polynesian lagoon showed that both intra- and interspecific competition strongly impact the growth of inferior *Porites* corals, but that intraspecific aggregation reduces this effect (Idjadi and Karlson 2007). Spatial aggregation by colonies of inferior coral competitors may be common on reefs, and may play an important role in allowing them to persist, thereby enhancing coral diversity.

Many of the early studies on coral competition did not reveal whether the effects led to fitness reductions (reviewed in Lang and Chornesky 1990). More recent experiments have remedied this deficit. Experiments over 1–2 years in Australia on competition among species of branching corals documented negative impacts on both growth and sexual reproduction (Tanner 1997). Van Veghel et al. (1996) transplanted several species of stony corals into contact with colonies of *Montastrea annularis* on Caribbean reefs, and observed that competitive outcomes changed over 60 days, in part due to the development of sweeper tentacles by some of the opponents. Field experiments conducted over 1–3 years in the Philippines showed that corals placed in monospecific stands (10 cm from other colonies) grow more slowly than those in multispecies stands, possibly in response to chemical cues released by neighbors (Dizon and Yap 2005). Peach and Hoegh-Guldberg (1999) transplanted colonies of the massive coral *Goniopora* into contact with branching corals *Pocillopora*, but detected no effects of competition, likely because their experiment extended only 17 days, shorter than the time required to fully develop aggressive organs in many corals (Lang and Chornesky 1990).

Field experiments also have revealed that soft-bodied cnidarians competitively damage many stony corals on tropical reefs. Several types of soft corals were transplanted adjacent to stony corals on the Great Barrier Reef, and observed to cause unilateral damage over 8 months (Sammarco et al. 1983, 1985). Soft corals on the Great Barrier Reef also have been shown to exude waterborne allelopathic chemicals that cause reduced recruitment of stony corals to adjacent settlement tiles (Maida et al. 1995a, b). Langmead and Chadwick-Furman (1999b) transplanted individuals of the corallimorpharian *Rhodactis rhodostoma* adjacent to stony corals at Eilat, northern Red Sea, and observed over 1.5 years that the corallimorpharians rapidly damaged and overgrew branching *Acropora* corals, but were killed by massive *Platygyra* brain corals. Kuguru et al. (2004) conducted the reverse experiment of transplanting *Acropora* coral fragments into contact with *R. rhodostoma* in Tanzania. Within 3 months, the corallimorpharians killed most of the corals they contacted, while the isolated control corals all survived. In a yearlong field experiment, Miles (1991) transplanted stony corals adjacent to the corallimorpharian *R. (Discosoma) sanctithomae*, and observed short-term damage to the corallimorpharians, but eventual long-term damage to the corals. Sebens and Miles (1988) conducted a field experiment on competition between the Caribbean encrusting octocoral *Erythropodium caribaeorum* and stony corals, which revealed that the octocorals damaged and overgrew the stony corals over 1 year. Competition between two Caribbean zoanthids (*Palythoa caribaeorum* and *Zoanthus sociatus*) was examined in 8–10 month experiments in Venezuela, resulting in standoffs due to possible growth inhibition, and faster growth by isolated *Palythoa* as a form of exploitation competition (Bastidas and Bone 1996).

A few field experiments on coral competition have been conducted on cold-water reefs. A 2.5-year field experiment in Central California showed that corallimorpharians unilaterally damaged and excluded corals from the tops of subtidal rocky reefs, but not from their lower edges (Chadwick 1991). A 3-year field experiment in Southern California demonstrated that both natural and artificial algae damaged the solitary coral *Balanophyllia elegans* by abrading the coral tissues and causing polyp retraction (Coyer et al. 1993). In contrast, individuals of *B. elegans* effectively defended themselves from encroachment by colonial ascidians, potentially via nematocyst release, during a 16-month experiment in Puget Sound, Washington (Bruno and Witman 1996). Morrow and Carpenter (2008) conducted a 1-year field experiment at Santa Catalina Island in Southern California, and demonstrated that foliose algae interfered with particle capture by corallimorpharians in *Eisenia* kelp forests.

In the only field experiment on competition between coral reef sea anemones and algae, glass slides were placed near the giant Caribbean anemone *Condylactis gigantea*. The growth of filamentous algae on the slides was inhibited, presumably due to allelopathic chemicals exuded by the anemone (Bak and Borsboom 1984). Two field experiments have been published on competition between corals and other sessile invertebrates, both on sponges. Live grafts of the bioeroding sponge *Cliona orientalis* that were affixed to stony corals invaded the coral tissues within 2–3 months on the Great Barrier Reef (Schönberg and Wilkinson 2001). Few species of bioeroding sponges have been described in detail, but their ability to resist coral defenses and persist in environments that are unfavorable to corals may give them a distinct competitive advantage under certain conditions. Crude extracts from the sponge *Agelas clathrodes* were incorporated into gels at natural volumetric concentrations and applied to Caribbean stony corals overnight (Pawlik et al. 2007). PAM fluorimetry revealed that both the baseline fluorescence (F_o) and potential quantum yield (Y) of zooxanthellae were reduced in coral tissues exposed to the sponge compounds for 18 h. Many benthic invertebrates are chemically defended (see Sect. 3.2, Glynn and Enochs 2010), and these experiments reveal that at least some sponges likely use allelopathy to gain a competitive advantage over corals.

Many field experiments have been conducted on competition between corals and macroalgae on tropical reefs, and often are necessary to reveal the sometimes subtle effects of interactions between these two competitors (see Sect. 2.1). These experiments often have lasted for only a few months, likely because many reef algae are short-lived and impact corals seasonally (Fong and Paul 2010). Tanner (1995) removed macroalgae on Australian reefs, and showed that over 2 years the percent cover of stony corals increased. A major cyclone obliterated the effects of competition in some of his experimental corals. A 9-month experiment in Mexico showed that contact with mixed turf algae can cause physiological stress to stony corals (Quan-Young and Espinoza-Avalos 2006). In Japan, placement of cyanobacteria and a brown alga in direct contact with live coral tissue for 2 months negatively impacted coral photosynthetic efficiency, chlorophyll concentration, and zooxanthellae density (Titlyanov et al. 2007). A second experiment revealed that red algae gain a competitive advantage over some corals under low light conditions due to their superior ability to photoacclimate to low irradiance (Titlyanov et al. 2009). A caging study on Florida reefs for 3 months demonstrated that reduced grazing pressure on macroalgae led to higher algal abundance, with varied impacts on adjacent stony corals (Lirman 2001). Littler and Littler (2006) manipulated nutrient levels and herbivory over 2 years on reefs in Belize, and combined their experimental results with those from field surveys to develop a model of the factors impacting coral–macroalgal competition (Fig. 1b).

On the Great Barrier Reef of Australia, Jamaluddin Jompa and Laurence McCook conducted a series of short-term experiments that revealed several aspects of coral–algal competition. They initially manipulated brown algae and corals to demonstrate mutual impacts of competition over 6 months (Jompa and McCook 2002a). They then enhanced algal growth via nutrient enhancement for 3 months, and caused more intense competition between algae and corals, but only when herbivory was insufficient to consume the excess algal biomass (Jompa and McCook 2002b). Further experiments lasting 6–18 months showed that, in sharp contrast to the relatively minor effects of mixed turf algae, filamentous red algae caused extensive damage to coral tissues, perhaps due to allelochemical production by the algae (Jompa and McCook 2003a,b). Finally, Diaz-Pulido and McCook (2004) conducted a 1-month study of algal recruitment onto several types of reef substratum, and documented that live coral tissue prevents algal settlement. Thus, algal colonization of reef surfaces appears to rely mainly on the previous death of coral tissues to open space for settlement.

Maggie Nugues and colleagues recently conducted several types of field experiments on coral–algal competition on Caribbean reefs. Effects of sedimentation on rates of macroalgal overgrowth of corals were examined near the mouths of river systems on St. Lucia for 15 months (Nugues and Roberts 2003). In a rare demonstration of competitive damage by corals to macroalgae, they also showed during a 3-week experiment in Curaçao that contact with some stony corals detrimentally impacts the common calcareous green alga *Halimeda opuntia* (Nugues et al. 2004a). Negative impacts were mutual: a subsequent 2-month experiment revealed that some individuals of *H. opuntia* may transfer a bacterial pathogen that potentially causes White Plague Type II, a disease that can lead to widespread coral mortality (Nugues et al. 2004b). However, there was no procedural control in this study, so it is unclear if the placement of an inert object on healthy coral tissue, or the role of the algae as a vector, caused the disease symptoms in the corals. A later experiment encapsulated coral larvae together with *H. opuntia* algae for 4 days and found that the coral planulae settled frequently on the algal blades (Nugues and Szmant 2006). Corals thus may suffer enhanced mortality on reefs where macroalgae are abundant, in part because the algae create an unstable, ephemeral substrate for coral attachment. Finally, Nugues and Bak (2006) employed a transplant design similar to that of Jompa and McCook (2002b) to determine mechanisms of competition between six Caribbean corals and the brown alga *Lobophora variegata* over 1 year.

Taken as a body, recent field experiments on coral–algal competition have employed a variety of transplant and caging methods, and revealed mainly negative impacts unilaterally on corals in terms of their recruitment, growth, survivorship, and physiological state. Some studies, however, found that the negative effects of macroalgae on coral recruitment and growth varied depending on the macroalgal species (Maypa and Raymundo 2004, Birrell et al. 2008a). Field experiments also have demonstrated that the ability of macroalgae to competitively damage stony corals appears to depend on levels of irradiance, sedimentation, dissolved nutrients, and herbivory on coral reefs.

2.4 Laboratory Experiments

Few laboratory manipulations have been conducted recently on competition among sessile reef organisms, due in part to the difficulties of maintaining many of these organisms over long periods under laboratory conditions. It remains a challenge to duplicate in the laboratory the often extreme conditions on tropical reefs of high irradiance and water flow, coupled with low dissolved and particulate nutrients. Thus, laboratory experiments have been limited mainly to outdoor aquaria supplied with natural irradiance and flow-through seawater, and underline the importance of well-equipped marine stations located near coral reefs. Advantages of laboratory manipulations are that they allow for greater control over environmental conditions, more frequent data collection, and more precise descriptions of competitive behavior than possible during limited dive times on coral reefs. Most laboratory experiments on competition have been conducted on stony corals, with a few on other cnidarians and algae.

Lapid et al. (2004) and Lapid and Chadwick (2006) used flow-through indoor tanks at a public seaside aquarium to determine the outcomes of contacts over 1 year between the brain coral *Platygyra daedalea* and the massive coral *Favites complanata* from the Red Sea. They found that laboratory experimental outcomes mimic the effects of both field experiments and field surveys. They also used laboratory experiments to quantify energetic costs of competition. Chadwick (1991) conducted a 1-year laboratory experiment on competition between temperate corals and corallimorpharians, and also found similar results among all three types of methods used to examine competition. Short-term laboratory experiments of 1 week on contacts between Caribbean sea anemones (corallimorpharians and actiniarians) and stony corals revealed similar outcomes to the early stages of an accompanying field experiment in Jamaica (Miles 1991) and a field survey in Panama (Sebens 1976). Chadwick and Adams (1991) used a 9-month laboratory experiment to show that rates of clonal replication and aggression correlate positively in corallimorpharians. Clones that replicate polyps rapidly also cause the most damage per individual polyp to stony corals, indicating a lack of trade-off between these two processes. Laboratory experiments also have quantified levels of heat-shock protein HSP70 expressed during competitive contacts between temperate corallimorpharians and actiniarian sea anemones (Rossi and Snyder 2001). Some sea anemones elevate their HSP70 expression within 3 days of competition, apparently as a mechanism to enhance repair of the cellular damage inflicted by opponents. Settlement tiles impregnated with extracts of the soft coral *Sinularia flexibilis* were used in larval choice experiments for stony corals in flow-through aquaria, and revealed that this soft coral uses allelopathy to inhibit adjacent stony coral settlement, corroborating field experiments on the nearby Great Barrier Reef (Maida et al. 1995a). Some soft corals have been shown to produce secondary metabolites (e.g., diterpines) that inhibit growth and cause tissue necrosis to stony coral colonies in laboratory dose experiments (Coll et al. 1982, Coll and Sammarco 1983). A 10-day laboratory experiment on competitive contacts between temperate corals and ascidians showed that outcomes were consistent with both field experiments and field surveys (Bruno and Witman 1996).

Outcomes of intraspecific competition in the branching stony coral *Stylophora pistillata* have been examined recently in laboratory experiments. Rinkevich and coworkers cultured newly settled individuals of *S. pistillata* to determine interactions among contacting colonies over 4 months (Frank et al. 1997) and 1 year (Amar et al. 2008). They found that fusion among nonidentical genotypes results in fitness costs in terms of the growth rates of individual genotypes, but not in terms of the overall size of fused chimeric colonies.

Most laboratory experiments on coral–algal competition have been brief, and have focused on impacts on the recruitment and survival of coral larvae. At Keys Marine Laboratory in Florida, 4-day experiments placed larvae of the stony coral *Porites astreoides* and the gorgonian *Briareum asbestinum* together in containers with macroalgae and cyanobacteria, and revealed deterrence of larval settlement (Kuffner et al. 2006). At Lizard Island Research Station on the Great Barrier Reef of Australia, Birrell et al. (2008b) examined the effects of water-soluble compounds from crustose coralline algae on coral settlement over 2 days. In a similar manner, Vermeij et al. (2009) conducted laboratory experiments over 1–14 days in Hawaii on planular survival rates of the stony coral *Montipora capitata* when exposed to macroalgae, and used the broad-spectrum antibiotic Ampicillin to reveal mediation of this effect by associated microbes. These recent laboratory studies at widely different sites have demonstrated how exposure to macroalgae and macroalgal compounds can reduce coral recruitment even before larvae contact the reef substratum.

Laboratory experiments also have been used to elucidate mechanisms of competition between macroalgae and stony coral colonies. Titlyanov et al. (2007) designed an outdoor water-table experiment at the Sesoko Marine Biological Station in Japan to examine how abrasion, shading, and allelopathy by cyanobacteria and brown algae impact the growth and physiology of the massive coral *Porites lutea*. In a laboratory experiment in the Philippines, Maypa and Raymundo (2004) measured coral settlement in the presence of macroalgae for 2 days, followed by 3 weeks of monitoring of the coral spat for survival and growth, and determined that both the physical and chemical impacts of macroalgae on young corals varied widely among algal species. A novel 4-day experiment aboard the OR/V White Holly during a Scripps Expedition to the Line Islands documented physiological damage to stony corals placed in close proximity to macroalgae, and the role of associated microbes through application of antibiotics (Smith et al. 2006).

In conclusion, researchers have utilized laboratory experiments to successfully determine some costs and mechanisms of competition among reef organisms, and to confirm field results on the directionality and outcomes of competition. Several studies have now combined field and laboratory methods and obtained similar results, providing powerful evidence that some competitive outcomes are consistent among environments (Chadwick 1991, Miles 1991, Bruno and Witman 1996, Langmead and Chadwick-Furman 1999a,b, Lapid et al. 2004).

2.5 Mathematical Modeling

Few researchers have applied mathematical models to describe or predict the outcomes of competition among sessile reef organisms, hindered in part by the spatial and temporal complexity of coral reefs (Lang and Chornesky 1990). Recently, Muko et al. (2001) developed a mathematical model to predict space use among competing corals based on variation in their life history traits. The authors explained the dominance of branching corals at protected sites by their rapid growth rates and thus exploitation and monopolization of space. At exposed sites, the dominance of tabular corals was explained by their relatively rapid recruitment and low mortality from physical disturbances. Connolly and Muko (2003) further explored modeling of size-dependent competition among corals, and concluded that patterns of disturbance, recruitment, and ontogenetic shifts in competitive ability all may strongly affect both transitive and hierarchical interactions. Competition among sessile reef organisms is difficult to model using traditional mathematical equations and isocline analysis (Lotka 1925, Volterra 1926, Tilman 1982, Connolly and Muko 2003), because outcomes vary spatially and often involve border effects along the edges of neighboring colonies. A fruitful area for future modeling might involve the application of spatial models of neighborhood competition to sessile reef organisms, as has been done for terrestrial plants (Pacala and Silander 1990, reviewed in Morin 1999).

3 Mechanisms of Competition

Recent studies have revealed few new mechanisms of competition among sessile reef organisms since extensive discoveries in the 1970s and 1980s of a wide array of competitive mechanisms, especially in cnidarians (reviewed by Lang and Chornesky 1990, Williams 1991). Recent work has added detail on how frequently each mechanism is used by each type of reef organism, and has greatly expanded the types of sessile organisms known to employ toxic chemicals to damage competitors. In addition, advances in molecular methods for microbial identification have revealed that alteration of microbial assemblages on corals is a potentially important pathway of competitive damage to the coral holobiont. Finally, our understanding of intraspecific mechanisms of competition among coral colonies has increased, and now rivals that of interspecific mechanisms. We review here recent advances in our understanding of the physical, chemical, and biological mechanisms of competition employed by each major taxonomic group of sessile reef organisms.

3.1 Cnidarians

In cnidarians, recent studies on mechanisms of exploitation competition have focused on the corallimorpharians, soft-bodied anthozoans often termed "corals without a skeleton." Some corallimorpharians preempt the settlement of stony corals and other organisms on reefs through both rapid asexual production of large clonal aggregations and sexual production of dispersive propagules (Chen et al. 1995a, b, Chadwick-Furman and Spiegel 2000, Chadwick-Furman et al. 2000). In the tropical corallimorpharian *Rhodactis rhodostoma*, up to 30% of body mass is devoted to gonads during the reproductive season, a large investment in the production of dispersive larvae (Chadwick-Furman et al. 2000). In a temperate corallimorpharian, rapid cloning correlates with aggressive ability, causing some clones to both aggressively attack corals and then quickly overgrow their skeletons (Chadwick and Adams 1991). Zoanthids also may replicate clonally at rapid rates and monopolize reef substratum in continuous mats. Members of the genera *Palythoa* and *Zoanthus* are particularly ubiquitous in coastal areas where nutrient concentrations are high, and thus may gain a competitive advantage over corals on nearshore reefs in Brazil (Costa et al. 2002, 2008) and Venezuela (Bastidas and Bone 1996).

Several recent studies have documented physical mechanisms of interference competition among cnidarians, and also during interactions with neighboring algae and sponges. Skeletal overgrowth, redirection of growth, locomotion away from competitors, and physical interception of food particles appear to be the main physical mechanisms used during interference competition among reef cnidarians. In terms of interspecific mechanisms, skeletal overgrowth is the main mechanism of competition among five species of *Porites* corals in Japan (Rinkevich and Sakai 2001), two species of *Porites* in Polynesia (Idjadi and Karlson 2007), and three types of corals and a sponge in the Red Sea (Rinkevich et al. 1992). Alino et al. (1992) observed soft corals to overgrow and smother stony corals, and Chadwick (1991) documented that the corallimorpharian *Corynactis californica* overgrows contacted stony corals. Interference competition also causes evasion responses by some corals. Small, mobile fungiid corals actively locomote away from overgrowing coral colonies (Chadwick 1987, Chadwick-Furman and Loya 1992). Rates of mobility decrease exponentially with fungiid coral size, and large individuals employ biological interference mechanisms to defend living space rather than the physical mechanism of avoidance. Some colonial corals in Hawaii also redirect their growth away from areas of contact with competing corals (Romano 1990). Stony corals evade contact with encroaching sponges through alteration of their angle of growth, leading to the formation of coral "domes" in response to contact with sponges (López-Victoria et al. 2006).

Corals employ a complex array of intraspecific competitive mechanisms that include several types of physical responses. Both skeletal overgrowth and retreat growth (avoidance) occur during intraspecific competition among colonies of octocorals (Frank et al. 1996), scleractinian corals (Chadwick-Furman and Rinkevich 1994), and hydrocorals (Frank and Rinkevich 1994, 2001). The formation of skeletal barriers and sutures can lead to competitive standoffs along the contact zones among genotypes of the common stony coral *Stylophora pistillata* (Chadwick-Furman and Rinkevich 1994, Rinkevich et al. 1994, Amar et al. 2008). Although described quantitatively in only a few species, many types of stony corals form skeletal barriers along intraspecific contact zones (Fig. 2). Some hydrocorals overgrow other colonies with their soft tissues only, in advancing zones that lack calcification (Frank et al. 1995). In a unique study, Kim and Lasker (1997) documented that gorgonian corals along the edges of conspecific aggregations appear to intercept enough food particles to significantly reduce the growth rates of colonies in the interior of aggregations. However, depletion rates of planktonic food were not quantified, and the observed changes in gorgonian growth could have been due in part to chemical mechanisms such as allelopathy among the colonies.

A wide diversity of cnidarians also employs chemical mechanisms of interference competition on coral reefs. Several types of bioactive compounds have been detected in scleractinians (Fusetani et al. 1986, Rashid et al. 1995, Koh 1997), some of which likely function as allelochemicals.

Fig. 2 Competitive interactions among massive scleractinian corals in the Caribbean Sea. (**a**) *Diploria clivosa*, (**b**) *Montastraea cavernosa*, (**c**) *D. clivosa* (*top*) and *D. strigosa* (*bottom*), (**d**) mixed species assemblage of *Diploria, Montastraea, Madracis,* and *Siderastrea.* Note the lack of clear damage along intraspecific contact zones, in contrast to gaps and areas of unilateral tissue damage along some interspecific contact zones. Photographs by E. Mueller

Aqueous-soluble toxic compounds occur in most of the 58 species of corals examined by Gunthorpe and Cameron (1990a, b). Waterborne extracts from the stony coral *Goniopora tenuidens* kill colonies of the competing coral *Galaxea astreata* (Gunthorpe and Cameron 1990c), inhibit larval metamorphosis and growth of the coral *Pocillopora damicornis*, and kill the swimming larvae of four other scleractinian corals. Koh and Sweatman (2000) examined the effects of natural products from the azooxanthellate coral *Tubastraea faulkneri* on its own larvae and the larvae of 11 other sympatric corals from seven genera and four families. They demonstrated that methanol extracts are toxic to all 11 species of heterospecific larvae tested, but not to conspecifics. The presence of such broad-spectrum activity suggests significant ecological function in mediating competitive interactions, in that the larvae of other species may be unable to settle in the vicinity of this coral. Three of the identified compounds (aplysinopsin, 6-bromoaplysinopsin, and 6-bromo-29-de-*N*-methylaplysinopsin) were isolated previously from other sponges and dendrophylliid corals, while the fourth, a dimer of 6-bromo-29-de-*N*-methylaplysinopsin, was a newly discovered toxin. Allelopathy against settling larvae appears to be widespread, as all 7 stony coral species examined thus far possess toxic compounds active against up to 13 species of scleractinian larvae (Fearon and Cameron 1996, Koh and Sweatman 2000). Colonies of the stony coral *Porites cylindrica* vary their growth rates depending on the identity of neighboring corals 5–10 cm away, indicating that waterborne chemicals may mediate this effect (Dizon and Yap 2005). The stony coral *Styophora pistillata* also appears to utilize allelopathy to damage colonies of the competing soft coral *Parerythropodium fulvum* in the Red Sea (Rinkevich et al. 1992). Thus, although allelopathy was believed to be employed largely by soft corals following initial discoveries in the 1980s (reviewed in Lang and Chornesky 1990), work since then has revealed that many stony corals also may produce halos of toxins in the water space around them, and thus reduce encroachment by coral competitors during larval settlement and afterwards.

Recent work on allelopathy by soft corals has expanded the number of species known to exude waterborne toxins. *Sinularia flexibilis* and *Lobophytum hedleyi* both exude terpenes, toxic allelomones that inhibit growth and produce tissue necrosis in neighboring scleractinian corals (Aceret et al. 1995). Diterpene concentrations of >5 ppm induce cytological damage in branchlets of the stony corals *Acropora formosa* and *Porites cylindrica*, manifested by expulsion of zooxanthellae, release of nematocysts, inhibition of polyp activity, necrosis, and ultimate death of the corals. Allelopathic compounds emitted by *S. flexibilis* also inhibit the recruitment of juvenile scleractinian corals located downstream from exuding colonies (Maida et al. 1995a, b). Other cnidarians may secrete allelopathic chemicals: the rosetip anemone *Condylactis gigantea* inhibits overgrowth by all major groups of macroalgae on Caribbean reefs, apparently via release of toxic compounds into the water column (Bak and

Borsboom 1984). Numerous zoanthids in the genus *Palythoa* are moderately toxic, and palytoxin, a neuromuscular poison isolated from the Hawaiian zoanthid *P. toxica*, was the most poisonous nonproteinaceous substance known in the early 1970s (Moore and Scheuer 1971). Zoanthids potentially use these chemical toxins during spatial competition. Field experiments on colonies of the zoanthids *Palythoa* and *Zoanthus* in Venezuela indicate that they inhibit the growth of contacted corals and create competitive standoffs, possibly via allelopathy (Bastidas and Bone 1996). At least some members of all major groups of cnidarians on reefs, including the hydrocorals and corallimorpharians which have not been tested, likely release toxins into the water to defend themselves from encroachment by competitors.

In contrast to chemical mechanisms of competition, the biological mechanisms employed during interspecific competition by cnidarians involve mostly behavioral and morphological modifications. Trade-offs may occur between investment in physical, chemical, and biological mechanisms of competition, but the energetic costs and major patterns of these trade-offs are not yet understood. The major biological mechanisms of competition employed by cnidarians mostly were described during the 1970s and 1980s (reviewed by Lang and Chornesky 1990, Williams 1991). Recent work has focused mainly on the ecological importance of elongated polyps or tentacles (sweeper polyps, sweeper tentacles, and bulbous marginal tentacles) developed by some species of reef cnidarians, and how they are used to defend space. Sweeper polyps in *Goniopora* corals appear to develop directionally toward colonies of soft corals (Dai 1990), and to possess a specialized cnidom and tentacle morphology compared to those of ordinary polyps (Peach and Hoegh-Guldberg 1999). Their scattered distribution on colonies of *Goniopora* led the latter authors to speculate that these elongated polyps serve functions in addition to competitive defense, such as deterrence of predators. Sweeper tentacles appear to develop and regress at random on some brain corals, serving as probes that detect and kill competitors that settle within the wide aggressive reach of these massive corals. They thus serve in part as a preemptive mechanism to detect and damage encroaching competitors at a distance (Lapid et al. 2004, Fig. 3). If nearby coral recruits are detected, then sweeper tentacles are deployed massively toward them, and also toward established coral colonies that grow within the reach of some brain corals (Lapid and Chadwick 2006). Sweeper tentacles are employed by brain corals to kill other types of encroaching cnidarians, such as corallimorpharians, and are one of the few coral competitive mechanisms that can damage polyps of the latter (Langmead and Chadwick-Furman 1999b).

The wide, competitor-free zones visible around colonies of *Platygyra daedalea* and many other brain corals likely are

Fig. 3 Competitive interactions between brain corals *Platygyra daedalea* and other stony corals at Eilat, northern Red Sea. (**a**) Colonies of *P. daedalea* (*center*) contacting those of *Favites* spp. Note the wide zones of unilateral damage to *Favites* along borders with the brain corals. Also note lack of damage along borders where several brain corals interact a center, and possible fusion among some colonies. (**b**) Unilateral damage to *Favites* sp. (*left*) by *P. daedalea* (*right*). (**c**) Exposed white skeleton on the branching coral *Acropora* sp. along the region of interaction with *P. daedalea*. Note the wide space between the two colonies, and distortion of the regular growth pattern of the branching coral on the side facing the brain coral. (**d**) Close-up of expanded tentacles on a brain coral at night. Note the elongated sweeper tentacles at center. Photographs by N. E. Chadwick (a, b, c) and R. Ates (d)

created by this inducible aggressive mechanism, which allows these corals to effectively defend space and to persist as large, long-lived colonies on shallow reefs (Fig. 3). Sweeper tentacles also occur on some encrusting soft corals and cause aggressive damage to contacted stony corals on Caribbean reefs (Sebens and Miles 1988). During the 1990s, corallimorpharians were discovered to develop bulbous marginal tentacles packed with large holotrich nematocysts that kill most contacted corals on reefs in the Caribbean (Miles 1991) and Red Sea (Langmead and Chadwick-Furman 1999a, b, Fig. 4). Similar to stony corals, corallimorpharians also directionally extrude their mesenterial filaments to digest the tissues of neighboring cnidarians (Chadwick 1987, 1991). These aggressive mechanisms, together with rapid rates of clonal replication, have allowed corallimorpharians to replace corals as dominants on coral reef flats in some parts of the Indian Ocean (Kuguru et al. 2004) and Red Sea (Chadwick-Furman and Spiegel 2000, Norström et al. 2009, Fig. 4).

A unique mechanism of mucus deployment to damage competitors was discovered in mobile mushroom corals during the 1980s (Chadwick 1988, Chadwick-Furman and Loya 1992), but the mucus components that caused damage to opponents were not determined. Recently, we observed that the mushroom coral *Fungia scutaria* is unable to damage competitors when separated by filters with 1–3 mm pores (N. E. Chadwick and A. Moss, 2006, unpublished data). Electron microscopy reveals large holotrich nematocysts in the mucus of *F. scutaria* in contact with the stony coral *Montipora verrucosa*, suggesting that these organelles are the agents of competitive damage (Fig. 5). Fungiids are the only corals, out of six Hawaiian species examined, that release large quantities of nematocysts into their mucus (Coles and Strathmann 1973). An early study with filters also indicated that "large, nondialyzable agents" in the mucus appear to cause competitive damage by these corals (Hildemann et al. 1977). Thus, fungiid mushroom corals actively release nematocysts into their surface mucus, and these nematocysts may be their major mechanism of competitive damage to neighboring stony corals on reefs.

In terms of biological mechanisms of competition within cnidarian species, recent studies have revealed nematocyst discharge along the borders between conspecific colonies, resulting in tissue damage to the scleractinian corals *Stylophora pistillata* (Chadwick-Furman and Rinkevich 1994) and *Acropora hemprichi* (Rinkevich et al. 1994). Intense competition also appears to occur between tissues of different genotypes within fused chimeric colonies of stony corals, leading to stunted growth (Amar et al. 2008). Because the larvae of many corals settle in aggregations, newly metamorphosed polyps often contact conspecifics, and if they are closely related, may fuse their tissues to form

Fig. 4 Competitive interactions between corallimorpharians (*Rhodactis* spp.) and stony corals. (**a**) Aggregation of corallimorpharians on a partially dead branching coral *Acropora* sp. in Yemen. (**b**) Close-up of a single corallimorpharian polyp and dead zone on adjacent stony coral in Yemen. (**c**) Large aggregation of *R. rhodostoma* on a reef flat at Eilat, northern Red Sea. Note that only macroalgae and the tip of a giant clam are not covered by corallimorpharians. (**d**) Aggregation (*right*) of *R. rhodostoma* interacting with a massive stony coral at Kenting Reefs, South Taiwan. Note the white bulbous marginal tentacles on the corallimorpharians facing contact with the stony coral. Photographs by F. Benzoni (a, b), N. E. Chadwick (c), and A. Chen (d)

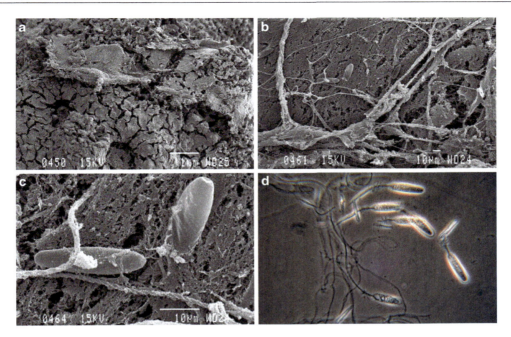

Fig. 5 (previous page). Electron micrographs of the surface of the colonial coral *Montipora verrucosa* in contact with the mushroom coral *Fungia scutaria*. (**a**) Intact coral polyps of *M. verrucosa* (*lower left*), advancing mucus sheet excreted by *F. scutaria* (*center*), and necrotic tissue and exposed skeleton of *M. verrucosa* behind the mucus front (*upper right*). (**b**) Mucus sheet secreted by *F. scutaria* and containing nematocyst capsules and threads. (**c**) Close-up of nematocyst threads in the mucus of *F. scutaria*. (**d**) Light micrograph of abundant nematocysts in the mucus of *F. scutaria*. Micrographs by A. Moss (a–c) and N. E. Chadwick (d)

chimeric colonies of more than one genotype coexisting within a single soma (reviewed in Amar et al. 2008, Fig. 3). The formation of chimeric coral colonies leads to within-soma competitive interactions at cellular and tissue levels, in which the somatic and/or germ cells of the dominant genotype may reduce the fitness of subordinates within a single morphological coral colony. Laboratory experiments have documented chimera formation and possible somatic and germ cell competition in the scleractinian coral *Stylophora pistillata* (Frank et al. 1997, Amar et al. 2008) and the hydrocoral *Millepora dichotoma* (Frank and Rinkevich 1994). These within-colony competitive mechanisms may be important drivers of population structure and evolution in stony corals, but their costs and benefits are not well understood.

3.2 Other Sessile Invertebrates

Recent work has elucidated a variety of competitive mechanisms in other sessile invertebrates, especially among major space occupiers on coral reefs such as sponges, colonial ascidians, and zoanthids (Glynn and Enochs 2010). Sponges that excavate and encrust carbonate substratum are particularly destructive competitors of stony corals (Glynn 1997, Rutzler 2002). Sponges in the family Clionaidae (Porifera, Hadromerida) can spread laterally at rates of 19 cm year^{-1} (Rutzler 2002, López-Victoria et al. 2006). Some sponges send out pioneering filaments that perforate coral skeletons and excavate below the live tissue layers, thus avoiding tissue-based defense mechanisms. Eventually, they can undermine coral skeletal support and induce polyp retraction and death (López-Victoria and Zea 2004, Schönberg and Wilkinson 2001). The bioeroding sponge *Cliona orientalis* may rapidly colonize recently grazed or broken coral skeletons (within 1–2 weeks), but contact with live coral tissue significantly reduces sponge survival and health (Schönberg and Wilkinson 2001). In contrast, the encrusting sponge *Diplastrella gardineri* rapidly overgrows the live tissues of some branching corals in the northern Red Sea (Rinkevich et al. 1992). Invasion by sponges may be enhanced under environmental conditions that are stressful to corals but tolerable for sponges, such as low irradiance or high levels of sedimentation, and following episodes of coral bleaching, disease, or attacks by predatory crown-of-thorns sea stars. Thus, mechanisms of coral competition with sponges and other sessile organisms may result in very different outcomes than for coral–coral competition. The competitive advantage gained by non-reef builders under the above conditions can lead to reduced coral cover and eventual replacement by these alternate dominants (Fig. 1a).

As with cnidarians, some sponges also produce potent allelochemicals that mediate competitive interactions through direct tissue contact, or via release in sponge mucus or directly into the water column (Becerra et al. 1997). Early studies with *Aplysinia fistularis* demonstrated that some sponges perform de novo biosynthesis of natural products (Thompson et al. 1983). Species of the bioeroding sponge *Siphonodictyon* (= *Aka*) bore deeply into living coral heads, leaving only the oscular chimneys exposed and surrounded by live coral tissue. This was the first sponge proposed to use allelochemicals to prevent growth and reduce respiration rates of adjacent coral polyps. *Siphonodictyon* sp. secretes toxic mucus that acts as a carrier for the secondary metabolite siphonodictidine, which inhibits coral growth around the base of the oscular chimneys (Sullivan and Webb 1983). The encrusting and excavating sponge *Cliona tenuis* produces the toxin clionapyrrolidine A, and can undermine and displace live coral tissue at a rate of 20 cm year^{-1}. However, live fragments of *C. tenuis* placed in direct contact with corals are killed by coral defenses, and the sponges do not appear to release allelochemicals (Chaves-Fonnegra et al. 2008). Corals can kill *C. tenuis* tissue within 24 h, thus allelopathy likely does not occur during external, but rather subdermal *C-tenuis*-coral contact (López-Victoria and Zea 2004). Chemical extracts of several Caribbean sponges damage the photosynthetic machinery of microalgal symbionts within the massive coral *Diploria labyrinthiformis*. PAM fluorimetry revealed that both baseline fluorescence (correlated with the production of chlorophyll a), and potential quantum yield (a measure of photosynthetic efficiency) were reduced in coral tissues exposed to extract from the sponge *Agelas clathrodes*. Thus, chemically active invertebrates such as sponges may compete with corals by undermining the skeletal integrity of corals and reducing their photosynthetic efficiency and respiration, leading to reduced coral growth (Sullivan and Webb 1983, Pawlik et al. 2007). Sponge chemicals can inhibit the settlement of fouling organisms and have been a consistent source of antifouling compounds; however, their effects from an ecological perspective have been examined only recently (Paul et al. 2007).

Ascidians generally constitute a minor component of the benthic community on coral reefs, and often occur in cryptic microhabitats such as caves, crevices, and the undersides of rubble and corals (Monniot et al. 1991). However, they potentially can compete with corals in exposed reef habitats due to their rapid growth rates (Bak et al. 1981), early sexual maturity, high fecundity (Millar 1971), and lack of successful predators (Lambert 2002). Some ascidians, such as *Trididemnum cyclops* and *T. miniatum*, also contain photosymbionts and UV-absorbing substances that protect them from harmful UV radiation (reviewed in Hirose 2009), thus enhancing their ability to compete with zooxanthellate corals. On deteriorating reefs, the biomass of hermatypic corals and filter-feeding organisms (e.g., sponges and ascidians) is negatively correlated (Bak et al. 1996, Aerts 1998). Recent evidence suggests that several species of ascidians are spreading rapidly in tropical regions of the world (Bak et al. 1996, Lambert 2002). A ninefold increase of the colonial ascidian *T. solidum* was reported along a fringing reef in Curaçao over a 15-year period (Bak et al. 1996). The newly described colonial ascidian *Botryllus eilatensis* rapidly overgrows reef-building corals in the Red Sea during periods of nutrient enrichment, and has become a potentially important competitor for space with corals (Shenkar et al. 2008). Zoanthids such as *Palythoa* and *Zoanthus* are ubiquitous in coastal areas near Brazil (Costa et al. 2002, 2008), Venezuela (Bastidas and Bone 1996), and the Florida Keys K. M. Morrow, pers. obs. (2009), where nutrient concentrations often are high. Thus, both ascidians and zoanthids can take advantage of nutrient loading on reefs and gain a competitive advantage over corals (Costa et al. 2002, Shenkar et al. 2008). Vermetids are sessile tube-dwelling gastropods that secrete a mucus net to entrap food and also possibly to competitively damage surrounding corals. The wide flat mucus nets of vermetids completely cover nearby corals on reef flats in the northern Red Sea. They appear to prevent coral feeding and also may allow the vermetids to consume photosynthetically derived coral metabolites (Zvuloni et al. 2008). In the Mediterranean Sea, bryozoans (*Watersipora* sp.) successfully compete for space with the introduced coral *Oculina patagonica* by overgrowing colonies during seasonal bleaching when the corals are physiologically weakened (Fine and Loya 2003). In non-bleached *O. patagonica* colonies, competition induces translocation of ^{14}C products to the interaction zone leading to rapid coral growth, but when corals become bleached (40–85%), they lack these translocated carbon resources (Fine et al. 2002) and are unable to prevent overgrowth by neighboring bryozoans. Thus, bryozoans may use overgrowth to effectively compete with some corals, especially when the latter become weakened by environmental stressors.

We conclude that sessile invertebrates such as sponges, ascidians, zoanthids, and bryozoans may compete successfully with stony corals using both physical and chemical mechanisms. The effectiveness of competitive mechanisms in many of these sessile invertebrates depends on environmental factors (e.g., levels of nutrients, temperature, and irradiance) that alter their growth rates relative to corals.

3.3 Macroalgae

Macroalgae are one of the major competitors with corals on tropical reefs, especially where rates of herbivory are low and/or dissolved nutrients are high, as on many anthropogenically impacted reefs (Fong and Paul 2010, Figs. 6 and 7). Algae can inflict competitive damage to stony corals using seven major types of mechanisms (reviewed by

Fig. 6 Competitive interactions between massive Caribbean corals *Montastraea* spp. and (**a**) filamentous red algae, (**b**) foliose brown algae *Dictyota* and cyanobacteria *Lyngbya*, (**c**) calcareous green algae *Halimeda*, and (**d**) unidentified sponge. Photographs by K. M. Morrow

Fig. 7 Competitive interactions between massive scleractinian corals and cyanobacterial mats: (**a**) *Montastraea* coral surrounded by cyanobacteria and *Dictyota* brown algae, (**b**) red cyanobacteria surrounding a small colony of *Diploria labyrinthiformis*, (**c**) colonies of *Montastraea* (*left*) and *Siderastrea* (*right*, bleached) separated by a brown cyanobacterial mat, (**d**) cyanobacteria in consortia with other marine bacteria forming black band disease on a stony coral colony. Newly dead coral skeleton (*left*) and non-diseased coral tissue (*right*). Photographs by R. Ritson-Williams

McCook et al. 2001): preemption by interfering with the settlement and survival of larval and juvenile corals (Maypa and Raymundo 2004, Birrell et al. 2005, Mumby 2006, Box and Mumby 2007), shading, allelopathy, attraction of settling larvae to emphemeral algal surfaces (Miller and Hay 1996, Littler and Littler 1997, Nugues and Szmant 2006, Vermeij et al. 2009), abrasion, basal encroachment (Coyer et al. 1993, Lirman 2001, Box and Mumby 2007, Titlyanov et al. 2009), and sedimentation due to reduced water flow (Nugues and Roberts 2003).

Mechanisms of competitive damage vary widely among algal species. The brown algae *Dictyota pulchella* and *Lobophora variegata* use shading and abrasion to damage juvenile corals (*Agaricia* sp.), and inert algal mimics of *D. pulchella* create a similar abrasive effect (Box and Mumby 2007). Individuals of *D. dichotoma* reduce the photosynthetic efficiency and growth rates of contacted stony corals *Porites lutea* through abrasion of coral tissues, but mats of the cyanobacteria *Lyngbya bouillonii* use both abrasion and unknown mechanisms such as allelopathy (Titlyanov et al. 2007). Individuals of *L. variegata* may exude a chemical deterrent that prevents swimming and metamorphosis of coral larvae in the vicinity of the alga (Morse et al. 1996, Baird and Morse 2004). In contrast, another study found that seawater collected from aquaria holding *L. variegata* thalli enhanced the settlement of larval corals by 40%, whereas the brown alga *Padina* sp. and the filamentous green alga *Chlorodesmis fastigata* both reduced coral settlement (Birrell et al. 2008a). Maypa and Raymundo (2004) showed that exudates from four species of macroalgae had either neutral or positive effects on coral settlement, illustrating the complex and highly species-specific mechanisms of interaction between algae and corals.

Some macroalgae release dissolved compounds that potentially mediate the diversity and abundance of microbial assemblages associated with stony corals (Nugues et al. 2004b, Smith et al. 2006), but these effects have not yet been quantified. Similar to the chemical defenses of sponges, macroalgae actively synthesize compounds that prevent microbial disease, biofouling, and herbivory (Boyd et al. 1999, Engel et al. 2006, Puglisi et al. 2007). Macroalgae respond to microbial challenges by releasing reactive oxygen species (ROS) and producing defensive secondary metabolites (Engel et al. 2006, Lane and Kubanek 2008). A survey of 42 seaweed species determined that mechanical damage to most algae induced activation of chemical defenses (Cetrulo and Hay 2000), thus abrasion from contact with adjacent corals may cause concentrated chemical release by macroalgae, similar to algal responses to herbivory. On reefs where rates of herbivory are high, the consequently enhanced release of harmful compounds from algae potentially results in chemical damage to corals and their associated microbes. The majority of crude extracts from 54 species of marine algae and 2 species of seagrasses examined from Indo-Pacific coral reefs are active against one or more types of ecologically relevant microorganisms, including saprophytic fungi, saprophytic stramenopiles, and a pathogenic bacterium (Puglisi et al. 2007). Broad-spectrum activity was demonstrated in 21–50% of the algae tested; extracts from the green alga *Bryopsis pennata* and the red alga *Portieria hornemannii* inhibited all assay microorganisms. Thus, many marine algae possess antimicrobial chemical defenses, and may use them to alter the diversity and abundance of coral-associated microbes along contact zones with competing corals.

Macroalgae also can stimulate the growth of microbes in coral mucus by releasing high levels of dissolved organics. Dissolved organic carbon (DOC) is a critical substrate for microbial growth, and is not readily available in coral mucus (Wild et al. 2004). Smith et al. (2006) showed that diffusible compounds released by macroalgae indirectly cause coral mortality by enhancing microbial activity on adjacent coral surfaces in low-flow environments. Elevated microbial growth can cause coral degradation through oxygen depletion, accumulation of poisons (e.g., hydrogen sulfide or secondary metabolites), and/or microbial predation on weakened coral polyps (Segel and Ducklow 1982). Thus, macroalgae might inhibit coral larval settlement via enhanced concentrations of noxious microbes or possibly by weakening larval resistance to microbial infection (Vermeij et al. 2009). In one case, a macroalga potentially served as a vector of a virulent disease (white plague type II) to adjacent corals, indicating that some algae might be reservoirs for coral pathogens (Nugues et al. 2004b). As such, the increased algal biomass on present-day reefs could account in part for elevated incidences of coral disease over the past few decades. Anthropogenic nutrient enrichment and encroaching macroalgae both also can increase the amount of labile DOC near corals, enabling mucus-associated microbes to break down complex and previously unavailable carbon sources via co-metabolism, and leading to uncontrolled and detrimental microbial growth on coral surfaces (Dinsdale and Rohwer 2010).

We conclude that mechanisms employed by macroalgae to compete with corals include diverse physical and chemical processes that impact all stages of the coral life cycle (Titlyanov et al. 2007) and may involve alteration of microbial assemblages on corals. As with sessile invertebrates, the competitive mechanisms used by macroalgae may be altered greatly by environmental conditions, especially seawater temperature and levels of water flow and dissolved nutrients, which affect their growth rates and the diffusion of compounds they release when interacting with corals.

4 Factors That Alter Competition Among Sessile Reef Organisms

Most recent studies on factors that alter outcomes of competition on reefs have focused on interactions between stony corals and other sessile organisms. These may be grouped into small-scale versus large-scale processes. In terms of small-scale local processes, recent evidence has confirmed earlier studies showing that both the relative age and time since contact between interacting organisms strongly influence competitive outcomes (reviewed in Lang and Chornesky 1990). Reversals in outcome may occur at days to months after initial contact (Romano 1990, Miles 1991, Van Veghel et al. 1996, Langmead and Chadwick-Furman 1999b). Outcomes can vary from colony fusion and subsequent chimera formation to tissue rejection and overgrowth, depending on the age and time since contact between conspecific stony corals (Frank and Rinkevich 1994, Frank et al. 1997). At least 5 months are needed to detect the long-term outcomes of some intraspecific coral contacts (Chadwick-Furman and Rinkevich 1994).

Water motion may alter competitive outcomes by deflecting aggressive organs and thus reducing the aggressive reach of corals, allowing them to grow closer together (Genin and Karp 1994), and also by reducing concentrations of dissolved chemicals released by competitors (Gunthorpe and Cameron 1990c, Smith et al. 2006, Chaves-Fonnegra et al. 2008). Levels of dissolved nutrients also can reverse the relative growth rates of corals in close proximity, and thus their patterns of competitive dominance (Dizon and Yap 2005). Reduced water flow may intensify the competitive ability of macroalgae (River and Edmunds 2001, Smith et al. 2006, Morrow and Carpenter 2008). Overgrowth and smothering of adjacent corals can be exacerbated by particle entrapment and sediment deposition within the low-flow environment under algal canopies (Richmond 1993). The morphology of macroalgae and other flexible benthic organisms can differentially modify water currents and alter the direction and deposition of suspended particles. Morrow and Carpenter (2008) demonstrated that the foliose macroalga *Dictyopteris undulata* inhibits particle capture by neighboring corallimorpharians to a greater extent than does the branching alga *Gelidium robustum*, through redirection of particles around polyps and induced contraction of the feeding tentacles of the corallimorpharian. Thus, the morphologies of flexible neighbors may in part determine competitive outcomes, by altering patterns of sedimentation and/or food particle capture in sessile reef organisms.

Frequencies of competition vary with depth on coral reef slopes; green macroalgae are more important competitors with corals in shallow well-lit areas, whereas ascidians compete more frequently with corals on the deeper reef slope (Van Veghel et al. 1996). Some red algae acclimate to extremely low levels of irradiance and can overgrow corals under these conditions, whereas the corals win at high irradiance (Titlyanov et al. 2009). The angle of encounter between individuals is important in sponge–coral competition (López-Victoria et al. 2006). Advancement of three species of *Cliona* sponges is more effective when sponge–coral tissues confront each other at 180° angles than when the sponges attack corals from below. Additionally, turfing and macroalgae that settle in the available space created by lateral sponge advancement, as well as increased sedimentation, may further weaken coral boundary polyps. A coral's ability to evade sponge advancement also appears to rely on coral defensive mechanisms such as upward growth away from the sponge–coral interface, overtopping of sponges in the case of foliose and plating corals, and allelochemical defense (López-Victoria et al. 2006).

Conspecific corals often aggregate due to co-settlement of their larvae (Amar et al. 2008), creating a spatial pattern that leads not only to chimera formation, but also to refuges from interspecific competition. Aggregations of inferior corals reduce the negative impacts of competition with superior corals, and can enhance the coexistence of diverse species on reefs (Idjadi and Karlson 2007). Coral percent cover also influences coral–sponge interactions. On reefs in Curaçao and Colombia, the proportion of aggressive sponge species increases with increasing coral cover, up to 25% coral cover (Aerts 1998). In addition, when algal biomass increases on reefs so does the intensity of overgrowth, shading, and chemical inhibition of corals by large macrophytes (e.g., *Turbinaria, Sargassum,* and *Dictyota*). Usually, corals can prevent algal settlement on their live tissues (Diaz-Pulido and McCook 2004), but newly settled recruits and juveniles are more vulnerable to algal overgrowth than are adult corals, due to their small size and greater sensitivity to physiological challenges (McCook et al. 2001, Fig. 19.7). The strength of algal competitive mechanisms varies with algal functional form (see Steneck and Dethier 1994), such that articulated calcareous and leathery macrophytes (e.g., *Halimeda* and *Turbinaria*, respectively) abrade coral tissues more than do foliose and crustose algae (e.g., *Dictyota* and *Peyssonnelia*, respectively). Filamentous turf and crust algae may impact only small coral recruits, but large algal mats can overgrow most recruits and some adult corals (Birrell et al. 2008b).

In terms of larger-scale processes at the level of entire coral reefs, disturbances that selectively remove certain types of organisms can shift the relative dominance of competing organisms. On some reefs, rates of colonization and spread by macroalgae depend primarily on the extent of prior damage to coral colonies from large-scale disturbances. For example, after a catastrophic bleaching event in the Maldives, a complete phase shift occurred within 1 year, in which algae spread to cover 60–90% of the reef benthos (Bianchi et al. 2006, Titlyanov and Titlyanov 2008). Bleaching and mortality of corals due to elevated sea temperatures and high UV radiation (global climate change) have allowed corallimorpharians to preempt space on polluted reefs in Tanzania (Muhando et al. 2002). Soft-bodied cnidarians may become dominant on coral reefs due to alteration of environmental conditions. Anthropogenic eutrophication on inshore areas of the Great Barrier Reef enhanced the growth rates of soft corals and allowed them to overgrow reef-building stony corals, resulting in their dominance on inshore but not offshore reefs (Alino et al. 1992). Eutrophication of coral reefs at Eilat in the northern Red Sea correlates with increased abundances of filter-feeding invertebrates that compete with corals, such as vermetid mollusks (Zvuloni et al. 2008), corallimorpharians (Chadwick-Furman and Spiegel 2000, Fig. 19.4), and ascidians (Shenkar et al. 2008). The presence of shipwrecks and other metal debris is associated with extensive overgrowth of corals by corallimorpharians at Palmira Atoll in the central Pacific, possibly due to stimulation of corallimorpharian growth by iron leaked from the rusting steel (Work et al. 2008).

Thus, human disturbances to corals reefs that alter conditions, including levels of dissolved nutrients, seawater temperature, sedimentation, and mechanical disturbance (e.g., ship grounding), may strongly alter competitive outcomes and contribute to large-scale phase shifts from coral-dominated reefs to alternate dominants that can rapidly take advantage of shifting conditions. These shifts are similar to those that occur in naturally marginal environments for coral reefs, such as near river mouths (Fabricius 2010), or near the limits of their latitudinal range where macroalgae outcompete corals due to lower temperatures and/or higher nutrient levels than near the equator (Bellwood et al. 2004, Bruno et al. 2007, Norström et al. 2009). We expect competitive exclusion of reef-building corals by non-reef builders to become increasingly frequent on the world's coral reefs, as anthropogenic impacts continue to impair the ability of corals to fend off competitors that once were kept in check through a combination of herbivory and slow growth constrained by the limited particulate and dissolved nutrients on pristine tropical reefs (Fig. 1). More than half the benthic cover currently on coral reefs worldwide consists of organisms other than hard corals and macroalgae (Bruno et al. 2009, Norström et al. 2009). Until some of the negative human impacts on coral reefs are reversed, the resulting alteration of competitive outcomes among sessile reef organisms is expected to continue as a major cause of the decline of coral reefs.

5 Effects of Competition Among Sessile Organisms

5.1 Effects on Individuals

In contrast to the lack of evidence 20 years ago, several recent studies have revealed impacts of intraspecific competition on sessile reef organisms at the level of the individual. All have been conducted on stony and soft corals, and most have shown reduced growth and/or survival of colonies due to competition with conspecifics, in terms of impacts of fusion among genetically different conspecific colonies. Chimera formation and intraspecific competition significantly reduce the growth rates of colonies of the stony coral *Stylophora pistillata*, but fusion also may enhance the survival of some chimeric colonies through larger body size (Amar et al. 2008). Thus, fusion of co-settled or contacting coral colonies may increase their survival through larger colony size and physical stability (Chornesky 1991, reviewed in Frank and Rinkevich 1994), but the fusion of genetically different colonies also may result in somatic or germ cell parasitism that ultimately decreases the growth and fitness of at least one of the partners (Amar et al. 2008). Intraspecific competition among solitary fungiid corals in aggregations also appears to decrease their growth rates (Elahi 2008). Colonies of the stony coral *Porites cylindrica* show complex responses to intraspecific competition: they grow more slowly but survive longer than when competing with other coral species, and these growth rates reverse under conditions of nutrient enrichment (Dizon and Yap 2005). Allogenic contacts between some corals can lead to net enhancement of colony survival due to physical support and stabilization in reef areas exposed to high water motion. Chornesky (1991) found that contacts among colonies of *Agaricia tenuifolia* in Belize anchored their skeletons and potentially allowed them to resist breakage and detachment better than did isolated colonies. However, this study consisted of comparative field surveys and did not quantify impacts of intraspecific contact on coral growth or survival.

Experimental work on soft corals indicates that gorgonian coral growth decreases with the presence of nearby conspecifics (Kim and Lasker 1997). The larvae of four species of soft corals in the Red Sea settle gregariously on reefs, leading to high rates of contact and fusion with allogeneic conspecifics, and the formation of chimeric colonies that have slower growth and more deformities relative to single-genotype colonies (Barki et al. 2002). Allogeneic fusion occurs only in young colonies <3 weeks old, and represents a window during development in which the alloimmune system is not fully matured. Thus, fusion among genetically different coral colonies may occur mainly during early post-metamorphosis, and lead to a combination of stunting effects due to mixing of genotypes within one colony, but also enhancement of colony size and stability via larger skeletons. Overall, recent evidence indicates mainly negative impacts on corals of within-species competition, with some possible "cooperation" among colonies leading to physical benefits. Because the larvae of many corals settle gregariously, effects of intraspecific competition may be widespread on reefs, but often are unseen because chimeric colonies are difficult to distinguish from single-genotype colonies.

To our knowledge, no published information exists on effects of intraspecific competition in other coral reef cnidarians such as actinarian sea anemones, corallimorpharians, or zoanthids, or among other sessile organisms including ascidians, sponges, or macroalgae. As these organisms also may settle gregariously, competition for space within species may be common (Figs. 2–4), and the impacts of this process need to be examined in future studies.

Effects of interspecific competition among sessile organisms on reefs are well studied, in part because the high biodiversity on coral reefs leads to high frequencies of interactions among species (Van Veghel et al. 1996, reviewed in Knowlton and Jackson 2001). Also, contact margins between species often are more obvious than those within species, and may involve unilateral overgrowth and wide damaged or cleared zones around individuals (Fig. 3). Several recent studies have documented negative impacts of interspecific competition on individual reef organisms, mostly among coral colonies, but in some cases between corals and other sessile organisms. Coral growth, fecundity, and survival all are reduced during competition with other corals (Romano 1990, Tanner 1997, Idjadi and Karlson 2007) and with macroalgae (Tanner 1995). Short-term growth rates of solitary fungiid corals are impaired, and their rates of mucus production and mobility enhanced, when they are surrounded experimentally by other species of fungiids (Elahi 2008). Polyps of corallimorpharians are significantly larger and contain more ovaries at the centers of aggregations than along the edges where they contact corals, on reefs in both the western Pacific (Chen et al. 1995a) and Red Sea (Chadwick-Furman et al. 2000). Thus, interspecific competition with corals appears to reduce both body size and sexual reproductive rates in these corallimorpharians. Conversely, competition with corallimorpharians inhibits the fecundity, recruitment, and survival of solitary temperate corals (Chadwick 1991). During competition between temperate corallimorpharians and actinian sea anemones, individuals express higher levels of the heat-shock protein HSP70 than when not in competition, indicating activation of mechanisms for the repair of cellular damage (Rossi and Snyder 2001). As such, interspecific competition appears to be costly for some cnidarians in terms of the production of energetically expensive compounds for the repair of cellular components.

Recent reviews also have concluded that when corals compete with macroalgae, the growth of both types of organisms can be inhibited along the contact zone (Jompa and McCook 2002a). Macroalgae may reduce coral reproductive success by inducing reallocation of energy from reproduction to repair of tissues damaged during competition, resulting in smaller egg sizes along coral–algal borders and in the center of coral patches (Foster et al. 2008). Competitive stress induced by contact with macroalgae likely diminishes the energy reserves that are critical to normal immune function and resilience of healthy corals, potentially contributing to chronic bleaching events and disease infection (Ritson-Williams et al. 2009). Macroalgae can impact coral replenishment through reduction of coral fecundity and larval survival, preemption of space for larval settlement, abrasion and overgrowth of new recruits, dislodgment of recruits settled on crustose algae, and changes to habitat conditions (reviewed in Birrell et al. 2008b). Some brown macroalgae cause recruitment inhibition or avoidance behavior of coral larvae, while others reduce the survival of new recruits of stony corals and octocorals (Kuffner et al. 2006).

As with the mechanisms employed by macroalgae to compete with stony corals, their competitive impacts are highly species-specific. Macroalgae cause significantly decreased growth of colonies of the stony corals *Porites astreoides* and *Montastraea faveolata*, but no clear effects on the coral *Siderastrea siderea* in caged treatments (Lirman 2001). Similarly, the red turf algae *Corallophila huysmansi* and *Anotrichium tenue* damage and overgrow live tissues of the branching coral *Porites cylindrica* and the massive coral *Porites* sp., but the corticated red alga *Hypnea pannosa* and the filamentous green alga *Chlorodesmis* cause no visible effects on coral tissues (Jompa and McCook 2003a,b). Four species of macroalgae tested by Maypa and Raymundo (2004) varied widely in their morphological and chemical impacts on the stony coral *Pocillopora damicornis*: some enhanced and others reduced coral settlement and survival, while some reduced juvenile coral growth rates more than others, illustrating the complex and species-specific effects of macroalgae on the early life-stages of corals. In field experiments, the brown alga *Lobophora variegata* did not cause tissue mortality to five species of Caribbean corals, but negatively impacted the leafy, fast-growing coral *Agaricia agaricites* (Nugues and Bak 2006). Subtle initial impacts in terms of physiological stress were detected in *M. faveolata*, in which contact with mixed turf algae caused reductions in coral zooxanthella densities, Chlorophyll a concentration, and tissue thickness, followed by complete coral overgrowth by the algae (Quan-Young and Espinoza-Avalos 2006). Physiological damage can be mutual, in that some corals extrude their mesenterial filaments onto macroalgae, causing algal discoloration, migration of cholorplasts away from the algal surface, and entry of coral nematocysts into the algal epidermis (Nugues et al. 2004a). These studies illustrate the complexity of competitive impacts on interacting corals and algae, and highlight the importance of variation among species and environments. Healthy corals that normally prevent the attachment and survival of algal recruits may become competitively inferior when stressed by environmental factors such as nutrient-enrichment, sedimentation, and disease (Diaz-Pulido and McCook 2004, Dinsdale and Rohwer 2010).

In some cases, interspecific competition appears to have no impact on coral growth rates. Connell et al. (2004) observed in a long-term study that competition does not appear to affect the long-term growth of corals on the Great Barrier Reef in Australia. They concluded that many mechanisms of coral competition require low energy investment, and thus may not alter rates of growth in dominant corals. A similar conclusion was reached by Lapid and Chadwick (2006), who detected no significant impacts on the growth of dominant brain corals when they developed sweeper tentacles to compete with other massive corals. However, subordinate massive corals that were damaged by the brain corals had severely impaired growth and survival over 1 year under laboratory conditions. Romano (1990) likewise demonstrated that competition reduces the growth rates of subordinate but not dominant coral species during field experiments in Hawaii. Thus, interspecific competition appears to severely reduce the growth, reproduction, and survival of some subordinate corals, but not necessarily that of competitive dominants. More studies need to be conducted on a wider range of species to determine the robustness of these patterns in fitness costs.

Few studies have examined the individual-level effects of competition on other types of space occupiers on reefs, such as sponges and ascidians. Future research should examine how competition, especially with corals, impacts the reproduction, growth, and survival of non-cnidarian sessile invertebrates on coral reefs.

5.2 Effects on Populations and Communities

Through the above-mentioned impacts on individual organisms, competition alters demographic patterns on coral reefs, in terms of population growth, decline, and turnover rates. For example, competitive impairment of sexual reproduction leads to fewer recruits on coral reefs, and eventually to smaller, older populations of impacted competitors. Slower individual growth rates likely delay ages at first reproduction, and could cause sexually mature individuals to revert to a nonreproductive state, thus slowing turnover rates of populations. Finally, increased mortality at all life stages reduces the population sizes of subordinate competitors and eventually may exclude them completely from reef areas where members of dominant species eventually kill all individuals. Competitive interactions with macroalgae impact all life stages of corals (see Sect. 5.1). High algal biomass on coral reefs can result in strong negative interactions with adult corals and decrease coral recruitment success by reducing larval settlement and post-settlement survival, thus potentially altering the population structure of corals. However, few studies have quantified impacts of competition on population-level processes in sessile reef organisms (Chadwick 1991, Tanner 1995, 1997). Analyses of population size structure can provide a wealth of information on patterns and rates of population decline or growth (Bak and Meesters 1999, Meesters et al. 2001, Goffredo and Chadwick-Furman 2003, Guzner et al. 2007), and need to be incorporated into studies on impacts of competitive interactions among sessile organisms on coral reefs. Competition also alters population genetics, in that nontransitive circular competitive networks among individuals or colonies likely promote the coexistence of diverse genotypes within populations (Chadwick-Furman and Rinkevich 1994, Frank and Rinkevich 1994, Rinkevich et al. 1994).

Effects of competition on whole reef communities have been more thoroughly studied, and a wide array of impacts has been documented. Recent studies have revealed alteration of species diversity on reefs, and in some cases, decreased diversity, due to competitive interactions mainly among stony corals. Competition significantly decreased stony coral diversity over 38 years at Heron Island on the Great Barrier Reef, and altered species composition on protected reef crests and in shallow pools (Connell et al. 2004). In contrast, on the exposed reef flat, storms disturbed the coral community and kept coral abundances low, reducing the importance of competition in this frequently disturbed intertidal habitat. In Taiwan, the most aggressive corals may grow rapidly and thus dominate the reefs (Dai 1990). Conversely, aggression allows slower-growing corals to persist in Hawaii (Romano 1990) and the Red Sea (Lapid et al. 2004). Abelson and Loya (1999) found no relationship between the level of aggressiveness and relative abundance of stony coral species at Eilat. They confirmed, as known for other reef areas (Dai 1990) that on Red Sea reefs an intransitive hierarchy occurs with circular networks among corals at intermediate aggressive levels. Rinkevich et al. (1992) showed that four major space occupiers at Eilat (three types of corals and a sponge) engage in a network of competitive interactions with no clear dominants, thus enhancing species diversity on these reefs.

Aggression allows corallimorpharians to dominate some reef flats in the Red Sea and Indian Ocean, to the exclusion of the once-dominant stony corals (Chadwick-Furman and Spiegel 2000, Langmead and Chadwick-Furman 1999a, b, Muhando et al. 2002, Kuguru et al. 2004, Fig. 4). Competition also alters patterns of dominance and vertical zonation among stony corals and corallimorpharians on temperate subtidal reefs in the northeastern Pacific (Chadwick 1991). Some macroalgae and sponges engage in competitive standoffs on Caribbean reefs, with neither advancing, but sponges lose during competition with tunicates, zoanthids, and gorgonians (López-Victoria et al. 2006). These competitive outcomes among sessile invertebrates likely alter their relative abundances on reefs. Damage to stony corals by macroalgae can result in algal dominance on some reefs and alter the overall structure of reef communities (Hughes and Tanner 2000, Kuffner et al. 2006, Fig. 1).

During the 1980s, studies showed that some coral reefs may undergo a phase shift from stony corals to macroalgae (Fong and Paul 2010, Fig. 1). Recent analyses demonstrate that coral reef phase shifts also occur from stony corals to a variety of organisms other than macroalgae (reviewed in Norström et al. 2009). Top-down processes such as changes in consumers and overfishing appear to cause phase shifts to urchins and macroalgae (Norström et al. 2009), while bottom-up processes such as changes in nutrient levels appear to drive phase shifts to sponges, soft corals, and corallimorpharians (including iron enrichment, Work et al. 2008). In both types of processes, the main drivers are human-induced changes to coral reefs in terms of overfishing and pollution. Thus, the impacts of competition on coral communities appear to be increasing, and are altered by a host of anthropogenic forces (Dinsdale and Rohwer 2010).

It is now clear that competition is a major driving force in many coral reef habitats, especially those that have been altered through human disturbance. On pristine reefs, natural physical disturbances may be more important in some reef zones than are biological interactions such as competition. Generally, on pristine reefs the community structure of intertidal reef flats tend to be controlled more by physical disturbances (low tides, storms, freshwater flooding), while that of subtidal reef slopes is controlled more by biological interactions (predation, competition for space, Connell et al. 2004). This variation in community structuring forces with depth below sea level and distance from terrestrial influences is similar to that documented a few decades earlier for temperate rocky shorelines (reviewed in Witman and Dayton 2001). It is also similar to the vertical gradient in physical versus biological factors controlling coral zonation on temperate rocky reefs, in which sand scouring impacts some corals along the bases of reefs, while competition with dominant corallimorpharians excludes them from the tops of the reefs, allowing them to become abundant only at intermediate heights above the sand–reef interface (Chadwick 1991).

6 Conclusions and Directions for Future Research

Much has been added to our understanding of competition among sessile organisms on coral reefs since the initial major review by Lang and Chornesky (1990). The inclusion of diverse methods into single integrated studies, such as long- and short-term surveys and experiments in a combination of laboratory and field settings, has revealed robust patterns and processes in reef competition, and should be continued in future studies. Foundational descriptions of competitive mechanisms and some of their effects on reefs were completed mostly during the 1970s and 1980s, with recent research focusing on intraspecific dynamics, costs and benefits of competition, and widespread effects on individuals and coral reef communities. Competition has been established as a major structuring force on coral reefs, with microscale to community-level impacts. Much recent work focuses on how human-induced changes on coral reefs are altering competitive outcomes, especially leading to phase shifts from reef-building corals to other organisms. These studies reveal how altered competitive outcomes can cause the demise of coral reefs and their replacement by alternate dominants with much lower biodiversity than the more structurally complex coral-built reefs they replace (Bellwood et al. 2004, Norström et al. 2009). Competitive interactions are destined to play an increasing role in the story of how reefs are affected by man, and are a model process for understanding how human interference affects the outcomes of interactions among organisms in communities.

Future research is needed especially on processes of competition in sessile space occupiers other than corals, such as the sea anemones, zoanthids, ascidians, sponges, and algae that may replace corals on some reefs. How competition impacts the diverse microbial assemblages that associate with sessile reef organisms is just beginning to be examined, as new methods are developed in molecular and microbial biology. Microbial studies need to be coordinated with those on the antibacterial and/or stimulatory nature of chemical compounds released by coral hosts, their symbionts, and adjacent competitors. Research on how ocean acidification affects calcium carbonate production by corals should be linked to examination of their changing ability to compete with non-reef-building organisms. Global climate change and ocean acidification also may promote increases in marine disease and the alteration of competitive networks due to compromised coral immune systems. A better understanding of how competition impacts the spread of disease among sessile organisms on reefs is especially needed, given ever-increasing levels of disease in corals and other reef organisms. Future research will benefit from interdisciplinary approaches that include both macro- and microscopic perspectives to reveal the complexity of competitive processes on coral reefs.

Acknowledgments We thank Zvy Dubinsky for inviting us to write this review. The manuscript was improved by comments from J. Bruno, A. Norstrom, and J. Szczebak. We dedicate this review to J. Lang and E. Chornesky, pioneering researchers on coral competition who inspired NEC to spend a lifetime investigating the complexity of interactions among cnidarians on coral reefs, and to T. Morrow and A. Morrow for their support of K. M. Morrow in her pursuit of a career in coral reef biology. Financial support was provided by grants from the Israel Science Foundation (ISF), National Science Foundation (NSF), and

National Oceanic and Atmospheric Administration (NOAA) to NEC for past and current fieldwork on coral reefs, and by a NOAA Nancy Foster Scholarship to KMM. This is contribution #65 of the Auburn University Marine Biology Program.

References

Abelson A, Loya Y (1999) Interespecific aggression among stony corals in Eilat, Red Sea: a hierarchy of aggression ability and related parameters. Bull Mar Sci 65:851–860

Aceret TL, Sammarco PW, Coll JC (1995) Effects of diterpines derived from the soft coral *Sinularia flexibilis* on the eggs, sperm and embryos of the scleractinian corals *Montipora digitata* and *Acropora tenuis*. Mar Biol 122:317–323

Aerts LAM (1998) Sponge/coral interactions in Caribbean reefs: analysis of overgrowth patterns in relation to species identity and cover. Mar Ecol Prog Ser 175:241–249

Aerts LAM, van Soest RWM (1997) Quantification of sponge/coral interactions in a physically stressed reef community, NE Colombia. Mar Ecol Prog Ser 148:125–134

Alino PM, Sammarco PW, Coll JC (1992) Competitive strategies in soft corals (Coelenterata, Octocoraliia) IV. Environmentally induced reversals in competitive superiority. Mar Ecol Prog Ser 81:129–145

Amar KO, Chadwick NE, Rinkevich B (2008) Coral kin aggregations exhibit mixed allogeneic reactions and enhanced fitness during early ontogeny. BMC Evol Biol 8:126

Baird AH, Morse ANC (2004) Unduction of metamorphosis in larvae of the brooding coral *Acropora* palifera and *Stylophora pistillata*. Mar Freshwater Res 55:469–472

Bak RPM, Borsboom JLA (1984) Allelopathic interaction between a reef coelenterate and benthic algae. Oecologia 63:194–198

Bak RPM, Meesters EH (1999) Population structure as a response of coral communities to global change. Am Zool 39:56–65

Bak RPM, Sybesma J, Van Duyl FC (1981) The ecology of the tropical compound ascidian *Trididemnum solidum*. 2. Abundance, growth and survival. Mar Ecol Prog Ser 6:43–52

Bak RPM, Lambrechts DYM, Joenje M et al (1996) Long-term changes on coral reefs in booming populations of a competitive colonial ascidian. Mar Ecol Prog Ser 133:303–306

Barki Y, Gateno D, Graur D et al (2002) Soft-coral natural chimerism: a window in ontogeny allows the creation of entities comprised of incongruous parts. Mar Ecol Prog Ser 231:91–99

Bastidas C, Bone D (1996) Competitive strategies between *Palythoa caribaeorum* and *Zoanthus sociatus* (Cnidaria: Anthozoa) at a reef flat environment in Venezuela. Biol Bull 59:543–555

Becerra MA, Turon X, Uriz M (1997) Chemically-mediated interactions in benthic organisms: the chemical ecology of *Crambe crambe* (Porifera: Poecilosclerida). Hydrobiologia 356:77–89

Bellwood DR, Hughes TP, Folke C et al (2004) Confronting the coral reef crisis. Nature 429:827–833

Bianchi CN, Morri C, Pichon M et al (2006) Dynamics and Pattern of Coral Recolonization following the 1998 Bleaching Event in the Reefs of the Maldives. Proc 10th Int Coral Reef Symp 1:30–37

Birrell CL, McCook LJ, Willis BL (2005) Effects of algal turfs and sediment on coral settlement. Mar Pollut Bull 51:408–414

Birrell CL, McCook LJ, Willis BL et al (2008a) Effects of benthic algae on the replenishment of corals and the implications for the resilience of coral reefs. Oceanogr Mar Biol Ann Rev 46:25–63

Birrell CL, McCook LJ, Willis BL et al (2008b) Allelochemical effects of macroalgae on larval settlement of the coral *Acropora millepora*. Mar Ecol Prog Ser 362:129–137

Box SJ, Mumby PJ (2007) Effects of macroalgal competition on growth and survival of juvenile Caribbean corals. Mar Ecol Prog Ser 342:139–149

Boyd KG, Adams DR, Burgess JG (1999) Antibacterial and repellent activities of marine bacteria associated with algal surfaces. Biofouling 14:227–236

Bruno JF, Selig ER (2007) Regional declines of coral cover in the Indo-Pacific: timing, extent, and subregional comparisons. PLoS ONE 2(8):e711

Bruno JF, Witman JD (1996) Defense mechanisms of scleractinian cup corals against overgrowth by colonial invertebrates. J Exp Mar Biol Ecol 207:229–241

Bruno JF, Selig ER, Casey KS et al (2007) Thermal stress and coral cover as drivers of coral disease outbreaks. PLoS Biol 5(6):1220–1227

Bruno JF, Sweatman H, Precht WF et al (2009) Assessing evidence of phase shifts from coral to macroalgal dominance on coral reefs. Ecology 90:1478–1484

Buss LW (1986) Competition and community organization on hard surfaces in the sea. In: Diamond J, Case TJ (eds) Community ecology. Harper and Row, New York

Carpenter RC (1990) Mass mortality of *Diadema antillarum* I Long-term effects on sea urchin population-dynamics and coral reef algal communities. Mar Biol 104:67–77

Cetrulo GL, Hay ME (2000) Activated chemical defenses in tropical versus temperate seaweeds. Mar Ecol Prog Ser 207:243–253

Chadwick NE (1987) Interspecific aggressive behavior of the corallimorpharian *Corynactis californica* (Cnidaria: Anthozoa): effects on sympatric corals and sea anemones. Biol Bull 173:110–125

Chadwick NE (1988) Competition and locomotion in a free-living fungiid coral. J Exp Mar Biol Ecol 123:189–200

Chadwick NE (1991) Spatial distribution and the effects of competition on some temperate Scleractinia and Corallimorpharia. Mar Ecol Prog Ser 70:39–48

Chadwick NE, Adams C (1991) Locomotion, asexual reproduction, and killing of corals by the corallimorpharian *Corynactis californica*. Hydrobiologia 216(217):263–269

Chadwick-Furman N, Loya Y (1992) Migration, habitat use, and competition among mobile corals (Scleractinia: Fungiidae) in the Gulf of Eilat, Red Sea. Mar Biol 114:617–623

Chadwick-Furman N, Rinkevich B (1994) A complex allorecognition system in a reef-building coral: delayed responses, reversals and nontransitive hierarchies. Coral Reefs 13:57–63

Chadwick-Furman NE, Spiegel M (2000) Abundance and clonal replication in the tropical corallimorpharian *Rhodactis rhodostoma*. Invertebr Biol 119:351–360

Chadwick-Furman NE, Spiegel M, Nir I (2000) Sexual reproduction in the tropical corallimorpharian *Rhodactis rhodostoma*. Invertebr Biol 119:361–369

Chaves-Fonnegra A, Castellanos L, Zea S et al (2008) Clionapyrrolidine A-A metabolite from the encrusting and excavating sponge *Cliona tenuis* that kills coral tissue upon contact. J Chem Ecol 34:1565–1574

Chen CA, Dai C-F (2004) Local phase shift from *Acropora*-dominant to *Condylactis*-dominant community in the Tiao-Shi Reef, Kenting National Park, southern Taiwan. Coral Reefs 23:508

Chen CA, Chen C, Chen I (1995a) Spatial variability of size and sex in the tropical corallimorpharian *Rhodactis* (=*Discosoma*) *indosinensis* (Cnidaria: Corallimorpharia) in Taiwan. Zool Stud 34(2):82–87

Chen CA, Chen C, Chen I (1995b) Sexual and asexual reproduction of the tropical corallimorpharian *Rhodactis* (=*Discosoma*) *indosinensis* (Cnidaria: Corallimorpharia) in Taiwan. Zool Stud 34(1):29–40

Chornesky EA (1991) The ties that bind: inter-clonal cooperation may help a fragile coral dominate shallow high-energy reefs. Mar Biol 109:41–51

Coles SL, Strathmann R (1973) Observations on coral mucus "flocs" and their potential trophic significance. Limnol Oceanogr 18:673–678

Coll JC, Sammarco PW (1983) Terpenoid toxins of soft corals (Cnidaria: Octocorallia): their nature, toxicity, and their ecological significance. Toxicon 21:69–72

Coll JC, La Barre S, Sammarco PW et al (1982) Chemical defences in soft corals (Coelenterata: Octocorallia) of the Great Barrier Reef: a study of comparative toxicities. Mar Ecol Prog Ser 8:271–278

Connell JH, Hughes TP, Wallace CC et al (2004) A long-term study of competition and diversity of corals. Ecol Monogr 74:179–210

Connolly SR, Muko S (2003) Space preemption, size-dependent competition, and the coexistence of clonal growth forms. Ecology 84:2979–2988

Costa OS Jr, Attrill M, Pedrini AG et al (2002) Spatial and seasonal distribution of seaweeds on coral reefs from Southern Bahia, Brazil. Bot Mar 45:346–355

Costa OS Jr, Nimmo M, Attrill MJ (2008) Coastal nutrification in Brazil: a review of the role of nutrient excess on coral reef demise. J South Am Earth Sci 25:257–270

Coyer JA, Ambrose RF, Engle JM et al (1993) Interactions between corals and algae on a temperate zone rocky reef: mediation by sea urchins. J Exp Mar Biol Ecol 167:21–37

Dai C-F (1990) Interspecific competition in Taiwanese corals with specific reference to interactions between alcyonaceans and scleractinians. Mar Ecol Prog Ser 60:291–297

Diaz-Pulido G, McCook LJ (2004) Algal recruitment and interactions. Coral Reefs 23:225–233

Dinsdale EA, Rohwer F (2010) Fish or germs? Microbial dynamics associated with changing trophic structures on coral reefs. In: Dubinsky Z, Stambler N (eds) Coral reefs: an ecosystem in transition Springer, Doedrecht

Dizon RM, Yap HT (2005) Coral responses in single- and mixed-species plots to nutrient disturbance. Mar Ecol Prog Ser 296:165–172

Done TJ (1992) Phase-shifts in coral reef communities and their ecological significance. Hydrobiologia 247:121–132

Elahi R (2008) Effects of aggregation and species identity on the growth and behavior of mushroom corals. Coral Reefs 27:881–885

Engel S, Puglisi MP, Jensen PR et al (2006) Antimicrobial activities of extracts from tropical Atlantic marine plants against marine pathogens and saprophytes. Mar Biol 149:991–1002

Fabricius K (2010) Factors determining the resilience of coral reefs to eutrophication: a review and conceptual model. In: Dubinsky Z, Stambler N (eds) Coral reefs: an ecosystem in transition Springer, Doedrecht

Fearon RJ, Cameron AM (1996) Larvotoxic extracts of the hard coral *Goniopora tenuidens*: allelochemicals that limit settlement of potential competitors? Toxicon 34:361–367

Ferrier-Pages C, Hoogenboom M, Houlbrèque F (2010) The role of plankton in coral trophodynamics. In: Dubinsky Z, Stambler N (eds) Coral reefs: an ecosystem in transition, Springer, Doedrecht

Fine M, Loya Y (2003) Alternate coral–bryozoan competitive superiority during coral bleaching. Mar Biol 142:989–996

Fine M, Oren U, Loya Y (2002) Bleaching effect on regeneration and resource translocation in the coral *Oculina patagonica*. Mar Ecol Prog Ser 234:119–125

Fong P, Paul JV (2010) Coral reef algae: the good, the bad, and the ugly. In: Dubinsky Z, Stambler N (eds) Coral reefs: an ecosystem in transition, Springer, Doedrecht

Fong P, Smith TB, Wartian MJ (2006) Epiphytic cyanobacteria maintain shifts to macroalgal dominance on coral reefs following ENSO disturbance. Ecology 87:1162–1168

Foster NL, Box SJ, Mumby PJ (2008) Competitive effects of macroalgae on the fecundity of the reef-building coral *Montastraea annularis*. Mar Ecol Prog Ser 367:143–152

Frank U, Rinkevich B (1994) Nontransitive patterns of historecognition phenomena in the Red Sea hydrocoral *Millepora dichotoma*. Mar Biol 118:723–729

Frank U, Rinkevich B (2001) Alloimmune memory is absent in the Red Sea hydrocoral *Millepora dichotoma*. J Exp Zool 291:25–29

Frank U, Brickner I, Rinkevich B, Loya Y et al (1995) Allogeneic and xenogeneic interactions in reef-building corals may induce tissue growth without calcification. Mar Ecol Prog Ser 124:181–188

Frank U, Bak RPM, Rinkevich B (1996) Allorecognition responses in the soft coral *Parerythropodium fulvum fulvum* from the R Sea. J Exp Mar Biol Ecol 197:191–201

Frank U, Oren U, Loya Y, Rinkevich B (1997) Alloimmune maturation in the coral *Stylophora pistillata* is achieved through three distinctive stages, 4 months post-metamorphosis. Proc Roy Soc Lond B 264:99–104

Fusetani N, Asano M, Matsunaga S et al (1986) Bioactive marine metabolites XV. Isolation of aplysinopsin from the scleractinian coral *Tubastrea aurea* as an inhibitor of development of fertilized sea urchin eggs. Comp Biochem Physiol 85B:845–846

Genin A, Karp L (1994) Effects of flow on competitive superiority in scleractinian corals. Limnol Oceanogr 39:913–924

Glynn PW (1997) Bioerosion and coral-reef growth. In: Birkeland C (ed) Life and death of coral reefs. Chapman & Hall, New York

Glynn P, Enochs I (2010) Invertebrates and their roles in coral reef ecosystems. In: Dubinsky Z, Stambler N (eds) Coral reefs: an ecosystem in transition Springer, Doedrecht

Goffredo S, Chadwick-Furman NE (2003) Comparative demography of mushroom corals (Scleractinia: Fungiidae) at Eilat, northern Red Sea. Mar Biol 142:411–418

Griffith JK (1997) Occurrence of aggressive mechanisms during interactions between soft corals (Octocorallia: Alcyoniidae) and other corals on the Great Barrier Reef, Australia. Mar Freshwater Res 48:129–135

Gunthorpe L, Cameron AM (1990a) Intracolonial variation in toxicity in scleractinian corals. Toxicon 28:1221–1227

Gunthorpe L, Cameron AM (1990b) Widespread but variable toxicity in scleractinian corals. Toxicon 28:1199–1219

Gunthorpe L, Cameron AM (1990c) Toxic exudate from the hard coral *Goniopora tenuidens*. Toxicon 28:1347–1350

Guzner B, Novoplansky A, Chadwick NE (2007) Population dynamics of the reef-building coral *Acropora hemprichii* as an indicator of reef condition. Mar Ecol Prog Ser 333:143–150

Harvell CD, Kim K, Burkholder JM et al (1999) Emerging marine diseases – climate links and anthropogenic factors. Science 285:1505–1510

Hildemann WH, Raison RL, Hull CJ et al (1977) Tissue transplantation immunity in corals. Proc 3rd Int Coral Reef Symp Miami 1:537–543

Hirose E (2009) Ascidian tunic cells: morphology and functional diversity of free cells outside the epidermis. Invertebr Biol 128:83–96

Hoegh-Guldberg O (2004) Coral reefs in a century of rapid environmental change. Symbiosis 37:1–31

Holland GJ, Webster PJ (2007) Heightened tropical cyclone activity in the North Atlantic: natural variability or climate trend? Philos Trans R Soc Ser A 365:2695–2716

Hughes TP, Tanner JE (2000) Recruitment failure, life histories, and long-term decline of Caribbean corals. Ecology 81:2250–2263

Hughes TP, Rodrigues MJ, Bellwood DR et al (2007) Phase shifts, herbivory, and the resilience of coral reefs to climate change. Curr Biol 17:360–365

Idjadi JA, Karlson RH (2007) Spatial arrangement of competitors influences coexistence of reef-building corals. Ecology 88:2449–2454

Jompa J, McCook LJ (2002a) The effect of herbivory on competition between a macroalga and a hard coral. J Exp Mar Biol Ecol 271:25–39

Jompa J, McCook LJ (2002b) The effect of nutrients and herbivory on competition between the hard coral (*Porites cylindrica*) and a brown alga (*Lobophora variegata*). Limnol Oceanogr 47:527–534

Jompa J, McCook LJ (2003a) Contrasting effects of turf algae on corals: massive *Porites* spp. are unaffected by mixed-species turfs, but killed by the red alga *Anotrichium tenue*. Mar Ecol Prog Ser 258:79–86

Jompa J, McCook LJ (2003b) Coral-algal competition: macroalgae with different properties have different effects on corals. Mar Ecol Prog Ser 258:87–95

Kim K, Lasker HR (1997) Flow-mediated resource competition in the suspension feeding gorgonian *Plexaura homomalla* (Esper). J Exp Mar Biol Ecol 215:49–64

Knowlton N (1992) Thresholds and multiple stable states in coral reef community dynamics. Am Zool 32:674–682

Knowlton N, Jackson JBC (2001) The ecology of coral reefs. In: Bertness MD, Gaines SD, Hay M (eds) Marine community ecology. Sinauer Associates, Sunderland

Koh EGL (1997) Do scleractinian corals engage in chemical warfare against marine microbes? J Chem Ecol 23:379–398

Koh EGL, Sweatman H (2000) Chemical warfare among scleractinians: bioactive natural products from *Tubastraea faulkneri* Wells kill larvae of potential competitors. J Exp Mar Biol Ecol 251:141–160

Kuffner IB, Walters LJ, Becerro MA et al (2006) Inhibition of coral recruitment by macroalgae and cyanobacteria. Mar Ecol Prog Ser 323:107–117

Kuguru BL, Mgaya YD, Ohman MC et al (2004) The reef environment and competitive success in the Corallimorpharia. Mar Biol 145:875–884

Lambert G (2002) Nonindigenous ascidians in tropical waters. Pac Sci 56:291–298

Lane AL, Kubanek J (2008) Secondary metabolite defenses against pathogens and biofoulers. In: Amsler CD (ed) Algal chemical ecology. Springer, Berlin

Lang JC, Chornesky EA (1990) Competition between scleractinian reef corals – a review of mechanisms and effects. In: Dubinsky Z (ed) Ecosystems of the world: coral reefs. Elsevier, Amsterdam, pp 209–252

Langmead O, Chadwick-Furman NE (1999a) Marginal tentacles of the corallimorpharian *Rhodactis rhodostoma*. 1. Role in competition for space. Mar Biol 134:479–489

Langmead O, Chadwick-Furman NE (1999b) Marginal tentacles of the corallimorpharian *Rhodactis rhodostoma*. 2. Induced development and long-term effects on coral competitors. Mar Biol 134:491–500

Lapid ED, Chadwick NE (2006) Long-term effects of competition on coral growth and sweeper tentacle development. Mar Ecol Prog Ser 313:115–123

Lapid ED, Wielgus J, Chadwick-Furman NE (2004) Sweeper tentacles of the brain coral *Platygyra daedalea*: induced development and effects on competitors. Mar Ecol Prog Ser 282:161–171

Lirman D (2001) Competition between macroalgae and corals: effects of herbivore exclusion and increased algal biomass on coral survivorship and growth. Coral Reefs 19:392–399

Littler M, Littler DS (1997) Disease induced mass mortality of crustose algae on coral reefs provides rationale for the conservation of herbivorous fish stocks. Proc 8th Int Coral Reef Symp 1:719–724

Littler MM, Littler DS (2006) Harmful algae on tropical reefs: bottom-up eutrophication and top-down herbivory. Harmful Algae 5:565–585

Littler MM, Littler DS, Brooks BL (2009) Herbivory, nutrients, stochastic events, and relative dominances of benthic indicator groups on coral reefs: a review and recommendations. Smithsonian Contrib Mar Sci 38:401–414

López-Victoria M, Zea S (2004) Storm-mediated coral colonization by an excavating Caribbean sponge. Mar Ecol Prog Ser 26:251–256

López-Victoria M, Zea S, Weil E (2006) Competition for space between encrusting excavating Caribbean sponges and other coral reef organisms. Mar Ecol Prog Ser 312:113–121

Lotka AJ (1925) Elements of physical biology. Williams & Wilkins, Baltimore (Reprinted as Elements of Mathematical Biology, Dover, New York, 1956.)

Maida M, Sammarco PW, Coll JC (1995a) Effects of soft corals on scleractinian coral recruitment: I Directional allelopathy and inhibition of settlement. Mar Ecol Prog Ser 121:191–202

Maida M, Sammarco PW, Coll JC (1995b) Preliminary evidence for directional allelopathic effects of the soft coral *Sinularia flexibilis* (Alcyonacea, Octocorallia) on scleractinian coral recruitment. Bull Mar Sci 56:303–311

Maliao RJ, Turingan RG, Lin J (2008) Phase-shift in coral reef communities in the Florida Keys National Marine Sanctuary (FKNMS), USA. Mar Biol 154:841–853

Maypa AP, Raymundo LJ (2004) Algal-coral interactions, mediation of coral settlement, early survival, and growth by macroalgae. Silliman J 45:76–95

McCook LJ, Jompa J, Diaz-Pulido G (2001) Competition between corals and algae on coral reefs: a review of evidence and mechanisms. Coral Reefs 19:400–417

Meesters EH, Hilterman M, Kardinaal E et al (2001) Colony size-frequency distributions of scleractinian coral populations: spatial and interespecific variation. Mar Ecol Prog Ser 209:43–54

Miles JS (1991) Inducible agonistic structures in the tropical corallimorphian, *Discosoma sanctithomae*. Biol Bull 180:406–415

Millar RH (1971) The biology of ascidians. Adv Mar Biol 9:1–100

Miller MW, Hay ME (1996) Coral-seaweed-grazer nutrient interactions on temperate reefs. Ecol Monogr 66:323–344

Monniot C, Monniot F, Laboute P (1991) Coral reef ascidians of New Caledonia. ORSTOM, Paris

Moore RE, Scheuer PJ (1971) Palytoxin: a new marine toxin from a coelenterate. Science 172:495–498

Morin PJ (1999) Community ecology. Blackwell, Oxford

Morrow KM, Carpenter RC (2008) Macroalgal morphology mediates particle capture by the corallimorpharian *Corynactis californica*. Mar Ecol Prog Ser 155:273–280

Morse ANC, Iwao K, Baba M et al (1996) An ancient chemosensory mechanism brings new life to coral reefs. Biol Bull 191:149–154

Muhando CA, Kuguru BL, Wagner GM et al (2002) Environmental effects on the distribution of corallimorpharians in Tanzania. Ambio 31:558–561

Muko S, Sakai K, Iwasa Y (2001) Dynamics of marine sessile organisms with space-limited growth and recruitment: application to corals. J Theor Biol 210:67–80

Mumby PJ (2006) The impact of exploiting grazers (Scaridae) on the dynamics of Caribbean coral reefs. Ecol Appl 16:747–769

Mumby PJ (2009) Phase shifts and the stability of macroalgal communities on Caribbean coral reefs. Coral Reefs 28:761–773

Mumby PJ, Steneck RS (2008) Coral reef management and conservation in light of rapidly-evolving ecological paradigms. Trends Ecol Evol 23:555–563

Norström AV, Nyström M, Lokrantz J et al (2009) Alternative states on coral reefs: beyond coral–macroalgal phase shifts. Mar Ecol Prog Ser 376:295–306

Nugues MM, Bak RPM (2006) Differential competitive abilities between Caribbean coral species and a brown alga: a year of experiments and a long-term perspective. Mar Ecol Prog Ser 315:75–86

Nugues MM, Roberts CM (2003) Coral mortality and interaction with algae in relation to sedimentation. Coral Reefs 22:507–516

Nugues MM, Szmant AM (2006) Coral settlement onto *Halimeda opuntia*: a fatal attraction to an ephemeral substrate? Coral Reefs 25:585–591

Nugues MM, Delvoye L, Bak RPM (2004a) Coral defence against macroalgae: differential effects of mesenterial filaments on the green alga *Halimeda opuntia*. Mar Ecol Prog Ser 278:103–114

Nugues MM, Smith GW, Hooidonk RJ et al (2004b) Algal contact as a trigger for coral disease. Ecol Lett 7:919–923

Pacala SW, Silander JA (1990) Field tests of neighborhood population dynamic models of two annual weed species. Ecol Monogr 60:113–134

Paine RT (1984) Ecological determinism in the competition for space. Ecology 65:1339–1348

Pandolfi JM, Bradbury RH, Sala E et al (2003) Global trajectories of the long-term decline of coral reef ecosystems. Science 301:955–958

Paul VJ, Arthur K, Ritson-Williams R (2007) Chemical defenses: from compounds to communities. Biol Bull 213:226–251

Pawlik JR, Steindler L, Henkel TP (2007) Chemical warfare on corals: sponge metabolites differentially affect coral in situ. Limnol Oceanogr 52:907–911

Peach MB, Hoegh-Guldberg O (1999) Sweeper polyps of the coral *Goniopora tenuidens* (Scleractinia: Poritidae). Invertebr Biol 118:1–7

Puglisi MP, Engel S, Jensen PR et al (2007) Antimicrobial activities of extracts from Indo-Pacific marine plants against marine pathogens and saprophytes. Mar Biol 150:531–540

Quan-Young, Espinoza-Avalos (2006) Reduction of zooxanthellae density, chlorophyll *a* concentration, and tissue thickness of the coral *Montastraea faveolata* (Scleractinia) when competing with mixed turf algae. Limnol Oceanogr 51:1159–1166

Rashid MA, Gustafson KR, Cardellina JH II et al (1995) Mycalolides D and E, new cytotoxic macrolides from a collection of the stony coral *Tubastrea faulkneri*. J Nat Prod 58:1120–1125

Raymundo LJ (2001) Mediation of growth by conspecific neighbors and the effect of site in transplanted fragments of the coral *Porites attenuata* Nemenzo in the central Philippines. Coral Reefs 20:263–272

Richmond RH (1993) Coral reefs – present problems and future concerns resulting from anthropogenic disturbance. Am Zool 33:524–536

Rinkevich B, Sakai K (2001) Interspecific interactions among species of the coral genus *Porites* from Okinawa, Japan. Zoology 104:1–7

Rinkevich B, Shashar N, Liberman T (1992) Nontransitive xenogenic interactions between four common red sea sessile invertebrates. Proc 7th Int Coral Reef Symp Guam 2:833–839

Rinkevich B, Frank U, Bak RPM et al (1994) Alloimmune responses between *Acropora hemprichi* conspecifics: nontransitive patterns of overgrowth and delayed cytotoxicity. Mar Biol 118:731–737

Ritson-Williams R, Arnold SN, Fogarty ND et al (2009) New perspectives on ecological mechanisms affecting coral recruitment on reefs. Smithsonian Contrib Mar Sci 38:437–458

River GF, Edmunds PJ (2001) Mechanisms of interaction between macroalgae and scleractinians on a coral reef in Jamaica. J Exp Mar Biol Ecol 261:159–172

Romano SL (1990) Long-term effects of interspecific aggression on growth of the reef-building corals *Cyphastrea ocellina* (Dana) and *Pocillopora damicornis* (Linnaeus). J Exp Mar Biol Ecol 140:135–146

Rossi S, Snyder MJ (2001) Competition for space among sessile marine invertebrates: changes in HSP70 expression in two pacific cnidarians. Biol Bull 201:308–393

Rutzler K (2002) Impact of crustose clionid sponges on Caribbean reef corals. Acta Geol Hisp 37:61–72

Sammarco PW, Coll JC, La Barre S et al (1983) Competitive strategies of soft corals (Coelenterata: Octocorallia): allelopathic effects on selected scleractinian corals. Coral Reefs 1:173–178

Sammarco PW, Coll JC, La Barre S (1985) Competitive strategies of soft corals (Coelenterata: Octocorallia). 2. Variable defensive responses and susceptibility to scleractinian corals. J Exp Mar Biol Ecol 91:199–215

Schönberg CGL, Wilkinson CR (2001) Induced colonization of corals by a clionid bioeroding sponge. Coral Reefs 20:69–76

Sebens KP (1976) The ecology of Caribbean sea anemones in Panama: utilization of space on a coral reef. In: Mackie GO (ed) Coelenterate ecology and behavior. Plenum Press, New York

Sebens KP, Miles JS (1988) Sweeper tentacles in a gorgonian octocoral: morphological modifications for interference competition. Biol Bull 179:378–387

Segel LA, Ducklow HW (1982) A theoretical investigation into the influence of sublethal stresses on coral-bacterial ecosystem dynamics. Bull Mar Sci 32:919–935

Shenkar N, Bronstein O, Loya Y (2008) Population dynamics of a coral reef ascidian in a deteriorating environment. Mar Ecol Prog Ser 367:163–171

Smith JE, Shaw M, Edwards RA et al (2006) Antifouling activity and microbial diversity of two congeneric sponges *Callyspongia* spp. from Hong Kong and Bahamas. Mar Ecol Prog Ser 324:151–165

Stambler N (2010) Zooxanthellae: the yellow symbionts inside animals. In: Dubinsky Z, Stambler N (eds) Coral reefs: an ecosystem in transition Springer, Doedrecht

Steneck RS, Dethier MN (1994) A functional group approach to the structure of algal dominated communities. Oikos 69:476–498

Sullivan B, Webb L (1983) Siphonodictine, a metabolite of the burrowing sponge *Siphonodictyon* sp. that inhibits coral growth. Science 221:1175–1176

Tanner JE (1995) Competition between scleractinian corals and macroalgae: an experimental investigation of coral growth, survival and reproduction. J Exp Mar Biol Ecol 190:151–168

Tanner JE (1997) Interspecific competition reduces fitness in scleractinian corals. J Exp Mar Biol Ecol 214:19–34

Thompson JE, Barrow KD, Faulkner DJ (1983) Localization of two brominated metabolites, aerothionin and homoaerothionin, in spherulus cells of the marine sponge *Aplysina fistularis* (=*Verongia thiona*). Acta Zoologica 64:199–210

Tilman D (1982) Resource competition and community structure. Princeton University Press, Princeton

Titlyanov EA, Titlyanov TV (2008) Coral–algal competition on damaged reefs. Russ J Mar Biol 34:199–219

Titlyanov EA, Titlyanov TV, Yakovleva IM et al (2007) Interaction between benthic algae (*Lyngbia bouillonii, Dictyota dichotoma*) and a scleractinian coral (*Porites lutea*) in direct contact. J Exp Mar Biol Ecol 342:282–291

Titlyanov EA, Titlyanov TV, Arvedlund M (2009) Finding the winners in competition for substratum between coral polyps and epilithic algae on damaged colonies of the coral *Porites lutea*. Mar Biodiv Rec 2(e85):1–4

Van Veghel MLJ, Cleary DFR, Bak RPM (1996) Interspecific interactions and competitive ability of the polymorphic reef-building coral *Montastrea annularis*. Bull Mar Sci 58:792–803

Vermeij MJA, Smith JE, Smith CM (2009) Survival and settlement success of coral planulae: independent and synergistic effects of macroalgae and microbes. Oecologia 159:325–336

Volterra V (1926) Variations and fluctuations in the numbers of individuals in animal species living together. Reprinted in 1931. In: Chapman RN (ed) Animal ecology. McGraw-Hill, New York

Wild C, Huettel M, Klueter A (2004) Coral mucus functions as an energy carrier and particle trap in the reef ecosystem. Nature 428:66–70

Williams RB (1991) Acrorhagi, catch tentacles and sweeper tentacles: a synopsis of "aggression" of actiniarian and scleractinian Cnidaria. Hydrobiologia 216(217):539–545

Witman JD, Dayton PK (2001) Rocky subtidal communities. In: Bertness MD, Gaines SD, Hay M (eds) Marine community ecology. Sinauer Associates, Sunderland

Work TM, Aeby GS, Maragos JE (2008) Phase shift from a coral to a corallimorph-dominated reef associated with a shipwreck on Palmyra Atoll. PLoS ONE 3(8):e2989:1–5

Zvuloni A, Armoza-Zvuloni R, Loya Y (2008) Structural deformation of branching corals associated with the vermetid gastropod *Dendropoma maxima*. Mar Ecol Prog Ser 363:103–108

Scaling Up Models of the Dynamics of Coral Reef Ecosystems: An Approach for Science-Based Management of Global Change

Jesús Ernesto Arias-González, Craig Johnson, Rob M. Seymour, Pascal Perez, and Porfirio Aliño

Abstract Coral reefs around the world and the populations who directly and indirectly depend on them are facing a multitude of global, regional, and local threats. In face of these unprecedented global changes, it is critical to understand how coral reef ecosystems and the goods and services they provide will evolve. The problem is complex and its solution is difficult because of the nature of biophysical connectivity of coral reef systems, and their connection with human social and economic systems. In order to increase our knowledge and the predictive capacities necessary to determine how coral reef ecosystems will respond to global change, it is necessary to employ a combination of data synthesis and numerical simulation. The increase in the knowledge base and predictive capacities regarding the influence of the drivers: ocean circulation, climate, ocean-acidification, terrestrial run-offs driven by enhanced human activities such as the clearing of native vegetation and its replacement with intensive agriculture and coastal development, pollution, overfishing, and invasive species on marine coral reef ecosystems is central for the effective management and conservation of coral reef ecosystem services. Modeling at multiple scales has revealed to be a vital tool in meeting this challenge by providing important technology that allows managers, other decision makers, and users to see the dynamics of the whole system – including ecological, biophysical, socioeconomic, and restoration aspects. The aim of this review is to provide information concerning the studies of local and regional coral reef ecosystem models, the coupling of ecological and social system models and models based on ecosystems as well as to provide suggestions for future development and use of models for science-based management of global change

Keywords Coral reefs • global change • local and regional models • coupling socio-ecological systems • models based on ecosystems • science-based management

1 Introduction

1.1 The Coral Reef Crisis

Coral reefs around the world and the populations who directly and indirectly depend on them are facing a multitude of global, regional, and local threats, including climate change; pollution from massive tourism, urban development, agriculture, and other sources; overfishing; and sedimentation (Bellwood and Hughes 2001; Hughes et al. 2003; Pandolfi et al. 2003; Hoegh-Guldberg et al. 2007; Carpenter et al. 2008; Sale 2008). The world has already effectively lost 19% of the original area of coral reefs in the past 30–50 years, 15% are seriously threatened and loss of these reefs is anticipated within the next 10–20 years, another 20% are under threat of loss in 20–40 years, while only 46% of the world's reefs are regarded as being relatively healthy and not under any immediate threats of destruction (Wilkinson 2008). In face of this unprecedented global deterioration and mass extinction, it is critical to understand how these ecosystems and the goods and services they provide will evolve. The

J.E. Arias-González (✉)
Laboratorio de Ecología de Ecosistemas de Arrecifes Coralinos
Departamento de Recursos del Mar, Centro de Investigación y de Estudios Avanzados del I.P.N.-Unidad Mérida, Antigua Carretera a Progreso Km 6, A.P. 73 CORDEMEX, 97310, Mérida, Yucatán, USA
e-mail: earias@mda.cinvestav.mx

C. Johnson
Marine and Anarctic Futures Centre,
Institute for Marine & Anarctic Studies, Private Bag 12,
7001 Hobart, TAS, Australia
e-mail: Craig.Johnson@utas.edu.au

R.M. Seymour
Department of Mathematics, University College London, Gower Street, WC1E 6BT, London, UK
e-mail: rms@math.ucl.ac.uk

P. Perez
CSIRO Marine & Atmospheric Research, Cleveland, QLD, Australia
e-mail: pascal.perez@anu.edu.au

P. Aliño
The Marine Science Institute, University of the Philippines, 1101 Diliman, Quezon City, Philippines
e-mail: pmalino2002@yahoo.com

problem is complex and its solution is difficult in part because of the nature of biophysical connectivity of coral reef systems, and their connectedness with human social and economic systems.

Addressing the complexity of coupled coral reef – human systems, including their interconnectedness at multiple scales, is an explicit goal of the Coral Reef Targeted Research (CRTR) project (www.gefcoral.org/Targetedresearch/Modelling). It urges researchers to "attack all problems simultaneously; understand how local problems affect global problems and vice versa; understand how social and economic problems affect biological and physical problems and vice versa; explore the effects of different management strategies on all problems; allow managers to learn and adapt" (Fig. 1).

Modeling emerges as a vital tool in rising to this challenge. It is an important technology that allows managers, other decision makers, and users to see the dynamics of the whole system – including ecological, biophysical and socio-economic, and restoration aspects (Hannon and Ruth 1997; Hall 2000; Hobbs and Suding 2009). Models can provide a means to capture and comprehend the complexity of the coral reef – human phenomenon at multiple scales. For example, they can simultaneously examine impacts of human activity on key trophic, competitive, and mutualistic relationships on coral reefs, and how changes in these biophysical characteristics feedback to influence human behavior. They can aid in the understanding of major shifts in the state of reef systems and the reversibility of these transitions (Mumby et al. 2007; Fung et al. in press). In particular, they can help to identify possible reef futures in any given context, and so can provide invaluable support to decision makers (Wooldridge et al. 2005; Melbourne-Thomas et al. 2011, in press). Ongoing development of modeling resources will enable reef managers to develop scenarios for their own areas, better understand the links between local, regional, and global processes, and facilitate access to relevant scientific and economic data over the Internet (i.e., www.gefcoral.org/Targetedresearch/Modelling).

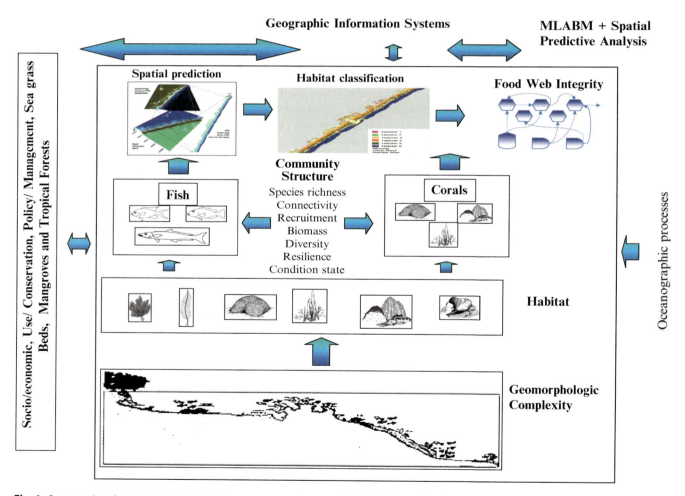

Fig. 1 Integrated reef ecosystem analysis with a management approach. MLABM : multilevel agent-based models (Modified from Arias et al. 2000)

1.2 The Rise of Modeling

The proceedings of the International Coral Reef Symposia give an idea of the increasing importance of modeling in coral reef research. Proceedings of the first eight ICRS conferences each contained between one and six papers focusing on modeling, while in the 11th ICRS (2008), there were ~50 papers with a primary focus on models. The increase in using models to better understand patterns and processes in coral reefs has considerably increased in the last decade.

Developments in software and hardware ICT technologies have facilitated integration of multiple processes in, and accessibility of, coral reef models. High-performance computing has made it possible to explore the consequences of relatively simple interactions in the context of complex emergent behaviors that can now be more easily understood. Improved capacity in areas of data availability (e.g. ReefBase, FishBase, SeaAroundUs, OBIS), remote sensing, information processing (i.e., http://typhon.rsmas.miami.edu/GIS), storing, and visualization have permitted the development of more complex and sophisticated models providing insights from local to global perspectives (Wooldridge et al. 2005; Cowen et al. 2006; Paris et al. 2007; Melbourne-Thomas et al. 2011, in press). These models enhance the capacity to realize spatially explicit prediction and provide a heuristic opportunity for learning while undertaking model formulation and applications for day-to-day decisions in reef management.

1.3 The Challenges of Complexity and Complicatedness

While modeling is a powerful tool for studying reef dynamics and making predictions useful to managers (McCook 2001; Lirman 2003), the complexity (sensu Miller and Page 2007) in behavior of reef systems nonetheless presents a challenge to modelers. Particular issues include that while a system will have inherent dynamic stabilities, it may also show signs of marked instability and thus demonstrate sudden phase shifts, metastable states, and hysteresis, and a system's trajectory may be inherently unpredictable.

Because of this inherent complexity, it is not possible to simultaneously optimize the *generality*, *realism*, and *precision* of a model (Levins 1970). Indeed, in practice, we are doing well if we can optimize any two of these. Generality means that a model created for a reef system in one region could be cloned readily for another. Realism refers to capturing specific features that are characteristic of coral reefs, while precision means that the detailed dynamics of individual components are faithfully captured by the model. A crucial decision is which of this triplet of characteristics to emphasize in constructing a model.

Achieving optimum complexity (*sensu* complicatedness) in building a model is another challenge (Fulton et al. 2003). Modelers typically use a strategy of "judicious simplification" of the real world to gain conceptual understanding (at some level of granularity) without trying to account for every twitch and flutter of the real-world output. The disadvantage here is that the modeler must often make ad hoc judgments in selecting the relevant key players and their interactions – it is not known, a priori, just how significant each "twitch and flutter" is. Such a strategy usually aims to preserve generality and realism at the expense of precision. Even when these decisions are made, modeling key processes that occur over various temporal and spatial scales is an inevitable challenge.

A related concern is that the complexity of the phenomena can get in the way of the models being properly verified and calibrated, or tested for sensitivity. When there are many (say, hundreds) of variables in a model, it is not possible to undertake the type of painstaking parameter sweep of each variable in combination with other variables in validating and verifying models that can be done with simpler models.

The intention of this review is to provide information about models of local and regional coral reef ecosystems and coupling of ecological and social system models, as well as to provide suggestions for future development and use of models for science-based management of global change.

2 Modeling Local-Scale Dynamics

Models of coral reef systems are mostly been developed and deployed on a *local* scale of the order 1–5 km, covering scenarios in which the modeled benthic habitat is relatively homogeneous, and can be conceived as dominated by interactions among a relatively small number of functional groups. At this scale, for example, it is usual to model habitats that are dominated by interactions between corals and algae separately from habitats dominated by sea grass. If fish groups are included, the major players are typically largely territorial within the modeled habitat.

However, it must be acknowledged that local habitats exist within a larger context, and are therefore mostly open systems. In particular, they are subject to both endogenous and exogenous recruitment. Since recruitment processes are poorly understood, this presents a challenge. Generally, it is assumed that a local model exists within a regional pool from which recruits are taken, i.e., within a metapopulation (Hanski and Gyllenberg 1993). Other issues that local models have been used to address, either individually or in combination, include:

- Coral-algae phase shifts in terms of competition for space
- The stability of coral-dominated and algal-dominated states
- Mechanisms to induce phase shifts in either direction, including external forcing and internal interactions
- Interactions between algae and fish, captured as "standard" plant-herbivore ecological dynamics
- Interactions between piscivorous and herbivorous fish, captured as "standard" prey-predator ecological dynamics
- Multiple interactions among a suite of functional groups, described in terms of standard prey-predator ecological interactions
- Effects of predation on other consumer interactions (herbivorous, piscivorous, carnivorous), and the consequences for reef dynamics

In this section, we will focus on models of benthic phase shifts between alternative stable states.

2.1 The Need for Local Models to Include Phase Shifts

Many coral reefs are being degraded rapidly. This can lead to reefs undergoing a "phase shift" – i.e., a rapid change (on ecological timescales) – from a coral- to an algal-dominated state, presenting a particular challenge for management (Fig. 2). Phase shifts to algal dominance typically result in significant loss of biodiversity and loss in ecosystem functioning with a concomitant decrease in the quality and/or quantity of ecosystem goods and services including, for example, lower fish stocks and tourism appeal (Done 1992; Hughes 1994; Costanza et al. 1997; McCook 1999; Hughes et al. 2003; Pandolfi et al. 2003; Bellwood et al. 2004). A distinction can be made between "continuous" phase shifts, which do not exhibit hysteresis, and "discontinuous" phase shifts, which do (Fig. 2). The distinction is critical because hysteresis is particularly problematical for management. Hysteresis arises as a region of multiple equilibria so that there is a range of conditions in which the system can exist in either of the two alternative stable states (Scheffer et al. 2001; Beisner et al. 2003; Collie et al. 2004), i.e., coral-dominated or algal-dominated. Thus, recovery of corals from the degraded algal-dominated state requires the causative agent(s) to be reduced to much lower levels than the threshold that triggered the shift in the first place.

It is clear that, at least for some reefs and reef systems, a challenge for managers is to clarify the underlying causes of phase shifts so that appropriate management options can be identified. Modeling is a valuable tool to help untangle the underlying mechanisms. However, many local models do not address the issues or mechanisms of phase shifts directly. For example, some models focus only on coral dynamics (e.g., Maguire and Porter 1977; Hughes 1984; Hughes and Tanner 2000; Lirman 2003; Langmead and Sheppard 2004; Wakeford et al. 2008). While these approaches can be useful in understanding changes in coral cover and, in some cases, changes to the resilience of reefs in the face of multiple stressors (e.g., Wakeford et al. 2008), they are not useful for investigating coral–algal interactions and the importance of algal grazing in any dynamic sense.

Models with explicit coral-algal dynamics have been more useful in evaluating the effects of anthropogenic stressors. For example, McClanahan (1995) used a mean-field model based on ordinary differential equations to show that fishing both herbivores and piscivores gave the highest maximum yield to Kenyan fishers. The model of McCook (2001)

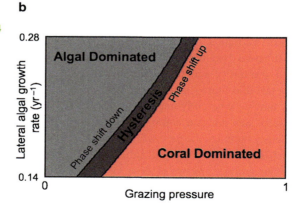

Fig. 2 Phase shift and hysteresis from reduced grazing and/or increased growth of macroalgae (through increased nutrients). (**a**) Discontinuous phase shifts with resultant hysteresis can arise either through reducing grazer biomass (*blue*) when there are no added nutrients, or increasing algal growth rate (by increasing nutrients) (*red*) when there is no reduction of herbivores (e.g., by fishing). (**b**) Combinations of algal growth rate (reflecting the amount of added nutrients) and abundances of grazers, which give rise to coral- and algal-dominated equilibria, and the hysteresis region where multiple stable equilibria occur

highlighted that long-term nutrification and sedimentation can result in phase shifts when combined with natural stressors. Mumby et al. (2006) constructed a stochastic cellular automaton model to explore the effects of different levels of grazing, hurricane disturbance, and nutrification on Caribbean reefs, and concluded that the presence of the sea urchin *Diadema antillarum* increased reef resilience to phase shifts. Further investigation of this model established the sensitivity of coral cover to algal grazing by scarid fish (Mumby, 2006). Mumby et al. (2007) investigated both phase shifts and hysteresis following reduction in grazing pressure using both a simple differential equation model and a stochastic cellular automaton model of a generalized Caribbean reef.

The models of Mumby et al. (2006, 2007) are important because they demonstrate the susceptibility of Caribbean reefs to discontinuous phase shifts as grazing pressure is reduced. The implications for management are clear; as coral cover declines in the hysteresis region, the level of grazing required to effect recovery of corals increases. However, management needs to identify the full range of drivers of phase shifts and hysteresis. Phase shift driven largely by "top-down" processes (removal of grazers) is well accepted, but it is also possible that "bottom-up" mechanisms, i.e., stimulation of algal growth by added nutrients, play an important role, although this is more contentious (McManus and Polsenberg 2004; Burkepile and Hay 2006). In particular, it is unclear how top-down and bottom-up processes might interact. For example, Pandolfi et al. (2003) suggest that the synergy between pollution and overfishing was partly responsible for historical coral reef decline. Similarly, McCook (1999) suggests that synergy exists between overfishing and nutrification in coral reef decline, while Knowlton (2001) highlights possible synergistic effects between sedimentation and high nutrient levels. Knowlton (2001) further observes that threshold responses are more likely if synergistic effects arise. Similarly, McClanahan et al. (2002) suggest that synergistic effects are often the cause of permanent ecological transitions on reefs, and McCook (1999) cites several studies with empirical evidence of synergy between acute disturbances and nutrification or overfishing in causing coral decline. It is clearly critical that models allow for the possibility of synergistic interactions, but it is vastly preferable that model structure allows for synergistic effects to arise as an emergent feature rather than coding them directly as "hard-wired" effects.

3 Connecting Processes at Local and Regional Scales

Nevertheless, the relative importance of the various anthropogenic stressors in relation to the existence of alternative stable states, and resultant phase shifts remains unresolved, both theoretically and empirically. The Modeling and Decision Support Working Group (MDSWG) of the CRTR project has developed a suite of mean-field biophysical models of varying complexity of a coral-algal system at a local scale that are also designed to function as building blocks for larger, regional-scale models. One of these models, focusing on benthic dynamics, has been parameterized generically using data from a range of reefs worldwide, and used to investigate the roles of, and possible synergies between, grazing, nutrification, and sedimentation in the induction of phase shifts on the benthos (Fung 2009; Fung et al. in press).

The local-scale model, representing behavior on patches of reef tens to hundreds of square meters, is nonspatial and uses ordinary differential equations to represent the interaction dynamics of benthic components represented as three functional groups:

- *Corals*, referring to live, hermatypic scleractinian corals, which can be differentiated as broadcast spawning and brooding
- *Algae*, referring to patches of fast growing "turf algae" (say~2–4 mm high), consisting mainly of simple filamentous green, red, and blue-green algae, but which may also include algal crusts and unicells
- *Macroalgae*, referring to all functional groups of algae apart from turfing, filamentous algae (included as the turf algae functional group), and microalgae and crustose coralline algae (included in the "space" category – see below)

These groups compete for *space*, which refers to everything else. Space need not be empty, but is generally inhabited by a variety of sponges, clams, molluscs, ascidians, bryozoans, and, in particular, nongeniculate coralline algae and dead coral skeletons. But the most important feature of *space* is that it is covered by fine algal turfs (microturfs, <~2–4 mm high). The low canopy height of these fine turfs is maintained by grazing, mostly by herbivorous fish, but also by sea urchins, which crop it as fast as it can grow. Examples of *macroalgae* include *Sargassum* spp., which are abundant on inshore reefs (Jompa and McCook 2002), *Lobophora variegata*, which is common on degraded Caribbean reefs (Jompa and McCook 2002) and the Great Barrier Reef (Diaz-Pulido and McCook 2004), and *Dictyota* spp., which are common on Caribbean reefs (McCook 2001).

In the stand-alone benthic model, grazing pressure on the algal groups is represented by an exogenously determined parameter. However, the benthic model can be extended by including fish and urchin functional groups, thereby rendering grazing pressure a dynamic process, which can respond, for example, to fishing pressure. Thus, grazing pressure can be represented as a function of grazer biomass, a dynamic variable. *Fish* refers to those fishes that are associated with reef habitat postsettlement (e.g., Chaetodontidae, Acanthuridae, Scaridae, Labridae, Serranidae, Haemulidae, Lutjanidae). In addition to *exogenous* recruitment, fishes are assumed to recruit *endogenously*, to take account of the process whereby eggs produced by the endogenous fish population hatch and are retained in their pelagic stage in a nearby area, with larvae subsequently recruiting back to the same reef area as their parents. Though the majority of reef fishes are broadcast spawners, endogenous recruitment can also occur when brooding species lay eggs and take care of them until hatching. The model considers three fish groups in order to attain reasonable trophic detail (Fung 2009). In addition, sea urchin (e.g., *Diadema antillarum*) dynamics can be included, as they are important grazers on some coral reefs. The *fish* groups are:

- Herbivorous fish – principally Scaridae, Acanthuridae, Pomacentridae, Siganidae
- Small, carnivorous fish – e.g., small Labridae, juveniles of large carnivorous species
- Large, carnivorous fish – e.g., large Labridae, Serranidae

Recruitment is to the herbivore and the small carnivore functional groups, the latter of which then have the potential to grow into large carnivores.

The influence of anthropogenic stressors in relation to phase shifts can be investigated using this model – either using the reduced benthic model (with a "passive" grazing pressure parameter, as in Fung et al. (in press), or with the fully coupled fish-benthic system model containing all nine functional groups (Fung 2009). Fishing of herbivores can be represented through reduced grazing pressure, and nutrification through increased algal growth rates. Sedimentation acts principally on coral characteristics, through a decrease in settlement and growth rates, and an increase in natural mortality. The principal conclusions from an analysis of this model are that the presence of significant macroalgae is necessary for discontinuous phase shifts to be possible; only the less dramatic continuous phase shifts appear to be possible without macroalgae. That the presence of macroalgae can result in discontinuous phase shifts supports the results of Mumby et al. (2007). Further, though each of the three stressors (reduced grazing pressure, nutrification, and sedimentation) is capable of inducing a phase shift on its own, there are significant synergistic effects between them. This is illustrated for overfishing and nutrification in Fig. 2. This means that management strategies for recovery of reefs with macroalgae need to take into account possible hysteresis effects. In addition, managing nutrients could increase the resilience of reefs to fishing, since added nutrients can make reefs more prone to fishing-induced discontinuous phase shifts and an increase in the size of the corresponding hysteresis region. This applies to many reefs in the Caribbean and Indo-Pacific reefs associated with significant land mass. Where a discontinuous phase shift does arise then, by definition, managers must reduce the magnitude of a stressor (fishing, nutrient, or sediment levels) to a level much less than the threshold at which the phase shift occurred in the first place. Thus, recovery targets and actions should not be based on these thresholds alone.

Since management is usually effected at regional scales, it follows that, if models are to be useful in the role of decision support for managers, at least some of them must capture dynamics at regional scales. To be most useful in a management context, models at a regional scale will attempt to capture the essential features of corals reefs as complex systems and, by parameterizing them for particular areas, enable some minimum level of both realism and precision. Ecologists have long recognized that the dynamics of coral reefs are influenced by a suite of interacting processes that operate over multiple scales in space and time (e.g., Hatcher et al. 1987; Hatcher 1997), and so the goal of most models that include a regional focus is to capture key processes across a range of scales from local to global. It is also essential to capture the emergent dynamics that arise from complex interactions among many elements, and for this reason it can be advantageous that regional models have an explicitly spatial component.

The regional model developed by the MDSWG (called CORSET) is spatially explicit, and divides a region into a large but finite number of cells, typically of dimensions 0.5–2 km per side (Melbourne-Thomas et al. 2011, in press). The local model of Fung (2009) runs independently in each cell characterized as a reef, and reef cells are connected by transport of larvae, sediment, and other pollutants. Connectivity of each reef cell with every other reef cell in the system is described by connectivity matrices, and there are separate matrices to describe connectivities for broadcast spawning corals, sea urchins, and fishes. The regional model demonstrates emergent subregional dynamics well in accordance with empirical observation (Melbourne-Thomas et al. 2011, in press).

Because of the complexity of the myriad processes that operate across local–regional scales, most models to date address only a particular subset of processes. Early heuristic models showed that recovery of reefs following coral decline depended on patterns of connectivity, the magnitude of self-seeding to natal reefs, life-history features, background levels of mortality and, at a system-wide scale, patterns of variance in reef state (e.g., Johnson and Preece 1993; Preece and Johnson 1993). Moreover, these simple models demonstrated strong interactions among these factors; the sensitivity of the system to one parameter depended on the levels of the others. More recently, a raft of models focused on specific regions have demonstrated the importance of connectivity among reefs to the maintenance of populations and communities. While these models suggest that some generalizations may be possible, e.g., that most reefs require some kind of external subsidy through larval input, in particular they serve to emphasize that because of particular current patterns and the layout of reefs in a region, individual reefs show great variation in their importance as sources or "sinks" of larvae (Bode et al. 2006; Cowen et al. 2000, 2006; Paris et al. 2007; Treml et al. 2008). Of great significance is the evidence from several models that larvae do not disperse as passive particles, and that larval dispersal patterns are as much influenced by the behavior of the larvae as the physics of water movement (e.g., Cowen et al. 2006; Paris et al. 2007). Importantly, these kinds of results are supported by empirical work (e.g., Jones et al. 1999; Stobutzki 2001; Swearer et al. 2002).

While spatially explicit models of metapopulation dynamics at a regional level have been employed to evaluate possible alternative management strategies in a fisheries context (e.g., Little et al. 2007), most regional-scale models have not attempted to model community dynamics and/or enable evaluation of strategies to manage the system as a whole. Mumby (2006) developed a useful spatially explicit model of community dynamics on a Caribbean reef, but his extension to multiple reefs was an abstract heuristic exercise and not intended to model any particular region. Spatially explicit deterministic biogeochemical "whole of ecosystem" models that incorporate physics show some promise to assist with management decisions (e.g., Fulton et al. 2003), but large instantiations of this kind of approach can be very difficult to parameterize and lack the flexibility to capture all the processes affecting reef dynamics that managers may wish to consider.

To adequately capture biophysical dynamics and the dynamics of humans interacting with coral reefs in a socioeconomic and management setting, and to deliver capacity for management strategy evaluation (MSE), agent-based models arguably show greatest promise. Depending on the availability of data to make them meaningful to particular settings, these kinds of models can be large and complex as with the In Vitro agent-based model developed for the tropical North-West Shelf of Australia (Gray et al. 2006). However, the complexity of In Vitro makes it difficult to apply to other systems without considerable tailoring of the code. The MDSWG have chosen a middle ground in building their regional model, developing an agent-based framework that is sufficiently generic to capture the essential local dynamics of coral reefs anywhere in the world, and to have reefs interacting at regional scales through transport of larvae, sediment, and other pollutants.

Parameterizations for the Mesoamerican Barrier Reef (Melbourne-Thomas et al. in press), and extensive tracts of reefs in the Philippines in the South China Sea (Melbourne-Thomas et al. 2011) are highly encouraging in that they validate well for the biophysical component (Fig. 3). The model is currently being expanded to include

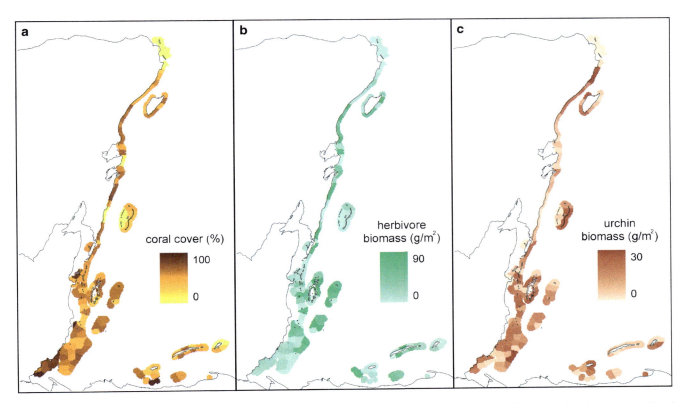

Fig. 3 Spatial variation in modeled coral cover (**a**), herbivorous fish biomass (**b**), and sea urchin biomass (**c**) across the MAR for the "healthy reef" scenario. For visualization purposes, values displayed are averages across cells within polygons at the final timestep of a 100-year model run. Spatial variability within the region is pronounced and emerges as a result of variability in larval supply, and how this affects "downstream" dynamics directly and indirectly (After Melbourne Thomas et al. in press)

a generic socioeconomic model. Notwithstanding the need for integration of the socioeconomic elements, the validated biophysical model is directly applicable to decision-support and management strategy evaluation through testing of specified management (and socioeconomic) scenarios.

4 Coupling of Ecological and Social Systems

Adams et al. (2004) argue that while biodiversity loss and poverty are linked problems, integrated policies targeting both poverty alleviation and biodiversity conservation together are in need of robust conceptual frameworks in order to be effective. Fortunately, over the last decade, a growing number of resource managers and researchers from diverse disciplines have embraced a complex systems approach to link ecological resilience to governance structures, economics, and society (Hughes et al. 2005). In their book "Exploring Resilience in Socio-Ecological Systems," Walker et al. (2006) lay the theoretical foundation for an integrative analytical framework. Lachica-Aliño et al. (2006) provide a remarkable account of the influence of these innovative ideas onto successive modeling attempts of fisheries in the Philippines. But the quest for better-integrated socioecological models creates its own impediments, as established earlier by Villa and Constanza (2000).

A report from the World Bank (2006) on coral reef ecosystem health states that useful environmental indicators must pass the test of: (i) relevance, (ii) availability of data, (iii) scientific soundness, (iv) management responsiveness, and (v) communicability (transparency). Likewise, models should meet these same requirements, especially when they aim to simulate socioeconomic aspects influencing coral reef health. While researchers have developed and refined local and regional ecological models (see sections above), it remains that local and regional socioeconomic models are much less well documented. Beyond fishing, several other industries and activities often dramatically influence coral reef health. For example, direct threats of tourism on coral reefs include exploitation (for building material), pollution (waste treatment), and degradation (carrying capacity), while indirect threats include increasing demand on reef products (seafood, material for jewelry) and pollution associated with urban sprawl around resort areas.

Unlike biophysical modeling of coral reefs for which experimental evidence and validating data sets remain limiting factors in the development of meaningful models, research on socioeconomics of coastal management has developed several standardized assessment frameworks like the SocMon framework (Bunce et al. 2000), the MPA evaluation framework (Pomeroy et al. 2004), and the EcoServices framework (Bulte et al. 2005). However, thus far very few examples of an evidence-based and integrative model for coral reefs exist (Lutz et al. 2000; McCausland et al. 2006; Gray et al. 2006). One reason, among others, is in the contextual nature of most of the socioeconomic interactions that are at stake. Unlike ecological processes that can be described by a set of generic equations for very different reefs around the world, socioeconomic processes often elude such an approach. For example, fishermen in Mexico and in the Philippines might be very different social entities, with unrelated decision-making processes and dissimilar sets of values and beliefs. Dambacher et al. (2007) provide a compelling example of such contextual intricacies with their modeling of Lihir island's coastal communities.

Due to the contextual nature of socioeconomic processes and, more importantly, to the overall objective of integrated models to better inform policies and contribute to poverty alleviation, the legitimacy of these models is of paramount importance (Perez 2008). These models are not neutral in that they carry social and economic views of their creators. As stated by Lynam et al. (2007), the most dangerous fate awaiting the so-called integrated models is to become a means to their own sake. In order to avoid such a predicament, a growing number of modelers advocate more holistic approaches to modeling. These approaches acknowledge the necessity for a greater transparency in the designing process. They often involve a transdisciplinary team of experts who contribute to the model by providing key drivers of the system, their operational scales, and related data sets where they exist. This information is then compiled and transformed into a preliminary intermediate object – a prototype – in order to structure and enhance further development. Altogether, the model aims to support a common semantic, to develop a collective metaphor, and to outline actual data limitations. Then, this prototype is turned into an interactive platform in order to engage with diverse groups of stakeholders who become model designers themselves (Barreteau et al. 2007). As stated by Gurung et al. (2006), even very simple models with a limited degree of realism can be very efficient at facilitating understanding, communication, and negotiation. This holistic approach differs from more traditional modeling by involving experts and stakeholders in the design process itself.

SimReef (Perez et al. 2009) was developed in the MDSWG according to these principles. The model describes the social, economic, and ecological evolution of Quintana Roo's coastline (Mexico). Successive causal

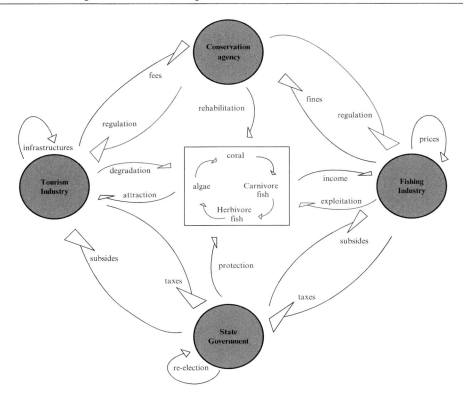

Fig. 4 Causal network of the coral reef–human ecosystem in Quintana Roo (Mexico). Arrows indicate directional relationships. Gray circles represent major socioeconomic drivers (After Perez et al. 2009. Reproduced by permission of Modelling and Simulation Society of Australia and New Zealand Inc.)

loop diagrams are used to collectively and iteratively build a holistic model among experts. The model includes major mechanisms through which regional drivers (tourism development, taxation, subsidies) influence local social and economic dynamics (immigration, fishing effort, pollution) and, ultimately, interactions at the level of coral reefs (Fig. 4; note that for readability, not all relevant relationships are shown).

This part of the Mesoamerican barrier reef is typical of a multilevel socioecological system (Hughes et al. 2005; World Bank 2006) where observed environmental harm concentrates on limited fragile areas, while primary social drivers (point-source pollution, reef exploitation, and management) are strongly conditioned by regional drivers of much larger magnitude. For example, the southern coast of Yucatan, known as Costa Maya, currently supports 5,000 people who rely mainly on traditional fishing activities, while the current government development plan for the region forecasts an increase of 175,000 people through migration by 2020 (Daltabuit et al. 2006). However, several key relationships in this diagram are not supported by robust quantitative evidence. For example, little is known yet about direct impact of coastal urbanization on the health of adjacent reefs, due to the complex hydro-geological processes in the karstic underground. Likewise, the relationship between regional growth of tourism and its local economic benefits is highly ambiguous (Juárez 2002). In addition, political and administrative decisions related to real-estate development are often prone to hidden and intense pressure from powerful lobbying groups.

These limitations had two interconnected consequences. First, it was decided to opt for an agent-based modeling structure with rule-based relationships in order to offer as much flexibility as possible to the design process (Shafer 2007). Second, the computer model, i.e., SimReef, was used as an interactive platform in order to engage with stakeholders in a participatory modeling approach (Barreteau et al. 2007; Lynam et al. 2007). Thus, throughout the design process, a careful balance was struck between complexity (i.e., greater realism) and the need for interactivity, keeping in mind that the model acts as a catalyst for discussion and the creation of new understanding of the real system (Fig. 5).

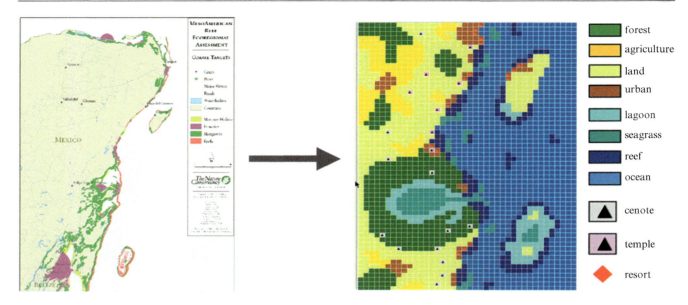

Fig. 5 SimReef's visual representation of Quintana Roo's coastline based on a 60 × 40 raster grid, compared with a standard map of the region (courtesy of The Nature Conservancy) (After Perez et al. 2009. Reproduced by permission of Modelling and Simulation Society of Australia and New Zealand Inc.)

5 Ecosystem-Based Models

From a management perspective, it is also important to understand how an ecosystem's functioning, and the goods and services it provides, will change. One of the ways for understanding the mechanisms of functioning of coral reef ecosystems is the generalization at the level of the models of trophic or energy relationships between its component communities, representing different links in the food webs (Sorokin 1993) (Figs. 6 and 7). Odum and Odum (1955) were the first to create such models by constructing a biomass pyramid of the components of a reef community at the Eniwetok atoll. From this classic work emerged the concept that coral reefs are mature systems with multiple sources of primary production and complex self-regulated food webs. Later, other models have exposed the relative functional importance of different components, such as the central role of microbial populations in the energy interconnection between separate reef communities (Sorokin 1973; but see Johnson et al. 1995). These early models were based in a bottom-up approaches where the basis of energy metabolism was the detritus pool (Sorokin 1993). Other approaches such as the top-down ECOPATH model of Polovina (1984) used functional groups of organisms as the systems components where bacterioplankton, microflora, detritus, and DOM (dissolved organic matter) were totally excluded from the food web. Later models based on network analysis began to integrate detritus, DOM, particle organic matter (POM), microflora and multiple sources of primary producers and bacteria (Aliño et al 1993; Johnson et al. 1995; Opitz 1996; Arias-González et al. 1997; Niquil et al. 1999) into the food web.

5.1 Bottom-Up Effects on Ecosystem Function

Not surprisingly perhaps, several studies based on network analysis models have identified several functional responses associated with changes in the cover of coral and algae. Modeling a steady-state network carbon exchange among 19 trophic compartments for the shallow front slope of a mid-shelf reef in the Great Barrier Reef (GBR), Johnson et al. (1995) examined the effect of the replacement of live coral cover by epilithic algae on patterns and magnitudes of carbon flux. These authors indicated that "the shift in structure to an algae-dominated system increases the proportion of the total flow that is recycled and transferred to the detritus pool (although the structure of recycling is not affected), and the balance of pathways in the network is changed: average path length increases, while the average trophic level of most of the second order consumers, and trophic efficiencies of most trophic categories, decreases. Also, there are marked changes in dependencies of particular trophic groups on others. The analysis shows that, in the coral-dominated state, carbon fixed by zooxanthellae is used indirectly by most organisms in the system, even those seemingly remotely connected."

A similar analysis carried out by Arias-Gonzalez (1993) and Arias-Gonzalez et al. (1997) in French Polynesia on fringing and barrier reef zones also suggested a change in ecosystem functioning with a change in coral cover. Relative to sites with low coral cover, in coral-dominated areas there is an increase in the number of fish functional groups, more of the primary production is processed directly by herbivores, the trophic structure more efficiently conserves energy

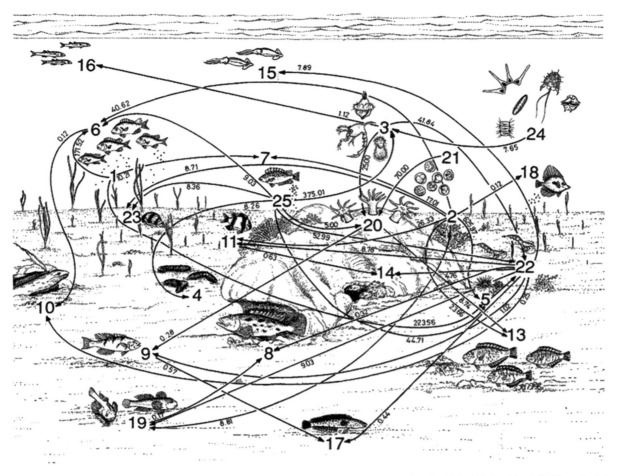

Fig. 6 Illustration of trophic interactions in one of the early Ecopath models showing major trophic flows (gm-2year-1) of the Bolinao reef flat ecosystem, Philippines (Alino et al. 1993). Trophic group number guide: (1) Seagrasses; (2) Seaweeds; (3) Zooplankton; (4) Sea cucumber; (5) Sea urchins; (6) Siganus fuscescens; (7) Siganus spinus; (8) Groupers; (9) Wrasses; (10) Moray; (11) Damselfishes; (12) Gobies; (13) Parrotfish; (14) Cardinalfish; (15) Squids; (16) Other planktivorous fish; (17) Other piscivorous fish; (18) Other herbivorous fish; (19) Other omnivorous fish; (20) Sessile invertebrate consumer; (21) Sessile invertebrate producers; (22) Other invertebrates; (23) Crustaceans; (24) Molluscs; (25) Phytoplankton

and materials, the absolute food intake of important functional groups increases (piscivorous, carnivorous, herbivorous fishes, and heterotrophic benthos), and there is a decrease in bacteria associated with sediments.

Even a small change in coral cover can produce important cascades in trophic structure and alters the cycle of organic material. A recent study on uninhabited atolls in the northern Line Islands (Sandin et al. 2008) shows empirical evidence of the trends suggested by the models developed by Johnson et al. (1995) and Arias-González et al. (1997). Sandin et al. (2008) indicated that apex predators and reef-building organisms dominated unpopulated Kingman and Palmyra, while small planktivorous fishes and fleshy algae dominated the populated atolls of Tabuaeran and Kiritimati. This is important because it shows that a shift in benthic structure can produce a concomitant shift up the trophic web. In the case of the models presented for French Polynesia by Arias-González et al. (1997), the number of trophic groups declined with a reduction in live coral cover, resulting in less complex and less efficient trophic webs.

These trends can affect humans who directly benefit from coral reefs. A small reduction in coral cover may affect services such as potential fishing production. A mass balance simulation based on wet weight for the Coral Triangle, Indonesia, has helped to identify that removal of refuge space is as harmful to juvenile reef fish populations as is direct mortality due to the fishery itself in that reef damage is cumulative due to the slow regrowth rate of corals (Ainsworth et al. 2008).

While these studies provide some indications, in general very little is known about the functional bottom-up mechanisms that underpin key elements of ecosystem functioning such as production of autotrophs and high-order consumers. Modeling the effect of reef degradation on patterns of trophic structure and flow is important if we are to properly understand the consequences of major shifts in the structure of reef systems, whether anthropogencially driven or not.

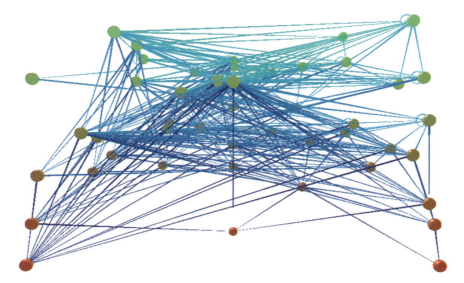

Fig. 7 Illustration of trophic interactions in one of the early Ecopath of the Moorea Island barrier reef complex, Indo-Pacific (Arias-González 1993). The web structure in the image is organized vertically, with node color representing trophic level. Red nodes represent basal species, such as primary producers and detritus, orange and brown nodes represent intermediate species, and green nodes represent top species or primary predators. Links characterize the interaction between two nodes, and the width of the link attenuates down the trophic cascade (i.e., a link is thicker at the predator end and thinner at the prey end). Image produced with FoodWeb3D, written by R. J. Williams and provided by the Pacific Ecoinformatics and Computational Ecology Lab (www.foodwebs.org, Yoon et al. 2004)

5.2 *Top-Down Effects on Ecosystem Function*

Compared to bottom-up effects, there are more studies of top-down effects on coral reefs. Profound indirect effects of overfishing have been suggested for coral reefs (Jackson et al. 2001; Pandolfi et al. 2003; Bellwood et al. 2004), and it seems that in a range of marine ecosystems the effect of fisheries typically extends well beyond the collapse of exploited fish stocks (Scheffer et al. 2005). A particular example is the role of fishing in the collapse of Caribbean coral reef ecosystems (Hughes 1994; Bellwood et al. 2004). The depletion of herbivorous fish in Jamaica apparently left sea urchins as the only grazer to control macroalgae (Hay 1984). When disease affected the sea urchins during the early 1980s, there was rapid overgrowth of corals by brown fleshy algae, inducing a change in the reef structure. Fisheries may produce notable impacts on coral reefs where predation is a major force regulating primary and/or secondary consumers, particularly if there is nutrient enrichment (Polovina 1984; Atkinson and Grigg 1984; Grigg et al. 1984).

Different impacts of fishing have been estimated depending on the type and intensity of fisheries on reefs. Some models indicate that despite high primary productivity, fishery yields are relatively low, apparently because of high levels of internal predation and low ecotrophic efficiencies at lower trophic levels (Polovina 1984; Opitz 1996; Arias-González et al. 2004; Gribble 2003; Wantiez and Chauvet 2003; Bozec et al. 2004; Tsehaye and Nagelkerke 2008). Polovina (1984) suggests that this is also a reason why there can be substantial variations in fishery yields depending on the harvest strategy. For example, in an energy-based simulation of Caribbean reefs, removing all fish results in a rapid and dramatic drop in fisheries yields; reduces algal abundance, coral biomass, and productivity; and realizes increased dominance by sea urchins once their predators are removed (McClanahan 1995). Fishing piscivores alone also results in low fisheries yields, but high reef accretion by indirectly releasing coral from competition with algae (McClanahan 1995). Similarly, based on a mass balance simulation model, Ainsworth et al. (2008) suggested that limiting commercial fishing of groupers in artisanal fisheries would slightly benefit overall catch rates, but that the improvement would be lost if the fisheries increased effort to compensate for missing catch of groupers. They suggested that limiting commercial net fisheries for reef fish would see a marginal increase in reef fish biomass and an unexpected benefit to large pelagic species due to reduced interception of their anchovy prey.

By performing a retrospective simulation of trends in Eritrea fisheries, Tsehaye and Nagelkerke (2008) found that a decline in fisheries yield was possible if the fishing level was fivefold the official estimate. Fishing only piscivores and herbivorous fishes results in the highest and most stable fisheries yields, but high levels of fishing can result in increased algae that competitively exclude coral, producing a reduction in calcium carbonate deposition (McClanahan 1995).

For many reefs there is evidence of fishing down the food web (Waddell 2005; Wilkinson 2008), which is a compelling reason to understand the potential impact of fishing on ecosystem functioning. Arias-González et al. (2004) examined the potential effect of fisheries on the functioning of

Caribbean coral reef ecosystems. Their mass balance models suggested that over-exploited fisheries on coral reefs could decrease by 20 fold the primary production required to sustain fish yield compared to partially protected reefs.

6 Conclusions and Outlook

There is a raft of different types of ecosystem models – a 'tool box' of models – that are potentially useful in managing coral reefs. These different models, at local and regional scales, have helped to advance understanding of coral reef ecosystems by including key processes operating across a range of scales in time and space. These efforts contribute to the growing body of knowledge of the dynamics of coral reef ecosystems and to ecosystem based management. They open new possibilities to support decision making in managing coral reefs at a range of scales because they can comprehensively include:

- competition, predation and other ecological interaction
- biogeochemical processes
- dispersal processes
- accretion and erosion processes
- effects of climate change
- impacts of human populations on coral reefs, e.g. fisheries and pollution
- ecosystem services to human populations
- patterns of biodiversity, including dynamics

Key areas to include in a modeling 'tool box' likely to be optimally useful in decision support include:

- integration of ecosystem, social and economic approaches in a spatially explicit framework across different time scales
- inclusion of dispersal and connectivity either directly or indirectly using 3D hydrodynamic models
- visualization of model behaviors and products to build a broad consensus among scientists, managers, politicians and other stakeholders

To improve predictive capacity to better resolve how coral reef ecosystems will respond to climate change and other anthropogenic drivers, it is necessary to employ a combination of data synthesis and numerical simulation. Better knowledge of effects of ocean circulation, climate, ocean acidification, terrestrial run-off stimulated by human activity (such as clearing native vegetation and its replacement with intensive agriculture), coastal development, pollution, overfishing and invasive species on coral reef ecosystems will help to improve predictive capacity, and ultimately underpins good management and conservation of coral reef ecosystems.

7 Summary

The imperatives brought about by the coral reef crisis focuses managers and modelers alike on the effects of overexploitation, degradation, pollution, and loss of ecosystem goods and services. These problems are complex and their solution is therefore challenging due to both the biophysical complexity of coral reef systems, and their connectedness with social and economic systems. Modeling these complexities is a useful approach to help understand the dynamics of coral reefs as coupled socio-ecological systems. Modeling as an approach can provide explanatory and predictive value, helping to realize the phenomenon of coral reefs and their spatial and temporal variability. The examples we present, of models from reef-specific zones to local reef-scale models to regional reef models of Mexico and the Philippines, reveal complex but realistic dynamics, including phase shifts, and show the importance of the synergistic effects of multiple anthropogenic stressors such as nutrient enrichment and reduction of predators and herbivores that can lead to overgrowth by algae. These models are helpful to identify thresholds in the system, and to predict community trajectories, providing managers with a way to visualize alternative futures for coral reef systems. In addition to providing novel ways of representing the complex nature of coral reefs, modeling is an effective way to communicate with and engage the participation of stakeholders, as occurred so successfully with the MDSWG activity in the World Bank Global Environment Facility Coral Reef Targeted Research project. Linking ecological and social systems at various scales using FilReef and SIMREEF in the MDSWG work proved useful, not only for decision-support but particularly for stakeholder education and communication. New emerging technologies to facilitate improved visualization, learning experiences and access to these kinds of models will broaden participation in and acceptance of the modeling process. Access to CORSET as a 'Web 2.0' technology, running on a 'supercomputer' but accessible over the web (see www.reefscenarios.org), is an example of the kind of development that should see modeling and its products utilized more widely by a range of stakeholders spanning general public through to specialist modelers. Furthermore, the science-based ecosystem management tools presented here open opportunities for greater synergies between the social and natural sciences. This review offers a glimpse of the way in which transdisciplinary modeling paves the way for models to attain greater relevance and wider utility, providing sound science and responsiveness and transparency in management in a way that can be readily communicated in a user-friendly manner.

References

Aliño PM, McManus LT, McManus JW, Nañola C, Fortes MD, Trono GC, Jacinto GS (1993) Initial parameter estimations of a coral reef flat ecosystem in Bolinao, Pangasinan, Northwestern Philippines. In: Christensen V, Pauly D (eds) Trophic models of aquatic ecosystems, vol 26, pp. 252–258. ICLARM Conference Proceedings

Adams WM, Aveling R, Brockington D, Dickson B, Elliott J, Hutton J, Roe D, Vira B, Wolmer W (2004) Biodiversity conservation and the eradication of poverty. Science 306:1146–1149

Ainsworth CH, Varkey DA, Pitcher TJ (2008) Ecosystem simulations supporting ecosystem-based fisheries management in the coral triangle. Indones Ecol Model 214:361–374

Arias-González JE (1993) Fonctionnement trophique d'un ècosystème récifal: secteur de Tiahura, île de Moorea, Polynésie francaise. Thèse de doctorat, EPHE, Perpignan

Arias-González JE, Salvat B, Delesalle B, Galzin R (1997) Trophic fonctioning of the Tiahura sector "Moorea Island, French Polynesia". Coral Reefs 16:231–246

Arias-González JE, Garza-Pérez JR, González-Salas CF, González-Gándara C, Hernández-Landa RC, Membrillo-Venegas N, Nuñez-Lara E, Pérez E, Ruíz-Zárate MA (2000) Coral reef ecosystem research: towards integrated coastal management on the Yucatan Peninsula, Mexico. In: Done T, Lloyd D (eds) Information management and decision support for marine biodiversity protection and human welfare: coral reefs. Australian Institute of Marine Science, Townsville, pp 39–50

Arias-González JE, González-Salas CF, Nuñez-Lara E, Galzin R (2004) Trophic models for investigation of fishing effect on coral reef ecosystems. Ecol Model 172:197–212

Atkinson MJ, Grigg RW (1984) Model of a coral reef ecosystem. II: gross and net primary production at French Frigate Shoals, Hawaii. Coral Reefs 3:313–322

Barreteau O, Le Page C, Perez P (2007) Contribution of simulation and gaming to natural resource management issues: an introduction. Simul Gam 38:185–194

Beisner BE, Haydon DT, Cuddington K (2003) Alternative stable states in ecology. Front Ecol Environ 1:376–382

Bellwood DR, Hughes TP (2001) Regional-scale assembly rules and biodiversity of coral reefs. Science 292:1532–1534

Bellwood DR, Hughes TP, Folke C, Nystrom M (2004) Confronting the coral reef crisis. Nature 429:827–833

Bode M, Bode L, Armsworth PR (2006) Larval dispersal reveals regional sources and sinks in the Great Barrier Reef. Mar Ecol Prog Ser 308:17–25

Bozec Y-M, Gascuel D, Kulbicki M (2004) Trophic model of lagoonal communities in a large open atoll (Uvea, Loyalty islands, New Caledonia). Aquat Living Resour 17:151–162

Bulte E, Hector A, Larigauderie A (2005) ecoSERVICES: Assessing the impacts of biodiversity changes on ecosystem functioning and services, p 40. DIVERSITAS Report 3

Bunce L, Townsley P, Pomeroy RS, Pollnac R (2000) Socioeconomic manual for coral reef management. AIMS, Townsville

Burkepile DE, Hay ME (2006) Herbivore vs. nutrient control of marine primary producers: context-dependent effects. Ecology 87:3128–3139

Carpenter KE, Abrar M, Aeby G, Aronson RB, Banks S, Bruckner A, Chiriboga A, Cortés J, Delbeek JCh, DeVantier L, Edgar GJ, Edwards AJ, Fenner D, Guzmán HM, Hoeksema BW, Hodgson G, Johan O, Licuanan WY, Livingstone SR, Lovell ER, Moore JA, Obura DO, Ochavillo D, Polidoro BA, Precht WF, Quibilan MC, Reboton C, Richards ZT, Rogers AD, Sanciangco J, Sheppard A, Sheppard Ch, Smith J, Stuart S, Turak E, Veron JEN, Wallace C, Weil E, Wood E (2008) One-third of reef-building corals face elevated extinction risk from climate change and local impacts. Science 321:560–563

Collie JS, Richardson K, Steele JH (2004) Regime shifts: can ecological theory illuminate the mechanisms? Progr Oceanogr 60:281–302

Costanza R, d'Arge R, de Groot R, Farberparallel S, Grasso M, Hannon B, Limburg K, Naeem S, O'Neill RV, Paruelo J, Raskin RG, Suttonparallelparallel P, van den Belt M (1997) The value of the world's ecosystem services and natural capital. Nature 19:253–260

Cowen RK, Paris CB, Srinivasan A (2000) Connectivity of marine populations: open or closed? Science 311:857–859

Cowen RK, Paris CB, Srinavasan AB (2006) Scaling of connectivity in marine populations. Science 311:522–527

Dambacher JM, Brewer DT, Dennis DM, McIntyre M, Foale S (2007) Qualitative modelling of gold mine impacts on Lihir Island's socio-economic system and reef-edge fish community. Environ Sci Technol 41:555–562

Daltabuit M, Vasquez LM, Cisneros H, Ruiz GA (2006) El Turismo en la Ecorregion del Sistema Arrecifal Mesoamericano. UNAM-CRIM, Cuernavaca, Mexico, WWF, Washington

Diaz-Pulido G, McCook LJ (2004) Effects of live coral, epilithic algal communities and substrate type on algal recruitment. Coral Reefs 23:225–233

Done TJ (1992) Phase shifts in coral reef communities and their ecological significance. Hydrobiologia 247:121–132

Fung TC (2009) Local scale models of coral reef ecosystems for scenario testing and decisions support. Ph.D. thesis, University College, London

Fung TC, Seymour RM, Johnson CR (in press). Alternative stable states and phase shifts in coral reefs under anthropogenic stress. Ecology

Fulton EA, Smith ADM, Johnson CR (2003) Effect of complexity on marine ecosystem models. Mar Ecol Prog Ser 253:1–16

Gray R, Fulton E, Little R, Scott R (2006) Ecosystem model specification with an agent based framework. Northwest Shelf joint environmental management study. Technical Report No. 16, CSIRO

Gribble NA (2003) GBR-prawn: modelling ecosystem impacts of changes in fisheries management of the commercial prawn (shrimp) trawl fishery in the far northern Great Barrier Reef. Fish Res 65:493–506

Grigg RW, Polovina JJ, Atkinson MJ (1984) Model of a coral reef ecosystem. III. Resource limitation, community regulation, fisheries yield and resource management. Coral Reefs 3:23–27

Gurung TR, Bousquet F, Trébuil G (2006) Companion modeling, conflict resolution, and institution building: sharing irrigation water in the Lingmuteychu Watershed, Bhutan. Ecol Soci 11:36. http://www.ecologyandsociety.org/vol11/iss2/art36/

Hall ChAS (ed) (2000) Quantifying sustainable development: the future of tropical economies. Academic, San Diego

Hannon B, Mathias R (1997) Modeling dynamics biological systems. Springer, New York

Hanski I, Gyllenberg M (1993) Two general metapopulation models and the core-satellite species hypothesis. Am Nat 142:17–41

Hatcher BG (1997) Coral reef ecosystems: how much greater is the whole then the sum of the parts? Proc 8th Int Coral Reef Symp 1:43–56

Hatcher BG, Imberger J, Smith SV (1987) Scaling analysis of coral reef systems: an approach to problems of scale. Coral Reefs 5:171–181

Hay ME (1984) Patterns of fish and urchin grazing on Caribbean coral reefs: are previous results typical? Ecology 65:446–454

Hobbs RJ, Suding KN (eds) (2009) New models for ecosystem dynamics and restoration. Island Press, Washington, DC

Hoegh-Guldberg O, Mumby PJ, Hooten AJ, Steneck RS, Greenfield P, Gomez E, Harvell CD, Sale PF, Edwards AJ, Caldeira K, Knowlton N, Eakin CM, Iglesias-Prieto R, Muthiga N, Bradbury RH, Dubi A, Hatziolos ME (2007) Coral reefs under rapid climate change and ocean acidification. Science 5857:1737–1742

Hughes TP (1984) Population-dynamics based on individual size rather than age – a general-model with a reef coral example. Am Nat 123:778–795

Hughes TP (1994) Catastrophes, phase shifts, and large-scale degradation of a Caribbean coral reef. Science 265:1547–1551

Hughes TP, Tanner JE (2000) Recruitment failure, life histories, and long-term decline of Caribbean corals. Ecology 81:2250–2263

Hughes TP, Baird AH, Bellwood DR, Card M, Connolly SR, Folke C, Grosberg R (2003) Climate change, human impacts, and the resilience of coral reefs. Science 301:929–933

Hughes TP, Bellwood DR, Folke C, Steneck RS, Wilson J (2005) New paradigms for supporting the resilience of marine ecosystems. Tree 20:380–386

Jackson JBC, Kirby MX, Berger WH, Bjorndal KA, Botsford LW, Bourque BJ, Bradbury RH, Cooke R, Erlandson J, Estes JA, Hughes TP, Kidwell S, Lange CB, Lenihan HS, Pandolfi JM, Peterson ChH, Steneck RS, Tegner MJ, Warner RR (2001) Historical overfishing and the recent collapse of coastal ecosystems. Science 293:629–638

Johnson CR, Preece AL (1993) Damage, scale and recovery in model coral communities: the importance of system state. Proc 7th Int Coral Reef Symp 1:606–615

Johnson C, Klumpp D, Field J, Bradbury R (1995) Carbon flux on coral reef: effects of large shifts in community structure. Mar Ecol Prog Ser 126:123–143

Jompa J, McCook LJ (2002) Effects of competition and herbivory on interactions between a hard coral and a brown alga. J Exp Mar Biol Ecol 271:25–39

Jones GP, Milicich MI, Emslie MJ, Lunow C (1999) Self-recruitment in a coral reef fish population. Nature 402:802–804

Juárez A (2002) Ecological degradation, global tourism, and inequality: Maya interpretations of the changing environment in Quintana Roo, Mexico. Hum Organiz 61:113–124

Knowlton N (2001) The future of coral reefs. Proc Nat Acad Sci U S A 98:5419–5425

Lachica-Aliño L, Wolff M, David LT (2006) Past and future fisheries modelling approaches in the Philippines. Rev Fish Biol Fish 16:201–2112

Langmead O, Sheppard C (2004) Coral reef community dynamics and disturbance: a simulation model. Ecol Model 175:271–290

Levins R (1970) Complex systems. In: Waddington CH (ed) Towards a theoretical biology, vol 3. Edinburgh University Press, Edinburgh, pp 73–88

Lirman D (2003) A simulation model of the population dynamics of the branching coral Acropora palmata – effects of storm intensity and frequency. Ecol Model 161:169–182

Little LR, Punt AE, Mapstone BD, Pantus F, Smith ADM, Davies CR, McDonald AD (2007) ELFSim – a model for evaluating management options for spatially structured reef populations: an illustration of the 'larval subsidy' effect. Ecol Model 205:381–396

Lutz W, Prieto L, Sanderson W (eds) (2000) Population, development, and environment on the Yucatan Peninsula: from ancient Maya to 2030. IIASA, Laxenburg

Lynam T, De Jong W, Sheil D, Kusumanto T, Evans K (2007) A review of tools for incorporating community knowledge, preferences, and values into decision making in natural resources management. Ecol Soc 12:5

Maguire LA, Porter JW (1977) A spatial model of growth and competition strategies in coral communities. Ecol Model 3:249–271

Melbourne-Thomas J, Johnson CR, Fung T, Seymour R, Chérubin M, Arias-González JE, Fulton EA (in press). Regional-scale scenario modeling for coral reefs: a decision support tool to inform management of a complex system. Ecol Appl doi:10.1890/09–1564.1

Melbourne-Thomas J, Johnson CR, Aliño PM, Geronimo RC, Villanoy CL, Gurney GG (2011). A multi-scale biophysical model to inform regional management of coral reefs in the western Philippines and South China Sea. Environ Model Soft 26:66–82

McCausland WD, Mente E, Pierce GJ, Theodossiou I (2006) A simulation model of sustainability of coastal communities: aquaculture, fishing, environment and labour markets. Ecol Model 193:271–294

McClanahan TR (1995) A coral reef ecosystem-fisheries model: impacts of fishing intensity and catch selection on reef structure and processes. Ecol Model 80:1–19

McClanahan T, Polunin N, Done T (2002) Ecological states and the resilience of coral reefs. Cons Ecol 6:18. (online) URL: http://www.consecol.org/vol6/iss2/art18

McCook LJ (1999) Macroalgae, nutrients and phase shifts on coral reefs: scientific issues and management consequences for the Great Barrier Reef. Coral Reefs 18:357–367

McCook LJ (2001) Competition between corals and algal turfs along a gradient of terrestrial influence in the nearshore central Great Barrier Reef. Coral Reefs 19:419–425

McManus JW, Polsenberg JF (2004) Coral-algal phase shifts on coral reefs: ecological and environmental aspects. Progr Oceanogr 60:263–279

Miller JH, Page SF (2007) Complex adaptive systems: an introduction to computational models of social life. Princeton University Press, Princeton, p 284

Mumby PJ (2006) The impact of exploiting grazers (scaridae) on the dynamics of Caribbean coral reefs. Ecol Appl 16:747–769

Mumby PJ, Hastings A (2006) Connectivity of reef fish between mangroves and coral reefs: algorithms for the design of marine reserves at seascape scales. Biol Consow 128:215–222

Mumby PJ, Hedley JD, Zychaluk K, Harborne AR, Blackwell PG (2006) Revisiting the catastrophic die-off of the urchin *Diadema antillarum* on Caribbean coral reefs: fresh insights on resilience from a simulation model. Ecol Model 196:131–148

Niquil N, Arias-González JE, Delesalle B, Ulanowicz R (1999) Characterization of the planktonic food web of Takapoto atoll, using network analysis. Oecologia 118:232–241

Odum HT, Odum EP (1955) Trophic structure and productivity of windward coral reef community on Eniwetok atoll. Ecol Monogr 25:291–320

Opitz S (1996) Trophic interactions in Caribbean coral reefs. ICLARM Technical Report 43, p. 341

Pandolfi JM, Bradbury RH, Sala E, Hughes TP, Bjorndal KA, Cooke RG, McArdle D, McClenachan L, Newman MJH, Paredes G, Warner RR, Jackson JBC (2003) Global trajectories of the long-term decline of coral reef ecosystems. Science 301:955–958

Paris CB, Cherubin LM, Cowen RK (2007) Surfing, spinning or diving from reef to reef: effects on population connectivity. Mar Ecol Prog Ser 347:285–300

Perez P (2008) Embracing social uncertainties with complex systems science. In: Smithson M, Bammer G (eds) Challenges of uncertainty. EarthScan Publisher, London, pp 147–155

Perez P, Dray A, Cleland D, Arias-González JE (2009) SimReef: an agent-based model to address coastal management issues in the Yucatan Peninsula, Mexico. In: Anderssen RS, Braddock RD, Newham LTH (eds) 18th World IMACS congress and MODSIM09 international congress on modelling and simulation, pp 72–79. MSSANZ/IMACS, Canberra, http://www.mssanz.org.au/modsim09/H6/perez.pdf

Polovina JJ (1984) Model of a coral reef ecosystem: I. The ECOPATH model and its application to French Frigate Shoals. Coral Reefs 3:1–11

Pomeroy RS, Parks JE, Watson LM (2004) How is your MPA doing? IUCN, Cambridge

Preece AL, Johnson CR (1993) Recovery of model coral communities: complex behaviours from interaction of parameters operating at different spatial scales. In: Green DG, Bossomaier T (eds) Complex systems: from biology to computation. IOS Press, Amsterdam, pp 69–81

Sale PF (2008) Management of coral reefs: where we have gone wrong and what we can do about it. Mar Poll Bull 56:805–809

Sandin SA, Smith JE, DeMartini EE, Dinsdale E, Donner SD, Friedlander AM, Konotchick T, Malay M, Maragos8 JE, Obura D, Pantos O, Paulay G, Richie M, Rohwer F, Schroeder RE,

Sheila Walsh1, Jackson JBC, Knowlton N, Sala E (2008) Baselines and degradation of coral reefs in the northern Line Islands. PLoS One 3: e1548. doi:10.1371/journal.pone.0001548

Shafer JL (2007, August) Agent based simulation of a recreational reef fishery: linking ecological and social dynamics. Ph.D. University of Hawaii, Hondulu, p 209

Scheffer M, Carpenter S, Foley JA, Folkes C, Walker B (2001) Catastrophic shifts in ecosystems. Nature 413:591–596

Scheffer M, Carpenter S, de Young B (2005) Cascading effects of overfishing marine systems. Trends Ecol Evol 20:579–581

Sorokin IY (1973) Microbial aspects of productivity of coral reefs. In: Jones O, Endean R (eds) Biology and geology of coral reefs, 2. Biology 1. Academic, New York, pp 17–45

Sorokin IY (1993) Coral reefs ecology. Springer, Berlin, Heidelberg, New York

Stobutzki I (2001) Marine reserves and the complexity of larval dispersal. Rev Fish Biol Fish 10:515–518

Swearer SE, Shima JS, Hellberg ME, Thorrold SR, Jones GP, Robertson DR, Morgan SG, Selkoe KA, Ruiz GM, Warner RR (2002) Evidence of self-recruitment in demersal marine populations. Bull Mar Sci 70(Suppl 1):251–271

Treml EA, Halpin PN, Urban DL, Pratson LF (2008) Modeling population connectivity by ocean currents, a graph-theoretic approach for marine conservation. Landscape Ecol 23:19–36

Tsehaye I, Nagelkerke LAJ (2008) Exploring optimal fishing scenarios for the multispecies artisanal fisheries of Eritrea using a trophic model. Ecol Model 212:319–333

Villa F, Constanza R (2000) Design of multi-paradigm integrating modelling tools for ecological research. Environ Modell Softw 15:169–177

Waddell J (ed) (2005) The state of coral reef ecosystems of the United States and Pacific freely associated states: 2005, p 522. NOAA Technical Memorandum NOS NCCOS 11. NOAA/NCCOS Center for Coastal Monitoring and Assessment's Biogeography Team. Silver Spring

Wakeford M, Done TJ, Johnson CR (2008) Decadal trends in a coral community and evidence of changed disturbance regime. Coral Reefs 27:1–13

Walker BH, Anderies JM, Kinzig AP, Ryan P (eds) (2006) Exploring resilience in social-ecological systems comparative studies and theory development. CSIRO Publishing, Canberra

Wantiez L, Chauvet C (2003) First data on community structure and trophic Networks of Uvea coral reef fish assemblages (Wallis and Futuna, South Pacific Ocean). Cybium 27:83–100

Wilkinson CR (ed) (2008) Status of coral reefs of the World: 2008. Australian Institute for Marine Science, Journsirlle

World Bank (2006) Measuring coral reef ecosystem health: integrating social dimensions. The World Bank, Washington, Report No. 36623-GLB, 65

Wooldridge S, Done T, Berkelmans R, Jones R, Marshall P (2005) Precursors for resilience in coral communities in a warming climate: a belief approach. Mar Ecol Progr Ser 295:157–169

Yoon I, Williams RJ, Levine E, Yoon S, Dunne JA, Martinez ND (2004) Webs on the Web (WoW): 3D visualization of ecological networks on the WWW for collaborative research and education. Proc IS&T/SPIE Symp Electr Imag Visual Data Anal 5295:124–132

Part V
Disturbances

The Impact of Climate Change on Coral Reef Ecosystems

Ove Hoegh-Guldberg

Abstract Human activities such as the burning of fossil fuels, deforestation and changing land use have dramatically altered the atmospheric concentration of greenhouse gases such as carbon dioxide and methane. These changes have resulted in global warming and ocean acidification, both of which pose serious threats to coral reef ecosystems through increased thermal stress and ocean acidity as well as declining carbonate ion concentrations. Observed impacts on coral reefs include increased mass coral bleaching, declining calcification rates, and a range of other changes to subtle yet fundamentally important physiological and ecological processes. There is little evidence that reef-building corals and other organisms will be able to adapt to these changes leading to the conclusion reef ecosystems will become rare globally by the middle of the current century. Constraining the growth of carbon dioxide in the atmosphere as well as reducing local stresses such as overfishing and declining water quality, however, holds considerable hope for avoiding this gloomy future for coral reefs. Given the importance of coral reefs to the livelihoods of millions of people, actions such as these must be pursued as a matter of extreme urgency.

Keywords Climate change • ocean acidification • mass coral bleaching • declining calcification • erosion • disruption of sensory systems • IPCC • carbon dioxide • cethane • green-house gases

1 Introduction

Climate change is the defining environmental, economic, and social issue of our time and there is no any longer reasonable doubt that the rapid increases in the atmospheric carbon dioxide and other greenhouse gases since the beginning of the Industrial Period are driving significant changes to the physical and chemical environment of the Earth (IPCC 2007). These changes are occurring at rates that are 2–3 orders of magnitude faster than the rapid shifts between glacial and interglacial periods seen over the past 740,000 years (Petit et al. 1999; Augustin et al. 2004; Hansson et al. 2006). There is also abundant evidence that biological systems are changing across the planet (Walther et al. 2002). Coral reef ecosystems have played a particularly key role in our understanding of how the earth's ecosystems may respond to rapid anthropogenic climate change. This chapter reviews our understanding of these changes and develops a series of projections for how important marine ecosystems such as coral reefs are likely to change over the next few decades and century. As will be discussed, there is now an even greater urgency for the international community to stabilize carbon dioxide at or below current concentrations as quickly as possible or risk losing these crucial ecosystems from the human experience.

2 The Coral Reef Environment

The environmental conditions that define where coral reefs exist today provide important insight into how they will change in the future under rapid environmental change. As several chapters have already outlined, coral reefs are distributed in the shallow, sunlit waters of the tropics and subtropics. Here, they capture the abundant sunlight, converting it into organic energy, which either flows directly through the ecosystem or is used to power important processes such as calcification (Muscatine 1990). Coral reefs occupy coastal areas in a band from roughly 30° north and south of the equator (Kleypas et al. 1999a). At higher latitudes, calcification decreases to a point where it decreases below the rate of erosion, reef accretion becomes negative, and carbonate coral reefs no longer persist. Instead, coral communities form relatively

O. Hoegh-Guldberg (✉)
Global Change Institute, The University of Queensland St Lucia, QLD 4072, Australia
e-mail: oveh@uq.edu.au

slow-growing colonies on rocky and sandy surfaces, and when they die, their skeletons are removed such that they do not result in a reef framework. At the lower latitudes, however, the rate at which calcium carbonate is deposited greatly exceeds physical and biological erosion leading to the net accumulation of the three-dimensional structure or framework of the reef. This framework is home to a relatively poorly documented biodiversity, which may involve millions of species of animal, plant, fungi, and protists (Reaka-Kudla 1997).

Kleypas et al. (1999a) were the first to provide a comprehensive and systematic examination of the conditions that correlate with the development of carbonate coral reef ecosystems. In their study of over 1,000 reef locations worldwide, coral reefs were found to occupy location is characterised by distinct temperature, salinity, nutrient, aragonite saturation, and light (depth, turbidity) regimes. The environmental factors under which present-day coral reefs have evolved are shown in Table 1. Unraveling the relative importance of each factor is not possible and it is unlikely that any one factor limits the distribution of coral reefs on its own. A key feature of this distribution is that exceeding these conditions will lead to physiological distress and mortality. For example, increases or decreases above a particular temperature at a particular geographic location will lead to the bleaching of coral communities *en masse* (Hoegh-Guldberg 1999), as described in Chapter 23. Equally, decreases in the carbonate ion concentration below values of around 200 μmol per kilogram water (inherent within the aragonite saturation state, Table 1) will result in reef calcification fall behind biological and physical erosion (Guinotte et al. 2003; Hoegh-Guldberg et al. 2007).

3 The Influence of Rising Atmospheric Carbon Dioxide and Other Greenhouse Gases

Solar radiation ultimately drives the temperature of the terrestrial, oceanic, and atmospheric components of the Earth. In addition to the chemical composition of the atmosphere, the reflectivity of key components of the earth such as ice, clouds, and vegetation work together to define the heat balance of the earth (IPCC 2007). Approximately 30% of incoming energy is eventually reradiated back out to space, with approximately 70% ultimately being absorbed resulting in an average global temperature of approximately 14°C. This effect is similar to the way greenhouses work in agriculture, with solar radiation entering through the glass panels of the greenhouse and warming the air, soil, and vegetation. But some is returning. The net effect is that heat is trapped within the greenhouse. The "greenhouse effect" is crucial to life on Earth – without it, the earth would global temperature of −18°C and carbon-based life forms would be unable to exist (Solomon et al 2008).

There is now considerable insight into how the key factors have driven the earth's temperature over the past million or so years. In this regard, using either direct measurements of trapped bubbles of ancient atmospheres (Petit et al. 1999; Augustin et al. 2004; Hansson et al. 2006) or a number of chemical proxies (Mann 2002), scientists have been able to derive long-term records of atmospheric conditions and their impact on global temperature and other factors such as desertification and ice volume. There is now abundant evidence from instrument and proxy records such as these that humans have dramatically changed the chemical composition of the earth's atmosphere. One of the most famous of these records

Table 1 Environmental averages and extremes associated with coral reefs (Kleypas et al. 1999 a, b)

Variable	Minimum	Maximum	Mean	Standard deviation
Temperature (°C)				
Average	21	29.5	27.6	1.1
Minimum	16	28.2	24.8	1.8
Maximum	24.7	34.4	30.2	0.6
Salinity (ppt)				
Minimum	23.3	40	34.3	1.2
Maximum	31.2	41.8	35.3	0.9
Nutrients (μmol L^{-1})				
Nitrate	0	3.34	0.25	0.28
Phosphate	0	0.54	0.13	0.08
Aragonite Saturation (Ω_{arag})				
Average	3.28	4.06	3.83	0.09
Maximum depth of light penetration (m) calculated from the monthly average depth at which average light decreases below the perceived minimum for reef development of 250 mmol m^{-2} s^{-1}				
Average	−9	−81	−53	13.5
Minimum	−7	−72	−40	13.5
Maximum	−10	−91	−65	13.4

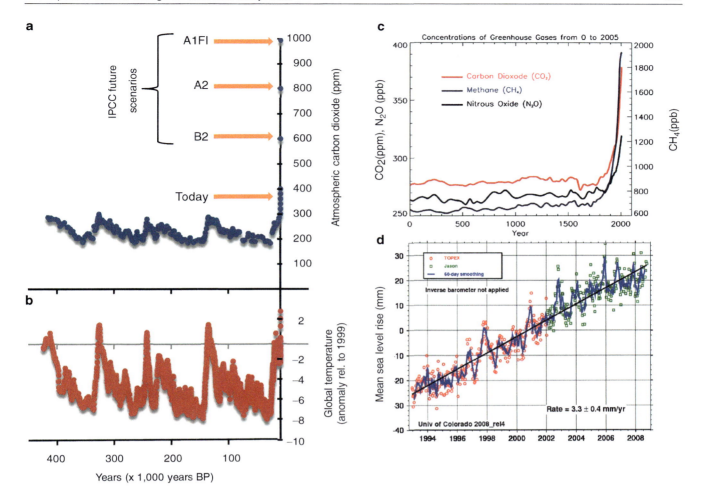

Fig. 1 Variation in planetary conditions as a result of natural and anthropogenic climate change. (**a**) Variation in atmospheric carbon dioxide over the past 420,000 years (Petit et al. 1999). *Arrows* indicate atmospheric concentrations today, as well as for three key scenarios from the fourth assessment report of the Intergovernmental Panel on Climate Change (IPCC 2007). (**b**) Anomalies in global temperature of relative to 1999 calculated from isotope proxies. (**c**) Changes in key greenhouse gases over the past 2,000 years (IPCC 2007). (**d**) Changes in sea level measured using satellite altimeters (Courtesy of Colorado Center for Astrodynamics Research, University of Colorado)

arises from the Vostok ice core (Fig. 1, (Petit et al. 1999), which chronicles the rise and fall of carbon dioxide in past atmospheres over a period stretching back 420,000 years. This record has been obtained from sampling tiny bubbles of ancient atmospheres trapped in layers of compacted snow that accumulated over this time in the relatively stable Antarctic climate. The record has been verified and extended by groups such as Augustin et al. (2004) and now stretches back to 740,000 years BP. This record documents the rise and fall of atmospheric carbon dioxide as a result of the heating and cooling of the planet over vast periods of time driven by the Milankovitch cycle. The Milankovitch cycle is a consequence of the variability in eccentricity, axial tilt, and precession of the Earth's orbit around the Sun, which has a periodicity of approximately 100,000 years. As a result of the Earth's distance from the Sun varying over time, the earth temperature decreases periodically by 5–10° (ice ages), which impacts atmospheric carbon dioxide (Petit et al. 1999; Augustin et al. 2004).

Importantly, carbon dioxide over this period has varied between 180 and 280 ppm, with the maximum value being dramatically less than the current atmospheric content called 387 ppm. During these changes in temperature and atmospheric composition, which took thousands of years to unfold, the biological characteristics of the planet changed dramatically (Augustin et al. 2004).

Instrument records from the last 130 years indicate a sharp departure from the conditions that have defined the earth for over 1 million years (IPCC 2007). Not only has global temperature increased, but carbon dioxide and other significant greenhouse gases have increased to a level not seen in the last million (Fig. 1a) if not 20 million years. These changes have been driven by the combustion of fossil fuels, which have entered the atmosphere in ever increasing rates since the advent of the Industrial Revolution during the second half of the nineteenth century. Changes have also been accompanied by the increased levels of atmospheric methane arising from agricultural and land-use sources, as well as a slew of other warming gases such as chlorinated fluorocarbons (CFCs) many of which have warming potentials that may be 100 times that of carbon dioxide (Fig. 1b).

Initially, most of the changes in global temperature arose from the changes to atmospheric chemistry. Increasingly, however, changes are now also being exerted by changes to the earth's albedo (or diffuse reflectivity), which is a consequence of the changes that have resulted from the increase in global temperature to such elements as the distribution of ice and snow relative to darker more absorptive components of our planet. The current dramatic collapse of Arctic summer sea ice over the past 5 years (Cressey 2007), for example, represents a major change in the amount of light reflected back into space on account of the greater absorptivity of the darker ocean that replaces the once highly reflective, icy seascape (Zhang et al. 2008).

One of the most important components inherent to the enhanced greenhouse effect is the influence of a modified planetary radiation budget on climate at a local scale. For example, changing the relative patterns of heating and cooling in the ocean, for example, will lead to dramatic changes in where and when water will flow (currents) as well as dramatic changes in storm strength, activity, and rainfall patterns (IPCC 2007). All these characteristics have the potential to affect biological systems in ways that are significant relative to those experienced over the last several million years. In this respect, there is now abundant evidence that the earth's biosphere has begun to change (Walther et al. 2002; IPCC 2007). Almost every aspect of biology is temperature dependent, and as a consequence there is now strong evidence that the distribution of plants and animals is being significantly modified by climate change (Walther et al. 2002). These changes, for example, have driven bird and butterfly populations in Europe and the United States 50–200 km north of where they were 50 years ago (Parmesan et al. 1999; Walther et al. 2002; Parmesan and Yohe 2003). Seasonal warming occurs earlier in many regions resulting in responses from the life cycles of many organisms. Changes have also been documented in the arrival time of migratory birds, flower opening times, and calling times of reproducing frogs and other amphibians, all occurring earlier as the start date of spring has advanced (Walther et al. 2002; Parmesan and Yohe 2003).

In addition to adjustments in the range and phenology of plants and animals, other more dramatic changes to ecosystems have occurred. Impacts from the catastrophic breakdown of summer sea ice in the Arctic, for example, have resulted in dramatic decreases in the health and number of polar bears and other Arctic life (Derocher et al. 2004; Schipper et al. 2008). In this case, changes to key resources (i.e. sea ice) and its absence elsewhere have undermined ice-dependent organisms. As the discussion below will outline, coral reefs face a somewhat similar situation, with rapid changes to key environmental variables that threatened to eliminate their very existence (Hoegh-Guldberg et al. 2008).

4 Changes to Tropical/Sub-tropical Oceans

Increasing atmospheric concentrations of greenhouse gas have driven an increase in the average temperature of the global ocean of 0.74°C, and have increased sea levels by an average of 17 cm over the twentieth century (IPCC 2007). At the same time as carbon dioxide has directly influenced global temperature, it has also been absorbed by the ocean where it has reacted with sea water to form a dilute acid (carbonic acid). Carbonic acid dissociates to produce a proton, which in turn converts carbonate ions to bicarbonate ion. This change has resulted in a decline of 0.1 pH units (Equivalent to an increase in hydrogen ions of around 26%), as well as a decrease in the typical carbonate ion concentration of typical coral reef environments of approximately 30 μmol kg^{-1} water (Hoegh-Guldberg et al. 2007). Comparing the changes that have occurred during the last century to those that have occurred over 420,000 years prior to the Industrial Revolution reveals that coral reefs are already facing novel combinations of sea temperatures, pH, and carbonate ion concentrations (Fig. 1a). These changes in sea temperature and carbonate ion concentration have occurred extremely rapidly. Within a century, we have essentially moved the same distance as would normally take over 20,000–40,000 years under the slow dynamics of the glacial cycle (Petit et al. 1999; Augustin et al. 2004). With further changes, the conditions with respect to the carbonate ion concentration that we know support coral reefs will contract rapidly to the equatorial region as we head beyond atmospheric concentrations of carbon dioxide of 450 ppm. The speed and endpoint of the changes represent enormous challenges for organisms living in and around coral reefs, especially in the light of the time required for biological responses such as migration and/or evolution.

Rising greenhouse gas concentrations are likely to change a range of other factors that are important for near-shore tropical ecosystems. Changing sea levels probably pose minimal threat if their current slow rate of change continues and organisms such as reef-building corals remained healthy enough to keep up with the changes in water depth. While there is considerable uncertainty around the estimates as sea-level rise, several leading research groups project that average sea-level rise could rise by 1.4 m or more by 2100 (Hansen 2007; Rahmstorf 2007). These estimates may in turn be conservative given recent evidence of the accelerated breakdown of the terrestrial ice sheets in Greenland (Gregory et al. 2004; Witze 2008) and Western Antarctica (Steig et al.

2009), which suggest that current IPCC projections of approximately 30–50 cm by 2100 (IPCC 2007) severely underestimate potential sea-level rise if the input of water and ice from these sources increases.

Changing weather patterns along coastlines fringed by coral reefs are also likely to play an important role in the health of coral reefs into the future. As has already been outlined in previous chapters, the flux of nutrients and sediments from coastal areas plays a strong influence on whether or not coral reefs flourish. Warmer seas are likely to drive more intense storms (Emanuel 2005; IPCC 2007), with the prospect that some coral reefs many receive increasing episodes of damage from destructive storms that arrive more frequently. In this respect, changing rainfall patterns and storm intensity may also lead to the destabilization and erosion of river catchments that ultimately exit into waters that bathe coastal ecosystems such as coral reefs. As will be discussed in the final section of this chapter, how we manage these coastal catchments may be critical to how much these global–local interactions affect coral reefs in the future.

Fig. 2 A normally pigmented (*left*) and bleached (*right*) coral on reefs around Great Keppel Island, on the Great Barrier Reef (GBR) in January 2006. Anomalously warm water covered reefs in this area for over 6 weeks, resulting in almost 90% of corals bleaching. Resulting coral mortalities were between 30% and 40%, depending on the species. Photographer: Ove Hoegh-Guldberg, Global Change Institute, University of Queensland

5 Impacts on Coral Reefs

5.1 Impacts of Thermal Stress

One of the first examples of how rapid anthropogenic climate change might severely impact biological systems is associated with coral reefs. Starting in the early 1980s, a new phenomenon called mass coral bleaching entered the scientific literature. As outlined in Chapter 23, coral bleaching is a phenomenon that has been known for over a 100 years. However, large-scale 'mass' coral bleaching began to occur in the early 1980s in Panama, Florida, and at other Caribbean sites (Glynn 1983; Lessios et al. 1983; Glynn 1996) with corals across large areas of reef literally turned white over a number of weeks ("bleached", Fig. 2). The whitening of coral tissues during bleaching events is primarily due to the loss of pigments and/or cells of their essential symbiotic dinoflagellates (*Symbiodinium*). It is triggered by a range of different phenomena, including elevated temperature and irradiance (for review, see Hoegh-Guldberg 1999). While early workers were uncertain of why mass coral bleaching events suddenly appeared on coral reefs, it was clear by the end of the 1980s that they were associated with periods of exceptionally warm water coupled with bright sunny still weather ("doldrums"; Lessios et al. 1983; Hoegh-Guldberg and Smith 1989; Glynn and D'Croz 1990). Alongside coral disease, which is also considered to be driven by warming seas (see Chapter 23), mass coral bleaching and mortality may affect thousands of square kilometers within a few weeks and hence contributes the greatest source of mortality for corals (Hoegh-Guldberg et al. 2007).

Since the first reports of mass coral bleaching almost 30 years ago, it has become a regular phenomenon on most of the world's coral reefs. Much is known about the mechanism, which involves perturbations to the light-capturing ability of *Symbiodinium* within the tissues of reef-building corals as well as impacts on other key cellular processes (Chapter 23). Satellite measurements of sea surface temperature have also played an important role in verifying the underlying driver associated with the appearance of coral bleaching, and in understanding future changes to coral reefs that might be expected under rapid warming of tropical/subtropical oceans. Satellite measurements have verified that mass bleaching is highly likely on a coral reef if sea temperatures rise by 1°C above the long-term summer maximum temperatures in a particular region (Strong et al. 1996; Toscano et al. 2000). These measurements have been further extended by incorporating the length of time that a coral reef region is exposed to a particular anomaly. In this respect, analysis of the "degree heating weeks" (DHW = anomaly size × length of exposure; Strong et al. (1996) has provided even greater insight into the impact of coral bleaching on a particular reef. If DHW values rise above 4, coral reefs will bleach but recover quickly with little or no lasting damage. If DHW values rise above 8, coral bleaching is likely to be extensive with significant mortality. DHW values of over 12 are associated with devastating impacts on coral reefs. In 1998, many sites across the Western Indian Ocean experienced DHW values of over 12, with as much as 95% of corals on some reef systems

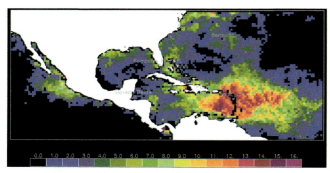

Fig. 3 Degree Heating Weeks (DHW) accumulating from anomalies at the end of October 2005 (Oct 28 2005). This warming event preceded the worst mass bleaching event on record for the Caribbean. DWH is calculated by multiplying the anomaly above the summer sea temperature maxima by the amount of time that a region has experienced the anomaly. This figure shows how much thermal stress has accumulated in an area over the preceding 12 weeks. Values that exceed 3–4 are associated with the first signs of bleaching on coral reefs while values that exceed 10–12 are usually associated with significant levels of coral mortality. Courtesy of Coral Watch, NOAA/NESDIS, Washington, DC

being killed (Hoegh-Guldberg 1999). Almost 50% of all corals died across the region by the end of the very long warm summers associated with this period (Hoegh-Guldberg 1999; Wilkinson and Hodgson 1999; Wilkinson et al. 1999).

More recently, the extent to which DHW values can be used to project the severity and outcome of thermal stress on coral reefs was tested during the widespread mass bleaching event in the Caribbean that occurred in late 2005 (Fig. 3). The highest mortality of corals during this event occurred in the eastern Caribbean, coinciding with DHW values that exceeded 12 (Donner et al. 2007). Other areas, such as those of Florida and Mexico, were affected by bleaching but recovered without a substantial increase in mortality.

The impact of extreme events like those that occurred in 1998 and on subsequent occasions provide important insight into how future stress may impact coral reefs over the next few decades and century as tropical/subtropical seas continue to warm. In this respect, taking the known thresholds of coral communities from past events, and comparing these to future sea temperatures estimated by the global climate modeling community, allows insight into the timing and eventual fate of coral reefs (Hoegh-Guldberg 1999) given assumptions about how effective human societies will be in reducing emissions of CO_2 and other greenhouse gases.

Understanding how coral bleaching might change under climate change has assumed considerable importance as we have struggle to understand what lies ahead for coral reefs and the many societies that depend on them. The combination of the current behavior of coral reefs under thermal stress with the output of climate models for how sea temperature is likely to change over the coming decades and century reveals that the conditions for coral bleaching are likely to become more common until they occur on a yearly basis by the middle of this century (Hoegh-Guldberg 1999). Calculating DHW values for future oceans reveals that thermal stress *on par* or greater than that which occur in the Western Indian Ocean in 1998 will occur on an annual basis within the next 30–50 years under the current upwardly spiraling carbon dioxide concentrations. Subsequent analysis done (Done et al. 2003; Donner et al. 2005) verified these conclusions, indicating the largely unsustainable nature of the changes will occur as ocean temperatures in tropical and temperate regions rise 2°C or more above preindustrial values.

5.2 Impacts of Ocean Acidification

At the same time as it was becoming clear that rising sea temperatures were a major threat to coral reef ecosystems, Kleypas and coworkers postulated that the acidification of oceans by rising atmospheric carbon dioxide represented another serious challenge to coral reefs in a high CO_2 world. This phenomenon later attracted the term "ocean acidification" (Caldeira and Wickett 2003) and is increasingly being seen as a major threat to marine calcifiers among which coral reefs are a prominent component. Ocean acidification is a consequence of the increasing flux of carbon dioxide into ocean waters as a result of increasing atmospheric carbon dioxide concentrations (Fig. 4). On entering the ocean, carbon dioxide reacts with water molecules to create carbonic acid. Carbonic acid subsequently dissociates releasing a proton, which reacts with carbonate ions, converting them into bicarbonate ions. Carbonate ion concentrations are often expressed relative to the saturation state of sea water with respect to aragonite, the principal crystal form of calcium carbonate crystals deposited by reef-building corals and many other marine calcifiers. In a seminal paper, Joanie Kleypas, Jean-Pierre Gattuso and colleagues (Kleypas et al. 1999b) proposed that the decrease in the carbonate ion as a result of a doubling of atmospheric carbon dioxide would be enough to significantly decrease (by up to 40%) the ability of corals and other marine calcifiers to form their calcium carbonate skeletons.

The first experiments done by Gattuso, Langdon, and others quickly verified the theoretical predictions of Kleypas et al (1999a). A large number of studies have shown that a doubling of pre-industrial atmospheric carbon dioxide concentrations will result in decreases of up to 40% in the calcification and growth of corals and other reef calcifiers

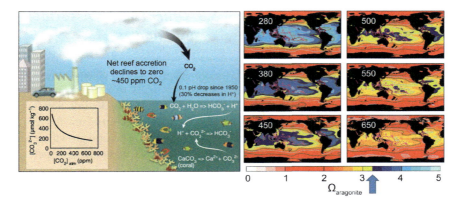

Fig. 4 Ocean acidification and its potential influence on the coral reefs. (**a**) Schematic showing link between anthropogenic sources of atmospheric carbon dioxide and acidification of tropical/subtropical oceans. Rising atmospheric concentrations of carbon dioxide leads to ocean acidification, which results in a decrease in the carbonate (CO_3^{2-}) ion concentration, limiting the amount available for calcifying organisms like reef-building corals and calcareous red algae. The inset graph depicts the typical relationship between atmospheric carbon dioxide and ocean carbonate ion concentration. (**b**) The aragonite saturation constant as a function of location and atmospheric carbon dioxide concentration (white numbers within each panel). Aragonite saturation constants are calculated by dividing the product of calcium and carbonate ion concentration by the solubility product of aragonite, which is the form of calcium carbonate that reef corals and many other organisms deposit to form their skeletons. The *blue arrow* indicates aragonite saturation constant below which carbonate coral reefs tend not to form. *Pink dots* represent the location of carbonate coral reef ecosystems (Adapted from Hoegh-Guldberg et al (2007) and used with the permission of Science Magazine)

such as red calcareous algae (Kleypas and Langdon 2006). Field studies have also suggested that net carbonate accretion on coral reefs is likely to decrease below zero when carbonate ion concentrations reach 200 μmol per kilogram water, which corresponds approximately to aragonite saturation states of 3.3 or less (Kleypas et al. 1999a; Guinotte et al. 2003, see legend Fig. 4). The aragonite saturation state (Ω_{arag}) is sensitive to the partial pressure of carbon dioxide above the ocean as well as ocean temperature. In the latter case, the aragonite saturation of warm tropical oceans is higher than that of polar oceans due to the fact that carbon dioxide is more soluble in cold water (Fig. 4b). As atmospheric carbon dioxide has increased, the aragonite saturation of the world's oceans has decreased. This trend will continue until the chemical conditions that are needed for the development and maintenance of carbonate coral reef ecosystems are all but restricted to the equator, eventually disappearing as we exceed concentrations above atmospheric carbon dioxide concentration of 550 ppm (Hoegh-Guldberg et al. 2007).

Demonstrations of declining reef accretion have been difficult to demonstrate in the field due to be difficulty of attributing observed changes to any one factor. A good example of this is the rigorous and compelling study (De'ath et al. 2009), which demonstrated that the calcification of reef-building corals over vast areas of the Great Barrier Reef (GBR) has decreased by around 14% since 1990. This study, which investigated the growth rate over the past several 100 years of 328 colonies of a massive coral *Porites* from 69 reefs across the GBR in Australia, also revealed that changes unprecedented in the 400 years of record examined. The observation that the decrease in calcification occurred both inshore and offshore sites down the entire length of the GBR seems to indicate that the general downturn of calcification has occurred. Given that bleaching events and indeed anthropogenic disturbances such as deteriorating water quality have been the most intense on the inshore reefs of the GBR, the explanation that this was due to general factors such as ocean acidification with/without the influence of thermal stress is highly plausible.

While most of the work that has been done so far on ocean acidification has concentrated on calcifying organisms, it is important to note that changes in pH can have significant direct effects on processes other than that of calcification (Portner 2008). These other impacts are associated with acidosis and are a direct result of the increased numbers of protons influencing membrane transport, enzyme activity, and a whole range of other key physiological processes (Portner and Knust 2007; Portner and Farrell 2008). So far, our understanding and documentation of these types of impacts is in its infancy. It is clear, however, that direct pH effects may be extremely important and many influence a wide range of processes. Species that have highly exposed respiratory surfaces may also experience changes to gas exchange, a consequence of perturbations to all-important proton gradients and blood pigments. Organisms such as squid, annelid worms, and bivalve molluscs have been reported to show metabolic depression as pH decreases in the water surrounding (Portner 2008). Equally subtle effects may be felt on reproductive behavior of organisms through changes to the ability to sense different habitats, as well as impacts on the ability of noncoral organisms to tolerate thermal extremes (Portner 2008; Portner and Farrell 2008).

As discussed above, this phenomenon has already been documented in corals, where acidification makes some species of corals more sensitive to coral bleaching (Anthony et al. 2008). These subtle physiological impacts of ocean acidification may have a wide range of effects over and above that of physiology of corals. Recent work by Munday and colleagues has shown that the ability of fish to navigate and avoid predators is disturbed at relatively low levels of acidity (Munday et al. 2007, 2009). These subtle yet fundamental impacts could have major implications for coral reef organisms and the ecosystems that they form. Understanding the ramifications of the full spectrum of these changes must be a priority for future research.

Without serious action on global warming and ocean acidification, coral reef environments will steadily move towards temperatures and carbonate ion concentrations in which carbonate reef systems are unlikely to form due to the excessive coral mortality rates and negative reef carbonate accretion (Hoegh-Guldberg et al. 2007). Under these circumstances, reefs that are currently coral dominated will be replaced by reef structures that will be dominated by other organisms (Fig. 5).

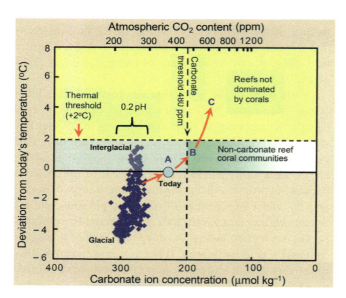

Fig. 5 Carbonate ion concentrations calculated from atmospheric carbon dioxide levels and sea temperature over the past 420,000 years for the coral reef with the average temperature of 25°C. Glacial and interglacial conditions are indicated, along with thresholds for temperature (2°C above today) and carbonate ion concentrations (200 μmol kg^{-1}). Conditions under which noncarbonate coral reefs will dominate and where reefs will potentially disappear as coral-dominated ecosystems are indicated, along with the relative position of today (**a**), and when atmospheric carbon dioxide reaches 480 ppm (**b**) and 600 ppm (**c**). Details of calculations can be obtained from Hoegh-Guldberg et al. (2007) from which this figure was adapted and used with the permission of Science Magazine

5.3 Other Factors Associated with Climate Change

As discussed above, enhanced greenhouse conditions on our planet are likely to affect many other factors in addition to sea temperature and acidity. Sea-level rise is generally seen as not posing a problem for coral reefs (IPCC 2007; Rahmstorf 2007) if coral growth remains vigorous. However, clear signs that other factors such as sea temperature and acidity are likely to reduce the growth of corals suggest that some reef communities may struggle to maintain themselves under even the most minimum changes in sea level. Dramatic changes in sea level in response to the disintegration of the Greenland and western Antarctic ice sheets (Ridley et al. 2005; Gregory and Huybrechts 2006; Hansen 2007) will further exacerbate this problem, causing to deeper sections of reef to experience lower light levels and consequent loss (Gregory and Huybrechts 2006; Steig et al. 2009). Other factors such as changing patterns of drought and storm activity are likely to also interact with coral reefs and influence their future. Many parts of the world are already experiencing droughts, leading to the excessive drying and destabilisation of sediments and increasing outflows from river catchments (e.g. Queensland coastline and the adjacent GBR). Given the projection of drought coupled with sporadic and more intense rainfall events, it is possible that the flux of sediments and nutrients from these catchments may increase sharply as we enter the coming decades and century. Given the crucial role that sediments and nutrients play in limiting reef development in many parts of the world today, the prospect of reduced reef health on account of these factors remains likely.

There are many other factors that are important to coral reefs, which may change as the climate changes. The presence of clouds along coastlines may change as the hydrological cycle shifts, affecting both the amount of photosynthetic and ultraviolet radiation. These changes can modify the impact of thermal stress, as seen around French Polynesia in 1998 when the impact of warmer than normal conditions was offset by excessive cloudiness (Mumby et al. 2001). Changes to ocean circulation may lead to dramatic changes to currents, leading to modification of the extent and direction with which coral reefs in particular regions are connected. One of the most important issues is how different factors are likely to interact. Anthony et al. (2008) revealed that the impacts of thermal stress are exacerbated by ocean acidification, with the observation that the thermal threshold at which corals will bleach is significantly reduced if thermal stress occurs within an acidified ocean. Interaction between global and local factors also assumes importance, not only in terms of understanding how global change will play out, but also in terms of potential opportunities for reef managers to respond to the challenges of climate change. Reducing fish grazing, for example, has a negative effect on the ability of bleached reef to recover (Hughes et al. 2003), and hence suggests that maintaining levels of herbivory on coral reefs may form an important strategy for buying time as we struggle to stabilize atmospheric concentrations of carbon dioxide and other greenhouse gases.

5.4 Ecosystem Responses and Outcomes

Most analyses that have examined how coral populations are likely to respond to climate change suggests that the abundance of corals on reefs will decrease sharply as mortality increases and growth slows (Baskett et al. 2009). The resulting decrease in the abundance and community structure of reef-building corals will have important implications for the role that corals play in forming the habitat for the many species that live in and around coral reefs. While the impact on a few groups such as coral reef fishes is becoming clear, the implications for the vast majority of coral reef organisms are poorly understood (Przeslawski et al. 2008). Studies of the fish on reefs with various amounts of damage from coral removed from them as a result of Crown of Thorns starfish (*Acanthaster planci*) infestations and mass coral bleaching have indicated that climate change is likely to drive major changes in the community composition of many reef-associated species (Wilson et al. 2006; Graham et al. 2007; Pratchett et al. 2008; Wilson et al. 2008). For example, Wilson et al. (2006) explored the data from 17 independent studies done in the past decade finding that 62% of fish species on coral-dominated reefs declined in abundance within 3 years of disturbances that resulted in >10% decline in coral cover. Inspection of the types of fish that disappeared as corals decreased in abundance revealed that those species that either eat coral directly or require live coral as a settlement queue are among the first to be lost from reefs affected by mass coral bleaching and mortality (Wilson et al. 2006; Graham et al. 2007; Pratchett et al. 2008). In some cases, those fish that either eat invertebrates or graze the epilithic algal matrix (EAM) showed either no change or an increase in abundance. Clearly, we need to learn more about how climate change will modify the ability of organisms such as corals and calcareous algae to maintain the three-dimensional structures that are so important as fish habitats. It is also clear that, like other organisms, there are also a series of direct impacts on fish of changing sea temperatures and pH. The early evidence (Munday et al. 2008; Portner and Farrell 2008) suggest that these changes will also play important roles in determining the health of coral reef fish, especially in the way factors and responses interact synergistically.

Our understanding of the ecological implications of global warming and ocean acidification remained to be fully understood. However, given the enormous range of processes that are likely to be impacted by the physical and chemical changes exerted on tropical/subtropical oceans by global warming and ocean acidification, it is highly likely that coral reefs will continue to be influenced in complex and often surprising ways. The next few decades of research will no doubt be focused on the enormous challenges that lie ahead for marine ecosystems like coral reefs.

6 The Role of Acclimation and Adaptation in Altering Projections of the State of Coral Reefs Under Climate Change

Organisms may respond to change in several fundamental ways. They may tune their physiological processes so that they operate more optimally under a set of new conditions ("acclimation"; see Chapters 8 and 24), or their populations may undergo genetic change via natural selection such that a greater proportion of these populations is represented over time by genetically tolerant individuals. In the latter case, new capabilities are introduced through mutation or recombination via sexual reproduction. Lastly, populations of organisms may also experience genetic change through the introduction of new capabilities via migration. The key consideration in terms of understanding the trajectory of any organism is whether or not these processes can keep up with the current rapid rate of anthropogenic climate change. As mentioned above, the current rate of climate change is several orders of magnitude greater than even the fastest rates of environmental change seen during the ice age transitions of the last million years or more. This should be of considerable concern given that ice age transitions, even with their slower rates of change, were characterized by enormous amounts of disruption and change to the earth's biosphere (Mayewski et al. 2004).

Understanding of how coral reefs will change over the next few decades and century hinges very much on our understanding of how thresholds to stress will change over time. Several authors (Hoegh-Guldberg 1999; Done et al. 2003; Hughes et al. 2003; Donner et al. 2005) have discussed how changes to tolerance thresholds would delay the impact of changing sea temperatures on coral populations. Donner et al. (2005) conclude that coral populations would need to show an increase in their thermal tolerance of 0.2–1.0°C per decade if they were to avoid the scenarios originally proposed by Hoegh-Guldberg (1999). After almost 30 years of data on sea temperature and corals, there is little or no evidence that corals have adjusted their thermal threshold by this amount. Other evidence that corals and their dinoflagellates symbionts can change rapidly enough is equivocal at best. Reef-building corals have long generation times with populations that generally have low genetic diversity, resulting in relatively slow rates of evolution. It is clear that changes in species composition (Hoegh-Guldberg and Salvat 1995; Marshall and Baird 2000) have occurred, which has led to more robust and less diverse communities of corals. It is also clear that adaptive optima within populations may be influenced by natural selection removing sensitive individuals from populations of reef-building corals (Maynard et al. 2008). These changes are likely to result in thermally more tolerant populations and communities, but which have all the negative characteristics of populations and communities in which genetic diversity has been reduced (e.g. greater vulnerability to other factors such as disease).

The prospect that corals can swap their symbionts for truly novel forms within an ecological timeframe as postulated within the adaptive bleaching hypothesis (Buddemeier and Fautin 1993) has never been demonstrated (Goulet and Coffroth 2003; Goulet 2006). Rather, studies have identified the interesting behavior of multicladal symbiosis in which coral–dinoflagellate associations that host two or more different types of dinoflagellates symbionts within the one association (Rowan et al. 1997; Knowlton and Rohwer 2003; Baker and Romanski 2007) may respond to environmental change by encouraging the dominance of one variety of symbiont over another (Baker 2001; Baker 2004; Jones et al. 2008). In some cases, this has been associated with increased thermal tolerance as one variety of symbiont dominates over another (Rowan et al. 1997; Sampayo et al. 2008). These examples are interesting but are only restricted to a small number of symbiotic associations (Goulet 2006), and are relatively fixed in time in terms of their other thermal tolerance. For example, the evolution of multicladal symbioses will ultimately be attuned to the environment surrounding them, much as the expressions of different forms of an enzyme are tuned to the conditions around an organism. For this reason, shifts in the dominance of one clade of symbiont over another are really a form of physiological acclimation and hence do not have the crucially important characteristic of being able to change endlessly as the climate increases into unknown "territory" (Hoegh-Guldberg et al. 2002).

There are a couple other reasons why varying the symbiotic makeup of a coral–dinoflagellate association is unlikely to provide the rapid rates of change in thermal threshold that are required to keep up with climate change. Firstly, acclimation to thermal stress is only one of the several factors and thus far there has been no evidence of acclimation to reduced carbonate ion concentrations (Dove 2004). Secondly, it would be highly unusual that the thermal tolerance of reef-building corals was solely determined by the physiological performance of the dinoflagellates symbionts when there are so many other processes that are likely to be affected by temperature, pH, and carbonate ion concentrations within the coral host (Kleypas et al. 1999b; Anthony et al. 2008). Thirdly, the very fact that the abundance of coral communities appear to be on an increasingly downward trend a opposed to the opposite (Bruno and Selig 2007) suggests that corals are not becoming more tolerant to the stress conditions increase. And finally, any consideration of adaptation as a mechanism has to appreciate the difficulty of adapting to a continuously changing environment. That is, the environment is not undergoing a step change (passing from one environmental condition to another) but rather continuously changing (and accelerating) as atmospheric carbon dioxide continues to build up in the Earth's atmosphere (IPCC 2007). This puts enormous and continuous pressure on coral communities and makes sufficient evolution in ecological time extremely unlikely.

7 Prospects for Coral Reefs and Dependent Societies in the Coming Decades and Century

On the balance of evidence, coral reefs are likely to face tough times over the coming decades and century if we continue to increase atmospheric carbon dioxide at the phenomenal rate of 2 ppm per year (IPCC 2007). As has been argued above, exceeding atmospheric carbon dioxide concentrations of more than 450 ppm will place coral reefs in conditions under which the thermal threshold for bleaching will be exceeded on a yearly basis, and where rates of calcification will fall behind the rates of physical and biological erosion (Hoegh-Guldberg et al. 2007). Under these conditions, coral reefs are likely to change from coral-dominated ecosystems into eroding seascapes in which biodiversity will be a fraction of what it is today (Fig. 6). Less clear are the impacts on processes such as primary productivity and nutrient recycling. While reefs occupied by other organisms such as seaweeds and cyanobacteria are less visually appealing, their ability to trap sunlight may equal or exceed that of healthy coral reefs. Productivity aside, however, it is very clear that the appeal of coral reefs for industries such as tourism is likely to be substantially reduced as reefs change. Given that tourism is a major export earner for many developing as well as developed countries, these changes are likely to affect the amount of international income generated by reefs for national economies that are in many cases highly dependent (Hoegh-Guldberg 2000; Hoegh-Guldberg and Hoegh-Guldberg 2004). Together with the other impacts that are likely to be felt on coastal infrastructure due to changing weather patterns, sea-level rise and the loss of coastal protection by carbonate reefs, the outcomes for many coral reef dependent societies of unrestrained climate change could be serious (IPCC 2007).

Fig. 6 Coral reefs face an uncertain future if their ability to form carbonate skeletons is diminished under high atmospheric carbon dioxide. This series of photographs depict a series of sites on the Great Barrier Reef (GBR) that are analogous to reef states expected under the rapid buildup of atmospheric CO_2, (**a**). Reef-building coral reefs like this one have healthy populations of reef-building corals and red coralline algae that are able to build and maintain the 3-D framework of the reef. This framework provides habitat for an estimated million species and ultimately provides food, livelihoods, and coastal protection to hundreds of millions of people worldwide. (**b**) As atmospheric concentrations of carbon dioxide cross thresholds for thermal stress and reef accretion (~450 ppm), corals are likely to be outcompeted by other benthic organisms such as seaweeds and soft corals. Reefs may not show net accretion of calcium carbonate this point and available habitats for coral reef species will diminish dramatically. (**c**) Beyond 450 ppm, corals and other calcifiers are likely to be rare on coral reefs, leading to circumstances in which reefs may begin to crumble and disappear (Adapted from Hoegh-Guldberg et al. (2007) and used with permission of Science Magazine. See this article for a full detail of locations photographed)

Under an ideal set of circumstances, the current failure of international climate treaty negotiations aside, international negotiations such as those in the post-Kyoto period will have to achieve drastic cuts in global carbon dioxide emissions (>80% by 2050) in order to stabilize levels at or below 450 ppm. If this is the case, then coral reefs have a good chance of persisting albeit in a rather different form to that of today. In this respect, coral-dominated areas such as those of the Indo-Pacific region may resemble more the algal-dominated (with a few corals present) condition of many reefs in the Caribbean or Eastern Pacific (Manzello et al. 2008). In this situation, reducing the impact of other stresses such as the overexploitation of key fish species and deteriorating water quality will become increasingly important as corals are weakened (Hughes et al. 2003). As has been discussed above, promoting healthy populations of grazing fishes on coral reefs that have been impacted by mass coral bleaching has been shown to aid the recovery of bleached reefs as compared to reefs where grazing fishes have been removed (Hughes et al. 2007).

8 Conclusions and Future Directions

Coral reef ecosystems are seriously threatened by current and future rates of change in ocean temperature and acidity. If current trends in atmospheric carbon dioxide concentrations continue, coral reefs will reach, and quickly exceed, critical thresholds associated with their thermal tolerance and carbonate ion requirements. At the same time, other factors such as sea-level rise, storm intensity, and rainfall intensity are likely to exacerbate g along tropical/subtropical coastlines. Under these conditions, the corals that build the framework of coral reefs will become rare organisms on tropical/subtropical reef systems, and the distribution and abundance of carbonate reef systems will undergo severe contraction. With this will go an enormous number of species that are directly dependent on reef-building corals and the carbonate structures that they build and maintain. Prospects for rapid evolution of corals and/or their dinoflagellate symbionts are largely unsupported and appear unlikely to keep pace with the rapid rate of anthropogenic change. Given the serious threat that global climate change poses for coral rests, there is an urgent imperative for governments everywhere to rapidly reduce carbon dioxide emissions while assisting developing nations reduce the impact of local stresses such as overfishing and unsustainable coastal development on coral reefs. If these circumstances can be achieved, then coral reef ecosystems will have some chance of surviving the coming century of extreme environmental stress. Let us hope that this is the case.

Acknowledgments The author is grateful to Jez Roff, Ken Anthony, and Sophie Dove for reading and commenting on the original manuscript, and for support through the Queensland Smart State Premier's Fellowship program.

References

Anthony KR, Kline DI, Diaz-Pulido G, Dove S, Hoegh-Guldberg O (2008) Ocean acidification causes bleaching and productivity loss in coral reef builders. Proc Natl Acad Sci U S A 105: 17442–17446

Augustin L, Barbante C, Barnes PRF, Barnola JM, Bigler M, Castellano E, Cattani O, Chappellaz J, DahlJensen D, Delmonte B et al (2004) Eight glacial cycles from an Antarctic ice core. Nature 429:623–628

Baker AC (2001) Reef corals bleach to survive change. Nature 411:765–766

Baker AC (2004) Symbiont diversity on coral reefs and its relationship to bleaching resistance and resilience. In: Rosenberg E, Loya Y (eds) Coral health and disease. Springer, Berlin, p 488

Baker AC, Romanski AM (2007) Multiple symbiotic partnerships are common in scleractinian corals, but not in octocorals: comment on Goulet (2006). Mar Ecol Prog Ser 335:237–242

Baskett M, Gaines S, Nisbet R (2009) Symbiont diversity may help coral reefs survive moderate climate change. Ecol Appl 19:3–17

Bruno JF, Selig ER (2007a) Regional decline of coral cover in the Indo-Pacific: timing, extent, and subregional comparisons. PLoS ONE 2(8):e711

Buddemeier RW, Fautin DG (1993) Coral bleaching as an adaptive mechanism. Bioscience 43:320–326

Caldeira k, Wickett ME (2003) Anthropogenic carbon and ocean pH. Nature 425:365

Cressey D (2007) Arctic melt opens Northwest passage. Nature 449:267–267

De'ath G, Lough JM, Fabricius KE (2009) Declining coral calcification on the Great Barrier Reef. Science 323:116–119

Derocher AE, Lunn NJ, Stirling I (2004) Polar bears in a warming climate. Integr Comp Biol 44:163–176

Done T, Whetton P, Jones R, Berkelmans R, Lough J, Skirving W, Wooldridge S (2003) Global climate change and coral bleaching on the Great Barrier Reef. State of Queensland Greenhouse Taskforce, Department of Natural Resources and Mines, Brisbane, p 54

Donner SD, Skirving WJ, Little CM, Oppenheimer M, Hoegh-Guldberg O (2005) Global assessment of coral bleaching and required rates of adaptation under climate change. Glob Change Biol 11:2251–2265

Donner SD, Knutson TR, Oppenheimer M (2007) Model-based assessment of the role of human-induced climate change in the 2005 Caribbean coral bleaching event. Proc Natl Acad Sci U S A 104:5483–5488

Dove S (2004) Scleractinian corals with photoprotective host pigments are hypersensitive to thermal bleaching. Mar Ecol Prog Ser 272:99–116

Emanuel K (2005) Increasing destructiveness of tropical cyclones over the past 30 years. Nature 436:686–688

Glynn PW (1983) Extensive bleaching and death of reef corals on the pacific coast of panama. Environ Conserv 10:149–154

Glynn PW (1996) Coral reef bleaching: Facts, hypotheses and implications. Global Change Biol 2:495–509

Glynn PW, D'Croz L (1990) Experimental-evidence for high-temperature stress as the cause of el-nino-coincident coral mortality. Coral Reefs 8:181–191

Goulet TL (2006) Most corals may not change their symbionts. Mar Ecol Prog Ser 327:1–7

Goulet TL, Coffroth MA (2003) Stability of an octocoral-algal symbiosis over time and space. Mar Ecol Prog Ser 250:117–124

Graham NA, Wilson SK, Jennings S, Polunin NV, Robinson J, Bijoux JP, Daw TM (2007) Lag effects in the impacts of mass coral bleaching on coral reef fish, fisheries, and ecosystems. Conserv Biol 21:1291–1300

Gregory JM, Huybrechts P (2006) Ice-sheet contributions to future sea-level change. Philos Trans R Soc A Math Phys Eng Sci 364:1709–1731

Gregory JM, Huybrechts P, Raper SCB (2004) Climatology – threatened loss of the Greenland ice-sheet. Nature 428:616–616

Guinotte JM, Buddemeier RW, Kleypas JA (2003) Future coral reef habitat marginality: temporal and spatial effects of climate change in the Pacific basin. Coral Reefs 22:551–558

Hansen JE (2007) Scientific reticence and sea level rise. Environ Res Lett 2:2

Hansson M, Hoffmann G, Hutterli MA, Huybrechts P, Isaksson E, Johnsen S, Jouzel J, Kaczmarska M, Karlin T, Kaufmann P et al (2006) One-to-one coupling of glacial climate variability in Greenland and Antarctica. Nature 444:195–198

Hoegh-Guldberg O (1999) Climate change, coral bleaching and the future of the world's coral reefs. Mar Freshw Res 50:839–866

Hoegh-Guldberg H, Hoegh-Guldberg O (2004) Great Barrier Reef 2050: implications of climate change for Australia's Great Barrier Reef, vol WWF-Australia February 2004, Brisbane, p 345

Hoegh-Guldberg O, Salvat B (1995) Periodic mass-bleaching and elevated sea temperatures – bleaching of outer reef slope communities in Moorea, French-Polynesia. Mar Ecol Prog Ser 121:181–190

Hoegh-Guldberg O, Smith GJ (1989) The effect of sudden changes in temperature, light and salinity on the population density and export of zooxanthellae from the reef corals *Stylophora pistillata* and *Seriatopra hystrix*. J Exp Mar Biol Ecol 129:279–303

Hoegh-Guldberg O, Hoegh-Guldberg H, Stout DK, Cesar H (2000) Pacific in peril: biological, economic and social impacts of climate change on Pacific coral reefs. Greenpeace, Suva, Fiji, p 72

Hoegh-Guldberg O, Jones RJ, Ward S, Loh WL (2002) Is coral bleaching really adaptive? Nature (London) 415:601–602

Hoegh-Guldberg O, Mumby PJ, Hooten AJ, Steneck RS, Greenfield P, Gomez E, Harvell CD, Sale PF, Edwards AJ, Caldeira K et al (14 Dec 2007) Coral reefs under rapid climate change and ocean acidification. Science 318(5857):1737–1742

Hoegh-Guldberg O, Hughes L, McIntyre S, Lindenmayer DB, Parmesan C, Possingham HP, Thomas CD (2008) Ecology. Assisted colonization and rapid climate change. Science 321:345–346

Hughes TP, Baird AH, Bellwood DR, Card M, Connolly SR, Folke C, Grosberg R, Hoegh-Guldberg O, Jackson JB, Kleypas J et al (2003) Climate change, human impacts, and the resilience of coral reefs. Science 301:929–933

Hughes TP, Rodrigues MJ, Bellwood DR, Ceccarelli D, Hoegh-Guldberg O, McCook L, Moltschaniwskyj N, Pratchett MS, Steneck RS, Willis B (20 Feb 2007) Phase shifts, herbivory, and the resilience of coral reefs to climate change. Curr Biol 17(4):360–365. Epub 2007 Feb 8

IPCC (2007) Synthesis report. Contribution of Working Groups I, II and III to the Fourth Assessment Report of the Intergovernmental Panel on Climate Change, eds. C. W. Team R. K. Pachauri and A. Reisinger. Geneva, Switzerland: Intergovernmental Panel on Climate Change, 2008, p 104

Jones AM, Berkelmans R, van Oppen MJH, Mieog JC, Sinclair W (2008) A community change in the algal endosymbionts of a scleractinian coral following a natural bleaching event: field evidence of acclimatization. Proc R Soc B Biol Sci 275:1359–1365

Kleypas JA, Langdon C (2006) Coral reefs and changing seawater chemistry, Chapter 5. In Phinney J, Hoegh-Guldberg O, Kleypas J, Skirving W, Strong AE (eds) Coral reefs and climate change: science and management, vol 61. AGU Monograph Series, Coastal and Estuarine Studies. Geophysical Union, Washington, DC

Kleypas JA, McManus JW, Menez LAB (1999a) Environmental limits to coral reef development: where do we draw the line? Am Zool 39:146–159

Kleypas JA, Buddemeier RW, Archer D, Gattuso JP, Langdon C, Opdyke BN (1999b) Geochemical consequences of increased atmospheric carbon dioxide on coral reefs. Science 284:118–120

Knowlton N, Rohwer F (2003) Multispecies microbial mutualisms on coral reefs: the host as a habitat. Am Nat 162:S51–S62

Lessios HA, Glynn PW, Robertson DR (1983) Mass mortalities of coral-reef organisms. Science 222:715–715

Mann ME (2002) The value of multiple proxies. Science 297:1481–1482

Manzello DP, Kleypas JA, Budd DA, Eakin CM, Glynn PW, Langdon C (2008) Poorly cemented coral reefs of the eastern tropical Pacific: possible insights into reef development in a high-CO2 world. Proc Natl Acad Sci U S A 105:10450–10455

Marshall PA, Baird AH (2000) Bleaching of corals on the Great Barrier Reef: differential susceptibilities among taxa. Coral Reefs 19:155–163

Mayewski PA, Rohling EE, Stager JC, Karlen W, Maasch KA, Meeker LD, Meyerson EA, Gasse F, van Kreveld S, Holmgren K et al (2004) Holocene climate variability. Quat Res 62:243–255

Maynard JA, Anthony KRN, Marshall PA, Masiri I (2008) Major bleaching events can lead to increased thermal tolerance in corals. Mar Biol 155:173–182

Mumby PJ, Chisholm JRM, Edwards AJ, Andrefouet S, Jaubert J (2001) Cloudy weather may have saved Society Island reef corals during the 1998 ENSO event. Mar Ecol Prog Ser 222:209–216

Munday P, Jones G, Sheaves M, Williams A, Goby G (2007) Vulnerability of fishes of the Great Barrier Reef to climate change. Climate Change Great Barrier Reef 357–391

Munday PL, Kingsford MJ, O'Callaghan M, Donelson JM (2008) Elevated temperature restricts growth potential of the coral reef fish Acanthochromis polyacanthus. Coral Reefs 27:927–931

Munday PL, Donelson JM, Dixson DL, Endo GGK (2009) Effects of ocean acidification on the early life history of a tropical marine fish. Proc R Soc Lond B Biol Sci 276(1671):3275–3283

Muscatine L (1990) The role of symbiotic algae in carbon and energy flux in reef corals. In: Dubinsky Z (ed) Ecosystems of the world, vol. 24. Coral reefs. Elsevier, Amsterdam, pp 75–87

Parmesan C, Yohe G (2003) A globally coherent fingerprint of climate change impacts across natural systems. Nature 421:37–42

Parmesan C, Ryrholm N, Stefanescu C, Hillk JK, Thomas CD, Descimon H, Huntley B, KailaI L, Kullberg J, Tammaru T et al (1999) Poleward shifts in geographical ranges of butterfly.pdf. Nature 399:579–783

Petit JR, Jouzel J, Raynaud D, Barkov NI, Barnola J-M, Basile I, Bender M, Chappellaz J, Davis M, Delaygue G et al (1999) Climate and atmospheric history of the past 420,000 years from the Vostok ice core, Antarctica. Nature 399:429–436

Portner HO (2008) Ecosystem effects of ocean acidification in times of ocean warming: a physiologist's view. Mar Ecol Prog Ser 373:203–217

Portner HO, Farrell AP (2008) Ecology physiology and climate change. Science 322:690–692

Portner HO, Knust R (2007) Climate change affects marine fishes through the oxygen limitation of thermal tolerance. Science 315:95–97

Pratchett MS, Munday PL, Wilson SK, Graham NAJ, Cinnerm JE, Bellwood DR, Jones GP, Polunin NVC, McClanahan TR (2008) Effects of climate-induced coral bleaching on coral-reef fishes – ecological and economic consequences. Oceanogr Mar Biol Annu Rev 46:251–296

Przeslawski R, Ahyong S, Byrne M, Worheide G, Hutchings P (2008) Beyond corals and fish: the effects of climate change on noncoral benthic invertebrates of tropical reefs. Global Change Biol 14:2773–2795

Rahmstorf S (2007) A semi-empirical approach to projecting future sea-level rise. Science 315:368–370

Reaka-Kudla ML (1997) Global biodiversity of coral reefs: a comparison with rainforests. In: Reaka-Kudla ML, Wilson DE (eds) Biodiversity II: understanding and protecting our biological resources, vol II. Joseph Henry Press, Washington, DC, p 551

Ridley JK, Huybrechts P, Gregory JM, Lowe JA (2005) Elimination of the Greenland ice sheet in a high CO_2 climate. J Clim 18:3409–3427

Rowan R, Knowlton N, Baker A, Jara J (1997) Landscape ecology of algal symbionts creates variation in episodes of coral bleaching. Nature 388:265–269

Sampayo EM, Ridgway T, Bongaerts P, Hoegh-Guldberg O (2008) Bleaching susceptibility and mortality of corals are determined by fine-scale differences in symbiont type. Proc Natl Acad Sci USA 105:10444–10449

Schipper J, Chanson JS, Chiozza F, Cox NA, Hoffmann M, Katariya V, Lamoreux J, Rodrigues ASL, Stuart SN, Temple HJ et al (2008) The status of the world's land and marine mammals: diversity, threat, and knowledge. Science 322:225–230

Solomon S, Qin D, Manning M et al (2008) Climate change 2007: the physical science basis. Cambridge University Press, Cambridge/New York/Melbourne/Madrid/Cape Town/Singapore/São Paulo/Delhi

Steig EJ, Schneider DP, Rutherford SD, Mann ME, Comiso JC, Shindell DT (2009) Warming of the Antarctic ice-sheet surface since the 1957 international geophysical year. Nature 457:459–462

Strong AE, Barrientos CS, Duda C, Sapper J (1996) Improved satellite technique for monitoring coral reef bleaching. Proceedings of the eighth international coral reef symposium, Panama, June 1996, pp 1495–1497

Toscano MA, Liu G, Guch IC, Casey KS, Strong AE, Meyer JE (2000) Improved prediction of coral bleaching using high-resolution HotSpot anomaly mapping. Proceedings of the ninth international coral reef symposium, Bali, 2000, vol 2, pp 1143–1147

Walther GR, Post E, Convey P, Menzel A, Parmesan C, Beebee TJ, Fromentin JM, Hoegh-Guldberg O, Bairlein F (2002) Ecological responses to recent climate change. Nature 416:389–395

Wilkinson C, Hodgson G (1999) Coral reefs and the 1997–1998 mass bleaching and mortality. Nat Resour 35:16–25

Wilkinson C, Linden O, Cesar H, Hodgson G, Rubens J, Strong AE (1999) Ecological and socioeconomic impacts of 1998 coral mortality in the Indian ocean: an ENSO impact and a warning of future change? Ambio 28:188–196

Wilson SK, Graham NAJ, Pratchett MS, Jones GP, Polunin NVC (2006) Multiple disturbances and the global degradation of coral reefs: are reef fishes at risk or resilient? Global Change Biol 12:2220–2234

Wilson SK, Fisher R, Pratchett MS, Graham NAJ, Dulvy NK, Turner RA, Cakacaka A, Polunin NVC, Rushton SP (2008) Exploitation and habitat degradation as agents of change within coral reef fish communities. Global Change Biol 14:2796–2809

Witze A (2008) Losing Greenland. Nature 452:798–802

Zhang J, Lindsay R, Steele M, Schweiger A (2008) What drove the dramatic retreat of arctic sea ice during summer 2007? Geophys Res Lett 35:L11505. doi:10.1029/2008GL034005

Coral Bleaching: Causes and Mechanisms

Michael P. Lesser

Abstract Unprecedented changes in coral reef systems have focused attention on a wide range of stressors on local, regional, and global spatial scales but global climate change resulting in elevated seawater temperatures is widely accepted as having contributed to the major declines in coral cover or phase shifts in community structure on time scales never previously observed or recorded in the geological record. The major mechanism of scleractinian mortality as a result of global climate change is "coral bleaching," the loss of the endosymbiotic dinoflagellates (=zooxanthellae) that occurs as part of the coral stress response to temperature perturbations in combination with several other synergistic factors. Over several years many studies have shown that the common mechanism underlying the stress response of corals to elevated temperatures is oxidative stress that is exacerbated when exposure to high irradiances of solar radiation accompanies the thermal insult. Oxidative stress, the production and accumulation of reduced oxygen intermediates such as superoxide radicals, singlet oxygen, hydrogen peroxide, and hydroxyl radicals can cause damage to lipids, proteins, and DNA. Reactive oxygen species are also important signal transduction molecules and mediators of damage in cellular processes, such as apoptosis, autophagy, and cell necrosis all of which are believed to have roles in coral bleaching depending on the intensity and duration of the environmental insult. This chapter examines the current evidence supporting the hypothesis that the production and accumulation of reactive oxygen species leads to oxidative stress and is the proximal cause of coral bleaching.

Keywords Corals • zooxanthellae • coral bleaching • seawater temperature • global climate change

1 Introduction

We now have ample evidence that global climate change, principally the emission and accumulation of greenhouse gases (e.g., CO_2, CH_4), has had multiple effects on coral reefs including increases in seawater temperature, changes in the calcium carbonate saturation point, large-scale changes in atmospheric/oceanic coupling (e.g., El Niño-Southern Oscillation [ENSO]) and changes in sea level (Hoegh-Guldberg 1999; Kleypas et al. 1999; Wilkinson 1999; Hoegh-Guldberg et al. 2007). Terrestrial, aquatic, and marine ecosystems are all affected by global climate change but coral reefs have become the "poster child" for ecosystems that will experience profound ecological changes in the next 50 years in a "business as usual" scenario where reefs as we currently know them will only exist in very isolated places (Hoegh-Guldberg et al. 2007). Zooxanthellate scleractinian corals are a major contributor to the productivity of coral reef ecosystems worldwide between 30°N and 30°S latitude and their prolific growth rates (3–15 cm year^{-1}) in optically clear, oligotrophic, tropical seas is responsible for the three-dimensional framework of coral reef systems and biodiversity rivaling tropical rain forests. Coral reefs are also a source of food and livelihood for at least 125 million people worldwide; they support major industries (fishing and tourism), and play a key role in stabilizing coastlines (Hoegh-Guldberg 1999).

Coral reefs are experiencing unparalleled levels of anthropogenically induced stress (Hoegh-Guldberg et al. 2007; Carpenter et al. 2008). Until recently, global climate change was seen as just one of the many factors (e.g., eutrophication, coastal development, sedimentation, overfishing) responsible for the decline in the health of coral reefs (Hughes 1994; Hughes and Connell 1999). The impact of anthropogenically induced stress on the percent cover of living coral worldwide and the projection of continued rising sea temperatures under greenhouse warming scenarios (Hoegh-Guldberg 1999) has changed research priorities towards understanding the potential impact of greenhouse gas driven climate change on the world's coral reefs. Additionally, in the last decade, another

---

M.P. Lesser (✉)
Department of Molecular, Cellular and Biomedical Sciences,
University of New Hampshire, Durham, NH 03824, USA
e-mail: mpl@unh.edu

effect of the accumulation of dissolved carbon dioxide in the world's oceans has been a significant change in seawater pH, or ocean acidification (Hoegh-Guldberg et al. 2007). The combined effects of elevated seawater temperature and ocean acidification on corals are only beginning to be addressed but early studies on calcification, metabolism, and coral mortality are revealing complex physiological interactions and poor outcomes for corals (Hoegh-Guldberg et al. 2007; Anthony et al. 2008; Crawley et al. 2009). Currently, it is believed that the most effective, and immediate, strategy to undertake is to reduce local effects such as industrial and agricultural pollution, eutrophication, and over-fishing as compounding stressors in order to provide more time to solve the longer timescale problems associated with global climate change (Hoegh-Guldberg et al. 2007).

The most profound response of corals to environmental stress is to expel their symbiotic dinoflagellates known as zooxanthellae from their tissues into the environment during a process known as "coral bleaching." Coral bleaching results in a paling or whitening of the affected coral with varying levels of coral mortality depending on the severity of the thermal stress and can be caused by the expulsion of zooxanthellae and/or the loss of photosynthetic pigment per cell (Hoegh-Guldberg 1999; Lesser 2004). Coral bleaching is distinct from the seasonal cycling of zooxanthellae densities in reef corals where the areal concentration of zooxanthellae, as well as the number of functional photosystem II (PSII) reaction centers, decreases to annual lows during the summer months (Fagoonee et al. 1999; Fitt et al. 2000; Warner et al. 2002) and then recovers during the Fall and Winter (Northern Hemisphere). The number and severity of coral bleaching events has been described as a "biological signal" (Hughes 2000) for the consequences of global climate change on coral reefs that is occurring worldwide, and it is predicted to continue if current scenarios of greenhouse gas accumulation persist (Hoegh-Guldberg 1999; Sheppard 2003; Hoegh-Guldberg et al. 2007). The principal concern regarding global climate change is that within the framework of evolutionary adaptation, scleractinian corals will not be able to physiologically adapt at the current rates of environmental change (Gates and Edmunds 1999). In fact, modeling studies have shown that severe bleaching will become common in the Caribbean basin in the next 20–30 years without changes in the rate of greenhouse gas emissions or changes in the physiological tolerances of corals and their zooxanthellae (Hoegh-Guldberg 1999; Donner et al. 2007; Lesser 2007).

Mass mortalities of corals have been reported as far back as the 1870s (Glynn 1993; Williams and Bunkley-Williams 1990), but these early events provide limited insight into the specific perturbation, such as elevated seawater temperatures, that these events were associated with. The earliest comprehensive report of temperature-related bleaching comes from studies on the Great Barrier Reef from 1928 to 1929 (Yonge and Nichols 1931). Yonge and Nichols (1931) recorded a mass bleaching event during the Austral Summer of 1929 on reef flats during a Spring tide where water temperatures reached 35.1°C. Many corals that were submerged or aerially exposed were killed. Yonge and Nicholls surmised that elevated seawater temperature, and not irradiance or aerial exposure, was the main cause of this mortality as the same reefs exposed to Spring tide conditions under similar irradiances of solar radiation did not exhibit high rates of mortality when the temperatures were normal. Additionally, some corals (e.g., *Favia* and *Goniastrea*) exhibited bleaching but survived under these harsh conditions. These bleached corals still had zooxanthellae when examined histologically and recovered quickly over a period of less than a month (Yonge and Nichols 1931). Yonge and Nichols (1931) then used samples of *Favia* in controlled experiments and reported that the duration and intensity of the thermal insult were critical factors affecting both bleaching and mortality. Many experiments have now shown that increases in seawater temperature are the primary cause of the unprecedented number of coral bleaching events since the early 1980s (Brown 1997a; Glynn 1993; Fitt et al. 2001; Lesser 2004). Several reviews on the causes, mechanisms, economic costs, and ecological outcomes of coral bleaching have also been published (Brown 1997a, b; Hoegh-Guldberg 1999; Loya et al. 2001; Coles and Brown 2003; Douglas 2003; Bellwood et al. 2004; Lesser 2004, 2006; Sotka and Thacker 2005; Smith et al. 2005; Hoegh-Guldberg et al. 2007; Weis 2008). Here, I review specifically the causes and cellular mechanisms of bleaching and examine the current evidence supporting the hypothesis that the production and accumulation of reactive oxygen species (ROS) leads to oxidative stress and is the proximal mechanism underlying the cellular phenomenon known as coral bleaching.

2 Causes of Coral Bleaching

Many field and laboratory studies on bleaching in corals and other symbiotic cnidarians have established a causal link between temperature stress and bleaching (Hoegh-Guldberg and Smith 1989; Jokiel and Coles 1990; Lesser et al. 1990; Glynn 1993; Fitt et al. 1993; Warner et al. 1999, 2002; Lesser 1997; Hoegh-Guldberg 1999; Fitt et al. 2001; Coles and Brown 2003; Lesser and Farrell 2004), while the extent of bleaching and subsequent mortality are related to the magnitude of temperature elevation and the duration of exposure. As is typical in the natural world, the occurrence and severity of coral bleaching events varies significantly in space and time. Similar variability is consistently observed in the high number of published experimental results employing different experimental protocols by investigators to understand the mechanisms and timeframes of cellular events leading to bleaching.

While thermal stress is viewed as the principal cause of coral bleaching, other environmental factors can cause bleaching independently, and act synergistically by effectively lowering the threshold temperature at which coral bleaching occurs (Lesser 2004, 2006). These other abiotic factors include salinity changes, sedimentation, exposure to supra-optimal irradiances of visible radiation, exposure to ultraviolet radiation, and low-temperature thermal stress (Brown 1997a, b; Lesser et al. 1990; Gleason and Wellington 1993; Kerswell and Jones 2003; Coles and Brown 2003; Lesser 2004; Mayfield and Gates 2007; Brown and Dunne 2008). The principal abiotic factor that has a significant influence on the severity of thermally induced coral bleaching is solar radiation, both its visible (photosynthetically active radiation, PAR: 400–700 nm, Hoegh-Guldberg and Smith 1989; Lesser et al. 1990; Shick et al. 1996; Dunne and Brown 2001; Jones and Hoegh-Guldberg 2001; Banaszak and Lesser 2009), and ultraviolet (UVR: 290–400 nm, UVB: 290–320 nm, UVA: 320–400 nm) components (Lesser et al. 1990; Gleason and Wellington 1993; Shick et al. 1996). The optical properties of most tropical waters results in low attenuation coefficients and allows UVR to penetrate to depths of 15 m or more (Gleason and Wellington 1993; Shick et al. 1996; Lesser 2000; Lesser and Gorbunov 2001). Ultraviolet radiation is known to have a detrimental effect on photosynthesis and growth in zooxanthellae (Lesser and Shick 1989; Kinzie 1993; Lesser 1996; Lesser and Lewis 1996; Shick et al. 1996; Banaszak and Lesser 2009) with the harmful effects of UVR involving damage to critical proteins that are the result of both the direct and indirect effects of UVR.

Evidence for bleaching caused by UVR in the field is anecdotal (Harriot 1985), but field experiments supporting UVR as the sole factor causing bleaching do exist (Gleason and Wellington 1993) despite experimental problems suggesting that differences in visible irradiance may have contributed to the observed effects (Dunne 1994). In the study by Gleason and Wellington (1993), the differences in PAR irradiance with UVR (488 µmol quanta m^{-2} s^{-1}) and PAR irradiances without UVR (442 µmol quanta m^{-2} s^{-1}) are physiologically insignificant and should not undermine the conclusion that UVR alone can induce coral bleaching under the right circumstances. For sessile corals, exposure to solar UVR in shallow tropical waters is unavoidable and exposure to UVR is particularly important during hyperoxic conditions (D'Aoust et al. 1976; Crossland and Barnes 1977; Dykens and Shick 1982; Shick 1990; Rands et al. 1992; Shashar et al. 1993; Kühl et al. 1995) that occur intracellularly in corals during photosynthesis and leads to both the biochemical and photodynamic production of reactive oxygen species (ROS) (Halliwell and Gutteridge 1999; Lesser 2006). The photoprotective processes discussed below act in concert to suppress oxidative damage to the photosynthetic apparatus but the damage incurred, and the energetic costs associated with repairing damaged proteins and synthesizing antioxidant enzymes ultimately leads to a decrease in the quantum yield of photosynthesis (Long et al. 1994; Niyogi 1999), and oxidative damage to key cellular components (Fridovich 1998; Asada 1999; Halliwell and Gutteridge 1999).

3 Mechanisms of Coral Bleaching

Much of the early work on coral bleaching took an "algal-centric" viewpoint with photoinhibition of photosynthesis, and damage at PSII specifically, as the primary suspect. Exposure to elevated temperatures (Iglesias-Prieto et al. 1992), visible radiation (Hoegh-Guldberg and Smith 1989), or UVR (Lesser and Shick 1989) alone, and in combination with thermal stress (Lesser 1996, 1997), can result in photoinhibition of photosynthesis in zooxanthellae defined as a decrease in maximum net photosynthesis. Photoinhibition occurs as a result of the reduction in photosynthetic electron transport, combined with the continued high absorption of excitation energy leading to damage at photosystem II (PSII) reaction centers (Long et al. 1994; Niyogi 1999). Our early guidance for studying the mechanisms of photoinhibition in zooxanthellae comes from the literature on higher plants and phytoplankton where studies on thermal and light stress, and their interactions, had already been conducted (Long et al. 1994; Niyogi 1999).

Based on studies from higher plants and phytoplankton, Lesser and Shick (1989) conducted culture experiments showing that exposing zooxanthellae to UVR caused a decrease of photosynthetic pigment per cell using flow cytometry, lowered growth rates, and a decrease in maximum productivity measured as carbon fixation. Combined with the observation of increasing levels of oxidative stress with increasing irradiances of PAR and exposure to UVR, Lesser and Shick (1989) postulated that these findings had direct relevance to bleaching in corals through the interactive effects of high temperature and exposure to high irradiances of solar radiation on the zooxanthellae of corals. Subsequently, a multifactorial experiment showed that for a shallow-water zooxanthellate zoanthid, there were significant effects on the expulsion of zooxanthellae associated with thermal stress and exposure to UVR, but not with changes in PAR (Lesser et al. 1990). These changes were, again, correlated with increased levels of antioxidant enzyme activity and therefore the production of ROS (Lesser et al. 1990). While these studies showed that zooxanthellae exhibited photoinhibition upon exposure to temperature and UVR, there were no data providing insight on the specific site of damage. Iglesias-Prieto et al. (1992) addressed this issue by using DCMU-induced chlorophyll fluorescence, a technique commonly used in studies on phytoplankton (e.g., Vincent 1980), combined with oxygen flux measurements of both photosynthesis and respiration. These measurements showed physiological stress above 30°C as declines in photosynthesis, respiration, and the quantum yield of PSII fluorescence. The authors suggested that changes in thylakoid membrane fluidity, and subsequent changes in photosynthetic electron transport capacity, lead to the observed photoinhibition of photosynthesis (Iglesias-Prieto et al. 1992). These authors were also keenly aware of the potential physiological diversity of zooxanthellae and how that might relate to the heterogeneous patterns of bleaching, even within a coral species, observed on many coral reefs.

As a result of this study, Lesser (1996) employed semi-continuous cultures of zooxanthellae isolated from the sea anemone *Aiptasia pallida* (Clade B) and simulated the *in hospite* nutrient conditions and visible light regime while exposing the cultures to high temperature stress with and without exposure to UVR. During this experiment, several parameters were monitored to assess the effects of thermal stress and UVR on photosynthesis by looking at effects on both the light and dark reactions as well as measuring the production of ROS and the activity of antioxidant enzymes. Photosynthesis, PSII function, growth rates, and the activities of ribulose 1,5-bisphosphate decarboxylase/oxygenase (Rubisco) were shown to be significantly affected by exposure to an increase in temperature from 25°C to 31°C. Additionally, photosynthesis, PSII function, growth rates, and Rubisco activities further declined, significantly, when the same cells were exposed to elevated temperature and UVR. These data showed for the first time that both the light and dark reactions were affected simultaneously by thermal stress with or without exposure to UVR. These observations were always accompanied by increases in the cellular concentration of superoxide radicals and hydrogen peroxide as well as increases in enzymatic antioxidant defenses. Exposing cultures to ascorbate, a nonenzymatic quencher of hydroxyl radicals, and catalase, which decomposes hydrogen peroxide to water and oxygen at the end of the experiment for only 1 h, improved photosynthetic performance (i.e., P_{max}) by 24% in cultures exposed to 31°C and by 37% in cultures exposed to elevated temperature and UVR (Lesser 1996). Subsequent studies on cultured zooxanthellae have supported these results and also showed that thermal tolerance and ROS production has a genetic component (Suggett et al. 2008; Saragosti et al. 2010) and is involved in programmed cell death or apoptosis in zooxanthellae (Franklin et al. 2004).

The study by Lesser (1996) showed that there are, in fact, multiple sites of damage in the photosynthetic apparatus of zooxanthellae during thermal stress that are significantly affected by the interaction of other abiotic factors such as exposure to high irradiances of visible, and UVR (Lesser et al. 1990; Bhagooli and Hidaka 2004; Lesser and Farrell 2004). Additionally, it was shown that ROS are the effecter molecules for these observations and are consistent with studies showing that PSII and Rubisco are damaged or inhibited by exposure to ROS (Richter et al. 1990; Asada 1999). The observation that ROS is directly involved in several aspects of photoinhibition is not limited to zooxanthellae in culture. Similar experiments on the coral *Agaricia tenuifolia* showed a significant improvement in the photosynthetic performance of thermally stressed corals, and a decrease in bleaching, when corals were exposed to exogenous antioxidants (Lesser 1997).

As can be surmised from the higher plant literature and the discussion above ROS formation is a pervasive theme in the stress response of all organisms (Halliwell and Gutteridge 1999) and corals in particular (Lesser 2006). While oxidative stress has long been an important area of research in the biomedical community, for many environmental physiologists this has been, until recently, an underappreciated facet of organismal physiology and biochemistry. It would, therefore, be informative, to highlight the basic principles of oxidative stress here. All photosynthetic and respiring cells produce ROS including superoxide radicals ($O_2^{\cdot-}$) via the univalent pathway (Eq. 1), and hydrogen peroxide (H_2O_2) with the continued reduction of O_2^- (Eq. 2), and the formation of hydroxyl radicals (HO·, Eq. 3), which is then reduced to the hydroxyl ion and water (Eq. 4) as a consequence of exposure to, and use of, molecular oxygen (Fridovich 1998; Asada 1999; Halliwell and Gutteridge 1999).

$$O_2 + e^- \longrightarrow O_2^- \quad \text{(superoxide radical)} \tag{1}$$

$$O_2^- + 2H^+ + e^- \longrightarrow H_2O_2 \quad \text{(hydrogen peroxide)} \tag{2}$$

$$H_2O_2 + e^- \longrightarrow OH^- + HO \text{(hydroxyl radical)} \tag{3}$$

$$HO + e^- \longrightarrow OH^- \quad \text{(hydroxyl ion)} \tag{4}$$

$$\text{(Net Reaction)} O_2 + 4H^+ + 4e^- \longrightarrow H_2O$$

The production of ROS is directly, and positively, related to the concentration or pO_2 of O_2 (Jamieson et al. 1986) and this has unique consequences for photosynthetic organisms, including corals. Oxidative stress, the production and accumulation of ROS beyond the capacity of an organism to quench these reactive species, can cause damage to lipids, proteins, and DNA, but can also act as important signal transduction molecules (Fridovich 1998; Asada 1999; Halliwell and Gutteridge 1999). The purpose of antioxidant defenses in biological systems is to quench singlet oxygen ($^1O_2$) at the site of production (e.g., PSII reaction centers in chloroplasts), and quench or reduce the flux of reduced oxygen intermediates such as $O_2^{·-}$ and H_2O_2 to prevent the production of HO·, the most damaging of the ROS (Fridovich 1998; Asada 1999; Halliwell and Gutteridge 1999). In addition to the production of $^1O_2$ within PS II (Macpherson et al. 1993), it has recently been shown that $O_2^{·-}$ and HO· are also produced in the PS II reaction center (Liu et al. 2004) and that ascorbate can react with $^1O_2$ to produce H_2O_2 in the chloroplast (Kramarenko et al. 2006). The enzymes superoxide dismutase (SOD), catalase, ascorbate peroxidase, and nonenzymatic antioxidants inactivate $O_2^{·-}$ and H_2O_2, thereby preventing the formation of HO·, and subsequent cellular damage (Fridovich 1998; Asada 1999; Halliwell and Gutteridge 1999). For many marine organisms in symbiosis with photoautotrophic symbionts, including corals, antioxidant defenses in the animal host occur in proportion to the potential for photooxidative damage, which is functionally correlated with the biomass of photosynthesizing symbionts (Dykens and Shick 1982; Dykens et al. 1992). In corals, the cnidarian host expresses a Cu/Zn and Mn SOD (Lesser and Farrell 2004; Plantivaux et al. 2004) while zooxanthellae also express Fe SOD (Matta et al. 1992) with additional evidence that they may also express a Cu/Zn SOD (Lesser and Shick 1989; Matta et al. 1992). It has also been demonstrated that green fluorescent protein (GFP) found in high concentrations in corals can quench $O_2^{·-}$ (Bou-Abdallah et al. 2006) and H_2O_2 (Palmer et al. 2009). GFP can improve survival in model systems (e.g., *Escherichia coli*) exposed to an $O_2^{·-}$ generating system directly demonstrating a positive effect of GFP expression that is not coupled to bioluminescence (Fig. 1a). The modest SOD-like activity of GFP may well be compensated by its high concentration in corals (Mazel et al. 2003; Dove 2004, Fig. 1b), making it a significant contributor to the overall antioxidant defenses of corals. GFP protein (Dove et al. 2006, Fig. 1c) also decreases in corals exposed to thermal stress. This is consistent with an *in hospite* environment where high rates of $O_2^{·-}$ production occurs during thermal stress and GFP quenches $O_2^{·-}$ but not without a decrease in GFP concentrations, which is caused by oxidative degradation of the protein (Bou-Abdallah et al. 2006) along with a decrease in transcription of the gene (Smith-Keune and Dove 2008). GFP is one of a suite of nonenzymatic antioxidants known to occur in symbiotic cnidarians that includes high concentrations of dimethylsulfide (DMS) and dimethylsulfoniopropionate (DMSP) (Broadbent et al. 2002), which have been shown to quench $^1O_2$ and HO·, respectively (Sunda et al. 2002), as well as ultraviolet absorbing compounds such as mycosporine glycine, which can also quench $^1O_2$ (Dunlap et al. 2000).

Many studies on coral bleaching use noninvasive techniques such as active fluorescence to assess the damage to PSII in the symbiotic zooxanthellae. Using more direct and quantitative techniques (e.g., immunoblots), we now also know that damage to PSII reaction centers in zooxanthellae occurs principally at the D1 protein of PSII and is correlated with changes in PSII fluorescence and oxidative stress during exposure to thermal stress and/or solar radiation (Warner et al. 1999; Lesser and Farrell 2004). There is also evidence that under high irradiances of solar radiation, a significant proportion of PSII reaction centers can be chronically damaged (30%) without exposure to thermal stress and without negative effects on productivity (Gorbunov et al. 2001). Several species of coral exposed to elevated temperatures (>2°C above seasonal highs) and saturating light (>350 µmol quanta $m^{-2} s^{-1}$) show that 14–35% of PSII reaction centers were damaged and may constitute a photoprotective mechanism to prevent damage to all PSII reaction centers and subsequent coral bleaching (Hill and Ralph 2006). The site of damage in these instances is also likely to be the D1 protein since the oxygen-evolving complex of PSII in zooxanthellae appears to be thermotolerant within the range of recorded bleaching temperatures (Hill and Ralph 2008), although photobleaching of antenna pigments may also be involved (Takahashi et al. 2008). Similar to higher plants, the zooxanthellae of corals can dissipate excess excitation energy through the xanthophyll cycle (Brown et al. 1999; Hoegh-Guldberg and Jones 1999; Gorbunov et al. 2001; Levy et al. 2006). Under conditions where the irradiance required to saturate photosynthesis is 2–5 times greater than the saturation constant (E_k) for corals, mid-day depressions in the quantum yield of PSII fluorescence are consistently observed and are not correlated with decreases in productivity (e.g., Lesser and Gorbunov 2001) but are photoprotective. This process of photoprotection is also known as dynamic photoinhibition, a regulatory process to prevent the overexcitation of the photosynthetic apparatus and damage to PSII but the capabilities of this photoprotective mechanism can be exceeded under high irradiances of solar radiation and thermal stress exposing zooxanthellae to oxidative stress and its consequences (Lesser and Farrell 2004; Lesser 2006).

Jones et al. (1998), using active fluorescent measurements, proposed that the observed collapse of thermally induced decreases in photosynthesis and the quantum yield of PSII fluorescence were secondary effects of sink limitation in the dark reactions of photosynthesis and subsequent overreduction of photosynthetic electron transport causing decreases in the quantum yield of PSII fluorescence. Using the kinetics of steady state, or effective, quantum yields of PSII fluorescence

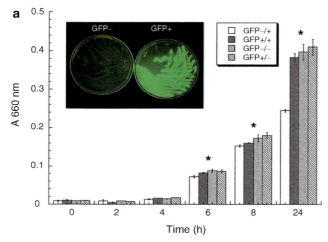

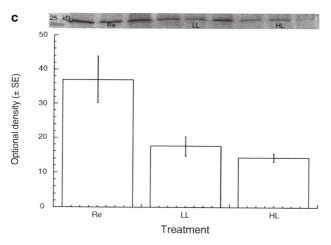

Fig. 1 (a) Growth (mean ± SE of absorption at 660 nm) of control (GFP−/+, GFP−/−, GFP+/−) and treatment (GFP+/+) *Escherichia coli* (pGLO plasmid, Bio-Rad Inc.) cultures exposed to 100 nM superoxide. (O_2^-) generated using the hypoxanthine/xanthine oxidase system in 50 mM phosphate buffer at pH 7.4. Cultures were grown in LB media with ampicillin (0.1 mg ml^{-1}) and with or without arabinose (0.167 mg ml^{-1}). Arabinose is required to activate the expression of GFP in this vector. Cultures of cells expressing GFP (GFP +) or not expressing GFP (GFP −) were grown to log phase and inoculated into tubes (15 ml LB media with ampicillin and arabinose (*N*=3), and in LB media with ampicillin minus arabinose (*N*=3) while control cells of GFP+ and GFP−, not exposed to O_2^- (GFP+/− and GFP−/−), were grown under identical

($\Delta F/F_m'$) and nonphotochemical quenching (*qN*), coral samples at 34°C exposed to irradiances between 500 and 1,500 µmol quanta m^{-2} s^{-1} showed no induction of the Calvin cycle as suggested by the decrease in $\Delta F/F_m'$ and an increase in *qN* while decreases in gross photosynthesis were also observed under similar experimental conditions (Jones et al. 1998). The work by Jones et al. (1998) clearly illustrates the importance of carbon sink limitation in exacerbating damage to PSII. This observation is significant because carbon limitation has been observed in shallow-water corals (Muscatine et al. 1989), and can be significantly affected by water flow (Lesser et al. 1994) that partially explains observed patterns of bleaching on coral reefs (Nakamura and van Woesik 2001). The Jones et al. (1998) model of sink limitation leading to overreduction of photosynthetic electron transport, oxidative stress, and damage to PSII is consistent with the data on damage to both photochemistry and carbon fixation (Lesser 1996) under conditions experienced by shallow-water corals (<10 m). Corals at depths deeper than 10 m experience decreasing amounts of solar radiation and less sink limitation although critical enzymes of the Calvin cycle (e.g., Rubisco activase) could still be affected by thermal stress (Crafts-Brandner and Salvucci 2000), with the result being decreased productivity and the potential for overreduction of photosynthetic electron transport and damage to PSII reaction centers. It should also be noted that Rubisco is itself sensitive to ROS, specifically H_2O_2, which would be formed by the dismutation of O_2^- in the chloroplast (Asada 1999). In either case, damage to PSII leads to enhanced ROS production due to the Mehler reaction on the reducing side of photosystem I (PSI) and is the most significant site of O_2 production in the chloroplast (Asada 1999). The Mehler reaction is often described as an alternative sink for electrons when sink limitation (e.g., carbon or nitrogen limitation) occurs, as is the reduction of molecular oxygen during photorespiration, using the oxygenase or C_2

Fig. 1 (continued) conditions at the same time. All cultures were incubated at 35°C. Significant treatment effects were detected (ANOVA: *P*=0.008), * indicates significant differences using *post hoc* SNK (*P*<0.05) multiple-comparisons. Beginning at 6 h in cells not expressing GFP and exposed to O_2^- significantly lower growth rates were observed while all other treatment groups were not statistically different. (b) Underwater fluorescence photograph of the coral, *Montastraea faveolata*, showing the uniform distribution of GFP among the polyps (Photograph by Charles Mazel). (c) Western blots of host-associated GFP (*N*=3 for each treatment) for *Montastraea faveolata* from experiments described in Lesser and Farrell (2004) and expressed as optical density (± SE) of immunoblots. A significant treatment effect was detected (ANOVA: *P*=0.019) and significant differences were observed using *post hoc* SNK (*P*<0.05) multiple-comparisons to show that colonies recovering from bleaching stress (Re) had higher concentrations of GFP when compared with corals exposed to thermal stress and either low irradiances (LL) or high irradiances (HL) of solar radiation

pathway for Rubisco that results in formation of H_2O_2 (Asada 1999). Under stressful conditions, ROS production at PSI and PSII would then overwhelm algal antioxidant defenses as the proximal series of events leading to the expulsion of zooxanthellae (Lesser 2006).

In higher plants, additional, and novel, mechanisms of reducing excitation pressure on PSII have been identified and are collectively called alternate electron transport pathways. One of these, chlororespiration, involves the oxygen-dependent reduction of the plastiquinone pool in the dark using a membrane-bound NAD(P)H-oxidoreductase in the thylakoid membrane (Peltier and Cournac 2002). These pathways are upregulated when plants are exposed to abiotic stress such as elevated temperatures and reduce the production of ROS while maintaining the production of ATP (Peltier and Cournac 2002). In corals, as well as cultured zooxanthellae, there is evidence of a functional chlororespiration pathway (Jones and Hoegh-Guldberg 2001; Hill and Ralph 2005; Reynolds et al. 2008), but evidence to the contrary also exists (Suggett et al. 2008). While corals are known to be severely hypoxic in the dark when chlororespiration would not be operating, chlororespiration may still help corals remain poised for efficient photosynthesis and ATP synthesis when corals are re-illuminated (Jones and Hoegh-Guldberg 2001).

As originally suggested by Iglesias-Prieto et al. (1992), recent studies have shown that differences in the lipid composition of thylakoid membranes in genetically distinct zooxanthellae has a significant effect on membrane fluidity during thermal stress, the uncoupling of electron transport, subsequent oxidative stress, and photoinhibition measured as a decrease in the maximum quantum yield of PSII fluorescence (Tchernov et al. 2004). Tchernov et al. (2004) acknowledge that their results are also consistent with light-driven ROS production and subsequent lipid peroxidation in the thylakoid membranes, which could then establish a positive feedback loop as membranes become more fluid further uncoupling photosynthetic electron transport from the reaction centers during exposure to elevated seawater temperatures. Other studies, however, suggest that at least in the early stages of bleaching, the thylakoid membranes are intact (Dove et al. 2006; Hill et al. 2009) as are host membranes (Sawyer and Muscatine 2001).

Oxidative stress has also been implicated in the inhibition of the repair of damage to PS II that includes initial damage to the oxygen-evolving complex (Nishiyama et al. 2001, 2006). From these results, a new model of photoinhibition of photosynthesis has been proposed (Murata et al. 2007; Takahashi and Murata 2008) and is believed to represent the mechanism of photoinhibition in zooxanthellae (Takahashi and Murata 2008). While providing an interesting basis to examine alternative mechanisms of photoinhibition of photosynthesis in zooxanthellae, much of the model requires damage to the oxygen-evolving complex as the primary event and is largely based on studies conducted on cyanobacteria. First, zooxanthellae apparently do not sustain damage to their oxygen-evolving complex at temperatures consistent with coral bleaching (Hill and Ralph 2008). Second, in cyanobacteria, both photosynthesis and respiration occur on the same membranes, and have common redox proteins used in both respiratory and photosynthetic electron transport, which allows for significant capabilities to remove excitation pressure away from PSII even at low irradiances (Bailey et al. 2008). Cyanobacteria also maintain a low ratio of PSII:PSI reaction centers, which guarantees that PSI turnover does not limit electron flow through PSII and the potential damage that would occur from overreduction of photosynthetic electron transport (Bailey et al. 2008). Additionally, fluorescence measurements can be problematic as a result of the presence of the intersystem electron transport and can overestimate maximum fluorescence (F_m) (Büchell and Wilhelm 1993). These significant physiological differences between eukaryotic and prokaryotic photoautotrophs, and the fact that the lone study on corals uses active fluorescence and inhibitors as the primary tools (Takahashi et al. 2004) suggests that more research in this area is required to assess whether the results from the cyanobacterial studies are applicable to zooxanthellae.

The cnidarian host also responds to thermal stress and the production of ROS (Lesser 2006; Flores-Ramírez and Liñán-Cabello 2007; Baird et al. 2008; Császár et al. 2009; Fitt et al. 2009). Several studies have shown that the antioxidant activity of the host increases during periods of photooxidative stress (Lesser et al. 1990; Levy et al. 2006; Flores-Ramírez and Liñán-Cabello 2007), but a recent study by Fitt et al. (2009) showed the importance of not only zooxanthellae genotype, but of both the constitutive expression of antioxidant proteins (e.g., SOD), as well as the ability to respond to thermal stress by synthesizing more SOD and heat shock proteins (HSPs) upon exposure to thermal stress. One of the most important studies showing the direct production of ROS in symbiotic cnidarians was done on the sea anemone *Anthopleura elegantissma* where direct measurements of the flux of ROS were accomplished (Dykens et al. 1992). This well-designed study clearly showed the simultaneous production of ROS in both the host tissues and zooxanthellae upon illumination. Dykens et al. (1992) showed conclusively the disproportional fluxes of photosynthesis-dependent ROS produced in zooxanthellae relative to their biomass in the symbiosis, and the direct photodynamic production of ROS in the tissues of azooxanthellate samples. While these experiments did not include temperature as an experimental factor, it is reasonable to assume that the flux of ROS would further increase in both the host and zooxanthellae as observed from indirect measurements of antioxidant enzyme activity in other photoautotrophic symbioses (Lesser et al. 1990; Levy et al. 2006). Other studies have suggested that oxidative stress is primarily a response of the animal host to hyperoxia imposed by the photosynthetic zooxanthellae (Nii and Muscatine 1997). In their study, Nii and Muscatine (1997) suggested that O_2^- was not released by intact, nonstressed, zooxanthellae, which has recently been shown not to be true (Saragosti et al. 2010). Hydrogen peroxide, with its significantly greater diffusion constants, had been proposed as the most likely species of active oxygen to be released from zooxanthellae whether stressed or not, and as membranes are compromised by processes such as lipid peroxidation O_2^- could also be released from damaged zooxanthellae (Lesser et al. 1990; Lesser 2006). Other studies have also shown the occurrence of oxidative stress occurring in both the host and zooxanthellae during thermal stress and exposure to solar radiation (Lesser and Farrell 2004; Levy et al. 2006), but some investigators still suggest that oxidative stress is primarily an animal response despite data showing a strong antioxidant response of both the host and zoxanthellae during exposure to thermal stress (Levy et al. 2006). One of the arguments that oxidative stress in the animal compartment is the primary cause of bleaching has been that while there are studies showing zooxanthellae produce H_2O_2 (Lesser 1996; Franklin et al. 2004; Suggett et al. 2008), the primary oxidant suspected of initiating the bleaching response (Lesser et al. 1990; Smith et al. 2005; Lesser 2006), there are no studies where the release of hydrogen peroxide by zooxanthellae has been demonstrated. This has now been shown to occur at significant rates that are dependent on the genotype of zooxanthellae (Suggett et al. 2008). Corals can be at a disadvantage during thermal stress and high solar irradiances because their skeletal elements scatter those photons not initially absorbed, which increases their residence time, and the possibility of being absorbed by the reaction centers in zooxanthellae (Enríquez et al. 2005). This amplification of the *in hospite* light field could exacerbate the response to thermal stress (Enríquez et al. 2005).

The host also responds to thermal stress in other ways that are related to the mechanism (s) of coral bleaching. In particular, HSPs are upregulated in response to thermal stress (Black et al. 1995; Fang et al. 1997; Sharp et al. 1997). Heat shock proteins are inducible by a number of environmental factors, including oxidative stress, and appear to be part of a generalized stress response that is evolutionarily conserved. Under stressful conditions, HSPs interact with proteins to maintain their conformation and function or in targeting damaged proteins for degradation. This function is also consistent with patterns of expression for HSP and markers of protein degradation observed in corals (Downs et al. 2000, 2002). One area in need of study on corals is the relationship between HSPs and apoptosis. It is known from other systems that HSP72 can regulate the response to stress by intervening in the mitochondrial apoptotic pathway downstream of the release of cytochrome *c* (Beere and Green 2001) and this should be an interesting area of work for environmentally induced apoptosis in sea anemones and corals. Studies on the effect of UVR and thermal stress on corals have also shown significant DNA damage in host tissues upon exposure to UVR (Anderson et al. 2001; Baruch et al. 2005) and thermal stress combined with exposure to solar radiation (Lesser and Farrell 2004). DNA damage can occur from the direct effects of UVR or indirectly by oxidative stress and can lead to apoptosis or programmed cell death if not repaired. One of the key cell cycle genes activated after DNA damage is *p*53 (Johnson et al. 1996). If DNA repair is not possible, then *p*53-mediated apoptosis may be initiated. The expression pattern of *p*53 protein in *Montastraea faveolata* after exposure to thermal stress and high irradiances of solar radiation was consistent with the observed pattern of DNA damage and subsequent coral bleaching (Lesser and Farrell 2004).

Nitric oxide, or nitrogen monoxide (NO·), is a molecule involved in signal transduction (e.g., neurotransmission), but is also involved in a diverse array of processes associated with oxidative stress. It is now known that the inducible enzyme nitric oxide synthase (NOS) produces NO· and reacts with O_2^- to form highly reactive nitrogen species (RNS) such as the peroxynitrite anion (ONOO$^-$). Because the solubility of NO· is similar to water, it can readily diffuse across biological membranes where it reacts at near diffusion-limited rates with O_2^- to form ONOO$^-$, which itself can diffuse across biological membranes at rates 400 times greater than O_2^- (Marla et al. 1997). It has been suggested that high concentrations of NO· creates significant competition between NO· and SOD for O_2^-, and that the outcome of this competition for O_2^- may be a major determinant of the level of oxidative stress in many organisms. Both the host cells (Perez and Weis 2006) and zooxanthellae can produce NO· (Bouchard and Yamasaki 2008) and the presence of nitric oxide synthase activity in corals and sea anemones (Trapido-Rosenthal et al. 2005; Morrall et al. 2000) suggests that NO·, and subsequently ONOO$^-$, production may also be important contributors to oxidative stress and apoptosis in corals (Lesser 2006; Weis 2008).

Two apoptotic pathways have been described and are known as the death-receptor pathway and the mitochondrial pathway. The mitochondrial pathway is commonly associated with DNA damage and upregulation or activation of the cell cycle gene *p*53 (Hengartner 2000). Exposure to UVR can also cause ROS production in the electron transport chain of mitochondria (Gniadecki et al. 2000) and can lead to apoptosis (Pourzand and Tyrell 1999). Both the death receptor and mitochondrial pathways converge at the mitochondria and the Bcl-2 family of genes. Bcl-2 can also be directly downregulated, and therefore promote apoptosis, by exposure to ROS (Hildeman et al. 2003). In the mitochondria, the release of proapoptotic effectors (e.g., cytochrome *c*, ROS, caspase 9) subsequently leads to the assembly of the apoptosome, which among other things activates caspase-dependent DNase (Green and Reed 1998; Rich et al. 2000). Caspases have been identified in *Hydra* sp. and are involved in apoptosis (Cikala et al. 1999), while caspases have also been identified in phytoplankton and are regulated in a similar fashion when compared to higher plant and metazoan caspases during apoptosis (Segovia et al. 2003; Bidle and Falkowski 2004).

Ultrastructural studies have shown that both apoptosis and cell necrosis are occurring in host and algal cells of thermally stressed symbiotic sea anemones (Dunn et al. 2002, 2004). Recent advances in the molecular genetics of cnidarians has shown that sea anemones have highly conserved homologues to Bcl-2 and caspase (Dunn et al. 2006; Richier et al. 2006), that caspase activity and apoptosis increase with thermal stress (Richier et al. 2006), and that RNA interference (RNAi) assays of sea anemone caspase can be used to control apoptosis in sea anemones (Dunn et al. 2007a). One of the other functions of Bcl-2 is to regulate Ca^{2+} concentrations, the intracellular level of which is an important signal for cells to undergo apoptosis (Rong and Distelhorst 2008) and originally proposed as one possible mechanism involved in coral bleaching (Gates et al. 1992). In fact, studies on corals and zooxanthellae have revealed that a rise in intracellular Ca^{2+} levels tracks thermal stress and is correlated with coral bleaching (Fang et al. 1997; Huang et al. 1998; Sandeman 2006), but strong evidence arguing against the involvement of intracellular Ca^{2+} levels in coral bleaching also exists (Sawyer and Muscatine 2001). Additionally, an enzyme belonging to the highly conserved family of cyclophilins has been implicated in the regulation of oxidative stress during exposure to stress in sea anemones and may also play a role in coral bleaching as cyclophilins have been described as mediators of apoptosis (Perez and Weis 2008).

Based on the ultrastructural evidence that apoptosis and necrosis both occur in thermally stressed symbiotic cnidarians (Dunn et al. 2002, 2004), that a putative *p53* protein is upregulated in response to DNA damage (Lesser and Farrell 2004), and that exogenous antioxidants improve photosynthesis and decrease bleaching (Lesser 1997), the most parsimonious conclusion is that coral bleaching occurs as a result of oxidative stress leading to apoptosis and cell necrosis in thermally stressed symbiotic cnidarians. Cellular necrosis and apoptosis can both result from oxidative stress, both lead to cell death, and both have features that overlap one another (Martindale and Holbrook 2002). Whereas high levels of oxidative stress cause cell necrosis, lower levels generally cause DNA damage and cell cycle arrest, or initiate apoptosis (Halliwell and Gutteridge 1999; Martindale and Holbrook 2002). As previously suggested, apoptosis and cell necrosis are the extreme cases in a range of likely cellular responses to thermal stress in corals (Gates et al. 1992).

Recently, new data have suggested a role for autophagy in coral bleaching and there is evidence that the apoptosis and autophagy pathways are interrelated (Dunn et al. 2007b). Autophagy has been described for a large number of taxonomically unrelated organisms and is known to be upregulated during metabolic stress such as starvation, hypoxia, or oxidative stress that leads to decreases in protein synthesis and low ATP/AMP ratios (Levine and Yuan 2005; Lum et al. 2005). Through the regulation of nutrient sensing by the TOR (target of rapamycin) gene, cells can be selectively slated to catabolize their own cytoplasmic components, known as "self-digestion," if irreversibly damaged or senescent, which is the hallmark signature of autophagy (Lum et al. 2005). Although the final phenotype of apoptotic and autophagic cells is very similar, there are significant differences in the genes involved and the time frame of events, which suggests that autophagy may be important during chronic stress (e.g., lower temperatures and lower irradiances) while apoptosis and necrosis occurs under acute stress (e.g., higher temperatures and high irradiances).

Oxidative stress has been proposed as a unifying mechanism for several environmental insults that cause bleaching (Lesser 1996, 2006) via exocytosis from coral host cells (Gates et al. 1992; Lesser 1997) or apoptosis (Gates et al. 1992; Dunn et al. 2002, 2004, 2007a, b; Lesser and Farrell 2004). A cellular model of bleaching in symbiotic cnidarians has been developed and includes oxidative stress, PSII damage, DNA damage, and apoptosis as underlying processes (Lesser 1996, 2006; Downs et al. 2002; Weis 2008, Fig. 2).

4 Acclimatization/Adaptation of Host and Zooxanthellae

Coral bleaching results in the breakdown of a mutualistic symbiosis that is essential for the survival of corals. There is growing evidence that the range of responses of corals to environmental stress (Fitt et al. 2001) is also a function of the genotype(s) of zooxanthellae within the host. The controversial notion that coral bleaching is an adaptive response (e.g., Baker 2001) is an area of exciting research and recent papers have shown that members of Clade D zooxanthellae either exhibit enhanced thermal tolerance or become the predominant genotype in corals after a bleaching event (Berkelmans et al. 2006; Jones et al. 2008; Sampayo et al. 2008). From a physiological perspective, we are beginning to understand that zooxanthellae from different clades exhibit differences in their ability to prevent overexcitation of the photosynthetic apparatus (Warner and Berry-Lowe 2006), in their production of ROS such as hydrogen peroxide (Suggett et al. 2008), and in their constitutive expression of antioxidant enzymes

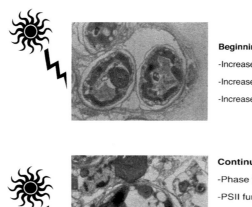

27-28°C

Beginning of Exposure to Elevated Temperatures

-Increase in metabolic rates

-Increase in cell division

-Increased turnover of zooxanthellae

29-30°C

Continued Exposure to Elevated Temperatures

-Phase transition of membranes

-PSII function decreases

-Decrease translocation to host

-Increased production of ROS/RNS

-Exocytosis of zooxanthellae

30-32°C

Chronic Exposure to Elevated Temperatures

-Excess ROS/RNS production

-Damage to membranes, proteins, DNA

-Damage to cell adhesion proteins

-Rapid decrease in PSII and Rubisco function

-Increase in intracellular calcium

-Increase in cell cycle and pro-apoptotic genes

-Decrease in ATP, decrease in protein sysnthesis

-Increase in auotophagy gene expression

-Continued exocytosis and detachment of gastrodermal cells

-Apoptosis/Autophagy/Cell Necrosis

Fig. 2 Model of coral bleaching modified from Lesser (2006) and including additional concepts from Dunn et al. (2007a, b), Desalvo et al. (2008), and Weis (2008). Early thermal stress results in an increase in metabolic rates and cell division (Beginning of Exposure to Elevated Temperature) but normal morphology while increasing thermal stress, especially in the presence of solar radiation, causes an increase in the production of ROS and RNS and begins to affect membranes and PSII function (Continued Exposure to Elevated Temperatures), which leads to mid-stage apoptosis (Dunn et al. 2004). These reactive molecules cause damage to lipids, proteins, and DNA damage that ultimately causes cellular damage that changes the expression of cell cycle and pro-apoptotic genes with the subsequent occurrence of apoptosis or cell necrosis (Chronic Exposure to Elevated Temperatures) and continued degradation consistent with late-stage apoptosis (Dunn et al. 2004). An increase in ROS, or a decrease in protein synthesis or decrease in ATP, can also result in autophagy pathways being upregulated. Electron micrographs provided by Simon Dunn

such as SOD where clades C and D have greater constitutive SOD activities than clades A and B (Fig. 3). The availability of molecular genetic data on zooxanthellae genotypes, their micro- and macroscale distributions, and the mapping of physiological capabilities on those genetic differences will play a significant role in determining who are the winners and losers (Loya et al. 2001) under any continuing scenario of global climate change. We should not, however, underestimate the potential for hosts to play a decisive role in their own fate during and after coral bleaching (Grottoli et al. 2007; Fitt et al. 2009).

The consensus opinion is that the rate of environmental change far exceeds the capabilities of the majority of corals to adapt in an evolutionary sense (Gates and Edmunds 1999). There have been several studies exploring the acclimatization capacities of corals to both high temperature stress and solar irradiance as the basis for observed and experimentally induced thermotolerances (Brown et al. 2002; Castillo and Helmuth 2005; Maynard et al. 2008; Middlebrook et al. 2008). Several explanations for acquired thermotolerance have been suggested and include enhanced photoprotective mechanisms, selective mortality, shuffling of different zooxanthellae genotypes, greater energy reserves, and rapid evolution of long-term physiological memory. We know from other well-studied systems that changes in the range of acclimatization capabilities of

5 Conclusions and Future Directions

The future for integrated studies on coral bleaching will continue to include molecular genetics, microarrays, proteonomics, RNAi assays, knockouts, and marine model organisms combined with a quantitative organismal approach (Hofmann et al. 2005; Weis et al. 2008). Methods used (e.g., electron paramagnetic resonance [EPR], enzyme assays, fluorochromes) to assess the level of oxidative stress should be routinely incorporated, and these techniques combined in an interdisciplinary manner with the physiological measurements at the organismal level (e.g. DNA damage or photosynthesis). Some research groups have established expressed sequence tag (EST) libraries for different genotypes of zooxanthellae and corals (Leggat et al. 2007; Desalvo et al. 2008) and have already showed that a large group of functionally related genes indicates that oxidative stress is a signature feature of coral bleaching (Desalvo et al. 2008). These EST libraries will facilitate progress on the development of microarrays using both stress and metabolic genes, which can then be used to simultaneously assess stress levels in corals exposed to a wide range of bleaching conditions (Desalvo et al. 2008; Richier et al. 2008). One should not forget, however, that proteins are the functional entity facilitating physiological changes and studies on genes alone, without their protein counterparts, may be limited in what information they can provide (Feder and Walser 2005).

Acknowledgments The author thanks numerous colleagues for engaging in many conversations on the subject. Many funding agencies, including NOAA, NSF, and ONR, have supported this work over the years. In particular, the Coral Reef Targeted Research (CRTR) Program provided funding to support this review.

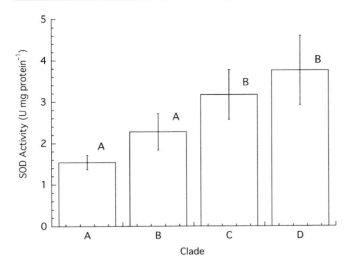

Fig. 3 Constitutive superoxide dismutase (SOD) activities for zooxanthellae grown in the same liquid culture media (ASP8) and under the same visible irradiances (~100 µmol quanta m^{-2} s^{-1}) from different clades (Clade A, $N=29$; Clade B, $N=19$; Clade C, $N=4$; Clade D, $N=6$) analyzed as described in Lesser and Shick (1989). Significant effects of Clade were detected (ANOVA: $P=0.005$) and the results of *post hoc* multiple comparison testing is indicated by the superscripts. Samples of different zooxanthellae clades were generously provided by Mark Warner

thermally sensitive traits can lead to successful shifts in thermal tolerances (Somero 2002). These changes in thermotolerance can be the result of more thermally tolerant genotypes, more efficient repair processes to replace damaged proteins, or the biosynthesis of alternate forms of the same protein (e.g., isozymes) with increased stability under the new temperature regime (Somero 2002). All of these processes affect the energy budget of the organism and subsequently other life history traits (e.g., reproduction). Since markers of physiological stress are commonly used to assess thermotolerance, it would seem appropriate to undertake specific studies on the turnover of critical proteins (Gates and Edmunds 1999), as has been done for other sentinel species in marine habitats (e.g., mussels, Bayne 2004), as well as detailed studies that include models of damage and repair of critical proteins under gradients of environmental stress (e.g., Lesser et al. 1994) and reaction norms for a variety of host and zooxanthellae genotypes (Angilletta et al. 2003; Edmunds and Gates 2008). These types of studies would provide the range of acclimatization potential required as biological input for models that integrate long-term monitoring of sea surface temperature and atmosphere–ocean coupled general circulation models, which then provide robust predictive capabilities under various scenarios of global climate change (Donner et al. 2007; Lesser 2007).

References

Anderson S, Zepp R, Machula J, Santavy D, Hansen L, Mueller D (2001) Indicators of UV exposure in corals and their relevance to global climate change and coral bleaching. Human Ecol Risk Assess 7:1271–1282

Angilletta MJ Jr, Wilson RS, Navas CA, James RS (2003) Tradeoffs and the evolution of thermal reaction norms. Trends Ecol Evol 18:234–240

Anthony KRN, Kline DI, Diaz-Pulido G, Dove S, Hoegh-Guldberg O (2008) Ocean acidification causes bleaching and productivity loss in coral reef builders. Proc Natl Acad Sci U S A 105:17442–17446

Asada K (1999) The water-water cycle in chloroplasts: scavenging of active oxygen and dissipation of excess photons. Ann Rev Plant Physiol Mol Biol 50:601–639

Bailey S, Melis A, Mackey KRM, Cardol P, Finazzi G, van Dijken G, Berg GM, Arrigo K, Shrager J, Grossman A (2008) Alternative photosynthetic electron flow to oxygen in marine *Synechococcus*. Biochim Biophys Acta 1777:269–276

Baird AH, Bhagooli R, Ralph PJ, Takahashi S (2008) Coral bleaching: the role of the host. Trends Ecol Evol 24:16–20

Baker AC (2001) Reef corals bleach to survive change. Nature 411:765–766

Banaszak AT, Lesser MP (2009) Effects of ultraviolet radiation on coral reef organisms. Photochem Photobiol Sci 8:1276–1294

Baruch R, Avishai N, Rabinowitz C (2005) UV incites diverse levels of DNA breaks in different cellular compartments of a branching coral species. J Exp Biol 208:843–848

Bayne BL (2004) Phenotypic flexibility and physiological tradeoffs in the feeding and growth of marine bivalve mollusks. Integr Comp Biol 44:425–432

Beere HM, Green DR (2001) Stress management-heat shock protein-70 and the regulation of apoptosis. Trends Cell Biol 11:6–10

Bellwood DR, Hughes TP, Folke C, Nystrom N (2004) Confronting the coral reef crisis. Nature 429:827–833

Berkelmans R, van Oppen MJH (2006) The role of zooxanthellae in the thermal tolerance of corals: a 'nugget of hope' for coral reefs in an era of climate change. Proc R Soc Lond B 273:2305–2312

Bhagooli R, Hidaka M (2004) Photoinhibition, bleaching susceptibility and mortality in two scleractinian corals, *Platygyra ryukyuensis* and *Stylophora pistillata*, in response to thermal and light stress. Comp Biochem Physiol A 137:547–555

Bidle KD, Falkowski PG (2004) Cell death in planktonic, photosynthetic microorganisms. Nat Rev Microbiol 2:643–655

Black NA, Voellmy R, Szmant AM (1995) Heat shock protein induction in *Montastraea faveolata* and *Aiptasia pallida* to elevated temperatures. Biol Bull 188:234–240

Bou-Abdallah F, Chasteen ND, Lesser MP (2006) Quenching of superoxide radicals by green fluorescent protein. Biochim et Biophys Acta (General Subjects) 1760:1690–1695

Bouchard JN, Yamasaki H (2008) Heat stress stimulates nitric oxide production in *Symbiodinium microadriaticum*: a possible linkage between nitric oxide and the coral bleaching phenomenon. Plant Cell Physiol 49:641–652

Broadbent AD, Jones GB, Jones RJ (2002) DMSP in corals and benthic algae from the Great Barrier Reef. East Coast Shelf Sci 55:547–555

Brown BE (1997a) Coral bleaching: causes and consequences. Coral Reefs 16(Suppl):S129–S138

Brown BE (1997b) Adaptations of reef corals to physical environmental stress. Adv Mar Biol 31:222–299

Brown BE, Dunne RP (2008) Solar radiation modulates bleaching and damage protection in a shallow water coral. Mar Ecol Prog Ser 362:99–107

Brown BE, Ambarsari I, Warner ME, Fitt WK, Dunne RP, Gibb SW, Cummings DG (1999) Diurnal changes in photochemical efficiency and xanthophylls concentrations in shallow water reef corals: evidence for photoinhibition and photoprotection. Coral Reefs 18:99–105

Brown BE, Downs CA, Dunne RP, Gibbs SW (2002) Exploring the basis of thermotolerance in the reef coral *Goniastrea aspera*. Mar Ecol Prog Ser 242:119–129

Büchell C, Wilhelm C (1993) In vivo analysis of slow chlorophyll fluorescence induction kinetics in algae: progress, problems, and perspectives. Photochem Photobiol 58:137–148

Carpenter KE, Abrar M, Aeby G et al (2008) One-third of reef building corals face elevated extinction risk from climate change and local impacts. Science 321:560–563

Castillo KD, Helmuth BST (2005) Influence of thermal history o the response of *Montastraea annularis* to short-term temperature exposure. Mar Biol 148:261–270

Cikala M, Wilm B, Hobmayer E, Böttger A, David CN (1999) Identification of caspases and apoptosis in the simple metazoan *Hydra*. Current Biol 9:959–962

Coles SL, Brown BE (2003) Coral bleaching-capacity for acclimatization and adaptation. Adv Mar Biol 46:184–223

Crafts-Brandner S, Salvucci ME (2000) Rubisco activase constrains the photosynthetic potential of leaves at high temperature and CO_2. Proc Natl Acad Sci U S A 97:13430–13435

Crawley A, Kline DI, Dunn S, Anthony K, Dove S (2010) The effect of ocean acidification on symbiont photorespiration and productivity in *Acropora Formosa*. Global Change Biol 16:851–863

Crossland CJ, Barnes DJ (1977) Gas-exchange studies with the staghorn coral *Acropora acuminata* and its zooxanthellae. Mar Biol 40:185–194

Császár NBM, Seneca FO, van Oppen MJH (2009) Variation in antioxidant gene expression in the scleractinian coral *Acropora millepora* under laboratory thermal stress. Mar Ecol Prog Ser 392:93–102

D'Aoust BG, White R, Wells JM, Olsen DA (1976) Coral-algal association: capacity for producing and sustaining elevated oxygen tensions *in situ*. Undersea Biomed Res 3:35–40

Desalvo MK, Voolstra CR, Sunagawa S, Schwarz JA, Stillman JH, Coffroth MA, Szmant AM, Medina M (2008) Differential gene expression during thermal stress and bleaching in the Caribbean coral *Montastraea faveolata*. Mol Ecol 17:3952–3971

Donner SD, Knutson TR, Oppenheimer M (2007) Model-based assessment of the role of human-induced climate change in the 2005 Caribbean coral bleaching event. Proc Natl Acad Sci U S A 104:5483–5488

Douglas AE (2003) Coral bleaching-how and why? Mar Poll Bull 46:385–392

Dove S (2004) Scleractinian corals with photoprotective host pigments are hypersensitive to thermal bleaching. Mar Ecol Prog Ser 272:99–116

Dove S, Ortiz JC, Enríquez S, Fine M, Fisher P, Iglesias-Prieto R, Thornhill D, Hoegh-Guldberg O (2006) Respone of holosymbiont pigments from the scleractinian coral *Montipora monasteriata* to short-term heat stress. Limnol Ocenogr 51:1149–1158

Downs CA, Mueller E, Phillips S, Fauth JE, Woodley CM (2000) A molecular biomarker system for assessing the health of coral (*Montastraea faveolata*) during heat stress. Mar Biotechnol 2:533–544

Downs CA, Fauth JE, Halas JC, Dustan P, Bemiss J, Woodley CM (2002) Oxidative stress and seasonal coral bleaching. Free Radic Biol Med 33:533–543

Dunlap WC, Shick JM, Yamamoto Y (2000) UV protection in marine organisms. I. sunscreens, oxidative stress and sntioxidants. In: Yoshikawa T, Toyokuni S, Yamamoto Y, Naito Y (eds) Free radicals in chemistry, biology and medicine. OICA International, London

Dunn SR, Bythell JC, Le Tessier DA, Burnett WJ, Thomason JC (2002) Programmed cell death and necrosis activity during hyperthermic stress-induced bleaching of the symbiotic sea anemone *Aiptasia* sp. J Exp Mar Biol Ecol 272:29–53

Dunn SR, Thomason JC, Le Tessier MDA, Bythell JC (2004) Heat stress induces different forms of cell death in sea anemones and their endosymbiotic algae depending on temperature and duration. Cell Death Differ 11:1213–1232

Dunn SR, Phillips WS, Spatafora JW, Green DR, Weis VM (2006) Highly conserved caspase and Bcl-2 homologues from the sea anemone *Aiptasia pallida*: lower metazoans as models for the study of apoptosis evolution. Mol Evol 63:95–107

Dunn SR, Philips WS, Green DR, Weis VM (2007a) Knockdown of actin and caspase gene expression by RNA interference in the symbiotic anemone *Aiptasia pallida*. Biol Bull 212:250–258

Dunn SR, Schnitzler CE, Weis VM (2007b) Apoptosis and autophagy as mechanisms of dinoflagellate symbiont release during cnidarian bleaching: every which way you lose. Proc R Soc Lond B 274:3079–3085

Dunne RP (1994) Radiation and coral bleaching. Nature 368:697

Dunne R, Brown B (2001) The influence of solar radiation on bleaching of shallow water reef corals in the Andaman sea, 1993–1998. Coral Reefs 20:201–210

Dykens JA, Shick JM (1982) Oxygen production by endosymbiotic algae controls superoxide dismutase activity in their animal host. Nature 297:579–580

Dykens JA, Shick JM, Benoit C, Buettner GR, Winston GW (1992) Oxygen radical production in the sea anemone *Anthopleura elegantissima*: and its symbiotic algae. J Exp Biol 168:219–241

Edmunds PJ, Gates RD (2008) Acclimatization in tropical reef corals. Mar Ecol Prog Ser 361:307–310

Enríquez S, Méndez ER, Iglesias-Prieto R (2005) Multiple scattering on coral skeletons enhances light absorption by symbiotic algae. Limnol Oceanogr 50:1025–1032

Fagoonee I, Wilson HB, Hassell MP, Turner JR (1999) The dynamics of zooxanthellae populations: a long-term study in the field. Science 283:843–845

Fang L, Huang S, Lin K (1997) High temperature induces the synthesis of heat–shock proteins and the elevation of intracellular calcium in the coral *Acropora grandis*. Coral Reefs 16:127–131

Feder ME, Walser J-C (2005) The biological limitations of transcriptomics in elucidating stress and stress responses. J Evol Biol 18:901–910

Fitt WK, Spero HJ, Halas J, White MW, Porter JW (1993) Recovery of the coral *Montastrea annularis* in the Florida Keys after the 1987 Caribbean "bleaching event". Coral Reefs 12:57–64

Fitt WK, McFarland FK, Warner ME, Chilcoat GC (2000) Seasonal patterns of tissue biomass and densities of symbiotic dinoflagellates in reef corals and relation to coral bleaching. Limnol Oceanogr 45:677–685

Fitt WK, Brown BE, Warner ME, Dunne RP (2001) Coral bleaching: interpretation of thermal tolerance limits and thermal thresholds in tropical corals. Coral Reefs 20:51–65

Fitt WK, Gates RD, Hoegh-Guldberg O, Bythell JC, Jatkar A, Grottoli AG, Gomez M, Fisher P, Lajuenesse TC, Pantos O, Iglesias-Prieto R, Franklin DJ, Rodrigues LJ, Torregiani JM, van Woesik R, Lesser MP (2009) Response of two species of Indo-Pacific corals, *Porites cylindrical* and *Stylophora pistillata*, to short-term thermal stress: the host does matter in determining the tolerance of corals to bleaching. J Exp Mar Biol Ecol 373:102–110

Flores-Ramírez LA, Liñán-Cabello MA (2007) Relationships among thermal stress, bleaching and oxidative damage in the hermatypic coral, *Pocillopora capitata*. Com Biochem Physiol C 146:194–202

Franklin DJ, Hoegh-Guldberg O, Jones RJ, Berges JA (2004) Cell death and degeneration in the symbiotic dinoflagellates of the coral *Stylophora pistillata* during bleaching. Mar Ecol Prog Ser 272:117–130

Fridovich I (1998) Oxygen toxicity: a radical explanation. J Exp Biol 201:1203–1209

Gates RD, Edmunds PJ (1999) The physiological mechanisms of acclimatization in tropical reef corals. Am Zool 39:30–43

Gates RD, Baghdasarian G, Muscatine L (1992) Temperature stress causes host cell detachment in symbiotic cnidarians: implications for coral bleaching. Biol Bull 182:324–332

Gleason DF, Wellington GM (1993) Ultraviolet radiation and coral bleaching. Nature 365:836–838

Glynn PW (1993) Coral reef bleaching: ecological perspectives. Coral Reefs 12:1–17

Gniadecki R, Thorn T, Vicanova J, Petersen A, Wulf HC (2000) Role of mitochondria in ultraviolet-induced oxidative stress. J Cell Biochem 80:216–222

Gorbunov M, Kolber ZS, Lesser MP, Falkowski PG (2001) Photosynthesis and photoprotection in symbiotic corals. Limnol Oceanogr 46:75–85

Green DR, Reed JC (1998) Mitochondria and apoptosis. Science 281:1309–1312

Grottoli AG, Rodrigues LJ, Palardy JE (2007) Heterotrophic plasticity and resilience in bleached corals. Nature 440:1186–1189

Halliwell B, Gutteridge JMC (1999) Free radicals in biology and medicine. Oxford University Press Inc., New York, p 936

Harriot VJ (1985) Mortality rates of scleractinian corals before and during a mass bleaching event. Mar Ecol Prog Ser 21:81–88

Hengartner MO (2000) The biochemistry of apoptosis. Nature 407:770–776

Hildeman DA, Mitchell T, Aronow B, Wojciechowski S, Kappler J (2003) Control of Bcl-2 expression by reactive oxygen species. Proc Natl Acad Sci U S A 100:15035–15040

Hill R, Ralph PJ (2005) Diel and seasonal changes in fluorescence rise kinetics of three scleractinian corals. Funct Plant Biol 32:549–559

Hill R, Ralph PJ (2006) Photosystem II heterogeneity of *in hospite* zooxanthellae in scleractinian corals exposed to bleaching condition. Photochem Photobiol 82:1577–1585

Hill R, Ralph PJ (2008) Impact of bleaching stress on the function of the oxygen evolving complex of zooxanthellae from scleractinian corals. J Phycol 44:299–310

Hill R, Ulstrup KE, Ralph PJ (2009) Temperature induced changes in thylakoid membrane thermostability of cultured, freshly isolated, and expelled zooxanthellae from scleractinian corals. Bull Mar Sci 85:223–244

Hoegh-Guldberg O (1999) Climate change, coral bleaching and the future of the world's coral reefs. Mar Freshwater Res 50:839–866

Hoegh-Guldberg O, Jones RJ (1999) Photoinhibition and photoprotection in symbiotic dinoflagellates from reef-building corals. Mar Ecol Prog Ser 183:73–86

Hoegh-Guldberg O, Smith GJ (1989) The effect of sudden changes in temperature, light, and salinity on the population density and export of zooxanthellae from the reef corals *Stylophora pistillata* Esper and *Seriatopora hystrix* Dana. J Exp Mar Biol Ecol 129:279–303

Hoegh-Guldberg O, Mumby PJ, Hooten AJ, Steneck RS, Greenfield P, Gomez E, Harvell CD, Sale PF, Edwards AJ, Caldeira K, Knowlton N, Eakin CM, Iglesias-Prieto R, Muthinga N, Bradbury RH, Dubi A, Hatziolos ME (2007) Coral reefs under rapid climate change and ocean acidification. Science 318:1737–1742

Hofmann GE, Burnaford JL, Fielman KT (2005) Genomics-fueled approaches to current challenges in marine ecology. Trends Ecol Evol 20:305–311

Huang S-P, Lin K-L, Fang L-S (1998) The involvement of calcium in heat-induced coral bleaching. Zool Stud 37:89–94

Hughes T (1994) Catastrophes, phase shifts, and large-scale degradation of a Caribbean coral reef. Science 265:1547–1551

Hughes L (2000) Biological consequences of global warming: is the signal already apparent. Trends Ecol Evol 15:56–61

Hughes TP, Connell JH (1999) Multiple stressors on coral reefs: a long-term perspective. Limnol Oceanogr 44:932–940

Iglesias-Prieto R, Matta JL, Robins WA, Trench RK (1992) Photosynthetic response to elevated temperature in the symbiotic dinoflagellate *Symbiodinium microadriaticum* in culture. Proc Natl Acad Sci U S A 89:10302–10305

Jamieson D, Chance B, Cadenas E, Boveris A (1986) The relation of free radical production to hyperoxia. Ann Rev Physiol 48:703–719

Johnson TM, Yu Z, Ferrans VJ, Lowenstein RA, Finkel T (1996) Reactive oxygen species are downstream mediators of p53-dependent apoptosis. Proc Natl Acad Sci U S A 93:11848–11852

Jokiel PL, Coles SL (1990) Responses of Hawaiian and other Indo-Pacific reef corals to elevated temperatures. Coral Reefs 8:155–162

Jones RJ, Hoegh-Guldberg O (2001) Diurnal changes in the photochemical efficiency of the symbiotic dinoflagellates (Dinophyceae) of corals: photoprotection, photoinactivation, and the relationship to coral bleaching. Plant Cell Environ 24:89–99

Jones RJ, Hoegh-Guldberg O, Larkum AWD, Schreiber U (1998) Temperature-induced bleaching of corals begins with impairment of the CO_2 fixation mechanism in zooxanthellae. Plant Cell Environ 21:1219–1230

Jones AM, Berkelmans R, van Oppen MJH, Mioeg JC, Sinclair W (2008) A community change in the algal endosymbionts of a scleractinian coral following a natural bleaching event: field evidence of acclimatization. Proc R Soc Lond B 275:1359–1365

Kerswell AP, Jones RJ (2003) Effects of hypo-osmosis on the coral *Stylophora pistillata*: nature and cause of 'low-salinity bleaching'. Mar Ecol Prog Ser 253:145–154

Kinzie RA III (1993) Effects of ambient levels of solar ultraviolet radiation on zooxanthellae and photosynthesis of the reef coral *Montipora verrucosa*. Mar Biol 116:319–327

Kleypas JA, Buddemeier RR, Archer D, Gattuso JP, Langdon C, Opdyke BN (1999) Geochemical consequences of increased atmospheric CO_2 on corals and coral reefs. Science 284:118–120

Kramarenko GG, Hummel SG, Martin SM, Buettner GR (2006) Ascorbate reacts with singlet oxygen to produce hydrogen peroxide. Photochem Photobiol 82:1634–1637

Kühl M, Cohen Y, Dalsgaard T, Jørgensen BB, Revsbech NP (1995) Microenvironment and photosynthesis of zooxanthellae in scleractinian corals studied with microsensors for O_2, pH, and light. Mar Ecol Prog Ser 117:159–172

Leggat W, Hoegh-Guldberg O, Dove S (2007) Analysis of an ESt library from the dinoflagellate (*Symbiodiniumi* sp.) symbiont of reef-building corals. J Phycol 43:1010–1021

Lesser MP (1996) Exposure of symbiotic dinoflagellates to elevated temperatures and ultraviolet radiation causes oxidative stress and photosynthesis. Limnol Oceanogr 41:271–283

Lesser MP (1997) Oxidative stress causes coral bleaching during exposure to elevated temperatures. Coral Reefs 16:187–192

Lesser MP (2000) Depth-dependent effects of ultraviolet radiation on photosynthesis in the Caribbean coral, *Montastraea faveolata*. Mar Ecol Prog Ser 192:137–151

Lesser MP (2004) Experimental coral reef biology. J Exp Mar Biol Ecol 300:217–252

Lesser MP (2006) Oxidative stress in marine environments: biochemistry and physiological ecology. Ann Rev Physiol 68:253–278

Lesser MP (2007) Coral reef bleaching and global climate change: can coral survive the next century? Proc Natl Acad Sci U S A 104:5259–5260

Lesser MP, Farrell J (2004) Solar radiation increases the damage to both host tissues and algal symbionts of corals exposed to thermal stress. Coral Reefs 23:367–377

Lesser MP, Gorbunov MY (2001) Diurnal and bathymetric changes in chlorophyll fluorescence yields of reef corals measured in situ with a fast repetition rate fluorometer. Mar Ecol Prog Ser 212:69–77

Lesser MP, Lewis S (1996) Action spectrum for the effects of UV radiation on photosynthesis in the hermatypic coral *Pocillopora damicornis*. Mar Ecol Prog Ser 134:171–177

Lesser MP, Shick JM (1989) Effects of irradiance and ultraviolet radiation on photoadaptation in the zooxanthellae of *Aiptasia pallida*: primary production, photoinhibition, and enzymic defenses against oxygen toxicity. Mar Biol 102:243–255

Lesser MP, Stochaj WR, Tapley DW, Shick JM (1990) Bleaching in coral reef anthozoans: effects of irradiance, ultraviolet radiation, and temperature on the activities of protective enzymes against active oxygen. Coral Reefs 8:225–232

Lesser MP, Cullen JJ, Neale PJ (1994) Photoinhibition of photosynthesis in the marine diaton *Thalassiosira pseudonana* during acute exposure to ultraviolet B radiation: relative importance of damage and repair. J Phycol 30:183–192

Levine B, Yuan J (2005) Autophagy in cell death: an innocent convict? J Clin Invest 115:2679–2688

Levy O, Achituv Y, Yacobi YZ, Dubinsky Z, Stambler N (2006) Diel 'tuning" of coral metabolism: physiological responses to light cues. J Exp Biol 209:273–283

Liu K, Sun J, Song Y, Liu B, Xu Y, Zhang S, Tian Q, Liu Y (2004) Superoxide, hydrogen peroxide, and hydroxyl radical in D1/D2/cytochrome *b-559* photosystem II reaction center complex. Photosynth Res 81:41–47

Long SP, Humphries S, Falkowski PG (1994) Photoinhibition of photosynthesis in nature. Ann Rev Plant Physiol Mol Biol 45:633–662

Loya Y, Sakai K, Yamazato K, Nakano Y, Sambali H, van Woesik R (2001) Coral bleaching: the winners and the losers. Ecol Lett 4:122–131

Lum JJ, DeBerardinis RJ, Thompson CB (2005) Autophagy in metazoans: cell survival in the land of plenty. Nat Rev Mol Cell Biol 6:439–448

Macpherson AN, Telfer A, Barber J, Truscott GT (1993) Direct detection of singlet oxygen from isolated photosystem II reaction centers. Biochim Biophys Acta 1143:301–309

Marla SS, Lee J, Groves JT (1997) Peroxynitrite rapidly permeates phopholipid membranes. Proc Natl Acad Sci U S A 94:14243–14248

Martindale JL, Holbrook NJ (2002) Cellular response to oxidative stress: signaling for suicide and survival. J Cell Physiol 192:1–15

Matta JL, Govind NS, Trench RK (1992) Polyclonal antibodies against iron-superoxide dismutase from *Escherichia coli* B cross react with superoxide dismutases from *Symbiodinium microadriaticum* (Dinophyceae). J Phycol 28:343–346

Mayfield AB, Gates RD (2007) Osmoregulation and osmotic stress in coral dinoflagellate symbiosis: role in coral bleaching. Comp Biochem Physiol A 147:1–10

Maynard JA, Anthony KRN, Marshall PA, Masiri I (2008) Major bleaching events can lead to increased thermal tolerance in corals. Mar Biol 155:173–182

Mazel C, Lesser MP, Gorbunov M, Barry T, Farrell J, Wyman K, Falkowski PG (2003) Green fluorescent proteins in Caribbean corals. Limnol Oceanogr 48:402–411

Middlebrook R, Hoegh-Guldberg O, Leggat W (2008) The effect of thermal history on the susceptibility of reef-building corals to thermal stress. J Exp Biol 211:1050–1056

Morrall CE, Galloway TS, Trapido-Rosenthal HG, Depledge MH (2000) Characterization of nitric oxide synthase activity in the tropical sea anemone *Aiptasia pallida*. Comp Biochem Physiol B 125:483–491

Murata N, Takahashi S, Nishyama Y, Allakhverdiev SI (2007) Photoinhibition of photosystem II under environmental stress. Biochim Biophys Acta 1767:414–421

Muscatine L, Porter JW, Kaplan IR (1989) Resource partitioning by reef corals as determined from stable isotope composition. I. $\delta^{13}C$ of zooxanthellae and animal tissue vs depth. Mar Biol 100:185–193

Nakamura T, van Woesik R (2001) Water-flow rates and passive diffusion partially explain differential survival of corals during the 1998 bleaching event. Mar Ecol Prog Ser 212:301–304

Nii CM, Muscatine L (1997) Oxidative stress in the symbiotic sea anemone *Aiptasia pulchella* (Calgren, 1943): contribution of the animal to superoxide ion production at elevated temperature. Biol Bull 192:444–456

Nishiyama Y, Yamamoto H, Allakhverdiev SI, Inaba M, Yokota A, Murata N (2001) Oxidative stress inhibits the repair of photodamage to the photosynthetic machinery. EMBO J 20:5587–5594

Nishiyama Y, Allakhverdiev SI, Murata N (2006) A new paradigm for the action of reactive oxygen species in the photoinhibition of photosystem II. Biochim Biophys Acta 1757:742–749

Niyogi KK (1999) Photoprotection revisted: genetic and molecular approaches. Ann Rev Plant Physiol Plant Mol Biol 50:333–359

Palmer CV, Modi CK, Mydlarz LD (2009) Coral fluorescent proteins as antioxidants. PLoS ONE 4:e7298

Peltier G, Cournac L (2002) Chororespiration. Ann Rev Plant Biol 53:523–550

Perez S, Weis V (2006) Nitric oxide and cnidarian bleaching: an eviction notice mediates breakdown of a symbiosis. J Exp Biol 209:2804–2810

Perez S, Weis V (2008) Cyclophyllin and the regulation of symbiosis in *Aiptasia pallida*. Biol Bull 215:63–72

Plantivaux A, Furla P, Zoccola D, Garello G, Forcioli D, Richier S, Merle P-L, Tambutté S, Alemand D (2004) Molecular characterization of two CuZn-superoxide dismutases in a sea anemone. Free Radic Biol Med 37:1170–1181

Pourzand C, Tyrell RM (1999) Apoptosis, the role of oxidative stress and the example of solar UV radiation. Photochem Photobiol 70:380–390

Rands ML, Douglas AE, Loughman BC, Ratcliff RG (1992) Avoidance of hypoxia in a cnidarian symbiosis by algal photosynthetic oxygen. Biol Bull 182:159–162

Reynolds JM, Bruns BU, Fitt WK, Schmidt GW (2008) Enhanced photoprotection pathways in symbiotic dinoflagellates of shallow-water corals and other cnidarians. Proc Natl Acad Sci U S A 105:13674–13678

Rich T, Allen RL, Wyllie AH (2000) Defying death after DNA damage. Nature 407:777–783

Richier S, Sabourault C, Courtiade J, Zucchini N, Allemand D, Furla P (2006) Oxidative stress and apoptotic events in the symbiotic sea anemone, *Anemonia viridis*. FEBS J 273:4186–4198

Richier S, Rodriguez-Lanetty M, Schnitzler CE, Weis VM (2008) Response of the symbiotic cnidarian *Anthopleura elegantissima* transcriptome to temperature and UV increase. Comp Biochem Physiol D 3:283–289

Richter M, Rüle W, Wild A (1990) Studies on the mechanism of photosystem II photoinhibition II. The involvement of toxic oxygen species. Photosynth Res 24:237–243

Rong Y, Distelhorst CW (2008) Bcl-2 protein family members: versatile regulators of calcium signaling in cell survival. Ann Rev Physiol 70:73–91

Sampayo EM, Ridgway T, Bongaerts P, Hoegh-Guldberg O (2008) Bleaching susceptibility and mortality of corals are determined by fine-scale differences in symbiont type. Proc Natl Acad Sci U S A 105:10444–10449

Sandeman I (2006) Fragmentation of the gastrodermis and detachment of zooxanthellae in symbiotic cnidarians: a role for hydrogen peroxide and Ca^{2+} in coral bleaching and algal density control. Rev Biol Trop (Int J Trop Biol) 54:79–96

Saragosti E, Tchernov D, Katsir A, Shaked Y (2010) Extracellular production and degradation of superoxide in the coral Stylophora pistillata and cultured Symbiodinium. PLoS ONE 9:e12508

Sawyer SJ, Muscatine L (2001) Cellular mechanisms underlying temperature-induced bleaching in the tropical sea anemone *Aiptasia pulchella*. J Exp Biol 204:3443–3456

Segovia M, Haramaty L, Berges JA, Falkowski PG (2003) Cell death in the unicellular chlorophyte *Dunaliella tertiolecta*. A hypothesis on the evolution of apoptosis in higher plants and metazoans. Plant Physiol 132:99–105

Sharp VA, Brown BE, Miller D (1997) Heat shock protein (HSP 70) expression in the tropical reef coral *Goniopora djiboutiensis*. J Therm Biol 22:11–19

Shashar N, Cohen Y, Loya Y (1993) Extreme diel fluctuations of oxygen in diffusive boundary layers surrounding stony corals. Biol Bull 185:455–461

Sheppard CRC (2003) Predicted recurrences of mass coral mortality in the Indian Ocean. Nature 425:294–297

Shick JM (1990) Diffusion limitation and hyperoxic enhancement of oxygen consumption in zooxanthellate sea anemones, zoanthids, and corals. Biol Bull 179:148–158

Shick JM, Lesser MP, Jokiel P (1996) Effects of ultraviolet radiation on corals and other coral reef organisms. Global Change Biol 2:527–545

Smith DJ, Suggett DJ, Baker NR (2005) Is photoinhibition of zooxanthellae photosynthesis the primary cause of thermal bleaching in corals? Global Change Biol 11:1–11

Smith-Keune C, Dove S (2008) Gene expression of a green fluorescent protein homolog as a host-specific biomarker of heat stress within a reef-building coral. Mar Biotechnol 10:166–180

Somero GN (2002) Thermal physiology and vertical zonation of intertidal animals: optima, limits, and costs of living. Integr Comp Biol 42:780–789

Sotka EE, Thacker RW (2005) Do some corals like it hot? Trends Ecol Evol 20:59–62

Suggett DJ, Warner ME, Smith DJ, Davey P, Hennige S, Baker NR (2008) Photosynthesis and production of hydrogen peroxide by *Symbiodinium* (Pyrrhophyta) phylotypes with different thermal tolerances. J Phycol 44:948–956

Sunda W, Kleber DJ, Klene RP, Huntsman S (2002) An antioxidant function for DMSP and DMS in marine algae. Nature 418:317–320

Takahashi S, Murata N (2008) How do environmental stresses accelerate photoinhibition. Trends Plant Sci 13:178–182

Takahashi S, Nakamura T, Sakamizu M, van Woesik R, Yamasaki H (2004) Repair machinery of symbiotic photosynthesis as the primary target of heat stress for reef-building corals. Plant Cell Physiol 45:251–255

Takahashi S, Whitney S, Itoh S, Maruyama T, Badger M (2008) Heat stress causes inhibition of the *de novo* synthesis of antenna proteins and photobleaching in cultured *Symbiodinium*. Proc Natl Acad Sci U S A 105:4203–4208

Tchernov D, Gorbunov MY, de Vargas C, Yadav SN, Milligan AJ, Häggblom M, Falkowski PG (2004) Membrane lipids of symbiotic algae are diagnostic of sensitivity to thermal bleaching in corals. Proc Natl Acad Sci U S A 101:13531–13535

Trapido-Rosenthal H, Zielke S, Owen R, Buxton L, Boeing B, Bhagooli R, Archer J (2005) Increased zooxanthellae nitric oxide synthase activity is associated with coral bleaching. Biol Bull 208:3–6

Vincent WF (1980) Mechanisms of rapid photosynthetic adaptation in natural phytoplankton communities. 2. Changes in photochemical capacity as measured by DCMU-induced chlorophyll fluorescence. J Phycol 20:201–211

Warner ME, Berry-Lowe S (2006) Xanthophyll cycling and photochemical activity in symbiotic dinoflgellates in multiple locations of three species of Caribbean coral. J Exp Mar Biol Ecol 339:86–95

Warner ME, Fitt WK, Schmidt GW (1999) Damage to photosystem II in symbiotic dinoflagellates: a determinant of coral bleaching. Proc Natl Acad Sci U S A 96:8007–80012

Warner ME, Chilcoat GC, McFarland FK, Fitt WK (2002) Seasonal fluctuations in the photosynthetic capacity of photosystem II in symbiotic dinoflagellates in the Caribbean reef-building coral *Montastraea*. Mar Biol 141:31–38

Weis VM (2008) Cellular mechanisms of cnidarian bleaching: stress causes the collapse of symbiosis. J Exp Biol 211:3059–3066

Weis VM, Davy SK, Hoegh-Guldber O, Rodriguez-Lanetty M, Pringle JR (2008) Cell Biology in model systems as the key to understanding corals. Trends Ecol Evol 23:369–376

Wilkinson CR (1999) Global and local threats to coral reef functioning and existence: review and predictions. Mar Freshwater Res 50:867–878

Williams EH Jr, Bunkley-Williams L (1990) The world-wide coral reef bleaching cycle and related sources of coral mortality. Atoll Res Bull 335:1–67

Yonge CM, Nichols AG (1931) The structure, distribution and physiology of the zooxanthellae. (Studies on the Physiology of Corals IV). Sci Rep Great Barrier Reef Exped 1928–29 1:135–176

The Potential for Temperature Acclimatisation of Reef Corals in the Face of Climate Change

Barbara E. Brown and Andrew R. Cossins

Keywords Temperature • acclimatisation • acclimation • epigenetics

Coral bleaching has taken centre-stage in the debate over the likely biological effects of global environmental change. Central to any judgements on this issue is the ability of corals to display increased tolerance of debilitating or lethal conditions through phenotypic adaptations, such as heat-hardening, longer-term acclimatisation responses or even trans-generational epigenetic effects. But the key question is whether the magnitude of such responses can match the predicted increases in sea temperatures over the period of global warming. In the recent literature, much has been said about the potential for acclimatisation in tropical reef corals and how it may, or may not, be significant in the context of the world's changing climate (Hughes et al. 2003; Hoegh-Guldberg 2004; Donner et al. 2007; Hoegh-Guldberg et al. 2007; Maynard et al. 2008a; Donner 2009). In fact, we know remarkably little about the potential for and extent of acclimatisation in corals, and the complex physiology and behaviour underlying the phenomenon (Edmunds and Gates 2008; Maynard et al. 2008a). It is important at this stage to define the terms used in this chapter following Bligh and Johnson (1973) since there has been, and continues to be, considerable confusion in their use in the literature together with established concepts in thermal biology (see Box 1).

B.E. Brown (✉)
School of Biology, University of Newcastle upon Tyne, Newcastle upon Tyne, NE1 7RU, United Kingdom
e-mail: profbarbarabrown@aol.com

A.R. Cossins (✉)
School of Biological Sciences, University of Liverpool, L69 3BX, Liverpool, United Kingdom
e-mail: cossins@liverpool.ac.uk

Box 1 – Terms and Concepts Used in Considering Resistance Adaptations of Corals

Terminology

Genotypic adaptation is a process whereby natural selection adjusts the frequency of genes that code for traits affecting fitness (Cossins and Bowler 1987; Willmer et al. 2004) and which typically occurs over many generations. In contrast, phenotypic adaptation refers to adaptive changes made by an organism within its lifetime. These phenotypic responses are usually short-medium term (minutes or hours for diurnal responses, or days and weeks for seasonal responses) and reversible, but are limited in extent by the organism's genotype (Coles and Brown 2003). They can also occur during development to generate altered morphologies usually with irreversible consequences. Acclimatisation refers to phenotypic adaptations in response to fluctuations in natural conditions, whilst acclimation is reserved for adaptations generated under controlled laboratory experiments when the effects of the factor of interest can be isolated. Since there is no useful adjective to be derived from acclimatisation, 'acclimatory' is frequently used in the literature to describe responses of organisms in their natural environment (Willmer et al. 2004).

Resistance Adaptations

Phenotypic responses manifest themselves either as responses induced over the 'normal' range of

(continued)

> **Box 1** (continued)
>
> temperatures that affect the rates of living ('capacity' adaptations, see Precht et al. (1973)) or as responses that enhance survival of extreme conditions ('resistance' adaptations). The latter are usually induced by prior short- or long-term exposure to mildly stressful conditions and they generally comprise rapidly occurring but transient increases in resistance to lethal stress ('hardening' responses), or to more slowly developing and more long-lasting resistance (as with seasonal resistance acclimatisation). Damage, debilitation and death are heavily dependent on the length and frequency of exposure to extreme environmental conditions since thermal damage is progressive and cumulative over time. The rate by which damage occurs is also very highly dependent on lethal temperature (i.e. the Q_{10} temperature coefficient is very high, see appendix in Schmidt-Nielsen (1997)), such that a small increase in exposure temperature (i.e. 0.1°C) can induce a disproportionately large increase in the rates of damaging processes. However, thermal damage can be mitigated by repair using a suite of mechanisms of which the best known are heat shock (stress) proteins and those which repair oxidative damage. Tolerance and resistance are often linked to the status of these mechanisms. All of these considerations relate directly to the concept of 'phenotypic plasticity', in which single genotypes can generate different phenotypes in response to exposure to different environmental conditions (Pigliucci et al. 2006). Some, but not all, phenotypic plasticity is adaptive.
>
> ### Measuring Thermal Resistance
>
> A widely used and more ecologically meaningful approach to the determination of thermal tolerance properties is the critical thermal maximum (CTM) in which animals are warmed or cooled at a constant rate (e.g. 1°C/h) and the temperature at which a loss of a critical locomotory, postural or behavioural attribute is observed (i.e. loss of righting response, or of orientation). However, CTM is negatively related to heating rate, and is generally affected by the prior thermal conditioning of the animals, and ecologically relevant experimental conditions need to be chosen (Terblanche et al. 2007).

Underlying our attempts to predict the status of corals in future years are two important considerations – first, to what extent can they acclimatise to changing environmental conditions, and second, how close are these organisms to their thermal limits? As highlighted by Edmunds and Gates (2008), it is not a case of whether corals can or cannot acclimatise to changing environmental conditions but rather to what degree they are able to acclimatise. According to Donner et al. (2005), who combined NOAA bleaching predictions with atmosphere-ocean general circulation models, the thermal tolerance of corals must increase by 0.2–1.0°C per decade over the next 30–50 years if bleaching is not to become an annual or biannual event on the world's coral reefs. The arguments highlighted by Hoegh-Guldberg (1999, 2004), Donner et al. (2005) and Hoegh-Guldberg et al. (2007) suggest that corals will be unable to physiologically adjust in the time frame required to match this rate of warming.

Closely related to the above are two claims that have become engrained in the recent coral literature. They are that corals are living close to their upper lethal temperatures (Mayer 1914; Edmondson 1928; Coles et al. 1976) and, second, that coral bleaching, in response to steadily rising sea temperatures, has increased in frequency and intensity in recent years (Hoegh-Guldberg 1999, 2004; Hughes et al. 2003; Donner et al. 2005; Hoegh-Guldberg et al. 2007) implying that thermal thresholds have changed little during the last 2 decades (Kleypas et al. 2008). Coral bleaching (i.e. the loss of zooxanthellae and/or their pigments) in response to stresses such as elevated temperature are discussed in detail in Chapter 23 Coral Bleaching: Causes and Mechanisms and the bleaching response, while not precisely defining the thermal limits of the coral, has regularly been used as a sensitive indicator of thermal sensitivity of the holobiont (see reviews by Jokiel and Coles 1990; Glynn 1993; Hoegh-Guldberg 1999).

In this Chapter we shall revisit the above claims in the context of evidence provided from other poikilotherms (organisms with a variable body temperature) on the possible limitations of living at temperatures close to lethal limits. We shall also review recent evidence for acclimatisation of corals to elevated temperature and solar radiation while briefly exploring the new and emerging science of epigenetics (the study of heritable changes in gene expression and function that cannot be explained by changes in DNA sequence) and the implications that this may have for a more thorough understanding of the acclimatisation potential of reef corals in future decades. Defining the magnitude of resistance adaptations is complicated by several factors, notably (i) complications arising from the symbiotic engagement of host and zooxanthellae, since they may have different individual sensitivities as well as possessing interacting effects on thermal sensitivity, (ii) the existence of stressors in addition to damaging high temperature, including solar radiation and desiccation, and (ii) the sheer complexity of the natural environment with the interplay of tidal, diurnal and other multiannual cycles.

1 Historical Perspectives on Coral Acclimatisation and Acclimation

Interestingly, physiological adjustments to environmental factors such as temperature and salinity were discussed by some of the earliest scientists working on corals (Mayer 1914; Edmondson 1928) while Yonge (1940) attributed much of the success of the stony corals to their considerable powers of 'adaptation'. In this context Yonge was using the term in a broad context to incorporate both long-term genotypic adaptation (as defined above) and short-term phenotypic acclimatisation. His views were strongly influenced by his field and experimental observations of corals living on shallow, inter-tidal reef flats during the 1928–1929 Great Barrier Reef Expedition. Although unsuccessful, because of methodological constraints, he was the first to attempt measurement of coral acclimatisation responses through the assessment of monthly zooxanthellae densities in reef corals (British Natural History Museum Archives 1928). Indeed, it was over 50 years later before we appreciated the significance of seasonal adjustments in the depths of coral tissue and zooxanthellae densities in response to increased temperature and solar radiation (Stimson 1997; Brown et al. 1999; Fagoonee et al. 1999; Fitt et al. 2000).

Subsequently, studies of the temperature physiology of corals and their powers of acclimation were carried out by Coles and Jokiel (1978) in their evaluation of the effects of a thermal effluent from a power-generating station in Hawaii. Their experiments not only established that *Montipora verrucosa* was capable of acclimation to temperatures of 1–2°C above the summer maximum temperature but that this species was able to tolerate a fluctuating temperature regime. Furthermore, they established that sub-tropical coral species appeared to have an upper lethal limit that was 2°C lower than their tropical counterparts (Coles et al. 1976). These studies and others are discussed in detail in Brown (1997) and Coles and Brown (2003) in their reviews of the acclimatisation potential of corals to rising sea temperatures.

By the late 1990s concern about the effects of climate change on coral reefs prompted renewed interest in acclimatory responses of corals with Berkelmans and Willis (1999) demonstrating that the winter bleaching threshold of *Pocillopora damicornis* on the Great Barrier Reef was 1°C lower than the summer threshold for this species. In addition, Berkelmans (2002) concluded that thermal adaptation occurred on scales of 10–100 km on the Great Barrier Reef with cross-shelf and latitudinal differences in bleaching thresholds corresponding to specific temperature regimes on mid and outer-shelf reefs.

Clearly, early workers were intrigued by the environmental rigours experienced particularly by inter-tidal corals and their ability to endure hours of aerial exposure, high temperatures and solar radiation. As coral reef science developed in the early 1970s, thermal thresholds of corals were a focus for those evaluating the effects of potentially polluting discharges. By the 1990s, questions over the extent of thermal acclimatisation by corals and the rate at which physiological adjustments could be made were frequently voiced. These questions continue to be asked as projections are attempted of coral reef status by the middle of the twenty-first century (Berkelmans et al. 2004; Donner et al. 2005; Hoegh-Guldberg et al. 2007; Baker et al. 2008; Donner 2009).

Before looking at the recent evidence for acclimatisation in reef corals it is valuable to review the latest thinking of thermal physiologists working on similar problems with other phyla and the idea that species living in tropical climates are likely to suffer disproportionately from small temperature increments.

2 Organisms Living Close to their Lethal Limits are more Vulnerable to the Effects of Climate Change

Janzen (1967) has argued that because tropical ectotherms had evolved in relatively benign aseasonal environments, they should be thermal specialists with a limited ability to acclimate compared with higher latitude, more generalist species. More recent work (Tewksbury et al. 2008) notes that while these conclusions are broadly supported by several studies on terrestrial ectotherms they do not necessarily hold in the marine environment where thermal specialists are found at both high and low latitudes with thermal generalists being common in the mid-latitudes. According to Tewksbury and his colleagues, such a pattern mirrors the seasonality of ocean temperatures leading to the conclusion that both tropical and high-latitude organisms live at near-stressful temperatures, which could make them particularly susceptible to global warming.

An elegant study by Stillman (2002) on the thermal tolerance limits of the porcelain crab across both latitudinal boundaries and vertical gradients in the inter-tidal/sub-tidal zone has shown that the upper thermal tolerance limits of a number of crab species mirrors their microhabitat conditions with a strong positive correlation between lethal temperatures and maximal habitat temperature. Interestingly, a porcelain crab species from the upper inter-tidal habitat, which had evolved the greatest tolerance to damaging high temperatures, appeared to have much reduced acclimation capacity when thermally conditioned to 8°C and 18°C, compared with lower inter-tidal and sub-tidal species. These results led Stillman (2003) to conclude that the upper inter-tidal species would be the most susceptible to the smallest increases in microhabitat temperatures and hence most vulnerable to climate warming. He argued that intertidal species might be living closer to their thermal maxima and may have reduced their abilities to increase their upper thermal tolerance limits when compared with sub-tidal species. Similar results were obtained for turban snails of the genus *Tegula*, which inhabit the intertidal and subtidal zone (Hellberg 1998) with thermal limits of protein synthesis reflecting their vertical distribution on the shore while inter-tidal species displayed reduced acclimation abilities compared to those in the sub-tidal zone. In reviewing this work Somero (2005) concluded that 'warm adapted intertidal species face current – and most likely, future-threats from high temperatures than less heat-tolerant, subtidal congeners'.

But how relevant are these conclusions to reef corals? Certainly it would appear from the existing literature that the thermal limits of corals are dictated, at least in part, by their thermal environment both on a latitudinal (Coles et al. 1976) and a seasonal basis (Berkelmans and Willis 1999). However, the temperature environment encountered by corals worldwide is very variable as shown in Fig. 1 with corals in low latitudes living at high temperatures with relatively little seasonal variation (~2–3°C) while those at high latitudes, such as in the Iki Islands of Japan, experience annual lows of 14°C and annual highs of 27.5°C (Nomura 2004). Considerable seasonal temperature ranges are also recorded for corals living in the Arabian region (Sheppard et al. 2002). Marked fluctuations (>6°C) in daily temperatures have been noted for corals living in the vicinity of upwelling areas in Oman (Coles 1997), as well in the Andaman Sea (Phongsuwan personal communication) and Chagos archipelago where reefs are subject to the effects of internal waves (Sheppard 2009). Such fluctuations are likely to have significant bearing on overall thermal sensitivity as suggested in field studies of thermally induced coral bleaching (McClanahan and Maina 2004; McClanahan et al. 2007). Furthermore, some of the most temperature-susceptible branching corals known, i.e. corals of the genus *Acropora* and genus *Pocillopora*, have been found in the vicinity of geothermal vents in the Banda Sea, Indonesia where temperatures of 34°C were recorded (Tomascik et al. 1997).

Clearly corals have adapted over time to a wide range of temperature scenarios but it seems likely that annual and daily variations in temperature might also play a role, alongside maximal habitat temperatures, in defining thermal tolerances. This is particularly likely for high latitude sites where corals experience low winter temperatures that lead to declines in photochemical efficiency (Suwa et al. 2008) or even bleaching (Nomura 2004; Hoegh-Guldberg and Fine 2005), factors which could have profound implications for the energy budgets of corals as they approach maximal summer temperatures later in the year. Indeed, it might be argued that temperature increases associated with climate change might provide some benefit to the corals of high latitude reefs since the negative effects of winter low temperatures might be eliminated and the physiological status of the coral improved to face the rigours of an increased summer maximum.

Unlike the porcelain crabs and turban snails described earlier, the thermal tolerances of corals are defined not only by their maximum temperature exposure but also by their experience of solar radiation (Fitt et al. 2001; Lesser and Farrell 2004; Brown and Dunne 2008) since the latter plays a key role in the bleaching process (see Chapter 23) as temperature rises. Because light is so central in either reducing or improving bleaching tolerance at elevated sea temperatures (Brown and Dunne 2008), the thermal limits of corals must be influenced by the combination of both the thermal and light regimes of their microhabitat.

Corals typically undergo thermal damage and bleaching at temperatures above 32°C, yet many other species, both vertebrate and invertebrate, including those inhabiting higher latitudes, can withstand much higher temperatures. For example, (i) goldfish can be conditioned indefinitely at temperatures up to 38°C and can display complex trained behaviours even above 40°C (Hoyland et al. 1979); (ii) some tropical marine fish species in Indonesia display upper critical thermal maxima of 40°C or above (Eme and Bennett 2009) (ii) European diving beetles of the genus *Agabus* display a mean upper lethal limit of 43–46°C (Calosi et al. 2008), (iii) the tropical prawn *Macrobrachium acanthurus* can be acclimated to 32°C and displays a CTM of 39.8°C (Diaz et al. 2002), and (iv) tropical bivalves typically display upper thermal limits at approx 36–40°C (Compton et al. 2007). Indeed, the upper tolerances of tropical corals are closer to that of freshwater crayfish inhabiting the cool temperate rivers of northern Europe (Cossins and Bowler 1976). Thus, it seems clear that corals are not as thermally resistant as might be

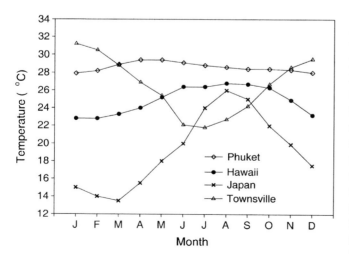

Fig. 1 Average monthly sea temperatures over the year for four different reef locations: Phuket, Thailand (After Brown et al. 1996); Hawaii (After Jokiel and Brown 2004); Iki Islands, Japan (After Nomura 2004) and Townsville, Australia (After Kenny 1974)

expected from their tropical distribution, which is perhaps surprising given their sessile nature as adults and their inability to avoid stress by moving in a thermal gradient.

This sensitivity might have two explanations. First, it might well be that the apparently high thermal sensitivity of corals is due to some fundamental physiological limitation which is not evident in other taxa. Corals certainly possess conventional heat shock proteins (HSP) responses (Sharp et al. 1997) and the cnidarian genome (see Section 5 Genomics Approaches to Stress Responses in Corals) certainly contains a broad range of stress-response genes. But they might lack some other critical protective response that is expressed in more complex organisms. Alternatively, their sensitivity might be linked to some property of the symbiotic relationship between zooxanthellae and host. Second, it may be that the microclimatic conditions experienced by corals occupies a lower range of temperatures than is currently appreciated, and that their tolerance states relate to this rather than the temperatures indicated by large geographic-scale models or remote sea-surface measurements. In any case, the thermal microenvironment of intertidal and sub-tidal zones is a complex mosaic over space and time due to the powerful influences of solar and tidal cycles, and the experiences of animals within that niche cannot be simply characterised by a single temperature. This applies particularly to the sea-surface temperatures generated by satellite imaging, which by estimating surface skin of the ocean does not reflect the degree of variability with depth, the occurrence of cold upwelling or other micro-environmental features. This mosaic presents opportunities for survival even in a globally warmed world, i.e. thermal refugia. A similar situation was recorded by Huey et al. (2009) who found that tropical species of forest lizard were active over a lower temperature range and displayed lower thermal tolerance limits than lizard species inhabiting more open, lowland tropical sites, and this was found to match the lower temperatures beneath the forest canopy. Thus, the thermal properties of forest and savannah lizards were linked to the microclimatic properties of their respective environments. Similarly, it may be that the corals inhabit waters with temperatures well below those causing damage and have evolved a resistance appropriate to the temperature variations more frequently experienced than for the occasional thermal bleaching event.

3 Has Coral Bleaching Increased in Intensity and Frequency in Recent Years?

One of the earliest attempts to collate the incidence of bleaching events over time was that of Glynn (1993) who showed that major bleaching had only been documented since the 1980s with very few records prior to this date. One explanation that Glynn suggested might account for this pattern was the lack of interest and accessibility to reefs prior to 1970 though he noted that in the 1960s and 1970s there were few bleaching reports despite an active and expanding reef research base.

However, one of the earliest documented examples of coral bleaching was made by Yonge on the Great Barrier Reef Expedition of 1928–1929 (Yonge and Nicholls 1931). In March 1929 Yonge and his colleagues noted extensive bleaching and mortality of reef flat corals at Low Isles (Fig. 2) during a period of calm conditions when seawater temperatures reached at least 35°C. Although these observations led Yonge to experimentally investigate the loss of zooxanthellae at elevated temperatures he never published photographs of the bleached reef in the extensively illustrated scientific reports of the expedition. Having witnessed recovery of many of the bleached corals within 3 months in the field (Yonge and Nicholls 1931), one can only assume that Yonge thought that there was nothing extraordinary about this bleaching episode and that it was a phenomenon that might be witnessed regularly by corals living on the shallow reef flat.

One of the first references to the fact that bleaching events were occurring more and more frequently was that of Hoegh-Guldberg (1999), though no quantitative evidence was provided in this paper. An earlier publication (Brown et al. 1996) documenting the steadily rising sea temperatures over a 50-year period in the eastern Indian Ocean certainly predicted an increased frequency of bleaching in this region but it was not until Oliver et al. (2009) analysed the comprehensive

Fig. 2 Coral bleaching at Low Isles, Great Barrier Reef on March 21st 1929 (Photograph from C.M. Yonge collection held at the British Museum of Natural History and reproduced with the permission of the latter)

ReefBase bleaching database that any attempt was made to quantify the incidence of bleaching events over recent time. These authors concluded that their data did not allow them to differentiate between true increases in bleaching frequency and increases in reporting effort. They did, however, identify four major bleaching peaks on a global level in the previous 20 years, but stressed that this number was too small quantitatively to establish that the frequency of severe events was increasing. Similarly, their data did not indicate any increases in the intensity of bleaching over this time. This result was in stark contrast to the findings of Eakin et al. (2009) who demonstrated clear increases in both the frequency and intensity of bleaching-level temperature stresses derived from instrumental observations of global SST and modern near real-time satellite data. Oliver et al. (2009) explained this mismatch by considering coral colonies that survived severe bleaching, such as that in 1998, as being more capable of surviving subsequent thermal stress. In other words, they suggested that corals have accommodated to the steadily rising sea temperatures over recent years. This explanation is also one of a number used by Berkelmans (2009) to account for the fact that bleaching did not occur at four sites on the Great Barrier Reef in 2004 despite the bleaching threshold temperatures far exceeding those eliciting bleaching in 1998. Other explanations invoked the modulating effects of light and the selection of more thermally resistant holobiont genotypes among surviving populations.

Reduced susceptibility to bleaching was also observed at sites around Phuket, Thailand, in 1997 and 1998 when environmental stresses (both temperature and solar radiation) were much higher than in 1991 and 1995 when extensive bleaching was witnessed (Dunne and Brown 2001). In this case, experience of unusually high solar radiation in the months preceding the seasonal maximum temperature, which leads to improved thermal tolerance (Brown et al. 2002a), was suggested as the explanation for reduced bleaching in 1997–1998. Interestingly, despite steadily rising sea temperatures over the last 60 years at this location, which is in the Indian Ocean warm pool, there have been no major bleaching events in recent years on the scale of those witnessed in 1991 and 1995. A more recent example of possible acclimatisation involves that of increased thermal tolerance of three major coral genera, namely *Acropora*, *Pocillopora* and *Porites,* on the Great Barrier Reef in 2002 following earlier thermal stress in 1998 when bleaching was extensive (Maynard et al. 2008b). In this example, bleaching was 30–100% lower in 2002 than that predicted from the relationship between bleaching severity and thermal stress in 1998, in spite of much higher solar irradiances in 2002. Prior experience of high solar radiation before the 2002 event and selective mortality of less tolerant genotypes, as a result of the 1998 bleaching, were not considered to be significant in explaining the observed increase in thermal tolerance of corals. However, symbiont shuffling, trophic plasticity and/or access to heterotrophic feeding and physiological acclimatisation were highlighted as possible mechanisms accounting for the improved coral tolerances observed in 2002.

We therefore conclude that despite the popular notion that global bleaching events are increasing in intensity and frequency there is at present no rigorous evidence to support such a statement and furthermore some suggestion that the thermal tolerances of corals in different parts of the world are adjusting to warmer scenarios. In the following section, we examine the existing evidence for resistance acclimatisation in reef corals.

4 Recent Work on Phenotypic Resistance Adaptations to Thermal/Irradiance Stresses in Reef Corals

The recent literature on responses of reef corals to thermal and irradiance stresses can be broadly divided into two categories. First, acclimatisation studies on the effects of experimentally elevated temperature and/or solar radiation on corals which have acclimatised to particular conditions (i.e. high or low sea temperatures and/or irradiance) in their natural environment (Brown et al. 2002a, b; Anthony and Hoegh-Guldberg 2003; D'Croz and Mate 2004; Castillo and Helmuth 2005: Berkelmans and van Oppen 2006; Dove et al. 2006; Griffin et al. 2006; Brown and Dunne 2008) and second, a number of acclimation studies where corals were experimentally pre-exposed to high temperatures and high or low solar radiation before subsequent evaluation of their physiology at a later date under stressful temperature conditions (Visram and Douglas 2007; Yakovleva and Hidaka 2004; Castillo and Helmuth 2005; Middlebrook et al. 2008).

Where acclimatisation had been demonstrated then it appears to involve several different processes – these include changes in physiological/biochemical traits of both the coral host and/or its zooxanthellae (Brown et al. 2000, 2002a,b; Brown and Dunne 2008); the replacement of bleaching susceptible zooxanthellae by genetically distinct, bleaching-resistant zooxanthellae (Rowan 2004; Baker et al. 2004) or by shifts in the dominant members of zooxanthellae populations in corals, which host multiple clades or types of algae (Berkelmans and van Oppen 2006).

As far as physiological traits are concerned, then the field experience of high irradiance on the western surfaces of *Goniastrea aspera* colonies from Phuket, Thailand was shown to confer subsequent temperature tolerance mediated, at least in part, by high levels of stress proteins and antioxidants in the coral host and improved xanthophyll cycling in the zooxanthellae (Brown et al. 2000, 2002b). Stress proteins have also been shown to be critical in seasonal acclimatisation of other marine invertebrates (Hoffman and Somero 1995; Roberts et al. 1997) with increased stress protein and antioxidant enzyme activity being noted in several studies on thermally stressed corals (Downs et al. 2000, 2002; DeSalvo et al. 2008). The overall thermal tolerance of corals is affected by both coral host and zooxanthellae (Baird et al. 2009) with publications highlighting properties of the host (Salih et al. 2000; Brown et al. 2002b; Bhagooli and Hidaka 2004; Dove 2004; Grottoli et al. 2006; Ainsworth et al. 2008) and the zooxanthellae (Baker et al. 2004; Rowan 2004; Tchernov et al. 2004; Berkelmans and van Oppen 2006) in influencing bleaching susceptibility at elevated temperatures.

The background thermal tolerance of *G. aspera*, discussed above, will be affected by the fact that it hosts Clade D zooxanthellae, which are recognised as being the most thermally tolerant zooxanthellae known to date (Rowan 2004; see also Chapter 23). More specifically, these corals harbour symbionts identified as type D1a (Pettay and LaJeunesse 2009) on both east and west surfaces of the colony though the zooxanthellae on the western surfaces appear to have improved photoacclimatory abilities compared with those on the east (Brown and Dunne 2008).

A clear example of corals, with mixed symbiont populations, acquiring improved thermal tolerance through changing the dominant symbiont type is that documented by Berkelmans and van Oppen (2006). These authors investigated the potential for acclimatisation of the branching coral *Acropora millepora*, which hosts both Clade C and Clade D zooxanthellae. Corals were able to gain an increased thermal tolerance in the order of 1–1.15°C by switching their dominant symbiont Clade to Clade D. The authors argue that although this capability is significant in the context of global climate change, in the absence of other mechanisms of thermal acclimatisation it would not be sufficient to meet the required tolerance increases of 0.2–1.0°C per decade demanded by some models (Donner et al. 2005).

It is thus clear that corals have a number of mechanisms by which they can respond to changes in their thermal and irradiance environments and that these can be effected within a relatively short time period (~9 months in the case of symbiont shuffling as described by Berkelmans and van Oppen (2006)). The timescale of acquisition of thermal tolerance in *G. aspera* is also relatively short being in the order of 1–3 years as inferred from bleaching patterns in corals of different heights in the field. Many of the corals displaying nonbleached western surfaces during bleaching events were 4–6 years old (Brown et al. 2000) and these colonies would not have been exposed to differential irradiance regimes on east and west faces until they were ~1–2 years of age. How long the irradiance 'memory', and the thermotolerance it confers, is retained in the absence of the environmental signal that first induced the tolerance is unknown though preliminary observations suggest that the 'memory' might be retained for several years (see Section 6 Epigenetics and Its Significance for Coral Acclimatisation to Elevated Temperature).

It is important to point out that not all corals have the ability to improve their thermal tolerance by shuffling their symbionts, with *A. millepora* from some sites on the Great Barrier Reef retaining their native dominant clade and showing greater bleaching susceptibility when manipulated in the same way as their congeners from other reefs (Berkelmans and van Oppen 2006). Similarly, while some laboratory studies show temperature acclimation of corals after pre-exposure to elevated seawater temperatures (Yakovleva and Hidaka 2004; Middlebrook et al. 2008), others do not (Visram and Douglas 2007) though as the latter authors point out it is often extremely difficult to mimic environmental conditions on the reef in laboratory manipulations. Nevertheless, it appears that there is overwhelming evidence for acclimatory ability in different coral species to increased sea temperatures and that this could very well account for the apparent increase in thermal tolerance of three major genera on the Great Barrier Reef between 1998 and 2002 (Maynard et al. 2008b).

Hoegh-Guldberg (2009) has recently dismissed phenotypic acquired tolerance responses as having any bearing on the ability of corals to tolerate future warmer climates, justifying this by pointing to the small scale of protective responses in relation to the challenges of the projected increase in environmental temperature. Hoegh-Guldberg et al. (2007) went further in suggesting that evidence of coral phenotypic resistance adaptation to bleaching is equivocal or nonexistent, the implication being that there was sufficient experimental exploration of the issue to make a proper judgement. Obura (2005) recognised that thermal tolerance properties of corals are particularly complex and argues for a more explicit definition of terms. This would permit the separation of adaptive responses so that the underpinning tolerance traits can be more properly quantified, and the status of coral reefs can be more precisely defined over time. The current literature is very limited in quantifying acquired tolerance responses in corals, including both rapid heat hardening and longer-term acclimation responses. Indeed, Hoegh-Guldberg and colleagues (Middlebrook et al. 2008) recognise this point, in stating that only a single study has examined thermal acclimation experimentally. Good examples of the systematic approaches required (conditioning and exposure regimes, CTM's, tolerance polygons, etc.) are those described by Cossins and Bowler (1987) and put into effect in respect of reef fish by Eme and Bennett (2009). It is therefore premature to reject a meaningful role for acquired tolerance adaptations in mitigating the effects of global warming. But being colonial corals are especially tractable and interesting for this kind of analysis since it is possible to replicate the analysis of single genotypes, and they can potentially display differentiated adaptive states across the colony.

An important issue in defining the survivable thermal limits of corals, and their ability to display resistance acclimation responses, is to establish an ecologically meaningful criterion of thermally induced debilitation or death. Conventional methods of quantifying tolerance properties for other organisms (see Cossins and Bowler 1987) involves exposing a sample of animals from a single population to a range of lethal temperatures over time and recording the time or temperature causing 50% mortality (LT_{50},). In order to determine the extent to which LT_{50} can be modified by prior thermal experience some batches of animals need to be pre-conditioned for at least 3 weeks to a range of temperatures over the normal, non-stressful range of temperatures. Their LT_{50} can then be determined as before. The resulting data can be used to construct a 'tolerance polygon', which defines the tolerable thermal niche of that population of animals (Fig. 3). At present a range of different measures of thermal debilitation have been used for corals by different investigators (see Table 1), which are mainly focused on the viability and performance of the photosynthetic apparatus and bleaching of zooxanthellae rather than mortality of the holobiont itself. But Obura (2005) argues that bleaching does not inevitably lead to the coral mortality and shows examples from the coast of Kenya of an inverse relationship in the branching coral *Pocillopora damicornis*. Moreover, measures of thermal injury and death vary considerably depending on the methods used, and this directly affects any conclusions drawn regarding the effects of global warming. Thus, to develop a more meaningful prediction of global effects on coral viability it is important first to identify the critical lesions of thermal damage leading not just to loss of photosynthetic performance but also to long-term viability of the holobiont. In addition to the existing ideas of bleaching susceptibility and photosystem damage, this could be related to other possible lesions of the host tissue, such as loss of energy supply due to debilitation of oxidative phosphorylation (El-Wadawi and Bowler 1995), or collapse of ion gradients across cellular plasma membranes (Gladwell 1975; Cossins and Bowler 1976), both of which have been documented in other taxa.

Of course, phenotypic resistance adaptations have their limits, particularly at the high end of the spectrum of predicted temperatures and when faced by a 6°C warmed planet. But there is good reason to expect that phenotypic resistance

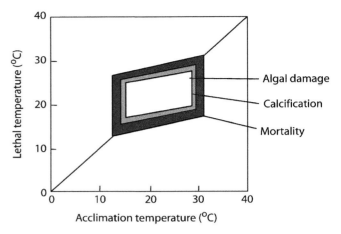

Fig. 3 Hypothetical tolerance polygons constructed for a coral holobiont. The polygons represent the effect of acclimation temperature upon the upper and lower thermal limits of a coral for three different measures of effect, namely holobiont mortality, inhibition of coral calcification and onset of algal damage. The upper line of each polygon is based on data for 50% debilitating effects of high temperature and the lower line for low temperatures, and the polygon is constructed for each with vertical construction lines. The polygons thus contains the combination of acclimation and lethal temperatures within which median effects are evident. Similar lines can be constructed for any other level of effect as a series of 'onion skins' around the median limits; thus the algal damage polygon lies within that for other measurements indicating that it has greater thermal constraints. The slope of the graphs indicating upper and lower thermal limits reflects the effect of prior thermal history upon the thermal tolerance; animals displaying large such responses typically have a slope of 0.3. The diagonal line indicates where lethal temperature equals the acclimation temperature

Table 1 Examples from the recent literature of parameters measured during experimental exposure of corals to elevated temperatures where P/R=photosynthesis/respiration measurements and calcif.=coral calcification

Authors	Coral			Symbiont			
	Mortality	Calcif.	HSPs/anti-oxidants	Algal density	Algal pigments	Fv/Fm	P/R
Downs et al. (2000)			X				
Downs et al. (2002)					X		
Brown et al. (2002b)			X	X	X	X	
Reynaud et al. (2003)		X		X			X
Bhagooli and Hidaka (2004)	X					X	
Yakovleva and Hidaka (2004)				X	X	X	
Lesser and Farrell (2004)			X		X	X	
Griffin et al. (2006)			X				
Grottoli et al. (2006)					X		X
Visram and Douglas (2007)					X		
Middlebrook et al. (2008)				X	X	X	
Anthony et al. (2008)		X		X[a]	X[a]		X

[a]Bleaching metric derived colorimetrically from digital photography as luminance reduction relative to maximal algal densities/chlorophyll concentration

adaptations can indeed provide meaningful protection, thereby enhancing survival. First, stress exposures in the field are highly variable both in space and time and whilst some corals will be damaged and may die, others will have time to mount protective responses. This variability means that a fraction of any population will survive based on resistance enhancements perhaps in thermal refugia. Second, the processes leading to thermal damage, and thus the extent of damage caused by a specified period of exposure, is very heavily dependent on temperature. Q_{10}s for damaging processes can exceed 50 (Cossins and Bowler 1987) rather than two to three for most biological processes, so that an increase of just 0.1°C in lethal limits can have a disproportionately large effect in slowing the rate of damage accumulation, mitigating damage for a given exposure, and this can make the difference between survival and death. Third, as already described, stress genes are as inducible in corals as in other taxa.

5 Genomics Approaches to Stress Responses in Corals

The recent publication of a draft genome sequence of the first cnidarian, the starlet sea anemone, *Nematostella vectensis* (Putnam et al. 2009), has demonstrated that the cnidarian genome possesses a surprising degree of conservation of gene content and even gene synteny with vertebrate animals. This represents the oldest conserved synteny in the eukaryotes and suggests that cnidarians are better models for understanding the vertebrate genome than the more widely used invertebrate model species, such as *Drosophila* and *Caenorhabditis*. This includes an almost complete list of genes known to be part of developmental signalling pathways including those involved in signal transduction, cell communication and adhesion pathways, as well as genes responding to environmental disturbance, such as heat shock proteins and molecular chaperones, superoxide dismutases, etc. Thus, there is no reason to expect that cnidarians lack any molecular apparatus involved in stress responses. The ability to identify coral genes using conventional homology-based methods favours the direct application of contemporary post-genomic technologies to the problem of acquired thermotolerance mechanisms in corals.

Recent years have seen the increasing application of transcript screening technologies in stress tolerance studies of corals (Forêt et al. 2007). All studies to date have generated cloned cDNA libraries whose PCR amplification generated hybridisation probes are arrayed on the surface of glass microscope slides. The number of genes represented on arrays is rapidly increasing as cDNA libraries expand; thus, Edge et al. (2005) created a microarray containing just 32 cDNA gene probes from *Acropora cervicornis* and *Montastraea faveolata* to test the effects of elevated temperature, salinity and UV light; Desalvo et al. (2008) generated 1310 cDNA probes from the latter species and most recently Bay et al. (2009) have constructed microarray for *Acropora millepora* composed of a 18,000 cDNA probes. However, the need for sequence data for emerging model species will be more easily met by the combination of new, very high throughput sequencing technologies. Thus, Meyer et al. (2009) have recently generated 600,000 sequence reads for *A. millepora* in just one run of the Roche 454 instrument, which on assembly generated a comprehensive list of 11,000 genes. This in combination with in situ synthesis of microarrays from *in silico* sequence data (e.g. Agilent) offers a rapid and low-cost route to microarray generation.

Given that to date most published studies have been constrained by lack of sequence data and microarray probes, what has been achieved so far in understanding thermal responses of corals? DeSalvo et al. (2008) have compared *M. faveolata* exposed to thermal stress and bleaching with non-stressed controls to find 392 differentially expressed genes. Of these, only 68 were identified by homology-searching, revealing roles in a range of stress-related processes such as those involved in HSP expression and oxidative stress. They also completed a simple time-course experiment thereby classifying genes into different groups or clusters according to speed and direction of response. Bay et al. (2009) have explored array differences in *A. millepora* from two locations in the Great Barrier Reef varying in turbidity, and responses to relocation to clean water conditions. Again, small proportions of genes displayed differential expression between sites but these differences were lost on transfer of specimens from both sites to common garden conditions for 10 days. The small array developed by Edge et al. (2005) revealed gene responses to increased temperature including carbonic anhydrase, thioredoxin, a urokinase plasminogen activator receptor (uPAR) and three ribosomal genes. Richier et al. (2008) employed microarrays containing 10 K anonymous (unsequenced or unidentified) probes predominantly from host tissues of the temperate sea anemone *Anthopleura elegantissima* to assess thermal responses. Of these 2.7% or 284 genes were differentially expressed due to thermal or UV treatment, and of these only a fraction of these probes possessed an identity by homology alignment when sequenced. Nevertheless, certain gene responses were discovered including a 18-fold up-regulation of ferritin, a protein involved in cytoprotection and immunity.

Whilst it is still early in the application of these advanced technologies, and gene coverage is incomplete, these papers show clear promise in defining in much greater detail the responses of corals to environmental stress. The technical approach has strong comparative properties allowing correlation of responses to acquired tolerance and thermal sensitivity between species, populations and sites, and comparing specimens exposed to experimental stress treatments. Knowledge of gene identity and function is obviously crucial for relating responses to biological processes and pathways, but arrays offer strong phenotyping or classification capability, which allows discrimination between specimens in relation to prior experience or genotype, etc.

6 Epigenetics and Its Significance for Coral Acclimatisation to Elevated Temperature

Being colonial and possessing indeterminate growth properties, corals may grow over time into sufficiently large structures such that different regions of the colony might be exposed to quite divergent kinds and amounts of stress. This includes the east- and west-facing surfaces or upper and lower parts of inter-tidal corals and light and shade adapted parts of sub-tidal coral colonies. These regions might well develop into quite different phenotypic forms, some of which might endow stress-adaptive properties in some parts of the colony relative to others. However, there is some anecdotal evidence that differentiated parts of a single colony, which possess a common genetic constitution, might retain divergent properties even when subsequently exposed to common garden conditions. Thus, the normally high-light-adjusted, west-facing sides of *Goniastrea aspera,* when subjected to shade conditions for 4 years, still appeared to retain a greater temperature tolerance when subjected to elevated temperatures compared to the east-facing sides (Brown, personal communication).

Seasonal and short-term acclimatisation are generally reversible, so these long lasting effects appear to result from mechanisms other than the conventional. This might relate to earlier concepts of genetic assimilation when inducible phenotypes become fixed at least for a period of time, or when developmental experiences have lifelong and even trans-generational effects, as with 'canalisation' (Pigliucci et al. 2006), genomic imprinting, gene silencing etc. Based largely on advances in genome science, these so-called epigenetic mechanisms have moved from vague ideas to suggest a range of specific mechanisms (Pal and Hurst 2004; Suzuki and Bird 2007) by which long-lasting, even multi-generational, resistance adaptations might become fixed. The ecological significance of long-lasting epigenetic influences is now becoming rather well established including the vernalisation response in plants, effects of carcinogens and teratogens and trans-generational effects of starvation and dietary effects on the *agouti* gene (Pal and Hurst 2004; Suzuki and Bird 2007; Bossdorf et al. 2008).

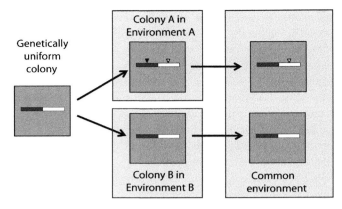

Fig. 4 Hypothetical involvement of acclimatisation and epigenetic modification in generating tolerant phenotypes in corals. A single coral (indicated by the box with a bar containing two genes) grows and splits into two parts one of which by virtue of its orientation to the sun is exposed to much more intense solar radiation (Environment A) than the other (Environment B) over several growing seasons. Both colonies possess exactly the same genotype. The two new corals develop divergent thermal phenotypes, linked to some modification in the expression of the two genes in Colony A relative to Colony B, as indicated by the arrowheads. Conventional phenotypic adaptation results in restoration of one gene to its pre-existing state when the colony is returned to its previous (common) environment with the loss of the black arrowhead; that is, the phenotypic change is fully reversible. However, the persistence of the white arrowhead in colony A under the same conditions is *prima facie* evidence of an epigenetic effect (Modified after Bossdorf et al. 2008)

Epigenetic changes can be induced by interaction of small non-coding RNA products with gene promoters or enzymatic modification of histones, but the most exciting mechanism, and which is currently the subject of intense interest in a variety of cancer, plant and animal model systems (Pal and Hurst 2004; Suzuki and Bird 2007; Bossdorf et al. 2008; Kronforst et al. 2008), is the stable modification of chromosomal regions by DNA methylation. The latter can be readily fingerprinted by modifying a standard AFLP protocol (Vos et al. 1995) with methylation-sensitive endonucleases (methylation-sensitive AFLP) (Xu et al. 2000). Restriction fragments that correlate with distinct phenotypes can be cloned subsequently and used as probes enabling a microarray-based approach to genome-wide identification of responding genes (Yamamoto and Yamamoto 2004). The power of this approach lies in the ability to compare phenotypically different forms of the same genomic DNA sequence, generated over the same time period and in the ability to explore multiple specimens from each of several different genotypes. Divergent patterns between corals or parts of corals subjected to different experiences would thus point directly at an adaptive role for DNA methylation and open up a new direction for future research in this model system (Fig. 4). If shown to exist in corals this would have a considerable impact on our understanding of how corals can maximise their survival in a globally warmed climate.

7 Summary and Conclusions

1. The potential for corals to display phenotypic resistance adaptations to the damaging or lethal environmental conditions in an era of warming climate should not be dismissed, particularly on the basis of a currently inadequate understanding of the primary lesions of thermal damage, and of coral resistance acclimatisation.
2. Prior experience of high solar radiation levels or elevated temperatures has been shown to increase the thermal tolerance of corals both in the field and in the laboratory. Also, there is now limited evidence from work in the field that some corals have increased their temperature tolerances and reduced their bleaching susceptibility in recent bleaching events where temperature elevations and solar radiation levels were comparable with earlier bleaching episodes. Coral mortality was low in earlier events and was not considered significant as an explanation of reduced bleaching susceptibility, which was attributed to acclimatisation and/or adaptation.
3. Mechanisms implicated in increasing thermal tolerance involve algal clade shuffling and switching, improved photoprotective defences by symbiotic algae and up-regulated stress protein and antioxidant enzyme responses in both symbiotic algae and coral host.
4. The role of epigenetics mechanisms in acclimatisation to warming sea temperatures is currently unexplored in corals, but is suggested by observations of the divergent long-term thermally tolerant phenotypes of split coral colonies. This offers an additional potential mechanism that might significantly contribute to the prolonged survival of corals in a warming climate.

References

Ainsworth TD, Hoegh-Guldberg O, Heron SF, Skirving WJ, Leggat W (2008) Early cellular changes are indicators of pre-bleaching thermal stress in the coral host. J Exp Mar Biol Ecol 364:63–67

Anthony KRN, Hoegh-Guldberg O (2003) Kinetics of photoacclimation in corals. Oecologia 134:23–31

Anthony KRN, Kline DI, Diaz-Pulido G, Dove S, Hoegh-Guldberg O (2008) Ocean acidification causes bleaching and productivity loss in coral builders. Proc Natl Acad Sci 105:17442–17446

Baird AH, Bhagooli R, Ralph P, Takahashi S (2009) Coral bleaching: the role of the host. Trends Ecol Evol 1:16–20

Baker AC, Starger CJ, McClanahan TR, Glynn PW (2004) Coral reefs: corals' adaptive response to climate change. Nature 430:741

Baker AC, Glynn PW, Riegl B (2008) Climate change and coral reef bleaching: an ecological assessment of long-term impacts, recovery trends and future outlook. Estuarine Coast Shelf Sci 80:435–471

Bay LK, Ulstrup KE, Bjørn Nielsen H, Jarmer H, Goffard N, Willis BL, Miller DJ, Van Oppen MJH (2009) Microarray analysis reveals transcriptional plasticity in the reef building coral *Acropora millepora*. Mol Ecol 18:3062–3075

Berkelmans R (2002) Time-integrated thermal bleaching thresholds of reefs and their variation on the Great Barrier Reef. Mar Ecol Prog Ser 237:309–310

Berkelmans R (2009) Bleaching and mortality thresholds: how much is too much? In: van Oppen MJH, Lough JM (eds) Coral bleaching. Springer, Heidelberg, pp 103–119

Berkelmans R, Willis BL (1999) Seasonal and local spatial patterns in the upper thermal limits of corals on the inshore central Great Barrier Reef. Coral Reefs 18:219–228

Berkelmans R, van Oppen MJH (2006) The role of zooxanthellae in the thermal tolerance of corals: a 'nugget of hope' for coral reefs in an era of climate change. Proc R Soc Lond B 272:29–38

Berkelmans R, De'ath G, Kininmonth S, Skirving WJ (2004) A comparison of the 1998 and 2002 coral bleaching events on the Great Barrier Reef: spatial correlation, patterns and predictions. Coral Reefs 23:74–83

Bhagooli R, Hidaka M (2004) Photoinhibition, bleaching susceptibility and mortality in two scleractinian corals, *Platygyra ryukyuensis* and *Stylophora pistillata*, in response to thermal and light stresses. Comp Biochem Physiol A 137:547–555

Bligh J, Johnson KG (1973) Glossary of terms for thermal physiology. J Appl Physiol 35:941–961

Bossdorf O, Richards CL, Pigliucci M (2008) Epigenetics for ecologists. Ecol Lett 11:106–115

British National History Museum Archives (1928) Sir Maurice Yonge collection: expedition progress reports DF214/7 for Aug 17th-November 14th 1928

Brown BE (1997) Coral bleaching: causes and consequences. Coral Reefs 16(Suppl):S129–S138

Brown BE, Dunne RP (2008) Solar radiation modulates bleaching and damage protection in a shallow water coral. Mar Ecol Prog Ser 362:99–107

Brown BE, Dunne RP, Chansang H (1996) Coral bleaching relative to elevated seawater temperature in the Andaman Sea (Indian Ocean) over the last 50 years. Coral Reefs 15:151–152

Brown BE, Dunne RP, Ambarsari I, Le Tissier MDA, Satapoomin U (1999) Seasonal fluctuations in environmental factors and variations in symbiotic algae and chlorophyll pigments in four Indo-Pacific coral species. Mar Ecol Prog Ser 191:53–69

Brown BE, Dunne RP, Warner ME, Ambarsari I, Fitt WK, Gibb SW, Cummings DG (2000) Damage and recovery of Photosystem II during a manipulative field experiment on solar bleaching in the coral *Goniastrea aspera*. Mar Ecol Prog Ser 195:117–124

Brown BE, Dunne RP, Goodson MS, Douglas AE (2002a) Experience shapes the susceptibility of a reef coral to bleaching. Coral Reefs 21:119–126

Brown BE, Downs CA, Dunne RP, Gibb SW (2002b) Exploring the basis of thermotolerance in the reef coral *Goniastrea aspera*. Mar Ecol Prog Ser 242:119–129

Calosi P, Bilton DT, Spicer JL, Atfield A (2008) Thermal tolerance and geographical range size in the *Agabus brunneus* group of European diving beetles. J Biogeogr 35:295–305

Castillo KD, Helmuth BST (2005) Influence of thermal history on the response of *Montastraea annularis* to short-term temperature exposure. Mar Biol 148:261–270

Compton TJ, Rijkenberg MJA, Drent J, Piersma T (2007) Thermal tolerance ranges and climate variability: a comparison between bivalves from differing climates. J Exp Mar Biol Ecol 352:200–211

Coles SL (1997) Reef corals occurring in a highly fluctuating temperature environment at Fahal Island, Gulf of Oman (Indian Ocean). Coral Reefs 16:269–272

Coles SL, Jokiel PL, Lewis CR (1976) Thermal tolerance in tropical versus subtropical Pacific reef corals. Pac Sci 30:159–166

Coles SL, Jokiel PL (1978) Synergistic effects of temperature, salinity and light on the hermatypic coral *Montipora verrucosa*. Mar Biol 49:187–195

Coles SL, Brown BE (2003) Coral bleaching – capacity for acclimatization and adaptation. Adv Mar Biol 46:183–223

Cossins AR, Bowler K (1976) Resistance adaptation of the freshwater crayfish and thermal inactivation of membrane-bound enzymes. J Comp Physiol B 111:15–24

Cossins AR, Bowler K (1987) Temperature biology of animals. Chapman and Hall, London

Desalvo MK, Voolstra CR, Sunagawa S, Schwarz JA, Stillman JH, Coffroth MA, Szmant AM, Medina M (2008) Differential gene expression during thermal stress and bleaching in the Caribbean coral *Montastraea faveolata*. Mol Ecol 17:3952–3971

Díaz F, Sierraa E, Reb AD, Rodríguez L (2002) Behavioural thermoregulation and critical thermal limits of *Macrobrachium acanthurus* (Wiegman). J Thermal Biol 27:423–428

D'Croz L, Mate JL (2004) Experimental responses to elevated water temperature in genotypes of the reef coral *Pocillopora damicornis* from upwelling and non-upwelling environments in Panama. Coral Reefs 23:473–483

Donner SD, Skirving WJ, Little CM, Oppenheimer M, Hoegh-Guldberg O (2005) Global assessment of coral bleaching and required rates of adaptation under climate change. Global Change Biol 11:2251–2265

Donner SD, Knutson TR, Oppenheimer M (2007) Model-based assessment of the role of human-induced climate change in the 2005 Caribbean bleaching event. Proc Natl Acad Sci 104:5483–5488

Donner SD (2009) Coping with commitment: projected thermal stress on coral reefs under different future scenarios. PLoS ONE 4:1–10

Dove S (2004) Scleractinian corals with photoprotective host pigments are hypersensitive to thermal bleaching. Mar Ecol Prog Ser 272:99–116

Dove S, Ortiz JC, Enríquez S, Fine M, Fisher P, Iglesias-Prieto R, Thornhill D, Hoegh-Guldberg O (2006) Response of holosymbiont pigments from the scleractinian coral *Montipora monasteriata* to short-term heat stress. Limnol Oceanogr 51:1149–1158

Downs CA, Mueller E, Phillips S, Fauth JE, Woodley CM (2000) A molecular biomarker system for assessing the health of coral (*Montastrea faveolata*) during heat stress. Mar Biotechnol 2:533–544

Downs CA, Fauth JE, Halas JC, Dustan P, Bemiss J, Woodley CM (2002) Oxidative stress and seasonal coral bleaching. Free Radic Biol Med 33:533–543

Dunne RP, Brown BE (2001) The influence of solar radiation on bleaching of shallow water reef corals in the Andaman Sea, 1993-1998. Coral Reefs 20:201–210

Eakin CM, Lough JM, Heron SF (2009) Climate variability and change: monitoring data and evidence for increased coral bleaching stress. In: van Oppen MJH, Lough JM (eds) Coral bleaching. Springer, Heidelberg, pp 41–67

Edge SE, Morgan MB, Gleason DF, Snell TW (2005) Development of a coral cDNA array to examine gene expression profiles in *Montastraea faveolata* exposed to environmental stress. Mar Pollut Bull 51:507–523

Edmondson CH (1928) The ecology of an Hawaiian coral reef. Bull Bernice P Bishop Mus 45:1–64

Edmunds PJ, Gates RD (2008) Acclimatization in tropical reef corals. Mar Ecol Prog Ser 361:307–310

El-Wadawi R, Bowler K (1995) The development of thermotolerance protects blowfly flight muscle mitochondrial function from heat damage. J Exp Biol 11:2413–2421

Eme J, Bennett WA (2009) Critical thermal tolerance polygons of tropical marine fishes from Sulawesi. Indones J Thermal Biol 3:220–225

Fagoonee I, Wilson HB, Hassell MP, Turner JR (1999) The dynamics of zooxanthellae populations: a long-term study in the field. Science 283:843–845

Fitt WK, McFarland FK, Warner ME, Chilcoat GC (2000) Seasonal patterns of tissue biomass and densities of symbiotic dinoflagellates in reef corals and relation to coral bleaching. Limnol Oceanogr 45:677–685

Fitt WK, Brown BE, Warner ME, Dunne RP (2001) Coral bleaching: interpretation of thermal tolerance limits and thermal thresholds in tropical corals. Coral Reefs 20:51–65

Forêt S, Kassahn KS, Grasso LC, Hayward DC, Iguchi A, Ball EE, Miller DJ (2007) Genomic and microarray approaches to coral reef conservation biology. Coral Reefs 26:475–486

Glynn PW (1993) Coral-reef bleaching – ecological perspectives. Coral Reefs 12:1–17

Griffin SP, Bhagooli R, Weil E (2006) Evaluation of thermal acclimation capacity in corals with different thermal histories based on catalase concentrations and antioxidant potentials. Comp Biochem Physiol C 144:155–162

Gladwell RT (1975) Heat death in the crayfish *Austropotamobius pallipes*: thermal inactivation of muscle-bound ATPase in warm and cold adapted animals. J Thermal Biol 1:95–100

Grottoli AG, Rodrigues LJ, Palardy JE (2006) Heterotrophic plasticity and resilience in bleached corals. Nature 440:1186–1189

Hellberg ME (1998) Sympatric sea shells along the sea's shore: the geography of speciation in the marine gastropod *Tegula*. Evolution 52:1311–1324

Hoegh-Guldberg O (1999) Climate change, coral bleaching and the future of the world's coral reefs. Mar Freshwat Res 50:839–866

Hoegh-Guldberg O (2004) Coral reefs in a century of rapid environmental change. Symbiosis 37:1–32

Hoegh-Guldberg O (2009) Climate change and coral reefs: Trojan horse or false prophecy? Coral Reefs 28:569

Hoegh-Guldberg O, Fine M (2005) Coral bleaching follows wintry weather. Limnol Oceanogr 50:256–271

Hoegh-Guldberg O, Mumby PJ, Hooten AJ, Steneck RS, Greenfield P, Gomez E, Harvell CD, Sale PF, Edwards AJ, Caldeira K, Knowlton N, Eakin CM, Iglesias-Prieto R, Muthiga N, Bradbury RH, Dubi A, Hatziolos ME (2007) Coral reefs under rapid climate change and ocean acidification. Science 318:1737–1742

Hoyland J, Cossins AR, Hill MW (1979) Thermal limits for behavioural function and resistance-adaptation of goldfish, *Carassius auratus* L. J Comp Physiol 129:241–246

Hoffman GE, Somero GN (1995) Evidence for protein damage at environmental temperatures: seasonal changes in levels of ubiquitin conjugates and hsp70 in the intertidal mussel *Mytilus trossulus*. J Exp Biol 198:1509–1518

Huey RB, Deutsch CA, Tewksbury JJ, Vitt LJ, Hertz PE, Pérez HJA, Garland T Jr (2009) Why tropical forest lizards are vulnerable to climate warming. Proc R Soc B 276:1939–1948

Hughes TP, Baird AH, Bellwood DR, Card M, Connolly SR, Folke C, Grosberg R, Hoegh-Guldberg O, Jackson JBC, Kleypas J, Lough JM, Marshall P, Nystrom N, Palumbi SR, Pandolfi JM, Rosen B, Roughgarden J (2003) Climate change, human impacts, and the resilience of coral reefs. Science 301:929–933

Janzen DH (1967) Why mountain passes are higher in tropics. Am Nat 101:233–249

Jokiel PL, Coles SL (1990) Response of Hawaiian and other Indo-Pacific reef corals to elevated temperature. Coral Reefs 8:155–162

Jokiel PL, Brown EK (2004) Global warming, regional trends and inshore environmental conditions influence coral bleaching in Hawaii. Global Change Biol 10:1627–1641

Kenny R (1974) Inshore surface sea temperatures at Townsville. Mar Freshwat Res 25:1–5

Kleypas J, Danabasoglu G, Lough JM (2008) Potential role of the ocean thermostat in determining regional differences in coral bleaching events. Geophys Res Lett 35:L03613

Kronforst MR, Gilley D, Strassmann J, Queller D (2008) DNA methylation is widespread across social Hymenoptera. Curr Biol 18:R287–R288

Lesser MP, Farrell JH (2004) Exposure to solar radiation increases damage to both host tissues and algal symbionts of corals during thermal stress. Coral Reefs 23:367–377

Mayer AG (1914) The effects of temperature on tropical marine animals. Carnegie Inst Washington Publ Dept Mar Biol Pap Tortugas Lab 183:1–24

Maynard JA, Baird AH, Pratchett MS (2008a) Revisiting the Cassandra syndrome; the changing climate of coral reef research. Coral Reefs 27:745–749

Maynard J, Anthony K, Marshall P, Masiri I (2008b) Major bleaching events can lead to increased thermal tolerance in corals. Mar Biol 155:173–182

McClanahan TR, Maina J (2004) Response of coral assemblages to the interaction between natural temperature variation and rare warm-water events in Kenyan reef lagoons. Ecosystems 6:551–563

McClanahan TR, Ateweberhan M, Muhando C, Maina J, Mohammed MS (2007) Effects of climate and seawater temperature variation on coral bleaching and mortality. Ecol Monogr 77:503–525

Middlebrook R, Hoegh-Guldberg O, Leggat W (2008) The effect of thermal history on the susceptibility of reef-building corals to thermal stress. J Exp Biol 211:1050–1056

Meyer E, Aglyamova GV, Wang S, Buchanan-Carter J, Abrego D (2009) Sequencing and *de novo* analysis of a coral larval transcriptome using 454 GSFlx. BMC Genomics 10:219

Nomura K (2004) The Ki Peninsula. In: Japanese Coral Reef Society and Ministry of Environment (eds) Coral reefs of Japan Ministry of the Environment, Tokyo, pp 252–256

Obura DO (2005) Resilience and climate change: lessons from coral reefs and bleaching in the Western Indian Ocean. Estuarine Coast Shelf Sci 63:353–372

Oliver JK, Berkelmans R, Eakin CM (2009) Coral bleaching in space and time. In: van Oppen MJH, Lough JM (eds) Coral bleaching. Springer, Heidelberg, pp 21–39

Pal C, Hurst LD (2004) Epigenetic inheritance and evolutionary adaptation. In: Hirt RP, Horner DS (eds) Organelles, genomes and eukaryote phylogeny. CDC Press, Boca Raton, pp 347–364

Pettay DT, LaJeunesse TC (2009) Microsatellite loci for assessing genetic diversity, dispersal and clonality of coral symbionts in 'stress-tolerant' clade D *Symbiodinium*. Mol Ecol Res 9:1022–1025

Pigliucci M, Murren CJ, Schlichting CD (2006) Phenotypic plasticity and evolution by genetic assimilation. J Exp Biol 209:2362–2367

Precht H, Christopherson J, Hensel H, Larcher W (1973) Temperature and life. Springer, Berlin

Putnam NH, Srivastava M, Hellsten U, Dirks B, Chapman J, Salamov A, Terry A, Shapiro H, Lindquist E, Kapitonov VV, Jurka J, Genikhovich G, Grigoriev IV, Lucas SM, Steele RE, Finnerty JR, Technau U, Martindale MQ, Rokhsar DS (2009) Sea anemone genome reveals ancestral eumetazoan gene repertoire and genomic organization. Science 317:86–94

Richier S, Rodriguez-Lanetty M, Schnitzler CE, Weis VM (2008) Response of the symbiotic cnidarian *Anthopleura elegantissima* transcriptome to temperature and UV increase. Comp Biochem Physiol D 3:283–289

Reynaud S, Leclerq N, Romaine-Lioud S, Ferrier-Pages C, Jaubert J, Gattuso J-P (2003) Interacting effects of CO2 partial pressure and temperature on photosynthesis and calcification in a scleractinian coral. Global Change Biol 9:1660–1668

Roberts DA, Hoffman GE, Somero GN (1997) Heat shock protein expression in *Mytilus californianus*: acclimatization (seasonal and tidal-height comparisons) and acclimation effects. Biol Bull 192:309–320

Rowan R (2004) Coral bleaching: thermal adaptation in reef coral symbionts. Nature 430:742

Salih A, Larkum A, Cox G, Kuhl M, Hoegh-Guldberg O (2000) Fluorescent pigments in corals are photoprotective. Nature 408:850–853

Schmidt-Nielsen K (1997) Animal physiology; principals and adaptations. Cambridge University Press, New York

Sharp V, Brown BE, Miller D (1997) Heat shock protein (HSP 70) expression in the tropical reef coral *Goniopora djiboutuensis*. J Therm Biol 22:11–19

Sheppard C (2009) Large temperature plunges recorded by data loggers at different depths on an Indian Ocean atoll: comparison with satellite data and relevance to coral refuges. Coral Reefs 28:399–403

Sheppard C, Price A, Roberts C (2002) Marine ecology of the Arabian Region. Academic Press, London

Somero GN (2005) Linking biogeography to physiology: evolutionary and acclimatory adjustments of thermal limits. Front Zool 2:1. doi:10.1186/1742-9994-2-1

Stillman JH (2002) Causes and consequences of thermal tolerance limits in rocky intertidal porcelain crabs, Genus *Petrolisthes*. Int Comp Biol 42:790–796

Stillman JH (2003) Acclimation capacity underlies susceptibility to climate change. Science 301:65

Stimson J (1997) The annual cycle of density of zooxanthellae in the tissues of field and laboratory-held *Pocillopora damicornis*. J Exp Mar Biol Ecol 214:35–48

Suwa R, Hirose M, Hidaka M (2008) Seasonal fluctuation in zooxanthellar genotype composition and photophysiology in the corals *Pavona divaricata* and *P. decussata*. Mar Ecol Prog Ser 361:129–137

Suzuki MM, Bird A (2007) DNA methylation landscapes: provocative insights from epigenomics. Nature 447:396–398

Tchernov D, Gorbunov MY, De Vargas C, Narayan Yadav S, Milligan AJ, Haggblom M, Falkowski PG (2004) Membrane lipids of symbiotic algae are diagnostic of sensitivity to thermal bleaching in corals. Proc Natl Acad Sci 101:13531–13535

Terblanche JS, Deere JA, Clusella-Trullas S, Janion C, Chown SL (2007) Critical thermal limits depend on methodological context. Proc R Soc B 274:2935–2943

Tewksbury JJ, Huey RB, Deutsch CA (2008) Putting the heat on tropical Animals. Science 320:1296–1297

Tomascik T, Mah AJ, Nontji A, Moosa MK (1997) *The Ecology of the Indonesian Seas*. Part Two. Periplus Editions (HK) Ltd, Hong Kong

Visram S, Douglas AE (2007) Resilience and acclimation to bleaching stressors in the scleractinian coral *Porites cylindrica*. J Exp Mar Biol Ecol 349:35–44

Vos P, Hogers R, Bleeker M, Reijans M, van de Lee T, Hornes M, Friters A, Pot J, Paleman J, Kuiper M, Zabeau M (1995) AFLP: a new technique for DNA fingerprinting. Nucleic Acids Res 23:4407–4414

Willmer P, Stone G, Johnstone IA (2004) Environmental physiology of animals. Wiley-Blackwell, London

Xu M, Li X, Korban SS (2000) AFLP-Based detection of DNA methylation. Plant Mol Biol 18:361–368

Yakovleva I, Hidaka M (2004) Different effects of high temperature acclimation on bleaching-susceptible and tolerant corals. Symbiosis 37:87–105

Yamamoto F, Yamamoto M (2004) A DNA microarray-based methylation-sensitive (MS)-AFLP hybridization method for genetic and epigenetic analyses. Mol Genet Genomics 271:678–686

Yonge CM (1940) The biology of reef building corals. Sci Rep Great Barrier Reef Exped 1928–1929 1:353–391

Yonge CM, Nicholls AG (1931) Studies of the physiology of corals. IV. The structure, distribution, and physiology of the zooxanthellae. Sci Rep Great Barrier Reef Exped 1928–1929 IV:135–176

Reef Bioerosion: Agents and Processes

Aline Tribollet and Stjepko Golubic

Keywords Reef bioerosion • Microborers • Euendoliths • Cyanobacteria • Microalgae • Fungi • Macroborers • Sponges • Bivalves • Grazers • Urchins • Parrotfishes • Carbonate dissolution • Sedimentation • Carbonate budget • Reef framework • Coral reefs • Ocean acidification • Anthropogenic factors

1 Introduction

Coral reef maintenance depends on the balance between constructive and destructive forces. Constructive forces are mainly calcification and growth of corals and encrusting coralline algae. Destructive forces comprise physical, chemical, and biological erosion. Bioerosion is considered as the main force of reef degradation because physical erosion (storms) is temporary and localized, and chemical erosion is considered as negligible due to the actual ocean chemistry (Scoffin et al. 1980). Reef bioerosion affects sedimentary and skeletal carbonate substrates. It plays an important role in reef sedimentation, diversity maintenance by creating habitats and by providing food resources, and in biogeochemical cycles (recycling of dissolved Ca^{2+} and C). Thus, bioerosion is an integral part of the coral reef carbonate balance. The concept of bioerosion was introduced by Neumann (1966). It includes biocorrosion, which refers to destruction of carbonates by chemical means, and bioabrasion which refers to mechanical removal of carbonates by organisms (Golubic and Schneider 1979; Schneider and Torunski 1983).

Most of the coral reef studies to date have focused on reef growth and accretion. In contrast, much less attention has been paid to bioerosion processes. Since the 1950s, the number of publications on bioerosion processes has increases exponentially (Wilson 2008). It is obvious, however, from Wilson's bibliographic review (http://www.wooster.edu/geology/bioerosion/BioerosionBiblio.pdf) that most of the studies on bioerosion concerned macroboring and grazing agents. Less attention has been given to microboring agents. Moreover, only a few reviews have been published on the different aspects of bioerosion. For instance, Otter (1937) reported on rock-destroying organisms related to coral reefs. Hutchings (1986) reviewed bioerosion processes in coral reefs in general. Schneider and Le Campion (1999) reviewed the roles of cyanobacteria in construction and destruction of carbonates in freshwater and marine environments including reefs. Golubic et al. (2005) reviewed the roles of boring fungi in marine ecosystems. Perry and Hepburn (2008) reviewed the main processes, including bioerosion, involved in the carbonate balance in coral reefs. Only recently Tribollet (2008a) reviewed the roles of microboring organisms in modern coral reef ecosystems. Finally, the most recent review on bioerosion processes and agents in all kinds of ecosystems (terrestrial, freshwater, and marine environments) was provided by Tribollet et al. (in press). Overall, very few reviews have focused on bioerosion processes in coral reefs and are now dated.

In this chapter, we review the various agents of bioerosion, their interactions, and their roles in coral reef ecosystems in the present context, and project potential changes with climate change combined with local anthropogenic stressors. In particular, we emphasize on the roles of microborers in bioerosion and coral reef maintenance in the context of climate change.

2 Agents of Bioerosion

Agents of bioerosion comprise internal and external bioeroders of different sizes and from a wide range of taxonomic affiliation. Internal bioeroders excavate carbonate substrates in search of shelter or food, while external bioeroders graze on both epilithic and endolithic organisms thereby abrading

A. Tribollet (✉)
Unité de Recherche CAMELIA, Institut de Recherche pour le Développement, BP A5 101 Promenade Laroque, 98848, Nouméa Cedex, Nouvelle-Calédonie
e-mail: aline.tribollet@ird.fr

S. Golubic (✉)
Department of Biology, Boston University, 5 Cummington Street, 02215, Boston, MA, USA
e-mail: golubic@bu.edu

the substrates. Epiliths are attached to the surface of hard substrates, whereas endoliths reside inside hard substrates. Endoliths include organisms that colonize existing fissures (chasmo-endoliths) or cavities in porous substrates (crypto-endoliths), as well as those which actively penetrate carbonate substrates, such as euendoliths (Golubic et al. 1981). Euendoliths comprise various organisms of different sizes; microbial euendoliths are often referred to as microborers.

2.1 Internal Agents

Internal agents comprise microborers (<100 µm) and macroborers (>100 µm). The microboring organisms penetrate actively into carbonate substrates by dissolving them, whereas macroborers, depending on the organisms, penetrate into the substrates using chemical and mechanical means. The mechanisms of penetration by those different organisms have not been completely resolved. Hutchings (1986) and Tribollet (2008a) reviewed different hypotheses referring to the mechanisms of penetration into substrates used by internal bioeroders, such as calcium pumps in boring cyanobacteria (Garcia-Pichel 2006) and etching cells in boring sponges (Pomponi 1980).

2.1.1 Microborers

Microborers are phototrophic and organotrophic microorganisms (Tribollet 2008a). Microboring phototrophs are prokaryotic cyanobacteria and eukaryotic chlorophytes and rhodophytes (Fig. 1). Organotrophs (heterotrophs) are fungi (Fig. 2), foraminifera, and other mostly unidentified prokaryotic and eukaryotic light-independent microorganisms. Microborers can colonize both live and dead calcareous substrates, including main framebuilders of coral reefs, corals, and encrusting coralline algae. In live corals, colonization of skeletons from outside is restricted by polyps. Microborers, however, are able to colonize the skeleton of live corals from their base as soon as the larvae settle, or enter through lateral fissures. Then, they keep up with coral growth. The colonization of live coral skeletons is a selective process requiring a positive phototropic growth orientation and a rate of growth and carbonate penetration equal to or exceeding that of skeletal calcification (Le Campion-Alsumard et al. 1995a; Tribollet and Payri 2001). Microborers with such ability comprise cyanobacteria, such as *Plectonema terebrans* and the chlorophyte *Ostreobium quekettii*, and less frequently, the Conchocelis stages of Bangial rhodophytes (Fig. 3) (Laborel and Le Campion-Alsumard 1979). Among light-independent organotrophs, fungi are common in skeletons of live corals (Kendrick et al. 1982), where they attack microboring algae as well as coral polyps (Le Campion-Alsumard et al. 1995b)

and may represent a major hazard to coral health (Bentis et al. 2000). Following the death of corals and encrusting coralline algae, colonization by different microborers starts at the surface of the substrates. Colonization occurs within a few days (Le Campion-Alsumard 1979) followed by a succession of microborer communities. Pioneer microborers include septate chlorophytes, such as *Phaeophila dendroides*, and cyanobacteria, such as *Hyella* spp. (Fig. 1a) and *Mastigocoleus testarum* (Fig. 1b) (Le Campion-Alsumard 1975; Chazottes et al. 1995). In mature communities, i.e., after about 6 months of exposure to colonization in tropical waters, dead substrates are dominated by the cyanobacterium *Plectonema terebrans* (Fig. 1c), the chlorophyte *Ostreobium quekettii* (Fig. 1d), and different species of fungi (Chazottes et al. 1995; Tribollet and Golubic 2005; Tribollet 2008b). Microborers can be very abundant in dead substrates, up to more than 250,000 filaments per square centimeters (Schneider and Campion-Alsumard 1999; Tribollet 2001).

Phototrophic microborers are distributed throughout the illuminated areas of the ocean floor, many of them exhibiting adaptation to very low light intensities (oligophotes; Magnusson et al. 2007; see review by Tribollet 2008a). Light reaching microborers vary qualitatively and quantitatively depending on the depth of penetration of their filaments into substrates, the depth in the ocean, and the nature of the epilithic cover. The euendolithic cyanobacterium *Plectonema terebrans* and the siphonal chlorophyte *Ostreobium quekettii* have been found in carbonate substrate as deep as 270 and 300 m, respectively, in clear oligotrophic seas (Lukas 1978; Le Campion-Alsumard et al. 1982; Radtke et al. 1996). At the upper end of the distribution range of marine phototrophic microborers, euendolithic cyanobacteria inhabit at high concentrations coastal limestone rocks and skeletal carbonates of intertidal and supratidal wave-sprayed ranges. These ecotone communities show sharply defined zonal distribution and are responsible for the dark coloration of coastal rocks (Ercegovic 1932). Most are cyanobacteria, e.g., chroococcalean *Hormathonema violaceo-nigrum*, *Hyella balani*, *Solentia foveolarum*, and filamentous heterocystous *Mastigocoleus testarum* and *Kyrtuthrix dalmatica*. In conjunction with grazers, supratidal and intertidal microborers are responsible for the formation of biokarst (Schneider and Torunski 1983) and bioerosional notches (Neumann 1966) (Fig. 4). In subtidal ranges, shoaling ooid sands are common in the shallow lagoons of carbonate platforms and back reefs. These sands are extensively bored by a special set of microborers. Being periodically buried and unburied, ooid sands impose a limitation and selective pressure to resident microborers. A new genus of coccoid pleurocapsalean cyanobacteria, *Cyanosaccus* (Lukas and Golubic 1981) and several species of the genus *Hyella* have been described from shoaling ooids, e.g., *H. immanis* (Al-Thukair and Golubic 1991a, b). Most common habitats of microboring organisms

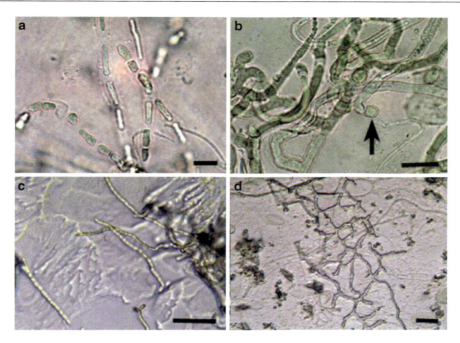

Fig. 1 Examples of microboring filaments. (**a**) Cyanobacterium *Hyella sp.* (**b**) *Mastigocoleus testarum* (heterocyst cell indicated by *arrow*), (**c**) Cyanobacterium *Plectonema terebrans* (isodiametric filament, distinctive cells) (**d**) Chlorophyte *Ostreobium quekettii* (polymorphic filaments). Scale bar is 10 μm

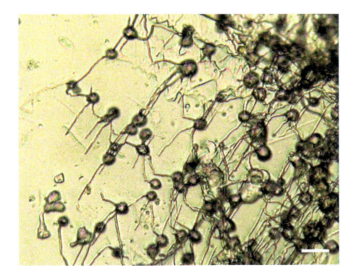

Fig. 2 Hyphae filaments and reproductive organs ("balls") of *Lithopythium gangliiforme* colonizing calcite. Scale bar is 10 μm

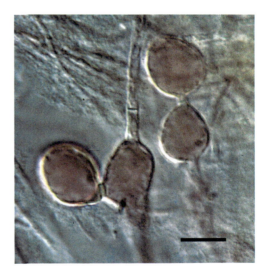

Fig. 3 Euendolithic Conchocelis stage of a rhodophyte *Porphyra*. Scale is 10 μm

in shallow subtidal ranges are relatively stable substrates, such as skeletons of dead corals, coral rubble, and bivalve shells. The distribution of microborers in these substrates mimics their bathymetrical distribution. In shallow tropical waters, *Ostreobium quekettii* and *Plectonema terebrans* penetrate the deepest into dead coral and coralline algal substrates, up to more than 1 cm depth (Chazottes et al. 1995; Tribollet and Payri 2001; Tribollet 2008b), whereas cyanobacteria *Mastigocoleus testarum* and *Hyella caespitosa*, and chlorophytes, such as *Phaeophila dendroides* and *Eugomontia sacculata* remain restricted to the surface layer of the carbonate substrates.

2.1.2 Macroborers

Macroborers are comprised of various organisms (Hutchings 1986 for review), including protists (foraminifera), sponges

Fig. 4 Bioerosional notch from New Caledonia (Lifou Island)

(e.g., clionids; Rützler 2002), bryozoans (Todd 2000), polychaetes (Wielgus et al. 2006), sipunculids (Gherardi and Bosence 2001), bivalves (e.g., lithophagid mussels; Kleemann 1996), and crustaceans (e.g., cirripeds; Kolbasov 2000). Organisms such as some foraminifera, bryozoans, and serpulids (polychaetes) only etch the surface of the substrates to anchor themselves (Fig. 5). In articles on bioerosion, the term "macroborer" usually refers to the fauna that actively penetrate into substrates. Among reef macroborers, sponges (Fig. 6) are probably the bioderoding animals that have received the most attention, especially within the past 10 years (see work of C.H.L. Schönberg; e.g., Schönberg and Tapanila 2006; Schönberg 2008). Macroborers can be divided into two categories based on their size: meiofauna (0.1–1 mm) and macrofauna (>1 mm). The meifauna includes mainly larvae of the macrofauna, and small foraminifera and polychaetes, such as sabellids and spionids.

Similar to microborers, macroborers are present in both live and dead carbonate substrates in coral reefs. In live corals, most of the macroborers penetrate from the dead parts or cracks in the skeleton. However, some macroborers, such as the bivalve *Lithophaga bisulcata* (Scott 1988) and the sponge *Cliona orientalis* (Schönberg and Wilkinson 2001), are able to penetrate into the coral skeletons through the polyp layers. In live bivalves, organic conchiolin layers composed of scleroproteins form natural barriers against penetration by macroborers, but at length, they are conquered (Fig. 7) (Cobb 1975). Colonization of dead substrates by macroborers starts with settlement of their planktonic larvae onto the substrate surface (Hutchings 1986; Lewis 1998), and occasionally by juveniles in the case of polychaetes (Hutchings and Murray 1982), or by extension of tissues in the case of sponges (Schönberg 2003). Colonization of newly exposed dead substrates by macroborers takes place within a few weeks (Hutchings 1986; Schönberg and Wilkinson 2001). The pioneer macroborers are short-lived polychaetes, such as cirratulids (Hutchings and Peyrot-Clausade 2002). They are replaced later by spionids and sabellids depending upon the reef environmental conditions (Davies and Hutchings 1983). Hutchings and Murray (1982) showed that the recruitment of polychaetes varies seasonally, as well from year to year depending on the environmental conditions and substrate availability. Sponges reproduce by sexual reproduction and by propagation through tissue fragments, making them effective in colonizing both live and dead coral colonies (Schönberg and Wilkinson 2001). Hutchings (1986) reviewed the variability in macroborer recruitment in detail. After 1 year of exposure, sipunculids such as *Aspidosiphon sp.* can become very abundant in dead corals (Peyrot-Clausade and Chazottes 2000). There are several evidences that colonization of dead substrates by boring bivalves and sponges takes more than 2–3 years (Kiene 1985; Peyrot-Clausade et al. 1992; Kiene and Hutchings 1994; Chazottes et al. 1995; Tribollet and Golubic 2005). These are long-lived boring organisms. The development of adult sponges and bivalves depends on the

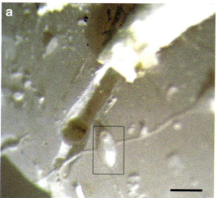 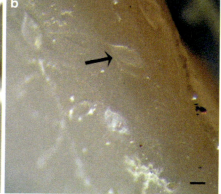

Fig. 5 Traces of (**a**) etching by serpulids (*the scare* indicates a preserved bryozoan) – scale bar is 2 mm, and (**b**) bryozoan after removal of the organism (*arrow*) – scale bar is 1 mm

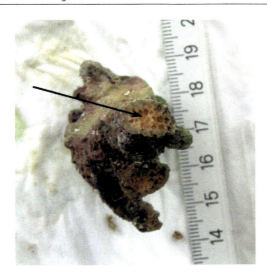

Fig. 6 Boring sponges (*Cliona* sp.) colonizing a dead rubble of a branching coral

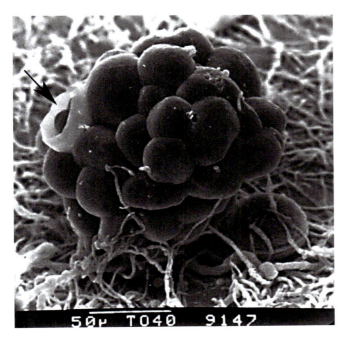

Fig. 7 Resin cast of the boring sponge *Entobia* (fossil trace from the Eocene, Paris Basin) showing a cell removing a carbonate "chip" (*arrow*). (Courtesy: Gudrun Radtke) Scale bar is 50 μm

grazing pressure and food availability (Risk and Sammarco 1982; Holmes et al. 2000).

Macroborers, similar to microborers in reefs, have a circumtropical distribution. They are present in the Red Sea (Hassan et al. 1996), Indo-Pacific waters (Holmes et al. 2000; Pari et al. 1998; Schönberg 2000; Tribollet and Golubic 2005), and Atlantic waters (Hein and Risk 1974; Rützler 2002; Zea and Weil 2003; Ward-Paige et al. 2005). Boring sponges are extremely abundant in Caribbean reefs contributing to high rates of bioerosion (MacGeachy 1977; Perry 1998; Mallela and Perry 2007). Geographical distribution of macroborers, which are filter feeders or detritivores, varies greatly depending on the water quality (Risk et al. 1995; Edinger et al. 2000; Tribollet and Golubic 2005). Filter-feeding macroboring communities are usually more abundant in eutrophic and turbid waters rich in organic particles than in oligotrophic, clear waters (Rose and Risk 1985; Hutchings and Peyrot-Clausade 2002; Tribollet and Golubic 2005; Ward-Paige et al. 2005). Perry (2000) highlighted the high variability in diversity, abundance, and distribution of macroborers, as well as the rates of macrobioerosion among the fore reefs, back reefs, and patch reefs in Jamaica. Perry (1998) noticed that bioerosion by macroborers, especially sponges, increases with depth on the fore reefs and that some species of corals are more susceptible than others to colonization by macroborers. Contrary to microborers, macroborers are not light-limited, and therefore, should be found at different depths (Perry 1998). Macroboring traces have been observed in deep-sea corals and shells (Freiwald et al. 1997; Wisshak 2006). A few species of boring foraminifera and sponges, which possess phototrophic symbiont zooxanthellae, similar to those of corals, are, however, only distributed within the photic zone (Schönberg and Loh 2005). Depth of penetration of macroborers into substrates varies depending on the organism and its size. Small polychaetes do not penetrate deep inside substrates, while other macroborers, such as sponges, can produce extensive branched and interconnected systems of galleries within carbonate substrates (Rützler 1974; Mariani et al. 2000), while communicating with the surface by numerous small openings. Other boring sponges are partially endolithic with bodies that bore and externally overgrow the substrate (Rützler 2002).

2.2 External Agents

External agents comprise grazers and predators. They are not considered endolithic, although some predators drill holes into calcareous exoskeletons of their prey.

Grazers involved in bioerosion comprise organisms that abrade or excavate carbonate substrates by using their radulas, teeth (beak-like jaws), or other hard buccal apparatus (Aristotle's lantern) in search of food. They range in size from mm-sized mollusks (gastropods, such as limpets and littorinas; Radtke et al. 1996) to polyplacophores (chitons; Rasmussen and Frankenberg 1990), large echinoderms (sea urchins, such as *Echinometra mathei* and *Eucidaris thouarsii*; Reaka-Kudla et al. 1996), and fishes (scarids, such as the parrotfish *Sparisoma viride*, and acanthurids or surgeonfish; see Hutchings 1986; Hoey and Bellwood 2008). Grazing takes place on both live and dead substrates (Fig. 8), but is more important on dead carbonate substrates (Bruggemann

Fig. 8 Grazing traces (*arrows*) on live (**a**) and dead (**b**) carbonate substrates due to parrotfishes and urchins, respectively

et al. 1994; Rotjan and Lewis 2006). Diet selection of grazers remains poorly studied. However, it was shown that some fish grazers, such as *Sparisoma viride* (Bonaire, Caribbean), feed preferentially on dead substrates covered by algal turfs and endoliths than on crustose coralline algae (Bruggemann et al. 1994). Similarly, Chazottes et al. (2002) showed that grazing urchins at Reunion Island preferentially feed on algal turfs than on crustose coralline algae and phaeophytes see also Mills et al. 2000. Foraging preferences are related to the nutritional quality of the food types and their yield. Bellwood and Choat (1990) showed that excavating fishes graze preferentially on convex surfaces of dead substrates and, consequently, tend to reduce rugosity of reef structure and topography. They also produce carbonate particles contributing to reef sedimentation and keep down macroalgal development. It has been shown by carrying cage exclusion experiments that the absence of important herbivores, such as parrotfishes, can induce shifts from coral-dominated reef to a macroalgal-dominated reef (Hughes et al. 2007). On live corals, grazers, such as parrotfishes, may exacerbate stress of bleached corals by reducing their fitness and altering coral-zooxanthellae symbiosis (Rotjan et al. 2006).

Although grazers do not leave boreholes corresponding perfectly to their body, like micro- and macroborers, their scars can be identified and thus used to indicate the main grazers of dead substrates in different reef areas (Fig. 8). Different buccal apparatus of parrotfish of the family Scaridae, and the corresponding scars, were reported by Bellwood and Choat (1990). Depending on their apparatus and size, grazers excavate more or less substrates. Gastropods play a relatively minor role in grazing (Trudgill 1976), while urchins are the most efficient grazers in coral reefs together with a variety of fish (Hutchings 1986). They can be very abundant with the highest densities reported on disturbed reefs in French Polynesia (210 ± 60 ind. m^{-2}, Pari et al. 1998) and the Galapagos (60 ± 5 ind. m^{-2}, Reaka-Kudla et al. 1996). In contrast, some of the lowest densities of urchins were reported on "healthy" reefs on the Great Barrier Reef (Australia) and in lagoons of French Polynesian atolls (Bak 1976, 1990). The species composition of grazers varies regionally and increases in diversity in lower latitudes. Specialized grazers are often arranged in zones parallel to the coastline, specializing to different regimes in water supply, as exemplified for the Bahamas in Radtke et al. (1996). Inshore–offshore differences have also been noticed for parrotfishes along the Great Barrier Reef by Bellwood and Choat (1990) and Hoey and Bellwood (2008). These authors, as well as Russ (1984), showed that parrotfishes are less abundant at inshore reefs than at offshore reefs, influencing the rates of grazing across the barrier reef (Tribollet and Golubic 2005).

Predators that successfully attack invertebrates protected by shells include snails of the families Naticidae, Muricidae, and five other families (Walker 2007), as well as foraminifera (Cedhagen 1994). Predatory snails use a combination of mechanical and chemical means to perforate into the shell of their preys and to kill them (Carriker 1969). Literature is abundant on fossil and modern bivalves that are attacked by predatory snails (e.g., Harper 1994; Hansen and Kelley 1995).

3 Geological History

The first questions that come in mind are why some organisms chose to bore into substrates? How and when did this habit evolve? One of the earliest hypotheses was that microbial euendoliths evolved seeking shelters to escape grazers when those evolved around 540 million years ago. This hypothesis should be rejected (Campbell 1982) since the oldest known microbial euendoliths are silicified remains of the coccoid cyanobacterium *Eohyella campbellii* that had penetrated the laminae of a stromatolite in Mesoproterozoic strata of China, some 1,400 million years ago (Zhang and Golubic 1987). Later, several species of *Eohyella* have been found penetrating Neoproterozoic ooids of East Greenland and Spitzbergen (ca. 800 My; Knoll et al. 1986; Green et al. 1988). The first organically preserved remains were of a Conchocelis stage of an ancient rhodophyte extracted from ca. 420 My old crinoid ossicles in cored Wenlockian (Silurian) rocks of Poland (Campbell et al. 1979; Campbell 1980).

The bulk of evidence of past microbioerosion is usually derived from the study of their microborings. Microbial euendoliths leave recognizable, often species-specific traces in the substrate that preserve well and constitute instant trace fossils. These traces contain important paleoecological information regarding bioerosion in the geological past, and their study constitutes a separate discipline with its own methods and rules of nomenclature. However, the paleoecological information that these fossil traces contain depends largely on their interpretation, which, in turn, rests on comparisons with modern counterparts Klein et al. 1991. In modern settings, microborings can be studied in parallel with the microborers that make them, so as to identify bioindicators for the interpretation of ancient environments (Golubic et al. 1970, 1975 1983; Vogel et al. 2000; Calcinai et al. 2003). The names of formally described euendolith traces and their identified makers are listed in Table 1.

Bioerosion evolved parallel to the evolution of reef-building organisms. In Middle Cambrian, both body and trace fossils of microbial endoliths were present (Stockfors and Peel 2005), and the entire microboring assemblages appear to be well established by the Ordovician times, with about 35% of fossil ichnotaxa that survived to the Present; this record also includes the first occurrence of borings attributable to euendolithic green alga *Ostreobium* (Glaub and Vogel 2004). Macroboring organisms diversified rapidly with representatives of various invertebrate phyla (James et al. 1977). Boring worms and boring sponges were documented in lower Cambrian (Kobluk and Risk 1977a). Macroborings increased significantly in abundance and diversity during the Ordovician, and by the Devonian, many macroborings became similar to the modern ones (Wilson 2007). In the course of the Mesozoic, the diversity of macroborers reached its peak in Jurassic (Wilson and Palmer 2006). Bioerosion by modern macroboring invertebrates has been reviewed by Hutchings (1986) and their fossil traces by Wilson (2007). Boreholes of macroborers are relatively easy to identify at the taxa level, because they fossilized well, like microborer boreholes, and thus, are also used in paleoecological studies (Bromley and D'Alessandro 1990; Tapanila et al. 2004; Glaub et al. 2007). As for the boring bivalves, their taxonomy and the description of their boreholes have been given by Evans (1970) and Warme (1975).

4 Interactions and Ecological Impact of Bioeroding Agents on Coral Reefs

Microborers and grazers act in synergy. The synergy between microborers and grazers was first described in detail in the formation of biokarst on limestone coasts (Schneider and Torunski 1983). Phototrophic microborers penetrate carbonate substrates to a depth of several mm, depending on the translucency of the substrate and the ability of microborers to use the reduced light for photosynthesis. On reaching their compensation depth (photosynthesis equals respiration), their penetration ceases. Thus, by itself, microbioerosion represents an activity limited to surface layers of the substrate, where the process would stabilize. However, microbioerosion continues and turns into a progressing process through the activity of grazers. In the process, grazers' sharp radulas and teeth remove the surface layers of the carbonate, loosened by the activity of microborers, thereby promoting light penetration deeper into the substrate. As a consequence, microboring activity continues as well. Due to this process, microborers with the other endoliths are important benthic primary producers in reefs (Tribollet et al. 2006a).

A positive correlation between microbioerosion and grazing rates was shown by Chazottes et al. (1995) and Tribollet and Golubic (2005) in different coral reefs. However, very low residual rates of microbioerosion reflect intensive grazing on carbonate substrates, and vice versa. Thus, microbioerosion and grazing, when intense, become negatively correlated. Very high grazing activity is usually due to sea urchins, such as the one at Faaa in French Polynesia (Chazottes et al. 1995), in the Galapagos (Reaka-Kudla et al. 1996), and at Reunion Island (Conand et al. 1998; Chazottes et al. 2002). Urchins can erode as much as 22.8 kg m^{-2} of exposed area per year (Table 2) (Reaka-Kudla et al. 1996). On reefs where dominant grazers are parrotfishes, such as at Osprey reef on the Great Barrier Reef, the rates of grazing are important (Table 2) but lower than those resulting from urchin activity (Tribollet and Golubic 2005). At high densities, sea urchins can cause destruction of the reef with loss of

Table 1 Euendolith traces and their identified makers

Group	Microboring organism	Trace fossil
Cyanobacteria	*Hyella caespitosa*	*Fascichnus dactylus*
	H. gigas	*F. frutex*
	Mastigocoleus testarum	*Eurigonum nodosum*
	Plectonema terebrans	*Scolecia filosa*
Chlorophytes	*Phaeophila dendroides*	*Rhopalia catenata*
	Gomontia polyrhiza	*Cavernula pediculata*
	Acetabularia rhizoid	*Fasciculus grandis*
	Ostreobium quekettii	*Ichnoreticulina elegans*
Heterotrophs	*Dodgella priscus*	*Saccomorpha clava*
	Lithopythium gangliiforme	*S. spherula*
	Ostracoblabe implexa	*Orthogonum fusiferum*

Table 2 Examples of rates of microbioerosion, macrobioerosion, grazing and total bioerosion quantified in coral reefs worldwide. Here, rates were recalculated to be expressed in kg of $CaCO_3$ removed per scare meter of planar reef area per year. A microdensity of 2.3 g cm^{-3} for the coral skeleton of *Porites lobata* was used in the calculation of microbioerosion rates when blocks of this coral species were used (as recommended by Tribollet et al. in press). The bulk density of *Porites lobata/lutea* (1.4 g cm^{-3} on average) was however used in the calculation of macrobioerosion and grazing rates when blocks of this coral species were used. Differences among rates result from type of substrate, techniques, length of exposure and environmental conditions

Location	Microbioerosion	Macrobioerosion	Grazing	Total bioerosion	Type of substrate	Type of reefs	Max length of experiment	Reference
French Polynesia (Moorea)	2	0-1	9-12	11-15	Blocks *Porites lobata*	2 patch reefs, Tiahura	2 yrs	Chazottes et al. (1995)
French Polynesia (islands & atolls)		0-4	1-28		Blocks *Porites lutea*	Fringing reefs, lagoon, pinnacles	5 yrs	Pari et al. (1998, 2002)
Hawaii (Kaneohe Bay)	2-3				Blocks *Porites lobata*	Inshore-offshore profile Eutrophic gradient	8 mo	Tribollet et al. (in press)
Galapagos Islands	3		22		Blocks *Porites lobata*	Reef at Champion Island	14 mo	Reaka-Kudla et al. (1996)
Great Barrier Reef		0-1	0-10		Blocks *Porites*	Leeward, lagoon, windward reefs at Lizard Island	5 yrs	Kiene & Hutchings (1994)
Great Barrier Reef	4				Sediments	Davis Reef lagoon	1 yr	Tudhope & Risk (1985)
Great Barrier Reef	2-10	0-2	0-8	2-11	Blocks *Porites*	Inshore-offshore profile Eutrophic and turbid gradient	3 yrs	Tribollet and Golubic (2005)
Belize	<0.5				Shell (*Strombus gigas*)	Patch reef at Glovers Reef Eutrophic experiment	2 mo	Carreiro-Silva et al. (2005)
Bahamas	<0.5				Limestone	Leeward/windward reefs	2 yrs	Vogel et al. (2000)
Jamaica	<0.5	<0.1	<0.05	<1	Natural dead colonies	Rio Bueno Embayment-offshore	Natural	Mallela & Perry (2007)
Reunion Island	<0.5	<0.2	6-16	0-16	Blocks *Porites lobata*	Fringing reef -Eutrophic gradient	1 yr	Chazottes et al. (2002)
Indonesia		1-10			Colonies of *Porites lobata*	South Sulawesi, Ambon and Java reefs Eutrophic gradient	Colonies of 8-11 yrs	Edinger et al. (2000)

reef framework (Hubbard et al. 1990; Glynn 1997), such as in the Galapagos (Reaka-Kudla et al. 1996). This is particularly true on reefs where fishing is intensive, with urchins becoming the main key herbivorous grazers (e.g., Kenyan reefs; McClanahan et al. 1994).

Measured rates of microbioerosion are considered "residual" because the amount of calcium carbonate dissolved by microborers in the piece of substrate removed by grazing herbivores cannot be quantified (see Chazottes et al. 1995). Thus, the rates of microbioerosion indicated in the literature are underestimated. In dead coral skeletons, the rates of microbioerosion (Table 2) range between 0.02 mol $CaCO_3$ m^{-2} of planar reef per year (0.002 kg m^{-2} of planar reef per year after conversion of rates per square meters of exposed surface area into per square meters planar reef area; >100 m depth in Bahamas; Vogel et al. 2000) and 102 mol $CaCO_3$ m^{-2} of planar reef per year (10.2 kg m^{-2} of planar reef per year; shallow oligotrophic waters on the Great Barrier Reef; Tribollet 2008b). In sediments, the rates have been estimated to be 3.5 mol $CaCO_3$ m^{-2} of planar reef per year on an average (0.35 kg m^{-2} of planar reef per year; Tudhope and Risk 1985). Recently, it was shown that the most active microborer is the chlorophyte *Ostreobium quekettii* (Tribollet 2008b). Although the major geological impact of microbioerosion is realized in conjunction with their grazers (Schneider 1976), microborers play various other roles in coral reefs. They contribute to cementation of sedimentary structures (MacIntyre et al. 2000) and micritization of carbonate grains (Bathurst 1966; Kobluk and Risk 1977b; Hook et al. 1984), thereby strengthening the reef framework. Phototrophic microborers also seem to play an important role in the survival of corals, including bleached corals, by providing photoassimilates (Fine et al. 2005) and nutrients (Ferrer and Szmant 1988) to their host (ectosymbiotic relationship). In contrast, boring fungi have a parasitic relationship with live corals (Golubic et al. 2005) and may be involved in some coral diseases (Alker et al. 2001). These relationships are described in details by Tribollet (2008a).

Macroborer activity also plays an important role in reef destruction as boreholes produced by adult macroborers can significantly weaken the base of live and dead coral colonies (Fig. 9), and thus render them susceptible to breakage by storms (Scott and Risk 1988). Despite the absence of significant direct interactions with the other agents of bioerosion, it is suggested that macroboring communities cannot develop when grazing is too intense as to remove settled larvae and juveniles (Risk and Sammarco 1982; Chazottes et al. 2002; Tribollet and Golubic 2005). Moreover, Chazottes et al. (2002) suggested that microborers and algal turfs may facilitate settlement of macroboring larvae. In return, macroborers may enhance the penetration of microborers into substrates especially once they die and vacate their borings (more light penetrates into substrates).

Fig. 9 Coral head of a *Porites* with a narrow base because of intense bioerosion (*arrow*). This kind of coral colony is highly susceptible to breakage during storms

The cumulative result of the synergistic activity between microborers and grazers as well as the activity of macroborers, which depends on the response of all actors to abiotic and biotic factors, is in selective removal of the substrate, which in turn strongly modifies the rock surfaces. For example, in the supratidal ranges, the preferable location for both microbial euendoliths and their grazers is in shaded microenvironments with good water retention (Schneider 1976). Such locations are bioeroded in preference to well drained and strongly insolated protuberances. As a consequence, the micro-landscape of the rock surface intensifies, deepening the depressions. The result is a sharp-edged and pointed rock surface called biokarst. The carbonate removed from the rock passes through the digestive tract of the bioeroders and contributes to the fine-grain fraction of the sediment deposited nearby (Schneider and Torunski 1983). Sediments produced by grazers can also contribute to reef cementation (Glynn 1997). This biogenic modification of the geomorphology of carbonate coasts and reefs operates at a centimeter-to-diameter scale. At a much larger meter-scale, the cumulative activity of several zonally arranged microboring communities and their similarly differentiated communities of grazers modify the entire coastal profiles by producing a bioerosional notch (Radtke et al. 1996). Bioerosional notches are observed on limestone coasts and elevated coral reefs (Fig. 4), marking extended positions of past and present sea levels (Neumann 1966), often displayed on submersed or elevated coral reef deposits. Bioerosional notches are especially conspicuous in protected coastal sections, whereas on high energy wave-exposed coasts, the overhangs they produce are vulnerable to wave impact and tend to collapse. In this way, the action of combined microbioerosion and grazing enhances coastal destruction by physical forces of erosion (see Radtke et al. 1996).

5 Effects of Anthropogenic and Climatic Changes on Bioerosion Processes

The number of studies attempting to assess the evolution of coral reef health and reef framework is increasing because coral reef ecosystems are increasingly threatened by anthropogenic activities and global climate change. With increasing awareness of these impacts and with the threat such changes pose to human populations that depend on coral reefs for protection of food, a number of studies are dedicated to projections and models of coral reef future developments.

Changes in coral reefs can be assessed if all processes involved are monitored (reef accretion, cementation, bioerosion, micritization, sediment deposition) under various environmental conditions and over time. A recent review of all these processes is given by Perry and Hepburn (2008). The problem is that most of the studies that tend to understand the factors affecting reef framework overlook the importance and complexity of bioerosion processes (see table 6 in Mallela and Perry 2007) or improperly used bioerosion rates. Consequently, the interpretations of rates and trends may be erroneous and the models may be inaccurate or incomplete. This is the case, for example, of early reef carbonate budget studies by Stearn et al. (1977) and Land (1979), which provided incomplete pictures of reef accretion in Barbados and Jamaica, respectively. Recently, Edinger et al. (2000) attempted to estimate reef net carbonate production in polluted versus unpolluted reefs (Indonesia) by simultaneously studying coral growth, coral calcification, live coral cover, skeleton density, and macrobioerosion rates in live corals. Their study revealed that polluted sites had net reef erosion while unpolluted sites had net reef accretion. However, Tribollet and Golubic (2005) demonstrated that bioerosion is driven mainly by the synergistic activity of grazers and microborers (see also Chazottes et al. 1995; Pari et al. 1998; Tribollet et al. 2002), and that combined rates of grazing and microbioerosion are higher at "healthy" oligotrophic reefs than at inshore turbid and eutrophic reefs. They further stressed that bioerosion is more intense on dead substrates than on live ones (Le Campion-Alsumard et al. 1995a; Tribollet and Payri 2001), and that it varies with the type of substrate (Risk et al. 1995; Perry and Hepburn 2008). The model of reef carbonate budget proposed by Edinger et al. (2000) is therefore incomplete and has to be considered with caution, especially regarding the state of the health of unpolluted reefs. Models created to predict coral reef evolution in the context of global change, such as the model proposed by Hoegh-Guldberg et al. (2007), also overlooked bioerosion processes. In the case of Manzello et al.'s (2008) study, bioerosion rates were improperly used. To better understand the effects of ocean acidification on reef framework, they studied in situ effects of this climatic factor on reef cementation at different sites, and used literature data on all kinds of bioerosion rates. They quantified that naturally acidic waters upwelling along the Galapagos have a negative effect on reef cementation and reported that bioerosion in general is more intense in the Galapagos than anywhere else, based on Reaka-Kudla et al.'s (1996) results. They thus concluded that cementation and bioerosion, and bioerosion and the saturation state of aragonite are negatively correlated (fig. 5 in Manzello et al. 2008). We suggest that these correlations are erroneous unless there are experimentally supported. Reaka-Kudla et al. (1996) reported only the rates of macrobioerosion and grazing, and not total bioerosion. Manzello et al.'s table 5 reports a combination of all kinds of rates of bioerosion (grazing, macrobioerosion, microbioerosion, total bioerosion), which are not comparable among the sites. Only rates due to the same agent of bioerosion or total bioerosion rates quantified with similar techniques can be compared among each other. Glynn (1984) also reported a dramatic increase in the number of dead substrates available to colonization by agents of bioerosion and an outbreak of grazing urchins (*Eucidaris thouarsii*) in the Galapagos following the 1982–1983 ENSO event (see Reaka-Kudla et al. 1996). It is thus more than probable that the high rates of bioerosion found by Reaka-Kudla et al. (1996) were due to the very high number of substrates available for colonization after massive coral mortality in 1982–1983. Bioerosion rates at the time of the study of Manzello et al.'s study may have been completely different.

Bioerosion as a whole can no longer be ignored in models and, especially, in biogeochemical models. The rates of bioerosion may dramatically increase in the near future for reasons discussed below, and greatly impact the coral reef framework and maintenance.

5.1 Direct Effects

Several natural abiotic factors can directly affect the agents of bioerosion. For instance, Le Campion-Alsumard (1979) reported effects of natural abiotic factors, such as salinity and water shortage on microborers living in supra- and infra-tidal environments. Species living in these environments show adaptations, such as well-developed sheaths allowing retention of water. In addition to abiotic factors natural environmental conditions, agents of bioerosion are exposed to anthropogenic and climatic influences. While eutrophication may have negative effects on live corals (Tomascik and Sanders 1987; Edinger et al. 2000; Fabricius et al. 2005), this abiotic factor increases the rates of macrobioerosion and microbioerosion. Rose and Risk (1985), Risk et al. (1995), Edinger et al. (2000), Holmes et al. (2000), Tribollet et al. (2002), and Tribollet and Golubic (2005) highlighted in situ that macroborers, especially filter feeders, such as boring sponges and bivalves, are particularly abundant at inshore eutrophic reefs. On the Great Barrier Reef, the rates of macrobioerosion are three to ten times higher at inshore eutrophic reefs than at offshore oligotrophic reefs (Australia) (Risk et al. 1995; Tribollet and Golubic 2005). Carreiro-Silva et al. (2005) showed that after adding nutrients at Glovers Reef (Belize), microbioerosion rates are five to ten times higher under eutrophic conditions (depending on the grazing pressure) than under oligotrophic conditions. Chazottes et al. (2002) showed a similar trend under natural conditions, while Tribollet and Golubic (2005) found the opposite trend along an inshore–offshore transect on the Great Barrier Reef, where inshore reefs were considered eutrophic. The rates of microbioerosion decrease from the coast of Queensland to the Coral Sea due to the impact of terrigenous inputs at the inshore reefs. Turbidity and sedimentation of fine particles on dead carbonate substrates certainly limited microborer settlement and development despite eutrophic conditions (Tribollet et al. 2002). The reefs studied by Chazottes et al. (2002) and Carreiro-Silva et al. (2005) were eutrophic but not turbid. These studies highlight that combined factors can have different effects on organisms and processes than a single factor.

Similarly, Fine et al. (2005) revealed that microborers are highly susceptible to photoinhibition when they are quickly exposed to high light intensity and elevated temperature, while a slow elevation of irradiance only stimulates the growth and development of filaments of *Ostreobium quekettii* (Fine and Loya 2002). Thus, the thermal stress appears to affect more microborer metabolism than irradiance, and its effects are enhanced when it is combined with solar stress. These abiotic factors are related to global warming. Other climatic factors can also greatly affect microborer metabolism. Recently, Tribollet et al. (2009) showed under controlled conditions, the positive effect of ocean acidification on *O. quekettii*'s growth. These authors highlighted that under a double pCO_2 (750 ppm predicted for the end of the century; IPCC 2007), which corresponds to a pH of 7.9 and Ω of 2.5, the depth of penetration of microboring filaments of *O. quekettii* and, thus, the rates of microbioerosion were increased by 48%.

Based on the available literature, we thus suggest that the effects of ocean acidification combined with other abiotic factors on microborers, such as eutrophication, elevated temperature, and irradiance, may greatly affect microbioerosion by the end of the century. We hypothesize that bioerosion due to sponges may increase under low pH. Their penetration into substrates might indeed be facilitated. In contrast, low pH and Ω might reduce the erosive activity of boring bivalves, grazing urchins, and fishes, as these organisms possess calcified parts, such as shell, test, radula, and beak. Very recently, studies showed that low pH and Ω can impact planktonic molluscs, sea urchins, and fishes in different ways (Fabry et al. 2008; Havenhand et al. 2008). A possible reduction in grazing pressure on dead substrates may mitigate the increase in microbioerosion under the acidic conditions predicted by the end of the century. This has to be investigated.

5.2 Indirect Effects

Bioerosion processes should increase significantly due to the increasing rate of coral mortality by the end of the century. The rate of coral mortality increases because of different abiotic factors, such as rising sea surface temperature. This factor is responsible for massive bleaching events and coral disease outbreaks (Mumby et al. 2001; Wilkinson 2002; Muller et al. 2008). Recently, Anthony et al. (2008) revealed that ocean acidification also causes coral bleaching under high irradiance, acting synergistically with elevated sea surface temperature. Corals show high mortality when the thermal stress last more than a few weeks and/or when it is combined with other abiotic stress-causing factors (Hoegh-Guldberg 1999; Wilkinson 2002; Mumby et al. 2001). Consequently, more and more dead substrates become available for colonization by agents of bioerosion. Bioerosion may also potentially increase because of the reduced calcification rates of reef framebuilders under low pH and Ω (Langdon and Atkinson 2005; Kleypas et al. 2006; Tribollet et al. 2006b; Anthony et al. 2008; Kuffner et al. 2008). Fine and Loya (2002) showed that when corals bleach and their growth slows down, phototrophic microboring communities bloom. Fine and Loya (2002) did not quantify the change in rates of microbioerosion, but it appears evident that if filaments of microborers grow and increase in abundance, more carbonate is dissolved. Boring fungi may also take advantage of the weakness of bleached, diseased, or stressed corals by attacking polyps. Normally, polyps defend themselves against such intrusions by precipitating pearl-like deposits of carbonate (or "cones") around hyphae filaments to prevent further penetration (Le Campion-Alsumard et al. 1995b; Bentis et al. 2000). Stressed corals may not be able to efficiently defend themselves against fungal attacks as it is energy demanding (Domart-Coulon et al. 2004). These latter ones may potentially bloom as the organic matter that polyps offer may be more accessible (Bentis et al. 2000). Thus, fungi may dissolve more carbonate. Similarly, it is hypothesized that larvae of boring bivalves and sponges might be more successful in penetrating the polyp layer and colonize coral skeleton as coral polyps are stressed and weakened by global warming and ocean acidification. Nevertheless, this has to be demonstrated.

6 Perspectives

Based on current knowledge, we would like to stress the urgency to investigate the combined effects of climatic and anthropogenic factors on bioeroding organisms and associated processes, by coupling different approaches (in situ and ex situ experiments) and by using different timescale sampling rates, if the state of the health of coral reefs and the maintenance of their framework in the context of global change has to be assessed, monitored, predicted, and managed properly. Some suggestions for research direction are as follows:

1. Continue to evaluate changes in species composition and abundance of bioeroding communities under different environmental conditions and over time (at different timescales).
2. Continue to determine interactions among the different agents of bioerosion and other reef organisms.
3. Determine the mechanism(s) of carbonate dissolution used by microborers and some macroborers, and how they may be affected by an acidic environment.
4. Compare, improve, and homogenize techniques and methods used to quantify the different rates of bioerosion for future comparisons.
5. Develop experiments under controlled conditions to investigate the effects of combined factors on the different agents of bioerosion, at both the organismal and process levels. This should allow identifying main factors influencing bioerosion.
6. Develop similar experiments as mentioned above, but in situ. In parallel, at several coral reefs with different saturation states, a program to evaluate the relationships between boring community metabolism and dissolution rates should be develop to test whether presently naturally varying saturation state is affecting bioerosion rates (dissolution rates).
7. Expand efforts in monitoring rates of bioerosion (microbioerosion, macrobioerosion, grazing, and total bioerosion) on a number of coral reefs worldwide.
8. Develop a model of bioerosion processes to improve biogeochemical models and highlight potential bioindicators of environmental changes among agents of bioerosion.

References

Alker AP, Smith GW, and Kim K (2001) Characterization of *Aspergillus sydowii* (Thom et Church), a fungal pathogen of Caribbean sea fan corals. Hydrobiologia 460:105–111

Al-Thukair AA, Golubic S (1991a) New endolithic cyanobacteria from the Arabian Gulf I. *Hyella immanis* sp. nov. J Phycol 27:766–780

Al-Thukair AA, Golubic S (1991b) Five new Hyella species from the Arabian Gulf. Algol Stud 64:167–197

Anthony KRN, Kline DI, Diaz-Pulido G, Dove S, Hoegh-Guldberg O (2008) Ocean acidification causes bleaching and productivity loss in coral reef builders. PNAS 105:17442–17446

Bathurst RGC (1966) Boring algae, micrite envelopes, and lithification of molluscan biosparites. Lpool Manchr Geol J 5:15–32

Bak RPM (1976) The growth of coral colonies and the importance of crustose coralline algae and burrowing sponges in relation with carbonate accumulation. Neth J Sea Res 10:285–337

Bak RPM (1990) Patterns of echinoid bioerosion in two Pacific coral reef lagoons. Mar Ecol Prog Ser 66:267–272

Bentis CJ, Kaufman L, Golubic S (2000) Endolithic fungi in reef-building corals (Order: Scleractinia) are common, cosmopolitan, and potentially pathogenic. Biol Bull 198:254–260

Bellwood DR, Choat JH (1990) A functional analysis of grazing in parrotfishes (family Scaridae): the ecological implications. Environ Biol Fish 28:189–214

Bromley RG, D'Alessandro A (1990) Comparative analysis of bioerosion in deep and shallow water, Pliocene to Recent, Mediterranean Sea. Ichnos 1:43–49

Bruggemann JH, van Oppen MJH, Breeman AN (1994) Foraging by the spotlight parrotfish *Sparisoma viride*. I. Food selection in different, socially determined habitats. Mar Ecol Prog Ser 106:41–55

Calcinai B, Arilla A, Cerrano C, Bavestrello G (2003) Taxonomy-related differences in the excavating micro-patterns of boring sponges. J Mar Biol Ass 83:37–39

Campbell SE (1980) *Palaeoconchocelis starmachii*, a carbonate boring microfossil from the Upper Silurian of Poland (425 million years old): implications for the evolution of the Bangiaceae (Rhodophyta). J Phycol 19:25–36

Campbell SE (1982) Precambrian endoliths discovered. Nature 299:429–431

Campbell SE, Kazmierczak J, Golubic S (1979) Palaeoconchocelis starmachii gen. n., sp. n., an endolithic rhodophyte (Bangiaceae) from the Silurian of Poland. Acta Palaeontol Pol 24:405–408

Carreiro-Silva M, McClanahan TR, Kiene WE (2005) The role of inorganic nutrients and herbivory in controlling microbioerosion of carbonate substratum. Coral Reefs 24:214–221

Carriker MR (1969) Excavation of boreholes by the gastropod, *Urosalpinx*: an analysis by light and scanning electron microscopy. Am Zool 9:917–933

Cedhagen T (1994) Taxonomy and biology of *Hirrokkin sarcophaga* gen. et sp. n., a parasitic foraminiferan (Rosalinidae). Sarsia 79:65–82

Chazottes V, Le Campion-Alsumard T, Peyrot-Clausade M (1995) Bioerosion rates on coral reefs: interaction between macroborers, microborers and grazers (Moorea, French Polynesia). Palaeo 113:189–198

Chazottes V, Le Campion-Alsumard T, Peyrot-Clausade M, Cuet P (2002) The effects of eutrophication-related alterations to coral reef communities on agents and rates of bioerosion (Reunion Island, Indian Ocean). Coral Reefs 21:375–390

Cobb WR (1975) Fine structural features of destruction of calcareous substrata by the burrowing sponge *Cliona celata*. Trans Am Microsc Soc 94:197–202

Conand C, Heeb M, Peyrot-Clausade M, Fontaine MF (1998) Bioerosion by the sea urchin Echinometra on La Reunion reefs (Indian Ocean) and comparison with Tiahura reefs (French Polynesia). In: Mooi R, Telford M (eds) Echinoderms: San Francisco. AA Balkema, Rotterdam, pp 609–615

Davies PJ, Hutchings PA (1983) Initial colonization, erosion and accretion on coral substrates. Experimental results, Lizard Island, Great Barrier Reef. Coral Reefs 2:27–35

Domart-Coulon IJ, Sinclair CS, Hill RT, Tambutté S, Puverel S, Ostrander GK (2004) A basidiomycete isolated from the skeleton of *Pocillopora damicornis* (Scleractinia) selectively stimulates short-term survival of coral skeletogenic cells. Mar Biol 144:583–592

Edinger EN, Limmon GV, Jompa J, Widjatmoko W, Heikoop JM, Risk MJ (2000) Normal coral growth rates on dying reefs: are coral growth rates good indicators of reef health? Mar Poll Bull 40:404–425

Ercegovic A (1932) Etudes écologiques et sociologiques des Cyanophycées lithophytes de la côte Yougoslave de l' Adriatique. Bull Int Acad Youg Sci Arts Cl Sc Math Nat 26:33–56

Evans JW (1970). Palaeontological implications of a biological study of rock boring clams (Family Pholadidae). In: Crimes TP, Harper JC (eds) Trace fossils, pp. 127–140. Geol J Special Issue 3. Seel House Press, Liverpool

Fabricius K, De'ath G, McCook L, Turak E, Williams DM (2005) Changes in algal, coral and fish assemblages along water quality gradients on the inshore Great Barrier Reef. Mar Poll Bull 51:384–398

Fabry VJ, Seibel BA, Feely RA, Orr JC (2008) Impacts of ocean acidification on marine fauna and ecosystem processes. ICES J Mar Sci 65:414–432

Ferrer LM, Szmant AM (1988) Nutrient regeneration by the endolithic communities in coral skeletons. Proc 6th Int Coral Reef Symp 3:1–4. Townsville, Australia

Fine M, Loya Y (2002) Endolithic algae: an alternative source of photoassimilates during coral bleaching. Proc R Soc Lond 269:1205–1210

Fine M, Meroz-Fine E, Hoegh-Guldberg O (2005) Tolerance of endolithic algae to elevated temperature and light in the coral *Montipora monasteriata* from the southern Great Barrier Reef. J Exp Biol 208:75–81

Freiwald A, Reitner J, and Krutschinna J (1997) Microbial alteration of the deep-water coral *Lophelia pertusa*: Early post-mortem processes. Facies 36:223–226

Garcia-Pichel F (2006) Plausible mechanisms for the boring on carbonates by microbial phototrophs. Sedim Geol 185:205–213

Gherardi DFM, Bosence DWJ (2001) Composition and community structure of the coralline algal reefs from Atol das Rocas, Suth Atlantic, Brazil. Coral Reefs 19:205–219

Glaub I, Golubic S, Gektidis M, Radtke G, Vogel K (2007) Microborings and microbial endoliths: geological implications. In: Miller W III (ed) Trace fossils: concepts, problems, prospects. Elsevier, Amsterdam-Oxford-New York, pp 368–381

Glaub I, Vogel K (2004) The stratigraphic record of microborings. Fossils Strata 51:126–135

Glynn PW (1984) Widespread coral mortality and the 1982-83 El Nino event. Environ Conserv 11:133–146.

Glynn PW (1997) Bioerosion and coral reef growth: a dynamic balance. In: Birkeland C (ed) Life and death of coral reefs. Chapman and Hall, New York, pp 68–95

Golubic S, Schneider J (1979) Carbonate dissolution. In: Trudinger PA, Swaine DJ (eds) Biogeochemical cycling of mineral-forming elements. Elsevier, Amsterdam-Oxford-New York, pp 107–129

Golubic S, Brent G, Le Campion-Alsumard T (1970) Scanning electron microscopy of endolithic algae and fungi using a multipurpose casting-embedding technique. Lethaia 3:203–209

Golubic S, Perkins RD, Lukas KJ (1975) Boring microorganisms and microborings in carbonate substrates. In: Frey RW (ed) The study of trace fossils. Springer, Heidelberg-Berlin-New York, pp 229–259

Golubic S, Friedmann I, Schneider J (1981) The lithobiontic ecological niche, with special reference to microorganisms. J Sedim Petrol 51:475–478

Golubic S, Campbell SE, Spaeth C (1983) Kunsharzausguesse fossiler Mikroben-Bohrgaenge (Resin-casting of fossil microbial borings). Der Praeparator, Bochum 29:197–200

Golubic S, Radtke G, Le Campion-Alsumard T (2005) Endolithic fungi in marine ecosystems. Trends Microbiol 13:229–235

Green JW, Knoll AH, Swett K (1988) Microfossils from oolites and pisolites of the Upper Proterozoic Eleonore Bay Group, central east Greenland. J Paleontol 62:835–852

Hansen TA, Kelley PH (1995) Spatial variation of naticid gastropod predation in the Eocene of North America. Palaios 10:168–278

Harper EM (1994) Are conchiolin sheets in corbulid bivalves primarily defensive? Palaeontology 37:551–578

Hassan M, Dullo W. -C, Fink A (1996). Assessment of boring activity in *Porites lutea* from Aqaba (Red Sea) using computed tomography. Proceedings of the 8th international coral reef symposium, Panama, 1996, 5

Havenhand JN, Buttler FR, Thorndyke MC, Williamson JE (2008) Near-future levels of ocean acidification reduce fertilization success in a sea urchin. Curr Biol 18:R651–R652

Hein FJ, Risk MJ (1975) Bioerosion of coral heads: inner patch reefs, Florida reef tract. Bull Mar Sci 25:133–137

Hoegh-Guldberg O (1999) Climate change, coral bleaching and the future of the world's coral reefs. Mar Fresh Res 50:839–866

Hoegh-Guldberg O, Mumby PJ, Hooten AJ, Steneck RS, Greenfield P, Gomez E, Harvell CD, Sale PF, Edwards AJ, Caldeira K, Knowlton N, Eakin CM, Iglesias-Prieto R, Muthiga N, Bradbury RH, Dubi A, Hatziolos ME (2007) Coral reefs under rapid climate change and ocean acidification. Science 318:1737–1742

Hoey AS, Bellwood DR (2008) Cross-shelf variation in the role of parrotfishes on the Great Barrier Reef. Coral Reefs 27:27–47

Holmes KE, Edinger EN, Hariyadi, Limmon GV, Risk MJ, Limmon GV, Risk MJ (2000) Bioerosion of live massive corals and branching coral rubble on Indonesian coral reefs. Mar Pollut Bull 40:606–617

Hook JE, Golubic S, Milliman JD (1984) Micritic cement in microborings is not necessarily a shallow-water indicator. J Sedim Petrol 54:425–431

Hubbard DK, Miller AI, Scaturo D (1990) Production and cycling of calcium carbonate in a shelf-edge reef system (St Croix, U.S.: Virgin Islands): applications to the nature of reef systems in the fossil record. J Sediment Petrol 60:335–360

Hughes TP, Rodrigues MJ, Bellwood DR, Ceccarelli D, Hoegh-Guldberg O, McCook L, Moltschaniwskyj N, Pratchett MS, Steneck RS, Willis B (2007) Phase shifts, herbivory, and the resilience of coral reefs to climate change. Curr Biol 17:360–365

Hutchings PA (1986) Biological destruction of coral reefs. Coral Reefs 4:239–252

Hutchings PA, Murray A (1982) Patterns of recruitment of polychaetes to coral substrates at Lizard Island, Great Barrier Reef – an experimental approach. Aust J Mar Fresh Res 33:1029–1037

Hutchings PA, Peyrot-Clausade M (2002) The distribution and abundance of boring species of polychaetes and sipunculans in coral substrates in French Polynesia. J Exp Mar Biol Ecol 269:101–121

IPCC, Climate Change (2007) The physical science basis. In: Solomon S et al. (eds) Contribution of working group I to the fourth assessment. Report of the Intergovernmental Panel on Climate Change. Cambridge University Press, Cambridge, UK, and New York

James NP, Kobluk DR, Pemberton SG (1977) The oldest macroborers: lower Cambrian of labrador. Science 197:980–983

Kendrick B, Risk MJ, Michaelides J, Bergman K (1982) Amphibious microborers: bioeroding fungi isolated from live corals. Bull Mar Sci 32:862–867

Kiene WE (1985) Biological destruction of experimental coral substrates at Lizard Island, Great Barrier Reef, Australia. Proc 5th Int Coral Reef Symp 5:339–344

Kiene WE, Hutchings PA (1994) Bioerosion experiments at Lizard Island, Great Barrier Reef. Coral Reefs 13:91–98

Kleemann KH (1996) Biocorrosion by bivalves. Mar Ecol 17:145–158

Klein R, Mokady O, Loya Y (1991) Bioerosion in ancient and comtemporary corals of the genus *Porites*: patterns and palaeoenvironmental implications. Mar Ecol Prog Ser 77:245–251

Kleypas JA, Feely RA, Fabry VJ, Langdon C, Sabine CL, Robbins LL (2006) Impacts of ocean acidification on coral reefs and other marine calcifyers. In: NSF, NOAA and U.S. Geological Survey (eds) A guide for future research, report of a workshop held 18-20 April 2005, St Petersburg, pp 1–88

Knoll AH, Golubic S, Green J, Swett K (1986) Organically preserved microbial endoliths from the Late Proterozoic of East Greenland. Nature 321:856–857

Kobluk DR, Risk MJ (1977a) Algal borings and fromboidal pyrite in Upper Ordovician brachiopods. Lethaia 10:135–143

Kobluk DR, Risk MJ (1977b) Calcification of exposed filaments of endolithic algae, micrite envelope formation and sediment production. J Sedim Petrol 47:517–528

Kolbasov GA (2000) *Lithoglyptes cornutes*, new species (Cirripedia: Acrothoracica), a boring barnacle from the Seychelles, with some data on its ultrastructure. Hydrobiology 438:185–191

Kuffner IB, Andersson AJ, Jokiel PL, Rodgers KS, Mackenzie FT (2008) Decrease abundance of crustose coralline algae due to ocean acidification. Nat Geosci 1:114–117

Laborel J, Le Campion-Alsumard T (1979) Infestation massive u squelette de coraux vivant par des Rhodophycees de type Conchocelis. C R Acad Sci Ser III 288:1575–1577

Land LS (1979) The fate of reef derived sediment on the North Jamaican Island slope. Mar Geol 29:55–71

Langdon C, Atkinson MJ (2005) Effect of elevated pCO_2 on photosynthesis and calcification of corals and interaction with seasonal change in temperature/radiance and nutrient enrichment. J Geophys Res Oceans 110:C09S07. doi:10.1029/2004JC002576

Langdon C, Takahashi T, Marubini F, Atkinson MJ, Sweeney C, Aceves H, Barnet H, Chipman D, Goddard J (2000) Effect of calcium carbonate saturation state on the calcification rate of an experimental coral reef. Global Biogeochem Cycles 14:639–654

Le Campion-Alsumard T (1975) Etude experimentale de la colonisation d'eclats de calcite par les cyanophycees endolithes marines. Cahiers Biol Mar 16:177–185

Le Campion-Alsumard T (1979) Les cyanophycées endolithes marines. Systématique, ultrastructure, écologie et biodestruction. Oceanol Acta 2:143–156

Le Campion-Alsumard T, Campbell SE, Golubic S (1982) Endoliths and the depth of the photic zone. Discussion. J Sed Petrol 52:1333–1338

Le Campion-Alsumard T, Golubic S, Hutchings P (1995a) Microbial endoliths in skeletons of live and dead corals: *Porites lobata* (Moorea, French Polynesia). Mar Ecol Prog Ser 117:149–157

Le Campion-Alsumard T, Golubic S, Priess K (1995b) Fungi in corals: symbiosis or disease? Interaction between polyps and fungi causes pearl-like skeleton biomineralization. Mar Ecol Prog Ser 117:137–147

Lewis JB (1998) Reproduction, larval development and functional relationships of the burrowing, spionid polychaete *Dipolydora armata* with the calcareous hydrozoan *Millepora complanata*. Mar Biol 130:651–662

Lukas KJ (1978) Depth distribution and form among common microboring algae from the Florida continental shelf. Geol Soc Am 10:448

Lukas KJ, Golubic S (1981) New endolithic cyanophytes from the North Atlantic Ocean: I. *Cyanosaccus piriformis* gen. et sp. nov. J Phycol 17:224–229

MacGeachy JK (1977) Factors controlling sponge boring in Barbados reef corals. Proc 3rd Int Coral Reef Symp, Miami 2:477–483

MacIntyre IG, Prufert-Bebout L, Reid RP (2000) The role of endolithic cyanobacteria in the formation of lithified laminae in Bahamian stromatolites. Sediment 47:915–921

Magnusson SH, Fine M, Kühl M (2007) Light microclimate of endolithic phototrophs in the scleractinian corals *Montipora monasteriata* and *Porites cylindrical*. Mar Ecol Prog Ser 332:119–128

Mallela J, Perry CT (2007) Calcium carbonate budgets for two coral reefs affected by different terrestrial runoff regimes, Rio Bueno, Jamaica. Coral Reefs 26:129–145

Manzello DP, Kleypas JA, Budd DA, Eakin CM, Glynn PW, Langdon C (2008) Poorly cemented coral reefs of the eastern tropical Pacific: possible insights into reef development in a high-CO_2 world. PNAS 105:10450–10455

Mariani S, Uriz M-J, Turon X (2000) Larval bloom of the oviparous sponge *Cliona viridis*: coupling of larval abundance and adult distribution. Mar Biol 137:783–790

McClanahan TR, Nugues M, Mwachireya S (1994) Fish and sea urchin herbivory and competition in Kenyan coral reef lagoons: the role of the reef management. J Exp Mar Biol Ecol 184:237–254

Mills SC, Peyrot-Clausade M, Fontaine MF (2000) Ingestion and transformation of algal turf by *Echinometra mathei* on Tiahura fringing reef (French Polynesia). J Exp Mar Biol Ecol 254:71–84

Muller EM, Rogers CS, Spitzack AS, van Woesik R (2008) Bleaching increases likelihood of disease on *Acropora palmate* (Lamarck) in Hawksnest Bay, St John, US Virgin Islands. Coral Reefs 27:191–195

Mumby PJ, Chisholm JRM, Edwards AJ, Clark CD, Roark EB, Andrefouet S, Jaubert J (2001) Unprecedented bleaching-induced mortality in *Porites* spp. At Rangiroa Atoll, French Polynesia. Mar Biol 139:183–189

Neumann AC (1966) Observations on coastal erosion in Bermuda and measurements of the boring rate of the sponge, *Cliona lampa*. Limnol Oceanogr 11:92–108

Otter GW (1937) Rock-destroying organisms in relation to coral reefs. Scientific Reports of the Great. Barr Reef Exped 1:323–352

Pari N, Peyrot-Clausade M, Le Campion-Alsumard T, Hutchings PA, Chazottes V, Golubic S, Le Campion J, Fontaine MF (1998) Bioerosion of experimental substrates on high islands and atoll lagoons (French Polynesia) after two years of exposure. Mar Ecol Prog Ser 166:119–130

Perry CT (1998) Grain susceptibility to the effects of microboring: implication for the preservation of skeletal carbonates. Sediment 45:39–51

Perry CT (2000) Macroboring of Pleistocene coral communities, Falmouth formation, Jamaica. Palaios 15:483–491

Perry CT, Hepburn LJ (2008) Syn-depositional alteration of coral reef framework through bioerosion, encrustation and cementation: taphonomic signatures of reef accretion and reef depositinal events. Earth Sci Rev 86:106–144

Peyrot-Clausade M, Hutchings PA, Richard G (1992) The distribution and successional patterns of macroborers in marine *Porites* at different stages of degradation on the barrier reef, Tiahura, Moorea, French Polynesia. Coral Reefs 11:161–166

Peyrot-Clausade M, Chazottes V (2000) La Bioérosion récifale et son rôle dans la sédimentogénèse à Moorea (Polynésie française) et à la Réunion. Océanis 26:275–309

Pomponi SA (1980) Cytological mechanisms of calcium carbonate excavation by boring sponges. Int Rev Cytol 65:301–319

Radtke G, Le Campion-Alsumard T, Golubic S (1996) Microbial assemblages of the bioerosional "notch" along tropical limestone coasts. Algol Stud 83:469–482

Rasmussen KA, Frankenberg EW (1990) Intertidal bioerosion by the chiton *Acanthopleura granulata*; San Salvador. Bahamas Bull Mar Sci 47:680–695

Reaka-Kudla ML, Feingold JS, Glynn W (1996) Experimental studies of rapid bioerosion of coral reefs in the Galapagos Islands. Coral Reefs 15:101–107

Risk MJ, Sammarco PW (1982) Bioerosion of corals and the influence of damselfish territoriality, a preliminary study. Oecologia 52:376–380

Risk MJ, Sammarco PW, Edinger EN (1995) Bioerosion in *Acropora* across the continental shelf of the Great Barrier Reef. Coral Reefs 14:79–86

Rotjan RD, Lewis SM (2006) Parrotfish abundance and corallivory on a Belizean coral reef. J Exp Mar Biol Ecol 335:292–301

Rotjan RD, Dimond JL, Thornhill DJ, Leichter JJ, Helmuth BST, Kemp DW, Lewis SM (2006) Chronic parrotfish grazing impedes coral recovery after bleaching. Coral Reefs 25:361–368

Rose CS, Risk MJ (1985) Increase in *Cliona delitrix* infestation of *Montastrea cavernosa* heads on an organically polluted portion of the Grand Cayman fringing reef. Mar Ecol 6:345–363

Russ GR (1984) The distribution and abundance of herbivorous grazing fishes in the central Great Barrier Reef. II. Levels of variability across the entire continental shelf. Mar Ecol Prog Ser 20:23–34

Rützler K (1974) The burrowing sponges of Bermuda. Smithson Contribut Zoo 165:1–32

Rützler K (2002) Impact of crustose Clionid sponges on Caribbean reef corals. Acta Geol Hisp 37:61–72

Schneider J (1976) Biological and inorganic factors in the destruction of limestone coasts. Contributions Sediment 6:1–112

Schneider J, Campion-Alsumard T (1999) Construction and destruction of carbonates by marine and freshwater cyanob. Eur J Phycol 34:417–426

Schneider J, Torunski H (1983) Biokarst on limestone coasts, morphogenesis and sediment production. Mar Ecol 4:45–63

Schönberg CHL (2000) Bioeroding sponges common to the Central Great Barrier Reef: descriptions of three new species, two new records, and additions to two previously described species. Senckenb Marit 30:161–221

Schönberg CHL (2003) Substrate effects on the bioeroding Desmosponge *Cliona orientalis*. 2. Substrate colonisation and tissue growth. Mar Ecol 24:59–74

Schönberg CHL, Loh WKW (2005) Molecular identity of the unique symbiotic dinoflagellates found in the bioeroding desmosponge *Cliona orientalis*. Mar Ecol Prog Ser 299:157–166

Schönberg CHL (2008) A history of sponge erosion: from past myths and hypotheses to recent approaches. In: Wisshak M, Tabanila L (eds) Current developments in bioerosion. Springer, Berlin, pp 165–202

Schönberg CHL, Tapanila L (2006) The bioeroding sponge *Aca paratypica*, a modern tracemaking analogue for the Paleozoic ichnogenus *Entobia devonica*. Ichnos 13:147–157

Schönberg CHL, Wilkinson CR (2001) Induced colonization of corals by a clionid bioeroding sponge. Coral Reefs 20:69–76

Scoffin TP, Stearn CW, Boucher D, Frydl P, Hawkins CM, Hunter IG, MacGeachy JK (1980) Calcium carbonate budget of a fringing reef on the West coast of Barbados. Part II. Erosion, sediments and internal structure. Bull Mar Sci 30:475–508

Scott PJB, Risk MJ (1988) The effect of *Lithophaga* (Bivalvia: Mytilidae) boreholes on the strength of the coral *Porites lobata*. Coral Reefs 7:145–151

Scott PJB (1988) Distribution, habitat and morphology of the Caribbean coral- and rock-boring bivalve, *Lithophaga bisulcata* (d'Orbigny) (Mytilidae: Lithophaginae). J Molluscan Stud 54:83–95

Stearn CW, Scoffin TP, Martindale W (1977) Calcium carbonate budget of a fringing reef on the west coast of Barbardos. Part 1: zonation and productivity. Bull Mar Sci 27:479–510

Stockfors M, Peel JS (2005) Euendolithic Cyanobacteria from the Middle Cambrian of North Greenland. Geologiska Föreningens i Stockholm Förhandlingar, IFF 127:179–185

Tapanila L, Copper P, Edinger E (2004) Environmental and substrate controls on Paleozoic bioerosion in corals and stromatoporoids, Anticosti Island, eastern Canada. Palaios 19:292–306

Todd JA (2000) The central role of ctenostomes in bryozoan phylogeny. Proceedings of the 11th international Bryozoology association conference, Lawrence pp 104–135

Tomascik T, Sander F (1987) Effects of eutrophication on reefbuilding corals. II. Structure of scleractinian coral communities on fringing reefs, Barbados, West Indies. Mar Biol 94:53–75

Tudhope AW, Risk MJ (1985) Rate of dissolution of carbonate sediments by microboring organisms, Davies Reef, Australia. J Sediment Petrol 55:440–447

Tribollet A (2001). Processus de bioérosion récifale (Grand Barrière de Corail, Australie). Importance du rôle joué par la microflore perforante, p 190. Ph.D. thesis. Université de la Méditerranée Aix-Marseille, II

Tribollet A (2008a) The boring microflora in modern coral reef ecosystems: a review of its roles. In: Wisshak M, Tapanila L (eds) Current developments in bioerosion. Springer, Berlin-Heiderlberg, pp 67–94

Tribollet A (2008b) Dissolution of dead corals by euendolithic microorganisms across the northern Great Barrier Reef (Australia). Microb Ecol 55:569–580

Tribollet A, Payri C (2001) Bioerosion of the crustose coralline alga *Hydrolithon onkodes* by microborers in the coral reefs of Moorea. French Polynesia. Oceanol Acta 24:329–342

Tribollet A, Golubic S (2005) Cross-shelf differences in the pattern and pace of bioerosion of experimental carbonate substrates exposed for 3 years on the northern Great Barrier Reef, Australia. Coral Reefs 24:422–434

Tribollet A, Decherf G, Hutchings PA, Peyrot-Clausade M (2002) Large-scale spatial variability in bioerosion of experimental coral substrates on the Great Barrier Reef (Australia): importance of microborers. Coral Reefs 21:424–432

Tribollet A, Radtke G, Golubic S Bioerosion. In: Encyclopedia of Geobiology. Springer, Berlin (in press)

Tribollet A, Langdon C, Golubic S, Atkinson M (2006a) Endolithic microflora are major primary producers in dead carbonate substrates of Hawaiian coral reefs. J Phycol 42:292–303

Tribollet A, Atkinson M, Langdon C (2006b) Effects of elevated pCO_2 on epilithic and endolithic metabolism of reef carbonates. Glob Change Biol 12:2200–2208

Tribollet A, Godinot C, Atkinson M, Langdon C (2009) Effects of elevated pCO_2 on dissolution of coral carbonates by microbial euendoliths. Glob Biogeoch Cycles 23:GB3008. doi:10.1029/2008GB003286

Trudgill ST (1976) The marine erosion of limestone on Aldabra Atoll, Indian Ocean. Z Geomorphol Suppl 26:164–200

Vogel K, Gektidis M, Golubic S, Kiene WE, Radtke G (2000) Experimental studies on microbial bioerosion at Lee Stocking Island, Bahamas and One Tree Island, Great Barrier Reef, Australia: implications for paleoecological reconstructions. Lethaia 33:190–204

Walker SE (2007) Traces of gastropod predation on molluscan prey in tropical reef environments. In: Miller W III (ed) Trace fossils: concepts, problems, prospects. Elsevier, Amsterdam, pp 324–344

Ward-Paige CA, Risk MJ, Sherwood OA, Jaap WC (2005) Clionid sponge surveys on the Florida Reef Tract suggest land-based nutrient inputs. Facies 51:570–579

Warme JE (1975) Borings as trace fossils, and the processes of marine bioerosion. In: Frey RW (ed) The study of trace fossils. Springer, Berlin Heidelberg, New York, pp 181–229

Wielgus J, Glassom D, Chadwick NE (2006) Patterns of polychaete worm infestation of stony corals in the northern Red Sea and relationships to water chemistry. Bull Mar Sci 78:377–388

Wilkinson C (2002) The status of the coral reefs of the world: 2002, Australian Institute of Marine Science and the Global Coral Reef Monitoring Network, Townsville, pp 1–378

Wilson MA (2007) Macroborings and the evolution of marine bioerosion. In: Miller W III (ed) Trace fossils: concepts, problems, prospects. Elsevier, Amsterdam-Oxford-New York, pp 356–367

Wilson MA, Palmer TJ (2006) Patterns and processes in the Ordovician bioerosion revolution. Ichnos 13:109–112

Wilson MA (2008) An online bibliography of bioerosion references. In: Wisshack M, Tapanila L (eds) Current development in bioerosion. Springer, Berlin-Heidelberg, pp 473–478, http://www.wooster.edu/geology/bioerosion/BioerosionBiblio.pdf

Wisshak M (2006). High-latitude bioerosion. In: S. Bhattachararji, H. J. Neugebauer, J. Reitner, and K.Stüwe (eds.) Lecture, Notes in Earth Sciences, Berlin-Heidelberg Springer, vol 109 pp 1–202

Zea S, Weil E (2003) Taxonomy of the Caribbean excavating sponge species complex *Cliona caribbaea – C. aprica – C. langae* (Porifera, Hadromerida, Clionaidae). Caribbean J Sci 39:348–370

Zhang Y, Golubic S (1987) Endolithic microfossils (cyanophyta) from early Proterozoic stromatolites, Hebei, China. Acta Micropaleontologica Sinica 4:1–12

Microbial Diseases of Corals: Pathology and Ecology

Eugene Rosenberg and Ariel Kushmaro

Abstract Microorganisms play a fundamental role in the health and disease of all corals. Disease can be caused by pathogenic microorganisms, environmental stress, and a weakening of the host's innate immunity system. Often these factors interact. Pathogens responsible for the following coral diseases are described: Bacterial bleaching, aspergillosis, black band, white band, white plague, white pox, yellow band, brown band, porites trematodiasis, and skeletal eroding band. The small amount of data that is available on the modes of transmission and mechanism of pathogenicity are presented. Several possible modes of coral innate immunity to disease are discussed.

Keywords Coral diseases • coral pathology • coral health • vibrio • coral bleaching • bacteria • bacterial bleaching • aspergillosis • black band • white band • white plague • white pox • yellow band • brown band • porites trematodiasis • skeletal eroding band

1 Introduction

Microorganisms play a fundamental role in the health and disease of all plants and animals (Zilber-Rosenberg and Rosenberg 2009), including corals (Rohwer et al. 2002). Disease is defined as a process resulting in tissue damage or alteration of physiological function, producing visible symptoms (Stedman 2005). Coral diseases can be caused by pathogenic microorganisms, environmental stress (discussed by Dindsdale and Rohwer, this volume), and a weakening of the host innate immunity system. Often, these factors interact (Fig. 1). This chapter will review the current knowledge of the pathology and ecology of coral pathogens. The worldwide distribution of coral diseases is discussed in another chapter (Weil, this book). Several reviews have been published on different aspects of microbial diseases of corals (Bourne et al. 2009; Harvell et al. 2007; Rosenberg et al. 2007; Richardson and Aronson 2002).

The causative agents or pathogens responsible for most of the coral diseases are not known. The classical way to prove that a particular microorganism is responsible for a particular disease is to satisfy all of Koch's postulates: (1) The microorganism is present in all diseased individuals and absent in healthy ones; (2) the presumed pathogen is isolated in pure culture; (3) the pathogen infects healthy individuals and causes the appropriate disease signs; and (4) the pathogen can be re-isolated from the tissues of the experimentally infected individuals. It is not always possible to demonstrate Koch's postulates because some pathogens cannot be grown as pure cultures (Postulate 2) and the method of infection may not be understood. Table 1 lists the best-studied infectious diseases of corals.

2 The Bacterial Bleaching Disease

2.1 *Vibrio shiloi*

Bleaching of the scleractinian coral *Oculina patagonica* (Fig. 2) in the eastern Mediterranean Sea was first recorded in the summer of 1993 (Fine and Loya, 1995).
Since that time, the annual pattern of bleaching has been monitored as a function of seawater temperature (Fig. 3). Bleaching begins during the early summer when seawater temperature reaches 25°C and increases until late in the summer, when the temperature approaches 30°C. At the peak of bleaching, 70–80% of the colonies showed signs of bleaching.

Kushmaro et al. (1996, 1997) reported that the bleaching of *O. patagonica* was the result of an infection by *Vibrio shiloi*. The demonstration that *V. shiloi* was the causative agent of the disease was established by rigorously satisfying all of Koch's postulates, including the fact that bleached coral in the sea contained the bacterium (Kushmaro et al. 1996,

E. Rosenberg (✉)
Department of Molecular Microbiology and Biotechnology, Tel Aviv University, Ramat Aviv, Israel
e-mail: eros@post.tau.ac.il

A. Kushmaro
Department of Biotechnology Engineering, Ben-Gurion University of the Negev, Beer-Sheva, Israel

Fig. 1 Coral diseases have multiple interacting causes, including the environment, pathogenic microorganisms, and host susceptibility

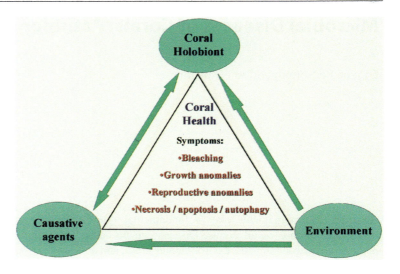

Table 1 Some of the better-studied infectious diseases of corals

Disease	Acronym	Species infected	Pathogen
Bacterial bleaching	BBL	*Oculina patagonica*	*Vibrio shiloi*
Bacterial bleaching	BBL	*Pocillpopora damicornis*	*Vibrio corallilyticus*
Aspergillosis	ASP	Octocorals (Gorgonians)	*Aspergillus sydowii*
Black band	BBD	Many	Consortium
White band I	WBD-I	Acroporids	Unknown
White band II	WBD-II	Many	*Vibrio carchariae (harveyi)*
White plague (Caribbean)	WPD	Many	*Aurantimonas coralicida*
White plague (Red Sea)	WPD	Mainly *Favia*, *Goniastrea*	*Thalassomonas loyana*
White pox	WPX	*Acropora palmata*	*Serratia marcescens*
Yellow band	YBS	Many	*Vibrio* sp.
Brown band	BrB	Many	Ciliate
Porites trematodiasis	PTR	Porites	*Podocotyloides stenometra*
Skeletal eroding band	SEB	Many	*Halofolliculina corallasia*

1997), whereas it was absent in healthy corals. Furthermore, Kushmaro et al. (1998) showed that the laboratory infection and subsequent bleaching only occurred at temperatures above 25°C. Thus, for bleaching to occur, both elevated temperature and the causative agent must be present (Table 2).

2.2 Infection of O. patagonica by V. shiloi

The specific steps in the infection of *O. patagonica* by *V. shiloi* have been studied extensively (Rosenberg and Falkovitz 2004). The bacteria are chemotactic to the coral mucus (Banin et al. 2001a), adhere to a β-galactoside-containing receptor on the coral surface (Toren et al. 1998), penetrate into the epidermal layer (Banin et al. 2000), and multiply intracellularly, reaching 10^8–10^9 cells cm^{-3}. The intracellular *V. shiloi* produces an extracellular linear peptide toxin (PYPVYPPPVVP) that inhibits algal photosynthesis (Banin et al. 2001b). Another important factor for the virulence of *V. shiloi* is the expression of superoxide dismutase (SOD) (Banin et al. 2003). Adhesion, production of the toxin, and expression of SOD are all temperature-dependent reactions, occurring at summer (25–30°C), but not winter (16–20°C) temperatures. Thus, *V. shiloi* cannot infect, multiply, or survive in the coral during winter. Sussman et al. (2003) demonstrated that the marine fireworm *Hermodice carunculata* is a winter reservoir and spring-summer vector for *V. shiloi*.

2.3 Development of Resistance of O. patagonica to V. shiloi

Sometime between 2002 and 2004, it was discovered that *O. patagonica* had become resistant to *V. shiloi*. The evidence for the development of resistance is based on the following (Reshef et al. 2006):

1. From 1995 to 2002, the pathogen *V. shiloi* was readily isolated from 46/50 bleached and bleaching corals collected

Fig. 2 Photograph of *Oculina patagonica* showing bleached (upper) and unbleached polyp (photo by A. Shoob)

10^8–10^9 cells cm^{-2}; now, *V. shiloi* adheres, penetrates the ectoderm, and is rapidly killed.

Not only can we now *not* isolate *V. shiloi* from bleached corals, but molecular techniques failed to recover the 16S rRNA gene from ca. 1,000 clones that were sequenced (Koren and Rosenberg 2006). Recently, Ainsworth et al. (2008) confirmed that *V. shiloi* is not currently present in bleached *O. patagonica*, using FISH technology. Bleaching of *O. patagonica* still occurs. However, it now begins in the spring when the temperature reaches ca. 25°C (well below the 29°C summer temperatures during the previous periods of bleaching in corals). In addition, the geographic distribution of bleached corals has changed and spread north along the coast (Y. Loya, personal communication). Furthermore, in the past, the bleached corals did not develop gonads with ova and sperm (Fine et al. 2002), whereas now they do (R. Armoza personal communication). These differences may be the result of an alternative pathogen or a change in the virulence factor, development of an immune response, such as the release of host cytokines (e.g., Mastroeni and Sheppard 2004), or some combination thereof, though this warrants further study.

from the wild; from 2004 to the present, it has not been possible to isolate *V. shiloi* from bleached or bleaching corals.
2. From 1995 to 2002, all laboratory strains of *V. shiloi* caused bleaching in controlled aquaria experiments; from 2004 to the present, none of the same strains bleach *O. patagonica* in the laboratory.
3. From 1995 to 2002, *V. shiloi* adhered to the corals, penetrated into the ectoderm and multiplied intracellularly to

2.4 Vibrio coralliilyticus

The coral *O. patagonica* is a temperate Mediterranean coral not found on corals reefs, and hence, it was important to test if bacterial pathogens could also cause bleaching of reef corals. A new species, *Vibrio coralliilyticus* (Fig. 4), was initially isolated from a bleached coral, *Pocillopora damicornis*, present on the Zanzibar coral reef (Ben-Haim and Rosenberg 2002; Ben-Haim et al. 2003). When bleaching of *P. damicornis* was observed on the Eilat coral

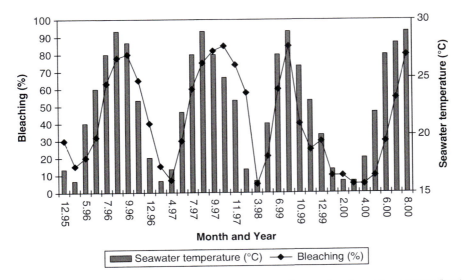

Fig. 3 Seasonal bleaching (diamonds) of *Oculina patagonica* in the Mediterranean Sea as a function of temperature (bars)

Table 2 Bleaching of *O. patagonica* as a function of temperature and inoculum size

Vibrio shiloi (Cells/ml)	Temperature (°C)	Bleaching (%)
0	29	0
120	29	95
0	16	0
10^2–10^8	16	0

Summarized from Kushmaro et al. 1998

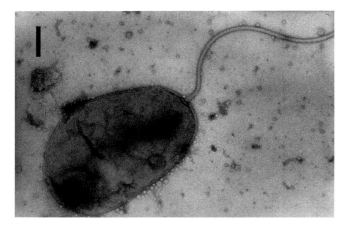

Fig. 4 Electron micrograph of *Vibrio coralliilyticus*, a causative agent of bleaching *Pocillopora damicornis*. Bar, 0.2 µm

reef, *V. coralliilyticus* was isolated from five different bleached coral colonies. It was absent in healthy corals. Using the different *V. coralliilyticus* strains, it was demonstrated that this *Vibrio* species is an etiological agent, bleaching *P. damicornis* in the Indian Ocean and Red Sea.

The infection of *P. damicornis* by *V. coralliilyticus* shows strong temperature dependence (Ben-Haim et al. 2003). Below 22°C, no signs of infection occurred. From 24°C to 26°C, the infection resulted in bleaching, whereas from 27°C to 29°C, the infection caused rapid tissue lysis. Previously, Jokiel and Coles (1990) reported that corals undergo tissue lysis following bleaching when the temperature is elevated. In the case of *V. coralliilyticus* infection, the tissue lysis was shown to be the result of the synthesis of a potent metalloproteinase by the pathogen at temperatures above 26°C.

2.5 Mass Bleaching

What is the etiology of mass bleaching that occurs periodically on coral reefs around the world? On this question, there are at least two different viewpoints. Most coral biologists take the position that high temperature and light act directly on the symbiotic algae to inhibit photosynthesis and produce reactive oxygen species, leading to bleaching (Jones et al. 1998). According to this hypothesis, microorganisms play no role in the bleaching process and that changes in the microbial community of bleached corals are a result, not a cause, of the process. The second viewpoint, taken by certain coral microbiologists (including the authors of this chapter) is that high temperature acts on the coral microorganisms as well as on the coral host causing a change in the microbial community that in some cases, contributes directly or indirectly to bleaching, i.e., the microbial hypothesis of coral bleaching (Rosenberg et al. 2008).

The two examples of bacterial bleaching of corals discussed earlier did not involve mass bleaching on coral reefs. The only evidence for the role of bacteria in mass bleaching events is indirect. Ritchie et al. (1994) enumerated the culturable heterotrophic bacteria of bleached and healthy *Montastraea annularis* coral colonies in the Caribbean. *Vibrio* spp. were never isolated from healthy corals, but represented 30% of the isolates from bleached corals. A similar shift in the bacterial community occurred on *Agaricia* sp. during the 1995/1996 and 1998/1999 bleaching events on the reefs of San Salvador Island, Bahamas (McGrath and Smith 1999). Prior to bleaching, *Vibrio* comprised ca. 20% of the bacterial population, whereas during bleaching, they rose to 40%, and at the height of bleaching, they represented over 60% of the culturable bacteria. When the corals recovered, the *Vibrio* population decreased to 20%. Clearly, these pioneering experiments could not distinguish between the *Vibrio* being the cause or result of the disease.

Bourne and coworkers (2008a) carried out a comprehensive study of changes in the microbial composition of the coral *Acropora millepora* over 2.5 years, which included a severe bleaching event on the Great Barrier Reef (January/February 2002). The data obtained by culture-independent techniques led to several important conclusions: (1) as corals bleached, the microbial community shifted, revealing a correlation between increasing temperature and the appearance of *Vibrio*-affiliated sequences; (2) this shift commenced prior to visual signs of bleaching; and (3) the coral microbial associations shifted again after the bleaching event, returning to a profile similar to the fingerprints obtained prior to bleaching. The authors suggest that microbial shifts can act as an indicator of stress prior to the appearance of visual signs of bleaching. They speculate further that the temperature-induced change in the microbial community prior to bleaching could result in a decrease in antibiotics secreted by symbiotic microorganisms, thereby causing the coral to become more susceptible to bacterial infection. This is in agreement with Ritchie (2006) who demonstrated that bacterial-produced antibiotics in coral mucus can regulate the coral microbial community.

Fig. 5 Black band disease of *Diploria strigosa* (We thank E. Weil for the photograph)

3 Black Band Disease

One of the first coral diseases to be reported in the literature was the black band disease (BBD) (Antonius 1981). It was described as a dark band that moved across apparently healthy corals (Fig. 5). As the band progressed, it left behind a bare coral skeleton. The band appeared to be composed of microbial biomass. The disease affects both hard and soft corals, primarily large reef-building corals with low growth rates. BBD has been reviewed by Richardson (2004). BBD is a widespread coral disease, which infects mainly the massive-framework-building corals (Frias-Lopez et al. 2004). It was first described on the reefs of Belize, the Florida Keys, and Bermuda (Antonius 1973; Garrett and Ducklow 1975). In the following decades, it has also been reported to occur in the Indo-Pacific and Red Sea (Antonius 1985; Al-Moghrabi 2001; Barneah et al. 2007) and in the Great Barrier Reef (Dinsdale 2002; Willis et al. 2004), indicating that this disease has a global distribution.

A seasonal pattern of the disease has been reported, coinciding with elevated water temperatures (Kuta and Richardson 2002). Symptoms of the disease are well described and include the black band (5–30 mm wide) that moves across the surface of the coral colony causing tissue lysis leading to either partial or complete death of the colony (Viehman et al. 2006). Although this disease affects a relatively low proportion of the susceptible coral community (approx 1%), its persistence in the environment and rate of progress across colonies (3 mm – 1 cm/day) make it an important threat to the reef community (Antonius 1981; Carlton and Richardson 1995; Edmunds 1991; Kuta and Richardson 1996).

Studies showed that similar to other stratified microbial communities (e.g. biofilms and microbial mats), the BBD microbial mat oxygen dynamics was dominated by the photosynthetic activity of the cyanobacterium (for review, see Carlton and Richardson 1995). It was found that changes in light intensity caused changes in O_2 concentration, in turn causing an opposite trend in H_2S concentration. This resulted in alterations between oxic and anoxic state inside the BBD mat (Carlton and Richardson 1995). In spite of the similarities between BBD and other microbial mats, BBD is unique in that the <1-mm-thick biofilm migrates horizontally across a living animal substratum that is subsequently killed by an anoxic, sulfide-rich microenvironment, created under the band (Carlton and Richardson 1995). The deadly sulfide is presumably generated by sulfate-reducing bacteria (SRB) (Richardson 2004).

BBD is believed to be caused by a consortium of microorganisms (Antonius 1981; Richardson et al. 1997; Richardson 1998; Dinsdale 2002; Barneah et al. 2007), and may not have a primary pathogen (Richardson 2004). The recent progress in molecular techniques has enabled molecular characterization of the bacterial community associated with BBD (Cooney et al. 2002; Frias-Lopez et al. 2002, 2004; Sekar et al. 2006; Barneah et al. 2007). It was found that the BBD microbial community is dominated, in terms of biomass, by filamentous cyanobacteria (Rützler and Santavy 1983; Bythell et al. 2002; Cooney et al. 2002; Frias-Lopez et al. 2003; Richardson 2004), numerous heterotrophic bacteria (Garrett and Ducklow 1975; Cooney et al. 2002), sulfide-oxidizing bacteria (Ducklow and Mitchell 1979; Viehman and Richardson 2002), SRB (Chet and Mitchell 1975; Garrett and Ducklow 1975; Cooney et al. 2002; Frias-Lopez et al. 2002; Viehman et al. 2006), and marine fungi (Ramos-Flores 1983). This diversity pattern differed according to geographic region, host coral species, and methodology used (Cooney et al. 2002; Frias-Lopez et al. 2004; Sekar et al. 2006). In spite of the fact that this disease was already identified in the 1970s, there is still a wide gap of knowledge pertaining to its etiology.

3.1 Environmental Factors

BBD is most prevalent during the summer when water temperature exceeds 28°C (Kuta and Richardson 1996) and is rarely observed at temperatures below 20°C. BBD is usually limited to shallow waters of 10 m or less (Antonius 1985). Kuta and Richardson (2002) found a positive correlation between elevated nitrite and the occurrence of BBD, but none with nitrate, ammonium, or phosphate. Furthermore, BBD is generally observed on pristine reefs far from obvious anthropogenic perturbations.

3.2 What is the Causative Agent(S) of BBD?

Although the microbiology of BBD has been studied extensively for the last 30 years, using both culturable and molecular methods, the causative agent of the disease remains obscure. The fact that the band contains numerous bacterial species and that no single isolate has yet been shown to cause the disease in healthy corals has led Richardson (2004) to propose that the disease is a result of a pathogenic microbial consortium.

4 Aspergillosis of Gorgonians

Aspergillosis is the only coral disease named after the causative agent of the disease, the fungus *Aspergillus sydowii*. The disease was first observed in the 1980s, causing mass mortalities of sea fans (*Gorgonia flabellum*) in certain areas of the Caribbean, including the coasts of Costa Rica (Guzman and Cortez 1984), Panamá (Garzón-Ferreira and Zea 1992), and Trinidad. In 1995, the disease reappeared near the island of Saba and spread throughout the Caribbean, affecting both *Gorgonia ventalina* and *G. flabellum* (Nagelkerken et al. 1997a). Disease incidence was positively correlated with water depth (Nagelkerken et al. 1997b) and total nitrogen concentration in the water (Baker et al. 2007). Kim and Harvell (2004) have evaluated the rise and fall of aspergillosis in the Florida Keys.

4.1 Disease Signs

Aspergillosis (Fig. 6) exhibits two characteristic pathological signs: (1) purpling of the tissue, together with the loss of polyps and (2) tissue loss where the bare skeleton is either exposed or covered with algae, cyanobacteria, and other organisms (Smith et al. 1996). Galls occur with some, but not all infected sea fans. Microscopically, fungal hyphae appear in all diseased samples.

4.2 Identification of the Pathogen

Taking affected tissues from diseased sea fans and grafting them onto healthy sea fans resulted in some of them developing the disease, demonstrating that the disease was transmissible (Smith and Weil 2004). The fact that not all exposures resulted in the disease indicated the presence of some resistance mechanisms. Resistance of corals to infectious disease will be discussed at the end of the chapter. When infected tissue, containing fungal hyphae, was plated on the growth

Fig. 6 Aspergillosis of *Gorgonia ventalina* (We thank L. Richardson and NOAA for the photograph)

media, a bluish colony developed. Phenotypically, the fungus appeared to be a species of the spore-forming genus *Aspergillus*. Sequential analysis of a 485-base pair fragment of the *trp* C gene demonstrated that the sea fan pathogen was *A. sydowii* (Geiser et al. 1998). Furthermore, the pathogen produced the secondary metabolites sydowinol, sydowinin A, and sydowii (Malmstrom et al. 2001), confirming that the pathogen was *A. sydowii*.

4.3 Source of the Pathogen

Aspergillus sydowii is considered to be a soil microorganism. This raised the question of how it became a pathogen in the marine environment. One of the interesting possibilities is that it arose from African and Asian soils that were carried by dust storms (Garrison et al. 2003). Spores of *A. sydowii* could survive the trip. In fact, Weir-Brush et al. (2004) isolated *A. sydowii* directly from African dust storms in the Caribbean and demonstrated that they caused aspergillosis. It has also been suggested that there are multiple origins of the pathogen, including terrestrial run-off (Rypien 2008).

5 White Plague Diseases

White plague in corals (Fig. 7) was first reported by Dustan (1977) in corals from the Florida Keys, and has

Fig. 7 White plague disease of *Diploria. strigosa* (We thank E. Weil for the photograph)

other coral genera. Although diseased *F. favus* contained 80–10 000 times more culturable bacteria than healthy specimens, none of the 25 isolates initially tested infected healthy corals. Filtration of aquarium water containing diseased *F. favus* indicated that the infectious agent was larger than 0.2 µm and smaller than 3 µm. Microorganisms retained by the 0.2-µm filter did not infect the corals; however, when the retentate was combined with the 0.2-µm filtrate, infection took place. This suggested that the infectious agent requires a filterable factor to cause the disease. Combining the 0.2-µm-filtered water from an aquarium containing a diseased coral with each of the 25 pure cultures, previously obtained from diseased *F. favus*, allowed for the recognition of one strain that caused rapid lysis of *F. favus*. The pathogen was identified as a new species *Thalassomonas loyana* sp. nov. (Barash et al. 2005; Thompson et al. 2006).

been subsequently reported from around the world (Green and Bruckner 2000). The historical basis of the name plague is from its epidemic character and the high rate of mortality. There are two types of plague affecting different nonacroporid coral species. Type I plague usually affects *Copophyllia* and *Mycetpphyllia,* and is characterized by patches of tissue loss on the coral surface at a rate of a few millimeters per day (Dustan 1977). Type II affects *Dichocoenia, Dendrogyra, Stephanocoenia,* and *Monastraea* (Richardson 1998), and is characterized by a higher rate of tissue loss, a few cm per day, and a pattern of infection that starts at the edge and base of the colony.

Dustan (1977) showed that tissue obtained from the diseased corals of the species *Mycetophyllia ferox, M. lamarkiana,* and *Copophyllia natans* could induce the disease in healthy corals of the same species, but not in other species. At that time, no microbiology was performed. Plague type II was first seen in Florida in 1995 (Richardson 1998) and was suggested to be distinct from the original plague by the rate of advancement and by the fact that it begins at the edge of the colony. Denner et al. (2003) identified the new genus and species *Aurantimonas coralicida* from the order *Rhizobiales* as the causative agent of the white plague type II.

A white plague-like disease has affected the Eilat, Israel, coral reefs. Two of the major reef-building coral genera, *Favia* and *Goniastrea*, were most affected by this disease. Approximately 10% of these corals showed progressive signs of the disease or were already dead as a result of the disease. Controlled aquarium experiments demonstrated that the disease is infectious and that transmission from diseased to healthy corals does not require direct contact (Barash et al. 2005). Infection was not genus-specific, because diseased *F. favus* infected three

6 White Band Diseases

White band disease (Fig. 8) is divided into two separate diseases. WBD-I, solely affecting Acroporids, is characterized by a white band that moves from the base of the coral to the tip, leaving behind a bare skeleton that is rapidly colonized by algae (Aronson and Precht 2001). This band advances up the coral branch at the rate of a few millimeters per day (Antonius 1981). The disease has been so devastating to *A. cervicornis* and *A. palmata* in the Caribbean that these two corals have been listed on the US Endangered Species Act (Hogarth 2006). WB1 can be transmitted directly and through vectors, such as snails

Fig. 8 White band disease of *Acropora cervicornis* (We thank E. Weil for the photograph)

(Williams and Miller 2005; Gil-Agudelo et al. 2006). However, a search for the causative agent of WB1, using molecular techniques, failed to reveal any bacterium present in the diseased corals and absent in healthy ones, suggesting that the disease pathogen may be non-bacterial (Casas et al. 2004).

WBD-II white band disease differs from Type I in that in Type II, there is a loss of zooxanthellae (bleaching) adjacent to the dying band of the tissue. Type II affects a variety of coral species (Richardson 1998). A bacterium resembling *Vibrio charcharina* was isolated from the surface mucopolysaccharide layer of the bleached margin of the affected *Acroporas* (Ritchie and Smith 1998). A microbial study of a more recent outbreak of WBD-II demonstrated that the causative agent was a bacterium closely related to *Vibrio harveyi*, a synonym of *V. charcharina* (Gil-Agudelo et al. 2006). It is interesting to note that the two known coral bleaching pathogens are *Vibrio* species.

7 White Pox Disease

White pox disease (Fig. 9), also termed *Acroporid serratiosis* (Patterson et al. 2002) and patchy necrosis (Bruckner et al. 1997), was first documented in 1996 on reefs off Key West, Florida (Holden 1996). White pox has since been observed throughout the Caribbean, exclusively affecting the elkhorn coral *Acropora palmata* (Patterson et al. 2002). The disease is characterized by irregularly shaped distinct white patches devoid of coral tissue. Lesions progress most rapidly, ~2.5 cm²/day, during periods of high seawater temperature. White pox is highly contagious, with nearest neighbors most susceptible to infection (Patterson-Sutherland and Ritchie 2004).

The causative agent of white pox disease is non-pigmented *Serratia marcescens* (Patterson et al. 2002), a member of the Enterobacteriaceae family. *S. marcescens* is an opportunistic pathogen of humans and other animals associated with both waterborne infections in tropical waters and hospital-acquired infections. As *S. marcescens* is common in human feces, it has been suggested that poor waste disposal practices may be the reason for disease outbreaks (Patterson-Sutherland and Ritchie 2004).

8 Yellow Band Disease

Yellow band (sometimes called yellow blotch) disease (Fig. 10) first appears as yellow blotches (= bleached areas) that suffer tissue mortality in the center and form yellow bands or rings all over the colony (Gil-Agudelo et al. 2004; Bruckner and Bruckner 2006). Histological studies revealed the presence of crystalline-like material in the gastric cavity (Santavy and Peters 1997). Outbreaks of yellow blotch disease have been recorded in Panamá in 1996 (Santavy et al. 1999), Netherlands Antilles in 1997 (Cervino and Smith 1997), Puerto Rico, Bermuda, and Grenada, and a similar syndrome has been reported from the southern Arabian Gulf in 1998 (Korrubel and Riegl 1998). Initially observed on *Montastraea faveolata* and *M. annularis* in the Caribbean, the disease signs have been found on many coral species. One of the characteristics of the disease is that it progresses slowly, ~ 0.7 cm/month. It is highly variable (Bruckner and Bruckner 2006).

Fig. 9 White pox disease of *Acropora palmata* (We thank L. Richardson and NOAA for the photograph)

Fig. 10 Yellow blotch disease of *Montastraea faveolata* (We thank E. Weil for the photograph)

Although the causative agent of yellow band disease is unknown, several bacterial strains metabolically related to *Vibrio* were found to be present in the mucus of the diseased corals and absent in healthy ones (Weil et al. 2006). The pathogen appears to attack not only the coral tissue, but also the endosymbiotic zooxanthellae, as demonstrated by the loss of 41–97% of the algae (Cervino and Smith 1997; Cervino et al. 2004b) and decrease in the photosynthetic quantum yield in the diseased parts of the coral. Unlike temperature-induced bleaching, there is no evidence of zooxanthellae in the coral mucus (Cervino et al. 2004a, b).

9 Brown Band Disease

The brown band disease (BrB) was first reported in 2004 on the Great Barrier Reef, Australia (Willis et al. 2004). The disease affected three coral families: Acroporidae, Pocilloporidae, and Faviidae. The macroscopic signs of BrB are a brown zone on the coral, preceded by healthy tissue and followed by exposed white skeleton (Fig. 11). Microscopically, the brown zone reveals large numbers of a protozoan ciliate. The brown color results from the fact that the ciliates contain zooxanthellae, > 50 dinoflagellates per ciliate (Ulstrup et al. 2007). Apparently, the ciliates ingest coral tissue at the lesion interface, accumulating the symbiotic zooxanthellae. The zooxanthellae remain photosynthetically active inside the protozoan (Ulstrup et al. 2007). The ciliate band migrates rapidly along the length of the branching corals from the base to the tip (>5 cm day^{-1}).

The BrB ciliate was classified as a new species belonging to the class Oligohymenophorea, subclass Scuticocilatia (Bourne et al. 2008b). The cells were typically 200–400 μm in length and 20–50 μm in diameter.

10 Porites Trematodiasis

Corals from the genus *Porites* are susceptible to infection by the parasitic worm *Podocotyloides stenometra*, a disease referred to as Porites trematodiasis (PTR). The sign of the disease is pink, swollen nodules on the coral colony (Fig. 12). The disease was first reported in Kaneohe Bay, Oahu (Aeby 2007). PTR was most prevalent in areas with intermediate coral cover, regardless of reef or zone. The primary vector for the disease is the coral-feeding butterfly fish *Chaetodon multicinctus*. Adult worms live in the guts of fish, which pass out the worm eggs with their feces (Aeby 2002).

11 Skeletal Eroding Band

Antonius (2000) described a coral disease that looked somewhat similar to black band disease, but was caused by the protozoan *Halofolliculina corallasia*. One of the characteristics of this disease was the apparent chemical etching of the coral skeleton by the protozoan, from which the name skeletal eroding band (SEB) was derived (Antonius and Lipscomb 2000). It has been suggested that skeleton etching would leave a pattern that, in principle, could be observed in the fossil record (Riegl and Antonius 2003). SEB disease has been observed on coral reefs in Papua New Guinea, Australia and the Red Sea's Gulf of Aqaba (Winkler et al. 2004).

Fig. 11 Brown band disease (We thank E. Weil for the photograph)

Fig. 12 Porites Trematodiasis disease of *Porites* (We thank E. Borneman for the photograph)

12 Coral Resistance to Disease

Corals are sessile organisms living mainly in temperate and warm oceans containing very low concentrations of organic material. To maximize uptake of nutrients, corals contain a high surface to volume ratio. The price that corals pay for their high surface to volume ratio is that they are highly exposed to potentially pathogenic microorganisms. In spite of their lack of a sophisticated immune system, corals are neither less healthy nor suffer higher morbidity than organisms possessing a combinational adaptive immune system. Previous attempts to explain these findings (e.g., Janeway and Medzhitov 2002) have been unsatisfactory (McFall-Ngai 2005).

Generally, animal resistance to infectious disease is discussed in terms of adaptive and innate immunity. As corals do not produce antibodies, it would suggest that they have little or no adaptive immune system. However, there are at least two documented cases where corals became resistant to specific pathogens, *O. patagonica* to *V. shiloi* (Reshef et al. 2006) and Caribbean corals to the causative agent of WPD *A. coralicida* (Richardson and Aronson 2002). The coral probiotic hypothesis was put forth to explain *O. patagonica* resistance to *V. shiloi* (Reshef et al. 2006). Recently, Vollmer and Kline (2008) have reported that ca. 6% of staghorn corals are currently resistant to WBD.

Coral resistance to disease has been reviewed by Mullen et al. 2004. Corals have a robust innate system, which includes (1) the surface mucus layer as a barrier to infection, (2) production of antibacterials, (3) circulating amoebocytes, (4) production of free radicals and other antimicrobial compounds, and (5) bacterial viruses (bacteriophages).

12.1 The Surface Mucus Layer as a Barrier to Infection

All corals secrete a layer of mucus over their surface (SML). Most of the carbon in SML originates from the symbiotic zooxanthellae (Patton et al. 1977), but is secreted by coral epidermal mucus cells as an insoluble complex gel-like layer over the coral surface (Meikle et al. 1988). The structure and function of the SML has been reviewed by Brown and Bythell (2005). Numerous publications indicate that SML contains an abundant and diverse symbiotic microbial community. It has been hypothesized that the microorganisms in the SML are spatially organized, each occupying a specific niche (Ritchie and Smith, 2004). Microorganisms in the SML can protect the coral against invasion by potential pathogens occupying potential sites of adhesion of the pathogen and producing antimicrobial compounds (see below). In addition, discontinuous sloughing off the SML, especially under stress, cleans the coral surface and helps prevent infection. Thus, the SML of corals acts like the mucus-containing bacteria on human skin, in helping to prevent infection by foreign microbes.

12.2 Coral Production of Antibacterials

Mechanical stress applied to the surface of scleratinian corals causes a rapid release of antibacterial material that kills a wide variety of bacterial species, including known coral pathogens (Geffen and Rosenberg 2005). Coral symbiotic bacteria present in the coral mucus are much more resistant to the antibacterial activity than bacteria isolated from the surrounding seawater. It has been suggested that release of the antibacterial material plays a role in preventing infection following mechanical injury, such as the bite of a predator. An active component of the stress-induced antibacterial activity has a molecular weight of 40 kDa and is sensitive to proteases (Y. Geffen, personal communication).

Although classical methods of extraction with organic solvents have yielded numerous antibiotics from sponges and soft corals, these methods have yielded little or no antibiotics from hard corals (Kelman 2004; Kelman et al. 2006). For example, Koh (1997) used extracts of 100 different hard corals and only a few yielded activity. Based on the studies of Geffen and Rosenberg (2005) described earlier, at least part of the reason for the failure to find antibiotics in hard corals is that they are released into the water (which is discharged) when the corals are handled during the harvesting process. Recently, several investigators have found that bacteria present in coral mucus produce antibiotics (Ritchie 2006; Nissimov et al. 2009; Shnit-Orland and Kushmaro 2009). These latter studies provide some support for the coral probiotic hypothesis (Reshef et al. 2006).

12.3 Circulating Amoebocytes

It has recently been shown that circulating granular amoebocytes in corals are not only involved in wound repair and histocompatibility, but also in defense against pathogens (Mydlarz et al. 2008). Both with regard to naturally occurring infections and experimental inoculations with the fungal pathogen *A. sydowii*, there was an inflammatory response characterized by a massive increase in amoebocytes. It was suggested that the ability to mount an inflammatory response may be a contributing factor in the survival of infected sea fans, even at times of elevated temperature stress.

12.4 Production of Free Radicals and Antibacterial Biochemicals

During daytime, the coral holobiont produces large amounts of oxygen, causing the coral tissue and SML to be supersaturated with oxygen (Kuhl et al. 1995), resulting in the formation of oxygen radicals. These radicals, as well as other highly oxidative compounds produced by living tissues, are toxic to microorganisms and are known to play a role in the innate defense of all animals. Bacteria that lack a potent superoxide dismutase are rapidly killed by corals (Banin et al. 2003). Thus, the symbiotic zooxanthellae not only provide energy and nutrients for coral growth, but also help protect corals against infection by potential pathogens by production of toxic oxygen radicals.

Mullen et al. (2004) have suggested that chitinases that are widely distributed in marine invertebrates could be an important source of induced antifungal resistance. Subsequently, it was shown that sea fans have exochitinase activity and release the activity upon injury, agitation, or manipulation of the tissue (Douglas et al. 2007). Furthermore, they showed that the chitinase was active against the sea fan pathogen *A. sydowii*.

Another compound that has been shown to protect sea fans against *A. sydowii* is melanin (Mydlarz et al. 2006). Infected corals build up melanin as a barrier to advancing fungal hyphae, thus preventing its spread (Petes et al. 2003). The melanin deposition results in the dark purple color that is associated with aspergillosis. These three examples of compounds produced by corals to help protect corals against infectious diseases are only the "tip of the iceberg." Just as in other areas of coral disease studies, much research remains to be performed.

12.5 Bacteriophages (Phages)

Although generally not considered a part of the immune system, phages can play a role in preventing transmission of bacterial pathogens and limiting the extent of damage caused by a pathogen. Marine phages are the most diverse and abundant form of life on the planet (Paul and Sullivan 2005) and for every pathogen, there is probably one or more phages that can destroy it. The bacteria–phage interaction is dynamic, so that when the bacterial pathogen increases in concentration, there is a greater chance for the phage to adsorb to its host, multiply, and produce more phages that can interact with additional pathogens, thereby reducing the severity of the infection. At least, in one case, where it was examined (Efrony et al. 2007), the phage remained in the coral mucus for several weeks. In a certain sense, this can lead to adaptive immunity in that a coral that recovers from a pathogen by phage will be immune to further attack of the pathogen for some time.

References

Aeby GS (2002) Trade-offs for the bufferfly fish, *Chaetodon multicinctus*, when feeding on coral prey infected with *Trematode metacercariae*. Behav Ecol Sociobiol 52:158–163

Aeby GS (2007) Spatial and temporal patterns of porites trematodiasis on the Reefs of Kaneohe Bay, Oahu, Hawaii. Bull Mar Sci 80:209–218

Ainsworth TD, Fine M, Roff G, Hoegh-Guldberg O (2008) Bacteria are not the primary cause of bleaching in the mediterranean coral *Oculina patagonica*. ISME J 2:67–73

Al-Moghrabi M (2001) Unusual Black Band Disease (BBD) outbreak in the northern tip of the Gulf of Aquaba (Jordan). Coral Reefs 19:330–331

Antonius A (1973) New observations on coral destruction in reefs. In: tenth meeting of the Association of Island Marine Laboratories of the Caribbean, University of Puerto Rico (Mayaguez), Puerto Rico, 1973, p 3

Antonius A (1981) The "band" diseases in coral reefs. In: Proceedings of the 4th international coral reef symposium, vol 2, Manila, 1981, pp 7–14

Antonius A (1985) Coral diseases in the Indo-Pacific: a first record. PSZNI Mar Ecol 6:197–218

Antonius A (2000) *Halofolliculina corallasia*, a new coral-killing ciliate on Indo-Pacific reefs. Coral Reefs 18:300

Antonius A, Lipscomb D (2000) First protozoan coral-killer identified in the Indo-Pacific. Attol Res Bull 48:1–21

Aronson RB, Precht WF (2001) White-band disease and the changing face of Caribbean Coral Reefs. Hydrobiologia 460:25–38

Baker DM, MacAvoy SE, Kim K (2007) Relationship between water quality, $\delta^{15}N$, and aspergillosis of Caribbean sea fan corals. Mar Ecol Prog Ser 343:123–130

Banin E, Israely T, Kushmaro A, Loya Y, Orr E, Rosenberg E (2000) Penetration of the coral-bleaching bacterium *Vibrio shiloi* into *Oculina patagonica*. Appl Environ Microbiol 66:3031–3036

Banin E, Israely T, Fine M, Loya Y, Rosenberg E (2001a) Role of endosymbiotic zooxanthellae and coral mucus in the adhesion of the coral-bleaching pathogen *Vibrio shiloi* to its host. FEMS Microbiol Lett 199:33–37

Banin E, Sanjay KH, Naider F, Rosenberg E (2001b) Proline-rich peptide from the coral pathogen *Vibrio shiloi* that inhibits photosynthesis of zooxanthellae. Appl Environ Microbiol 67:1536–1541

Banin E, Vassilakos D, Orr E, Martinez RJ, Rosenberg E (2003) Superoxide dismutase is a virulence factor produced by the coral bleaching pathogen *Vibrio shiloi*. Curr Microbiol 46:418–422

Barash Y, Sulam R, Loya R, Rosenberg E (2005) Bacterial strain BA-3 and a filterable factor cause a white plague-like disease in corals from the Eilat Coral Reef. Aquat Micro Ecol 40:183–189

Barneah O, Ben-Dov E, Kramarsky-Winter E, Kushmaro A (2007) Characterization of black band disease in Red Sea stony corals. Environ Microbiol 9:1995–2006

Ben-Haim Y, Rosenberg E (2002) A novel *Vibrio* sp. pathogen of the coral *Pocillopora damicornis*. Mar Biol 141:47–55

Ben-Haim Y, Zicherman-Keren M, Rosenberg E (2003) Temperature-regulated bleaching and lysis of the coral *Pocillopora damicornis* by the novel pathogen *Vibrio coralliilyticus*. Appl Environ Microbiol 69:4236–4242

Bourne D, Iida Y, Uthicke S, Smith-Keune C (2008a) Changes in coral-associated microbial communities during a bleaching event. ISME 2:350–363

Bourne DG, Boyett HV, Henderson ME, Muirhead A, Willis BL (2008b) Identification of a ciliate (Oligohymenophorea: Scuticociliatia) associated with brown band disease on corals of the Great Barrier Reef. Appl Environ Microbiol 74:883–888

Bourne DG, Garren M, Work MT, Rosenberg E, Smith GW, Harvell CD (2009) Microbial disease and the coral holobiont. Trends Microbiol 17:554–562

Brown BE, Bythell JC (2005) Perspectives on mucus secretion in reef corals. Mar Ecol Prog Ser 296:291–309

Bruckner AW, Bruckner RJ (2006) Consequences of Yellow Band Disease (YBD) on *Montastraea annularis* (species complex) populations on remote reefs off Mona Island, Puerto Rico Dis Aquat Organ 69:67–73

Bruckner AW, Bruckner RJ, Williams EH (1997) Spread of a blackband disease epizootic through the reef system in St. Ann's Bay, Jamaica. Bull Mar Sci 61:919–928

Bythell JC, Barer MR, Cooney RP, Guest JR, O'Donnell AG, Pantos O et al (2002) Histopathological methods for the investigation of microbial communities associated with disease lesions in reef corals. Lett Appl Microbiol 34:359–364

Carlton RG, Richardson LL (1995) Oxygen and sulfide dynamics in a horizontally migrating cyanobacterial mat: black band disease of corals. FEMS Microbiol Ecol 18:155–162

Casas V, Kline DI, Wegley L, Yu Y, Breitbart M, Rohwer F (2004) Widespread association of a rickettsiales-like bacterium with reef-building corals. Environ Microbiol 6:1137–1148

Cervino JM, Smith G (1997) Corals in Peril. Ocean Realm 2:33–34

Cervino JM, Hayes RL, Polson SW, Polson SC, Goreau TJ, Martinez RJ, Smith GW (2004a) Relationship of *Vibrio* species infection and elevated temperatures to yellow blotch/band disease in Caribbean corals. Appl Environ Microbiol 70:6855–6864

Cervino JM, Hayes RL, Goreau TJ, Smith GW (2004b) Zooxanthellae regulation in yellow bloch/band and other coral diseases contrasted with temperature related bleaching: in situ destruction vs expulsion. Symbiosis 37:63–85

Chet I, Mitchell R (1975) Bacterial attack on corals in polluted seawater. Microbiol Ecol 2:227–233

Cooney RP, Pantos O, Le Tissier MDA, Barer MR, O'Donnell AG, Bythell JC (2002) Characterization of the bacterial consortium associated with black band disease in coral using molecular microbiological techniques. Environ Microbiol 4:401–413

Denner EBM, Smith G, Busse HJ, Schumann P, Narzt T, Polson SW, LubitzW RLL (2003) *Aurantimonas coralicida* gen. nov., sp. nov., the causative agent of white plague type II on Caribbean scleractinian corals. Int J Syst Evol Microbiol 53:1115–1122

Dinsdale EA (2002) Abundance of black-band disease on corals from one location on the Great Barrier Reef: a comparison with abundance in the Caribbean region. In: Proceedings 9th international coral reef symposium, Vol 2, Bali, Indonesia 23–27 October 2000, pp 1239–1243

Douglas NL, Mullen KM, Talmage SC, Harvell CD (2007) Exploring the role of chitinolytic enzymes in the sea fan coral, *Gorgonia ventalina*. Mar Biol 150:1137–1144

Ducklow HW, Mitchell R (1979) Observations on naturally and artificially diseased tropical corals: a scanning electron microscopy study. Microb Ecol 5:215–223

Dustan P (1977) Vitality of reef coral populations off Key Largo, Florida: recruitment and mortality. Environ Geol 2:51–58

Edmunds PJ (1991) Extent and effect of black band disease on a Caribbean reef. Coral Reefs 10:161–165

Efrony R, Loya Y, Bacharach E, Rosenberg E (2007) Phage therapy of coral disease. Coral Reefs 26:7–13

Fine M, Loya Y (1995) The coral *Oculina patagonica*: a new immigrant to the Mediterranean coast of Israel. Isr J Zool 41:81

Fine M, Banin E, Israely T, Rosenberg E, Loya Y (2002) Ultraviolet (UV) radiation prevents bacterial bleaching of the Mediterranean coral *Oculina patagonica*. Mar Ecol Progr Ser 226:249–254

Frias-Lopez J, Zerkle AL, Boneheyo GT, Fouke BW (2002) Partitioning of bacterial communities between sea water and healthy black band diseased, and dead coral surfaces. Appl Environ Microbiol 68:2214–2228

Frias-Lopez J, Bonheyo GT, Jin Q, Fouke BW (2003) Cyanobacteria associated with coral black band disease in Caribbean and Indo-Pacific reefs. Appl Environ Microbiol 69:2409–2413

Frias-Lopez J, Klaus JS, Bonheyo GT, Fouke BW (2004) Bacterial community associated with black band disease in corals. Appl Environ Microbiol 70:5955–5962

Garrett P, Ducklow H (1975) Coral diseases in Bermuda. Nature 253:349–350

Garrison VH, Shinn EA, Foreman WT, Griffin DW, Holmes CW, Kellogg CA, Majewski MS, Richardson LL, Ritchie KB, Smith GW (2003) African and Asian dust: from desert soils to coral reefs. Bioscience 53:469–480

Garzón-Ferreira J, Zea S (1992) A mass mortality of *Gorgonia ventalina* (Cnidaria: Gorgoniidae) in the Santa Marta area, Caribbean coast of Columbia. Bull Mar Sci 50:522–526

Gil-Agudelo DL, Smith GW, Garzon-Ferreira J, Weil E, Petersen D (2004) Dark spots disease and yellow band disease, two poorly known coral diseases with high incidence in Caribbean Reefs. In: Rosenberg E, Loya Y (eds) Coral health and disease. Springer, Berlin, pp 337–349

Gil-Agudelo DL, Smith GW, Weil E (2006) The white band disease type II pathogen in Puerto Rico. Rev Biol Trop 54:59–67

Geffen Y, Rosenberg E (2005) Stress-induced rapid release of antibacterials by Scleractinian corals. Mar Biol 146:931–935

Geiser DM, Taylor JW, Ritchie KB, Smith GW (1998) Cause of sea fan death in the West Indies. Nature 394:137–138

Green EP, Bruckner AW (2000) The significance of coral disease pizootiology for coral reef conservation. Biol Conserv 96:347–361

Guzman HM, Cortez J (1984) Mortand de *Gorgonia flabellum* Linnaeus (Octocorallia: Gorgoniidae) en la Costa Caribe de Costa Rica. Rev Biol Trop 32:305–308

Harvell CD, Jordan-Dalgreen E, Merkel S, Rosenberg E, Raymundo L, Smith G, Weil E, Willis B (2007) Coral diseases, environmental drivers and the balance between coral and microbial associates. Oceanography 20:172–195

Hogarth WT (2006) Endangered and threatened species: final listing determinations for the Elkhorn coral and Staghorn coral. Federal Register, pp 26852–26861

Holden C (1996) Coral disease hot spot in the Florida keys. Science 274:2017

Janeway CA, Medzhitov R (2002) Innate immune recognition. Annu Rev Immunol 20:197–216

Jokiel PL, Coles SL (1990) Response of Hawaiian and othe Indo-Pacific reef corals to elevated temperature. Coral Reefs 8:155–162

Jones RJ, Hoegh-Guldberg O, Larkum AWD, Schreiber U (1998) Temperature-induced bleaching of corals begins with impairment of the CO_2 fixation mechanism in zooxanthellae. Plant Cell Environ 21:1219–1230

Kelman D (2004) Antimicrobial activity of sponges. In: Rosenberg E, Loya Y (eds) Coral health and disease. Springer, Berlin, pp 243–258

Kelman D, Kashman Y, Rosenberg E, Kushmaro A, Loya Y (2006) Antimicrobial activity of Red Sea corals. Mar Biol 149:357–363

Kim H, Harvell CD (2004) The rise and fall of a six-year coral-fungal epizootic. Am Nat 164:552–563

Koh EGL (1997) Do scleractinian corals engage in chemical warfare against microbes? J Chem Ecol 23:379–398

Koren O, Rosenberg E (2006) Bacteria associated with mucus and tissues of the coral *Oculina patagonica* in summer and winter. Appl Environ Microbiol 72:5254–5259

Korrubel JL, Riegl BR (1998) A new disease from the southern Arabian Gulf. Coral Reefs 17:22

Kuhl M, Cohen Y, Dalsgard T, Jorgensen BB, Revsbech NP (1995) Microenvironment and photosynthesis of zooxanthellae in scleractinian corals studied with microsensors for O_2, pHand light. Mar Ecol Prog Ser 117:159–172

Kushmaro A, Loya Y, Fine M, Rosenberg E (1996) Bacterial infection and coral bleaching. Nature 380:396

Kushmaro A, Rosenberg E, Fine M, Loya Y (1997) Bleaching of the coral *Oculina patagonica* by *Vibrio* AK-1. Mar Ecol Prog Ser 147:159–165

Kushmaro A, Rosenberg E, Fine M, Ben-Haim Y, Loya Y (1998) Effect of temperature on bleaching of the coral *Oculina patagonica* by *Vibrio shiloi* AK-1. Mar Ecol Prog Ser 171:131–137

Kuta KG, Richardson LL (1996) Abundance and distribution of black-band disease on Coral Reefs in the Northern Florida Keys. Coral Reefs 15:219–223

Kuta K, Richardson L (2002) Ecological aspects of black band disease of corals: relationships between disease incidence and environmental factors. Coral Reefs 21:393–398

Malmstrom J, James SC, Polson SW, Smith GW, Frisvad JC (2001) Study of secondary metabolites associated with virulent and non-virulent strains of *Aspergillus sydowii*: sea fan pathogen. In: Proceedings of the 8th symposium on the natural history of the Bahamas, vol 8, San Salvador Island, 2001, pp 48–51

Mastroeni P, Sheppard M (2004) Salmonella infections in the mouse model: host resistance factors and in vivo dynamics of bacterial spread and distribution in the tissue. Microbes Infect 6:398–405

McFall-Ngai M (2005) The interface of microbiology and immunology: a comparative analysis of the animal kingdom. In: McFall-Ngai M, Henderson B, Ruby ED (eds) The Influence of Cooperative Bacteria on Animal Host Biology. Cambridge University Press, Cambridge, pp 35–57

McGrath TA, Smith GW (1999) Community shifts in the surface mucopolysaccharide layer microbiota of Agaricia sp. during the 1995/6 and 1998/9 bleaching events on patch reefs of San Salvador Island, Bahamas. In: Cortès JN, Fonseca AG (eds). Proceedings of the 29th meeting of the Association of Marine Laboratories of the Caribbean, 2000 Cumana, Venezuela. CIMAR, Universidad de Costa Rica: San Jose, Costa Rica

Meikle P, Richards GN, Yellowlees D (1988) Structural investigations on the mucus from six species of coral. Mar Biol 99:187–193

Mullen K, Peters E, Harvell CD (2004) Coral resistance to disease. In: Rosenberg E, Loya Y (eds) Coral health and disease. Springer, Berlin, pp 377–399

Mydlarz LD, Jones LE, Harvell CD (2006) Innate immunity, environmental drivers and disease ecology of marine and freshwater invertebrates. Ann Rev Ecol Evol Syst 37:251–288

Mydlarz LD, Holthouse SF, Peters EC, Harvell CD (2008) Cellular responses in sea fan corals: granular amoebocytes react to pathogen and climate stressors. PLoS ONE 4(10):e7298

Nagelkerken I, Buchan K, Smith GW, Bonair K, Bush P, Garzon-Ferreira J, Botero L, Gayle P, Herberer C, Petrovic C, Pors L and Yoshioka P (1997a) Widespread disease in Caribbean sea fans: I. Spreading and general characteristics. Proceedings of the. 8th international coral reef symposium, vol 1, Panama pp 679–682

Nagelkerken I, Buchan K, Smith GW, Bonair K, Bush P, Garzon-Ferreira J, Botero L, Gayle P, Harvell CD, Herberer C, Kim K, Petrovic C, Pors L, Yoshioka P (1997b) Widespread disease in Caribbean sea fans: II. Patterns of infection and tissue loss. Mar Ecol Prog Ser 160:255–263

Nissimov J, Rosenberg E, Munn C (2009) Antimicrobial properties of resident coral mucus bacteria of *Oculina patagonica*. FEMS Microb Lett 292:210–215

Patterson KL, Porter JW, Ritchie KB, Polson SW, Mueller E, Peters EC, Santavy DL, Smith GW (2002) The etiology of white pox, a lethal disease of the Caribbean elkhorn coral, *Acropora palmata*. Proc Natl Acad Sci U S A 99:8725–8730

Patterson-Sutherland K, Ritchie KB (2004) White pox disease of the Caribbean elkhorn coral, *Acropora palmate*. In: Rosenberg E, Loya Y (eds) Coral health and disease. Springer, Berlin, pp 289–300

Patton JR, Abraham S, Benson AA (1977) Lipogenesis in the intact coral *Pocillopora capitata* and its isolated zooxanthellae: evidence for a light-driven carbon cycle between symbiont and host. Mar Biol 44:235–247

Paul JH, Sullivan MB (2005) Marine phage genomics: what have we learned? Curr Opin Biotech 3:299–307

Petes LE, Harvell CD, Peters EC, Webb MAH, Mullen KM (2003) Pathogens compromise reproduction and induce melanization in Caribbean sea fans. Mar Ecol Prog Ser 264:167–171

Ramos-Flores T (1983) Lower marine fungus associated with black line disease in star corals (*Montastrea annularis* E. & S.). Biol Bull 165:429–435

Reshef L, Koren O, Loya Y, Zilber-Rosenberg I, Rosenberg E (2006) The coral probiotic hypothesis. Environ Microbiol 8:2068–2073

Richardson LL (1998) Coral diseases: what is really known? Trends Ecol Conserv 13:438–443

Richardson LL (2004) Black band disease. In: Rosenberg E, Loya Y (eds) Coral health and disease. Springer, Berlin, pp 325–336

Richardson LL, Aronson RB (2002) Infectious diseases of reef corals. In: Proceedings of the 9th international coral reef symposium, Indonesia, 2002

Richardson LL, Kuta KG, Schnell S, and Carlton RG (1997) Ecology of the black band disease microbial consortium. Proc 8th Int Coral Reef Symp, Smithsonian Tropical Research Institute, Balboa, Panama 1:597–600

Riegl B, Antonius A (2003) *Halofolliculina corallasia* eroding band (SEB): a coral disease with fossilization potential. Coral Reefs 22:48

Ritchie KB (2006) Regulation of microbial populations by coral surface mucus and mucus-associated bacteria. Mar Ecol Prog Ser 322:1–14

Ritchie KB, Smith GW (1998) Type II white-band disease. Revista de Biol Trop 46:199–203

Ritchie KB, Smith GW (2004) Microbial communities of coral surface mucopolysaccharide layers. In: Rosenberg E, Loya Y (eds) Coral health and disease. Springer, Berlin, pp 259–264

Ritchie KB, Dennis JH, McGrath T, Smith GW (1994) In: Kass LB (ed). Proceedings of the 5th symposium of the national history of the bahamas. vol 5. Bahamian field station, San Salvador, pp 75–80

Rohwer F, Seguritan V, Azam F, Knowlton N (2002) Diversity and distribution of coral-associated bacteria. Mar Ecol Prog Ser 243:1–10

Rosenberg E, Falkovitz L (2004) The *Vibrio shiloi/Oculina patagonica* model system of coral bleaching. Ann Rev Microbiol 58:143–159

Rosenberg E, Koren O, Reshef L, Efrony R, Zilber-Rosenberg I (2007) The role of microorganisms in coral health, disease and evolution. Nat Rev Microbiol 5:355–362

Rosenberg E, Kushmaro A, Kramarsky-Winter E, Banin E, Loya Y (2009) The role of microorganisms in coral bleaching, ISME J 3:139–146

Rützler K, Santavy DL (1983) The black band disease of Atlantic reef corals. I. Description of the cyanophyte pathogen. Mar Ecol 4:301–319

Rypien KL (2008) African dust is an unlikely source of *Aspergillis sydowii*, the causative agent of sea fan disease. Mar Ecol Prog Ser 367:125–131

Santavy DL, Peters EC (1997) Microbial pests: coral disease in the Western Atlantic. Proceeding 8th international coral reef symposium, vol 1, pp 607–612

Santavy DL, Peters EC, Quirolo C, Porter JW, Bianchi CN (1999) Yellow-Blotch disease outbreak on reefs of the San Blas Islands. Panama. Coral Reefs 18:97

Sekar R, Mills DK, Remily ER, Voss JD, Richardson LL (2006) Microbial communities in the surface mucopolysaccharide layer and the black band microbial mat of black band-diseased *Siderastrea siderea*. Appl Environ Microbiol 72:5963–5973

Shnit-Orland M, Kushmaro A (2009) Coral mucus-associated bacteria: a possible first line of defense. FEMS Microbiol Ecol 67:371–380

Smith GW, Ives LD, Nagelkerken IA, Ritchie KB (1996) Caribbean sea fan mortalities. Nature 383:487

Smith GW, Weil E (2004) Aspergillosis of gorgonians. In: Rosenberg E, Loya Y (eds). Coral health and disease. Springer, New York, pp 279–288

Stedman TL (2005) Stedmans' medical dictionary, 27th edn. Lippencott Williams & Wilkins, Philadelphia

Sussman M, Loya Y, Fine M, Rosenberg E (2003) The marine fireworm *Hermodice carunculata* is a winter reservoir and spring-summer vector for the coral-bleaching pathogen *Vibrio shiloi*. Environ Microbiol 5:250–55

Thompson FL, Barash Y, Sawabe T, Sharon G, Swings J, Rosenberg E (2006) *Thalassomonas loyana* sp. nov., a causative agent of the white plague-like disease of corals on the Eilat coral reef. IJSEM 56:365–368

Toren A, Landau L, Kushmaro A, Loya Y, Rosenberg E (1998) Effect of temperature on adhesion of *Vibrio* strain AK-1 to *Oculina patagonica* and on coral bleaching. Appl Environ Microb 64:1379–1384

Ulstrup KE, Kühl M, Bourne DG (2007) Zooxanthellae harvested by ciliates associated with brown band syndrome of corals remain photosynthetically competent. Appl Environ Microbiol 73:1968–75

Viehman S, Richardson LL (2002) Motility patterns of *Beggiatoa* and *Phormidium corallyticum* in black band disease. Proc 9th Int Coral Reef Symp, Bali, Indonesia 2:1251–1255

Viehman S, Mills DK, Meichel GW, Richardson LL (2006) Culture and identification of *Desulfovibrio* spp. from corals infected by black band disease on Dominican and Florida Keys reefs. Dis Aquat Organ 69:119–127

Vollmer SV and Kline DI (2008) Natural disease resistance in threatened staghorn corals. PloS ONE 3: doi:10.1371/journal.pone 0003718

Weil E, Smith GW, Gil-Agudelo DL (2006) Status and progress in coral reef disease research. Dis Aquat Organ 69:1–7

Weir-Brush JF, Garrison VH, Smith GW, Shinn EA (2004) The relationship between Gorgonian coral (Cnidaria: Gorgonacea) diseases and African dust storms. Aerobiologia 20:119–126

Williams DE, Miller MW (2005) Coral disease outbreak: pattern, prevalence and transmission in *Acropora cervicornis*. Mar Ecol Prog Ser 301:119–128

Willis B, Page CA, Dinsdale EA (2004) Coral disease on the Great Barrier Reef. In: Rosenberg E, Loya Y (eds) Coral health and disease. Springer, Berlin, pp 69–104

Winkler R, Antonius A, Renegar A (2004) The skeleton eroding band disease on Coral Reefs of Aqaba, Red Sea. Mar Ecol 25:129–144

Zilber-Rosenberg I, Rosenberg E (2008) Role of microorganisms in the evolution of animals and plants: the hologenome theory of evolution. FEMS Microbiol Rev 32:723–735

Coral Reef Diseases in the Atlantic-Caribbean

Ernesto Weil and Caroline S. Rogers

1 Introduction

Coral reefs are the jewels of the tropical oceans. They boast the highest diversity of all marine ecosystems, aid in the development and protection of other important, productive coastal marine communities, and have provided millions of people with food, building materials, protection from storms, recreation and social stability over thousands of years, and more recently, income, active pharmacological compounds and other benefits. These communities have been deteriorating rapidly in recent times. The continuous emergence of coral reef diseases and increase in bleaching events caused in part by high water temperatures among other factors underscore the need for intensive assessments of their ecological status and causes and their impact on coral reefs.

In the last few decades, coral reefs around the world have experienced significant declines with changes in composition, structure, and function attributable to one or more natural and anthropogenic interacting factors (Harvell et al. 1999, 2005, 2007; Hoegh-Guldberg 1999; Ostrander et al. 2000; Hayes et al. 2001; Jackson et al. 2001; Gardner et al. 2003; Hughes et al. 2003; Pandolfi et al. 2003; Weil et al. 2003; Willis et al. 2004; Weil 2004; Sutherland et al. 2004; Wilkinson 2006; Rogers and Miller 2006; Hoegh-Guldberg et al. 2007; Lesser et al. 2007; Rogers et al. 2008a, b; Miller et al. 2009; Cróquer and Weil 2009b). The effect of each factor, or combination of factors, varies within regions and across time. A recent report indicated that 32% of zooxanthellate scleractinian corals face an elevated risk of extinction due mainly to bleaching and disease that seem positively and significantly correlated with elevated sea water temperature and further exacerbated by local anthropogenic stressors (Carpenter et al. 2008).

E. Weil (✉)
Department of Marine Sciences, University of Puerto Rico,
P.O. Box 9000, Mayagüez, PR 00680, USA
e-mail: eweil@caribe.net

C.S. Rogers
USGS Caribbean Field Station, 1300 Cruz Bay Creek, St. John,
VI, 00830, USA

A significant die-off of acroporids and other corals in the Florida Keys and the Dry Tortugas occurred during severe cold weather in the winter of 1977–1978 (Roberts et al. 1982). Some presumably minor restricted disease outbreaks that occurred in the 1970s in the Florida Keys and the Virgin Islands were followed by two apparently concurrent biotic, wide-geographic epizootic events during the late 1970s and early 1980s. [An epizootic or disease outbreak is defined as "an unexpected increase in disease or mortality in a time or place where it does not normally occur or at a frequency greater than previously observed" (Wobeser 1994; Work et al. 2008a, b; Woodley et al. 2008)]. The white band disease (WBD) outbreak affected *Acropora palmata* and *A. cervicornis*, two of the most abundant and important reef-building corals in the region (Gladfelter 1982), and an unknown pathogen produced the mass-mortality of the black sea urchin *Diadema antillarum*, an important and abundant species in all tropical and subtropical shallow marine habitats of the western Atlantic and Caribbean (Lessios et al. 1984a, b). Coral and sea urchin populations experienced over 90% mortalities over their geographic range, resulting in significant losses in genetic diversity, coral cover, and spatial heterogeneity of coral reefs across the Caribbean.

Almost 30 years after these events, the affected coral and urchin species have not recovered to their former densities and populations structures in reefs off Puerto Rico (Weil et al. 2003, 2005). Furthermore, hurricanes, storms, and two major widespread bleaching events in 1998 and 2005 led to further localized mortalities of surviving acroporids and other major reef-building species (Miller et al. 2006, 2009; Wilkinson and Souter 2008; McClanahan et al. 2009; Cróquer and Weil 2009a).

Disease is considered here as "any impairment to health resulting in physiological dysfunction," involving an interaction between a host, an agent i.e., pathogen, environment, genetics, and the environment (Martin et al. 1987; Wobeser 1994). These three components must interact in a precise way for disease to occur. This definition includes both non-infectious (produced by genetic mutations, malnutrition, and/or environmental factors), and infectious diseases (produced by pathogens). The host is the organism affected by the disease (e.g., coral, octocoral), the agent(s) is/are the factor(s) that directly or indirectly cause(s) disease. Infectious agents are capable of causing infection and may be transmissible between hosts (Stedman 1976; Wobeser 2006). The environment is considered to be the third factor of the disease triad and provides the stage where host–agent interactions occur (Wobeser 2006; Work et al. 2008a, b).

In a recent controversial report, Lesser et al. (2007) emphasized the importance of the environmental drivers causing disease outbreaks and questioned the generalized conclusion that diseases of corals are caused by a primary pathogen and are infectious in nature. They suggested that coral diseases are most often a secondary phenomenon caused by opportunistic pathogens after physiological stress produced by changing environmental conditions. Although it is important to understand the role of environmental co-factors, which in some cases could render corals more susceptible to disease, it is important to exercise a balanced approach that would increase our understanding of the interactions among host, agent, and environment (Work et al. 2008b). By definition, "if an organism develops an infectious disease, then there has to have been some breakdown in host defenses to allow pathogens to establish. By this logic, all diseases (e.g., common cold, TB, AIDS) are opportunist. The distinction that needs to be made (and that we can make in other animals but not corals yet) is whether we can measure or quantify the decrease in host response prior to development of disease, and thus make the case that this animal had a quantifiable suppression in immune status before development of disease" (T. Work, personal communication 2010; Work et al. 2008a).

Studying diseases in the marine environment has proven to be challenging. For example, it is difficult to collect samples without contamination and variability in sample collection methods may confound comparative results. Additionally, most marine bacteria, possibly including pathogens, are difficult to culture or are unable to be cultured today, making their identification, laboratory manipulation, and testing of Koch's postulates difficult. Even if a putative pathogen is identified, testing which environmental variable or driver is responsible for its emergence is extremely difficult. Furthermore, recent evidence indicates that bacterial and fungal communities living in association with coral tissues are highly dynamic and different bacteria and fungi may produce similar physiological responses (i.e., disease signs) (Ritchie 2006; Voss et al. 2007; Toledo-Hernandez et al. 2008; Sunagawa et al. 2009).

Corals are "ecological communities" (holobionts), harboring high diversities and abundances of bacteria, zooxanthellae, endolithic algae, fungi, and other boring invertebrates interacting in complex ways (Knowlton and Rohwer 2003; Ritchie 2006; Kimes et al. 2010). Changes in environmental conditions would presumably affect this physiological equilibrium by changing the resident microbial community, which could enhance susceptibility to infectious agents and/or weakening of the host immune system, which could render corals more susceptible to infection, or loss of zooxanthellae (Harvell et al. 1999, 2002, 2007; Ritchie 2006; Mydlarz et al. 2009; Thurber et al. 2009). The compromised-host hypothesis suggests that rising ocean temperatures may increase the number and prevalence of coral diseases by making corals more susceptible to ubiquitous pathogens or by causing shifts in microbial communities making some of them pathogenic (Rosenberg and Ben-Haim 2002).

Few quantitative studies have attempted to relate the emergence, prevalence, and incidence of coral reef diseases with deterioration/change in environmental quality. This requires either large spatial and/or long temporal scales to produce reliable results. Short-term studies, however, have established significant correlations between increasing sea water temperatures and increases in prevalence of white syndrome (WS) and black band disease (BBD) in the Great Barrier Reef (GBR), and prevalence and virulence of Caribbean yellow band disease (YBD) and white patches (Boyett et al. 2007; Bruno et al. 2007; Weil 2008; Harvell et al. 2009; Muller et al. 2008; Weil et al., in press).

Very little is known about the composition and dynamics of the natural microbial communities living in association with most reef organisms (but see Rohwer et al. 2001, 2002; Rosenberg 2004; Ritchie 2006; Rosenberg et al. 2007; Lesser et al. 2007; Sunagawa et al. 2009; Thurber et al. 2009). A clear determination of causality, therefore, is difficult to accomplish. This would require controlled experiments, which is a problem when moving holobiont colonies or fragments from the field into laboratory conditions. Natural changes in composition of bacterial populations may follow the initial infection by a disease-causing agent. Other bacteria could become dominant and pathogenic producing similar signs (Bourne et al. 2007; Voss et al. 2007; Toledo-Hernandez et al. 2008; Sunagawa et al. 2009). The fact that some coral epizootic events occurred over large geographic scales within short periods of time, or simultaneously, suggests a response of already present bacteria (or other pathogens) to similar changes in environmental conditions favoring the infectious disease outbreak. We are currently unable to fully explain the source(s) and sudden emergence of the majority of diseases and/or the outbreaks in coral reef organisms.

Fig. 1 (continued) aspergillosis and red band disease in *G. ventalina* (**j, k**), other compromised health conditions in *E. caribbaeorum* (**i**) and *P. nutans* (**l**), and purple spots produced by an unknown protozoan (Labyrinthulomycote) in *G. ventalina* (**m**). Disease conditions in the hydrocoral *M. complanata* (**n**), the vase sponge *X. muta* (**o**), the tubular sponge *C. vaginalis* (**p**) and the zoanthid *P. caribbaeorum* (**q**). Caribbean coralline lethal orange disease (**r**) and crustose coralline white band disease in *N. accretum* (**s**) (Photos E. Weil)

Fig. 1 Photographs of other conditions affecting Caribbean corals and other reef organisms. Growth anomalies in *D. strigosa* (**a**, **b**) (hyperplasia) and *A. palmata* (**c**) (neoplasia) (photo courtesy of E. Peters), pigmentation responses from unknown conditions producing tissue mortality in *M. faveolata* (**d**), *S. siderea* (**e**) and *D. labyrinthiformis* (**f**), white syndromes in *S. siderea* (**g**) and *M. faveolata* (**h**). Diseases in other Cnidarians include

The goal of this chapter is to present a historical perspective of diseases in coral reefs and a summary of the current distribution and status of coral diseases in the Atlantic-Caribbean along with recommendations for future research. There are several other important biological members of the coral reef community such as hydrocorals, sponges, zoanthids, sea urchins, and crustose coralline algae (CCA) that are affected by diseases (see Fig. 1) that will not be discussed in this chapter for lack of space. They will be discussed in a future publication. We have adopted the coral reef disease nomenclature recently updated in Work et al. (2008a), Raymundo et al. (2008), Beeden et al. (2008), and Weil and Hooten (2008b).

2 Historical Perspective

Diseases of coral reef organisms have likely been around for millions of years and may have produced significant population mortalities in the past, but this cannot be confirmed in the fossil record. The emergence of coral reef diseases in the Caribbean in the past few decades appears to be unprecedented in the geological record. Limited paleontological evidence suggests that the white band disease (WBD) outbreak in the late 1970s, which killed acroporid corals throughout their geographic distribution, was unparalleled on a timescale of at least three millennia (Aronson and Precht 2001a, b). Moreover, in recent years, large colonies (many over 500 years old) of the other main reef-building species in the region have succumbed to single virulent diseases such as white plague disease (WPD) and Caribbean yellow band disease (YBD), or the combination of these and bleaching in short periods of time, further indicating that this seems to be a recent and expanding problem (Weil et al. 2006; Bruckner and Hill 2009; Weil et al. in press). However, a hiatus in *A. palmata* accretion occurred 3,000 years ago and again 800 years ago over a wide geographic area. "Understanding the causes of such large-scale community shifts provides both opportunities and challenges with respect to unraveling both natural and anthropogenic change" (Hubbard et al. 2008).

2.1 Black Band Disease

The first scleractinian infectious disease reported in the Caribbean was black band disease (BBD). It was first observed in Belize, Bermuda, and Florida in the early 1970s (Antonius 1973; Garrett and Ducklow 1975), but has since been found throughout the wider Caribbean and the Indo-Pacific (Antonius 1985; Willis et al. 2004; Galloway et al. 2007).

It is characterized by a dark bacterial mat of varying composition forming a band separating healthy-looking tissue from the clean skeletal matrix (Table 1, Fig. 2a, b). Little etiological work was done in the early days but information on host range, mortality rates, and depth distribution was provided. It is the better-known coral disease with a significant number of studies expanding pathogeneses, etiology, and epizootiology (Rützler and Santavy 1983; Rützler et al. 1983; Richardson 1997; Aeby and Santavy 2006; Richardson et al. 2007), and including recent debates concerning the variability of the microbial community composition (Voss et al. 2007).

2.2 White Plague Diseases

The first report of a disease outbreak producing significant coral mortalities occurred in Florida in 1975 and was referred to as white plague type I. (WPD-I). It mainly affected the plating coral *Mycetophyllia ferox* (Dustan 1977) leading to fears of its disappearing from some areas in the Florida Keys. During a second WPD-I epizootic event, colonies of *M. ferox* were unaffected whereas massive *Montastraea* were affected (Dustan 1999), suggesting resistant *Mycetophyllia* colonies or a different causative agent. A third and more virulent outbreak of WPD occurred in the Florida Keys in 1995 mainly affecting *Dichocoenia stokesi* and 16 additional species over the next 2 years (Richardson 1998; Richardson et al. 1998a) (Table 1, Fig. 2e, f). In this case, *Aurantimonas coralicida* was isolated from WPD lesions and the disease termed white plague disease type-II (WPD-II) (Richardson et al. 1998b; Miller et al. 2001; Denner et al. 2003). In the late 1990s and early 2000s, a fourth more virulent epizootic was termed WPD-III and affected mostly *Montastraea* spp in Florida, the Virgin Islands, Puerto Rico, and Venezuela (Richardson and Aronson 2002; Weil 2002; Croquer et al. 2005). However, the disease agent was not identified (Richardson and Aronson 2002) and the term WPD-III was discarded.

Another condition called shut down reaction (SDR) in which tissue sloughed off quickly from coral colonies was described in the mid 1970s (Antonius 1977), but was never further studied and there have been no reports of this condition in the region in the last 2 decades.

2.3 White Band Disease and Diadema

Two years after the initial white plague epizootic, an outbreak of a disease with similar signs but affecting only acroporids called white band disease type I (WBD-I) devastated high proportions of *A. palmata* and *A. cervicornis* (Fig. 2c, d) pop-

Table 1 Common coral reef diseases in the western Atlantic and acronym (ACR), year reported/observed (year), pathogen/agent (P/A) identified = Y, No = N, number of taxa showing disease signs in corals (CO), octocorals (OC), hydrocorals (HY), sponges (SP), zoanthids (ZO) and crustose coralline algae (CCA) (number in parenthesis are Brasilian spp), depth distribution (DE), average community prevalence (PR), tissue mortality rate (TM), and their geographic distribution (GD) (WA = western Atlantic, WC = wider Caribbean, VI = Virgin Islands, FL = Florida, BE = Bermuda, CA = Caribbean, BA = Bahamas, ME = Mexico, PR = Puerto Rico), Updated from Weil et al. (2006) and other sources

Disease	ACR	Year	P/A	CO	OC	HY	SP	ZO	CCA	DE (m)	PR (%)	TM (mm/day)	GD
Bleaching	BL	1911	N	62	29	5	8	2		0–100	.2–85	?	WA
Growth anomalies	GA	1965	N	10	8	1				0–25	–	–	WC
Black band disease	BBD*	1973	Y	19(4)	6					0–25	.3–6	3–10	WA
White band disease-I	WBD-I	1977	N	2						0–10	0.1	?	VI,WC?
White plague disease-I	WPD-I	1977	N	12						10–21	3.6	3.1	FL
Shut Down reaction	SDR	1977	N	6						5–12	–	–	FL
White band disease-II	WBD*	1982	Y	3						1–25	.1–25	3–30	WC not BE
Red band disease	RBD	1984	Y	13(1)	5					2–20	–	1	WA
White patch disease[1]	WPA*	1992	Y	1						0–5	.002	15	CA,FL,BA
Caribbean yellow band[a]	YBD*	1994	Y	11						3–20	1–24	0.1–0.4	WC
White plague disease-II	WPD*	1995	Y	41(5)						3–30	.9–18	3–30	WA
Aspergillosis	ASP*	1996	Y		9(1)					1–25	1.9	.1–2.5	WA
Dark spots disease	DSD	2001	N	11(1)						1–25	1.1	–	WA
Crustose-Coralline white b.	CCWB	2004	N						3	1–20	1–6	.1–2	WC[a]
Caribbean white syndromes[2]	CWS	2004	N	15		2	3	1		2–25	–	–	WC[a]
Caribbean ciliate infection	CCI	2006	Y	21						2–25	–	–	WC[a]
Sea fan purple spots[3]	SFPS	2008	Y		1					2–18	–	–	ME,FL,PR
Coralline lethal orange disea.	CCLOD	2008	N						1	20	–	–	PR,CY,ME
Other coral health conditions[4]	CCH	–		15	8					1–25	–	–	WA
Other octocoral health condi.[4]	OCH	–			8					3–20	–	–	WA

* = Koch's postulates fulfilled. 1 = White patch disease is also termed white pox and patchy necrosis, 2 = White syndromes include several patterns of tissue loss exposing bands, stripes, blotches, or irregular shapes of clean skeleton (different from the other "white" diseases) with very low prevalence. 3 = PS produced by an unknown protozoan (Labyrinthulomycote). 4 = Other coral and octocorals health conditions include unhealthy-looking tissues with some degree of mortality, low prevalence, and limited geographic distribution with no pathological or etiological information. a = Including Flower Gardens, north Gulf of Mexico. Western Atlantic distribution includes the wider Caribbean and Brazil. Bleaching-affected species from Brazil have not being included in this list

Fig. 2 Photographs of the most common diseases in Caribbean corals. Active black band disease in *M. faveolata* (**a**) and *D. strigosa* (**b**), white band disease in *A. palmata* (**c**) and *A. cervicornis* (**d**), fast moving white plague disease in *D strigosa* (**e**) and *D. labyrinthiformis* (**f**), two different

ulations throughout their geographic range (Aronson and Precht 2001b). A different pattern or phase of this disease in *A. cervicornis* was described in the late 1990s and was named white band disease type II (WBD-II) (Ritchie and Smith 1998). It differed from WBD-I in having a bleaching band leading the necrotic edge of living tissue. These signs have only been observed in *A. cervicornis*, and it is not clear if these two patterns were caused by different pathogens, if the disease is expressed differently in the different species, or if the two etiologies represent different phases of the same syndrome (Weil 2004; Bythell et al. 2004).

Mortality of the surviving colonies/populations of acroporids have continued over the years due to recurrent WBD events, hurricanes and storms, bleaching, predation, and local environmental deterioration (sedimentation, turbidity, untreated sewer outflow, etc.) (Bruckner 2003; Weil et al. 2002, 2003; Wilkinson and Souter 2008; McClanahan et al. 2009), which together with the slow recovery of populations, led to these two corals being listed as threatened under the US Endangered Species Act (Hogarth 2006).

Almost concurrently with the WBD outbreak, although occurring over a shorter time span, a widespread and highly virulent infectious disease wiped out up to 99% of the populations of the black sea urchin *D. antillarum* throughout the wider Caribbean, including Bermuda (Lessios et al. 1984b; Lessios 1988). This urchin was at the time a keystone species regulating algae and coral community structure (Carpenter 1981, 1985, 1990a; Hughes et al. 1987). The causes of these events were never determined (but see Peters et al. (1983) and Rosenberg and Kushmaro, in this book). The consequence of these two epizootics was a significant change in the structure and morphology of most shallow-water coral reef communities throughout the wider Caribbean. These two outbreaks followed increasing water temperatures during an intense El Niño event, which produced limited bleaching in 1983.

2.4 White Patches and Octocoral Mortalities

Other localized invertebrate mass mortalities and disease outbreaks reported during the 1980s included: a die-off of the sea-fan *Gorgonia flabellum* in Panamá (Guzmán and Cortés 1984), and an outbreak of a thin red cyanobacteria mat in corals and octocorals termed red band disease (RBD) in Florida reefs (Rützler et al. 1983), later redescribed by Santavy and Peters (1997) and Richardson (1998) (Fig. 1k). The putative pathogens were not identified.

Several new coral and octocoral diseases were reported during the 1990s with more frequent and virulent epizootic events. White patches of clean skeletal tissue were observed in *A. palmata* (Porter and Meier 1992) which were clearly different from WBD signs, and were termed patchy necrosis (Bruckner and Bruckner 1997) and later, white pox (Rodriguez-Martinez et al. 2001; Patterson et al. 2002) (Fig. 2m). An early photograph from the USVI suggests this disease or a similar one could have been affecting *A. palmata* in the 1970s (Rogers et al. 2005). Several outbreaks of diseases with similar signs were observed in Florida, Puerto Rico, USVI, Mexico, and elsewhere in the late 1990s and early 2000s (Rodriguez-Martinez et al. 2001; Weil and Ruiz 2003; Rogers et al. 2008a) (Fig. 1h). All these terms and disease signs have now been pooled as white patch disease (Raymundo et al. 2008)

2.5 Dark Spots Disease

Dark spots disease (DSD) was first documented in the early 1990s in the Islas del Rosario archipelago, Colombia, as a type of bleaching that affected ca. 16% of *Montastraea annularis* colonies. It was called "Medallones Mostaza" ("mustard rings") (Solano et al. 1993). In 1994, similar signs were observed in other islands off Colombia mainly affecting *M. annularis*, *Siderastrea siderea*, and *Stephanocoenia intersepta*, and it was called "enfermedad de los lunares oscuros" (Diaz et al. 1995) or dark spots disease (DSD) (Figs. 2g, h). Dark spot lesions were characterized as "small, round, dark areas that apparently grow in size over time, some of which can be associated with a depression of the coral surface and others expand into a dark ring surrounding dead coral" (Garzón-Ferreira and Gil 1998). Other names characterizing different manifestations of the disease include "Dark Spots type II" in *S. intersepta*, *Colpophyllia natans*, and *Montastraea cavernosa*, and "Dark Bands" in *M. annularis*, *M. faveolata*, *S. siderea*, and *C. natans* (Weil 2004; Weil et al. 2006) (Fig. 2h). These conditions were all pooled as DSD (Raymundo et al. 2008; Weil and Hooten 2008); however, they could represent different diseases since their etiologies have not been resolved. Dark spots disease has now been found throughout the Caribbean basin (Cervino et al. 2001; Weil et al. 2002; Weil and Croquer 2009; Cróquer and Weil 2009a).

Fig. 2 (continued) etiologies of dark spots disease in *S. siderea* (**g**) and *S. intersepta* (**h**), bleached colonies of *M. faveolata* and *C. natans* (**i**), Caribbean yellow band disease in *M. franksi* (**j**) and *M. faveolata* (**k**), white patches in *A. palmata* (**l, m**), and Caribbean ciliate infections in *A. tenuifolia* (**n**) and *D. labyrinthiformis* (**o**) (Photos E. Weil)

Sixteen important scleractinian species have been reported to show signs corresponding to those characteristic of the disease (Table 1), and the disease appears less prevalent in the more northern portions of the Caribbean (Weil et al. 2002; Gil-Agudelo et al. 2004; Cróquer and Weil 2009a). Recently, a disease with lesions resembling DSD was described for Brazil where it was affecting *Siderastrea* sp. (Francini-Filho et al. 2008). In the Indo-Pacific, DSD has been documented in *Pavona varians* and *P. maldivensis* from Kahoolawe, Hawaii and in *P. varians*, *Psammocora nierstrazi* and *Montipora* sp. from Tutuila, American Samoa (Work et al. 2008c).

2.6 Caribbean Yellow Band Disease

Caribbean yellow band disease (YBD) (Fig. 1k) was first reported in the Florida Keys in 1997 by C. Quirolo in *Montastraea* colonies (Santavy and Peters 1997); however, Brown and Ogden (1993) published a photo of a large colony of *M. faveolata* from the Florida Keys with clear signs of YBD in a National Geographic article about bleaching, indicating that the disease could have been around in the 1980s. As with other diseases, different terms have been used for this disease (i.e., yellow blotch disease [Santavy et al. 1999], yellow band disease [Green and Bruckner 2000], and yellow blotch syndrome [Weil 2004]). Here we use Caribbean yellow band disease (YBD) following the original name and the geographic location to differentiate it from a similar syndrome described in the Red Sea with the same name (Korrubel and Riegl 1998). Signs of YBD were observed throughout the Caribbean and north to Bermuda in 1999 (Weil et al. 2002), and the disease is now widely distributed. Outbreaks of YBD were observed in Panamá in 1996 (Santavy et al. 1999), in the Netherland Antilles and Puerto Rico in 1997 (Cervino and Smith 1997; Bruckner and Bruckner 2006) and in Grenada, Mexico, Bermuda, and Puerto Rico between 2005 and 2009 (Cróquer and Weil 2009a, Weil et al., in press; Weil, unpublished data). This disease has become one of the major causes of tissue and colony mortality in three species of *Montastraea* (Fig. 2j, k), the most important reef-building genus in the region (Weil et al. 2006; Bruckner and Hill 2009; Weil et al., in press).

2.7 Caribbean Ciliate Infection

In 2004, ten coral species were observed with dead areas preceded by a dark band different from BBD in reefs off Venezuela. This band was formed by dense populations of a ciliate protozoan (*Halofoliculina* sp.) (Cróquer et al. 2006a) (Fig. 2n, o). Further surveys found the same ciliate infecting up to 22 coral species throughout the Caribbean, and the condition was termed Caribbean ciliate infection (Cróquer et al. 2006b; Weil et al. 2006; Weil and Hooten 2008; Weil and Croquer 2009; Cróquer and Weil 2009a; Weil et al., in press).

2.8 Aspergillosis and Purple Spots

A widespread epizootic of a fungal infection producing wide areas of tissue mortality surrounded by purple pigmentation (an immune response by the host) in the early 1990s, affected thousands of colonies of the sea fan *Gorgonia ventalina* (Fig. 1j) in many reef localities (Nagelkerken et al. 2007a, b). It was suggested that this was the same problem responsible for widespread *Gorgonia* mortalities in 1984 in Central America (Guzmán and Cortés 1984; Garzón-Ferreira and Zea 1992). The putative pathogen was later identified as the common terrestrial fungus *Aspergillus sydowii* and the disease was called aspergillosis (ASP) (Smith et al. 1996). At least eight other abundant octocoral species throughout the wider Caribbean have been reported to be affected by ASP (Weil 2001, 2002; Harvell et al. 2001; Smith and Weil 2004; Weil et al. 2006). Two independent reports have confirmed the impact of aspergillosis on the reproductive output of *G. ventalina* colonies, essentially reducing fitness and potential population recovery (Petes et al. 2003; Flynn 2008; Flynn and Weil 2008). Other disease signs have been observed in other octocoral species such as the common and abundant encrusting *Briareum asbestinum* and *Erythropodium caribaeorum*, which have been affected throughout their geographic range by "necrotic-like" lesions, which progress rapidly, sometimes killing large areas in a short time (Harvell et al. 1999; Weil 2004; Weil et al. 2006) (Fig. 1i).

In the last 4–5 years, colonies of the sea fan *G. ventalina* have been observed with small purple spots in Mexico and Florida (Harvell et al., 2008), and more recently, in Puerto Rico (Weil and Hooten 2008) (Fig. 1m). These purple spots are caused by a protozoan (Labyrinthulomycote) that infects colonies mostly during the summer (C.D. Harvell, personal communication). Prevalence has been increasing in several reefs off the southwest coast of Puerto Rico in the last few years (Weil, unpublished data 2009).

2.9 Other Diseases

Several other signs of presumed diseases such as white spots, white bands, white stripes and rings, pigmentation responses, dark bands, tissue loss and tissue "necrosis," that are usually found in a few colonies of a wide range of coral and octocoral species have been observed throughout the region in recent years (Fig. 1d, e, f, g and h). Furthermore, many other important members of the coral reef community such as hydrocorals, sponges, zoanthids, and other important calcifying organisms have been affected by diseases in the Caribbean for some time (Fig. 1n–s). In the late 1990s and early 2000s, the common crustose coralline alga (CCA) *Neogoniolithon accretum* and at least two other species were observed with an advancing, thin, white band separating healthy-looking tissues from dead areas (Weil 2004; Weil and Hooten 2008). The condition was termed crustose coralline white syndrome (CCWB) (Fig. 1s) and was found in high prevalence in deep reef habitats (18–23 m) of Puerto Rico and Grenada (Ballantine et al. 2005; Weil et al. in press; Weil, unpublished data 2007). Since then, signs of the condition have been observed throughout the Caribbean and in many locations in the Indo-Pacific and Indian Ocean (Weil, unpublished data 2008). More recently, signs similar to coralline orange lethal disease (CLOD) described for the Pacific (Littler and Littler 1995) have been observed on deep (20 m) CCA in Puerto Rico, the Cayman Islands, and Mexico and the condition termed Caribbean CLOD (CCLOD) (Weil et al., in press) (Fig. 1r).

3 Current Status of Coral Diseases

The Caribbean has been dubbed a "disease hot spot" due to the fast emergence and high virulence of coral reef diseases, their widespread geographic distribution, wide host ranges, and frequent epizootic events with significant coral mortalities. The Wider Caribbean includes the Gulf of Mexico, Florida, the Bahamas, and Bermuda and only about 8% of the coral reef area worldwide (Spalding and Greenfeld 1997), yet over 60% of all disease reports up to 2000 came from this region (Green and Bruckner 2000).

Local environmental and anthropogenic stresses in combination with global warming trends have been proposed as factors that could affect species susceptibility/resistance to pathogens, as well as enhance bacterial growth and virulence favoring local disease outbreaks, which can then be dispersed by the rapid currents in the basin (Peters 1997; Epstein et al. 1998; Goreau et al. 1998; Richardson 1998; Richardson and Aronson 2002; Weil 2004; Weil et al. 2006; Harvell et al. 2007).

Besides the Caribbean, the south coast of Brazil is the only other area with significant coral reef formations in the Western Atlantic, and until recently, no coral diseases had been reported for this region. Besides distance, two major dispersion barriers, the outflows of the Amazon and Orinoco Rivers, separate Caribbean and Brazilian coral reefs, which could also be effective barriers to dispersing pathogens. Nevertheless, several diseases with similar signs to those in the Caribbean were recently described for this region (Acosta 2001; Francini-Filho et al. 2008).

At least eighteen disease conditions affecting corals and other important reef organisms have been described for the Wider Caribbean (Table 1, Figs. 1 and 2). Most do not have defined pathologies nor have they been well characterized (Bythell et al. 2004; Weil et al. 2006). Of these, ten diseases affecting corals show consistent signs that allow their recurrent identification, black band disease (BBD), white plague disease (WPD), Caribbean yellow band disease (YBD), white band disease (WBD), white patches (WPA) (formerly called patchy necrosis, *Acropora* serriatosis, and white pox), dark spots disease (DSD), red band disease (RBD), Caribbean ciliate infection (CCI), growth anomalies (GA), and bleaching (BL). Other "white" diffuse/inconsistent signs (bands, spots, stripes, etc.) producing minor tissue loss affect several corals and have been grouped as Caribbean white syndromes (CWS). Unhealthy-looking conditions such as dark areas, bands, pigmentation responses, etc. have been pooled into "Other compromised health" conditions (OCH) until their etiologies and pathologies are clarified to avoid further confusion (Table 1, Fig. 1). Most "white" diseases are characterized by the recently exposed white skeleton after the tissue died, so identification is based on the appearance of the skeleton in contrast with the edge of live tissue and not on any pathology of the tissues (Bythell et al. 2004; Lesser et al. 2007; Work et al. 2008a).

Four diseases with consistent signs [aspergillosis (ASP), red band disease (RBD), growth anomalies (GA), and sea fan purple spots (PS)], and several other conditions affect common and abundant octocoral species (Table 1, Fig. 1). Little histopathological work has been done other than in corals and a few octocorals (Weil et al. 2006), a significant gap in our current approaches to the study of cnidarians diseases. Descriptions of the common Caribbean and Indo-Pacific coral-octocoral diseases can be found in Richardson (1998), Rosenberg and Loya (2004), Raymundo et al. (2008), Beeden et al. 2008; and Weil and Hooten (2008).

Compared to the Caribbean, only a few coral diseases and diseases in other organisms have been reported for the Indo-Pacific (Littler and Littler 1995; Korrubel and Riegl 1998; Willis et al. 2004; Galloway et al. 2007) and the Red Sea (Loya 2004). The first reported coral disease for the Indo-Pacific was black band disease affecting two massive faviid species in the Philippines and later, seven other species in the Red Sea (Antonius 1985). The disease showed similar signs to the Caribbean BBD, but the bacterial mat seemed to have a different species composition (see later). Besides BBD, brown band disease (BrB), skeletal eroding band (SEB), ulcerative white spots (UWS), atramentous necrosis (AtN), growth anomalies (GA), and white syndromes (WS) are the most commonly found diseases with low and variable prevalence and limited geographic distribution (Willis et al. 2004; Raymundo et al. 2008; Galloway et al. 2007). The number, distribution, and prevalence of diseases has been increasing across the Indo-Pacific and the Red Sea as more research is being done, with several reports of epizootic events across the region (Green and Bruckner 2000; Willis et al. 2004; Rosenberg and Loya 2004; Weil and Jordán-Dahlgren 2005; Page and Willis 2006; Raymundo et al. 2003, 2008; Work and Aeby 2006; Aeby 2006a, b; Galloway et al. 2007; McClanahan et al. 2009).

3.1 Pathogenesis

Descriptions of many coral diseases are limited and often confounded by the lack of clear diagnostic criteria and the absence of pathological observations, so that similar disease signs may reflect multiple conditions in one or more coral species (Bythell et al. 2004; Weil 2004; Work and Aeby 2006; Raymundo et al. 2008; Work et al. 2008a). There is evidence that different pathogenic bacteria and fungi can produce similar signs in the same and/or in different species (Toledo-Hernandez et al. 2008; Sunagawa et al. 2009).

Because the BBD microbial mat consistently contained dominant populations of the same microorganisms, Carlton and Richardson (1995) proposed that it was caused by a microbial consortium instead of a single pathogen. The three mayor players were a cyanobacterium (*Phormidium corallyticum*), a sulfide-oxidizing bacterium (*Beggiatoa* sp.), and a sulfate-reducing bacterium (*Desulfovibrio* sp.). The dark coloration during the day is provided by the red cyanobacterial pigment phycoerythrin. Richardson (1997) demonstrated that BBD sulfate-reducing bacteria are functionally specific to BBD pathogenicity, and suggested that they may be species-specific. Recent molecular studies, however, suggest that the primary pathogen may be a nonphotosynthetic, eubacterial heterotroph (Cooney et al. 2002; Frias-Lopez et al. 2002). Furthermore, these results also showed that *P. corallyticum*, originally identified as the cyanobacterial component of BBD (Rützler and Santavy 1983; Taylor 1983) may not be the cyanobacterium associated with BBD. Recent studies indicated that the BBD mat is dominated by an unidentified cyanobacterium most closely related to the genus *Oscillatoria* (Cooney et al. 2002; Frias-Lopez et al. 2003). At least three different taxa of cyanobacteria associated with BBD were identified by Frias-Lopez et al. (2003), who showed that they vary between the Caribbean and Indo-Pacific.

Furthermore, differences in composition of the bacterial community have been reported for BBD affecting corals in Florida, the Bahamas, and the US Virgin Islands (Voss et al. 2007). The mat composition seems to be variable spatially and/or temporally, or the components may have been misidentified originally (Voss et al. 2007; Sekar et al. 2008). Most recently, a single cyanobacteria ribotype was found to be associated with both red band disease (RBD) and BBD in corals from Palau, having a 99% sequence identity with a Caribbean strain (Sussman et al. 2006). Further research is needed to clarify whether RBD and BBD are the same.

Denner et al. (2003) identified the bacterium *A. coralicida* as the putative pathogen of WPD-II in the coral *D. stokesi* in the Florida Keys. Another 40 coral species have been reported to be susceptible to WPD (Weil 2004; Sutherland et al. 2004) because they showed signs similar to those described for *D. stokesi* colonies infected with *A. coralicida* (Richardson et al. 1998a). However, *A. coralicida* has not been consistently found in corals with signs of WPD, and probes developed and used to identify the pathogen are insufficiently specific to consistently incriminate this bacterium (Bythell et al. 2004; Polson et al. 2009). Even though *A. coralicida* has been found in a few other coral species, Koch's postulates have only been verified for *D. stokesi* (Richardson et al. 1998b; Denner et al. 2003; Pantos et al. 2003). Recent analyses of several diseased tissue samples from *Montastraea faveolata* colonies with typical WPD-II signs failed to find *A. coralicida* (Sunagawa et al. 2009), and no other experimental data show that any of the other species reported with WPD signs have been actually infected by *A. coralicida*. Therefore, WPD signs in these colonies and other species of corals might be caused by a different agent.

More than 20 years after the WBD epizootic, most populations of acroporids have not recovered and the disease agent associated with this epizootic has not been identified. A potential pathogen named *Vibrio charchariae* was identified but Koch's postulates were never fulfilled until recently. Results from controlled isolation/inoculation experiments in Puerto Rico showed that the potential cause of WBD type II is possibly a *Vibrio* species very close to *Vibrio harveyi*, a synonym of *V. charchariae* (Gil-Agudelo et al. 2006). This study also reported that other *Vibrio* species tested produced similar WBD signs, but their virulence was lower than *V. harveyi*.

The putative pathogen associated with white pox (WPX) signs on *A. palmata* in the Florida Keys was identified as the bacterium *Serratia marcescens*, a common gut bacterium in sheep, other mammals, and fish (Patterson et al. 2002). The disease was then called *Acropora* serratiosis; however, all conditions with similar signs (white patches, patchy "necrosis") in *A. palmata* have been recently pooled as "white patches" (WPA) (Raymundo et al. 2008; Weil and Hooten 2008). The disease agent in WPA-infected colonies outside the Florida Keys has never been identified, and preliminary results from a few samples from St. John did not show any correlation between signs of "white pox" and presence of *S. marcescens* (Polson et al. 2009).

The cause of dark spots disease is still unknown. Gil-Agudelo et al. (2004) examined bacterial flora of mucus in corals affected with DSD and found that the corals were infected with *Vibrio charchariae* whereas this bacterium was absent in normal corals. Experimental infections of corals in the field using this bacterium failed to replicate DSD signs. Corals maintained in aquaria incidentally developed DSD, and treatment with antibiotics led to further tissue loss, thereby arguing against a bacterial etiology for DSD (Gil-Agudelo et al. 2004). In Florida, fungi have been associated with *S. siderea* affected with DSD (Galloway et al. 2007). Chemical analyses of colonies of *S. siderea* resistant and susceptible to DSD in Puerto Rico suggested that phenotypic plasticity in antimicrobial activity may affect microbial infection and survival in the host colonies (Gotchfeld et al. 2006).

Recent experimental evidence suggests that a particular combination of four *Vibrio* species infects and kills zooxanthellae in the coral endoderm producing the characteristic signs of yellow band disease in both Caribbean and Indo-Pacific corals (Cervino et al. 2004a, b, 2008). However, the mechanisms by which the zooxanthellae and the coral tissue are killed are unclear. The onset of infection seemed to be temperature-dependent (Weil et al. 2009b), and prevalence increased under high-nutrient conditions (Bruno et al. 2003) and high water temperatures (Harvell et al. 2009; Weil et al., in press).

Further evidence indicates that the dynamics of bacterial communities in corals are more complicated and more responsive to changes in environmental conditions than previously thought (Ritchie 2006; Gil-Agudelo et al. 2006, 2007; Voss et al. 2007; Toledo-Hernandez et al. 2008). Similar to other coral diseases, recent findings have revealed that *A. sydowii*, the pathogen causing aspergillosis in sea fans and other octocorals (Smith et al. 1996; Smith and Weil 2004) has been found in sea fans without disease signs. Furthermore, ASP signs could be produced by other *Aspergillus* species and fungi in other groups (Toledo-Hernandez et al. 2008).

Changes in the composition and dynamics of the bacterial community after environmental or biological changes in the coral host could be related to different bacteria producing similar disease signs over time, sometimes indicating (although not conclusively proving) a potential development of resistance to the initial pathogenic agent (Reshef et al. 2006). In a recent study of *Montastraea* corals showing typical white plague signs in Puerto Rico, the primary WPD pathogen *A. coralicida* was not found after screening mucus and tissue samples (Sunagawa et al. 2009). Similar results were reported for the fungal disease ASP (Toledo-Hernandez et al. 2008) and for bacterial bleaching in *O. patagonica* in the Mediterranean (Ainsworth et al. 2008). A suite of other fungal species can produce similar aspergillosis signs in sea fans (Smith and Weil 2004; Toledo-Hernandez et al. 2008). These results emphasize the need to examine tissue microscopically in attempts to identify potential causative agents, to follow up with appropriate laboratory confirmation, and to be cautious about naming and describing diseases without a clear pathogenesis.

Many cnidarian diseases have yet to be characterized. Their etiologies have not been properly described and their putative pathogens have not been identified (Ritchie et al. 2001; Weil et al. 2006; Work et al. 2008a). As researchers become more familiar with disease signs and more pathological studies are carried out, the number of described diseases affecting corals and other reef organisms could grow. Koch's postulates have only been verified for five diseases and in most cases, for only one species in each condition (Table 1); so there is ample room for new pathologies and changes in our understanding of these diseases.

3.2 Geographic Distribution

Most coral diseases in the Caribbean have spread throughout the region (Table 1; Fig. 3). White plague, YBD, DSD, ASP, BBD, CCI, GA, and bleaching show the widest geographic distribution, from Bermuda to Trinidad and Tobago, the northern coast of Venezuela and Colombia, Central America, and the southern region of the Gulf of Mexico (Weil 2004; Weil and Croquer 2009). With the exception of the white band disease outbreak and the mass mortality of *D. antillarum*, no correlations between dispersion patterns of most of the recent epizootic events (ASP, WPD, YBD, etc.) with current patterns in the region have been reported. There is some evidence, however, that WBD spread against the predominant current in St. Croix in the early 1980s (Gladfelter 1982). A dispersion pattern following local and/or regional water currents is expected for new putative, infectious agents introduced or "activated" in one particular location, as was the case of the agent killing the sea urchins in the early 1980s (Lessios et al. 1984a,b Lessios 1988; Carpenter 1990a, b).

Most reefs in the northern Gulf of Mexico are far from the US mainland and deeper than 22 m where water temperatures are cooler and light conditions are reduced, conditions that might limit the development of diseases. A short-lived outbreak of a white syndrome was observed in 2005 in the deep coral reef areas of the Flower Garden Banks (Hickerson and Schmahl 2006). Other conditions such as growth anomalies, "mottling syndrome" and "pale ring syndrome" were reported by Borneman and Wellington (2005), and more

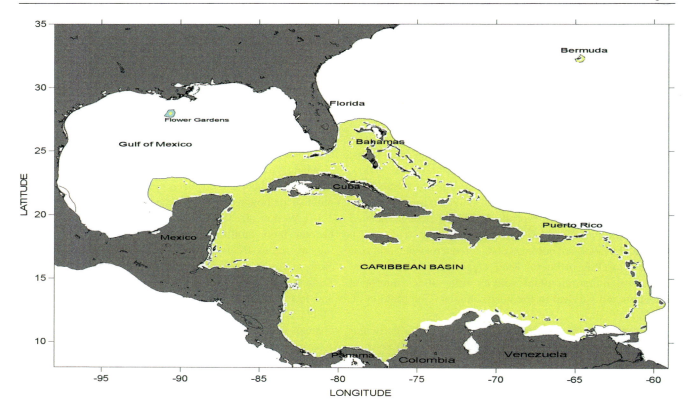

Fig. 3 Geographic distribution of the most common coral and octocoral diseases in the wider Caribbean: BBD, WPD, YBD, DSD, GAN, CCI, ASP, white syndromes and other compromise health conditions. Crustose coralline white band is also distributed throughout the wider Caribbean. A virulent white syndrome was observed in 2005 in the Flower Gardens in the north area of the Gulf of Mexico and bleaching has affected all reefs across the region

recently, Caribbean ciliate infections and GA were observed during disease surveys in the two major banks (Zimmer et al. 2009). Caribbean ciliate infection (CCI) was first observed in Venezuela in 2005 (Cróquer et al. 2006a, b), but recent surveys have reported this condition in corals in at least six other countries in the region (Bermuda, Puerto Rico, Grenada, Caymans, Mexico, and Panamá) (Cróquer and Weil 2009a; Weil and Croquer, unpublished data 2009), but this is the deepest report so far.

Until recently, only one disease, affecting zoanthids, was reported from Brazil (Acosta 2001). The first signs of scleractinian coral and octocoral diseases were observed in 2005 in the Abrolhos Bank, the largest reef system in Brazil. Conditions with signs similar to WPD, BBD, RBD, ASP, GA, and octocoral compromised health conditions ("tissue necrosis") were recently reported from this area (Francini-Filho et al. 2008). From 2005 to 2007, the distribution of these diseases has widened and their prevalence and virulence have increased producing significant coral and octocoral mortalities in the Abrolhos Bank. Based on estimates of disease prevalence and progression rates, as well as on the growth rates of a major reef-building coral species (the Brazilian-endemic *Mussismilia braziliensis*), it is predicted that eastern Brazilian reefs will suffer a massive coral cover decline in the next 50 years, and that *M. braziliensis* will be nearly extinct in less than a century if the current rate of disease mortality continues (Francini-Filho et al. 2008).

Limited connectivity between the Caribbean and the Brazilian reefs suggests that either these diseases are produced by different agents, possibly triggered by similar environmental changes (increase in water temperatures), or they have similar etiologies to their Caribbean counterparts. If pathogens are different, this shows the limited responses (signs) cnidarians can develop when affected by infectious diseases and other agents, and the importance of identifying putative pathogens.

3.3 Depth Distribution

Caribbean wide surveys indicate that WPD, YBD, DSD, and ASP have the widest depth distribution, from 1 to 25 m (Weil 2004; Cróquer and Weil 2009a, b; Weil et al., in press). If the pathogens are species-specific, disease infections would be limited to the depth distribution of the host species. Most of the species affected by the major diseases (*M. faveolata, M. franksi, M. cavernosa, M. ferox, Agaricia lamarcki*, etc.) have a wide depth distribution, some down to 90 m. However, some diseases affecting these species have limited depth distribution. White plague disease has been more prevalent in deeper habitats (10–25 m) in Puerto Rico (Weil et al., in press); however, recently, a colony of *M. ferox* was observed with signs of WPD at 50 m off the southwest coast of Puerto Rico, the deepest record so far (H. Ruiz, personal communication). YBD has only recently been observed below 20 m in the Caymans and Puerto Rico (Weil, unpublished), which could be related to different zooxanthellae composition in deeper corals. Bleaching has the deepest distribution of all diseases with pale or white corals (of a few different species) observed down to 100 m in some reefs in the Caymans in 2009 (McCoy C, personal communication 2009). Aspergillosis has the widest depth distribution among octocoral diseases, with sea fan colonies showing signs of this condition from 1 to 25 m (Jolles et al. 2002; Kim and Harvell 2002; Flynn and Weil 2008; Flynn and Weil, in press).

3.4 Prevalence, Incidence, and Virulence

Prevalence is the proportion of infected colonies in a population or a community. It is expressed by absence/presence per individual and usually does not give any indication of the severity (virulence) of the disease, which could include the number (and size) of the lesions that are present, the rate of tissue mortality, the proportion of the colony that is affected, or the rate of spread of the disease in the population (Work et al. 2008a, b; Weil et al. 2008). Most surveys are done yearly or only during "outbreaks" and, prevalence can vary greatly even over short periods of time. Prevalence of white pox disease (= white patches) on *A. palmata* colonies off St. John ranged from 0% to 52% (Rogers et al. 2008b). Disease incidence is a rate expressing the number of newly infected colonies over time. It requires temporal monitoring of the same reef area with mapping, tagging, and photographing of colonies along the sampled area (Weil et al. 2008; Work et al. 2008b).

Average disease prevalence at the coral community level for major coral diseases in the Caribbean remains low (<6%) and has not changed significantly in the last 10 years (Weil et al. 2002; Weil and Croquer 2009). However, in some localities, prevalence of some chronic diseases within populations could be much higher and, even low levels of disease over long periods of time can produce significant mortalities in reef communities. Frequent monitoring is needed to address the spatial and temporal variability in prevalence and virulence (number of lesions and rate of disease advance) and to assess disease incidence (new cases of infected colonies over time) at population and/or community levels and their cumulative effects.

One-time surveys are limited in that they only reflect the disease status at a particular time. Data for different reefs generated with different methods could be difficult to compare, especially if the areal extent of the surveys are not the same and the data are collected by different people without standardization of the disease identification. Similarly, data generated with the same methods but taken in different seasons and/or different years could also generate problems of interpretation. Prevalence will go down when the disease has run its course, and most susceptible colonies have died, so their proportion relative to the resistant survivors drops. A recent disease manual with two sets of underwater disease identification cards, one for the Caribbean and one for the Indo-Pacific, were published with the goal of standardizing the disease identifications, nomenclature used to describe and characterize them, and methodology to estimate prevalence, incidence, virulence, and their variability (Raymundo et al. 2008; Beeden et al. 2008; Weil and Hooten 2008).

Prevalence of WPD, YBD, WBD, BBD, and DSD showed high seasonal variability in Caribbean localities with usually higher prevalence and frequent outbreaks during summer's high water temperatures (Borger 2003; Gil-Agudelo et al. 2004; Borger and Steiner 2005; Bruckner and Bruckner 2006; Weil et al., in press; Weil, unpublished). In the late 1990s and early 2000s, YBD was a seasonal disease in La Parguera, Puerto Rico, active and highly visible during Summer--Fall, nearly disappearing from some colonies and completely from others during the Winter--Spring. In the last 6 years however, this seasonality disappeared. Prevalence of YBD increased every year with corresponding increase in severity (number of lesions and rate of disease advance) to epizootic levels in many reefs in Puerto Rico and other Caribbean localities (Bruckner and Bruckner 2006; Weil and Croquer 2009, Cróquer and Weil 2009a, b; Weil et al., in press). Moreover, increase in prevalence and severity over time covaried significantly with increasing average winter and yearly water temperatures (Fig. 4a, see below) (Weil 2008; Harvell et al. 2009; Weil et al., in press), thus warmer winters seem to have affected disease dynamics. However, co-variation does not mean causality, so the increase in prevalence could also result from increased incidence (number of new infected colonies per unit time) until most susceptible colonies were infected in the population. Prevalence could stay high due to the warmer temperatures until all susceptible colonies die from the disease and the relative proportion of diseased colonies is reduced over time (Bruckner and Hill 2009).

Rates of tissue loss are also highly variable and depend on the virulence of the pathogen, the susceptibility/resistance of the host, synergistic environmental conditions, and the duration of the infection (Bruckner 2002; Bruckner and Bruckner

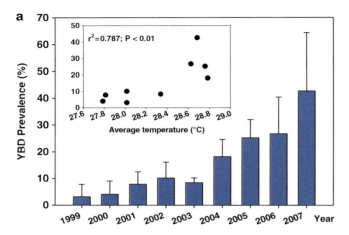

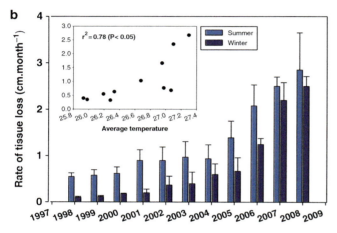

Fig. 4 Temporal changes in the dynamics of Caribbean yellow band disease. (**a**) Increase in YBD prevalence in the coral genus *Montastraea* in reefs off the south-west coast of Puerto Rico from 1999 to 2007 and the positive and significant ($r^2 = 0.787$, $P \leq 0.01$) correlation with average yearly surface water temperature (*inset*). (**b**) Seasonal variability in YBD lesion growth rates (virulence) measured in over 100 tagged colonies of *M. faveolata* in La Parguera, Puerto Rico from 1999 to 2008, and the significant positive co-variation ($r^2 = 0.54$, $P \leq 0.05$) between linear YBD lesion growth rates and the average seasonal surface seawater temperature for the same period (*inset*). (Modified from Weil et al., in press)

2006; Harvell et al. 2007). For example, rates of tissue loss were significantly higher during the fourth WPD outbreak even though the same pathogen seemed to have been the cause (Richardson and Aronson 2002). In Puerto Rico, the number of YBD disease lesions in *M. faveolata* colonies increased significantly over time with some very large colonies showing over 32 lesions at once. This led to a significant increase in rate of tissue loss, killing the colonies faster (Weil et al., in press).

The average rate of tissue loss (advance of the disease edge) estimated from *M. faveolata* colonies over the years in Mona and Desecheo islands was generally low (0.5–1.0 cm/month) and variable across colonies, months, and localities

(Bruckner and Bruckner 2006). Similar results from over 200 tagged colonies of the same species that were checked bi-annually were obtained in La Parguera. However, the rates of tissue loss increased with time, from 0.2 to 3.6 cm month^{-1} (Fig. 4b) and the seasonality observed in prevalence and virulence tended to decline over the years. Both prevalence and rates of tissue loss significantly increased from 2003 to 2008 and were correlated with increasing water temperatures (Weil et al., in press; Weil, unpublished) (Fig. 4). In Colombia, *M. annularis* and *S. siderea* had the highest prevalence of dark spot disease (10% and 5%, respectively) whereas the disease was much less common in *M. faveolata*, *M. franksi*, *S. intersepta*, and *M. cavernosa* (Gil-Agudelo 1998). Recent wide geographic surveys in the Caribbean showed lower prevalence values of DSD than those reported for Colombia and in general, lower prevalence in northern compared to southern localities (Weil and Croquer 2009).

3.5 Host Ranges

Overall, host ranges for most Caribbean coral reef diseases have remained stable or have increased over the years with many other reef organisms observed with similar disease signs in the region (Table 1) (Weil 2004; Bruckner 2009). Corals and other reef invertebrates are relatively simple organisms with a limited range of signs or visible "responses" to infections. Unless we do histology, most of what we can see (or characterize) are the external manifestation of responses of the diseased tissues or, just the pattern of tissue mortality. These include, but are not limited to, different patterns of tissue loss, pigmentation changes, general tissue conditions, mucus release, and growth anomalies. There has been no consistent histopathological research that could provide reliable descriptors for the different pathologies (Work and Aeby 2006; Work et al. 2008a).

Bleaching (presumably due to elevated temperature) has the widest host range affecting at least 62 corals, 29 octocorals, eight sponges, five hydrocorals, and two zoanthids in the Caribbean (McClanahan et al. 2009; Prada et al. 2009) (Table 1, Fig. 5). Bleached corals are still alive, and if conditions return to normal quickly enough, most colonies can fully recover. Four coral diseases, WPD, CCI, BBD, and RBD had the widest host ranges in the Caribbean with 41, 21, 19, and 13 susceptible scleractinian coral species, respectively (Table 1). Caribbean yellow band has been reported in 11 species of important reef-building corals. The total number of coral species affected by DSD has increased over time possibly due to the expansion of surveys. Sixteen important scleractinian species have been reported to show signs corresponding to those characteristic of the disease (Table 1)

Fig. 5 Bleached colonies of important reef-building species during the 2005 event. In many reefs of Puerto Rico and the US Virgin Islands, up to 90% of all colonies of important reef-building species were fully or significantly bleached [*Montastraea faveolata* (**a, b**), *A. palmata* (**c**), *D. cylindrus* (**d**), *C. natans* and *S. intersepta* (**e**), and *D. strigosa* and *S. siderea* (**j**)]. Bleached colonies of species that never been observed bleached in these reefs included *Scolymia cubensis* (**f**) and *Mycetophyllia ferox* (**g**) among others. The event produced significant mortalities in the agaricids and acroporids (**h, i**). Adjacent colonies of the same species showed significant differences in bleaching intensity in some localities (i.e., *M. faveolata* and *C. natans* (**k, l**), and some species did not bleach in certain areas (i.e., *Meandrina meandrites*) (**m**). Several species of octocorals bleached in many localities (**n**) (Photos E. Weil)

Fig. 6 Potential invertebrate vectors (reservoirs?) of coral diseases in the western Atlantic include the snails *Coralliophila abbreviata* and *C. caribaea*, here seen preying on *D. labyrinthiformis* (**a**), *M. faveolata* (**b**) and *A. palmata* (**c**); the flamingo tongue *Cyphoma gibossum*, a common predator of sea fans (**d**) and other octocorals (**e**); sea urchins which are "omnivorous" and can pick up pathogens from sediments or turf algae and move them to corals (**f**); and the fireworm *Hermodice carunculata* which preys on both octocorals (**g**) and corals. This fireworm has been frequently observed eating at the edges of diseased and healthy areas of corals with BBD (**h**, **i**), WPD (**j**) (Photo by C. Rogers), and white band disease (Photos E. Weil)

(Gil-Agudelo et al. 2004; Weil et al. 2002; Cróquer and Weil 2009a).

White plague disease, BBD, RBD, and DSD have been reported to affect five, four, one, and one coral species, respectively, in Brazil. Nine of the most common and abundant octocorals in the Caribbean seem to be susceptible to ASP, six to BBD, at least five to RDB, and several to other compromised health conditions and growth anomalies. Overall, at least six different diseases affect five of the most important and most common reef-building genera (including *Montastraea, Diploria, Colpophyllia, Acropora,* and *Agaricia*), and four different diseases affect 11 of the main reef-building genera (Fig. 7).

Most of the information used to compile lists of susceptible species came from single observations in space and time. A new host was added to the list if a single, or just a few colonies of a species was (were) observed with the disease signs without any verification of the pathology and/or etiology. As mentioned above, Koch's postulates have only been verified for a few pathogens in a few species, usually one for each condition [i.e. *Diploria strigosa* (BBD) (Rützler and Santavy 1983, *A. palmata* (white patches) (Patterson et al. 2002), *D. stokesi* (WPD) (Denner et al. 2003); *A. cervicornis* (WBD) (Gil-Agudelo et al. 2006); *G. ventalina,* and *G. flabellum* (ASP) (Smith et al. 1996; Geiser et al. 1998)]. Confirmation of the pathology of all the potentially susceptible species for each disease has never been done. The temporal dynamics of these infections (species could become resistant after the first infection, or susceptible colonies could be quickly eliminated from the population) has never been properly investigated.

Overall, only a fraction of the listed host species for each disease is usually observed with the disease signs when conducting typically annual field surveys, and the actual number of susceptible species (host range) for each disease will only be determined when the same pathogen is not only found in each species, but is shown to be the cause of the disease signs.

3.6 Vectors and Reservoirs

There are only two reports with experimental evidence of invertebrates acting as reservoirs (organism, substrate, or other media where the pathogen spends some time and completes part of its life cycle) and vectors (organisms or other media that act as a carrier and delivery medium for a pathogen) of a coral disease. The fireworm *Hermodice carunculata* in the Mediterranean is a vector for *Vibrio shiloi,* the pathogen that causes bacterial bleaching in the coral *Oculina patagonica* (Sussman et al. 2003), and the predatory Caribbean snail *Coralliophila abbreviata,* harbors *S. marcescens,*

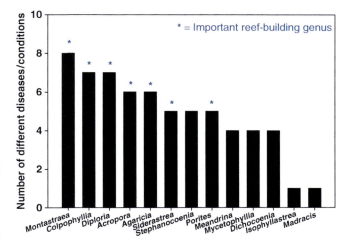

Fig. 7 The number of different diseases affecting the most important Caribbean reef-building scleractinian coral genera (*). Eleven genera of the 26 reported for the Caribbean are affected by at least four or more different diseases/conditions

the pathogen responsible for WPA in *A. palmata* in the Florida Keys (Williams and Miller 2005) (Fig. 6c).

Hermodice carunculata is a coral predator, ingesting *Vibrio shiloi* and keeping it alive in its gut (Sussman et al. 2003). This worm (or a similar species) is very common in coral reef communities across the Caribbean and is one of the main predators of acroporid and massive corals, hydrocorals, and octocorals. It is frequently observed feeding on the edges of WPD, YBD, and BBD active lesions in colonies of *M. faveolata, D. strigosa,* and *C. natans,* and ASP-infected sea fans (Fig. 6g – j). It is then possible that this fireworm could act as a vector and a reservoir for one or several of these Caribbean diseases as well. To date, disease reservoirs have only been identified for BBD (biofilms in reef sediments, which contain nonpathogenic versions of the BBD consortium) (Richardson 1997), and WPD (*Halimeda opuntia* mats, which seem to harbor the WPD pathogen) (Nugues et al. 2004).

Other potential disease vectors and possible reservoirs include parrotfishes (i.e., *Sparisoma viride*), damsel fishes (*Stegastes planifrons* and *Microspathodon chrysurus*), and the butterfly fish *Chaetodon capistratus,* a common coral predator capable of moving the BBD pathogen from diseased to healthy corals in experimental settings (Aeby and Santavy 2006), and the snail *Cyphoma gibossum,* a predator of sea fans and other octocorals (Fig. 6d, e). Fishes tend to directly bite diseased and healthy colonies of important reef-building species in the Caribbean potentially moving pathogens around. Vectors may be involved in disease transmission and spread at both local and regional scales. This is an important aspect of the dynamics of coral reef diseases, particularly as populations of many of these potential vectors have been significantly increasing (mostly as a consequence of overfishing of their main predators or recent successful reproduction) for which there is little information.

4 Environmental Drivers

The multiple and complex biological associations within the coral holobiont and the currently changing environmental conditions complicate attempts to isolate individual drivers/causes and to make predictions and/or extrapolations based on short-term and single-locality studies. Of many potential factors, increasing sea water temperature seems to be one that may have favored the emergence of coral diseases (Harvell et al. 1999, 2002, 2007, 2009). Evidence of this includes the following:

1. Bleaching events and most of the early disease outbreaks affecting coral reef organisms and other marine animals occurred during the warm Summer and early Fall seasons (Harvell et al. 1999; Weil et al. 2006; Van Oppen and Lough 2009).
2. The first Caribbean-wide surveys of coral diseases conducted during the Summer–Fall season in 1999 showed an increase in disease prevalence at the community level from Bermuda in the north-west Atlantic to the more tropical southern Caribbean (Venezuela and Colombia), suggesting a potential relationship with warmer temperatures (Weil et al. 2002).
3. The infection of *Pocillopora damicornis* by *V. coralliilyticus* (bacteria producing bleaching in this species) showed no signs of infection below 22°C, tissue bleaching from infection between 24°C and 26°C, and rapid tissue lyses from 27°C to 29°C (Ben-Haim et al. 2003a, b).
4. Recent evidence indicates that high water temperatures compromise host susceptibility and increase virulence (Harvell et al. 2002; Ritchie 2006; Bruno et al. 2007; Weil et al. in press, which would presumably increase prevalence and incidence over time.
5. Functional gene analysis of samples from experimental *Porites compressa* colonies subjected to environmental stressors (increased temperature, elevated nutrients and CO_2, and lower pH) showed increased abundance of microbial genes involved in virulence, stress resistance, sulfur and nitrogen metabolism, and coral-associated microbiota (Archaea, Bacteria, protists) shifted from a healthy community (e.g., Cyanobacteria, Proteobacteria, and zooxanthellae) to a community of microbes often found on diseased corals (Thurber et al. 2009)
6. Water temperature and disease prevalence in *A. palmata* colonies in St. John, USVI, were positively correlated, but bleached colonies exhibited a stronger relationship than unbleached colonies. In addition, a positive relationship between the severity of disease, as estimated by the area of the disease lesions, was apparent only for bleached corals (Muller et al. 2008).
7. Bacteria populations in corals change in composition, abundance, and possibly virulence when temperature increases (Ritchie 2006; Bourne et al. 2007; Weil et al. 2009b; Harvell et al. 2009; Sunagawa et al. 2009). Similar correlations of increased prevalence with increasing water temperatures have been found in a long-term study in the Great Barrier Reef (Selig et al. 2006; Bruno et al. 2007).

The WPD and YBD outbreaks in the eastern Caribbean have been associated with higher than normal water temperatures over the years, which have also produced widespread bleaching in the area (Bruckner and Bruckner 2006; Miller et al. 2006, 2009; Rogers et al. 2008a, b; McClanahan et al. 2009). Furthermore, in contrast to previous WPD outbreaks in La Parguera, Puerto Rico, the WPD outbreak after the 2005 bleaching event peaked during the unusually warmer winter season (February–March) (Cróquer and Weil 2009b, Weil et al., in press).

A positive correlation between increasing water temperatures and increasing prevalence of YBD in *Montastraea* colonies was found over a 9-year study in Puerto Rico (Fig. 4a, b) (Weil 2008; Weil et al., in press.). Furthermore, the seasonal rates of YBD-induced tissue loss in *M. faveolata* were significantly different between 1999 and 2004, with higher rates of tissue loss during the Summer–Fall compared to the Winter–Spring. These differences, however, disappeared as winter water temperatures became warmer after 2004 (Fig. 4b), and YBD has remained active all year long with similar rates of tissue loss throughout the year (Weil 2008; Bruckner and Hill 2009; Harvell et al. 2009;

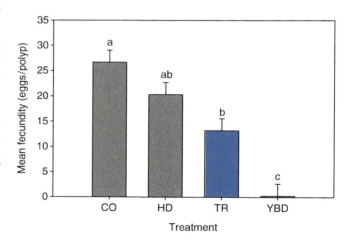

Fig. 8 Significant decline in reproductive output within diseased and healthy-looking areas of YBD-infected and control colonies of the important reef-building coral *Montastraea faveolata* in La Parguera, Puerto Rico. CO = control colonies with no signs of disease, HD healthy looking areas of diseased colonies, TR = transition areas (area between the YBD pale tissue and the healthy looking tissue), and YBD = disease area in colony (Modified from Weil et al. 2009a)

Weil et al., in press). Warmer winters favor the higher annual advance rates of lesions, significantly increasing the overall tissue and colony mortality, reducing fecundity (Fig. 8) and potentially affecting the short- and long-term recovery of populations (Weil et al. 2009a, b, in press).

High prevalence of dark spots disease seemed to be related to high water temperatures and specific depths in some localities but not in others (Gil-Agudelo and Garzón-Ferreira 2001; Borger 2003; Gotchfeld et al. 2006). Cróquer and Weil (2009a) found that populations of *S. siderea* exhibited a higher prevalence of DSD at intermediate (10 m) depths (25–40%), whereas *Stephanocoenia* populations were significantly more affected by DSD in deeper (>15 m) habitats (21–26%). In South Florida, prevalence of DSD increased during April-July but decreased during winter months (Borger and Steiner 2005). After 2 years of monitoring, Borger and Steiner (2005) suggested that DSD may be a general stress response of *S. siderea* that is exacerbated by an increase in water temperature, thereby illustrating geographic differences in environmental conditions conducive to development of DSD.

Other factors such as nutrient concentration might also affect the dynamics of some coral diseases. High nutrient exposure doubled rates of tissue loss in YBD diseased colonies of *M. faveolata* in Mexico (Bruno et al. 2003), showing response of the disease to changing nutrient conditions in surrounding waters (dissolved nutrients were artificially added). However, in similar nutrient experiments to those in Mexico, results from Puerto Rico showed no significant increase in number of lesions or rates of advance in YBD-infected colonies of *M. faveolata* when compared with controls (Bruno and Weil, unpublished). Another study showed higher prevalence of BBD in reefs closer to sewage effluents (Kaczmarsky et al. 2005).

Recent studies have shown that *A. palmata* colonies that were physically damaged by heavy swells had higher disease prevalence than undamaged colonies, with statistically greater prevalence when average monthly water temperature exceeded 28°C (Bright, personal communication 2010). Fragmentation of *A. palmata* and *A. cervicornis*, as well as other branching or columnar species during storms, leads to increases in number of colonies in the populations. However, during this process, open wounds and damaged tissues are presumably more susceptible to infection by opportunistic bacteria; thus, more intense storms predicted to occur with climate change could lead to more disease and higher mortality of these fragments.

5 Consequences and Management Implications

The potential of disease outbreaks to significantly change coral reefs was first shown by the massive mortalities of the acroporids and the black sea urchin *D. antillarum* in the Caribbean in the early 1980s. In a relatively short time and over a wide geographic region, populations of these species suffered up to 95% mortality (Gladfelter 1982; Lessios et al. 1984a, b; Carpenter 1990a, b) producing a cascade of significant ecological changes in the dynamics, function, and structure of coral reefs at local and geographic scales (Hughes 1994; Harvell et al. 1999; Aronson and Precht 2001a, b; Weil et al. 2003).

Disease etiology and dynamics seem to be highly variable across different spatial (populations, depth gradients, and reefs) and temporal scales (Weil et al. 2002; Bruckner and Bruckner 2006; Weil and Croquer 2009; Cróquer and Weil 2009a, b). Several infectious diseases have been persistent over the years throughout the Caribbean. Six of these have wide geographic distributions (BBD, WPD, YBD, DSD, ASP, and CCI) and could be considered chronic in many localities (Table 1) (Weil and Croquer 2009; Cróquer and Weil 2009a). Several more epizootics have occurred since the widespread WBD outbreak of the early 1980s, three of which have had wide geographic distributions and significant impact on the host species, WPD, ASP, and YBD (Bruckner and Bruckner 2006; Bruckner and Hill 2009; Weil and Croquer 2009; Cróquer and Weil 2009a; Weil et al. 2009a; Flynn and Weil, in press; Weil et al., in press).

Some of the most important reef-building coral genera (i.e., *Montastraea*, *Diploria*, *Siderastrea*, *Colpophyllia*) in the Caribbean are susceptible to at least five of the most prevalent and virulent diseases and several other compromised health problems (Fig. 7). Local populations of the three species of *Montastraea* and other important reef-building species in reefs off the Virgin Islands and Puerto Rico have been devastated by WPD epizootics, pervasive YBD, two intensive bleaching events, and their synergistic impact. For example, WPD from 2005 to 2007 caused more coral loss (over 60% loss in live tissue cover) than any other factor up until 2005/2006 in the USVI (Miller et al. 2009), some reefs off the east coast of Puerto Rico (García-Sais et al. 2008), and Curacao and Grenada (Weil and Cróquer 2009).

Similarly, reefs off La Parguera, on the southwest coast of Puerto Rico, showed an average loss of coral cover of 53.7% over 4 years (2004–2007), but as a consequence of two WPD epizootics, a persistent YBD outbreak and the bleaching event of 2005 (Weil et al., in press) (Fig. 9). Similar coral tissue losses have been reported for Mona and Desecheo islands west of Puerto Rico where YBD and bleaching have been the major problems (Bruckner and Bruckner 2006; Bruckner and Hill 2009).

As average water temperatures increase, bleaching events have become more frequent and intense over wider geographic scales, affecting most zooxanthellae-bearing reef organisms to increasing depths. Before 2005, bleaching in the Caribbean caused variable, but generally low coral

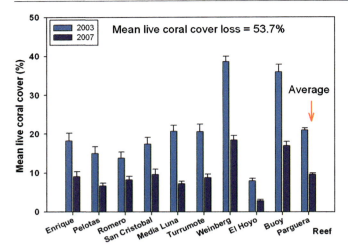

Fig. 9 Average loss in live coral cover in nine reefs off La Parguera between 2003 and 2004, and 2007. The average loss in live coral tissue for the area was 53.7% (±7.2%) (Modified from Weil et al., in press)

mortality, nothing like the mass mortalities of the scale observed in the Indo-Pacific. Only the recent extensive bleaching of 2005, the worst ever recorded for the region, produced widespread mortality in several reefs of the eastern Caribbean, which was compounded with disease outbreaks in several localities (Miller et al. 2009; Rogers et al. 2008a, b; McClanahan et al. 2009; Cróquer and Weil 2009a, b).

Significant coral mortality due to bleaching was observed for the first time in Puerto Rico during this event; however, this mortality was not widespread across all species, suggesting differential susceptibility and specific resistance to bleaching. Most agaricids, *M. ferox*, and *A. palmata* showed the highest mortality due to bleaching. A high proportion of other reef organisms (octocorals, hydrocorals, and zoanthids, and a few sponges) bleached in many localities (Fig. 5), with some of these showing significant population mortalities (i.e. *Millepora* spp. *Erythropodium caribbaeorum*, and *Palythoa caribbaeorum*) (McClanahan et al. 2009; Prada et al. 2009; Weil et al., in press).

At the peak of the bleaching, outbreaks of WPD started to be observed in Puerto Rico, the Virgin Islands, and Grenada (Hernández-Delgado et al. 2006; Cróquer and Weil 2009b; Miller et al. 2009; Rogers et al. 2008a, b; Weil et al., in press). Increasing water temperature trends in the region, possibly associated with global climate change, and the frequency and intensity of bleaching events could have affected and will probably affect the emergence, dispersion, and virulence of coral reef diseases and their consequences in the recent past and the near future. An increasing number of colonies and species are being infected by more than one disease at the same time. Some colonies of *M. faveolata*, for example, have been observed with four different pathologies simultaneously (Cróquer and Weil 2009a, b; Weil and Croquer, unpublished data), which significantly increases rates of tissue and colony mortality over time.

Furthermore, recent studies showed that in addition to the general tissue loss, some diseases could significantly reduce the fitness (reproductive output) in important reef species, similar to what was reported for bleaching almost 20 years ago (Szmant and Gassman 1990). Sexual reproduction is critical to coral population dynamics and the long-term regeneration of coral reefs. Recurrent recruitment failure and low reproductive output in corals have been highlighted as explanations as to why reefs are not recovering from major coral losses (Hughes and Connell 1999; Hughes and Tanner 2000). Recent data showed that ASP significantly reduced fecundity (fitness) in *G. ventalina* (Flynn and Weil 2008) and YBD significantly affected the reproductive output of *M. faveolata* (Weil et al. 2009). Furthermore, Cervino et al. (2001, 2004a, 2008) indicated that all the pale, yellowish areas in the yellow band lesions are depleted of zooxanthellae, which would reduce local energy supply and potential energy available for the rest of the colony.

Thirty two percent of reef-building scleractinian corals around the world face an elevated risk of extinction due mainly to bleaching and disease driven presumably by elevated sea water temperature and further exacerbated by local anthropogenic stressors (Carpenter et al. 2008). The proportion of threatened (not including Near Threatened coral species) recognized by IUCN exceeds that of most terrestrial animal groups apart from amphibians, particularly because of corals' apparent susceptibility to climate change (particularly high sea water temperatures) and local anthropogenic factors (Carpenter et al. 2008). This plus all the recent reports of high mortality rates due to local and/or extensive disease outbreaks affecting coral reefs worldwide is a major cause of concern for the future of these important tropical marine communities.

Progress in coral disease research requires collaboration among experts across many different disciplines, including genetics, physiology, cell biology, ecology, pathology, microbiology, and epidemiology. We need innovative, preferably nondestructive techniques to diagnose diseases. One promising new approach involves the use of custom-designed microarrays to characterize microbial patterns (Kellogg and Zawada 2009). Sequential, frequent sampling of coral colonies with and without disease will reveal changes in microbial communities over time. Additional histological analysis of healthy and diseased corals should provide further clues. Further understanding will come from laboratory experiments that test the effects of temperature, irradiance, sediments, nutrients, and other pollutants on the development and progression of disease in corals. Currently, the tools for genotyping coral colonies are available for only a few species, and developing these tools for other major reef-building species would help us to evaluate whether or not certain genotypes are more resistant to diseases.

Even after more than 35 years since the first report of a coral disease in the Caribbean, researchers keep finding new diseases affecting corals and other important invertebrates and CCA groups responsible for building and maintaining these important tropical communities. Diseases of coral reef organisms have become one of the most, if not the most, important factors accelerating the decline of coral reefs and the potential loss of biodiversity (when the food and energy sources and the three-dimensional limestone framework in these communities disappear), compromising the future integrity of many coral reefs in the western Atlantic. However, to date there is no report of any coral species that has been extirpated from disease, even locally.

Reefs in the Caribbean have gone through significant changes in community structure with decreases in coral cover and increases in macroalgal cover (Edmunds 1991; Carpenter 1990a, b; Hughes 1994; Bellwood et al. 2004; Rogers and Miller 2006; Weil et al., in press). The impact of diseases on the reproductive output of corals and octocorals further hinder their potential future recovery. Ongoing monitoring programs and yearly surveys in many localities have failed to show any significant recovery following the recent outbreaks (Weil, unpublished data; C.S. Rogers, personal communication). Even after 25 years, the acroporids and sea urchin populations have not been able to recover from the epizootic events of the early 1980s. Collapsed coral populations would produce significantly fewer larvae, so recruitment and juvenile survivorship would be potentially very low making it difficult, or slow, for populations to recover, even if environmental conditions and other factors are favorable. In Puerto Rico, surviving populations and new recruits and juveniles of *A. palmata* were frequently affected by storms, hurricanes, sedimentation, and anthropogenic impacts after the mass mortalities of the 1980s (Weil et al. 2003).

Even though some positive results indicating lower numbers and prevalence of coral diseases inside marine protected areas (MPAs) and/or marine reserves have been reported (Raymundo et al. 2008), more information is needed to generalize the potential "protection" these areas provide to coral populations. Coral reefs within and outside of MPAs in the US Virgin Islands had losses of over 60% of the coral cover following the disease outbreak associated with the 2005 bleaching episode (Miller et al. 2009; Rogers et al. 2008a, b). In other localities, disease prevalence inside MPAs was not lower than outside MPAs (Coelho and Manfrino 2007: Page et al. 2009). It is challenging to test the hypothesis that coral reefs within MPAs will be less susceptible to diseases than those outside protected areas. A variety of complex factors has to be controlled for, for example:

- The actual level of protection that the area receives (reflecting compliance with regulations and effectiveness of enforcement)
- The length of time that protective measures have been in place
- The type of protection (prohibition of fishing, anchoring, etc.)
- The history of fishing and any other extractive uses before the area was effectively protected [some areas might have had very little fishing to begin with, for example]
- Other stressors like runoff, pollution that could affect the condition of the marine resources
- Prevalence of diseases before the MPA was established
- The validity of the control ("reference") areas used for comparison with protected areas

Our current limited knowledge of the pathogenesis and other important aspects of most coral reef diseases (Lesser et al. 2007) undermines our ability to develop strategies to solve the problem in the near future. MPA boundaries will not prevent disease outbreaks or bleaching events, but, protecting large, genetically variable fecund populations of the most important reef-building groups could increase the species survivorship by protecting potentially resistant genotypes that could then reseed other degraded populations outside the reserves (Vollmer and Kline 2008).

6 Summary

Coral reefs are declining around the world due to natural and/or human-produced stressors, including global climate change. Recently, diseases of coral reef organisms have become increasingly important in the deterioration of these important marine communities. The Caribbean has been dubbed a "disease hot spot" due to the fast emergence of diseases and frequent epizootic events with significant coral mortalities in the last 30 years. Fifteen disease conditions affecting corals and other important reef organisms have been described, but most do not have defined pathologies nor have they been well characterized. Of these, ten conditions with consistent signs are affecting most of the important reef-building coral species, five are affecting at least 12 species of octocorals, two at least three species of crustose coralline algae (CCA), and two a single zoanthid species, while several uncharacterized conditions affect sponges and other reef organisms. Variability in disease manifestation and distribution complicates their characterization and identification. Most of these diseases have a Wider Caribbean distribution with a few, like black band disease, showing worldwide distributions.

Three widespread and several local outbreaks produced significant mortalities over a wide geographic range, with a cascade of significant changes in community structure, decreases in coral cover, and increases in macroalgal cover. The impact of diseases on the reproductive output of corals and octocorals further hinders their potential for recovery. Our limited understanding of the pathogenesis of most coral reef diseases undermines our ability to develop strategies to reduce their effects in the near future. Further advances in coral disease research require collaboration among experts across many different disciplines, development and use of new techniques, and controlled laboratory experiments. Caribbean coral reefs will benefit from protection of resistant coral genotypes along with greater efforts to reduce the manageable stresses caused by humans. Protecting large, genetically variable fecund populations of the most important reef-building species could increase the survivorship of resistant genotypes that could then reseed degraded populations. However, this has to come together with greater efforts to manage human activities that stress coral reefs, and the reestablishment of former environmental quality for the survival of coral reefs in the region.

Acknowledgments We thank Zvy and Maya Dubinsky for their kind invitation to be part of this book. Thierry Work, Kim Ritchie, Ken Sulak, Noga Stambler, and an anonymous reviewer made important comments and suggestions that helped improve this contribution. Some results presented here come from research funded by the GEF-World Bank CRTR program through the disease working group and NOAA-CRES Grant (NA170P2919) to E. Weil.

References

Acosta A (2001) Disease in zoanthids: dynamics in space and time. Hydrobiologia 460:113–130

Aeby GS (2006a) Outbreak of coral disease in the Northwestern Hawaiian islands. Coral Reefs 24:481

Aeby GS (2006b) Baseline levels of coral disease in the Northwestern Hawaiian islands. Atoll Res Bull 543:471–488

Aeby GS, Santavy DL (2006) Factors affecting susceptibility of the coral *Montastraea faveolata* to black-band disease. Mar Ecol Prog Ser 318:103–110

Ainsworth TD, Fine M, Roff G, Hoegh-Guldberg O (2008) Bacteria are not the primary cause of bleaching in the Mediterranean coral *Oculina patagonica*. ISME J 2:67–73

Antonius A (1973) New observations on coral destruction in reefs. In: Tenth Meeting of the Association of Island Marine Laboratories of the Caribbean (abstract). University of Puerto Rico, Mayaguez, p 3

Antonius A (1977) Coral mortality in reefs: a problem for science and management. In: Proceedings of 3rd international coral reef symposium, vol 1, Miami, 1977, pp 617–623

Antonius A (1985) Coral diseases in the Indo-Pacific: a first record. PSZNI Mar Ecol 6:197–218

Aronson RB, Precht WF (2001a) Evolutionary palaeoecology of Caribbean coral reefs. In: Allmon WD, Bottjer DJ (eds) Evolutionary paleoecology: the ecological context of macro-evolutionary change. Columbia University Press, New York, pp 171–233

Aronson RB, Precht WF (2001b) White-band disease and the changing face of Caribbean coral reefs. In: Porter JW (ed) The ecology and etiology of newly emerging marine diseases, developments in hydrobiology. Kluwer, Dordrecht, pp 23–35

Ballantine D, Weil E, Ruiz H (2005) Coralline white band syndrome: a coralline algal affliction in the tropical Atlantic. Coral Reefs 24:117

Beeden R, Willis BL, Raymundo LJ, Page C, Weil E (2008) Underwater cards for assessing Coral health on Indo-Pacific reefs. GEF-CRTR Program, Center for Marine Sciences, University of Queensland, Brisbane, 18 pp

Bellwood DR, Hughes TP, Folke C, Nystrom M (2004) Confronting the coral reef crisis. Nature 429:827–833

Ben-Haim Y, Thompson FL, Thompson CC, Cnockaert MC, Hoste B, Swings J, Rosenberg E (2003a) *Vibrio coralliilyticus* sp. nov., a temperature-dependent pathogen of the coral Pocillopora damicornis. Int J Syst Evol Microbiol 53:309–315

Ben-Haim Y, Zicherman-Keren M, Rosenberg E (2003b) Temperature-regulated bleaching and lysis of the coral *Pocillopora damicornis* by the novel pathogen *Vibrio coralliilyticus*. Appl Environ Microbiol 69:4236–4242

Borneman EH, Wellington GW (2005) Pathologies affecting reef corals at the Flower Garden Banks, Northwestern Gulf of Mexico. Gulf Mexico Sci 1:95–106

Borger JL (2003) Three scleractinian coral diseases in Dominica, West Indies: distribution, infection patterns and contribution to coral tissue mortality. Rev Biol Trop 51(suppl 4):25–38

Borger JL, Steiner S (2005) The spatial and temporal dynamics of coral diseases in Dominica, West Indies. Bull Mar Sci 77:137–154

Bourne D, Iida Y, Uthicke S, Smith-Keune C (2007) Changes in coral-associated microbial communities during a bleaching event. ISME J 2:350–363

Boyett HV, Bourne DG, Willis BL (2007) Elevated temperature and light enhance progression and spread of black band disease on staghorn corals of the Great Barrier Reef. Mar Biol 151:1711–1720

Brown BE, Ogden JC (1993) Coral bleaching. Sci Am 268:64–70

Bruckner AW (2002) Priorities for effective management of coral diseases. NOAA Technical Memorandum, NMFS-OPR-22, 54 pp

Bruckner AW (2009) The global perspective of incidence and prevalence of coral diseases. In: Galloway SB, Bruckner AW, Woodley CM (eds) (2007) Coral health and disease in the Pacific: vision for action. NOAA Technical Memorandum, NOS NCCOS 97 and CRCP 7 National Oceanic and Atmospheric Administration, Silver Spring, pp 90–290

Bruckner AW, Bruckner RJ (1997) Outbreak of coral disease in Puerto Rico. Coral Reefs 16:260

Bruckner AW, Bruckner R (2006) Consequences of yellow band disease on *Montastraea annularis* (species complex) populations on remote reefs off Mona Island, Puerto Rico. Dis Aquat Org 69:67–73

Bruckner AW, Hill R (2009) Ten years of change to coral communities off Mona and Desecheo islands, Puerto Rico, from disease and bleaching. Dis Aquat Org 87:19–31

Bruno JF, Selig ER (2007) Regional decline of coral cover in the Indo-Pacific: timing, extent, and sub-regional comparisons. PLoS ONE 2:e711. doi:10.1371/journal.pone.0000711

Bruno JF, Peters LE, Harvell CD, Hettinger A (2003) Nutrient enrichment can increase the severity of coral diseases. Ecol Lett 6:1056–1061

Bruno JF, Selig ER, Casey KS, Page CA, Willis BL (2007) Thermal stress and coral cover as drivers of coral disease outbreaks. PLoS Biol 5:e124. doi:10.1371/journal.pbio.0050124

Bythell J, Pantos O, Richardson L (2004) White plague, white band and other "white" diseases. In: Rosenberg E, Loya Y (eds) Coral health and disease. Springer, New York, pp 351–366

Carlton R, Richardson LL (1995) Oxygen and sulfide dynamics in an horizontally migrating cyanobacterial mat: black-band disease of corals. FEMS Microbiol Ecol 18:144–162

Carpenter RC (1981) grazing by *Diadema antillarum* (Philippi) and its effects on the benthic algal community. J Mar Res 39:749–765

Carpenter RC (1985) Sea urchin mass mortality: effects on reef algal abundance, species composition, and metabolism and other coral reef herbivores. In: Proceedings of 5th international coral reef congress, vol 4, Tahiti, 1985, pp 53–60

Carpenter RC (1990a) Mass mortality of *Diadema antillarum*. I. Long-term effects on sea urchin population-dynamics and coral reef algal communities. Mar Biol 104:67–77

Carpenter RC (1990b) Mass mortality of *Diadema antillarum*. II. Effects on population densities and grazing intensity of parrotfishes and surgeonfishes. Mar Biol 104:79–86

Carpenter K, Livingston S et al (2008) One third of reef-building corals face elevated extinction risk from climate change and local impacts. Science 321:560–563

Cervino JM, Smith G (1997) Corals in peril. Ocean Realm 2:33–34

Cervino J, Goreau TJ, Nagelkerken I, Smith GW, Hayes R (2001) Yellow band and dark spots syndrome in Caribbean corals: distribution, rate of spread, cytology, and effects on abundance and division of zooxanthellae. In: Porter JW (ed) The ecology and etiology of newly emerging marine diseases, developments in hydrobiology. Kluwer, Dordrecht, pp 53–63

Cervino JM, Hayes RL, Polson SW, Polson SC, Goreau TJ, Martinez RJ, Smith GW (2004a) Relationship of *Vibrio* species infection and elevated temperatures to yellow blotch/band disease in Caribbean corals. Appl Environ Microbiol 70:6855–6864

Cervino JM, Hayes RL, Goreau TJ, Smith GW (2004b) Zooxanthellae regulation in yellow blotch/band and other coral diseases contrasted with temperature related bleaching: In situ destruction vs. expulsion. In: 4th international symbiosis congress, vol 37, Nova Scotia, 2004, pp 63–85

Cervino JM, Thompson FL, Gómez-Gil B, Lorence EA, Goreau TJ, Hayes RL, Winiarski KB, Smith GW, Hughen K, Bartells E (2008) *Vibrio* pathogens induce yellow band disease in Caribbean and Indo-Pacific reef-building corals. J Appl Microbiol. doi:10.1111/J.1365-2672.2008.03871x

Coelho VR, Manfrino C (2007) Coral community decline at a remote Caribbean island: marine no-take reserves are not enough. Aquat Conserv Mar Freshw Ecosyst 17:666–685

Cooney RP, Pantos O, Le Tissier MDA, Barer MR, O'Donnell AG, Bythell JC (2002) Characterization of the bacterial consortium associated with black band disease in coral using molecular microbiological techniques. Environ Microbiol 47:401–413

Cróquer A, Weil E (2009a) Local and geographic variability in distribution and prevalence of coral and octocoral diseases in the Caribbean II: genera-level analysis. Dis Aquat Org 83:209–222

Cróquer A, Weil E (2009b) Changes in Caribbean coral disease prevalence after the 2005 bleaching event. Dis Aquat Org 87:33–43

Cróquer A, Bastidas C, Lipscomb D, Rodríguez-Martínez RE, Jordan-Dahlgren E, Guzmán HM (2006a) First report of folliculinid ciliates affecting Caribbean corals. Coral Reefs 25:187–191

Cróquer A, Bastidas C, Lipscomb D (2006b) Folliculinid ciliates: a new threat to Caribbean corals? Dis Aquat Org 69:75–78

Cróquer A, Weil E, Zubillaga AL, Pauls SM (2005) Impact of an outbreak of white plague II on a coral reef in the Archipelago Los Roques National Park, Venezuela. Carib J Sci 41:815–823

Denner EBM, Smith GW, Busse HJ, Schumann P, Nartz T, Polson SW, Lubitz W, Richar-son LL (2003) *Aurantimonas coralicida* gen. nov., sp. nov., the causative agent of white plague type II on Caribbean scleractinian corals. Int J Syst Evol Microbiol 53:1115–1122

Diaz JM, Garzon-Ferreira J, Zea S (1995) Los arrecifes coralinos de la isla de San Andrés, Colombia: estado actual y perspectivas para su conservación. Academia Colombiana de Ciencias Exactas, Físicas, y Naturales, Colección Jorge Alvarez Lleras, 95 pp

Dustan P (1977) Vitality of reef coral populations off Key Largo, Florida: recruitment and mortality. Environ Geol 2:51–58

Dustan P (1999) Coral reefs under stress: sources of mortality in the Florida Keys. Nat Resour Forum 23:147–155

Edmunds PJ (1991) Extent and effect of black band disease on Caribbean reefs. Coral Reefs 10:161–165

Epstein PR, Sherman K, Spanger-Siegfried E, Langston A, Prasad S, McKay B (1998) Marine ecosystems: emerging diseases as indicators of change. Health Ecological and Economic Dimensions (HEED), NOAA Global Change Program, 85 pp

Francini-Filho RB, Moura RL, Thompson FL, Reis RM, Kaufman L, Kikuchi RKP, Leão ZMAN (2008) Diseases leading to accelerated decline of reef corals in the largest South Atlantic reef complex (Abrolhos Bank, eastern Brazil). Mar Poll Bull 56:1008–1014

Flynn K (2008) Impact of the fungal disease aspergillosis on populations of the sea fan *Gorgonia ventalina* (Octocorallia, Gorgonacea) in La Parguera, Puerto Rico. MS thesis, University of Puerto Rico, Puerto Rico, 85 pp

Flynn K, Weil E (2008) Impact of aspergillosis on the reproduction of the sea fan *Gorgonia ventalina*. Abstract. In: 11th international coral reef symposium, Ft. Lauderdale, 2008, p 307

Flynn K, Weil E (2010) Variability of Aspergillosis in *Gorgonia ventalina* in La Parguera, Puerto Rico. Caribb J Sci (in press)

Frias-Lopez J, Zerkle AL, Bonheyo GT, Fouke BW (2002) Partitioning of bacterial communities between seawater and healthy black band diseased and dead coral surfaces. Appl Environ Microbiol 68:2214–2228

Frias-Lopez J, Bonheyo GT, Jin Q, Fouke BW (2003) Cyanobacteria associated with coral black band disease in Caribbean and Indo-Pacific reefs. Appl Environ Microbiol 69:2409–2413

Galloway SB, Work TM, Bochsler VS et al (2007) Coral disease and health workshop: coral histopathology II. NOAA Technical Memorandum NOS NCCOS 56 and NOAA Technical Memorandum CRCP 4. National Oceanic and Atmospheric Administration, Silver Spring, 84 p

García-Sais J, Appeldoorn R, Battista T, Bauer L, Bruckner A, Caldow C, Carrubba L, Corredor J, Diaz E, Lilyestrom C, García-Moliner G, Hernández-Delgado E, Menza C, Morell J, Pait A, Sabater J, Weil E, Williams E, Williams S (2008) The state of coral reef ecosystems in Puerto Rico. In: Waddell JE, Clarke AM (eds) The state of coral reef ecosystems of the United States and Pacific freely associated states. NOAA Technical Memorandum, NOS NCCOS 73, pp 75–116

Gardner TA, Cote IM, Gill JA, Grant A, Watkinson AR (2003) Long-term region-wide declines in Caribbean corals. Science 301:958–960

Garrett P, Ducklow H (1975) Coral disease in Bermuda. Nature 253:349–350

Garzón-Ferreira J, Gil D (1998) Another unknown Caribbean coral phenomenon? Reef Encounters 24:10–13

Garzón-Ferreira J, Zea S (1992) A mass mortality of *Gorgonia ventalina* (Cnidaria: Gorgoniidae) in the Santa Marta area, Caribbean coast of Colombia. Bull Mar Sci 50:522–526

Geiser DM, Taylor JW, Ritchie KM, Smith GW (1998) Cause of sea-fan death in the West Indies. Nature 394:137–138

Gil-Agudelo DL (1998) Caracteristicas, incidencia y distribución de la enfermedad de "Lunares oscuros": en corales pétreos del Parque Nacional Natural Tayrona, Caribe colombiano. Marine biology. Universidad de Bogota Jorge Tadeo Lozano, Bogota, 90 p

Gil-Agudelo DL, Garzón-Ferreira J (2001) Spatial and seasonal variation of dark spots disease in coral communities of the Santa Marta area (Colombian Caribbean). Bull Mar Sci 69:619–629

Gil-Agudelo DL, Smith GW, Weil E (2006) The white band disease type II pathogen in Puerto Rico. Rev Biol Trop 54:59–67

Gil-Agudelo D, Smith GW, Garzón-Ferreira J, Weil E, Peterson D (2004) Dark spots disease and yellow band disease, two poorly known coral diseases with high incidence in Caribbean reefs. In: Rosenberg E, Loya Y (eds) Coral health and disease. Springer, New York, pp 337–350

Gil-Agudelo D, Fonseca DP, Weil E, Garzón-Ferreira J, Smith GW (2007) Bacterial communities associated with the mucopolysaccharide layers of three coral species affected and unaffected with dark spots disease. Can J Microbiol 53:465–471

Gladfelter WB (1982) White band disease in *Acropora palmata*: implications for the structure and growth of shallow reefs. Bull Mar Sci 32:639–643

Goreau TJ, Cervino J, Goreau M, Hayes R, Hayes M, Richardson L, Smith GW, DeMeyer G, Nagelkerken I, Garzon-Ferreira J, Gill D, Peters EC, Garrison G, Williams EH, Bunkley-Williams L, Quirolo C, Patterson K (1998) Rapid spread of diseases in Caribbean coral reefs. Rev Biol Trop 46:157–171

Gotchfeld D, Olson J, Slattery M (2006) Colony versus population variation in susceptibility and resistance to dark spot syndrome in the Caribbean coral *Siderastrea siderea*. Dis Aquat Org 69:53–65

Green EP, Bruckner AW (2000) The significance of coral disease epizootiology for coral reef conservation. Biol Conserv 96:347–461

Guzmán HM, Cortés J (1984) Mass death of *Gorgonia flabellum* L. (Octocorallia: Gorgonidae) in the Caribbean coast of Costa Rica. Rev Biol Trop 32:305–308

Harvell CD, Kim K, Burkholder JM, Colwell RR, Epstein PR, Grimes DJ, Hofmann EE, Lipp EK, Osterhaus ADME, Overstreet AM, Porter JW, Smith GW, Vasta GR (1999) Emerging marine diseases–climate links and anthropogenic factors. Science 285:1505–1510

Harvell CD, Kim K, Quirolo C, Weir J, Smith GW (2001) Coral bleaching and disease: contribution to 1998 mass mortality of *Briaeorum asbestinum* (Octocorallia, Gorgonacea). In: Porter JW (ed) The ecology and etiology of newly emerging marine diseases, Developments in hydrobiology. Kluwer, Dordrecht, pp 97–104

Harvell CD, Mitchell CE, Ward JR, Altizer S, Dobson AP, Ostfeld RS, Samuel MD (2002) Climate warming and disease risk for terrestrial and marine biota. Science 296:2158–2162

Harvell CD et al (2005) The rising tide of ocean diseases: unsolved problems and research priorities. Front Ecol Environ 2:375–382

Harvell CD, Markel S, Jordan-Dahlgren E, Raymundo LJ, Rosenberg E, Smith GW, Willis BL, Weil E (2007) Coral disease, environmental drivers and the balance between coral and microbial associates. Oceanography 20:36–59

Harvell CD, Altize S, Cattadori IM, Harrington L, Weil E (2009) Climate change and wildlife diseases: when does the host matter the most? Ecology 90:912–920

Hayes ML, Bonaventura J, Mitchell TP, Prospero JM, Shinn EA, Van Dolah F, Barber RT (2001) How are climate and marine biological outbreaks functionally linked? Hydrobiologia 460:213–220

Hickerson E, Schmahl GP (2006) Flower Garden Banks National Marine Sanctuary coral disease monitoring. Cruise Summary, 9 pp

Hernández-Delgado EA, Toledo CG, Claudio H, Lassus J, Lucking MA, Fonseca J, Hall K, Rafols J, Horta H, Sabat AM (2006) Spatial and taxonomic patterns of coral bleaching and mortality in Puerto Rico during year 2005. Coral bleaching response workshop, NOAA, St. Croix, USVI, p 16

Hoegh-Guldberg O (1999) Climate change, coral bleaching and the future of the world's coral reefs. Mar Freshw Res 50:839–866

Hoegh-Guldberg O, Mumby PJ, Hooten AJ, Steneck RS, Greenfield P, Gomez E, Harvell CD, Sale PF, Edwards AJ, Caldeira K, Knowlton N, Eakin CM, Iglesias-Prieto R, Muthiga N, Bradbury H, Dubi A, Hatziolos ME (2007) Coral reefs under rapid climate change and ocean acidification. Science 318:1737–1742

Hogarth WT (2006) Endangered and threatened species: final listing determinations for the Elkhorn coral and Staghorn coral. Federal Register, pp 26852–26861

Hubbard DK, Burke RB, Gill IP, Ramirez WR, Sherman C (2008) Coral reef geology: Puerto Rico and the US Virgin Islands. In: Riegl B, Dodge R (eds) Coral reefs of the USA. Springer, Dordrecht, pp 263–302

Hughes TP (1994) Catastrophes, phase shifts and large scale degradation of a Caribbean coral reef. Science 265:1547–1549

Hughes TP, Connell JH (1999) Multiple stressors on coral reefs: a long-term perspective. Limnol Oceanogr 44:932–940

Hughes TP, Tanner JE (2000) Recruitment failure, life histories, and long-term decline of Caribbean corals. Ecology 81:2250–2263

Hughes TP, Reed DC, Boyle MJ (1987) Herbivory on coral reefs: community structure following mass mortalities of sea urchins. J Exp Mar Biol Ecol 113:39–59

Hughes TP, Baird AH, Bellwood DR, Card M, Connolly SR, Folke C, Grosberg R, Hoegh-Guldberg O, Jackson JBC, Kleypas J, Lough JM, Marshall P, Nystrom M, Palumbi SR, Pandolfi JM, Rosen B, Roughgarden J (2003) Climate change, human impacts and the resilience of coral reefs. Science 301:929–933

Jackson JBC, Kirby MX, Berger WH, Bjorndal KA, Botsford LW, Bourque BJ, Bradbury RH, Cooke R, Erlandson J, Estes JA, Hughes TP, Kidwell S, Lange CB, Lenihan HS, Pandolfi JM, Peterson CH, Steneck RS, Tegner MJ, Warner RR (2001) Historical overfishing and the recent collapse of coastal ecosystems. Science 293:629–638

Jolles A, Sullivan PK, Alker A, Harvell CD (2002) Disease transmission of aspergillosis in sea fans: inferring process from spatial patterns. Ecology 83:2373–2378

Jonas RB, Gillevet P, Peters E (2006) Comparison of bacterial communities among corals exhibiting white-syndrome (disease) signs in the Flower Garden Banks National Marine Sanctuary. Progress report, 13 pp

Kaczmarsky LT, Draud M, Williams EH (2005) Is there a relationship between proximity to sewage effluent and the prevalence of coral disease? Caribb J Sci 41:124–137

Kellogg CA, Zawada DG (2009) Applying new methods to diagnose coral diseases. USGS Fact Sheet 2009–3113

Kim K, Harvell CD (2002) Aspergillosis of sea fan corals: disease dynamics in the Florida Keys. In: Porter J, Porter K (eds) The Everglades, Florida Bay, and coral reefs of the Florida Keys, an ecosystem sourcebook. CRC Press, Boca Raton, pp 813–824

Kimes NE, Van Nostrand JD, Weil E, Zhou J, Morris PJ (2010) Microbial functional structure of *Montastraea faveolata*, an important Caribbean reef-building coral, differs between healthy and Caribbean yellow-band diseased colonies. Environ Microbiol 12:541–561. doi:10.1111/j.1462-2920. 2009.02113.x

Knowlton N, Rohwer F (2003) Multispecies microbial mutualisms on coral reefs: the host as a habitat. Am Nat 162:51–62

Korrubel JL, Riegl BR (1998) A new disease from the southern Arabian Gulf. Coral Reefs 17:22

Lesser MP, Bythell JC, Gates RD, Johnstone RW, Hoegh-Guldberg O (2007) Are infectious diseases really killing corals? Alternative interpretations of the experimental and ecological data. J Exp Mar Biol Ecol 346:36–44

Lessios HA (1988) Population dynamics of *Diadema antillarum* (Echinodermata: Echinodea) following mass mortality in Panamá. Mar Biol 95:515–526

Lessios HA, Cubit JD, Robertson RD, Shulman MJ, Parker MR, Garrity SD, Levings SC (1984a) Mass mortality of *Diadema antillarum* on the Caribbean coast of Panama. Coral Reefs 3:173–182

Lessios HA, Robertson DR, Cubit JD (1984b) Spread of *Diadema* mass mortality throughout the Caribbean. Science 226:335–337

Littler M, Littler D (1995) Impact of CLOD pathogen on Pacific coral reefs. Science 267:1356–1360

Loya Y (2004) The coral reefs of Eilat, past, present and future: three decades of coral community structure studies. In: Rosenberg E, Loya Y (eds) Coral health and disease. Springer, New York, pp 1–34

Martin SW, Meek AH, Willerberg P (1987) Veterinary epidemiology, principles and methods. Iowa State University Press, Ames, p 343

McClanahan TR, Weil E, Cortés J, Baird A, Ateweberhan M (2009) Consequences of coral bleaching for sessile organisms. In: Van Oppen M, Lough J (eds) Coral bleaching: patterns, processes, causes and consequences, Springer Ecological Studies. Springer, Berlin, pp 121–138

Miller J, Rogers C, Waara R (2001) Monitoring the coral disease plague type II on coral reefs of St. John, US Virgin Islands. In: Proceedings of the 30th Sientific Meeting of the Association of Marine Laboratories of the Caribbean, Puerto Rico. Rev Biol Trop 51(Suppl 4):47–56

Miller J, Waara R, Muller E, Rogers CS (2006) Coral bleaching and disease combine to cause extensive mortality on reefs in US Virgin Islands. Coral Reefs 25:418

Miller J, Muller E, Rogers CS, Waara R, Atkinson A, Whelan KRT, Patterson M, Witcher B (2009) Coral disease following massive bleaching in 2005 causes 60% decline in coral cover on reefs in the US Virgin Islands. Coral Reefs 28:925–937

Muller EM, Rogers CS, Spitzack AS, van Woesik R (2008) Water temperature influences disease prevalence and severity on *Acropora palmata* (Lamarck) at Hawksnest Bay, St. John, US Virgin Islands. Coral Reefs 27:191–195

Mydlarz LD, Couch CS, Weil E, Smith GW, Harvell CD (2009) Immune defenses of healthy, bleached and diseased *Montastraea faveolata* during a natural bleaching event. Dis Aquat Org 87:67–71

Nagelkerken I, Buchan K, Smith GW, Bonair K, Bush P, Garzon-Ferreira J, Botero L, Gayle P, Harvell CD, Heberer C, Kim K, Petrovic C, Pots L, Yoshioka P (2007a) Widespread disease in Caribbean sea fans: I. Spreading and general characteristics. In: Proceedings of 8th international coral reef symposium, vol 1, Panama, 1997, pp 679–682

Nagelkerken I, Buchan K, Smith GW, Bonair K, Bush P, Garzon-Ferreira J, Botero L, Gayle P, Harvell CD, Heberer C, Kim K, Petrovic C, Pots L, Yoshioka P (2007b) Widespread disease in Caribbean sea fans: II. Pattern of infection, and tissue loss. Mar Ecol Prog Ser 160:255–263

Nugues MN, Smith GW, van Hooidonk RJ, Seabra MI, Bak RPM (2004) Algal contact as a trigger for coral disease. Ecol Lett 7:919–923

Ostrander GK, Meyer-Armstrong K, Knobbe ET, Gerace D, Scully EP (2000) Rapid transition in the structure of a coral reef community: the effects of coral bleaching and physical disturbance. Proc Natl Acad Sci U S A 97:5294–5302

Page C, Willis B (2006) Distribution, host range and large-scale spatial variability in black band disease prevalence on the Great Barrier Reef, Australia. Dis Aquat Org 69:41–51

Page CA, Baker DA, Harvell CD, Golbuu Y, Raymundo L, Neale SJ, Rosell KB, Rypien KL, Andras JP, Willis BL (2009) Influence of marine reserves on coral disease prevalence. Dis Aquat Org 87:135–150

Pandolfi JM, Bradbury RH, Sala E, Hughes TP, Bjorndal KA, Cooke RG, McArdle D, McClanahan L, Newman MJH, Paredes G, Warner RR, Jackson JBC (2003) Global trajectories of the long-term decline of coral reef ecosystems. Science 301:955–958

Pantos O, Cooney RP, Le Tissier MDA, Barer MR, O'Donnell AG, Bythell JC (2003) The bacterial ecology of a plague-like disease affecting the Caribbean coral *Montastrea annularis*. Environ Microbiol 5:370–382

Patterson KL, Porter JW, Ritchie KB, Polson SW, Mueller E, Peters EC, Santavy DL, Smith GW (2002) The etiology of white pox, a lethal disease of the Caribbean elkhorn coral, *Acropora palmata*. PNAS 99:8725–8730

Peters E (1997) Diseases of coral reef organisms. In: Birkeland C (ed) Life and death of coral reefs. Kluwer, Boston, pp 114–136

Peters EC, Oprandy JJ, Yevich PP (1983) Possible cause of "white band disease" in Caribbean corals. J Invert Pathol 41:394–396

Petes LE, Harvell CD, Peters EC, Webb MAH, Mullen KM (2003) Pathogens compromise reproduction and induce melanization in Caribbean sea fans. Mar Ecol Prog Ser 264:167–171

Porter JW, Meier O (1992) Quantification of loss and change in Floridian reef coral populations. Am Zool 23:625–640

Polson SW, Higgins JL, Woodley CM (2009) PCR-based assay for detection of four coral pathogens. In: Proceedings of 11th International coral reef symposium session, vol 8, Ft. Lauderdale, 2008, pp 247–251

Prada C, Weil E, Yoshioka P (2009) Octocoral bleaching under unusual thermal stress. Coral Reefs. doi:10.1007/s00338-009-0547-z

Raymundo LJ, Harvell CD, Reynolds TL (2003) *Porites* ulcerative white spot disease: description, prevalence, and host range of a new coral disease affecting Indo-Pacific reefs. Dis Aquat Org 56:95–104

Raymundo LJ, Couch CS, Harvell CD (eds) (2008) Coral disease handbook. Guidelines for assessment, monitoring and managing. GEF-CRTR program. Currie Communications, Australia, 121 pp

Reshef L, Koren O, Loya Y, Zilber-Rosenberg I, Rosenberg E (2006) The coral probiotic hypothesis. Environ Microbiol 8:2068–2073

Richardson LL (1997) Occurrence of the black band disease cyanobacterium on healthy corals of the Florida Keys. Bull Mar Sci 61:485–490

Richardson LL (1998) Coral diseases: what is really known? Trends Ecol Evol 13:438–443

Richardson LL, Aronson RB (2002) Infectious diseases of reef corals. In: Proceedings of 9th international coral reef symposium, vol 2, Bali, 2000, pp 1225–1230

Richardson LL, Goldberg WM, Carlton RG, Halas JC (1998a) Coral disease outbreak in the Florida Keys: plague type II. Rev Biol Trop 46:187–198

Richardson LL, Goldberg WM, Kuta KG, Aronson RB, Smith GW, Ritchie KB, Halas JC, Feingold JS, Miller M (1998b) Florida's mystery coral killer identified. Nature 392:557–558

Richardson LL, Sekar R, Myers JL, Gantar M, Voss JD, Kaczmarsky L, Remily ER, Boyer GL, Zimba PV (2007) The presence of the cyanobacterial toxin microcystin in black band disease of corals. FEMS Microbiol Lett 272:182–187

Ritchie KB (2006) Regulation of microbial populations by mucus-associated bacteria. Mar Ecol Prog Ser 322:1–14

Ritchie KB, Smith GW (1998) Description of type II white band disease in acroporid corals. Rev Biol Trop 46:199–203

Ritchie KB, Polson SW, Smith GW (2001) Microbial disease causation in marine invertebrates: problems, practices and future prospects. In: Porter JW (ed) The ecology and etiology of newly emerging marine diseases, developments in hydrobiology. Kluwer, Dordrecht, pp 131–139

Roberts HH, Rouse LJ, Walker ND, Hudson JH (1982) Cold-water stress in Florida bay and northern Bahamas: a product of winter cold-air outbreaks. J Sed Petrol 52(1):0145–0155

Rodriguez-Martinez RE, Benaszak AT, Jordán-Dahlgren E (2001) Necrotic patches affect *Acropora palmata* (Scleractinia: Acroporidae) in the Mexican Caribbean. Dis Aquat Org 47:229–234

Rohwer F, Seguritan V, Azam F, Knowlton N (2002) Diversity of coral-associated bacteria. Mar Ecol Prog Ser 243:1–10

Rohwer F, Breitbar M, Jara J, Azam F, Knowlton N (2001) Diversity of bacteria associated with the Caribbean coral *Montastraea franksi*. Coral Reefs 120:85–91

Rogers C, Patterson K, Porter J (2005) Has white pox been affecting elkhorn coral for over 30 years? Coral Reefs 24:194

Rogers CS, Miller J (2006) Permanent "phase shifts" or reversible declines in coral cover? Lack of recovery of two coral reefs in St. John, USVI. Mar Ecol Prog Ser 306:103–114

Rogers CS, Miller J, Muller EM (2008a) Coral diseases following massive bleaching in 2005 cause 60 percent decline in coral cover and mortality of the threatened species, *Acropora palmata*, on reefs in the U.S. Virgin Islands. USGS Fact Sheet 2008–3058

Rogers CS, Miller J, Muller EM, Edmunds P, Nemeth RS, Beets J, Friedlander AM, Smith TB, Boulon R, Jeffrey CFG, Menza C, Caldow C, Idrisi N, Kojis B, Monaco ME, Spitzack A, Gladfelter EH, Ogden JC, Hillis-Starr Z, Lundgren I, Schill WB, Kuffner IB, Richardson LL, Devine BE, Voss JD (2008b) Ecology of coral reefs in the US Virgin Islands. In: Riegl BM, Dodge RE (eds) Coral reefs of the USA. Springer, Dordrecht, pp 303–373

Rosenberg E (2004) The bacterial disease hypothesis of coral bleaching. In: Rosenberg E, Loya Y (eds) Coral health and disease. Springer, Berlin, pp 445–461

Rosenberg E, Ben-Haim Y (2002) Microbial diseases of corals and global warming. Environ Microbiol 4:318–326

Rosenberg E, Loya Y (2004) Coral health and disease. Springer, Berlin, p 484

Rosenberg E, Kushamaro A (2010) Microbial diseases of corals: pathology and ecology. In: Dubinsky S, Stambler N (eds) Corals and reefs: their life and death. Springer

Rosenberg E, Koren O, Reshef L, Efrony R, Zilber-Rosenberg I (2007) The role of microorganisms in coral health, disease and evolution. Nat Rev Microbiol 5:355–362

Rützler K, Santavy DL (1983) The black band disease of Atlantic reef corals. I. Description of the cyanophyte pathogen. PSZNI Mar Ecol 4:301–319

Rützler K, Santavy DL, Antonius A (1983) The black band disease of Atlantic reef corals III. Distribution, ecology and development. PSZNI Mar Ecol 4:329–335

Santavy DL, Peters EC (1997) Microbial pests: coral disease in the Western Atlantic. In: Proceedings of 8th international coral reef symposium, vol 1, Panama, 1997, pp 607–612

Santavy DL, Peters EC, Quirolo C, Porter JW, Bianchi CN (1999) Yellow-blotch disease outbreak on reefs of the San Blas islands, Panamá. Coral Reefs 18:97

Sekar R, Kaczmarsky LT, Richardson LL (2008) Microbial community composition of black band disease on the coral host *Siderastrea siderea* from three regions of the wider Caribbean. Mar Ecol Prog Ser 362:85–98

Selig ER, Harvell CD, Bruno JF, Willis BL, Page CA et al (2006) Analyzing the relationship between ocean temperature anomalies and coral disease outbreaks at broad spatial scales. In: Phinney J, Hoegh-Guldberg O, Kleypas J, Skirving W, Strong A (eds) Coral reefs and climate change: science and management. American Geophysical Union, Washington, DC, pp 111–128

Smith GW, Ives ID, Nagelkerken IA, Ritchie KB (1996) Aspergillosis associated with Caribbean sea fan mortalities. Nature 382:487

Smith GW, Weil E (2004) Aspergillosis of gorgonians. In: Rosenberg E, Loya Y (eds) Coral health and disease. Springer, New York, pp 279–286

Spalding DL, Greenfeld A (1997) New estimates of global and regional coral reef areas. Coral Reefs 16:225–230

Solano OD, Navas G, Moreno-Forero SK (1993) Blanqueamiento coralino de 1990 en el Parque Nacional Natural Corales del Rosario (Caribe colombiano). Anales del Instituto de Investigación del Mar, Punta de Betin 22:97–111

Stedman TL (1976) Stedman's medical dictionary. Williams and Wilkin Company, Baltimore, 1678 pp. http://www.stedmans.com

Sunagawa S, DeSantis TZ, Piceno YM, Brodie EL, DeSalvo MK, Voolstra CR, Weil E, Andersen GL, Medina M (2009) Bacterial diversity and white plague disease-associated community changes in the Caribbean coral *Montastraea faveolata* assessed by 16S rRNA microarray analysis and clone library sequencing. ISME J 3:1–10

Sussman M, Loya Y, Fine M, Rosenberg E (2003) The marine fireworm *Hermodice carunculata* is a winter reservoir and spring-summer vector for the coral-bleaching pathogen *Vibrio shiloi*. Environ Microbiol 5(4):250–255

Sussman M, Bourne D, Willis B (2006) A single cyanobacterial ribotype is associated with both black and red bands on diseased coral from Palau. Dis Aquat Org 69:111–118

Sutherland KP, Porter JW, Torres C (2004) Disease and immunity in Caribbean and Indo-Pacific zooxanthellate corals. Mar Ecol Prog Ser 266:273–302

Szmant AM, Gassman NJ (1990) The effects of prolonged "bleaching" on the tissue biomass and reproduction of the reef coral *Montastraea annularis*. Coral Reefs 8:217–224

Taylor D (1983) The black band disease of Atlantic reef corals. II. Isolation, cultivation, and growth of *Phormidium corallyticum*. PSZNI Mar Ecol 4:320–328

Thurber RV, Willner-Hall D, Rodriguez-Mueller B, Desnues C, Edwards RA, Angly F, Dinsdale D, Kelly L, Rohwer F (2009) Metagenomic analysis of stressed coral holobionts. Environ Microbiol 11:2148–2163

Toledo-Hernandez C, Zuluaga-Montero A, Bones-Gonzalez A, Rodriguez JA, Sabat AM, Bayman P (2008) Fungi in healthy and diseased sea fans (*Gorgonia ventalina*): is *Aspergillus sydowii* always the pathogen? Coral Reefs 27:707–714

Van Oppen M, Lough J (2009) Coral bleaching: patterns, processes, causes and consequences, ecological studies. Springer, Berlin, 350 pp

Vollmer SV, Kline DI (2008) Natural disease resistance in threatened staghorn corals. PLoS ONE 11:33718

Voss JD, Mills DK, Myers JL, Remily ER, Richardson LL (2007) Black band disease microbial community variation on corals in three regions of the wider Caribbean. Microb Ecol 54:730–739

Weil E (2001) Caribbean coral reef diseases. Status and research needs. In: McManus J (ed) Priorities for Caribbean coral reef research. National Center for Caribbean Coral Reef Research. RSMAS, University of Miami, coral gables, p 10

Weil E (2002) Coral disease epizootiology: status and research needs. Coral health and disease: developing a national research plan. Coral Health and Disease Consortium, Charleston, South Carolina, 14 pp

Weil E (2004) Coral reef diseases in the wider Caribbean. In: Rosenberg E, Loya Y (eds) Coral health and disease. Springer, New York, pp 35–68

Weil E (2008) Temporal dynamics of the ongoing Caribbean yellow band epizootic event: potential link to increasing water temperatures. In: 11th international coral reef symposium (Abstract), Ft. Lauderdale, p 57

Weil E, Ruiz H (2003) Tissue mortality and recovery in *Acropora palmata* (Scleractinia, Acroporidae) after a patchy necrosis outbreak in southwest Puerto Rico. In: 31th scientific meeting of the association of marine laboratories of the Caribbean (Abstract), Port of Spain, Trinidad, p 22

Weil E, Jordán-Dahlgren E (2005) Status of coral reef diseases in Zanzibar and Kenya, western Indian ocean. Progress report, GEF-WB Coral Reef Targeted Research and Capacity Building-Coral Disease Working Group, 16 pp

Weil E, Hooten AJ (2008) Underwater cards for assessing coral health on Caribbean reefs. GEF-CRTR Program, Center for Marine Sciences, University of Queensland, Bruisbane, 24 pp

Weil E, Croquer A (2009) Spatial variability in distribution and prevalence of Caribbean coral and octocoral diseases I: community level analysis. Dis Aquat Org 83:195–208

Weil E, Urreiztieta I, Garzón-Ferreira J (2002) Geographic variability in the incidence of coral and octocoral diseases in the wider Caribbean. In: Proceedings of 9th international coral reef symposium, vol 2, Bali, pp 1231–1238

Weil E, Torres JL, Ashton M (2005) Population characteristics of the black sea urchin *Diadema antillarum* (Philippi) in La Parguera, Puerto Rico, 17 years after the mass mortality event. Rev Biol Trop 53:219–231

Weil E, Smith GW, Gil-Agudelo DL (2006) Status and progress in coral reef disease research. Dis Aquat Org 69:1–7

Weil E, Croquer A, Urreiztieta I (2009a) Caribbean Yellow Band Disease compromises the reproductive output of the reef-building coral *Montastraea faveolata* (Anthozoa, Scleractinia). Dis Aquat Org 87:45–55

Weil E, Croquer A, Urreiztieta I (2010) Temporal variability and consequences of coral diseases and bleaching in La Parguera, Puerto Rico from 2003–2007. Carib J Sci (in press)

Weil E, Smith GW, Ritchie KB, Croquer A (2009b) Inoculation of *Vibrio spp*. onto *Montastraea faveolata* fragments to determine potential pathogenicity. In: Proceedings of 11th international coral reef symposium session, vol 7, Ft. Lauderdale, pp 202–205

Weil E, Hernandez-Delgado EA, Bruckner AW, Ortiz A, Nemeth M, Ruiz H (2003) Distribution and status of Acroporid (scleractinia) populations in Puerto Rico. In: Bruckner AW (ed) Proceedings of the Caribbean *Acropora* workshop: potential application of the

US Endangered Species Act (ESA) as a conservation strategy. NOAA Technical Memorandum, NMFS-OPR-24, Silver Spring, 199 pp

Weil E, Jordan-Dahlgren E, Bruckner AW, Raymundo LJ (2008) Assessment and monitoring protocols of coral reef diseases. In: Raymundo L, Couch CS, Harvell D (eds) Coral disease handbook: guidelines for assessment, monitoring and management. GEF-CRTR-Currie Communications, Australia, pp 48–64

Wilkinson C (2006) Status of coral reefs of the world: summary of threats and remedial action. In: Coté IM, Reynolds JM (eds) Coral reef conservation. Cambridge University Press, Cambridge, pp 3–39

Wilkinson C, Souter D (2008) Status of Caribbean coral reefs after bleaching and hurricanes in 2005. Global Coral Reef Monitoring Network, and Reef and Rainforest Research Centre, Townsville, 152 p

Williams DE, Miller MW (2005) Coral disease outbreak: patterns, prevalence and transmission in *Acropora cervicornis*. Mar Ecol Prog Ser 301:119–128

Willis BL, Page CA, Dinsdale EA (2004) Coral disease on the Great Barrier Reef. In: Rosenberg E, Loya Y (eds) Coral health and disease. Springer, Berlin, pp 69–104

Wobeser GA (1994) Investigation of management of diseases in wild animals. Plenum, New York, p 265

Wobeser GA (2006) Essentials of disease in wild animals. Blackwell, Oxford, p 243

Woodley CM, Bruckner AW, McLenon AL, Higgins JL, Galloway SG, Nicholson JH (2008) Field manual for investigating coral disease outbreaks. NOAA Technical Memorandum, NOS NCCOS 80 and CRCP 6, 85 pp

Work TM, Aeby GS (2006) Systematically describing gross lesions in corals. Dis Aquat Org 70:155–160

Work TM, Richardson LL, Reynolds TL, Willis BL (2008a) Biomedical and veterinary science can increase our understanding of coral disease. J Exp Mar Biol Ecol 362:63–70

Work TM, Woodley C, Raymundo L (2008b) Confirming field assessments and measuring disease impacts. In: Raymundo L, Couch SS, Harvell D (eds) Coral disease handbook: guidelines for assessment, monitoring and management. GEF-CRTR – Currie Communication, Australia, pp 48–64

Work TM, Aeby GS, Stanton FG et al (2008c) Overgrowth of fungi (endolithic hypermycosis) associated with multifocal to diffuse distinct dark discoloration of corals in the Indo-Pacific. Coral Reefs 27:663

Zimmer B, Duncan L, Aronson RB, Deslarzes KJP, Deis D, Robbart M, Precht WF, Kaufman L, Shank B, Weil E, Field J, Evans DJ, Clift L (2009) Long-term monitoring at the East and West Flower Garden Banks National Marine Sanctuary, 2004–2008. Volume I: Technical report. U.S. Dept. of the Interior, Minerals Management Service, Gulf of Mexico OCS Region, New Orleans, Louisiana. OCS Study MMS 2009, 233 pp

Factors Determining the Resilience of Coral Reefs to Eutrophication: A Review and Conceptual Model

Katharina E. Fabricius

Keywords Nutrients • particulate organic matter • turbidity • sedimentation • coral reef • calcification • recruitment • competition

1 Introduction

Eutrophication and increased sedimentation have severely degraded many coastal coral reefs around the world. This chapter reviews the main impacts of eutrophication on the ecology of coral reefs and the properties of reefs that determine their exposure, resistance, and resilience to it. It shows that eutrophication affects coral reefs by way of nutrient enrichment, light loss from turbidity, and the smothering and alteration of surface properties from sedimentation. These changes lead to changes in trophic structures, reduced coral recruitment and diversity, the replacement of corals by macroalgae, and more frequent outbreaks of coral-eating crown-of-thorns starfish. The reefs and areas most susceptible to degradation from pollution are deeper reef slopes, reefs located in poorly flushed locations and surrounded by a shallow sea floor, frequently disturbed reefs, and reefs with low abundances of herbivorous fishes. The chapter concludes with a conceptual model of the main links between water quality and the condition of inshore coral reefs.

The term "eutrophication" defines the increase in nutrient concentrations (especially nitrogen or phosphorus) in a water body, which can increase the production of algae, turbidity, sedimentation of particulate matter, and in severe cases hypoxia. A "contaminant" is a substance (including nutrients and sediments) that occurs at above "natural" concentrations, while a "pollutant" is defined as a substance that occurs at a concentration causing environmental harm (GESAMP 2001). Minor nutrient enrichment may therefore be considered contamination, while major eutrophication constitutes pollution.

Eutrophication degrades reefs through three mechanisms (Rogers 1990; Dubinsky and Stambler 1996; Szmant 2002; Fabricius 2005):

(a) Nutrient enrichment causes trophic changes
(b) Turbidity causes light loss, which affects photosynthesis in deeper water
(c) Sedimentation causes reduced larval settlement and mortality

The main source of new dissolved and particulate nutrients and sediments that enter coastal marine systems is rivers. How much nutrient and sediment rivers carry is determined by geography (e.g., rainfall, soil type, and slope) and land management (e.g., degree of vegetation cover and fertilizer application) (Bourke et al. 2002). Other sources of nutrients and particulate matter are aquaculture and sewage outfall sites (Loya et al. 2004), diffuse coastal runoff, groundwater seepage, the upwelling of nutrient-rich water from greater depths, nitrogen fixation by benthic and pelagic cyanobacteria and algae, and the atmosphere [e.g., rain, settling dust (Dumont et al. 2005; Duce et al. 2008)].

Marine coastal ecosystems, including coastal coral reefs, are exposed to increasing amounts of soil, fertilizer, and pesticide washed from cleared land, and discharged from sewage and fish farms and other point sources (Vitousek et al. 1997; Tilman et al. 2001; Smith et al. 2003). Globally, an additional 1% of the earth's surface is being cleared every year; nitrogen fertilizer use has increased more than sixfold since 1960 (Matson et al. 1997); and human population along the coast is growing even faster than elsewhere. All this contributes to a rapidly intensifying exploitation of coastal resources and increasing losses of nutrients from the land (Crossland et al. 2005). Oxygen-depleted seafloor zones, attributable to runoff of agricultural nitrogen and phosphorus, have doubled in area in the last few decades, indicating that many marine water bodies are becoming more eutrophic (Diaz 2001; GESAMP 2001; Vaquer-Sunye and Duarte 2008).

New nutrients that enter coral reefs cycle rapidly and in extremely complex ways. It is useful to differentiate between dissolved and particulate, and between organic and inorganic

K.E. Fabricius (✉)
Australian Institute of Marine Science, PMB No. 3, Townsville, MC, QLD 4810, Australia
e-mail: k.fabricius@aims.gov.au

forms of nutrients, because the biological effects of the four differ fundamentally. *Dissolved inorganic nutrients* (especially phosphate, nitrate, or ammonium) are highly bioavailable; they are incorporated into benthic and pelagic food webs within hours to days, and, thereafter, are found as particulate organic matter (especially detritus, bacteria, phytoplankton, and marine snow). *Dissolved organic nutrients* often occur in relatively high concentrations, but a large proportion cannot be used as food by any organism, including bacteria. *Particulate organic matter* (POM) is dominated by detritus and plankton, while particulate inorganic nutrients are often bound to suspended sediment grains. Concentrations of particulates are typically determined by measuring proxies–chlorophyll, particulate nitrogen, particulate phosphorus, and total suspended solids. Uptake, excretion, and decomposition in benthic and pelagic food webs ensure continuous conversions between dissolved and particulate, and between organic and inorganic forms of nutrients.

Both the concentration and type of suspended particles strongly determine water clarity, which in turn determines *benthic irradiance* at a given water depth. While coastal coral reefs can flourish at relatively high levels of turbidity (Kleypas 1996; DeVantier et al. 2006), in turbid water, they tend to be restricted to the upper 4–10 m because of reduced coral photosynthesis and growth at greater depth. In comparison, in clear oceanic waters, coral reefs can be found at depths of over 40 m (Yentsch et al. 2002). Water clarity is strongly governed by the degree to which waves resuspend sediments in shallow shelf seas (Larcombe and Woolfe 1999). Water clarity can also be affected by increased phytoplankton productivity, such as recorded in the past around a sewage outfall site in Kanehoe Bay, Hawaii (Hunter and Evans 1995), and fish farms in the Northern Red Sea (Loya et al. 2004). Historic data on water clarity in marine systems are, however, sparse, and the conditions leading to long-term changes in water clarity in tropical coastal systems are poorly understood.

Altered *sedimentation* regimes are often linked to eutrophication, as particulate matter eventually settles onto the seafloor and onto benthic organisms. Larger grain sizes are deposited within a few kilometers of the source, but the smallest grain fractions (clay and silt particles) remain suspended for prolonged periods of time and are often distributed over tens to hundreds of kilometers, undergoing several cycles of deposition and resuspension. Such small particles carry more nutrients and pesticides (Gibbs et al. 1971), absorb more light (Moody et al. 1987), and cause greater stress and damage to corals than do sediments that are coarse and poor in organic matter (Weber et al. 2006). Not only the amount of sedimentation but also its type and organic contents thus determine the extent of damage caused by sedimentation to coral reefs.

Although degradation or damage from elevated nutrients and sediments has been reported from reefs around the world, it has often been difficult to distinguish whether human activities have contributed to cause the damage. Comprehensive reports documenting damage include studies from Hawaii (Hunter and Evans 1995), Indonesia (Edinger et al. 1998), Costa Rica (Cortes and Risk 1985; Hands et al. 1993), Barbados (Tomascik and Sander 1987; Wittenberg and Hunte 1992), St Croix (Hubbard and Scaturo 1985), Kenya (McClanahan and Obura 1997), and the Great Barrier Reef (van Woesik et al. 1999; Koop et al. 2001; McCulloch et al. 2003; Fabricius et al. 2005). Most of these studies assessed spatial and/or temporal variation at control and impact sites. By their nature, such assessments based on associations cannot assert causal relationships between pollution and reef degradation, because historical data are often missing and other disturbances (e.g., overfishing, coral bleaching, storms, and floods) tend to co-occur. In shallow nearshore waters, nutrient concentrations and water clarity can vary – both spatially and temporally – by up to two orders of magnitude. Even without increased terrestrial runoff and eutrophication, reef communities change along water quality gradients: from terrestrially influenced conditions (characterized by fluctuating salinity, more variable or higher nutrient load, siltation, and turbidity) to oceanic conditions (where nutrients, siltation, and turbidity are typically low) (Fig. 1). Epidemiological tools originally developed to assess the weight of evidence that smoking causes lung cancer (U.S. Department of Health, Education, and Welfare 1964) have now been applied to coral reef studies to assess causal links between reef degradation and increased levels of nutrients and sediments in the central Great Barrier Reef (Fabricius and De'ath 2004).

2 Responses of Reef Organisms to Eutrophication

The following section of this chapter reviews the available information on two things: first, how eutrophication directly and indirectly effects corals and other reef-associated organisms; and, second, how the ecological balances of coral reef ecosystems changes with nutrient enrichment, light loss, and sedimentation.

2.1 Hard Corals

Hard corals are competitive in low-nutrient environments for three reasons. First, they are remarkably efficient in internally recycling nutrients between the host and its endosymbiotic

Fig. 1 Shallow-water coral reef communities can change visibly along water quality gradients, as shown by these photographs from reefs near the mouths of the Proserpine and O'Connell Rivers, central Great Barrier Reef. (**a**) 20 km from the rivers: reef assemblages in the shallow water are dominated by green and red macroalgae and have few coral colonies; assemblages at greater depths are dominated by heterotrophic filter-feeders (not shown); (**b**) 40 km from the rivers: communities are dominated by macroalgae and low coral cover; (**c**) 60 km from the rivers: octocoral and hard coral dominate communities; (**d**) 70 km from the rivers: diverse *Acropora*-dominated communities (Photos: K. Fabricius)

unicellular algae (zooxanthellae; Stambler 2010 in this book). Second, they have the capacity to occupy most trophic levels simultaneously. This is because they are efficient phototrophs (due to their endosymbiotic algae), they take up dissolved inorganic and organic nutrients, graze on primary producers (such as large phytoplankton), capture and prey upon herbivorous and predatory zooplankton, and feed on decompositional material such as detritus (Rosenfeld et al. 1999). Third, many hard coral species are able to phenotypically adapt to varying light and food availability within days, and such plasticity helps maximizing energy gains throughout their lifetime (Anthony and Fabricius 2000).

Studies of the impacts of eutrophication have often focused on *dissolved inorganic nutrients* (e.g., Stambler et al. 1994; Dubinsky and Stambler 1996; Koop et al. 2001; Szmant 2002). Most studies show that high levels of dissolved inorganic nitrogen and phosphorus can cause significant physiological changes in corals, but do not kill or greatly harm individual coral colonies (reviewed in Fabricius 2005). However, exposure to dissolved inorganic nitrogen can lead to declining calcification, higher concentrations of photopigments (affecting the energy and nutrient transfer between zooxanthellae and host; Marubini and Davies 1996), and potentially higher rates of coral diseases (Bruno et al. 2003). In areas of nutrient upwelling or in heavily polluted locations, chronically elevated levels of dissolved inorganic nutrients may so alter the coral physiology and calcification as to cause noticeable changes in coral communities (Birkeland 1997). However, dissolved inorganic nutrients are generally quickly removed from the water column by way of biological uptake by bacteria, phytoplankton, and the benthos; hence, concentrations will often not increase greatly along pollution gradients. The main way in which dissolved inorganic nutrients affect corals appears to be by enriching organic matter in the plankton and in sediments.

Particulate organic matter (POM) is used by corals as food, and tissue thickness, photosynthetic pigment concentrations, and calcification in corals increase in response to POM feeding. At higher levels of POM, some hard coral species can increase rates of heterotrophy – and, thus, partly or fully compensate for energy losses resulting from light attenuation – while other species show feeding saturation and are unable to compensate for light loss (Anthony and Fabricius 2000). At even higher levels of POM, gross photosynthesis and respiration, tissue thickness and calcification start to decline in all species as light attenuation outweighs further energy gains from POM feeding. Photosynthesis, tissue thickness, and calcification therefore change in a modal fashion along eutrophication gradients (Tomascik and Sander 1985; Marubini and Davies 1996). However, photosynthetic pigment concentrations further increase with increasing POM and have, therefore, been suggested as one of the most useful early-warning indicators of eutrophication (Marubini 1996; Cooper et al. 2009). While corals are not greatly harmed by dissolved inorganic nutrients – and may even benefit from particulate organic matter – macroalgae and heterotrophic filter-feeders benefit more from dissolved inorganic and particulate organic nutrients than do corals (see below). As a result, corals that can grow at extremely low food concentrations may be out-competed by macroalgae and/or more heterotrophic communities that grow best in high nutrient environments.

Benthic irradiance is a crucial factor for reef corals. Shading reduces photosynthesis, leading to slower calcification and thinner tissues (Anthony and Hoegh-Guldberg 2003; Allemand et al. 2010 in this book). Few corals can grow when surface irradiance is less than about 4% – as encountered at ~40 m in clear water or at ~4 m in highly turbid water – and reef development tends to cease at around this light level (Yentsch et al. 2002; Cooper et al. 2009). Phototrophic species are more severely affected by light limitation than are other species, with the result that species richness declines at high turbidity in deeper depths, due to the loss of sensitive species. On the other hand, slower-growing species may be out-competed at high irradiance by fast-growing phototrophic species and by macroalgae; hence, species richness is often highest at intermediate light levels (Cornell and Karlson 2000). Corals acclimatize to low irradiance by increasing their pigment density. However, full photoacclimation takes about 5–10 days and is too slow to compensate for energy losses if turbidity fluctuates (Anthony and Hoegh-Guldberg 2003).

Sedimentation reduces coral recruitment rates and coral biodiversity, with many sensitive species being under-represented or absent in sediment-exposed communities. Small colonies and species with thin tissues and flat morphologies are often more sensitive to sedimentation than are large colonies or those with thick tissues or branching growth forms (Rogers 1990; Stafford-Smith and Ormond 1992). In coral colonies, sedimentation stress increases linearly with the duration and amount of sedimentation: for example, a certain amount of sediment deposited on the coral for one time unit exerts the same measurable photophysiological stress as twice the amount deposited for half the time (Philipp and Fabricius 2003). High sedimentation rates (up to >100 mg dry weight cm^{-2}) can kill exposed coral tissue within a few days, while lower rates reduce photosynthetic yields in corals within ~24 h (Philipp and Fabricius 2003). Exposure to a few days of sedimentation can cause long-term damage to coral populations, by removing whole cohorts of small and sensitive corals. Some suggest a sedimentation threshold of 10 mg cm^{-2} day^{-1}, with reefs being severely damaged at higher sedimentation rates (Rogers 1990).The damage to tissue under a layer of sediment further increases with increasing organic content and bacterial activity and with decreasing grain size of the sediment (Hodgson 1990; Weber et al. 2006). Levels of ~12 mg cm^{-2} day^{-1} can kill newly settled corals with <48 h exposure if sediments are rich in organic contents, but such levels can be tolerated if the organic content is low (Fabricius et al. 2003). These and similar data demonstrate the critical, but as yet poorly understood, effects of organic enrichment of sediments.

2.2 Coral Recruitment

It is during their recruitment stages that corals are the most sensitive to pollution. While adult corals can tolerate prolonged periods of low light, competition with macroalgae, and moderate levels of sedimentation, the settlement of coral larvae and the survival of newly settled young and small colonies are extremely sensitive to them (Fabricius 2005). Indeed, very little settlement occurs on sediment-covered surfaces, and the tolerance of coral recruits to sediment is at least one order of magnitude lower than that of adult corals. Settlement of coral larvae is also controlled by light intensity and spectral composition; reduced light reduces the depth at which larvae settle (Baird et al. 2003). Because successful coral recruitment is essential if reefs are to recover from bleaching, storms, or other disturbances – and because recruitment is one of the main factors determining speed of recovery from disturbances – reduced coral recruitment is one of the most deleterious effects of eutrophication on coral reefs.

2.3 Crustose Coralline Algae

Certain species of crustose coralline algae are essential for coral settlement (Harrington et al. 2004). Experiments and field data suggest that high nutrient levels do not greatly alter the physiology of crustose coralline algae, but high sedimentation rates are related to low abundances of corallines in coral reefs (Kendrick 1991; Fabricius and De'ath 2001). Some crustose coralline algae associated with coral reefs survive burial under coarse inorganic sediments for days to weeks, but their survival rates rapidly decline if the sediments contain traces of herbicides (Harrington et al. 2005). Turf algae often out-compete coralline algae (Steneck 1997), and, because turf algae can trap large quantities of sediments (Purcell 2000), they make the surrounding substratum less suitable for coralline algae and for coral settlement (Birrell et al. 2005).

2.4 Macroalgae

Macroalgal communities are an integral and often diverse component of inshore reef systems. They use photosynthesis to satisfy their carbon demand, while their nutrient demand is met by uptake of dissolved inorganic nutrients plus, in some species, by demineralization of particulate organic matter deposited on their fronds (Schaffelke 1999). In the absence of grazing, the growth and productivity of some, but not all, groups of macroalgae are nutrient-limited and increase with minute increases in dissolved inorganic nutrients and POM (Littler and Littler 2007). Macroalgae can be dominant around point sources of nutrients (Smith et al. 1981; Lapointe et al. 2004; Lapointe and Bedford 2007). On the Great Barrier Reef, total macroalgal cover increases several-fold along gradients of declining water clarity and increasing nutrients (van Woesik et al. 1999; Fabricius and De'ath 2004; De'ath and Fabricius 2010). Long-term data have also shown that expansion (Cuet et al. 1988) or decline (Smith et al. 1981) in macroalgal cover over time coincide with increasing and declining nutrients, respectively. These time series data indicate a causal link between macroalgal abundances and nutrient availability. Macroalgae also tend to flourish in areas of nutrient upwelling, on eastern sides of continents or large islands, where more rivers originate than in the west (Birkeland 1988), and with latitude, as do nutrient concentrations (Johannes et al. 1983). These large-scale geographic data all add evidence that the availability of nutrients controls macroalgal biomass. However, increases in macroalgal biomass are only observed in areas where grazing by herbivorous fishes or invertebrates is low (McCook 1999; Littler and Littler 2007), and where light is not limiting.

Macroalgae that form low ephemeral mats tend to overgrow, damage, or even kill understorey corals by shading, restricting gas exchange, and creating hypoxia when mats collapse (Loya et al. 2004). In contrast, tall perennial species, such as *Sargassum* spp., do not usually kill corals, but they can reduce coral growth by shading (Littler and Littler 2007). All dense macroalgal assemblages (low ephemeral mats and tall perennial stands) suppress coral recruitment through space occupancy, allelopathy, silt-trapping, or shading (Connell et al. 1997; Szmant 2002).

2.5 Crown-of-Thorns Starfish (Acanthaster planci)

Another severe consequence of eutrophication is the apparent increase in frequencies of outbreaks of crown-of-thorns starfish (Birkeland 1982; Brodie et al. 2005; Fabricius et al. 2010). *A. planci* is corallivore – it feeds on coral tissue – and has been the most common cause of coral mortality throughout many tropical Indo-Pacific regions in the last 5 decades. The pelagic larvae of this starfish filter-feed on large phytoplankton, and experiments suggest that these larvae are food limited: their survivorship in the laboratory increases steeply with increasing availability of suitable food at environmentally relevant concentrations (Fabricius et al. 2010). In the field, increased nutrient availability can increase the abundance of large phytoplankton cells: a strong temporal and spatial relationship exists between drought-breaking floods from high continental islands and outbreaks of this starfish (Birkeland 1982). New research further strengthens the evidence that more frequent outbreaks of *A. planci* are linked to high nutrient levels, while acknowledging that the removal of predators of *A. planci* can further increase the likelihood of outbreaks (Brodie et al. 2005; Houk et al. 2007). After primary *A. planci* outbreaks have formed in a region with high phytoplankton concentrations, many of their numerous larvae may be transported by currents to remote regions, hence secondary *A. planci* outbreaks may form far away from areas of eutrophication.

2.6 Filter-Feeders, Macrobioeroders, and Suspension Feeders

Benthic filter feeders, such as sponges, bryozoans, bivalves, barnacles, and ascidians, are important components of reef ecosystems. Most of them live hidden in crevices or drill or etch their own holes within coral skeletons (so-called bioeroders). Far fewer live on the reef surface where they would compete with hard corals and with algae for space. Most filter-feeders are heterotrophic (i.e., not associated with photosynthetic symbionts) and are specialized to feed on a narrow spectrum of planktonic particles. Because of these trophic constraints, they are often unable to obtain a positive carbon balance in oligotrophic waters (Birkeland 1988), and their densities increase in response to nutrient enrichment (Smith et al. 1981; Costa Jr et al. 2000). Literature suggests that, unlike macroalgae – which directly compete with corals for well-lit habitats – surface-inhabiting heterotrophic filter-feeders tend to monopolize space in poorly lit, highly productive environments, i.e., conditions that are marginal for corals. Filter-feeders therefore rarely outcompete corals directly (Aerts and Van Soest 1997), with the few observed exceptions all occurring in areas of low light, high plankton productivity, and organic enrichment (Smith et al. 1981; Brock and Smith 1983). The demise of corals and the establishment of filter-feeders therefore appear largely as independent symptoms of eutrophication, with the fate of each group determined by altered trophic conditions rather than by altered balances in their competition for space.

Many internal macrobioeroders are filter-feeders that actively bore into or chemically erode the calcium carbonate skeletons of live corals and dead reef substrata. The main groups of macrobioeroders are boring sponges (e.g., *Cliona* spp.) and bivalves (e.g., *Lithophaga* spp.). In high densities, they can substantially weaken the structure of coral reefs and increase their susceptibility to storm damage. Several studies have documented abundances of internal macrobioeroders increasing in response to increased nutrient availability (Hallock 1988; Cuet et al. 1988). For example, abundances of the boring sponge *Cliona delitrix* increased fivefold in an area exposed to untreated sewage (Rose and Risk 1985), and abundances of most internal macrobioeroders are higher in productive inshore environments than they are offshore. Increased bioerosion in areas of nutrient enrichment, combined with reduced coral growth, diminished skeletal densities, and lower recruitment rates, can lead to conditions in which reef erosion exceeds calcium carbonate accretion (Pari et al. 2002).

Octocorals (soft corals and sea fans) are passive suspension feeders rather than internal filter-feeders, with the more abundant genera, all containing photosynthetic endosymbionts (Fabricius and Alderslade 2001). Very few studies report photosynthetic soft corals monopolizing space after hard coral disturbance, and a shift from hard corals to octocorals appears to be a rare occurrence and restricted to productive and high-irradiance, high-current, and wave-protected waters. Indeed, octocorals appear to be more strongly affected by declining water quality than are hard corals: octocoral species richness declines by up to 60% along a gradient of increasing turbidity, due to the disappearance of zooxanthellate octocorals (De'ath and Fabricius 2010).

2.6.1 Fishes

Fishes play an essential role in the ecology of coral reefs. The abundances of many taxa are related to the structural complexity of reefs (Wilson et al. 2006). High coral cover – a proxy for structural complexity – promotes local fish abundances, thus sustaining grazing pressure on algae. Reduced abundances of herbivorous fish lead to proliferation of macroalgae (McCook 1999). However, it is unclear whether the abundances of herbivorous fish on coral reefs decline with increasing turbidity (Wolanski et al. 2004), as they do in some estuarine areas (Mallela et al. 2007). It has also been argued that increased fishing pressure may lead to complex and largely unpredictable trophic cascades. For example, the overfishing of large carnivores could lead to higher densities of small carnivores, which in turn may reduce the number of invertebrates that feed on juvenile crown-of-thorns starfish (*Acanthaster planci*), and therefore an increased survival of this coral-eating starfish (Sweatman 2008).

3 Factors Influencing the Susceptibility of Reefs to Eutrophication

Reefs differ significantly in their exposure, resistance, and resilience to pollution, and the identification of coral reef areas with low exposure and high resistance and resilience is a high management priority (West and Salm 2003).

Exposure to a pollutant is typically a function of the amount (concentration or load) of pollutant and the length of time it is in contact with a coral (Fig. 2). A coral that is exposed to a high concentration of a pollutant for a short period of time may suffer a similar fate to one that is exposed to lower concentrations for longer periods (Philipp and Fabricius 2003). Peak concentrations are only a relevant measure of pollutants if the onset of damage or mortality is rapid (Fig. 2). For example, dredging may kill benthic organisms by acute sedimentation within a few hours to days, but prolonged dredging can also cause damage through cumulative physiological changes due to light loss from turbidity (Bak 1978).

Resistance of an ecosystem is defined as the ease or difficulty with which it can change in response to a disturbance (Pimm 1984). Resistance can arise from environmental (extrinsic) or biological (intrinsic) factors (Done 1999; West and Salm 2003). For example, a coral reef composed of coral communities adapted to naturally turbid settings may be low in species richness and dominated by the turbidity-tolerant genera *Turbinaria* and massive *Porites*. Such communities are more resistant to pollution than communities adapted to clearwater environment that contain many sensitive species.

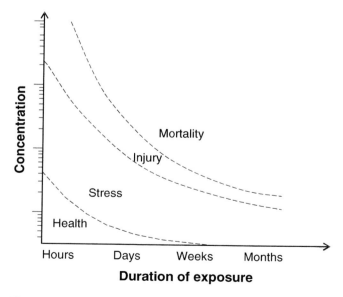

Fig. 2 Schematic representation of responses of coral exposed to a pollutant. The severity of response is typically a combined function of the duration and amount (concentration or load) of exposure

Ecological *resilience* is defined here as the time it takes an ecosystem to return to a stable equilibrium point after being disturbed (Pimm 1984; Tilman and Downing 1994). Note that several definitions of resilience are in use (Nyström et al. 2008). Notably, resilience has sometimes been defined as the *capacity* of an ecosystem to recover after some acute disturbance without undergoing a phase shift into an alternate state (West and Salm 2003), a definition of resilience that is more difficult to quantify and hence not employed here.

The factors that predict exposure, resistance, and resilience of reefs to degradation are listed in Table 1, derived from a qualitative review of some of the better-described case studies in the literature (Fabricius 2005). A formal risk analysis will be required to quantify the relative contributions of these properties.

Currents, waves, and tides are extremely important properties of coral reefs, and key predictors of exposure, resistance, and resilience at both local and regional scales. At local scales, current-swept reef fronts, flanks, and channels, as well as reef crests with moderate wave exposure, are the least likely to retain pollutants; they support high recruitment and have the fastest coral growth (Sebens 1991). Ideally, wave exposure is sufficient to remove sediment deposits without causing frequent coral breakage (in inshore waters, coral skeletons are, typically, very brittle). As fast currents and waves also facilitate macroalgal growth, competition between corals and macroalgae is intense if nutrient levels are high. At regional scales, currents determine where and how far pollutants are being transported. Where currents are predominantly tidal, very low and very high tidal ranges affect transport and resuspension and the ability of reefs to withstand pollution. A strong connectivity to upstream populations that produce pelagic larvae also decreases the time for recovery from a disturbance and is a strong predictor of resilience.

Spatial factors are the simplest predictors of exposure: the closer downstream a coral reef is to a pollution source and the higher the mean annual load of this source, the greater is its exposure to a pollutant. Geospatial models of the global scale of pollution around coral reefs have been developed based on the distance of reefs to pollution sources (Bryant et al. 1998). These models estimate that 22% of all coral reefs worldwide are classified as at high (12%) or medium (10%) threat from inland pollution and soil erosion. They also classify 12% of reefs at threat from marine pollution (distance from ports, oil tanks, oil wells, and shipping areas), and 30% of reefs as threatened from coastal development, such as cities, mines, and resorts (Bryant et al. 1998). At regional scales, the percentage of reefs at risk is a direct function of the extent of land clearing, and up to 50% of reefs are at risk in the countries with the most widespread land clearing (Bryant et al. 1998). At local scales, such as downstream from well-defined point sources, or in coastal reefs fringing eroding land, terrestrial runoff can be the single most significant pressure for selected

Table 1 Spatial, geophysical, and biological properties that predict the exposure, resistance, and resilience of coral reefs to degradation by eutrophication

Factor	Property	Highest risk, lowest resistance and resilience
Hydrodynamics	Currents and waves determine exposure to pollutants (through dilution, mixing, and removal), and resistance of organisms (growth rates and photosynthesis are high at strong currents)	Weak currents, weak or extreme wave exposure
Connectivity	Connectivity to region with large brood stock determines larval supply and rates of self-seeding	Low connectivity to region with large brood stock, and low self-seeding
Location	Downstream distance from source, and discharge load of source, determines exposure	Near to and downstream of point of discharge
Local topography	Local: steepness of reef slope determines the accumulation or downward transport of settled pollutants	Terraces, gradual slopes
Geomorphology	Geomorphology determines sediment retention versus flushing: retention is greatest in lagoons, embayments, and on leeward reefs sides, while sediments and pollutants are flushed away from headlands, channels, or reef flanks	Lagoons, semi-enclosed embayments, leeward areas
Bathymetry	Depth of area, and depth and nature of surrounding seafloor determine the rate of wave resuspension and removal of sediments, and light limitation in turbid water	Lower reef slopes, areas surrounded by shallow sea floor
History	Exposure history facilitates adaptation to local conditions, and determines successional stage	Large and fast changes from historical to present conditions
Ecology	Substrata suitable for coral recruitment	Low coral recruitment success
	Fish abundances balance community structures (herbivores controlling macroalgal biomass, predators controlling invertebrate populations)	High fishing pressure, low abundances of herbivores and predators
	Biodiversity (functional redundancy, presence of more tolerant taxa)	Low biodiversity
	Additional coral disturbances (coral predators, diseases, bleaching, storms) determine cumulative stress	Frequent and severe additional disturbances, synergistic stressors

coral reefs (Bourke et al. 2002). However, such geospatial models do not factor in additional indirect effects that may occur hundreds of kilometers away from areas of eutrophication, such as increasing outbreak frequencies of crown-of-thorns starfish populations.

Topography/geomorphology and bathymetry can modify exposure enormously: pollutants are retained for prolonged periods in poorly flushed embayments and lagoons, but may rapidly disappear from a well-flushed headland surrounded by a deep water body (Hopley et al. 2007). Around the world, most severely polluted and hypoxic marine sites appear to be located in shallow and poorly flushed semienclosed water bodies such as gulfs, bays, fjords, or bights (e.g., Kaneohe Bay, the Baltic Sea, Chesapeake Bay, Irish Sea; Diaz 2001). Bathymetry also modifies exposure by affecting the balance between sedimentation and resuspension of materials. Materials are easily washed away from shallow, upper reef slopes, but accumulate below the reach of surface waves in deeper areas (Wolanski et al. 2005). Upper reef slopes are also less affected by turbidity than are deeper areas, where light becomes limiting for photosynthetic organisms. Repeated wave resuspension and deposition of materials is worst at sites surrounded by a wide, shallow continental shelf, whereas pollutants are flushed away into deeper waters if no shelf retains them. For example, the Great Barrier Reef is located on a 50–200 km-wide shallow continental shelf with >2,000 barrier reefs between the land and the open ocean. Retention times here are unknown, but some estimate them to be up to 300 days for dissolved materials (Luick et al. 2007). Particulate materials are likely to be retained for even longer periods of time, as they are repeatedly deposited and resuspended from the shallow sea floor. So, although nutrient enrichment is less severe on the Great Barrier Reef than in many other more densely populated regions, symptoms such as macroalgal dominance and low coral diversity on numerous inshore reefs (Fig. 1) have been attributed to enhanced terrestrial runoff (van Woesik et al. 1999; Fabricius and De'ath 2004; De'ath and Fabricius 2010).

Last, *biological processes* are believed to modify the resistance and resilience of coral reefs, but many of these processes are as yet poorly understood. For example, abundant herbivorous fish strongly control macroalgal abundances, thereby promoting resilience (Littler and Littler 2007). It is also still unresolved to what extent resistance and resilience are codetermined by biodiversity; regions of low biodiversity have fewer species to replace the loss of sensitive species and may be more likely to undergo structural and functional changes in their communities (Bellwood et al. 2004). For example, the loss of the dominant coral species *Acropora palmata* and *Acropora cervicornis* and the one remaining important algal grazer, *Diadema antillarum*, in the Caribbean, has led to a widespread collapse of reef ecosystems there. It is also unknown whether or not the resistance and resilience of reefs varies along latitudinal gradients, as reefs in higher latitudes naturally have lower calcification rates, higher macroalgal biomass, and lower coral biodiversity than do low-latitude reefs. Last, regions that are prone to severe or frequent disturbances (e.g., from coral bleaching, storms, cold water upwelling, or outbreaks of crown-of-thorns starfish) are more likely to be prone to degradation than rarely disturbed regions. This is because poor water quality often does not directly kill the adult coral populations, but retards coral recruitment and, hence, the speed of recovery from unrelated disturbances. It has also been shown that exposure to one form of stress may decrease the resilience of an ecosystem to another stressor (Hughes et al. 2003; Wooldridge et al. 2005).

In summary, Table 1 suggests that degradation from poor water quality is most likely to occur in poorly flushed locations with weak currents, on deeper reef slopes, in places where fish abundances are low, and in regions that are frequently affected by other forms of disturbance. In contrast, reefs with strong currents, well-flushed locations, shallow reef crests surrounded by a deep water body, and reefs inhabited by healthy populations of fishes are likely to have the highest levels of resistance and resilience.

4 The Conceptual Model

So far, we have seen that corals, and many groups that interact with corals, are either inhibited or promoted by eutrophication, especially through (a) trophic shifts resulting from greater availability of dissolved and particulate inorganic and organic nutrients, (b) light limitation in deeper water, and (c) sedimentation. We have also seen that the severity of response to eutrophication varies spatially – with reefs located in poorly flushed locations at greatest risk of damage – and depends on a number of environmental and biological conditions at a site.

A qualitative model (Fig. 3) may help to summarize the numerous links between water quality (blue) and the condition of inshore coral reefs (yellow) in the context of external environmental processes and parameters that determine exposure, resistance, and resilience (grey). At first sight, the model could appear daunting, but it is easily deconstructed into a set of causal links, many of which are explained in the text above, and briefly summarized here:

(a) *Pollutant loads* are typically determined by geographic conditions (vegetation cover, rainfall, soil type, slope, etc.) and land management.
(b) *The main sources of new dissolved nutrients* are rivers, point sources, upwelling, and the atmosphere via rain; de/mineralization and biological uptake and release (see below at d, e) affect both nutrient gains and losses.

(c) *The main sources of new particulate matter, particulate nutrients, and sediments* entering coastal marine systems are rivers and point sources; burial is the main pathway of removal, while biological uptake and release (see below at d, f) affect both gains and losses.

(d) *Cycling:* Dissolved and particulate matter enter into complex cycles of conversions by way of biological uptake and release, chemical absorption and desorption, and sedimentation and resuspension. Dissolved inorganic nutrients can increase concentrations of phytoplankton and other forms of POM, leading to higher turbidity, reduced light, and increasing rates of sedimentation.

(e) *Dissolved inorganic nutrients* release nutrient limitation in some *macroalgae*, which may thus gain a competitive advantage over corals.

(f) High levels of *particulate organic matter* in reef waters favor the growth of some macroalgae; some species of coral also gain advantage at moderate levels of particulate organic matter. High phytoplankton loads are also linked to an increased survival of filter-feeders that thrive when particle loads are high. Crown-of-thorns starfish (COTs) have filter-feeding pelagic larvae that appear to be limited by the availability of large phytoplankton. Circumstantial and experimental evidence suggest large terrestrial runoff events and/or increased oceanic productivity resulting from phytoplankton blooms stimulate outbreaks of this starfish. Higher fish biomass is also often associated with increasing marine productivity.

(g) *Light reduction* resulting from turbidity leads to greatly reduced photosynthesis and recruitment in corals, and a shallower depth limit for reef development. Stress from light limitation varies greatly between species, and after light-dependent species disappear a reef will have fewer species and suffer a decline in biodiversity. Settlement of coral larvae is also controlled by light intensity and spectral composition. It also decreases the depth range for some species of macroalgae. In contrast, many heterotrophic filter-feeders can only live at low-light conditions. The main symptoms of light limitation in the field are, therefore, reduced coral recruitment and biodiversity, a shallower depth limit for reef growth, and a shift from phototrophic to heterotrophic processes.

(h) Increased *sedimentation* severely disturbs most aspects of for coral reefs. Settlement rates are low, and young corals have high mortality rates on sediment-covered surfaces. The effects of increased levels of sedimentation are, therefore, slower recovery from disturbance, altered species composition, and reduced coral diversity. Sediment coated or mixed with organic matter is more difficult to remove than clean calcareous sediments and is particularly detrimental to small organisms including newly settled corals. Sedimentation also inhibits the growth of some species of crustose coralline algae.

(i) Some *macroalgae* compete for space with corals, and space occupied by macroalgae is often unavailable for the settlement of coral larvae.

(j) *Fishes* directly interact with corals and macroalgae. Grazing fishes can control macroalgae, and a reduction in grazing by fishes can leads to their proliferation at high nutrient levels. High coral cover leads to high structural complexity, promoting local fish abundances and sustaining grazing pressure on algae. Predatory and omnivorous fish have also been associated with controlling abundances of juvenile crown-of-thorns starfish and numerous other invertebrates, playing an immensely important role to protect the integrity of numerous ecological functions. It remains to be investigated to what extent herbivorous fishes are directly affected by turbidity.

(k) *Crown-of-thorns starfish (COTS)*: coral loss from predation by crown-of-thorns starfish is greater than coral loss through any other cause of mortality. However, adult crown-of-thorns starfish are also controlled by food availability as outbreaks collapse at less than 5% coral cover. Crown-of-thorns starfish also require crustose coralline algae for settlement and as food in their first 6 months after metamorphosis, but it is unknown whether their survival is food limited at this stage.

(l) Successful coral recruitment is a prerequisite for the recovery of corals from disturbances. Coral recruitment depends on the availability of larvae, either from external larval sources upstream or from local brood stock (self-seeding). Coral larvae also need crustose coralline algae on which to settle.

Last, models need to include the factors that determine exposure, resistance, and resilience to changing water quality. As has been shown above, the main factors are hydrodynamics, spatial location (distance to a point source, to upwelling, etc.), bathymetry, and the propensity to disturbance; all of which strongly determine the fate of a specific reef system exposed to eutrophication.

5 Discussion

This study has shown that increased abundances of macroalgae, reduced recruitment success in corals, and increased frequencies of outbreaks of *A. planci* are arguably the most significant effects of eutrophication on coral reefs. While variations in nutrients, light, and sediment are naturally found in pristine conditions, they are often exacerbated by human activity, such as land clearing, agriculture and urban activities, and aquaculture. With severe exposure, reefs will suffer shallower reef development, changed coral community structure, and greatly

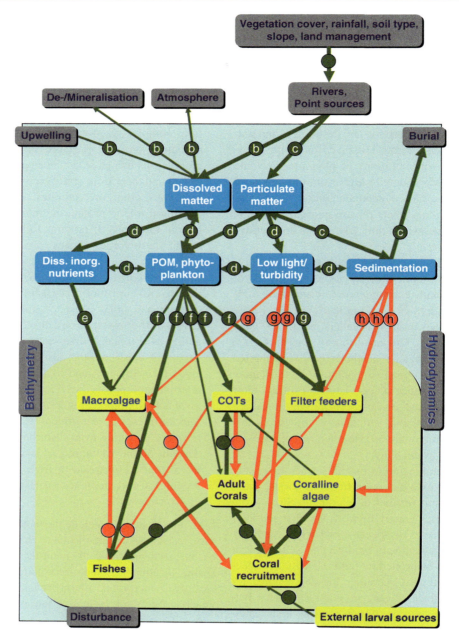

Fig. 3 Conceptual model of the relationships between the main water quality constituents (blue) and biotic responses (yellow) on coral reefs (*pale yellow box*) within their marine setting (*light blue box*). External and geophysical factors (*gray boxes*) are additional drivers further shaping the relationships. *Thick arrows* indicate strong effects, *thin arrows*, weaker effects. *Arrow* colors indicate the direction of change (green: promote, red: reduce), *round* letter symbols identify the main processes and relationships explained in the text. Minor links are omitted for clarity. Abbreviations: POM, particulate organic matter; COTS, crown-of-thorns starfish

reduced species richness. Hence, increasing exposure to terrestrial runoff causes reef ecosystems to become less diverse and compromises their ability to maintain essential ecosystem functions and to recover from disturbances.

The study has also shown that the severity of eutrophication effects is strongly determined by a relatively small number of environmental properties and biotic conditions. Several of these factors are identical to those that determine the resistance and resilience of coral to bleaching caused by warming oceans. For example, fast currents, topography, proximity to deep water, and a diverse community with abundant herbivores are considered reliable factors in predicting the likelihood of coral communities dying as a result of bleaching (West and Salm 2003) – as they are in predicting death arising from eutrophication. However, reefs in shallow waters are – relative to deeper reef slopes – tolerant of the turbidity associated with eutrophication but sensitive to bleaching. A better understanding of the additive or interactive effects between eutrophication and climate change is clearly needed.

The conceptual model presented here incorporates the main factors associated with coral reefs exposed to changing water quality. It may serve as a starting point to develop quantitative models to predict how specific coral reefs would respond to environmental change. Such models could be used to assess degradation and recovery prospects due to deteriorating or improving water quality, and to identify research gaps.

Although disturbances are a normal and important aspect of their environment, coral reefs are inherently quite stable over time, with coral cover and composition often not changing for many years, or even decades (Connell 1997). After severe acute and short-term disturbances, coral in shallow, well-lit, and well flushed windward reefs can recover within 10–15 years if larvae are plentiful (Connell 1997). Recovery takes longer (possibly up to 50 years) on deeper reef slopes and in poorly flushed settings (such as lagoons) and areas with poor connectivity. Recovery from chronic and human-induced disturbances that alter the physical environment is also slower and less commonly observed than recovery from fast and acute disturbances (Connell et al. 1997).

In most Indo-Pacific coral reefs including the Great Barrier Reef (GBR), coral cover has been declining at a rate of 0.2–1.5% per year since the 1960s (Bruno and Selig 2007). To avoid further reef degradation, it is essential that the disturbance frequency does not exceed the average recovery time in anyone location, and that chronic disturbances are minimized. In recent times, the frequency and nature of major disturbances such as outbreaks of crown-of-thorns starfish, coral bleaching, and severe storms have exceeded the capacity of many reefs around the world to recover (Bruno and Selig 2007; Wilkinson 2004). Such large-scale events are typically not controllable by management action. In contrast, eutrophication is much more manageable and can often be prevented by preserving vegetation cover on land, reducing fertilizer loss into the sea, and restricting aquaculture facilities to well-flushed locations where dilution is rapid and the resistance and resilience of reefs is greatest. With increasing reef disturbances as a result of global warming and seawater acidification, management of water quality and healthy fish abundances will be critically important to the future of coral reef ecosystems (Wooldridge et al. 2005).

Acknowledgments This study was funded by the Marine and Tropical Sciences Research Facility (MTSRF), a part of the Australian Government's Commonwealth Environment Research Facilities Program, and the Australian Institute of Marine Science (AIMS).

References

Aerts LAM, Van Soest RWM (1997) Quantification of sponge/coral interactions in a physically stressed reef community, NE Colombia. Mar Ecol Prog Ser 148:125–134

Anthony KRN, Fabricius KE (2000) Shifting roles of heterotrophy and autotrophy in coral energetics under varying turbidity. J Exp Mar Biol Ecol 252:221–253

Anthony KRN, Hoegh-Guldberg O (2003) Variation in coral photosynthesis, respiration and growth characteristics in contrasting light microhabitats: an analogue to plants in forest gaps and understoreys? Funct Ecol 17:246–259

Baird AH, Babcock RC, Mundy CP (2003) Habitat selection by larvae influences the depth distribution of six common coral species. Mar Ecol Prog Ser 252:289–293

Bak RPM (1978) Lethal and sublethal effects of dredging on reef corals. Mar Pollut Bull 9:14–16

Bellwood DR, Hughes TP, Folke C, Nyström M (2004) Confronting the coral reef crisis. Nature 429:827–833

Birkeland C (1982) Terrestrial runoff as a cause of outbreaks of *Acanthaster planci* (Echinodermata: Asteroidea). Mar Biol 69:175–185

Birkeland C (1988) Geographic comparisons of coral-reef community processes. In: Choat JH, Barnes D, Borowitzka M, Coll JC, Davies PJ, Flood P, Hatcher BG, Hopley D, Hutchings PA, Kinsey DW, Orme GR, Pichon M, Sale PF, Sammarco PW, Wallace CC, Wilkinson C, Wolanski E, Bellwood O (eds) Proceedings of the 6th international coral reef symposium, Townsville, 1988, pp 211–220

Birkeland C (1997) Geographic differences in ecological processes on coral reefs. In: Birkeland C (ed) Life and death of coral reefs. Chapman & Hall, New York, pp 273–287

Birrell CL, McCook LJ, Willis BL (2005) Effects of algal turfs and sediment on coral settlement. Mar Pollut Bull 51:408–414

Bourke L, Selig E, Spalding M (2002) Reefs at risk in Southeast Asia. World Resources Institute, Cambridge

Brock RE, Smith SV (1983) Response of coral reef cryptofaunal communities to food and space. Coral Reefs 1:179–183

Brodie J, Fabricius K, De'ath G, Okaji K (2005) Are increased nutrient inputs responsible for more outbreaks of crown-of-thorns starfish? An appraisal of the evidence. Mar Pollut Bull 51:266–278

Bruno J, Selig E (2007) Regional decline of coral cover in the Indo-Pacific: timing, extent, and subregional comparisons. Public Libr Sci ONE 2:e711. doi:710.1371/journal.pone.0000711

Bruno J, Petes LE, Harvell D, Hettinger A (2003) Nutrient enrichment can increase the severity of coral diseases. Ecol Lett 6:1056–1061

Bryant DG, Burke L, McManus J, Spalding M (1998) Reefs at risk: a map-based indicator of threats to the world's coral reefs. World Resources Institute, Washington, DC

Connell JH (1997) Disturbance and recovery of coral assemblages. Coral Reefs 16(Suppl):101–113

Connell JH, Hughes TP, Wallace CC (1997) A 30-year study of coral abundance, recruitment, and disturbance at several scales in space and time. Ecol Monogr 67:461–488

Cooper T, Gilmour J, KE F (2009) Coral-based bioindicators of changes in water quality on coastal coral reefs: a review and recommendations for monitoring programs. Coral Reefs 28:589–606

Cornell HV, Karlson RH (2000) Coral species richness: ecological versus biogeographical influences. Coral Reefs 19:37–49

Cortes JN, Risk MJ (1985) A reef under siltation stress: Cahuita, Costa Rica. Bull Mar Sci 36:339–356

Costa OS Jr, Leao ZM, Nimmo M, Attrill MJ (2000) Nutrification impacts on coral reefs from northern Bahia, Brazil. Hydrobiologia 440:370–415

Crossland CJ, Bairn D, Ducrotoy JP (2005) The coastal zone: a domain of global interactions. In: Crossland CJ (ed) Coastal fluxes in the anthropocene. Springer, Berlin, pp 1–37

Cuet P, Naim O, Faure G, Conan JY (1988) Nutrient-rich groundwater impact on benthic communities of La Saline fringing reef (Reunion Island, Indian Ocean): preliminary results. In: Choat JH, Barnes D, Borowitzka M, Coll JC, Davies PJ, Flood P, Hatcher BG, Hopley D, Hutchings PA, Kinsey DW, Orme GR, Pichon M, Sale PF, Sammarco PW, Wallace CC, Wilkinson C, Wolanski E, Bellwood O (eds) Proceedings of the 6th international coral reef symposium, Townsville, 1988, pp 207–212

De'ath G, Fabricius KE (2010) Water quality as a regional driver of coral biodiversity and macroalgae on the Great Barrier Reef. Ecol Appl 20:840–850

DeVantier L, De'ath G, Done T, Turak E, Fabricius K (2006) Species richness and community structure of reef-building corals on the nearshore Great Barrier Reef. Coral Reefs 25:329–340

Diaz RJ (2001) Overview of hypoxia around the world. J Environ Qual 30:275–281

Done TJ (1999) Coral community adaptability to environmental change at the scales of regions, reefs and reef zones. Am Zool 39:66–79

Dubinsky Z, Stambler N (1996) Marine pollution and coral reefs. Glob Change Biol 2:511–526

Duce RA, LaRoche J, Altieri K, Arrigo K, Baker A et al (2008) Impacts of atmospheric anthropogenic nitrogen on the open ocean. Science 320:893–897

Dumont E, Harrison JA, Kroeze C, Bakker EJ, Seitzinger SP (2005) Global distribution and sources of DIN export to the coastal zone: results from a spatially explicit, global model (NEWS-DIN). Glob Biogeochem Cycles 19:GB4S02. doi:10.1029/2005GB002488: 1–14

Edinger EN, Jompa J, Limmon GV, Widjatmoko W, Risk MJ (1998) Reef degradation and coral biodiversity in Indonesia: Effects of land-based pollution, destructive fishing practices and changes over time. Mar Pollut Bull 36:617–630

Fabricius K, Alderslade P (2001) Soft corals and sea fans: a comprehensive guide to the tropical shallow water genera of the central-west Pacific, the Indian Ocean and the Red Sea. Australian Institute of Marine Science, Townsville

Fabricius K, De'ath G (2001) Environmental factors associated with the spatial distribution of crustose coralline algae on the Great Barrier Reef. Coral Reefs 19:303–309

Fabricius K, Wild C, Wolanski E, Abele D (2003) Effects of transparent exopolymer particles (TEP) and muddy terrigenous sediments on the survival of hard coral recruits. Estuar Coast Shelf Sci 57: 613–621

Fabricius K, De'ath G, McCook L, Turak E, Williams DM (2005) Changes in algal, coral and fish assemblages along water quality gradients on the inshore Great Barrier Reef. Mar Pollut Bull 51: 384–398

Fabricius KE (2005) Effects of terrestrial runoff on the ecology of corals and coral reefs: review and synthesis. Mar Pollut Bull 50:125–146

Fabricius KE, De'ath G (2004) Identifying ecological change and its causes: a case study on coral reefs. Ecol Appl 14:1448–1465

Fabricius KE, Okaji K, De'ath G (2010) Three lines of evidence to link outbreaks of the crown-ofthorns seastar Acanthaster planci to the release of larval food limitation. Coral Reefs 29:593–605

GESAMP (2001) Protecting the oceans from land-based activities. Land-based sources and activities affecting the quality and uses of the marine, coastal and associated freshwater environment. United Nations Environment Program, 71, Nairobi

Gibbs RJ, Matthew MD, Link DA (1971) The relation between sphere size and settling velocity. J Sediment Petrol 41:7–18

Hallock P (1988) The role of nutrient availability in bioerosion: consequences to carbonate buildups. Palaeogeogr Palaeoclimatol Palaeoecol 63:275–291

Hands MR, French JR, O'Neill A (1993) Reef stress at Cahuita point, Costa Rica: anthropogenically enhanced sediment influx or natural geomorphic change. J Coastal Res 9:11–25

Harrington L, Fabricius K, De'ath G, Negri A (2004) Fine-tuned recognition and selection of settlement substrata determines post-settlement survival in corals. Ecology 85:3428–3437

Harrington L, Fabricius K, Eaglesham G, Negri A (2005) Synergistic effects of diuron and sedimentation on photosynthetic yields and survival of crustose coralline algae. Mar Pollut Bull 51:415–427

Hodgson G (1990) Tetracycline reduces sedimentation damage to corals. Mar Biol 104:493–496

Hopley D, Smithers SG, Parnell K (2007) The geomorphology of the Great Barrier Reef. Cambridge University Press, Cambridge

Houk P, Bograd S, van Woesik R (2007) The transition zone chlorophyll front can trigger Acanthaster planci outbreaks in the Pacific Ocean: Historical confirmation. J Oceanogr 63:149–154

Hubbard DK, Scaturo D (1985) Growth rates of seven species of scleractinian corals from Cane Bay and Salt River, St. Croix, USVI. Bull Mar Sci 36:325–338

Hughes TP, Baird AH, Bellwood DR, Card M, Connolly SR, Folke C, Grosberg R, Hoegh-Guldberg O, Jackson JBC, Kleypas J, Lough JM, Marshall P, Nystroem M, Palumbi SR, Pandolfi JM, Rosen B, Roughgarden J (2003) Climate change, human impacts, and the resilience of coral reefs. Science 301:929–933

Hunter CL, Evans CW (1995) Coral reefs in Kaneohe Bay, Hawaii: two centuries of western influence and two decades of data. Bull Mar Sci 57:501–515

Johannes RE, Wiebe WJ, Crossland CJ, Rimmer DW, Smith SV (1983) Latitudinal limits of coral reef growth. Mar Ecol Prog Ser 11:105–111

Kendrick GA (1991) Recruitment of coralline crusts and filamentous turf algae in the Galapagos archipelago: effect of simulated scour, erosion and accretion. J Exp Mar Biol Ecol 147:47–63

Kleypas JA (1996) Coral reef development under naturally turbid conditions: fringing reefs near Broad Sound, Australia. Coral Reefs 15:153–167

Koop K, Booth D, Broadbent A, Brodie J, Bucher D, Capone D, Coll J, Dennison W, Erdmann M, Harrison P, Hoegh-Guldberg O, Hutchings P, Jones GB, Larkum AWD, O'Neil J, Steven A, Tentori E, Ward S, Williamson J, Yellowlees D (2001) ENCORE: the effect of nutrient enrichment on coral reefs. Synthesis of results and conclusions. Mar Pollut Bull 42:91–120

Lapointe BE, Bedford BJ (2007) Drift rhodophyte blooms emerge in Lee County, Florida, USA: evidence of escalating coastal eutrophication. Harmful Algae 6:421–437

Lapointe BE, Barile PJ, Yentsch CS, Littler MM, Littler DS, Kakuk B (2004) The relative importance of nutrient enrichment and herbivory on macroalgal communities near Norman's Pond Cay, Exumas Cays, Bahamas: a 'natural' enrichment experiment. J Exp Mar Biol Ecol 298:275–301

Larcombe P, Woolfe K (1999) Increased sediment supply to the Great Barrier Reef will not increase sediment accumulation at most coral reefs. Coral Reefs 18:163–169

Littler MM, Littler DS (2007) Assessment of coral reefs using herbivory/nutrient assays and indicator groups of benthic primary producers: a critical synthesis, proposed protocols, and critique of management strategies. Aquat Conserv Mar Freshw Ecosyst 17:195–215

Loya Y, Lubinevsky H, Rosenfeld M, Kramarsky-Winter E (2004) Nutrient enrichment caused by in situ fish farms at Eilat, Red Sea is detrimental to coral reproduction. Mar Pollut Bull 49:344–353

Luick JL, Mason L, Hardy T, Furnas MJ (2007) Circulation in the Great Barrier Reef Lagoon using numerical tracers and in situ data. Cont Shelf Res 27:757–778

Mallela J, Roberts C, Harrod C, Goldspink CR (2007) Distributional patterns and community structure of Caribbean coral reef fishes within a river-impacted bay. J Fish Biol 70:523–537

Marubini F (1996) The physiological response of hermatypic corals to nutrient enrichment. Ph.D. thesis, Faculty of Science, Glasgow

Marubini F, Davies PS (1996) Nitrate increases zooxanthellae population density and reduces skeletogenesis in corals. Mar Biol 127:319–328

Matson PA, Parton WJ, Power AG, Swift MJ (1997) Agricultural intensification and ecosystem properties. Science 277:504–509

McClanahan TR, Obura D (1997) Sedimentation effects on shallow coral communities in Kenya. J Exp Mar Biol Ecol 209:103–122

McCook LJ (1999) Macroalgae, nutrients and phase shifts on coral reefs: scientific issues and management consequences for the Great Barrier Reef. Coral Reefs 18:357–367

McCulloch M, Fallon S, Wyndham T, Hendy E, Lough J, Barnes D (2003) Coral record of increased sediment flux to the inner Great Barrier Reef since European settlement. Nature 421:727–730

Moody J, Butman B, Bothner M (1987) Near-bottom suspended matter concentration on the continental shelf during storms: estimates based on in situ observations of light transmission and a particle size dependent transmissometer calibration. Cont Shelf Res 7:609–628

Nyström M, Graham NAJ, Lokrantz J, Norström AV (2008) Capturing the cornerstones of coral reef resilience: linking theory to practice. Coral Reefs 27:795–809

Pari N, Peyrot-Clausade M, Hutchings PA (2002) Bioerosion of experimental substrates on high islands and atoll lagoons (French Polynesia) during 5 years of exposure. J Exp Mar Biol Ecol 276:1–2

Philipp E, Fabricius K (2003) Photophysiological stress in scleractinian corals in response to short-term sedimentation. J Exp Mar Biol Ecol 287:57–78

Pimm LP (1984) The complexity and stability of ecosystems. Nature 307:321–326

Purcell SW (2000) Association of epilithic algae with sediment distribution on a windward reef in the northern Great Barrier Reef, Australia. Bull Mar Sci 66:199–214

Rogers CS (1990) Responses of coral reefs and reef organisms to sedimentation. Mar Ecol Prog Ser 62:185–202

Rose CS, Risk MJ (1985) Increase in *Cliona delitrix* infestation of *Montastrea cavernosa* heads on an organically polluted portion of the Grand Cayman fringing reef. Pubblicazioni della Stazione Zoologica di Napoli I: Mar Ecol 6:345–362

Rosenfeld M, Bresler V, Abelson A (1999) Sediment as a possible source of food for corals. Ecol Lett 2:345–348

Schaffelke B (1999) Particulate organic matter as an alternative nutrient source for tropical *Sargassum* species (Fucales, Phaeophyceae). J Phycol 35:1150–1157

Sebens KP (1991) Effects of water flow on coral growth and prey capture. Am Zool 31:59A

Smith SV, Kimmener WJ, Laws EA, Brock RE, Walsh TW (1981) Kaneohe Bay sewerage diversion experiment: perspectives on ecosystem response to nutritional perturbation. Pac Sci 35:279–395

Smith SV, Swaney DP, Talaue-Mcmanus L, Bartley JD, Sandhei PT, McLaughlin CJ, Dupra VC, Crossland CJ, Buddemeier RW, Maxwell BA, Wulff F (2003) Humans, hydrology and the distribution of inorganic nutrient loading to the ocean. Bioscience 53:235–245

Stafford-Smith MG, Ormond RFG (1992) Sediment-rejection mechanisms of 42 species of Australian scleractinian corals. Aust J Mar Freshw Res 43:683–705

Stambler N, Jokiel PL, Dubinsky Z (1994) Nutrient limitation in the symbiotic association between zooxanthellae and reef-building corals: the experimental design. Pac Sci 48:219–223

Steneck R (1997) Crustose corallines, other algal functional groups, herbivore and sediments: complex interactions along reef productivity gradients. In: Proceedings of the 8th international coral reef symposium, Panama, pp 695–700

Szmant AM (2002) Nutrient enrichment on coral reefs: is it a major cause of coral reef decline? Estuaries 25:743–766

Tilman D, Downing J (1994) Biodiversity and stability in grasslands. Nature 367:363–365

Tilman D, Fargione J, Wolff B, D'Antonio C, Dobson A, Howarth R, Schindler D, Schlesinger WH, Simberloff D, Swackhamer D (2001) Forecasting agriculturally driven global environmental change. Science 292:281–284

Tomascik T, Sander F (1985) Effects of eutrophication on reef-building corals. 1. Growth rate of the reef-building coral *Montastrea annularis*. Mar Biol 87:143–155

Tomascik T, Sander F (1987) Effects of eutrophication on reef-building corals. 2. Structure of scleractinian coral communities on fringing reefs, Barbados, West Indies. Mar Biol 94:53–75

U.S. Department of Health, Education, and Welfare (1964) Smoking and health: report of the advisory committee to the Surgeon General of the Public Health Service. Public Health Service Publication No 1103, Washington, DC

van Woesik R, Tomascik T, Blake S (1999) Coral assemblages and physico-chemical characteristics of the Whitsunday Islands: evidence of recent community changes. Mar Freshw Res 50:427–440

Vaquer-Sunye R, Duarte CM (2008) Thresholds of hypoxia for marine biodiversity. Proc Natl Acad Sci U S A 105:15452–15457

Vitousek PM, Aber JD, Howarth RW, Likens GE, Matson PA, Schindler DW, Schlesinger WH, Tilman D (1997) Human alteration of the global nitrogen cycle: sources and consequences. Ecol Appl 7:737–750

Weber M, Lott C, Fabricius K (2006) Different levels of sedimentation stress in a scleractinian coral exposed to terrestrial and marine sediments with contrasting physical, geochemical and organic properties. J Exp Mar Biol Ecol 336:18–32

West JM, Salm RV (2003) Resistance and resilience to coral bleaching: implications for coral reef conservation and management. Conserv Biol 17:956–967

Wilkinson C (2004) Status of coral reefs of the world: 2004. Australian Institute of Marine Science, Townsville

Wilson SK, Graham NAJ, Pratchett MS, Jones GP, Polunin NVC (2006) Multiple disturbances and the global degradation of coral reefs: are reef fishes at risk or resilient? Glob Change Biol 12:2220–2234

Wittenberg M, Hunte W (1992) Effects of eutrophication and sedimentation on juvenile corals. 1. Abundance, mortality and community structure. Mar Biol 116:131–138

Wolanski E, Richmond R, McCook L (2004) A model of the effects of land-based, human activities on the health of coral reefs in the Great Barrier Reef and in Fouha Bay, Guam, Micronesia. J Mar Syst 46:133–144

Wolanski E, Fabricius K, Spagnol S, Brinkman R (2005) Fine sediment budget on an inner-shelf coral-fringed island, Great Barrier Reef of Australia. Estuar Coast Shelf Sci 65:153–158

Wooldridge S, Done T, Berkelmans R, Jones R, Marshall P (2005) Precursors for resilience in coral communities in a warming climate: a belief network approach. Mar Ecol Prog Ser 295:157–169

Yentsch CS, Yentsch CM, Cullen JJ, Lapointe B, Phinney DA, Yentsch SW (2002) Sunlight and water transparency: cornerstones in coral research. J Exp Mar Biol Ecol 268:171–183

Part VI
Conservation and Management

The Resilience of Coral Reefs and Its Implications for Reef Management

Peter J. Mumby and Robert S. Steneck

Abstract Our view of ecosystems has evolved from one emphasizing determinism to an understanding that systems can exhibit dramatic, and often surprising, shifts in state. Perhaps the most well-known shift is the replacement of corals by macroalgae, but others occur when systems experience overwhelming bioerosion or heavy sedimentation. Preventing undesirable shifts in ecosystem state is a key goal of management, particularly given the need to stem the loss of ecosystem services. However, ecosystem shifts have proved difficult to predict because they can occur with little warning. Worse, the symptoms, such as loss of coral, and may be difficult to reverse because ecological feedback processes can constrain recovery. Thus, it is important to understand the factors that drive shifts in ecosystem state and the stability of such states. This is the study of resilience. A resilient reef is usually considered to be one that absorbs disturbances and recovers to a coral-rich state (though other states are also possible). We describe methods to quantify explicitly the resilience of a reef by combining models of a reef's equilibrial dynamics with its stochastic disturbance regime. In this case, resilience can be calculated as the probability that a reef will avoid shifting to an alternate stable state in a prescribed period of time, given its current state and anticipated disturbance regime. We then discuss the opportunities to "manage for resilience." Because many acute disturbances, such as coral bleaching, cannot be mitigated directly, the emphasis for management is to enhance processes of coral recovery through the management of watersheds, nutrient-runoff, and grazers. In addition, scientists are beginning to understand spatial patterns of the response of corals to disturbance. Although such research is at an embryonic stage, it promises to play an important role in helping to stratify the interventions of managers across the seascape.

P.J. Mumby (✉)
School of Biological Sciences, University of Queensland,
St. Lucia Brisbane, Qld 4072, Australia
e-mail: p.j.mumby@uq.edu.au

R.S. Steneck
School of Marine Sciences, University of Maine, Darling Marine Center, 193 Clarks Cove Road, Walpole, Maine 04573, USA
e-mail: steneck@maine.edu

1 Introduction

While it has long been appreciated that coral reefs experience periodic cyclone disturbance (Stoddart 1963), there had, until recently, been little evidence that processes of recovery may be impaired. This assumption is reflected in the types of community-level studies that were carried out in the 1970s. For example, the high density of corals led to studies of competition (Lang 1973; Benayahu and Loya 1977; Maguire and Porter 1977) and the study of disturbance focused on its influence on species diversity (Connell 1978). In the early 1980s, several catastrophic events hinted at the potentially fragile nature of coral domination. First was the regional outbreak of white band disease in the Caribbean that reduced the abundance of the major, reef-building acroporid corals in much of the region (Gladfelter 1982; Bythell and Sheppard 1993). Large-scale coral mortality was also observed in the Pacific Ocean after the severe ENSO event of 1983/4 that caused mass coral bleaching, particularly in the West Pacific (Glynn 1984). These two events revealed that coral populations can be acutely disturbed at regional scales. Moreover, evidence began to emerge that macroalgae can become an important, possibly even dominant, component of the reef community. A shipwreck on the Great Barrier Reef resulted in a small-scale but persistent shift from coral to macroalgal domination (Hatcher 1984). In 1983, another disease began to sweep through the Caribbean, this time eradicating the long-spined sea urchin *Diadema antillarum* (Lessios et al. 1984). Shortly after the loss of this urchin, macroalgae began to bloom on many Caribbean reefs (Carpenter 1990; Hughes 1994; Steneck 1994).

In 1997, Connell published a global review of coral recovery from disturbance and arrived at two conclusions: (a) recovery is generally poorer following chronic sources of stress (e.g., loss of herbivores) than acute disturbance (e.g., hurricanes) and (b) recovery had not been reported for the Caribbean whereas many Indo-Pacific reefs were either in stasis or exhibited strong recovery (Connell 1997). A year later, reefs were impacted by the first truly global disturbance phenomenon: the 1998 ENSO, which triggered mass coral bleaching worldwide and resulted in high levels of coral mortality at many sites in the Indian and Pacific Oceans (Wilkinson 1998). Given that this event had such dire impacts on much of the Indo-Pacific – an area for which mass coral mortality had not previously been observed – it became clear that coral recovery was an issue of global concern that warranted urgent research.

This chapter begins by reviewing the concept of resilience. We then describe a recent means of quantifying resilience and use this framework to describe some key ecological properties of coral reef ecosystems. Finally, we consider the ecological processes driving resilience and implications for reef management.

2 The Concept of Resilience: Definitions and History

In the 1970s, ecologists began to realize that ecosystems can have their own "behavior," which is distinct from the properties of the populations and individuals comprising them. In the extreme, ecosystems were considered almost a superorganism that is born, grows, matures, and can die (Whittaker 1970). It was about that time when terrestrial insect ecologist C.S. Holling (1973) observed that changes in the structure and functioning of ecosystems was "*profoundly influenced by changes external to* [the ecosystem] *and continually confronted by the unexpected.*" Further, he observed that drivers of ecosystem change were often nonlinear. He recognized the management challenges associated with unexpected and nonlinear changes to ecosystems so he developed the ecosystem concept of "*resilience*" as a way of describing the stability of ecosystems when perturbed.

Over time, two definitions of resilience evolved in the ecological literature reflecting two different aspects of stability (Holling 1996). These are often described as "ecological" or "engineering" resilience (Holling 1996). Holling's (1973) initial definition, now called "ecological resilience" emphasized the amount of disturbance an ecosystem can withstand without changing its self-organizing processes to alternative stable states (Gunderson 2000). In contrast, "engineering resilience" focuses on the stability around an equilibrium steady state. It is measured as the ability to both resist departing the steady state and the time required to return to equilibrium (Pimm 1984; O'Neill et al. 1986).

An important distinction between the two definitions is that ecological resilience does not assume ecosystems rebound to their previous state. They can flip to a new alternate stable state. While most people are familiar with engineering resilience such as how pressure on a spring deforms it but when the pressure is relaxed, the spring returns exactly to its shape before the deformation. However, with the same analogy, ecological resilience not only may not revert to the initial shape of the spring, it may suddenly change into something that looks and behaves in completely different ways from its initial state. For that reason, it is possible that the ecosystem will not or possibly cannot "flip back" to its original state (Holling 1996). Adding to those inconvenient truths, Holling (1973) observed that change is often unpredictable in cases of ecological resilience.

The theoretical underpinning to resilience theory evolved rapidly over about a 10-year period with much (if not most) of the empirical support coming from studies in the marine realm (often from coral reef ecosystems). Arguably, it was Lewontin's paper (1969) entitled "The meaning of stability" that sensitized the scientific community to the idea that populations, species, and communities can exist at "multiple stable points." To explain how this might be possible Lewontin emphasized the importance of antecedent history and "certain fixed forces" that would be called "drivers" today. That theoretical work set the stage for Sutherland's (1974) classic paper: "Multiple stable points in natural communities." Sutherland had observed that that ecological dominance among several species of encrusting bryozoans depended upon which larva arrived first. He confirmed Lewontin's theory "that history often has an important effect on the observed structure of many communities." A few years later working on the Great Barrier Reef, Sale (1979) observed the same patterns among damselfishes. Sale (1979) concluded that the dominant damselfishes resulted from "the chance allocation" of larvae to his study reefs. Sale's "lottery hypothesis" argued that since the species of larvae reaching the reef is unknown, the change in reef fish community structure will be unpredictable at the species level.

The idea, central to ecological resilience, that community development can be unpredictable at the species level was antithetical to the earlier deterministic concepts advanced by G. E. Hutchinson. Hutchinson (1957) believed that evolution shapes each species to possess unique ecological niches. Competition among species resulting in evolved resource partitioning should result in communities predictably comprised of competitively dominant species. This was mathematically modelled and empirically illustrated in an ornithological study by Hutchinson's student Robert MacArthur (1958). Over the next several decades, most community ecologists followed the lead of Hutchinson and MacArthur by stressing the *uniqueness* of species. However, the conclusions of Sutherland, Sale, and others suggested a different paradigm; one that departed from crisp determinism to one in which multiple stable states may describe the structure of natural communities.

In the 1980s and 1990s, a new "functional group approach" was developed (Steneck and Watling 1982; Steneck and Dethier 1994; Steneck 2001). Functional groups are based on ecologically important similarities among unrelated species. Thus, functional convergence such as molluscan limpets, echinoderm sea urchins, and vertebrate parrotfish can be treated together as herbivores on coral reefs because they all scrape substrates as they feed (Steneck and Dethier

1994). This distinguishes them in functional ways from other types of herbivores. Similarly, red, green, and brown algae that have similar canopy heights exert functionally similar shading and water flow effects to the benthos even though they evolved separately as distinct algal divisions (i.e., phyla) during the Precambrian. In short, functional similarities may be more important than phyletic relatedness.

The key to seeing how these two different world views (uniqueness versus similarity of species) coexist centers on the question being asked and the units being used. The early studies illustrating multiple stable states focused on functionally similar organisms. One encrusting bryozoan replaced another encrusting bryozoan on a settlement plate or one small territorial damselfish replaced another territorial damselfish on an Australian reef in the examples by Sutherland and Sale, respectively. So the *history* of which larva arrives first was paramount for the question of which species dominated the system. However, in Sutherland's (1974) discussion, most examples were fundamentally different from his research. His discussion often focused on functional changes driven by changes in ecological processes. In those examples, the alternate stable states were highly polyphyletic. For instance, Sutherland described an experiment in Washington State by Paine and Vadas (1969). They removed sea urchins from a tide pool dominated by a "mixture of [encrusting] calcareous and small fleshy algae" (Paine and Vadas 1969). As a result of the reduction in herbivory, the algal community structure flipped to one dominated by large, competitively dominant kelp. Another example, this time from coral reefs, described research of Stephenson and Searles (1960) showing how wire cages that exclude numerous species of herbivorous fishes increased both the diversity and biomass of benthic algae. Both the algal flora and the herbivores were phyletically diverse but functionally similar. Sutherland (1974) was describing fundamental changes to the structure and functioning of the ecosystem in ways that Lewontin (1969) would say were the result of "certain fixed forces" or drivers.

We learn from those early studies and from subsequent research that Holling's tenants of resilience are largely supported. We understand that ecosystems are profoundly influenced by external drivers. However, we have little or no capacity to predict what will happen at the species level because we will never know which larva will be drawn in the recruitment lottery.

3 Resistance and Recovery

Most theoretical analyses of resilience include consideration of how stable the ecosystem is relative to perturbations or disturbances. This is often depicted using "topography analogies" (e.g., Holling 1996). These analogies illustrate the structure of an ecosystem as a conceptual ball on a topographic landscape of peaks and valleys. The ball is most stable in this landscape when it resides in a steeply sided valley. Perturbations are anything capable of moving the ball within this landscape. Engineering resilience, by definition, requires a single basin of attraction because it must return to its past configuration. This contrasts with ecological resilience that may have multiple basins of attraction or, in other words, have multiple stable states but each may be functionally entirely different from each other (Fig. 1). Thus, the "two faces of resilience" (Holling 1996) today integrates both the capacity of an ecosystem to *resist change* and its capacity to *recover to its previous state*.

4 Calculating Resilience by Combining Disturbance and Recovery into a Single Framework

The topography analogies depicted in Fig. 1 can be placed into a formal framework by modeling the dynamics of the ecosystem (Ives and Carpenter 2007). Specifically, Mumby et al. (2007a) provided a method to integrate the disturbance (resistance) and recovery aspects of resilience into a single framework that estimates the probability that a reef will avoid being pushed beyond a threshold and entrained toward a coral-depauperate state. In other words, resilience is calculated as the probability that a reef will shift toward an alternate stable state. Two models are needed to derive this measure of resilience. First is an understanding of the equilibrial properties of the system in the absence of acute disturbance. While the reef may, in reality, never reach an equilibrium, an analysis of equilibrial properties indicates whether a reef will tend to exhibit a trajectory of recovery, degradation, or stasis. Clearly, recovery is the preferred trajectory to maintain resilience. However, to assess whether a reef will be able to maintain a recovery trajectory, a second model is needed to simulate the impact of the appropriate disturbance regime.

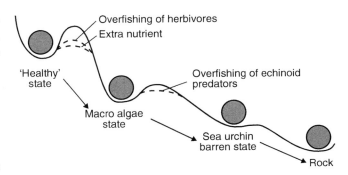

Fig. 1 Topographic analogy of ecosystem stability (depth of the valleys) and change (lateral movement of the ball) (From Bellwood et al. 2004). Note that each of the four locations of the ecosystem (i.e., conceptual ball) represents a very different ecosystem

4.1 Equilibrial Dynamics

The example given here uses a simulation model of a Caribbean forereef that has been tested against an independent time series of empirical data from Jamaica (Mumby et al. 2007a). The model is not described in detail but further information can be found in associated papers (Mumby et al. 2007a; Mumby and Hastings 2008). The processes simulated include coral recruitment, growth, competition, reproduction, background partial-colony mortality, background whole-colony, parrotfish corallivory, parrotfish herbivory, parrotfish exploitation, urchin herbivory, nutrification, algal colonization, algal growth, and coral–algal competition.

To illustrate the key dynamics of corals and macroalgae, a simple simulation was implemented in which coral cover was set initially at 35% (with an equal mix of brooders and spawners) and macroalgal cover at 10%. A total of 36 reefs were created, each having a unique but fixed level of grazing ranging from 5% of the reef being maintained in a grazed state per 6 months to 40% 6 month^{-1}. Sources of acute disturbance were removed although chronic levels of background partial-colony mortality, whole-colony mortality, and algal overgrowth of coral were free to occur. In essence, the simulation represents the natural trajectory of reefs between acute disturbance events; i.e., when free of further acute disturbance does the reef degrade, remain at equilibrium, or become healthier?

A striking pattern in reef trajectories developed over time as a function of grazing (Fig. 2). Those reefs with grazing of less than 20% 6 month^{-1} exhibited a gradual decline in coral cover and the rate of decline increased at lower levels of grazing impact (Fig. 2). A contrasting pattern of coral cover occurred once grazing levels exceeded 20% 6 month^{-1}. Here, coral cover increased and did so at an increasingly greater rate as grazing levels rose (Fig. 2). Thus, reefs with higher grazing exhibited recovery (coral colonization and growth outweighing mortality) whereas reefs at lower grazing exhibited degradation (mortality outweighing processes of colonization and growth). The reef at the fulcrum or threshold level of grazing (around 20% 6 month^{-1}) remained at the initial coral cover of 35% and was therefore at equilibrium. Viewing these dynamics allows some terminology to be assigned. Transient dynamics are those that occur as the reef moves toward a stable equilibrium (transients are shown in circles and equilibrial dynamics as squares in Fig. 2). The equilibrium at 35% coral cover and grazing of 20% 6 month^{-1} is considered to be *unstable* because reefs slightly to either side of this point are attracted toward either high or low coral cover. Indeed, most reefs eventually reach one of two possible *stable* equilibria in the absence of acute disturbance; a high coral cover state or a coral-depauperate state.

The example given here identified an unstable equilibrium (threshold, or tipping point) occurring at a coral cover of 35% and a grazing level of 20% 6 month^{-1}. To locate other thresholds, the simulation model was run for 200 years for every combination of coral cover and grazing. For many levels of grazing, three equilibria emerged; a stable coral-depauperate state, an unstable equilibrium at intermediate coral cover, and a stable coral-rich state (shown in detail in Fig. 3, which identifies two additional unstable equilibria to that identified in Fig. 2). Unstable equilibria were identified where coral cover remained within two percentage units of the initial value during the 200 year simulation (Fig. 2). For the range of grazing associated with herbivorous fish on exposed forereefs (i.e., 5–40% 6 month^{-1}), unstable equilibria fell along a diagonal line from an upper left bifurcation (>70% coral cover and grazing 5% 6 month^{-1}) to a bottom right bifurcation, located at low coral cover (<5%) and a relatively high grazing level of 30% 6 month^{-1} (all equilibria shown in Fig. 4). The two bifurcations delimit a zone of system instability, termed the "bifurcation fold," such that corals can either exhibit recovery trajectories, remain constant, or exhibit decline; the outcome depends on whether the reef sits above, on, or below the threshold respectively. Thus, a single level of grazing (e.g., 15% 6 month^{-1}) can be associated with reef recovery, stasis, or decline, the direction being determined by coral cover (Fig. 4). Such bistability, or the existence of two possible community states from a single level of a process (grazing), is the very essence of alternate states (Petraitis and Dudgeon 2004).

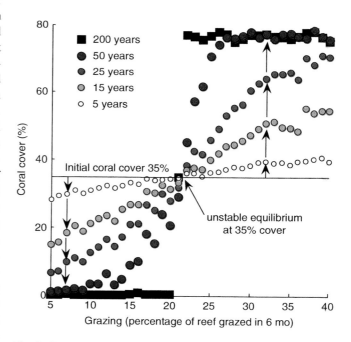

Fig. 2 Temporal trajectories of reefs under chronic disturbance and varying fish grazing starting from an initial coral cover of 35%. Transient cover is represented using *circular markers* and stable equilibria by *squares* (From Mumby 2009)

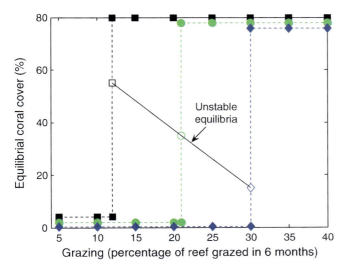

Fig. 3 Identification of system equilibria for three levels of starting coral cover; 55% (*black squares*), 35% (*green circles*), and 15% (*blue diamonds*). Stable equilibria are denoted using *solid markers* and unstable equilibria using *open markers*. Unstable equilibria lie between two alternate stable states. To aid inspection of the graph, the stable equilibria for each initial coral cover have been offset artificially above one another whereas in reality all stable equilibria fall at the level indicated for 15% cover (From Mumby 2009)

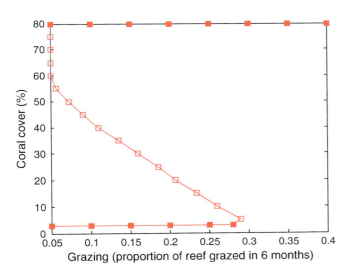

Fig. 4 Positions of stable (*solid squares*) and unstable (*open circles*) equilibria for *Montastraea* reefs at 7–15 m depth with high productivity and little sediment input. Equilibria determined after 200 year simulations

4.2 Disturbance Dynamics

If a reef lies just above a threshold today, it is logical to ask whether an acute disturbance will push the system over the tipping point toward a trajectory of coral decline. Whether this is likely to occur will depend upon the magnitude of the disturbance, the resistance of the coral community to the disturbance, and the distance of the reef from the threshold when the disturbance occurs. The first factor, the magnitude of the disturbance, is usually a physical process such as hurricane strength or a sustained warm anomaly of sea temperature. The resistance of the coral community to such stress will depend upon factors like coral morphology (Done 1992; Madin and Connolly 2006), the complement of zooxanthellae housed within the coral's tissues (Little et al. 2004), and the local physical conditions to which the corals are acclimated (McClanahan et al. 2007). Last, the distance of the reef from the threshold will depend on its current state and the rate at which the community is "recovering" or moving away from the tipping point (which is determined from the analysis of equilibrial dynamics seeing as it constitutes a trajectory between disturbance events). Thus, for a reef at any given state today, it is possible to simulate the processes of recovery, resistance, and stochastic disturbance. As the simulation proceeds, the coral community will undergo periods of recovery, disturbance, recovery, and so on. The simulation runs for a timeline of relevance to management (e.g., 20 years) and determines whether the reef was pushed beyond a tipping point (clearly, management would wish to prevent this occurring). Given that most acute disturbances are stochastic events, the entire simulation process is repeated many times, allowing the frequency of simulations in which the reef was driven beyond a threshold to be determined. Resilience is then expressed as the probability that the reef remains above a tipping point. An example is given in Fig. 5 in which resilience was determined for windward reefs in Belize that experience hurricanes at an overall return time of 17 years (Mumby et al. 2007a).

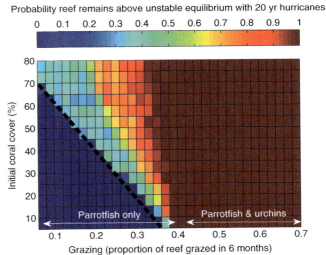

Fig. 5 Probability that reefs of given initial state will remain above the unstable equilibrium during a 25-year period. The physical disturbance regime includes stochastic hurricanes with a 20-year periodicity and the algal–coral overgrowth rate is 8 cm^2 year^{-1}. The unstable equilibrium is denoted (---) (From Mumby et al. 2007b)

Although this approach to quantifying resilience makes heavy use of models, and is therefore limited to systems for which such models are available (e.g., the Caribbean), it does have some desirable properties. First, it allows many of the issues of relevance to reef management to be integrated within a single analysis. These might include monitoring data on the current state of the reefs, the impacts of fishing, nutrification, sedimentation, coral bleaching, ocean acidification, and hurricane damage. Impacts like nutrification and sedimentation enter the analysis through their effects on algal community dynamics and coral population dynamics; both of which will influence the equilibrial properties of the system and change the location of the tipping points. Second, with appropriate testing and validation, the approach may help managers compare the resilience of reefs among locations, identify priority sites for conservation, and estimate the potential benefits of management interventions in terms of the increases expected in resilience. Third, as we show below, even a simple analysis of system dynamics has an important bearing on reef management.

5 Hysteresis in Reef Dynamics and the Urgency for Reef Management

The equilibrial dynamics of a Caribbean coral reef community can be used to revisit the trajectory of some Jamaican reefs over a 20-year period (Fig. 6). By 1979, forereefs reefs had not experienced a severe hurricane for 36 years and coral cover was high at ~75% (Hughes 1994). In 1980, a combination of coral disease and Hurricane Allen reduced coral cover to around 38% but since urchins were present, the reef began to recover. When the urchins died out in 1983, grazing levels were decimated in part because long-term overfishing had removed larger parrotfishes. With a coral cover of approximately 44% and a grazing intensity of only 0.05–0.1 (Mumby 2006), the reef began a negative trajectory toward algal domination that was exacerbated by further acute disturbance. By 1993, coral cover had fallen to less than 5%. A key feature of this graph is that reversing reef decline becomes ever more difficult as the cover of corals declines; as coral cover drops, the level of grazing needed to place the reef on the reverse trajectory (to the right of the unstable equilibrium) increases. Continuing the example of a simulated Jamaican reef, conservation action in the mid-1990s would require grazing levels to be elevated at least fourfold to the maximum observed levels for fishes in the Caribbean. In contrast, if action were taken a decade earlier when coral cover was still around 30%, target grazing levels would be more achievable, requiring only a two- to threefold increase. The term "hysteresis" is used to describe a situation where the trajectories of system decline and recovery differ. Here, it becomes increasingly difficult to help a system recover as its health declines, and there is therefore a pressing need for urgent reef management to prevent decline rather than waiting to attempt future restorative action once the system has degraded further.

The example from Jamaica nicely illustrates another concept of resilience; that of a "brittle" ecosystem. A brittle system is one that may suddenly break, often unexpectedly (as predicted by Holling 1973). Although the reefs of Jamaica were among the most intensively studied in the world in the 1970s, the brittle nature of the system went unrecognized; few herbivorous fishes were left because of intense fishing but because coral cover was high the system appeared to be healthy. The sudden loss of the urchin *Diadema antillarum* in 1983 led to a rapid phase shift that persists at many sites to this day. Thus, brittle systems often lull managers into a false sense of security. More generally, management for the resilience of coral reefs is hampered by the fact that there is usually little warning that a phase shift is about to occur.

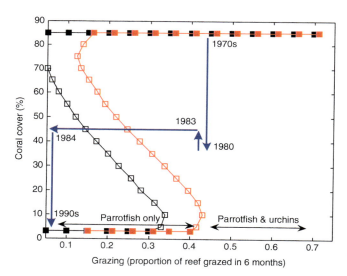

Fig. 6 Stable and unstable equilibria for Caribbean coral reefs at two levels of algal--coral overgrowth. Stable and unstable equilibria denoted (■) and (□), respectively. Black denotes 8 cm^2 year^{-1}, red denotes 14 cm^2 year^{-1}. *Blue lines*, marked with appropriate dates, represent model predictions of the trajectory of reefs in Jamaica (From Hughes 1994)

6 Ecological Feedbacks Drive Hysteresis

Coral reefs are complex ecosystems that are characterized by ecological feedback mechanisms (Table 1; Fig. 7). This can be illustrated by considering a diagrammatic representation of a coral reef (Fig. 7). A number of state variables, such as "corals" and "algae," are connected by multiple ecological processes such that they form a circle, or feedback, acting in both directions. Some examples of feedbacks are listed in Table 1 and illustrated in Fig. 7. Feedbacks can be positive or

Table 1 Feedback mechanisms causing threshold dynamics. Reef ecosystems are particularly susceptible to the emergence of alternative stable states of the ecosystem (Knowlton 1992; Hughes et al. 2005; Mumby et al. 2007a). Stable states are reinforced by ecological feedbacks that "attract" or drive a reef toward a particular state and then maintain the ecosystem within a specific state. The following table lists some of the feedback mechanisms that are suspected to occur on coral reefs and highlights how they are exacerbated by climate change

Feedback mechanism	Exacerbated by climate change
Competitive interactions between macroalgae and corals: Macroalgae pre-empt settlement space and therefore inhibit coral recruitment thereby constraining coral cover and facilitating further algal colonization (Diaz-Pulido and McCook 2004).	Frequent **mass coral mortality** events (bleaching, disease, and hurricanes) facilitate algal colonization because grazing intensity decreases (Kramer 2003). Note, the reverse process also occurs; coral growth and recruitment **reduce the area available to grazers**, which intensifies grazing and can reduce macroalgae.
Competitive interactions between macroalgae and corals: Algal competition causes increased postsettlement mortality in coral due to reduced light, flow, or growth rate (Steneck 1994; Box and Mumby 2007).	Frequent **mass coral mortality** events (bleaching, disease, hurricanes) facilitate algal colonization.
Competitive interactions between macroalgae and corals: Macroalgae overgrow adult corals causing direct reductions in **coral fecundity** because of absent coral (Lirman 2001; Jompa and McCook 2002; Hughes et al. 2007) and indirect chronic reductions in fecundity because of competition (Tanner 1995; Hughes et al. 2007). Reduced fecundity reduces demographic rates of colonization in corals, reinforcing shift toward algae. Note that rate of algal – coral overgrowth is poorly understood and varies dramatically among the taxa involved (Nugues and Bak 2006).	Frequent mass coral mortality events (bleaching, disease, hurricanes) reduce larval output of reefs further. **Calcification rates** of corals are slowed by increased ocean acidification. Results in greater competitive effectiveness of macroalgae relative to corals. Thermal stress also reduces **fecundity** in corals (Nugues and Bak 2006), which may ultimately reduce **larval supply** and coral recruitment (Hughes et al. 2000).
Competitive interactions between macroalgae and corals: Macroalgae may act as vectors of organisms that cause coral disease (Nugues et al. 2004), thereby promoting losses of corals.	Coral mortality events promote **algal colonization** and rising temperature may enhance efficacy of disease organisms (Harvell et al. 2002).
Competitive interactions between macroalgae and corals: Macroalgae exude polysaccharides that may stimulate bacterial growth near corals causing local hypoxia and coral mortality (Smith et al. 2006). Note, mechanism not demonstrated in situ and probably highly dependent on flow regime.	Coral mortality events promote **algal colonization**.
Reductions in coral colonization and survival (**coral loss**) lead to a reduction in **reef accretion** and therefore a reduction in reef **rugosity** (structural complexity). This in turn reduces the carrying capacity of reefs for herbivores, which require high rugosity to provide **shelter from predators** and sustain high densities (van Rooij et al. 1996). A reduction in herbivory continues to enhance the colonization of algae (Mumby et al. 2006b; Hughes et al. 2007).	Rate of rugosity loss may be exacerbated by acidification, which leads to elevated rates of **bioerosion** and physical erosion because coral **skeletons become weaker** (less densely calcified). Bleaching damage and slower rates of coral recovery will cause **habitat loss** and increase the average **distance among patches of high-quality habitat**. This may in turn **reduce the population connectivity** of reef organisms and reduce recruitment (Sale et al. 2005).
Certain species of coralline red algae act as **inducers to coral settlement** (Harrington et al. 2004). Reductions in the cover of encrusting coralline red algae caused by increases in the cover of carpeting macroalgae that trap sediments reduces the availability of settlement substratum for corals, thereby facilitating proliferation of algae (Steneck 1997).	Acidification increases energetic cost of calcification in coralline algae reducing their growth rate and increasing susceptibility to disease. Increase in macroalgal **competitors** after coral mortality events exacerbates process further by making the benthos increasingly hostile to encrusting corallines.
Failure of recovery of the urchin, *Diadema antillarum*, in much of the Caribbean may be driven by feedbacks. Hostile, macroalgal-dominated reefs possess high densities of microinvertebrates that prey upon settling urchin spat causing a bottleneck in urchin colonization because of high **postsettlement mortality** (i.e. macroalgae are a **predator refuge** for juvenile urchins). In contrast, macroalgae are scarce at high densities of adult urchins (Edmunds and Carpenter 2001). Thus, urchins can maintain high-quality habitat for urchin survival but only once grazing levels are high. Modest urchin recovery would enhance the health of many Caribbean reefs (Mumby et al. 2006a).	Acidification may further reduce urchin survival by reducing test strength and/or enhancing vulnerability to disease because of increased **energetic requirements of calcification**. Reductions in rugosity, which are exacerbated by climate change (above), increase the postsettlement mortality of urchins and increases in macroalgae after bleaching-induced coral mortality also add to density of urchin predators.
Recruitment of corals declines because of **Allee effects**, which reduce **fertilization success** and reduce levels of larval supply (Edmunds and Elahi 2007). The problem is then exacerbated because Allee effects may become more severe as **coral density declines** because of reduced recruitment and elevated adult mortality.	Frequent mass coral mortality events reduce the density of adult corals and enhance the severity of Allee effects further. **Chronic stress** caused by bleaching also causes **reduced fecundity** in corals (Nugues and Bak 2006).

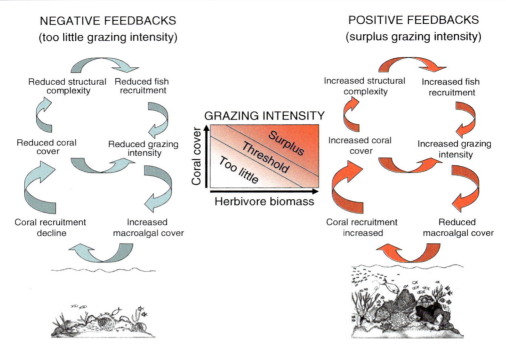

Fig. 7 Schematic representation of positive and negative feedback loops on coral reefs (From Mumby and Steneck 2008. With permission)

negative depending on which side of a threshold the system sits (Fig. 7). An example of a negative feedback might start with inadequate grazing intensity. This could be caused by one or more of the following circumstances: depleted herbivorous fish biomass (because of excessive fishing), high algal productivity (high wave power or elevated nutrient concentration), and/or low coral cover (after a recent bleaching event). Macroalgal cover begins to increase (Fig. 7), which reduces the settlement space available for corals. Furthermore, the increase in macroalgal cover enhances the frequency and duration of competitive interactions with coral recruits, which serves to increase their postsettlement mortality rate. The combined effects of reduced settlement space and enhanced mortality reduce the density of recruits on the reef. If this impact on recruitment is large enough, a bottleneck forms in the coral population such that natural losses of adult corals are not replaced, thereby liberating additional space for rapid macroalgal colonization. The resulting algal bloom further reduces the density of coral recruits, thereby intensifying the bottleneck in coral population dynamics, and the weakening of grazing intensity. Moreover, the continued loss of corals reduces the structural complexity of the reef. Lower habitat complexity then exerts deleterious impacts on the recruitment of corals (reduced availability of refugia from algae) and the recruitment of herbivorous fishes (because of increased predator efficiency). These mechanisms exacerbate the failure of coral recruitment and reduce grazing intensity even more, thus reinforcing the feedback.

7 Managing the Resilience of Reefs

Many of the disturbances facing coral reefs, such as coral bleaching, ocean acidification, hurricanes, eutrophication, and disease are described in detail in other chapters of the book. Some of these disturbances are preventable by appropriate management whereas many others are not. Here, we briefly summarize the management opportunities for addressing each type of disturbance to reef ecosystems. We focus on maintaining healthy corals because corals underpin so many ecosystem services on reefs (Done et al. 1996). Of course, the management of sustainable fisheries is just as important, and we direct readers to reviews on this topic elsewhere (Jennings et al. 2001).

7.1 Management of Preventable Disturbances

Not surprisingly, many local disturbances can be mitigated through local action. For example, sedimentation is usually caused by poor watershed management, particularly on high islands (Richmond et al. 2007). Action to reduce the clearance of vegetation, stabilize soils, and reduce the sediment load of runoff will help reduce the level of sediment reaching coral reefs. Though similar arguments can be made for nutrient runoff, some aspects of nutrification are more difficult to tackle because the effluent is difficult to detect and may diffuse onto the reef over a large area. The latter problem is likely to be particularly acute in karst environments that allow contaminated groundwater to seep onto the reef through multiple, and often inaccessibly deep, locations. A shift toward secondary and tertiary sewage treatment would reduce the contamination of groundwater in the first place.

The process of grazing can be managed using fishing regulations or no-take marine reserves. Unfortunately, few fisheries regulations exist to manage ecological processes explicitly. A shift away from using fish traps would benefit the management of herbivorous fish, which are highly susceptible to trap capture (Hawkins and Roberts 2004). However, specific recommendations on gear and catch limitations are not yet available for most herbivorous fish species. This is particularly problematic because achieving an outright ban on their harvest is either economically or culturally infeasible in many locations. For example, parrotfish are much sought after in Palau and in fact preferred to larger-bodied piscivores like grouper. The challenge, therefore, is to identify ecologically sustainable levels of herbivore exploitation, including urchins that are widely consumed in Asia. This is no easy task as it is constrained by the use of indiscriminate fishing methods, and limited data are available on species' life histories, demographic rates, and ecological functions. Indeed, the ecological function of herbivorous fishes is continually being reviewed. For example, it was only recently discovered that the relatively unpalatable fleshy macroalga, *Sargassum*, could be consumed by a species of batfish (Bellwood et al. 2006). Any plan to manage herbivorous fish species should ensure that a high diversity of species are included (Burkepile and Hay 2008) and that large body size is promoted because large-bodied individuals undertake a disproportionately high level of grazing (Bruggemann et al. 1994; Mumby et al. 2006b). Nonetheless, high levels of grazing have been found to improve coral recruitment (Edmunds and Carpenter 2001; Mumby et al. 2007b) and systems with intense herbivory, such as Palau, have recovered from severe disturbance without phase shifting to macroalgal dominance (Golbuu et al. 2007). In contrast, some systems exhibiting trophic level dysfunction (sensu Steneck et al. 2004), in which herbivory is severely constrained, have exhibited little recovery from disturbance (Cote et al. 2005).

Sea urchins also require explicit management because they have a remarkable ability to adapt to limited food availability (Levitan 1988). Thus, little prevents urchin populations from exploding if their predators, which include balistids and sparids, are heavily depleted because of fishing (McClanahan and Shafir 1990). Plagues of urchins in heavily fished environments, such as Kenya or Jamaica (before the urchin die-off in 1983), have caused extensive reef erosion and "urchin barrens" (Fig. 2, Bellwood et al. 2004).

7.2 Management of Unpreventable Disturbances

From a management perspective, there will always be a limited capacity to reduce perturbations or disturbances (Mumby and Steneck 2008). The best documented of these have resulted from climate-driven warming resulting in massive coral bleaching (Wilkinson 1998) and widespread coral disease (Aronson and Precht 2001). Such large-scale reef disturbances cannot be managed directly, but there are two complementary approaches to meeting such challenges. The first is to enhance coral recovery by managing for high recruitment potential. Conditions hostile to coral recruitment such as high algal biomass (Steneck 1988; Mumby et al. 2007b), high sediment loads (Birrell et al. 2005), and extensive erosion of the substrate by urchins (Sammarco 1980) should be minimized whereas conditions that facilitate coral settlement such as some encrusting coralline algae (Harrington et al. 2004), and subcryptic nursery habitats (high structural complexity) should be maximized. In short, actions to assist coral recovery will include managing the fishery (particularly herbivores and invertivores), the watershed, and nutrient runoff.

Actions that promote coral recovery can potentially be stratified by the level of physical disturbance a reef is subjected to. It is widely appreciated that not all reefs are equal and that some experience more intense and/or frequent physical disturbance than others (West and Salm 2003; McClanahan et al. 2007). For example, reefs situated in a hydrodynamic regime that includes strong vertical mixing of the water column will tend to be subjected to less extreme temperatures during a bleaching event (Skirving et al. 2006), which may result in less severe coral mortality. Managers might capitalize on such natural differences in stress among reefs and either target resources to areas that experience relatively benign conditions for corals or, if the beneficial impacts of management interventions on coral recovery are large enough, target areas that experience particularly harsh physical conditions. The stratification of management activities across the seascape is currently in its infancy but may help managers target the use of their precious resources wisely and begin to meet the challenge of incorporating the anticipated effects of climate change into decision-making.

Perhaps the greatest challenge for reef management is the need to scale up conservation activities toward national scales (The World Bank 2006). At present, many conservation programs are heavily focused on the implementation of marine reserves, yet it is also vital to maintain the quality of reef habitats in fished areas; which, after all, usually represent the majority of the seascape. Failure to manage habitat quality explicitly in fished areas may lead to a progressive erosion of habitat quality and a reduction in the value of ecosystem services provided to people including fisheries and coastal defence

(Hoegh-Guldberg et al. 2007). Ultimately, an ecosystem-based approach to managing reefs is required that locates marine reserves in strategically desirable sites and combines the benefits of reserves with a national policy to manage key ecosystem processes like coral recruitment across the entire seascape.

References

Aronson RB, Precht WF (2001) White-band disease and the changing face of Caribbean coral reefs. Hydrobiologia 460:25–38
Bellwood DR, Hughes TP, Folke C, Nystrom M (2004) Confronting the coral reef crisis. Nature 429:827–833
Bellwood DR, Hughes TP, Hoey AS (2006) Sleeping functional group drives coral reef recovery. Curr Biol 16:2434–2439
Benayahu Y, Loya Y (1977) Space partitioning by stony corals soft corals and benthic algae on the coral reefs of the northern gulf of eilat (Red Sea). Helgoländer wiss Meeresunters 30:362–382
Birrell CL, McCook LJ, Willis BL (2005) Effects of algal turfs and sediment on coral settlement. Mar Pollut Bull 51:408–414
Box SJ, Mumby PJ (2007) The effect of macroalgal competition on the growth and survival of juvenile Caribbean corals. Mar Ecol Prog Ser 342:139–149
Bruggemann JH, Kuyper MWM, Breeman AM (1994) Comparative analysis of foraging and habitat use by the sympatric Caribbean parrotfish *Scarus vetula* and *Sparisoma viride* (Scaridae). Mar Ecol Prog Ser 112:51–66
Burkepile DE, Hay ME (2008) Herbivore species richness and feeding complementarity affect community structure and function of a coral reef. Proc Natl Acad Sci U S A 42:16201–16206
Bythell JC, Sheppard CRC (1993) Mass mortality of Caribbean shallow corals. Mar Pollut Bull 26:296–297
Carpenter RC (1990) Mass mortality of *Diadema antillarum*. I. Long-term effects on sea urchin population dynamics and coral reef algal communities. Mar Biol 104:67–77
Connell JH (1978) Diversity in tropical rain forests and coral reefs. Science 199:1302–1309
Connell JH (1997) Disturbance and recovery of coral assemblages. Coral Reefs 16:S101–S113
Cote IM, Gill JA, Gardner TA, Watkinson AR (2005) Measuring coral reef decline through meta-analyses. Philos Trans R Soc B Biol Sci 360:385–395
Diaz-Pulido G, McCook LJ (2004) Effects of live coral, epilithic algal communities and substrate type on algal recruitment. Coral Reefs 23:225–233
Done TJ (1992) Effects of tropical cyclone waves on ecological and geomorphological structures on the Great Barrier Reef. Cont Shelf Res 12:859–872
Done TJ, Ogden JC, Wiebe WJ, Rosen BR (1996) Biodiversity and ecosystem function of coral reefs. In: Mooney HA, Cushman JH, Medina E, Sala OE, Schulze E-D (eds) Functional roles of biodiversity: a global perspective. Wiley, Chichester, pp 393–429
Edmunds PJ, Carpenter RC (2001) Recovery of *Diadema antillarum* reduces macroalgal cover and increases abundance of juvenile corals on a Caribbean reef. Proc Natl Acad Sci U S A 98:5067–5071
Edmunds PJ, Elahi R (2007) The demographics of a 15-year decline in cover of the Caribbean reef coral *Montastraea annularis*. Ecol Monogr 77:3–18
Gladfelter EH (1982) White-band disease in *Acropora palmata*: Implications for the structure and growth of shallow reefs. Bull Mar Sci 32:639–643
Glynn PW (1984) Widespread coral mortality and the 1982–83 el niño warming event. Environ Conserv 11:133–146
Golbuu Y, Victor S, Penland L, Idip D, Emaurois C, Okaji K, Yukihira H, Iwase A, van Woesik R (2007) Palau's coral reefs show differential habitat recovery following the 1998-bleaching event. Coral Reefs 26:319–332
Gunderson LH (2000) Ecological resilience – in theory and application. Annu Rev Ecol Syst 31:425–439
Harrington L, Fabricius K, De'Ath G, Negri A (2004) Recognition and selection of settlement substrata determine post-settlement survival in corals. Ecology 85:3428–3437
Harvell CD, Mitchell CE, Ward JR, Altizer S, Dobson AP, Ostfeld RS, Samuel MD (2002) Climate warming and disease risks for terrestrial and marine biota. Science 296:2158–2162
Hatcher BG (1984) A maritime accident provides evidence for alternate stable states in benthic communities on coral reefs. Coral Reefs 3:199–204
Hawkins JP, Roberts CM (2004) Effects of fishing on sex-changing Caribbean parrot fishes. Biol Conserv 115:213–226
Hoegh-Guldberg O, Mumby PJ, Hooten AJ, Steneck RS, Greenfield P, Gomez E, Harvell CD, Sale PF, Edwards AJ, Caldeira K, Knowlton N, Eakin CM, Iglesias-Prieto R, Muthiga N, Bradbury RH, Dubi A, Hatziolos ME (2007) Coral reefs under rapid climate change and ocean acidification. Science 318:1737–1742
Holling CS (1973) Resilience and stability of ecological systems. Annu Rev Ecol Syst 4:1–23
Holling CS (1996) Engineering resilience versus ecological resilience. In: Schulze P (ed) Engineering within ecological constraints. National Academy Press, Washington, DC, pp 31–44
Hughes TP (1994) Catastrophes, phase shifts, and large-scale degradation of a Caribbean coral reef. Science 265:1547–1551
Hughes TP, Baird AH, Dinsdale EA, Moltschaniwskyj NA, Pratchett MS, Tanner JE, Willis BL (2000) Supply-side ecology works both ways: The link between benthic adults, fecundity, and larval recruits. Ecology 81:2241–2249
Hughes TP, Bellwood DR, Folke C, Steneck RS, Wilson J (2005) New paradigms for supporting the resilience of marine ecosystems. Trends Ecol Evol 20:380–386
Hughes TP, Rodrigues MJ, Bellwood DR, Ceccarelli D, Hoegh-Guldberg O, McCook L, Moltschaniwskyj N, Pratchett MS, Steneck RS, Willis BL (2007) Phase shifts, herbivory, and the resilience of coral reefs to climate change. Curr Biol 17:1–6
Hutchinson GE (1957) Concluding remarks. Cold Spring Harb Symp Quant Biol 22:415–427
Ives AR, Carpenter SR (2007) Stability and diversity of ecosystems. Science 317:58–62
Jennings S, Kaiser MJ, Reynolds JD (2001) Marine fisheries ecology. Blackwell, Malden, p 417
Jompa J, McCook LJ (2002) Effects of competition and herbivory on interactions between a hard coral and a brown alga. J Exp Mar Biol Ecol 271:25–39
Knowlton N (1992) Thresholds and multiple stable states in coral reef community dynamics. Am Zool 32:674–682
Kramer PA (2003) Synthesis of coral reef health indicators for the western Atlantic: results of the AGRRA program (1997–2000). Atoll Res Bull 496:1–58
Lang JC (1973) Interspecific aggression by scleractinian corals. 2. Why the race is not only to the swift. Bull Mar Sci 23:260–279
Lessios HA, Robertson DR, Cubit JD (1984) Spread of *Diadema* mass mortality through the Caribbean. Science 226:335–337
Levitan DR (1988) Density-dependent size regulation and negative growth in the sea urchin *Diadema antillarum* Philippi. Oecologia 76:627–629
Lewontin RC (1969) The meaning of stability. Brookhaven Symp Biol 22:12–23
Lirman D (2001) Competition between macroalgae and corals: effects of herbivore exclusion and increased algal biomass on coral survivorship and growth. Coral Reefs 19:392–399

Little AF, van Oppen MJH, Willis BL (2004) Flexibility in algal endosymbioses shapes growth in reef corals. Science 304:1492–1494

MacArthur RH (1958) Population ecology of some warblers of northeastern coniferous forests. Ecology 39:599–619

Madin JS, Connolly SR (2006) Ecological consequences of major hydrodynamic disturbances on coral reefs. Nature 444:477–480

Maguire LA, Porter JW (1977) A spatial model of growth and competition strategies in coral communities. Ecol Model 3:249–271

McClanahan TR, Shafir SH (1990) Causes and consequences of sea urchin abundance and diversity in Kenyan coral reef lagoons. Oecologia 83:362–370

McClanahan TR, Ateweberhan M, Muhando CA, Maina J, Mohammed MS (2007) Effects of climate and seawater temperature variation on coral bleaching and mortality. Ecol Monogr 77:503–525

Mumby PJ (2006) The impact of exploiting grazers (Scaridae) on the dynamics of Caribbean coral reefs. Ecol Appl 16:747–769

Mumby PJ (2009) Phase shifts and the stability of macroalgal communities on Caribbean coral reefs. Coral Reefs 28:761–773

Mumby PJ, Hastings A (2008) The impact of ecosystem connectivity on coral reef resilience. J Appl Ecol 45:854–862

Mumby PJ, Steneck RS (2008) Coral reef management and conservation in light of rapidly-evolving ecological paradigms. Trends Ecol Evol 23:555–563

Mumby PJ, Harborne AR, Hedley JD, Zychaluk K, Blackwell PG (2006a) Revisiting the catastrophic die-off of the urchin *Diadema antillarum* on Caribbean coral reefs: fresh insights on resilience from a simulation model. Ecol Model 196:131–148

Mumby PJ, Dahlgren CP, Harborne AR, Kappel CV, Micheli F, Brumbaugh DR, Holmes KE, Mendes JM, Box S, Broad K, Sanchirico JN, Buch K, Stoffle RW, Gill AB (2006b) Fishing, trophic cascades, and the process of grazing on coral reefs. Science 311:98–101

Mumby PJ, Hastings A, Edwards HJ (2007a) Thresholds and the resilience of Caribbean coral reefs. Nature 450:98–101

Mumby PJ, Harborne AR, Williams J, Kappel CV, Brumbaugh DR, Micheli F, Holmes KE, Dahlgren CP, Paris CB, Blackwell PG (2007b) Trophic cascade facilitates coral recruitment in a marine reserve. Proc Natl Acad Sci U S A 104:8362–8367

Nugues MM, Bak RPM (2006) Differential competitive abilities between Caribbean coral species and a brown alga: a year of experiments and a long-term perspective. Mar Ecol Prog Ser 315:75–86

Nugues MM, Smith GW, Hooidonk RJ, Seabra MI, Bak RPM (2004) Algal contact as a trigger for coral disease. Ecol Lett 7:919–923

O'Neill RV, DeAngelis DL, Waide JB, Allen TFH (1986) A hierarchical concept of ecosystems. Princeton University Press, Princeton

Paine RT, Vadas RL (1969) The effects of grazing by sea urchins, *strongylocentrotus* spp. on benthic algal populations. Limnol Oceanogr 14:710–719

Petraitis PS, Dudgeon SR (2004) Detection of alternative stable states in marine communities. J Exp Mar Biol Ecol 300:343–371

Pimm SL (1984) The complexity and stability of ecosystems. Nature 307:321–326

Richmond RH, Rongo T, Golbuu Y, Victor S, Idechong N, Davis G, Kostka W, Neth L, Hamnett M, Wolanski E (2007) Watersheds and coral reefs: conservation science, policy, and implementation. Bioscience 57:598–607

Sale PF (1979) Recruitment, loss and coexistence in a guild of territorial coral reef fishes. Oecologia 42:159–177

Sale PF, Cowen RK, Danilowicz BS, Jones GP, Kritzer JP, Lindeman KC, Planes S, Polunin NVC, Russ GR, Sadovy YJ, Steneck RS (2005) Critical science gaps impede use of no-take fishery reserves. Trends Ecol Evol 20:74–80

Sammarco PW (1980) *Diadema* and its relationship to coral spat mortality: grazing, competition, and biological disturbance. J Exp Mar Biol Ecol 45:245–272

Skirving W, Heron ML, Heron SF (2006) The hydrodynamics of a bleaching event: implications for management and monitoring. In: Hoegh-Guldberg O, Kleypas J, Phinney JT, Skirving W, Strong A (eds) Corals and climate change. Coastal and Estuarine Series, 61. American Geophysical Union, Washington, DC, pp 145–161

Smith JE, Shaw M, Edwards RA, Obura D, Pantos O, Sala E, Sandin SA, Smriga S, Hatay M, Rohwer FL (2006) Indirect effects of algae on coral: algae-mediated microbe-induced coral mortality. Ecol Lett 9:835–845

Steneck RS (1988) Herbivory on coral reefs: a synthesis. Proc 6th Int Coral Reef Symp 1:37–49

Steneck RS (1994) Is herbivore loss more damaging to reefs than hurricanes? Case studies from two Caribbean reef systems (1978–1988). In: Global aspects of coral reefs: health, hazards, and history, Miami, 1993, pp C32–C37

Steneck RS (1997) Crustose corallines, other algal functional groups, herbivores and sediments: Complex interactions along reef productivity gradients. In: Proceedings of 8th international coral reef symposium, vol 1, Panama, 1997, pp 695–700

Steneck RS (2001) Functional groups. In: Levin SA (ed) Encyclopedia of biodiversity, vol 1. Academic, New York, pp 121–139

Steneck RS, Dethier MN (1994) A functional group approach to the structure of algal-dominated communities. Oikos 69:476–498

Steneck RS, Watling L (1982) Feeding capabilities and limitations of herbivorous molluscs: a functional group approach. Mar Biol 68:299–319

Steneck RS, Vavrinec J, Leland AV (2004) Accelerating trophic level dysfunction in kelp forest ecosystems of the western north Atlantic. Ecosystems 7:323–331

Stephenson W, Searles RB (1960) Experimental studies on the ecology of intertidal environments at Heron Island, i. Exclusion of fish from beach rock. Aust J Mar Freshw Res 11:241–267

Stoddart DR (1963) Effects of hurricane hattie on the British honduras reefs and cays, October 30–31, 1961. Atoll Res Bull 95:1–142

Sutherland JP (1974) Multiple stable points in natural communities. Am Nat 108:859–873

Tanner JE (1995) Competition between scleractinian corals and macroalgae: an experimental investigation of coral growth, survival and reproduction. J Exp Mar Biol Ecol 190:151–168

The World Bank (2006) Scaling up marine management: the role of marine protected areas. The World Bank, Washington, DC

van Rooij JM, Kok JP, Videler JJ (1996) Local variability in population structure and density of the protogynous reef herbivore *Sparisoma viride*. Environ Biol Fishes 47:65–80

West JM, Salm RV (2003) Resistance and resilience to coral bleaching: implications for coral reef conservation and management. Conserv Biol 17:956–957

Whittaker RH (1970) Communities and ecosystems. Macmillan, London, p 158

Wilkinson CR (1998) Status of coral reefs of the world 1998. Global Coral Reef Monitoring Network and Australian Institute of Marine Science, Townsville, p 184

Index

Bold page numbers refer to entire chapter

A
a*. *See* In-vivo absorption, 98
Abiotic, 128, 245
Abiotic factors, 245, 407, 408, 444,
Abiotic process, 263
Abiotic stress, 411, 445
Abrasion, 255, 281, 291, 347, 355, 361, 362, 365, 435
Abrolhos Bank, 476
Absorb, 29, 107, 108, 110–113, 188, 189, 191, 246, 251, 494, 509
Absorption, 29–31, 53, 69, 98, 99, 112, 123, 138, 179, 182, 189, 192, 202, 245, 246, 252, 306, 338, 407, 410, 501
Abudefduf, 333
Abundance, 7, 13, 17, 19, 21, 32, 89, 93, 100, 109, 133, 134, 234, 241, 243–245, 247, 251–253, 255, 258, 259, 264, 273, 274, 276, 278, 279, 282, 288–290, 293–295, 300, 302, 303, 305, 307–309, 311, 312, 347–351, 362, 364, 366, 377, 384, 399–401, 439, 441, 445, 446, 466, 482, 493, 496–498, 500, 509
Acanthaster, 5, 275, 286–288, 299, 300, 302, 304, 305, 307,
    A. *brevispinus*, 299
    A. *ellisii*, 299
    A. *planci*, 3, 5, 273, 274, 288, 297. 299–301, 309, 312, 399, 497, 498, 501
    A. *plancii*, 3, 5
Acanthochromis, 333
Acanthochromis polyacanthus, 332, 334, 336
Acanthophora, 257
Acanthophora spicifera, 244, 256, 257, 262
Acanthuridae, 329, 330, 377, 378
Acanthurus nigrofuscus, 333
Acanthurus nigroris, 333
ACC. *See* Amorphous calcium carbonate
Accessory pigments, 31, 246
Acclimation, 95, 99, 100, 115, 399–400, 421, 423, 424, 426–428
Acclimatisation, 413–415, 421–431
Acclimatization/adaptation, 413–415
Acclimatory ability, 427
Accretive growth model, 179, 181–183
Accuracy assessment, 29
Acetabularia rhizoid, 441
Acetogenins, 251, 252
Acidic proteins, 133, 134
Acidification, 151–172, 312, 503
Acoelomorph worms, 299
Acronym, 452
Acrophora cytherea, 94, 95
Acropora sp., 15, 21, 63, 65, 95, 100, 143, 154, 160, 254, 295, 302, 357, 358
    A. *arabensis*, 61
    A. *cervicornis*, 69, 122, 127, 128, 140, 159, 429, 457, 465, 468, 470, 471, 483, 500
    A. *eurystoma*, 154, 163, 166

    A. *formosa*, 99, 126–128
    A. *hemprichi*, 351, 358
    A. *hyacinthus*, 97
    A. *intermedia*, 165
    A. *longicyathus*, 93
    A. *millepora*, 93, 124, 128, 129, 131, 133, 254, 255, 427, 429, 454
    A. *nobilis*, 98
    A. *palifera*, 92, 254
    A. *palmata*, 69, 183, 219, 457, 458, 465, 468, 471, 474, 477, 481–485, 500
    A. *serriatosis*, 473
    A. *squamosa*, 128
    A. *surculosa*, 254
    A. *tenuis*, 61, 92, 93, 100, 254
    A. *valid*, 97
    A. *verweyi*, 159
Acroporidae, 70, 72, 75, 459
Acroporids, 92, 254, 294, 299, 301, 307, 450, 458, 468, 479, 481, 509
Acroporid serratiosis, 458, 474
Actinarian, 347, 348, 349, 354, 365
Actinia schmidti, 190
Active fluorescence, 409, 411
Active/passive, 27, 126–129
Active remote sensing, 27
Active swimming species, 303
Active transport of ions, 127
Activity cycle, 341
Act synergistically, 116, 399, 407
Acute disturbance, 377, 503, 509, 512–514
Acute stress, 411
Adapt, 88, 101, 168, 171, 189, 236, 246, 247, 274, 406, 414, 495, 517
Adaptations, 8, 20, 87, 100, 101, 140, 168, 187–192, 246, 286, 304, 309, 311, 332, 339–400, 406, 413–415, 421–423, 426–431, 436
Adaptive, 6, 257, 259, 311, 400, 413, 421, 422, 427, 430, 460, 461
Adaptive bleaching hypothesis, 400
Adaptive response, 6, 413, 427
Adenosine monophosphate, 413
Adenosine triphosphate, 127, 128, 135, 139, 222, 411, 413
Adhesion, 215, 430, 452, 460
Adult habitat, 332, 338
Adults, 31, 60, 71, 75, 93, 131, 135, 286, 287, 291, 297, 300, 302, 303, 306, 332, 336, 338–340, 363, 366, 425, 438, 443, 459, 496, 500, 501, 516
Advection-diffusion, 179, 181–183
Advection-diffusion simulations, 181, 182
Advective flux, 202
Aeolidia edmondsoni, 301
Aeolid nudibranchs, 308
Aequorea victoria, 109, 191, 192
AFLP. *See* Amplified fragment length polymorphism
AFM. *See* Atomic force microscopyatomic force microscopy

Agabus, 424
Agaricia sp., 254, 302, 362, 454, 481
    A. agaricites, 67, 255, 365
    A. humilis, 67
    A. tenuifolia, 232, 348, 364, 408
Agariciidae, 66, 67, 73
Agelas clathrodes, 352, 360
Agelas wiedenmayeri, 242
Aggregating behavior, 342
Aggregation, 5, 288, 300, 307, 329, 333, 339, 341, 342, 351, 354, 358, 363–365
Aggression, 347, 353, 366
Aggressive, 283, 303, 307, 348, 351, 355, 357, 358, 363, 366
Agricultural effluents, 154
Agricultural pollution, 406
Agriculture, 21, 232, 257, 385, 501
Ahermatypic, 59–61, 141
Aiolochroia crassa, 242
Aiptasia pallida, 88, 220, 408
Aiptasia pulchella, 90, 91, 97, 191
Aiptasia tagetes, 91
Aka, 360
α. See Efficiency of photosynthesis
Albedo, 394
Albulidae, 330
Alcohols, 217
Alcyonaria, 307
Alcyoniina, 279, 280, 294
Algal growth rate, 98, 376
Algae, **6**, 31, 59, 87, **120**, **153–154**, 187, **200**, **210**, **231**, 233–236, **241–264**, 273, 340, 347, 375, 377, **436**, **456**
Algae expulsion, 96
Algae survival, 87
Algal, **6**, 28, 88, 110, **152**, 191, **208**, **219**, **241**, 245, 250–251, 254–256, 261, **291**, **307**, **340**, **348**, **376**, **428**, **440**, **511**
Algal abundance, 256, 308, 348, 352, 384
Algal biomass, 250, 256, 307, 340, 353, 362, 363, 366, 497, 500, 517,
Algal bloom, 249, 260, 261, 516
Algal canopies, 363
Algal domination, 261–264
Algal pavement, 202
Algal pigments, 193, 246, 428
Algal recruits, 365
Algal standing crops, 340
Algal tissue, 191, 248, 249
Algal turfs, 242–245, 247, 253, 255, 259, 260, 264, 338, 348, 377, 440, 443
Algologists, 9
Algorithms, 177
Alizarin red, 221, 335
ALK. See Total alkanity
Alkaline vacuoles, 164
Allee effects, 515
Allelochemical defense, 363
Allelopathic chemicals, 347, 352, 356
Allelopathy, 255, 256, 307, 347, 352, 354–357, 360–362, 497,
Allochthonous, 201, 247–248
Allochthonous sources, 247
Allogeny, 364
Alternate dominant, 347, 349, 359, 364, 367
Alternative states, 350
Alveopora, 73
Amazon, 473
Ambassidae, 330
Amblyglyphidodon, 333
Amblypomacentrus, 333

Ambushing, 309
American Samoa, 328, 472
Amgalaxin, 133
Amino acids, 93–95, 109, 131, 133, 134, 138, 167, 207, 209–211, 216, 218, 220–222, 303
Ammonia, 202, 204, 205, 233
Ammonium, 87, 94–96, 203, 207, 208, 211, 455
Amoebocytes, 460
Amorphous, 97, 119, 122, 129–130, 137, 142, 143
Amorphous calcium carbonate (ACC), 130, 142
AMP. See Adenosine monophosphate
Amphidinium, 90
Amphinomid, 302, 309
Amphinomidae, 281, 302
Amphinomid polychaete worm, 302
Amphipods, 251, 282, 283, 287, 289, 294, 296, 297, 303, 308
Amphiprion percula, 336
Amphiprion spp., 294, 333, 334
Amphistegina lobifera, 124
Ampicillin, 354, 410
Amplified fragment length polymorphism (AFLP), 430
Anatomically diverse group, 331
Anchovy prey, 384
Ancient heritage hypothesis, 135
Andaman Sea, 424
Anemonefish, 295, 334, 336, 337
Anemones, 189, 191, 220, 280, 283, 294–296, 302, 304, 334, 348, 349, 352, 353, 365, 367, 412, 413
Anemones identification, 334
Anemonia viridis, 167, 190
Angelfish, 330, 333, 337, 340
Anglerfish, 330
Anguilla, 328
Aniculus elegans, 302
Anilocra apogonae, 306
Animal-algal symbiosis, 6
Animal intracellular pH, 188
Annelida, 277, 280–282, 303
Annelids, 273, 275, 287, 292, 294, 296, 303, 306, 309
Annelid worms, 281, 289, 302, 312, 357
Annual, 3, 6, 53, 141, 155, 220, 331
Anodonta cygnea, 129
Anotrichium tenue, 365
Anoxia, 168
Anoxic, 201, 205, 455
Antarctic, 393, 398
Antecedent history, 510
Antenna pigments, 409
Antennariidae, 330
Anthopleura elegantissima, 91, 192, 409, 429
Anthozoa, 59, 74
Anthozoans, 101, 123, 191, 192, 215, 220, 279, 292, 355
Anthropogenic disturbances, 4, 18, 347, 351, 397
Anthropogenic eutrophication, 363
Anthropogenic factor, 260, 445, 446, 484
Anthropogenic modification, 341
Anthropogenic stressors, 3, 68, 76, 253, 378, 385, 435, 465, 484
Anthropogeny, 3, 4, 16, 18, 21, 25, 53, 152, 247, 257, 259, 261, 263, 311, 327, 340, 341, 349, 351, 360, 362–364, 367, 376–378, 383, 385, 391, 393, 394, 397, 399, 401, 404, 405, 444, 455, 468, 485
Antibacterial material, 460
Antibacterials, 367, 460, 461
Antibiotic, 232, 234, 254, 347, 354, 454, 460, 475
Antifeedant activity, 251
Antifouling activity, 251

Index

Antigua and Barbuda, 328
Antimicrobial activity, 251, 475
Antioxidant defenses, 190–192, 408, 409, 411
Antioxidant enzyme activity, 407, 411, 427
Antioxidant enzymes, 101, 191, 407, 431
Antioxidative defenses, 190–191
Antipatharia, 279, 280, 295–296
APEC. *See* Asia Pacific Economic Cooperation
Apex, 139, 231, 234–236, 384
Apical and basolateral ectodermal plasma membranes, 189
Aploactinidae, 330
Aplodactylidae, 330
Aplysinia fistularis, 360
Aplysinopsin, 356
Apogon, 306, 332
Apogonidae, 311, 330, 336
Apogon nigrofasciatus, 306
Apogon trimaculatus, 306
Apoptosis, 96, 97, 405, 408, 410–414
Apoptotic, 89, 97, 412–414
Apoptotic stage, 89
Aposomatic function, 308
Aposymbiotic, 94, 171
Aposymbiotic colonies, 94
Applied research, 120
Aquaculture, 6, 257–260, 493
Aquarium, 141, 257, 304, 341, 342, 353, 457
Aquatic ecosystem, 403
Aquatic environments, 201
Arabian Gulf, 139, 458
Arabian region, 47, 424
Arabian Sea, 49
Aragonite, 6, 14, 19, 59, 69, 120, 124, 128–130, 139, 140, 143, 152, 155, 158, 166, 169, 181, 241, 278, 280, 392, 396, 397
  crystals, 124
  skeletons, 59
Aragonitic tablets, 134
Archaea, 16, 212, 231, 233, 482
Archaeological sites, 21
Archipelago, 40, 47, 51, 235, 328, 336, 424, 471,
Areschougiageae, 256
Arginase, 167
Armorheads, 330
Artemia salina, 217, 218, 220–222
Arthropoda, 276, 282, 303
Artificial algae, 352
Artificial habitat, 337
Artificial reefs, 253, 311
Aruba, 328
Ascidiacea, 290, 296
Ascidians, 129, 275, 282, 284, 290, 296, 307, 347–350, 352, 359, 360, 362, 364–367, 377, 497
Ascidians season, 350
Ascorbate peroxidase, 101
Asexual production, 59–61, 75, 355
Asexual reproduction, 59, 60, 62, 92, 257, 297
Asia, 5, 50, 51, 515
Asia Pacific Economic Cooperation (APEC), 48
Asparagopsis, 243
ASP. *See* Aspergillosis
Aspartic acid (Asp), 94, 133
Aspergillosis (ASP), 7, 451, 452, 456, 461, 466, 469, 472, 473, 475, 477, 481, 483,484
Aspergillosis and purple spots, 466, 472
Aspergillus, 456, 475
Aspergillus sydowii, 452, 456, 472

Aspidosiphon sp., 281, 438
Assemblage, 14, 15, 17, 21, 28, 37, 63–65, 87, 242, 255, 280, 293, 300, 309, 337, 362, 367, 441, 495, 497
Association, 110, 247
Asteroidea, 287, 297
Asteroids, 287, 297, 303
Astrangia poculata, 96, 101
Astrocoeniidae, 66, 72
Astroides, 62, 70
Astrophyton muricatum, 274, 297
Astropyga radiata, 296
Atlantic, 19, 20, 49, 50, 62, 65, 66, 90, 91, 93, 264, 279, 280, 290, 301, 302, 312, 341, 439, 465–486
Atlantic/Mediterranean, 302
Atmosphere, 26, 29, 31, 107, 156–158, 259, 391–393, 400, 415, 422, 493, 501
Atmospheric, 6, 14, 17, 27, 30, 32, 151–153, 156–158, 162, 165, 167–171, 247, 321, 391–394, 396–401, 405,
Atmospheric correction, 28, 29, 31
AtN. *See* Atramentous necrosis
Atoa, 5
Atolls, 3, 151, 247
Atomic force microscopyatomic force microscopy (AFM), 141
  phase images, 134
ATP. *See* Adenosine triphosphate
ATP/AMP ratios, 413
ATPase, 119, 127–129, 131, 135, 138, 154, 155, 162, 188, 189
Atramentous necrosis (AtN), 474
Attenuation, 7, 28, 53, 110, 245–247, 407, 495
Attenuation coefficient, 28, 245, 407
Aulostomidae, 330
Aulostomus chinensis, 333
Aurantimonas coralicida, 452, 457
Australia, 3, 47, 154, 259, 300, 304, 312, 313, 328, 331, 349, 351, 353, 354, 366, 379
Australian, 18, 20, 49, 52, 65
Autecological studies, 15
Autochthonous sources, 247–248
Autophagic cells, 413
Autophagy, 405, 413
Autotrophic, 202, 218, 223
Autotrophs, 152, 202, 203, 205, 383
Autrotrophy, 202
AZ Acetazolamide (Diamox), 130
Azooxanthellate, 51, 59, 60, 75, 279, 312, 356, 411

B

Back reefs, 202, 241, 244, 245, 300, 436, 439
Bacteria, 6, 7, 16, 29, 95, 96, 108, 130, 192, 207, 209, 211, 224, 231, 233, 236, 242, 254, 278, 292, 303, 341, 347, 361, 382, 383, 452, 454, 455, 457, 460, 461, 466, 474, 480, 482, 483, 492, 494, 495
Bacterial activity, 460
Bacterial bleaching, 234, 451–454, 475, 481
Bacterial bleaching disease, 234, 451–454
Bacterial farm, 215
Bacterial viruses (bacteriophages), 460
Bacteriophages (Phages), 461
Bacterioplankton, 382
Bahamas, 39, 246, 287, 337, 440, 443, 454, 469, 473, 474
Bahrain, 328
Balanophyllia elegans, 352
Balistes, 339
Balistidae, 311, 330
Baltic Sea, 500

Banda Sea, 424
Banding, 141, 155, 221
Bands, 109, 112, 141, 181, 458, 471–473
Bangial rhodophytes, 436
Bangladesh, 328
Barbados, 21, 281, 302, 337, 444, 494
Barnacles, 283, 284, 296, 297, 299, 301, 305, 306, 497
Barracudas, 329, 330
Barrier, 44, 123, 127, 133, 247, 307, 460, 461
Barrier reefs, 5, **17, 40, 47, 92,** 165, **209, 215, 232, 242, 274, 333, 349, 377, 395, 406, 423, 440, 454, 494, 509**
Baseline fluorescence (F_o), 260
Basolateral cell, 123, 128
Batfish, 517
Bathymetry, 500, 501
Batrachoididae, 330
Bays, 30, 32, 37, 204, 208, 244, 247, 248, 264, 274, 305, 459, 500
BBD. See Black band disease
BBL. See Bacterial bleaching
Bcl-2, 412, 413
Beaches, 241, 257, 274, 336
Beer–Lambert law, 245
Beetles, 424
Beggiatoa sp., 474
Behavioral diversification, 340
Behaviorally, 329, 334
Behavioral response, 254, 337
Behavioral traits, 300, 331
Behavior/behaviour, 3, 61, 108, 124, 141, 166, 208–210, 215, 254, 255, 283, 294, 299, 303–305, 307–309, 328, 333, 334, 339, 340, 353, 365, 374, 375, 377, 379, 396, 397, 400, 421, 510
Belize, 17, 21, 242, 255, 348, 349, 352, 364, 445, 455, 468, 513
Beneficial effect, 304
Benthic, 16, 17, 20, 25, 26, 28, 31, 60, 69, 107, 152, 204, 211, 232, 235, 236, 241, 245, 249, 251–261, 279, 286, 288, 291, 292, 297, 298, 303, 309, 311, 331–334, 337, 340, 347, 350–352, 360, 363, 364, 375–378, 383, 401, 441, 493, 494, 496–498, 511
Benthic boundary layer, 249
Benthic community, 25, 28, 31, 232, 235, 245, 253–258
Benthic community structure, 25, 31, 253
Benthic complexity, 311
Benthic organisms, 107, 251, 347, 350, 363, 401, 494
Benthic-spawning, 332
Benthos, 61, 200–204, 243, 283, 290, 291, 307, 309, 349, 363, 377, 383, 495, 511
Bermuda, 5, 65, 292, 455, 458, 468, 469, 471–473, 475, 476, 482
Berycidae, 330
Berycids, 330
Bicarbonate (HCO_3^-), 89, 94, 114, 124, 125, 127, 129, 130, 133, 137, 138, 142, 152, 155–162, 165, 166, 189, 221, 259, 394, 396
Bigeyes, 330
Bights, 500
Bilirubin, 190
Biodiversity, 14, 16, 17, 19–21, 44, 47–53, 63, 207, 273, 293, 304, 309, 312, 313, 341, 367, 376, 380, 385, 392, 400, 405, 485, 496, 500, 501
Bioeroding, 275, 292, 297, 352, 359, 360, 441–444, 446
Bioeroding sponge, 352, 359, 360
Bioerosion, 16, 169, 171, 261, 276, 286, 297, 435–446, 498
Bioerosion processes, 435–446
Biofilm, 245
Bio-filters, 151
Biofouling, 362
Biogeochemical cycles, 203, 207, 212, 435
Biogeochemistry, 6, 27, 199–205
Biogeographic theory, 47, 50

Biogeography, 17, 43–44, 47, 48, 52, 53, 332, 341
Biogeography/zoogeograph, 329
Bioherm, 59, 151
Biological adaptation, 8
Biological conservation, 342
Biological control, 42, 131, 139–141
Biological processes, 15, 167, 171, 207, 429, 430, 500
Biological response, 107, 220, 394
Biomass, **6, 61, 96, 136,** 199, **209, 215, 231, 244, 278, 331, 353, 376, 455, 497**
Biomineralization, 119, 120, 122, 130, 131, 133, 135, 153–155, 162, 164
Biophysical research, 3
Biosphere, 165, 394, 399
Biosynthesis, 192, 251, 252, 360, 415
Biota, 14, 18, 19, 201, 246, 247, 276, 291, 294, 303, 312, 341
Biotic factor, 15, 245, 443
Biotic interactions, 16, 245, 273, 277, 299–308
Biotic process, 241, 263
Bird-nose wrasse, 333
Bivalve molluscs, 14, 296, 302, 304, 397
Bivalves, 15, 16, 169, 286, 290, 292–294, 297, 304, 306, 424, 438, 440, 441, 445, 497, 498
Bivalve shells, 437
Bivalvia, 284, 286, 303
BL See Bleaching
BL sec^{-1}. See Body lengths per second
Black band, 7, 234, 361, 455–456, 459, 466, 468, 470, 473, 485
Black band disease (BBD), 7, 234, 361, 455–456, 459, 466, 468, 470, 472–478, 480, 481, 483, 485
Black coral, 280, 295
Blast fishing, 341, 342
Bleached coral colonies, 28, 454
Bleached corals, 137, 217–219, 223, 225, 406, 425, 440, 443, 453, 454, 478
Bleaching (BL), 6, 28, 53, 69, 87, 116, 127, 152, 183, 191, 200, 234, 245, 302, 347, 392, 405, 421, 445, 451, 465, 494, 509
Bleaching phenomenon, 6
Blennies, 296, 332, 334, 336, 337, 339
Blenniidae, 334, 336
Blooms, 210, 211, 220, 235, 244, 247, 249, 250, 256, 257, 259–262, 501
Bluehead wrasse, 338
BMP. See Bone morphogenetic protein
Body lengths per second (BL sec^{-1}), 337
Body size, 342, 364, 365, 517
Bolinao reef, 383
Bombyx mori, 123
Bonaire, 75, 292, 440
Bonefish, 330
Bone morphogenetic protein (BMP), 135
Bonnemaisoniaceae, 252
Border, 354, 357, 358
Boring bivalves, 169, 304, 438, 441, 445
Boring fungi, 435, 443, 445
Boring organisms, 152, 170, 438
Boring sponges, 436, 439, 441, 445, 498
Boring worms, 441
Botryllus eilatensis, 360
Bottlenecks limiting, 260
Bottom-up effects, 382–384
Bottom-up force, 261
Bottom-up processes, 367, 377
Boundary, 6, 89, 164, 177, 179, 180, 182, 183, 203, 205, 249, 253, 363
Boundary layers, 6, 177, 179, 182, 183, 203, 205, 249, 253
Boxfish, 330, 339
Brachiopoda, 291, 303
Brachiopods, 135, 296, 312

Index 525

Brachyurans, 273, 282, 283, 287, 292, 294, 297, 302, 303, 305
Brain corals, 71, 74, 348, 352, 353, 357, 366
Branch, 41, 98, 115, 116, 154, 178–180, 183, 282, 295, 299, 301, 302, 309, 311, 351, 457
Branching, 15, 70, 98, 115, 116, 126, 130, 136, 177–181, 183, 216, 218, 243, 244, 275, 280, 282, 292, 294, 298, 302, 340, 351, 352, 354, 357, 359, 363, 365, 424, 427, 428, 439, 459, 483, 496
Branch-spacing, 178–180, 183, 294
Brazil, 50, 66, 355, 360, 472, 473, 476, 481
Brazilian, 71, 312, 473, 476
BrB. *See* Brown band disease
Breakage, 153, 364, 443, 499
Breakers, 30
Breeding, 64, 65, 67, 71
Briareum asbestinum, 280, 354, 472
British Virgin Islands, 328
Brittle system, 514
Broadband, 26
Broadcast spawn, 62, 65, 68–71, 74, 75, 92, 377
Broadcast spawners, 60, 62, 70, 74, 75, 91, 378
Broken coral skeletons, 359
Bromine, 251
6-Bromoaplysinopsin, 356
6-Bromo-29-de-N-methylaplysinopsin, 356
Bromoperoxidases, 251
Brooders, 5, 60, 62, 70, 74, 91, 512
Brood planula, 65, 74
Brown algae, 242–244, 246, 251, 252, 254, 353, 354, 361, 362, 511
Brown band, 312, 459, 473
Brown band disease (BrB), 459, 473
Brown macroalgae, 365
Brunei, 328
Bryopsis, 257, 362
Bryopsis pennata, 362
Bryozoans, 152, 287, 292, 294, 296, 298, 312, 360, 377, 438, 497, 510, 511
Buffer factor, 158
Bulbous marginal tentacles, 357, 358
Business as usual scenario, 405
Butterfish, 330
Butterflyfish, 295, 305, 337, 339, 459, 481
Bythitidae, 330, 332

C
^{14}C, 125, 164, 224, 360
δ^{13}C, 138, 155, 178, 217, 220, 339
CA. *See* Carbonic anhydrase
Ca^{2+}, 124, 125, 127–131, 138, 140, 142, 143, 154, 155, 164, 188, 221, 413, 435
^{45}Ca, 125–127, 129, 130, 153, 164, 221
Ca^{2+}-ATPase, 127, 128, 131, 138, 154, 155, 162
Ca^{2+}-binding protein (CaBP), 128
Ca^{2+} channels, 127, 128
Caenorhabditis, 429
Caesionidae, 330
Calcareous, 13, 119, 124, 243–245, 278–280, 282, 286–288, 363, 436, 439, 501
Calcareous algae, 151, 153, 243, 257, 259, 397, 399
Calcareous green alga, 244, 353, 361
Calcification, 6, 25, 32, 53, 59, 90, 119–143, 151–157, 162–171, 181, 199, 221, 222, 251, 311, 312, 340, 355, 391, 392, 396, 397, 400, 406, 428, 435, 436, 444, 445, 495, 496, 500
Calcification rates, 6, 126–128, 137–139, 141, 152–154, 162–165, 168, 169, 221, 445, 500, 515
Calcifiers, 154, 167, 396, 401

Calcinus obscurus, 302
Calcite, 19, 129, 140, 241, 278
Calcitic prisms, 134
Calcium carbonate (CaCO$_3$), 14, 60, 125, 133–134, 136, 180, 220, 221, 241, 251, 252, 278, 367, 384, 392, 396, 397, 401, 405, 443, 498
 crystal exoskeleton, 59
 precipitation, 137, 152, 156–158
Calcium/proton pump, 222
Calcium pumps, 436
Calicoblastic cells, 121–126, 128, 129, 131–133, 138, 140, 142, 162
Calicodermis, 121–124, 126–128, 130, 131, 135, 138, 139, 141–143
California, 75, 349, 352
Callionymidea, 330
Callyspongia vaginalis, 294
Calvin cycle, 99, 410
Cambodia, 328
Cambrian, 14, 441, 511
Capacity, 3, 5, 9, 15, 19, 44, 52, 87, 99, 114, 133, 162, 167, 168, 171, 219–222, 225, 248–250, 259, 261, 290, 375, 379, 380, 385, 408, 409, 422, 423, 495, 499, 503, 511, 517
Caracanthidae, 330
Carangidae, 311, 330
Carapidae, 297, 330
Carbohydrates, 216, 233, 303, 340
Carbon, 6, 53, 87, 110, 124, 151, 178, 199–202, 204, 207, 215, 216, 218, 220–225, 256, 290, 360, 382, 391, 406, 460, 497
 budget, 222–224, 444
 fixation, 94, 188, 205, 232, 407, 410
 metabolism, 233
 translocation, 89, 220, 224
Carbonate, 14, 16, 19, 59, 125, 151–158, 163, 165, 166, 168, 169, 180, 220, 241, 278, 311, 359, 384, 391, 405, 435, 498
 budget, 444
 dissolution, 446
 species, 156, 162
Carbon-concentrating mechanism (CCM), 125, 131, 154
Carbon dioxide (CO$_2$), 53, 69, 130, 156, 234, 391–394, 396–401, 406
Carbonic acid (H$_2$CO$_3$), 130, 131, 137, 155, 189, 394, 396
Carbonic anhydrase (CA), 94, 129–131, 133–135, 138, 142, 154, 167, 189, 192, 429
Carbonylcyanide m-chlorophenylhydrazone (CCCP), 127
Carcasses, 303
Cardinalfish, 330, 332, 335–337, 340, 383
Caribbean, 5, 15–17, 19, 32, 40, 50, 51, 63, 65, 69, 74, 90, 152, 154, 178, 231, 232, 251, 254, 256, 264, 273, 278, 279, 281, 288, 298, 302–304, 306–308, 310, 329, 337, 338, 341, 348–353, 360, 361, 365, 377, 378, 385, 395, 396, 401, 439, 454, 456–458, 460, 465–486, 509, 512, 514
 basin, 406, 471
 ciliate infection, 472, 473, 476
 reefs, 253, 287, 290, 302, 308, 349, 351, 353, 356, 358, 366, 377, 379, 384, 437, 439, 481, 505, 515
Caribbean ciliate infections, 471, 476
Caribbean crustose lethal orange disease (CCLOD), 473
Caribbean sea, 14, 16, 279, 281, 287, 348, 350, 353, 356
Caribbean–Western Atlantic, 68
Caribbean white syndromes (CWS), 473
Caribbean yellow band disease, 466, 468, 470–473, 478
Carideans, 287, 292, 294, 296, 297
Carnivores, 282, 283, 286, 293, 302, 303, 309–311, 331, 339, 378, 498
Carnivorous, 7, 286, 311, 376, 378, 383
Carnivorous fish, 7, 311, 378
β-Carotene, 98, 112, 191
Carotenoid peridinin, 112
Carotenoids, 98, 121, 191

Carrying capacity, 3, 9, 114, 380, 515
Caryophyllia, 71
Caryophyllia smithii, 68, 123
Caryophylliidae, 66, 68
Caspase, 97, 412, 413
Cassiopeia xamachana, 91
Catalases, 101, 113, 167, 191, 408, 409
Catastrophic, 8, 9, 263, 264, 363, 394, 509
Caulerpa, 245, 251, 252, 255–258
Caulerpaceae, 243, 256
Caulerpales, 252
Caulerpa taxifolia, 256, 258
Causes and mechanisms, 405–415
Cavernula pediculata, 414
Caves, 278, 279, 309, 333, 348, 360
Cavities, 279, 281, 282, 292, 293, 297, 306, 309, 332, 436
Cayman Islands, 328, 473
^{13}C:^{12}C, 339
CCA, 256. See also Crustose coralline algae
CCCP. See Carbonylcyanide m-chlorophenylhydrazone
CCH. See Coral compromise health
CCI. See Caribbean ciliate infections
CCLOD. See Caribbean crustose lethal orange disease
CCM. See Carbon-concentrating mechanism; CO_2-concentrating mechanism
CCWB. See Crustose coralline white band disease
cDNA, 429
cDNA probes, 429
Cell communication, 429
Cell cycle, 89, 90, 95, 96, 412–414
Cell necrosis, 413, 414
Cell-specific density (CSD), 89, 219
Cellular anatomy, 88–89
Cellular-automata-based particle model, 182
Cellular biology, 5, 332
Cenozoic, 14, 16, 19, 88
Centers of calcifications (COCs), 133, 140
Centriscidae, 330
Centropomidae, 330
Centropyge bicolor, 333
Cephalaspideans, 251
Cephalopod, 303, 309
Cerithium, 302
CFC. See Chlorinated fluorocarbon
Chaenopsidae, 285, 286, 291, 292, 296, 302, 303, 336
Chaetodon, 336, 339
    C. fasciatus, 295
    C. multicinctus, 459
Chaetodontidae, 311, 330, 377
Chaetopteridae, 296
Chagos, 49
Chagos archipelago, 32, 328, 424
Channel, 127, 128, 138, 288, 336, 341, 499
CHAR. See Contribution of heterotrophically acquired carbon to daily animal respiration
Charybdis, 303
Chasmo-endoliths, 436
Cheilodactylidae, 330
Cheilodipterus, 332
Cheilodipterus quinquelineatus, 306
Cheiloprion, 333
Chemical bioerosion, 435
Chemical defenses, 241, 243, 245, 251–253, 264, 290, 308, 311, 362, 363
Chemical factor, 335
Chemical mechanisms of competition, 355, 357, 360

Chemical pollutant, 341
Chemical pollution, 341
Chemical stress, 7
Cherax quadricarinatus, 124
Chesapeake Bay, 500
Chimera, 61, 187–188, 359, 363, 364
Chimeric, 188
Chimeric colonies, 354, 358, 359, 364
China, 328, 379, 440
Chitons, 251, 284, 286, 297, 439
Chlorinated fluorocarbon (CFC), 394
Chlorine, 251
Chlorodesmis, 365
Chlorodesmis fastigata, 362
Chloroperoxidases, 251
Chlorophyll, 97, 99, 100, 109–112, 161, 191, 192, 219, 352, 408, 428, 494
Chlorophyll *a*, 98, 111, 112, 220, 246, 337, 360, 365
Chlorophyll *b*, 246
Chlorophyll *c*, 98, 111, 112
Chlorophyll concentration, 98, 100, 110, 111, 191, 352, 428
Chlorophyta, 242–244, 256
Chlorophyte, 31, 130, 244, 434, 436, 437, 441, 443
Chloroplast, 88, 90, 94, 121, 409, 410
Chlororespiration pathway, 411
Chlorurus sordidus, 333
Chromis, 333, 337, 338
    C. atripectoralis, 336
    C. viridis, 337
Chromophores, 108, 109
Chromophyta, 112
Chronic, 168, 350, 393, 477, 483, 497, 503, 509, 512
Chronic bleaching, 365
Chronic photoinhibition, 99, 101
Chronic stress, 413, 515
Chrysiptera, 333
Cidaroid, 287, 300
Ciliate, 287, 291, 312, 452, 459, 472
Cirratulid, 281, 297
Cirrhitidae, 330
Cirripedia, 282, 283, 303
Cirriped species (*Pyrgoma*), 302
Clades, 6, 66–68, 70–73, 75, 76, 87, 88, 90–95, 97, 99–101, 110, 187, 190, 300, 301, 400, 408, 413–415, 427, 431
Cladocora caespitosa, 73, 161, 166, 168
Cladophora, 257
Clams, 129, 189, 209, 210, 286, 358, 377
Classification, 27–31, 37–38, 70, 73, 74, 87, 273, 276, 278, 279, 281, 285, 287, 292–293, 313
Cleaning, 132, 274, 282, 294, 309, 331
Cleaning behavior, 304, 308
Clear water, 110, 246, 439, 496, 498
Cl-/HCO_3-antiport, 189
Climate, 18, 225, 260, 385, 393, 394, 396, 400, 401, 424, 427, 430, 517
Climate changes, 4, 6–9, 14, 18, 19, 21, 22, 43, 53, 69, 78, 139, 258–261, 311, 312, 341, 373, 385, 391–401, 406, 421–431, 435, 483, 486, 503, 515, 517
Clingfish, 330, 334, 339
Clinidae, 330
Clinids, 330
Clionaid sponges, 297
Cliona sp., 439, 498
    C. delitrix, 498
    C. orientalis, 352, 359, 438
    C. tenuis, 360
Clipperton, 274, 328

Index   527

"Clock" genes, 107
CLOD. *See* Coralline orange lethal disease
Clone-mates, 177
Closed system, 92, 156, 187
Closed-system transmission, 91
Clostridium, 341
Clouds, 29, 30, 107, 247, 331, 333, 392, 398
Cnidaria, 276, 282, 303
Cnidarian, 62, 87, 95, 101, 129, 167, 187, 191, 192, 279, 358, 409, 411,
   diseases, 475
   genome, 425, 429
C:N:P ratios, 200, 202, 204, 205
CNP stoichiometery, 200
C/N ratio, 209, 219
Coastal, 7–9, 18, 19, 21, 53, 64, 110, 209, 249, 250, 253, 256,
   258–260, 355, 360, 380, 381, 385, 391, 395, 400, 405, 436,
   443, 444, 465, 493, 494, 499, 501, 517
Coastal ocean, 7
Coastal protection, 400, 401
Coastal zone color scanner, 29
Coastlines, 13, 51, 241, 331, 395, 398, 405, 440
Coccoid, 88, 436
Coccolithophores, 125, 135, 155
CO_2-concentrating mechanisms (CCMs), 94, 125, 189
Coconut Island, 4, 305
COCs. *See* Centers of calcifications
Codium, 256, 257
Coelenteric cavity, 121, 123, 124, 190
Coelosmilia sp, 140
Coevolution, 93, 341
Cold fronts, 349
Cold upwelling, 425
Cold-water corals, 59
Cold water upwelling, 66, 259, 500
Collaboration, 5, 9, 276, 313, 342, 484, 486
Collapse, 19, 263, 264, 282, 302, 384, 394, 409, 428, 443, 497, 500, 501
Colloidal matrix, 122
Colombia, 349, 363, 471, 475, 478, 482
Colombian reefs, 350
Colonization, 15, 75, 245, 247, 258–260, 311, 338, 347, 348, 353,
   363, 436, 438, 444, 445, 512, 515, 516
Colonize, 233, 243, 245, 257, 260, 296, 359, 436, 445, 457
Colony, 5, 15, 28, 38, 60–65, 67, 70–75, 89, 108, 126, 151, 169,
   177, 192, 216, 217, 219–223, 233, 243, 275, 337, 348, 351,
   354–361, 363, 365, 366, 392, 426, 438, 451, 466, 495, 512
   architecture, 115–116
   branch, 311
   survival, 61, 100, 364
Color vision, 338
Colouring, 192
Colpophyllia, 481, 483
Colpophyllia natans, 471
Columbia, 328
Commensalism, 16, 297, 303–305, 340
Commensal relationships, 304
Commensals, 281–283, 296, 329
Communication partnership for science and the sea (COMPASS), 8
Communit, 329
Community, 8, 93, 151, 200, 202, 204, 205, 222, 231–234, 237,
   243, 338, 339, 342, 360, 366, 379
   assembly, 17
   state, 263, 264, 512
   structure, 14, 17, 21, 25, 26, 29, 31, 199, 253, 255, 256, 258, 273,
     275, 288, 293, 299, 312, 341, 347, 367, 485, 486, 499, 501, 511
Comoros, 328
COMPASS. *See* Communication partnership for science and the sea

Compensation light intensity (E_c), 98
Competition, 6, 14, 16, 20, 42, 43, 61, 99, 154, 245, 255–256, 275,
   290, 298, 299, 307, 341, 347–367, 376, 384, 385, 412, 496,
   497, 499, 510, 512, 515
Competitive, 16, 43, 120, 255, 261, 278, 307, 311, 348–367, 374, 384,
   494, 501, 510, 511, 515, 516
Competitive network, 348, 349, 366, 367
Competitors, 129, 253, 308, 312, 348, 351, 352, 355–360, 363, 364,
   366, 367, 515
Computed tomography (CT), 178
Computed tomography scanning, 180
Concfocal microscopy, 164, 167
Conchocelis, 436, 437, 440
Condylactis, 91
Condylactis gigantea, 89, 91, 188, 352, 356
Condylactis gigantean (Actiniaria), 91
Connectivity, 3, 6, 42, 43, 69, 247, 294, 310, 338, 374, 378, 385, 476,
   499, 503, 515
Consequences and management implications, 483–485
Conservation, 8, 19–21, 32, 47–51, 53, 60, 264, 327, 328, 331,
   341–343, 380, 387, 514, 517
Conservationists, 47, 331, 341
Consumption, 96, 101, 167, 200, 202, 207, 208, 232, 233, 236, 250,
   299, 300, 302, 307, 309, 310, 347
Continental shelves, 13
Contribution of heterotrophically acquired carbon to daily animal
   respiration (CHAR), 218
Contribution of zooxanthellae to animal respiration (CZAR), 5, 87, 224
Control, **5, 14, 27, 39, 92, 108, 120, 158, 188, 203, 232, 245, 275,
   330, 347, 384, 406, 429, 482, 497**
Conus, 285, 286, 303
Cook Islands, 328
Copepoda, 282, 283, 292, 303
Copepods, 216, 251, 283, 287, 290, 296, 297, 299, 305, 306
Copophyllia, 457
Copophyllia natans, 457
   Coral, (the entire book) bleaching, 6, 8, 69, 89, 100, 152, 165,
     245, 261, 312, 359, 395, 396, 398, 399, 401, 405–415, 421,
     422, 424–426, 445, 452–454, 458, 494, 500, 503, 509, 514,
     516, 517
   breakage, 153
   cover, 7–9, 26, 152, 170, 232, 236, 261, 273, 300, 307, 311, 347,
     349, 359, 363, 376, 377, 379, 382, 383, 399, 444, 459, 465,
     476, 483–486, 495, 499, 501, 503, 512–516
   disease outbreaks, 445
   diseases, 7, 233–236, 307, 362, 395, 443, 445, 451, 452, 455–461,
     466, 468, 471–486, 495, 514, 515
   fecundity, 365, 515
   growth, 6, 14, 16, 21, 136, 159, 171, 181, 217, 225, 234, 255, 360,
     364, 365, 398, 436, 444, 461, 497–499, 515
   host, 87, 92, 95, 97, 99, 111, 113, 116, 152, 189, 219, 220, 224,
     233, 281, 283, 295, 298, 299, 303–307, 367, 400, 413, 426,
     427, 431, 454, 455, 475
   larvae, 76, 92, 135, 254–255, 353, 354, 362, 365, 496, 501
   mortality, 152, 153, 231, 256, 260, 275, 298, 309, 312, 350, 353,
     362, 396, 398, 406, 428, 431, 444, 445, 484, 497, 509, 510,
     515, 517
   physiology, 140, 159, 167, 216–224, 495
   planulae settled, 353
   rubble, 210, 211, 245, 254, 281, 297, 338, 437
   settlement, 354, 362, 365, 496, 515, 517
   size, 72, 355
   skeletons, 119, 131, 134, 140–143, 155, 247, 278, 281, 283, 297,
     299, 304, 306, 359, 377, 436, 438, 443, 497, 499, 515
   slab, 338
   surface, 123, 125, 234, 298, 362, 452, 457, 460, 471

Copophyllia natans (cont.)
    survival, 60, 101, 233, 354, 364–366, 413, 430, 431, 443, 486, 496, 498
        taxonomy, 37–45, 53
        zonation, 367
Coral–algal interaction, 254, 348, 349, 376
Coral-ascidian interactions, 350
Coral compromise health 476
Coral-depauperate state, 349, 511, 512
Coral life-stage, 60–62, 255, 362
Corallimorpharia, 279, 280, 303
Corallimorpharian, 280, 347–349, 351–355, 357, 358, 363–367
Corallinaceae, 243, 256
Coralline algae, 6, 14, 60, 202, 242, 244, 254, 354, 377, 496, 515
Coralline orange lethal disease (CLOD), 466, 473
Coralline red algae, 241, 242, 246, 401, 515
Coralliophila, 285–287, 301
Coralliophila abbreviata, 301, 307, 480, 481
Corallites, 39, 180–183, 292, 300
Corallite structure, 38
Corallivore fishes, 311
Corallivores, 287, 288, 299–302, 305, 307, 311, 312, 497
Corallivorous starfish, 5
Corallophila huysmansi, 365
Corallovexiidae, 284, 285, 306
Coral reefs, (the entire book)
    algae, 241–264
    crisis, 373–374, 385
    diseases, 465, 473
    diseases in the Atlantic-Caribbean, 465–486
    ecosystems, 7, 9, 18, 25, 153, 171, 207, 210, 211, 233, 250, 273–313, 327, 331, 373–385, 391–401, 405, 435, 444, 494, 503, 510
Coral reef targeted research (CRTR), 374, 377, 385
Coral resistance to disease, 460–461
Coral triangle (CT), 47–53, 383
Coral triangle initiative (CTI), 48, 50, 53
Corculum cardissa (Bivalvia), 91
Cornetfish, 329, 330, 339
Correlation, 14, 19, 26, 32, 53, 67, 92, 108, 137, 154, 164, 167, 191, 211, 220, 236, 294, 307, 348, 423, 429, 441, 444, 455, 466, 475, 478, 482
Cortisol, 336
Corynactis californica, 355
Cost, 25, 26, 28, 53, 135–136, 171, 187, 429, 515
Costa Rica, 156, 328, 494
COTS. *See* Crown-of-thorns starfish
C_2 pathway, 410–411
Crabs, 251, 273, 281–283, 293, 294, 296–299, 302–309, 424
Crabs and shrimps, 299
Crayfish, 124, 424
Creediidae, 330
Cretaceous, 15, 16, 140
Cretaceous period, 15, 16, 140
Cretaceous-tertiary (K-T) extinction event, 16
Crevices, 115, 233, 293, 333, 348, 360, 497
Crinoidea, 287, 296, 303
Crinoids, 152, 281, 287, 292, 296
Critical enzymes, 410
Critical factors, 406
Critical proteins, 407, 415
Critical thermal maxima, 424
Critical threshold, 263, 401
Crown-of-thorns starfish (COTS), 7, 8, 497, 498, 501–503
CRTR. *See* Coral reef targeted research
Crustacea, 282, 289, 297, 303, 311, 313

Crustaceans, 51, 129, 275, 277, 282, 286, 287, 290–294, 296, 299, 302–304, 307, 308, 311, 312, 336, 383, 438
Crustose, 243, 249, 250, 257, 363, 365,
Crustose calcareous algae, 257, 259
Crustose coralline, 6, 254, 466, 469
Crustose coralline algae (CCA), 6, 14, 59, 60, 242, 244, 245, 254, 255, 259, 297, 377, 440, 468, 469, 473, 485, 496, 501
Crustose coralline white band disease (CCWB), 473
cry1, 108
cry2, 108
Cryptic, 244, 276, 278, 279, 281, 286, 287, 290, 292–294, 297, 298, 302, 303, 307, 309, 311, 312, 329, 339, 348, 360
Critical thermal maximum, 422, 424, 427
Cryptic fauna, 281, 292, 309
Cryptic species, 51, 72, 75, 76, 278, 288, 294
Cryptochirus coralliodytes, 298
Cryptochromes, 68, 108, 137
Crypto-endoliths, 436
Cryptofauna, 292, 293, 296, 303
Crystals, 120, 124, 131, 134, 140, 142
CSD. *See* Cell-specific density
Ctenactis, 71
    C. crassa, 71
    C. echinata, 71
Ctenophores, 297, 308
CT. *See* Computed tomography
CTI. *See* Coral triangle initiative
CTM. *See* Critical thermal maximum
Cuba, 280, 328
Culcita, 287, 288, 300
Culture, 3, 6, 9, 88–90, 95, 99, 101, 191, 408, 415, 451, 454, 466
Curaçao, 67, 75, 246, 285, 353, 360, 363
Currents, 6, 14, 39, 41–43, 53, 202, 205, 247, 287, 288, 294, 334, 337, 338, 347, 363, 394, 473, 475, 497, 499, 500, 502
Current speed, 337
Cuthona, 285, 301
Cuthona poritophages, 301
$CuZn^-$, 190
Cu/Zn, 409
CuZnSOD, 190
CWS. *See* Caribbean white syndromes
Cyanide, 127, 341, 342
Cyanobacteria, 6, 87, 107, 132, 210, 211, 233, 242–244, 247, 251–254, 256, 259, 261, 262, 264, 349, 352, 354, 361, 362, 400, 411, 436, 437, 441, 455, 456, 471, 474, 482
Cyanobacterial algal mats, 348
Cyanobacterial blooms, 259
Cyanobacterial mat, 361
Cyanobacterium, 254, 436, 437, 440, 455, 474
Cyanophyta, 242
Cyanosaccus, 436
Cycles per second. *See* Hertz
Cyclic peptides, 252
Cycling, 99, 109, 209, 212, 213, 328, 329, 406, 426, 501
Cyclophilins, 413
Cymothoid isopod, 304, 306
Cynoglossidae, 330
Cyphastrea serailia, 98
Cyphoma gibbosum, 307, 480, 481
Cystocloniaceae, 256
Cytochrome *c*, 412
Cytochromes, 167
Cytoplasm, 88, 89, 97, 128, 131, 189, 413
Cytoprotection, 429
CZAR. *See* Contribution of zooxanthellae to animal respiration (CZAR),

D

D. *See* Dissolution
D1, 92, 99, 100, 409, 427
Dactylopteridae, 330
Damage, 6, 8, 61, 97, 99, 101, 116, 190, 192, 220, 245, 247, 299, 300, 302, 304–306, 339, 341, 342, 348, 351–358, 360, 362, 365, 366, 383, 395, 399, 405, 407–415, 422, 424, 425, 428, 429, 431, 451, 461, 483, 494, 496–498, 500, 514
Damselfish, 15, 289, 290, 311, 329–334, 336, 337, 339, 340, 383, 481, 510, 511
Dark, 67, 89, 90, 97–99, 114, 125, 127, 129–131, 136–139, 154, 158, 163, 169, 171, 181, 190, 201, 217, 221, 222, 280, 301, 335, 408, 409, 411, 436, 455, 461, 468, 471–474
Dark calcification, 127, 128, 139, 153–154, 163, 170, 221
Dark respiration, 99, 162, 167, 202, 259
Dark spots disease (DSD), 469, 471–473, 475–478, 481, 483
Darwin, 5, 43, 44, 53
Darwinian order, 37
Darwin's centres of origin, 43, 53
Dascyllus, 333, 338
Dascyllus aruanus, 337
DCAA. *See* Dissolved combined amino acids
DCMU. *See* 3-(3,4-dichlorophenyl)-1,1-dimethylurea(Diuron)
DD. *See* Dinoxanthin
DDAMDOC disease, fleshy algae, and microbes (DDAMed Model), 236, 237
Dead, 127, 208, 245, 275, 283, 286, 291, 293, 296, 300, 358, 436, 438–440, 444, 445, 457, 472, 473, 498
Dead coral, 170, 245, 261, 288, 300, 301, 312, 338, 361, 377, 437, 438, 443, 471
Death, 7–9, 20, 21, 61, 96, 97, 152, 192, 231–235, 250, 305, 312, 353, 356, 359, 408, 412, 413, 422, 428, 429, 436, 455, 502
Decapod, 290, 292, 297
Decapoda, 282, 303
Decapod crustaceans, 277, 291, 299, 304, 309
Declines, 3, 7, 9, 14, 18, 21, 25, 32, 51, 53, 97, 99, 165, 169, 200, 220, 232, 236, 273, 276, 303, 307, 312, 313, 327, 336, 347, 364, 366, 377, 378, 383, 384, 394, 399, 405, 408, 424, 465, 476, 478, 482, 485, 495–498, 501, 512–515
Declining coral reef ecosystem health, 6
Decompositional material, 495
Deeper-water, 59, 92, 99, 100, 244, 286, 493, 500
Deep-sea, 59, 60, 71, 439
Defend, 252, 286, 305, 332, 348, 355, 357, 358, 445
Defense mechanism, 359
Defensive mechanism, 251, 292, 309, 363
Defensive strategies, 251
Degradation, 18, 95, 96, 99, 190, 207, 209–211, 235, 261, 303, 313, 362, 383, 385, 409, 412, 414, 435, 493, 494, 499, 500, 503, 511, 512
Degree heating weeks (DHW), 395, 396
Demograph, 329
Demographic variables, 17
Demography, 328
Dendrogyra, 457
Dendrophylliid, 356
Dendrophylliidae, 66, 72
Denitrification, 200, 205
Deoxyribonucleic acid (DNA) 7, 88, 90, 190, 341, 405, 409, 412
    damage, 61, 412–415
    methylation, 430
    sequence, 7, 422, 430
Depsipeptides, 252
Depth, 27, 30, 31, 60, 92, 93, 98–101, 107, 109, 110, 113, 115, 116, 136, 159, 178, 191, 192, 201, 202, 207, 211, 217, 242, 245, 246, 255, 258–260, 273, 278–280, 286, 287, 290, 297–299, 303, 338, 339, 342, 363, 367, 392, 394, 407, 410, 423, 425, 436, 437, 439, 441, 443, 445, 456, 468, 483, 493–496, 499, 501, 511, 513
    distribution, 107, 246, 468, 469, 476–477
    gradient, 192, 246, 483
    zonation, 246, 258
Desulfovibrio sp., 474
Detectors, 27, 338
Detrital particles, 246
Detritivores, 277, 281, 283, 308–310, 439
Detritus, 205, 250, 288, 291, 294, 303, 339, 382, 384, 494, 495
Development, 3, 5–9, 14–19, 21, 25–27, 31, 59–63, 66–71, 74–76, 93, 109, 110, 130, 135, 248, 258, 260, 273, 286, 299, 307, 312, 313, 327, 329, 331, 332, 334, 336, 341, 342, 349, 351, 364, 373–375, 380, 381, 385, 392, 397, 401, 405, 415, 421, 438, 445, 452–453, 465, 466, 475, 483, 484, 486, 496, 499, 501, 510
Developmental stages, 292, 306, 307
Devonian, 441
Dewatering, 341
DFAA. *See* Dissolved free amino acids
DHW. *See* Degree heating weeks
Diadema, 256, 286, 287, 297, 300, 303, 307, 468–471
Diadema antillarum, 32, 273, 289, 301, 377, 378, 465, 500, 509, 514, 515
Diadematid, 297, 311
Diadinoxanthin (D_n), 98, 191
Diamox, 130
Diatoxanthin(Dt), 98, 191
DIC. *See* Dissolved inorganic carbon
3-(3,4-dichlorophenyl)-1,1-dimethylurea(Diuron), 127, 137, 154, 408
Dichocoenia, 457
Dichocoenia stokesi, 468
DICOM format, 180
Dictyopteris undulata, 363
Dictyotaceae, 244
Dictyota pulchella, 242, 254, 255, 362
Dictyota spp., 242, 244, 252, 261, 377
Dicytospheria cavernosa, 244
DIDS, 129, 189
Died, 113, 152, 295, 396, 473, 477, 514
Diel cycles, 107, 113
Diet, 217, 252, 253, 291, 309, 339, 340, 440
Diffuse, 28, 126, 129, 164, 189, 394, 412, 473, 493, 517
Diffusion, 120, 123, 126–129, 131, 143, 177, 179, 181–183, 189, 203, 216, 362, 412
Diffusive boundary layer, 179, 203
Digestion, 95–97, 110, 222, 413
Digital image, 28
Dimethylsulphide (DMS), 409
Dimethylsulphoniopropionate (DMSP), 409
DIN. *See* Dissolved inorganic nitrogen
2,4-Dinitrophenol, 127
Dinoflagellates, 59, 87–90, 94, 98, 101, 110, 112, 187, 188, 192, 215, 216, 219, 287, 291, 297, 395, 399–401, 405, 406, 459
Dinoxanthin, 98, 191
Diodon, 339
Diodontidae, 330
Diplastrella gardineri, 359
Diploastrea heliopora, 67, 71
Diploid, 89, 90
Diploprion bifasciatum, 338
Diploria, 356, 481, 483
    D. clivosa, 356
    D. labyrinthiformis, 360, 361, 467, 470, 471, 480
    D. strigosa, 356, 455, 457, 467, 470, 479, 481
Direct effect of CO_2, 158, 166–167
Discarded gastropod, 296

Dischistodus, 333
Disease, 231, 233, 234, 236, 241, 259, 264, 302, 304, 305, 312, 341, 347–349, 353, 359, 362, 365, 367, 384, 395, 400, 443, 445, 451–460, 465–468, 470–478, 481–486, 499, 509, 515, 516
  hot spot, 473, 485
  infection, 365, 476
  morphology, 233
  outbreaks, 231, 445, 458, 465, 466, 468, 471, 473, 475, 482–485
  reservoirs, 481
  symptoms, 353
Display complex orientation, 342
Dissolution (D), 132, 141, 151, 152, 154, 156–158, 169–171, 259, 446
Dissolved and particulate nutrient fluxes, 200
Dissolved carbon dioxide, 155, 162, 165, 188, 406
Dissolved chemical signals, 351
Dissolved combined amino acids (DCAA), 210
Dissolved free amino acids (DFAA), 209–210, 216
Dissolved inorganic carbon (DIC), 124, 125, 128–129, 131, 143, 154–158, 162–164, 178, 181, 188–190, 221
Dissolved inorganic carbon transport, 128, 189
Dissolved inorganic nitrogen (DIN), 202, 208, 209, 219, 231–235, 495
Dissolved inorganic nutrients, 216, 495–497
Dissolved material, 178, 500
Dissolved nutrients, 203, 353, 360, 362–364, 483, 500
Dissolved organic carbon (DOC), 202, 207–209, 211, 222–224, 232–236, 256, 362
Dissolved organic matter (DOM), 209, 215–217, 224, 246, 278, 382
Dissolved organic nitrogen (DON), 204, 207–213, 224
Distribution, 13, 14, 17, 18, 25, 26, 28, 29, 31, 38, 41–43, 48–51, 89, 91, 97–98, 100, 107, 110, 112, 123, 124, 133, 139, 152, 156, 178, 179, 183, 191, 192, 207, 208, 244–246, 251, 256, 259, 279, 282, 298, 299, 305, 309, 311, 334, 341, 357, 392, 394, 401, 410, 424, 425, 436, 437, 439, 451, 453, 455, 468, 469, 473–477, 485
Disturbance, 4, 18, 243, 245, 246, 250, 257, 258, 261, 263, 264, 302, 304, 311–313, 347, 349, 350, 354, 363, 364, 367, 377, 397, 399, 429, 494, 496, 498–503, 509–514, 516–518
Disturbance dynamics, 513–514
Diterpenoids, 252, 308
Diverse, 14, 15, 49–53, 64, 65, 76, 87, 90, 92, 110, 171, 187, 199, 210, 216, 241–243, 251, 253, 254, 276–279, 282, 283, 286, 288, 290, 292, 293, 296, 299, 303, 304, 309, 311, 312, 329, 331, 332, 334, 337, 341, 343, 348, 362, 363, 366, 367, 380, 400, 412, 460, 461, 495, 497, 502, 511
Diversity, 6, 7, 15, 17–20, 31, 32, 43, 44, 47–53, 60, 61, 64, 75, 87, 88, 92, 93, 100, 140–143, 151, 152, 168, 190–192, 216, 241–244, 247, 251, 253, 255, 258, 263, 273–275, 280, 281, 286–289, 291, 305, 306, 311, 312, 330, 338–340, 347, 349, 351, 355, 362, 366, 400, 408, 435, 439–441, 455, 465, 493, 500, 501, 509, 511, 517
Djibouti, 328
DMS. *See* Dimethylsulphide
DMSP. *See* Dimethylsulphoniopropionate
D$_n$. *See* Diadinoxanthin
DNA. *See* Deoxyribonucleic acid (DNA)
DNase, 413
DOC. *See* Dissolved organic carbon
DOC, disease, fleshy algae, and microbes (DDAM), 236
Dodgella priscus, 441
Dolabella auricularia, 252
DOM. *See* Dissolved organic matter
Domecia acanthophora, 299
Dominance, 16, 74, 90, 93, 219, 237, 241, 244, 256, 260, 261, 264, 290, 347–351, 354, 363, 364, 366, 376, 384, 400, 500, 510, 517
Dominated, 14–16, 30, 43, 75, 93, 100, 110, 162, 179, 183, 202, 205, 211, 234, 236, 241, 244, 245, 249, 250, 255, 260–264, 311, 336, 349, 350, 375, 376, 382, 383, 398–401, 436, 440, 455, 474, 494, 495, 498, 511

Domination, 15, 241, 261–264, 349, 509
Dominica, 328
Dominican Republic, 328
DON. *See* Dissolved organic nitrogen
Doridomorpha gardineri, 301
Dottiebacks, 330
Dottyback, 333
Doubling rate, 113, 331
Drag, 204, 340
Dragonets, 330
Drosophila, 123, 429
Drupella, 285, 286, 300, 301, 307
DSD. *See* Dark spots disease
Dt. *See* Diatoxanthin
Dynamics, 7, 17–20, 44, 45, 95–97, 142, 154, 156, 162, 199, 200, 205, 212, 219, 231–237, 248–250, 273, 282, 297, 299, 309, 328, 329, 331, 342, 347, 351, 367, 373–385, 394, 409, 455, 461, 466, 475, 477, 478, 481, 483, 484, 509, 511–516

E

EAM. *See* Epilithic algal matrix
Early mineralization zones (EMZ), 140, 141
Eastern Pacific, 4, 19, 50, 67, 68, 75, 92, 288–291, 293, 301, 302, 307, 309, 401
E_c. *See* Compensation light intensity
ECF. *See* Extracellular calcifying fluid
Echeneidae, 330
Echinodermata, 276, 282, 287, 289, 297, 303
Echinoderms, 129, 131, 135, 187, 275, 279, 281–283, 287, 288, 290, 293, 294, 296, 299, 305, 306, 439, 510
Echinoecus pentagonus, 306
Echinoids, 275, 288, 292, 296, 297, 300, 303, 309, 311, 312
Echinometra, 287, 300
Echinometra mathaei, 311, 439
Echinostrephus, 287, 300
Echinothrix, 287, 297, 300
ECM. *See* Extracellular calcifying medium
Ecological, 5, 6, 8, 9, 14–19, 21, 22, 28, 29, 32, 44, 53, 59, 61, 64, 69, 76, 93, 151–152, 200, 207, 241, 244–261, 264, 275, 276, 278, 281, 290, 297, 298, 300, 303, 305, 308, 309, 313, 328, 329, 331, 332, 337–340, 342, 348, 349, 356, 357, 360, 362, 374–377, 380–382, 385, 399, 400, 405, 406, 422, 428, 430, 441–444, 465, 466, 483, 494, 509–511, 514–517
Ecological characters, 41
Ecological data, 313
Ecological diversification, 340
Ecological issue, 51, 341
Ecological resilience, 8, 380, 499, 510, 511
Ecological succession, 15
Ecological zones, 26
Ecology, 4–6, 8, 13, 16–22, 27, 60, 62, 100–101, 110, 199, 207–213, 273, 275, 282, 300, 311, 328, 329, 340, 451–461, 484, 493, 498, 499
Economic, 6, 8, 9, 21, 53, 258, 374, 380, 381, 385, 391, 406, 517
Economic consequences, 8
ECOPATH model, 310, 382, 383
Ecosystem, 13, 15, 152, 242, 328, 329, 331, 373–385
  responses, 7, 309, 399
  shifts, 509
Ecotoxicologists, 9
Ecotoxicology, 5
Ectoderm, 108, 120–122, 125
Ectoprocta, 277, 303
Ectotherms, 423
Ecuador, 301, 328

Index

Eeltail catfish, 330
EEP. *See* Equatorial Eastern Pacific region
EEZ. *See* Exclusive economic zones
Efficiency of photosynthesis (α), 98
Efficiently conserves energy, 382
Egesta method, 222
Egg, 5, 61, 67, 68, 71–75, 216, 331–334, 336–339, 365, 378, 459
    mortality, 336
    quality, 332, 336
    size, 332, 333, 365
Egypt, 4, 328
Egyptian, 63, 64
Eilat, 4, 63, 67, 108, 109, 113, 169, 287, 305, 351, 352, 357, 358, 364, 366, 453, 457
Eisenia kelp forests, 352
E_k. *See* Light intensity of incipient saturation
Electron microscopy, 122, 358
Electron paramagnetic resonance (EPR), 415
Eleotridae, 330
Elevated seawater temperature, 363, 405, 406, 411
El Niño, 6, 53, 152, 471
El Niño-La Nina fluctuations, 7
El Niño-Southern Oscillation 264, 349, 405, 444, 509
Elysia halimedae, 308
Embryo development, 60
Emission spectra, 192
Emperors, 330, 336, 337, 342
Emydocephalus annulatus, 334
EMZ. *See* Early mineralization zones
ENCORE, 204
Encrusting coralline algae, 435, 436, 515, 517
Encrusting corals, 15
Encrusting sponge, 359
Endemism, 43, 44, 48–52
Endocrine, 328
Endocrine disruptors, 7, 9
Endodermal, 139, 188, 189
Endodermal plasma membrane, 189
Endogenous digestive enzyme, 340
Endogenously, 107, 109, 136, 340, 375, 377, 378
Endolithic, 210, 278, 281, 289, 292–294, 435, 439
Endolithic algae, 16, 210, 211, 466
Endolithic upwelling, 200
Endolymph, 124, 129
Endoplasmic, 128, 328, 329
Endosymbionts, 16, 87, 101
Endosymbiotic algae, 494, 495
Energy, 15, 27, 71, 107–116, 127, 135, 136, 151, 154, 164, 178, 199, 200, 203, 204, 210, 215–220, 222–225, 233, 236, 246, 305, 308, 309, 328, 340, 347, 365, 366, 382, 384, 391, 392, 407, 409, 414, 428, 445, 461, 484, 485, 495, 496
    budget, 136, 215, 222, 223, 415, 424
    flow, 15, 328, 329
    fluxes, 111, 113–115, 199
Enewetak, 4, 68, 288–290, 309
Engineering, 27, 120, 132
Engineering resilience, 510, 511
England, 68
Enhanced nutrient, 249, 258
Enrichment, 20, 99, 165, 166, 217, 220, 248, 253, 496, 497
ENSO. *See* El Niño-Southern Oscillation
Entacmea quadricolor, 191
Enterobacteria, 236
Enterobacteriaceae, 458
Entobia, 439
Entoprocta, 277, 291, 303

Environment, 19, 25, 38, 39, 47, 51, 87–89, 91, 92, 99, 120, 123, 138, 153, 158, 171, 177–179, 181–183, 187, 188, 191, 201, 205, 212, 225, 234, 253, 258, 263, 293, 297, 311, 342, 363, 391–392, 400, 406, 409, 421–424, 426, 430, 446, 452, 455, 465, 466, 498, 503
Environmental changes, 15, 16, 20–22, 43, 93, 96, 152–153, 263, 391, 399, 400, 406, 414, 421, 446, 503
Environmental conditions, 69, 71, 76, 93, 97, 101, 108, 189–191, 224, 225, 234, 243, 261, 263, 264, 298, 304, 353, 359, 362, 363, 391, 400, 422, 427, 438, 442, 446, 466, 475, 477, 482, 485
Environmental drivers, 258, 263, 466, 482–483
Environmental lawyers, 9
Environmental variation, 20, 38–40
Enzymatic, 131, 134, 190–191, 287, 408, 430
Enzymatic mechanisms, 190
Enzyme, 87, 94, 99, 101, 113, 129–131, 151, 154, 162, 167, 188–192, 203, 233, 251, 300, 400, 412, 413
    activity, 113, 140, 167, 397
    assays, 415
Enzyme carbonic anhydrase (CA), 129–131, 154, 167, 189
Enzyme ribulose-1,5-bisphosphate carboxylase/oxygenase (Rubisco), 94
Eocene, Paris Basin, 439
Eohyella campbellii, 440
Ephippidae, 330
Epidemiological, 494
Epidermis, 120, 121, 365
Epifaunal, 311
Epigenetics, 421, 422, 427, 430, 431
Epilithic algae, 382
Epilithic algal matrix (EAM), 399
Epiliths, 436
Epinephelus maculates, 340
Epinephelus striatus, 341
Epiphytes, 245, 257, 258, 349
Epiphytic cyanobacteria, 256, 262, 349
Epithelial, 120, 123, 127, 306
Epizoites, 304, 305
Epizootic, 280, 465, 466, 468, 471–475, 477, 483, 485
EPR. *See* Electron paramagnetic resonance
Epulopiscium, 341
Epulopiscium fishelsoni, 341
Equatorial, 20, 49, 52, 53, 63, 65, 110, 168, 394
Equatorial Eastern Pacific region (EEP), 65
Equilibria, 349, 376, 509, 511–514
Equilibrium, 125, 129, 156, 157, 263, 264, 349, 350, 499, 510–514
    constants, 156
    theory, 44
    thermodynamics, 133
Eritrea, 49, 328, 384
Erosion, 152, 241, 258–260, 385, 390–392, 395, 400, 435, 515, 517
Erythropodium caribaeorum, 280, 352, 472, 484
Escherichia coli, 236, 409, 410
EST libraries. *See* Expressed sequence tag libraries
Estuaries, 248, 249
Ethacrynic acid, 127, 128
Ethoxyzolamide (EZ), 130
Eucheuma, 256–258
Eucidaris, 287, 300
    E. galapagensis, 300, 301
    E. thouarsii, 439, 444
Euendolithic green alga, 441
Eugomontia sacculata, 437
Eukaryotes, 108, 429
Eukaryotic algae, 107, 243
Eukaryotic microorganism, 436
Eukaryotic photoautotrophs, 411

Eunicid, 297, 303
Euphyllidae, 66, 67
Eupomacentrus planifrons, 15
Eurigonum nodosum, 441
Europe, 5, 37, 394, 424
Eurythoe complanata, 302
Eusmilia fastigiata, 75
Eutrophic, 232, 248, 305, 439, 442, 444, 445, 493
Eutrophication, 89, 96, 97, 152–153, 234–236, 264, 349, 363, 364, 405, 406, 444, 445, 493–503, 516
Eutrophic water, 439
Evolution, 37–45, 52–53, 69, 70, 73, 74, 107, 113, 243, 251, 307, 311, 329, 332, 338, 339, 359, 380, 394, 400, 414, 441, 444, 510
Evolutionary, 16, 19–20, 22, 38, 41–45, 51, 53, 60, 61, 66, 67, 76, 88, 93, 100, 101, 152, 243, 278, 288, 312, 332, 334, 414
Evolutionary adaptation, 406
Evolutionary biology, 5
Evolutionary controls, 67
Exclusive economic zones (EEZ), 330
Exogenous antioxidants, 408, 413
Exogenous recruitment, 375, 377
Exoskeleton, 59, 61, 119, 129, 134, 279, 312, 439
Exotic marine species, 256
Expedition, 5, 215, 236, 274, 283, 354, 423, 425
Exploitation, 3, 21, 307, 311, 331, 354, 380, 381, 493, 512, 517
Exploitation competition, 352
Exports, 38, 128, 200, 201, 205, 208, 232, 250, 400
Exposure, 3, 7, 9, 39, 75, 97, 134, 168, 179, 192, 244, 277, 298, 308, 338, 354, 395, 405–409, 411–414, 422–424, 427–429, 436, 438, 442, 456, 493, 495, 496, 498–502
Expressed sequence tag libraries 123, 128, 129, 141, 415
Expulsion, 61, 95–97, 192, 356, 406, 407, 411
Extinction, 13, 14, 16, 19, 41–44, 273, 276, 312, 338, 373, 465, 484
Extracellular calcifying fluid (ECF), 123
Extracellular calcifying medium (ECM), 120–125, 127, 129–131, 141–143, 154, 162, 164, 167, 168
Extracellular pH (pHe), 167, 168
EZ. See Ethoxyzolamide

F
Faaa, 441
Factors limiting growth, 247–250
Factors limiting settlement, 247–250
FADH, 108
Fall, 406, 482
Fascichnus dactylus, 441
Fasciculus grandis, 441
Fascichnus frutex, 441
Fatty acids, 217, 251
Faunas, 5, 17, 19, 51, 65, 90, 151, 153, 273, 278, 279, 281, 286–288, 292, 294, 296, 309, 438
Fauna turnover, 19, 20
Favia sp., 124, 406, 452, 457
  F. favus, 113, 457
  F. fragum, 68, 74, 90, 161, 254, 285
Faviids, 65, 70, 73–75, 299, 348, 473
Faviidae, 66, 73, 459
Favites spp., 71, 357
  F. abdita, 92
  F. complanata, 353
Fe-, 190
Feces, 208, 303, 309, 458, 459
Fed, 89, 215, 217–222, 224, 310, 332, 339
Federated States of Micronesia, 3
Feed, 19, 232, 281, 283, 286, 287, 291, 292, 294, 296, 299–302, 305, 306, 308–312, 329, 333, 334, 339, 340, 440, 495, 497, 498, 510

Feedback loop, 236, 349, 411, 516
Feeding, **17, 95, 108, 191, 203, 215, 250, 274, 329, 360, 426, 439, 481, 495**
Feeding ecology, 339
Feeding mode, 216, 222–224, 277, 299, 301
Ferritin, 167, 429
Fertilization, 5, 60, 66, 71, 72, 74, 75, 109, 165, 312, 515
Fertilize eggs, 74, 332
FeSOD, 190, 409
Field Emission Scanning Electron Microscopy (FESEM), 122
Fiji, 50, 51, 64, 152, 297, 307, 328, 330
Filamentous algae, 244, 307
Filamentous heterocystous, 436
Filamentous red algae, 245, 353, 361
Filefish, 330
Filters, 151, 234, 309, 358, 457
  feeders, 152, 233–235, 283, 309, 439, 445
  feeding, 177, 291, 293, 296–298, 309, 364
Filter-feeding organism, 360
Fins, 338, 339
Fireworm, 302, 452, 480, 481
Fish, **5, 20, 25, 38, 47, 122, 152, 225, 231, 251, 281, 327, 375, 398, 440, 459, 474, 510**
  abundances, 499–501, 503
  biomass, 231, 236, 379, 384, 501, 516
  cage farming, 153
  conservation, 327, 331
  distributions, 47
  farms, 493, 494
  grazing, 38, 399, 512
  health issue, 341
  hosts, 282, 304, 306, 308
  traps, 517
Fisheries, 7, 53, 263, 307, 308, 329, 331, 379, 380, 384, 385, 516, 517
Fisheries yields, 384
Fishers, 329
Fishing, 8, 232, 234–236, 241, 300, 307, 329, 331, 334, 341, 342, 376–378, 380, 381, 383–385, 405, 443, 485, 514, 516, 517
Fishing pressure, 236, 341, 377
Fissures, 436
Fistularia, 339
Fistulariidae, 330
FITC-Dextran. See Fluorescein isothiocyanate-Dextran
Fitness, 71, 259, 305, 339, 351, 359, 364, 421, 440, 472, 484
Fitness costs, 347, 354
Fix nitrogen, 247, 252
FIZ. See Freshly isolated zooxanthellae
Fjords, 500
Flabellidae, 66, 67, 72
Flagella, 88, 89, 188
Flatworms, 275, 290, 292, 299, 305, 308
Fleshy algae, 6, 233–236
Flexibility, 91, 92, 249, 379, 381
Flooding, 258, 259
Flora, 5, 51, 234, 475, 511
Florida, 4, 32, 287, 348, 350, 354, 395, 396, 457, 458, 468, 469, 471–475, 483
Florida Keys, 31, 32, 242, 243, 245, 254, 255, 274, 301, 337, 360, 455, 456, 465, 468, 472, 474, 481
Florida reefs, 21, 260, 352, 471
Flow cytometry, 210, 407
Flow regimes, 178, 256, 297, 341, 515
Flow velocities, 177–179
Fluid, 122, 124, 125, 129, 131, 140, 154, 164, 170, 178, 203, 411
Fluorescence, 99, 109, 127, 134, 352, 360, 408–411
Fluorescent pigments, 101
Fluorescein isothiocyanate-Dextran, 154

Fluorescent protein (FPs), 109, 191
Fluorochromes, 415
Flying gurnards, 330
F_m. See Maximum fluorescence
F_o. See Baseline fluorescence
Foliose, 243, 244, 257, 363
Foliose algae, 243, 352, 361, 363
Food, 15, 96, 125, 166, 178, 200, 215, 216, 221, 222, 225, 241, 251, 252, 282, 288, 291, 293, 296, 299, 303, 304, 306–311, 329, 332, 335, 339–341, 347, 351, 355, 360, 363, 383, 401, 405, 435, 439, 440, 444, 465, 485, 494–497, 501, 517
Food available, 222
Food resources, 339, 435
Food webs, 199, 200, 234, 235, 250, 309–310, 339, 382, 384, 494
Food-web theory, 309
Foraging, 282, 283, 287, 303, 304, 309, 440
Foraminifera, 20, 92, 124, 125, 127, 129, 151, 154, 164, 296, 436–440
Foraminiferans, 292, 296
Forereef, 244, 274, 278, 279, 286, 512, 514
Form, 14, 28, 37, 59, 88, 109, 120, 151, 177, 188, 202, 207, 215, 231, 241, 279, 334, 352, 391, 412, 430, 438, 458, 494, 514
Fossils, 13–19, 22, 152, 153, 243, 286, 287, 440, 441
    fuels, 152, 153, 391
    record, 13–18, 20, 43, 44, 51, 140, 243, 459, 468
    trace, 16, 439, 441
FPs See Fluorescent protein
Fragmentation, 61, 91, 245, 257, 258, 297, 483
Fragments, 126, 177, 297, 430, 438, 456, 483
Free amino acids, 303
Free-living, 88, 89, 91, 188, 189, 215, 286, 288, 351
French Frigate Shoals, 312
French Polynesia, 64, 247, 274, 282, 303, 328, 336, 337, 341, 382, 383, 398, 440–442
Freshly isolated zooxanthellae (FIZ), 98, 99, 101, 114, 188
Freshwater flooding, 351, 367
Freshwater marine species, 334
Fringing, 208, 210, 247, 499
Fringing reef flats, 259
Fringing reefs, 4, 151, 247, 281, 289, 290, 360, 382, 442
Frogfish, 330
Fucoxanthins, 246
Functional group, 110, 243, 308, 310, 349, 350, 375–378, 382, 510
Fungal hyphae, 456, 461
Fungi, 7, 16, 120, 132, 233, 276, 362, 392, 435, 436, 443, 445, 455, 466, 474, 475
Fungiacyathidae, 66, 72
Fungia fungites, 68
Fungia repanda, 71
Fungia scruposa, 71
Fungia scutaria (Hawaii), 92
Fungia spp., 160, 168
Fungiid, 61, 71, 72, 74, 351, 355, 358, 364, 365
Fungiidae, 66, 67, 72, 73, 75
Fusiliers, 330, 333
Fusion, 61, 89, 90, 290, 354, 357, 363, 364, 473
Future, 3–9, 14, 18–21, 31, 53, 64, 69, 76, 120, 140, 156, 158, 162, 165, 169, 171, 188, 202, 215, 224–225, 236–237, 249, 256, 258, 263, 264, 276, 311–313, 327, 329, 331, 341, 348, 354, 365–368, 373–375, 385, 391, 395, 396, 398, 401, 415, 422, 424, 427, 430, 444, 446, 468, 484–486, 503, 514
Fv/Fm, See Photochemical efficiency

G
G. See Gross calcification
GA. See Growth anomalies
Galápagos Islands, 171, 260, 273, 300, 301, 305, 440–444

Galaxaura, 243, 244
Galaxea, 72, 301
    G. *astreata*, 67, 72, 356
    G. *fascicularis*, 67, 72, 73, 123, 124, 128, 133, 142, 143, 159, 160, 165, 188, 217, 285
Galaxin, 133, 134
Gall crabs, 283, 298
Gametes, 60–72, 74, 75, 107–109, 245, 310, 333
Gametogenesis, 64, 65, 68, 75, 109
GAN. See Growth anomalies
Gardineroseris planulata, 71
Gastrodermis, 120, 121
Gastropod, 243, 251, 282, 286, 287, 289, 290, 292, 294, 296, 297, 299–301, 303, 307, 349, 360, 439, 440
Gastropoda, 285, 286, 303
GBR (Australia), 4, 5, 7, 47, 164, 165, 209, 259, 278, 312, 342, 349, 353, 354, 366, 397, 440, 459
Gelbstoff, 110
Gelidium robustum, 363
Gene, 87, 131, 135, 138, 143, 191, 409, 412, 413, 422, 429, 430, 453, 456, 482
    flow, 38, 41, 44, 69, 76
    regulation, 183
Genetic, 5, 6, 14, 38–44, 52, 53, 60, 61, 69, 87, 90, 136, 259, 294, 312, 313, 328, 329, 338, 400, 408, 414, 430, 465
Genetic analysis, 44, 334
Genetic barcoding, 313
Genetic changes, 41, 399
Genetic concept, 37
Genetic data, 313, 414
Geneticists, 9
Genetic study, 19, 43, 75, 76, 338
Genomes, 90, 130, 135, 139, 189, 192, 341, 425, 429, 430
Genotypes, 39, 60, 61, 75, 88–93, 95, 99–101, 177, 351, 354, 355, 358, 359, 364, 366, 411–415, 421, 422, 426, 427, 430, 485, 486
Genotypic diversity, 87, 312
Genuine progress indicator (GPI), 9
Genus-specific *Symbiodinium*, 92
Geochemists, 9
Geographic, 14, 16, 18, 20, 32, 37–41, 43, 44, 48–50, 59, 62, 70, 76, 92, 93, 100, 108, 110, 247, 259, 276, 299, 308, 313, 327, 339, 342, 392, 453, 455, 465, 466, 468, 469, 471–476, 478, 483, 486, 497, 500
Geographical limits, 245
Geographical regions, 349
Geographic distance, 38, 49
Geographic distributions, 14, 18, 93, 453, 468, 469, 473–478, 483
Geographic information system (GIS), 48, 375
Geographic location, 38, 39, 59, 92, 93, 247, 392, 472
Geographic occurrence, 276, 299
Geographic-scale model, 425
Geographic variations, 37–41
Geological, 15, 17, 19, 21, 22, 42, 47, 50–52, 107, 199, 405, 441, 443, 468
Geological history, 16, 47, 50, 51, 88, 440–441
Geological scale, 338
Geologic history, 21
Geologists, 5, 8
Geomorphology, 25, 26, 31, 499, 500
Geospatial models, 499, 500
Germ cell parasitism, 347, 364
GFP. See Green-fluorescent protein
Ghost pipefish, 330
Giant, 189, 209, 210, 304, 352, 358
Gilvin, 110
GIS. See Geographic information system
Glacial, 398

Glacial cycle, 394
Glassfish, 330
Glass slides, 352
Glint, 29, 30
Global, 3, 13, 47, 59, 119, 151, 199, 242, 276, 327, 347, 373, 391, 405, 421, 444, 455, 473, 499, 509
Global bleaching events, 426
Global change, 13, 17–19, 151–171, 258, 260, 373–385, 398, 444, 446
Global climate change, 3, 6–9, 21, 311, 312, 341, 347, 363, 367, 401, 405, 406, 414, 415, 427, 444, 484, 485
Global warming, 6, 8, 18, 151, 152, 164, 200, 261, 341, 391, 398, 399, 421, 423, 427, 428, 445, 473, 503
Glutamine/glutamate, 87, 94, 220
Glutamine synthetase (GS), 87, 94
Glycerol, 93, 94, 139, 220
Glycolysis, 167
Glycosaminoglycans, 123, 132, 133
Gnathiid isopod larvae, 304
Goatfish, 311, 330
Gobies, 282, 330, 331, 334, 336, 383
Gobiesocidae, 330, 334
Gobiidae, 330, 336
Goldfish, 424
Golfe du Lion, 68
Gomontia polyrhiza, 441
Gomphosis varius, 333
Gonads, 355, 453
Goniastrea, 406, 452, 457
    G. aspera, 68, 74, 426, 430
    G. australiensis, 98
    G. favulus, 75, 92
    G. retiformis, 221, 225, 254
Goniopora, 61, 351, 357
    G. lobata, 113
    G. tenuidens, 356
Gonochoric, 5, 60, 62, 63, 65, 66, 70–76
Gonodactylid, 293
Gonodactylus bredini, 302
Gonodactylus falcatus, 312
Gorgonacea, 279, 280, 289, 294
Gorgonia flabellum, 456, 471, 481
Gorgonians, 123, 130, 274, 279, 280, 292, 294, 354, 355, 364, 366, 452, 456
Gorgonia ventalina, 456, 472
GPP. See Gross primary production
Gracilariaceae, 256
Gracilaria salicornia, 256, 257
Gracilaria spp., 243, 244, 256, 257, 259, 260
Gradients, 5, 18, 38, 92, 100, 116, 126–129, 138, 166, 178, 179, 182, 183, 189, 192, 235, 236, 246, 248, 281, 315, 367, 397, 425, 428, 442, 494, 495
Graz, 329
Graze, 277, 399, 435, 440, 495
Grazed coral skeletons, 359
Grazers, 205, 209, 236, 250, 252, 257, 260, 277, 347, 376–378, 384, 436, 439–441, 443, 444, 500, 509, 515
Grazing, 16, 38, 243, 245, 246, 250, 252, 256, 283, 286, 291, 301, 307, 376–378, 399, 401, 439–446, 497, 501, 512–517
Grazing agents, 435
Grazing pressure, 352, 377, 378, 439, 445, 498, 501
Grazing rates, 216, 218, 225, 442
Great barrier reef (GBR), 4, 17, 40, 47, 61, 92, 165, 209, 215, 232, 259, 274, 333, 349, 377, 395, 406, 423, 440, 454, 466, 494, 509
Great Barrier Reef Marine Park Authority (GBRMPA), 7
Green algae, 241–246, 251, 252, 256, 308, 353, 361, 362, 365, 441

Green-fluorescent protein (GFP), 109, 191, 409, 410
Greenhouse gases, 6, 53, 391–394, 399, 405, 406
Greenhouse warming scenarios, 405
Green macroalgae, 257, 348, 363
Grenada, 328, 458, 472, 473, 476, 483, 484
Gross calcification (G), 169
Gross domestic product (GDP), 8, 9
Gross primary production (GPP), 32, 200–202, 205, 208
Groundtruth, 28
Groundwater, 200, 247, 493, 517
Groupers, 274, 329, 331, 332, 336, 341, 342, 383, 384, 517
Growth, 6, 14, 38, 59, 87, 111, 119, 152, 177, 200, 215, 232, 245, 276, 327, 347, 376, 391, 405, 430, 435, 455, 467, 494, 512
    of corals, 133, 366, 396, 398, 435
    rates, 6, 14, 16, 21, 90, 95, 96, 98, 113, 116, 126, 160, 178, 215, 218, 221, 222, 250, 256, 258, 259, 303, 304, 331, 335, 354–356, 360, 362–366, 376, 378, 397, 405, 407, 408, 410, 455, 476, 478, 499, 515
Growth anomalies (GA), 467, 469, 473–476, 478, 481
Grunts, 304, 330
GS. See Glutamine synthetase
Guadeloupe, 328
Guam, 4, 64, 253, 254, 274, 279, 282, 312, 328, 341
Gulf of California, 68, 75
Gulf of Eilat, 63, 92, 95, 152, 169
Gulf of Mexico, 63, 65, 329, 473, 475, 476
Gulf of Panamá, 260
Gulf of Suez, 300
Gulfs, 92, 500
Gut microfloras, 340

H
Habitat, 15, 25, 37, 48, 139, 201, 222, 244, 273, 327, 365, 375, 399, 423, 515
    architecture, 294
    degradation, 18, 21
    loss, 261, 341, 515
    selection, 339
    tracking, 17
    zones, 338
Haemulidae, 330, 377
Haimesiastraea Conferta, 16
Haiti, 328
Halimeda, 241, 242, 244–246, 252, 253, 255, 260, 261, 308, 348, 361, 363
    crown-of-thorns-starfish, 7
    H. opuntia, 254, 353, 481
Halimedia spp., 236
Halofoliculina sp., 472
Halofolliculina corallasia, 452, 459
Halymenia, 243, 244
Haploid, 89, 90
Haplosyllis spongicola, 294
Hard corals, 59, 351, 364, 460, 494–498
Harmful effects, 171, 407
Hatch, 332, 333, 378
Hatchings, 282, 332, 334, 336, 378
Hatchlings, 332, 334, 336
H^+-ATPase, 129, 189
Hawaii, 3–6, 28, 30–32, 50, 67, 68, 92, 100, 115, 152, 204, 208, 246–248, 256, 259, 289, 295, 303, 305, 333, 341, 349, 351, 354, 355, 366, 423, 424, 442, 472, 494
Hawaiian, 3, 65, 92, 94, 95, 115, 257, 287, 312, 339, 357, 358
Hawaiian Islands, 231
Hawkfish, 329, 330

HCl, 158–161
HCO₃⁻. *See* Bicarbonate
Health, 6, 7, 9, 14, 21, 32, 225, 232, 236, 304, 341, 359, 380, 381, 394, 395, 398, 399, 405, 436, 444, 446, 451, 465, 466, 469, 473, 476, 481, 483, 514, 515
Healthy, 217–219, 235, 250, 330, 339, 373, 379, 394, 401, 440, 444, 451, 454, 456–460, 480–482, 484, 500, 503, 514
Healthy coral reef, 6, 234, 400
Healthy corals, 218, 223, 245, 365, 452, 454–457, 481, 516
Healthy coral tissue, 353
Heat shock protein (HSP70), 353, 354, 365, 411, 412, 425, 429
Heat tolerant, 100, 424
Heavy metals, 7, 236
Helcogramma obtusirostris, 334
Helicostoma nonatum, 312
Heliofungia actiniformis, 67, 68, 71, 74, 126
Heliopora coerulea, 280, 301
Hemiglyphidodon, 333
Hemispherical, 178
Herbicides, 69, 496
Herbivores, 166, 231–232, 235, 236, 242–244, 250–253, 257, 258, 261, 286, 287, 300, 303, 308–311, 331, 339, 340, 376, 378, 385, 440, 443, 499, 502, 509–511, 515, 517
Herbivorous, 273, 286, 337, 340, 443, 495
Herbivorous fish, 7, 8, 231, 250, 253, 311, 340, 376–379, 383, 384, 493, 497, 498, 500, 501, 511, 512, 514, 516, 517
Herbivory, 16, 241, 243, 245, 250–251, 253, 257, 260–262, 264, 283, 286, 288, 299, 303, 349, 352, 353, 360, 362, 364, 399, 511, 512, 515, 517
Hermaphroditic, 5, 60
Hermatypic, 49, 59, 109, 151, 377
Hermatypic scleractinians, 60, 295, 377
Hermit crabs, 282, 283, 296, 297, 302
Hermodice, 281, 302
Hermodice carunculata, 301, 302, 307, 312, 452, 480, 481
Heron Island, 289, 366
Herring, 337
Hertz, 336
Heterocapsa triquetra, 192
Heterokontophyta, 242
Heterotrophic, 101, 152, 178, 202, 215, 220, 223–225, 232, 237, 293, 454, 495–497, 501
Heterotrophic benthos, 383
Heterotrophic feeding, 215, 216, 219, 220, 222–225, 426
Heterotrophic microbes, 231, 236
Heterotrophs, 152, 215, 225, 236, 436, 441, 474
Heterotrophy, 215–225, 295, 495
Hexacorallia, 75, 279, 313
H. gigas, 441
Higher temperature, 140, 152, 164–166, 210, 234, 424
High irradiance, 116, 244, 247, 353, 405, 407–410, 412, 413, 426, 445, 498
High light (HL), 30, 87, 95, 97–99, 101, 111–116, 178, 192, 215, 217, 218, 221, 222, 225, 245, 247, 410, 430, 445
High-light colonies, 112–114
High light (HL) *vs.* low light (LL), 95
High/low solar radiation, 426
Histidine, 94, 167
Histology, 121, 478
Historical ecology, 20–21
HL. *See* High light
3H₂O, 200
HO. *See* Hydroxyl radicals
Holobiont genotypes, 426
Holobionts, 16, 87–89, 92, 97, 99–101, 107, 111, 113, 187–192, 212, 233, 422, 428, 461, 466, 482

Holocene, 14, 17, 22, 155
Holocentrid, 337
Holocentridae, 311, 330
Holothurians, 288, 292, 294, 297, 306, 309, 311
Honduras, 328
Hong Kong, 68, 75, 296, 300
H₂O₂. *See* Hydrogen peroxide
H₂O₂ production, 101, 138, 190, 405, 408, 412, 413
Horizontal gene transfer, 191
Horizontal transmission, 91–93, 97, 187
Hormathonema violaceo-nigrum, 436
Host factor, 93–95, 114–116
Host ranges, 468, 473, 478–481
Host-release factor (HRF), 89, 94
Hosts, **6, 22, 47, 87, 107, 136, 152, 187, 215, 233, 274, 334, 367, 400, 409, 422, 443, 451, 465, 494, 515**
Host tissue, 89, 94, 96, 97, 99, 114, 191, 192, 218, 225, 305, 411, 412, 428, 429
Hotspots, 20, 32, 47, 49–50, 473, 485
HRF. *See* Host-release factor
HSP70. *See* Heat shock protein
HSPs/anti-oxidants, 428
Human, 3, 7–9, 18, 20, 21, 28, 37, 41, 123, 128, 129, 131, 311, 327, 331, 337, 343, 349, 367, 373, 374, 391, 396, 458, 460
Human activity, 231, 236, 374, 385, 501
Human disturbances, 364, 367
Human impacts, 18, 20, 21, 50, 53, 347, 350, 364
Human metallo-proteases, 167
Human populations, 21, 47, 76, 327, 328, 331, 342, 385, 444, 493
Human socioeconomic issues, 342
Huon Peninsula, 17
Hurricane, 3, 4, 8, 9, 245, 264, 347, 377, 465, 471, 485, 509, 513–516
Hybridisation, 429
Hybridization, 5, 44, 60, 69, 76, 341
Hybrids, 42, 69, 143
Hydnophora, 306
Hydra sp., 413
Hydrocorals, 279, 283, 284, 287, 299, 305, 348, 349, 351, 355, 357, 466, 468, 469, 473, 478
Hydrodynamics, 177–179, 183, 201, 205, 298, 385, 499, 517
Hydrogel, 123, 132, 133, 142, 143
Hydrogen peroxide (H₂O₂), 138, 190, 191, 405, 408, 412, 413
Hydroida, 279, 303
Hydroids, 279, 287, 296
Hydrolithon spp., 254
  H. onkodes, 254
  H. reinboldii, 254, 255
Hydrological cycle, 398
Hydrophiidae, 334
Hydroxyl radicals (HO), 190, 405, 408, 409
Hyella sp., 436, 437
  H. balani, 436
  H. caespitosa, 437, 441
  H. immanis, 436
Hymenocera picta, 301, 302, 309
Hypercalcified sponges, 278, 279, 299
Hyperoxia, 188, 190–192, 412
Hypnea, 256
  H. musciformis, 256, 257
  H. pannosa, 365
Hypoxia, 167, 234, 413, 493, 497, 515
Hypoxic, 234, 333, 411, 500
Hypsipops, 333
HyspIRI (Hyperspectral infrared imager), 31
Hysteresis, 137, 263, 375–378, 514–516
Hz. *See* Hertz

I

IAA. See Indo-Australian Archipelago
Ichnoreticulina elegans, 441
Ichthyocides, 341
Ichthyologists, 9
Iki Islands, 424
IKONOS, 26, 31, 32
Illumination, 98, 125, 221, 293, 411
Images, 26–32, 134, 179, 180, 202, 384
Immigration, 44, 282, 381
Immune function, 365
Immune response, 93, 97, 192, 236, 453, 472
Immune system, 367, 460, 461, 466
Immunity, 429, 451, 460, 461
Immunoblots, 409, 410
Immunorecognition, 192
Incidence, 72, 233, 256, 425, 426, 456, 466, 477–478, 482
In culture, 88–90, 95, 99, 101, 191, 408
Independent Na$^+$-ATPase, 188
India, 256, 275, 296, 328
Indian Ocean, 49, 50, 52, 152, 210, 211, 275, 280, 283, 289, 300, 303, 358, 364, 366, 395, 396, 425, 426, 454, 473
Indo-Australian Archipelago (IAA), 20, 47
Indonesia, 18, 40, 47–53, 64, 152, 295, 328, 331, 383, 424, 442, 444, 494
Indonesian/Philippines Archipelago, 40, 47
Indo-Pacific, 17, 19, 20, 47, 49–53, 62, 64, 65, 68, 90, 91, 93, 100, 254, 273–275, 278–280, 283, 286–288, 290, 300–303, 306, 307, 309, 312, 313, 351, 362, 378, 384, 401, 439, 455, 469, 472–475, 477, 484, 497, 503, 509, 510
Indo-Pacific reefs, 53, 279, 303, 307, 378, 509
Indo-Pacific waters, 439
Indo-West Pacific (IWP), 19, 280, 281, 283, 287, 288
Industrial agriculture runoff, 232
Infect, 91, 97, 291, 451, 452, 455, 457, 472, 475
Infectious diseases, 451, 452, 456, 460, 465, 466, 468, 471, 476, 483
Influence of microbes, 231
Inhibition, 90, 95, 128–131, 137, 139, 154, 164, 189, 264, 349, 352, 356, 363, 365, 407, 411, 428
Inhibitor, 138, 190, 191, 411
    of anion transport, DIDS, 189
    of photosynthesis, 127, 137, 154
Inhibit research, 330
In hospite, 88–90, 97–99, 101, 138, 188, 189, 408, 409, 412
Inorganic carbon, 124, 129, 133, 137–138, 155, 156, 164, 179, 181, 189
Insects, 123, 166, 168, 510
Inshore, 363, 364, 397, 444, 498
    reefs, 377, 397, 440, 445, 493, 497, 500
    water, 217, 338, 498
Insoluble organic matter/matrix (IOM), 132, 133
Interactions, 101, 133–134, 156, 159, 160, 177, 190, 217–221, 225, 234, 260, 296, 303, 304, 307, 309, 357, 360, 362, 384, 385, 398, 408, 430, 461
Intercolony damage, 351
Interference competition, 351, 355
Interglacial, 18, 391
Intergovernmental Panel on Climate Change (IPCC), 8, 53, 152, 157, 171, 258–260, 392–395, 398, 400, 445
Internal clocks, 107
Internal waves, 200, 424
International coral reef symposia (ICRS), 5, 375
Interspecific competition, 245, 298, 351, 363, 365, 366
Intertidal ridges, 241
Intra-and interspecific contact, 351
Intracellular pH (pHi), 125, 162, 166, 167, 188, 189
Intraspecific competition, 245, 351, 354, 355, 364

Invertebrate larvae, 254
Invertebrates, 60, 131, 134, 251, 252, 273, 275–299, 301–305, 307–313, 336, 339, 383, 424, 429, 441, 471, 480, 499
Inverted biomass pyramid, 231, 234
Invertivores, 307, 311
In-vivo absorption, 98
IOM. See Insoluble organic matter/matrix
Ionic conditions, 188
IPCC. See Intergovernmental Panel on Climate Change
Iran, 328
Irish Sea, 500
Iron enrichment, 367
Irradiance, 67, 92, 99, 100, 107, 109, 111, 112, 115, 116, 244, 245, 247, 347, 352, 353, 359, 360, 363, 395, 407, 409, 414, 426–429, 445, 484, 496
Island biogeography, 44
Islands, 4, 8, 31, 44, 49, 51, 53, 93, 208, 209, 211, 231, 236, 247, 248, 258, 262, 280, 289, 290, 301, 305–307, 313, 328, 336–338, 342, 351, 352, 354, 366, 380, 384, 395, 438, 440–442, 454, 456
Isodiametric filament, 437
Isophyllia sinuosa, 73
Isopods, 251, 283, 297, 303–306, 308, 309, 340
Isopora cuneata, 60
Isotope ratio, 339
Isotopic, 18, 21, 124, 127, 128, 217, 220, 306, 339
Isotopic equilibrium, 155
Israel, 4, 259, 328, 457
Istiblennius edentulous, 334
IWP. See Indo-West Pacific

J

Jacks, 330, 337
Jamaica, 4, 5, 67, 68, 72, 274, 278, 302, 328, 348, 353, 384, 439, 442, 444, 512, 514, 517
Japan, 4, 37, 48, 50–52, 64, 68, 74, 75, 91, 93, 208–210, 300, 328, 335, 341, 348, 351, 352, 354, 355, 424
Jawfish, 330, 332
Jelly fish, 109
Johnston Atoll, 50, 333
Jordan, 188, 248, 301, 328, 339
Junctions, 120, 123, 126, 129, 141, 143
Jurassic, 441
Juvenile corals, 61, 93, 100, 255, 256, 297, 298, 300, 361, 362, 365
Juveniles, 60, 61, 75, 93, 100, 160, 254, 255, 287, 302, 304, 306, 308, 329, 332, 334, 336–340, 356, 363, 378, 383, 438, 443, 485, 498, 501, 515

K

Ka. See Thousand years ago Kahoolawe, 472
Kaneohe Bay, 244, 264, 305, 500
Kaneohe Bay, Hawaii, 27, 30, 32, 204, 208, 248, 289, 442, 459
Kappaphycus, 6, 256–258
Kawaguti, 5, 64, 68, 119, 124, 136
Kenya, 15, 65, 289, 307, 328, 428, 494, 517
Kenyan, 376, 443
Key West, 458
Kill, 205, 232, 234–236, 261, 305, 340, 356–358, 360, 366, 440, 495–498, 500
Kinetic effect, 155
Kinetic fractionation, 220
Kinetics of calcification, 138
Kingman, 236, 383
Kingman Reef, 231
Kiribati, 328

Kiritimati, 231, 236, 383
Kleptocnidae, 308
Knockouts, 415
Koch's postulates, 451, 466, 474, 475, 481
Kumulipo, 3
Kuwait, 65, 328
Kyphosidae, 330
Kyrtuthrix dalmatica, 436

L
Labridae, 329, 330, 377, 378
Labrisomidae, 330
Labrisomids, 330
Lagoons, 31, 116, 202, 205, 208, 210, 241, 244, 245, 250, 258, 260, 261, 280, 283, 337, 338, 351, 436, 440, 442, 499, 500, 503
Lampreys, 341
Landsat, 26, 32
La Parguera, 477, 478, 482–484
Larva, 91, 135, 254, 329, 332, 333, 337, 510, 511
Larvae, 5, 43, 51, 60–62, 74–76, 91–93, 97, 133, 135, 168, 225, 254–255, 282, 291, 304, 306, 332, 334–339, 347, 353–356, 358, 361–365, 378, 379, 436, 438, 443, 445, 485, 496, 497, 499, 501, 503, 510
Larvae settle, 254, 436, 496
Larval, 5, 6, 60, 67, 69, 70, 74, 76, 92, 108, 254, 256, 290, 291, 306, 329, 332–338, 354, 361, 362, 378, 501, 515
Larval dispersal, 14, 70, 379
Larval distribution, 334
Larval metamorphosis, 356
Larval settlement, 69, 76, 121, 254, 255, 336, 354, 356, 362, 365, 366, 493
Larval stages, 303, 306, 334
Larval supply, 379, 499, 515
Larval survival, 300, 336, 365
Larva survives, 60
Laser ablation inductively coupled plasma mass spectrometry (LA-ICPMS), 338
Lateral branch, 351
Latitude, 14, 39, 52, 53, 65, 66, 91, 92, 100, 107, 152, 166, 168, 258–260, 287, 288, 290, 302, 391, 392, 405, 423, 424, 440, 497, 500
Latitudinal gradient, 236, 500
Laurencia, 242–244, 252, 255
Law of uniformitarianism, 14
Layers, 6, 41, 89, 120, 121, 123, 125, 126, 128, 129, 141–143, 164, 177, 179, 181–183, 189, 201, 203, 205, 215, 221, 233, 243, 244, 249, 253, 256, 259, 305, 393, 437, 438, 441, 445, 452, 458, 460, 496
Lead, 338
LEC. *See* Light-enhanced calcification
Leeuwin current, 18, 49, 52
Lepidozygus, 333
Leptastrea purpurea, 68
Leptoseris fragilis, 110
LER. *See* Linear extension rate
Lethal, 115, 191, 339, 421–425, 428, 429, 431, 466, 469, 473
Lethal temperatures over time and recording the time or temperature causing 50% mortality (LT$_{50}$), 428
Lethrinidae, 330, 336
Lidar (Light detection and ranging), 27, 28, 31, 32
Life cycles, 60–62, 89, 243, 245, 255, 256, 260, 292, 305, 336, 341, 362, 394, 481
Life history, 60, 62, 75, 92, 244, 291, 306, 312, 329, 332–339, 354, 378, 415
Life history stages, 291, 306, 312, 329, 332, 339

Light, **5–7**, **14**, **27**, **38**, **59**, **87**, 89–92, 94–101, **107–116**, **119**, 125–131, 139, 141, 142, **153–154**, 158, 162, 163, 165, 167, 169–171, **177**, **188**, **201**, **209**, **216**, **244–247**, 257, 260, **273**, 280, 293, 298, 301, 312, 313, **335**, **352**, **392**, **407**, **424**, 426, 429, 430, **436**, **454**, **475**, **493**, **515**
  absorption, 99, 202
  availability, 14, 38, 247
  calcification, 126–128, 221
  limitation, 100, 496, 499–501
  quality, 245, 246
  quantity, 245, 246
  reflected, 394
  traps, 336, 337
Light-adapted colonies, 223
Light and dark calcification, 153–154, 163, 170
Light and dark reactions, 408
Light/dark regulation, 138
Light-enhanced calcification (LEC), 6, 127, 131, 136–139, 153, 154, 162, 165
Light harvesting pigments, 111
Light intensity of incipient saturation (E_k), 98
Light microscope, 335
Light-saturated rate of photosynthesis (P_{max}), 98
Lihir Island, 380
Limitation, 15, 90, 95, 114, 135, 180, 202, 204, 205, 220, 235, 247, 248, 253, 260, 351, 381, 409, 410, 422, 425, 436, 496, 499–501, 517
Limiting, 8, 52, 91, 96, 127, 133, 138, 143, 165, 178, 179, 181, 207, 224, 245–248, 250–251, 258, 260, 300, 303, 305, 347, 384, 397, 398, 461, 497, 500
Limiting nutrients, 224
Limiting resource, 178, 347
Limits, 3, 14, 53, 65, 96, 100, 107, 113, 115, 158, 170, 171, 178, 199, 203–205, 245, 248, 249, 251, 252, 258, 261, 290, 307, 308, 342, 364, 392, 411, 422–425, 428, 429, 475, 501
Limpets, 251, 303, 439, 510
Linear extension rate (LER), 16, 182, 183, 221
Line Islands, 64, 231, 235, 236, 287, 354, 383
Linuche unguiculata (Coronatae), 91
Lionfish, 329, 341
Lipid peroxidation, 138, 190, 411, 412
Lipids, 72, 93, 94, 132, 135, 138, 189–191, 217, 218, 225, 233, 409, 411, 412, 414
Lithophaga bisulcata, 438
Lithophaga spp., 498
Lithophage, 293, 294, 297
Lithopythium gangliiforme, 437, 441
Live and dead calcareous substrates, 436
Livebearing brotulas, 332
Live coral, 152, 153, 283, 300, 306, 338, 352, 353, 359, 360, 399, 436, 438, 440, 443, 444, 484, 498
Live coral cover, 152, 169, 170, 261, 273, 300, 382, 383, 444, 484
Live food fish trade, 341
Live organic matter (LOM), 215, 216, 225
Live young, 332
Lizardfish, 330
Lizard Island, 8, 306, 336, 337, 354, 442
LL. *See* Low light
LMW. *See* Low-molecular weight
Lobophora, 253
Lobophora variegata, 244, 246, 254, 255, 353, 362, 365, 377
Lobophyllia corymbosa, 92
Lobophytum hedleyi, 356
Lobsters, 49, 282, 283, 297, 312
Local to global scales, 327
LOM. *See* Live organic matter

Longer-term acclimatisation, 421
Long-term, 8, 14, 15, 17, 18, 32, 53, 65, 71, 72, 125, 126, 152, 159, 171, 247, 347, 349–352, 363, 365, 366, 377, 392, 395, 414, 422, 423, 428, 482–484, 494, 497, 514
Long-term monitoring, 32, 349–351, 415
Lottery hypothesis, 510
Low, 17, 29, 39, 50, 65, 87, 111, 121, 152, 178, 188, 199, 207, 217, 232, 242, 280, 336, 349, 352–354, 359, 360, 363, 366, 367, 377, 398, 407, 423, 436, 455, 469, 493, 512
Lower irradiance, 413
Lower temperature, 166, 413, 425
Low irradiance, 352, 359, 411, 496
Low Isles, 425
Low light (LL), 89, 95, 97, 99, 111–116, 178, 192, 217–222, 225, 244, 246, 247, 338, 352, 436, 496, 497, 501
Low-molecular weight (LMW), 190, 191, 208, 209
Low-nutrient water, 199, 200, 205
Low tides, 208, 292, 367
LT_{50}. *See* Lethal temperatures over time and recording the time or temperature causing 50% mortality
Luminescent, 109, 288
Lunar, 64, 65, 108, 109
Lunar cycles, 3, 64–66, 107–109, 141
Lutjanidae, 329, 330, 377
Lyngbya spp., 252, 349, 361
    L. bouillonii, 362
    L. majuscula, 252, 254
    L. polychroa, 254

M

MAAs. *See* Mycosporine-like amino acids
Macroalgae, 7, 14, 189, 203, 205, 231, 541–251, 253–257, 259–261, 334, 338, 348, 349, 351–354, 356, 358, 360–367, 376–378, 384, 495–497, 499, 501, 509, 512, 515, 517
Macroalgal, 247, 249–251, 255, 256, 261, 348–350, 352–354, 440, 485, 486, 497, 499, 500, 509, 512, 515–517
Macroalgal-dominated state, 350
Macrobioeroders, 497–498
Macrobioerosion, 439, 442, 444–446
Macroborers, 437–441, 443, 445, 446
Macroboring agents, 435
Macrobrachium acanthurus, 424
Macrofauna, 274, 275, 293, 438
Macrophytes, 25, 204, 208, 243, 244, 254, 363
Macroscale distribution, 414
Madagascar, 50, 328
Madracis, 75, 92, 100, 356
Madracis mirabilis, 126, 129, 160, 161, 178–180, 183, 219
Malacanthidae, 330
Malaysia, 50, 51, 93, 328
Maldives, 65, 67, 152, 274, 328, 363
Males brood fertilized eggs, 332
Management, 5–8, 21, 27, 33, 53, 171, 256, 263, 264, 327, 331, 341–343, 373–385, 483–485, 493, 498, 500, 503, 509–518
Management issue, 341, 342
Management strategy evaluation (MSE), 379, 380
Managers, 5, 7–9, 27, 263, 329, 331, 341, 374–376, 378–380, 385, 398, 509, 514, 517
Mangroves, 7, 20, 51, 244, 245, 258, 277
Manicina areolata, 68, 74
Marblefish, 330
Marianas Islands, 303, 312
Marine bacteria, 361, 466
Marine biodiversity hotspots, 20
Marine ecosystems, 16, 19, 21, 210, 256, 296, 384, 391, 399, 405, 435, 465

Marine environment, 53, 171, 233, 241, 243, 327, 423, 435, 456, 466
Marine protected areas (MPAs), 3, 6, 31, 32, 48, 53, 250, 342, 380, 485
Marine spatial planning (MSP), 6
Marshall Islands, 4, 288, 302, 328
Martinique, 328
Mass, 16, 179, 183, 201–205, 249, 251, 302, 309, 310, 332, 338, 340, 355, 373, 378, 383–385, 395,
    bleaching, 6, 8, 32, 53, 347, 395, 396, 399, 401, 406, 454, 509
    mortalities, 406, 456, 465, 471, 475, 484, 485, 510, 515
    spawning, 5, 62–65, 76, 254, 310, 333
Mass coral bleaching, 6, 8, 32, 395, 399, 401, 509–510
Massive bleaching events, 152, 445
Mass transfer limitation, 202, 204, 205
Mastigocoleus testarum, 436, 437, 441
Maternal stress hormones, 336
Mathematical model, 354, 510
Mats, 210, 211, 243, 245, 249, 252, 280, 340, 348, 355, 361–363, 455, 468, 471, 473, 474, 481, 497
Maui, 256, 257, 295
Mauritius, 328
Maximum fluorescence (F_m), 411
Mayotte, 328
MDSWG. *See* Modeling and Decision Support Working Group
Meandrina meandrites (Scleractinaria), 91, 95, 285, 479
Meandrinidae, 66, 72
Mechanical damage, 304, 362
Mechanism, 6, 20, 38, 48, 69, 93, 108, 119, 153, 183, 189, 221, 234, 245, 273, 342, 350, 351, 355–363, 365, 367, 376, 395, 405, 422, 436, 456, 475, 493, 514
Mechanisms of competition, 255, 307, 354–362
Medial axis, 180
Mediterranean, 50, 62, 168, 256, 258, 260, 302, 360, 451, 453, 475, 481
Mehler reaction, 101, 410
Meifauna, 438
Melanin, 190, 461
Melatonin, 190
Melichthys niger, 333
Melichthys vidua, 333
Membran, 329
Membranes, 88, 89, 94, 97, 99, 101, 112, 122, 123, 127, 128, 130, 132, 134, 139, 162, 188, 189, 249, 328, 397, 408, 411, 412, 414, 428
Membrane transport, 397
Meroplankton, 303
Mesenterial filaments, 307, 358, 365
Mesoamerican barrier reef, 379, 381
Mesocosm, 154, 158–162, 165, 169
Mesophotic, 59
Mesoproterozoic strata, 440
Mesozoic, 19, 441
Metabacterium, 341
Metabolic, 9, 87, 94, 97, 100, 107, 112, 113, 116, 125, 131, 138, 152, 167, 168, 170, 183, 188, 189, 199, 202, 203, 301, 338, 339, 397, 413
Metabolic demand, 223, 225, 249
Metabolic energy, 107, 123, 127
Metabolic fractionation, 220
Metabolic genes, 415
Metabolic inhibitors, 127
Metabolic rate, 128, 166, 201, 203, 414
Metabolic requirements, 87, 218, 311
Metabolism, 87, 107, 125, 138, 139, 166, 167, 169, 171, 191, 202, 207, 209, 216, 219, 224, 233, 382, 406, 445, 446, 482
Metabolites, 137, 251–253, 308, 354, 360, 362, 456
Metagenomic analysis, 233
Metal debris, 364
Metamorphosed polyps, 358
Metamorphosis, 60, 131, 254, 255, 338, 347, 356, 362, 501

Index 539

Metapopulation, 328, 375, 379
Metazoan caspases, 413
Metazoans, 14, 107, 135, 183, 273, 274, 276, 283, 288, 291–294, 296, 297, 299, 302, 303, 305, 308–310
Methods, 25–27, 32, 43, 44, 89, 120, 124–126, 128, 133, 158–161, 166, 171, 177, 179, 180, 182, 189, 190, 207, 211, 216, 221, 222, 224, 248, 250, 252, 253, 263, 308, 309, 312, 327, 335, 337, 338, 340, 342, 348–355, 367, 415, 428, 429, 441, 446, 451, 456, 460, 466, 477, 511, 517
Mexico, 63, 65, 328, 352, 380, 381, 385, 396, 469, 471–473, 475, 476, 483
MI. *See* Mitotic index
Michaelis-Menten relationship, 129
Michaelis-Menten uptake kinetics, 248
Michaelis-Menten kinetics, 128, 200
Microarray analysis, 131, 135
Micrabaciidae, 66
Micritization, 443, 444
Microarrays, 415, 429, 430, 484
Micro-atolls, 204
Microbes, 231–237, 256, 277, 341, 354, 362, 460, 482
Microbial activity, 234–236, 256, 362
Microbial communities, 170, 234, 237, 256, 454, 455, 460, 466, 468, 484
Microbial disease, 362, 451–461
Microbial dynamics, 231–237
Microbial growth, 233, 236, 362
Microbial infection, 362, 475
Microbioerosion, 441–445
Microbiota, 482
Microcosm, 248, 253, 259
Microdesmidae, 330
Microelectrode, 154, 162
Micro-environments, 38, 120, 154, 297, 425, 455
Microflora, 277, 340, 382
Microhabitats, 100, 288, 291, 296–298, 360, 423, 424
Micronesia, 3, 4, 9, 51, 328
Microorganisms, 87, 278, 288, 302, 362, 436, 451, 452, 454, 456, 457, 460, 461, 474
Microplankton, 29
Micropredators, 277, 299, 303, 306, 307, 312, 340
Microsatellite alleles, 90
Microscale distribution, 414
Microscope, 429
Microspathodon, 333
Microspathodon chrysurus, 481
Microstructure, 136, 140, 143
Migration, 17, 18, 133, 292, 294, 381, 394, 399
Milankovitch cycle, 393
Millepora dichotoma, 351, 359
Millepora spp., 279, 284, 299, 305, 484
Milleporid, 279, 296, 298, 299
Million years, 3, 14, 17–20, 51, 393, 394, 399
Million years ago, 14, 19, 51, 140, 187, 243, 440
Mimic cleaner, 339
Mineralogy, 19, 120, 140
Miocene, 19, 20
Miocene Ryukyu limestones, 51
Mitochondria, 88, 94, 121, 122, 125, 128, 190, 328, 329, 412
Mitochondrial, 90, 412
Mitotic index (MI), 90, 96
MLABM. *See* Multilevel agent-based models
MnSOD, 190, 409
Modelers, 6, 9, 375, 380, 385
Modeling and Decision Support Working Group (MDSWG), 377
Modeling community, 396

Modelling, 6, 7, 28, 182, 183, 308, 310–311, 354, 374–377, 380–383, 385, 396, 406, 511
Models, 7–9, 15, 17, 29, 30, 32, 60, 75, 94, 95, 114, 115, 124, 125, 131, 134, 135, 142, 166–168, 171, 177, 179, 181–183, 200, 204, 222, 225, 234–237, 253, 259, 310, 311, 331, 338, 349, 350, 352, 354, 367, 373–385, 396, 409–411, 413–415, 422, 425, 427, 429, 430, 444, 446, 493–503, 511, 512, 514
Modification, 16, 112, 129, 135, 138, 179, 188, 192, 298–299, 341, 357, 398, 430, 443
Molecular data, 20, 73, 299
Molecular biomarkers, 7
Molecular genetics, 413, 415
Molecular level, 93, 131, 207
Mollusca, 135, 276, 282, 284, 289, 297, 303, 312, 313
Molluscan, 20, 134, 273, 286, 296, 301, 302, 312, 313, 510
Molluscs, 14, 20, 49, 51, 119, 122–125, 129, 131, 134, 135, 142, 275, 281, 283, 285, 286, 288, 290, 292–294, 296, 297, 299, 302, 304, 306, 307, 309, 311, 312, 377, 383, 397, 445
Mollusks, 92, 152, 153, 364, 439
Mona and Desecheo islands, 478, 483
Monacanthidae, 311, 330
Monad kinetics, 200, 203
Monastraea, 457
Monitoring, 7, 9, 25, 32, 222, 331, 349–351, 446, 477, 483, 485, 514
Monk seals, 231
Monomyces rubrum, 67, 71, 73
Montastrea spp., 69, 73, 95, 97, 100, 181, 255, 351, 356, 361, 468, 472, 475, 478, 481–483, 513
  M. *annularis*, 16, 69, 94, 177, 232, 348, 454, 458, 471
  M. *cavernosa*, 72, 73, 356, 471
  M. *faveolata*, 365, 410, 412, 429, 458, 474, 479, 482
  M. *franksi*, 222
  M. *nancyi*, 16
Montipora sp., 93, 100, 300, 472
  M. *capitata*, 87, 160, 161, 218, 219, 225, 349, 354
  M. *digitata*, 92, 100, 140, 210
  M. *monasteriata*, 99
  M. *verrucosa*, 92, 96, 115, 232, 305, 358, 359
  M. *verrucosa* (Scleractinaria), 91
Moorea, 247, 310, 312, 337, 338, 442
Moorea Island barrier reef, 384
Moorish idol, 329, 330
Moray, 339, 383
Moray eels, 311
Morphogenesis, 119, 177, 178, 183
Morphological data, 313
Morphological diversification, 340
Morphologically, 41, 60, 72, 76, 243, 257
Morphologically visible, 38
Morphological modifications, 357
Morphological plasticity, 177–179
Morphological variation, 38, 180
Morphology, 16, 17, 43, 90–91, 115, 116, 119, 120, 124, 133, 136, 140, 178, 179, 243244, 249, 250, 256, 275, 279, 280, 292, 294, 295, 298, 299, 357, 363, 414, 471, 513
Morphometrics, 5, 179, 180, 183, 279
Morphospecies, 60
Morphotype, 339, 348
Mortality, 7, 9, 32, 61, 152, 153, 183, 232, 234, 245, 254–256, 259–261, 275, 289, 290, 298, 300, 303, 306, 308–310, 312, 334–338, 348–350, 353, 354, 362, 366, 378, 383, 392, 395, 396, 398, 399, 406, 414, 425, 426, 428, 431, 444, 445, 457, 458, 465, 467–469, 471, 472, 475–479, 483–486, 493, 497, 498, 501, 509, 510, 512, 515–517
Morwongs, 330
Motile, 88, 287, 292, 294, 296, 297, 302, 312

Mouthbrooding, 332
Movement, 18, 20, 43, 51, 177–179, 234, 236, 291–293, 298, 305, 336, 338, 342, 379, 511
Mozambique, 328
MPA. *See* Marine protected area
MSE. *See* Management strategy evaluation
μ. *See* Algal growth rate
Mucopolysaccharide, 256, 458
Mucus, 208, 209, 211, 216, 225, 233, 280, 303, 305, 309, 349, 358–360, 362, 365, 452, 459, 460, 475, 478
    adhesion, 215
    and fat bodies, 283, 299
    matrix, 75
    spawn, 232
Mugilidae, 330
Mullets, 329, 330
Mullidae, 311, 330
Multilevel agent-based models (MLABM), 374
Multiple cycles, 64, 71
Multispectral, 26–28, 31
Muraena, 339
Muraenidae, 311
Muricidae, 285, 300, 440
Musculus, 296
Mushroom corals, 50, 71, 72, 358, 359
Musid, 299
Mussidae, 66
Mussismilia braziliensis, 476
Mussismilia hispida, 71, 312
Mutualisms, 282, 297, 304–305, 308, 340
Mutualistic, 233, 282, 294, 304, 374
Mutualistic association, 188, 304
my. *See* Million years
mya. *See* Million years ago
Myanmar, 328
Mycetophyllia, 457
    M. ferox, 457, 468, 479
    M. lamarkiana, 457
    M. reesi, 122, 132
Mycosporine-glycine, 101, 191, 409
Mycosporine-like amino acids (MAAs), 101, 190, 191, 333
Mysida, 282, 283, 292, 303

N
N. *See* Nitrogen
^{15}N, 217, 340
δ^{15}N, 138, 217, 339, 340
Na$^+$, 128, 135, 167, 188
Nacrein, 135
NAD(P)H-oxidoreductase, 411
NaHCO$_3$, 158
Nanograins, 134, 141–143
Nanoplankton, 215, 216, 222–225, 298
NaOH, 158
Naso lituratus, 252
Natural algae, 352
Natural products, 251, 347, 356, 360
Natural selection, 37, 43–45, 100, 399, 400, 421
Natural stressors, 377
Nauru, 328
Near infrared (NIR), 28
Necrosis, 96, 97, 312, 354, 356, 405, 413, 414, 458, 471–474
Net calcification (N), 169
Neighborhood competition, 354
Nematocysts, 215, 280, 294, 308, 352, 356, 358, 359, 365

Nematostella vectensis, 192, 429
Nemipteridae, 330
Neogoniolithon accretum, 473
Neopomacentrus, 333
Neoproterozoic ooids, 440
Netherlands, 328, 458, 472
Netherlands Antilles, 328, 458
Neurophys, 328, 329
New Caledonia, 31, 50, 51, 210, 274, 283, 296, 301, 312, 328, 337–340
New Guinea, 17, 50, 51, 53, 64, 272, 328, 336, 459
New Zealand, 67, 328, 381, 382
N-fixation, 200, 204, 207, 210–211, 242, 247, 264, 493
NH$_4$+, 203, 210, 249
Nicaragua, 328
Niches, 15–18, 52, 274, 277, 291, 303, 309, 425, 428, 460, 510
Night, 64, 65, 67, 108, 109, 113, 123, 129–131, 138, 139, 169, 208, 211, 222, 283, 287, 288, 294, 300, 311, 336, 337
Ningaloo Reef, 300
NIR. *See* Near infrared
Nitrate, 95, 170, 202–205, 207, 211, 233, 249, 392, 455
Nitric oxide, or nitrogen monoxide (NO•), 412
Nitric oxide synthase (NOS), 412
Nitrification, 200, 205
Nitrogen(N), 87, 114, 200, 216, 249
    fixation, 200, 210–211, 242, 247
    fixers, 205
    limited, 219
Nitrogen-and phosphorus-limitation, 114, 248
Nitrogenase, 167, 211
Nitrogenous, 87, 209, 251
Niue, 328
^{15}N:^{14}N, 339
NO$_3$-, 154, 249
NO. *See* Reactive nitrogen species nitric oxide
NO•. *See* Nitric oxide, or nitrogen monoxide
NOS. *See* Nitric oxide synthase
Nocturnal, 63, 282, 287, 292, 294, 309, 311, 336, 337
Non-enzymatic antioxidants, 191, 409
Non-enzymatic mechanisms, 190
Nonindigenous, 256, 312, 341
Non-indigenous organisms, 256, 341
Non-invasive techniques, 409
Non-photochemical quenching (NPQ), 99, 100, 410
Non-reef builders, 347, 350, 359, 364
Non-symbiotic, 130, 138, 139, 190–192, 220, 221, 313
North America, 327, 341
Noxious microbes, 362
NPQ. *See* Non-photochemical quenching
N:P ratio, 200, 202, 204, 205, 248
N-remineralization, 204, 205, 212, 250, 303
Nucleus, 88, 89, 121
Nudibranch, 282, 285, 299, 301, 303, 308
Nudibranch gastropods, 299
N uptake, 96, 192, 204, 205, 224, 229
Nutrients, 5, 7, 87, 113, 154, 188, 199, 215, 216, 232, 241, 242, 247–248, 264, 347, 443, 482, 493–495
    additions, 234, 235, 253, 262
    assimilation, 222
    cycling, 328, 329
    enrichment, 154, 243, 249, 258, 360, 362, 364, 384, 493, 494, 497, 498, 500
    fluxes, 111, 113–115, 201
    levels, 350, 352, 364, 367, 377, 496, 497
    limitation, 95, 205, 220, 235, 248, 253, 260, 501
    recycling, 200, 494
Nutrification, 377, 378, 512, 514, 517

Nutrition, 6, 90, 95–97, 111, 136, 216, 217, 220, 225, 298
Nutritional benefit, 295, 298, 304
N vs. P limitation, 248

O

O$_2$, 188, 408, 409
O$_2^-$ (superoxide ion), 190
δ^{18}O, 155, 178
Oahu, 208, 351, 459
Obligate shrimp *(Alpheus lottini)*, 302
Ocean acidification, 6,-8, 18, 53, 69, 151–171, 258, 311, 312, 367, 385, 396–399, 406, 444, 445, 514, 516
Ocean circulation, 51, 385, 398
Oceanic water, 32, 242, 246, 247, 250, 494
Oceanographic principle, 342
Oceanographic process, 334, 337
Oceanographic research vessel, 354
Oceanography, 19, 199
OCH. *See* Other compromised health conditions
O'Connell Rivers, 495
Octocoral, 279, 280, 297, 348, 352, 466, 471–473, 476, 495, 498
Octocorallia, 279, 303
Octocorals, 280, 298, 352, 355, 365, 469, 471, 473, 475, 479–481, 484, 485, 498
Octocoral species, 279, 297, 472, 473, 498
Octopus cyanea, 303
Octopuses, 286, 291, 296, 303
Ophiocanops fugiens, 295
Oculina arbuscula, 101, 218, 219
Oculina diffusa, 126, 130
Oculina patagonica, 171, 360, 451, 453, 481
Oculinidae, 66
Offshore, 31, 397, 498
    reefs, 64, 364, 440, 445
    water, 208, 336, 338
Ogcocephalidae, 330
OH- (hydroxyl radical), 190
Oil globules, 332
Okinawa, 91, 168, 169, 208–211
Oligohymenophorea, 459
Oligotrophic, 16, 87, 95, 110, 113, 152, 188, 207, 213, 242, 405, 436, 439, 443–445
Oligotrophic conditions, 95, 445
Oligotrophic sea, 405, 436
Oligotrophic waters, 87, 113, 188, 443
OM. *See* Organic matrix
Oman, 328, 424
ONOO-. *See* Peroxynitrite anion
Ooids, 436
Open ocean, 154, 169, 207, 209–211, 233, 500
Open system, 91, 158, 187
Open-system transmission, 91
Ophiocanops fugiens, 295
Ophiothrix lineata, 294
Ophiuroidea, 288, 303
Ophiuroids, 288, 294, 295
Opisthobranchia, 301
Opisthobranch molluscs, 297, 309
Opisthobranchs, 286, 303
Opisthognathidae, 330
Opportunistical, 234, 243, 286
Opportunistic macroalgae, 249
Opportunistic pathogen, 233, 236, 458
O$_2$ production, 137
Optimum temperature, 139, 140, 166
O$_2$-radicals, 192

Orb. velvetfish, 330
Ordovician, 15, 16, 441
Ordovician period, 16
Organic acids, 93, 94, 210
Organic compounds, 132, 142, 188, 204, 209, 210, 216, 249, 250
Organic matrix (OM), 120, 122, 127, 131–136, 138–143, 154, 167, 168, 212, 220–222
Organic matter, 107, 154, 201, 204, 205, 207–211, 215–217, 224, 232, 235, 246, 247, 287, 288, 291, 293, 299, 303, 306, 309, 382, 445, 494–497, 501, 502
Organic nitrogen, 204, 205, 207, 219, 220, 224, 231, 232, 235, 495
Organotrophic microorganism, 436
Organotrophs, 436
Orientation, 116, 221, 336, 342, 422, 430, 436
Orinoco Rivers, 473
Orthogonum fusiferum, 441
OR/V. *See* Oceanographic research vessel
Oscillatoria spp., 252, 474
Osmoregulat, 329
Osmosis, 89
Osmotic, 127
Ostraciidae, 330
Ostracion, 333, 339
Ostracion meleagris, 333
Ostracoblabe implexa, 441
Ostracoda, 283, 303
Ostreobium, 441
Ostreobium quekettii, 436, 437, 443, 445
Other compromised health (OCH) conditions, 466, 473, 481, 483
Otoliths, 122, 124, 138, 168, 334–336, 338
Oulastrea crispata, 62, 68
Over-exploitation, 7, 307, 341, 385, 401
Overfishing, 200, 235, 331, 385, 406, 494, 499
Overgrow, 261, 274, 349, 355, 363, 365, 439
Overgrowth, 96, 290, 306, 307, 351, 353, 355, 356, 360, 363–365, 384, 385, 512, 513
Oxidative damage, 407, 422
Oxidative phosphorylation, 127, 428
Oxidative stress, 190, 406–413, 415, 429
Oxygen, 97, 99, 101, 109, 113, 139, 162, 189–191, 205, 234, 235, 243, 333, 362, 406–412, 454, 455, 461
Oxygenase, 94, 188, 189, 408, 410
Oxygen-evolving complex, 409, 411
Oxypora glabra, 232
Oxyrrhis marina, 192

P

P. *See* Phosphorus
^{32}P, 200
p53, 412, 413
Pachyseris, 72
Pacific, 18, 19, 47–50, 52, 53, 62, 64, 65, 71, 75, 92, 93, 100, 231, 254, 256, 262, 264, 280, 281, 283, 287, 288, 290, 291, 293, 296, 300–302, 307, 309, 313, 330, 340, 341, 349, 364–366, 384, 401, 473, 509
Pacific coral reefs, 231, 288, 302, 312, 349
Pacific endemic sea, 300
Pacific islands, 3, 5, 342
Pacific Ocean, 211, 235, 300, 348, 509, 510
Padina sp., 244, 260, 362
Pagurites sp, 302
Palaemonid shrimps, 294
Palao tropical research station, 5
Palau, 3, 64, 67, 68, 74, 274, 474, 517
Paleoceanographic, 155

Paleoecological, 19, 22, 441
Paleoecology, 13–22
Paleontologists, 15, 19, 47, 119
Paleontology, 120
Paleoproductivity, 16
Paleotemperatures, 155
Palisada (Laurencia) poiteaui, 242
Palmira Atoll, 364
Palmyra, 383
Palythoa, 279, 352, 355, 357
    P. caribaeorum, 352, 484
    P. caribbea, 242
    P. toxica, 357
Palytoxin, 357
PAM fluorimetry. *See* Pulse amplitude modulated fluorimetry
Panamá, 17, 21, 260, 262, 277, 280, 288, 293, 298, 301, 302, 312, 456, 458, 471, 472, 476
Panbiogeography, 43, 44
Papua, 50
Papuan Birds Head Peninsula, 50
Papua New Guinea, 17, 50, 51, 53, 64, 459
PAR. *See* Photosynthetic available/active radiation
Paracellular, 123, 127, 129, 189
Paraglyphidodon, 333
Parapercis hexopthalma, 333
Parasites, 166, 281–283, 287, 291, 297, 299, 302, 304–309, 312, 339, 340
Parasitic, 283, 291, 297, 304–306, 340, 341, 459
Parasitic relationship, 443
Parasitisms, 187, 283, 286, 297, 303–308, 341, 364
Parasitoids, 309
Paratypton siebenrocki, 299
Parerythropodium fulvum, 356
Parma, 333
Parrotfish, 252, 329–331, 338, 340, 342, 383, 439–441, 481, 510, 512, 514, 517
    corallivory, 512
    herbivory, 512
Particle, 178, 179, 182, 204, 205, 215, 246, 260, 281, 288, 296, 298, 303, 309, 352, 355, 363, 379, 439, 440, 445, 494, 497, 501
Particulate matter, 216, 224, 245, 493, 494, 501
Particulate organic matter (POM), 107, 201, 208, 216, 235, 278, 287, 288, 291, 299, 303, 494, 495, 497, 501
Particulate organic nitrogen (PON), 208
Parupeneus multifasciatus, 333
Passive airborne, 28
Passive diffusional, 189
Passive remote sensing system, 27
Pathogenesis, 474–475, 485, 486
Pathogenic bacterium, 362
Pathogens, 7, 233, 234, 236, 251, 256, 308, 312, 353, 362, 451–461, 465, 466, 469, 471–478, 480, 481
Pathology, 451–461, 473, 481, 484
Pattern recognition receptor (PRR), 93
Pavona sp., 72, 92, 222
    P. cactus, 132, 159, 222
    P. chiriquiensis, 71
    P. clavus, 221
    P. decussata, 98
    P. gigantea, 71, 100, 222
    P. varians, 65, 67, 71, 472
Pb, 338
PCD. *See* Programmed cell death
pCO$_2$, 99, 136, 153, 156–160, 162, 163, 165–168, 170, 171, 445
PCR. *See* Polymerase chain reaction
Pearlfish, 297, 330

Pearl oyster, 134
Pegasidae, 330
Pegasus, 339
Pelagic, 235, 292, 307, 311, 333, 334, 336, 337, 384, 493, 494
Pelagic larva, 254, 332, 333
Pelagic larvae, 254, 334–337, 497, 499, 501
Pelagic phytoplankton, 232
Pelagic predators, 336
Pelagic spawners, 332
Pelagic stage, 245, 378
Pocillopora elegans, 68
Pempheridae, 330
Penicillus, 245
Pentacerotidae, 330
Peptide toxin, 452
Percent cover, 7, 31, 349, 350, 352, 363, 405
Periclimenes pedersoni, 304
Periclimenes perryae, 297
Peridinin, 31, 98, 191
Peroxidase, 101, 113, 409
Peroxydases, 167, 190
Peroxynitrite anion (ONOO$^-$)., 412
Persistence, 9, 17, 19, 261, 297, 307, 338, 341, 455
Pesticides, 7, 494
Petrochirus diogenes, 302
PGPase. *See* Phosphoglycolate phosphatase
pH, 6, 7, 21, 69, 93, 94, 123–125, 127, 129–131, 136–139, 142, 143, 152–158, 162–164, 166–168, 170, 188, 189, 259, 394, 397, 399, 400, 406, 410, 445, 482
Phaeophila dendroides, 436, 437, 441
Phaeophyceae, 242–244
Phaeophytes, 31, 440
Phanerozoic Eon, 14, 16, 19
Pharmaceuticals, 7
Phase shifts, 200, 231, 232, 234, 249, 253, 261–264, 311, 349, 350, 363, 364, 367, 375–378, 385, 499, 514
Phenol, 232
Phenotype, 430, 431
Phenotypic adaptations, 421, 430
Phenotypic plasticity, 422, 475
Pherallodichthys meshimaensis, 334
Pherecardia striata, 301, 302, 309
Phestilla lugubris, 301
Phestilla minor, 301
Phestilla sibogae, 303
Philippines, 40, 47, 50, 51, 64, 300, 328, 331, 351, 354, 379, 380, 383, 385, 473
Phloroglucinol (1,3,5-trihydroxybenzene), 252
Phormidium corallyticum, 474
Phoronida, 277, 291, 303
Phosphates, 94, 96, 138, 167, 202, 204, 205, 231, 295, 410, 455
Phosphate uptake, 94, 200
Phosphofructokinase, 167
Phosphoglycolate phosphatase (PGPase), 99
Phosphorus (P), 87, 95, 110, 113, 114, 116, 200, 202, 208, 215, 216, 224, 225, 232, 248, 493–495
Phosphorylase kinase, 167
Phosphorylation, 108, 127, 428
Photic zone, 98, 110, 154, 259, 260, 439
Photoacclimate, 98, 352
Photoacclimation, 90, 98, 99, 111–113, 116, 220, 496
Photoacclimative responses, 107
Photo-adaptation, 220
Photoautotrophs, 101, 409, 411
Photochemical efficiency (Fv/Fm), 98, 99, 424, 428
Photographic survey, 348

Index 543

Photoinhibition, 98, 99, 101, 247, 407, 408, 411, 445
Photooxidative damage, 409
Photoprotection, 99, 409
Photoprotective agent, 109
Photo quadrats, 349
Photorespiration, 162, 165, 410
Photosymbionts, 360
Photosymbiotic, 87
Photosynthates, 89, 94, 107, 112, 114, 116, 215, 216, 218, 220, 231
Photosynthesis, 89, 90, 94, 95, 97–101, 109, 110, 112, 115, 116, 125, 127, 131, 136–139, 153, 154, 156–158, 162, 165, 166, 169, 178, 181, 188–191, 200, 204, 207, 216–220, 222, 223, 225, 232–234, 236, 243, 245, 246, 253, 260, 407–411, 413, 428, 441, 452, 454, 493–497, 501
Photosynthesis inhibiting factor (PIF), 89, 94
Photosynthetic available/active radiation (PAR), 98, 192, 246, 407
Photosynthetic efficiency, 202, 220
Photosynthetic pigment, 112, 406, 407, 495
Photosynthetic units, 98
Photosystem I (PSI), 410
Photosystem II (PSII), 99, 406, 407
Phototrophic, 178, 188, 191, 436, 439, 441, 443, 445, 501
Phototrophic dissolved inorganic carbon, 178
Phototrophic microorganism, 436
Phototrophic species, 496
Phototrophs, 189, 293, 436
Phuket, Thailand, 4, 424, 426
Phycobilin, 246
Phycoerthyrin, 246
Phylogeny, 42, 43, 70, 73, 92, 340
Physical bioerosion, 435
Physical disturbances, 245, 250, 257, 258, 354, 367, 513, 517
Physical factor, 335
Physical habitat loss, 341
Physical oceanographers, 8, 9
Physiolog, 329
Physiological, 6, 51, 87, 97, 99, 100, 115, 119, 125, 127, 130, 133, 139, 143, 162, 167, 168, 171, 187, 212, 220, 225, 246, 254, 257, 338, 340–342, 353, 392, 397–400, 406–408, 411, 413–415, 423–426, 451, 465, 466, 495, 498
Physiological adaptation, 187–192
Physiological characters, 41
Physiological damage, 365
Physiological effect, 253, 340, 341
Physiological issue, 341
Physiological response, 212, 332, 341, 466
Physiological stress, 365, 408, 415, 466
Physiologists, 9, 408, 423
Physiology, 4, 97, 100, 119, 125–136, 140, 143, 158, 159, 166–168, 216–222, 328, 332, 342, 354, 398, 408, 421, 423, 426, 484, 495, 496
Phytoplankton, 25, 110, 207, 216, 232, 248, 251, 300, 303, 383, 407, 408, 413, 495, 497, 501
Phytoplankton blooms, 501
Picoplankton, 215, 216, 222–225
φ See Quantum yield
PIF. See Photosynthesis inhibiting factor
Pigmentation, 98, 112, 113, 192, 467, 472, 473, 478
Pigments, 31, 97, 99, 111, 112, 191, 192, 220, 246, 247, 338, 395, 397, 406, 407, 409, 422, 474, 495, 496
Pinctada fucata, 134
Pinguipedidae, 330
Pinnotheridae (pea crabs), 283, 306
Pinufius, 285, 301
Pinufius rebus, 301
Pipefish, 329, 330, 336

Piscivores, 231, 309, 339, 376, 384, 517
Piscivorous, 376,
Piscivorous fish, 311, 376, 383
Pistol shrimps, 297, 305
Pitcairn, 328
Planktivores, 235, 294, 309, 311
Planktivorous, 333, 337
Planktivorous fish, 311, 383
Plankton, 110, 151, 171, 200, 203, 215–225, 232, 234, 247, 260, 279, 283, 287, 288, 290–292, 303, 309, 336, 494, 495, 497
Planktonic, 60, 69, 74, 208, 222, 292, 294, 309, 355, 438, 445, 497
Planktonivorous, 152
Plants, 6, 32, 38–40, 43, 44, 71, 129, 165–168, 187–190, 200, 204, 208, 243, 244, 246, 247, 251, 258, 308, 354, 392, 394, 407–409, 411, 413, 430, 451
Planulae, 61, 62, 64–66, 69, 70, 74–76, 108, 109, 125, 130, 353
Planula larvae, 60, 62, 74
Planula larval, 60
Plasma, 124, 338
Plasma membrane, 94, 97, 127, 128, 131, 189, 428
Plate tectonics, 20, 51, 53
Platygyra, 352
Platygyra daedalea, 353, 357
Plectonema terebrans, 436, 437, 441
Plectroglyphidodon, 333
Plectropomus leopardus, 336
Pleistocene, 14–20, 22, 51, 155
Plerogyra sinuosa, 113, 191
Plesiastrea versipora, 94, 140, 220
Plesiopidae, 330
Plexaura homomalla, 351
P limitation, 247, 248
Plotosidae, 330
P$_{max}$. See Light-saturated rate of photosynthesis
Pocillopora spp., 74, 93, 100, 262, 279, 283, 300, 302, 351, 424, 426
  P. *capitata*, 92
  P. *damicornis*, 39, 41, 61, 65, 68, 74, 75, 92, 96, 98, 109, 178, 220, 222, 289, 293, 301, 356, 365, 423, 428, 453, 454, 457, 482
  P. *elegans*, 68
  P. *meandrina*, 68, 92, 232
  P. *verrucosa*, 68, 92, 100, 178, 191
Pocilloporid, 108, 254, 283, 288, 298, 300, 305, 307
Pocilloporidae, 66, 70, 75, 282, 459
Pocilloporins, 192
Podocotyloides stenometra, 305, 452, 459
Poecilostomatoida, 283, 299, 306
Poikilotherms, 422
Poland, 440
Policy, 6–9, 171, 331, 518
Pollutants, 69, 341, 378, 379, 493, 498–500
Pollution, 7, 8, 21, 305, 341, 367, 373, 377, 380, 381, 385, 406, 485, 493, 495, 496, 498, 499
Poly-Asp, 134
Polychaeta, 280, 289, 296, 297
Polychaetes, 251, 281, 283, 292, 293, 296, 297, 299, 302, 304, 309, 438, 439
  annelids, 287, 303
  worms, 279, 286, 291, 294, 296, 297, 302, 304, 307, 309
Polycrystalline, 141
Polycyclic aromatic hydrocarbons (PAH), 7
Polygons, 181–183, 379, 427, 428
Polymerase chain reaction (PCR), 131
  amplification, 429
Polymorphic filament, 437
Polymorphic species, 40
Polynesia, 64, 247, 274, 282, 303, 337, 341, 355, 382, 383, 398, 440, 441

Polynesian lagoon, 351
Polyphenolics, 252
Polyplacophores, 439
Polyp layers, 438, 445
Polyps, 3, 59–62, 69–72, 74–76, 97–98, 119, 121, 127, 130, 151, 154, 180, 183, 216, 222, 261, 280, 283, 291, 300, 305–307, 311, 347, 348, 352, 353, 356–360, 362, 363, 365, 410, 436, 438, 445, 453, 456
Polysiphonia, 257
Poly-unsaturated fatty acids, 93, 217, 218
POM. *See* Particulate organic matter
Pomacanthidae, 330
Pomacentridae, 311, 329, 330, 333, 336, 339, 378
Pomacentrus, 333
    P. amboinensis, 332, 333, 336
    P. lepidogenys, 336
Pomachromis, 333
PON. *See* Particulate organic nitrogen
Population biology, 5
Populations, 3, 5, 7–9, 21, 38–40, 44, 47, 53, 61–66, 69, 72–76, 89, 92, 93, 95–97, 100, 168, 200, 219, 225, 234, 244, 245, 253, 256, 259, 261, 273, 276, 279, 283, 287, 288, 293, 294, 296, 300, 302, 303, 305, 307, 310–312, 328, 331, 332, 337, 338, 341, 342, 359, 366–367, 373, 378, 382, 383, 385, 399–401, 426, 428, 429, 444, 454, 465, 466, 471, 472, 474, 477, 481–483, 485, 486, 493, 496, 500, 510, 516
    dynamics, 7, 95–97, 282, 299, 328, 329, 331, 484, 514, 516
    genetics, 5, 76
    mortalities, 468, 484
    size, 9, 16, 231, 366
Porcelanid crab, 294
Porcellio scaber, 124, 129
Porgies, 329, 330
Porifera, 129, 276, 278, 282, 289, 293–294, 297, 303, 359
Porites polyps, 305
Porites sp., 72, 93, 141, 164–166, 220, 279, 301–305, 312, 337, 348, 351, 355, 365, 397, 426, 443, 459, 498
    P. astreoides, 67, 72, 160, 220, 254, 255, 289, 354, 365
    P. attenuata, 351
    P. brighami, 67
    P. compressa, 87, 92, 96, 159, 160, 219–221, 225, 305, 482
    P. cylindrica, 225, 356, 364, 365
    P. divaritaca, 130
    P. furcata, 232
    P. lobata, 87, 160, 166, 225, 232, 442
    P. lutea, 39, 160, 166, 354, 362, 442
    P. panamensis, 92
    P. porites, 125, 130, 136, 159, 160, 181
    P. rus, 160, 274, 337
    P. sillimaniani, 178
    P. solida, 97
Porites trematodiasis (PTR), 452, 459
Poritidae, 66, 67, 72, 73
Porphyra, 259, 437
Porphyra-334, 101
Portieria hornemannii, 362
Portunid crabs, 302, 309
Portunus, 282, 303
Positive feedback, 235, 264, 349, 350, 411
Potential quantum yield (Y), 352, 360
Poverty, 331, 380
$p53$ protein, 412, 413
P/R, 428
Prawn, 424
Precambrian, 511

Predation, 16, 20, 91, 112–114, 222, 223, 235, 274, 282, 283, 286, 288, 290, 292–294, 296, 299–305, 307–311, 341, 362, 367, 376, 384, 385, 471
Predator outbreaks, 348, 349
Predators, 5, 231, 232, 234–236, 252, 253, 275, 282, 283, 286, 288, 294, 297, 299, 300, 302–304, 307–309, 312, 333, 336, 339, 340, 348, 349, 357, 360, 383–385, 398, 439, 440, 460, 481, 482, 497, 516, 517
Predatory crown-of-thorns sea stars, 359
Predatory threat, 336
Predictive capacities, 15, 385
Premnas, 294, 333
Premnas biaculeatus, 334
Presettlement, 337
Pressure, 45, 66, 94, 129, 156, 166, 190, 232, 236, 290, 292–294, 296, 302, 307, 308, 311, 330, 334, 341, 352, 377, 378, 381, 397, 400, 411, 436, 439, 445, 498, 499, 501, 510
Prevalence, 234, 236, 307, 309–311, 466, 469, 472–478, 482, 483, 485
Prevalence, incidence, and virulence, 477–478
Prey, 110, 116, 216, 218–222, 224, 225, 232, 282, 283, 286–288, 293, 294, 300–303, 307, 308, 310–312, 336, 339, 384, 439, 440, 480, 495
    selection, 339
    species, 232, 286, 293, 299, 303
Prey-predator, 376
Priacanthidae, 330
Primary producers, 170, 243, 245–247, 250, 310, 382, 384, 441, 495
Primary production, 32, 192, 200–202, 204, 205, 207, 208, 210, 211, 234, 241, 250, 382, 385
Primary productivity, 16, 19, 189, 199, 202, 205, 241, 242, 250, 259, 384, 400
Prionocidaris baculosa, 300
Pristotus, 333
Prochordates, 135
Production, 16, 19, 32, 60, 61, 66, 75, 94, 101, 125, 127, 137, 139, 151, 153, 165, 169, 189, 191, 192, 200–202, 204, 205, 207, 208, 210–212, 218, 219, 233–236, 241, 242, 247, 250, 252, 259, 290, 292, 300, 303, 305, 310, 338, 351, 353, 355, 360, 365, 367, 382, 383, 385, 406–412, 414, 444, 452, 460, 461, 493
Productivity, 16, 19, 25, 32, 53, 99, 165, 166, 189, 199–203, 205, 241–244, 247, 248, 250, 257–260, 384, 400, 405, 407, 409, 410, 494, 497, 501, 513, 516
Programmed cell death (PCD), 96, 408, 412
Prokaryotes, 120, 310
Prokaryotic blue-green algae, 242, 243
Prokaryotic cells, 243
Prokaryotic genes, 130
Prokaryotic microorganism, 436
Prokaryotic photoautotrophs, 411
Proliferation, 16, 90, 152, 246, 250–251, 256, 261, 273, 348, 498, 501
Proserpine, 495
Protein carbonylation, 190
Proteins, 7, 53, 93, 94, 99, 101, 108, 109, 113, 123, 128, 132–135, 138, 140, 141, 143, 167, 189–192, 217–219, 221, 249, 252, 280, 353, 365, 407, 409, 411–415, 422, 424, 426, 429, 431
Proteobacteria, 482
Proteonomics, 415
Protists, 287, 437
Proton pump, 222
Protons, 125, 137, 138, 143, 154, 155, 222, 394, 396, 397
Protozoan, 459, 466, 472
PRR. *See* Pattern recognition receptor
Psammocora nierstrazi, 472
Psammocora stellata, 68
Pseudoalteromonas sp., 254
Pseudoamphithoides incurvaria, 308

Pseudobrooding, 75
Pseudochromidae, 330
Pseudochromis fuscus, 333
Pseudosquilla ciliata, 312
PSI. *See* Photosystem I
PSII. *See* Photosystem II
PSII:PSI, 411
PSUs. *See* Photosynthetic units
Pterois spp., 341
PTR. *See* Porites trematodiasis
Puerto Rico, 67, 72, 75, 248, 249, 273, 274, 458, 465, 468, 469, 471–479, 482–485
Puffer, 329, 339
Pufferfish, 330
Puget Sound, 352
Pulse amplitude modulated (PAM) fluorimetry, 352, 360
Pyrenoid, 88, 112, 121, 189
Pyrgomatids, 306
Pyruvate dehydrogenase, 167

Q

Q_{10}, 422
Qatar, 328
Q_{10}s, 429
Quantitative techniques, 409
Quantum yield (φ), 98
Quantum yield of photosynthesis, 407
Quaternary period, 13, 15, 18
Quickbird, 26, 27, 32
Quintana Roo, 380, 381
Quintana Roo's coastline, 382
Quoyula, 285, 300, 301

R

Rabbitfish, 252, 330, 337, 340, 342
Radiation, 69, 98, 107–109, 116, 191–192, 246, 286, 293, 360, 363, 392, 394, 398, 406, 407, 409, 410, 412, 414, 422–424, 426, 430, 431
Radiative transfer effects, 29
Rain, 200, 405, 493, 500
Rainfall, 21, 67, 247, 394, 395, 398, 401, 493, 500
Raman microscopy, 130
Raspailia inaequalis, 180
RBD. *See* Red band disease
RDM. *See* Relative Dominance Model
Reaction centers, 99, 112, 406, 407, 409–412
Reactive nitrogen species (RNS), 101, 412, 414
Reactive nitrogen species nitric oxide (NO), 101
Reactive oxygen species (ROS), 97, 99, 101, 109, 190–192, 362, 406–414, 454
Real-time satellite, 426
Recognize, 27, 93, 299, 303, 340, 341
Recovers, 264, 395, 399, 453, 478, 485, 496, 499, 502, 503, 511, 514
Recovery, 18, 90, 96, 218, 256, 264, 300, 304, 349, 376–378, 401, 425, 471, 472, 483, 485, 486, 496, 499–501, 503, 509–514, 517
Recruitment, **5**, **14**, **69**, **236**, **245**, **288**, **338**, **352**, **375**, **438**, **484**, **493**, **511**
Recruits, 245, 256, 377
Recycling, 87, 199–201, 204, 205, 207, 211, 212, 241, 242, 247, 250, 303, 382, 435, 494
Red algae, 241–246, 251, 252, 256–258, 348, 352, 353, 361, 363, 397
Red band disease (RBD), 466, 471, 473, 474, 476, 478, 481
Redfield equation, 158
Redfield ratios, 110, 113, 200, 204
Red macroalgae, 14, 349, 495
Redox reactions, 107
Red Sea, **39**, **49**, **63**, **95**, **108**, **168**, **221**, **251**, **273**, **333**, **348**, **439**, **454**, **472**, **494**
Reduction, 6, 8, 9, 65, 94, 129, 152, 154, 159, 167, 169–171, 211, 235, 236, 255, 256, 259–262, 273, 341, 351, 365, 376, 377, 383–385, 407, 408, 410, 411, 501, 511, 517
Reef, 5, 188, 331
  accretion, 16, 258, 260, 297, 384, 391, 397, 401, 444, 515
  bioerosion, 286, 297, 435–446
  carbonate budget, 444
  cave, 309
  crest water, 336
  ecosystem, **6**, **13**, **25**, **59**, **152**, 200, **207**, **225**, **233**, **244**, **273**, **327**, **373**, **391**, **405**, **494**, **510**
  erosion, 391, 444, 498, 517
  flats, 423, 425
  framework, 15, 59, 110, 152, 170, 171, 261, 273, 278, 279, 286, 297, 311, 392, 443, 444,
  function, 199, 205, 340
  management, 8, 171, 256, 331, 375, 509–518
  zones, 241, 243, 280, 286, 292, 295, 367, 382
Reef-building corals, 14, 16, 59, 107, 139, 191, 215, 225, 276, 284, 299, 308, 347, 349, 364, 367, 394–397, 399–401, 455, 457, 476, 478, 482, 483
Reef fish biology, 327, 339, 340
Reef-hugging, 337
Reflection, 29
Regions, 8, **18**, **28**, **39**, **47**, **59**, **90**, **108**, **126**, **153**, **178**, **203**, **242**, **274**, **328**, **348**, **375**, **394**, **405**, **424**, **455**, **465**, **497**, **509**
Relative Dominance Model (RDM), 349, 350
Remineralization, 204, 205, 250
Re-mineralization, 303
Remoras, 330
Remote sensing, 21, 25, 202, 375
Remote sensing application, 29, 31–33
Removal, 94, 123, 126–129, 132, 137, 138, 153, 154, 157, 204, 231, 235, 236, 245, 250–251, 282, 284, 295, 304, 309, 333, 337, 377, 383, 435, 438, 443, 497, 501
Renal, 189, 328
Repair, 136, 171, 354, 365, 411, 412, 415, 422, 460
Replication, 353, 358
Reproduc/spawn, 329
Reproduction, **5**, **14**, **59**, **90**, **107**, **135**, **215**, **297**, **328**, **347**, **399**, **415**, **438**, **482**, **512**
Reproduction/spawning, 328
Reproductive, **5**, **37**, **59**, **108**, **243**, **300**, **331**, **355**, **397**, **437**, **472**
Reproductive behavior, 332, 397
Reproductive biology/larval ecology, 5
Reproductive characters, 41
Reproductive cycle, 66–69, 108, 341
Reproductive investment, 332
Reproductive output, 61, 339, 472, 482, 484–486
Reproductive season, 355
Reproductive success, 66, 300, 307, 332
Reproductive system, 331
Reservoirs, 155, 250, 256, 312, 362, 452, 480–482
Resilience, 6, 8, 9, 18, 21, 218, 220, 264, 365, 376–378, 380, 493–503, 509–518
Resilience, structure and function, 6
Resistance, 7, 190, 192, 243, 256, 362, 421, 422, 425–431, 452–453, 456, 460–461, 473, 475, 477, 482, 484, 493, 498–503, 511, 513
Resistance adaptations, 421–422, 426–431
Resisting herbivory, 251

Respiration, 5, 87, 94, 97–99, 110, 125, 127, 137, 139, 152, 154, 156–158, 162, 164, 166, 167, 169, 170, 179, 202, 205, 215, 217, 218, 220, 224, 245, 259, 360, 408, 411, 428, 441, 495
Restriction, 44, 51, 341, 430
Retention, 6, 67, 69, 70, 109, 210, 220, 241, 242, 247, 250, 443, 444, 500
Reticulate evolution, 41–45, 52, 53
Retinomotor, 338
Returning larvae, 336
Reunion, 210, 211, 328, 440, 441
Reunion Island, 211, 440, 441
Revelle factor, 158
Rhizangiidae, 66, 72
Rhizobiales, 457
Rhizostomeae, 91
Rhizophyllidaceae, 252
Rhodactis spp., 358
    R. lucida (Corallimorph), 91
    R. rhodostoma, 348, 352, 355
    (*Discosoma*) sanctithomae, 348, 352
Rhodomelaceae, 252, 256
Rhodophyta, 242–244, 256
Rhodophyte *Porphyra.*, 437
Rhodophytes, 31, 436, 437, 440
Rhopalia catenata, 441
Ribulose 1, 5-bisphosphate decarboxylase/oxygenase (Rubisco), 94, 99, 188, 189, 408, 410, 411
Rising sea surface temperature, 445
Rising sea temperature, 396, 405, 423, 425, 426
Risk of extinction, 465, 484
RNAi assays, 413, 415
RNA interference (RNAi), 413
RNS. *See* Reactive nitrogen species
Rock/Rocky, 13, 170, 241, 245, 281, 307, 348, 352, 367, 392, 435, 443
ROS. *See* Reactive oxygen species
Rotifera, 291, 303
Rubisco, *See* Ribulose 1, 5-bisphosphate decarboxylase/oxygenase
    Rubisco activase, 410
RuBP. *See* Ribulose 1, 5-bisphosphate decarboxylase/oxygenase
Rudderfish, 329, 330, 340, 342
Rudist bivalves, 15, 16
Rugosa, 19
Rugosity, 31, 32, 235, 440, 515
Runoff, 5, 8, 69, 232, 485, 493, 499–502, 516, 517

S
Sabellariidae, 281, 296
Sabellidae, 281, 292, 296
Sabellids, 297, 438
Saccomorpha clava, 441
Saccomorpha spherule, 441
Sacoglossans, 251
Saculinid barnacle, 297
Saint Kitts and Nevis, 328
Saint Lucia, 328
Saint Vincent and the Grenadines, 328
Salinity, 14, 18, 21, 69, 91, 94, 136, 156–158, 171, 392, 423, 429, 444, 494
Salinity changes, 407
Samoan reefs, 300
San Blas, 298
Sand, 28, 30, 31, 91, 201, 241, 242, 244, 259, 282, 288, 338, 367, *Sandalolitha robusta*, 67, 71
Sandburrowers, 330
Sand divers, 330
Sandperch, 330, 333
Sands, 91, 241, 436
Sandy, 210, 211, 241, 247, 392
Sandy areas, 202
Santa Catalina Island, 352
Saprophytic fungi, 362
Saprophytic stramenopiles, 362
Sardine, 337
Sargassum spp., 252, 377, 497
Satellite, 6, 26–29, 31–33, 201, 393, 395, 426
Satellite Pour l'Observation de la Terre, 26
Saturated, 94, 128, 137, 165, 203, 217
Saturation, 6, 14, 69, 124, 128, 129, 140, 141, 152, 158, 166, 169, 178, 219, 248, 392, 405, 446,
Saturation constant (E_k), 397
Saudi Arabia, 49, 328
Scales, 14, 17, 19–22, 25, 28, 32, 43, 63, 64, 76, 100, 122, 123, 129, 139, 141, 157, 178, 183, 202, 203, 248, 253, 292, 308, 309, 327, 334, 339, 349, 373–380, 385, 394, 423, 427, 443, 446, 468, 484, 499
Scaridae, 329, 377, 378, 440
Scarids, 377, 439
Scarus sordidus, 252
Scattering, 29–31, 107, 192, 245
Scavengers, 192, 283, 303, 308, 309
Scavenging, 191, 281, 283, 284, 288, 309
Schooling clupeids, 337
Scientists, 5, 8, 9, 21, 26–28, 200, 313, 331, 342, 385, 392, 423, 509
Scleractinia, 19, 59, 73, 74, 125, 216, 279, 294, 303
Scleractinian, 16, 19, 50, 59, 60, 62–64, 66, 67, 69–71, 73–76, 130, 178, 183, 283, 284, 295, 297, 305, 347, 355, 468, 472, 478
Scleractinian corals, 5, 14, 16, 18, 19, 59–76, 87, 88, 119, 126, 136, 140, 141, 177–180, 188, 209, 223, 225, 279, 280, 282–285, 287, 294, 299, 306, 309, 349, 355, 356, 358, 359, 361, 377, 405, 406, 461, 465, 478, 481, 484
Sclerosponges, 298
Scolecia filosa, 441
Scorpaena, 339
Scorpaenidae, 330
Scorpionfish, 329, 339
SCUBA, 4, 153, 342,
Scuba diving, 37
Scuticocilatia, 459
SDR. *See* Shut down reaction
Sea anemone, 88, 89, 96, 97, 135, 167, 188–192, 220, 280, 283, 304, 348, 349, 352–354, 365, 367, 408, 411–413, 429
Sea basses, 329, 336, 340, 352
Sea cucumber, 288
Sea fan purple spots, 473
Sea fans, 280, 456, 460, 461, 471–473, 475, 476, 480, 481, 498
Sea floor, 493, 494, 500
Seafloor composition, 31
Seagrass (es), 7, 31, 51, 116, 243, 245, 362, 383
Sea hares, 251, 252
Sea horse, 332, 336, 339
Sea level rise, 152, 258–260, 394, 395, 398, 400, 401
Sea levels, 3, 7, 14, 15, 17–19, 43, 51, 52, 199, 260, 367, 393, 394, 398, 405, 443
Sea moths, 330, 339
Season, 52, 64, 65, 71, 96, 202, 482
Seasonal disease, 477
Seasonal variability, 247, 477, 478
Seastars, 273, 274, 281, 287, 288, 297, 299–301, 307, 309, 359
Sea-surface temperature (SST), 14, 18, 21, 32, 183, 236, 259, 261, 264, 395, 415, 426, 445
Sea temperatures, 363, 394–396, 398, 399, 405, 423–427, 431

Sea urchins, 168, 231, 232, 250–252, 273, 283, 288, 296, 297, 300, 306, 307, 311, 312, 349, 377, 378, 384, 439, 441, 445, 465, 468, 471, 475, 480, 483, 485, 509–511, 517
Seawater, 14, 29, 69, 76, 94, 114, 119, 123–130, 136, 137, 140, 151–164, 167, 168, 170, 171, 188, 189, 192, 204, 207, 216, 220, 221, 224, 233, 236, 248, 251, 255, 259, 353, 362, 406, 460, 503
   temperature, 6, 141, 362, 405, 406, 411, 425, 427, 451, 458
   vacuoles, 164
Seaweb, 8
Seaweeds, 242, 243, 245, 251–253, 256, 258, 362, 383, 401
SEB. *See* Skeletal eroding band
Secondary metabolites, 251–253, 308, 354, 362, 456
Sedimentary record, 22
Sedimentation, 38, 69, 141, 171, 200, 255, 260, 264, 297, 298, 304, 341, 353, 359, 361, 363–365, 373, 377, 378, 405, 407, 440, 445, 471, 485, 493, 494, 496, 498, 500, 501, 514, 516
Sediments, 6–8, 29, 154, 170, 200–202, 205, 208, 209, 216, 224, 225, 232, 233, 241, 242, 245–248, 250, 253–255, 258, 260, 264, 280–283, 286, 288, 291, 293, 296–298, 303, 305, 311, 378, 379, 383, 398, 443, 480, 484, 493–496, 499, 501, 513, 516
Segmentation, 281, 284, 306
Selective pressure, 44, 66, 436
Self digestion, 413
Sensitive, 68, 76, 97, 100, 101, 126, 151, 152, 158, 162, 164, 166–168, 171, 189, 259, 260, 334, 397, 398, 400, 410, 415, 422, 460, 496, 503
Sensitive species, 496, 498,
Sensitivity, 100, 129, 141, 151, 154–171, 363, 375, 377, 422, 425, 430
Seriatopora, 283, 300
Seriatopora hystrix, 92, 93, 178
Serpulidae, 281, 292, 296
Serpulids, 296, 438
Serranid, 338
Serranidae, 329, 336, 377, 388
Serratia marcescens, 458, 474
Sesquinoids, 252
Sessile, 152, 180, 275, 279, 288, 292, 294, 300, 304, 308, 347–368, 407, 425, 460
Sessile invertebrate, 299, 348, 352, 359–360, 362, 383
Sessile reef organisms, 348, 349, 353–355, 362–364, 366, 367
Settlement, 69, 245, 336, 338, 347, 354, 378, 397, 513, 514
Settles, 60, 69, 70, 245, 254, 293, 296, 298, 332, 334, 338, 356–358, 363, 494
Settling, 247, 255, 334, 337, 338, 356, 493
Sewage, 152, 483, 493, 494, 498, 517
Sewer, 471
Sex change, 71, 72
Sexual, 62, 66, 67, 70–74
Sexual production, 355
Sexual reproduction, 59–76, 366, 399, 438, 484
Seychelles, 8, 116, 328
Shade-adapted colonies, 223
Shade corals, 113
Shading, 99, 245, 247, 255, 354, 361–363, 496, 497, 511
Shallow, 51, 53, 59, 60, 93, 110, 152, 192, 211, 243, 244, 258, 279, 286, 288, 292, 312, 347, 358, 363, 382, 391, 407, 423, 425, 437, 443, 494, 500, 501, 503
Shallow lagoon, 116, 436
Shallow waters, 16, 27, 29–31, 92, 95, 98–100, 111–113, 115, 116, 188, 191, 217, 244, 246, 258, 280, 282, 336, 407, 410, 455, 471, 495, 502
Sharks, 231, 235
Shells, 124, 129, 133, 136, 155, 158, 245, 282, 286, 292, 293, 296, 302, 437, 439, 440, 445
Shelter, 152, 171, 280, 282, 293, 294, 296–298, 304, 305, 312, 435, 440

Sheltering, 116, 287, 293, 295, 332
Shift, 255
Shifting baseline syndrome, 14, 21
Shinorine, 101
Shipwrecks, 364, 509
Shore, 7, 110, 247, 282, 336, 424,
Short-wave infrared (SWIR), 28, 29
Shrimp and fish cleaners, 308
Shrimpfish, 282
Shrimps, 218, 220, 251, 282, 291, 294, 296, 297, 299, 302, 304, 305, 307–309, 331, 336
SHF. *See* Synthetic host factor
Shut down reaction (SDR), 468
Sibling species, 40, 43, 276
Siderastrea, 356, 361, 483
   S. radians, 222
   S. siderea, 365, 471
Siganidae, 378
Siganus fuscescens, 383
Siganus spinus, 252, 383
Signal transduction molecules, 412
Silica, 130, 200, 278
Silico sequence, 429
Simulate/simulating, 177, 179
Simulation models, 177–179, 181, 183, 349, 350, 384, 512
Sinai Peninsula, 49
Singapore, 64, 67, 71, 294
Sinularia flexibilis, 354, 356
Siphamia argentea, 296
Siphon, 296
Siphonaceous, 241, 257
Siphonaceous Chlorophyta, 244
Siphonal, 434
Siphonodictyon sp., 360
Sipuncula, 276, 281, 297, 303
Size, 6, 26, 28, 29, 41, 49–51, 61, 62, 71, 72, 88, 89, 98, 123, 126, 179, 182, 183, 201, 208, 209, 215, 232, 276, 279, 280, 282, 288, 291–293, 310, 311, 334, 341, 342, 354, 364, 378, 395, 435, 436, 438–440, 477
Skeletal, 122, 355
Skeletal architecture, 140, 298, 299
Skeletal carbonate substrates, 435, 436
Skeletal eroding band (SEB), 459, 473
Skeletal growth, 136, 140, 217, 220–222, 225
Skeletal surface area, 217, 219
Skeletonization, 180
Skeletons, 19, 60, 61, 70, 99, 111, 112, 114, 119–125, 127, 129, 130, 132, 134, 138–140, 178, 180, 182, 183, 211, 217, 219–221, 280, 292, 300, 305, 355, 357, 364, 396, 436–438, 456, 457, 459, 473
Skeleton/tissue, 119
Skylight, 29
Sleepers, 330
Smell, 334, 336
SML. *See* Surface mucus layer
Snails, 123, 251, 287, 300, 424, 440, 457, 480, 481
Snappers, 329, 331, 337, 342
Snooks, 330
Social sciences, 8, 9
Sodium azide, 127
Sodium cynanide, 127
SOD. *See* Superoxide dismutases
Soft-bodied cnidarians, 352, 363
Soft corals, 5, 59, 216, 279, 280, 283, 348, 349, 351, 352, 354–358, 363, 364, 367, 401, 455, 460, 498,
Soft tissues, 132, 243, 300, 355

Soil, 247, 392, 456, 493, 499, 500, 516
Solar irradiance, 412, 414, 426
Solar irradiation, 191
Solar radiation, 107, 109, 191–192, 392, 406, 407, 409, 410, 412, 414, 422–424, 426, 430, 431
Soldierfish, 330
Solenostomidae, 330
Solentia foveolarum, 436
Solitary, 60, 61, 65, 70–72, 135, 280, 288, 290, 349, 352, 364, 365
Solomon Islands, 50, 53, 64
Soluble organic matrix (SOM), 132
Soluble reactive phosphorus (SRP), 232
Somalia, 328
Somatic parasitism, 364
SOM. *See* Soluble organic matrix
Sound, 8, 44, 334, 336, 343, 385
Sources, 6–8, 32, 43, 48, 53, 87, 88, 95, 107–116, 125, 127, 128, 137, 162, 164, 178, 181, 182, 189, 200, 203, 204, 207–210, 212, 220, 224, 225, 232, 241, 247–249, 253, 254, 256, 260, 276, 277, 279, 282, 331, 333, 334, 336, 338, 360, 373, 378, 382, 456, 461, 485, 493, 497, 499–501, 512
South Africa, 341
Southern California, 352
sp..*See* Species
Space, 3, 4, 6, 7, 17, 38, 40–42, 61, 62, 76, 88, 90, 95, 107, 110, 122, 123, 125, 128, 129, 154, 162, 188, 235, 236, 243, 245, 255, 259, 260, 278, 280, 293, 294, 298, 307, 333, 347–351, 353–360, 363, 365–367, 376–378, 383, 385, 392, 394, 406, 425, 429, 468, 481, 497, 498, 501, 516
Space availability, 89
Spadefish, 329, 330
Spanning, 19, 332, 341, 385
Sparidae, 330
Sparisoma viride, 439, 440, 481
Spatial, 255, 495
Spatial aggregation, 351
Spatial resolution, 26, 28, 31
Spatial scales, 25, 30, 51, 264, 298, 375
Spawn, 60, 64–66, 72–76, 108, 232, 333,
Spawners, 5, 74, 91, 512
Spawning, 5, 61–69, 71, 72, 74–76, 107, 108, 275, 328, 329, 332–334, 341, 342
Spawning behavior, 3
Spawning events, 5, 63–65, 74, 76, 254, 310, 333
Spearfishing, 342
Specialized, 52, 120, 121, 280, 287, 300, 339, 357, 440, 497
Species (sp.), 37–44
    distribution, 17, 18, 38, 43, 48, 258
    diversity, 7, 15, 44, 50, 251, 275, 339, 349, 366, 509
    infected, 452
    richness, 32, 48–52, 59, 60, 276, 279, 287, 290, 294, 295, 311, 496, 498, 502
Species-specific mechanism, 362
Spectral composition, 496
Spectral reflectances, 28, 29, 31
Sperm, 5, 61, 62, 70–75, 453
Spermatogenic, 72
Sphaeramia ordicularis, 335
Spheciospongia vesparia, 294
Sphyraenidae, 330
Spiny basslets, 330
Spiny puffers, 329, 330, 339
Spionids, 296, 299, 438
Spionid worms, 297
Spirobranchus, 281, 304
Spirobranchus giganteus, 242, 304

Spitzbergen, 440
SPM. *See* Suspended particulate matter
Sponge chemicals, 360
Sponge-coral competition, 363
Sponges, 152, 153, 180, 187, 232–234, 242, 245, 275, 278–283, 287, 293, 294, 296–299, 304, 307–309, 312, 347–352, 355, 356, 359–363, 365, 367, 436–439, 441, 445, 458, 466, 469, 473, 478, 484, 485, 497, 498
Spores, 243, 245, 456
SPOT. *See* Satellite Pour l'Observation de la Terre
Spratly Islands, 328
Spring tide, 406
Squid, 286, 292, 302, 383, 397
Squirrelfish (Holocentridae), 311, 337
SR-B1 expression, 192
Sri Lanka, 50, 328
SRP. *See* Soluble reactive phosphorus
SSPS. *See* Sea fan purple spots
SST. *See* Sea-surface temperature
Stability, 17, 247, 261–264, 282, 293, 364, 376, 415, 465, 510, 511
Stabilizing coastlines, 405
Stable isotope analysis, 218, 339
Stable isotopes, 125, 128, 138, 218, 339
Stable states, 6, 235, 260–264, 376, 377, 509–511, 513, 515
Staphylococcus, 236
Starfish, 5, 123
Starved, 89, 217–221
Statisticians, 9
St Croix, 311, 475, 494
Stegastes, 333
    S. leucostictus, 333
    S. planifrons, 481
Stephanocoenia, 457, 483
Stephanocoenia intersepta, 75, 471
Sterols, 217
Stichodactyla haddoni, 336
Stoichiometery, 204
Stomatopod crustaceans, 309, 312
Stomatopods, 282, 293, 294, 302
Stony corals, 280, 348–356, 358–367
Storms, 61, 69, 171, 200, 245, 259, 260, 297, 302, 348, 366, 367, 394, 395, 398, 401, 435, 443, 456, 465, 471, 483, 485, 494, 496, 498, 500, 503
STPCA, 131, 138
Streamlined fish, 340
Streptococcus, 236
Stress, 7, 9, 68, 69, 88, 93, 97, 99, 101, 191, 218, 256, 262, 336, 352, 365, 399–401, 405, 408, 410–412, 415, 422, 425, 426, 429–430, 445, 454, 460, 483, 496, 500, 501, 509, 513, 517
Stress at sublethal levels, 7
Stressors, 5–8, 153, 225, 263, 264, 311, 360, 376, 377, 406, 422, 435, 465, 482
Stress-tolerant, 100
Stromatolites, 16, 440
Structural defenses, 250–252
Structures, 6, 14, 15, 39, 59, 62, 88, 93, 115, 119, 120, 130, 133–135, 140, 142, 151, 152, 171, 179, 180, 203, 243, 250, 253, 273, 279, 280, 291, 292, 294, 296, 332, 366, 383, 392, 399, 443, 510, 511
Stylaraea punctata, 254
Stylocheilus, 252
Stylocheilus longicauda, 252
Stylocheilus striatus, 252
Stylophora, 92, 93, 154, 283, 300
Stylophora pistillata, 71, 89, 92, 95, 96, 100, 108, 109, 111–113, 115, 122, 128–132, 135, 138, 143, 154, 165, 167, 189, 191, 209, 217, 218, 220–224, 254, 301, 351, 354, 355, 358, 359, 364

Stypopodium zonale, 244
Subsistence, 341, 342
Substrates, 15, 170, 178, 182, 244, 245, 247–249, 254, 261, 282, 287, 290, 293, 296–298, 300, 303, 306, 307, 332, 337, 338, 362, 436–443, 445, 481, 517
Subtidal rocky reefs, 352
Sub-tidal zones, 423–425
Subtropical, 18, 20, 59, 63–65, 152, 243, 258, 259, 395, 397, 399, 401
Sudan, 328
Sugars, 93, 132, 133, 138, 232, 233
Sulawezi, 335
Sulfur, 233, 482
Summer, 8, 52, 64, 65, 96, 108, 202, 394–397, 423, 424, 451–453, 455, 472, 477, 483
Sunlight, 29, 98, 110, 115, 136, 391, 400
Superoxide dismutases (SODs), 7, 113, 167, 190, 191, 409, 411, 412, 414, 415, 429, 452, 461
Superoxide radicals, 408
Supervised classification, 28, 29
Surface mucus layer (SML), 305, 460
Surgeonfish, 252, 329–333, 337, 340–342, 439
Survival, 60, 71, 75, 92, 100, 101, 113, 220, 233, 302, 304, 306, 336–339, 354, 359, 364–366, 409, 425, 429, 430, 460, 475, 501
    of corals, 354, 413, 431, 443, 486
    rates, 306, 354, 496
Survive/surviving, 60, 61, 70, 87, 97, 99, 101, 171, 236, 260, 286, 306, 401, 426, 452, 456, 471, 496
Survivorship, 6, 254, 288, 297, 304, 305, 307, 353, 485, 486
Suspended particulate matter (SPM), 215, 216, 223–225
Suspended sediment, 246, 494
Suspension, 6
Suspension feeding, 281, 283, 286, 288, 291, 303, 309
Sweeper polyps, 357
Sweepers, 329, 330
Sweeper tentacles, 307, 351, 357, 358
Swimming, 204, 245, 281, 282, 287, 292, 303, 336, 337
Swimming larvae, 356
Swimming speed, 337
SWIR. *See* Short-wave infrared
Syllid polychaetes, 303
Syllid worms, 299
Symbiodinium, 87–93, 97–101, 395
    S. bermudense, 88, 90, 91
    S. californium, 90, 91
    S. cariborum, 90, 91
    S. corculorum, 90, 91
    S. goreauii, 90, 91
    S. kawagutii, 90, 91
    S. linucheae, 91
    S. meandrinae, 90, 91
    S. microadriaticum, 90, 91, 187
    S. muscatinei, 90, 91
    S. pilosum, 90, 91
    S. pulchroru, 90, 91
    S. pulchrorum, 90
    S. trenchii, 90, 91
    S. varians, 91
Symbionts, 6, 87–101, 107, 110, 112, 114–116, 131, 153, 154, 162, 165, 170, 187–192, 216, 219, 220, 223–225, 275, 276, 283, 288, 290, 295, 297, 298, 301, 305, 307, 341, 360, 400, 401, 409, 426, 427
Symbionts shuffle, 93
Symbiophagy, 6
Symbiosis, 6, 16, 59, 87, 88, 92, 93, 96, 101, 110, 139, 152, 187–192, 210, 212, 215, 218, 219, 222, 247, 299, 303–308, 341, 409, 411, 413, 440

Symbiosome membrane, 88, 89, 94, 188, 189
Symbiosomes, 88–89, 94, 97, 188, 189
Symbiotic dinoflagellates, 94, 406
Synchronization, 107, 108
Synchronous, 63–66, 76
Synchronous coral spawning, 332
Synchronous reproductive timing, 5
Synergistic effects, 377, 378, 385
Syngameons, 38, 40, 44, 45
Syngnathidae, 330
Syngnathids, 336
Syngnathus, 339
Synodontidae, 330
Synthetic host factor (SHF), 94
Systematic data, 313

T

Tabuaeran, 383
Tabulata, 19
Taiwan, 64, 348, 349, 358, 366
Tanaids, 283, 296, 303
Tannin, 232, 252
Tanzania, 352, 363
Target of rapamycin (TOR), 413
Taxonomic, 5, 19, 38, 40–41, 44, 60, 72, 112, 119, 211, 242, 256, 273, 280, 282, 299, 312, 313, 327, 328, 339, 349, 355, 435
Tectonic events, 19, 20
Tegula, 424
TEK. *See* Traditional ecological knowledge
Temperate, 65, 87, 91, 94, 96, 140, 164, 166, 168, 190, 191, 219, 243, 245, 247–253, 256, 259, 288, 302, 327, 334, 340, 341, 348, 353–355, 365–367, 396, 424, 429, 453, 460
Temperate marine fish, 341
Temperate marine species, 334
Temperate rocky reefs, 348, 367
Temperate rocky shorelines, 367
Temperate zones, 245, 247, 250
Temperature, 6, **14**, **32**, **39**, 67, **88**, **108**, **129**, **152**, **183**, **207**, **219**, **234**, **245**, **312**, 337, **392**, **405**, **421**, **445**, **451**, **465**, **517**
Temperature stress, 101, 261, 406, 408, 414, 426, 460
Temporal, 15, 17–19, 30, 51, 63, 65, 76, 119, 208, 246, 248, 253, 263, 264, 308, 354, 375, 385, 466, 474, 477, 478, 483, 494, 497, 512
Temporal scales, 15, 17, 18, 30, 246, 263, 466, 483
Tentacles, 75, 108, 111, 154, 189, 215, 280, 281, 286, 288, 294, 304, 307, 351, 357, 358, 363
Tentacular activity, 107
Terapontidae, 330
Terebellidae, 296
Terpenoids, 251, 252
Terrestrial, 5, 32, 43, 44, 124, 129, 217, 242, 246, 247, 251, 252, 258–260, 327, 340, 354, 367, 385, 394, 405, 423, 435, 456, 472, 484, 494, 499–502, 510
Terrestrial ecosystem, 242, 405
Terrestrial run-off, 5, 385, 456, 494, 499–502
Terrestrial sediments, 258, 260
Tetiaroa Atoll, 247
Tetralia spp., 283, 294, 302
Tetraodon, 339
Tetraodontidae, 330
Thailand, 51, 64, 286, 296, 424, 426
Thalamita, 303
Thalassoma bifasciatum, 338
Thalassoma duperrey, 333
Thalassoma lunare, 333
Thalassomonas loyana, 457

Thalli, 208, 243, 248–250, 257, 262, 362
Thallus, 249, 257, 340
Theonella swinhoei, 294
Thermal, **52, 69, 100, 110, 152, 208, 259, 395, 406, 421, 445**
Thermal bleaching, 152, 153, 218, 425
Thermally tolerant, 101, 415, 427, 431
Thermal sensitivity, 422, 425, 430
Thermal stress, 69, 208, 210, 259, 395–398, 400, 401, 406–414, 426, 429, 445
Thermal threshold, 398–400
Thermal tolerance, 100, 101, 152, 399–401, 408, 413, 422, 423, 425–428, 431
Thermoclines, 259, 260
Thermodynamic equilibrium, 133, 156
Thermotolerance, 414, 415, 427, 429
Thioredoxin, 429
Thousand years ago, 16, 21
Threadfin bream, 330, 337
Threats, 6, 8, 328, 331–333, 336, 341, 342, 373, 380, 424
Threshold responses, 377
Threshold temperature, 407, 426
Thylakoid membrane, 99, 112, 408, 411
Thylakoids, 88, 89, 98, 99, 112, 408, 411
Tidal, 14, 109, 246, 292, 352, 422, 425, 499
Tidal regime, 14
Tigerperch, 330
Tight recycling, 199, 200, 205
Tilefish, 330
Tissues, 60, 87, 111, 119, 152, 181, 187, 209, 216, 243, 283, 332, 348, 395, 406, 423, 438, 451, 466, 495
  damage, 61, 97, 306, 339, 352, 353, 356, 358, 365, 412, 451, 483
  layers, 120, 121, 164, 181, 359
  loss, 456–458, 472, 473, 475, 477, 478, 482–484
  mortality, 303, 348, 467, 469, 472, 477, 478, 483, 484
Titanoderma prototypum, 254
Toadfish, 330
Tocopherols, 191
Tokelau, 328
Tolerance, 52, 69, 88, 91, 100, 101, 152, 168, 171, 257, 399–401, 406, 408, 413–415, 421–431, 496
Tolerant, 15, 53, 100, 101, 168, 257, 259–261, 399, 400, 409, 415, 424, 426, 427, 430, 431, 498, 502
Tolerate, 53, 87, 191, 244, 308, 397, 423, 427, 496
Tolerating herbivory, 251
Tonga, 328
Tongue soles, 330
Top-down, 261, 311, 367, 377, 382, 384–385
Top-down effects, 384–385
Top-down force, 261
Top-down processes, 367, 377
Topical, 327, 328
Topography, 205, 242, 298, 440, 500, 502, 511
TOR. *See* Target of rapamycin
Total alkanity (ALK), 155, 157
Total dissolved inorganic carbon (CT), 157
Tourism, 153, 373, 376, 380, 381, 400, 405
Tourists, 153
Toxic compounds, 292, 356
Toxicity, 128, 167, 249, 253, 338
Toxin, 302, 356, 357, 360, 452
Trace elements
  Ba, 155
  $\delta^{13}C$, 155
  $\delta^{18}O$, 155
  Mg, 155
  Sr, 155

Traditional ecological knowledge (TEK), 3
Transcellular, 126–129, 189
Transect, 25, 26, 29, 31, 204, 303, 445
Transfer, 18, 28–31, 92, 95, 112, 126, 138, 139, 141, 183, 188, 189, 191, 192, 201–205, 215, 216, 219, 220, 234, 353, 429, 495
Translocated, 87, 93–96, 110, 113, 114, 116, 224, 290, 360
Translocated carbon resources, 360
Translocation, 89, 95, 99, 107, 114, 218–220, 224, 233, 290, 360
Transmitted, 27, 93, 187, 457
Transparent body, 338
Transplant, 340, 353
Transplantation, 113, 177, 178
Trap capture, 517
Trapezia spp., 283, 302, 305, 307
Trap fishing, 341
Triassic, 3, 14, 88, 134
Trichinotidae, 330
Tridacnid, 286, 297
Trididemnum cyclops, 360
Trididemnum miniatum, 360
Trididemnum solidum, 290, 360
Trigger, 234, 311
Triggerfish, 311, 329, 330, 333, 339
Trinidad and Tobago, 328, 475
Triplefins, 330, 334
Tripterygiidae, 330
Trizopagurus magnificus, 302
Trophic, **7, 188, 217, 231, 241, 273, 339, 374, 426, 493, 517**
Trophic interactions, 225, 291, 303, 307–311, 383, 384
Trophic levels, 200, 217, 234–236, 309, 310, 339, 340, 382, 384, 495, 517
Trophic modeling, 308
Trophic position, 339
Trophic structures, 231–237, 382, 383, 493
Trophodynamics, 215–225, 273, 308
Tropical, **5, 13, 37, 53, 59, 87, 107, 152, 188, 210, 218, 241, 274, 340, 347, 394, 405, 421, 436, 458, 465, 494**
Tropical-adapted taxa, 18
Tropical affinities, 14
Tropical cyclone, 8, 258
Tropical sea, 59, 282, 396, 405
Tropical storms, 349
Tropical waters, 5, 243, 245, 251, 252, 288, 407, 436, 437, 458
Tropiometra afra-macrodiscus, 296
Trumpetfish, 329, 330
Tsehay, I., 384
Tubastraea coccinea, 312
Tubastraea faulkneri, 356
Tubastraea tagusensis, 312
Tubastrea, 61, 62, 75, 130
  T. aurea, 130
  T. coccinea, 61, 68
  T. diaphana, 62
Tube blennies, 330
Tube worms, 152
Tunicates, 287, 290, 292, 294, 306, 308, 312
Turbid, 39, 110, 246, 439, 444, 445, 494, 496, 498
Turbidity, 7, 8, 14, 21, 69, 136, 142, 225, 392, 429, 445, 471, 493, 494, 496, 498, 500, 501, 503
Turbid waters, 246, 439, 494, 496
Turbinaria, 72, 217, 218, 242, 243, 250, 363, 498
  T. mesenterina, 72
  T. reniformis, 217, 218
Turf algae, 231, 236, 244, 245, 250, 261, 291, 338, 352, 353, 365, 377, 480, 496
Turfing, 363

Index

Turks and Caicos Islands, 328
Turnover of carbon, 199
Turnover rates, 243, 250, 366
Tutuila, 272
Tuvalu, 328, 341

U

Udotea, 245
Udoteaceae, 243
Ulcerative white spots (UWS), 473
Ultrastructural, 94, 413
Ultra-violet radiations (UVR), 69, 191, 293, 398, 407
Ultraviolet screens, 191–192
Ultra-violet. *See* UV
Unfed, 217, 337
United Arab Emirates, 328
Unsupervised classification, 28
Upar. *See* Urokinase plasminogen activator receptor
Uptake and storage, 247, 248
Uptake rate, 95, 200, 202–204, 209, 216, 243, 248, 249
Upwelling, 19, 30, 53, 65, 66, 110, 200, 259, 260, 424, 425, 444, 493, 495, 497, 500, 501
Urban development, 373
Urbanization, 341
Urchins, 231, 232, 250–252, 283, 288, 297, 300, 307, 311, 349, 367, 377, 378, 383, 384, 439–441, 443–445, 468, 475, 480, 510, 511, 514, 517
Urease, 167
Urochordata, 290, 303
Urokinase plasminogen activator receptor (Upar), 429
USVI, 471, 482, 483
US Virgin Islands, 328, 474, 479, 485,
UV, 99, 101, 107, 109, 110, 116, 191, 192, 247, 252, 333, 360, 363, 429
    absorption, 252, 360
    light, 109, 429
    screen, 191
    stress, 247
UV-absorbing mycosporine-like amino acids (MAAs), 333
Uva Island, 262, 301
Uva Reef, 293
UVR. *See* Ultra-violet radiations
UWS. *See* Ulcerative white spots

V

Vagile, 152, 282, 292, 311
Vagile richness, 311
Vanuatu, 50, 51, 328
Vectors, 7, 157, 158, 179, 181, 257, 307, 312, 353, 362, 410, 452, 457, 459, 480–482
Vectors and reservoirs, 481–482
Velvetfish, 330
Venezuela, 75, 328, 352, 355, 357, 360, 468, 472, 475, 476, 482
Vermetids, 349, 360, 364
Vertebrate, 135, 310, 341, 424, 429, 510
Vertical gradients, 367, 423
Vertical system transmission, 91
Vibrio sp., 450
    V. carchariae (harveyi), 452
    V. charchariae, 474, 475
    V. charcharina, 458
    V. coralliilyticus, 453–454
    V. harveyi, 458, 474
    V. shiloi, 451–453, 460, 481
Vicariance, 20, 43–44, 52, 53

Viet Nam, 51, 297, 328
Vietnam, 51, 297, 328
Virgin Islands, 273, 302, 328, 465, 468, 469, 474, 479, 483–485
Virulence, 187, 305, 452, 453, 466, 473, 474, 477–478, 482, 484
Virulent diseases, 362, 468, 483
Viruses, 16, 29, 233, 234, 460
VIS. *See* Wavebands in the visible
Visible radiation, 407
Visible spectrum (VIS), 26, 28, 110
Viviparous. brotulas, 330

W

Ω, 445
Wallis & Futuna Islands, 328
Warming events, 7, 8, 349, 396
Washington, 274, 352, 396, 511
Washington State, 511
Water currents, 294, 347, 363, 475
Water flow, 178, 203, 225, 249, 256, 294, 309, 333, 353, 361–363, 394, 410, 511
Water molecules, 29, 245, 396
Water motion, 5, 6, 250, 363, 364
Water optical properties, 30, 31
Water quality, 8, 14, 250, 397, 401, 439, 493–495, 498, 500–503
Watersheds, 7, 247, 260, 516
Watersipora sp., 360
Water-table, 109, 354
Water temperatures, 14, 92, 259, 337, 455, 465, 466, 471, 475–478, 482, 483
Water velocity, 178, 202–204
Wavebands in the visible 28, 29
Wavelengths, 27, 29, 30, 192, 246
Waves, 6, 14, 16, 28, 39, 61, 152, 171, 202, 203, 241, 242, 244, 246, 247, 250, 259, 293, 298, 338, 436, 443, 498–500, 516
    action, 39, 241, 244, 250, 259, 293, 298
    energy, 14, 242, 246, 443
WBD. *See* White band disease
WBD-I. *See* White band I
WBD-II. *See* White band disease type II
Weight, 94, 126, 132, 133, 158, 190, 191, 208–210, 218, 221, 224, 250, 259, 303, 310, 340, 383, 460, 494, 496
Weight loss, 340
Wenlockian (Silurian) rocks, 440
Western Atlantic, 19, 65, 264, 279, 280, 287, 301, 312, 341, 465, 469, 473, 480, 485
Western Australian (WA), 18, 63, 64, 300
Western Pacific, 47, 53, 100, 280, 281, 290, 296, 300, 312, 365
Western Samoa, 64, 328
White band disease (WBD), 234, 457–458, 460, 465, 466, 468–471, 473, 475, 480, 481, 483, 509
White band disease and diadema, 468–471
White band disease type II (WBD-II), 458, 471
White band I, 452
White band II, 452
White patch disease (WPA), 471
White patches and octocoral mortalities, 471
White plague, 7, 456, 468, 475
White plague (Red Sea), 452
White plague disease (WPD), 456–457, 460, 468, 470, 473, 47–47, 480–484
White plague type, 353, 362, 457, 468
White pox, 452, 458, 471, 473–475, 477
White pox disease(WPX), 458, 477
White syndrome (WS), 466, 467, 473–476
Whole reef communities, 366
Winter, 110, 202, 406, 423, 452, 454, 465, 477, 482, 483

Within-species interactions, 347, 348
Wormfish, 330
Worms, 152, 167, 242, 275, 279, 281, 283, 286, 290–292, 294, 296, 297, 299, 301, 302, 304, 305, 307–309, 312, 397, 441, 452, 459, 480, 481
WPA. *See* White patch disease
WPD. *See* White plague disease
WPX. *See* White pox disease
Wrasses, 329–333, 338, 340, 342, 383
Wrigglers, 330
WS. *See* White syndrome

X

XANES spectroscopy. *See* X-ray Absorption Near Edge Structure (XANES) spectroscopy
Xanthid crab, 303
Xarifia, 274, 283, 285, 306
Xenisthmidae, 330
X-ray, 126, 128, 142, 180, 181
X-ray Absorption Near Edge Structure (XANES) spectroscopy, 134

Y

Y. *See* Potential quantum yield
Yap, 3
YBD. *See* Caribbean yellow band disease
Yellow band (YBS), 458, 484
Yellow band disease (YBD), 7, 458–459, 466, 468, 471–473, 475–478, 481–484
Yemen, 328, 358

Z

Zanclidae, 330
Zoanthidea, 303
Zoanthids, 280, 352, 355, 357, 359, 360, 365–367, 407, 466, 468, 469, 473, 476, 478, 484, 485
Zoanthus, 352, 355, 357, 360
  Z. robustusas, 94
  Z. sociatus (Zoantharia), 91, 352
Zonation, 15, 246, 258, 274, 309, 366, 367
Zooanthid, 242
Zooplanktivores, 339
Zooplankton, 96, 112, 113, 116, 191, 207, 216–218, 220–225, 275, 283, 299, 303, 311, 347, 383, 495
Zoospore, 89, 245
Zooxanthellae, **5, 59, 87, 107, 119, 187, 209, 215, 233, 286, 347, 406, 422, 440, 458, 466, 495**
  density, 96, 97, 110, 113, 189, 219, 352, 406, 423
  genotype, 89–93, 95, 99–101, 413–415
  recovery, 90, 96, 425
  release, 89, 91, 93–97
Zooxanthellate, 44, 48–51, 59, 60, 75, 88, 107, 178, 187, 220, 279, 280, 283–285, 300–302, 305–307, 312, 360, 405, 407, 465, 498
Zooxanthellate corals, 300, 305, 306
Zygotes, 243

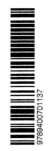